Light Metals 1984

Edited By
J. P. McGeer
Alcan International Limited
Kingston Laboratories

Proceedings of the technical sessions
sponsored by the
TMS Light Metals Committee
at the 113th Annual Meeting,
Los Angeles, California,
February 27 - March 1, 1984.

A Publication of The Metallurgical Society of AIME

A Publication of The Metallurgical Society of AIME
420 Commonwealth Drive
Warrendale, Pennsylvania 15086
(412) 776-9000

Printed in the United States of America.
Library of Congress Card Catalogue Number **72-623660**
ISBN Number **0-89520-470-3**

Reprinted in 1985

Foreword

Light Metals 1984 continues the tradition of publishing current advances in science and technology of primary processes in the aluminum industry. The quantity in and quality of this volume and the companion conference, at a time when the industry continues to be depressed, is a tribute to the significance which academia and industry attach to this exchange of information on aluminum smelting. Continued growth of *Light Metals* and the conference is a credit to the foresight of those industry members who started this tradition in 1962. Many of them have since retired, but our thanks and congratulations go to them once again for having started a most important activity.

Our thanks must also go to those who, over the years, have contributed to the continued success of this venture. Their names are too numerous to list, but a survey of the membership of the Light Metals Committee, of the authors who have contributed to various volumes, and certainly not least the staff of The Metallurgical Society, signifies the value which is attached to the activity. Their efforts have benefited us all, and have undoubtedly contributed to the growth and competitiveness of the aluminum industry.

Our industry is changing rapidly; the recession over the recent years has resulted in a significant restructuring. It is therefore heartening to see that once again our activity has grown and this year we have more sessions, more papers, and more entries participating than ever before.

Another recognition of the work of the Light Metals Committee of The Metallurgical Society is the creation of an annual award which will be given for the first time in 1985. This award will be for the best paper appearing in *Light Metals* each year. It constitutes recognition of the contribution which *Light Metals* is making to the literature of aluminum production. Details regarding the criteria for the award, and of the award itself, will be available in 1984.

The preparation of the program which is represented by the papers in this volume could not possibly occur without the enthusiastic cooperation of the subject chairmen. They are:

D. Donaldson	Alumina and Bauxite Production
E.L. Cambridge	Aluminum Reduction Technology
L.G. Boxall	Carbon Technology
F.R. Mollard	Cast Shop Technology
L.C. Blayden	Environmental Control and Health

On behalf of all the subject chairmen, and from the bottom of my heart, I would like to thank the TMS headquarters staff, who have cooperated so well throughout the year to ensure an outstanding program and its publication in *Light Metals 1984*. These people are:

John Ballance	Director of Publications and Editor, Journal of Metals
Marilyn Zabel	Meetings Manager
Barbara Kamperman	Program Coordinator
Elizabeth Luzar	Book Publishing Manager

J.P. McGeer
Alcan International Limited
Kingston Laboratories
P.O. Box 8400
Kingston, Ontario
Canada K7L 4Z4

December 1983

Light Metals Committee

<u>Chairman</u>
J.P. McGeer
Alcan International Limited
Kingston Laboratories
Kingston, Ontario, Canada

<u>Vice Chairman</u>
H.O. Bohner
Swiss Aluminum Limited
Zurich, Switzerland

<u>Past Chairman</u>
E. Max Adkins
Dubai Aluminum Company, Limited
Dubai, United Arab Emirates

<u>Committee Members</u>

J.C. Agarwal
Amax Specialty Metals Corporation
Parsippany, New Jersey

H.P. Armintrout
Intalco Aluminum Corporation
Ferndale, Washington

G.M. Bell
Aluminum Company of America
Alcoa Laboratories
New Kensington, Pennsylvania

K. Bengtson
Kaiser Aluminum and Chemical Corporation
Pleasanton, California

L.G. Boxall
Martin Marietta Corporation
Baltimore, Maryland

V. L. Bullough
Reynolds Metals Company
Sheffield, Alabama

H. Forberg
National Southwire Aluminum Company
Hawesville, Kentucky

S.D. Hill
U.S. Bureau of Mines
Salt Lake City, Utah

C.J. McMinn
Reynolds Metals Company
Sheffield, Alabama

F.R. Mollard
Cabot Corporation
Reading, Pennsylvania

H.S. Monahan
Alcan Smelters and Chemicals
Montreal, Quebec, Canada

T. Naterstad
Ardal og Sunndal Verk
Oslo, Norway

C.B. Wilson
Dow Chemical Company
Freeport, Texas

R. Zabreznik
Kaiser Aluminum and Chemical
Tacoma, Washington

Associate Members

A. Aksnes
Granges Aluminum Company
Sundsvall, Sweden

E. Dunn
Kaiser Aluminum and Chemical Corporation
Pleasanton, California

T.G. Edgeworth
Alcan International Limited
Kingston, Ontario, Canada

K. Higbie
Stroudsburg, Pennsylvania

S.R. Leavitt
Alcan East Asia Limited

J.J. Miller
Alumax, Inc.
San Mateo, California

R.E. Miller
Aluminum Company of America
Alcoa Center, Pennsylvania

P.G. Thaure
Aluminum Pechiney
Paris, France

J.B. Todd
Kaiser Aluminum and Chemical Corporation
Oakland, California

Contents

Foreword iii

Light Metals Committee v

ALUMINA AND BAUXITE I
Process Design and Control

Optimisation of a Bayer Flowsheet by Successive Simulations Automatically Generated 3
A. Adjemian and P. Colombe

Steady State Simulation of the Bayer Process 13
D.R. Chinloy and R.K. Holzwarth

Optimum Control of Bauxite Charge to High Temperature Digestion Units at QAL 27
P. McIntosh

Improved Particle Size Analysis Method for Precipitation Control 39
J.T. Creehan

Control of Process Liquor Impurities at San Ciprian 53
J.L. Monita Benito

Application of Optimal Control to Lurgi Fluidized Bed Calciner for Sandy Alumina 63
T. Kohno and M. Itoh

The Use of Small Nuclear Magnetic Resonance Spectrometers as On-Line Analyzers for Rotary Kiln Control 81
R.M. Pearson and T.L. Parker

ALUMINA AND BAUXITE II
Analytical Technology

Automated Laboratory Liquor Analyzer--ALLA--for Analysis of Bayer Liquor 101
C.K. Matocha

Application of Ion Chromatography to the Analysis of Bayer and Aluminum Process Liquors 115
K. The and R. Roussel

Characterization of Organics in Bayer Liquor 127
J.D. Guthrie, P.J. The, and W.D. Imbrogno

Equilibrium Solubility of Sodium Sulfate in Bayer Liquor 147
J.T. Malito

Dehydration Products of Gibbsite by Rotaly Kiln and Stational Calciner.......... 157
K. Yamada, T. Harato, S. Hamano, and K. Horinouchi

Particle Size Analysis in Bayer Plant Control.......... 173
J.D. Zwicker

ALUMINA AND BAUXITE III
Bayer Process Technology

Flocculation and Thickening of Red Mud.......... 197
C. Sato, Y. Shibue, N. Arakawa, and Y. Kawai

Hydrothermal Recovery of Soda and Alumina from Red Mud: Tests in a Continuous Flow Reactor.......... 211
P.J. Cresswell and D.J. Milne

A Graphical Construction for Predicting the Yield from Continuous Precipitator Trains.......... 223
E.T. White, M. Steemson, and D. Milne

Mathematical Model of the Precipitation Section of a Bayer Plant.......... 237
M.L. Steemson, E.T. White, and R.J. Marshall

ALUMINA AND BAUXITE IV

Attrition of Aluminas.......... 257
D.J. Braun

Measurement of Sandy Alumina Dustiness.......... 269
S. Perra

Study of Alumina Behavior in Smelting Plant Storage Tanks.......... 287
H.P. Hsieh

ALUMINA AND BAUXITE V
Plant Design and Equipment

Corrosion in Caustic Environment Application to Alumina Plants.......... 307
M.V. Chaubal

New Mechanical Seal for Abrasive Suspensions.......... 325
Ph. Marchessaux

Hydraulic Actuator for Angle Valves.......... 343
F. Lavastre and R. Saulnier

Development of a Dustless Alumina Shiploader.......... 351
T. Tanabe, S. Assakura, O. Yoshida, and T. Kitahara

New Coal Based Alumina Plant Energy Concept.......... 357
H. Beisswenger and H.W. Schmidt

ALUMINA AND BAUXITE VI

Advances in the Filtration of Alumina Trihydrate........................ 373
R.F. Nunn and R.M. Talcott

Utilization of Horizontal Belt Filters for Washing Fine Alumina Seed and Kiln Feed Product........................ 385
R.F. Crawford, D.A. Dahlstrom, and G. Minear

The Ammonoalunite Process for Production of Alumina from Clay............ 397
R.W. Bartlett, R.J. Wesely, and T.R. Bolles

Alumina Via Hydrochloric Acid Leaching of High Silica Bauxites - Process Development........................ 411
D. Elsner, D.H. Jenkins, and H.N. Sinha

ALUMINUM REDUCTION TECHNOLOGY I
New Developments

Start-Up of First Pre-Bake Potline in Brasil........................ 433
J.E. Tauscher and D.S. Ratcliffe

Development of a High Amperage End-to-End Reduction Cell Design.......... 443
C.J. Rogers and D.W. Macmillan

Development of the Kaiser Aluminum 195 kA Cell........................ 455
R.D. Zabreznik and E.D. Tarapore

Reduction of Power Consumption in 1940/1950 Designed Pots............... 461
O.S. Chaudhry

Energy Savings Through Improved Cathode Construction.................... 475
P. Aeschbach and K. Etzel

A New Approach to Aluminum Cathode Design and Relining.................. 483
J.J. Vadla

ALUMINUM REDUCTION TECHNOLOGY II
Measurement and Control

Bath Ratio Control at the Mead Works........................ 493
R.C. Jeltsch

Routine Analysis of Bath Ratio: A New Method........................ 501
K. Nagy, R. Bjørgo, and D. Fremstad

Alumina Dissolution and Sludge Formation........................ 513
R. Keller

Metal Pad Velocity Measurements in Prebake and Soderberg Reduction Cells........................ 519
A.T. Tabereaux and R.B. Hester

Metal Pad Velocity by the Iron Rod Method........................ 541
B.F. Bradley, E.W. Dewing, and J.N. Rogers

ALUMINUM REDUCTION TECHNOLOGY III
Alternative Processes

TiB_2 Coatings Prepared by Plasma Spraying and Electroplating.............. 555
D.W. Townsend and L.G. Boxall

Use of TiB_2 Cathode Material Application and Benefits in Coventional VSS Cells.. 573
L.G. Boxall and A.V. Cooke

Carbon Partition Between Carbothermically Reduced Al and Al_2O_3-Al_4C_3 Melts... 589
L.M. Ruch, A.F. Saavedra, and R.M. Kibby

The Production of Raw Al/Si Alloys from Italian Leucitic Minerals by a Direct Carbothermic Process. Experiments and Results on a Pilot Plant.. 601
M.F. Landi, S. Da Roit, S. Montesi, and L. Piras

Microscopic Investigation of Clay Chlorination............................ 619
E.S. Martin and K. Wefers

ALUMINUM REDUCTION TECHNOLOGY IV

Alumina Transportation To Cells.. 635
I. Stankovich

Crusting Behavior of Smelter Aluminas...................................... 649
D.W. Townsend and L.G. Boxall

Aluminum Pechiney Specifications for Optimising the Aluminas Used in Sidebreak and Point Feeding Reduction Pots............. 667
Y. Bertaud and A. Lectard

ALUMINUM REDUCTION TECHNOLOGY V
Modelling and Analysis

Thermal Analysis of Soederberg Pots... 689
T. Ohto and T. Matsushima

Investigation of the Dynamic Behavior of Aluminum Pots.................... 701
P. Entner, W. Schmidt-Hatting, Z. Mitrovic, U. Gruber, D. Rufer, and B. Fjørnes

The Current Efficiency of a Shorted Anode in a Prebake Cell............... 711
W.W. Hyland

Physical Modelling of Bubble Behaviour and Gas Release from Aluminum Reduction Cell Anodes... 721
S. Fortin, M. Gerhardt, and A.J. Gesing

Potroom Ventilation System by Segregated Air Layers....................... 743
O. Lid and Ø. Haugen

ALUMINUM REDUCTION TECHNOLOGY VI
Fundamental Studies

Thermodynamics of Metal Oxide Reactions in Hall Cells.................. 757
A.J. Becker and D.R. Careatti

Interactions of Gases in Molten Salts: Carbon Dioxide and Oxygen in Cryolite-Alumina Melts.................................. 759
H. Numata and J. O'M. Bockris

The Rate Determining Step of Metal Loss in Cryolite-Alumina Melts........ 789
Z. Qiu and L. Fan

Model Calculations of the Electrical Conductivity of Cryolite Melts.. 805
P. Fellner, K. Grjotheim, and K. Kvande

On the Anode Effect in Industrial Aluminum Cells......................... 825
J. Thonstad, F. Nordmo, A.H. Husøy, K.Ø. Vee, and O.G. Austrheim

Determination of Melting Temperatures and Al_2O_3 Solubilities for Hall Cell Electrolyte Compositions.............................. 841
S.S. Lee, K.S. Lei, P. Xu, and J.J. Brown, Jr.

CARBON TECHNOLOGY I

Coker Feedstock Characteristics and Calcined Coke Properties............. 859
P.J. Rhedey and S.K. Nadkarni

Influence of Maximum Temperature and Heat-Soak Times on the Properties of Calcined Coke.................................... 869
A.F. Bopp, G.B. Groff, and B.H. Howard

Desulfurization of Petroleum Coke.. 883
H.H. Brandt and R.S. Kapner

New Applications to Upgrade Coal Tar Products............................ 889
H. Kremer and S. Cukier

CARBON TECHNOLOGY II

Correlation of Raw Material Properties with Binder Content of HS Soderberg Paste... 901
L. Castonguay and S.K. Nadkarni

Temperature Stability of Soderberg Anode Pitch........................... 909
M. Sørlie

Evaluation of Volumetric Anode Mix Formulation........................... 923
D. Belitskus

Anode Consumption Test Cell.. 933
P.P. Russell, K.R. Weisbrod, and S.S. Jones

Interrelation Between Composition and Properties of Coal Tar Pitches and Properties of Anode.................................. 943
M.L. Itskov, R.V. Svoboda, V.I. Denisenko, E.A. Koledin, and G.F. Vedernokov

CARBON TECHNOLOGY III

Factors in the Design of Reduction Cell Anodes.......................... 961
D.G. Brooks and V.L. Bullough

Temperature and Stresses in Carbon Anodes in an Aluminum Reduction Cell.. 977
T.L. Weng and M.B. Hsu

A Referee Method for Evaluating Anode Coke Density...................... 989
V.L. Bullough, A.L. George, and E.R. Russell

Effect of Butt Particle Size and Butt Fraction on Prebake Anode Properties.. 1005
T.L. Weng

Removal of Solid Bath from Spent Anodes................................. 1015
J.C. Broad and B.D. Farmer

CARBON TECHNOLOGY IV

Noranda Anode Baking Furnace - Design and Performance.................. 1027
J. Auton, W.F. Crowell, R.R.Ebersole, W.W. Ellison, J. Kaiser, J.G. Peacy, and E.J. Chin

Mathematical Simulation of a Horizontal Flue Ring Furnace............... 1033
R.T. Bui, E. Dernedde, A. Charette, and T. Bourgeois

Operational Experience with Cold Ramming Paste Rheinfelden 1AK.......... 1041
A.J. Franke, L. Förster, and A. Sturm

New Development in Pot Wall Linings..................................... 1053
K. Etzel and P. Aeschbach

Chemical Resistance of Cathode Carbon Materials During Electrolysis..... 1059
M. Sørlie and H.A. Øye

Electrochemically Induced Crystallographic Changes in Graphite in Molten Salts.. 1071
A. Hafeez

CAST SHOP TECHNOLOGY I
Casting Technology

Pipeline for Transferring Molten Aluminum Alloys........................ 1085
O.G. Sivilotti and J. Sulzer

Automation of Slab Casting.. 1103
J. Moriceau and P. Heuillard

Twin Roll Casting of Aluminum: The Occurrence of Structure Inhomogeneities and Defects in As-Cast Strip......................... 1111
H. Westengen and K. Nes

Rolling Aspects of Twin Roll Casting.................................... 1129
R. Iricibar and I. Jin

New Casting Method for Improving Billet Quality........................ 1145
J.P. Faunce, F.E. Wagstaff, and H. Shaw

Mould Temperatures During DC-Casting of Eight-Inch Diameter Extrusion Ingots in Alloy 6063.. 1159
E.K. Jensen

Production of Continuously Quenched Wire Rod of Al-Mg-Si Alloys and Properties of Rods and Wires.............................. 1177
P. Fiorini, L. Gasparotto, A. Lama, and G. Zago

CAST SHOP TECHNOLOGY II
Melt Treatment and Ingot Structure

Coreless Induction Versus Rotary Furnace Melting of Ball-Milled Drosses.. 1193
P.J. Bamji

Electrolytic Removal of Magnesium from Scrap Aluminum.................. 1205
B.L. Tiwari and R.A. Sharma

Improved Grain Refining with TiBAl Alloys Containing Three Percent Titanium... 1217
J. Pearson and M.E.J. Birch

The Use of Undercooling Measurements to Evaluate Grain Size of Wrought Aluminum Alloys...................................... 1231
S.A. Levy and J.L. Kirby

Role of Ingot Structure in Structural Streaking....................... 1249
D.A. Granger and C.L. Jensen

CAST SHOP TECHNOLOGY III
Filtration and Degassing

The TAC Process: A Proven Technology.................................. 1267
B. Gariépy, G. Dubé, C. Simoneau, and G. Leblanc

Molten Aluminum Filtration: Fundamentals and Models.................... 1281
C.E. Eckert, R.E. Miller, D. Apelian, and R. Mutharasan

Molten Aluminum Quality Measured with Reynolds 4M™ System.............. 1305
T.L. Mansfield

Removal of Hydrogen from Molten Aluminum by Gas Purging................ 1329
T.A. Engh and T. Pedersen

Determination of Hydrogen in Aluminum and Its Alloys by the Fusion/Carrier Gas Method Using the LECO RH-3 Determinator............ 1345
W.D. Lamb

Segregation Effects in Aluminum Metal Samples Effect of Sample Mould Design.. 1373
J.L. Fortier and C. Tremblay

ENVIRONMENTAL CONTROL I
Disposal and/or Recovery of Spent Potlining

Presentations and Panel Discussion.................................... 1389

ENVIRONMENTAL CONTROL II

Hydrocarbon Emissions from Soderberg Pots.............................. 1393
D.R. Augood and H.J. Seim

Mathematical Model of Fluoride Evolution from Hall-Heroult Cells........ 1429
W.E. Haupin

Methods of Monitoring Occupational Exposure in Aluminum Smelters....... 1441
A. Steinegger

Reduction Cell Hooding Efficiencies with Various Cell, Superstructure, and Shield Design Factors............................. 1449
R.J. Schlager and R.K. Ng

Control of Gas Collection Efficiency in VSS-Potlines by Monitoring CO_2 and CO in the Primary Pot Gases................................... 1457
L. Wikström

ENVIRONMENTAL CONTROL III

The Effluent Limitations Guidelines for the Primary Aluminum Industry.. 1475
J.H. Goldman

St. Jude Industrial Park's Sanitary Wastewater Zero Discharge System... 1507
R.E. Hurt

Control of Pitch Volatiles in Green Electrode Production................ 1521
T.J. Robare and M.W. Wei

Gradients of Fluoride Accumulation in Southern Pine Forests Near a New Primary Aluminum Smelter................................... 1531
R.C. Dickie, D.C. McCune, D.C. MacLean, and J.A. Laurence

Effects of Staged Combustion on NO_x Formation During Coke Calcination... 1541
A.F. Bopp and J.F. Bent

Contribution to the Interpretation of Coal Tar Pitch Measurements at the Workplace.. 1553
R. Glaus and A. Steinegger

ADDENDUM

Overview of Alcoa Direct Reduction Process Technology.................. 1571
M.J. Bruno

Feasibility of an Aluminum-Silicon Blast Furnace....................... 1591
R.L. Troup

Feasibility of an Aluminum-Silicon Blast-Arc Process.................. 1613
D.T. Stevenson

The Life Span and Market Demands of Some Bauxite Mines................ 1633
V.G. Hill and S. Ostojic

The Use of Synthetic Flocculant Polymers in Settling Red Muds Derived from High Goethite Bauxite Ores............................. 1653
S.E. Sankey and R.J. Schwarz

Flash Calciner By Fives-Cail Babcock.................................. 1669
J.P. Henin

The Use of Refractory Lifters in Coke Calcination..................... 1697
M.F. Vogt, G.R. Jones, and G.A. Tyler

Loss of Weight Systems in Continuous Carbon Production - A Competitor to Traditional Belt Scales.............................. 1715
S. Frosta and T. Naterstad

Recrystallization of Commercial Beryllium Sheet....................... 1727
J.M. Marder

Author Index.. 1747

CHAIRMEN

F. N. Newchurch
Reynolds Metals Company
Little Rock, Arkansas

B. J. Foster
Kaiser Aluminum and Chemical Corp.
Gramercy, Louisiana

OPTIMISATION OF A BAYER FLOWSHEET

BY SUCCESSIVE SIMULATIONS AUTOMATICALLY GENERATED

A. ADJEMIAN and P. COLOMBE

ALUMINIUM PECHINEY
Centre d'Etude de l'Alumine
13120 GARDANNE
FRANCE

A computer program package has been developped to obtain, for a given production rate, optimal values of the operating parameters which significantly affect the economy of the plant, taking into account all the process and technological constraints of the system.

Operation of the plant at a stationary state is simulated with a sequential modular approach flowsheeting system composed of all classical unit operations suited to represent a Bayer process. Optimisation is carried out using a direct search method - namely the Simplex method of Nelder an Mead - and combining the objective function and inequality constraints in a single modified function by the use of penalty functions. A simple example is described in the paper.

Introduction

Operation of a Bayer plant under optimal conditions whith respect to operating cost implies a multi-level hierarchical control system, part of which is a supervisory program to determine the optimal values of the operating parameters which contribute to the economy of the plant (1).

In this respect, a supervisory program which consists essentially of a flowsheeting system coupled to an optimisation routine, is described below. It computes, by simulation, the material and energy balance of the whole plant and determines the optimal values of the operating parameters or set points, taking into account any process and/or technological constraint.

Flowsheeting System

General Features

Our flowsheeting system (2) is a sequential modular flowsheeting program for any Bayer-process and is essentially made up of two parts :

- the unit modules
- the system executive.

The input to and ouput of our package are handled by the user through special programs.

Every stream in the process is a sixteen-dimensional vector providing all mass components that are important in the Bayer-process, namely alumina, caustic soda, carbonate, calcium oxide, silica, water and other compounds contents and diverse properties like temperature, volume, specific enthalpy, density and weight ratio of alumina over caustic soda called RP.

Unit Modules

A Bayer-cycle consists of several plant-units (digestion, hydrate precipitation, evaporation, steam power plant, etc ...) which can be further divided into different pieces of equipment.

Presently, the library of unit modules contains 22 items grouped as follows :

- mixers, splitters
- unit representing a thickener or a washer
- units with temperature, or pressure change
- reactor units
- unit representing a heat-exchanger
- units representing the operations of filtration and washing on filters
- evaporation unit
- power plant unit
- controller for regulation

Each unit module computes unit output stream values given unit input stream values, specific operating data and equipment parameters. A Newton-Raphson type iterative procedure is used whenever the heat and material balances are represented by a collection of nonlinear equations to be solved simultaneously.

The evaporation unit calls for some comments ; a black-box model of a multi-effect evaporator with condensate and liquor flash tanks, liquid mixing/division vessels and vapor mixing/division vessels is complicated and time consuming to solve. Therefore a black-box model was run to produce data so that a central-composite orthogonal design (3) in seven dimensions might be used to find quadratic representations of the data.

The System Executive

It is the leading part of the program. After reading the data which describe the functional block diagram of the flowsheet, it checks the coherence of all the conventional input and tear stream values. Then it asks for the information concerning the optimisation, namely

- the objective function to minimize
- the decision variables and their limits
- the constraints.

Computation starts, then, calling sequentially the routines of the unit module in a sequence order selected by the user at the elaboration of the functional block diagram. Moreover the system executive provides the means to drive the tear streams to convergence.

The last function of the system executive is to call the optimisation routine for getting a new set of values of the decision variables to restart the calculations with, whenever a complete solution of the flowsheet is obtained, unless the optimum is reached. All the results are saved in a data file.

This system, very simple in fact, gives facilities for joining new user-supplied unit modules making our flowsheeting package quite evolutionary.

Control Blocks and Tear Streams Convergence

To be able to handle specifications on intermediate and/or output stream variable values (for example a fixed production rate or the caustic soda concentration of an intermediate flow) control blocks with information streams can be inserted as particular unit modules (4). An information stream is simply a means to transfer information from one unit to another where the information is not necessarily stream-related data. This procedure leads to creating a recycle in the flow diagram which is treated in the same manner as other tear streams. Fig. 1 shows for example, how the caustic soda concentration of a stream can be fixed by controlling the flow rate of another stream.

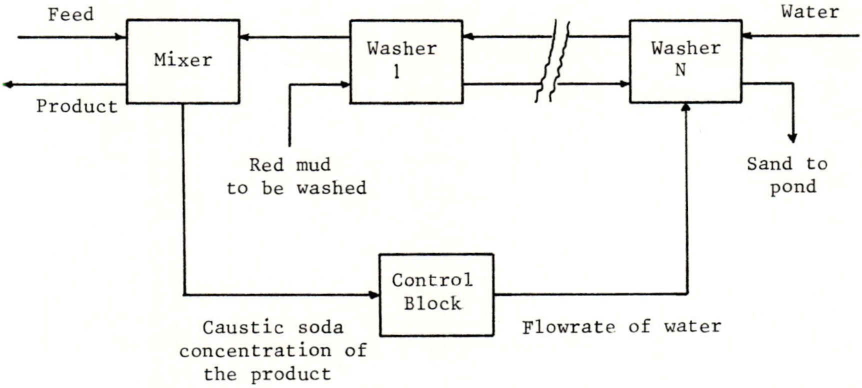

Fig. 1 : Example of control block usage. The caustic soda concentration of the product is controlled by acting on the flowrate of water.

Three well known convergence methods are provided :

- successive subtitutions
- Wegstein
- generalized secant.

In practice the Wegstein's method is exclusively used when the successive substitutions method appears to be somewhat slow (lets say, more than five iterations to get convergence) while the last one is mainly used for implicit iteration loops which is generally the case of loops associated to control blocks.

Our experience indicates that more sophisticated methods are not more efficient to solve our type of simulation problems.

Optimisation

The Optimisation Problem

A formal presentation of the problem is to minimize the objective function Φ (X, Y) subject to the following constraints :

$$x_{iL} \leq x_i \leq x_i^L \quad \text{for } i = 1, 2, \ldots\ldots, n \tag{1}$$

$$g_k (X, Y') \leq 0 \quad \text{for } k = 1, 2, \ldots\ldots, 1 \tag{2}$$

where X is the set of the decision variables x_i,
Y and Y' are sets of intermediate stream variables,
x_{iL} is the lower bound on the i-th decision variable,
x_i^L is the upper bound on the i-th decision variable,
g is a process and/or technological linear and/or non-linear inequality constraint.

The objective function is the operating cost of a working plant ; it can be written as below :

$$\Phi\ (X, Y) = \sum_{j=1}^{m} y_j\ c_j \tag{3}$$

with c_j representing the cost of raw materials and/or power and utilities expressed in Francs/unit of measure and y_j being the flowrate of the item of cost c_j.

The values of y_j for j = 1, 2, ..., m, which depend on the values of the decision variables x_i, are the result of a flowsheet calculation initiated from the corresponding set of x_i values.

The algorithm

The simulated flowsheet may be highly constrained and discontinuous, because the unit models used may contain logical switches rather than mathematically continuous relationships. An optimisation expert would probably select, among recent and complex techniques, the one leading to the minimum computer time while our criteria are simplicity of application, robustness and reliability. Therefore, in agreement with several authors (5, 6), we have chosen a direct search method, namely the version of the Sequential Simplex method modified by Nelder and Mead (7). In this method, the simplex is permitted to alter in shape and thus no longer stays a simplex. So a function of n independant variables is minimized using (n+1) vertices of a flexible polyhedron. By means of either expansions or contractions the size and shape of the flexible polyhedron is scaled to fit into the particular topology of the problem being solved.

The transformation

$$x_i = x_{iL} + (x_i^L - x_{iL})\ \sin^2 z_i \tag{4}$$

allows a search in z_i to be made, all the x_i being bounded automatically. Z becomes the set of the transformed decision variables z_i.

Constraints given by equation (2) are handled by including them within a modified objective function F using penalty functions, converting our initial restricted problem into one which is unrestricted.

The problem is reduced to minimizing the modified objective function

$$F = \Phi (Z, Y) + \sum_{k=1}^{l} P_k \, g_k^2 (Z, Y') \, H (g_k) \qquad (5)$$

where $H (g_k)$ is the unit step function defined so that

$$H (g_k) = 1 \text{ for } g_k > 0$$
$$H (g_k) = 0 \text{ for } g_k \leq 0 \qquad (6)$$

and the P_k are large positive penalty constants.

Example for plant operating cost optimisation

Description of the Bayer plant simulated

The flowsheet corresponds to conventional European process technology for production of alumina.

Bauxite is digested at high temperature in a series of autoclaves. The bauxite charge, in agreement with digestion kinetics data, is set to obtain a desired weight ratio RP after digestion.

After diluting to a set value of caustic soda concentration the digested slurry, the red mud is settled and washed in successive stages.

The solid concentration in the washer underflows goes from about 400 to 300 g/l.

After a security filtration, the settler overflow is cooled in successive stages of exchangers to an appropriate temperature before entering the trihydrate precipitation unit. The cooling mediums can be a washer overflow, spent liquor and water when necessary.

Feed to the alumina trihydrate precipitation is the aluminate liquor mixed with precipitated alumina trihydrate to a solid concentration of 400 g/l of mixture. Precipitation is achieved in a series of crystallizers at decreasing temperatures.

The product hydrate wash water ratio is equal to 1 t/t Al_2O_3.

Part of the spent liquor is concentraced in a 6-effect falling- film evaporator with live steam available at a pressure equal to 5 kg/cm2.

Caustic carbonate concentrations of liquors in the cycle are kept to reasonable values by treating a washer overflow with calcium hydroxide added in such a quantity that an excess expressed in CaO be present for reacting.

The power plant consists of three vapor-generation plants in operation as follows :

- two of them work at fixed load to produce high pressure steam, one burning coal the other burning natural gas. Only a small fraction of that steam is expanded to a pressure of 5 kg/cm2.

- the third one is operated at variable load to produce high pressure steam which is expanded to a pressure of 5 kg/cm2 through a steam-turbine generating power. The boiler furnace is fired with natural gas and fuel oil.

Optimisation problem

The operating cost function to be minimized for the Bayer plant described above is given by :

$$\Phi (X, Y) = \left(\sum_{j=1}^{6} y_j c_j \right) /Prod \qquad (7)$$

with Φ the operating cost function expressed in Francs/Mton of alumina produced,
y_j the consumption of
- coal for j = 1
- fuel oil for j = 2
- natural gas for j = 3
- kWh for j = 4
- CaO for j = 5
- NaOH for j = 6

c_j the corresponding prices,
Prod the alumina production rate in Mton/h.

Three decision variables are selected :
x_1 : the caustic soda concentration of the aluminate liquor, in g Na_2O/l
x_2 : the caustic soda concentration of the liquor used for bauxite digestion, in g Na_2O/l
x_3 : the feed flow rate to the evaporation plant, in m3/h with the following limit values :

$$180 \le x_1 \le 220 \qquad (8)$$

$$140 \le x_2 \le 170 \qquad (9)$$

$$225 \le x_3 \le 325 \qquad (10)$$

Technological and process constraints to be satisfied in our example are related to the following parameters :

y'_1 : the slurry flow rate fed to the digestion plant in m3/h,

y'_2 : the caustic soda concentration of the sand slurry to pond, in g Na_2O/l

y'_3 : the temperature of the liquor leaving the first effect of the evaporation plant, in deg. Celsius
y'_4 : the evaporator live steam pressure, in kg/cm2.

Thus requirements are :

$$g_1 = y'_1 - 450 \leq 0 \qquad (11)$$

$$g_2 = y'_2 - 5 \leq 0 \qquad (12)$$

$$g_3 = y'_3 - 135 \leq 0 \qquad (13)$$

$$g_4 = y'_4 - 5 \leq 0 \qquad (14)$$

Our optimisation problem is so reduced to minimizing the function :

$$F = (\sum_{i=1}^{6} c_i y_i) / Prod + 100 \sum_{k=1}^{4} H(g_k) g_k^2 \qquad (15)$$

Results

The plant simulated in our example led to a functional block diagram containing 46 functional modules with more than 130 interconnecting streams, 18 of them being tear streams.

Results, given for an alumina production rate fixed at 35 Mtons/h, are displayed in Table I.

Table I. Search Stages

Point j	x_{1j}	x_{2j}	x_{3j}	Relative values of the function to minimize	Points in Simplex
1	155.7	190.2	263	1	
2	163.9	192.9	270	1.0265	
3	159.0	196.4	268	1.0124	
4	157.9	192.9	292	1.0061	1 2 3 4
5	160.9	193.0	272	1.0139	1 5 3 4
6	154.0	193.2	276	1.0010	1 6 3 4
7	149.7	184.8	294	.9865	1 6 7 4
8	148.5	185.9	264	.9878	1 6 7 8
9	146.1	180.1	268	.9793	1 9 7 8
10	142.3	180.0	289	.9801	10 9 7 8
11	143.4	180.6	303	.9798	10 9 7 11
12	146.5	181.1	291	.9789	10 9 12 11
13	143.7	180.0	288	.9778	13 9 12 11
14	144.3	180.0	293	.9779	13 9 12 14
15	142.5	180.1	322	.9766	13 15 12 14
16	142.2	180.1	310	.9775	13 15 16 14
17	140.6	180.2	325	.9753	13 15 16 17
18	142.6	180.0	308	.9772	18 15 16 17
19	142.0	180.0	316	.9766	18 15 19 17

Values of the decision variables corresponding to point 0 represent the base point of the initial polyhedron. That base point has to be a feasible one. The next three points correspond to the other vertices of the initial polyhedron which must be located in the feasible region. Coordinates of those points are obtained from the coordinates of the base point. The initial polyhedron is not a regular one to avoid violation of constraints g_1 and g_3.

The minimum of the operating cost - .9745 - is found at the 26th stage of the search for x_1 = 140.4, x_2 = 180.0 and x_3 = 325. It comes out that the only constraints being operative are g_3 and the constraints on the decision variables x_2, x_3.

Conclusion

In addition to all common applications enabled by any process flowsheeting package, our flowsheeting system developped for the simulation of the Bayer process and related processes, combined with a simple direct search optimisation technique has demonstrated its ability, efficiency and reliability for solving complex and highly constrained problems such as the minimization of the operating cost of an existing Bayer plant, hereby the operation of a plant, taking into account all constraints which are, in most cases, time-dependent.

References

(1) G. Lang, M. Matula, B. Nemeth and F. Sitkei, "Process Control System of Bayer Plant", paper presented at TMS-AIME Annual Meeting, Chicago, Ill., Feb. 1981.

(2) A. Adjémian, "Simulation des Traitements Chimiques et Hydrométallurgiques des Minerais", Industrie Minérale - Série les Techniques, 1(1983) pp. 53-56.

(3) G.E.P. Box and K.B. Wilson, "On the Experimental Attainment of Optimum Conditions", Journal of the Royal Statistical Society, 13B(1951) pp. 1-45.

(4) A.W. Westerberg, H.P. Hutchison, R.L. Motard an P. Winter, Process Flowsheeting, Cambridge University Press, Cambridge, England, 1979.

(5) L.M. Rose, "Some Comments on the Choise of Optimization Algorithms for Engineering Optimization Problems", paper presented at the Third International Meeting "Informatique et Génie Chimique", Paris, France, April 1983.

(6) Private communication, D.C. Stocker, "A Comparative Study of Nonlinear Programming Codes", M.S. Thesis, University of Texas, Austin, Texas, 1969.

(7) J.A. Nelder and R. Mead, "A Simplex Method for Function Minimization", Computer Journal, 7(1964) pp. 308-313.

STEADY STATE SIMULATION OF THE BAYER PROCESS

D.R. Chinloy

Alcan International Limited
Kingston Laboratories
P. O. Box 8400
Kingston, Ontario, K7L 4Z4
CANADA

R.K. Holzwarth

Alcan International Limited
Alcan Engineering Division
P. O. Box 6090
Montreal, Quebec, H3C 3H2
CANADA

FLOW I is a sequential modular flow sheet program designed to assist the process engineer in computing steady state heat and mass balances for processes involving Bayer liquor, steam and solids. By being Bayer process specific a high degree of computational efficiency is achieved.

This paper describes the general capabilities and limitations of FLOW I and demonstrates its usage through a specific example. Convergence techniques to handle the large number of recycle streams typical of a Bayer plant are also discussed.

This simulation package is being widely used for conceptual designs, and modification or optimization of existing operations.

Introduction

The Bayer process, the caustic digestion of bauxite, has been and continues to be the dominant method for the production of alumina, accounting for essentially all of the Western world's production. In the seventies, sharp increases in energy and bauxite costs, and in capital equipment costs and interest rates gave incentive to the industry to seek ways of reducing alumina production costs. In parallel, the steady-state simulation packages (eg. FLOWTRAN, PACER, CHESS...) were becoming commercially available, after proven success in the multi-component multi-product petrochemical industry. The question was naturally posed, "could the Bayer process be modelled accurately? and, if so, could the resulting model lead to process optimization of existing processes, and improved design of new plants?"

The Bayer Process

A simplified schematic of the Bayer process is given in Figure 1. Caustic liquor is pumped through a series of regenerative heat exchange stages and finally through a live steam-fed heat exchange stage. Ground bauxite is contacted with the heated liquor in digesters, and the hydrated alumina species are dissolved. The digested slurry is cooled, the heat being transferred to the incoming spent caustic liquor. The residue of the digestion process, red mud, is separated from the now super-saturated liquor, is washed with water in counter-current settling stages, and is ultimately disposed of. The remaining traces of red mud are removed from the super-saturated liquor by security filtration. This liquor is further cooled, then seeded with gibbsite crystals, the precipitate separated into product and new seed, and the product calcined. The now spent liquor is returned to digestion after caustic soda has been added to make-up for losses. Prior to caustic soda addition, part or all of the spent liquor may pass through an evaporator. Heat is supplied to the digestion circuit and to the evaporation circuit, usually from a captive boiler-house. Electric power must be supplied to all parts of the process for pumping. Co-generation of steam and electricity may be practised and increases the level of alumina plant/energy source integration.

With regard to the steady-state modelling of this process, the following points are noteworthy :

(1) The number of recycle loops

(2) The close interaction between the process areas, more especially if there is co-generation of steam and power.

System Design

For a company such as ours, which employs one chemical process, the following limitations could be applied to any computer model without compromising its usefulness :

- the model could deal only with that one process, i.e. the Bayer process

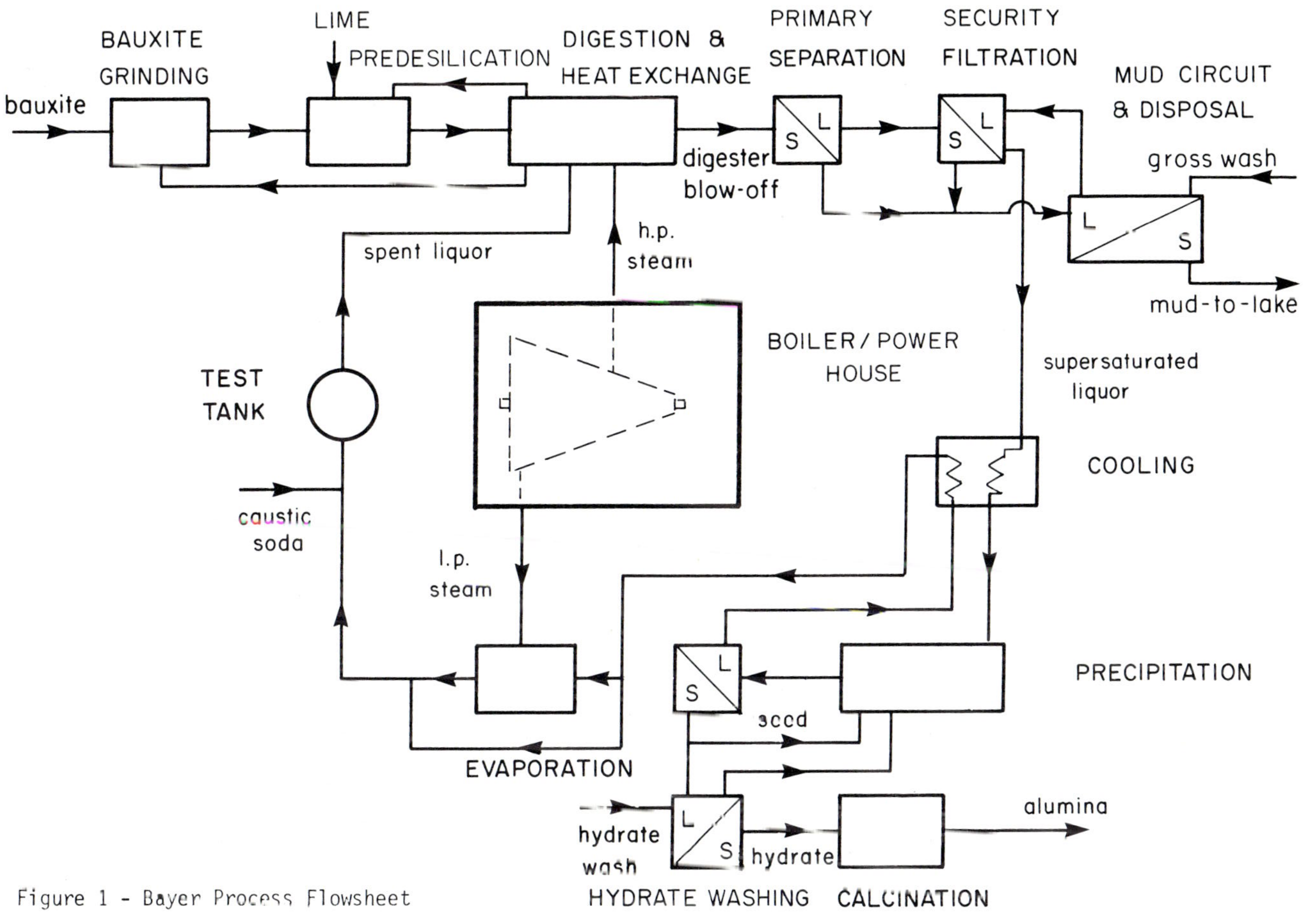

Figure 1 - Bayer Process Flowsheet

- the model should be flexible

- the user would be presumed to be an experienced process engineer (with little modelling knowledge), or a novice guided by an experienced process engineer

Further, it would be desirable if:

- the keywords in the model approximated that of the process engineer

- model convergence required process intuition rather than mathematical knowledge

- the model was easy-to-use, and interactive.

FLOW I

FLOW I is a sequential modular flexible flowsheet program, which has been implemented in FORTRAN. Modules are provided which perform heat and mass balance calculations for unit operations normally encountered in the Bayer process. To use the program, an operation list is written which specifies how the unit operations are connected through process streams and the order of unit operation calculation.

Streams

Two stream types are supported, slurry and saturated steam.

Slurry Streams: Streams of liquor and/or solids are represented by slurry streams. The stream parameters stored are,

- Temperature
- Vapor pressure
- Solids density
- Solids specific heat
- Enthalpy flowrate - Total stream
- Mass flowrate - Solids
- Mass flowrate - Solvent H_2O
- Mass flowrate - Solute $NaAlO_2$
- Mass flowrate - Solute NaOH
- Mass flowrate - Solute Na_2CO_3

Solid and water streams are special cases of the general slurry stream.

Vapor Streams: Vapor streams are represented by saturated steam. The stream parameters stored are,

- Temperature
- Pressure
- Enthalpy flowrate
- Mass flowrate

Since super-heated steam has a negligible effect on the heat and mass balances in the Bayer process, the FLOW I assumption of saturated steam is reasonable. However, steam tables have been used for the companion program, SPOW, which models the Power House.

General Assumptions. The following general principles have been adopted for performing process calculations, although specific operations may over-ride these principles:

Solids: Stream solids are assumed to be a single species characterized by density and specific heat. Solid properties are assumed to be invariant with temperature and pressure, and vapor pressure is taken as zero.

Liquor: Empirical physical property routines relate liquor composition, temperature, density, enthalpy, and vapor pressure. The specific heat is assumed to be a function of composition only; the vapor pressure of the solutes is taken as zero.

Vapor: Property routines relating temperature, pressure and enthalpy of saturated steam are built into FLOW I.

Flowrates: Negative flowrates are, in general, valid and represent a reversal in flow direction.

Initialization: All stream parameters are initialized to zero. This eliminates the need for a separate input file, and reduces the number of keystrokes (and the chances of error) required to perform a simulation.

External Stream Representation. All stream input and output is in terms of external parameters which are dictated by North American industry convention. FLOW I automatically handles the translation between internal and external stream representations.

Operations

The operations available in FLOW I are summarized in Appendix 1. New process operations can be coded, as FORTRAN subroutines, and incorporated. In addition a number of important general purpose and calculation control facilities are included.

Interactive Operation. FLOW I operations may be read into computer memory from a disk file or, they may be entered and executed from the user's terminal. The latter facility allows the effect of process variables on a flowsheet to be investigated interactively.

FLOW I Subroutines. A subroutine facility is provided which allows mini-plants, such as the digestion area or the mud circuit area, to be written and separately de-bugged, before inclusion in the main flowsheet.

Reactions. A general purpose reaction operation, REAC, has been provided. The stoichiometric coefficients and the heat of reaction, based on an arbitrary key component, must be specified. The coefficients which are implemented for some common reactions are given in Table I.

Convergence

A typical Bayer flowsheet may have about three hundred streams. Due to the presence of many counter-current unit operations, about a

Table I. Stoichiometric Coefficients of some Common Reactions Implemented in REAC Operation

Reaction	Stoichiometric Coefficient in REAC				
	Solids	H_2O	$NaAlO_2$	NaOH	Na_2CO_3
dissolution of gibbsite $Al_2O_3 \cdot 3H_2O + 2NaOH \to 2NaAlO_2 + 4H_2O$	-1.529	0.706	1.608	0	-0.785
dissolution of boehmite $Al_2O_3 \cdot H_2O + 2NaOH \to 2NaAlO_2 + 2H_2O$	-1.176	0.353	1.608	0	-0.785
lime causticization $CaO + H_2O + Na_2CO_3 \to CaCO_3 + 2NaOH$	0.786	-0.321	0	1.429	-1.894
calcium aluminate formation $3CaO + 2NaAlO_2 + 7H_2O \to 3CaO \cdot Al_2O_3 \cdot 6H_2O + 2NaOH$	1.250	-0.750	-0.976	+0.476	0
desilication of kaolin $Al_2O_3 \cdot 2SiO_2 \cdot 2H_2O \to$ bayer sodalite	0.906	0.049	0	-0.755	-0.200

third of these streams have to be initialized. These characteristics may lead to computational difficulties. However, a series of heuristics have been developed which assure stability and convergence:

(1) The sequence of calculation of the operation modules in the FLOW I program follows the path of a major slurry stream through the process unit operations in the plant being modelled.

(2) The initial values of all unspecified streams are set at start-up values.

(3) In a model consisting of more than one mini-plant the whole model is closed, rather than closing the mini-plants individually and then closing the whole plant.

Convergence is rapid (less than fifty iterations). Consequently, only one convergence technique, successive substitution, is provided - which relieves the user of the responsibility of having to choose a convergence technique.

The above heuristics are rooted in Bayer process experience - plants can be started-up smoothly, and reach steady-state without danger of over-shoot.

Example

To illustrate the ease with which FLOW I handles the Bayer process, a typical mud-washing circuit problem is solved.

Multi-Stage Counter-Current Washers

10,000 kg/min. of digester blow-off, with 10 weight % solids, is washed in a series of four counter-current washers, each with an underflow of 10 weight % solids, 100 % washing efficiency, with a gross wash of 3 kg gross wash : 1 kg solids. The process flow diagram is shown in Figure 2 and the FLOW I flow diagram in Figure 3.

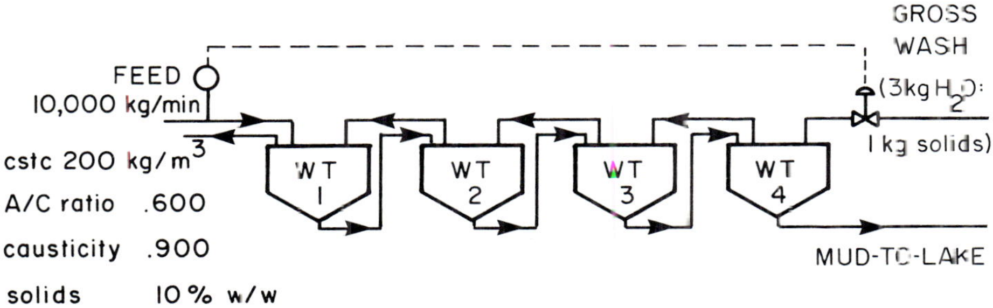

Figure 2. Multi-Stage Counter-Current Process Process Flow Diagram

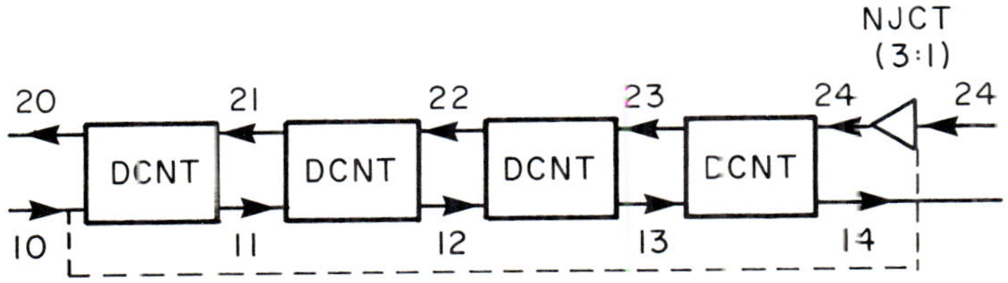

Figure 3. Multi-Stage Counter-Current Process FLOW I Flow Diagram

The actual terminal dialogue, showing the application of the heuristics, is given below:

```
.RUN FLOW01
ENTER OPERATION FILE NAME---->LIVE
***   specify feed streams
OP    1---->   LSTR 10 100,200,.6,.9,10,3,.25
OP    2---->   FLOW 10 10000.
OP    3---->   LSTR 24 25.
OP    4---->   NJCT 10,24,1 7 3.0
***          heuristic (2)
***   initialize recycle streams to start-up
***   values (gross wash stream 24)
OP    5---->   COPY 24 21,22,23
***          heuristic (1)
```

```
***    calculate mud circuit following the
***    mud-containing stream
OP   6---->010DCNT 10,21 20,11 10.
OP   7---->   DCNT 11,22 21,12 10.
OP   8---->   DCNT 12,23 22,13 10.
OP   9---->   DCNT 13,24 23,14 10.
***         heuristic (3)
***    test for convergence of mud circuit
***    rather than individual stages
OP  10---->   TEST 010 20,90 21,91
***    display streams 10 to 29
OP  11---->   SHOW 0 10 -29
```

STRM NO	LIQUOR KG/MIN	SOLIDS KG/MIN	CAUSTIC KG/M**3	TEMP. DEG C	V.PRES KPA
10	9000.0	1000.0	200.00	100.0	84.8
11	9000.0	1000.0	198.00	99.4	83.2
12	9000.0	1000.0	192.05	97.7	78.5
13	9000.0	1000.0	174.55	92.1	65.2
14	9000.0	1000.0	124.98	74.8	34.8
20	3000.0	0.0	198.00	99.4	83.2
21	3000.0	0.0	192.05	97.7	78.5
22	3000.0	0.0	174.55	92.1	65.2
23	3000.0	0.0	124.98	74.8	34.8
24	3000.0	0.0	0.00	25.0	3.2

Plant Simulation

About three hundred streams, and slightly fewer operations are required for an alumina plant quantitative flow diagram, and this is well within the capacity of most mainframe computers. FLOW I is currently implemented on a DEC-10KI and a VAX 11/780. This flowsheet program has been matched to existing plant conditions. Calibration is necessary as we have no *a priori* means of predicting parameters such as the overall heat transfer coefficients and stage washing efficiencies. The memory requirement for the Bayer process-specific FLOW I is a modest 120 kilo-bytes. A quantitative flowsheet for a Bayer plant with a low temperature digestion circuit requires 86 C.P.U. seconds (DEC-10KI) for convergence; while a high temperature digestion circuit converges in 193 C.P.U. seconds (see Table II).

Table II. Alumina Plant Simulation with FLOW I

	Low Temp. Digestion	High Temp. Digestion
No. of digester regenerative heater stages	3	10
Approx. No. of Streams	200	230
Approx. No. of Operations	270	280
Memory (kilo-bytes)	120	120
C.P.U. time (sec.)*	86	193
Approx. cost ($ U.S.)*	13	30

* DEC-10KI

The heat and material balance programs have been used to optimise raw material consumption rates, as well as the more mundane aspects such as indicating excess parasitic water.

Process Synthesis

FLOW I is part of the larger strategy to rapidly analyse the total cost of alumina production - the sum of the raw material, labour and capital costs. The logic diagram for conceptual alumina plant/power house synthesis is given in Figure 4.

The process is conceptualized in general terms, taking into consideration the mineralogical composition of the bauxite, the environmental constraints on red mud discharge, the type of alumina to be produced (viz. sandy or floury) etc. Process flow diagrams and their corresponding FLOW I flow diagrams are drawn. Design specifications consistent with good Bayer plant practise and with the particular physical and chemical characteristics of the bauxite and red mud are made (e.g. slurry feed temperature to liquor decanters should be less than the atmospheric boiling point, the bauxite mill discharge must be pumpable, the lime:bauxite charge must be adequate for phosphorous control etc). A FLOW I heat and material balance is made and the flow diagrams modified to meet the design specifications if necessary, and the heat and material balance recalculated until satisfactory (e.g. digester regenerative flashing/heating stage(s) may be added/deleted to adjust the liquor decanter feed temperature).

The electricity demand is estimated by comparison with other plants. A co-generating power house and a steam-generating boiler house (with purchased electricity) are designed with the help of SPOW to meet the process steam demands from FLOW I and the estimated electricity demand.

Knowing the required residence time in the various process vessels from laboratory data, and the flowrates and intensive properties of the streams from FLOW I and SPOW, a major equipment list is prepared. The capital costs for the major equipment is estimated from internal correlations, or by using external software houses. A battery-limit capital cost estimate of the process area is made using internally-developed installation factors, which take account of the complexity of the given process area. The capital costs of the individual process areas are summed, and infrastructure and overhead charges are factored in to give a study estimate of the capital cost of the plant.

From the raw material prices, the unit consumption rates from FLOW I and the energy consumption rates from SPOW, the raw material costs for alumina production are calculated. Estimating the labour costs and the capital charges, the total alumina production cost is calculated. The case study approach is used to estimate the minimum production cost.

At any juncture, the sequence of events in the logic diagram may be interrupted to allow the alumina plant/power house conceptual design to be modified, or expenditure-to-date compared to budget.

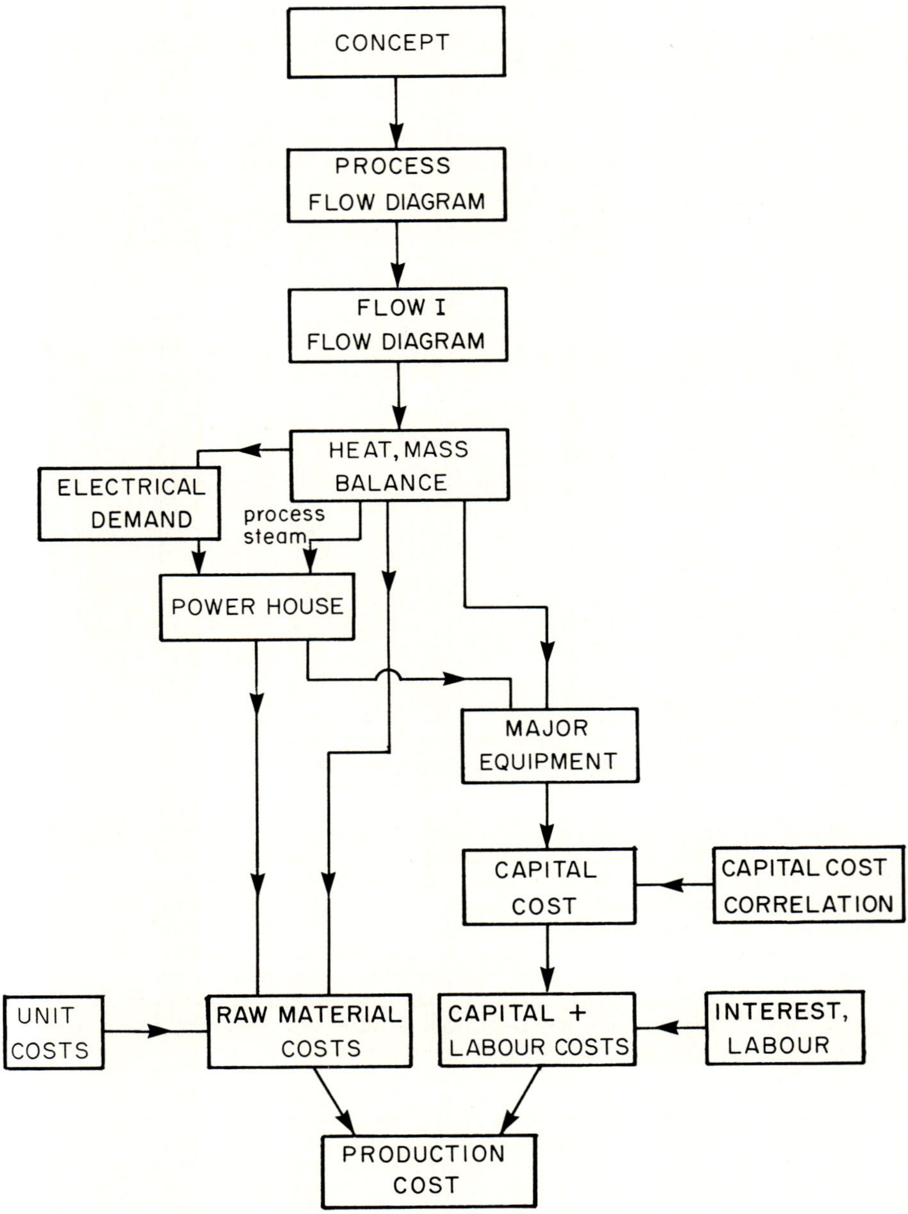

Figure 4. Logic Diagram for Conceptual Alumina Plant/Power House Synthesis

Discussion

Until recently, the commercially-available simulation packages were not suited to the Bayer process, as they did not handle solid streams nor solutions of electrolytes. Consequenly in-house Bayer process-specific models have been coded. A fixed flowsheet program was developed in the late sixties (3). As with all fixed flowsheet programs, the effect of changes in stream parameters on the overall process may be easily evaluated; however, the effect of changes to the flowsheet configuration are not readily evaluated. A flexible, Bayer process-specific model (BUOP), which overcomes the latter problem, was developed in the late seventies (4). Like FLOW I, BUOP represents vapor as saturated steam; however, unlike FLOW I, a companion Power House program is not available with BUOP.

More recently, the general-purpose simulation package, ASPEN, which is capable of handling solid streams and solutions of electrolytes, has been applied to the Bayer process (1,2). ASPEN is a very large program, requiring of the order of 2-5 mega-bytes of memory (1). For Bayer plants, convergence is slow due to the counter-current nature of the major process areas, and in particular the mud washing circuit. The slow closure prohibits this simulator from being run interactively.

On the other hand, with the application of the heuristics for convergence, FLOW I closes rapidly. Further an "interpretive" language is provided. Coding and running a FLOW I operation list is similar to coding and running a BASIC program. Precoded portions of the list may be called into memory, altered and executed; similarly new operations may be entered interactively, altered and executed. Thus, the effect of changes to process parameters and changes to flowsheet configurations can be examined speedily and interactively.

Conclusions

By being Bayer process-specific, FLOW I is a computationally efficient simulator that allows interactive use of the heat and material balance package.

Acknowledgments

The authors wish to acknowledge the role of Dr. J.-P. Riffaud and co-workers in the earlier development of Bayer process simulators and the helpful advice given by Dr. G.D. Fulford.

References

1. Chen, C.-C., et al, "Bayer Process Simulation by ASPEN", Light Metals, (1983) pp. 295-312

2. Shah, V.B., Gacka, P., and Langa, J.M., "The Application of ASPEN Flowsheet Simulator at Alcoa", A.I.Ch.E. Symp. Ser. No. 214, Vol. 78, (1982) pp 56-65

3. Siklosi, P., "The Mathematical Model of Material Flow of Alumina Production According to the Bayer Process", Proc. Second Intl. Symp. I.C.S.O.B.A., Vol. 3, (1971) pp 291-300

4. Steinemann, P., and Tschamper, O., "The Bayer Unit Operation Program (BUOP) and its Application for the Modification of the American Bayer Process", Light Metals, (1982) pp 187-198

Appendix 1. Operation Summary

The following is a brief description of the operations available in FLOW I.

Define Streams.

LSTR	Define liquor, solid or slurry stream
VSTR	Define saturated steam stream
COPY	Make copies of a stream
LOAD	Load streams from disk file created by operation SAVE

Modify Streams.

FLOW	Set stream flowrate
TEMP	Set stream temperature
PRES	Set stream vapor pressure

Combine Streams.

JUNC	Simulate adiabatic junction point
MAKE	Make up a stream of known flowrate from a prioritized list of input streams
NJCT	Control flow of injection stream to meet target specification post-injection

Separate Streams.

SPLT	Simulate flow splitter
DCNT	Simulate washed decanter
CLAS	Simulate classifier
FLSH	Simulate adiabatic flash

Heat Exchange.

COND	Simulate steam condenser
EXCH	Simulate counter-current heat exchanger
RGEN	Simulate flashing regenerative heat recovery stage

Chemical Reaction.

REAC	Simulate reaction specifying stoichiometric coefficients

Output.

SHOW	Display selected streams on user's terminal
LIST	List selected streams
SUMx	List summary report for area "x"
DETx	List detail report for area "x"
DISK	Route listings to disk file
TERM	Route listings to user's terminal
SAVE	Save selected streams on disk

Calculation flow control.

TEST	Test for convergence of recycle calculations
GOTO	Transfer control to a specified statement number
CALL	Transfer control to a specified subroutine
EXIT	Return control from a CALL'd subroutine
LIVE	Accept input statements from the user's terminal
FILE	Accept input statements from a disk file

Auxiliary Operations.

PARM	Set system parameter
SETP	Set operation parameter

OPTIMUM CONTROL OF BAUXITE CHARGE TO HIGH TEMPERATURE

DIGESTION UNITS AT QAL

P. McIntosh

Queensland Alumina Limited
Gladstone, 4680
Queensland, Australia

Summary

The current expansion of the Queensland Alumina Limited plant will increase maximum annual capacity from 2 438 000 to 2 736 000 metric tonnes. Previous expansions were based on increased plant flow and evaporation capacity. This 298 000 t/y increment relies on improved liquor productivity. A microprocessor-based instrumentation and control system has been installed to interface process computers to the digestion and bauxite grinding areas. Parameters, such as digester temperature, liquor and bauxite slurry flows, flash steam have been stabilised. Alumina/caustic (A/C) ratio and caustic concentrations of liquor entering and leaving each digestion unit are measured on-line. Mass balance calculations are performed to determine alumina equilibrium in the digesters, and bauxite charge adjusted to achieve the desired approach to equilibrium. This paper reviews the control hardware installed, sampling techniques, hierarchical control philosophy and productivity benefits achieved.

Introduction

The Queensland Alumina Limited (QAL) alumina refinery is sited at Gladstone, Queensland, Australia.

Alumina is produced on behalf of four shareholder companies, for smelters in Australia and overseas. Companies participating in the consortium are Alcan Aluminium Limited, Comalco Limited, Kaiser Aluminum and Chemical Corporation, and Pechiney Ugine Kuhlmann. QAL functions as a tolling operation, supplying alumina to the owners in proportion to their equity in the company for a charge per tonne which covers the cost of processing. Bauxite is sourced by Comalco from extensive deposits at Weipa in Northern Australia.

The initial plant was commissioned in 1967 with a capacity of 732 000 tonnes per annum. Since commissioning, the plant has undergone 3 expansions (Table I) with the fourth due to be commissioned March 1984.

Table I. Capacity Increase at QAL

	Maximum Capacity* Metric t/y	Comments
Initial Plant (1967)	732 000	Initial Digestion Unit
First (1968)	1 097 000	2nd Digestion Unit, 50% Flow Capability
Second (1971)	1 554 000	Upgrade 2nd Digestion Unit to Full Flow
Third (1973)	2 438 000	3rd Digestion Unit
Fourth (1984)	2 736 000	Improved Liquor Productivity

*Capacity figures rounded from original design in long tons (2240 lb).

The first three expansions were based on installation of additional equipment to provide increased plant flow and evaporation capacity and the attendant increased production. The current expansion relies on liquor productivity, with the 298 000 t/y capacity increase gained from improvements in precipitation yield, digestion yield and liquor purity.

This paper reviews the digestion yield component where the aim has been to maximise bauxite charge by improving approach to equilibrium in the digesters while maintaining high alumina extraction from bauxite.

Digestion Process at QAL

Bauxite Grinding

Weipa bauxite is a mixture of gibbsite and boehmite. A typical analysis is shown in Table II. The plant is designed to extract both alumina phases. Bauxite usage rate is approximately 2.2 tonnes of dry bauxite per tonne of alumina produced.

Table II. Typical Bauxite Analysis

Constituent	%
Total Alumina	55
Alumina Available in Process	49-50
$Al_2O_3.3H_2O$, Gibbsite	37
$Al_2O_3.H_2O$, Boehmite	13
Fe_2O_3	13
Total SiO_2	5-6
TiO_2	3
Moisture	13

Bauxite received from Weipa is pisolitic and sized to a maximum of 20 mm. It is stockpiled in the open, reclaimed as required and wet ground with spent liquor in two-compartment rod and ball mills to reduce particle size and provide greater surface area for extraction of alumina during digestion. Slaked lime is added for mud conditioning and phosphate control. Mill discharge slurry is heated and pumped to slurry holding tanks prior to charging to the digesters.

Digestion

Three parallel digestion units are installed. Design of each unit is typical of the Kaiser high temperature digestion process.

Spent liquor is pumped through 9 stages of heating in parallel trains of tubular heat exchangers and a further stage of heating by direct steam injection, following which it is pumped into the digesters.

Two vertical, agitated digesters operate in series on each unit. Holding time is typically less than 10 minutes. During the digestion process, liquor A/C ratio (mass ratio of alumina as Al_2O_3 to caustic as Na_2CO_3) increases from approximately 0.360 to 0.680 as alumina dissolves.

After leaving the digester vessels, slurry is flash cooled to atmospheric pressure by 10 successive flash stages, each operating at successively lower pressures. Flash steam is used to preheat incoming spent liquor prior to entry to the digesters.

Flash tank effluent is diluted with mud washer overflow liquor to control caustic concentration to the level required for the precipitation step, prior to being pumped to the mud settlers.

Historical Bauxite Charge Control

For given conditions, bauxite charge rates are constrained by the need to maintain a safe margin below the boehmite equilibrium solubility in the digesters. When overcharging occurs boehmite extraction is incomplete, leaving undissolved alumina to act as seed and precipitate alumina from solution, which results in increased bauxite usage, increased alumina loss in red mud and potential production losses due to filter press blinding.

Historically, a safety margin of 0.035 - 0.045 A/C ratio points from equilibrium was employed to specify the digester target A/C ratio so that risk of overcharging was minimised. This was necessary because there was no practical means of continuously monitoring or controlling normal process fluctuations which affect bauxite charge requirements. The short holding time in the digesters also contributed to poor extraction at low safety margins. The result was that bauxite charge, and thus alumina production rates, were significantly less than optimum.

Under the historical QAL system, spent liquor to each digestion unit was analysed 9 times per day for caustic and alumina. Analyses and relevant process data were used to calculate the bauxite charge rate required to meet target conditions. Digester effluent (pregnant) liquor was also analysed to check approach to target A/C ratio. Bauxite charge was then adjusted to the new calculated rate, generally 1 - 1.5 hours after the samples were first taken.

This type of control required many simplifying assumptions. For example, it was assumed that conditions at sample time were representative of operations until the next sample, i.e. steady state operation was assumed for the 2-3 hour time period between samples.

The inadequacy of this approach to control is illustrated by examining typical time based data for spent liquor caustic concentration. Figure 1 compares an actual caustic concentration profile and ideal bauxite charge rate (broken lines) with those obtained using the historical philosophy (solid lines). It can be seen there is a period of undercharging of bauxite followed by considerable overcharging once caustic concentration begins to decrease, compounded by the time delay between sampling and bauxite charge adjustment. Fluctuations of other variables such as spent liquor A/C ratio, digester temperature, and spent liquor flow cause similar problems. Collectively, the fluctuations limit the approach to equilibrium.

Development of techniques for on-line measurement of caustic concentration and A/C ratio in spent and pregnant liquors have made it possible to continuously monitor liquor analysis. The on-line analysers, developed and proven over a number of years by Kaiser and QAL, employ electrodeless conductivity measurement to determine alumina content, along with density and sonic velocity to determine caustic content of the liquor.

On-line digital computers make it practical to closely monitor relevant process conditions and to match bauxite charge to the optimum rate.

Computer Based Bauxite Charge Control System

Following the decision to incorporate bauxite charge as a component of the plant expansion, the Foxboro Company was selected to supply the process management and control system.

System hardware and software design commenced late 1981 with development of functional specifications which detailed data base design and preparation criteria, operator displays, supervisory control programs, reports to be printed either routinely or on demand, and system security requirements.

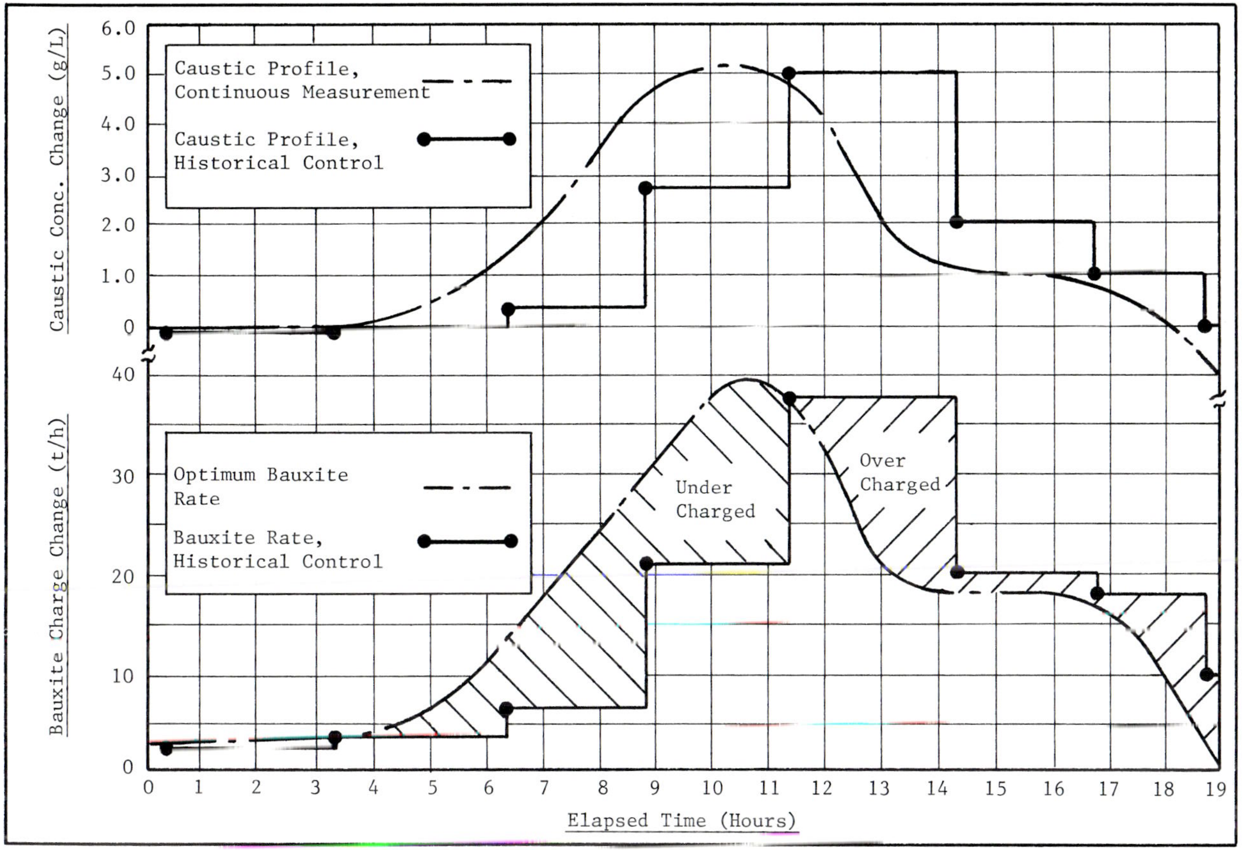

Figure 1. Historical Versus Optimum Bauxite Charge Control Systems

The existing panel mounted instrumentation, 60% pneumatic (installed 1966-71) and 40% electronic (1972-73), was no longer supported by the supplier, and could not be interfaced to the process computer system. It was therefore replaced.

Pneumatic, electronic and thermocouple signal transmissions were determined to be generally of acceptable standard. Extensive use of P/I and I/P transducers enabled pneumatic signal systems to be retained. Some critical loops required upgrading to modern electronic.

The final system, illustrated schematically in Figure 2, interfaced 110 control loops and 965 inputs into the new Foxboro system.

The system design provides 3 levels of control hierarchy.

Regulatory Control. This incorporates dedicated single-loop electronic analog controllers (Foxboro SPEC 200) and multiloop microprocessor-based distributed controllers (Microspec). Each of these units access information from the process directly. Set-point information is accessed via Foxnet communications links. SPEC 200 loops are interfaced through Communication Control Modules (CCM), while Microspec interfaces directly to Foxnet. Operators communicate with loops via Videospec. Additional data acquisition is via Universal Field Multiplexers (UFM), which also interface directly to Foxnet.

Videospec consoles provide all the display, command, alarm and trending functions previously available on the analog control panel. Videospec consoles are incorporated into 2 operator workstations which also incorporate alarm annunciators, motor stop/start controls, computer consoles and 3-pen trend recorders.

Supervisory Control. The Fox 1/A computers access process information via the Foxnet communications links using the standard Foxboro proprietary software package IMPAC (Industrial Multilevel Process Analysis and Control), which integrates the functions of data acquisition, monitoring, control and man/machine communication. Standard features used in the QAL system include scan blocks for acquiring process data, computational blocks for performing arithmetic operations, and for initiating programs written in FORTRAN, and control blocks for performing standard control algorithms.

Process data is scanned across Foxnet at intervals of 5 to 120 seconds, depending on duty.

Set points for field controllers are calculated by linking standard IMPAC blocks to execute the required calculation sequence and/or control functions.

Alternatively user-written FORTRAN programs are used where calculation complexity exceeds IMPAC capability.

Set point update information is transferred to field controllers across Foxnet at intervals of 5 to 120 seconds, depending on the duty.

Fox 1/A color consoles provide all display, command, alarm and graphic functions for supervisory control.

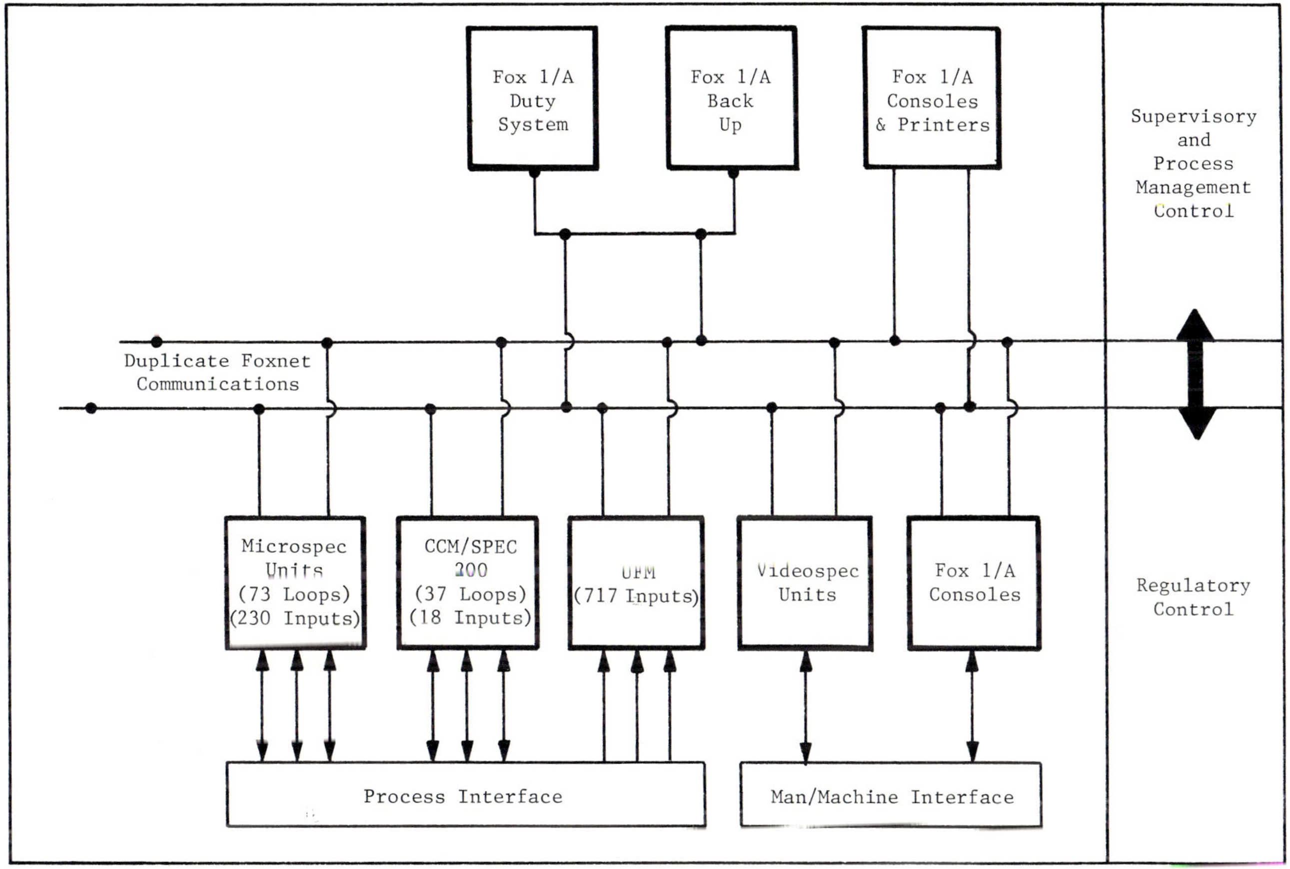

Figure 2. Schematic Outline of Computer Based Digestion Control System, Illustrating Hierarchical Control

Process Management. The Fox 1/A computers generate a hierarchy of reports and graphic displays from current and stored data. These are designed to assist operational, process engineering and management decision making. Technical support of operations by process engineers is enhanced by ready access to current and stored long term data and the ability to graphically and statistically analyse process interactions.

Software packages, to be implemented in the next phase of development, provide linear and non-linear optimisation techniques. These will be used to minimise total energy consumption across the 3 digestion units.

Installation and Commissioning

System hardware installation was completed May, 1983. The 3 digestion units have been running under supervisory control since early July, 1983.

Implementation of the computer-based system with its extensive data collection, storage, graphics and reporting capability, offered an immediate improvement in knowledge about control of the process. The system offered the opportunity to analyse in detail a large range of process variables, a facility not readily available within the previous analog control system. This revealed dynamic information about the process which showed it necessary to improve control and stability of process variables other than those directly related to bauxite charge.

For example, a study of the interaction between digester temperature, operating level in first flash tank, flow and temperature of spent liquor entering the digesters revealed each of these variables to be highly interdependent. This lead to the development of the multi-variable control strategies subsequently implemented.

Similarly, continuous measurement of pregnant liquor A/C ratio enabled the response of this variable to upstream disturbances to be measured.

Instabilities and interactions in regulatory control loops also became apparent. Control of flow of both spent liquor and bauxite slurry was improved, minimising variations in digester temperature and pregnant liquor A/C ratio which had previously been attributed to noise.

Supervisory Control Strategies

Spent Liquor Flow Control

Spent liquor flow rate is a prime variable in determining alumina production rate and must be held steady for good charge control and heat recovery on the units. The flow rate is a combination of spent liquor injected directly into the digesters and the spent liquor component of the bauxite slurry charged to the digesters. The latter figure is calculated using measured bauxite slurry flow rate and density plus calculated liquor density values. Subtraction of this value from the required total spent liquor flow defines the charge pump flow rate required to maintain steady conditions as bauxite rate is altered in response to other process changes.

Temperature Control

Flow rate and temperature of spent liquor exiting digestion heaters, flow rate and temperature of bauxite slurry, and digester temperature aim are used to predict, as a feed forward component, steam required to meet digester temperature aim. Comparison of actual temperature with aim temperature, as measured at the exit of the second digester vessel (dead-time approximately 10 minutes), enables feed back compensation to be applied to the feed forward prediction.

Weighfeeder Control

Nine bauxite grinding mills fed by 14 weighfeeders supply bauxite slurry. Bauxite rates across the weighfeeders are set by demand for bauxite as determined by the charge control programs and level trends in the slurry holding tanks which are used to buffer the process from mill fluctuations. A computer program distributes this demand evenly between the in-service grinding mills. Mill liquor is varied with bauxite rate and measured slurry density from each mill.

Flash Tank Level, Injection Heater and Injection Pump Control

Level in the first flash tank is regulated by varying steam supply to the injection heaters. This is achieved in two ways, regulating steam flow from the flash tank to control injection heater temperature and by varying injection pump suction pressure, thereby changing steam flow driving force.

Bauxite Charge Control

Bauxite charge rate is regulated to achieve a target approach to equilibrium conditions within the digesters. Since equilibrium A/C ratio varies with changes in temperature, caustic concentration, liquor contaminants etc., target A/C ratio must be varied to maintain a constant safety margin.

Factors which are taken into account to determine instantaneous bauxite charge requirements are:

- Mass flow of caustic soda into the digester.
- Mass flow of diluents such as water of crystallisation and free moisture in bauxite, steam injected into liquor for heating, lime slurry.
- Caustic losses within the digester, which primarily occur through reaction with silica contained in the bauxite.
- Mass flow of alumina in spent liquor.
- Digester temperature.
- Bauxite composition and expected alumina extraction.

Because of the number of inputs necessary to calculate mass and energy balances around the digester, and complexity of the balances themselves, a FORTRAN program has been developed to perform the balances on a repetitive basis. Since many of the variables are functions of bauxite charge rate, an iterative solution is necessary to solve the balances. A feed forward set point for bauxite mass charge rate is obtained from solution of the balances.

Feed back is necessary to correct for factors such as instrument drift and errors of assumption arising from changes in process conditions which can not be measured directly. Actual pregnant liquor A/C ratio, as measured by the on-line analyser is compared with the expected liquor A/C ratio, the latter being passed through dead-time (20-25 minutes) and filtering to simulate holding time in the digesters and flash tanks. Generation of a correction factor enables parameters in the mass balance to be modified so that actual liquor A/C ratio closely follows target A/C ratio as process conditioning change.

Benefits To Date

At time of writing (September, 1983), the system has been controlling the 3 digestion units for 3 months. Development of applications software, operator displays and reports is continuing in parallel with commissioning of new ancillary process equipment installed with the expansion project.

Significant improvement in process stability is evident, particularly pregnant liquor A/C ratio, digester temperatures, spent liquor flows and flash tank levels.

The feed forward/feed back bauxite charge strategy has proved particularly successful in stabilising pregnant liquor A/C ratio within tight limits (Table III). Since July, control of pregnant liquor A/C ratio, measured as standard deviation (aim A/C - actual A/C), has improved from ±0.013 to ±0.005 ratio points. Bauxite charge safety margins have been reduced to 0.010 - 0.015 as a result.

Table III. Improvements from Computer Based Charge Control

Parameter	Historical	Current
Safety Margin (A/C Ratio)	0.035 - 0.045	0.010 - 0.015
Pregnant Liquor (A/C Ratio)	Aim ± 0.013	Aim ± 0.005
Spent Liquor Flow (m^3/h)	Aim ± Unknown	Aim ± 6
Digester Temperature (^oC)	Aim ± 2.5	Aim ± 0.5
Heater Outlet Temperature (^oC)	219.0	224.2

The reduction in safety margin ensures that the required increase in alumina capacity will be met. Despite the reduced safety margin and the short holding time, alumina extraction from bauxite has improved due to elimination of excessive overcharging.

The combination of digester temperature, flash tank level, injection heater temperature and injection pump control strategies has dramatically improved operational stability and heat recovery on the digestion units. Standard deviation of digester temperature (aim - actual) has reduced from $\pm 2.5^{o}$ to $\pm 0.5^{o}$. Spent liquor heater outlet temperatures have increased 5°C on average compared with the best performance under analog control.

Energy savings resulting from increased yield and heat recovery exceed 700 MJ/t.

An added benefit is the maintenance of digester temperature and bauxite charge at aim conditions during major process upsets. Figure 3 presents actual plant data recorded during restoration of normal spent liquor flow following a bauxite outage on one digestion unit. Based on results from the on-line analyser (solid line), pregnant liquor A/C ratio was maintained within 0.003 of aim and digester temperature within 1°C of aim. This was in spite of changes in spent liquor caustic concentration of 25 g/L, changes in spent liquor A/C ratio of 0.017 and flow increases (not shown) totalling 850 m^3/h during a 14 hour period. Based on laboratory analyses every 2-3 hours (circles), A/C ratio was maintained within 0.006 of aim.

Other benefits are becoming increasingly evident with time. The data collection and disk based storage capability, coupled with graphics, history of operator induced changes and reporting provides a "window" on the process for tracking and solution of process and operational problems. It is now possible for Operations and Process Engineering personnel to correlate "cause" and "effect", removing much of the guesswork necessary under the previous analog control system.

Following a unit shutdown for maintenance, the rate of increase of digester temperature during start-up is now controlled from the computer using the temperature control strategy developed for normal operation. This generates a smooth temperature profile up to operating temperatures, reducing the risk of digesters shedding scale and plugging the flash train, and generally promoting a more rapid start-up.

Future Work

The control system offers the potential of optimising total raw material and energy costs associated with the digestion area utilizing standard optimisation packages resident in the computer. It is anticipated this work will commence early 1984.

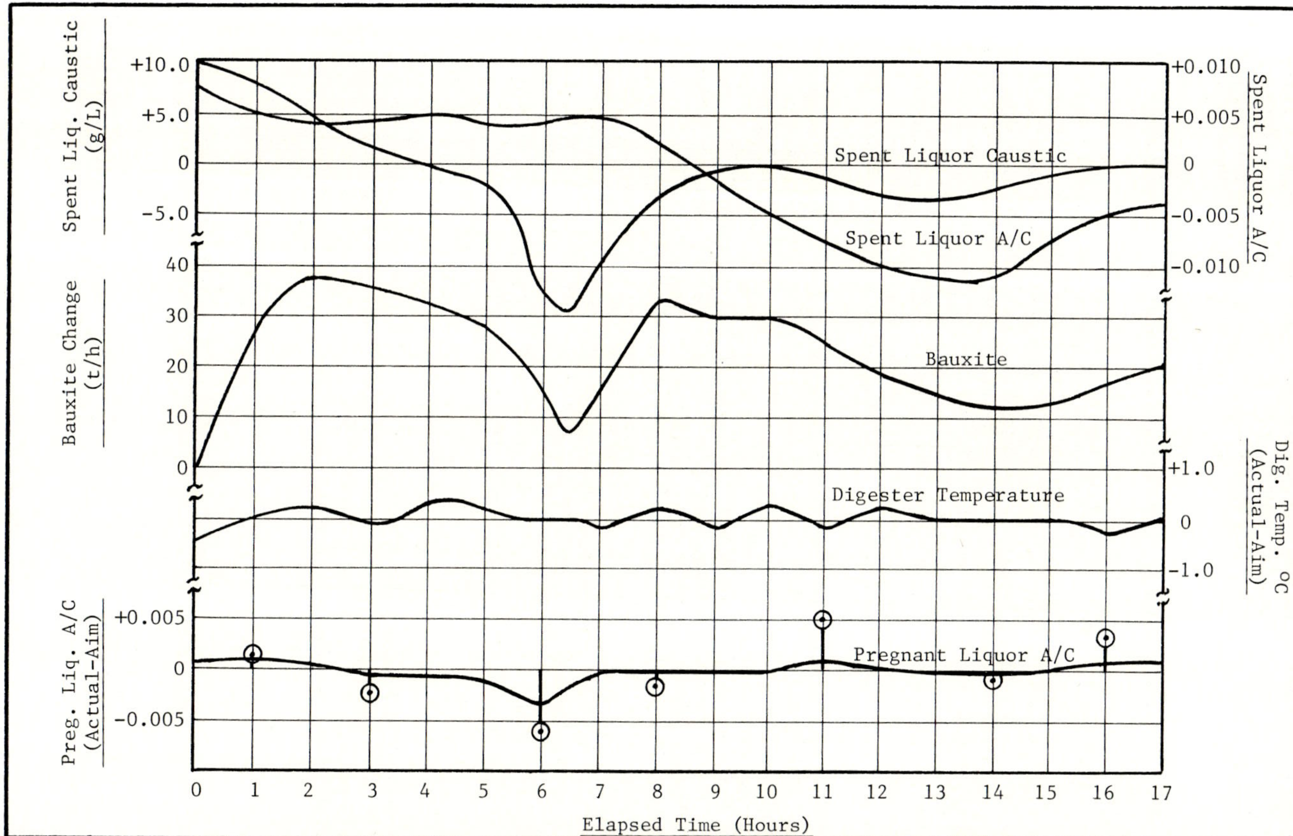

Figure 3. Digester Temperature and Liquor A/C Ratio Control During Major Plant Upsets.
⊙ Represents Laboratory Analysis of Pregnant Liquor Samples

IMPROVED PARTICLE SIZE ANALYSIS METHOD FOR PRECIPITATION CONTROL

John T. Creehan

Kaiser Aluminum and Chemical Corporation
P. O. Box 1031
Baton Rouge, Louisiana 70821
USA

Conventional methods for particle characterization of Bayer process aluminum trihydrate are time consuming and limited in completeness of information. A high degree of analytical skill is required for various instruments available for particle measurements, eliminating most from routine application in the control laboratory. One instrument that has qualified as a suitable replacement for the conventional methods is the extended range Microtrac particle-size-analyzer. Its advantage of operational simplicity and analytical ability to measure samples in native slurry form makes the instrument suitable for routine analyses of size distribution and specific surface. The instrument has replaced sieving and specific surface area methods to provide data for precipitation control at the Gramercy, Louisiana plant.

Introduction

Alumina producers worldwide are reacting to energy costs and environmental requirements by changing the Bayer process. Responses have been mainly to increase process liquor yield, to shift the product particle size to a coarser distribution, and to improve its quality for smelting (1).

Increasing productivity per unit volume of liquor depends upon a number of factors. One important factor affecting yield is the surface area of alumina trihydrate (ATH) seed made available for precipitation of soluble alumina in the saturated Bayer liquor (2).

Floury alumina conversion provides an incentive to retain high liquor productivity and effect physical properties characteristic of a sandy product (3). Quality product is alumina that meets minimum and maximum particle size specifications. Another quality factor is the amount of subsieve (<44 micron) alumina. The consensus on which size limit (22, 10, or 1 micron) defines dustiness is yet to be decided (1). Other quality parameters are adsorptivity of HF fumes and purity; such as, soda and calcium content (4).

The achievement of higher productivity and better quality depends upon the ability to monitor what is happening in the precipitation process, to interpret this data, and make proper control responses to produce economically a quality product. Conventional laboratory methods of particle size and surface area analysis are tedious and slow. This study results in the replacement of the conventional methods with Microtrac (5) which quickly provides data for control of precipitation yield and particle size.

Background and Objective

Certain parameters are used by the precipitation control engineer to maximize yield and maintain product quality. Decisions are influenced by the specific surface area (SSA) and size distribution of ATH seed and its precipitated product.

1. Seed Surface

The total SSA (cm^2/g) of seed recycled to the process must be known in order to maintain a target ratio of area to saturated liquor volume.

2. Precipitated Size of Hydrate Product

Greater than 74 micron (+200 mesh) cumulative percentage is an excellent indicator of precipitation trending for process control.

Greater than 44 micron (+325 mesh) cumulative percentage is also an excellent indicator of trends to coarser or finer distributions of precipitated ATH.

3. Precipitated Area of ATH

The fractional SSA of ATH representing 44 microns (-325 mesh) and below is an excellent indicator of nucleation or excessive production of fine particles in precipitation. The subsieve SSA may also be used as an indirect measurement of fines production by calculation of the mean particle diameter of that fraction.

Kaiser has provided this data in the past with a set of U. S. Standard sieves (100, 200, and 325 mesh) and the Fisher Subsieve Size Analyzer (FSSA). Samples presented for analysis by these methods must be deliquored, washed, and dried for accurate measurement. Although the actual analytical time is reasonable, the lapsed time from sampling to analysis completion is at least one and one-half hours.

Many instrumental methods are available for particle measurements, but most require highly skilled operators for effective use (6). Additionally, particles of ATH are not regularly shaped (Figure I) and are friable when sieved or packed for FSSA analyses. For these reasons and varying principles applied by instrumentation or equipment, the particle size distribution and SSA will be represented differently depending upon the particular method chosen for analysis (7). The method selected to replace sieves and FSSA must be precise, sensitively relate changes in particle size distributions and specific surface areas, and surpass the conventional methods in simplicity, speed, and scope of data production.

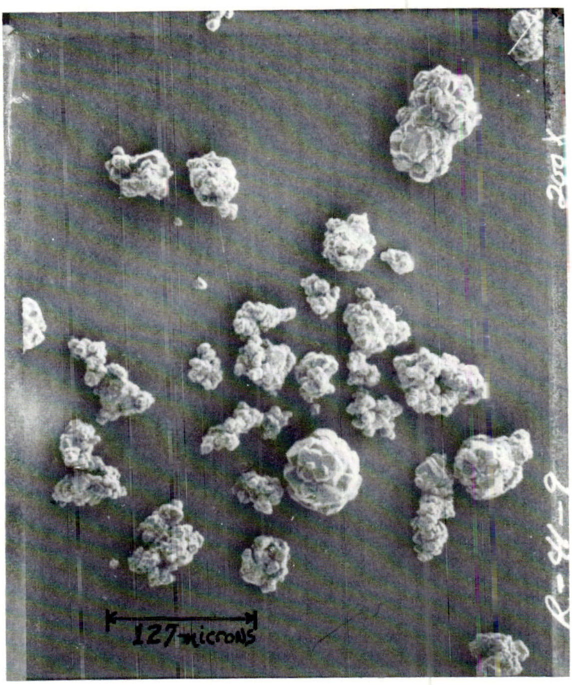

Figure I. Varied Shapes and Sizes of Alumina Trihydrate Precipitated in the Bayer Process.

Experimental Procedure

Leeds and Northrup Instruments manufactures an extended range (3.3 - 300 micron) Microtrac particle-size-analyzer. From forward scattering of its laser beam, passing through a flowing particle suspension, the instrument sizes particles and computes a distribution and overall surface area. Thirteen sensors, placed at a fixed distance and perpendicular to the axis of the laser, detect scattered light from the particles. Software converts the sensed light scatter into a cummulative percentage particle distribution and incremental and total surface areas (5). Evaluated by various screening procedures, Microtrac appears to have the greatest potential for replacing sieves and FSSA.

Suitability

Its suitability depends upon its capability to:

1. Measure the total and subsieve specific surface for seed charge and nucleation control.
2. Measure the complete particle distribution, including subsieve sizes, for product size control and also amplify nucleation trends.
3. Reproducibly measure and correlate well with certified sieves and the Fisher Subsieve Sizer to establish data credibility and benchmarks for the new methods.
4. Reduce the elapsed time of analysis greatly by making measurements directly on slurry samples from precipitation.
5. Produce reliable data on an uninterrupted basis for the process control engineer.

Discussion of Experimental Results

Subsieve (5 - 30 microns) Size Range - Precision

Reproducibility of the extended range Microtrac analyzer is compared with data on calibrated glass spheres used generally to calibrate equipment and evaluate methods for measuring small particles. The glass spheres are microscopically measured reference samples certified by the U. S. Department of Commerce.

Table I demonstrates Microtrac's reproducibility, which is actually better in this size range than that given by the tedious microscopic method used to characterize the glass spheres. The overall reproducibility expressed as percent relative standard deviation (% RSD) is 0.88% for Microtrac and 2.8% for the microscopic method.

Table I. Analysis Reproducibility Comparison*: Particle Sizing

Micron Size	NBS Certified Value			Extended Range Microtrac		
	Vol./%**	Std. Dev.	RSD	Vol./%**	Std. Dev.	RSD
27	-	-	-	89.4	0.36	0.40
25	94.5	1.0	1.1	-	-	-
20	79.5	1.2	1.5	-	-	-
19	-	-	-	63.2	0.24	0.38
17	58.0	1.8	3.2	-	-	-
15	39.5	1.0	2.5	-	-	-
13	23.0	1.1	4.7	33.4	0.10	0.30
10	9.0	0.3	3.7	-	-	-
9.4	-	-	-	16.2	0.14	0.86
6.6	-	-	-	7.98	0.06	0.75
4.7	-	-	-	3.12	0.08	2.6
Overall Precision			2.8%			0.88%

* Sample: National Bureau of Standards, SRM 1003, Glass Spheres
** Cumulative Percent Below Size

Table II. Analysis Reproducibility Comparison*: Particle Sizing

Micron Size	NBS Certified Value			Extended Range Microtrac		
	Wt./%**	Std. Dev.	RSD	Vol./%**	Std. Dev.	RSD
105	80.0	0.20	0.25	75.1	0.31	0.41
74	-	-	-	44.2	0.65	1.5
71	46.0	0.27	0.58	-	-	-
53	24.0	0.40	1.7	21.8	0.40	1.8
38	10.0	0.40	4.0	8.98	0.59	6.6
28	2.0	0.26	13.3	-	-	-
27	-	-	-	2.25	0.54	24.0
Overall Precision			4.0%			6.9%

* Sample: National Bureau of Standards, SRM 1004, Glass Beads
** Cumulative Percent Above Size

Sieve (34 - 120 microns) Size Range - Precision

In this range, the reproducibility is again compared and shown in Table II. Calibrated glass beads also certified by the Department of Commerce by microscopic measurement at the Bureau of Standards were also measured by Microtrac for comparison. Microtrac's reproducibility is similar to that given in the certificate of analysis; 6.9% RSD overall versus 4.0% for the certified analysis.

Certified Sieve Comparison - Precision

A sample of ATH was riffle-split into a large number of samples. Ten randomly selected samples were analyzed by the extended range Microtrac and sieves. The sieves were certified to meet ASTM E11-70 specifications. The overall reproducibility of the analysis of the Bayer hydrate sample was respectively 4.0% and 4.2% RSD for sieving and Microtrac.

Specific Surface Area - Precision

The SSA of ten randomly selected samples from an ATH composite was determined by the FSSA and Microtrac. FSSA is measured by the pressure drop across a packed sample bed. Reproducibility of the Microtrac is generally better than the FSSA for the entire sample or its subsieve fraction. See Table III. The means are expressed as cm^2/g.

Table III. Analytical Comparison of Reproducibility: Specific Surface Area

Portion \ Analysis	FSSA			Microtrac		
	cm^2/g	Std. Dev.	RSD	cm^2/g	Std. Dev.	RSD
-325 Fraction*	931	46	4.9%	1308	56	4.3
As-Is Sample	519	60	11.6%	314	10	3.2

* Sample: Alumina Trihydrate Analyzed after Passage Through a 325 Mesh Sieve.

Microtrac and Sieve Relationships

Sixty-eight precipitation slurry samples were analyzed by conventional sieving of a washed and dried portion of the slurry analyzed directly by Microtrac. As shown in Figure 2, the corresponding 325, 200, and 100 mesh measurements made by Microtrac correlate very well with the sieve data. The probability of correlation between the two methods was above 99.9% with an overall standard error of $\pm$ 2.2% absolute. The distribution of particles determined by Microtrac is slightly finer and coarser at those particular distribution ends than that characterized by sieving.

Microtrac and FSSA Relationships

Ninety-seven slurry samples were analyzed by both instruments. Microtrac determined the surface area directly on the slurry sample from precipitation. Figure 3 indicates that the Microtrac SSA is lower than FSSA; e.g., 646 versus 718 cm^2/g. Correlation between the two different methods is at the 98% probability level with a standard error of $\pm$ 95 cm^2/g. A number of samples in the comparison group analyzed had total areas below 496 cm^2/g, the measurement limit for the Fisher instrument. This bias certainly affected the standard error of the relationship since Fisher indicates no change for areas under this limit, while Microtrac is not limited.

Figure 3 also relates the specific surface area of the subsieve fraction of precipitated ATH product. Thirty slurry samples were analyzed by both methods to develop a correlation and linear regression equation relating them. The probability of correlation is at the 95% level with a standard error of $\pm$ 74 cm^2/g.

Range of Microtrac Particle Measurement

Microtrac analysis spans the regions covered by sieving, Celloscope, and FSSA. Figure 4 documents that conventional sieve analysis is limited routinely to sizes above 44 microns. Comparative data in Figure 4 is also plotted for Electrozone - Celloscope (Coulter principle) and the Microtrac particle size analyzers. The Electrozone - Celloscope is also known as the Elzone or Celloscope.

To produce a particle distribution of a similar range as Microtrac's, two Electrozone measurements, a blending procedure, and one hour are required. In addition to the long analytical time, a probable sampling error from less than 50 milligrams of sample must be considered. In ten minutes, Microtrac provides complete analysis, including complete particle size distribution, total SSA, and subsieve SSA from a forty gram sample.

Reliability of Microtrac Performance

To insure reliable data from day-to-day and over long periods of time, the extended range Microtrac is checked each day with an in-house reference ATH. When any deviation from the reference analysis is present, system checks are made; such as, laser alignment. This procedure has been followed since November, 1981. The instrument has been in continuous daily service from that time. Only three days were lost in the replacement of a faulty mechanical printer with an electronic one. Minor problems related to condensation on the sample cell lens (fogging) and background contamination have been eliminated. The instrument has provided data in an uninterrupted manner for over two years.

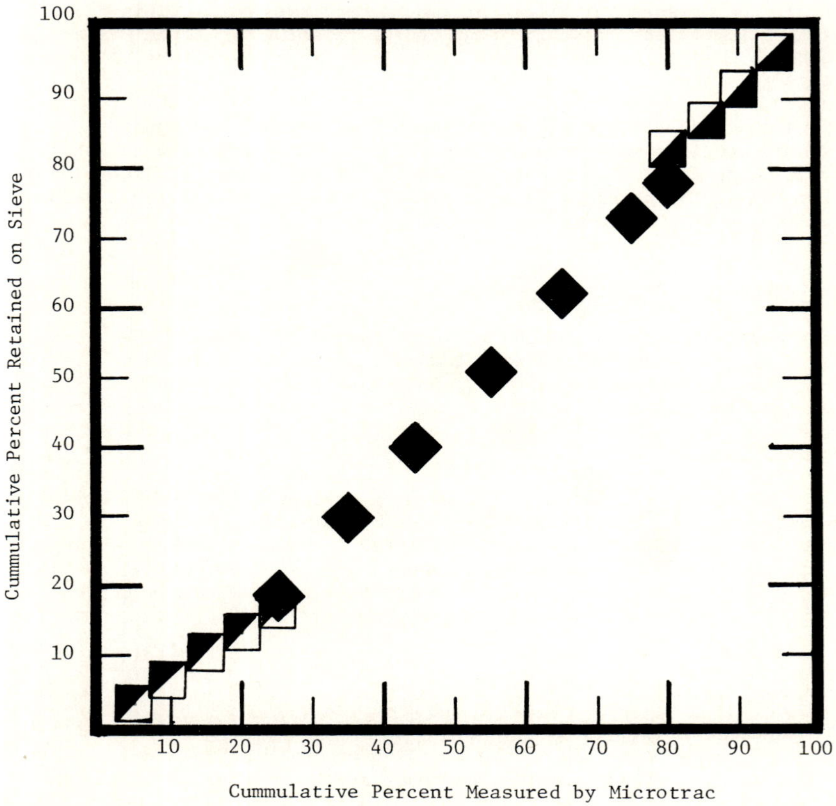

Figure 2. Ability of Microtrac to Predict Certified Sieve Data by Direct Analysis of Slurry Samples from Precipitation.

 Linear Regression Equation for 100 Mesh (149 micron) Relationship

$$y = 0.6501 \times MT + 0.1120$$

 Linear Regression Equation for 200 Mesh (74 micron) Relationship

$$y = 1.072 \times MT - 7.807$$

 Linear Regression Equation for 325 mesh (44 micron) Relationship

$$y = 0.9163 \times MT + 9.366$$

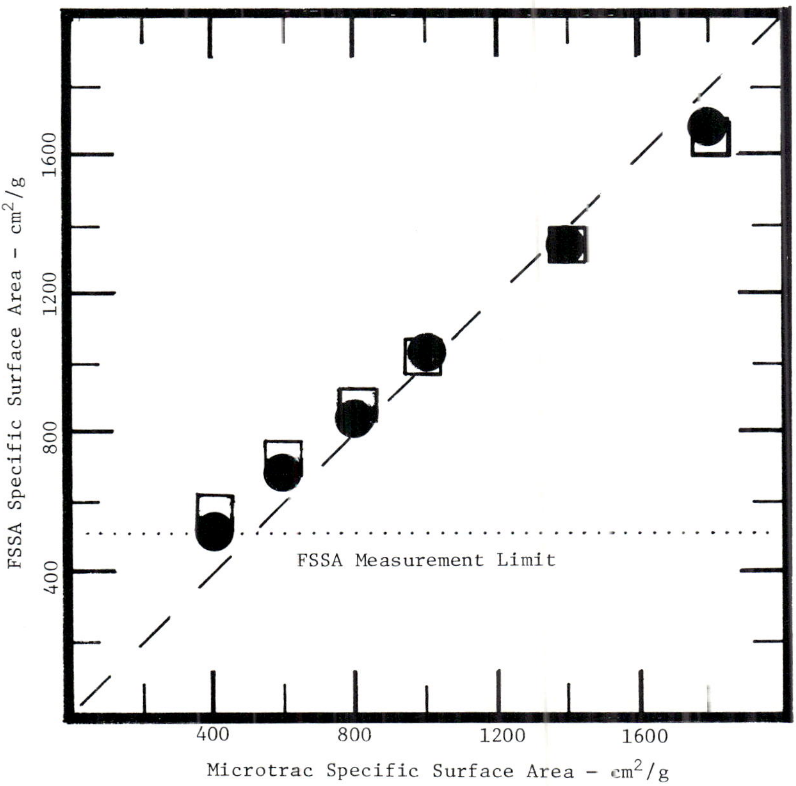

Figure 3. Ability of Microtrac to Predict Fisher Subsieve Sizer - Specific Surface Area.

□ Linear Regression Equation for Precipitation Slurry Particles Passing through 325 Mesh (<44 microns).

$$FSSA = 0.7742 \times MTSSA + 257$$

● Linear Regression Equation for FSSA Relationship on the Complete Sample from the Precipitation Unit.

$$FSSA = 0.8359 \times MTSSA + 178$$

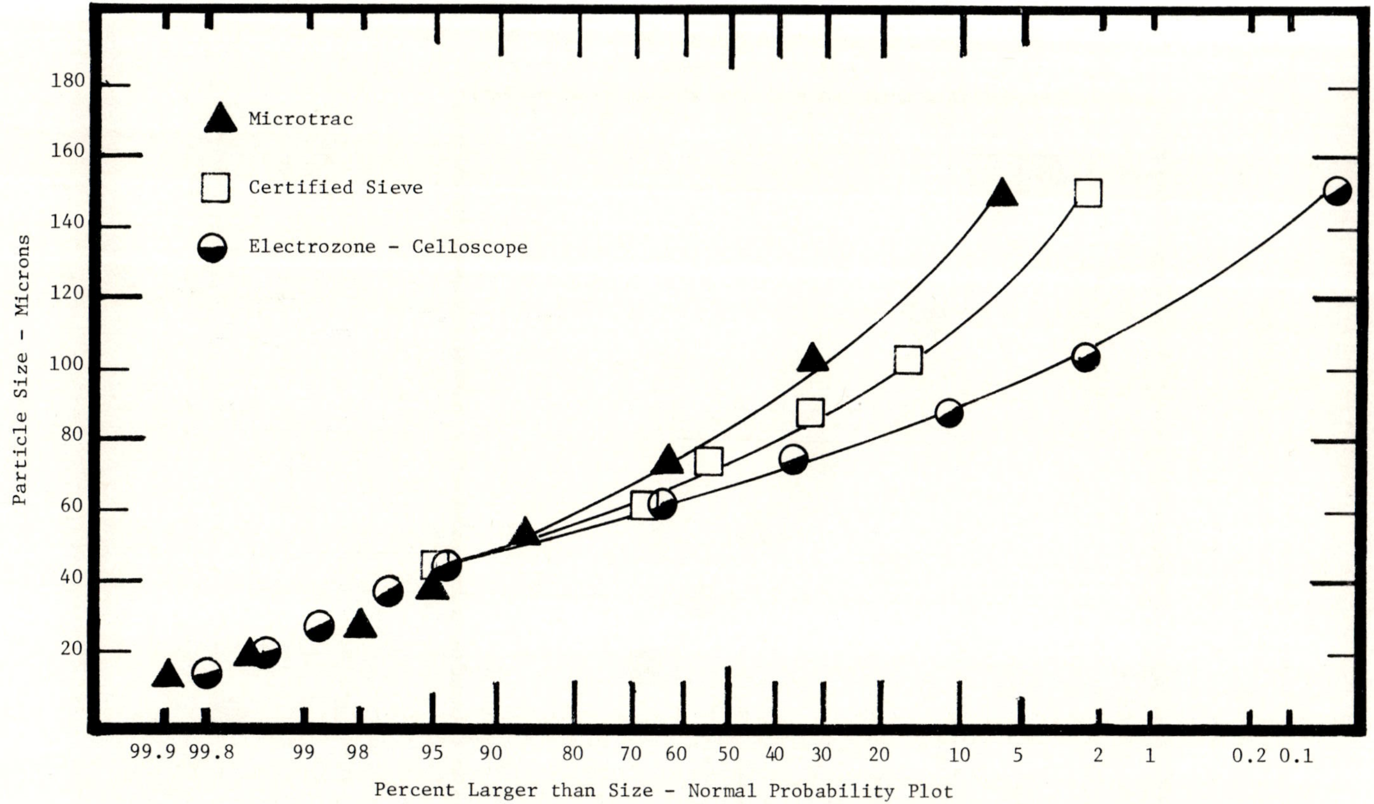

Figure 4. Comparison of Sizing Data from Extended Range Microtrac, Certified Sieves, and Electrozone - Celloscope for Precipitated Hydrate Sample (TT 52 TOP, TK 37 - 2200 Hours - 10/20/80).

Conclusion

Microtrac quickly provides precipitation control data in a reliable manner from direct analysis of precipitation slurry samples. Its reproducibility is comparable to sieving, the Fisher Subsieve Sizer, and the microscopic method. Its data relates to sieve data, but is not limited to sieve sizes. It measures particles of subsieve size as well as the Electrozone - Celloscope, but in a fraction of the time required for complete analysis of size distribution. Microtrac measures specific surface area of the total sample and its subsieve fraction, but is not limited to SSAs above 496 cm^2/g. It is a dependable instrument with a very low maintenance demand, easy to operate, and capable of fast sample analysis, providing data quickly to the process control engineer.

Nomenclature

ASTM E11-70	:	American Society for Testing and Materials. Specification for sieve (mesh) quality established by committee E11 in 1970.
ATH	:	$Al_2O_3 \cdot 3H_2O$, Bayer Hydrate, or Gibbsite.
cm^2/g	:	Centimeters squared per gram or the specific surface area.
Coulter Principle	:	Size measurement based upon displacement of conductive liquid between two conductors in a manner proportional to particle volume by measurement of the resultant voltage increase.
Cumulative Percentage	:	The fraction of sample above or below a mesh or certain micron size reported as a wt./% or vol./% of total sample.
FSSA	:	Fisher Subsieve Size Analyzer/Area.
HF	:	Hydrogen fluoride gas.
NBS	:	National Bureau of Standards/U. S. Department of Commerce.
MT	:	Microtrac cumulatively measured percentage of sample above or below a micron size.
MTSSA	:	Microtrac measured specific surface area.
% RSD	:	Measurement of precision as calculated from the standard deviation of repetitive analyses divided by the quantity measured and converted into a relative percentage. It is also known as a coefficient of variation.
Probability of Correlation	:	Based upon the correlation coefficient, it is probability of a relationship between variables. The closer it is to 100%, the less chance of being by chance not related to the variables.
SSA	:	Specific surface area in cm^2/g.
Standard Error	:	One standard deviation from the estimated or predicted value from the linear regression equation. It is expressed usually as an absolute percentage variation from the estimated value.
Vol./%	:	Percentage of sample based upon the ratio of fractional to total particle volume of a sample. If particle density is constant, it equals wt./%.
Wt./%	:	Percentage of sample based upon the ratio of fractional to total particle weight of a sample.
y	:	Cumulatively measured sieve wt./% of sample.

References

1. Errol A. Kirke, "Recent Trends in Bayer Precipitation Practice", Chemical Engineering in Australia, 7 (1) (1982), pp. 35-39.

2. T. G. Pearson, The Chemical Background of the Aluminum Industry, 3rd. ed., p. 28, The Royal Institute of Chemistry, London, 1957.

3. C. Sato and A. Furukawa, "Comparative Evaluation of the Economics of Floury and Sandy Type Bayer Alumina", presented at the Second International Symposium of ISCOBA, Volume 3, pp. 227-236, 1971.

4. M. Landi and R. Barchiega, "Alumina in Aluminum Smelting - Present and Future", presented at ISCOBA, Hungary, pp. 127-157, October, 1981.

5. E. C. Muly and H. N. Frock, "Industrial Particle Size Measurement Using Light Scattering", Optical Engineering, 19 (6) (1980), pp. 861-869.

6. Terence Allen, Particle Size Measurement, 3rd. ed., Chapman and Hall, London, 1981.

7. Wolfgang B. Pietsch, "An Evaluation of Techniques for Particle Size Analysis - Parts 1 and 2", Minerals Processing, November and December, 1968, pp. 6-11, and pp. 12-14, 24.

CONTROL OF PROCESS LIQUOR IMPURITIES AT SAN CIPRIAN

J.L. Monita Benito

Alumina Espanola, S.A.
San Ciprian
Cervo, Lugo, Spain

The San Ciprian Alumina Refinery is designed to process West African (Boké) bauxite with very low chemical losses. An effective concentration/crystallization system is installed to control the build-up of impurities in the liquor. This paper outlines the main characteristics of the system, and discusses the results over the first three years operation.

Introduction

The San Ciprian alumina plant is designed to process low silica West African (Boké) bauxite, using a high temperature low caustic concentration version of the Bayer process. The problem of impurity build-up in the circulating liquor was faced during the plant design period.

Boké bauxite contains about 0.1% organic carbon, in the form of humic acids and organic materials which partially react with Bayer liquor to form a series of sodium salts. Smaller quantities of organics are introduced into the process as floculating agents, anti-foam agents, etc. As the process liquor is recycled through the digestion system the impurities accumulate until an equilibrium is reached between input and losses. If the overall liquor losses are low, as they are in a well designed plant, the equilibrium values can be very high.

The long chain organic compounds introduced into the system degrade to produce the carbonate, formate, acetate and oxalate salts of sodium together with intermediate compounds. Some carbonate is also formed by reaction of atmospheric carbon dioxide.

The major problems which arise from the presence of these organic degradation products in the process liquors are:

a) Sodium oxalate precipitates in the coolest part of the circuit, increases the generation of fine particles, and interferes with the settling of the fines. The fines balance of the circuit is upset, potentially affecting product quality, and hydrate overflows the tertiary settlers. This, together with actions to correct product quality, decreases liquor productivity, and may cause problems in the heat exchange systems.

b) The organic materials inhibit the precipitation process, decreasing the precipitation rate, and causing a further loss in productivity.

c) Increasing quantities of sodium carbonate in the liquor (falling liquor causticity) may enforce a decrease in the caustic concentration to maintain the TTS within acceptable values, further decreasing productivity.

d) The high and medium molecular weight materials may cause foaming in precipitators and flashtanks.

e) The physical properties of Bayer liquor are affected by a high level of impurities:

Boiling point elevation
Specific gravity
Viscosity
Specific heat

Changes in these values will affect the performance of the plant.

There are many methods of impurity control available for industrial use, individually or in combination:

Oxalate crystallization columns
Oxalate precipitation by liquor destabilization
Oxalate removal by seed washing
Liquor calcination with bauxite or hydrate
Wetox high temperature oxidation
Carbonate removal by cooling
Carbonate causticisation in a dilute stream
Evaporative crystallization of spent liquor.

There is no general best system. The optimum process(es) for a particular plant will depend on the details of the process used and on the bauxite to be processed. The plant designer must choose an impurity removal system which is compatible with the plant design and operation.

For San Ciprian an evaporative crystallization process was chosen to control the build-up of oxalate, carbonate, and other process impurities.

Basic Process

The plant design called for a process liquor at 90% causticity, 3 g/l sodium oxalate and 200 g/l caustic expressed as carbonate. To maintain these conditions the removal rates required in excess of natural outputs are approximately 2 MTD Sodium Oxalate and 10 MTD Sodium Carbonate when operating at an alumina production rate of 800,000 MTY.

The principle adopted to achieve removal of these compounds is one of salting out the impurities by evaporating the process liquor to high concentration. Concentration takes place in a backward feed forced circulation multiple effect evaporator unit to a level just below that at which sodium carbonate begins to crystallize, and then the operation is transferred to a crystallizing battery where crystallization can take place in a controlled manner without risk of fouling the heating elements. A preliminary separation of solids is made by decantation, followed by a drying operation using solid bowl centrifuges.

Under design conditions the evaporator feed rate is determined by the need to control oxalate: excess carbonate is carried to the unit by this flow of liquor.

Detailed Design

The conceptual design and general specification of the unit were carried out in Montreal. To ensure maximum Spanish content the detailed engineering and equipment procurement were completed in Madrid. This caused some problems, which we will discuss later.

A diagram showing the major features of the unit is given as Fig. 1.

The evaporator unit consists of seven effects arranged in series for backward feed. Each effect is a forced circulation system, consisting of one flash tank, one external shell and tube heat exchanger, one axial flow pump for liquor recirculation, and one liquor transfer pump, together with the appropriate valving and piping.

The flash tanks are used to separate vapour from the entering superheated stream. Each consists of a cylindrical pressure vessel with dished head, 45° conical bottom, and tangential liquor entry. The vapour outlet of each tank is coupled to a vapour header running the length of the unit by an appropriately sized line. The tanks are designed to produce clean vapour without the aid of external separators or internal demister pads. Their elevation is determined by the need to provide adequate NPSH for the axial flow circulation pump.

The heaters are arranged horizontally, with fixed tube plates and shell expansion bellows. Tube velocity is set at 1.6m/sec. to minimize scale formation and secure good heat transfer coefficients. The heaters are located below the flash tanks, to prevent boiling in the tubes. They are connected to the vapour/live steam header on the shell side, and regenerative condensate is flashed from heater to heater. There are three separate flash stages for live steam condensate.

In operation the feed flash tank is fed with spent liquor, which flashes to the shell side of the last heater. The liquor flows by gravity into the last-effect, and is then pumped through the unit under level control.

Hot concentrated liquor passes from the first evaporator effect to the crystallizers by pressure differential, and is put through a temperature/concentration/time regime designed to promote coarse easily separable crystals.

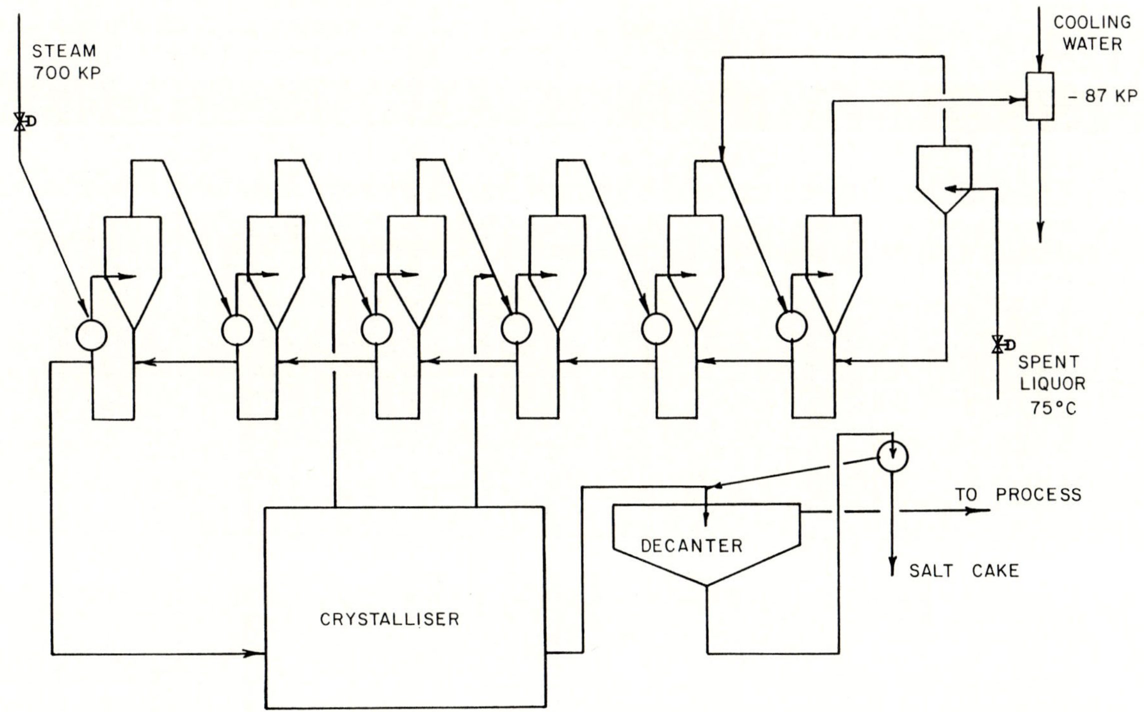

FIG.I 6 STAGE FORCED CIRCULATION BACKWARD FEED EVAPORATOR - CRYSTALLISER UNIT

The unit is stress relieved to resist caustic corrosion, and nickel is used in the areas of highest temperature.

Problems

The axial flow circulation pumps were purchased from a local supplier, licensed by a European manufacturer. Unfortunately the licensee did not understand the behaviour of the pumps he was selling, and received no support from his principals. The pumps could not satisfy the flow/head/NPSH specification, and cavitated badly in service.

After less than three months operation on caustic liquor, severe corrosion/erosion damage was found in pumps and pipework at the hot end of the unit. An unfounded fear of caustic embrittlement resulted in a four month shutdown of the unit in late 1981, during which the causticity of the plant liquor slipped to 83%, and the oxalate concentration rose to a little over 4 g/l.

A test programme was carried out at the pump manufacturers factory, which showed that the cavitation resulted from an NPSH problem, and that the pump characteristics changed significantly as the direction of flow was reversed. By reversing the pump installation, changing the blade angle, lowering the pump by 1 metre, and reducing the pump speed it was possible to obtain satisfactory operation at 90% of the designed circulating flow. The problem is illustrated by the typical axial flow pump curves shown in Fig. 2.

It was orginally intended to commission the unit six months after plant start-up, and to run it to maintain plant liquor causticity at 90%. The problem with the circulation pumps delayed unit operation, and allowed the causticity to fall, so changing the design feed conditions.

The sodium carbonate concentration in the feed liquor has now doubled. As a result the caustic concentration at which the carbonate solubility is exceeded is reduced by about 60 g/l.

It is a property of solutions that as the concentration rises the evaporation required to produce a given further concentration decreases. In the San Ciprian system the evaporation across the crystallizer battery is pre-set, and hence the concentration change across the crystallizers is also lower than the design value.

The concentration achievable is then controlled by fouling considerations in the live steam effect, and by reduced concentration obtainable across the crystallization battery.

The crystals are fragile, and it was necessary to control the degree of agitation in the crystallizers to obtain an easily separable product. It was further necessary to reduce the G forces at the centrifuges to obtain a free-flowing product which did not compact inside the centrifuges and cause blockages.

After tuning the unit to overcome the problems outlined above it became possible to produce saltcake in adequate quantity. Because of the change in feed quality, the control of the unit is more sensitive than envisaged in the design. It is necessary to control the concentration

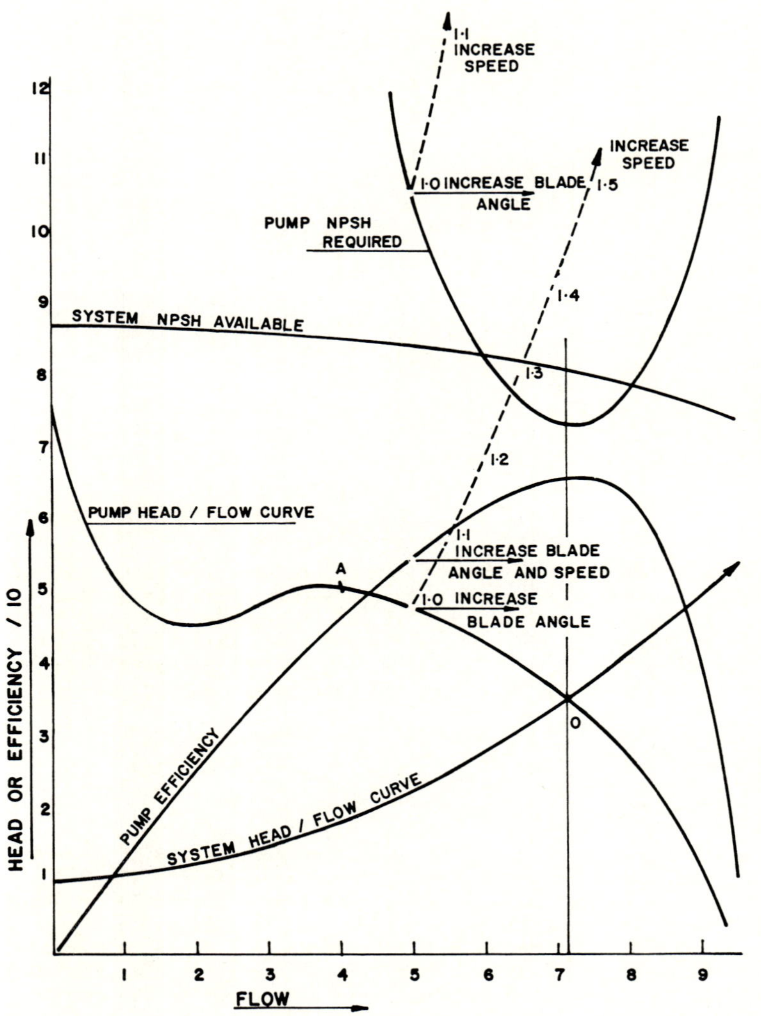

FIGURE: 11 AXIAL FLOW PUMP CURVES – SPEED: n RPM
– BLADE ANGLE: $\alpha°$

entering the crystallization section to ±5 g/l caustic for optimum results.

Results

The unit can be run at low concentration and low temperature to produce an oxalate/vanadate cake which is obtained as a dry brown granular material. Alternatively, at high concentration and moderate temperature a carbonate/oxalate/organic cake is produced. This is moist, pasty, and almost black in colour. The analysis of the cake obtained is presented in Fig. 3.

Operation is now steady, with a clear improvement in performance, particularly from October 1982, when most of the operating and mechanical problems were solved. Table I shows the quantity of solids removed per month, and the rising level of saltcake production.

Fig. 4 shows the evolution of the impurities in the plant liquor from the start-up to the present time. It should be noted that in the period Aug. 1981 - March 1982 the oxalate calibration of the analysis was lost, and hence the reported results are high. The path of the oxalate concentration is better represented by the dotted line.

It can be seen that the oxalate and organic carbon are being controlled at the desired levels, and that the carbonate is being controlled, but requires a long-term effort to bring down to the desired value.

Salt Cake Disposal

The original plant design anticipated the calcination of the saltcake with hydrate to recover alumina and caustic values. This concept was overtaken by rising energy prices, and is not economic at the present time. We find that entombment in the impermeable mud stack is a simple and effective technique for saltcake disposal. About 1000 MT/month of saltcake is buried in the 20,000 MT of dry mud sent to the mud stack.

As climatic conditions are sometimes unfavorable for burial of saltcake for extended periods, and the burial is better managed as a discontinuous once per month operation than as a daily routine, we established an intermediate covered storage station in the plant.

The mud is deposited in the stack area, and the feed point is moved several times so that a number of layers of mud accumulate. The depth of the densified mud is at least 4 meters in the area chosen for disposal. Using a back-hoe a trench 2 m deep, 4 m wide, and as long as necessary, is cut in the dried mud. The spoil from this trench is stacked on the upslope side. Once per month, during dry weather conditions the saltcake is trucked from the intermediate storage area and dumped in the trench. The spoil is pushed back to cover the saltcake and the red mud feed point is then located to cover the area with fresh red mud.

FIGURE: III CAKE ANALYSIS

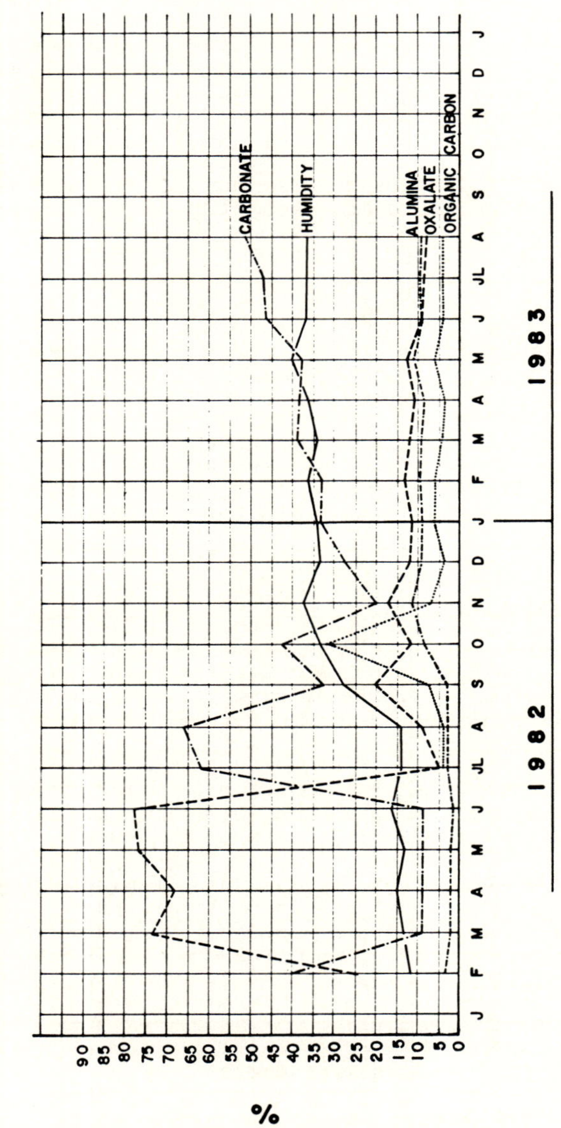

FIGURE IV PLANT LIQUOR IMPURITIES (200 g/l. CAUSTIC)

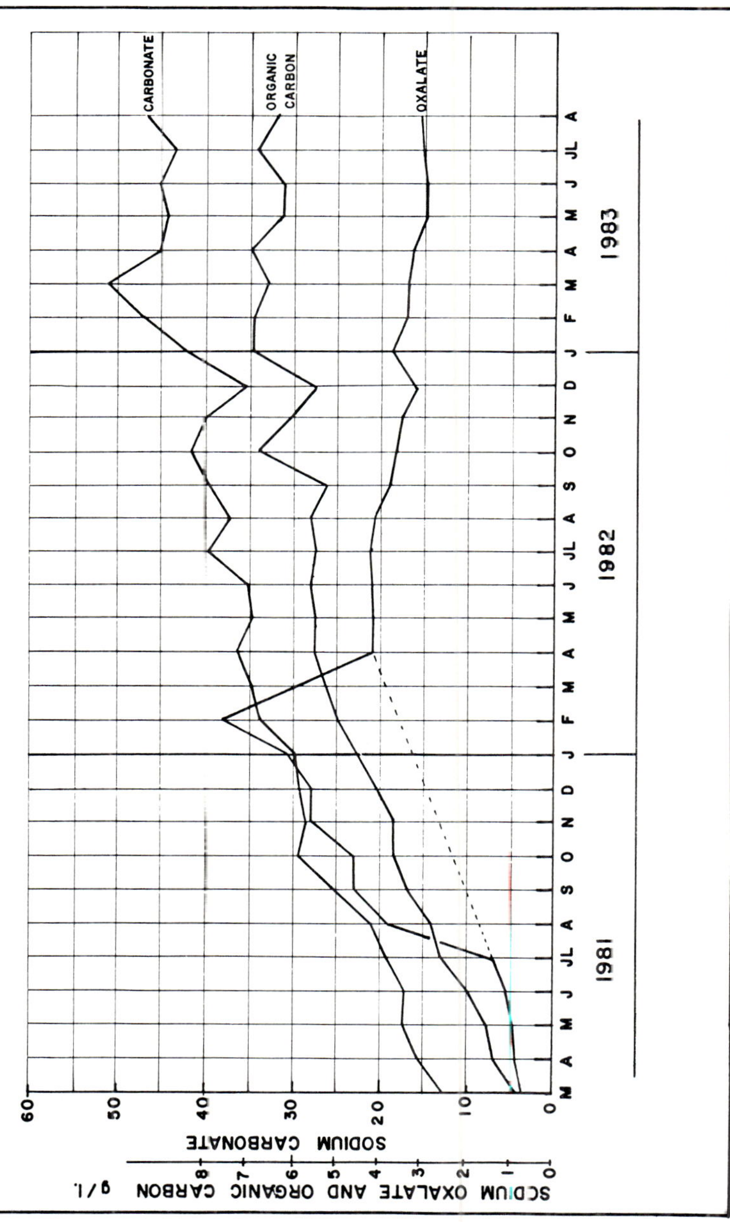

Acknowledgements

Special thanks are extended to Messrs. Bryan Hiscox, from Alcan International and E. Alvarez-Miranda, Tomas Guil and Juan Gutiérrez from Alumina Espanola for their help in the preparation of this paper.

I also thank Alumina Espanola and Alcan International for permission to publish this paper.

TABLE I

MONTHLY PRODUCTION (METRIC TONS)

	Wet cake	Dry cake	Oxalate	Carbonate
1.982				
Jan.	-	-	-	-
Feb.	167	103	26	41
March	130	113	83	9
April	68	58	39	5
May	66	58	45	5
June	129	109	75	9
July	247	212	11	131
Aug.	475	408	37	268
Sept.	202	147	29	46
Oct.	512	343	37	146
Nov.	719	451	78	91
Dec.	759	508	63	142
1.983				
Jan	656	421	47	142
Feb.	830	533	68	177
March	1084	696	76	310
April	1145	735	73	328
May	1252	769	86	311
June	1153	724	70	341
July	708	444	35	209
August	985	597	43	310

APPLICATION OF OPTIMAL CONTROL TO LURGI FLUIDIZED BED CALCINER FOR SANDY ALUMINA

TETSUYA KOHNO
WAKAMATSU WORKS
MITSUI ALUMINIUM CO., LTD.
JAPAN

MASAYASU ITOH
HEAD OFFICE
MITSUI MINING & SMELTING CO., LTD.
JAPAN

Summary

In industrial processes, classic control theory is generally used, of which PID control is typical. However, the theory sometimes fails to work for such processes which involve large time lag or strong interferences. In such cases, application of modern control theory is significantly effective. But this theory often poses severe mathematical problems and requires the use of sophisticated digital techniques. Therefore, it is seldom put to practical use.

Mitsui Aluminium has developed an advanced computer control system using color CRT (Cathod Ray Tube) display and has applied it to the alumina refinery process since 1972. Our Wakamatsu Works uses this most advanced computer control system.

On the basis of the experience we had with the above computer control system, we have recently applied modern control theory to Lurgi circulating fluidized bed calciner for sandy alumina. An exclusive control system for the fluidized bed calciner has been developed. This development would have been impossible using only classic control theory.

A statistical method and exclusive digital controllers as the control device were adopted in this new control system. It took 3 years to develop this new control system and it has been working satisfactorily since May 1981.

In this paper, we report on the development and the configuration of the new control system. We also discuss the computer control system for the entire aluminium hydrate process at Wakamatsu Works.

Introduction

Wakamatsu Works operates on the basis of the Bayer process. Two alumina production lines were constructed separately as follows ;

First stage construction

Product : Floury Alumina
Production Capacity : 200,000 TPY
Operation Start : October, 1972

Second stage construction

Product : Sandy Alumina
Production Capacity : 200,000 TPY
Operation Start : September, 1976

The final step in the Bayer process for extracting alumina from bauxite, is to calcine aluminium hydrate into anhydrous alumina. Usually, a rotary kiln is used as a calciner.

In order to calcine sandy-type alumina, we decided to introduce the fluidized bed calciner in 1974, developed by Lurgi/VAW in West Germany. Its operation was started in 1976. Its main characteristics include low energy consumption, low NOX emission in waste gas, low maintenance cost and high utilization factor.

The factors which mostly influence alumina quality are the temperature in the furnace and the mean retention time, represented by the differential pressure between the top of the furnace and the bottom. The set points of both the temperature and the differential pressure should be changed according to the quality of product alumina which is required by smelters. Once these parameters are set, they must be maintained steadily for continuous production with uniform quality. Therefore, it is not too much to say that the controllability of both the temperature and the differential pressure determine the quality of the alumina product.

The control system originally supplied by Lurgi was of the constant-value control type, using analog devices, which was put into operation in as early as 1976. Thanks to rapid and remarkable development on control devices, systems, and concepts in the mean time, we could apply digital devices backed up with the modern control theory.

While control of the temperature and the differential pressure by means of analog devices in a strict and narrow range may be possible, it takes more time and labour to do so as compared with the newly developed method, which even gives most practical approach to more stable and more accurate control of such parameters.

We have succeeded in developing a new control system, in which modern control theory and digital devices are adopted. It is said that modern control theory is seldom put to practical use, because it is very difficult to identify a state transition function which expresses a dynamic characteristic of a plant. Neverthless, we solved the above problem by using an auto-regressive model (hereinafter called " AR model ") and prepared serial software packages for data sampling, analyzing, synthesizing and etc, putting special emphasis on the inter-relations with such items as

1) Easy control of process to meet versatile quality requirements of smelters

2) Lower energy consumption

3) etc

Explanation of Process

The main features of the process are explained by a simplified flow sheet (Figure 1).

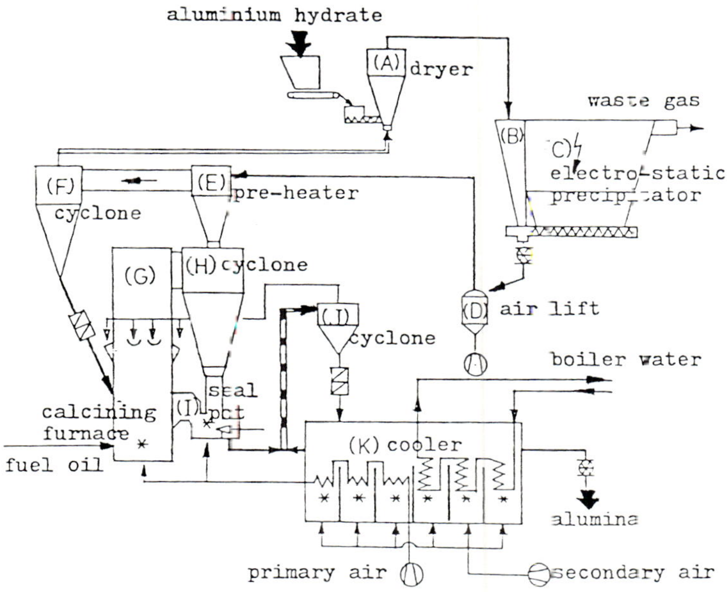

Figure 1 flow sheet of fluidized bed calciner

Damp aluminium hydrate, $Al_2O_3 \cdot 3H_2O$, is fed into dryer(A) to be mixed with the waste gas stream from cyclone(F). In the dryer, the hydrate having about 10% of surface moisture, is dried and exchanges heat with the gas. After being separated from the gas by mechanical pre-duster(B), and electro-static precipitator(C), it is conveyed to pre-heater(E) by airlift(D), and is mixed again with hot gas from cyclone(H). In the pre-heater, the hydrate is dried completely and also partially calcined. After being separated from gas again by cyclone(F), it is fed into calcining furnace(G).

The calcining furnace is the core of this process and has a fluidized bed with air nozzles. The air, which is pre-heated, is supplied through these nozzles for fluidization and for combustion.

Heat energy for endothermic calcination is generated by direct combustion of fuel oil in the fluidized bed of the calcining furnace. Fuel oil is supplied into the furnace through six oil lances which are located above the furnace floor. Secondary air for combustion is supplied into the furnace through four inlets at a point of middle height.

Alumina particles are lifted up by upward gas flow in the furnace and transferred into cyclone(H), which is directly connected to the calcining furnace. Alumina is separated from combustion gas by this cyclone and settles in the seal-pot(I). Alumina, built up in the seal-pot, is also fluidized by air and is recycled to the bottom of the furnace again. During recycling between the calcining furnace and the cyclone via the seal-pot, obtained alumina has the quality of the desired specifications.

In Lurgi's original control system, the temperature in the furnace was controlled by changing the feed rate of damp alumina. But we remodelled it so as to be controlled by adjusting fuel oil in order to realize more rapid response.

From the seal-pot, alumina is extracted continuously through a discharging device. The mean retention time is controlled by adjusting the discharging device. The differential pressure between the top of the furnace and the bottom expresses the inventory of alumina. In other words, it means the mean retention time of alumina.

Alumina product extracted through the discharging device is mixed with air from cooler(K), and is lifted up into a cyclone(J). Direct heat exchange is applied to transfer heat from the solids to air. After being separated from the solids, pre-heated air is sent into the calcining furnace as secondary air for combustion.

On the other hand, alumina is sent into a cooler(K) for further cooling. The cooler has six chambers and each chamber has a fluidized bed with air nozzles. Being fluidized, alumina flows over the walls which separate chambers. The first three chambers have pipe-assemblies in which primary air is pre-heated indirectly. The last three chambers have also pipe-assemblies for pre-heating boiler water to recover heat from the alumina product.

Development

PID control is useful for a controlled system having low order lag. But it is often not suitable for a controlled system which has large time lag or multi-variable interaction. The calcining furnace has large time lag and multi-variable interaction. Therefore, we tried to apply modern control theory to our calcining furnace.

The most difficult problems which we had to solve in successfully applying the modern control theory to our development were ;

(1) We must perfectly grasp the dynamic characteristic of a controlled system and express it in a state equation.

(2) PID control is easy to understand. But the modern control theory often poses severe mathematical problems.

(3) Though the modern control theory is going to be systematized with various techniques, its premises have to be

always simulated in the best way to confirm with actual industrial process requirement.

(4) Both hardware and software digital techniques are necessary for applying the modern control theory to actual processes.

It took 3 years to overcome the above difficulties and develop our new control system.

Dec., 1977 ∿ March, 1978	Study of control method
April, 1978 ∿ Aug., 1978	Programming for data sampling and simulation
Sep., 1978 ∿ June, 1979	Programming for analyzing the process and synthesizing the control loop
July, 1979 ∿ Nov., 1980	Field test and modifying
Dec., 1980 ∿ April, 1981	Installing the back-up device
May, 1981 ∿	Continuous actual operation

Computer Control System

Our computer control system for the whole alumina hydrate process, including the calcining furnace, is shown in figure 2.

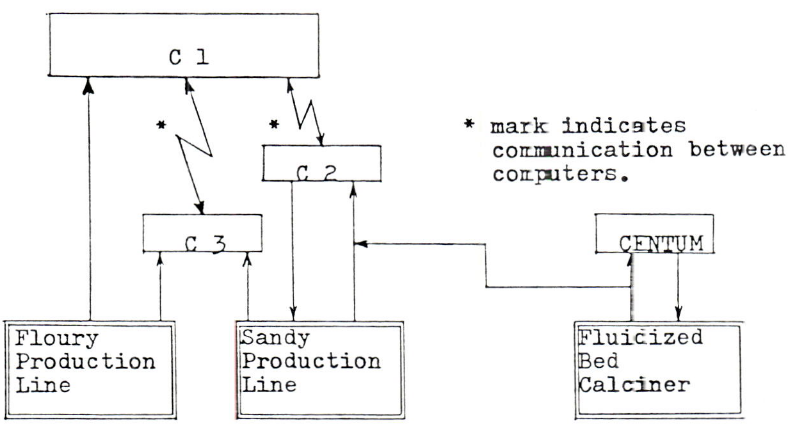

Figure 2 Configuration of computer control system

The computer control system consists of three computers C1, C2, and C3. The computer C1 is in communication with computers C2 and C3. It is a hierarchal system in which the computer C1 is a host.

The computer C2 is used exclusively for DDC (Direct Digital Control) and the computer C3 for monitoring the condition of electric motors. The computer C1 is especially designed to treat off-line jobs as well as on-line jobs at the same time. Various data of the whole process are finally communicated to the computer C1 and stored in its bulk memory. Therefore, we can analyze the whole plant by utilizing this stored data.

This system is primarily designed to promote advancements in the technological level of Mitsui Aluminium which newly joined the alumina business in 1972 by starting operation of its Wakamatsu Works. Therefore, this has the following outstanding features.

1) The whole plant with two process lines is monitored, controlled and briefed by one or two operators.

2) The flexibility of processing the data is increased as much as possible by developing the software by Mitsui Aluminium itself.

3) Communications among the computers within the system can be carried out as if they hold their memories and data in common.

4) All input data are represented by parameters. This makes the system very flexible in case of modifying processes or plants.

The CENTUM is a digital control device for the fluidized bed calciner. The data of this plant are gathered into the CENTUM. At the same time, they are communicated to the computer C1 through I/O devices in the computer C2. If necessary, by using actual data, it is possible to analyze data, to simulate, to synthesize control loop or etc. All software packages for developing the new control system are prepared in the computer C1.

Configuration of the New Control System

The CENTUM is a microprocessor-based distributed control device which is used in the new control system for our fluidized bed calciner.
The CENTUM is powerful enough for plant operation, and other devices are not necessary.
The configuration of the CENTUM is shown in figure 3.

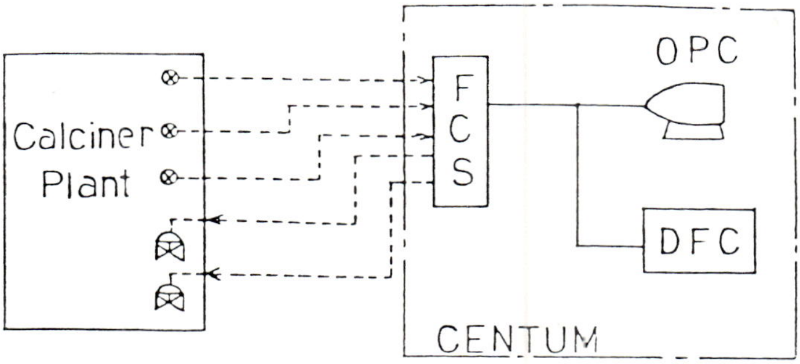

Figure 3 CENTUM System Configuration

The CENTUM consists of FCS, DFC and OPC. The FCS is an abbreviation of " Field Control Station ". It interfaces with the field equipments such as 2-wire transmitters and control valves in the fluidized bed calciner, and performs control computation and directly out-puts results to the plant. It incorporates a microprocessor and performs all signal linearization, alarm checking, indirect calculation, loop connection and etc. (Picture 1)

The DFC is an abbreviation of " Data File Console ". It contains three main subunits : a data file nest which can accommodate up to two flexible disk drives, a station control nest incorporating a 16-bit processor and 48K-words of core memory, and a system maintenance keyboard. A flexible disk has a capacity of 154K 16bit words and stores not only process data records and CRT display frames but also system programs and parameters. (Picture 1)

The OPC is an abbreviation of " Operators' Console ". It is used as a means of man/machine interface. One feature of an Operators' Console is virtual panel display. The size of the screen is 20 inches with 7 colors and it has a human-engineered operator keyboard. It permits us to monitor and tune the plant almost exactly like a conventional analog control instrument. (Picture 2)

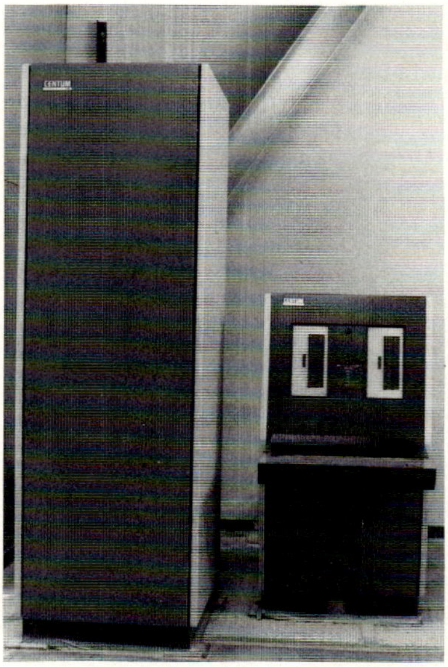

We can monitor and control the fluidized bed calciner with various CRT virtual display panels. For example, the control panels display simulated fronts of analog instruments in groups of 8 instruments per panel. (Picture 3) The graphic panels display graphic diagrams of the fluidized bed calciner. (Picture 4) The recording panels simulate 2-point analog trend recorders from data stored in memory. (Picture 5)

Picture 1 FCS and DFC

Picture 2 OPC

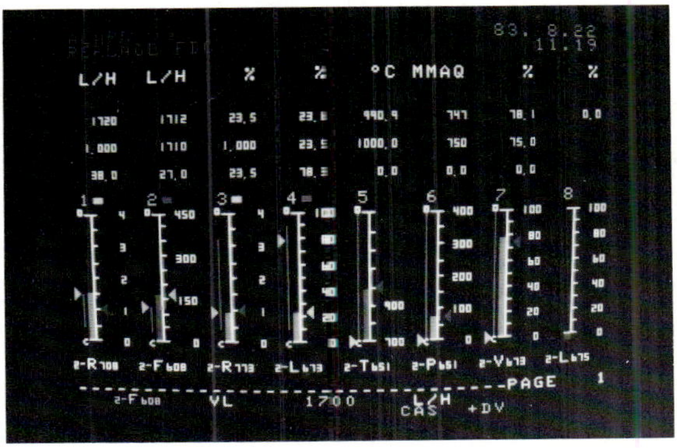

Picture 3

Control Panel

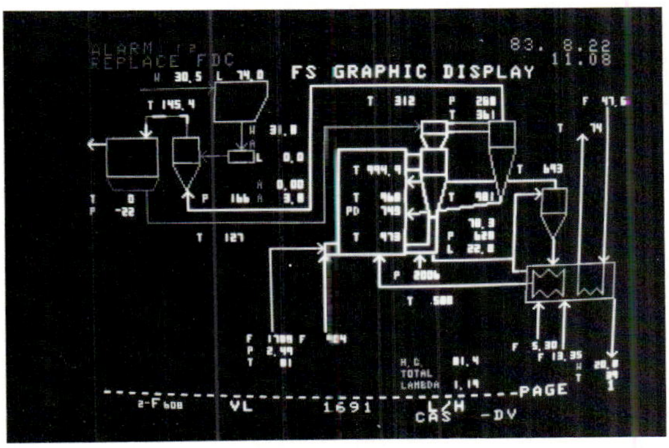

Picture 4

Graphic Panel

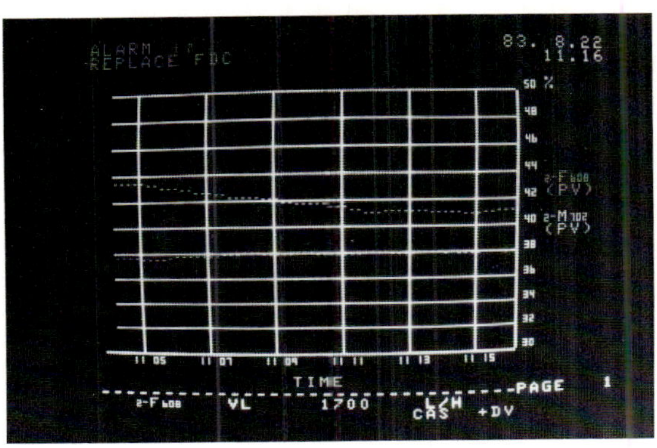

Picture 5

Recording Panel

Synthesis of the New Control System

We succeeded in identifying the state transition function by adopting the AR model, which uses statistical methods.
The principle of the AR model is based on the fact that predictive value of each variable is mathematically estimated from value of observed variables.
The basic equation of the AR model is given as Eq.(1)

$$Zs = A\, Zs_{-1} + B\, Ys_{-1} + Ws \quad \text{-------} \quad (1)$$

Where Zs : State variable vector at time s

Zs_{-1} : State variable vector at time $s-1$

Ys_{-1} : Manipulated variable vector at time $s-1$

Ws : Noise vector added to state variable at time s

A : State transition matrix (A matrix)

B : Manipulated transition matrix (B matrix)

After the state transition function is identified, an optimal feedback control system is automatically designed by the computer C1, which solves the DP (Dynamic Programming) problem so as to minimize the performance index with quadratic form.

Our procedure for synthesizing the new control system is shown in figure 4.

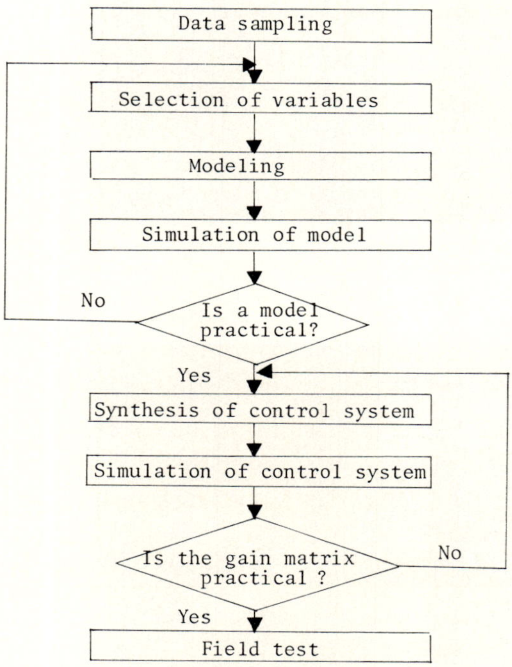

Figure 4 Procedure for synthesizing the new control system

Data Sampling

10 variables in the calcining furnace were scanned every 30 seconds, and these data were stored in the computer C1 through the computer C2. The term of scanning was about 8 hours. The variables which represent the characteristic of the calcining furnace have to be selected. We changed the value of manipulated variables every 20 minutes so as to identify a more useful AR model. We selected variables which are listed in Table I.

Name of variable and its function	Adoption
< Control variables >	
1. Temperature at the furnace (G) - the calcining temperature of alumina	**
2. Differential pressure at the furnace (G) - the alumina inventory of furnace	**
3. Temperature at the outlet of the pre-heater (E) - the heat exchanging condition of hydrate fed into the cyclone (F)	**
4. Temperature of off-gas at the outlet of the dryer (A) - the fluctuation of surface moisture in hydrate	**
5. Temperature of primary air at the inlet of the furnace (G) - the fluctuation of both volume and temperature of discharged alumina	**
6. Pressure at the outlet of the seal pot (I) - discharging condition of alumina	
7. Weight of damp alumina at the inlet of the dryer (A) - the fluctuation of hydrate feed into the dryer (A)	
< Manipulated variables >	
8. Flow rate of fuel oil at the inlet of the furnace (G)	**
9. Out put value to the discharging device at the outlet of the seal pot (I)	
10. Position of the discharging device at the outlet of the seal pot (I)	**

Table I Variables

Selection of variables

The variables which influenced the main state variables (temperature and differential pressure at the furnace) had to be selected by means of our experience and statistical analysis. The variables which are selected, are marked with ** in table I.

Modeling

After selecting variables, we must determine optimal order (LL), A-matrix and B-matrix. An example of a calculation result is shown in Table II. In this case, the optimal order is 4. Therefore, each value for 30 seconds from now, is calculated according to the present actual value and to the last 3 actual values.

Simulation of model

The above model must be examined and tested. The results of simulation of both the temperature and the differential pressure are shown in figures 5,6. The simulation was performed with computer C1. In each figure, the upper line indicates actual plant data and the lower line indicates the results of simulation.

It can be said that the predictive value in simulation bears a close resemblance to the actual value. This is a very good state transition function and predicts actual temperature and differential pressure very closely. If we can't obtain a practical function at this step, we must go back once more to the step of selection of variables.

Synthesis of control system

The control system has to be synthesized on the basis of the above-cited state transition function. A feedback control system is used as our basic principle. A control gain G which minimizes the quadratic performance index J1 (See Eq. 2) is determined with dynamic programming (D. P.).

$$J1 = \int_0^\infty \{ Q\, x^2(t) + R\, y^2(t) \}\, dt \quad \text{--------} \qquad (2)$$

where
- Q : Weighting factor of state variable
- R : Weighting factor of manipulated variable
- x(t) : Deviation of state variable
- y(t) : Deviation of manipulated variable

Simulation of control system

The simulation of the control system should be performed with a closed loop including both the state transition function and control gain G. This simulation is done with the computer C1 and its general flow is shown in fugure 7.

(A(I)B(I))

LL= 4 (Optimal Order)

I= 1

MATRIX 5 x 7

	1	2	3	4	5	6	7
1	0.11636D+01	-0.20862D+00	0.64064D-02	0.12708D+00	-0.62539D-01	-0.94211D-01	0.15405D-01
2	0.14699D-01	0.57089D+00	0.61808D-01	-0.72385D-01	0.13275D+00	0.68261D-01	-0.33265D-01
3	-0.41614D-01	-0.52096D-01	0.10798D+01	-0.13549D-01	-0.88496D-02	0.79331D-01	-0.15298D-01
4	-0.91155D-01	0.17606D-01	-0.74260D-02	0.46170D+00	-0.32026D-01	-0.10436D-01	-0.20358D-02
5	0.66235D-01	0.52196D-01	0.28711D-01	-0.31083D-01	0.29366D+00	0.62165D-01	-0.12120D-01

← A- matrix (columns 1–5) → ← B- matrix (columns 6–7) →

I= 2

MATRIX 5 x 7

	1	2	3	4	5	6	7
1	0.23140D-01	0.13261D-01	-0.27120D-01	-0.19028D-01	0.49554D-01	0.17249D+00	-0.49361D-01
2	-0.23007D-01	0.21186D+00	-0.67290D-01	0.77328D-02	-0.29217D-01	-0.92517D-01	0.13472D-01
3	-0.47692D-01	-0.85566D-01	0.71472D-01	-0.10554D+00	0.51114D-01	-0.11950D+00	0.72315D-02
4	0.58610D-01	-0.23908D-01	0.31642D-01	0.29013D+00	-0.42287D-01	0.18296D-01	-0.34991D-02
5	-0.56782D-01	0.25936D-01	-0.18189D-01	0.16462D-01	0.27793D+00	-0.10246D+00	0.18545D-01

I= 3

MATRIX 5 x 7

	1	2	3	4	5	6	7
1	-0.13407D+00	0.10525D+00	0.13275D-01	-0.11232D+00	0.17733D-01	-0.57074D-01	0.20233D-01
2	-0.46447D-01	0.12188D+00	0.21369D-01	-0.10760D+00	-0.13957D-01	-0.84423D-02	-0.15925D-01
3	0.82760D-01	0.14336D-01	-0.10827D+00	0.64750D-01	0.44079D-01	0.61749D-01	0.10243D-01
4	0.15652D-01	-0.27184D-02	-0.37563D-01	0.13110D+00	0.50181D-01	0.11657D-01	-0.44384D-02
5	0.23315D-02	-0.22545D-02	0.36615D-02	0.30595D-01	0.17829D+00	0.50284D-01	-0.14952D-01

I= 4

MATRIX 5 x 7

	1	2	3	4	5	6	7
1	-0.70360D-01	0.10744D+00	-0.49176D-02	0.84150D-02	0.17015D-01	0.35802D-01	0.13192D-01
2	0.55267D-01	0.85638D-01	-0.91022D-02	0.16979D+00	-0.91963D-01	0.25945D-01	0.19119D-01
3	0.50029D-02	0.13302D+00	-0.65411D-01	0.47340D-01	-0.83516D-01	0.77634D-02	0.26709D-02
4	0.21916D-01	0.12089D-01	0.23273D-02	0.85891D-01	0.20462D-01	0.53865D-02	0.17272D-01
5	0.33416D-02	-0.77883D-01	-0.11038D-01	-0.18796D-01	0.21289D+00	-0.17157D-01	0.25158D-01

Table II Optimal Order, A- matrix and D- matrix

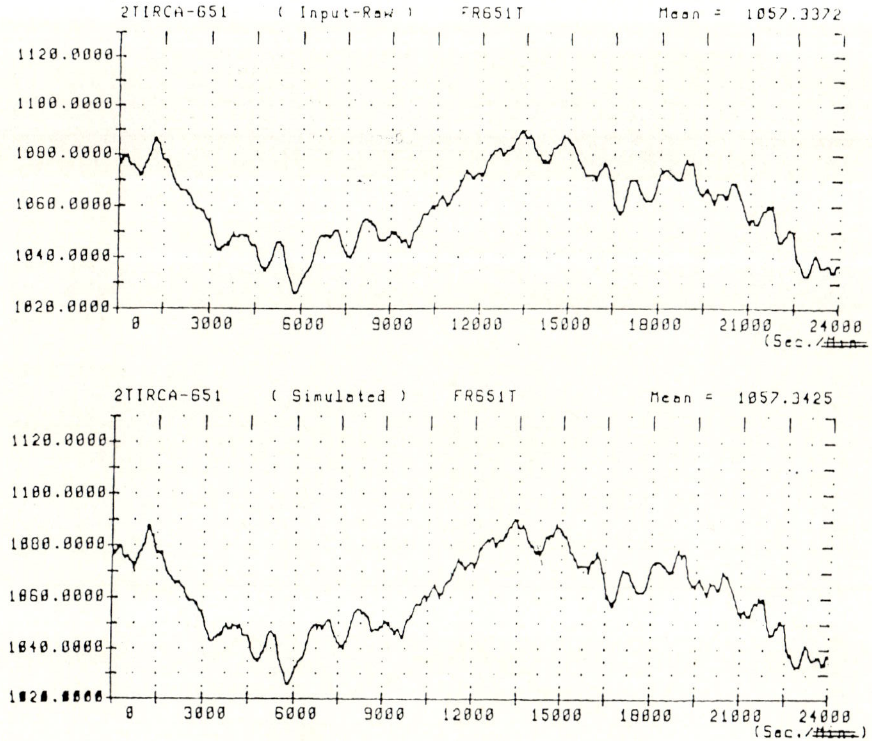

Figure 5 Simulation of Temperature

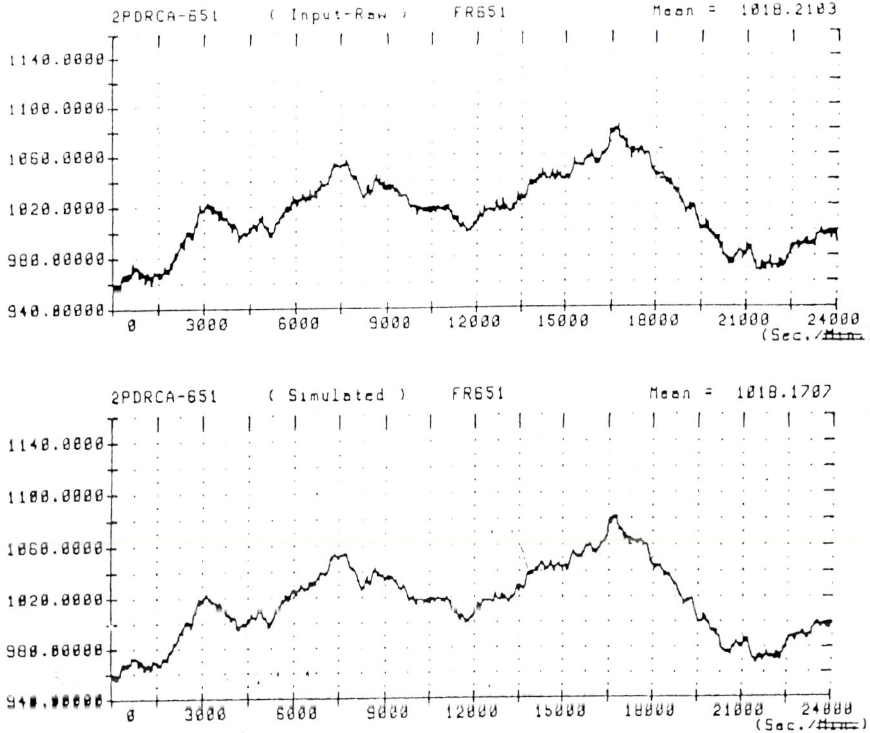

Figure 6 Simulation of Differential Pressure

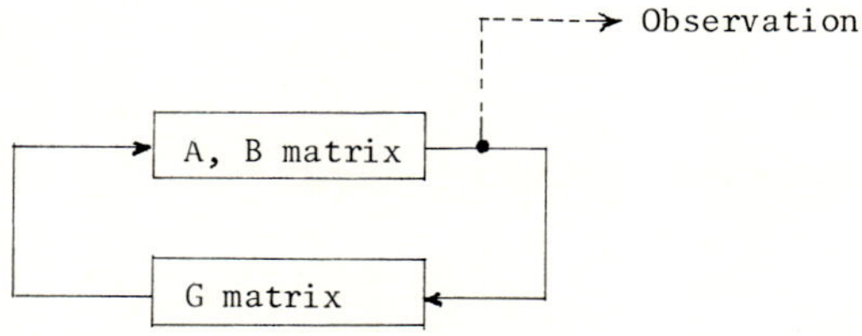

Figure 7 Simulation of control system

A practical control gain G should be determined by adjusting the weighting factors Q and R.

Field test

After matrices A, B and control gain G obtained by the above procedure are registered in the CENTUM's memory, field tests are performed in actual plant.

Conclusion of Actual Operation

We started the operation with the new control system in May, 1981 and it has been working satisfactorily. Our experiences of operation have led us to the following conclusions.

1) Temperature is controlled within the range of ± 20°C for the set point, and the differential pressure within ± 25mmAq constantly in spite of any disturbances. (See Figure 8)

2) A single AR model can cover the production range of 300 ∿ 850 T/D of alumina and adjustment of parameters due to such changes are continuously and smoothly controlled.

3) Only what an operator has to do to change the load rate of the plant etc, is to input new parameters through the operator console.

4) Actual operating conditions can be observed at a glance on the CRT display, which releases operators from their conventional heavy burdens.

In the initial stage of our attempt to introduce the modern control theory to the operation of the fluidized bed calciner, though we already had had experiences in application of computers by that time, we felt as if we were trying to thrust our hand into a new black box.

We started with the study of the theory and collecting relevant information. In further advanced stage, we often doubted that the application of the theory to an actual plant is possible, because of the difficulty of the theory itself for one reason. One of the main reasons why we succeeded lies in our modification of state transition function applying the AR model.

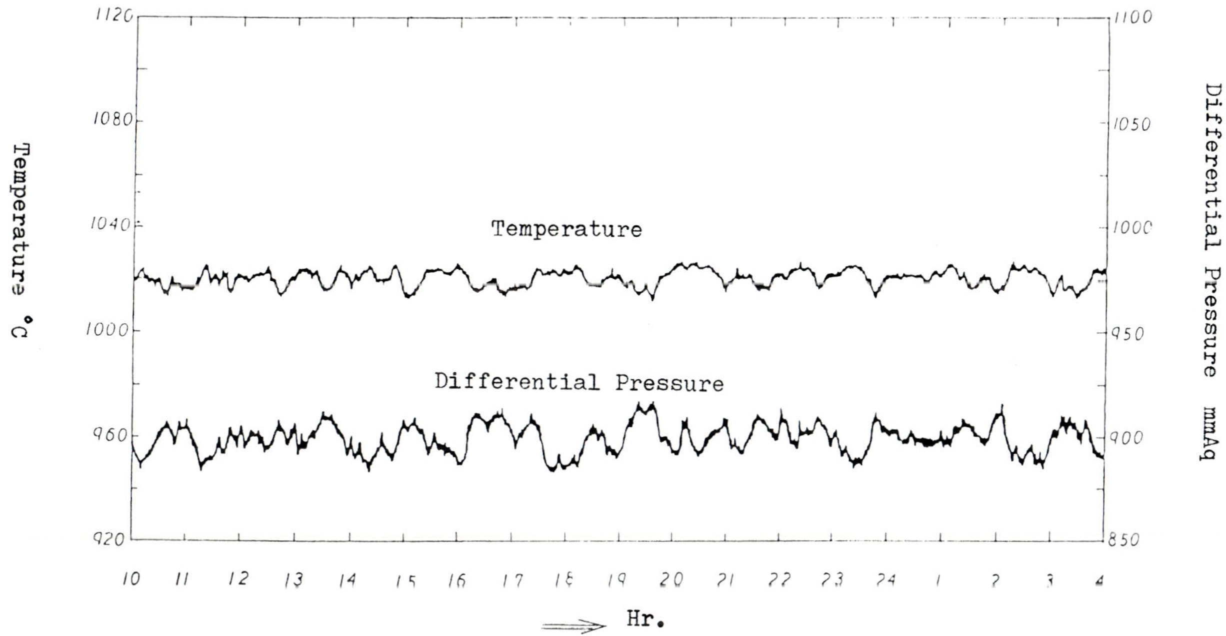

Figure 8 New Control System

It has been proved that the AR model can be applied not only to the control of a plant but also to prediction of states. We are now planning to apply this system to controls of other processes and to prediction of different states.

Acknowledgments

It is a pleasure to acknowledge the guidance of ex-Prof. Masaru Takada of Kyushu University, Prof. Kiyoshi Katoh of Kyushu Institute of Technology and Minoru Iseda of Mitsui Mining & Smelting Co..

References

1) H. Akaike, T. Nakagawa ; Statistical Analysis and Control of Dynamic System, Science Co. (1972)

2) H. W. Schmidt, H. Beisswenger ; Pratical Experience with the Operation of Lurgi/VAW Fluid Bed Calciners, 105th AIME Annual Meeting, Las Vegas, Nevada (1976)

3) T. Furuya ; Abstract of Statistical Control Theory, Paper presented at Kyushu Conference of the Society of Instrument and Control Engineers (1976)

4) H. Nakamura, M. Uchida, T. Kitami, Y. Kondo ; Application of Optimal Control Theory to Boiler Control, Journal of the Society of Instrument and Control Engineers, Vol. 18, No. 4 (1979)

5) S. Shimano ; Lurgi Fluidized Bed Calciner for Sandy-Type Alumina, Fourth International Conference on Fluidization (1983)

THE USE OF SMALL NUCLEAR MAGNETIC RESONANCE SPECTROMETERS AS ON-LINE ANALYZERS FOR ROTARY KILN CONTROL

R. M. Pearson
Kaiser Aluminum & Chemical Corporation
Center for Technology
Pleasanton, California 94566

T. L. Parker
Kaiser Aluminum & Chemical Corporation
Chemical Plant, Airline Highway
Baton Rouge, Louisiana 70821

Kaiser Aluminum & Chemical Corporation is currently using small nuclear magnetic resonance (nmr) spectrometers to measure the combined water content of reduction grade aluminum oxide as it leaves a rotary kiln. This paper discusses modifications made to a commercially available laboratory instrument for use as an on-line analyzer and problems encountered when used under actual plant conditions. NMR results are compared to gravimetric loss-on-ignition measurements and plant control data is discussed.

Introduction

Kaiser has been doing serious nmr since we acquired a Varian DA100-15D nuclear magnetic resonance (nmr) spectrometer in 1969. This spectrometer was capable of doing continuous wave (CW) nmr on a wide variety of solid samples of interest to the aluminum industry. These included both aluminum oxides and carbonaceous materials. The aluminum oxide work included both high surface area aluminas (1, 2) and the complete range of calcined samples. Both 'H nmr and ^{27}Al nmr were studied on these materials. Anodic coatings were also studied (3, 4).

This work laid the background which was used for our application of nmr to the control of Bayer plant rotary kilns. It will be shown how small nmr spectrometers have been modified and used as on-line analyzers to measure the hydrogen (water) content of the reduction grade ore as it leaves a rotary kiln. The cooperation of IBM Instruments in the work is acknowledged.

There are several commercially available moisture analyzers which are, at least in theory, capable of accurately determining the water content of a solid. Of those available, the ones most applicable to our problem would probably be infrared or slow neutron devices.

Infrared devices measure surface moisture of a solid. We had no reason to believe that the moisture on the surface of an alumina particle would be related to the material's total water content and could, therefore, be compared to a gravimetric loss on ignition value. In theory, a slow neutron device could probably be made to function in our application, but these devices are at least as expensive as nmr and would require a large (650 lb) sample. They also could not differentiate between physically absorbed and chemically bound water as can nmr.

Nuclear magnetic resonance is not a familiar technique in the aluminum industry. It will, therefore, be discussed in more detail than would normally be the case in a paper such as this. We shall see that the nmr method complements x-ray spectroscopy. X-ray is insensitive for the light elements such as ^{1}H, ^{19}F, ^{23}Na, and ^{27}Al. It turns out that nmr is at its best for those elements which are most important in the aluminum industry.

Theory

Because the following is an oversimplified version of nmr theory, it is not totally correct. It will, however, give the reader a qualitative feeling for how an nmr spectrometer works in our application. For a more complete discussion of magnetic resonance, I would suggest the reader obtain a copy of Pulse and Fourier Transform nmr by Farrar and Becker for a brief introduction to nmr principles and instrumentation.

If we place a small bar magnet in a coil and hook the coil to a volt meter (Figure 1), the meter would not register a voltage unless we were to spin the magnet. The idea of a magnet moving in a conductor is the basic principle on which the nmr experiment is based. In fact, a simple coil of wire placed in a magnetic field is used as the detector in an nmr spectrometer (Figure 2).

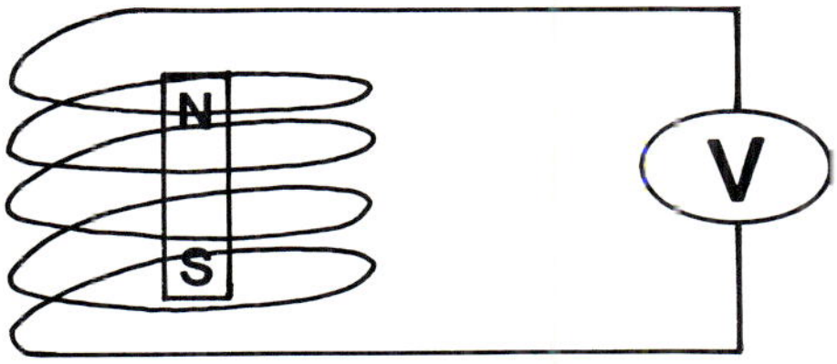

Figure 1. Magnet In Coil

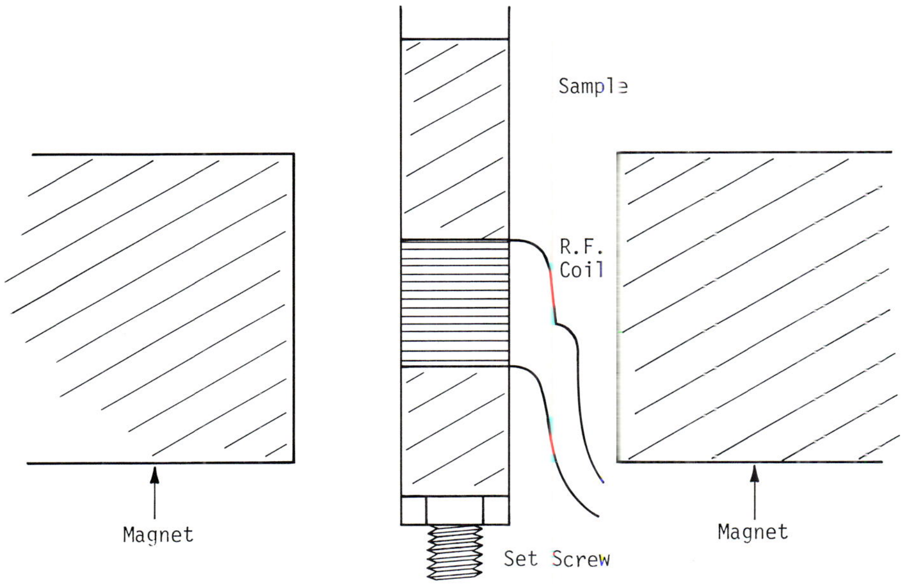

Figure 2. nmr Sample In Magnet

The above arrangement allows one to easily isolate the magnet and its electronics from the exterior atmosphere so that the system can be used under plant conditions. This is done by removing the set screw, drilling out the bottom of the probe, and inserting a sample tube as shown in Figure 3.

The theory of nmr is very similar to that of optical spectra. Just as electrons exhibit a property known as "spin", the nuclear particles, such as protons and neutrons, show the same "spin" behavior. We can discuss nuclear "Zeeman levels" just as we can discuss the more familiar atomic Zeeman levels well known in optical spectra.

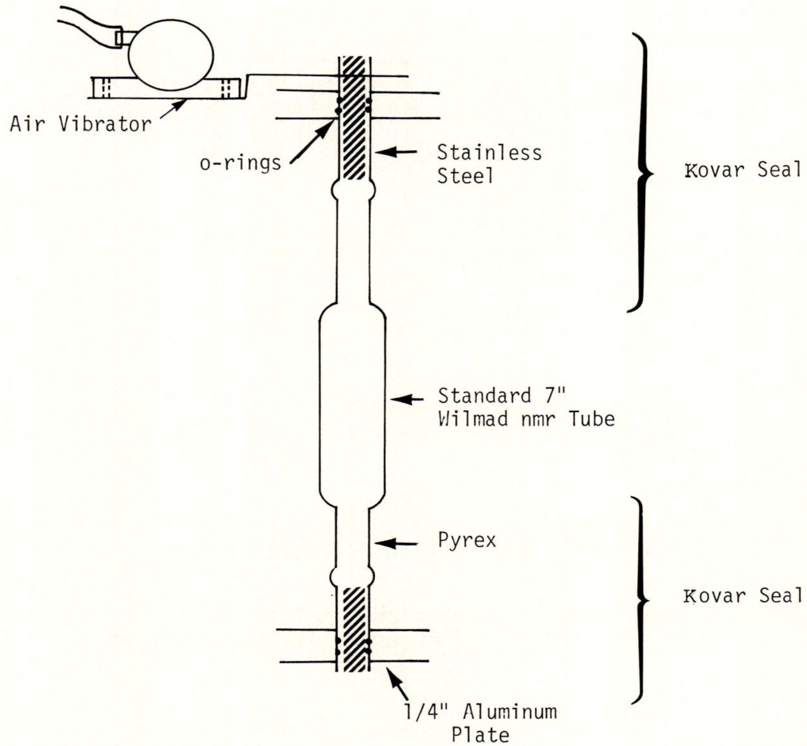

Figure 3. On-Line Analyzer Sample Tube

A hydrogen nucleus has one unpaired nucleon, a proton, and a spin $I = 1/2$. This means that the proton behaves as though it were a sphere spinning in space, i.e.,

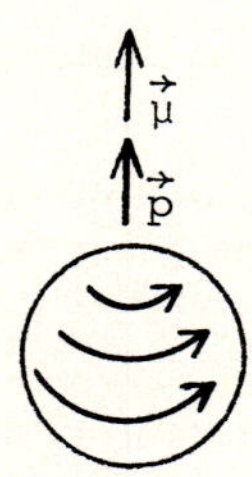

Figure 4. The Spinning Proton

Since the proton has a mass, it will have an angular momentum which can be represented by the vector $\vec{p}$, shown in Figure 4. Since the proton is a spinning electric charge, it will generate a magnetic field which is represented in Figure 4 by the magnetic moment vector $\vec{\mu}$.

The above picture of the proton shows that it will behave as a tiny bar magnet and line up with an external magnetic field thus:

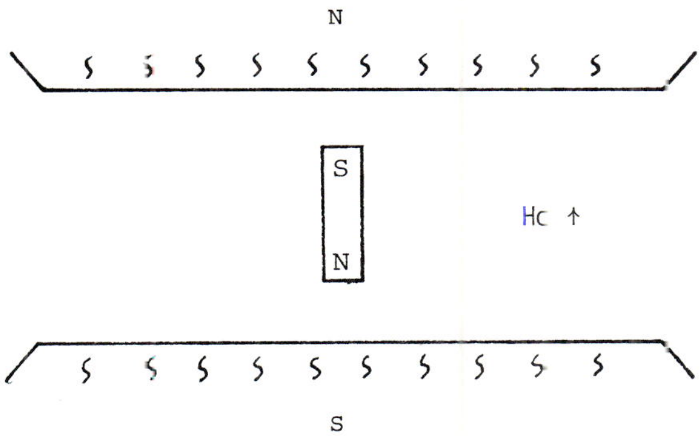

Figure 5. Bar Magnet In Magnetic Field, Ho

If we represent the total magnetization of the sample by the vector M and place a coordinate frame in the laboratory magnet, we obtain:

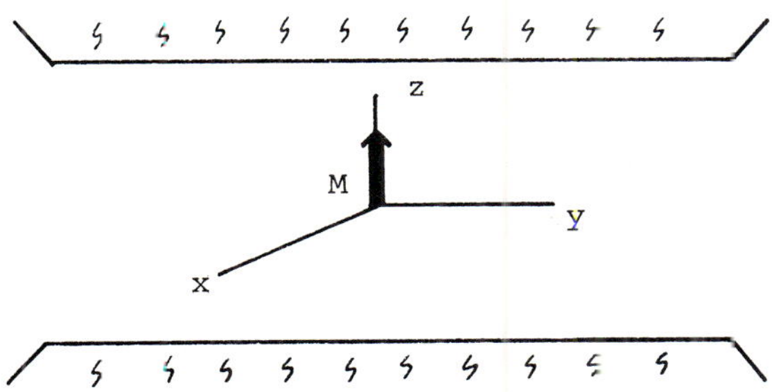

Figure 6. Magnetization Of Sample In Magnetic Field

Where: M represents the total magnetization of the sample.
x, y, and z are the laboratory coordinates.

The vector M, shown in Figure 6, represents the low energy position of the sample. The magnetization (M) is aligned with the z axis of the external magnet and is rotating about z at a rate proportional to the strength of the external field, Ho. If we tip the magnetization, M, onto the x-y plane of the magnet, M will then be a high energy state and will rotate in the x-y plane and about the z axis. The total magnetization of the sample in the magnetic field can be measured if the magnetization of the sample is changing with time. This change induces an electric current in a conductor surrounding the sample. It is this current which is the signal in a nuclear magnetic resonance spectrometer.

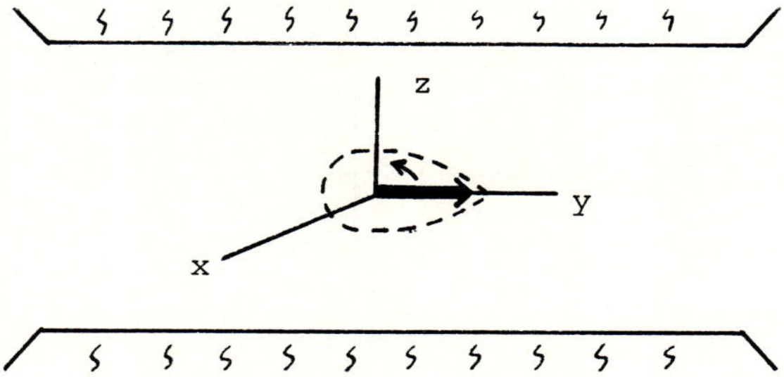

Figure 7. M - In A High Energy State

The ratio of the magnetic moment vector to the angular momentum vector times $\hbar$ ($\hbar = h/2\Pi$, where h = Planck's constant) is defined as the magnetogyric ratio, i.e., $\gamma = \vec{\mu}/\hbar\vec{p}$. The magnetogyric ratio (γ) is a fundamental property of the nucleus. We know that when $\vec{p} = 0$, then $\vec{\mu} = 0$ and when $p \neq 0$, then $\mu \neq 0$ and that nuclei of spin I have 2I + 1 energy levels when placed in a magnetic field. These energy levels are separated by an energy:

$$\Delta E = \frac{\vec{\mu}\ Ho}{I} \tag{1}$$

Where: I = the nuclear spin (1/2 for protons)
Ho = the strength of the external magnetic field
$\vec{\mu}$ = magnetic moment

For protons (hydrogen), the separation of the two energy levels is:

$$\Delta E = 2\ \vec{\mu}\ Ho \tag{2}$$

We know that for electromagnetic radiation

$$\Delta E = h\nu \tag{3}$$

Where: h = Planck's constant (6.62×10^{-27} erg sec) and
ν = the frequency

Equating the right sides of equations 2 and 3 gives:

$$h\nu = 2\ \mu\ Ho \tag{4}$$

$$\nu = \frac{2\ \mu}{h}\ Ho \tag{5}$$

and substituting γ for $2\ \mu/h$ gives:

$$\nu_0 = \frac{\gamma}{2\Pi}\ Ho \tag{6}$$

Where: ν is frequency in Hz (cycles per sec)
γ for protons = $2.67 \times 10^4\ sec^{-1}g^{-1}$

Equation 6 is the fundamental relationship for the nmr experiment. It says that a hydrogen atom placed in a magnetic field of Ho strength will absorb electromagnetic radiation of wavelength ν_0 Hz.

Instrumentation

Both pulsed and continuous wave nmr spectrometers are commercially available. In our application we use pulsed instruments from IBM. This instrument operates at a 4.69 kilogauss magnetic field and uses 20 megahertz R. F. pulses to excite the hydrogen atoms in the sample. Details of this technique are given by Farrar and Becker (5).

Modification of the IBM nmr for Plant Use

Description of the nmr. The IBM nmr is manufactured by Bruker in West Germany. This instrument consists of a console, which contains a micro-processor control unit and a magnet contained in a separate case. The magnet contains the probe which holds the sample. It is thus possible, with minor modification, to separate the magnet (i.e., the sample) from the console for at least the distance from the control room to the air slide on the kilns at Gramercy (∿100 feet).

Modifications of the Spectrometer. Our first task was to become familiar with the operation of the nmr so that we could use it to measure the water content of reduction grade alumina. The water content (∿1% by weight) of reduction grade alumina is low relative to other materials the spectrometer is designed to study.

Unacceptable "ringing" of the probe was observed when the spectrometer gain was increased so as to see the nmr signal of water in reduction grade alumina. Willie Shee of IBM suggested that this was "acoustical ringing" and could be corrected by substituting stainless steel for the aluminum probe sides. This was done and good nmr signals of water in reduction grade ore were obtained.

Dust-proof cases were built for both magnet and console units of the nmr. Both enclosures are similar in design. Each consists of an aluminum container fitted with an additional compartment to attach the necessary wiring, as shown below.

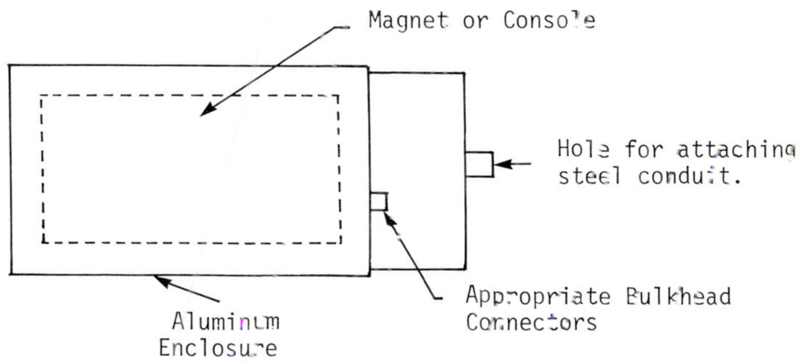

Figure 8. Aluminum Dust-Proof Enclosures

All connecting wires between the two units were enclosed in steel conduit, a 2 inch conduit for all lines except the 220 volt power cable between the two units and a 1 inch conduit for the power line. This separation was deemed necessary to minimize "cross talk" between the 220 volt line and the low voltage cables which connect the two units. The units enclosed in their cases are shown in Figures 9 and 10.

Figure 9. Magnet Enclosure

Figure 10. Console Enclosure

Calibration

The determination of water in a reduction grade alumina amounts to comparing the intensity of the nmr signal from a sample to the intensity of the nmr signal of a suitable known standard. The concentration of water in the sample can then be calculated from:

$$\% H_2O \text{ (sample)} = \% H_2O \text{ (Std)} \times \frac{\text{Intensity (sample)}}{\text{Intensity (Std)}} \times \frac{\text{Density (Std)}}{\text{Density (sample)}} \quad (7)$$

In practice, the bulk density of reduction grade alumina remains constant enough to drop the density term from equation (7).

The IBM calibration procedure for the nmr uses three known standard samples to calculate the slope and intercept of the straight line equation relating spectrometer reading to water content (i.e., y = mx + B). It would be difficult to obtain and keep three known alumina standards at a plant. It was easier to calibrate the spectrometer with one stable secondary standard. We chose a delrin (DuPont plastic) rod as this stable secondary standard. A 1/8 inch diameter delrin rod was placed in a special nmr tube as shown in Figure 11.

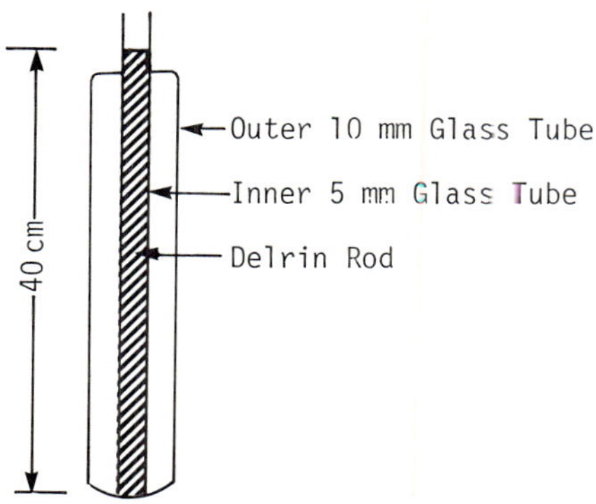

Figure 11. Delrin Rod In nmr Tube

The above delrin rod was chosen because it gave an nmr signal of about the same intensity and decay rate as a typical reduction grade alumina. An accurate calibration depends on knowing what the delrin rod must read in order to obtain the correct value of the combined water content of a reduction grade alumina. This was done by setting the slope and intercept of the spectrometer so that it will give the correct value for a known ore sample. The sample was then removed from the spectrometer and the delrin rod inserted and read. The rod is then used as a secondary standard for subsequent calibration of the instrument.

In actual practice the nmr has been found to be very stable and only minor changes in calibration have been required in the approximately two years we have been using them in our plants. Routine maintenance has consisted of replacing faulty electrical components in both the magnet and the console.

Results

Nine reduction grade alumina samples containing a broad range of loss-on-ignition (LOI) values were obtained from our Gramercy, LA, plant. The nmr free induction decays (FID) of these samples showed that some had picked up free (physical) water in transit. Figure 12 shows how combined water can be measured in the presence of free water in these samples.

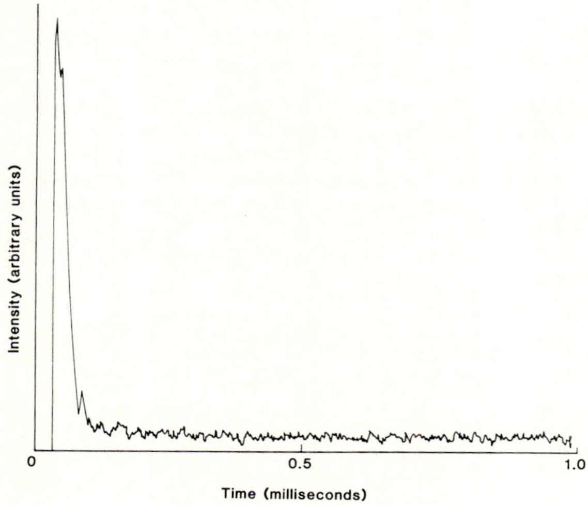

Figure 12A. 'H Free Induction Decay Due To Chemically Combined Water In Reduction Grade Alumina

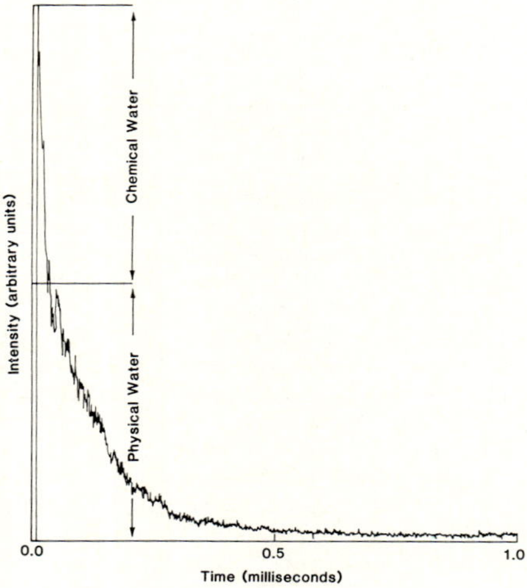

Figure 12B. Free Induction Decay For Alumina Containing Both Physically Adsorbed and Chemically Bound Water

The nmr and LOI methods agree quite well for the total water content of these samples (see Figure 13 and Table I). The distinction between the concentrations of physically absorbed and chemically bound water requires some discussion.

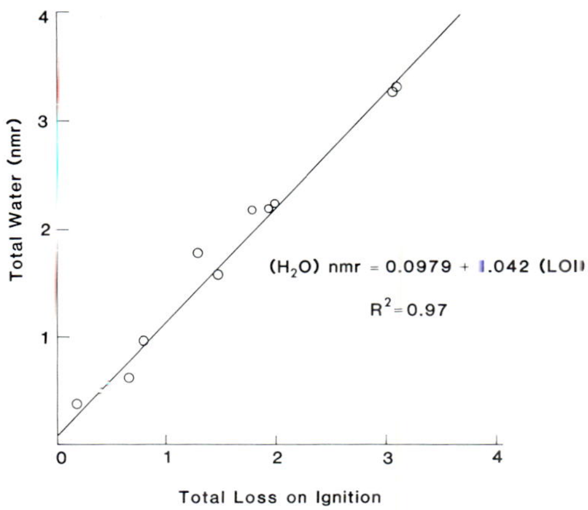

Figure 13. Total Water Content By nmr vs. Total Loss-On-Ignition For Nine Gramercy Alumina Samples

Table I. LOI and nmr Results For Nine Gramercy Aluminas

Sample	LOI					nmr				
No.	300°C		1000°C		Total	Physical		Chemical		Total
1A	0.28	0.28[1]	0.17	0.18	0.45	0.08	0.07	0.32	0.31	0.38
1B	0.28		0.18			0.06		0.30		
2A	0.27	0.25	0.79	0.80	1.45	0.09	0.09	0.89	0.87	0.97
2B	0.23		0.82			0.09		0.85		
3A	0.70	0.71	1.08	1.07	1.77	0.30	0.29	1.77	1.90	2.18
3B	0.72		1.06			0.27		2.02		
4A	-0.01	0.00	0.67	0.66	0.67	N.D.[2]		0.67	0.62	0.62
4B	0.01		0.66					0.57		
5A	0.04	0.07	1.29	1.27	1.34	N.D.		1.50	1.45	1.45
5B	0.09		1.26					1.40		
6A	2.11	2.12	1.50	1.49	3.57	2.99	2.51	1.50	1.60	4.11
6B	2.12		1.48			2.02		1.70		
7A	0.14	0.17	2.01	1.95	2.12	N.D.		2.24	2.18	2.13
7B	0.20		1.89					2.12		
8A	0.23	0.21	3.07	3.05	3.26	N.D.		3.26	3.28	3.23
8B	0.19		3.04					3.30		
9A	0.26	0.25	3.10	3.08	3.31	N.D.		3.23	3.31	3.31
9B	0.23		3.07					3.39		

[1]Average of duplicate samples underlined; [2]Not detected.

The fast decay is due to protons shown in bold type of Figure 14, i.e., chemically bound (combined water). The smaller protons in physically adsorbed water contribute to the slow decay (narrow nmr line) found in the nmr spectrum of these materials. It is important to note that the relative amounts of chemically bound to physically adsorbed water can be measured directly from the free induction decay of an aluminum oxide.

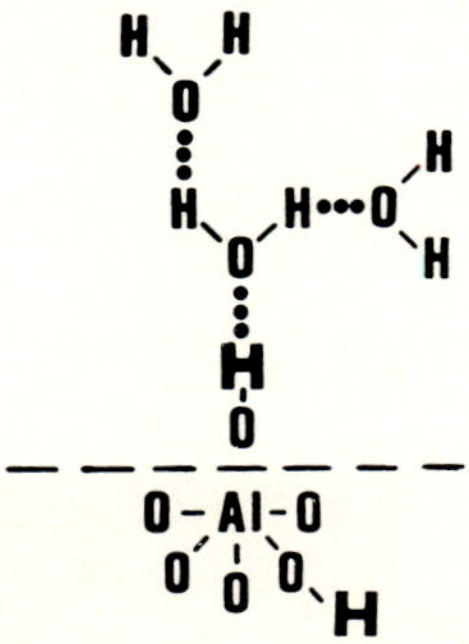

Figure 14. Types Of Protons In Reduction Grade Alumina

The gravimetric (LOI) methods attempt to make this distinction by treating the alumina at some intermediate temperature to measure the physically adsorbed water, then increasing the treatment temperature to the 1000°C to 1200°C range to measure the combined water. The intermediate temperature is arbitrary and considerable chemical water is lost at this temperature as can be seen from Table II.

Table II. Physically Adsorbed and Chemically Bound Water In Gramercy Aluminas

	Physical Water		Chemical Water		Total Water	
Sample	LOI, %	nmr, %	LOI, %	nmr, %	LOI, %	nmr, %
1	0.28	0.07	0.18	0.31	0.45	0.38
2	0.25	0.09	0.80	0.87	1.05	0.97
3	0.71	0.29	1.07	1.90	1.77	2.18
4	0.00	0.00	0.66	0.62	0.67	0.62
5	0.07	0.00	1.27	1.45	1.34	1.45
6	2.12	2.51	1.49	1.60	3.57	4.11
7	0.17	0.00	1.95	2.18	2.12	2.18
8	0.21	0.00	3.05	3.28	3.26	3.28
9	0.25	0.00	3.08	3.31	3.31	3.31

The nmr method will, on average, give smaller amounts of physically bound water because the sample is not heated to some intermediate temperature which will drive off chemically bound water. The nmr method will give higher chemically bound water so that the total water values for the two methods will agree quite well. (Figure 13)

Analysis Of Alumina From Kilns

The above problem can be avoided in analyzing ore from a kiln by taking the sample at a place close enough to the kiln so that no physical water is found. We have found that for distances up to 50 feet down the air slide, the physical water is not a problem.

Plant Installation

Figure 15A shows the magnet on a kiln at our Baton Rouge plant, while 15B shows the console in the control room.

Figure 15A. Spectrometer Magnet Installed At Baton Rouge

Figure 15B. Console In Control Room

Plant Data

The spectrometer was first installed at Gramercy without an automatic sampler. The magnet was located 25 feet down the air slide from the rotary coolers on the end of the kiln. Samples were dipped out of the air slide and poured into the sample tube in the magnet. The console was located in the control room about 100 feet away. Sixty consecutive samples were taken about four hours apart; combined water was determined by nmr; lower plug was removed from the tube, sample caught in a jar, and sent to the lab for LOI determination. Table III shows the results obtained. These samples cover a relatively narrow range of water concentrations and the usual linear regression analysis is probably not a very good indication of how well the data agree. It can be seen from Table IV that nmr data are more repeatable than LOI results.

It is important to note that although individual pairs of points may not agree, the time average of the results are very close, i.e., in Table III average LOI = 0.59%, whereas average nmr = 0.62%. This means that the lab results are less consistent than nmr values, but that the lab error is random.

Figure 16 compares nmr results to lab LOI measurements as a function of time.

Table III. Lab LOI vs. nmr LOI (%)

Lab	nmr	Lab	nmr	Lab	nmr	Lab	nmr	Lab	nmr
.70	.64	.77	.79	.53	.66	.40	.45	.14	.45
.48	.47	.30	.37	.22	.59	.53	.61	.91	.81
.86	.80	.19	.46	.36	.54	.77	.68	.95	.89
.88	.89	.53	.54	.92	.79	.76	.69	.48	.48
.51	.53	.53	.51	.31	.62	.49	.49	.58	.62
.49	.47	.67	.68	.50	.64	.80	.75	.61	.62
.71	.62	.80	.77	.73	.80	.79	.69	.84	.86
.58	.66	.69	.71	.66	.51	.70	.60	.45	.60
.32	.41	.56	.59	.66	.70	.50	.50	.52	.56
.92	.59	.74	.76	.66	.67	.54	.56	.47	.47
.26	.35	.53	.71	.58	.52	.81	.74	.59	.56
.28	.44	.47	.57	.67	.60	.76	.66	.53	.51

Lab Average = 0.59
nmr Average = 0.62

Table IV. Repeatability of nmr and LOI Data (%)

Analysis Performed By nmr (3 analyses)			Analysis Performed By Lab (2 analyses)		
Low	High	Range	Low	High	Range
.39	.45	.06	1.53	1.83	.30
.26	.36	.10	.30	.61	.31
.38	.46	.08	.43	.49	.06
.37	.39	.02	1.02	1.11	.09
.63	.67	.04	1.71	1.83	.12
.20	.24	.04	.45	.56	.11
.29	.35	.06	1.18	1.44	.26
.69	.71	.02	.34	.69	.35
.64	.70	.06	.33	.48	.15
.32	.39	.07	.44	.57	.13
.48	.55	.07	.47	.69	.22
.65	.70	.05	.79	1.13	.34
.31	.37	.06	.33	1.29	.96
.94	.98	.04	.45	.77	.32
.70	.77	.07	1.12	1.37	.25
.49	.56	.07	.76	1.05	.29
.35	.39	.04	1.19	1.42	.23
	Avg.	.06		Avg.	.26

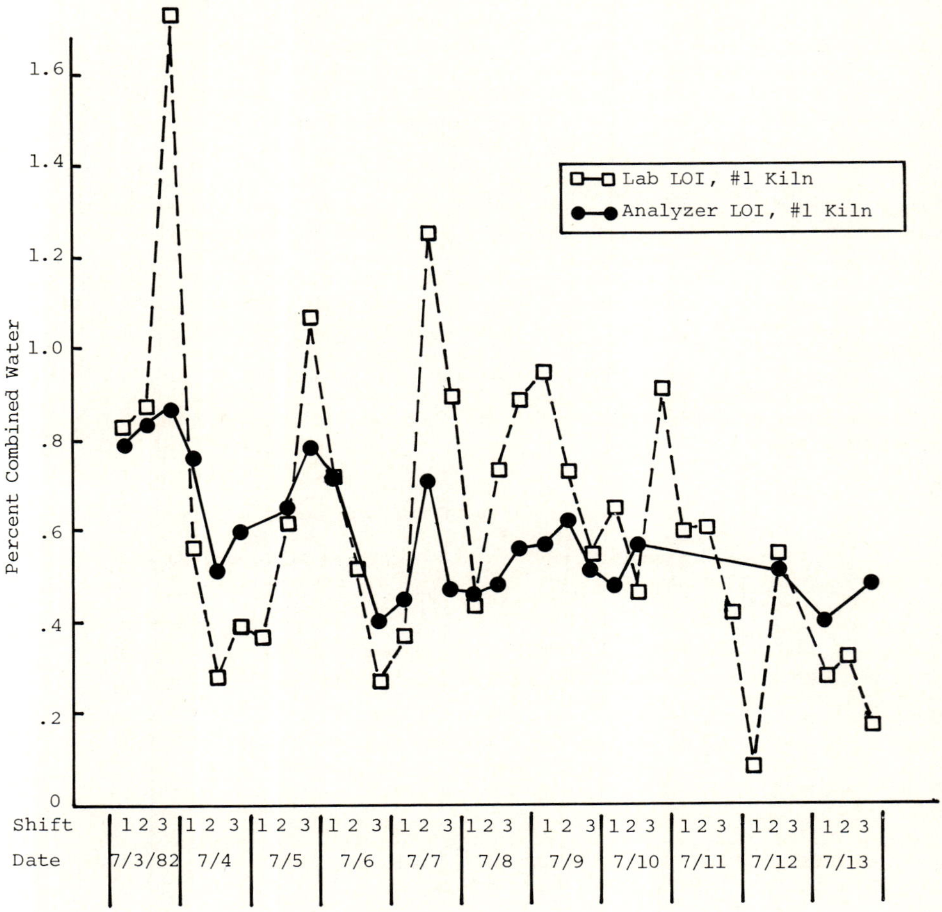

Figure 16. Lab LOI vs. nmr LOI Gramercy Plant

Conclusions

The data presented in this paper show how a commercially available nmr spectrometer can be refitted and used to measure the combined water content of reduction grade alumina as it exists in a rotary kiln. The nmr is probably more reliable than conventional gravimetric results. The spectrometer is much more rapid. An analysis every five minutes could be made. This allows one to monitor the kiln essentially on a continuous basis. On-line analysis of the ore from the kiln makes automatic feedback control of the kiln possible. This could result in large savings of energy required to calcine reduction grade alumina.

Although these modified IBM nmr spectrometers have functioned fairly well under plant conditions, they do present some rather serious problems. These instruments are made to operate under controlled laboratory conditions and will not operate on a year-round basis in a plant unless extensive additional temperature control is provided in the magnet unit. The modifications necessary to make these laboratory instruments function under plant conditions are not trivial. They tend to be expensive and difficult to maintain.

What is needed is an instrument designed as an on-line analyzer to operate under plant conditions with easier operation and maintenance capabilities.

References

1. R. M. Pearson, "Wide Line Nuclear Magnetic Resonance Studies on Transition Aluminas--Distribution of Protons Between Surface and Bulk Phases," Journal of Catalysis, 23 (1971) pp. 388-394.

2. B. R. Baker and R. M. Pearson, "Water Content of Pseudoboehmite: A New Model for Its Structure," Journal of Catalysis, 33 (1974) pp. 265-278.

3. B. R. Baker and R. M. Pearson, "Sealing Studies of Anodic Oxides by Wide-Line NMR Spectroscopy," Journal of Electrochem Soc., 119 (2) (1972) pp. 160-165.

4. B. R. Baker, J. d. Balser, R. M. Pearson, and E. O. Strahl, "Sealing of Anodic Oxide Films on Aluminum," Aluminium, 52 (1976) pp. 241-245.

5. T. C. Farrar and E. D. Becker, Pulse and Fourier Transform nmr - Introduction To Theory and Methods, Academic Press, New York, 1971.

CHAIRMEN

J.C. Winch
Alcan International Limited
Montreal, Quebec, Canada

F.S. Williams
Alcoa Laboratories
Alcoa Center, Pennsylvania

AUTOMATED LABORATORY LIQUOR ANALYZER--ALLA--FOR ANALYSES OF BAYER LIQUOR

Charles K. Matocha

Analytical Chemistry Division
Alcoa Laboratories
Alcoa Center, Pennsylvania 15069
USA

ALLA--the "Automated Liquor Analyzer" completely automates the gluconate-fluoride method for determination of total caustic carbonate and alumina in Bayer liquors. The advantages obtained include a better utilization of manpower, elimination of operator variability, and a significant improvement in accuracy and precision. The system is described as well as some of the technical aspects of the analytical procedure.

Introduction

Among the most important analyses required by a refining plant are the determinations of total caustic, carbonate, and alumina. All three components can be determined on a single sample using a method developed by Watts and Utley (1) that has become known as the "gluconate-fluoride" method (2). In this method, sodium gluconate is added to the solution containing sodium aluminate, sodium carbonate, and free sodium hydroxide. The gluconate complexes the aluminum, and titration with hydrochloric acid (HCl) determines the free and combined, or total caustic (TC) plus one-half the carbonate. The mixture is acidified, and the carbonate which is released is purged with nitrogen gas. After neutralizing the excess acid, the remaining half carbonate is titrated with HCl. The gluconate complex is broken by addition of potassium fluoride causing the formation of K_2NaAlF_6 and the release of three moles of hydroxide for each mole of aluminum. The hydroxyl ion release is titrated with HCl and equated to the alumina content. It is Alcoa's practice to report in g/L: Total Caustic (TC) as equivalent Na_2CO_3, Total Alkali (TA) as the sum of Total Caustic plus the determined sodium carbonate, and alumina (Al_2O_3). Additionally, the ratios of TC to TA and Al_2O_3 to TC are calculated and reported.

The "Automated Laboratory Liquor Analyzer"--ALLA--developed by Alcoa, completely automates this method. This system was developed to take advantage of state-of-the-art automation. Advantages obtained through the use of ALLA include:

1. Improved accuracy and precision.
2. Elimination of operator variability.
3. Improved manpower utilization.
4. Reliable operation.

This report describes the equipment used and some of the technical details investigated in the development of the system.

ALLA--The "Automated Laboratory Liquor Analyzer"

The Brinkmann/Metrohm hardware used for this system is shown in Figure 1. The system consists of a titrator (HCl) with single end-point detection, two dispensers (HCl and NaOH), valves for potassium fluoride and water addition and nitrogen purge, a control module, an 8-position turntable, and a system computer (H-P 85). Additionally, lids protect the sample from atmospheric carbonation prior to analysis, and a de-capping unit was designed and built by Alcoa. All sequencing for titrations, reagent dispensing, sample presentation, timing, and value operation is coordinated through the Brinkmann/Metrohm Control Module. This, importantly, allows access to the computer concurrent with analysis.

The system computer is dedicated to sample identification, calculations and data input/output. Initially, up to eight samples can be placed on the turntable, Figure 2. As samples are completed they can be replaced and new identifiers added to the sample queue in the computer. The system will continue with analyses as long as samples are identified into the array. When a beaker with water is positioned after the last sample and so identified, the analyzer stops with the electrode and dispenser tips in the water beaker.

FIGURE 1: ALLA--AUTOMATED LABORATORY LIQUOR ANALYZER

FIGURE 2: 8-POSITION TURNTABLE

All titration volumes are automatically transmitted to the system computer, which then calculates the constituent concentrations. At the conclusion of each sample analysis all values are reported on the printer and filed on the cassette tape of the H-P 85. The analysis log generated on the tape can be used as the data base when reporting to another computer, or to produce, for example, a shift report.

Potassium fluoride (KF) is added by a timed gravity feed through a control valve. All components are washed with water after being removed from the sample. This water addition is also a timed gravity feed through a control valve. The nitrogen purge and control valves are timed by the Control Module. Sodium gluconate is not added by the system; it is added during the sample preparation procedure.

Analytical time varies with the sample type, but a typical green liquor can be analyzed in about 10 minutes. When processing assorted samples, about 140 samples can be analyzed over a three-shift period. Eight analyzers are currently in operation in Alcoa's laboratories; five units are operated on a three-shift basis.

Analytical

Automated Titrations

In addition to analyzing typical Bayer liquors, ALLA is capable of analyzing other samples containing varying concentrations of TC, carbonate and alumina. Since some samples may not require complete analysis, the control module is programmed to select, upon demand, sequencing to determine:

A: TC, TA, Al_2O_3, Al_2O_3/TC, TC/TA

B: TC, TA, TC/TA

C: Al_2O_3

Sequence "C" is used to determine "Available Alumina" in the laboratory bauxite digest test or "Bomb Digest" method.

The titrator must, therefore, provide rapid titrations and accurate end-point detection for a variety of constituent concentrations and solution buffering--all without operator intervention. This required some changes in the titration procedure and use of the titrator/dispenser units. The original method was modified so that 3.00 mL of excess NaOH is present before the start of each titration. The excess NaOH is treated as a simple back titration and deducted as a blank value in the calculations. The use of currently determined blanks:

1. Improves the accuracy of analysis.
2. Eliminates the need for equivalency of NaOH and HCl solutions.
3. Eliminates adjustment of gluconate and potassium fluoride solutions to match the end-point.
4. Minimizes solution preparation time.

With excess NaOH stipulated, Brinkmann installed an additional circuit to allow all titrations to start and continue at maximum delivery rate until the appropriate control band is reached. This produces a fast titration. The excess NaOH also eliminates possible over-titration for low or missing concentrations.

The original investigation by Watts and Utley (1) specifies that after addition of potassium fluoride, HCl be rapidly added to the end-point and then a small excess for the quantitative formation of the fluoroaluminate. Brinkmann programmed the Control Module to over-ride the titrator control band at this point and allow unrestricted HCl addition which is terminated at the end-point; then the excess HCl is added. The ALLA procedure eliminates the previously required NaOH back-titration for the alumina determination (2). It should also be noted that the end-point for the alumina determination is sharper when determined by HCl titration.

The Brinkmann anti-diffusion tip is used on the titrators and dispensers. Its effectiveness was verified by leaving the tip in the solution after the end-point is established and monitoring the stability of the end-point with time. This stability is not observed for other tips without diffusion control. Diffusion is most apparent during the nitrogen gas purge. Elimination of diffusion improves accuracy.

Selection of End-Point

Watts and Utley (1), discussed the selection of end-point for the three titrations, and pH 8.1 is accepted as the common end-point (2). It was noted in this investigation that after potassium fluoride addition the solution is buffered to a higher pH which elevates the inflection point for the last titration (alumina determination). The tiration curves were reinvestigated with respect to the inflection point and the rate of change, ΔpH/Δ volume, as the inflection point is approached. In order to compensate for the end-point shift and the highly buffered titration for the alumina determination, the common end-point was changed to pH 8.2. Although buffered, the TC and carbonate titrations are sharp, and a change in these end-points is not as critical.

Calibration of the End-Point

Prior to using an automated titrator, the pH meter is calibrated with buffers, and the pH value of the end-point is set on the titrator. However, the titration proceeds under control of a null-point detector, that is, the titration is terminated when the measured and sought pH values coincide. Taking advantage of this, the ALLA procedure uses a more direct and accurate way of establishing the end-point; the null-meter is zeroed using a pH buffer of the desired end-point. A standard buffer at pH 8.20 has been established and is used to calibrate the end-point.

The next generation titrator should have the capability to off-set the isopotential point from the usual pH 7.0 to the end-point. Temperature effects would then be minimized.

Standardization of HCl

The ALLA procedure uses HCl and NaOH of approximately 0.5N, and it is not essential that they be equivalent normalities. Also, only the HCl needs to be standardized. The normality of the HCl is determined on the titrator. This insures accurate titrations and the same end-point. Potassium acid phthalate is used in the standardization routine and a procedure similar to analyzing a sample is followed. After the potassium acid phthalate (KHP) is dissolved in water that has been adjusted to the end-point, twenty-one mL of NaOH is dispensed into the solution and the excess is back-titrated with the HCl being standardized; its volume being T_1. An additional twenty-one mL of NaOH is then dispensed and the titrated volume of HCl is recorded as T_2. The normality is calculated as:

$$N\ HCl = \frac{\text{Weight KHP}}{0.2042\ (T_2 - T_1)}$$

Elimination of Atmospheric Carbonation

Carbonization of the caustic is very evident when diluted solutions are exposed to the atmosphere. To illustrate the effect of atmospheric carbonation on the analysis, eight samples were placed on the turntable and remained exposed to the atmosphere while awaiting analysis. The results are shown in Figure 3. The decrease in TC is off-set by an equal increase in Na_2CO_3, thus maintaining TA constant. The change in Al_2O_3/TC and TC/TA are also shown.

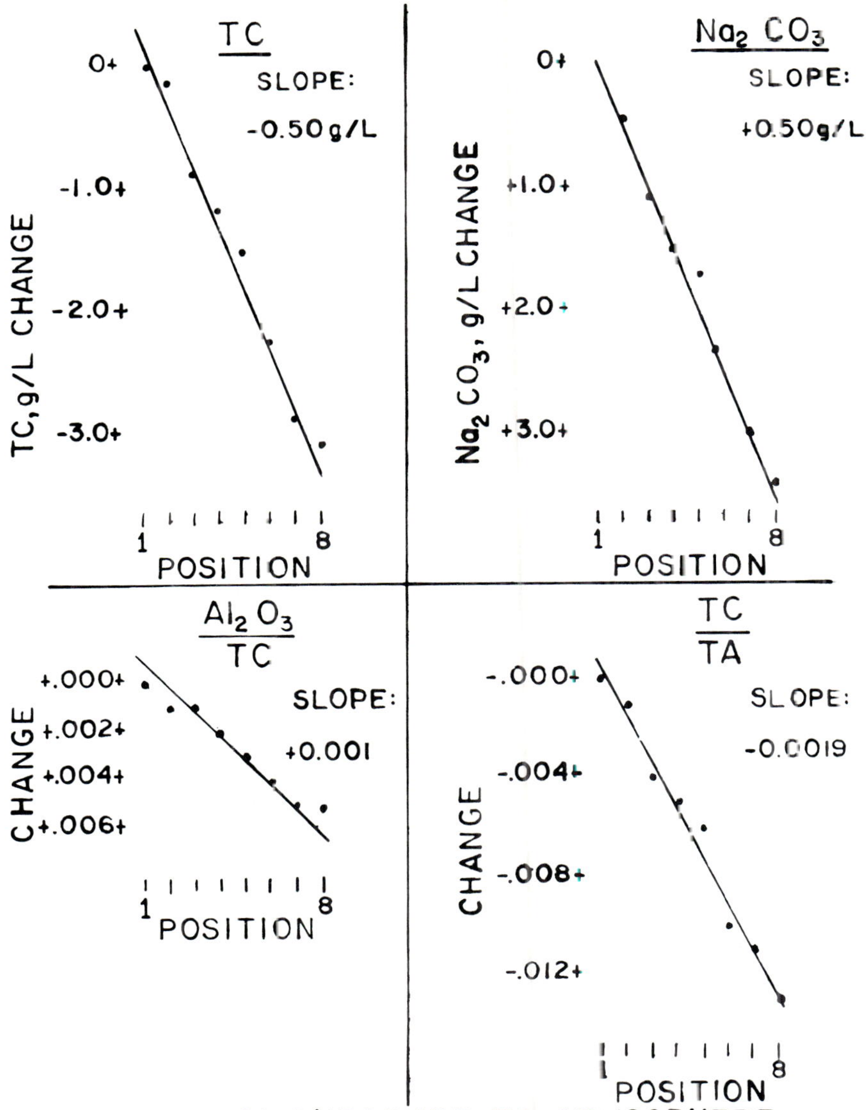

EFFECT OF EXPOSURE TO ATMOSPHERE
PRIOR TO ANALYSIS

FIG. 3 ATMOSPHERIC CARBONATION

Carbonation is eliminated by covering the solution as soon as the analytical sample is prepared. In the ALLA procedure, a magnetic stainless steel cover is used. A de-capping unit, Figure 4, designed at Alcoa Technical Center, uses an electromagnet to remove the cover prior to analysis and a sensor verifies its removal.

FIGURE 4: COVER LID AND DE-CAPPING UNIT

The carbonation test was repeated using the protecting cover. The analytical data and precision are given in Table I. The elimination of carbonation is evident in the stability of the carbonate determination. Additionally, three samples remained covered for three hours prior to analyses. Their analyses are also given in Table I. Only a slight degradation in TC and Na_2CO_3 is detected.

TABLE I

ELIMINATION OF ATMOSPHERIC CARBONATION

- PROTECTING COVER PRIOR TO ANALYSIS -

Position	TC, g/L	Na_2CO_3, g/L	TA, g/L	Al_2O_3, g/L	Al_2O_3/TC	TC/TA
1	191.5	56.1	247.6	52.11	.272	.773
2	191.7	56.1	247.8	52.14	.272	.774
3	191.4	56.1	247.5	52.04	.272	.773
4	191.8	55.3	247.7	52.11	.272	.774
5	191.5	56.1	247.6	52.08	.272	.773
6	191.2	56.7	247.9	52.15	.273	.771
7	191.2	56.3	247.5	52.04	.272	.772
8	191.6	56.3	247.9	52.11	.272	.773
AVG., $\bar{X}$	191.5	56.2	247.7	52.10	.272	.773
s*	±0.2	±0.2	±0.2	±0.04	±.0004	±.001
- COVERED 3 HOURS BEFORE ANALYSIS -						
A	190.9	56.7	247.6	52.11	.273	.771
B	191.3	56.5	247.8	52.14	.273	.772
C	191.3	56.3	247.6	52.08	.272	.773
AVG., $\bar{X}$	191.2	56.5	247.7	52.11	.273	.772

*Standard Deviation

The same sample is used for the three tests reported in Table I and Figure 3. The pooled standard deviations for TA and Al_2O_3 determinations are calculated and reported in Table II. This statistic includes three independent calibrations and a comparison to Table I shows that calibration is maintained.

TABLE II

PRECISION - POOLED STANDARD DEVISION, s
DATA FROM FIGURE 3 AND TABLE I

	n	$\bar{X}$	s
Al_2O_3,g/L	19	52.10	0.035
TA,g/L	19	247.7	0.14

Interferences

The interferences reported by Watts and Utley (1) should be accounted in reporting data. The principal interference is silica which titrates in the alumina determination as an equal amount of Al_2O_3 (2).

Analytical Procedure

The ALLA procedure is outlined in Table III. The operator involvement includes preparation of the samples, their placement and removal from the turntable and entry of corresponding identifiers into the computer.

The use of blanks has been previously discussed. Blanks are the V1, V2, V3 and V4 values determined by analyzing, as a sample, 100-mL of sodium gluconate plus 25-mL water. Following their determinations:

V1, Blank Solution = GB
V2, Blank Solution = CB
V3 + V4, Blank Solution = AB

These values (GB, CB, AB) along with the normality of the HCl, N, are stored in the computer and used to calculate the analysis.

The calculation routine is given in Table IV.

TABLE III

ALLA PROCEDURE OF ANALYSIS

Manual Preparation

Sample

100 mL of 150 g/L sodium gluconate
Aliquot of sample
Cover

Identification

Identifiers to computer
Samples to turntable
Remove 1st cover
Start

Automated Sequence - ALLA

Advance turntable, elevate container

Stirrer on, 15-sec. delay

HCl titration to pH 8.20 as V1

V1 to computer, refill

Dispense 21.00 mL HCl (7 x 3.00 mL)

4-min. nitrogen purge

Dispense 24.00 mL NaOH (8 x 3.00 mL)

HCl titration to pH 8.20 as V2

V2 to computer, refill

Add 35 ml 50% (w/w) $KF \cdot 2H_2O$.

10-sec. delay

Fast HCl titration to pH 8.20 as V3

V3 to computer, refill

Dispense 3.00 ml HCl (1 x 3.00 mL)

60-sec. delay

Dispense 6.00 mL NaOH (2 x 3.00 mL)

15-sec. delay

HCl titration to pH 8.2 as V4

V4 to computer, refill

Stirrer off, lower container

Rinse titration head assembly

Report analysis

Remove next cover

Next

TABLE IV

CALCULATION ROUTINE

TC, g/L as Na_2CO_3 =

[(V1 - V2) - (GB + CB)] * N * 53/mL Sample

CARBONATE, g/L Na_2CO_3 =

2 * (V2 - CB) * N * 53/mL Sample

TA, g/L Na_2CO_3 =

(TC + CARBONATE)

Al_2O_3, g/L =

(V3 + V4 - AB) * N * 17/mL Sample

TC/TA

Al_2O_3/TC

Where: V1, V2, V3 and V4 are titration volumes, mL, determined on a sample;

GB, CB and AB are blank values; and

N is the HCl normality.

Accuracy and Quality Control

Standardization of the 0.5 N hydrochloric acid using the titrator and a procedure similar to analysis as well as the use of current blank values has been previously discussed. The reliability of the standardization is significant to the accuracy. Current blank values account for any minor solution changes and insure long-term accuracy.

The alumina determination is standardized using a solution of high-purity aluminum in excess hydrochloric acid. This can be, for example, an aluminum standard as used for Atomic Absorption analysis containing 10 mg/mL. An appropriate aliquot is taken, and sufficient NaOH is added to provide excess after the acid is neutralized. This is then analyzed as a sample. Additionally, an aliquot of the aluminum standard is added to an analyzed spent liquor and the recovery is verified.

The bicarbonate titrations are verified first using a standard solution of sodium carbonate. To obtain accuracy the procedure should determine one-half carbonate in the first titrations. Following the acid treatment, the CO_2 should be completely purged and the remaining half carbonate should be determined in the second titration. An analyzed spent liquor is then spiked with standard carbonate solution. In addition to verifying Na_2CO_3 recovery, the TC and Al_2O_3 concentrations should remain constant. Data reported in Table V shows the accuracy of the qualifying tests which were obtained during the installation of one of the systems. A known volume of 0.5 N NaOH can also be dispensed into an analyzed liquor and the recovery verified.

Unfortunately, no single liquor, production or synthetic, is stable enough to qualify all determinations. The above tests, however, use stable solutions and practices that can be duplicated under operating conditions. This is important in developing a Q.C. program that can be implemented in a routine laboratory.

TABLE V

QUALIFYING TESTS

A. Determination of Na_2CO_3

1. Solution Analyzed: 96.0 g/L Na_2CO_3

	g/L Na_2CO_3
1st Titration	47.8
2nd Titration	47.9
Total	95.7

2. Solution Analyzed: 48.0 g/L Na_2CO_3 Added to Liquor:

	g/L		
	TC	Na_2CO_3	Al_2O_3
Starting Analysis	184.5	44.5	72.2
After Addition	184.0	92.2	72.2
Net Na_2CO_3 Recovered	--	47.7	--

B. Determination of Al_2O_3

1. Solution Analyzed: 50.0 g/L Al_2O_3 Added to Liquor

	g/L Al_2O_3
Starting Analysis	72.1
After Addition	121.9
Net Al_2O_3 Recovered	49.8

Conclusion

The use of an existing method and chemistry is significant in that the data base is pre-established. In addition to improved precision, the automated hardware eliminates operator variability and the completely automated ALLA frees analysts' time for other tasks. High performance precision obtained under research conditions was reported in Table II; precision under routine conditions can be maintained at:

	Determination				
	Al_2O_3 g/L	TC, g/L	TA, g/L	$\frac{Al_2O_3}{TC}$	$\frac{TC}{TA}$
Standard Deviation of Analysis - Routine	±0.15	±0.25	±0.35	±0.001	±0.002

References

1. Watts, H. L. and Utley, D. W., Analytical Chemistry, Vol. 28, p. 1731-1735, Nov. 1956.

2. Alcoa Standard Method No. 151.8; Analysis of Sodium Aluminate Liquor Determination of Hydroxide, Carbonate, and Alumina (Gluconate-Fluoride Procedure).

APPLICATION OF ION CHROMATOGRAPHY TO THE ANALYSIS OF BAYER AND ALUMINIUM PROCESS LIQUORS

Kwat The   Roger Roussel

Alcan International Limited
Arvida Laboratories
CP 1250, Jonquiere
Quebec, Canada G7S 4K8

Ion Chromatography (IC) has been successfully applied as an analytical technique for the simultaneous determination of inorganic anions in Bayer liquor, scrubber liquor, filter cake, alumina trihydrate, boiler water, as well as the determination of sulphur in Carbonaceous materials. Results were in good agreement with those obtained by the conventional methods or published values. These techniques make possible a complete analysis of several anions of the prepared sample within one hour.

Introduction

The determination of anions by conventional methods usually requires labour intensive and time consuming analytical techniques. The determination of oxalate in Bayer liquor by the conventional ether extraction procedure, for example, requires the extraction of the oxalate with ether for 6 to 12 hours and its isolation as calcium oxalate precipitate. This solid has to be filtered off, washed and dissolved. The resulting solution is then titrated with an excess of an oxidizing agent, and then back titrated with a reducing agent to obtain the value of the oxalate present in the Bayer liquor. It is a complicated procedure which takes over 12 hours for completion, and consequently is prone to error.

Thus the introduction of a commercial ion chromatograph in the market has been timely for our purpose (1). This instrument uses low capacity ion exchange columns to separate the anions, a suppressor column to transform the eluent into carbonic acid and the anions of interest into their respective acid forms. The Dionex Ion Chromatograph with a conductivity detector was found to be very sensitive, accurate and capable of analyzing simultaneously several anions at trace levels. In addition, little or no pretreatement of the sample is required, the equipment is easy to operate, and less than 30 minutes is required for most analyses. In the case of Bayer liquor, the only sample pretreatment step consists of a 500 fold dilution with the eluent electrolyte prior to the injection of 3 mL of the diluted sample. The resulting chromatogram gives us not only the concentration of oxalate but also those of chloride, phosphate and sulphate.

The purpose of this communication is to describe the successful application of ion chromatography to the determination of anions in Bayer solutions, boiler water (2), oxalate in alumina trihydrate, S in carbonaceous materials (3), and some anions in filter cakes and scrubber liquors.

Experimental

Apparatus

Dionex ion chromatograph system 10 equipped with the fast run S3 anion separator, the S3 precolumn and packed bed suppressor column; all columns from Dionex.

Sample Description and Preparation

Bayer Liquor. Twenty two samples of Bayer liquor from various sources, including those from Kirkvine, Belgaum and San Ciprian, were obtained for this experiment. They were prepared for analysis by diluting 500X with the matrix solution of the eluent. Three mL of the diluted solutions were used for analysis.

Alumina Trihydrate. For this experiment, several samples of washed and unwashed samples from our Vaudreuil, San Ciprian, Kirkvine and Ewarton plants were used. They were prepared for analysis by leaching a weighed amount of well dried and mixed samples with hot water and diluting the resulting leachate to 100 mL.

Boiler Water. Six samples of boiler water from the Vaudreuil (4), Kirkvine (1) and San Ciprian (1) plants were used. When necessary, the samples were filtered or diluted with water. Three mL's of the original, or the filtered and diluted samples were used for analysis.

Filter Cake, Scrubber Liquor and Carbonaceous Materials. Pretreatment of scrubber liquor involves dilution only, while that of filter cake requires weighing (0.5 g) dissolution in hot water and dilution to 1000 mL. Samples of carbonaceous materials are combusted in a Schöniger oxygen flask in the presence of an absorbing solution consisting of 30 mL of H_2O, 1 mL of 0.1 NaOH and 3 drops of H_2O_2 (30%).

Experimental Procedure

An aliquot of the prepared sample solution is injected into the chromatograph equipped with an S3 fast run separator and a packed-bed suppressor. The ions are eluted with either a solution of 0.00225 M $NaHCO_3$/0.0018 M Na_2CO_3 or 0.0024 M $NaHCO_3$/0.0030 M Na_2CO_3, depending on the nature of the sample to be analysed. The pressure is set at 500 psi and the attenuation set according to the ions and samples used. The height of the peaks of the ions of interest are measured, and their concentrations read off from their respective calibration curves. The time required for one analysis is about 20 minutes.

The regeneration of the suppressor is carried out with 1N H_2SO_4 after 6 hours of continuous use by means of the built-in automatic regeneration system.

Results

Bayer Liquors

A total of 22 samples of Bayer liquor for various plants were used for the experiment. Thirteen for the analysis of $PO_4^{\equiv}$ and Cl^- and 9 for $SO_4^=$ and $C_2O_4^=$. A typical chromatogram of the anions Cl^-, $PO_4^{\equiv}$, $SO_4^=$ and $CO_4^=$ in a Bayer solution is shown in Figure 1. Different attenuation settings were used to take into account the different sensitivities and concentrations of those ions. Their peaks are nevertheless sharp, well resolved and free from interferences.

Repeatability of the I.C. measurements for these ions is shown in Table I. The data reported here were the average of 7 analyses over a period of several days. The coefficients of variation are quite good, even at concentrations as low as 0.2 g/L of $PO_4^{\equiv}$.

The accuracy of the ion chromatographic method was tested against the conventional methods: Volhard titration for chloride, molybdenum blue colorimetric determination for phosphate, gravimetric analysis for sulphate and extraction procedure described above for oxalate. Typical results are shown in Table I for Bayer liquors having the approximate concentration ranges: Cl^- 0.4 to 7 g/L, $PO_4^{\equiv}$ 0.03 to 0.3 g/L, $SO_4^=$ 0.4 to 6 g/L and $C_2O_4^=$ 2 to 4 g/L.

Alumina Trihydrate

Figure 2 shows a typical chromatogram of the anions leached from unwashed sample of alumina trihydrate. The chloride and sulphate peaks are also well defined and free from interference and could be used for analytical determination, should the need arise. The reproducibility of the oxalate determination is estimated as $\sigma \sim 2\%$ and the results of a comparative study with the gas chromatographic procedure are shown in Table II. The accuracy of the method is also assessed with the recovery test, involving 29 spiked samples. The % recovery obtained was 96.9 ± 2.4%.

Boiler Water

Figure 3 shows a chromatogram of synthetic solution containing 8 anions that are commonly found in the environment and could conceivably be found in the water used for the boilers. However, only Cl^-, $PO_4^{\equiv}$ and $SO_4^=$ were present in appreciable amounts, and NO_3^-, $SO_3^=$ in smaller quantities in the samples analyzed. The presence of a minute amount of oxalate was also observed in one of the samples.

The results of the comparative study with the conventional methods are shown in Table III. No attempts were made to do similar experiments with samples from Kirkvine and San Ciprian, but the accuracy of the method was tested by spiking these samples with known amounts of Cl^-,

$SO_4^{=}$ and $PO_4^{\equiv}$, giving a recovery of 103% with a standard deviation of 0.4% for 18 determinations.

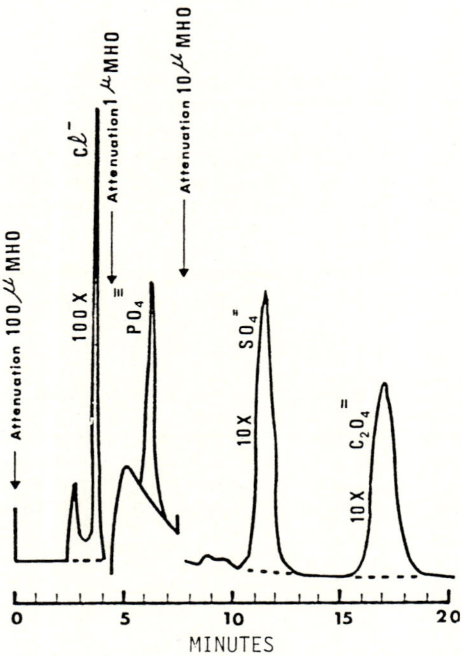

Figure 1 - Chromatogram of Anions in Bayer Solution

Table I. Comparison of Accuracy of Analysis of Cl^-, $PO_4^{\equiv}$, $SO_4^=$ And $C_2O_4^=$ in Bayer Solutions Using the Standard Methods and the Ion Chromatographic Procedure

SAMPLE #	Cl^- S.M./I.C.	$PO_4^{\equiv}$ S.M./I.C.	$SO_4^=$ S.M./I.C.	$C_2O_4^=$ S.M./I.C.
1	1.012	0.958	1.178	0.971
2	0.988	1.043	0.930	0.975
3	0.982	0.923	0.975	1.021
4	1.000	1.043	1.000	0.972
5	0.978	1.000	1.000	1.000
Average	0.992	0.993	1.017	0.988
C.V.	0.4%	1.15%	0.2%	1.0%

S.M. = Standard method
I.C. = Ion Chromatography
C.V. = Coefficient of variation from 7 determinations

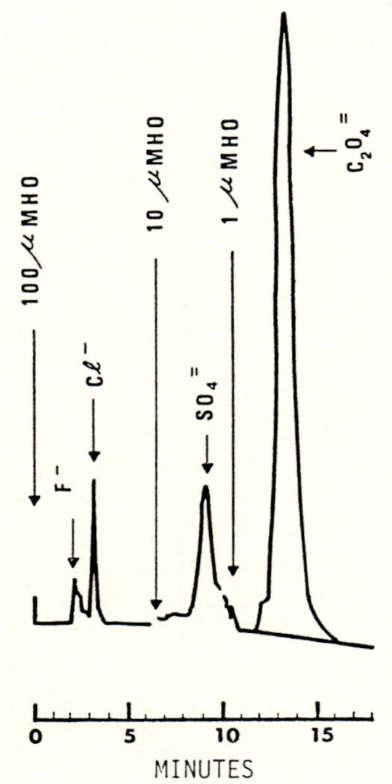

Figure 2 - Chromatogram of a Leachate of Alumina Trihydrate.

Table II. Results of the Comparative Study in the Determination of Sodium Oxalate in Alumina Trihydates by Ion and Gas Chromatographic Procedures

SAMPLE	SODIUM OXALATE, %	
	I.C.	G.C.
Plant #1, washed	0.020	0.016
Plant #1, washed	0.044	0.040
Plant #1, unwashed	0.15	0.11
Plant #1, unwashed	0.67	0.71
Plant #2, washed	0.009	N.D.
Plant #2, unwashed	0.034	N.D.
Plant #3, washed	0.0006	N.D.
Plant #3, unwashed	0.040	0.030

I.C. = Ion Chromatography
G.C. = Gas Chromatography
N.D. = Not detected

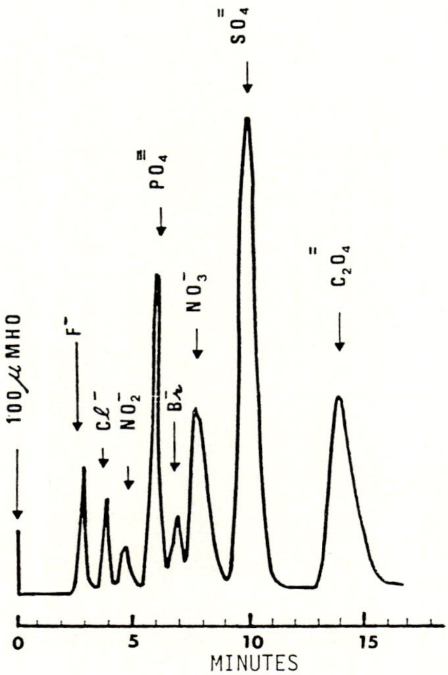

Figure 3 - Chromatogram of a Synthetic Solution of Mixed Anions

Table III. Comparison of Results of Analysis of Boiler Water of Vaudreuil Plant by the Standard Method and the Ion Chromatographic Procedure

SAMPLE #	$PO_4^{=}$, g/L		$SO_4^{=}$, g/L		Cl^-, g/l	
	S.M.	I.C.	S.M.	I.C.	S.M.*	I.C.
4	.02	.01	.10	.11		.24
5	.13	.14	.06	.07		-
6	.11	.10	.12	.13		.23
7	.10	.09	.15	.16		.18
Average	.09	.09	.11	.12		.22

* Unable to get consistent results. See text.

Filter Cake, Scrubber Liquor and Carbonaceous Materials

The procedure of analysis for scrubber liquor is straight forward since the sample is a liquid and requires dilution as the only pretreatment step. Sample of filter cake, on the other hand, is generally wet, but cannot be dried because drying will result in the unwanted formation of insoluble solids that contain sulphate and oxalate. The procedure adopted involves weighing the wet, thoroughly mixed sample and dissolving in hot water. Typical relative standard deviation is 1 to 2% for both the sulphate and oxalate determinations. In contrast, the method for the determination of sulphur in carbonaceous materials requires combustion of the sample in a Schöniger oxygen flask in the presence of an absorbing solution. The results agree quite well with those obtained by the Leco technique (Table IV).

Table IV. Results of a Comparitive Study on the Determination of Sulphur in Carbonaceous Materials by Ion Chromatography and Leco Technique

SAMPLE	SULPHUR, %	
	ION CHROMATOGRAPHY	LECO
ISO Pitch #93	0.69	0.73
ISO Pitch #94	0.40	0.38
ISO Pitch #95	0.57	0.56
ISO Pitch #96	3.02	3.06

Discussion

Table I also shows excellent accuracy and precision of the ion chromatographic procedure for the determination of Cl^-, $PO_4^{\equiv}$, $SO_4^=$ and $C_2O_4^=$ ions in Bayer solutions even for concentrations as low as 0.2 g/L $PO_4^{\equiv}$. These findings are further confirmed by the data of the recovery tests (25 determinations) showing essentially 100% recovery for all the four ions investigated.

In addition, the ion chromatographic procedure has a far higher sensitivity, suffers less interferences than the conventional method and requires less sample for analysis. Five mL of boiler water is more than sufficient for the determination of Cl^-, $PO_4^{\equiv}$, NO_3^-, $SO_3^=$ and $SO_4^=$ by the ion chromatographic method. In contrast, 300 mL of the sample is required for one determination of the $SO_4^=$ ion by the conventional gravimetric technique. This makes it prohibitive and sometimes impossible to carry out frequent analyses of anions. The conventional method of analysis for anions is prone to interference. A case in point was the determination of chloride in some samples of boiler water. Because of the dark brown coloration and the presence of a relatively large amount of suspended material in the samples, we were unable to obtain consistent results with the Volhard titration, but encountered no such difficulty with the ion chromatographic procedure.

The ion chromatographic method has also been successfully applied to the analysis of oxalate in alumina trihydrate. Pretreatment of the samples involves only leaching the sample with hot water and making up the leachate to 100 mL with a volumetric flask. The reproducibility is good and its accuracy as assessed by the recovery test was excellent (99% recovery). Our results indicate that this method can also be applied to the analysis of washed alumina trihydrate, containing as little as 0.001% oxalate. This is about 10 to 20 times more sensitive than the gas chromatographic method formerly used by Alcan.

The successful application of ion chromatography for the determination of $SO_4^=$ and $C_2O_4^=$ in filter cake and $SO_4^=$ (= sulphur) in carbonaceous materials further demonstrate the flexibility and versatility of the technique.

Conclusions

The ion chromatographic procedure has been found to be as accurate, more sensitive and more rapid than the conventional chemical methods for the determination of anions in Bayer liquor, boiler waters, scrubber liquor, filter cake, oxalate in alumina trihydrate and sulphur in carbonaceous materials.

References

(1) Dionex Corporation, 1228 Titan Way
Sunnyvale, C.A. 94086, U.S.A.

(2) Law, J.J., Power Engineering, 1981, June 94.

(3) Sawicki, E., Mulik, J.D. and Wittgenstein, E.,
Ion Chromatographic Analysis of Environmental Polluants,
Chapter 13, page 149.

Useful Reviews and Books:

1. Small, H., Anal. Chem. 1983, Vol. 55 (2), 235 A.

2. Johnson, E.L., American Laboratory, 1982, February, 98.

3. Girard, J.E. and Glatz, J.A., American Laboratory 1981, October, 26.

4. Fritz, J.S., Gjerde, D.T. and Pohlandt, C., Ion Chromatography, Dr. Alfred Hüthig Verlag Hudelberg, Basel, New York 1982

5. Mulik, J.D. and Sawicki, Ion Chromatographic Analysis of Environmental Polluants, Vol. 1 and 2, Ann Arbor Science Publishers Inc., Ann Arbor, Michigan 1979.

CHARACTERIZATION OF ORGANICS IN BAYER LIQUOR

J. D. Guthrie, P. J. The, and W. D. Imbrogno

Aluminum Company of America
Alcoa Technical Center
Alcoa Center, Pennsylvania 15069

Routine analytical methods have been developed for characterizing Bayer liquor. The methods consist of instrumental techniques, including gas chromatography for the low and intermediate molecular weight compounds and gel permeation chromatography for the high molecular weight material. Gas chromatography/mass spectrometry was used to identify several compounds in addition to those previously reported in the literature. Two spent liquor samples, together with corresponding lake water samples, were characterized to demonstrate applicability of the methods. Results of the analyses provided information for better understanding the nature of Bayer liquor.

Introduction

Organic matter found in bauxite is a composite of many substances described by such terms as humic, lignin, cellulose and protein. In the Bayer process, a portion of this organic matter is extracted into the liquor during digestion of the bauxite, and is decomposed to form soluble sodium organic compounds. To promote red mud settling, synthetic or natural flocculant, such as copolymers of acrylamide and sodium acrylate or starch is added to the liquor stream. Other materials which contribute small amounts to the organic content of liquor include anti-foaming agents and lubricants. With recycling of the liquor, the concentration of organics and their degradation products increases until an equilibrium concentration is reached.

It has been shown that the presence of a significant amount of organic material in Bayer liquor causes numerous process problems and lowers liquor productivity. Difficulties caused by organic impurities include lower alumina yield, generation of excessive fine aluminum trihydroxide particles, a higher impurities content in the alumina, colored liquor and aluminum trihydroxide, lower red mud settling rate, loss of caustic due to the formation of sodium organic compounds, increased liquor density, viscosity, boiling point and foaming of the liquor.

Numerous studies concerning the identification of organics in Bayer liquor have been reported (1-5). In general, the analytical methods employed were relatively long, thus limiting the number of samples which could be studied in detail.

The objective of the work reported here was to develop analytical techniques which were simple enough to allow characterization of organics in liquor on a routine basis. These techniques would facilitate studies related to control of organics in the Bayer process stream.

Methods incorporating three instrumental techniques were developed to characterize the organics. Gas chromatography (GC) under different sets of conditions was used to determine the low and intermediate molecular weight compounds. The high molecular weight materials were characterized by gel permeation chromatography (GPC). During development of the analytical methods, a gas chromatography/mass spectrometry (GC/MS) study was undertaken to identify compounds in the liquor in addition to those already reported in the literature.

To demonstrate the applicability of these methods, two spent Bayer liquor samples were characterized, along with the corresponding lake water for each sample. These samples were chosen because the organics in them were found to be different, even though the Bayer refining plants are quite similar.

Experimental Methods

Low Molecular Weight Compounds

To prepare a spent Bayer liquor sample for analysis, 1 ml was added to a 24 ml screw cap vial containing ~15mg of nonanoic acid (internal standard), and the mixture was cooled in an ice bath. To this, 1.5 ml of concentrated hydrochloric acid was carefully added with swirling until all of the alumina redissolved. Two milliliters of 1-butanol was added and the mixture was vigorously shaken for 2 minutes. The sample was heated for 1 hour at approximately 70°C in a water bath. After cooling, the butanol layer was removed with a disposable Pasteur pipet, without removing any of the aqueous phase, and 1 µl of the solvent containing the butyl esters was injected into the GC.

To prepare a lake water sample for analysis, 20 ml was placed in a Petri dish and allowed to evaporate in a hood. One milliliter of 12% sodium hydroxide was used to rinse the air dried material into a 24 ml vial containing ~15mg of nonanoic acid as an internal standard. The remainder of the sample preparation procedure was the same as described for low molecular weight compounds in a liquor sample

Samples were run on a Hewlett-Packard 5880 GC equipped with a flame ionization detector. The column was 1/8 in. O.D. x 10 ft. stainless steel packed with 10% SP-1000 on 100/120 mesh Chromosorb WAW (Supelco, Inc.). Chromatographic conditions were the following:

Injection port temp.	250°C
Detector temp.	350°C
Initial oven temp.	100°C
Time at initial temp.	5 min.
Program rate	10°/min.
Final oven temp.	250°C
Time at final temp.	5 min.
Carrier flow	30 ml/min.
Carrier	Helium
Attenuation	X512
Chart speed	0.5 cm/min.

A typical chromatogram of the low molecular weight compounds is shown in Figure 1. The compounds were quantified using an internal standard technique.

Organic acids were named as carboxylic acids, even though the compounds actually analyzed were butyl esters, silyl derivatives or a mixture of both, depending on the particular technique used. However, these compounds exist as sodium salts in the liquor.

Intermediate Molecular Weight Compounds

The liquor sample preparation procedure outlined for low molecular weight compounds was followed down to removal of the butanol layer from the 24 ml vial. At this point, the butanol layer was placed in a 30 ml beaker and allowed to evaporate overnight at room temperature in a hood. One milliliter of TRI-SIL silyl derivatizing agent (Pierce Chemical Company) was added to the dried sample. The sample and TRI-SIL were mixed, transferred to a 4 ml screw cap vial and centrifuged. One microliter of derivatized sample was injected into the GC.

To prepare a lake water sample for analysis, 20 ml was placed in a Petri dish and allowed to evaporate in a hood. One milliliter of 12% sodium hydroxide was used to rinse the dried material into a 24 ml vial. The remainder of the sample preparation procedure was the same as described for intermediate molecular weight compounds in a liquor sample.

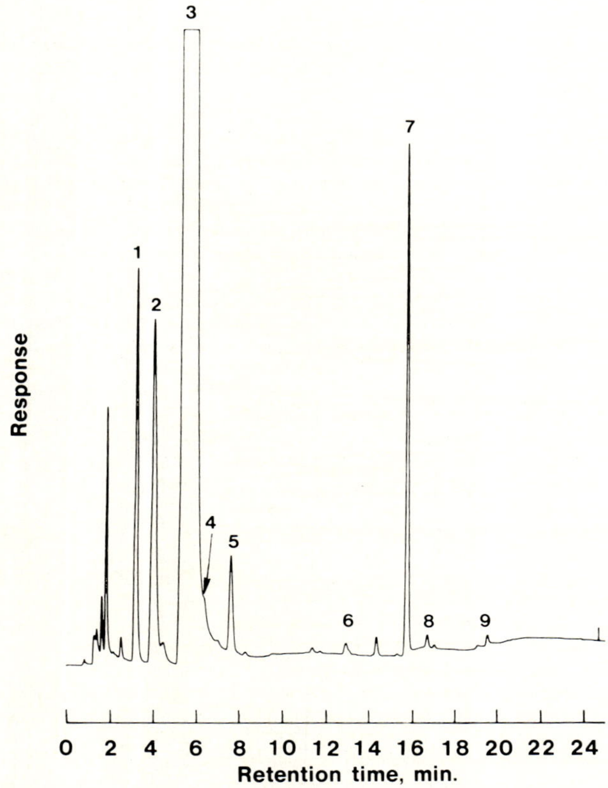

Figure 1 Gas Chromatogram of Low Molecular Weight Compounds in Bayer Liquor. Peak 1 = Formic, 2 = Acetic, 3 = Butanol, 4 = Propionic, 5 = Butyric, 6 = Lactic, 7 = Nonanoic (IS), 8 = Oxalic, 9 = Succinic.

Samples were run on a Hewlett-Packard 5880 GC equipped with a capillary column inlet and a flame ionization detector. The column was a 0.25 mm I.D. x 30 m DB-5 fused silica capillary (J&W Scientific, Inc.). Chromatographic conditions were the following:

Injection port temp.	300°C
Detector temp.	350°C
Initial oven temp.	50°C
Final oven temp.	325°C
Program rate	10°/min.
Time at final temp.	5 min.
Injection type	Splitless
Carrier flow	30 cm/sec.
Carrier	Helium
Attenuation	X32
Chart speed	1 cm/min.

A typical capillary GC chromatogram is shown in Figure 2.

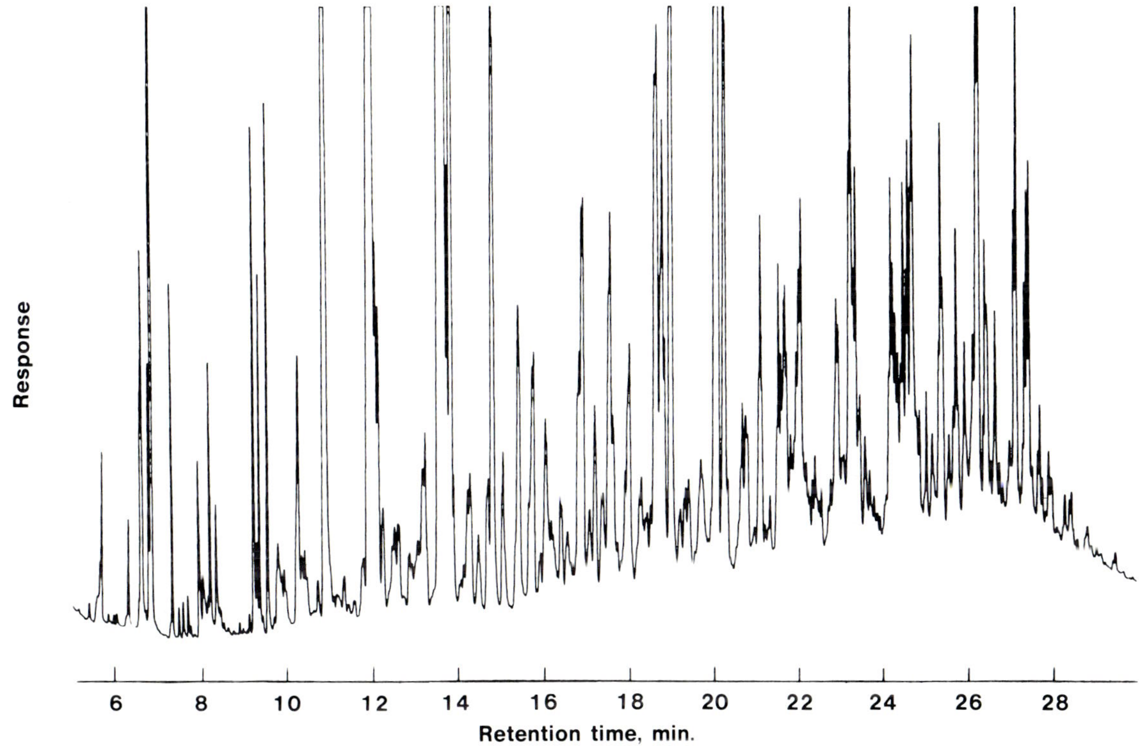

Figure 2 Gas Chromatogram (Capillary Column) of Intermediate Molecular Weight Compounds In Bayer Liquor

The same liquor sample shown in Figure 2 was run on a packed column for comparison (see Figure 3). The column used was 1/8 in. O.D. x 6 ft. stainless steel packed with 3% SP-2250 on 100/120 mesh Supelcoport (Supelco, Inc.). A 2 μl injection was used with the following conditions:

Injection port temp.	300°C
Detector temp.	350°C
Initial oven temp.	50°C
Final oven temp.	325°C
Program rate	10°/min.
Time at final temp.	5 min.
Carrier flow rate	20 ml/min.
Carrier	Helium
Attenuation	X64
Chart speed	1 cm/min.

Individual compounds can be quantified using an internal standard technique. However, when comparing two or more samples, visual inspection of the chromatograms for comparison of peak areas or peak heights may provide sufficient information and be more practical than attempting to quantify a large number of peaks.

High Molecular Weight Compounds

Sample preparation for GPC analysis was the same as described for GC analysis for intermediate molecular weight compounds down to the addition of TRI-SIL. In place of the TRI-SIL, 1 ml of uninhibited tetrahydrofuran (THF) was added, the solution was transferred to a 4 ml screw cap vial and centrifuged. Ten microliters of solvent containing butyl esters was injected into the GPC. The samples were run at two wavelengths, 254 nm and 400 nm. For the 254 nm run, the sample was diluted 1:20.

The lake waters were prepared by placing 20 ml of sample in a Petri dish and allowing them to evaporate in a hood. One milliliter of 12% sodium hydroxide was used to rinse the dried sample into a 24 ml vial. The remainder of the GPC sample preparation procedure was the same as described for the liquor sample above.

The GPC analysis was performed on a Hewlett-Packard 1080B liquid chromatograph equipped with a variable wavelength detector. The columns were 7.8 mm I.D. x 30 cm, 500A and 7.8 mm I.D. x 30 cm, 100A μ Styragel (Waters Associates) in series. Chromatographic conditions were as follows:

Eluent	Tetrahydrofuran (uninhibited)
Flow rate	1 ml/min.
Detector	400 nm and 254 nm
Chart speed	0.5 cm/min.
Attenuation	X128 for 254 nm run
	X64 for 400 nm run

Chromatograms resulting from runs at 254 nm and 400 nm are shown in Figures 4 and 5.

A calibration standard for the humic material determined at 400 nm was prepared in the following way. Five hundred milliliters of Bayer liquor was mixed with 500 ml of water and 30 g of powdered activated carbon (Nuchar, Kodak) for 16 hours at 50°C. The mixture was then filtered

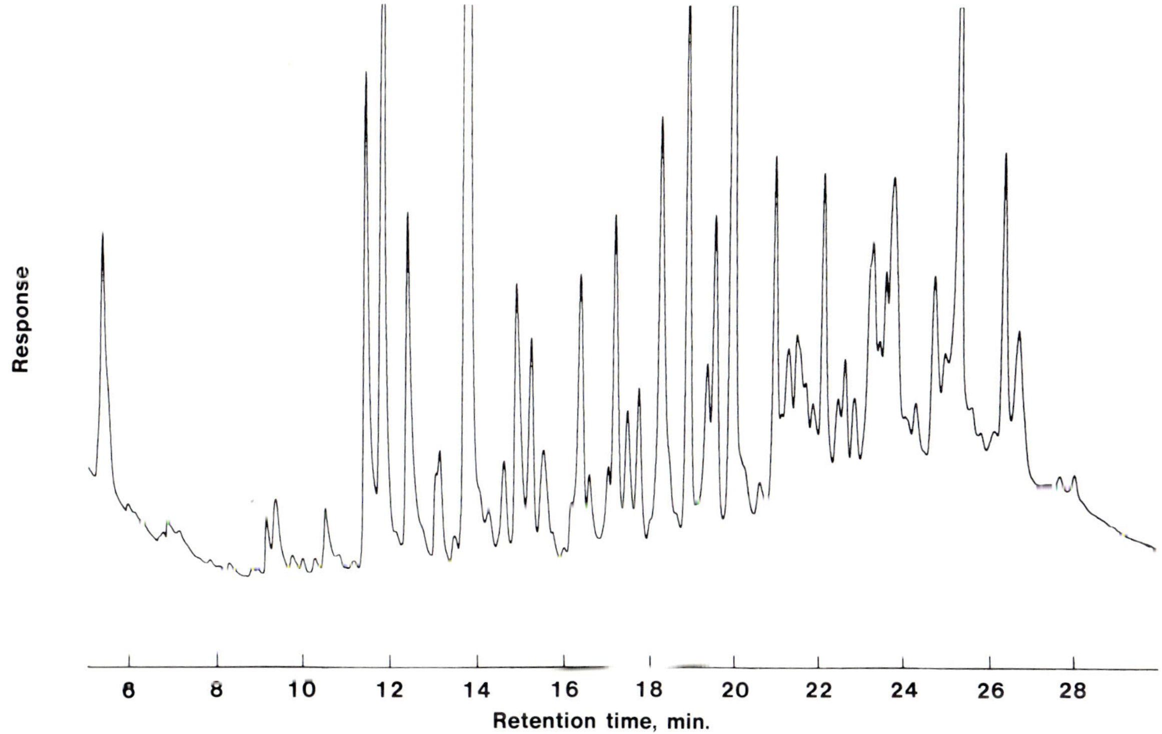

Figure 3 Gas Chromatogram (Packed Column) of Intermediate Molecular Weight Compounds in Bayer Liquor

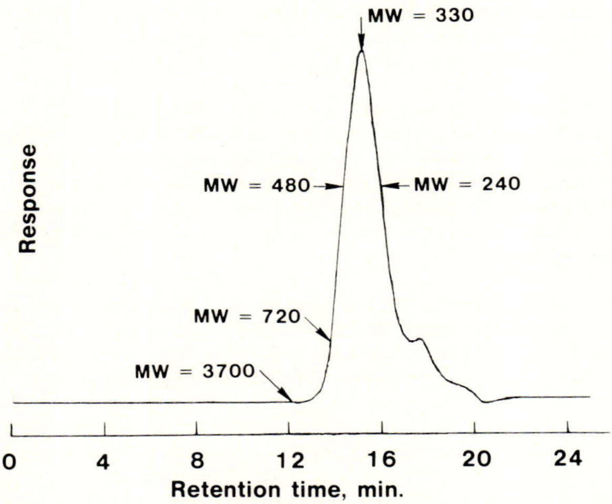

Figure 4 GPC Chromatogram of Organics in Bayer Liquor, Detector Set at 254 nm

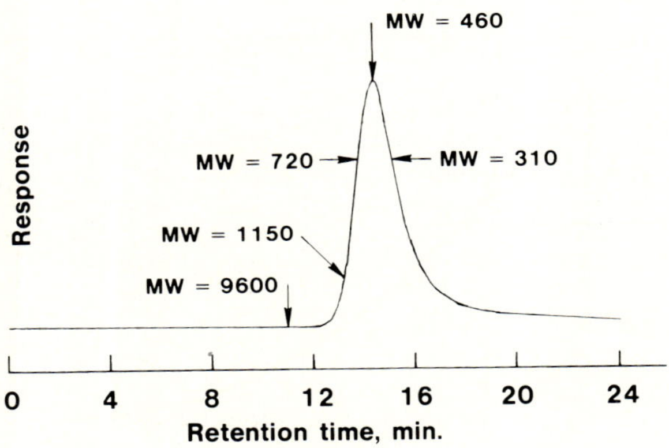

Figure 5 GPC Chromatogram of Organics in Bayer Liquor, Detector Set at 400 nm

through number 1 paper and washed with 100 ml of cold water. The organics were desorbed by mixing the carbon with 100 ml of 50% sodium hydroxide for 4 hours at 80°C. The carbon was filtered and washed with 500 ml of hot water, and the filtrate was acidified with concentrated hydrochloric acid to pH 1. The solution was allowed to stand overnight and the humics were filtered on a 1.2 micron membrane filter.

The organic carbon content of the dried humic material was 45%. A sample was prepared for GC (intermediate molecular weight compounds) and GPC analysis as described above. The gel permeation chromatogram showed a peak similar in molecular weight range to the peak obtained from liquor samples. The GC analysis showed a relatively clean chromatogram, indicating the sample separated by this technique was essentially high molecular weight material.

A molecular weight versus retention time calibration curve was prepared by running polystyrene standards with molecular weights of 9000, 4000, 2000, and 800. A fifth standard, di-n-butyl succinate (MW = 230), was also run. The calibration curve is plotted in Figure 6.

It should be pointed out that the standards used for the molecular weight calibration curve are structurally different from the humic type substances being analyzed. Therefore, the molecular weights reported for the samples should be used only to compare samples and should not be taken as absolute.

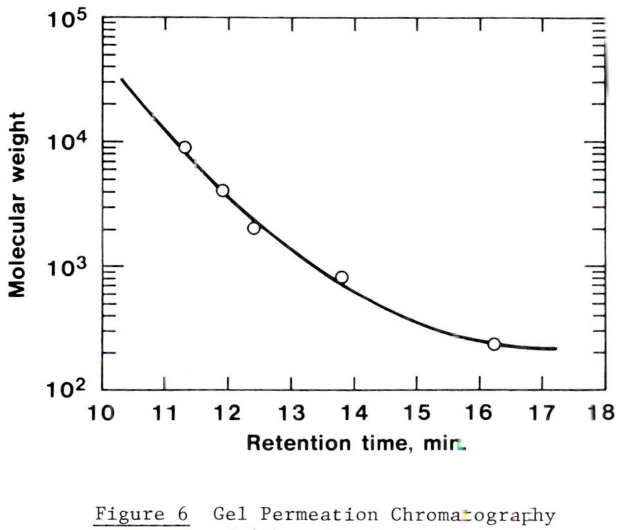

Figure 6 Gel Permeation Chromatography Calibration Curve

Gas Chromatography/Mass Spectrometry

A sample for GC/MS analysis was prepared by placing 1 ml of liquor in a 24 ml screw cap vial and carefully adding 1.5 ml of concentrated hydrochloric acid with swirling to redissolve the alumina. Two milliliters of 1-butanol was added to the acidified sample, followed by shaking. The lower aqueous phase was removed and 5% sodium hydroxide was added dropwise with shaking until the color transferred from the butanol back to the aqueous phase. This aqueous phase was separated and acidified to a pH <2 with concentrated hydrochloric acid and the water was allowed to evaporate overnight at room temperature. One milliliter of TRI-SIL was added to the dried sample, followed by centrifuging.

The sample was run on a Hewlett-Packard 5985 GC/MS, and the resulting chromatogram identifying the compounds by number is shown in Figure 7. Table I lists the compounds identified. The GC conditions were the same as those described for the intermediate molecular weight compounds using a capillary column. The mass spectrometer was equipped with an electron impact source set at 70 electron volts.

Results

Spent Bayer liquor and lake water samples from two low temperature digestion plants (A and B) were analyzed by the techniques described in the experimental methods section. The bauxite used by the plants was mined from different sites of the same deposit. Table II lists typical compositions for these four samples.

TABLE II

Compositions of Samples from Plants A and B

	Plant A		Plant B	
	Liquor	Lake Water	Liquor	Lake Water
Al_2O_3 (g/L)	72.0	4.8	70.7	3.7
Total Caustic (g/L Na_2CO_3)	192.9	8.0	205.6	4.4
Total Alkali (g/L Na_2CO_3)	241.7	14.5	248.0	10.0
Total Organic Carbon (g/L)	17.1	1.3	17.9	0.8
NaF (g/L)	3.3	0.3	2.4	0.2
NaCl (g/L)	13.5	1.0	10.3	0.7
Na_2SO_4 (g/L)	6.3	0.6	5.3	0.5

The gas chromatograms for the low molecular weight compounds were similar to the chromatogram in Figure 1. The comparison of formic and acetic acids are listed in Table III.

Gas chromatograms representing the intermediate molecular weight compounds in the four plant samples are shown in Figures 8 through 11. Selected peaks in samples A and B were compared by calculating the concentration of a component in B relative to the same component in A (peak areas in sample A were assigned a concentration of 100%). The results are listed in Table IV.

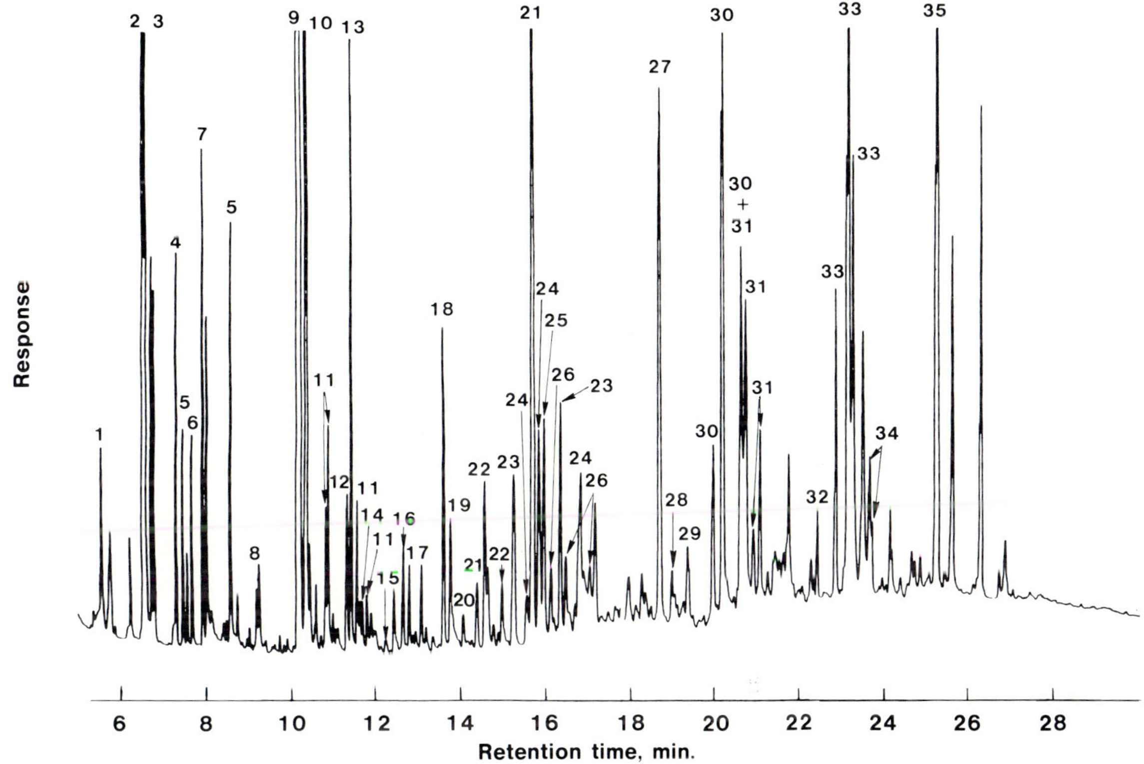

Figure / GC/MS Chromatogram Identifying Compounds in Bayer Liquor

Table I

Compounds Identified by GC/MS

1. Hydroxypropanoic (lactic)*
2. Ethanedioic (oxalic)*
3. Ethanolamine
4. Butanolamine
5. Hydroxybutanoic
6. Hydroxypropanoic
7. Propanedioic (malonic)*
8. Benzenecarboxylic (benzoic)
9. Butanedioic (succinic)*
10. Methylbutanedioic (methyl succinic)
11. Dimethylbutanedioic (dimethyl succinic)
12. Methylbutenedioic
13. Pentanedioic (glutaric)*
14. Dimethylpentanedioic (dimethyl glutaric)
15. Hydroxypropanedioic (hydroxy malonic)
16. Hydroxybutanedioic (hydroxy succinic, malic)
17. Hexanedioic (adipic)
18. Hydroxybenzoic (salicylic)
19. Hydroxypentanedioic (hydroxy glutaric)
20. Heptanedioic (pimelic)
21. Propanetricarboxylic (tricarballylic)*
22. Ethanetricarboxylic
23. Benzenedicarboxylic
24. Butanetricarboxylic
25. Propenetricarboxylic
26. Methylbenzenedicarboxylic
27. Hydroxybenzenedicarboxylic
28. Hexadecanoic (palmitic)
29. Butanetetracarboxylic
30. Benzenetricarboxylic (trimersic/trimellitic/hemimellitic)
31. Methylbenzenetricarboxylic
32. Hydroxydiphenylethanoic (benzilic)
33. Benzenetetracarboxylic (pyromellitic/prehnitic/mellophanic)
34. Methylbenzenetetracarboxylic
35. Benzenepentacarboxylic

*Indicates which compounds were confirmed by comparing GC retention time and mass spectral data with a known compound.

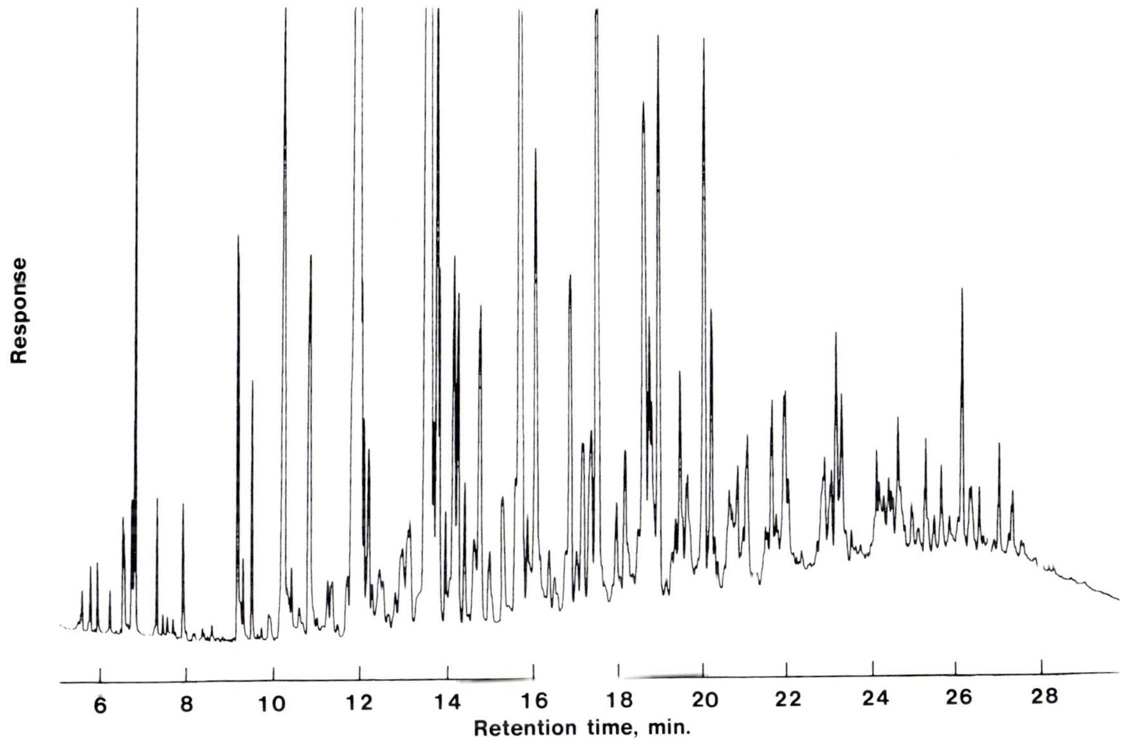

Figure 8 Intermediate Molecular Weight Compounds from Liquor A

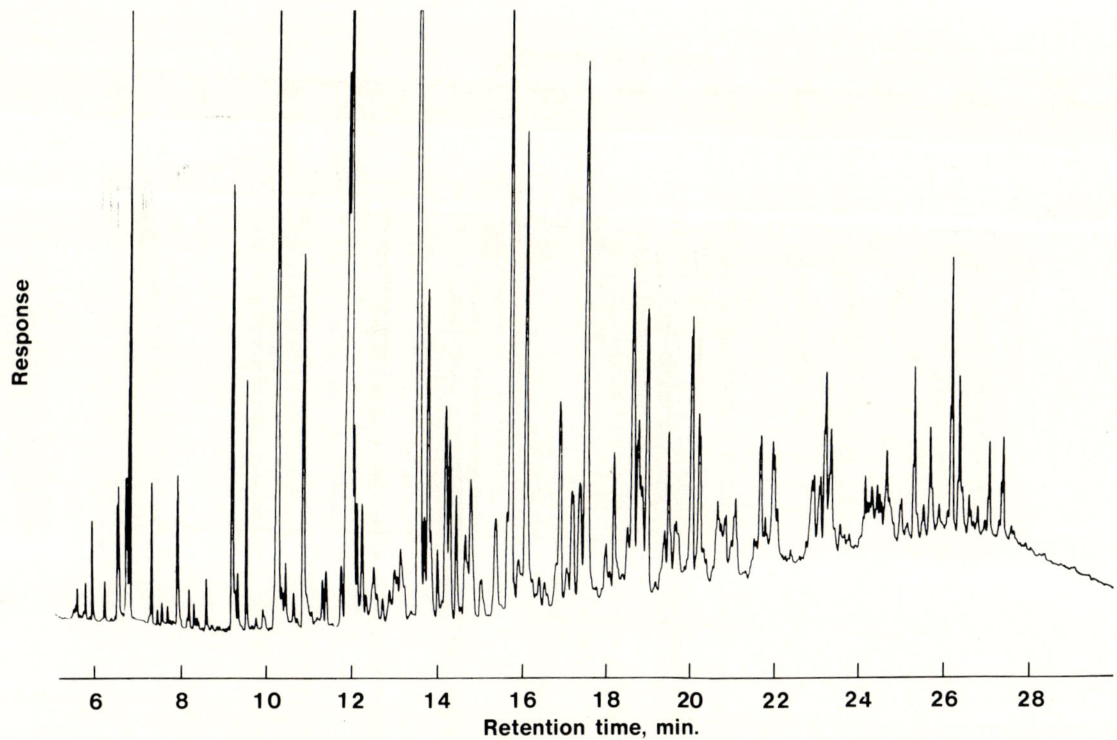

Figure 9 Intermediate Molecular Weight Compounds from Liquor B

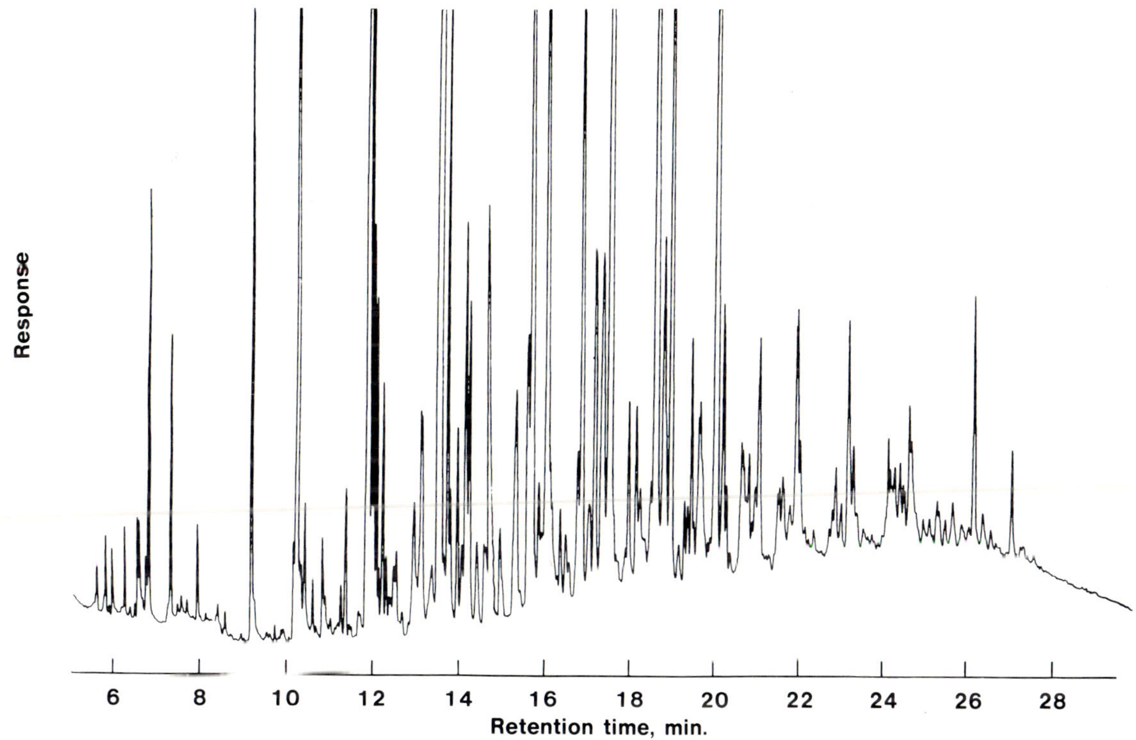

Figure 10 Intermediate Molecular Weight Compounds from Lake Water A

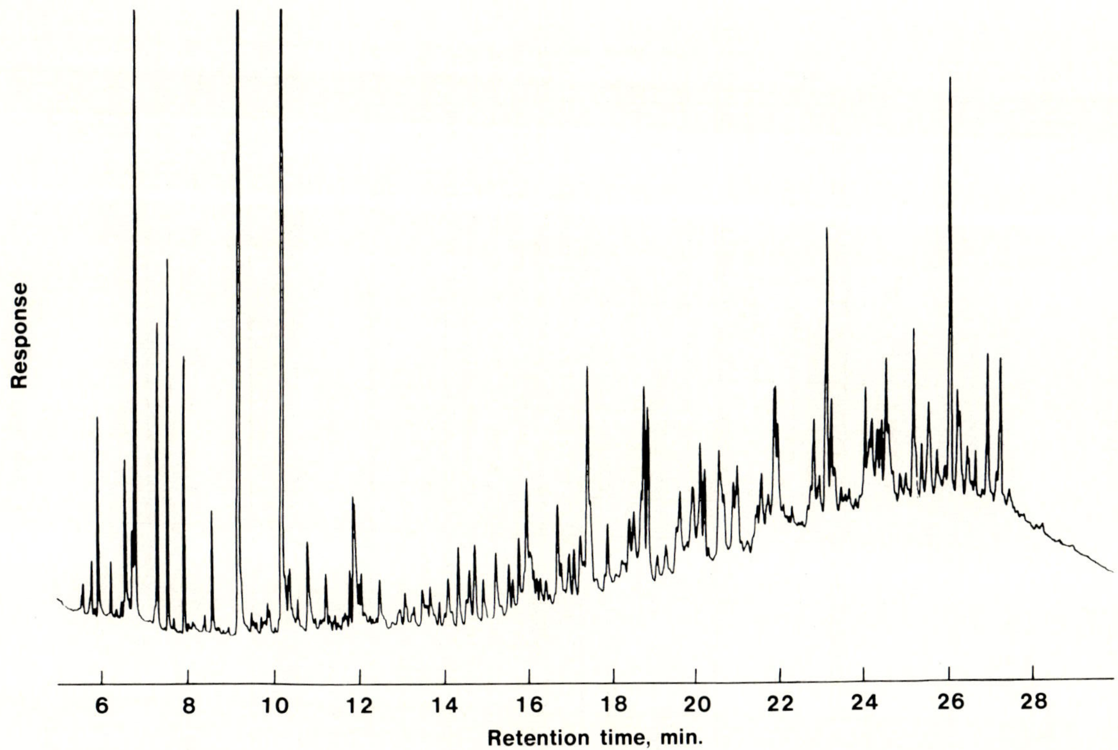

Figure 11 Intermediate Molecular Weight Compounds from Lake Water B

TABLE III

Comparison of Low Molecular Weight Compounds in Plant Samples

Sample	Formic (g/L)	Acetic (g/L)
Liquor A	5.9	11.0
Liquor B	3.4	6.1
Lake Water A	0.2	1.1
Lake Water B	None Detected	0.1

TABLE IV

Comparison of Intermediate Molecular Weight Compounds in Plant Samples

Sample	Oxalic	Succinic	Propane Tricarboxylic	Benzene Tetracarboxylic
Liquor A	100%	100%	100%	100%
Liquor B	80	32	41	54
Lake Water A	100	100	100	100
Lake Water B	70	1	17	79

The GPC peaks of the high molecular weight compounds resulting from the analysis of samples A and B were similar in shape to those in Figures 3 and 4. Table V lists the concentration of humic material, along with the molecular weights at the peak maxima for the 400 nm and 254 nm runs. An appropriate standard was not available for the peak at 254 nm, since this peak includes all compounds which absorb UV light at this wavelength. However, a column has been included in Table V showing the concentration of material in B absorbing at 254 nm relative to A.

TABLE V

Comparison of High Molecular Weight Material in Plant Samples

Sample	400 nm (Humic)	Molec. wt. at peak max.	254 nm	Molec. wt. at peak max.
Liquor A	3.1 g/L	604	100%	395
Liquor B	6.3	631	154	404
Lake Water A	0.06	571	100	394
Lake Water B	0.10	665	109	410

Discussion

Analytical Methods

The complexity of the mixture of organic material in Bayer liquor dictates the use of a combination of instrumental techniques to obtain a complete picture of the organics present. Figure 12 summarizes the sample preparation procedure, instrumental method and the approximate molecular weight range covered by each technique.

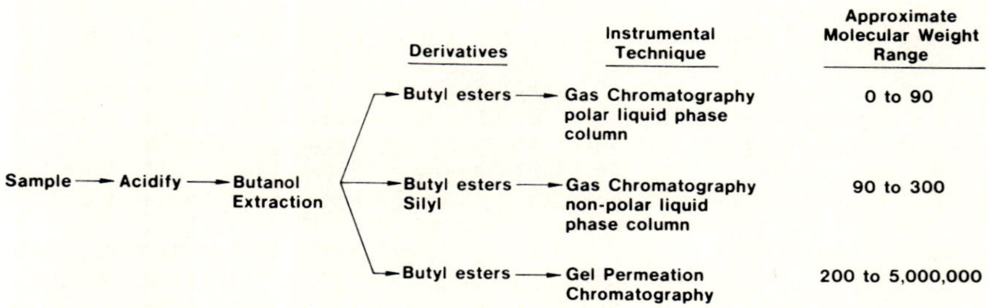

Figure 12 Analysis Scheme for Characterizing Organics in Liquor

Extraction of acidified Bayer liquor with 1-butanol served the dual purpose of separating the organics from the inorganics in the sample and esterifying the carboxylic acids as shown in reaction (1), making the compounds easier to analyze by gas chromatography.

$$RCOOH + R'OH \xrightarrow[\Delta]{H^+} RCOOR' + H_2O \qquad (1)$$

However, for the intermediate molecular weight compounds, a second derivatizing step was needed to convert the remaining polar groups (OH, SH and NH) into non-polar functionalities, again to make the chromatography easier. This was accomplished by preparing silyl derivatives as shown in reaction (2).

$$ROH + R'_3SiX \longrightarrow ROSiR'_3 + HX \qquad (2)$$

Thus, the GC chromatograms obtained by this sample preparation technique are mixtures of butylated and silylated compounds. Compounds which are only partially esterfied, and would not normally chromatograph, are also silylated, making the chromatograms (e.g. Figure 2) appear more complex than is really the case. The chromatogram in Figure 8, which consists of only silyl derivatives of the compounds, represents a more realistic picture of the complexity of the intermediate molecular weight compounds.

Although capillary GC is more difficult than packed column GC, comparison of Figures 2 and 3 shows the additional effort is beneficial in the case of the intermediate molecular weight compounds. Several peaks on the packed column were found to be more than one compound through the additional resolution provided by the capillary column.

The high molecular weight organic material present in Bayer liquor suggests the application of gel permeation chromatography as a means of characterizing this portion of the organics. Initially, porous silica size exclusion columns were used which were compatible with an aqueous system. However, several problems developed, including short column life and difficulty with reproducibility. The second approach to this problem was to employ a sample preparation technique similar to the one developed for the gas chromatographic samples. This method allowed the use of μ Styragel (styrene divinylbenzene copolymer) columns with the sample dissolved in solvent. In addition, the butanol extraction preparation technique removed inorganics from the sample, resulting in a cleaner sample for introduction into the GPC instrument. The esterification step also eliminated the problem of adsorption of the carboxylic acids on the column packing.

The identification of organic compounds by GC/MS is often facilitated by the preparation of silyl derivatives. From the mass spectra of these derivatives, the molecular weight of the compounds can usually be determined (6). To simplify the GC/MS identification, the butyl esters, resulting from the butanol extraction, were hydrolyzed back to the acids before preparing the silyl derivatives. The compounds were identified using a combination of GC retention time, molecular weight and mass spectral library searches. The compounds in Table I marked with an asterisk were confirmed by comparison of GC and MS data with reference compounds. The mass spectra of isomers is usually quite similar, requiring GC retention data to identify which isomer is present. When several isomers of a compound were found, as indicated by the same molecular weight, the parent compound was listed in Table I, and the isomers shown in the chromatogram were all given the same number, corresponding to the parent compound.

Results From Plant Samples

Although plants A and B operate under similar conditions, comparison of liquor and corresponding lake water samples by the methods outlined in the experimental methods section indicates that the organics are quite different.

Table III shows that liquor A has approximately twice the concentration of low molecular weight compounds as liquor B. Lake water A also has a higher concentration of these compounds than lake water B. A similar trend was observed for the intermediate molecular weight compounds, with a lower concentration of organics in plant B than in plant A samples. Comparison of GC retention times in Figures 8 through 11 shows that the same intermediate molecular weight compounds are present in all four samples. The results for the high molecular weight material shown in Table V are opposite those of the low and intermediate molecular weight compounds. Here, the organic material in the samples from plant B is approximately twice the concentration found in the samples from plant A. In addition, the molecular weight of the humic type material in plant B is higher than in plant A.

The results from characterization of these four samples indicate that the high molecular weight material in liquor A has degraded to a greater extent to low and intermediate molecular weight compounds than has the high molecular weight material in liquor B. The GC profile of the intermediate molecular weight compounds is very similar for both liquors, indicating no depletion or production of specific compounds in one liquor versus the other. However, this is not true for the lake water samples. The GC profile of the intermediate molecular weight compounds in lake water A is different from liquor A and the GC profile of lake water B is different from liquor B. These observations, together with comparison of GC retention times, indicate that the organics are undergoing changes in the lakes, and that these changes are primarily in relative concentrations of compounds, not in types of compounds. The shift in molecular weight distribution of the high molecular weight material between the liquor and lake water samples suggests that a change in the humic material is also occurring in the lakes. It is also possible that changes may occur between the time the organics leave the liquor and arrive at the lake. This would provide a second possible explanation for the results reported here. Further study is needed to determine if this is the case.

Acknowledgements

The authors would like to thank W. C. McGrew for his GC/MS work, J. P. Auses for his assistance in developing the GPC technique and D. J. Binet and M. E. Sumich for their helpful consultation throughout this project.

References

1. Gordon Lever, "Identification of Organics in Bayer Liquor," Light Metals, 2 (1978) pp. 71-83.

2. Pierre Salomon, "Contribution a l'analyse des substances organiques contenues das les bauxites etles liqueurs d'aluminate du cycle 'Bayer.' Oxydation par l'oxygene," Ph.D. thesis presented at L'Institute National Polytechnique De Grenoble, July, 1982.

3. Z. G. Szabo, M. Orban and I. Perl, "Some Fundamental Research for Alumina Industry in Hungary," ICSOBA Conference, Athens, Greece, (1978) pp. 422-436.

4. P. Roumeliotis and K. K. Unger, "Possible Uses of HPLC for the Separation, Isolation, Identification and Quantitative Determination of Substances from Complex Mixtures - Illustrated by Carboxylic Acid and Sodium Aluminate Liquors from the Bayer Process of Aluminum Recovery," Chromatographia, 15 (2) (1982) pp. 107-116.

5. K. Yamada, T. Harato and H. Kato, "Oxidation of Organic Substances in the Bayer Process," Light Metals, (1980) pp. 117-128.

6. Alan E. Pierce, Silylation of Organic Compounds, Pierce Chemical Company, Rockford, Illinois, 1968.

EQUILIBRIUM SOLUBILITY OF SODIUM SULFATE IN BAYER LIQUOR

John T. Malito

Kaiser Aluminum and Chemical Corporation
P. O. Box 1031
Baton Rouge, Louisiana 70821
USA

The equilibrium solubility of sodium sulfate in Bayer liquor (at constant alumina/caustic soda ratio of 0.357) over ranges of caustic soda concentration (245-315 g/l), caustic soda/total soda ratio (0.75-0.95) and temperature (150-450 ^{o}F) was determined. Using multivariate linear regression analysis an empirical mathematical model relating sodium sulfate solubility to caustic soda concentration, caustic soda/total soda ratio and temperature was derived. The model demonstrates how the sodium sulfate solubility decreases with increasing caustic soda and carbonate soda concentrations. The most characteristic feature of the solubility of sodium sulfate in Bayer liquor is the minimum which occurs in the temperature range 230-280 ^{o}F, and the temperature at the minimum solubility depends only on the caustic soda/total soda ratio. The model is useful in determining maximum allowable caustic and carbonate soda concentrations in avoiding formation of sulfate scale in the spent liquor evaporators and in the digestion heaters.

Introduction

A major problem encountered in the processing of Jamaican bauxite is the introduction of zinc into the circulating Bayer process liquor. Unchecked, the zinc coprecipitates with alumina and ultimately reports to the aluminum metal as an undesirable impurity. With current technology the amount of zinc in the liquor is controlled by precipitation as zinc sulfide using sodium sulfide or sodium hydrosulfide. However, because the zinc in Bayer liquor occurs mainly as the zincate anion rather than the free (or aquated) zinc cation, a large stoichiometric excess of the soluble sulfide salt is required to maintain a sufficiently low equilibrium zinc (zincate) concentration. Hence, the great majority of the sodium sulfide added for the control of zinc is eventually oxidized to sulfate, accounting for approximately one-half the total sodium sulfate in the liquor.

The most objectionable effect of sodium sulfate in Bayer liquor is the precipitation, under suitable conditions, of the double salt, Burkeite ($Na_2CO_3 \cdot 2Na_2SO_4$). Burkeite scaling in spent liquor evaporators and in digestion heaters, with attendent reduction in heat transfer rates, is all too common in plants processing Jamaican bauxite. In a high-caustic process it is especially important to monitor and control sodium sulfate levels because a) the amount of excess sodium sulfide needed for zinc control increases very rapidly with the concentration of caustic soda, and b) the solubility of sodium sulfate is reduced at higher caustic soda levels.

A model for predicting the maximum solubility of sodium sulfate under a wide range of Bayer process conditions would be desirable. At KACC an empirical equation was previously developed for predicting sodium sulfate solubility in Bayer liquor in which the total soda concentration does not exceed 400 g/l. The model relates the sulfate solubility to caustic soda, C, and thiosulfate soda concentrations, and to the caustic soda/total soda ratio, C/S. However, the relation holds only at the atmospheric boiling point of the liquor. Further attempts to correlate equilibrium sulfate and carbonate levels with the total sodium in solution failed (1). The solubilities of sodium sulfate and carbonate during evaporative desalting have been empirically related to the free sodium ion concentration (3). In these previous studies the effect of temperature upon sodium sulfate solubility was ignored, and no data were obtained at temperatures exceeding the atmospheric boiling point of the liquor.

Herein are reported new experimental data obtained on liquor taken from the KACC Baton Rouge plant. An empirical equation relating sodium sulfate equilibrium solubility to C (245-315 g/l), C/S (0.75-0.95) and temperature (150-450°F) was derived, and the general applicability of the model was tested on Gramercy plant liquor.

Experimental

About 20 liters of liquor having the composition C 320.4 g/l, A/C 0.357 and C/S 0.922 were obtained by the evaporative desalting of Baton Rouge test-tank liquor. For each series of variable-temperature runs, 1250 ml of head liquor, with the required caustic concentration and C/S, were prepared by dilution with water and by the addition of solid Na_2CO_3. Experiments were performed at nominal caustic concentrations of 250, 275 and 305 g/l, each at initial C/S levels of 0.75, 0.80, 0.85, and 0.90. The temperature was varied from 150 to 450°F in 50°F increments.

The prepared head liquor was charged to a 1500 ml bomb equipped with a discharge valve. Solid Na_2SO_4 in excess of its ambient-temperature solubility was added. The bomb was heated to 150°F and rotated horizontally about its axis for 1 hour. Rotation was discontinued, the bomb was oriented with the discharge valve in the horizontal position and the solids were allowed to settle for 10 min. without allowing the bomb to cool. A sample of the supernatant liquor was collected, at temperature, by connecting to the discharge valve a 40-ml stainless-steel bomb equipped with a needle valve. Using the same bomb contents, the above procedure was repeated sequentially at 50°F increments until 450°F. At higher temperatures the runs were sometimes interrupted to charge more Na_2SO_4. The sampling bomb was allowed to cool to room temperature, its entire contents were washed out and the concentrations of caustic, total soda and Na_2SO_4 were determined using alumina as a tie element.

Results and Discussion

The equilibrium solubilities of sodium sulfate in Baton Rouge test-tank liquor at constant A/C and at various caustic soda concentrations, C/S and temperatures are listed in Table I. In all experiments an equilibration time of one hour was used, since longer times produced identical results. In most cases, solid precipitated out of the equilibrium liquor sample as the sampling bomb cooled to ambient temperature. Hence, analyses were performed on the entire sampling bomb contents, and an alumina tie, referenced to each head liquor, was used to determine all concentrations. This method of analyses assumed that the liquor volume is independent of the quantities of dissolved salts. The iterative application of a KACC correlation for calculating the density of liquor at different solute concentrations indicated the error in the analyses was less than 2%. It is important to note that all concentrations given in Table I were obtained at room temperature, and that the actual concentrations at the elevated temperature would be somewhat lower due to a decrease in liquor density. Though correlations exist for estimating Bayer liquor densities at various temperatures, it is more convenient to express concentration as measured at room temperature.

An examination of the data in Table I suggested that the influence of each parameter (caustic soda concentration, C/S, and temperature) upon sulfate solubility could be adequately described by a 2nd order polynomial. The application of multivariate linear regression analysis gave the following empirical model:

$$\text{Sulfate} = A_o + A_1C + A_2C^2 + A_3(C/S) + A_4(C/S)^2 + A_5(C/S)T + A_6(C/S)^2T^2 \qquad (1)$$

Where:

Sulfate = sodium sulfate concentration, g/1 as Na_2SO_4

C = caustic soda concentration, g/1 as Na_2CO_3

C/S = caustic/soda ratio

T = temperature, °F

$A_o = 923 \quad A_1 = -1.67 \quad A_2 = 2.29 \times 10^{-3} \quad A_3 = -1486$

$A_4 = 998 \quad A_5 = -0.628 \quad A_6 = 1.51 \times 10^{-3}$

The R-squared value for this model was 0.951, and a comparison of the predicted and actual sulfate solubilities is shown graphically in Figure 1. A preliminary plot of the sulfate concentration against C/S (at constant caustic concentration and temperature) showed that some of the data were scattered. These data, marked by * in Table I, were not included in the model.

The behavior of the model is illustrated in Figures 2-4. The most characteristic feature is the solubility minimum which is a function of both temperature and C/S, Figure 2. The temperature of minimum solubility occurs at

$$T_{min.} = \frac{208.4}{(C/S)} \qquad (2)$$

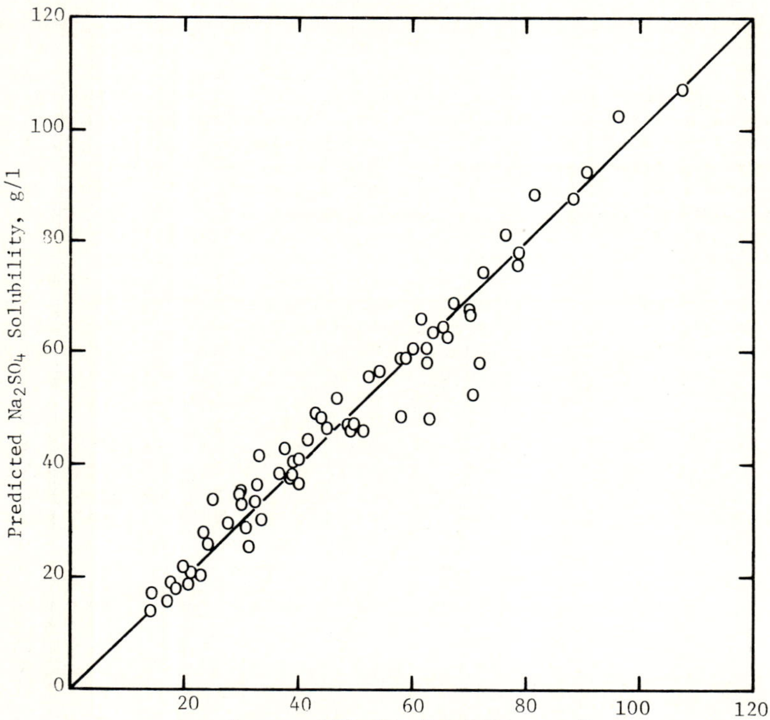

Figure 1. Measured vs. predicted (equation 1) Na_2SO_4 solubility in Bayer liquor at constant A/C, 0.357

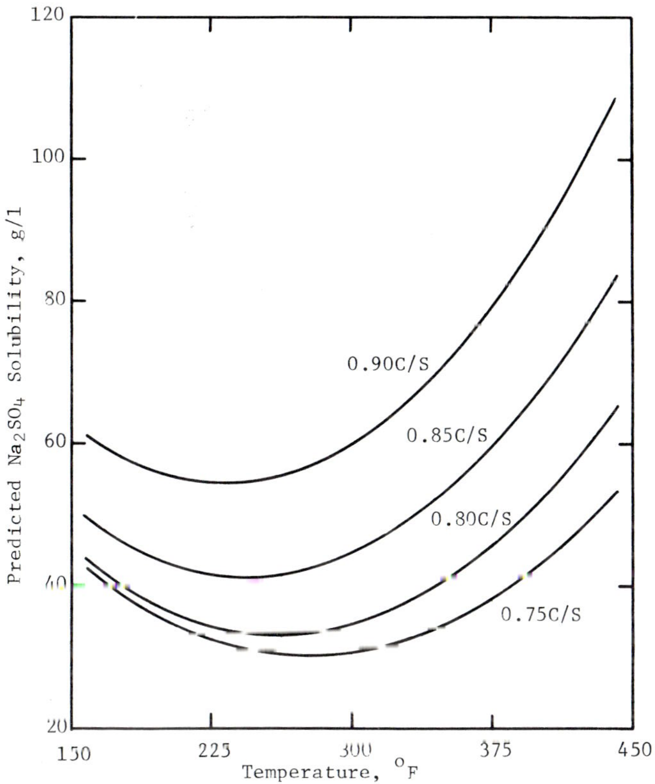

Figure 2. Plot of equation 1 at constant caustic soda concentration, 250 g/l, and constant A/C, 0.357.

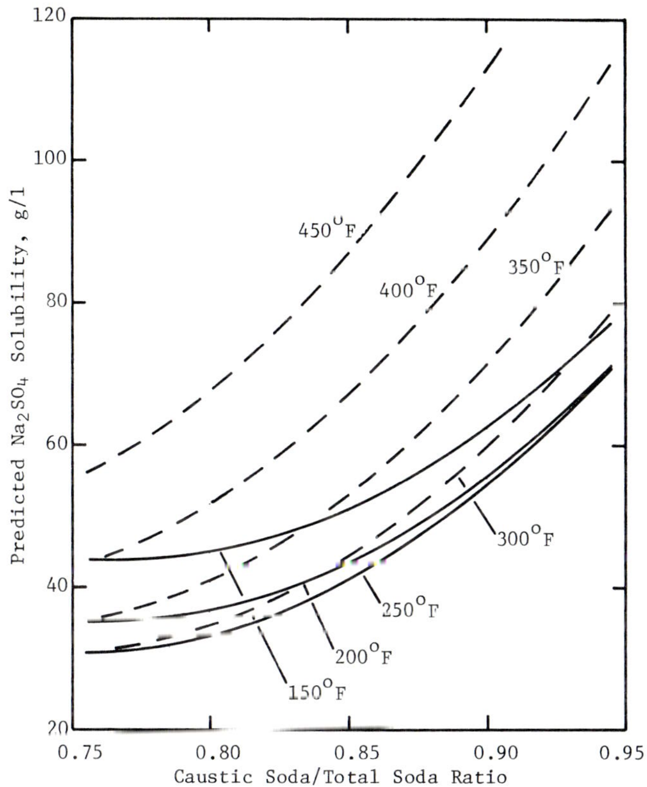

Figure 3. Plot of equation 1 at constant caustic soda concentration, 250 g/l, and constant A/C, 0.357.

and the sulfate solubility at the minimum is given by

$$\text{Sulfate}_{\text{min.}} = 857.4 + A_1C + A_2C^2 + A_3(C/S) + A_4(C/S)^2 \quad (3)$$

In aqueous sodium hydroxide and sodium chloride, sodium sulfate also displays a minimum solubility at about 250°F (2).

In a similar model

$$S = a_o + a_1C + a_2C^2 + a_3(C/S) + a_4(C/S)^2 + a_5T + a_6T^2 \quad (4)$$

C/S and temperature are not coupled, as they are in equation 1, and the temperature at minimum sulfate solubility would be independent of C/S. Equation 4, with an R-squared of 0.920, correlated the data almost as well as did equation 1. As shown in Figure 2, the minimum sulfate solubility occurs in the temperature range 230-290°F, similar to the operating temperatures found in the spent liquor evaporators and digestion heaters.

The following conditions and simplifying assumptions may limit the general applicability of the model to all Bayer liquors:

1.) The model assumes that the equilibrium sulfate solubility depends only on caustic concentration, C/S and temperature. According to the common-ion effect, all soluble sodium salts influence the sulfate solubility. However, attempts to correlate solubility with total soda were not very successful (1). A complete model would be quite complex, also taking into account sulfide, thiosulfate, chloride, and organic sodium. The effect of the A/C upon sulfate solubility may also be significant.

2.) In concentrated solutions, such as Bayer liquors, equilibrium expressions should be expressed in terms of activities rather than concentrations. The activity coefficients of the various ionic species in Bayer liquors are unknown and concentration dependent.

3.) More than one solid phase is in equilibrium with the liquid, i.e., the solid may consist of mixtures of Burkeite, sodium carbonate, and sodium sulfate. Previous work indicated that both Burkeite and excess carbonate are precipitated by addition of Na_2SO_4.

Because of the assumptions and conditions given above, the general applicability of the model to liquors from plants using different processes and bauxites may be limited. To assess the degree of limitation of general applicability, equation 1 was tested on a liquor from a different plant (Gramercy). The predicted and observed sulfate solubilities are shown in Table II. As suspected the agreement was poor. However, at temperatures of 200°F and above the predicted values were nearly a consistent 11.4 g/L higher than the observed values. This indicates that the model may be generally applied to Bayer liquors once the constant correction factor is found from a single solubility determination.

Table I. The Effects of Caustic Soda Concentration, C/S and Temperature on the Equilibrium Solubility of Na_2SO_4 in KACC, Baton Rouge, Bayer Liquor.

Observation	Caustic Concentration, g/l	C/S	Temperature °F	Na_2SO_4 Concentration, g/l
Head Liquor	248.7	0.752	-	56.4
1	248.4	0.767	150	41.8
2	247.8	0.779	200	32.8
3	247.7	0.781	250	30.1
4	248.0	0.782	300	32.6
5	247.3	0.774	350	38.8
6	247.3	0.767	400	46.0
7	246.5	0.767	450	60.4
Head Liquor	249.0	0.801	-	63.6
8	248.3	0.817	150	49.7
9	248.4	0.826	200	39.3
10	248.1	0.829	250	36.4
11	247.7	0.830	300	40.0
12	249.1	0.822	350	49.0
13	247.9	0.814	400	54.3
14	247.8	0.801	450	67.3
Head Liquor	253.1	0.852	-	76.8
15	252.1	0.862	150	70.6
16*	251.3	0.906	200	55.0
17	251.3	0.916	250	58.8
18	252.1	0.915	300	65.3
19	251.6	0.910	350	72.6
20	251.7	0.901	400	81.6
21	252.9	0.885	450	96.3
Head Liquor	246.0	0.894	-	84.8
22	245.0	0.915	200	65.9
23	244.9	0.918	250	63.5
24	246.2	0.913	300	70.2
25	245.9	0.905	350	78.5
26	244.9	0.892	400	88.3
27*	244.0	0.890	450	89.3
Head Liquor	283.3	0.768	-	45.0
28	282.0	0.786	150	27.5
29	281.1	0.795	200	19.7
30	281.4	0.798	250	20.5
31*	282.2	0.794	300	31.5
32	280.6	0.791	350	31.2
33*	280.6	0.779	400	47.7
34	279.3	0.770	450	51.2
Head Liquor	280.3	0.798	-	49.7
35	280.2	0.834	150	29.7
36	279.8	0.839	200	23.3
37	279.4	0.841	250	24.2
38	279.5	0.841	300	30.5
39	279.7	0.843	350	38.3
40	279.2	0.830	400	49.3
41	278.5	0.818	450	60.0

Table I - continued

Observation	Caustic Concentration, g/l	C/S	Temperature °F	Na_2SO_4 Concentration, g/l
Head Liquor	280.5	0.851	-	78.3
42	278.9	0.899	200	33.0
43	278.5	0.904	250	37.6
44*	278.5	0.987	300	48.9
45	277.2	0.892	350	52.3
46	278.5	0.883	400	61.5
47	277.9	0.866	450	76.8
Head Liquor	282.7	0.900	-	74.3
48	281.5	0.899	150	62.7
49	280.5	0.948	200	62.4
50	280.0	0.952	250	63.9
51	280.4	0.951	300	69.9
52	281.2	0.944	350	78.8
53	280.9	0.933	400	90.4
54	280.8	0.915	450	107.4
Head Liquor	299.2	0.751	-	32.6
55*	299.6	0.766	150	26.1
56	297.1	0.803	200	14.4
57	297.2	0.807	250	14.2
58	297.9	0.806	300	17.1
59	298.2	0.801	350	21.1
60	297.6	0.791	400	33.2
61*	296.8	0.778	450	29.2
Head Liquor	310.9	0.804	-	32.2
62*	309.3	0.833	150	26.2
63	309.5	0.844	200	17.5
64	309.3	0.849	250	18.2
65	308.8	0.845	300	22.7
66*	309.6	0.834	350	35.9
67	308.2	0.828	400	40.1
68*	306.6	0.813	450	55.4
Head Liquor	310.8	0.857	-	37.8
69*	309.2	0.898	150	29.5
70	309.7	0.905	200	25.1
71	309.0	0.910	250	29.8
72*	308.3	0.894	300	55.4
73	310.1	0.900	350	44.1
74	309.1	0.883	400	71.9
75*	307.8	0.872	450	82.0
Head Liquor	312.7	0.898	-	47.5
76*	312.4	0.897	150	50.0
77	311.4	0.950	200	42.8
78	311.3	0.954	250	46.7
79	312.3	0.953	300	58.0
80	311.8	0.944	350	78.0
81*	311.5	0.944	400	83.2
82	310.5	0.933	450	97.6

* These data were not included in the model.

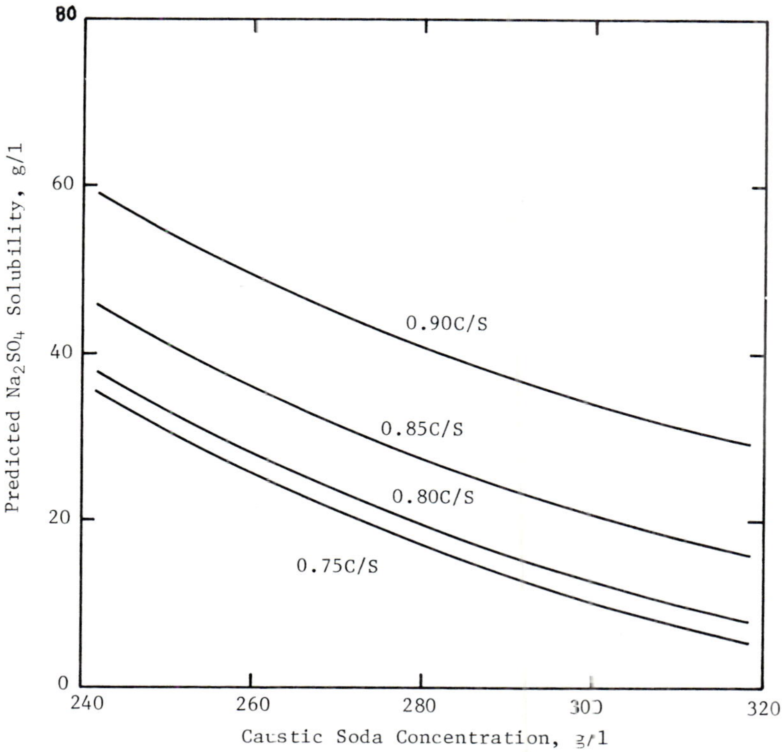

Figure 4. Plot of equation 1 at constant temperature, 250°F, and constant A/C, 0.357.

Table II. Predicted and Observed Na_2SO_4 Solubility in KACC, Gramercy, Bayer Liquor

Run	Caustic Concentration, g/1	C/S	Temperature °F	Na_2SO_4 Concentration, g/1	Predicted Na_2SO_4	Predicted-Observed
Head	226.2	0.704	-	68.6	-	-
1	226.3	0.707	150	65.0	60.9	-4.1
2	225.6	0.733	200	38.3	50.0	11.7
3	223.8	0.742	250	32.6	46.1	13.5
4	224.9	0.742	300	33.5	44.9	11.4
5	225.5	0.739	350	38.2	48.0	9.8
6	225.5	0.733	400	43.8	55.2	11.4
7	225.3	0.721	450	54.6	65.2	10.6

Conclusion

A seven-parameter empirical model for predicting the equilibrium solubility of sodium sulfate in Bayer liquor was developed. The model takes into account the caustic soda concentration, C/S and temperature. The general application to Bayer liquors requires the determination of a correction factor, obtained from a single sulfate solubility measurement on the liquor of interest. The model is useful in determining the maximum allowable caustic and carbonate soda concentration in avoiding the precipitation of sodium sulfate from Bayer process liquors.

References

(1) Private communication, R. G. Breuer, Kaiser Aluminum and Chemical Corporation, Center for Technology, October, 1982.

(2) W. C. Schroeder, Alton Gabriel, and Everett P. Partridge, "Solubility Equilibria of Sodium Sulfate at Temperatures of 150 to 350°. I. Effect of Sodium Hydroxide and Sodium Chloride", Journal of the American Chemical Society, 57 (9) (1935) pp. 1539-1546.

(3) Eddie L. Schiller, "Solubility of Bayer Liquor Impurities in Evaporative Crystallization", paper presented at TMS-AIME Spring Meeting, Atlanta, Georgia, March, 1983.

DEHYDRATION PRODUCTS OF GIBBSITE BY ROTALY KILN AND STATIONAL CALCINER

K. Yamada, T. Harato, S. Hamano and K. Horinouchi

Sumitomo Aluminium Smelting Co., Ltd.

Kikumoto Works, Niihama

Japan

Gibbsite was dehydrated in a fluid calciner with suspension preheaters (SPF).

Crystal forms of the dehydration products were determined and compared with those obtained by a conventional rotary kiln.

The former products were similar to those obtained by a suspension heater of lab scale whose heating velocity was of an order of 10^4 °C/S. The latter products were similar to those calcined in an electric furnace.

The surface area of products obtained by SPF is higher than that by the rotary kiln at a same α-Al_2O_3 content.

At a same surface area level, both products show the same reactivity with fluorine.

1. Introduction

Hydrated aluminas change to α-Al_2O_3 forming transition aluminas via various routes by the successive calcination.

Since Stumpf et al (1) reported the existence of several transition aluminas with different crystal structures, a variety of investigations have been reported with respect to their structures, the mechanisms and conditions of formation. Among the various hydrated aluminas, such as crystalline alumina monohydrates, crystalline alumina trihydrates and amorphous alumina hydrates, gibbsite obtained by the Bayer process is the most widely used alumina trihydrate.

Many reports have been presented on the dehydration process of gibbsite. Especially Brown et al (2) proposed two series of transition aluminas and studied the effect of the particle size of starting material and that of the calcination atmosphere on the dehydration products. In their studies heating velocity was relatively low, that is less than 100 °C/min.

Recently saving energy processes are very important in the Bayer process and suspension type calciners or fluidized bed type calciners with high thermal efficiency instead of a conventional rotary kiln have been developed. In these type calciners, heating velocity is very high and it achieves over 1000 °C/sec at near the dehydration temperature of gibbsite.

In this paper, the kinds and the properties of dehydration products obtained by the rapid heating process are described in relation to the slowly heated dehydration products. And the rates of various aluminas to α-phase transformation are also discussed.

2. Experimental

2.1 Raw material

Gibbsite was used for the starting material of the test. The mean diameter of the gibbsite was 55~90 μm. It contained 0.20 % of Na_2O, 0.02 % of SiO_2 and 0.01 % of Fe_2O_3.

2.2 Procedure

Four types of apparatus were used for the calcination of gibbsite. Two types (1), (2) were lab apparatus, and other two types (3), (4) were plant equipments.

(1) Electric furnace (lab apparatus)

10 g of gibbsite was heated at 5 °C/min to 300~1300 °C, and soaked for 2 hours.

(2) Rapid-heating apparatus (suspension heater, lab apparatus)

The apparatus was composed of a furnace, a gibbsite feeder, a heating zone (25 mm$^{\phi}$ x 3000 mm^{L}), a cyclone for separation. The furnace was heated by LPG. The gas velocity in the heating zone was typically 6.9 Nm/sec. Gibbsite was supplied typically at the rate of 0.1 kg/Nm^3. The elevation rate of the temperature was estimated about 30000 °C/sec. The residence time of gibbsite in the heating zone was less than 0.2 sec.

(3) Rotary kiln (plant equipment)

A conventional rotary kiln was used. The inlet temperature was 200~300 °C, and the outlet temperature was 1300~1400 °C. The residence time of gibbsite was 2~3 h.

(4) Fluid calciner with suspension preheaters (SPF, plant equipment)

SPF (3) was used. The inlet temperature of the first preheater was typically 40 °C, and the outlet temperature of the lost preheater was 1100~1200 °C. The residence time of gibbsite was approximately 2~3 minutes, its exact measurement was not possible. The calcined gibbsite was sampled from the nozzles attached to the under part of the SPF cyclones.

3. Results and Discussions

3.1 Dehydration products by electric furnace

Dehydration products by an electric furnace were investigated in order to compare with results of rapid-heating apparatus. The properties of dehydration products are shown in Table I.

As Brown has already reported, the transformation route of gibbsite is as follows : (2)

Table I. Dehydration product and surface area of calcined gibbsite in electric furnace

No.	Calci- tion temp. (°C)	Dehydration product (%) Gib.	Bo.	χ	γ	Surface area (m^2/g)
E300	300	32	16	30	20	190
E400	400	0	15	50	35	180
E500	500	0	6	50	40	171
E600	600	0	0	50	50	151
E700	700	0	0	50	50	110
E800	800	0	0	50	50	95

Calcination time : 120 min
Gib. : gibbsite
Bo. : boehmite
χ : χ-Al_2O_3
γ : γ-Al_2O_3

3.2 Dehydration products by rapid-heating apparatus

The properties of dehydration products are shown in Table II. The pore volume distributions of them are shown in Fig. 1. These figures give the following results :

(1) The dehydration product contained only 5 % of boehmite (calcination temp., 385 °C).

(2) The dehydration product did not contain boehmite or γ-Al_2O_3. It was composed of about 80 % of χ-Al_2O_3 and 20 % of ρ-Al_2O_3. (calcination temp., 520~705 °C)

(3) The dehydration product was composed of pseudo γ-Al_2O_3 alone. (calcination temp., 1100 °C)

In the case of rapid heating above 520 °C, gibbsite did not transform to boehmite, because of lack of hydrothermal conditions during heating. It is known that the transformation to boehmite occurs under hydrothermal conditions in gibbsite particel (4) (5), and gibbsite transforms to ρ-Al_2O_3 under vacuum heating conditions (6) (7).

The above experimental results suggest the following transformation route of gibbsite.

$$\text{Gibbsite} \begin{cases} (\rho) \text{ — pseudo } \gamma \text{ — } \delta \text{ — } \theta \text{ — } \alpha & \text{High temp.} \\ \rho \text{ — } \chi & \text{Low temp.} \end{cases}$$

Dehydration products by the rapid heating have high surface area, such as 300 m^2/g, compared with dehydration products in an electric furnace. This high surface area is due to the large micro-pore volume (radius < 20 Å) shown as in Fig. 1.

3.3 Dehydration products by rotary kiln

Next, we investigated the properties of calcined gibbsite sampled from different places of the rotary kiln, in order to clarify the transformation route in a plant equipment. The results are shown in Table III.

The samples at the inlet of the kiln (for example RK-1 and RK-2) contained 35~40 wt % of α-Al_2O_3. These high contents of α-Al_2O_3 were caused by mixing of α-Al_2O_3 powders which were accompanied by combution gas and recycled.

Table II. Dehydration product and surface area of rapid-heated gibbsite in suspension heater

No.	Calcination temp. (°C)			Dehydration product (%)						Surface area
	T_1	T_2	T_3	Gib.	Bo.	χ	ρ	γ		(m^2/g)
R385	500	415	385	83	5	5	0	0		23
R440	600	480	440	7	0	60	30	0		206
R520	700	565	520	0	0	80	20	0		270
R600	800	645	600	0	0	80	20	0		302
R675	900	725	675	0	0	80	20	0		312
R705	950	755	705	0	0	80	20	0		277
						P-γ	δ	θ	α	
R1100	1360		1100	0	0	120	0	0	0	144
R1300	1580		1300	0	0	15	15	12	40	

P-γ : pseudo γ-Al_2O_3

Flow velocity $\approx$ 7 Nm/sec

Residence time : 0.12~0.19 sec ($T_3 \leq$ 705)

0.04~0.05 sec ($T_3 \geq$ 1100)

Table III. Dehydration product of calcined gibbsite in rotary kiln

No.	Temperature °C	Gib.	Bo.	χ	$\gamma+\delta$	θ	κ	α
					%			
RK1	600	1	7	20	30	0	0	37
RK2	720	0	4	20	50	0	0	38
RK3	850	0	0	20	60	0	0	36
RK4	970	0	0	10	70	2	1	49
RK5	1100	0	0	tr.	20	11	4	49
RK6	1200	0	0	tr.	tr.	5	1	76
RK7	1300	0	0	0	0	0	0	100

Gib. : gibbsite tr. : trace

Bo. : boehmite

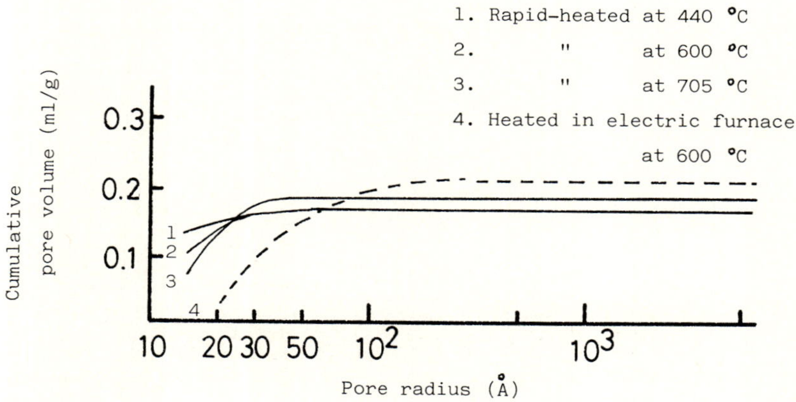

Fig. 1 Pore size distribution of dehydration products.

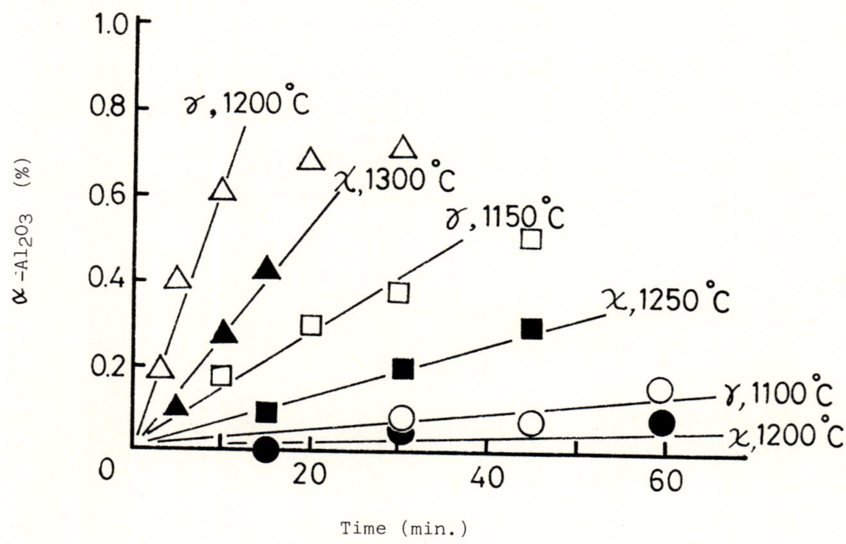

Fig. 2 Rate of α-Al_2O_3 phase transformation of χ, γ-Al_2O_3 in electric furnace.

Observed boehmite content of sample RK-1 in Table III was 7 %, but corrected content obtained after subtraction of α-Al_2O_3 was 10~20 %. This value was the same as for the dehydration product in the electric furnace.

These experimental results indicate the following transformation route of gibbsite as Brown proposed (2).

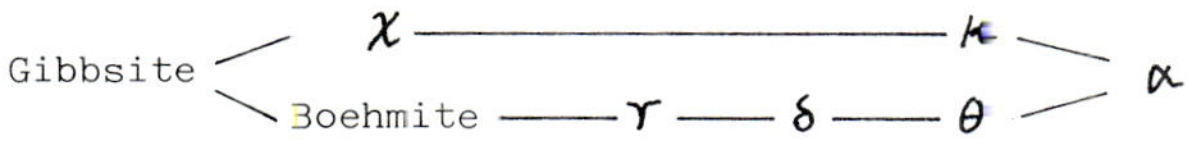

3.4 Dehydration products by SPF

The properties of dehydration products by SPF are shown in Table IV. The samples analysed were sieved and fine α-Al_2O_3 powder (under 44 μm) was removed.

The results were as follows :

(1) Boehmite content was less than 3 %.

(2) κ-Al_2O_3 was not observe, even though χ-Al_2O_3 was present.

(3) The sample of SPF 2 rehydrated easily in water at 100 °C. This fact demonstrated the presence of ρ-Al_2O_3.

From these experimental results, it is concluded that the transformation route in SPF is identified as that in rapid-heating apparatus.

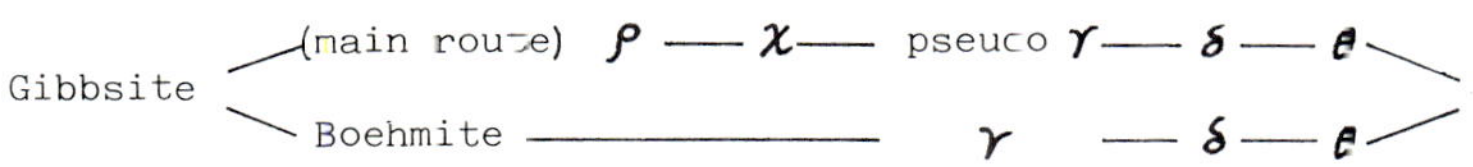

3.5 Rates of rapid-heated dehydrates and the transition aluminas to α-Al_2O_3 phase transformation

The rates of χ, γ-Al_2O_3, R600 and R1100 to α-Al_2O_3 phase transformation were measured in the electric furnace, as well as that of gibbsite in the suspension heater. Experimental results are shown in Fig.2.

α-Al_2O_3 phase transformation of these aluminas obeys zero-order kinetics ($\frac{dx}{dt}=k$) at the initial stage of the reaction, but deviates in the final stage (α-Al_2O_3 content $X > 0.6$) as reported by Yamagida (8).

Table IV. Dehydration product of calcined gibbsite in SPF

	Temperature °C	Gib.	Bo.	ρ	χ	$\gamma+\delta$	θ	κ	α
							%		
SPF1 (1st cyclon)	170	100	0	0	0	0	0	0	1
SPF2 (2nd ")	380	0	3	20	tr.	20	0	0	2
SPF3 (3rd ")	1100	0	0		tr.	60	7	0	15
SPF4 (Fluid bed)	1100	0	0		tr.	70	11	0	19

Before measurement, fine powder (under 44 μm) were removed by sieving.

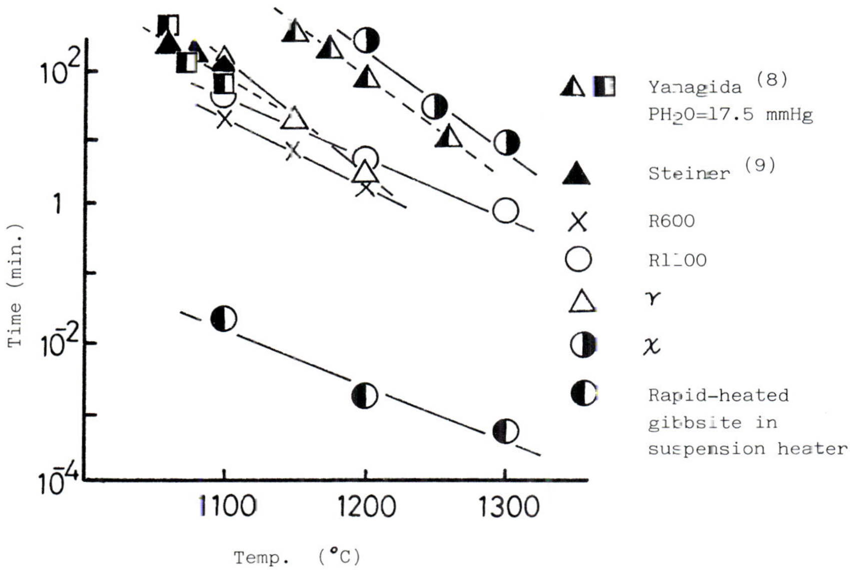

Fig.3 Relationship between time and temperature for formation of 20 % α-Al_2O_3 from various aluminas

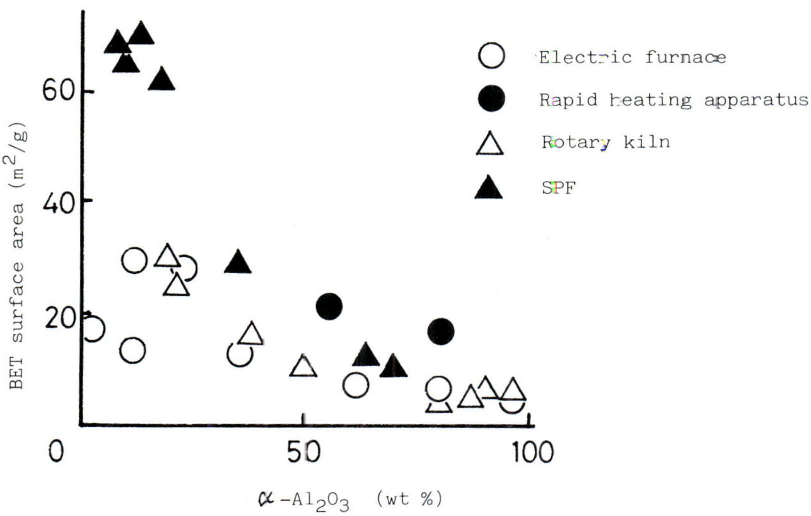

Fig. 4 BET surface area vs. α-Al_2O_3 content.

The rate constant (α-Al_2O_3 content/min) in the range of less than 0.6 is shown in Eq.(1) ~ Eq.(4) as a function of the reaction temperature, T (°K).

$$\chi\text{-}Al_2O_3 : K=10^{18.43} \exp (-143.7 \times 10^3/RT) \quad (1)$$

$$\gamma\text{-}Al_2O_3 : K=10^{16.63} \exp (-120.0 \times 10^3/RT) \quad (2)$$

$$R600 : K=10^{11.50} \exp (-84.3 \times 10^3 RT) \quad (3)$$

$$R1100 : K=10^{12.84} \exp (-92.3 \times 10^3 RT) \quad (4)$$

Activation energies to α-Al_2O_3 transformation are calculated to be 143.7, 120.0, 84.3 and 92.3 Kcal/mol for χ, γ-Al_2O_3, R600 and R1100, respectively. The dehydration products by rapid heating have lower activation energies than χ or γ-Al_2O_3. These lower activation energies can be due to the highly defective crystal structure of the starting material.

Usuallly alumina with ca. 20 wt % crystalization are used for aluminum smelting. Fig. 3 shows the relationship between the reaction time to attain X=0.2 (α-Al_2O_3 20 %) and the reaction temperature.

The rate of gibbsite to α-Al_2O_3 phase transformation in the suspension heaters is 10^4 as high as that of R600, which most easily transformes to α-phase in the electric furnace among other transition aluminas. This is caused mainly from the steam produced by fuel combution and from highly deformed crystal structure formed by instataneous dehydration.

The crystalization temperature of χ-Al_2O_3 at X=0.2 is 100~200 °C higher than that of R600 or R1100, but that of γ-Al_2O_3 is approximately the same temperature for R600 or R1100.

Table V. Properties of calcined gibbsite

Apparatus	α-Al_2O_3 %	Surface area m^2/g	Density g/cm^3	Reactivity with F F-g/100g Al_2O_3	Angle of repose	Dehydration product (%) χ	$\gamma+\delta$	ρ-$\gamma+\delta$	κ	θ	α
Electric furnace	12	30	3.55	2.3	36	tr.	20	0	30	15	10
	23	28	3.57	2.1	37	tr.	20	0	17	11	23
Rapid-heating	0	144	3.54	10		0	0	120	0	0	0
apparatus	40					0	0	15	0	12	40
Rotary kiln	3	73	3.45	6.0	32	20	50	0	1	8	3
	21	26	3.63	2.1	34	tr.	tr.	0	7	24	21
SPF	10	65	3.54	5.1	36	tr.	0	70	0	4	10
	18	62	3.60	4.5	35	tr.	0	70	0	10	18
	42	25	3.74	1.0	37	0	0	20	tr.	32	42

Before measurement, fine powder (under 44 μm) were removed by sieving.

Raw material gibbsite : 90 μm (average particle size)

3.6 Properties of α-Al_2O_3 obtained by rapid-heating

As described above, the transition alumina obtained from gibbsite heated rapidly differs from that heated slowly in the electric furnace in its crystal structure. Hereafter, the relationship between surface area and α-Al_2O_3 content and reactivity with fluorine gas are shown.

(1) BET surface area

Fig. 4 shows the relationship between the surface area and α-Al_2O_3 content for aluminas obtained by four different calcination apparatus, which are the electric furnace, rapid-heating apparatus, rotary kiln and SPF.

Fig. 4 indicates that rapid heated alumina calcined in SPF gives higher surface area at a same α-Al_2O_3 content as compared with those calcined in the electric furnace or in the rotary kiln. Because the rapid-heated product contains such transition aluminas as χ, ρ or pseudo γ-Al_2O_3 which have much higher surface areas than that of κ or θ -Al_2O_3 which are contained in the rotary kiln product.

(2) Reactivity with fluorine gas

In aluminum smelting, it is desired for alumina to adsorb fluorine containing gas. The results shown in Table V indicates that the reactivity of alumina obtained by rapid heating with fluorine is approximately twice as high as that of alumina obtained by the calcination in the rotary kiln at the α-Al_2O_3 content of 20 %. At a same BET level, all aluminas show the same reactivity with fluorine.

4. Conclusion

A different thermal decomposition process of gibbsite in SPF was found compared with the process in a conventional rotary kiln.

The surface area of alumina obtained by SPF is higher than that by the rotary kiln at a same α-Al_2O_3 content.

At a same surface area level, both aluminas shown the same reactivity with fluorine.

References

1. H. C. Stumpf et al. : Ind. Eng. Chem., 42, 1398 (1950)

2. J. F. Brown et al. : J. Chem. Soc. London, 13, 4 (1953)

3. K. Yamada et al. : Light Metal, 1983, PP. 159-171
"Development of a Fluid Calciner with Suspension Preheaters"
: Keikinzoku, 31, 583 (1981)

4. J. H. de Boer et al.: Pror, Kon. Ned. Akad. Wetersch. Ser. B, 57, 170 (1954)

5. D. Papee and R. Tertian : Bull. Soc. Chim., 983 (1955)

6. R. Tertian and D. Papee : J. Chim. Phys., 55, 341 (1958)

7. J. Fouguerol et al. : J. Catal., 57, 222 (1979)

8. H. Yanagida et al. : Bull Ceram. Soc. Japan, 74, 371 (1966)

9. C. J. P. Steiner et al. : J. Am. Ceram. Soc., 54, 412 (1971)

PARTICLE SIZE ANALYSIS IN BAYER PLANT CONTROL

J.D. Zwicker

ALCAN International Limited
Kingston Laboratories
P.O. Box 8400
Kingston, Ontario
Canada
K7L 4Z4

Abstract

Particle size analysis, with particular reference to Bayer plant materials, has evolved over the years from vibratory sieving to the sophisticated electric sensing-zone technique. Various methods have been tried and generally found wanting in one respect or another. These have included various sieving techniques, sedimentation, elutriation, permeametry, electric sensing-zone and microscopy. Other methods we have investigated include ultrasonics, β-ray attenuation, and laser diffraction. Reasons for our decision on the direction we are taking for Bayer plant control are discussed.

Introduction

For plant control purposes the general request is for a particle size analysis method which is fast and simple while being sensitive to change in the circuit. "Accuracy" is normally felt to be of secondary, if any, consequence, the argument being that only the sensing of relative change is necessary. This is contrary to the requirements for chemical control where absolute values are requested. This is a difficult argument to counter since it is not generally appreciated that the size which is reported is a function of the technique. There is no absolute method for most industrial materials. A particular instrument only reports the amount of light extinction or the settling rate or the change in electrical resistance as a particle goes through an orifice, or the amount of diffracted light, or the quantity of material passing a mesh of holes approximately equal in size and shape, or etc, etc. After various levels of mathematical manipulation a "size analysis" is reported. Obviously there are other parameters operating in these physical methods which must remain constant to permit the extraction of size related information. For most methods to give the same result with approximately the same sensitivity, the material must be spherical, opaque, each particle having the same and preferably low density, and the size distribution well within the capability of the method. All of this does not occur in Bayer plant materials.

The introduction of a new technique which is sensitive to change in its particular parameter(s) may over or under emphasize a particular section of the size distribution (e.g., fines) and thus provide signals which are at variance with those historically followed. Without continuous cross-checking the circuit may move to a different level. In addition, the product specifications will probably remain the same (i.e., the old technique) thus creating a potential conflict. However, if a bias is found and it remains constant and predictable this conflict may be resolved.

The best counter to the latest "fast and simple" instrument what ever it might be is to aim for the control technique which would be capable of analyzing all samples, intermediates and product, in terms of those parameters which have meaning to the customer, i.e., the reduction cell. It would permit following and interpreting changes in the circuit without the problem of finding correlations with different techniques, and would not be over or under sensitive to particular size fractions. An overview of the historical progression of our advances in this field indicates that we have been moving in this direction for some time. A review will be made in some detail of our work in sieving and electro-sensing zone and why we feel these, for the moment, are providing the best information for control purposes. Our investigations into the laser diffraction technique are detailed in an effort to indicate why we have not moved in this direction.

Historical Perspective

Over the last 30 years we have progressed in our control techniques from the Ro-tap (relegated to a far corner) for coarse material and hydrometers for fine, to the computerized Elzone for all material. A summary is given in Appendix I. Over the interval we have investigated many techniques only some of which we have actually purchased. More detail is available in the quoted references in the table but in general most of the techniques were not adopted due to the non-agreement with the entrenched sieving methods or the very difficult problems of handling slurries.

Our present control method employs the computerized Elzone for all materials while the product, for external purposes, is qualified by Ro-tap using calibrated sieves. Due to the agreement between the techniques, as discussed below, the net result is that we are controlling and shipping on sieve values but obtaining them by a faster, more precise, and more reliable technique.

Sieving

Today sieving is still probably the most widely used particle size analysis technique in the aluminum industry. Due to its long history and the simple equipment required it is usually assumed that the results obtained must be correct. In fact, any new technique is generally compared to sieving and if the results are different it is the new technique which is believed to be out of line. The main impetus to change has been the time requirement. Our move to Alpine was primarily due to savings in sieving time but the retention of the Alpine was due to precision and perceived "accuracy". With considerable effort this latter concept has been shown to be in error.

Kaye (6) enumerates some of the problems inherent in sieving, the most prevalent and important one being the non-uniformity of the sieves due to both manufacture and usage. In theory this can be overcome by calibration provided the calibration is done from time to time to overcome the changes due to usage. The effective cut-size or calibrated size is a function of loading and sieving action. Provided the latter is held constant the variable for individual samples is the sample load remaining on the sieve at the end of the process. This changes with the size distribution thus influencing the opportunity of particles to "see" an opening and thereby pass through. The Alpine uses a fresh sample for each sieve point thereby exaggerating this effect. This was not fully considered in our calibration procedure (3) which uses an alumina characterized with electroformed sieves. However, our system maintained several Alpine Air-Jet sievers and interchangeably used sieves within calibration, i.e., we were able to obtain the same results on whatever combination of sievers and sieves was used. These results led us to believe we had absolute calibration. Very little comparative work was done with the Ro-tap since the precision was better on the Alpine and the reporting of less fines for the alumina was assumed to be due to an apparent gentler sieving action.

The first indications that the Alpine results may not be dependable were noted during an investigation of the Attrition Index by the modified Forsythe-Hertwig Method (7). The attrition appeared to be less by Alpine than by Ro-tap. Careful sieving by a wet and dry method using electroformed sieves (3) indicated that fines were being generated and passed by the Alpine on unattrited hydrate. On the other hand the efficiency of sieving of alumina was inversely proportional to the amount of fines. Since the Rotap was the recognized method for attrition studies this investigation into alternative sieving method ceased.

During the investigation of the electrozone method very careful wet and dry electroformed sieving was shown to agree with it and that the Alpine results were erratic, especially on the alumina samples. No good explanation was found although static charging played a role both in the amount of breakage and in the sieving efficiency. Anti-static agents were of assistance at times in improving the efficiency while at other times they did nothing. Changes in fragility also caused variations in the amount of breakage of hydrate intermediates.

With the realization that the Alpine was providing unreliable results the Ro-tap was re-examined. Provided that the sample was sieved to completion using calibrated sieves the curve agreed very well with that obtained by using electroformed sieves. Examinations of what was retained on the sieves indicated that the completion point was achieved. The agreement with Elzone curves indicated that little or no breakage was occurring. A 20 minute sieving with a standardized set of sieves gave results for calcined alumina which were in agreement with the electroformed sieves and the Elzone.

Presently, only the Ro-tap is used for Bayer plant materials and only on calcined alumina for reference purposes. The main argument for maintaining this analysis is that it is the ISO referee method (9) for alumina and as such we must ensure that we know what our levels are by this method.

Electric Sensing Zone

This technique, based on what is known as the Coulter principle, is usually said to measure the volume of a particle. The particles to be characterized are suspended in an electrolyte and caused to pass through an orifice which is also carrying a current between two electrodes. The partial blocking of the orifice by the particle, which normally acts as a resistor, causes a voltage pulse. For our purposes, since our particles are reasonably spherical, this voltage pulse is essentially proportional to the volume of the particle. Through calibration using well-characterized latex or other spheres these voltage pulses are related directly to the particle diameter. By interfacing with a small computer it is possible to very rapidly analyse a sample across the complete range of any orifice tube. This range is normally assumed to be limited to approximately 2-40% of the diameter of the orifice. There is now evidence that the top size can be extended to at least 60% (9) with only a small error and that the lower limit can be as low as 1% (10). For routine control work the range on any one orifice tube can easily be from 4 to 50% (e.g. a 380 μm orifice can analyse samples falling within the range of 16 to 190 μm). By having tubes installed on switchable stands it is possible to handle most samples immediately without tube change or recalibration. Hard copy of the whole distributions, or whatever part is desired, would be available within ten minutes or less of putting the sample on the instrument. The main draw-backs to the instrument are its relative fragility and the use of mercury although for the computerized versions mercury can now be eliminated. The user must therefore be a bit more sophisticated analytically than most plant operators. Care must be taken to have clean electrolyte so that false counts are not obtained but this is not difficult for the regularly used orifices.

There has been much discussion in the literature over the last 15 years regarding the theoretical aspects of this technique, which has been summarized by Allen (11) and Kaye (6). During some work on calcined alumina it became necessary to verify the results being obtained by the Elzone since if it was correct we had a new insight into the calcining process. To verify if the response was as we assumed a variety of samples of alumina and one hydrate were wet and dry sieved using electroformed sieves. The median found was compared to that found by the electrozone technique after careful calibration. The results are shown in Table I.

Table I. Comparison of Analyses by Elzone and Sieving

Sample		Median Size, μm	
		Elzone*	Sieve**
Smelter-grade alumina	A	68	70
	B	70	73
	C	72	75
	D	73	77
	E	81	83
Alumina Tri-Hydrate		93	92

* Electrozone Technique (Computerized Elzone) ** Electroformed, used wet and dry

There is an obvious bias of about two micrometers between the two techniques. Upon close examination under the microscope it is apparent that the sieve aperture which would just pass an alumina particle is slightly greater than a volume diameter. Two micrometers is a reasonable estimate of that difference.

A sample of laboratory calcined alumina closely cut with electroformed sieves at -90 +70 μm was found to have an effective particle density of 2.0 by Hg porosimetry. By Elzone analysis with the computer the median size was found to be 79.5 μm and by manually counting the sample was found to contain $1.93x10^6$ particles per gram. Using this latter value and the density, the average particle size is found to be 79 μm. By microscope the average particle size was 80.3 μm. With these data we feel quite confident that for Bayer materials the electrozone technique provides a result which is as close to "truth" as is possible to attain today, of course, assuming that it is kept in calibration with good standards. Maintenance of calibration is relatively simple since our experience has been that it changes very little over extended periods (e.g. six months). Standards which at one time were a problem are now available from the equipment manufacturers (12,13) and distributors such as Duke Standards (14), and in general are quite reliable.

The instrumental precision is very high. If the same beaker of sample and electrolyte is repeatedly analyzed the same curve is generated, normally within 1%. The problem is the sample presented to the instrument. The quantity is quite small therefore segregation during sample extraction of dry material makes obtaining a representative sample by manual means very difficult. For alumina, riffling is required, preferably with a spinning riffler. The precision for routine work over the long term, then is about what is found for Ro-tap using calibrated sieves. The standard deviation about the median for both techniques is ±0.5 µm and at 45 µm, ±0.5 percentage points. Damp material does not present this segregation problem and is therefore easy to handle (i.e. hydrates should not be dried but just mixed and sampled). To avoid most of these manipulations which all introduce errors the slurry itself can be used directly provided a small representative portion can be extracted. The use of slurry permits immediate analysis without the necessity of filtering and washing and drying of hydrate which requires considerable time as well as introducing extra sampling errors.

A great advantage of this instrument is that it has a second use as a research tool. Discussion of this aspect is beyond the scope of this paper but such things as population balances can be performed beginning somewhat below 0.4 µm. This plus other studies can be performed and the results can be directly compared to routine data without transposition or discussion on the merits of two or more techniques.

Present Control and Characterization System

There are two manufacturers of electrozone equipment (12,13) but we have opted for the Elzone (13) because we feel that it tends to be state of the art. We obtained the prototype version of the computerized Elzone in 1969 and have grown with it. We have recently held discussions with the manufacturer regarding making more rugged instruments which hopefully will allow less sophisticated users.

The computerized Elzone now has replaced sieving for all hydrates and alumina except that the official method for alumina is the Rotap. Analyses are run periodically using sieves calibrated with NBS glass beads (15). Our experience on routine work is that there is very little difference in the long term average for the two methods, the bias being three micrometres, about what we found for very careful work (see Table I). However, the Ro-tap has been retained since it is the most widely accepted technique and tends to be the referee method for commercial purposes. It also ensures that the Elzone does not drift over the long term.

These Elzone analyses are run in a laboratory seperate from the plant which thus means a reporting lag. As well there is presently no easy way for plant personnel to obtain a fast, rough analysis. On the other hand we have a self-consistent technique for all materials which avoids long discussions on which technique is correct when two different methods are used for coarse and fine material or for intermediate vs product. Full size distributions from precise and reasonably "accurate" data give better control partly because there is more confidence in the results on the part of plant personnel.

Investigations of a Laser-Based Alternative Method

The technique which at the moment has the most appeal utilizes laser diffraction. The particle size data are calculated from the light intensity of various portions of the Fraunhofer diffraction pattern generated by passing a laser through a sample cell containing a dilute slurry. More complete outlines of the technique are available from the manufacturers (16,17,18) but the assumptions upon which the involved mathematics are based have not been revealed (19). We have examined two instrumental versions, one in some detail.

These instruments are advertised as being very fast, simple, and trouble-free. Such instruments would be welcomed for control purposes provided their precision was good and their bias, if any, with an acceptable control method was constant. We did not acquire these instruments but over the last several years samples of hydrate and calcined alumina have been submitted by various people to the manufacturers. In addition some samples were analyzed in our presence by a knowledgeable user and in another instance by a user who employed the instrument to control Bayer circuits. Data from eleven samples of hydrate and alumina in our normal size range and three samples of somewhat coarser hydrates are available from one version of the diffraction technique for which we have Elzone and/or sieve analysis. These are shown in Table II. Figure 1 presents the data in cumulative form for one sample from two commercial laser diffraction instruments and the curve produced by the Elzone. Figure 2 presents the data in differential form from only one diffraction instrument vs the Elzone for another sample. It is apparent that for our materials very different results are obtained. For control purposes this can be acceptable provided that the bias is constant and the instrument is sensitive to real changes in the distribution. An attempt was made to assess this. From all the data we have for one instrument (Table II), the differences (or biases) between the Elzone/sieve curves and the diffraction curves, in micrometres were determined and are reported in Table III. The Elzone/sieve results, the assumed correct value based on our experience and studies, were subtracted from the diffraction data.

Samples 1 to 11 in Table II, are characteristic of our normal smelter-grade alumina. The bias data for these samples, as reported in Table III, have been plotted in Figure 3. A linear regression yields the straight line shown which has an intercept at 55 µm and a slope of 1.24. The correlation coefficient is 0.828. It is apparent that there is considerable scatter about this line which is very marked in the coarse end. To determine if this scatter in micrometres has pertinence when translated into percent (Table IV) the bias was again calculated as discussed above and shown in Table V.

The distribution of the bias values versus the Elzone size is shown in Figure 4. There appears to be a tendency for the bias to be a function of the "true" percentage level but the scatter is rather large. The bias peaks, for these particular materials, in the region of 20% by Elzone, although one sample has quite a different bias. This plot is a prediction of what might be expected from the laser instrument based on an Elzone or sieve analysis. However, if the Laser Diffraction instrument were in use the requirement would be to predict the actual size distribution of the sample. To assess this the same bias but with reversed sign was plotted versus the level found by the Diffraction instrument and shown in Figure 5. A peak is again apparent but now displaced to about the 50% level. Two linear regression lines were drawn in as shown by using all

Table II. Particle Size Distribution Data Extracted From Laser Diffraction and Elzone Curves

Micrometres

Percentage Level	10		20		30		40		50		60		70		80		90	
Sample	LD[1]	E[2]	LD	E	LD	E	LD	E	LD	E	LD	E	LD	E	LD	E	LD	E
1	125	110	114	92	102	85	93	77	84	73	75	67	67	62	58	55	46	46
2	135	102	118	89	105	81	94	75	88	69	75	64	67	64	58	52	47	45
3	122	92	107	82	96	77	85	72	78	67	72	63	65	58	57	53	48	48
4	148	121	128	105	114	96	104	88	95	80	87	74	77	68	64	61	45	54
5	148	107	124	94	106	85	94	78	84	72	74	66	66	61	57	55	46	48
6	154	105	125	92	107	83	94	76	84	70	74	64	65	59	56	53	41	46
7	133	105	110	84	94	76	84	70	74	64	66	60	59	55	51	50	39	44
8	130	90	107	79	92	72	80	67	72	62	64	58	56	53	48	50	38	44
9	130	90	107	78	92	72	80	66	72	62	64	57	56	53	48	48	39	43
10	130	107	122	93	118	83	102	75	90	70	77	64	65	58	53	53	38	45
11	120	108	104	94	93	84	84	77	76	70	69	64	60	60	48	52	30	42
12	140	130	122	115	111	102	102	92	90	84	80	76	69	69	58	61	47	52
13	150	125	123	112	108	103	96	95	86	88	77	82	69	73	56	65	30	53
14	165	160	140	143	125	131	111	120	100	110	90	100	78	90	64	77	44	63

(1) LD - Laser Diffraction
(2) E - Elzone/Sieve

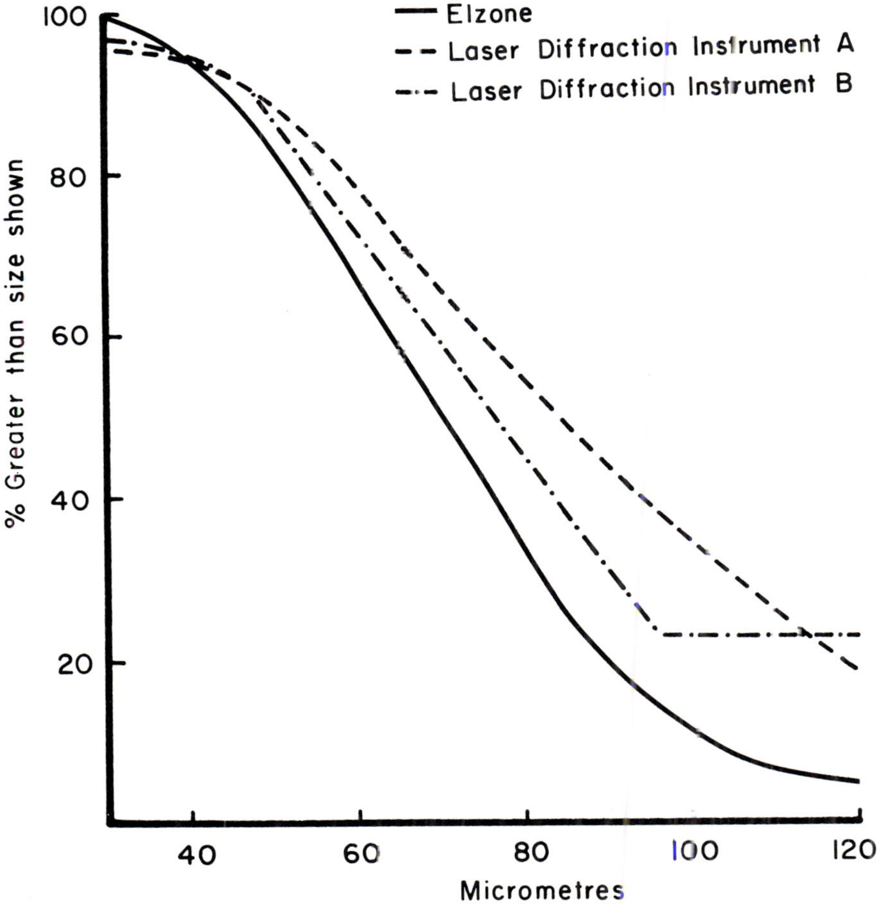

Figure 1 - Comparison of Reported Particle Size Distribution Curves for Smelter-Grade Alumina by Elzone and Two Laser Diffraction Instruments

(Cumulative Curves)

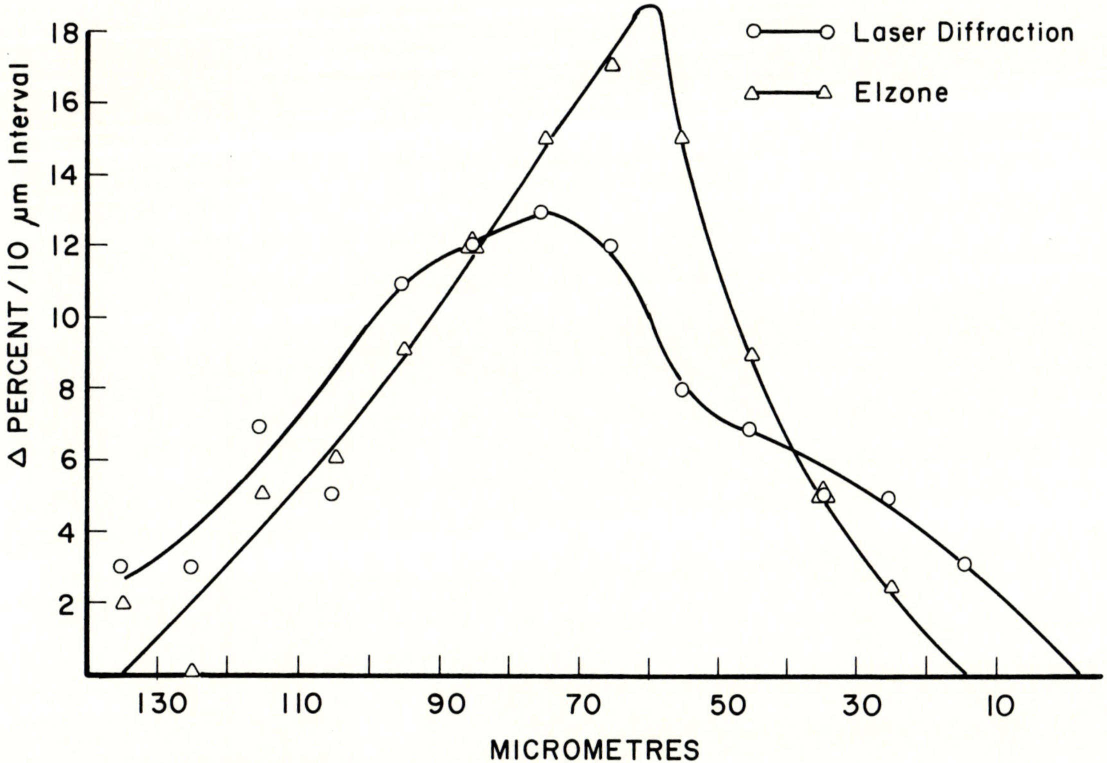

Figure 2 - Comparison of Reported Particle Size Distribution Curves for Smelter-Grade Alumina by Elzone and a Laser Diffraction Unit

(Differential Curves)

Table III. Bias Between Laser Diffraction and Elzone Data Vs. Elzone Data

Percentage Level Sample	10 E[1]	10 B[2]	20 E	20 B	30 E	30 B	40 E	40 B	50 E	50 B	60 E	60 B	70 E	70 B	80 E	80 B	90 E	90 B
	Micrometres																	
1	110	15	92	22	85	17	77	16	73	11	67	8	62	5	55	3	46	0
2	102	33	89	29	81	24	75	19	69	19	64	11	64	3	52	6	45	2
3	92	30	82	25	77	19	72	13	67	11	63	9	58	7	53	4	48	0
4	121	28	105	23	96	18	88	16	80	15	74	13	68	9	61	3	54	-9
5	107	41	94	30	85	21	78	16	72	12	66	8	61	5	55	2	48	-2
6	105	49	92	33	83	24	76	18	70	14	64	10	59	6	53	3	46	-5
7	105	28	84	26	76	18	70	14	64	10	60	6	55	4	50	1	44	-5
8	90	40	79	28	72	20	67	13	62	10	58	6	53	3	50	-2	44	-6
9	90	40	78	29	72	20	66	14	62	10	57	7	53	3	48	0	43	-4
10	107	23	93	29	83	35	75	27	70	20	64	13	58	7	53	0	45	-7
11	108	12	94	10	84	9	77	7	70	16	64	5	60	0	52	-4	42	-12
12	130	10	115	7	102	9	92	10	84	6	76	4	69	0	61	-3	52	-5
13	125	25	112	11	103	5	95	1	88	-4	82	-5	73	-4	65	0	53	-13
14	160	5	143	-3	131	-6	120	-9	110	-10	100	-10	90	-12	77	-13	63	-19

(1) E - Elzone
(2) B - Bias (Laser Diffraction - Elzone)

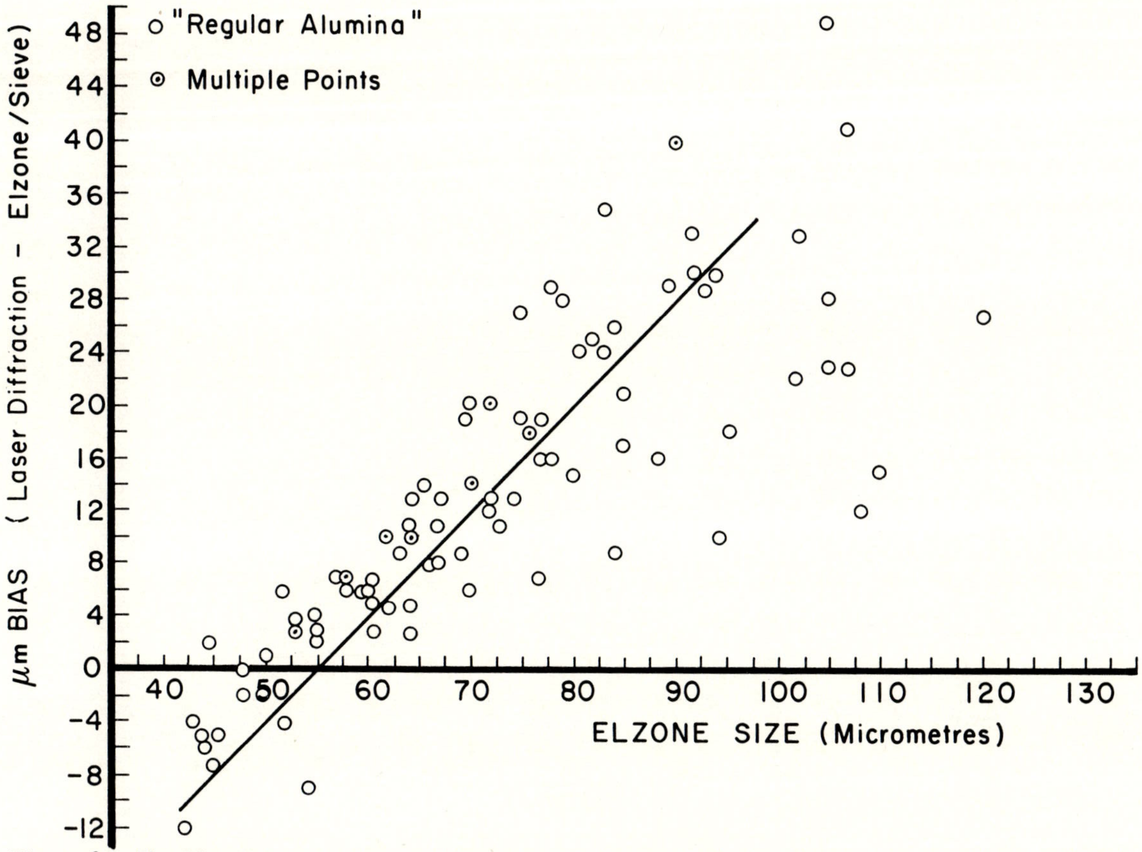

Figure 3 - The Bias Between a Laser Diffraction Instrument and the Elzone as a Function of Size for Smelter-Grade Alumina

Table IV. Particle Size Distribution Data Extracted From Laser Diffraction and Elzone Curves

Size Level-μm	Percent																	
	150		125		100		90		80		70		60		50		45	
Sample	LD[1]	E[2]	LD	E	LD	E	LD	E	LD	E	LD	E	LD	E	LD	E	LD	E
1	9	3	22	8	44	25	57	36	66	50	74	67	83	82	88	93	90	97
2	11	1	19	4	35	14	44	22	54	34	64	50	75	68	86	85	88	91
3	9	1	19	4	35	15	44	24	54	36	65	52	76	70	86	87	91	93
4	5	3	13	6	25	12	32	16	43	24	55	38	69	59	81	80	87	89
5	5	0	11	1	23	5	32	9	40	18	52	33	65	55	78	78	84	88
6	5	0	12	1	23	5	32	9	40	18	51	32	64	53	77	77	84	87
7	0	0	15	3	32	14	43	22	55	38	67	55	77	60	88	87	94	94
8	0	0	16	2	34	12	44	19	55	32	67	49	77	68	87	83	92	90
9	0	0	9	1	22	8	35	11	48	23	63	44	76	67	88	86	93	94
10	0	0	18	3	43	15	51	22	58	34	66	50	74	68	82	85	85	90
11	0	0	9	3	24	15	33	24	44	36	58	50	70	68	79	83	84	88
12	7	3	18	12	42	32	50	42	59	54	69	69	78	81	87	92	91	95
13	10	3	20	10	37	33	46	48	57	63	68	75	77	85	84	92	86	94
14	15	15	29	35	50	60	60	70	68	78	75	85	82	92	88	96	90	97

(1) LD - Laser Diffraction
(2) E - Elzone/Sieve

Table V. Bias Between Laser Diffraction and Elzone Data vs. Elzone Data

Size Level-μm	Percent																	
	150		125		100		90		80		70		60		50		45	
Sample	E[1]	B[2]	E	B	E	B	E	B	E	B	E	B	E	B	E	B	E	B
1	3	6	8	14	25	19	36	21	50	16	67	7	82	1	93	-5	97	-7
2	1	10	4	15	14	11	22	22	34	20	50	14	68	7	85	1	91	-3
3	1	8	4	15	15	20	24	20	36	18	52	13	70	6	87	-1	93	-2
4	3	2	6	7	12	13	16	16	24	19	38	17	59	10	80	1	89	-2
5	0	5	1	10	5	18	9	23	18	22	33	19	55	10	78	0	88	-4
6	0	5	1	11	5	18	9	23	18	22	32	18	53	11	77	0	87	-3
7	0	0	3	12	14	18	22	21	38	17	55	12	60	17	87	1	94	0
8	0	0	2	14	12	22	19	25	32	23	49	18	68	9	83	4	90	2
9	0	0	1	8	8	14	11	24	23	25	44	19	67	9	86	2	94	-1
10	0	0	3	15	15	28	22	29	34	24	50	16	68	6	85	-3	90	-5
11	0	0	3	6	15	9	24	9	36	8	50	8	68	2	83	-4	88	-4
12	3	4	12	6	32	10	42	8	54	5	69	0	81	-3	92	-5	95	-4
13	3	7	10	10	33	4	48	-2	63	-6	75	-7	85	-8	92	-8	94	-8
14	15	0	35	-6	60	-10	70	-10	78	-10	85	-10	92	-10	96	-8	97	-7

(1) E - Elzone
(2) B - Bias (Laser Diffraction - Elzone)

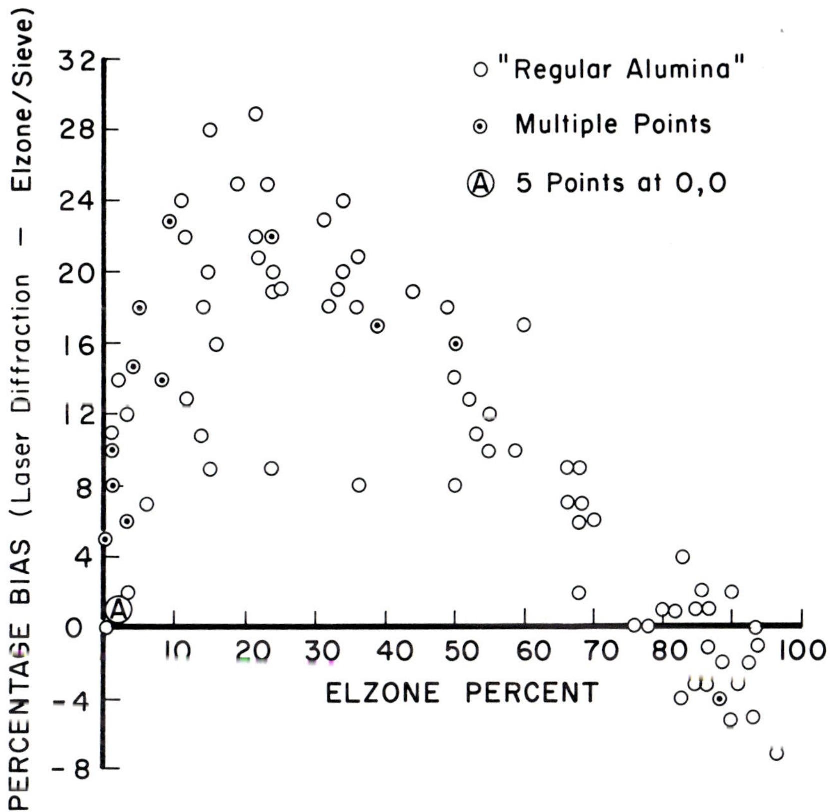

Figure 4 - Bias in Percent Between Laser Diffraction and Elzone Analyses Versus Elzone Percent for "Regular Alumina" Size Range

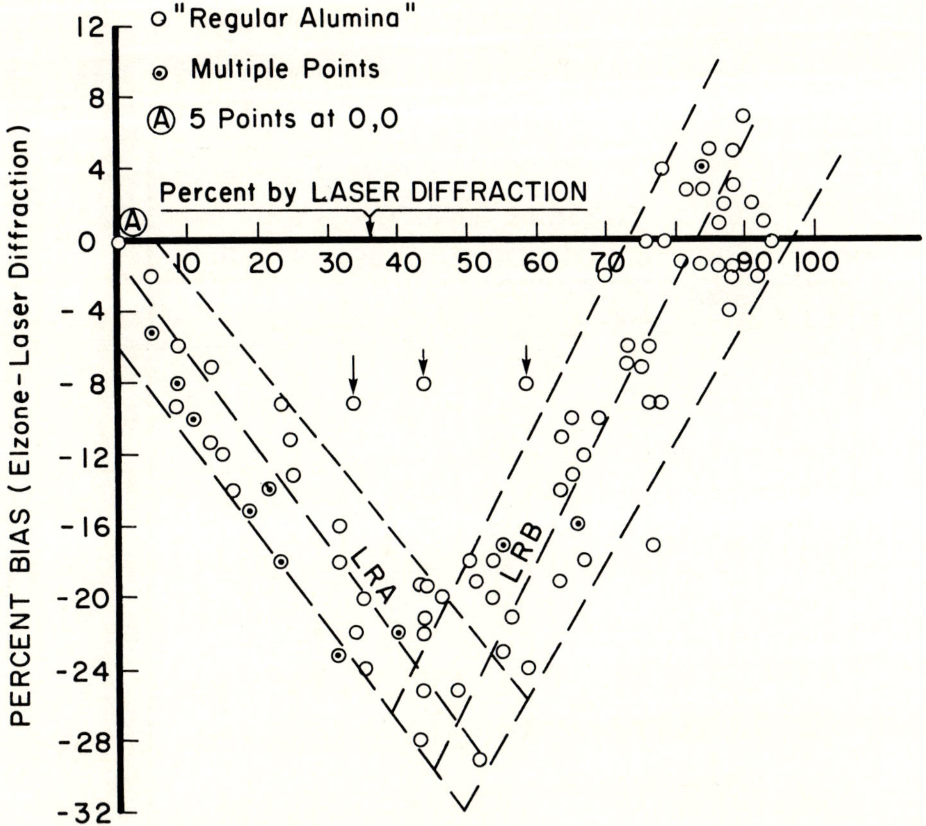

Figure 5 - Bias in Percent Between Elzone and Laser Diffraction Analyses Versus Percent by Laser Diffraction for Regular Smelter-Grade Alumina

the points, except the three indicated with arrows, from zero to 50% for LRA and from 45% to 95% for LRB. The limiting lines were drawn in to enclose essentially all the points used in the regression analyses. This provides an estimate of the range of the bias for any given point. For instance, at the important 90% point the bias is estimated to extend from about minus 5 to about plus 10. That is, the "true" value would be predicted to lie somewhere between 86% and 100%. If this were a control point, compared to the Elzone which would predict the "true" value within one percentage point, the diffraction range would be intolerable. The same applies to the rest of the distribution curve. In addition the amount of bias required for correction varies with the level found, even for materials in the same size range.

The question arises as to whether the bias curve shown in Figure 5 will apply to distributions outside the range of our regular smelter-grade aluminas. Fortunately a change in our circuit produced some coarser material, the median size shifting about 20 micrometres. Three samples of this were analyzed by the same knowledgeable analysts, the results being shown as samples 12, 13 and 14 in Tables II to V inclusive. Their bias data are shown in Figure 6 with heavy dots. A linear regression on these points alone gave a line of slope 1.26 but the coefficient was only 0.46. The relationship is so weak that the line was not drawn but it is interesting that the line for the regular alumina has similar slope. The effect of these erratic sizes on the percentages is shown in Figure 7, again with solid dots.

Although we only have three samples it would appear that there has been a shift in the bias which must be applied to find the "real" value. Since the only information available would be that from the diffraction instrument, how much bias to apply would be a difficult question indeed. In addition, the correctness of the response to changes in distribution is unclear. If a particular sampling location became finer would this be detected or would the unknowingly changing bias possibly obscure it? Interestingly the points not used in the regression analysis (with arrows) are now in the range of the new points.

In the Bayer circuit intermediates and products tend to have truncated distributions due to classification. Austin and Shaw (2) showed that the diffraction technique is very insensitive to such distributions (it gives wider than actual curves). Our data confirms this since the diffraction technique yielded wider distributions than the Elzone/sieve, as shown in Figure 1 and 2. Johnston and Swanson (21) found for AC Fine Test Dust that the technique "sees considerably less small particles, and somewhat more large particles" than a sedimentation instrument which had excellent agreement with an electrosensing zone instrument. This was after calibration of the diffraction unit with latex beads. The mis-match at the fine end was to be expected because some of the fines were below the capability of the diffraction instrument. The coarse-end however, is truncated but well within the instrumental capabilities. At the 45 μm point the accepted level was 7% and the bias was +10%, within the range of what we found for the coarse end of our much coarser distributions. We have no data for samples, such as a fine seed, which are in this range but see no evidence that they would respond differently from what was found by Johnston and Swanson. Two of our alumina samples were analyzed further into the fine end. By Elzone we found less than 0.5% smaller than 20 μm whereas the laser diffraction instrument indicated about 5%. The latter instrument found the 0.5% point at 4 μm for one sample and below its

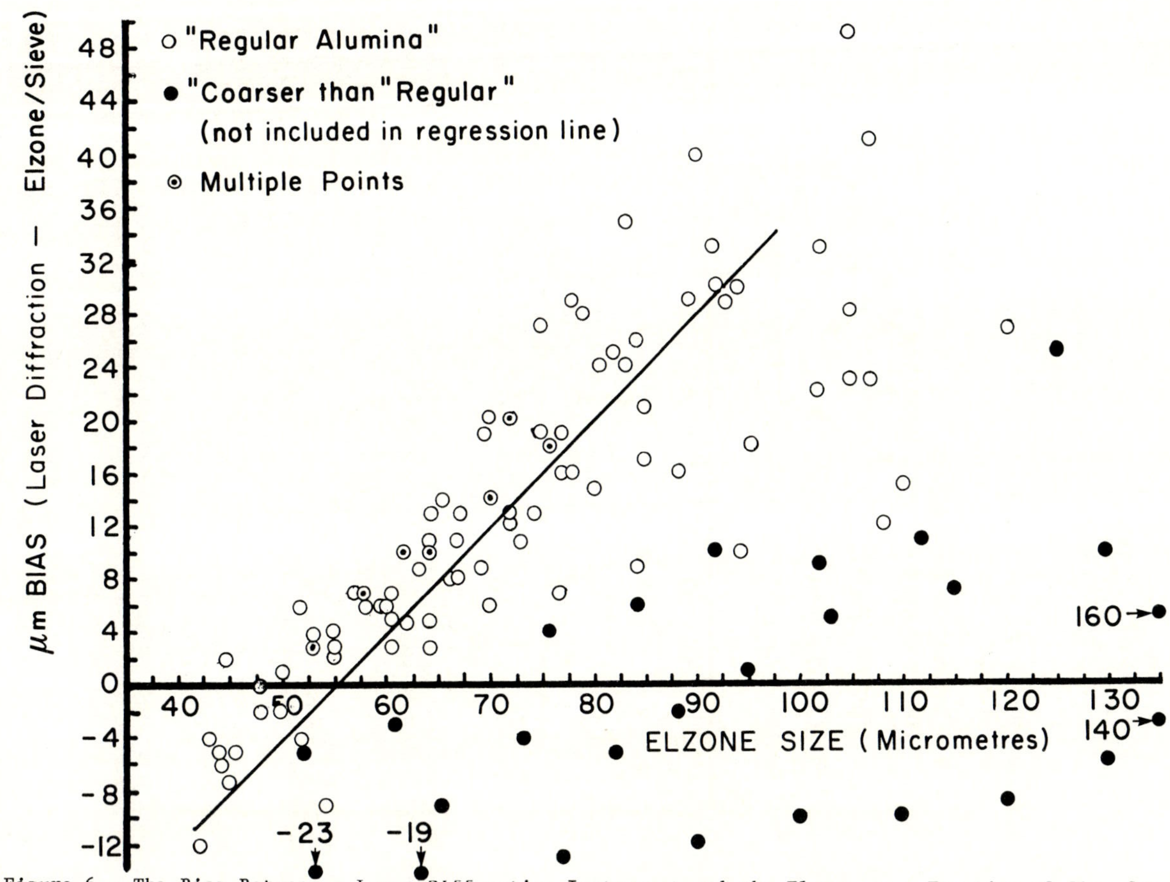

Figure 6 - The Bias Between a Laser Diffraction Instrument and the Elzone as a Function of Size for Smelter-Grade Alumina and Coarser Alumina

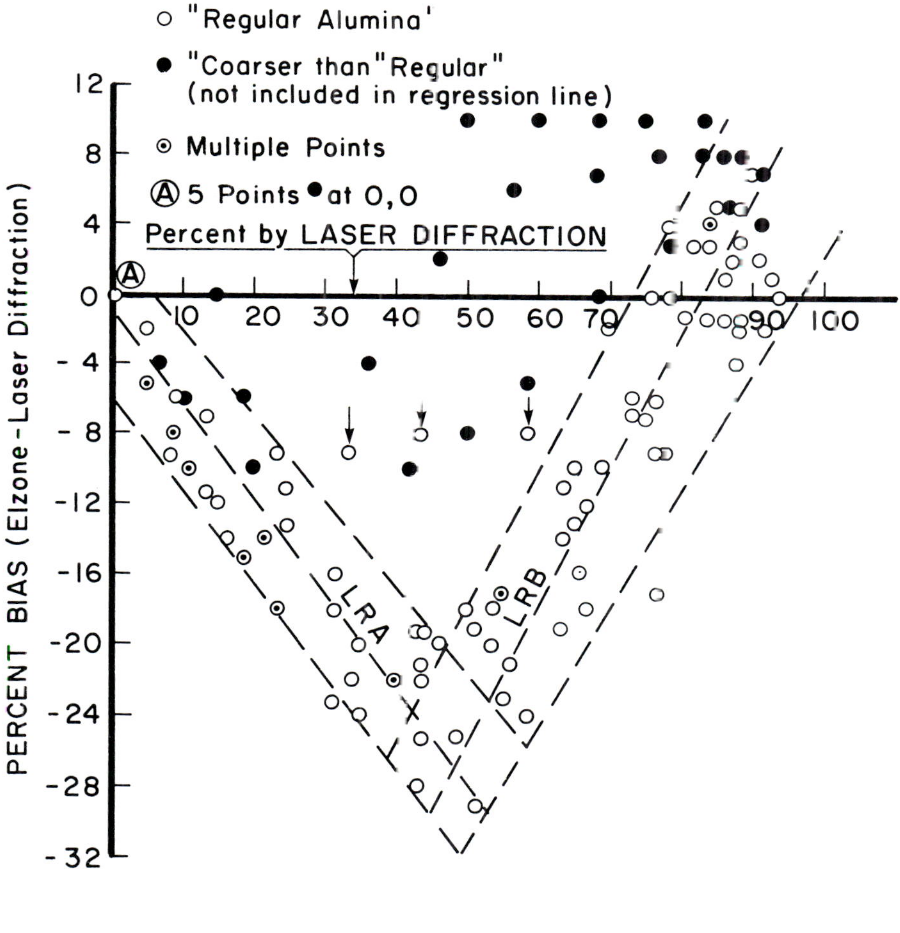

Figure 7 - Bias in Percent Between Elzone and Laser Diffraction Analyses Versus Percent by Laser Diffraction for Regular and Coarse Smelter-Grade Alumina

capability for the other. This also confirms the Austin and Shaw findings for truncated distributions.

The consequence of the variable bias and its wide range for any given point, due in part to the truncated distributions which have been cut at different sizes, is that the laser diffraction technique will not provide data which can be translated into "real" values. On this basis the technique cannot be recommended for control of the Bayer plant. It would probably mean that our control would be weakened.

Summary

Our present control technique involving the use of the computerized Elzone has been shown to provide data which is close to reality. In addition its range is such that it can be used for all intermediates and products. Thus changes in the circuit can be traced readily and cannot be attributed to mixed analytical methods. The technique is sophisticated but its output is such that a high level of control is possible.

Our experience with mixed techniques causes us to be very cautious about introducing this again even though the new laser diffraction technique is touted to be fast and simple. Although we have had very little hands-on experience with the technique data on our material is such that we are very hesitant about its introduction. Compared to our sophisticated system it has very questionable advantages. Its great disadvantage being its variable and unpredictable bias.

References

1. Wahl, B.J., and Larouche, P., Accuracy and Precision in Industrial Screen Analysis, Am. Ceram. Soc. Bull, 43 (5), 1964, pp. 377-379.

2. Zwicker, J.D., Sieve Analysis to Below Two Microns Using Micromesh Sieves, Am. Ceram. Soc. Bull, 45 (8), 1966, pp. 716-9.

3. Zwicker, J.D., Comparison of Particle Size Analysis Methods, Am. Ceram. Soc. Bull 46 (3), 1967, pp. 303-306.

4. Zwicker, J.D., On-Stream Particle Size Analysis Requirement for Industry, National Science Foundation Proceedings of the Workshop on Particle Technology, 1975, pp. 41-46.

5. Dubé, G., and Zwicker, J.D., Development of an On-Stream Particle Size Analyzer for Slurries, Light Metals, Vol. 1, 1978, pp. 85-94.

6. Kaye, Brian, H., Direct Characterization of Fineparticles 1981, John Wiley & Sons.

7. Forsythe, W.L., and Hertwig, W.R., Attrition Characteristics of Fluid Cracking Catalysts, Ind. Eng. Chem., 41 (6), 1949, pp. 1200-06.

8. International Standards Organization, Method 2926-1974(E), Aluminium Oxide Used Primarily for the Production of Aluminum-Particle Size Analysis, Sieving Method, International Organization for Standardization, Case Postale 56, CH-1121, Geneva 20, Switzerland.

9. Harfield, J.G., and Knight, P., Experimental Evidence of the Linear Response of the Coulter Counter, Particle Size Analysis 1981, pp. 151-164, John Wiley and Sons.

10. Berg, R.H., Particle Data Inc., July 1983, Private Communication.

11. Allen, T., Particle Size Measurement, Third Ed. 1981, Chapman and Hall.

12. Coulter Electronics Inc., 590 West 20th Street, Hialeah, Fl. 33010.

13. Particle Data Inc., PO Box 265, Elmhurst, Il. 60126.

14. Duke Scientific Corp., 445 Sherman Ave., Palo Alta, Ca. 94306.

15. Standard Reference Material 1004. Calibrated Glass Beads. National Bureau of Standards, Washington, D.C., 20234.

16. Microtrac Particle Size Analyser, Leeds & Northrup Co., Microtrac Division, Largo, Fl. 33543.

17. Laser Granulometer Model 715, Cilas Campagnie Industrielle des Lasers - U.S.Representative Marco Scientific Inc., 1031 H East Duane Ave., Sunnyvale, Ca. 94086.

18. Malvern 3300 Powder Particle Sizer, Malvern Instruments Ltd., Spring Lane, Malvern, Worcestershire, WR14 1AL, England.

19. Leschonski, K., Review of the Conference, Particle Size Analysis 1981, pp. 437-445, John Wiley and Sons.

20. Austin, L.G., and Shah, I., A Method for Inter-Conversion of Microtrac and Sieve Size Distributions, Powder Tech. 35 (1983) pp. 271-278.

21. Johnston, R.R. and Swanson, R., A Correlation Between the Results of Different Instruments Used to Determine the Particle Size Distributions in AC Fine Test Dust, Powder Tech. 32 (1982), pp. 119-124.

Appendix 1. Progression of Improvements in Particle Size Analysis for Control of Bayer Plants and Concurrent Investigations

	Control		
Date	Coarse	Fine	Research and Development
Before 1957	Ro-tap sieving	Hydrometers	Infrasizer
		Wagner Turbidimetry	Bahco
			Micromerograph
1957	Alpine air jet siever (1) calibrated sieves, complete size distributions		
1959			Coulter Counter, Model A fine seeds, special samples slurry samples
1961		Electroformed sieves to 2 μm(2)	
1965	A wide range sieving method to cover complete range(3)		
1969			Computerized Elzone - Coulter discarded
			On-stream size analysis investigations (4) Procedyne, Autometrics PSM, Royal School of Mines slurry sizer, Alcan On-Stream Sizer(5)
			Fine particle analysers investigated: Alcan Air Permeability Apparatus, MSA Particle Sizer, Sonic Sifter
			Broad-range techniques investigated Hiac Microtrac Cilas
1982	Computerized Elzone Ro-tap (for product reference)	Computerized Elzone	

CHAIRMEN

F.S. Williams
Alcoa Laboratories
Alcoa Center, Pennsylvania

J.C. Winch
Alcan International Limited
Montreal, Quebec, Canada

FLOCCULATION AND THICKENING OF RED MUD

C. Sato, Y. Shibue, N. Arakawa and Y. Kawai

Showa Aluminum Industries K.K.
Yokohama, Japan

In the red mud separation process, improvement of the clarification of the thickener overflow and of the thickening of the underflow is highly important.

Showa Aluminum Industries extensively studied flocculation and thickening of red mud and, as a result, succeeded in concentrating the underflow to about 700 - 800 grams/liter, which is considered a highest limit for transfer by the centrifugal pump.

It has been achieved by utilization of synthetic polymer and improvement of the thickener.

The above-mentioned improvement of the red mud flocculation and concentration has enabled the unit consumptions of the flocculation aid and caustic soda to be substantially reduced.

Introduction

Showa Alumiinum Industries (SAL) extracts alumina at its Yokohama Works from bauxite of three different origins, namely, Weipa and Gove in Australia and Bintain in Indonesia. For settling and washing of red mud, the Dorr type thickener is used.

Settling of the red mud is affected by factors such as types of material ores, alumina extractions, ground sizes of bauxite, solid concentration, types of flocculants used, construction of the thickener and properties of the aluminate liquor. In actual plant operation, these factors affect the mud settling in a most complicated manner[1)].

For efficient mud sedimentation, it is essential that the clarity of the aluminate liquor and the thickness of the mud slurry should be improved. SAL has been working on this since 1971[2)].

The slurry concentration of the thickener underflow has been increased from 400 - 600 grams/liter to 700 - 800 grams/liter through active utilization of flocculant and improvement of the thickener itself.

As the concentration of the underflow became higher, unit consumptions of flocculant and caustic soda were substantially improved.

Reported below are technical information based on SAL's experience in improvement of mud sedimentation through the use of flocculants and modification of the thickener.

Experiments

The experiments on settling of the mud were performed by using glass cylinders (10 cmø x 40 cmH) placed in a thermostatic vessel. An agitating plate having a number of holes in it, enabling the plate to travel up and down in the glass cylinder at a fixed rate of 10 cm/sec, was used to mix the red-mud slurry and a flocculant. The settling test was performed after agitating the mixture of the mud slurry and a flocculant three times.

Use of Synthetic Polymers as Settling Aid

Starch, lime and synthetic polymers are generally used to accelerate the mud sedimentation[3), 4), 5), 6), 7), 8), 9), 10), 11), 12), 13), 14)].

SAL's Yokohama Works had used a varieties of settling aids but, at present, it is using only sodium acrylate and acrylamide based synthetic polymers.

Switching from Powder to Emulsion Polymers

Further, SAL is switching polymers from powder to emulsion, due to the following reasons:

1) Cost of the emulsion polymer is lower than that of the powder polymer.

2) The intrinsic viscosity obtainable with respect to the powder polymer was not more than 10 dl/g (30°C, 1-N Na_2CO_3) due to the limitations on preparation of a solution. In case of the emulsion, a polymer having two times higher degree of polymerization is obtainable. Since powder polymer is manufactured by means of aqueous solution polymerization, it is impossible to make its molecular weight in excess of 10,000,000.

3) In the case of a powder polymer, an aqueous solution of 0.1 - 1% must be prepared in a normal practice and a resultant dissolving loss must be accepted. An emulsion polymer, on the other hand, does not require such troublesome work or causes such loss. The polymer content of an emulsion polymer is 70%, much higher than that of a powder polymer.

Experiments were performed to compare the efficiencies of polymers, emulsion and powder. The emulsion and powder polymers were used in the same amount, in terms of their effective ingredients, for treating the same amount of a mud slurry.

Thirty minutes after commencement of the experiment, a small part of the overflow was subjected to chemical analysis to determine the total iron content for the purpose of establishing the turbidity of the overflow.

The experiment was performed by adding the polymer to the slurry. After being well agitated, the slurry containing the polymer were allowed to stand for 30 minutes and then the height of the interface was measured. To study the tightness of the flocs, the test liquor was once again agitated, this time without adding any polymer, and then the height of the interface was measured after 30 minutes. The results are shown in Fig. 1.

These experiments proved that the emulsion polymer is superior to the powder polymer by far in respect of the settling rate, slurry compression, the turbidity of overflow and the tightness of flocs.

This is considered attributable to the orientation of the high molecule chains and the increase in number of bonds with suspended particles, both of which is obtainable in case of the emulsion polymer through high polymerization.

Feeding of Emulsion Polymer

Since the synthetic polymer is used by a small amount, it is charged into the thickener after being diluted in the process liquor or in the condensate so that it will disperse well in the slurry. As the feeding points, the feed launder, feed well and the feed pipe have been proposed[15), 16)].

The emulsion polymer becomes less active, as shown in Fig. 2, when it is diluted or held for a long time after being diluted. A contrivance has, however, been successfully introduced at SAL's Yokohama Works to feed the emulsion polymer undiluted at the feed well and disperse it well into the slurry[17)].

This arrangement has enabled the polymer additive to be saved by more than 30%.

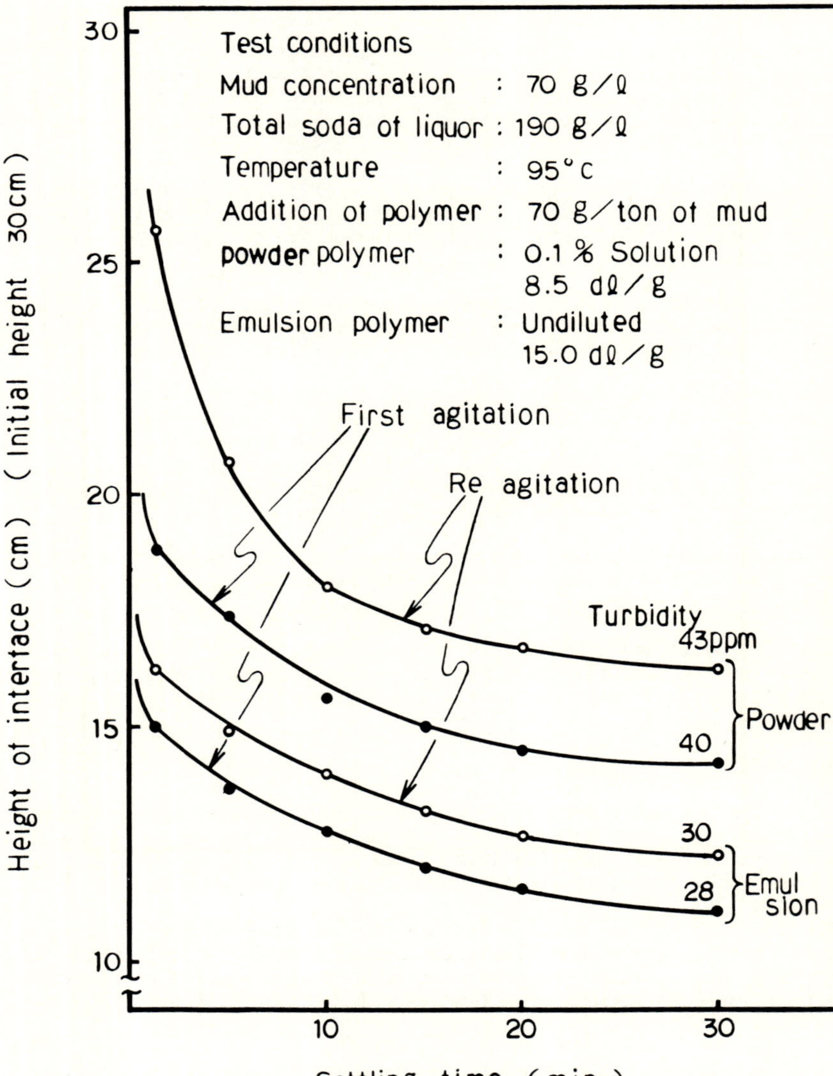

Fig. 1 Comparison with powder and emulsion polymer in mud settling

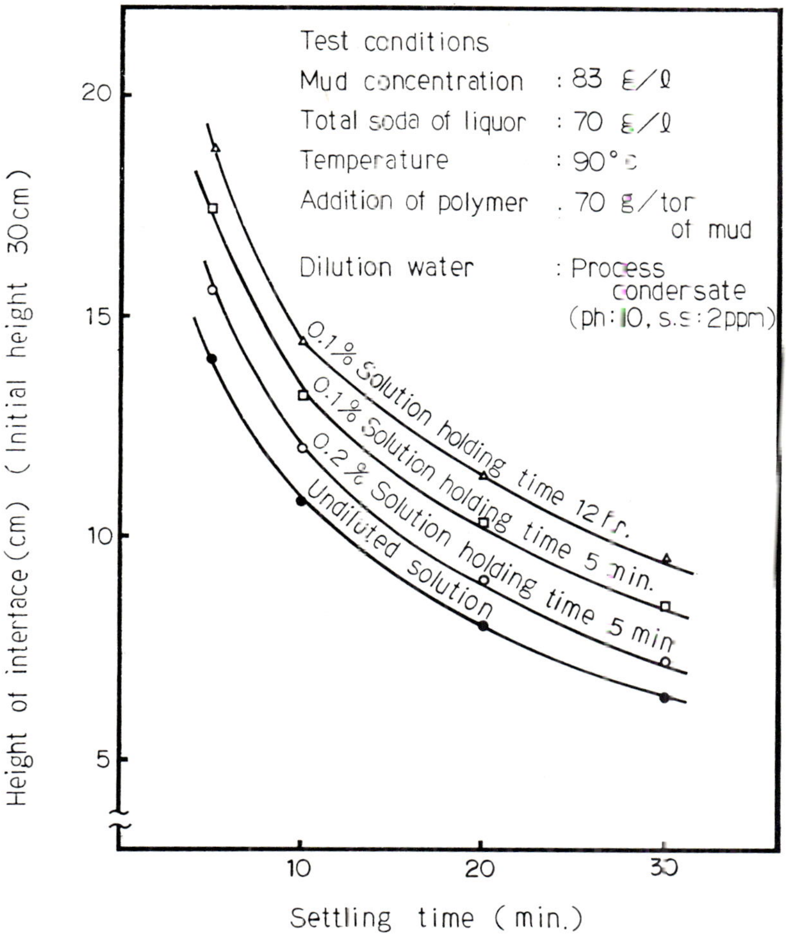

Fig. 2 Effect of dilution and holding time on the emulsion polymer

Flocculant for Washer

Since the temperature and alkalinity of the washer vary by stages, particles have different levels of electric charges. SAL has, therefore, studied the feasibility of adjusting the electric charges of the polymer by changing the composition of sodium polyacrylate and polyacrylamide as it has been found that the higher the amide ratio, the lower the anion ratio.

SAL's experience shows that the flocculants shown in Fig. 3 are most desirable for each washer. A flocculant having a high amide ratio is, therefore, suitable for the washer with low alkalinity.

At present, SAL uses 3 - 4 types of emulsion polymers, each having different amide ratio and molecular weight of 10 million to 25 million, undiluted for its six-stage washing operation.

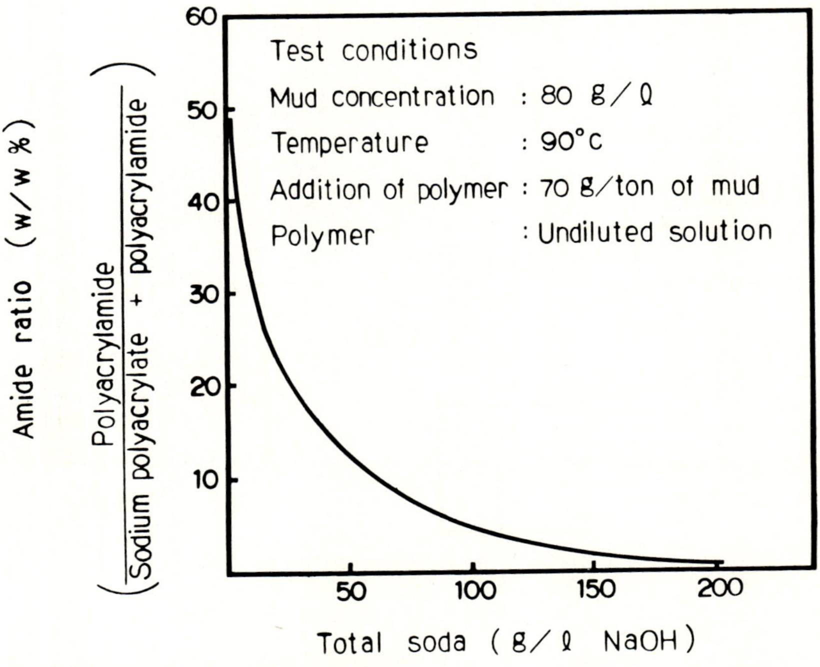

Fig. 3 Amide ratio v.s. total soda

Improvement of Cable Torque Thickener[18]

In 1974, SAL changed the type of its thickener from the conventional type to the cable torque type.

The cable torque thickener features a longer life as compared with the conventional type thickener because scale deposits less on the former.

Also, use of the cable in place of the conventional thixo-arm was believed, at least theoretically, to automatically lift the pipe-formed rake arm when the torque increases, enabling the torque trouble to be avoided. In reality, however, the torque trouble was unavoidable and, therefore, the specific gravity of the underflow had to be held low.

Studies were, therefore, made to see whether or not the lifting of the rake actually takes place as easily as it had been claimed. A resistance "R" required to push back the mud slurry was measured with different test pieces and different concentration of the slurry, using an apparatus shown in Fig. 4.

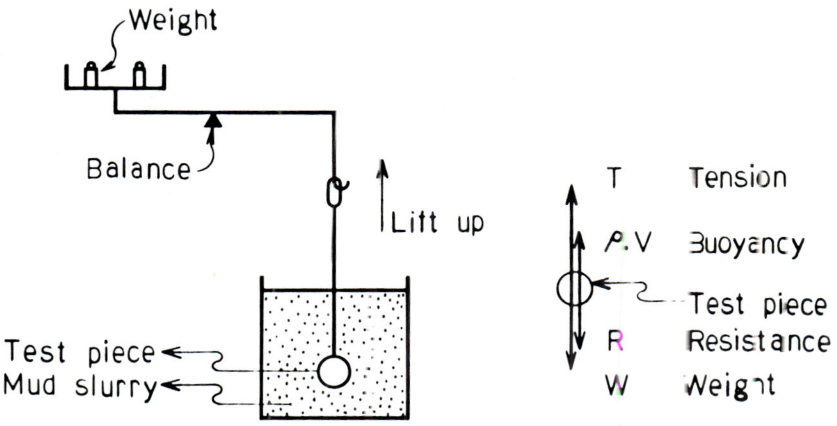

Fig. 4 Method for a measurement of resistance "R"

A test piece was hang from one end of a balance into a mud slurry and the tension required to move the test piece through the mud slurry was measured by placing weights on the other end of the balance.

If a test piece moves at the tension "T":

$$T = W - \rho \cdot V + R \quad \ldots\ldots\ldots \quad (1)$$

In this case "R" will be:

$$R = Cp \cdot S \quad \ldots\ldots\ldots\ldots\ldots \quad (2)$$

Where W = Weight of test piece
ρ = Specific gravity of the mud slurry
V = Volume of test piece
$\rho \cdot V$ = Buoyancy
R = Resistance required to thrust back red mud
S = Projected area of a test piece which is perpendicular to the normal slurry flow
Cp = A resistance coefficient of test piece, namely, the function of the shape of test piece and the concentration of the slurry.

On the basis of the test results, the relationship of Cp (=R/S) and ($\rho - \rho_o$) as the concentration of the slurry is shown in Fig. 5 (ρ is the specific gravity of the mud slurry and ρ_o, the specific gravity of the aluminate solution contained in the mud slurry).

Fig. 5 shows that shape of a test piece has little bearing on the results. Since it became possible to obtain the Cp value of each slurry concentration, the tension of the drag cable of the commercial thickener was obtained from the formulae (1) and (2) above.

In case of SAL's commercial thickener (18 mø), the tension was 88 kg when no scale was deposited on the rake and 120 - 150 kg, when scale was deposited on the rake after three months' operation. This applies, however, to the case in which ($\rho - \rho_o$) is 0.50. A test with a commercial thickener confirmed that, at this level of tension, lifting of the rake did not take place and the slurry was discharged sufficiently.

The above findings have lead the authors to believe that, under the normal circumstances, no automatic lifting of the rake arm takes place and that the torque alone increases with the increase in volume of deposited scale.

On the basis of the above findings, the cable torque thickener was modified to reduce the resistance "R". Fig. 6 compares the torque values and the specific gravity of the underflow, before and after the modification. It is clear from this chart that, after the modification, the torque value became stable and the specific gravity of the underflow became higher.

Thus, it has become possible to discharge a slurry stably at a concentration level of 700 - 800 grams/liter.

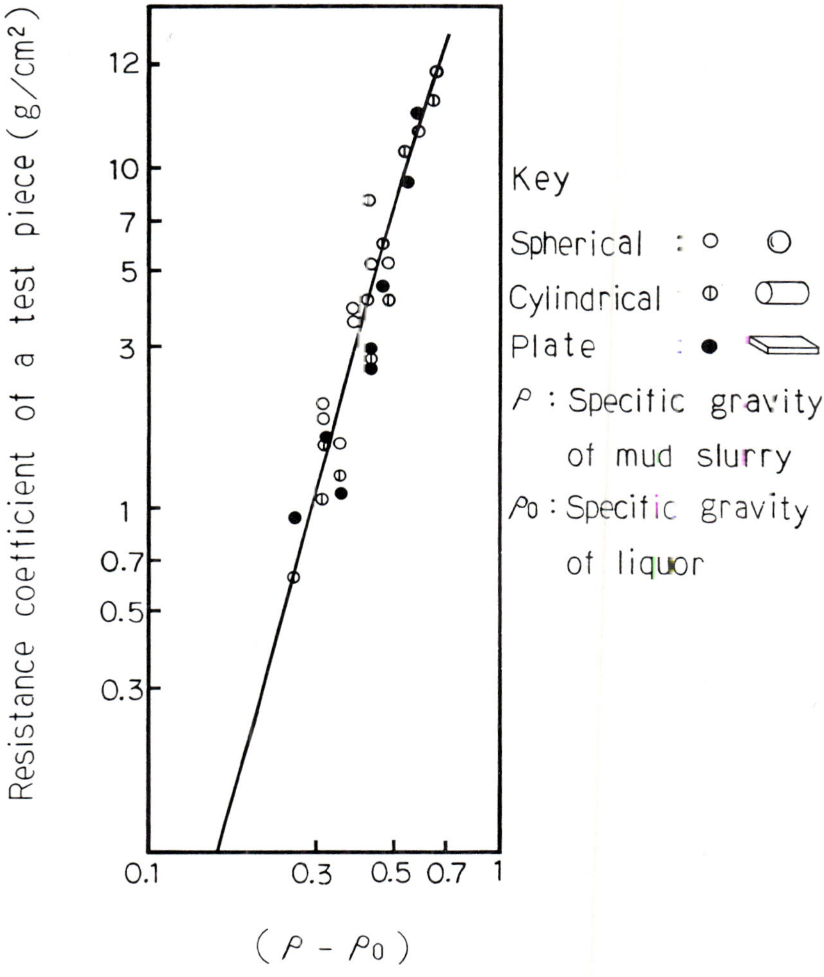

Fig. 5 Correlation between resistance coefficient of a test piece and $(\rho - \rho_0)$

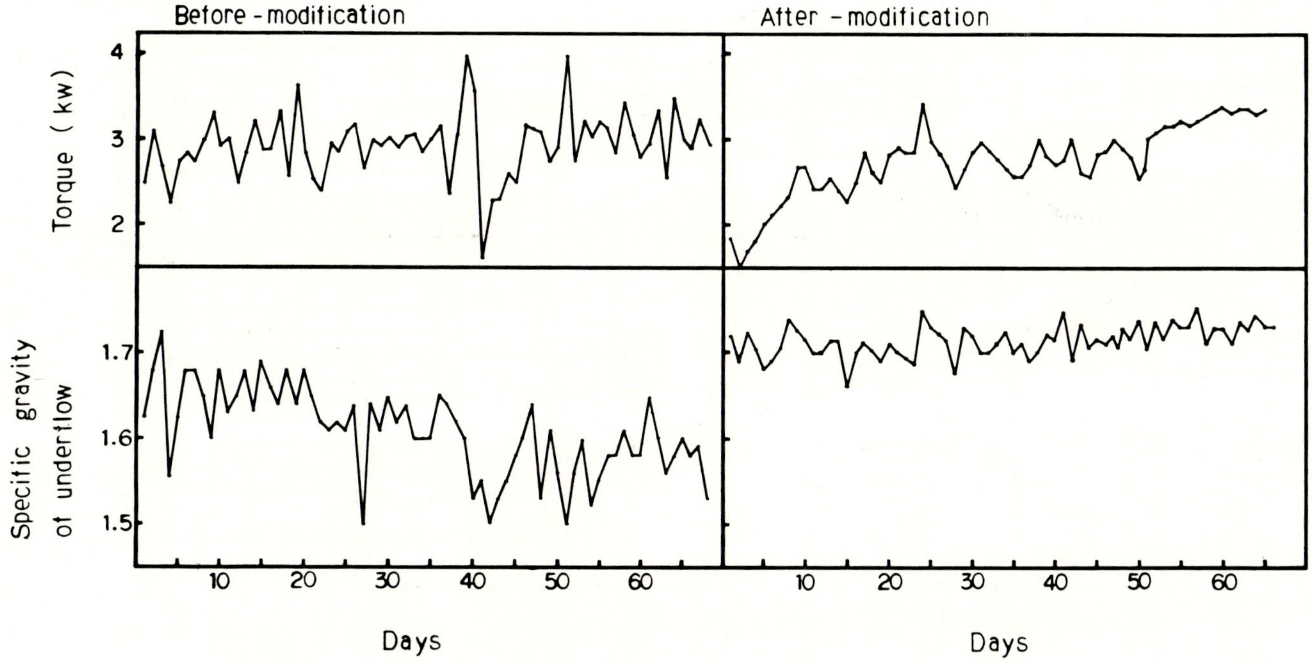

Fig. 6 Changing on standing of the torque and the specific gravity of the underflow of cable torque thickener's operation

Development of Low Torque Thickener

The conventional thickeners having the so-called "thixo-arms[19]" and "thixo-posts[20], [21]" have been developed. The maximum concentration of red mud that these thickeners were able to achieve was not more than 400 - 600 g/liter.

The low-torque thickener developed by SAL, on the other hand, is capable of concentrating the red mud slurry to 700 - 800 g/liter[22].

The low-torque thickener, the construction of which is shown in Fig. 7, has the following features:

1) Simple design with a number of vertically installed long posts. The apparatus has no rake blade and the thickner bottom is flat.

2) Extremely slow revolution of the long posts (0.005 - 0.01 rpm as compared to 0.05 - 5 rpm of the conventional thickener). Dehydration of the slurry is achieved slowly by these long posts taking sufficient time. The motive power requirement is one fourth of that of the conventional thickener.

3) Easy to build and reduced torque trouble resulting from deposition of scale.

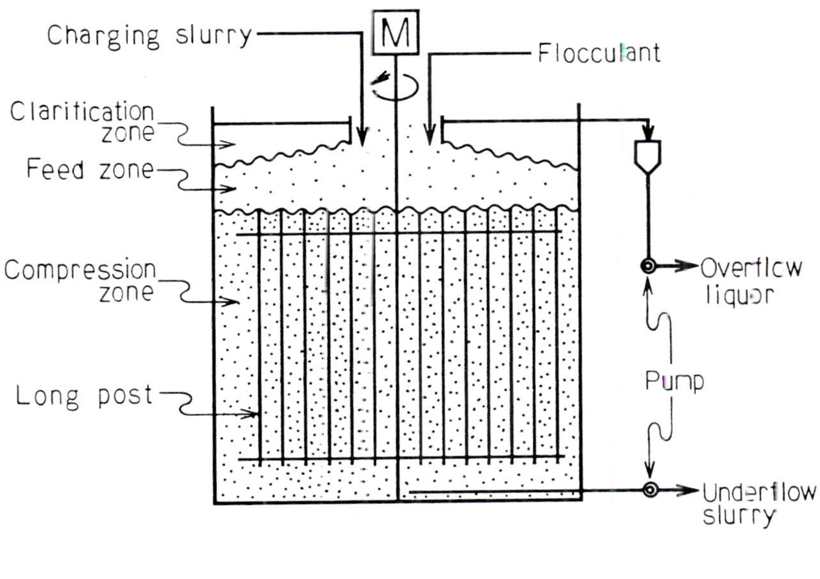

Fig. 7 Construction of low torque thickener

Tests were performed to study the relationship between the peripheral velocity of the long post and concentration of the mud slurry, using a 3-liter glass tube (10 cmø x 40 cmH) and a 450 g/liter slurry. As Fig. 8 shows, the slower the peripheral velocity, extremely higher the concentration effects.

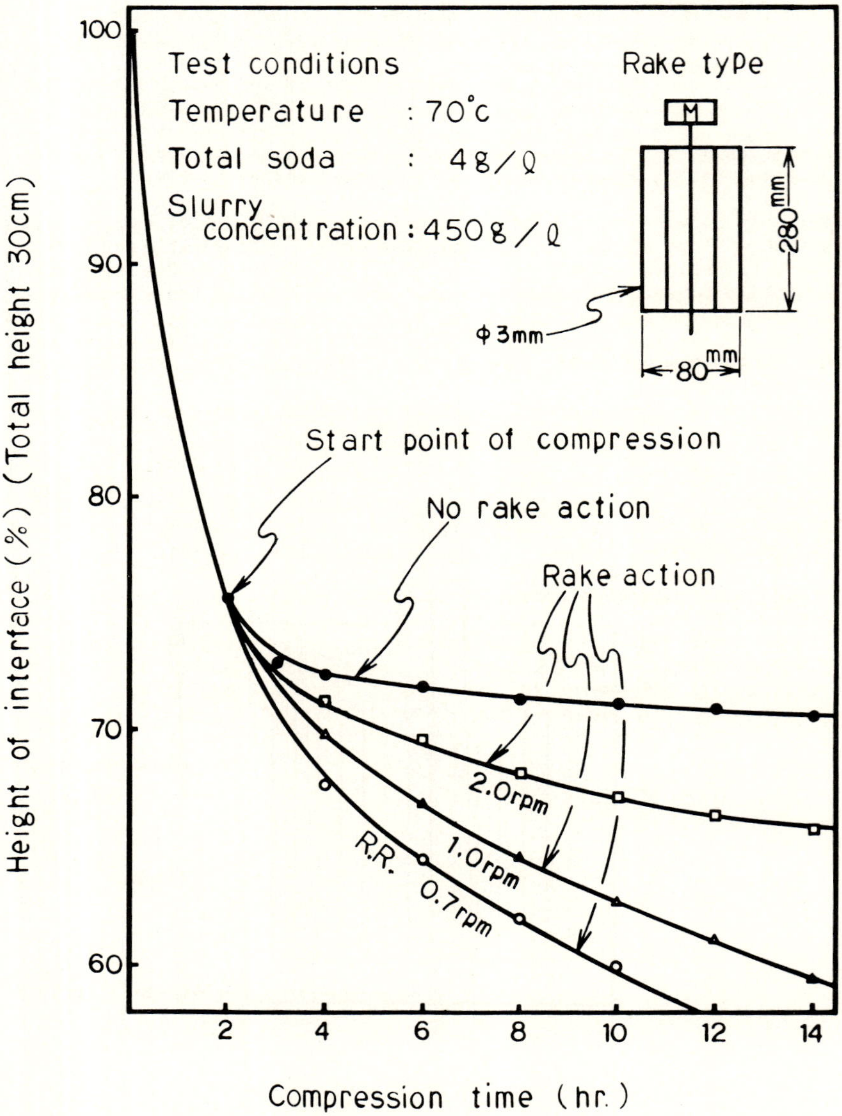

Fig. 8 Effect of rake action

Important factors in designing a low-torque thickener are as follows:

1) Distribution of long posts

2) Post diameter and interval between posts

3) Peripheral velocity of long posts

Three years since completion of the construction, SAL is successfully operating the low-torque thickener.

Although many problems had to be overcome in transportation of the high-concentration slurry, they have been studied from the viewpoint of the pseudo-Bingham flow and have been successfully resolved. Today, SAL finds no problems in transportation of the high-concentration slurry.

Conclusion

SAL has succeeded in reducing the unit consumption of water-soluble soda contained in the mud slurry and that of flocculant, by thickening the underflow of the mud thickener to 700 - 800 g/liter. Thickening of the underflow has been achieved through:

1) Selection of appropriate flocculants and establishment of their use method.

   SAL has decided to use emulsion polymers and selected polymers which are considered to give the optimum results both one for the settler and the other for washer. To ensure a long life of the polymers, they should be charged without dilution.

2) Modification and development of thickener design.

   It was found that reducing of the torque load enables the thickener to be operated stably at a high load. On the basis of this finding, SAL has modified its thickeners and is operating them successfully.

References

1) Y. Yamada, "Slurry Processing in Hydrometallurgy", Chemical Equipment No. 9 (1978) pp 24-31

2) C. Sato and T. Sato, "Control Method for Red Mud Thickeners", paper presented at Annual Meeting of AIME, New York, 1971

3) Kao Soap Co.: Japan Patent Pub. No. 17897 of 1960

4) Nalco Chemical Co.: U.S. Patent Pub. No. 3390959 of 1968

5) Nalko Chemical Co.: U.S. Patent Pub. No. 3445187 of 1969

6) Nalco Chemical Co.: U.S. Patent Pub. No. 3681012 of 1972

7) Atlas Chemical Co.: U.S. Patent Pub. No. 3397953 of 1968

8) Kao Soap Co.: Japan Patent Publ. No. 29083

9) Nalco Chemical Co.: Japan Patent Re. No. 15979 of 1977

10) Nalco Chemical Co.: Japan Patent Re. No. 39853 of 1977

11) American Cyanamid Co.: Japan Patent Pub. No. 51117 of 1980

12) American Cyanamid Co.: Japan Patent Pub. No. 120517 of 1981

13) Kurita Water Ind.: Japan Patent Pub. No. 22938 of 1976

14) Kurita Water Ind.: Japan Patent Pub. No. 78911 of 1982

15) J. L. Chandler, "Advances in the Use of Synthetic Flocculants", paper presented at Annual Meeting of AIME, 1976

16) Exxon Prod. Res. Co.: U.S. Patent No. 4273658 of 1981

17) Patent pending

18) K. D. Gruener, E. Martini and E. F. Porter, "The 'Cable Torque' a new thickener design of special value to the alumina industries", paper presented at I.C.S.O.B.A., 1973

19) R. C. Emmett, Jr., "Thickening in the Production of Alumina", paper presented at International Symposium on Unit Processes and Hydrometallurgy of AIME, Dallas, 1963

20) Metllgesellschaft AG: Deutsches Pat. Pub. No. 1557154 of 1967

21) I. U. Conversion Systems, Inc.: U.S. Pat. No. 4120791 of 1978

22) Patent pending

HYDROTHERMAL RECOVERY OF SODA AND ALUMINA FROM RED MUD:

TESTS IN A CONTINUOUS FLOW REACTOR

P.J. Cresswell & D.J. Milne

Comalco Research Centre
Thomastown, 3074
Victoria, Australia.

Further development of the Comalco hydrothermal process for recovery of soda and alumina from red mud has been undertaken with the aid of a laboratory scale (20kg/hr) continuous flow high temperature reactor. The apparatus is described. Under continuous flow conditions in backmixed reactors the kinetics of the process are found to be more favourable than those previously observed in batch experimentation. Good recoveries of soda (90%) and alumina (70%) are achieved using temperatures in the range 260-300°C with residence times of only 10-15 minutes. Liquor A/C limits are however, found to be strict.

Other tests show that an additional benefit of liquor purification may be achieved in the evaporation stage of the red mud treatment process such that the overall purification requirements of the Bayer mainstream are satisfied. Organic contamination is reduced by 20% on a single pass of deep evaporation and a broad spectrum of contaminants is removed.

Development of a process flowsheet is discussed.

Introduction

In a previous communication (1) a process was designed to recover soda and alumina from red mud. The process was developed essentially as a hydrothermal treatment with lime at 280-300°C. The mud is treated under medium caustic conditions (i.e. CS, 200-300g/ℓ) with addition of lime in the molar ratio to SiO_2 of 1.95-2.1. The process differs from higher caustic (500g/ℓ CS) processes described by others (2) (3) in that soda and alumina are recovered in a single digestion stage, at lower caustic levels, and with a different lime to silica ratio.

The hydrothermal treatment was initially aimed at the treatment of higher silica bauxites but in principle it may be applied to all grades. The sidestream treatment of mud offers advantages in using essentially well established technology similar to that used in the Bayer mainstream. It may be relatively easily integrated into the Bayer stream and, in fact, certain configurations might be adopted which could lead to high productivity in the integrated plant.

In this work, further improvements to the medium caustic process are described which were developed using a small continuously operating digester unit. The equipment was specially constructed to allow investigation of the reaction kinetics of digestion with the aim of reducing the overall processing costs. The ability of the sidestream to achieve liquor purification is also described together with further flowsheet development work.

Continuous Flow Digester Unit

A disadvantage of the previous batch digestion studies was the slow heat-up rates for the autoclaves which were available for digestions at around 300°C. In most experiments residence times of at least 1 hour were involved at 300°C and greater than 3 hours at lower temperatures; only approximate reaction times could therefore be estimated. The continuous flow reactor system allows more precise determination of reaction kinetics under conditions closer to those prevailing in a full scale plant.

An outline of the system is given in Figure 1. One or two monel autoclaves (Parr) in series are used with capacities of 1 or 2 litres each. The total reactor capacity is therefore 1 to 4 litres and residence times may be varied from around 5 minutes to more than 60 minutes by varying the flowrate up to 20 litres per hour. Slurry densities are variable up to 15% solids. Thus solids feed rates may be up to 2-3kg/hr or 10-20kg per day. The metering pump (Duraqual) is a double diaphragm slurry pump with a variable stroke adjustment. The pump is set to feed the slurry at a pressure slightly above the vapour pressure corresponding to the digest temperature. This is possible through an internal relief valve which keeps a constant diaphragm pressure on the process side; a pressure dampener reduces pulsations within the system. A gas fired preheater results in heating of the slurry stream to about 5°C below digest temperature in less than 2 minutes.

In order to obtain controlled pressure let down, after digestion, the slurry is first cooled to about 50-70°C in a water cooled heat exchanger. Flash evaporation of the liquor does not, therefore, take place in this system. A vibrating sensor mass flowmeter (Micromotion) measures flowrates, in kg/min, while flow is controlled by a motor operated regulating valve. The regulating valve incorporates a silicone rubber bush with an orifice

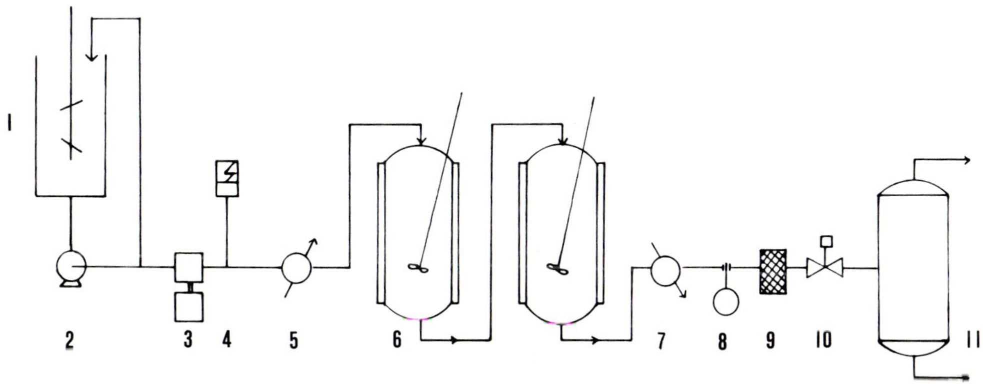

1. Mixing Tank
2. Recirculation Pump
3. Metering Pump
4. Dampener
5. Preheater
6. Autoclaves
7. Cooler
8. Flowmeter
9. Sandtrap
10. Flow regulating valve
11. Settling Tank

Figure 1: Continuous digestor system flowsheet

which is altered by compressing the bush. Where bauxite is being digested, predesilication may be used in a preliminary stage to minimize scale formation while a further scale trap minimizes the chance of blocking the small orifice in the regulating valve.

The performance of the system has been tested for its ability to reproduce plant conditions by carrying out Bayer style digestions on Weipa bauxite. Typically, a 15 minute digest at 180-185gpℓ CS, 235-240°C and a starting A/C of 0.35 gave an alumina extraction of 99% of TAA and an exit A/C of about 0.65. Final silica level in solution was 2.49×10^{-3} SiO_2/CS.

Studies of Lime Digest of Red Mud

The kinetics of the treatment of red mud with lime were investigated under continuous conditions in the apparatus described above. A plant red mud was used together with synthetic liquors and technical grade lime. The results of a series of tests are summarized in Table I. Conditions were chosen to study the effect of digest temperature, residence time and liquor composition on the extraction efficiencies of both soda and alumina.

Effect of Residence Time

Tests showed that good recoveries of both soda and alumina were achievable using residence times as low as 10-15 minutes in the continuous digester (Table I). Previous results under batch conditions (Ref. 1) indicated much longer times, of the order of 20-60 minutes, because of the obscuring effects of the slow heat-up time. In the current work, variation of residence time within the range 10-30 minutes appeared to have little effect on the extraction efficiency except under extreme conditions of low temperature or high A/C.

Effect of Digest Temperature

Results on the effect of varying digester temperature in the range 250-300°C are summarized in Figure 2. A minimum temperature of 260°C is required to achieve good extraction (95% of Na_2O, 70% $A\ell_2O_3$) but there is no improvement with higher temperatures. This finding greatly reduces the severity of the conditions required for the treatment as indicated in earlier batch tests. Below 280°C the batch tests indicated that a reaction time in excess of 2 hours was necessary to attain reasonable extractions.

Effect of Liquor Composition

Caustic concentration in the range 150-300g/ℓ CS was found to have only minor effects on extraction efficiency. This is in general agreement with earlier work which also indicated reduction in yields at very high concentrations because of the formation of different silicate products.

In the preliminary investigations, liquor A/C was determined to be an important factor and this has been further investigated. Results are summarized in Figure 3. This shows extraction efficiency to be rather sensitive to A/C. Soda recovery is high for A/C <0.15 but falls rapidly for higher values. Alumina recovery suffers a more gradual decline with increasing A/C but also decreases sharply for A/C $>$ 0.15. Productivity of the treatment is very dependent on A/C and so choice of an operating

Table I: Results of Tests in Continuous Digester

Temp. (°C)	Av.* Flow-rate (kg/min)	Residence Time (min)	Mud/ Liquor Ratio (w/w)	CaO/ SiO_2 Ratio (mole)	Exit Liquor		Extraction,%	
					CS	A/C	Na_2O	Al_2O_3
300	.32	11	.161	1.97	281	.097	90.7	64.0
300	.26	14	.161	1.97	280	.099	91.8	65.9
300	.32	11	.161	1.97	156	.120	96.0	57.6
300	.26	14	.161	1.97	140	.120	92.3	59.8
300	.32	11	.131	1.97	219	.098	92.6	71.0
300	.26	14	.131	1.97	219	.099	93.0	68.5
300	.32	11	.116	1.97	189	.090	89.3	62.1
300	.26	14	.116	1.97	190	.100	92.6	67.4
280	.22	17	.13	2.10	204	091	95.1	64.2
280	.16	23	.13	2.10	192	099	95.1	64.9
280	.26	14	.057	2.10	187	.106	76.1	45.3
280	.18	21	.057	2.10	193	.149	83.2	51.8
280	.19	19	.057	2.10	215	.166	41.9	21.1
280	.28	12	.057	2.10	213	.166	45.6	31.2
260	.19	19	.057	2.10	204	.049	95.1	72.0
270	.19	19	.057	2.10	204	.049	95.3	75.3
280	.18	21	.057	2.10	214	.047	86.3	68.5
280	.13	28	.057	2.10	216	.048	87.5	68.6
250	.12	31	.057	2.10	177	.152	45.4	10.9
270	.17	22	.057	2.10	187	.153	52.6	16.4

* Liquor plus solids.

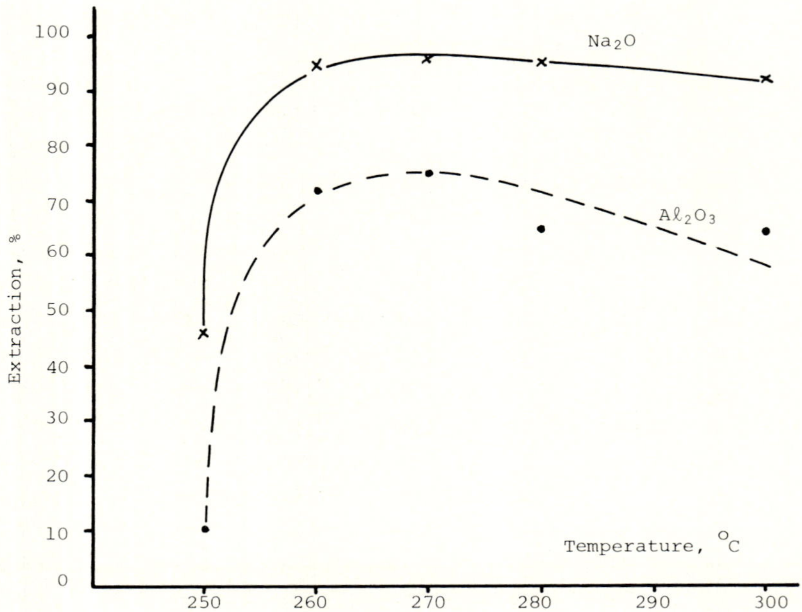

Figure 2: Effect of digestion temperature on extraction of soda and alumina from red mud.

optimum, in the range 0.10-0.15, must be a compromise based on extraction efficiency, flow rates and overall productivity. Clearly the starting A/C should be as low as possible to allow a sufficient margin.

Liquor Purification in the Evaporation Stage

The presence of a deep evaporation stage in the process circuit presents the possibility of substantial liquor purification by precipitating the impurities and consequent improvements in plant productivity. Table II shows the results of a test in which a sample of highly contaminated spent liquor containing 19g/ℓ organic carbon (as C) was subjected to deep evaporation. Salting out of sodium carbonate commenced at a concentration of about 400g/ℓ CS and continued up to 570g/ℓ CS where a liquor purity (C/S ratio) of 0.96 was obtained. Above 570g/ℓ CS crystallization of sodium aluminate ($Na_2O.A\ell_2O_3.2.5H_2O$) commenced as evidenced by the reduced liquor A/C at 590g/ℓ CS.

The progress of purification of organic components can be gauged by determining the ratio: Org.C/CS. This indicates no removal of organics at 430g/ℓ CS, 20% removal at 565g/ℓ CS and 22% removal at 589g/ℓ CS. The organic compounds coprecipitate with sodium carbonate and other salts.

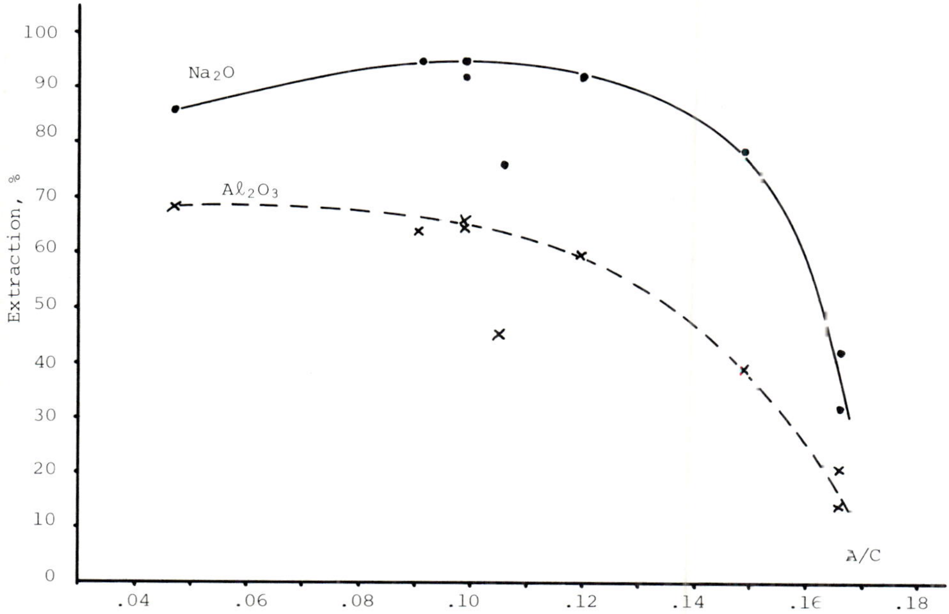

Figure 3: Relationship between extraction of soda and alumina and final liquor A/C.

Table II: Deep Evaporation: Analysis of Liquor Samples

CS g/ℓ	TS g/ℓ	Al_2O_3 g/ℓ	A/C	C/S	Na_2CO_3 g/ℓ (1)	Na_2CO_3 g/ℓ (2)	Org.C g/ℓ (3)	Org.C/CS
208.6	230.5	38.4	.184	.833	41.9	46.9	19.0	.091
430.7	478.5	73.8	.183	.844	74.8	71.5	36.6	.091
565.6	586.6	103.3	.183	.964	21.2	21.7	41.1	.073
588.9	ND	84.7	.144			25.5	41.6	.071

(1) Na_2CO_3 determined as (TS-CS).

(2) Determined by acid evolution of CO_2.

(3) Determined by oxidation/evolution of CO_2.

ND Not Determined.

The nature of the organic components precipitated was investigated using HPLC. Figure 4 shows the HPLC trace of an extract of the organic component from solids precipitated after evaporation to 570-590g/ℓ CS with, for comparison, an equivalent trace derived from the original spent liquor.

Comparison of the two traces in Figure 4 shows that both mixtures largely contain similar components. The extract obtained from the precipitated solids does not contain a concentration of any particular component (relative to the spent liquor) though a few components (e.g. A, B, C in Figure 4) are markedly reduced. This indicates that the evaporative crystallization is not specific for any particular component or class of components but results in a broad spectrum reduction in organic impurity levels. In this respect it is similar to wet oxidation, though the latter can be more complete.

Purification of this order would markedly reduce organic contamination of the circuit. Mass balance calculations predict an operating level of about 5g/ℓ. In the integrated circuit this would permit improvements in liquor productivity to be made relative to orthodox Bayer processing by allowing an increase in pregnant liquor concentration of at least 10%.

Process Description

Development of an overall process flowsheet covering both the Bayer process and red mud treatment with lime has been undertaken with the aid of a flexible computer mass balance programme. This enables rapid comparison of process options.

The process is outlined in Figure 5 and selected mass flows, relative to an orthodox plant treating a high silica bauxite, are outlined in Table III. It is apparent that a considerable increase in productivity in precipitation may be attained (ca. 15%) while bauxite digester efficiency is also increased (ca. 5%) by virtue of recycling a lower A/C liquor than in the orthodox process. The sidestream flow is relatively small at about 22% of the Bayer mainstream, while soda recoveries of about 90% are attained.

These advantages are offset by a relatively high evaporative load. The economics of the process will therefore be very dependent on local conditions. Estimates for Australian conditions (east coast) suggest that the hydrothermal process would certainly be preferable over orthodox treatment for bauxites containing 6-8% or more soluble silica. For higher grade bauxites containing less soluble silica, the economics will be sensitive to fuel costs but they can be markedly influenced by factors such as the availability of unused capacity in an existing refinery.

Discussion

The latest phase of development of the hydrothermal treatment of red mud with lime has resulted in a marked improvement in digestion kinetics. Reaction times to attain greater than 90% recovery of soda and greater than 65% recovery of alumina from the mud are now estimated to be in the region of 10 minutes. Equally importantly the improved reaction kinetics are apparent at lower temperatures (260°C) than those previously envisaged (300°C). The improvement under the partially backmixed conditions in the two-stage continuous reactor system suggests that the reaction is dependent upon the availability of a seed surface to catalyze formation of the

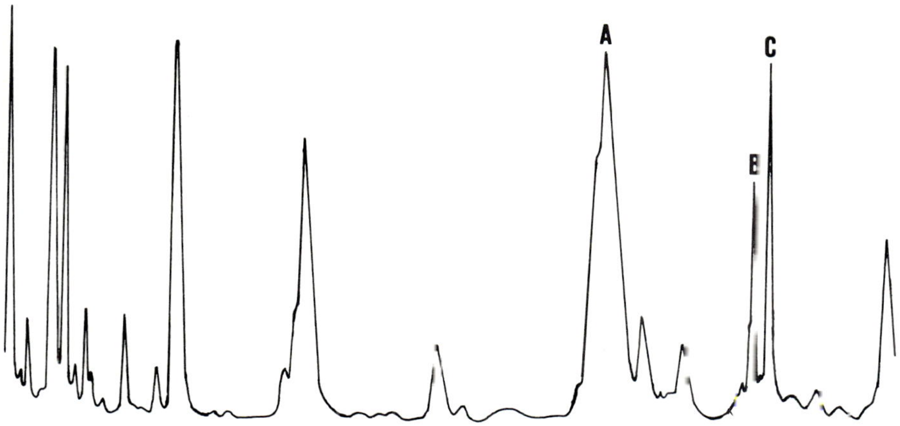

(a) Spent Liquor

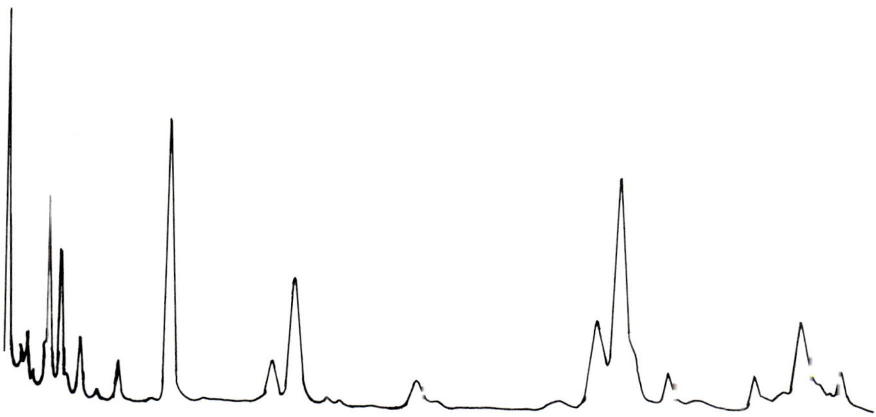

(b) Solids precipitated after evaporation of spent liquor to 570-590g/ℓ CS

Figure 4: HPLC traces of organic impurity extracts.

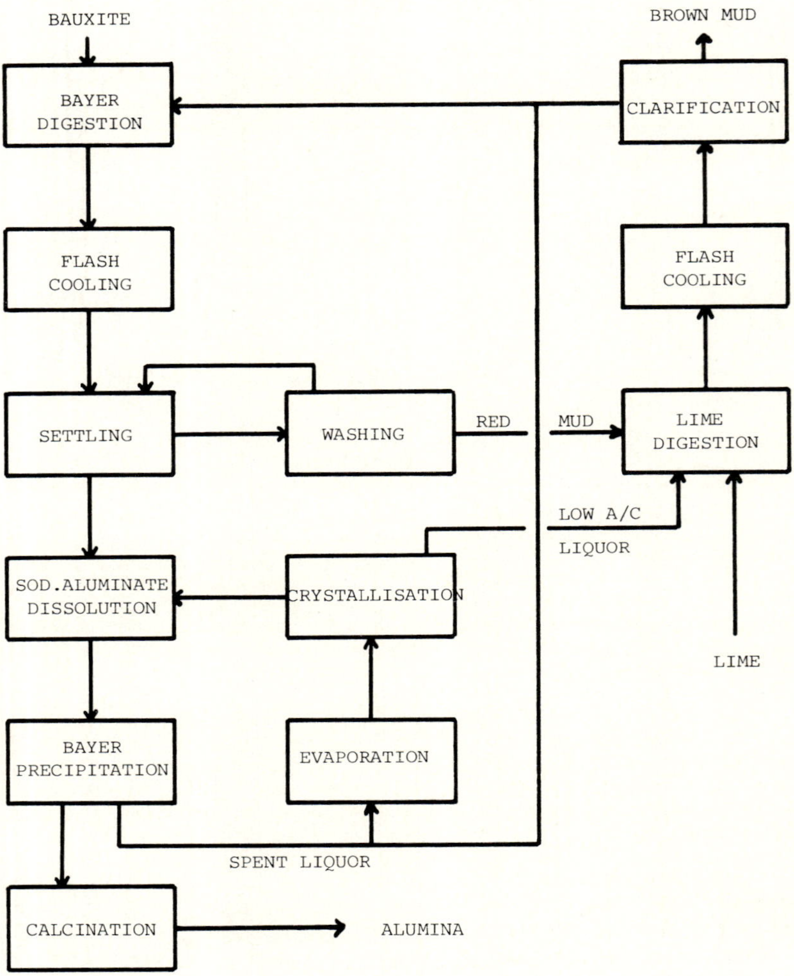

Figure 5: Process Flowsheet for combined Bayer process and Red Mud treatment.

Table III: Relative Mass Flows in Bayer and Hydrothermal Process

(Bauxite: 56% Al_2O_3, 8% SiO_2)

Stream	Bayer	Bayer + Hydrothermal Process
Bauxite digester flow	11,750	11,780
Bauxite	1,090	1,149
Liquor to lime digest	-	2,645
Evaporation	-	1,651
Make up NaOH (50% solution)	166	15
Liquor to precipitation	10,137	10 860
Product alumina	506	617

substituted hydrogarnets which are the main reaction products (e.g. 3CaO. $Fe_2O_3.2SiO_2$). Batch experiments, which simulate plug flow, minimize such effects and result in reaction times measured in hours at the lower digestion temperatures. These results have important implications with respect to final reactor design and demonstrate the importance of carrying out process investigations under continuous flow conditions.

The ability of the process to achieve a high degree of Bayer liquor purification enables a high caustic concentration to be maintained in the Bayer mainstream with a consequent high productivity. At the same time higher than normal productivity is attained for bauxite digestion because of the low A/C high strength liquor that is diverted from the lime treatment stream.

While a relatively high evaporative load is unavoidable, preliminary estimates show that this can be justified because of the reduced usage of high priced caustic soda and the overall improvement in plant productivity.

References

(1) Cresswell, P.J. and Milne, D.J., "A Hydrothermal Process for Recovery of Soda and Alumina from Red Mud", Light Metals 1982, pp.227-238.

(2) Hudson, L.K., Savage, K.I. and Stinson, J.M., U.S. Patent 4,044,095.

(3) Ni, L.P. and Romanov, L.G., "The Physical Chemistry of Hydrochemical Alkaline Processes for Alumina Extraction", Nauka (Alma-Ata), 1975 (Translated by ASM).

A GRAPHICAL CONSTRUCTION FOR PREDICTING THE YIELD FROM CONTINUOUS PRECIPITATOR TRAINS

E.T. White*, M. Steemson* and D. Milne#

*Department of Chemical Engineering, University of Queensland, Brisbane, Qld, 4067

#Comalco Research Ltd, 15 Edgars Road, Thomastown, Vic, 3074

The yield from a single precipitator can be related to the inlet a/c ratio, the product crystal surface area, the residence time and the operating temperature. This relation is presented as a graphical procedure which allows the performance of a crystallizer train, operated at optimal temperatures, to be estimated. Reasonable assumptions have to be made for the change in crystal surface area. The method allows good visualisation of the effect of process changes.

Complete models to predict the performance of the precipitation section of Bayer plants are complex. One such model is given in an accompanying paper [4]. Because of the interactions between yield, the surface area of particles and the changes in surface area, an involved computer solution is usually necessary. Considerable time is required to obtain, set up and run such models, so that solutions are not usually immediately available.

However, direct calculation of the yield in multi-tank trains is possible using simpler graphical techniques, if the user can provide certain information (eg. seed charge levels) from experience.

This paper outlines such a graphical method for the calculation of yield from continuous precipitation trains.

Yield from a Single Continuous Precipitator

Consider the continuous well mixed precipitator with representative overflow shown in Figure 1. This is fed with liquor of alumina concentration a_{in}.

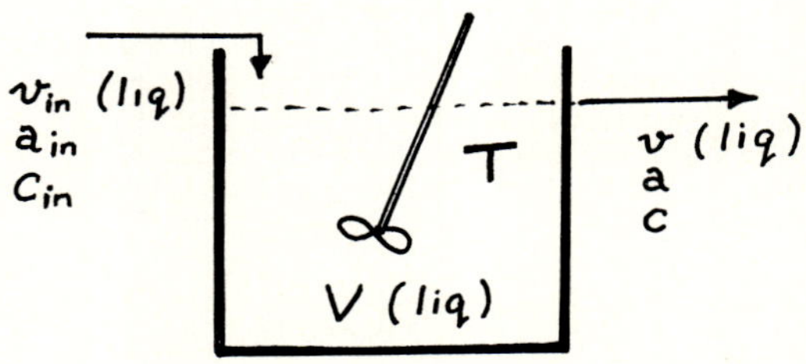

Figure 1 - Diagram of continuous precipitator

The only mechanism which removes alumina from solution is growth onto the suspended particles. The growth rate, G, expressed as a rate of change of linear size is known to be unaffected by particle size [3]. Therefore as the vessel is well mixed, all particles will experience the same growth rate. The mass rate of solid deposition in the vessel, w is then given by

$$w = \rho_c \; S \; V \; G/2 \tag{1}$$

where S is the particle surface area per unit volume of liquor, V is the volume of liquor in the vessel and ρ_c is the particle density. The factor of 2 is necessary since G refers to particle size (diameter) and thus two growing surfaces contribute to the increase in size.

If the volume flow rates of liquor into and leaving the precipitator are v_{in} and v, then the mass rate of deposition is also given by

$$w = (v_{in} \; a_{in} - v \; a)/\xi \tag{2}$$

where a is the exit liquor concentration and ξ = 0.654 is the conversion factor between $Al(OH)_3$, the crystal material and Al_2O_3, used for analyses.

A relationship between v_{in} and v can be obtained if it is assumed that as precipitation proceeds at a fixed temperature, the volume change of the slurry is negligible. Then,

$$v_{in} = v + w/\rho_c \qquad (3)$$

Combining these equations gives,

$$G = 2(a_{in} - a)/S\tau(\xi\rho_c - a_{in}) \qquad (4)$$

where $\tau = V/v$ is the mean residence time of liquor in the precipitator. Since the vessel is well mixed with representative overflow, this is the same as the mean residence time of the slurry. Thus the mean residence time is the working volume of the vessel divided by the slurry volumetric overflow rate.

The growth rate is taken to vary with concentration and temperature by a relation of the form [1,3]

$$G = k_G \exp(-E/T)(a - a^*)^2/c^2 \qquad (5)$$

where a^* is the equilibrium alumina concentration, c the caustic concentration and k_G and E kinetic parameters for growth.

Combining with (4) this gives,

$$[a/c - a^*/c]^2 = 2\beta[a_{in}/c - a/c] \qquad (6)$$

where $$1/\beta = (S\tau/c)(\xi\rho_c - a_{in})\, k_G \exp(-E/T) \qquad (7)$$

Relation (6) is a quadratic in a, which has the solution,

$$a/c = (a^*/c - \beta) + \sqrt{2\beta(a_{in}/c - a^*/c) + \beta^2} \qquad (8)$$

This gives the exit alumina concentration, a, from a continuous precipitator with mean residence τ fed with liquor of concentration a_{in} and with seed such that the <u>exit</u> surface area per unit volume of liquor is S.

The solution of equation (8) depends on values for $S\tau/c$, a_{in} and temperature, T. For example, if at 80°C, $a_{in}/c = 1.10$, $a^*/c = 0.59$ (both as Al_2O_3/Na_2O) and $\beta = 0.1$ then by substitution in (8), the exit a/c from the vessel will be 0.82, a drop of 0.28. For $\beta = 0.1$ at 80°C, with the growth kinetics for pure solutions, substitution in (7) gives about 240 h $m^2/kgNa_2O$. That is, for c = 100 gpl Na_2O and τ = 3h, the surface area is about 8 m^2/l of liquor. For a solid with surface area of 80 m^2/kg of solid (800cm^2/g) this would correspond to a solids content of 100 gpl. Methods of evaluating the surface area from measured size distributions are discussed in Appendix 1.

From equation 8, as β becomes large ($\beta > 20$), $S\tau$ is small and a/c approaches a_{in}/c, as would be expected. Similarly when β becomes small ($\beta < 0.005$), $S\tau$ is large and a/c approaches a^*/c.

Graphical Presentation

Figure 2 shows the solution of equation (8) as a function of temperature for c = 100 gpl (Na_2O) and $a_{in}/c = 1.1$ (as Al_2O_3/Na_2O). The parameter is $S\tau/c$, with S in m^2 of particle surface area <u>in tank</u> per m^3 of <u>liquor</u>, τ in hours and c as caustic in gpl (kg m^{-3}) of Na_2O.

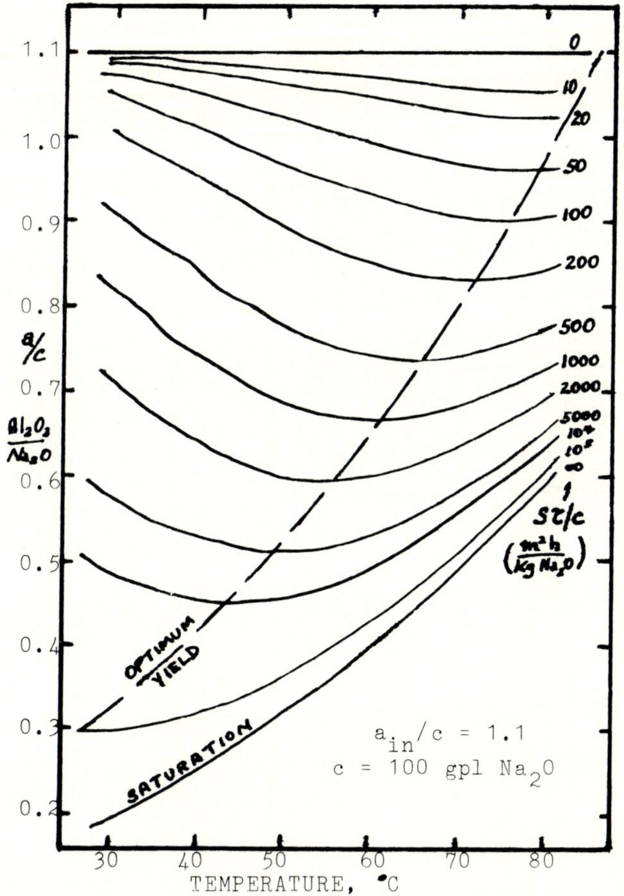

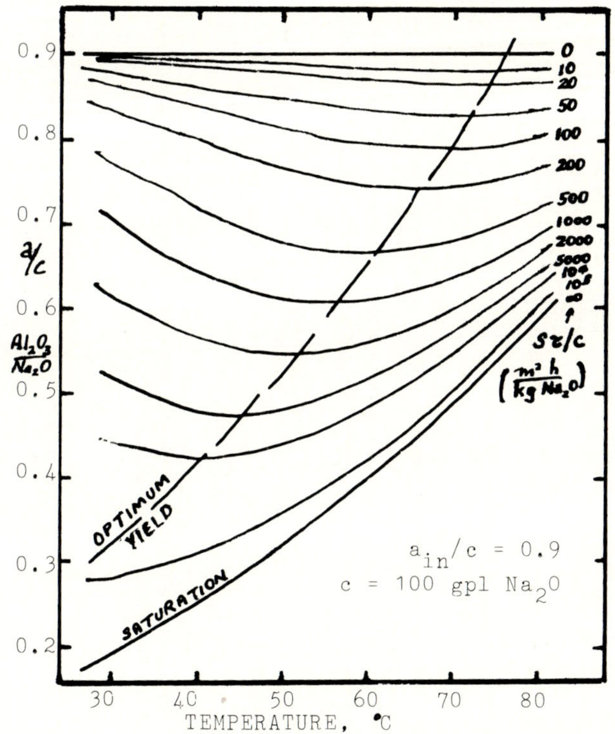

↑
Figure 3 - Exit a/c for a continuous precipitator $a_{in}/c = 0.9$

←Figure 2 - Exit a/c for a continuous precipitator $a_{in}/c = 1.1$.

The kinetic values obtained by Misra and White [3] for pure liquors of k_G = 1.96 x 10^{10} μm h^{-1} $(Al_2O_3/Na_2O)^{-2}$ and E = 7200 K^{-1} were used. Similarly the equilibrium solubility relation used was that for pure liquors obtained by Misra [2], viz.

$$\ln (a^*/c) = 6.211 - (2487 - 1.09c)/T \quad (9)$$

The graph can be used to determine the values used above. At 80°C, the saturation a^*/c can be read from the figure as 0.59. For particles of 80 m^2/kg surface area, at 100 gpl concentration and with a mean residence time of 3 hours, $S\tau/c$ = 240 h m^2/kg Na_2O. From the chart it can be seen that the corresponding exit a/c = 0.82.

For other caustic concentrations the plot changes near the botton since the saturation a^*/c increases slightly as the caustic level is increased. The changes however, will be small in the upper part of the figure.

For other inlet concentrations (a_{in}/c), similar charts can be drawn. Figure 3 is for a_{in}/c = 0.9 and c = 100 gpl Na_2O.

Once the a/c in the tank is known, the growth rate is also known. Figures 4 and 5 are plots of the growth rates (on a log scale) corresponding to Figures 2 and 3.

Implications of Graphs

For a continuous precipitator, as the yield increases the solution becomes more depleted, so the deposition rate decreases since the 'driving force' is less, which tends to reduce the yield. At steady state, a balance is reached which is represented by the yields and growth rates shown in the figures.

To increase the yield, it is necessary to increase the deposition rate by other means to compensate for the resulting drop in 'driving force'. This can be done by increasing the available surface area for deposition or by altering the temperature.

As seen from the above figures, as the surface area is increased, the yield increases and the growth rate decreases. A particular example showing the effect of surface area is shown in Figure 6. The dramatic drop in growth rate, and the large changes in surface area required to give significant increases in yield can be seen. It appears that high growth rates are essential for agglomeration [5], so if agglomeration is desired in a particular vessel, low surface areas and low yields may have to be tolerated.

The second variable which may be altered to increase the deposition rate is temperature. However, its effects are more complex A particular example showing the effect of temperature is given in Figure 7. There is indeed an optimum temperature for operating a continuous precipitator fed with a given liquor and containing a given solids surface area.

Provided it can be achieved, there seems no reason why precipitators should not always be operated at this optimum temperature (unless agglomeration or nucleation are adversely affected).

The presence of the optimum temperature can be seen in Figures 2 and 3 as minima in the exit a/c and in Figures 4 and 5 as maxima in growth rates. The locus of the optimum is marked on these figures.

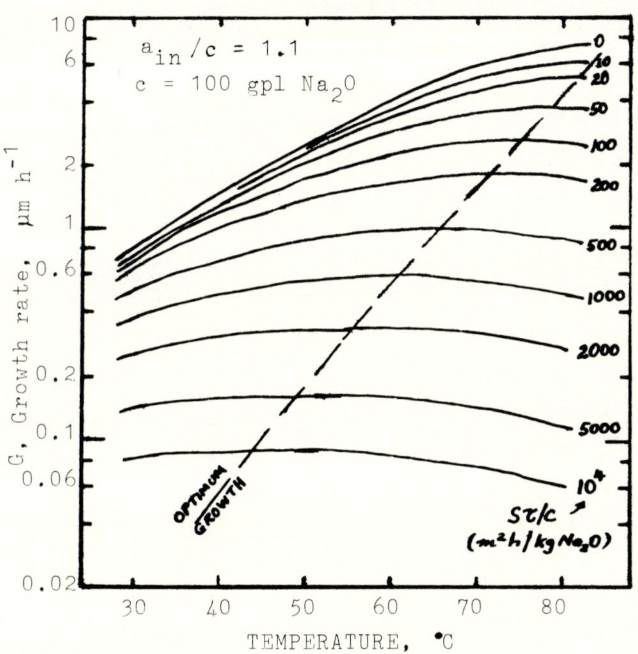

Figure 4 - Growth rate in continuous precipitator, $a_{in}/c = 1.1$

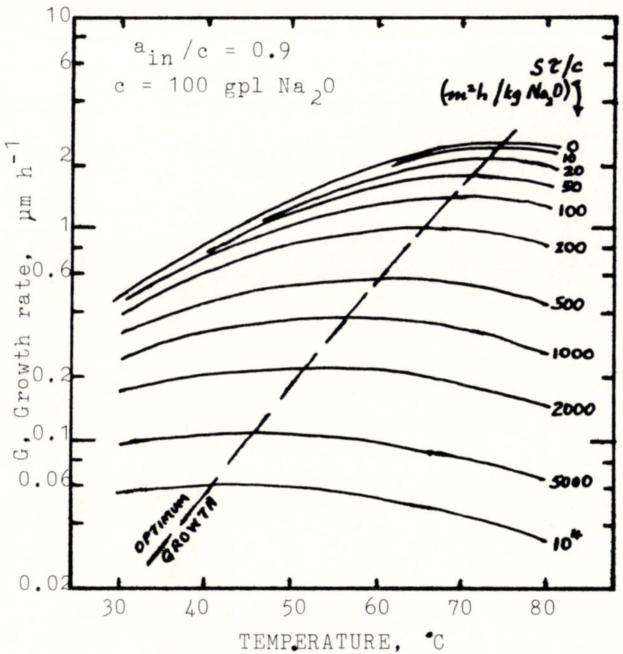

Figure 5 - Growth rate in continuous precipitator, $a_{in}/c = 0.9$

The reason for an optimum operating temperature can be explained as follows. For a given solution, when the temperature is very high, the equilibrium alumina concentration is also high, so the driving force (the difference between solution and equilibrium alumina concentrations) is small and a low growth rate results. The growth rate is the product of a kinetic term with a driving force term (see equation 5). As the temperature is reduced the driving force increases rapidly, and although the reaction kinetics slow because of the lower temperature, the net effect is a higher growth rate. Eventually at low temperatures the effect of the slower kinetics overwhelms the increase in the driving force and low growth rates are again observed.

Figure 8 shows a plot of growth rate against temperature for liquors of various constant concentrations. The maximum is clearly evident. This plot depends only on solution conditions and is equally true for any precipitation mode (batch or continuous).

The optimum temperature condition can be obtained solely from the growth (5) and solubility (9) relations by differentiation. Details are given in Appendix 2. The result obtained is that, to a good approximation, the optimum temperature is that at which the saturation a^*/c is 0.6 of the solution a/c. From this, the optimum temperature and optimum growth plots can be constructed (Figure 9). The effect of caustic concentration is shown.

As can be seen from Figures 2 to 5, the change in yield (or growth rate) is reasonably flat over a range of temperatures. Thus the optimum need only be attained within ± 7 °C of the values given in Figure 9.

To restate the point made earlier. Unless there are strong reasons for not doing so (cooling rate, effect on agglomeration or nucleation), precipitators should be designed to operate at these optimum temperature conditions.

Multi-Tank Calculation at Optimum Growth

Figure 10 was obtained from plots such as Figures 2 and 3, by plotting the exit value of a/c against the inlet concentration, a_{in}/c, with Sτ/c as parameter for continuous precipitators operating in the optimum temperature range. Since optimal temperature conditions are chosen, once a/c is selected T and G are known and these are also shown.

The 45° line (Sτ/c = 0) gives the condition $a/c = a_{in}/c$. Since the exit of one tank is the feed to the next, this line can be used to link tanks in a multi-tank analysis.

Design of a multi-tank continuous precipitation unit then becomes a stepping procedure. Starting with an initial feed concentration (A_{1i} on Figure 10) and a selected Sτ/c for the first tank, the exit concentration (point A_1) is obtained. Transferring horizontally to the 45° line gives A_{2i} the inlet concentration to the second tank. Knowing the value of Sτ/c for the second tank, (also operating at its optimum temperature) the exit concentration from the second tank, point A_2, may be determined. The procedure can then be repeated for further tanks.

The construction giving the A points on Figure 10 gives the results for a five tank train. The first three are low solids (to assist agglomeration) and the last two are high solids (to enhance yield). The construction gives the operating conditions shown in Table I.

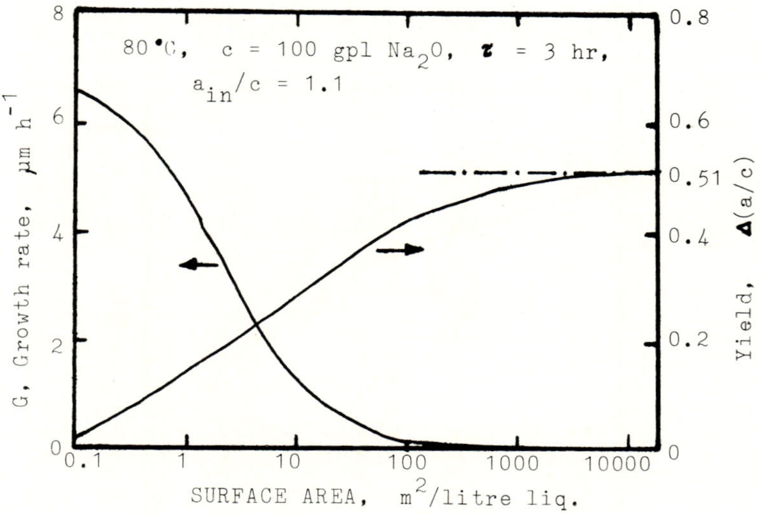

Figure 6 - Example showing effect of particle surface area on performance of continuous precipitator.

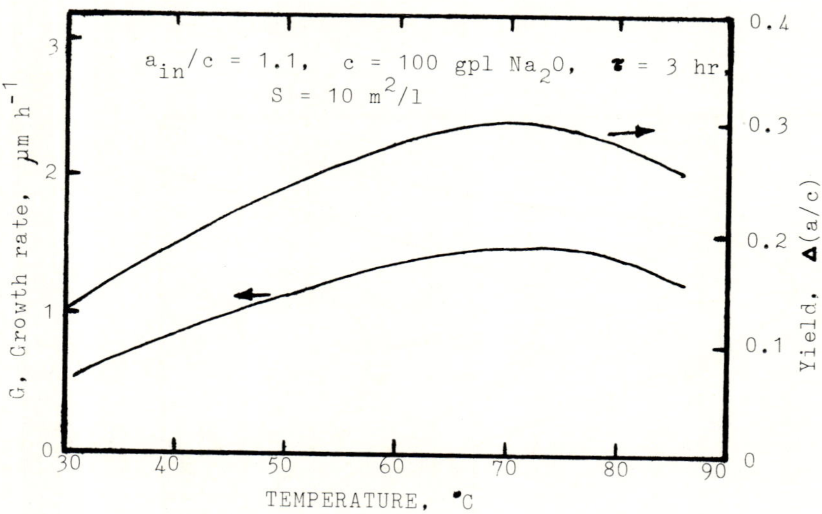

Figure 7 - Example showing effect of temperature on performance of continuous precipitator.

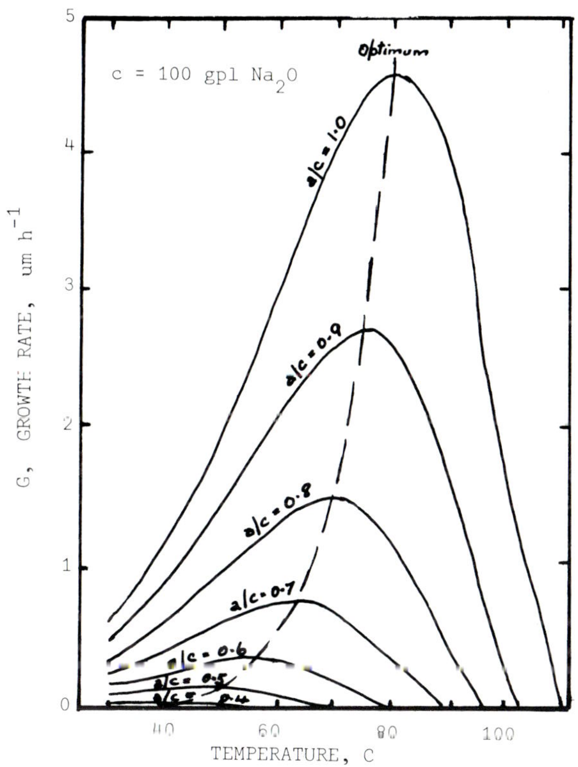

Figure 8 - Growth rate as a function of temperature showing maxima.

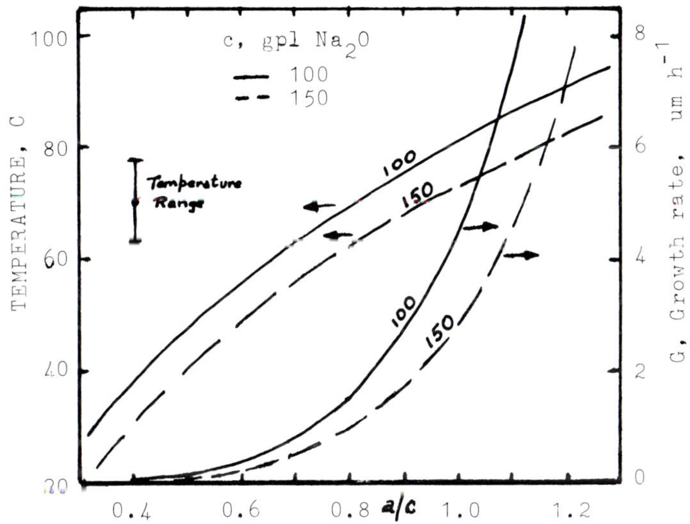

Figure 9 - Optimum temperature and growth rates for various a/c

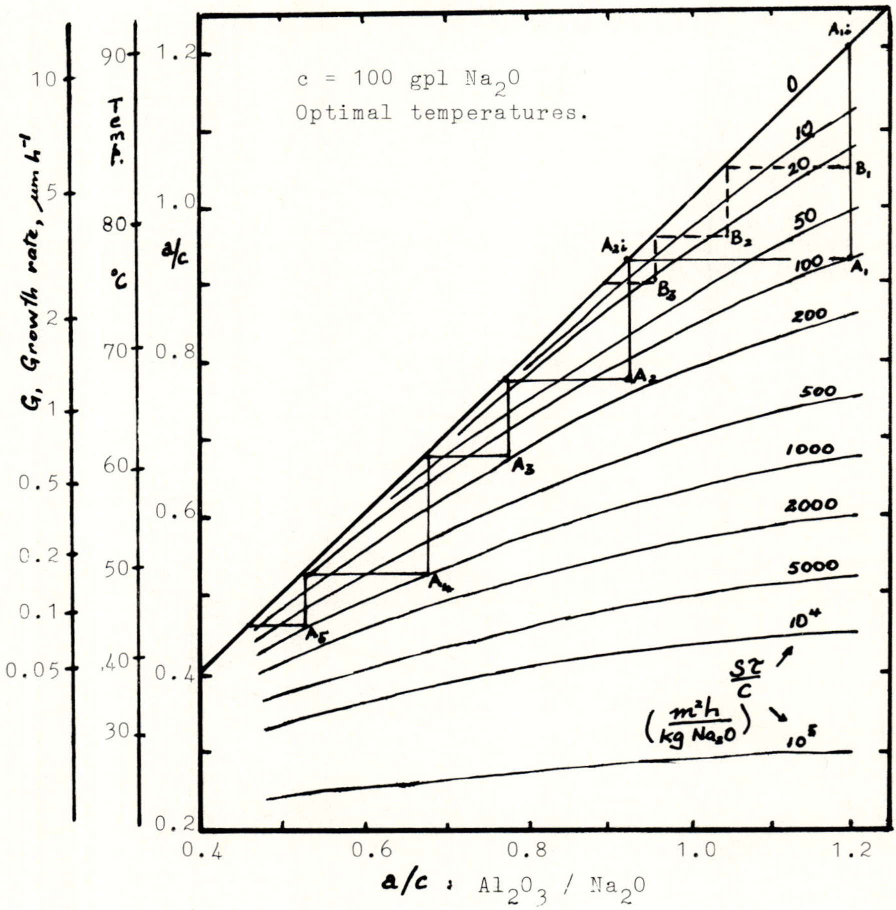

Figure 10 - Design chart for multi-tank continuous precipitator operating at optimum temperature.

Table I. Operating Conditions for 5 Precipitator Example shown on Figure 10.

Vessel No.	Temp* °C	Sτ/c † $m^2h/kgNa_2O$	G $\mu m\ h^{-1}$	Δ(a/c) Al_2O_3/Na_2O
1	77	100	3.2	0.27
2	68	150	1.3	0.15
3	63	200	0.77	0.10
4	50	1000	0.18	0.15
5	44	1000	0.09	0.07

* Optimum temperature (±7°C)

† Assumed

To demonstrate process alterations which could be undertaken, consider the result of dividing the first tank into three smaller tanks of one third the volume. As τ is one third of the previous value so is Sτ/c. This construction is shown on Figure 10 as the points B_1, B_2, B_3. According to these calculations, the yield would increase by 0.03 a/c units and there would be higher growth rates in the first and second of the new vessels, promoting increased agglomeration.

Limitations of Technique

The technique is easy to use and it allows visualisation of the effects of process changes on yield.

The calculations have been undertaken with the tanks at the optimum temperature. As this is a reasonably broad temperature range (±7°C) practical value should be reasonably close.

The charts given here have been constructed from the relations for pure liquors. The equilibrium solubility will differ for plant liquors as may the growth relation. However it is relatively easy to generate new plots for specific plant relationships.

The greatest limitation on the method is the need to select, or guess, the particle surface area in a tank. This will depend on the seed addition, the amount of growth and the effects of agglomeration and nucleation. This method will not indicate what seed is required to obtain the indicated exit surface area, what the size distributions are or whether the correct amount of recycle is available to provide the seed. Answers to these questions are generated by the more complex computer models.

Never-the-less, based on past experience, the process engineer can make a reasonable estimate of the particle surface areas expected (or that the tank will be run at by external control). In which case the graphical method described should be of considerable use in screening alternative process changes.

Conclusions

1. A material balance over a well mixed continuous precipitator gives a relation which allows its performance to be predicted, given growth kinetics and solubility data and the particle surface area in the tank.

2. There exists an optimum temperature at which to operate any tank, which depends on the liquor a/c in the tank.

3. For a multi-tank precipitator train running at the optimum temperatures, a simple graphical technique allows the yield in each vessel to be evaluated.

4. This technique should be useful for process designers to evaluate various flowsheets for product yield.

5. The technique gives no information about the quality of the product or the amount of recycle.

Acknowledgement

The assistance of a Comalco Research Laboratory Scholarship to M.L. Steemson is gratefully acknowledged.

Appendix 1

Evaluation of Surface Area from Size Distribution

The surface area per unit volume of solid, S_V, is given by $S_V = 6/\overline{\psi}\,\overline{L}_{32}$ where $\overline{L}_{32}$ is the volume-surface (or Sauter) mean size of the particles and $\overline{\psi}$ is the average sphericity, a measure of particle shape.

The volume surface mean size is calculated from a number size distribution by

$$\overline{L}_{32} = \Sigma\, n_i\, L_i^{3} / \Sigma\, n_i\, L_i^{2} \quad \text{if discrete, or}$$

$$= \int L^3\, f(L)\, dL / \int L^2\, f(L)\, dL, \text{if continuous}$$

where n_i is the number of particles of size L_i and f(L) is the number distribution density function.

If a mass size distribution is available, the estimate of $\overline{L}_{32}$ is given by

$$\overline{L}_{32} = \Sigma\, m_i / \Sigma\, (m_i / L_i) \quad \text{if discrete, or}$$

$$= \int f_M(L)\, dL / \int (f_M(L)/L)\, dL \quad \text{if continuous}$$

where m_i is the mass (strictly volume) of particles of size L_i and $f_M(L)$ is the mass distribution density function.

The sphericity of a particle, ψ, is the surface area of a sphere with the same volume as the particle divided by the actual surface area. This depends on particle shape. For hydrate particles, values of ψ range from 0.6 to 0.75.

Given a crystal content, X, as mass of crystal per unit volume of liquor, the surface area per unit volume of liquor, S, them becomes

$$S = 6\ X/\overline{\psi}\ \overline{L}_{32}\ \rho_c$$

where ρ_c is the crystal density.

Appendix 2

Optimum Growth Temperature

For a solution of given alumina concentration, a, there exists an optimum temperature giving a maximum growth rate (refer Figure 7).

The growth rate relation (equation 5) is of the form,

$$G = k \exp(-E/T)\ (a - a^*)^2 \tag{5}$$

where $k = k_G/c^2$.

Substituting in the equilibrium solubility relation (equation 7) of the form.

$$a^* = c \exp(b_1 - b_2/T) \tag{7}$$

where $b_1 = 6.211$, $b_2 = 2487 - 1.09c$, gives

$$G = k \exp(-E/T)\ [a - c \exp(b_1 - b_2/T)]^2$$

Equating the differential of G with respect to T (for constant a) to zero gives the expression

$$a/c = (2b_2/E + 1) \exp(b_1 - b_2/T_{opt})$$

Recognising the last term as the solubility expression gives

$$a/c = (2b_2/E + 1)\ (a^*/c)_{T_{opt}}$$

Thus the optimum temperature is that where

$$(a^*/c)_{T_{opt}} = (a/c)/(2b_2/E + 1)$$

For caustic concentrations from 100 to 150 gpl Na_2O, the value of $1/(2b_2/E + 1)$ ranges from 0.602 to 0.607, so a value of 0.6 will be chosen, ie.

$$(a^*/c)_{T_{opt}} = 0.6\ (a/c)$$

that is, $T_{opt} = (2487 - 1.09c)/[6.211 - \ln(0.6\ a/c)]$.

G_{opt} is obtained by substitution of this value of T_{opt} into the growth relation.

Nomenclature

a	alumina concentration, gpl Al_2O_3
a^*	equilibrium alumina concentration, gpl Al_2O_3
b_i	parameters used in Appendix 2
c	caustic concentration, gpl Na_2O
E	term in growth rate expression (equation 5), K^{-1}
$f(L)$	number distribution density function, μm^{-1}
$f_M(L)$	mass distribution density function, kg^{-1}
G	particle linear growth rate, $\mu m\ h^{-1}$
k_G	growth rate factor (equation 5), $\mu m\ h^{-1}\ (Al_2O_3/Na_2O)^{-2}$
L	particle size, μm
$\bar{L}_{32}$	volume-surface (Sauter) mean size, μm
m_i	mass of particles
n_i	number of particles
S	crystal surface area per unit volume of suspending liquor, m^2/m^3
S_v	crystal surface area per unit volume of solid, m^2/m^3
T	temperature, K; T_{opt} temperature giving maximum growth
v	volume flow rate of liquor, m^3h^{-1}
V	volume of liquor in tank
w	mass deposition rate of hydrate, $kg\ h^{-1}$
β	defined by equation 7, (Al_2O_3/Na_2O)
ρ_c	crystal density = 2420 $kg\ m^{-3}$
ξ	conversion between Al_2O_3 and $Al(OH)_3$ = 0.654
τ	mean residence time in vessel, h
ψ	sphericity of a particle

References

1. Kirke, E.A., "Recent Trends in Bayer Precipitation Practice", Univ of NSW Symposium on Alumina, Nov. 1981.

2. Misra, D., "Solubility of Aluminium Trihydroxide (Hydrargillite) in Sodium Hydroxide Solutions", Chemistry and Industry, (London) Vol 20 (1970), p 619-623.

3. Misra, C. and White, E.T., "Kinetics of Crystallization of Aluminium Trihydroxide from Seeded Caustic Aluminate Solutions", Chem. Eng. Progr. Sympos. Ser., No. 110, Vol 67, (1970), pp 63-65.

4. Steemson, M.L., White, E.T. and Marshall, R.J., "Mathematical Model of the Precipitation Section of a Bayer Plant", Light Metals, (1984).

5. Tschampfer, O., "Improvements by the New Alusuisse Process for Producing Coarse Aluminium Hydrate in the Bayer Process", Light Metals, (1982), pp 103-115.

MATHEMATICAL MODEL OF THE PRECIPITATION SECTION OF A BAYER PLANT

M.L. Steemson, E.T. White and R.J. Marshall

Department of Chemical Engineering
University of Queensland
St. Lucia, Queensland 4067
Australia

A mathematical model has been developed to predict the performance of the precipitation section of the Bayer process. For given feed conditions and imposed precipitation and classifier performance criteria, the model will give steady state plant predictions. The model is written in modular form, so that various flow sheet configurations can be considered.

Introduction

This project is concerned with the development of a mathematical model of the precipitation section in a Bayer plant. This is probably the least understood portion of the process and thus often the critical section controlling production from an operating alumina plant. The precipitation section consists of two major equipment components: crystallizers (precipitators) and classifers (thickeners). Figure 1 shows a typical continuous precipitation flowsheet.

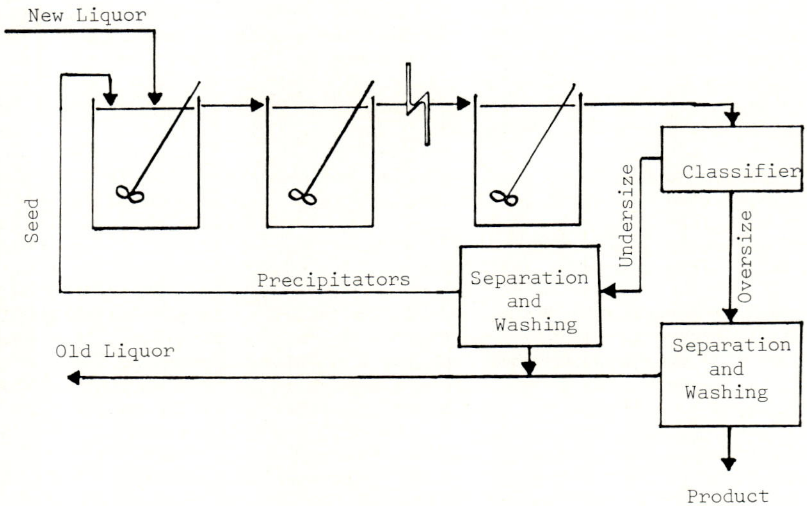

Figure 1 - Continuous industrial precipitation system

In the crystallizers, aluminium trihydroxide ($Al(OH)_3$) particles are grown by deposition from supersaturated liquor. The classifiers separate the coarser product fraction from the finer seed fraction which is recycled for further growth.

In a continuous precipitation system two operating criteria are involved: maximum recovery and a product of suitable quality. Quality covers impurity carryover, particle strength and particularly the particle size distribution. The size distribution of the product hydrate effects the subsequent handling characteristics of the calcined alumina. In practise these two criteria of yield and quality conflict. For example, increasing recovery often means an excess of fines. To aid in the choice of an optimum precipitation operating policy the use of a realistic process model is vital. The use of such a model would allow the effect of modifications to be predicted and an optimum design or operating policy to be determined.

Modelling Background

The concept of using a mechanistic population balance concept in crystallization work was initiated by Randolph and Larson (14). Since then considerable effort has been devoted to develop such models for Bayer precipitation. Misra and White (9) having established kinetics for growth and nucleation rates, developed a batch model for an alumina crystallizer based on population balance theory. This was a mechanistic model based on growth, nucleation and a very simple agglomeration mechanism. Extension to continuous systems was discussed (7). Later work by Low and White (5) resulted in a model for batch crystallization incorporating a more complex mechanistic collision-rupture agglomeration model. This however is not directly applicable to continuous precipitation systems.

Randolph and Tan (15) developed a simulation of a staged classified recycle precipitation system. This was based on the population balance applied to continuous systems and used moments to assist in the solution of the combined mass and population balances over a single crystallizer. The simulation also included the concept of a simple classifier performance curve in describing classifier behaviour. However for alumina studies the simulation is restrictive in that only a single recycle was considered and agglomeration effects were neglected.

A similar approach was used by Muller and Johnson (12). They developed a matrix solution technique for continuous classified recycle precipitation systems but, like Randolph and Tan, only considered single recycle systems without agglomeration effects. Their simulation however has the advantage that dynamic effects can be included.

Recently considerable interest has been shown by industrial corporations in developing mathematical models for Bayer precipitation. Overbey and Scott (13) developed an empirical batch model incorporating the use of seed surface area. This model is very restrictive in that nucleation and agglomeration effects are not considered. Another model for continuous alumina precipitators was developed by Mordini and Cristol (11). This is a mechanistic model but does not consider agglomeration. A simulation of the complete Bayer process was developed by Steinemann and Tschampfer (20) as a Bayer Unit Operation Program (BUOP) and was applied to the problem of conversion from floury to sandy alumina operation. It is not clear whether this contains a comprehensive precipitator-classifier model. Chen et al (1) applied ASPEN, one of the general process simulation packages to the Bayer process. They claim that a rigorous precipitation/classification model with rigorous kinetics is incorporated but details are not given. An alternative application of ASPEN is discussed by Shah et al (18). This simulation however is largely confined to energy usage and concentrates mainly on areas other than the precipitation section.

In conjunction with an interest with modelling work there has been considerable recent emphasis placed on process improvements. This has been largely due to rising energy costs and increasing environmental constraints. An excellent review of these improvements was presented by Kirke (3b) who highlighted the following key aspects -

(1) a trend to higher caustic concentrations. However particle strength and caustic losses can be a problem here
(2) higher seeding densities
(3) minimizing agitation power input
(4) the use of impurity removal schemes
(5) seeding strategies. Multiple seeding strategies can be used to enhance agglomeration and overall performance. Several industrial plants are currently using separate agglomeration and growth stages. An example of this dual process has been presented by Tschampfer (21)
(6) quality control. Two important aspects are evident: impurity removal systems and multiple seeding strategies. The aim of the latter

is to encourage seed toughening by recycling through multiple growth stages and to limit nucleation

(7) process control. An emphasis here has been the development of automated analyzers. With accurate kinetics and a good mathematical model, this will allow vastly improved control of the process.

Modelling Objective

The objective of this work is to develop a steady state model of a continuous Bayer precipitation plant. In order for the model to be applicable to handling current process trends (as outlined in the previous section) the following aspects have been incorporated.

(1) A mechanistic model of a continuous crystallizer which includes growth, nucleation and agglomeration mechanisms
(2) A direct model of a continuous classifier
(3) The capacity for considering multiple seeding and bypass strategies
(4) A versatile modular design which allows ease of flowsheet changes.

Crystallization Mechanisms

In order to develop a phenomenological model of the crystallization section of the precipitation plant, a knowledge of the crystallization mechanisms is necessary. For the aluminium trihydroxide system, these mechanisms are growth, nucleation, agglomeration and breakage.

Growth Mechanism

The growth rate of a crystal is defined as the rate of change of its size with time. This is usually correlated against process conditions such as supersaturation, temperature, impurity concentration, particle size and agitation conditions. Misra and White (10) concluded that for aluminium trihydroxide precipitation, growth rate is independent of both agitation conditions and particle size. They found that supersaturation is the driving force for growth and could be correlated by

$$G = k_G \, S^2 \qquad (1)$$

where G = linear growth rate (μm/hr)
k_G = rate constant dependent on solution temperature (and perhaps caustic and impurity concentrations, though there is little published data in this regard)
S = liquor supersaturation (Al_2O_3/Na_2O)

Following the work of King (3a) the supersaturation is taken as the solution a/c ratio minus that a/c ratio at equilibrium. For pure solutions in the range 0.8 - 1.2 (Al_2O_3/Na_2O) k_G can be related by

$$k_G = 1.96 \times 10^{10} \exp(-14.3 \times 10^3/RT) \qquad (2)$$

where RT has the units of cal/gmol.

Nucleation Mechanism

Nucleation is the mechanism by which new crystals (nuclei) are formed.

The rate of nucleation, B_o, is defined as the rate of increase in the number of crystals per unit volume of slurry due to nucleation, ie.

$$B_o = \frac{dN}{dt} \tag{3}$$

where B_o = nucleation rate
N = total number of crystals per unit volume of slurry.

For industrial alumina systems, the nucleation mechanism is that of secondary nucleation. Misra and White (10) in a study of pure aluminium trihydroxide solutions correlated the limited nucleation rate data by,

$$B_o = k_N A_S S^2 \tag{4}$$

where A_S = surface area of the particles per unit volume of precipitation
k_N = nucleation rate constant which depends on process conditions.

Misra determined k_N as a function solely of temperature for pure aluminate solutions. In industrial situations, correlations for nucleation rate under actual plant conditions will be required. Impurities may affect the value of k_N as could agitation conditions, the presence of intercoolers, scaled surfaces on the vessels, etc.

Agglomeration Processes

Agglomeration, the process by which small particles collide, adhere and eventually form a new larger particle (or agglomerate), results in a considerable reduction in fines and a coarsening of the final product.

The behaviour of agglomerates was observed by Scott (17) and Misra (7). Misra (7) and Sakamoto et al (16) suggest that agglomeration occurs as a two stage process:

(1) Collisions between the particles lead to the initial formation of loose aggregates
(2) Further growth cements these aggregates together eventually forming a single firm agglomerate. If insufficient growth occurs, these aggregates may break down back into individual particles.

Both of these workers proposed a simple batch model. Low and White (5) extended this concept by developing a quantitative collision-rupture model applicable to batch systems. They also proposed the concept of a critical diameter above which agglomeration does not occur. Halfon and Kaliaguine (2) developed a model based on binary collisions applicable to batch systems. They proposed an agglomeration rate which resulted from a fraction of the theoretical number of collisions between particles. This fraction was related to the fourth power of supersaturation. More recent work by Tschampfer (21) related agglomeration to a linear deposition rate. Although no model was presented, useful mechanistic observations were made.

A review of the literature indicates that while several models are available, none are considered applicable to continuous precipitators. For initial modelling purposes a simple agglomeration mechanism based on previous batch studies is proposed. Experimental work is currently being undertaken in this area.

An Agglomeration Model

The simple agglomeration model proposed for continuous aluminium trihydroxide processes involves the following:

(1) Agglomeration does not occur for particles above a certain critical size (D_{crit}). This critical size may be correlated with process conditions. Based on the work of Low (4) with batch experiments, the following correlation is obtained:

$$D_{crit} = G^{2/3} \; [2.1 - 3.5 \log_{10}(\varepsilon) + 1.5 \log_{10}(M_T)] \qquad (5)$$

where D_{crit} is the critical diameter (μm)
M_T = gpl solids in slurry
G = growth rate (μm/hr)
ε = power input per unit volume (kW/m^3)

(2) For particles less than D_{crit} a fraction (constant for all sizes) will be converted to agglomerates. The agglomerated particles so formed will lie in a narrow range of sizes about D_{crit}. The fraction agglomerated will depend on there being sufficient cementation to form an agglomerate. While the remaining particles may form loose aggregates, it will be assumed that these are ruptured by concentration changes in passing to the next vessel. The fraction (B) that agglomerates is assumed to be of the form:

$$B = \exp\left(\frac{-K \, \Delta L_{max}}{G\tau} \right) \qquad (6)$$

where ΔL_{max} = amount of growth (cementation) necessary for an aggregate to be converted to a rigid agglomerate. This has been assumed to be proportional to the mean size of the particles forming the agglomerate
G = growth rate
τ = mean residence time of particles within the crystallizer tank
K = proportionalities constant

The fraction of particles which agglomerate can be seen to be heavily dependent on the growth rate of particles. As G drops, B decreases and hence the amount of agglomeration decreases. In practise this implies that the majority of agglomeration will occur in the initial vessels of the train where the growth rate is high.

(3) Agglomeration effects can be individually evaluated. This assumes that the effects of agglomeration can be considered before growth and nucleation are considered. That is, in a precipitator the fraction agglomerated can first be evaluated, the distribution rearranged, and finally the effects of nucleation and growth can be determined.

Breakage of Particles

Particle breakage is a result of particle-particle and particle-equipment collisions, and results in an increase in the number of particles. In industrial precipitators, the level of agitation is probably insufficient to cause gross fracture. Particle breakage is therefore not considered further. Experimentally Misra (7) found gross breakage was only significant at very high agitation levels.

Crystallizer Model

Having established crystallization kinetics, it is possible to evaluate the steady-state behaviour of a continuous crystallizer (precipitator). The model we have used is mechanistic and based on a population balance approach. Figure 2 shows a schematic diagram of a typical continuous seeded precipitator. Seed and liquor from precipitator K-1 are fed into precipitator K.

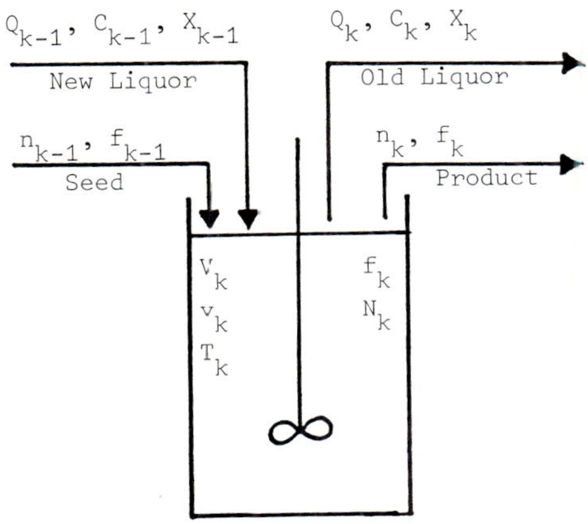

Figure 2 - Schematic diagram of a seeded continuous precipitator

The following assumptions are made,

(1) steady state operation
(2) crystallizers are well-mixed
(3) the overflow is representative of the bulk contents
(4) no crystal breakage occurs
(5) size independent growth ie. McCabes ΔL law of growth is valid. Misra (10) showed this to be a valid assumption.
(6) nuclei are generated at a negligably small size (close to zero).

Based on these assumptions the following steady state relations give the model: -

Mass Balances

Mass balances can be written for overall and individual components. There are two phases to be considered - solid particles and liquor. A brief summary of the detailed relations are given in the Appendix.

Equilibrium Relationship

A knowledge of aluminium trihydroxide solubility in the liquor is essential. For pure aluminate solutions Misra (8) developed the correlation given in the Appendix. This should be modified for plant liquors.

Kinetics

Kinetic relationships for growth, nucleation and agglomeration are required. Those described in the previous section were used.

Population Balance

Application of the population balance, as described by Randalph and Larson (14) leads to the moments equations for a seeded continuous crystallizer (Figure 2) as described in the Appendix. The relation for the third moment gives the material balance which includes the crystal deposition.

Solution of Model

It is possible to solve the equations for a single crystallizer by an iterative technique. A FORTRAN subroutine (CRYSTR) was developed. It is organised into an agglomeration section and a non-agglomerating section. Once the moments relations are solved iteratively the full size distribution is calculated using cubic spline interpolation. This has significant advantages over previous numerical techniques (eg Runge Kutta) in that it is very fast and provides an explicit solution. Also it allows the user by varying knot spacings to give greater accuracy to different portions of the distribution. Details of this technique are described elsewhere (19a).

Classifier Model

The second unit to be considered is the continuous classifier. The overall behaviour of the precipitation system is markedly affected by classification since this determines the quantity and size distribution of the recycle seed. The classifier cut size and sharpness of classification are key parameters. A classifier is modelled in terms of a classifier efficiency with the fraction of seed in the overflow related to the feed by:

$$n_o f_o(L) = \eta(L)\ n_i f_i(L) \tag{7}$$

where n_i and n_o are the inlet and overflow crystal rates (no/sec)
$f_i(L)$ and $f_o(L)$ are the inlet and overflow crystal size distributions
and $\eta(L)$ is the classifier efficiency.

For this model the classifier efficiency relation developed by Lynch (6) is used.

$$(L) = \frac{\beta(\exp(\alpha) - 1)}{(\exp(\alpha) + \exp(\frac{\alpha L}{d_{50}}) - 2)} \tag{8}$$

where β = the fraction of liquor carried to the overflow. This allows for the carry over of fines with the liquor
d_{50} = cut size of the classifier
α = parameter representing sharpness of classification.

Figure 3 shows a typical classifier efficiency curve with

α = 5
β = 0.92
d_{50} = 50 μm

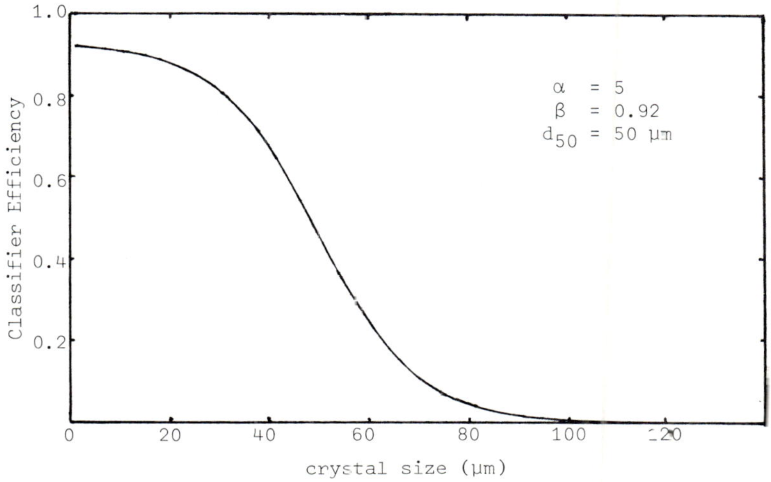

Figure 3 - Typical classifier efficiency curve

For a particular classifier d_{50}, α and β have to be determined experimentally. With classifier efficiency information it is then possible to perform mass and population balances over the classifier by the subroutine CLASSF. Underflow from the first classifier usually forms the aluminium trihydroxide product. The final classifier is taken as a clarifier and in the model is assumed to have clear overflow (no crystal content).

Precipitation System Configuration

The precipitation section consists of a number of crystallizers and a number of classifiers. Some of the streams from the classifiers are recycled as seed for the crystallizers. Since we have separate modules for crystallizers and classifiers, a variety of configurations are possible. Subroutine MIXER undertakes the mass and population balances for the mixing of process streams. The underflow and overflow from any classifier can be divided and sent to any crystallizer provided the underflow from the first is taken as product and the overflow from the last as spent liquor. Partial bypass of crystallizer feeds is allowable but only to crystallizers further down the train. Crystallizer recycle is not allowable.

The flowsheet (interconnection between crystallizers and classifiers) is input by numbering the units and indicating interconnects by giving the fraction of each outflow stream going to other vessels. Changing the flowsheet just requires changing unit numbers. The program is limited to a maximum of 12 crystallizers and 5 classifiers at present.

The total program called PRECIP is written in FORTRAN 77 and comprises about 2000 lines of code. Input data is all file based and takes several minutes of running time on a PDP 11-44 to evaluate each set of data. The time depends upon the closeness of the initial estimates to the final values.

Table I. Typical Simulation Run Using Model

Crystallizer	Temp (^oC)	A/C	G (μm/hr)	B^o $x10^{-7}no/m^3sec$	% agglomerates	D_{crit} (μm)	gpl solids	Surface Area (m^2/l)	$\bar{L}_{10}$ (μm)	$\bar{L}_{32}$ (μm)	$\bar{L}_{30}$ (μm)
Feed	83	1.2	-	-	-	-	-	-	-	-	-
1	82.2	0.86	1.84	-	0.53	14.0	118	11.8	22.4	41.9	31.9
2	81.1	0.76	0.62	-	0.37	7.7	138	13.3	25.3	43.1	34.1
3	80.0	0.70	0.29	-	0.19	5.0	147	14.0	26.9	43.6	34.9
4	79.4	0.67	0.16	-	0.12	3.6	153	14.4	27.3	43.9	35.3
5	68.3	0.62	0.28	8.31	0.21	4.8	166	15.4	21.3	44.4	32.7
6	67.8	0.58	0.16	5.72	0.23	3.6	170	15.5	19.2	44.7	31.7
7	67.8	0.56	0.10	3.66	0.16	2.8	174	15.8	18.5	44.9	31.4
8	67.2	0.55	0.08	3.16	0.12	2.3	177	16.0	17.8	45.0	31.0
9	67.2	0.53	0.05	2.24	0.08	1.9	178	16.1	17.3	45.1	30.7
10	66.7	0.52	0.04	2.09	0.05	1.7	180	16.2	16.8	45.2	30.4
11	66.7	0.52	0.03	1.58	0.06	1.5	180	16.2	16.4	45.3	30.2
12	66.1	0.51	0.03	1.67	0.04	1.4	182	16.3	16.1	45.3	30.0
C1U*	66.1	0.51	-	-	-	-	1024	-	30.9	53.7	43.0
C10	66.1	0.51	-	-	-	-	88	-	12.5	37.8	24.4
C2U	66.1	0.51	-	-	-	-	1384	-	31.1	43.0	37.5
C20	66.1	0.51	-	-	-	-	20	-	7.3	26.2	15.8
C3U	66.1	0.51	-	-	-	-	499	-	7.3	26.2	15.8
C30	66.1	0.51	-	-	-	-	-	-	-	-	-

* product

Some Results

Simulation Results

Several results have been obtained using PRECIP. These results were obtained using data for pure aluminate solutions (as previously outlined). This may need modification for actual Bayer precipitation plants.

Table 1 shows the steady state solution for a typical simulation using the model. Twelve precipitators in series followed by three classifiers are used. Each precipitator has a residence time of 3 hours. The cut size of the classifiers are 50, 30 and 0 microns. Secondary and tertiary seed are returned to the first precipitator. Growth, nucleation rates and percentage agglomerated as well as various mean sizes of the solid particles are presented for each precipitator and classifier stream.

It can be seen that for this simulation the bulk of the crystallization occurs during the first few precipitation tanks. In operation the program calculates the full size distributions for each stream, which may be plotted out.

Process Modifications

Various modifications to the above simulation may be investigated. For the same feed and same temperature profile, four changes were made (cases B to E), namely -

Case A: standard reference case (above)
Case B: running at high solids in the classifier underflow stream
Case C: returning secondary underflow to crystallizer 5 and tertiary underflow to crystallizer 1
Case D: returning secondary to 1 and tertiary to 7
Case E: bypassing, sending 50% of the feed to crystallizer 1 to crystallizer 5.

The effects of these modifications are summarized in Table II. Shown are the change in a/c ratio across the 12 crystallizers the recycle load (defined as the total crystal recycle mass divided by the mass of product) and various product mean sizes.

Table II. Results of Several Modifications

Case	$\Delta(a/c)$	Recycle Load (%)	Product			
			gpl solids	$\bar{L}_{10}$ (µm)	$\bar{L}_{30}$ (µm)	$\bar{L}_{32}$ (µm)
*A	0.689	77	1024	30.9	43.0	53.7
B	0.692	89	1308	32.3	43.2	52.5
C	0.688	85	1020	31.2	42.8	52.5
D	0.697	65	1025	17.1	34.3	57.2
E	0.708	219	1060	20.1	28.9	37.4

*reference case

Effect of Recirculation Rate

Thickening the classifier underflow (comparing case B with case A) causes little change to recovery of product, although the recycle load increases.

Effect of Recycle Strategy

Cases C and D show the effect of changing seeding strategy. In case C little change occurs. This is surprising, in that by introducing only tertiary seed to crystallizer 1 a coarser product would be expected. Table III shows details of each vessel. Higher growth rates are observed before the secondary seed is added to crystallizer 5, thus resulting in a coarser size fraction. However after the secondary seed addition much larger nucleation rates were observed leading to a similar sized product finally. In practise nucleation rates may be lower than those used here for pure aluminate solutions, in which case, a significantly coarser product would be expected.

Table III. Effect of Recycle Strategy

Crystallizer No	G (μm/hr)			B_o $\times 10^{-7}$ no/m^{3}sec			$\overline{L}_{10}$ (μm)		
	A	C	D	A	C	D	A	C	D
1	1.84	3.07	2.67	-	-	-	22.4	20.3	31.5
2	0.62	1.21	1.04	-	-	-	25.3	26.6	35.4
3	0.29	0.60	0.52	-	-	-	26.9	28.7	37.0
4	0.16	0.34	0.30	-	-	-	27.3	29.7	37.9
5	0.28	0.33	0.40	8.31	10.11	7.67	21.3	*22.1	21.6
6	0.16	0.19	0.25	5.72	6.64	5.67	19.2	19.8	18.3
7	0.10	0.11	0.14	3.66	4.11	5.08	18.5	18.9	*9.2
8	0.08	0.08	0.10	3.16	3.46	4.17	17.8	18.1	9.4
9	0.05	0.06	0.07	2.24	2.41	2.85	17.3	17.5	9.4
10	0.04	0.05	0.05	2.09	2.21	2.55	16.8	17.0	9.5
11	0.03	0.03	0.04	1.58	1.66	1.89	16.4	16.4	9.4
12	0.03	0.03	0.03	1.67	1.74	1.94	16.1	16.1	9.4

*mixing occurs

By changing tertiary seeding to crystallizer 7 (case D) we see an adverse effect of a much finer product distribution although the recycle load is lower. Table III indicates that higher growth rates (as compared to A) and similar nucleation rates result. However even with a lower nucleation rate we would expect to find a finer product. In general the results of C and D indicate that seeding strategy is one of the main methods available for modifying and optimizing the process.

Effect of Bypass

Case E illustrates the effect of bypass on the crystallization process. Significant changes occur. A 50% bypass of the feed to crystallizer 1 to crystallizer 5 causes an increase in recovery but at the expense of a much finer distribution. The recycle load also increases significantly owing to the large amount of fines. Bypass also appears to be a significant tool in process optimization.

Effect of Change in Input Flowrate

The final change investigated is a change in inlet flowrates (ie. residence times). Table IV summarizes the results found. Higher inlet flowrates can be seen to result in a slightly lower recovery and a finer product distribution. In addition the recycle load increases with increasing residence time. This indicates an advantage is gained at higher residence times.

Table IV. Effect of Changes in Inlet Flowrate

τ (hrs)	Δ(a/c)	Recycle Load (%)	Product			
			gpl solids	$\overline{L}_{10}$ (μm)	$\overline{L}_{30}$ (μm)	$\overline{L}_{32}$ (μm)
6	0.690	34	1002	34.5	52.0	68.1
4	0.694	68	1029	31.9	44.6	55.8
3	0.639	77	1024	30.9	43.0	53.7
2.4	0.685	92	1028	29.4	39.2	51.0
2	0.684	107	1043	28.2	39.1	48.8

Acknowledgements

The assistance of a Comalco Research Laboratory Scholarship is gratefully acknowledged.

Appendix

1. Steady State Mass Balance for an Industrial Seeded Precipitator

2 $Al(OH)_3$ Balance.

$$\frac{156}{102} X_{K-1} Q_{K-1} + n_{K-1} \int_0^\infty \rho_s \alpha_v L^3 f_{K-1}(L)\, dL = \frac{156}{102} X_K Q_K + n_K \int_0^\infty \rho_s \alpha_v L^3 f_K(L)\, dL$$

Na_2O Balance. $C_{K-1} Q_{K-1} = C_K Q_K$

Constant Volume Equation. $V_K = v_K + N_K V_K \int_0^\infty \alpha_v L^3 f_K(L)\, dL$

2. Solubility Relation for Aluminium Trihydroxide in Caustic Aluminate Solutions

$$X^* = C \exp\left(6.2106 - \frac{2486.7}{t + 273} + \frac{1.0876\ C}{t + 273}\right)$$

range of equation

$c = 30 - 230$ gpl Na_2O
$t = 25 - 100^oC$

3. Steady State Population Balance

The population balance applied to a seeded continuous crystallizer (Figure 2) gives

$$\frac{df_K}{dL} = \frac{-n_K}{N_K V_K G_K} f_K + \frac{n_{K-1}}{N_K V_K G_K} f_{K-1} + \frac{B^o{}_K\, \delta(L-L_o)}{N_K G_K}$$

Deriving a moments form by multiplying by L^γ and integrating from $-\infty$ to ∞.

$$\gamma = \emptyset \qquad n_K = n_{K-1} + B^o{}_K V_K$$

$$\gamma \neq \emptyset \qquad \mu_{\gamma,K} = \frac{\gamma v_K G_K}{Q_K} \mu_{\gamma-1,K} + \frac{n_{K-1}}{n_K} \mu_{\gamma,K-1}$$

where $\mu_{\gamma,K}$ is the γth moment of the Kth crystallizer and is defined by

$$\mu_{\gamma,K} = \int_0^\infty L^\gamma f_K(L)\, dL$$

Symbols Used

A_s	surface area of seed per unit folume of crystallizer
B	fraction of crystal agglomerating less than the critical diameter
B_o	nucleation rate (no per unit volume per unit time)
C	caustic concentration

d_{50}	classifier cut size
$f(L)$	number frequency density distribution
G	growth rate
k_G	growth rate constant
k_N	nucleation rate constant
K	parameter in agglomeration model
L	particle size
$\overline{L}_{10}$	number mean size
$\overline{L}_{30}$	volumetric mean size
$\overline{L}_{32}$	sauter mean size
M_T	gpl crystals
n	number rate of crystals
N	number of crystals per unit volume of crystallizer
Q	flow rate of liquor
R	gas constant
S	supersaturation (= $X-X^*$)
t	time
T	temperature
v	volume of liquor in crystallizer
V	volume of slurry in crystallizer
X	alumina concentration
X^*	equilibrium alumina concentration
α	sharpness of classification
α_v	crystal shape factor
β	classifier splitting fraction
δ	Dirac delta function
τ	crystallizer residence time
μ_γ	γth moment of seed distribution
ρ_s	density of crystals
$\psi(L)$	angularity factor
$\eta(L)$	efficiency of classification

Subscripts

k	crystallizer
i	inlet
o	outlet
s	solid

References

1. Chen, C.C., Boston, J.F., Mock, B., Galloway, T.J. and Loe, K.Y., "Bayer Process Simulation by Aspen", Light Metals, Oct (1982), pp 1-17.

2. Halfon, A. and Kaliaguine, S., "Alumina Trihydrate Crystallisation Part 2. A Model of Agglomeration", Can. Journal Chem. Eng., Vol 54, (1976), pp 168-172.

3a. King, W.R., "Some Studies in Alumina Trihydrate Precipitation Kinetics", Light Metals, (1973), pp 551-563.

3b. Kirke, E.A., "Recent Trends in Bayer Precipitation Practise". Univ. NSW Symposium on Alumina, Nov (1981).

4. Low, G., Ph.D. Thesis, "Agglomeration Effects in Aluminium Trihydroxide Precipitation", University of Queensland (1975).

5. Low, G.C. and White, E.T., "Agglomeration Effects in Alumina Precipit-

ation", Extractive Metallurgy Symposium, Melbourne, Nov (1975).

6. Lynch, A.J., "Mineral Crushing and Grinding Circuits: Their Simulation, Optimization, Design and Control", Elsevier Scientific Publishing Co., New York, (1977).

7. Misra, C., Ph.D. Thesis, "The Precipitation of Bayer Aluminium Trihydroxide", University of Queensland, (1970).

8. Misra, C., "Solubility of Alumina Trihydrate (Hydragillite) in Sodium Hydroxide Solutions", Chemistry and Industry (London), Vol. 20, (1970), pp 19 -.

9. Misra, C. and White, E.T., "A Mathematical Model for the Bayer Precipitation Process for Alumina Production", Proceedings of Chemeca 70, Butterworths (1971), pp 52-76.

10. Misra, C. and White, E.T., "Kinetics of Crystallization of Aluminium Trihydroxide from Seeded Caustic Aluminate Solutions", Chemical Engineering Progress, Symposium Series, Vol 67, No 110 (1970).

11. Mordini J. and Cristol, B., "Mathematical Model of Alumina Trihydrate Precipitation from Bayer Aluminate Liquors", paper presented at 4th Yugoslav International Symposium on Aluminium, Titrograd, Apr (1982).

12. Muller, T.B. and Johnson, T.L., "Simulation of Mass, Heat and Particulate Balances in the Bayer Precipitation", TMS-AIME, January Meeting, (1979), pp 3-29.

13. Overbey, T.L. and Scott, C.E., "Characterization of Bayer Plant Liquors and Seeds Utilizing a Mathematical Model for Precipitation", Light Metals, Vol 2, (1978), pp 163-178.

14. Randolph, A.D. and Larson, M.A., A.I.Ch.E.Jl., 8, 639 (1962).

15. Randolph, A.D. and Tan, C., "Numerical Design Techniques for Staged Classified Recycle Crystallizers: Examples of Continuous Alumina and Sucrose Crystallizers", Proc. Des. and Dev., Vol 17, April, (1978), pp189-200.

16. Sakamoto, K., Kanehara, M. and Matsushita, K., "Agglomeration of Crystalline Particles of Gibbsite during the Precipitation in Sodium Aluminate Solution", Chem. Eng. Soc. of Japan, 35, 481 (1971).

17. Scott, J., "Extractive Metallurgy of Aluminium", Vol 1, Alumina, 203, John Wiley & Sons, New York (1962).

18. Shah, V.B., Gacka, P. and Langa, J.M., "The Application of Aspen Flowsheet Simulator at Alcoa", A.I.Ch.E. Symposium Series, No 214, (1982), pp56-65.

19a. Steemson, M.L., White, E.T. and Marshall, R., "Modelling of Crystallizer-Classifier Recycle Systems", (in preparation).

19b. Steemson, M.L. and White, E.T., "Modelling of the Bayer Alumina Precipitation System", Proceedings of Chemeca 83, (1983), pp 473-483.

20. Steinmann, P. and Tschampfer, O., "The Bayer Unit Operation Program (BUOP) and its Application for the Modification of the American Bayer

Process by the Alusuisse Precipitation Process", Light Metals, (1982), pp 187-198.

21. Tschampfer, O., "Improvements by the New Alusuisse Process for Producing Coarse Aluminium Hydrate in the Bayer Process", Light Metals, (1982) pp 103-115.

CHAIRMEN

E.L. Cambridge
ARCO Metals Company
Tucson, Arizona

D.J. Donaldson
Kaiser Aluminum and Chemical Corp.
Oakland, California

ATTRITION OF ALUMINAS

D.J. Braun

Vereinigte Aluminium-Werke AG
Leichtmetall-Forschungsinstitut
Postfach 24 68, D-5300 Bonn 1

Mechanical stability constitutes an important characteristic of aluminas. Unstable aluminas disintegrate during transport from Bayer-Plant to smelters or - especially in modern plants - in dry gas cleaning. This grain-size reduction leads to operational difficulties in electrolysis.

Laboratory investigations had indicated, that the mode of calcination significantly influences the strength of alumina. Therefore, the aim of the present investigation was to test mechanical strength of aluminas from different plants in order to correlate strength to mode of calcination, BET-surface and to grain size of the products. A modified FORSYTHE-HERTWIG apparatus was used in these tests.

Introduction

The grains of commercial aluminas exhibit widely varying attrition rates when exposed to mechanical strains. In the present paper the attempt is made to relate the attrition rates of aluminas to the mode of calcination and also to measurable physical properties, such as BET-surface area or grain size distribution irrespective of the mode of precipitation of hydrate, which is quite different for the aluminas studied.

Experimental Methods

All samples investigated were tested in a modified FORSYTHE-HERTWIG apparatus. In the experimental set-up solid particles are exposed to an intensive mechanical strain caused by a high-velocity air jet. The test equipment employed in the experiments corresponds to the apparatus now widely used in the aluminum industry (1). Compared to the original FORSYTHE-HERTWIG apparatus (2), the testing time is shorter.

In this work the sample quantity used in each test run was 50 g, the air rate 0.436 cu.m (STP) and the duration of the test 15 min. The grain size distribution of the samples prior and after the test was determined by using the Coulter-Counter apparatus. The grain size reduction was characterized by the value of Δ 45 µm (before and after attrition) and by the attrition index AI:

$$AI\ [wt\%] = \frac{> 45\ \mu m_{before} - > 45\ \mu m_{after}}{> 45\ \mu m_{before}}$$

However, both data give only an indication of the attrition of particles > 45 µm into fractions < 45 µm. For a more exact determination of the extent of grain disintegration, it is necessary to determine for each size fraction the fraction in the sample prior and after the test. The sum of differences determined in this way indicates the extent of grain disintegration in a sample independently of any arbitrary separation cut. In the present study differences of 10 grain size fractions (< 22 µm, 22 - 28 µm, 28 - 35 µm, 35 - 45 µm, 45 - 57 µm, 57 - 71 µm, 71-90 µm, 90 - 115 µm, 115 - 140 µm and > 140 µm) were summed up.

For further characterization of grain disintegration scanning electron micrographs of samples were taken before and after the tests, and specific surfaces as well as α-Al_2O_3 contents of the aluminas (by X-ray diffraction) were determined.

Results

Table 1 shows the commercial aluminas studied. With low α-content, only fluid bed aluminas are listed. This is due to the higher α/BET-ratio of rotary kiln aluminas in comparison to the fluid bed aluminas, which is caused by inhomogeneous calcination in rotary kilns (3). Even at high BET-surfaces (around 50 m²/g) more than 20 % α-content are present in rotary kiln aluminas.

Table 1 Tested Aluminas

Description	Origin	Calcination Mode	Calcination Degree (% α-content)	BET-Surface m^2/g	Medium Grain Size (μm)
A	European	rotary kiln	24	55	70
B	European	rotary kiln	85	3	48
C	European	rotary kiln	71	5	43
D	European	rotary kiln	22	53	74
E	European	rotary kiln	91	5	40
F	European	rotary kiln	95	2	41
G	European	fluid bed	52	19	86
H	European	fluid bed	24	34	86
I	European	rotary kiln	93	2	40
J	European	fluid bed	26	34	82
K	European	fluid bed	16	38	82
L	European	fluid bed	46	21	50
M	European	fluid bed	42	24	54
N	European	fluid bed	25	36	34
O	American	rotary kiln	50	26	48
P	American	fluid bed	3	50	77
Q	Australian	rotary kiln	25	42	70
R	African	rotary kiln	35	50	67
S	African	rotary kiln	29	50	70
T	African	rotary kiln	71	5	48

Dependence of Attrition on Degree of Calcination

The attrition behavior of aluminas from rotary kilns and from fluid bed calciners is markedly different as is shown in figures 1 and 2. In general, attrition is higher with aluminas from rotary kilns. Whereas the strength of fluid bed products shows almost no dependence on the degree of calcination as characterized by their BET-surface, rotary kiln aluminas are increasingly unstable with increasing BET-surface, i.e. with lower degree of calcination. This holds only for calcines with BET-surfaces in the range > 20 m^2/g. Very high α-aluminas (samples B and C) again exhibit high attrition rates.

The markedly different attrition behavior of rotary kiln and fluid bed products becomes still more evident if the change in the different grain fractions after the attrition test is looked upon. This is shown in figures 3 and 4, where grain size distributions before and after the test are given for the fluid bed sample P and the rotary kiln sample D, resp. Both samples have nearly the same BET-surface and the same medium grain size. The latter is particularly important, because grain size also affects the attrition characteristics.

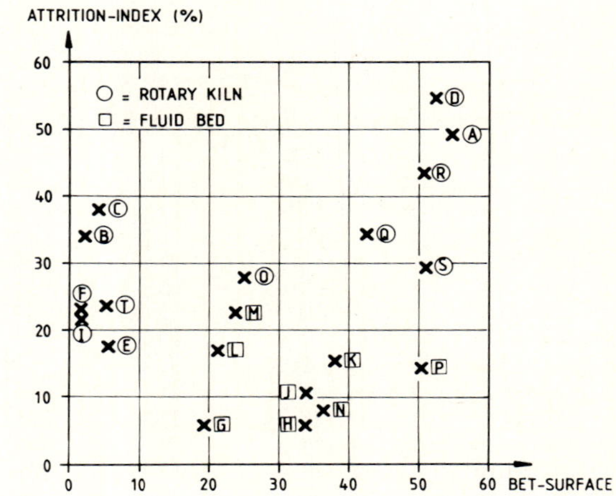

Figure 1 - Correlation between BET-surface and attrition index of aluminas

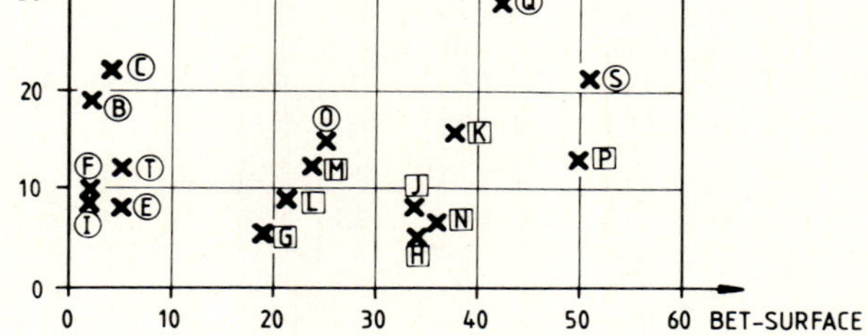

Figure 2 - Correlation between BET-surface and grain size reduction expressed as Δ 45 μm

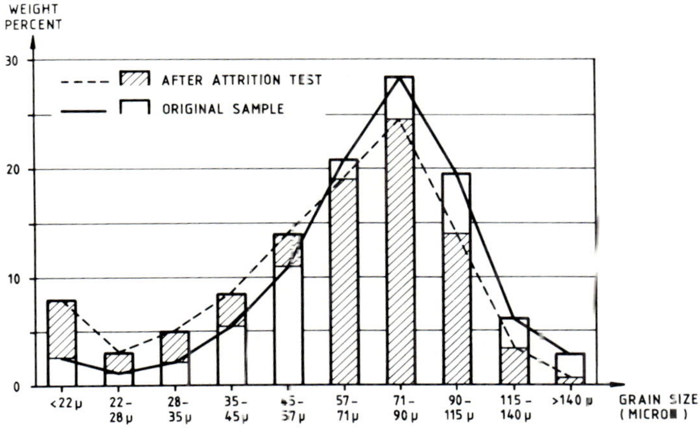

Figure 3 - Grain size distribution of fluid bed alumina sample P (BET 50 m²/g, d_{50} 77 μm) prior and after attrition test (436 l/h, 15 min)

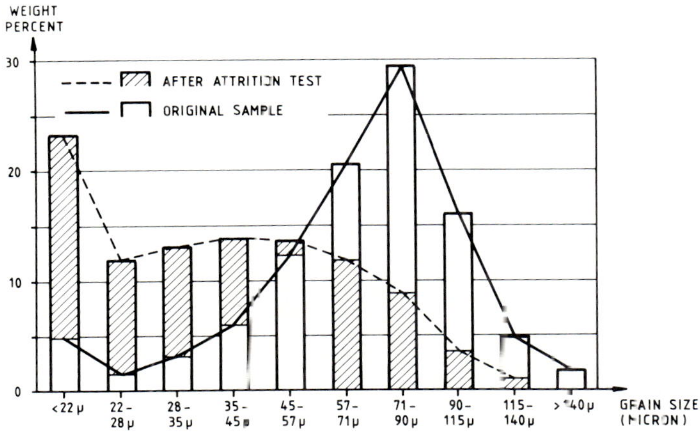

Figure 4 - Grain size distribution of rotary kiln alumina sample D (BET 53 m²/g, d_{50} 74 μm) prior and after attrition test (436 l/h, 15 min)

Original sample

100 μm

Sample after attrition test with an air jet of 200 l/h
Test time: 15 min

100 μm

Sample after attrition test with an air jet of 436 l/h
Test time: 15 min

100 μm

Figure 5 - Micrographs (SEM) of fluid bed alumina sample P (BET 50 m^2/g, d_{50} 77 μm)

Original sample

100 μm

Sample after attrition test with an air jet of 200 l/h
Test time: 15 min

100 μm

Sample after attrition test with an air jet of 436 l/h
Test time: 15 min

100 μm

Figure 6 - Micrographs (SEM) of rotary kiln alumina sample D (BET 53 m²/g, d_{50} 74 μm)

The majority of fine particles formed during the attrition test of the rotary kiln alumina D is in the range below 22 μm. The maximum of the grain size distribution of sample D initially found in the fraction 71 - 90 μm shifts to < 22 μm. The grain disintegration of the fluid bed alumina P follows another pattern: the newly formed grain fractions fall into the range < 22 μm to 57 μm, the distribution maximum remaining in the fraction 71 to 90 μm. This different behavior is also illustrated by the scanning electron micrographs given in figures 5 and 6. The photographs show the samples before and after the exposition to air jets of different velocities. The grain structure of the rotary kiln alumina (figure 6) changed completely after exposure for 15 minutes to an air jet of 436 l/h, while in the fluidized bed alumina (figure 5) rather slight increase of the finer fraction is registered.

The grain disintegration of a high α-alumina from a rotary kiln (sample B) is seen in figures 7 and 8. Again, the maximum of grain size distribution curve (fraction 45 - 57 μm) is shifted to the finest grain fraction.

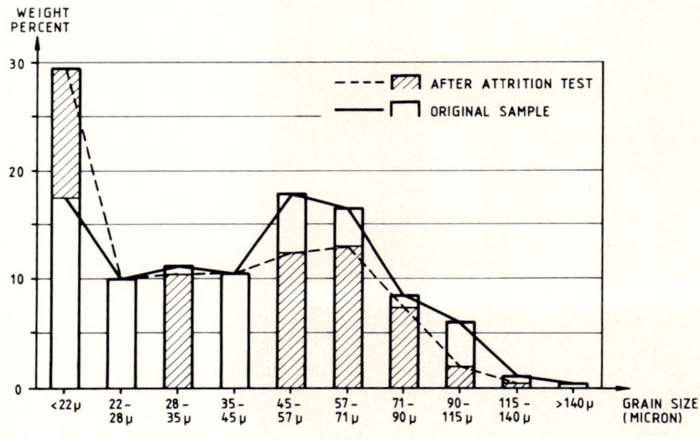

Figure 7 - Grain size distribution of a high α-alumina from a rotary kiln (sample B, BET 3 m²/g, d_{50} 48 μm) before and after attrition test

The apparent difference in mechanical stability of fluid bed aluminas compared to rotary kiln products can be explained by the fact that in fluid bed calcination due to the higher mechanical strains - compared to rotary kiln calcining - some attrition has already occured in the technical process, the remaining alumina particles being mechanically rather stable.

The difference found between rotary kiln and fluid bed products with respect to the dependence of mechanical strength

Original sample

100 µm

Sample after attrition test with an air jet of 200 l/h
Test time: 15 min

100 µm

Sample after attrition test with an air jet of 436 l/h
Test time: 15 min

100 µm

Figure 8 - Micrographs (SEM) of an α-alumina from a rotary kiln (sample B; BET 3 m²/g, d_{50} 48 µm)

on the degree of calcination (BET-surface) may be explained by the more homogeneous calcination in a fluid bed reactor. The instability of the high α-aluminas (samples B and C) is due to disintegration of secondary agglomerates composed of tiny α-alumina crystals at high calcination temperatures.

The above-mentioned correlations are valid for European aluminas as well as for American, Australian and African samples. This indicates, that the effect of the calcination mode is much greater than the the influence of the different precipitation modes used in the different plants.

Attrition and Grain Size

The medium grain size of the aluminas studied was in the range between 40 to 90 μm. In the range above 80 μm only fluidized bed aluminas were tested. Increase in grain size leads to a marked decrease in grain stability with rotary kiln oxides. This is especially evident by comparison of samples B and F, samples C and E or samples D and R. These pairs of rotary kiln calcined samples show different disintegration behavior having nearly the same BET-surface but different grain sizes. Fluid bed aluminas however, show practically no dependence of attrition values on grain size (figure 9).

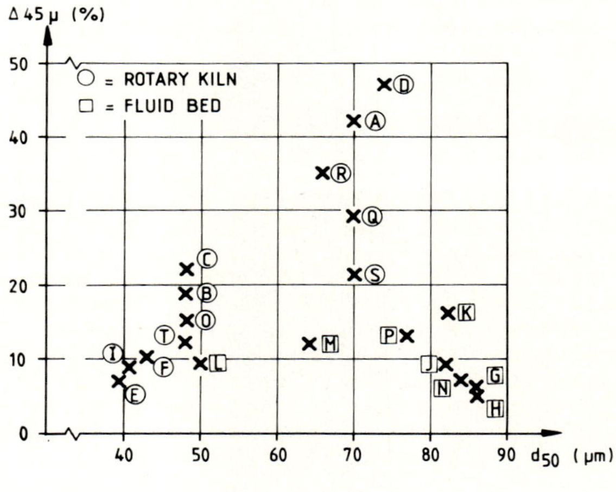

Figure 9 - Correlation between grain sizes and grain size reduction expressed as Δ 45 μm

A similar picture emerges if the overall grain disintegration expressed as the sum of all grain size fraction differences due to testing is plotted against grain size expressed as d_{50} (figure 10).

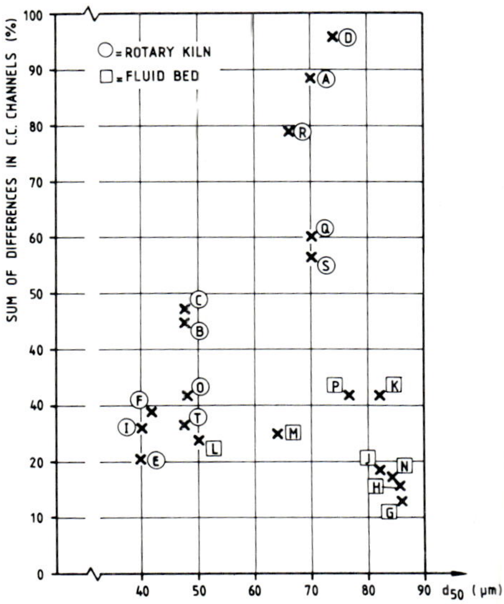

Figure 10 - Correlation between grain size and the sum of differences in Coulter-Counter channels caused by attrition

Attrition of Alumina Hydrates

A number of alumina hydrates from different American and European plants was also tested in the FORSYTHE-HERTWIG apparatus. Apart from the result that the overall disintegration of the hydrates - as studied by comparing grain fractions before and after the test - increases with increasing medium grain sizes of the hydrates, we were not able to demonstrate conclusive relations between attrition and grain size or crystallite sizes. These relations are complex and can only be studied by systematic variation of the precipitation mode in the laboratory. This is beyond the purpose of the present work. Previous work by other authors on this topis has to be mentioned (1).

Summary

Aluminas from fluid bed calciners, in general, are mechanically more stable than products from rotary kilns. The attrition behavior of these two types of commercial aluminas expressed by an attrition index as well as demonstrated by the overall change in the grain size distribution before and after attrition test differs markedly. This fact has to be taken into account in discussions on the grain size of aluminas.

Literature

1. J.L. Anjier and D.F.G. Marten,
"Particle Strength of Bayer Hydrate"
Light Metals 1982, S. 199 - 209

2. Forsythe, W.L. and Hertwig, W.R.,
"Attrition Characteristics of Fluid Cracking Catalysts"
Ind. and Eng. Chem. 41 (1949) 1200 - 1206

3. E. Barrillon,
"The merits and demerits of various types of Bayer aluminas"
Erzmetall 31 (1978) Heft 11, S. 519 - 522

"MEASUREMENT OF SANDY ALUMINA DUSTINESS"

S. Perra

Eurallumina S.p.A.
09010 Portoscuso - (Cagliari)
ITALY

The paper intends to offer a contribution for the quantitative determination of such intrinsic characteristic of the alumina.

Simple apparatus (pulvimeter) has been designed and built to measure on both weighted and numerical basis, the tendency of the alumina to produce dust (primary pulvation).

The pulvimeter enables to weight the free dust produced by an alumina after diffusion as settled in the measurement cell.

The alumina dust is then analyzed for the numerical counting (via Celloscope) (1) and for the weighted and numerical determination of the granulometric distribution.

The reproducibility of the method as realized is better than 5%.

Applying the above procedure to various alumina samples, the following conclusions have been reached:

- Alumina dustiness depends marginally from its -325 mesh content. It is on the contrary more related to the quality of -325 mesh fraction.

- On weighted terms, alumina dustiness in function of -20 microns content, even tough it has been observed that by increasing 5 times the content of -20 microns fraction, the weighted primary pulvation index increases only by 60%.

- In numerical terms, the alumina primary pulvation index is directly proportional to its -20 microns content.

Introduction

The study of dust levels in alumina is a problem which is so sensitive with end-users of alumina as to influence their choice of producers with whom they deal.

Producers, for their part, have responded to the problem by taking extreme care to control product fineness and by allocating a substantial part of their production costs and investment money to the problem.

However, it is authoritative opinion that both end-users and producers are subject mainly to emotional factors rather than hard facts. This makes it imperative to create a sound basis on which to work, such as the direct measurement of the tendency of a given alumina to produce dust, the identification of parameters which among alumina's physical characteristics are the catalysts of the phenomenon, and a criterium to establish the alumina specification based on facts, not emotions.

Conclusions

An instrument (the dust-meter) has been designed, constructed and tested to provide direct measurement of the "dust factor" of all types of alumina ; the instrument does this simply, reliably, and precisely.

The dust meter eliminates the need to extrapolate this important feature of alumina from other physical properties, since such results are not always realistic.

It has been demonstrated that the specification which requires maximum 10% on the -325 mesh is soundly based only to the extent that, generally speaking, the superfines increase as the -325 mesh increases.

On a weighted basis, the dust level index of alumina is correlated with its content of -20μ or -10μ (r^2 = 0.82) in such a way that, however, when there is a fivefold increase of superfine content, the weighted dust level index (Weighted Primary Dust Content Index = W.P.D.C.I.) increases by barely 60%.

This signifies that whatever means (operational or plant equipment based) are used to reduce superfines in the finished product, this will never produce a return on investment if the objective is to reduce losses of alumina during handling operations.

Table I summarizes the results of weighted dust content tests conducted on our alumina having -20μ content of between 0.95 and 5%.

The table also gives the results of weighted dust content tests conducted on our aluminas from which, by wet seiving, the -45, -30 and -20 μ have been removed, respectively.

Table I.

Weighted Primary Dust Content Index Versus -20μContent in Alumina

- 20μ (%)	W.P.D.C.I. (Kg/t Allumina)
5	1.78
1.68	1.15
0.95	1.04
2.03	1.36
1.5	1.27
2.4	1.49
1.0	1.36
1.6	1.20
3.2	1.62
4.01	1.82

CORRELATION INDEX

$r^2 = 0.82$

$$W.P.D.C.I. = 0.18 \times (\% -20\mu) + 0.98$$

Weighted Primary Dust Content Index of +45, +30, +20 Microns

SAMPLES	W.P.D.C.I.
AS IS	1.29
+45μ	0.18
+30μ	0.52
+20μ	0.80

Reasoning in numerical terms, which is the most correct approach from the standpoint of industrial hygiene (in the United States, for example, the inert dust limit in work environments is expressed as the number of suspended particles per unit volume), the numerical dust content index (Numerical Primary Dust Content Index = N.P.D.C.I.) is directly proportional to the -20μcontent in alumina.

By regressing linearly the data given in Table II it has been found that:

$$N.P.D.C.I. = \frac{\text{(millions of particles)}}{\text{gr of alumina}} = 1.235 (\% -20\mu) - 0.08$$

with a CORRELATION INDEX (r^2) of 0.99.

This happens because as superfines increase in the product, even with a slight increase of weighted dust content, the grain size distribution of the generated dust is increasingly finer (i.e. more particles for same weight).

Table II. Pulvation Test on Aluminas with Different Contents of Superfines

$-20\ \mu$ on the alumina sample (% weight)	N.P.D.C.I. millions of particles / gr. of alumina	NUMERICAL DISTRIBUTION (%)										DISTRIBUTION BY WEIGHT (%)								Nº of part. Alumina as is millions / gr
		Dust Deposited					Alumina as is					Dust Deposited				Alumina as is				
		$+45\ \mu$	$+20\ \mu$	$+10\ \mu$	$+5\ \mu$	$+2\ \mu$	$+45\ \mu$	$+20\ \mu$	$+10\ \mu$	$+5\ \mu$	$+2\ \mu$	$+45\ \mu$	$+20\ \mu$	$+10\ \mu$	$+5\ \mu$	$+45\ \mu$	$+20\ \mu$	$+10\ \mu$	$+5\ \mu$	
3.9	4.77	0.005	0.32	2.15	22.87	94.8	1.26	3.28	5.75	20.57	92.77	3.64	36.12	55.27	83.61	74.7	96.13	97.93	99.13	134
1.5	1.73	0.02	0.95	5.30	27.78	93.44	5.05	12.39	16.74	31.35	84.26	10.15	57.71	78.92	94.03	79.17	98.54	99.48	99.8	31
2.1	2.63	0.01	0.61	4.34	25.58	93.57	4.56	10.37	17.24	37.91	91.71	5.15	45.01	70.09	91.06	81.32	97.94	99.29	99.85	38
2.0	2.16	0.021	0.93	6.03	28.49	94.14	6.18	14.97	24.79	43.49	93.79	6.45	51.2	76.72	93.77	80.06	98.03	99.5	99.88	28.5
1.4	1.79	0.02	1.05	18.52	40.51	100	7.96	17.9	25.1	42.67	92.79	6.13	58	77.77	93.45	80.62	98.64	99.59	99.88	22.8

NOTE: All size distribution analysis are measured with the Celloscope.

CORRELATION INDEX

$r^2 = 0.99$

N.P.D.C.I. = $1.235 \times (\% - 20\ \mu) - 0.08$

In any case, if there is no threshold of superfines above zero for which with lower values we can reasonable state that the alumina is not "dusty" (since it has been widely demonstrated that the function is continuous) we can still fix as a criterium for superfines specification an alignment with the best products on the market.

The best approach, however, would be a large-scale use of a test making direct measurement of dust indexes.

Alumina Dust and Industrial Hygiene

The following concentration limits (TVL) of alumina dust are recommended for occupational environments:

SOURCE		TVL
I. SAX (2)	:	50 m.p.p.c.f.
Italian Chemical industry contract	:	30 m.p.p.c.f. or 10 mg/m^3

REMARKS: TVL is the maximum concentration to prevent damage to normal individuals exposed to alumina dust 8 hours per day.

m.p.p.c.f. = millions of particles per cubic foot.

m.p.p.c.m. = millions of particles per cubic meter.

Wishing to express the above limits in m.p.p.c.m., we have:

SOURCE		TVL
I. SAX	:	$\frac{50 \times 1000}{28.3168}$ = 1766 m.p.p.c.m.
Italian Chemical industry contract	:	$\frac{30 \times 1000}{28.3168}$ = 1059 m.p.p.c.m.

Since the Italian Chemical Industry Contract sets concentration limits in terms of both number of particles and weight, we can calculate particle diameter (in μ) which satisfies the following equation:

$$1059 \text{ m.p.p.c.m.} = 30 \text{ m.p.p.c.f.} = 10 \text{ mg/m}^3$$

$$\text{EQUIV. DIAMETER} = \sqrt[3]{\frac{(\text{mg/m}^3)}{(\text{m.p.p.c.m.}) \times \frac{4}{3} \times \pi \times \gamma}} \times 2 \times 10^4 =$$

$$= \sqrt[3]{\frac{10}{1059 \times 10^6 \times \frac{4}{3} \times 3.14 \times 3.56 \times 1000}} \times 2 \times 10^4 = 1.72\ (\mu)$$

This means that with a particle concentration of 10 mg/m^3, we also have a particle numerical concentration equal to 1059 m.p.p.c.m. provided that average numerical diameter is 1.72 μ.

If average numerical diameter is lower, the number of particles per cubic meter definitely surpasses the chemical contract limit. Likewise, the number of particles per cubic meter will be below the chemical contract limit whenever the average numerical diameter surpasses 1.72 μ.

In order to obtain reference values corresponding to our case in practice, on 3 February 1983 plant personnel took a sample of air in the tunnel of a belt conveyor during the loading of a ship from a storage silo.

The sample was taken 80 cm above the belt surface corresponding to the vertical gangway on the belt's outer edge.

The results were as follows:

a) particle concentration = 44 mg/m^3

b) properties of the dust:

- number of particles per mg of dust = 3.5×10^6

- grain size distribution	in numbers	in weight
+ 100 μ (%)	0.0000	0.0000
+ 45 μ "	0.0002	0.144
+ 20 μ "	0.028	2.34
+ 10 μ "	3.71	31.92
+ 5 μ "	27.66	81.41
+ 3 μ "	68.95	97.16
+ 2 μ "	93.58	99.74

In the case of particle matter generated by our sandy alumina the dust concentration in weight required to reach the numerical limit set by the chemical contract, is rather high:

Concentration of particle matter equivalent to 1059×10^6 particles per m^3, in the case of dust generated by our sandy alumina =

$$= \frac{1059 \times 10^6}{3.5 \times 10^5} = 303 \quad mg/m^3$$

as Table III indicates, this concentration is about four times greater than the visual perception limit (for approx. 4 μ of average numerical diameter, the visibility threshold is about 80 mg/m^3) and thirty times the maximum weighted limit provided for by the chemical contract.

Experiences with the dust meter on our alumina allows us to evaluate the pollution potential of our sandy product.

It should be borne in mind that the values to be computed are only potential figures, but still do represent the maximum reachable as the effect of air diffusion and/or transport from the environment in which the dust is freed into the atmosphere.

Table III.

Visual Perception Limit in mg/m^3 in Function of Particles Diameter in Micron

Diameter of Particles in Micron

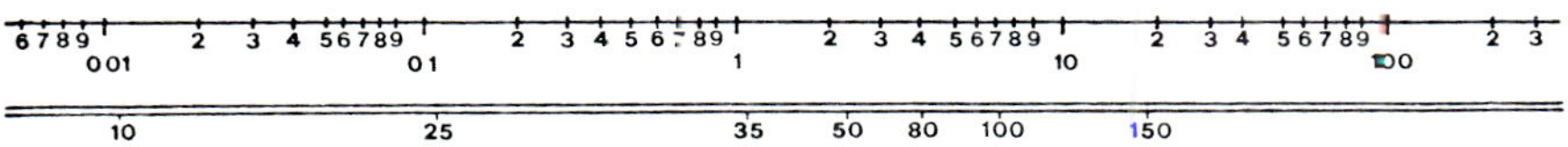

Visual Perception Limit in mg/m^3

These pollution potentials are as follows, for each phase of alumina handling (jump):

In weight : $\frac{1.4 \times 10^6 (\overline{W.P.D.C.I.})}{10} = \frac{140{,}000 \text{ m3 of air}}{\text{ton of alumina handled}}$

In numbers : $\frac{2.1 \times 10^6 (\overline{N.P.D.C.I.}) \times 10^6}{1059 \times 10^6} = \frac{2{,}000 \text{ m3 of air}}{\text{ton of alumina handled}}$

Of course the diffusion of alumina dust in the environment at its maximum potential value will be localized in the emission point.

From the emission point the dust is then subjected to a depositing phenomenon which reduces the concentration as the dust moves further away from said emission point.

The depositing phenomenon follows Stokes' law:

$$V = \frac{2 \times g \times r^2 \times (d_s - d_a)}{9 \times \mu \times 10^8}$$

where:

V = settling rate (cm/sec)

g = acceleration due to gravity (978 cm/sec^2)

r = radius of particle matter (μ)

d_s = density of particle matter (3.56 g/cm^3)

d_a = density of air (0.00121 g/cm^3 at 18°C)

μ = viscosity of air (182.7×10^{-6} poises at 18°C)

By computing the speed of sedimentation for several diameters, we have the following situation:

dia.	2 μ	V = 0.042 cm/sec
"	4 μ	V = 0.169 "
"	10 μ	V = 1.058 "
"	20 μ	V = 4.233 "
"	45 μ	V = 21.43 "

Therefore, the spreading distance from emission point, for each meter of vertical drop, depending on air speed which moves horizontally, is as follows:

DIAMETER OF PARTICLES (μ)	AIR SPEED (m/sec) 0	0.5	1	5	10
	Spreading Distances (meters)				
2	0	1190	2381	11905	23810
4	0	296	592	2959	5917
10	0	47	95	473	945
20	0	12	24	118	236
45	0	2	5	23	47

On the basis of analysis made with our dust-meter and through grain size control in the belt tunnel during ship loading, we may conclude that our interest is towards the behaviour of dust having diameter of 10 μ and less.

Method used to determine Primary Dust Content of Sandy Alumina

Basis of Method

A weighed sample of alumina is dropped through an orifice plate onto the base of a funnel fitted onto a conical diffuser, and from there onto a flat surface inside a closed chamber.

Part of the dust thus generated (primary dust content) settles onto a calibrated measuring cell. The resulting increase in cell weight constitutes the weighted primary dust content index (W.P.C.D.I.). The dust which has settled on the measuring cell is then analysed by CELLOSCOPE to determine its numerical primary dust content index (N.P.D.C.I.).

Field of Application

The method can be used with all sandy alumina samples and with all dust matter having diameter not greater than 1 millimeter and angle of repose not greater than 35°.

When matter has an angle of repose greater than 35° (such as floury alumina), measurement is difficult because the matter does not 'flow" with natural regularity through the orifice plate of the intake funnel.

Precision of method

5 percent.

Procedures

- Using a precision balance, weight out 250 grams +/- 0.1 grams of alumina taken by reduction of the whole sample by riffles.

- Assemble the dust meter apparatus, bearing the reference points in mind in order to guarantee the same relative position of the parts for each analysis (see attached figure 1 showing the dust meter and its dimensions).

- Using an analytical balance with tolerance range of +/- 0.0001 grams (P_1), weight the sedimentation cell which has been carefully cleaned beforehand.

- Position the sedimentation cell on its support bracket inside the dust meter.

- Using a suitable rubber plug, shut the funnel orifice plate.

- Pour 250 grams of weighed alumina into the funnel.

- Agitate the funnel to level its contents.

- Free the intake cone orifice by raising the rigid rod connected withe the rubber plug.

- Allow the sample to flow naturally through the orifice.

- Re-plug the orifice, start the timer.

- After 5 minutes have been timed, open the door provided (during all previous work, the door has been kept tightly shut).

- Remove the sedimentation cell.

- Brush-clean the outside surfaces of the sedimentation cell.

- Using an analytical balance with tolerance of +/- 0.0001 gram (P_2), weigh the sedimentation cell containing the dust.

- By means of electrolyte wash, transfer the settled dust inside the sedimentation cell into a special beaker for use with Celloscope; increase the total electrolyte volume to 100 ml.

- Analyze beaker contents using the Celloscope assembled with the 300 μ orifice, following the same general standard procedure used for hydrate of aluminum controls.

  The sole variation will consist in the procedure to correct the figures computed; therefore in response to the compute question: "FINAL DILUTED VOLUME", insert the result of the following equation:

  $$\text{FINAL DILUTED VOLUME (300 micron orifice)} = \frac{100}{(P_2-P_1) \times 1000} = \frac{0.1}{(P_2-P_1)}$$

- Record the following data produced by computer:

  - correct counts of particles greater than 15.79 μ (C + 15.79 expressed in counts/mg)

  - percentages in numbers at + 104, + 75, + 45, + 30, + 20 microns

  - total volume of particles greater than 15.79 microns (V + 15.79)

  - percentages in volume (weight) at +104, +75, +45, +30, +20 microns

- Repeat the dust meter analysis using a portion of the same sample, this in order to carry out the CELLOSCOPE analysis of settled dust using the 48 micron orifice.

- Transfer the dust from the sedimentation cell into a 200 ml flask, passing the dust through a 20 micron micro-sieve using electrolyte.

- In order to correct the computations, answer the computer's question: "FINAL DILUTED VOLUME" by inserting the result of the following equation;

$$\text{FINAL DILUTED VOLUME (48 micron orifice)} = \frac{200}{(P_2-P_1)\times 1000} = \frac{0.2}{(P_2-P_1)}$$

- Using the CELLOSCOPE, analyze the contents of the 200 ml flask using the 48 micron orifice.

- Record the following data produced by computer:

  - correct counts of particles smaller than 15.79 μ (C-15.79 expressed in counts/mg)

  - percentages in number to +15, +10, +5, +3, +2 microns

  - total volume of particles smaller than 15.79 μ (V-15.79)

  - percentages in volume (weight) to +15, +10, +5, +3, +2 microns.

Calculations

Weighted Primary Dust Content Index (W.P.D.C.I.)

$$\text{W.P.D.C.I.} = \overline{(P_2-P_1)} \times \frac{D_p^2}{D_C^2} \times \frac{1000}{250} = \overline{(P_2-P_1)} \times \frac{220^2}{39^2} \times \frac{1000}{250} =$$

$$= \overline{(P_2-P_1)} \times 127.28$$

Where:

D_p = diameter of dust meter (= 220 mm)

D_C = diameter of measuring cell (= 39 mm)

P_2 = weight of measuring cell, with dust, expressed in grams (average of two determinations)

P_1 = weight of measuring cell expressed in grams.

Numerical Primary Dust Content Index (N.P.D.C.I.)

$$N.P.D.C.I. = \left[C_{+15.79} \times (P_2 - P_1) + C_{-15.79} \times (P_2 - P_1)_2 \right] \times 1000 \times \frac{D_p^2}{D_C^2} \times \frac{1}{250 \times 10^6} =$$

$$= (C_{+15.79} + C_{-15.79}) \times (P_2 - P_1) \times 1000 \times \frac{220^2}{39^2} \times \frac{1}{250 \times 10^6} =$$

$$= (C_{+15.79} + C_{-15.79}) \times (P_2 - P_1) \times 1.273 \times 10^{-4}$$

Where:

$C_{+15.79}$ = correct counts obtained with 300 microns orifice plate

$C_{-15.79}$ = correct counts obtained with 48 microns orifice plate

Remark: the difference $(P_2 - P_1)$ of each of the two test must be used in combination with the relative value C±15.79 to have a more accurate result.

Grain Size Distribution by Number

- Normalizing factor (F.N. + 15.79)
  (300 micron orifice)

$$F.N._{+15.79} = \frac{C_{+15.79}}{\left[(C_{+15.79}) + (C_{-15.79}) \right]}$$

- Normalizing factor (F.N. -15.79)
  (48 micron orifice)

$$F.N._{-15.79} = \frac{C_{-15.79}}{\left[(C_{+15.79}) + (C_{-15.79}) \right]}$$

- Gran Size Distribution by Number

+ 104 μ = (percentage in the CELLOSCOPE + 104 μ) x F.N. +15.79

+ 75 μ = (" " + 75 μ) " "

+ 45 μ = (" " + 45 μ) " "

+ 30 μ = (" " + 30 μ " "

+ 20 μ = (" " + 20 μ) " "

$+ 15\mu = \left[(\text{percentages in the CELLOSCOPE} + 15\mu) \times F.N._{-15.79}\right] + F.N._{+15.79}$

$+ 10\mu = [($ " " $+ 10\mu)$ " "

$+ 5\mu = [($ " " $+ 5\mu)$ " "

$+ 3\mu = [($ " " $+ 3\mu)$ " "

$+ 2\mu = [($ " " $+ 2\mu)$ " "

Gran Size Distribution by Weight

- Normalizing factor ($FV_{+15.79}$)
  (300 micron orifice)

$$FV_{+15.79} = \frac{V_{+15.79}}{V_{+15.79} + V_{-15.79}}$$

- Normalizing factor ($FV_{-15.79}$)
  (48 micron orifice)

$$FV_{-15.79} = \frac{V_{-15.79}}{V_{+15.79} + V_{-15.79}}$$

- Gran Size Distribution by Weight

Use with a computation structure similar to the one described in gran size distribution by number but replacing:

- $F.N._{+15.79}$ with $FV_{+15.79}$

- $FN_{-15.79}$ with $FV_{-15.79}$

- the numerical percentages in the Celloscope with the volume (weight) percentages.

Apparatus

- Dust meter as per attached figure n° 1

- Precision balance; tolerance: 0.1 gram.

- Analytical balance: tolerance: 0.0001 gram.

- Soft bristle brush

FIG. 1 - DUST METER APPARATUS

(all dimensions in mm.)

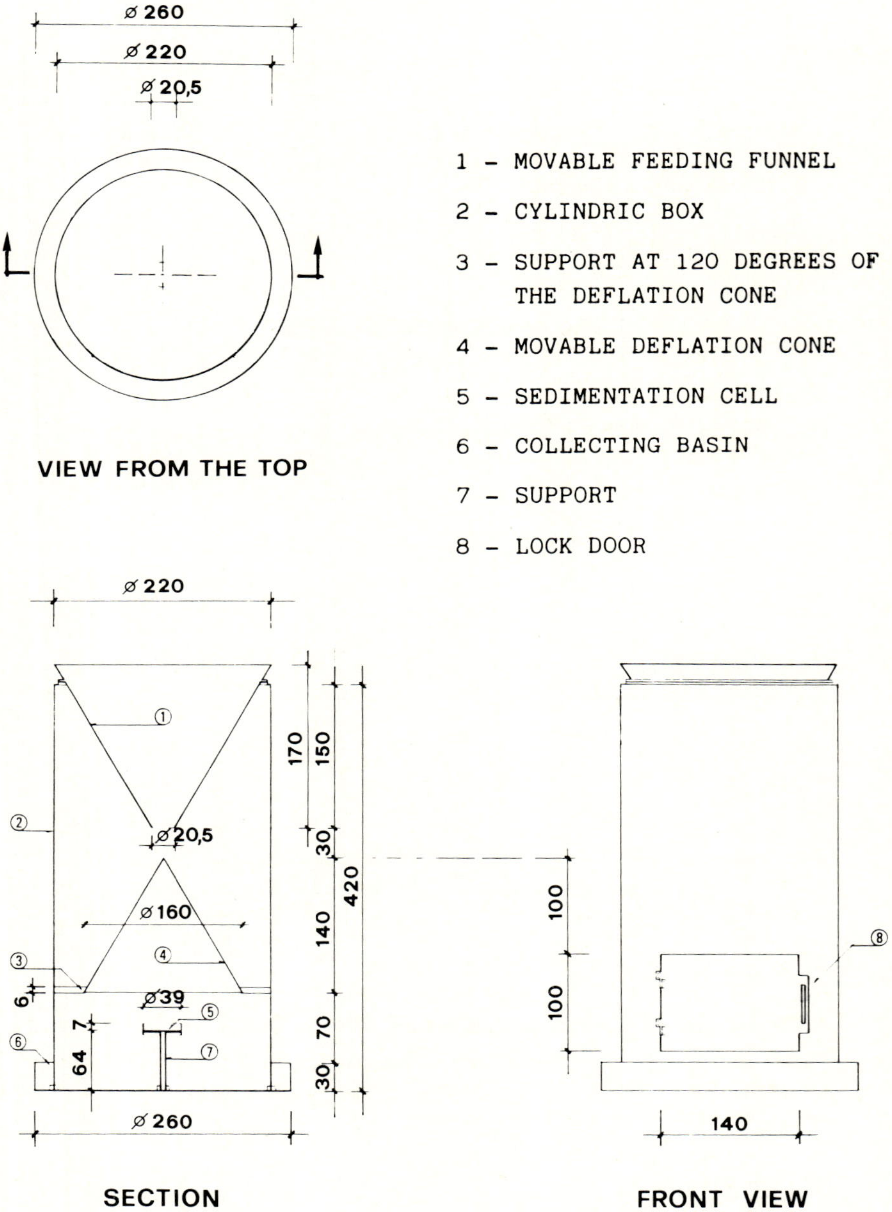

- Celloscope with 300 and 48 micron orifice, volume cross-sections: 200 and 5000 micron litre; beaker: 100 ml.
- Calibrated flask: 200 ml.
- 20 micron micro-sieve

Chemicals

- Distillate water
- Electrolyte for Celloscope (solution of trisodium phosphate in water with conductibility range of 16,500 to 20,000 microsimens/cm, filtered to 0.2 micron.
- Calibration materials for Celloscope: polystyrene micro-spheres; diameter : 10 and 60 microns.

Checking the Dust-Meter for Trouble-Free Operation

Reproducibility Check

The same sample (250 grams) was analyzed 10 times and gave the following W.P.D.C.I. results:

Test N°	W.P.D.C.I. (Kg/ton alumina)	
1	1.76	
2	1.85	
3	1.91	
4	1.93	W.P.D.C.I. = 1.80 av.
5	1.86	standard error =
6	1.78	=0.09 (5%)
7	1.76	
8	1.63	
9	1.76	
10	1.71	

W.P.D.C.I. and N.P.D.C.I. on basis of sample weight

Three series of tests (each series consisting of four tests) were made on another sample. 100, 250 and 500 grams of sample of alumina were used for each series, respectively.

Average test results were as follows:

Sample weight	W.P.D.C.I. Kg/T Al	N.P.D.C.I. million of particles/ gr Al
100 gr	1.03	1.64
250 gr	0.96	1.43
500 gr	1.37	1.52

As the data given in Table IV shows, in the 100 to 250 range, in addition to the relative consistency of the weight primary dust content index, there is also a well comparable grain size quality, both numerical and weighted.

With 500 grams of sample weight, the dust meter reacts with an increase in the weighted dust content index coupled with an increase of the granulometry of the deposited dust.

However, it is interesting to note that in spite of the increase in sample weight, the numerical dust content index does not increase appreciably, not even with 500 grams of sample; additionally, the numerical grain size distribution remains fairly constant.

There is a remarkable affinity between the quality of dust observed inside the dust meter and the dust wich naturally develops during the handling operations of our plant's alumina.

If we accept that the sample weight to be used in tests must be standard and such as to use the meter within its linear range (250 gr), then the weight of dust deposited in the measuring cell in function with sample weight varies according to the following law:

$$\text{Weight of deposited dust (g)} = A \times EXP^{B \times \text{(weight of sample)}}$$
$$(P_2 - P_1)$$

Where:

A is a constant wich varies from one alumina to another depending on its dust properties

B is a constant which depends on the instrument.

In the case reported above, A equals 2.14×10^{-3}, and B equals 4.676×10^{-3}.

Table IV. Pulvation Test on Different Quantities of the same Sample

Quantity taken from the sample	NPDCI million/gr	MILLION PARTICLES/mg			FV 48	FV 300	CUMULATIVE GRANULOMETRIC DISTRIBUTION (%)							
		N 48	N 300	NT			+70 μ	+45 μ	+30 μ	+20 μ	+10 μ	+5 μ	+3 μ	+2 μ
100 gr	1.64	1.583	0.012	1.595	0.992	0.008	0.000	0.01	0.112	0.47	5.89	33.11	72.86	95.06
250 gr	1.43	1.479	0.015	1.494	0.990	0.010	0.000	0.02	0.17	0.63	5.37	32.25	72.52	94.40
500 gr	1.52	1.095	0.016	1.991	0.986	0.014	0.001	0.03	0.30	0.98	6.87	33.03	71.17	94.40
	WPDCI Kg/t	μ^3/mg			FN 48	FN 300								
		V 48	V 300	VT										
100 gr	1.03	3.2665×10^8	1.91192×10^8	5.1784×10^8	0.631	0.369	0.96	4.37	17.29	31.53	64.17	91.55	98.78	99.91
250 gr	0.96	2.83238×10^8	2.51722×10^8	5.3496×10^8	0.529	0.471	0.3	5.93	23.66	41.28	68.18	92.28	98.91	99.90
500 gr	1.37	2.35287×10^8	3.19079×10^8	5.5437×10^8	0.424	0.576	1.00	8.72	32.36	52.26	77.48	94.67	99.17	99.93

References

1. Celloscope furnished by Particle Data, Inc., P.O. Box 265 Elmhurst, Ill - 60126.

2. SAX " Dangerous Properties of Industrial Materials" Radiological Sciences Laboratory, New York State Health Department, Albany N.Y. Third Edition - 1968 Page 389.

STUDY OF ALUMINA BEHAVIOR IN SMELTING PLANT STORAGE TANKS

H. P. Hsieh

Alcoa Laboratories
Aluminum Company of America
Alcoa Center, Pennsylvania 15069
U.S.A.

Field samples from inside the alumina storage tanks clearly indicate that significant size segregation can occur and the segregated fines can accumulate at the valleys between piles and at the periphery of a tank. This mode of segregation, caused by the airborne conditions, is a sharp contrast to the classical segregation. The segregated fines show profoundly different handling characteristics from the bulk of alumina particles. It is also demonstrated that other physical and chemical properties vary considerably among the segregated fractions. Possible mechanisms for the size segregation of alumina in storage tanks are discussed in terms of pulvation, secondary air currents, air slide feeding, etc.

Introduction

As the demand for better alumina handling and cell control is increasing, the need to optimize alumina characteristics becomes more urgent. It is then important that the handling properties of alumina be quantified and subsequently various mechanisms responsible for handling problems be studied to develop economic solutions. One of the most cited handling problems has been size segregation.

Size segregation of alumina can occur during belt conveying, pneumatic transport, etc., but most severely during filling of storage bins. Segregation of size fractions results in problems such as flushing or inability to convey, dustiness, irregular pot feeding and unbalanced distribution of impurities.

Most of the studies on particle segregation in the literature concern segregation due to sifting or percolation of fine particles through the voids formed by larger particles. In this mode of segregation, coarser particles tend to gather at the periphery of a storage tank. For this mode to be dominant, feeding of the powder into the storage tank is evenly distributed or the particles are large enough not to be affected by turbulent air flow inside the tank. When there is significant airborne movement, interaction of a cloud of fine particles with the surrounding air cannot be assumed to be that of individual particles and segregation can occur by a different mechanism (1).

Size segregation of alumina in storage tanks has been widely known to exist in many systems. The extent of the problem varies and depends on many factors. However, there has been only one published study of size segregation inside alumina storage tanks. Connally recently reported tests conducted on silos at an alumina plant (2). He attributed the slight segregation of coarse particles at the edge of a tank to fine particles sifting as the predominant mechanism. The maximum segregation mentioned in that study was represented by only 3% -45 μm due to multiple feeding ports used.

It is the purpose of this study to report on field samples collected from alumina storage bins of various configurations and infer the possible segregation mechanisms inside the bins. Segregated fractions of alumina will be characterized for handling behavior and other properties potentially affecting smelting operations.

Experimental

Aluminas used in Alcoa smelting plants originate from within as well as outside the company through exchange arrangements. The configuration of storage systems at smelting plants also varies. For a preliminary investigation of the nature of segregation phenomenon, two smelting plants were selected for sampling. Storage tanks of various configurations at these plants were studied. The configurations of the tanks tested are given in Figure 1. Some tanks were sampled once. Others were sampled twice at two different stages of a filling cycle so that two levels of alumina piles were sampled.

Samples were collected at different positions inside the tanks using a 0.9-m long powder sampler with three large openings across the length.

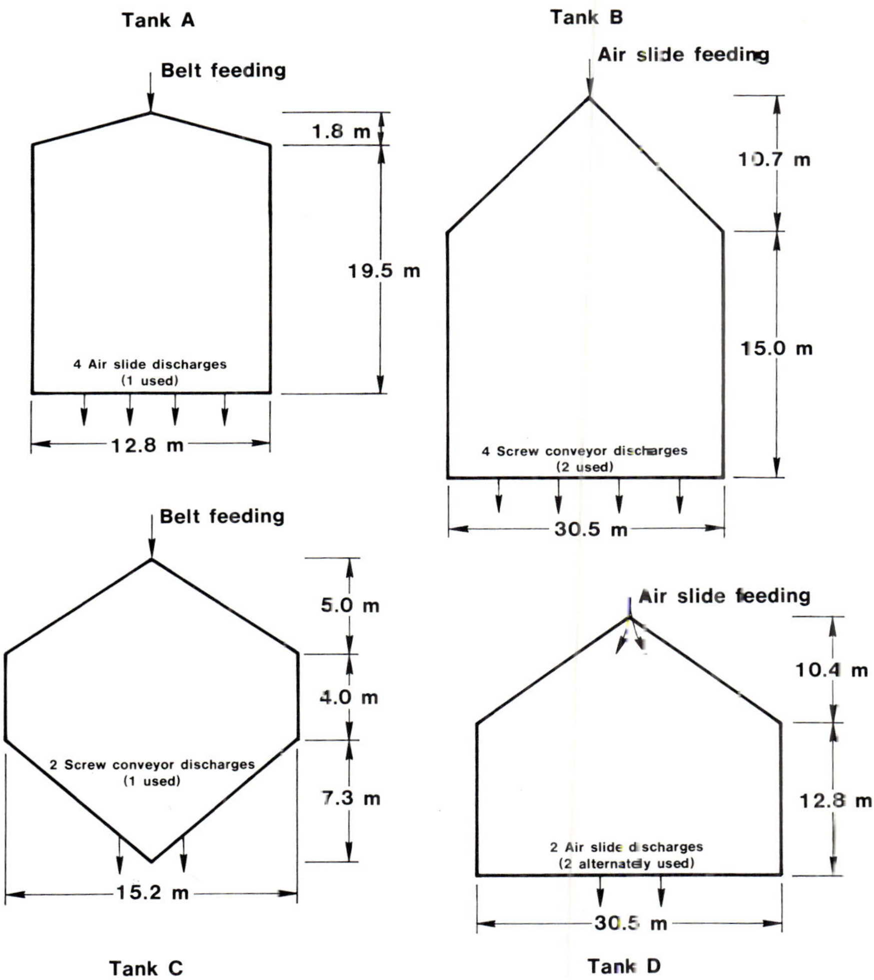

Figure 1 - Configurations of storage tanks studied

The sampler was driven into the pile beneath the surface to obtain a composite sample in the vertical direction. For Tanks C and D, samples were also collected from the top layers of piles within 2 cm of the surface.

The alumina samples thus obtained were than analyzed for particle size distribution by a Microtrac Particle Size Analyzer (Leeds & Northrup). The Powder Characteristics Tester (by Hosokawa Micromeritics Laboratory, Japan) was used to determine various handling properties such as angles of flow (repose, fall, and spatula), bulk densities, compressibility and cohesion. The tester was designed based on the test methods published by R. L. Carr, Jr. (3). A flow funnel test was also conducted in a 10-degree copper cone with a 0.24-cm orifice at the bottom and a 100-gram sample.

Results and Discussion

Size Segregation

For larger tanks (30.5 m in diameter) with air slide feeding, there is size segregation the degree of which depends on the feeding and discharge configuration. But in all cases, the segregation seems to occur at the low points (or "valleys") between piles of alumina and near the periphery of tanks where alumina particles were subject to least amount of disturbance and where finer particles or dust would settle and accumulate.

Listed in Table I is data for Tank A (12.8 m in diameter, fed by a conveyor belt and sampled at positions given in Figure 2) which did not show any segregation at any sampling points in the single pile. The particle size median was essentially the same everywhere sampled. In contrast, samples from Tank B (30.5 m in diameter and fed by an air slide) did show segregation with the same shipment of alumina as in the case of Tank A. Tank B was sampled twice: once midway of the filling cycle and secondly after the filling was completed. It is noted that valleys and multiple piles formed as a result of the feeding and discharge configuration. Both times, as shown in Table I, significant segregation occurred at the valleys between large and small piles and near the wall, but not on the slope of the piles. The bulk of the material in Tank B had a size median of 95-102 μm; however, portions with a size median as small as 18 μm could be found in the least disturbed low areas. The segregated fines had much higher percentages of -44 μm and -11 μm (39-82% and 16-33%, respectively) compared to the bulk material (9-16% and 2-5%, respectively).

Similar observations, although not as pronounced, were made in Tank C and Tank D at a separate smelting plant. A single pile was present in the smaller Tank C (15-m diameter and belt fed) while two large piles were observed in Tank D (30-m diameter and fed by an air slide). Because Tank D could not be safely entered, no samples could be collected from the "valleys" between piles. For both tanks, two rows of samples were collected along two slopes of the same pile representing two different angles relative to the manhole ladder. As shown in Table II, data from the collected samples (see Figure 3) indicates that practically no segregation was present in Tank C. Segregation of fines, even though not as severe as in Tank B, did exist at the wall in Tank D with the fines having a size median of 83 μm versus 88-91 μm in the bulk. Again, samples

TABLE I. PARTICLE SIZE MEDIANS OF ALUMINA SAMPLES FROM TANK A AND TANK B IN PLANT NO. 1

Sample Designation	Tank Designation	Distance From Wall (m)	Radial Angle From Manhole (°)	Particle Size Median (μm)	Remark
A-1	A	0	0	102	At the manhole ladder
A-2	A	3.2	0	100	
A-3	A	6.4	0	101	At top of the main pile and center of the cross-section.
B-1*	B	0	0	94.7	At the manhole ladder.
B-2*	B	3.0	0	33.5	At a "valley".
B-3*	B	7.0	0	100	
B-4*	B	10.7	0	99.9	
B-5*	B	0	25	17.5	At the wall.
B-6+	B	0	0	73.4	At the manhole ladder.
B-7+	B	3.0	0	97.8	
B-8+	B	7.0	0	102	
B-9+	B	10.7	0	98.5	
B-10+	B	1.0	25	99.3	At top of a small "ridge".

NOTE:

*Collected at midway of a filling cycle.
+Collected at end of a filling cycle.

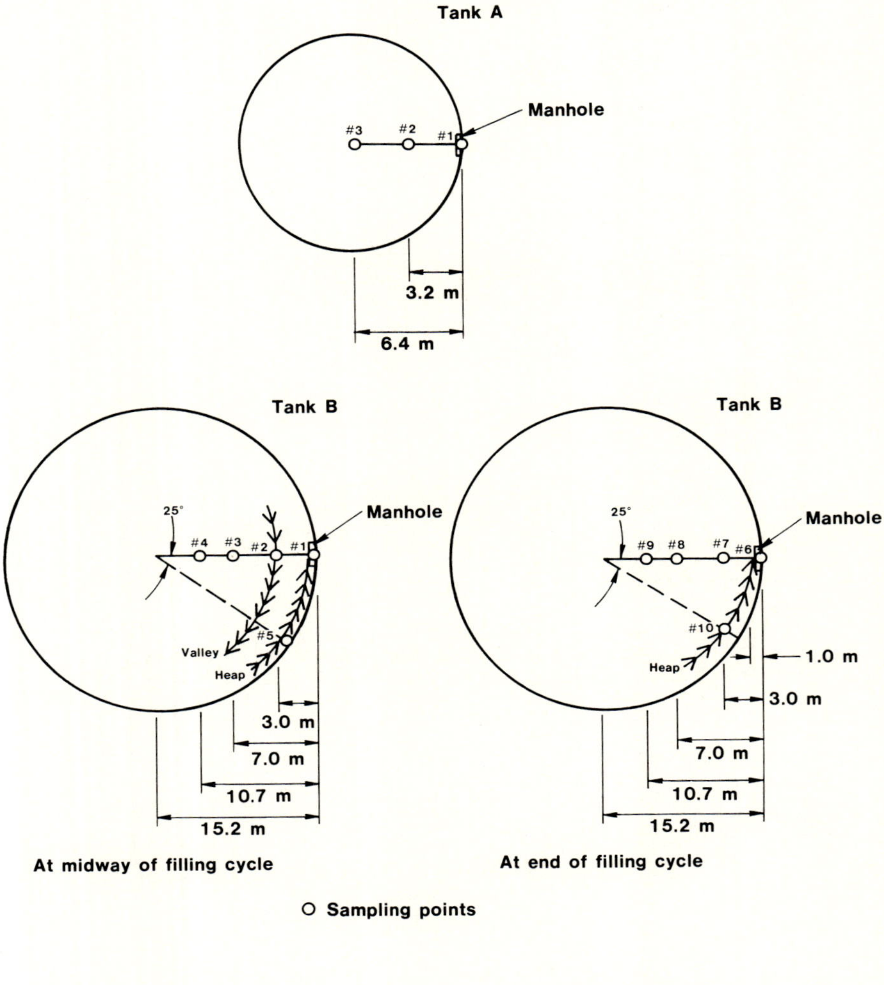

Figure 2 - Sampling positions inside Tank A and Tank B

TABLE II. PARTICLE SIZE MEDIANS OF ALUMINA SAMPLES FROM TANK C AND TANK D IN PLANT NO. 2

Sample Designation	Tank Designation	Distance From Wall (m)	Radial Angle From Manhole (°)	Particle Size Median (μm)	Remark
C-1	C	0.3	0	85.1	Near the wall.
C-2	C	2.8	0	86.7	
C-3	C	5.8	0	88.7	
C-4	C	7.5	0	87.2	Near top of the pile.
C-5	C	0.3	90	87.7	Near the wall.
C-6	C	2.8	90	86.8	
C-7	C	5.8	90	89.9	
C-8	C	7.5	90	88.2	Near top of the pile.
D-1	D	0.3	~25	82.9	Near the wall.
D-2	D	2.8	~25	89.0	
D-3	D	5.8	~25	88.3	
D-4	D	9.1	~25	91.3	
D-5	D	10.5	~25	91.0	At top of a pile.
D-6	D	0.3	~65	83.1	Near the wall.
D-7	D	2.8	~65	90.3	
D-8	D	5.8	~65	90.6	
D-9	D	9.1	~65	89.2	

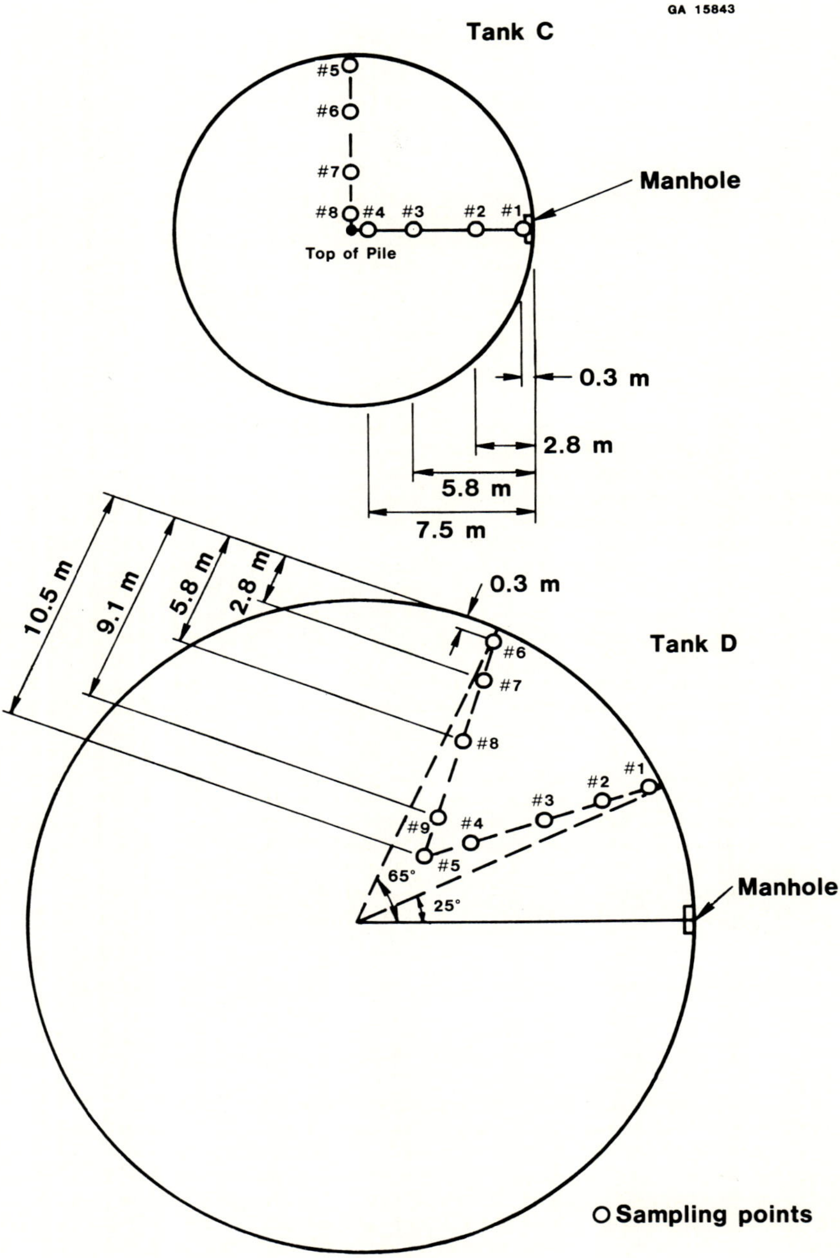

Figure 3 - Sampling positions inside Tank C and Tank D

along the slopes and at the top of piles were approximately the same in size.

The samples described above were collected beneath the pile surface as composite samples in the vertical direction. During sample collection from Tank C and Tank D, samples were also obtained from the pile surface (less than 2 cm deep). The results for Tank D are summarized in Table III. In all cases, the surface samples were finer than the corresponding samples collected beneath the surface. It appears that the finer particles resting on the pile surface after filling is completed can be airborne again and redispersed in the air upon the next vigorous air movement (such as filling) inside the tank. Fines eventually become accumulated in areas that are least disturbed by the air currents in the tank, namely, the wall and the low points between piles.

Handling Behavior

Quantifying handling behavior is an essential step in solving handling problems and predicting relative handling characteristics of a given material/equipment combination. We have previously studied other alumina samples actually used in smelting plants and compared bench test data with the field responses. A wide variety of samples with "good" or "poor" handleability have been examined. Some results are summarized in Table IV. It is apparent that in general the angle of spatula, compressibility, cohesion and flow funnel time are relatively good indicators for handling characteristics in practice. Particularly the flow funnel time seems to closely relate to the general handleability.

The aforementioned handling properties for the alumina samples collected from Tank A and Tank B are listed in Table V for illustrating the impact of size segregation. The angle of repose data is also included as it is a commonly referenced property, even though not very well related to practical general handleability. The angle of repose for the normal particle size distribution fractions is about 36-38° while for the segregated fines in Tank B it is significantly higher, 43-49°. The angle of spatula gives a relative angle of internal friction or angle of rupture for a dry material. Normal particles show an angle of spatula of 42-47° while segregated fines show 53-59°. It clearly indicates the poor flowability of segregated fines as they are discharged from the storage tanks.

To determine compressibility, one needs the bulk densities as follows:

$$C = 100\ (P-L)/P \qquad (1)$$

where L and P represent the loose and the packed bulk densities, respectively. The normal bulk particles have a loose and packed bulk density of 0.98-0.99 and 1.19-1.24 g/cm^3, respectively. For the segregated fines, the corresponding values are 0.82-0.89 and 1.33-1.41 g/cm^3 with the only exception of the slightly segregated sample B-6 from Tank B. The segregated fines have lower air permeability than the normal particles. Consequently once entrapped during handling, air cannot be easily released from the segregated fines, under the low-stress at-rest conditions, thus leading to a lower loose bulk density. When subject to vibration as used in the packed bulk density measurement, the deaerated

TABLE III. COMPARISON OF SURFACE AND COMPOSITE SAMPLES FROM TANK D IN PLANT NO. 2

Sample Designation	Method of Sampling[(1)]	Distance from Wall (m)	Radial Angle from Manhole (°)	Particle Size Median (μm)
D-1S	surface	0.3	~25	82.2
D-1	composite	0.3	~25	82.9
D-2S	surface	2.8	~25	81.7
D-2	composite	2.8	~25	89.0
D-3S	surface	5.8	~25	81.9
D-3	composite	5.8	~25	88.3
D-4S	surface	9.1	~25	81.0
D-4	composite	9.1	~25	91.3
D-5S	surface	10.5	~25	80.1
D-5	composite	10.5	~25	91.0
D-6S	surface	0.3	~65	74.3
D-6	composite	0.3	~65	83.1
D-7S	surface	2.8	~65	80.4
D-7	composite	2.8	~65	90.3
D-8S	surface	5.8	~65	84.5
D-8	composite	5.8	~65	90.6
D-9S	surface	9.1	~65	80.9
D-9	composite	9.1	~65	89.2

NOTE:

(1) Surface: sample collected less than 2 cm deep from surface.
Composite: sample collected as a composite over a distance of 0.9 m below surface.

TABLE IV. COMPARISON OF ALUMINA HANDLING PROPERTIES BY BENCH TESTS AND BY FIELD RESPONSES

Sample No.	S1	S2	S3	S4	S5	S6	S7	S8
Handling Characteristics[(1)]	good	good	good	good	good	poor	poor	poor
Angle of Repose (°)	37	31	35	32	35	39	36	41
Angle of Fall (°)	21	18	21	16	20	21	23	26
Angle of Difference (°)	16	13	14	16	15	18	13	15
Angle of Spatula (°)	45	43	46	39	37	46	47	52
Loose Bulk Density (g/cm^3)	1.05	1.02	1.04	1.02	1.01	1.02	0.97	1.03
Packed Bulk Density (g/cm^3)	1.30	1.24	1.25	1.20	1.19	1.31	1.19	1.30
Compressibility (%)	19.2	17.7	16.8	15.0	15.1	22.1	18.2	20.3
Cohesion (%)	18	14	10	5	4	34	15	20
Flow Funnel Time (min.)	5.6	5.2	7.2	4.4	4.5	13.5	8.1	12.4
Attrition Index (%)	14	12	11	19	14	9	14	13

NOTE:

(1) Field responses (remarks made by smelting personnel).

TABLE V. BULK HANDLING PROPERTIES OF SEGREGATED ALUMINA FRACTIONS IN TANK A AND TANK B

Sample Designation	Size Median (μm)	Angle of Repose (°)	Angle of Spatula (°)	Loose Bulk Density (g/cm^3)	Packed Bulk Density (g/cm^3)	Compressibility (%)	Cohesion (%)	Flow(1) Funnel Time (min.)
A-1	102	36	43	0.99	1.20	17.5	16.8	5.2
A-2	100	37	42	0.99	1.19	16.8	16.2	4.5
A-3	101	37	42	0.98	1.19	17.6	19.5	4.7
B-8	102	37	43	0.99	1.20	17.5	24.0	5.2
B-3	100	37	47	0.99	1.21	18.2	21.3	6.7
B-4	100	38	46	0.99	1.21	18.2	24.5	6.3
B-10	99.3	38	46	0.99	1.24	20.0	16.4	7.5
B-9	98.5	38	44	0.99	1.21	18.0	23.1	6.0
B-7	97.8	37	47	0.99	1.21	18.0	18.8	6.1
B-1	94.7	40	51	0.99	1.24	20.2	21.5	8.6
B-6	73.4	43	53	0.99	1.33	26.0	25.8	*
B-2	33.5	48	54	0.89	1.40	36.0	44.4	*
B-5	17.5	47	57	0.82	1.35	39.3	42.0	*

NOTE: (1)*flow funnel time indicating that sample would not flow through the funnel.

segregated fines achieve a higher packed bulk density due to better particle packing than the normal bulk particles. Packing volume fraction has been calculated from the particle size distribution data for the samples according to a mechanical packing model proposed by Lee (4). Table VI shows that the segregated fines have higher values of packing volume fraction than normal bulk particles, thus having higher packed bulk densities.

It has been suggested that when the compressibility is above 20%, the material would not be free-flowing. Particularly, when the compressibility is close to 40% or higher, it becomes very difficult to discharge the material from a hopper once the material has been stored in it for an extended period of time. The normal particles have a compressibility of 17-20% while the segregated fines have 26-39% as indicated in Table V.

The cohesion is a direct measurement of the amount of energy required to pull apart aggregates of cohesive particles in a specified time. A powder has less flowability when it has a high cohesion. Segregated fines have a cohesion greater than 25%, while normal bulk particles have a lower value (Table V).

The flow funnel test provides a simple and quick method of accessing the relative flow property of a powder, particularly suitable for implementation in the field. It is apparent from Table V that all the segregated fines smaller than 75 μm in size median would not flow through the funnel and the normal bulk particles flow through in less than 7.5 minutes.

Chemical Properties

Most of the samples collected from Tank A and Tank B were analyzed for LOI, water adsorption at 44% relative humidity, alpha-alumina and impurities. The impurity of particular interest and significantly different between the segregated fractions is silica. As apparent from Table VII, the segregated fines have appreciably higher values of LOI, water adsorption, alpha-alumina and silica. This was confirmed in a separate, more detailed study by analyzing ten narrow cut fractions (ranging from 4 to 127 μm in size median) prepared from a given batch of alumina. The surface area was found in that study to increase with the size median as a result of lower amounts of alpha-alumina at higher size medians.

Proposed Segregation Mechanisms in Alumina Storage Tanks

Fine particles move about in a storage tank as a result of the motion of the air in which they are suspended particularly during filling. Basically there are three modes of air motion inside a storage tank that can disperse the material: pulvation, secondary air currents by induction and through exhaust or openings (5). Pulvation occurs when a mass of falling alumina strikes existing piles in the tank, thus suddenly compacting the loose bulk material with a violent escape of air which carries fine particles with it by the shearing force. The fine particles then become airborne. The pulvation in this case is multi-directional.

TABLE VI. PACKED BULK DENSITY AND CALCULATED PACKING VOLUME FRACTION OF ALUMINA SAMPLES FROM TANK A AND TANK B

Sample Designation	Packed Bulk Density (g/cm^3)	Packing Volume Fraction
A-1	1.20	0.701
A-2	1.19	0.696
A-3	1.19	0.695
B-1	1.24	0.726
B-2	1.40	0.778
B-3	1.21	0.712
B-4	1.21	0.718
B-5	1.35	0.739
B-6	1.33	0.778
B-7	1.21	0.717
B-8	1.20	0.700
B-9	1.21	0.689
B-10	1.24	0.716

TABLE VII. CHEMICAL CHARACTERISTICS OF SEGREGATED ALUMINA FRACTIONS IN TANK A AND TANK B

Sample Designation	Size Median (μm)	LOI (%)[1]	Water[2] Adsorption (%)	α-Al_2O_3 (%)	SiO_2 (%)
A-1	102	0.61	2.89	7.3	0.018
A-3	101	0.56	2.93	7.2	0.018
B-1	94.7	0.57	2.97	8.5	0.018
B-2	33.5	0.89	4.01	23	0.021
B-4	100	0.64	2.97	8.4	0.017
B-5	17.5	0.91	4.17	27	0.021
B-6	73.4	0.71	3.32	14	0.018
B-10	99.3	0.62	2.94	8.4	0.017

NOTE:

(1) Loss on ignition (300-1200°C)
(2) Water adsorption at 44% relative humidity.

Induction occurs when a mass of alumina falls with each particle imparting some momentum to the surrounding air. This results in the induction of an air stream along with the alumina. The induced air flow moves some fine particles away from the column of alumina into the air inside the storage tank. Secondary air currents can also come from exhaust or openings. Both secondary air flows, by induction and through exhaust or openings, transport the localized dusty air, mostly formed by pulvation, away from the site of formation.

It appears from the field data that many airborne fine particles, carried around by the aforementioned secondary air currents, settle and accumulate in areas subject to least disturbance from air currents during filling, namely the areas near the periphery of a tank or "valleys" between piles. When filling or discharging stops, the remaining airborne fine particles will fall on the pile surface rather uniformly. Some of these fine particles may sift through voids between coarse particles and the remaining fine particles on the surface can become airborne again when there is sufficient air current inside the tank. Therefore fine particles may keep accumulating and concentrating in sizable zones, particularly near the wall. When the withdrawal from the tank is sufficient to include these peripheral and "valley" zones, slugs of fines are in the discharge.

As stated earlier, a potentially primary driving force for size segregation in a storage tank is the air currents inside. The secondary air currents carry the airborne particles away from the center into outer areas. To illustrate the significance of induced air flow in affecting the air movement and consequently the size segregation of alumina in a storage tank, an estimate was made based on a realistic set of conditions for Tank B: 120 t/hr alumina feed rate, filling cross section of 0.29 m^2 and a distance of fall of 10 m. According to an equation developed by Hemeon (5), the rate of the induced air flow for particles falling at terminal velocity is given by

$$q = 16.57 \sqrt[3]{R\,(\Delta S) A^2} \tag{2}$$

where q = induced air flow, m^3/min.
R = solid feed rate, t/hr
A = cross-sectional area of stream of particles, m^2
ΔS = distance of fall by particles, m

The calculated induced air flow for Tank B under the conditions specified above is 77 m^3/min. (or 4620 m^3/hr). The rate thus calculated tends to be high when compared with field data due to the underlying assumptions. Nevertheless, the calculated result can serve as a good indication. The fact that the calculated induced air flow can be about an order of magnitude higher than other combined air flows through the storage tank illustrates the importance of induced air flow.

A comparison of the magnitude of induced air flow rate was made between Tank A and Tank B in Plant No. 1. A typical set of conditions for Tank A is: 120 t/hr alumina feed rate, filling cross section of 0.073 m^2 and a distance of fall of 7 m. Applying Equation (2) results in a calculated induced air flow of 27 m^3/min. which is only about 1/3 of that for Tank B. This may partially explain why segregation occurred in Tank B but not in Tank A.

Another potentially important factor affecting size segregation is the mode of filling the storage tank. In a separate study, it was observed that air slide feeding has the tendency to separate the fine particles from the coarser ones even before the alumina particles reach the storage tank. The easy-to-separate particles, once transferred into the tank from the air slide, can easily segregate into fractions under the influence of primary pulvation and secondary air currents, especially the latter. In contrast, feeding the storage tanks by conveyor belt does not appear to promote particle separation. A more definitive study of the relative effects of air slide feeding and secondary air current (such as induced air flow) on size segregation in an alumina storage tank is planned to be conducted under controlled conditions in a pilot scale test using a given size tank.

Three of the four tanks cited in this study have single-entry feeding and one uses a splitter to generate two feeding streams. One approach to reduce size segregation is changing the storage system configuration. Well known methods in this category include multiple-entry feeding (2), multiple discharge points, use of a cone under the entry point to spread the particles more evenly to various positions, the use of an internal spreader pipe, placing a cone-shaped obstruction above the discharge orifice, the use of baffles, etc. Another approach is changing the particle size distribution of the alumina.

Conclusions

Size segregation of alumina particles in storage tanks has been widely known to exist but not been explored in a systematic fashion. This study analyzed the problem by collecting field samples from inside tanks of various configurations and characterizing the samples for important handling properties. Two case studies were made and, in both cases, segregation occurred in large tanks (30 m in diameter) with air slide feeding but not in smaller tanks (12.8-15 m in diameter) with conveyor belt feeding. Segregated fines were formed in "valleys" between piles and at the periphery of the tank while samples collected at other points along the slopes of piles had normal particle size distribution. The segregated fractions exhibited markedly different handling properties such as angles of flow (repose and spatula), compressibility, cohesion and flow funnel time. The fines had appreciably higher values of LOI, water adsorption, alpha-alumina and silica. The field data appears to indicate that fine alumina particles are separated from the bulk when an air slide is used to feed to the storage tanks. As the alumina feed strikes existing piles in a tank, pulvation makes the fine particles become airborne. Secondary air currents, particularly the induced air flow caused by the falling streams of alumina, carry the fines away from the pulvation sites. Fine particles can eventually settle and accumulate in areas subject to least disturbance from air currents, namely the areas near the periphery of a tank or "valleys" between piles. A more definitive study of the relative effects of air slide feeding and secondary air flow on size segregation in an alumina storage tank is needed and will be conducted under controlled conditions using a given size tank.

References

1. J. C. William, "The Segregation of Particulate Materials. A Review," Powder Technology, 15 (1976), pp. 245-251.

2. K. E. Connally, "Particle Segregation in Alumina Silos," Light Metals (1983), pp. 241-252.

3. R. L. Carr, Jr., "Evaluating Flow Properties of Solids," Chem. Eng. (1965 January 18), pp. 163-168.

4. D. I. Lee, "Packing of Spheres and Its Effect on the Viscosity of Suspension," J. Paint Technology, 42 (1970), pp. 579-587.

5. W. C. L. Hemeon, Plant and Process Ventilation, 2nd ed., pp. 46-157, The Industrial Press, Inc., New York, N. Y., 1963.

ALUMINA AND BAUXITE V

Plant Design and Equipment

CHAIRMEN

B.J. Foster
Kaiser Aluminum and Chemical Corp.
Gramercy, Louisiana

F.N. Newchurch
Reynolds Metals Company
Little Rock, Arkansas

CORROSION IN CAUSTIC ENVIRONMENT

APPLICATION TO ALUMINA PLANTS

M. V. Chaubal

Alumina Division Technology
Reynolds Metals Company
Union National Plaza
Little Rock, Arkansas 72201
USA

This study is a result of a comprehensive literature survey on corrosion of mild steel, nickel and nickel alloys in caustic environment as it relates to alumina plants. The literature indicates that the primary type of corrosion in caustic environment is stress corrosion cracking and that presence of sulfide and other oxidizable sulfur species enhance the corrosion propensity of materials of construction used in alumina plants. Available data suggest that presence of alumina in caustic solution permit operation at higher temperature than normally accepted in caustic plants. A failure-nonfailure criteria is suggested and should be tested as more data become available. Literature suggests that stress relieving of welds and minimizing concentration of oxidizable sulfur species, reduces probability of failure by stress corrosion cracking.

Introduction

This study is a result of a comprehensive survey of literature available on the subject of corrosion of mild steel, nickel and high nickel alloys in caustic environment. Extensive literature is available on corrosion in caustic solutions as experienced in pulp and paper industry, while very limited information exists on corrosion in aluminate solutions. However, the available information can be merged to understand corrosion of mild steel, nickel and high nickel alloys in alumina plants.

The results of literature survey reported herein firstly identifies the type of corrosion in caustic environment; secondly, it explains the possible mechanisms of corrosion and thirdly, it explains how the corrosion can be minimized.

It is emphasized again that this report is not a complete treatise on corrosion in alumina plants, but a comprehensive literature survey.

Discussion

Metals and alloys have a tendency to react with surrounding media to form stable compounds. Such reactions leading to metal loss are termed as corrosion. The science of corrosion encompasses a very broad field and has therefore been segmented by the type of corrosion attack into the following categories(1): (1) intergranular corrosion, (2) pitting corrosion, (3) stress corrosion cracking, (4) galvanic corrosion, (5) crevice corrosion, (6) oxygen concentration cell (7) corrosion fatigue, (8) cavitation corrosion, (9) fretting corrosion and (10) hydrogen embrittlement. These corrosion processes can occur in a piece of equipment constructed of mild steel, nickel or nickel alloys. However, the literature indicates that only a few of these are important sources of corrosion related failures in caustic environment and only these will be addressed in that which follows.

Types of Corrosion Applicable to Caustic Environment

Available literature indicates that the major type of corrosion seen in caustic environment is stress corrosion cracking also termed as caustic embrittlement or caustic cracking. Hydrogen embrittlement has also been mentioned in caustic environment. Thus it is necessary to understand first, these two types of corrosion attacks.

Stress Corrosion Cracking. Shreir states(2) that "occurrence of stress corrosion in a susceptible material requires the conjoint action of a tensile stress and a dissolution action." In other words, the mechanism of failure by stress corrosion involves a reaction of material with its environment and the presence of tensile stress, which by themselves could not have led to failure. A simple example can be described as follows: A chemical reaction occurs between a material and its environment. Reaction products are formed (normally metal oxides) which deposit at the reaction site and thus prevent further reaction. However, due to the presence of tensile stress, this protective film fractures and makes available fresh metal surface for further reaction. Then the cycle starts again. This is schematically shown in Figure 1. This process leads to crack propagation and eventual failure. Thus, both a dissolution reaction and tensile stress are required for failure by stress corrosion.

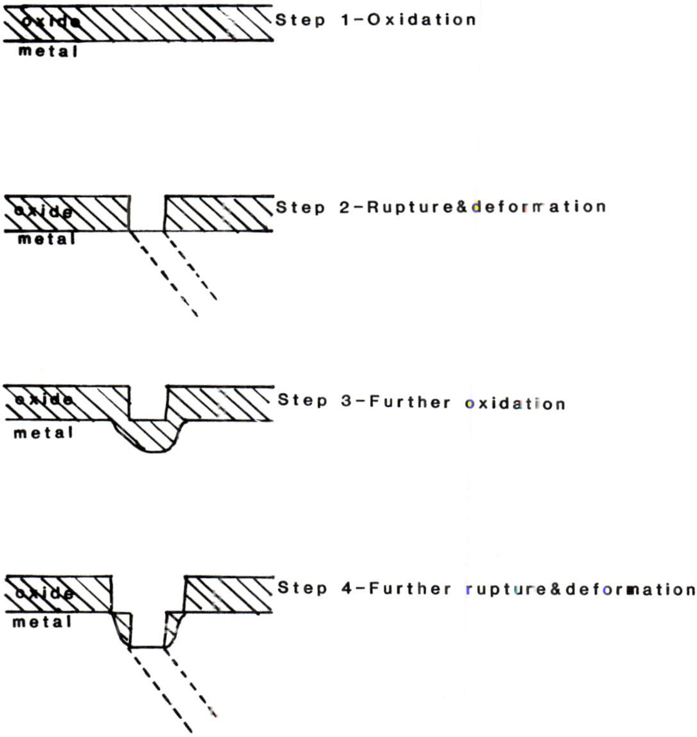

Figure 1: Strain Generated Active Path Mechanism for Stress Corrosion Cracking - A Film Rupture Model (2)

Champion[3] reports that caustic embrittlement, commonly known as stress corrosion, does not seem to occur in the absence of stress. He states further that corrosion cracks are intercrystalline. He identifies four criteria that serve as evidence of caustic embrittlement and they are: intercrystalline cracking, crack width decreasing with penetration to a fine point, crack direction is transverse to the stress and history of metal being exposed to high concentration caustic.

Hydrogen Embrittlement. In certain cases, the chemical reactions occurring between the metal and its environment can produce elemental hydrogen. Depending on the system, this elemental hydrogen can be either adsorbed into the metal lattice or form a metal hydride. Either of these can lead to loss of ductility in a crack and crack propagation due to internal stress can occur; leading to failure. Thus, this also is a form of stress corrosion cracking. Singbeil and Tromans[4] have shown that in caustic solutions, it is not thermodynamically possible to achieve the reversible electrode potential necessary for hydrogen evolution. Therefore, the possibility of hydrogen embrittlement leading to failure is very unlikely.

Thus, in dealing with corrosion of mild steel, nickel, and nickel alloys in caustic environment, it is necessary only to consider stress corrosion as a mode of failure of metal.

Mechanisms of Corrosion. The material of construction predominantly used in alumina plants is mild steel, but in some instances nickel and high nickel alloys have been used. Therefore, the discussion necessarily needs to deal with both these materials.

There exists a consensus amongst the various investigators that mild steel corrosion in caustic environment is essentially stress corrosion cracking[3,4,5,6,7,8,9,10,11,12,13,14,15]. These investigations have essentially tried to answer the questions why corrosion occurs and how various constituents in caustic solutions affect the rate of corrosion. The following is a summary of these investigations.

The evidence for caustic embrittlement (caustic cracking or stress corrosion cracking in caustic environment) can be isolated in either the pulp and paper industry or the alumina industry. Literature available in the pulp and paper industry is vast, while that for the alumina industry is very limited. However, parallels exist in the corrosive behavior of mild steel in both these applications and therefore an understanding of corrosion in aluminate solutions can be arrived at logically.

The investigations in 1953 for the pulp and paper industry found that the steel corrosion rate "increases rapidly with increasing concentrations of Na_2S, $Na_2S_2O_3$ and NaOH"[10]. Mueller's[10] data indicate that the liquor corrosivity is proportional to the thiosulfate concentration and this relationship is shown in Figure 2. Laliberte[11] indicates that mild steel corrosion rate is proportional to the product of concentrations of (S^{-2}) and ($S_2O_3^{-2}$) and he believes that the specie (S_2^{-2}) is responsible for corrosion. Laliberte explains his data, namely corrosion rate being proportioned to product of concentrations of S^{-2} and $S_2O_3^{-2}$ species, by the following reaction:

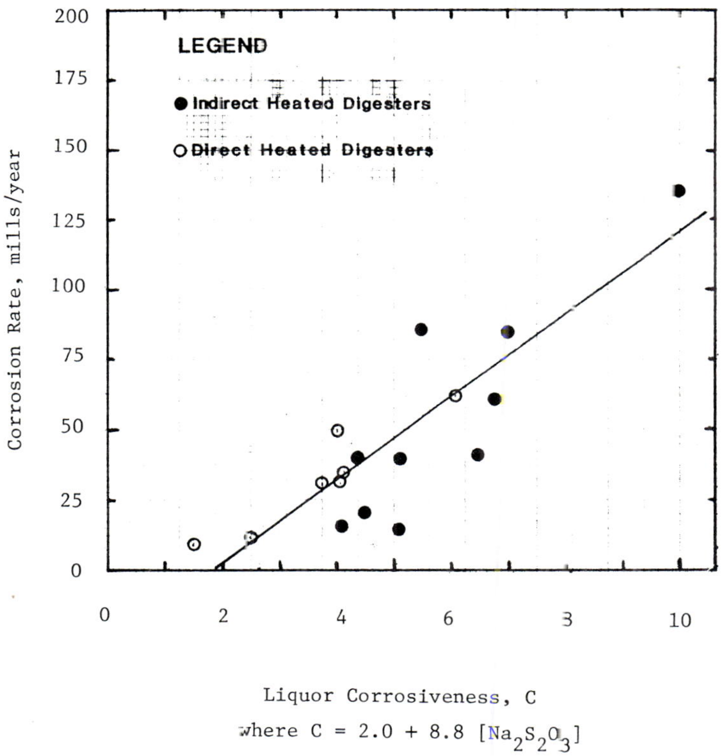

Figure 2: Effect of Liquor Corrosiveness on Corrosion Rate in Pulp and Paper Plants [10]

$S^{-2}+S_2O_3^{-2} \longrightarrow S_2^{-2}+SO_3^{-2}$

Higher the $[S^{-2}]*[S_2O_3^{-2}]$, higher the concentration of $[S_2^{-2}]$, a highly oxidizing specie, higher the corrosion rate. Laliberte further postulates that if partial oxidation of S^{-2} in solution to $S_2O_3^{-2}$ occurs, then higher corrosion rate will result.

A number of electrochemical studies by Singbeil and Tromans(4), Wensley and Charlton(9,16) and Staehle(17) have shown that the presence of oxidizable sulfur species, like (S^{-2}), (S_2^{-2}), lead to higher peak current densities at corrosion potential. This indicates a higher corrosion rate when compared to solutions without the presence of these sulfur compounds. The slow strain rate technique used by Singbeil and Tromans(4) also showed that a higher percent reduction in area occurred in the presence of sulfide. This indicates a higher susceptibility to stress corrosion failure. These data are shown in Table I.

Wenseley and Charlton(9,16) conducted electrochemical investigations to determine the effect of caustic concentration and also the effect of sulfide and thiosulfate in caustic solutions. Their data is presented in Figures 3 and 4. These data lead to the conclusion that susceptibility of stress corrosion cracking increases with increasing caustic concentation and also with increasing concentrations of sulfide and thiosulfate-oxidizable sulfur species –.

A comprehensive survey of corrosion behavior in alumina plants was reported in 1957 by Champion(3). The major conclusion drawn by Champion was that "the most important factors in caustic cracking appear to be tensile stress, sodium hydroxide concentration and temperature." "Within the ranges experienced in the Bayer process a rise in any of these accelerates stress corrosion." Champion also observed that cracking was seen with welded seams and not riveted seams; in tanks, cracks became less severe from the bottom to the top of the tank; the tendency to crack was higher at isolated welds than at welded seams; a tank containing boiling liquor at 200 gpl free soda (324 gpl caustic as Na_2CO_3) showed no signs of cracking after seven years but as the concentration was increased to 320 - 500 gpl free soda (550-850 gpl caustic as Na_2CO_3) and at higher temperatures, cracking developed. "Comparison of experience with caustic dissolving tanks and strong liquor tanks suggests that neutralization by alumina of part of the caustic soda in liquors does not destroy, though it may reduce effectiveness of that part of the soda in caustic cracking"; at concentration of free soda (Na_2O) below 250 gpl (425 gpl Caustic as Na_2CO_3) with little alumina serious cracking occurs and finally "serious cracking can probably be avoided by keeping the uncombined soda concentration below 200 - 250 gpl Na_2O (342-427 gpl Caustic as Na_2CO_3) or the temperature below 85°C."

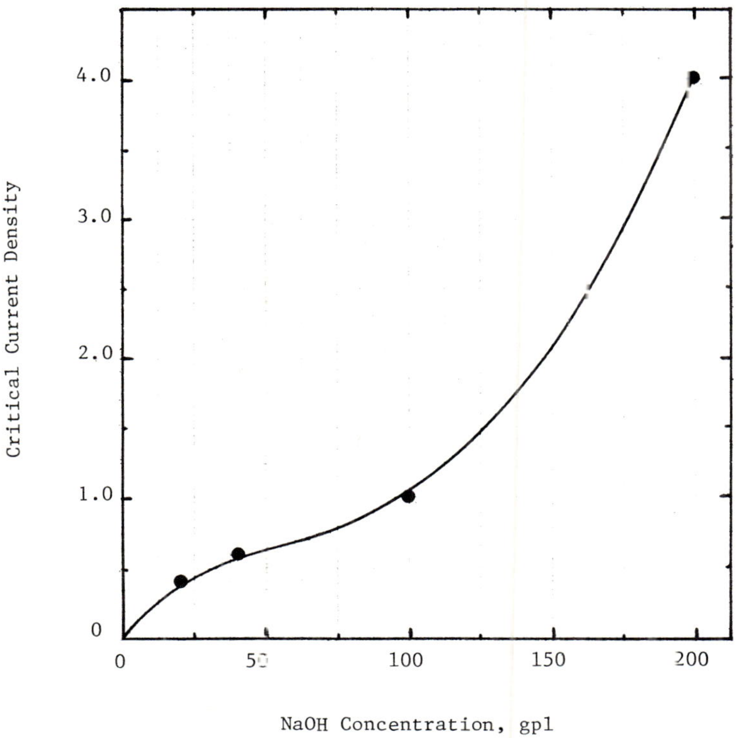

Figure 3: Effect of NaOH Concentration on Critical Current Density for Mild Steel at 90° C (4)

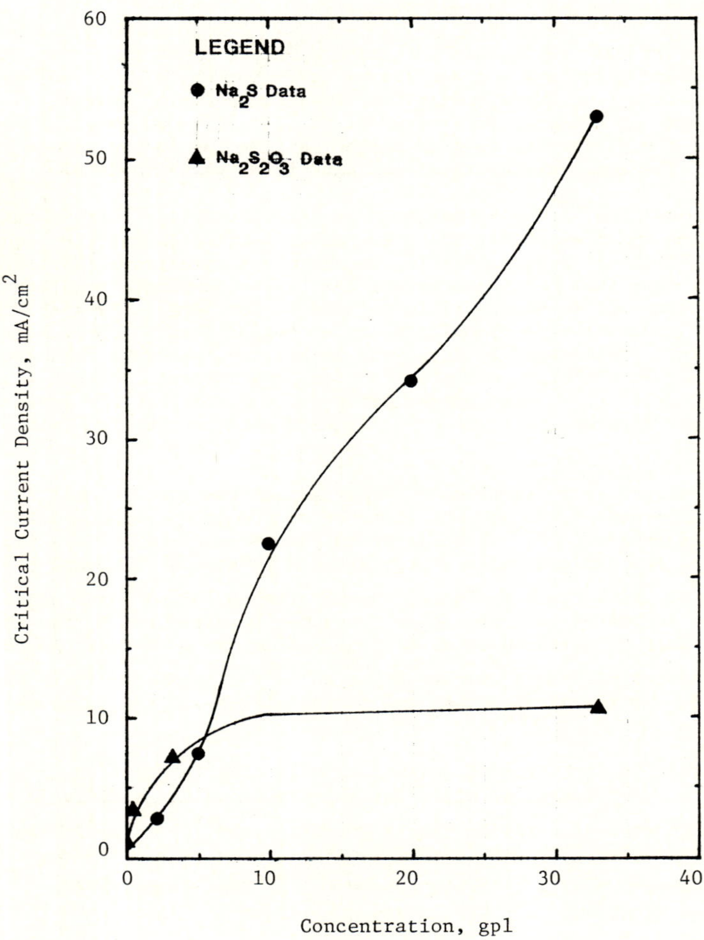

Figure 4: Effect of Na_2S and $Na_2S_2O_3$ on Critical Current Density of Mild Steel at 100 gpl NaOH and 90° C. (16)

Champions' data also indicate that for similar strong liquor concentrations and temperatures, the strong liquor tanks showed different corrosion behavior in different plants.

A similar observation has been made by Artem'ev et al. [5]. These authors found that in one plant stress corrosion cracking of decomposers (precipitators?) occurred during the first few years of operation. In a second plant no evidence of cracks was found in 12 - 20 years. They attribute the difference to presence of sulfide ion in the second plant liquors. The comparative liquor analysis is shown in Table II.

Artem'ev et.al[5]. also report data on laboratory electro-chemical experiments which show that the presence of sulfide in caustic liquors increases critical current density at corrosion potential. The data in Table III show that critical current density in aluminate solutions, which also contain sulfide ions, is higher than in pure NaOH. Further, the effect of concentration of sulfide ions is significant. These authors believe that presence of sulfide ions in solution prevents passivation of mild steel and therefore promotes "corrosion-mechanical damage" - stress corrosion cracking. These investigators also found that the presence of thiosulfate ions and sulfite ions also maintains the "active" state of the metal.

PPG Industries publication[18] has defined the process conditions for suitable use of various materials of construction. These data are reproduced in Figure 5. These data indicate that nickel alloys should be considered for use generally at temperature in excess of normal boiling point. Corrosion tests conducted in operating pulp and paper plants [19] have also indicated that corrosion rates can be minimized by the use of nickel alloys as compared to mild steel. Also use of nickel and nickel alloys in alumina plants is also reported[19]. Therefore, it is prudent to consider corrosion of nickel and nickel alloys and this discussion follows.

The rate of corrosion of nickel in pure caustic solution is low, but presence of impurities, specifically oxidizable sulfur species in caustic solution increases corrosion rate significantly[19]. In fact, higher the concentration of the oxidizable sulfur specie, higher the corrosion rate, as shown by data in Table IV. Data[19] also indicate that presence of chromium in nickel steel stabilizes the alloy from the point of view of corrosion. Hill and Johnson[20] reached the same conclusion and conclude that ferritic stainless steel are less prone to attack compared to austenite stainless steels. Literature in pulp and paper industry indicates that low nickel stainless steel perform well in the liquor heating stages of pulp digestion[21,22] and that the stainless steels are of the type 304, 304L, 316 and 316L [19,20,23,24,25,26,27,28].

An interesting ramification results when nickel steels and mild steels are used in the same liquor circuit. As a result of corrosion of nickel or its alloys, nickel ions are contributed to the caustic solutions containing sulfur ionic species. Investigators[11,22] have found that the presence of nickel ions increases the oxidation rate of S^{-2} ion and contributes to increased concentration of partially oxidized sulfur species to the solution and hence can increase the corrosion rate of mild steels.

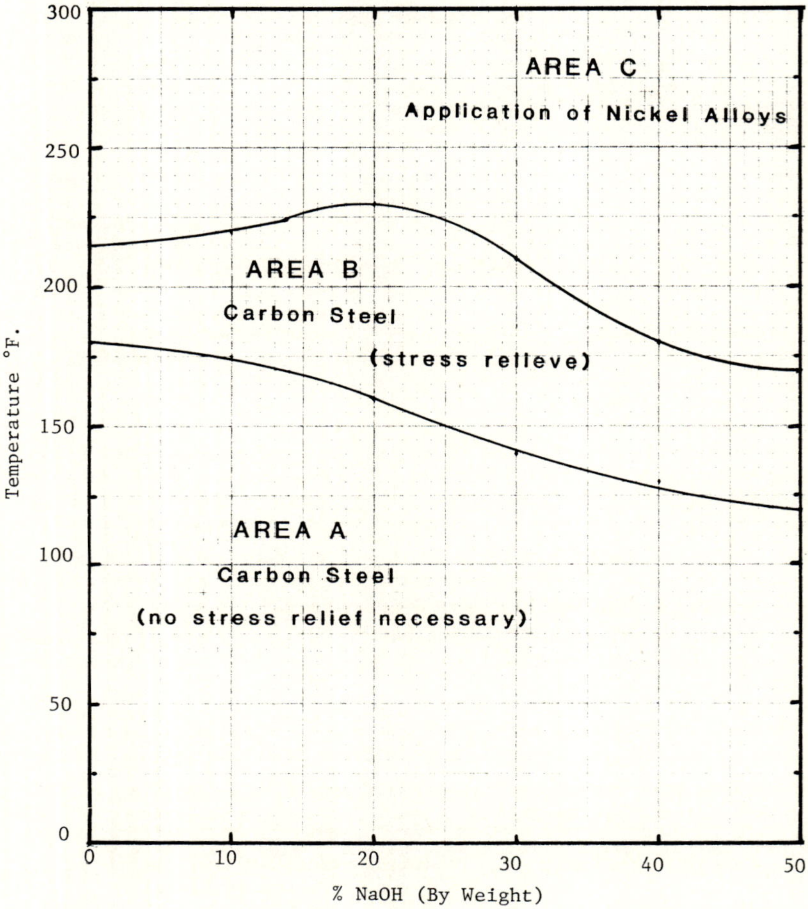

Figure 5: Caustic Soda Service Chart (18)

Table I[4]

Effect of Sulfide in Caustic Medium on Corrosion Rate at 92°C

Corrosive Media	Maximum Current Density MA/cm^2	% Reduction in Area
3.35M NaOH	8.0	29.0
2.5M NaOH+0.423M Na_2S	400	57.0

Table II. Liquor Analyses of Russian Alumina Plants [5]

Chemical Specie	Concentration in gpl Plant A	Plant B
Caustic as Na_2CO_3	239	239
Total S as Na_2CO_3	273	256
Al_2O_3	139	135
SiO_2	0.45	0.59
S, Total	3.5	0.08
$*S^{-2}$	0.022	ND
$*SO_4^{-2}$	0.0135	0.008

* These concentrations are in g ion/litre

Table III. Effect of Corrosive Medium on Corrosion Rate at 60°C[5]

Corrosive Medium	Maximum Current Density, mA/cm^2 Corrosion Potential -750 MV	-825MV
20% NaOH	0.1	0.8
20% NaOH+0.022 g ion/l S^{-2}	0.6	1.4
Aluminate Solution (0.01 g ion/l S^{-2}	1.0	4.3

Table IV. Effect of Oxidizable Sulfur Compounds on Corrosion of Nickel 200 in Caustic Soda at 130°C ± 5°C[20]

Test No.	Corrosive Medium	Corrosion Rate, Mils per year
1	Commercial Sodium Hydroxide being concentrated from 50 to 75% NaOH (Sulfur content at start, calculated as H_2S 0.009%	1.7
2	75% C.P.* Sodium Hydroxide	0.6
3	75% C.P. Sodium Hydroxide plus 0.75% Sodium Sulfide	22.8
4	75% C.P. Sodium Hydroxide plus 0.75% Sodium Thiosulfate	7.9
5	75% C.P. Sodium Hydroxide plus 0.75% Sodium Sulfite	5.2
6	75% C.P. Sodium Hydroxide plus 0.75% Sodium Sulfate	0.6

* Chemically pure

Failure-Nonfailure Criteria for Alumina Plants. The data available in the published literature and other unpublished data on material failure-nonfailure due to caustic embrittlement in alumina plants is summarized in Figure 6. Failure-nonfailure criteria for caustic plants as defined by PPG[18] publication is also shown. The Artem'ev data is not used since specific concentration and temperatures for equipment operation were not defined. These data indicate that presence of alumina in caustic solutions tends to passivate carbon steel, in that higher operating temperature at a given caustic concentration can be tolerated in the presence of alumina.

Given the failure-no failure data, a possible demarkation of failure-no failure zone can be constructed and is also shown in Figure 6. Admittedly, the number of data are few, but it is hoped more data will become available from alumina manufacturers to define this demarkation more accurately.

It is noted that the operating data used in Figure 6 are from plants operating in both low and high pressure digestion mode. British Aluminum data from Champion[3] does not indicate whether oxidizable sulfur species were present in the liquors, however some no failure data above the proposed demarkation line are from this publication and from a plant practicing low pressure digestion.

Summary

This study has indicated that:

1. Stress corrosion cracking is the primary type of corrosion in caustic environment encountered in caustic plants, pulp and paper mills, and alumina plants.

2. In caustic containing solutions, the possibility of stress corrosion cracking of carbon still increases as the caustic concentration and/or temperature are increased.

3. The possibility of stress corrosion cracking is enhanced due to the presence of oxidizable sulfur compounds, such as residual sulfide and thiosulfate, in a caustic containing solution.

4. The possibility of stress corrosion cracking can be minimized by firstly, stress relieving all welds during construction, and secondly by minimizing the concentration of oxidizable sulfur compounds.

5. A demarkation to define corrosion related failure-no failure in alumina plants with respect to operating temperature and caustic concentration is proposed.

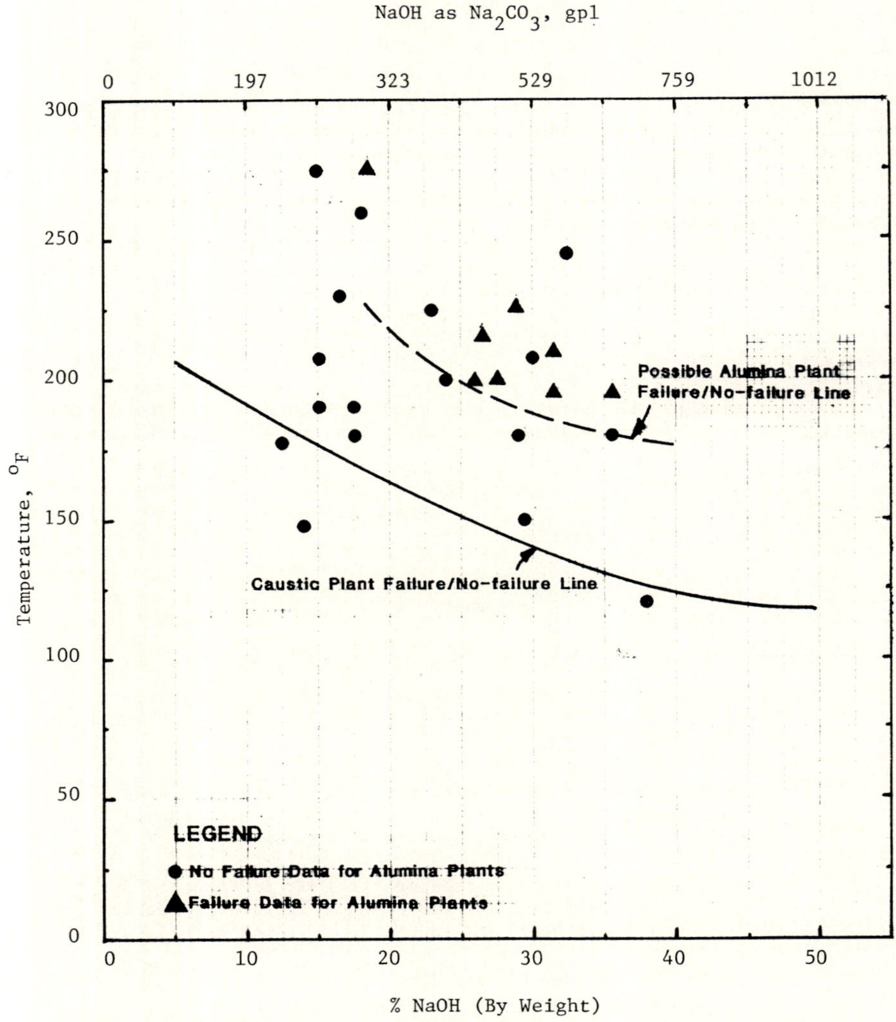

Figure 6: Corrosion Failure/No Failure Data for Alumina Plants (Adapted from Ref. 20)

Passivation of Caustic Embrittlement. A common denominator in the literature, to minimize caustic embrittlement, is minimization of residual stresses in equipment which comes into contact with caustic containing solutions. This technique is recommended in literature pertaining to both the pulp and paper industry[4,8,9,10,12,16,22] and the alumina industry[3,5,6]. In other words these investigators emphasize stress relieving of equipment after welded construction.

In plant practice, however, complete stress relieving is often not possible. So the problem reduces to minimizing the corrosive effects. In order to minimize corrosive effects, use of antenitic and ferritic stainless steels is suggested [13,19,22].

Passivation of caustic embrittlement is also achieved by presence of coatings on steel surface, thus preventing contact of corrosive caustic with the metal surface. It is reported [15] that in alumina plants presence of strongly adherent scale in digesters and precipitation, make them immune to stress corrosion cracking.

Since it has been established that the presence of oxidizable sulfur compounds lead to an increased rate of corrosion, it follows that if the concentration of oxidizable sulfur species in solution is minimized or avoided, the corrosion rate can be decreased substantially. This remedy is not feasible in the pulp and paper industry. However, in the alumina industry it is feasible to a certain extent. In the alumina industry, presence of oxidizable sulfur specie in solution is due to addition of sodium sulfide or sodium hydrosulfide to solutions for control of zinc contamination of alumina. In plants where such a control is practiced corrosion problems are significant. It is not possible to add a stoichiometric amount of sulfide to precipitate zinc sulfide, and some excess is always required which leads to presence of residual sulfide and partially oxidized sulfur compounds, such as thiosulfates in the plant solutions. As a result, this aggravates stress corrosion implications.

Bird and Vance[29] indicate that, significant improvements in reaction efficiency for zinc sulfide precipitation can reduce sulfide addition requirement significantly. It can then be presumed that as a result, residual oxidizable sulfide compound concentration in the Bayer liquors can be reduced. This in turn means that, corrosion rate of mild steel, nickel or nickel alloys can be lower than presently being encountered in alumina plants practicing zinc concentration control.

Artem'ev et.al.[5] have investigated the electro-chemical response of mild steel in aluminate plant solutions with additions of various compounds. Their aim was to find whether addition of these compounds passivated the corrosion reaction. These investigators found that "addition of 0.15 mole/lit. $K_2S_2O_8$ or 0.01 mole/lit. of H_2O_2 leads to a rapid shift in the potential of the metal into the passive region and thus should prevent cracking." They believe that oxidation of the sulfur compounds by these additives is responsible for metal passivation. However, at the same time these investigators caution that though this technique may passivate corrosion, the stress corrosion cracking that has occurred prior to passivation can lead to failure of the structure.

LIST OF REFERENCES

1. Perry, R. H., Chemical Engineers Handbook, Fifth Edition, McGraw Hill Book Company, New York, N. Y., 1973

2. Shrier, L. L., Corrosion - Volume 1, Newnes-Butterworths, London, 1979.

3. Champion, F. A., "Some Aspects of Stress Corrosion of Steel in Caustic Soda Solutions," Chemistry and Industry, 13 (July 1957) pp. 967-974.

4. Singbeil, D. and Tromans, D., "Effect of Sulfide Ion on Caustic Cracking of Mild Steel," Journal of Electro-Chemical Society, 128 (1981) pp. 2065-2070.

5. Arte'mev, V. I., Seregin, V. I., Zholobova, E. P. and Belyrev, V. P., "Stress Corrosion Cracking of Carbon Steel in Alkali Aluminate Solutions," Prot. Met., v. 15 (1) 1979 pp. 45-48.

6. Kjar, A. R., "Corrosion Evaluation and Control in Caustic Environment," Australian Corrosion Eng., 18 (7) (1974) pp. 13-16.

7. Humphries, M. J. and Parkins, R. N., "Stress Corrosion Cracking of Mild Steels in Sodium Hydroxide Solutions Containing Various Additional Substances," Corrosion Science, 7 (1967) pp. 747-761.

8. Bates, J. F., "*______________," Industrial and Engineering Chemistry, 58 (2) (Feb., 1966).

9. Wensley, D. A., and Charlton, R. S., "Corrosion Studies in Kraft White Liquor-Potentiostatic Polarization of Mild Steel in Caustic Solutions Containing Sulfur Species," Corrosion-NACE, 36 (8) (1980).

10. Mueller, W. A., "Mechanism and Prevention of Corrosion of Steels Exposed to Kraft Liquors," pp. 109-116 in " Pulp and Paper Industry Corrosion Problems International Seminar, National Association of Corrosion Engineers, 1974.

11. Laliberte, L. H., "Corrosion Problems in Pulp and Paper Industry," pp. 1-11 in Volume 2 Pulp and Paper Industry Corrosion Problems International Seminar, National Association of Corrosion Engineers, 1977.

12. Komp, M. E. and Mathay, W. L., "Steels for Pulp and Paper Industry," Paper #169 presented at NACE Annual Meeting, San Francisco, Mar. 1977.

13. Rozov, M. N., "Prospective Metals for Apparatus Used to Intensify Alumina Production," Soviet Journal of Non-Ferrous Metals, (Dec. 1960) pp. 28-32.

14. Uhlig, H. H., Corrosion and Corrosion Control, John Wiley and Sons, New York, N. Y., 1971.

15. Roach, G. I. D. and Clever, T. R., "Materials of Construction in Bayer Plants: Chemical Considerations," Journal of Geological Society of Jamaica. Proceedings of Bauxite Symposium, No. V, June 1982.

16. Wensley, D. A., and Charlton, R. S., "Corrosion Study in Kraft Liquor: Potentiostatic Polorization of Mild Steel in Caustic Solutions Containing Sulfur Species," Corrosion '79, NACE Publication, 1979.

17. Staehle, R. W. and Agraval, A. K., "Corrosion, Stress Corrosion Cracking and Electrochemistry of Iron and Nichel Alloys in Caustic Environments," Progress Report for March 1977 to February 1978 in ERDA Contract No. FY-76-5-02-2421, Ohio State University, Columbus, Ohio, p. 63.

18. *__________, NaOH: Caustic Soda, P.P.G. Industries Inc. Publication 1969.

19. *______________, "Corrosion Resistance of Nickel and Nickel-Containing alloys in Caustic Soda and other Alkalies," Corrosion Engineering Bulletin CEB-2, International Nickel Company, New York, N. Y. 1973, pp. 8-33.

20. Hill, J. B. and Johnson, M. J., "New Stainless Steels for the Pulp and Paper Industry," pp. 493-499, Book 3 in TAPPI Engineering Conference Proceedings, 1978. book 3, pp. 493-499.

21. Remean, J. J. and Barthelemy, H., "Quelques Problems deCorrosion dans la Fabrication de la pate a Papier par le Procede Kraft," Materiaux Et Techniques, November-December 1978.

22. Mueller, W. A.; 'Reduction and Oxidation Reactions in Kraft Liquor Recovery: Sources, Effects and Preventions," Pulp and Paper Magazine of Canada, 74 (4) (1973) pp. 69-76.

23. Chakrapani, O. G. and Czyzewski, H., "Improved Corrosion Control in Pulp Mills Can Reduce Costly Damage," Pulp and Paper (July 1979) pp. 146-150.

24. Lankenan, H. G., Engle, C. P. and Flores, A. R., "Corrosion of Carbon and Stainless Steel in Evaporators by Sulfate Black Liquor," pp. 117-124 in Pulp and Paper Corrosion Problems; NANCE Publication 1974, pp. 117-124.

25. *____________, Nickel Topics, 33 (2) (1980). Published by International Nickel Company, New York, N. Y.

26. Moller, G. E., "Designing with Stainless Steels for Service in Stress Corrosion Environments," paper presented at Corrosion' 76, Houston, Texas, 1976.

27. Komp, M. E. and Mathay, W. L., "Steels for Pulp and Paper Industry," Paper presented at Corrosion' 77, San Francisco, California, 1977.

28. Lankenan, H. G. and Bohnenberger, B., "Service Experience of Materials for Pulping Spent Liquor Evaporators," TAPPI (1978) pp. 501-507.

29. Bird, R. D. and Vance, H. R., "Zinc Removal From Aluminate Solutions," U.S. Patent Numbers 4, 282, 191, August 4, 1981.

* Unknown at press time.

NEW MECHANICAL SEAL FOR ABRASIVE SUSPENSIONS

Ph. MARCHESSAUX

ALUMINIUM PECHINEY - DIRECTION DES VENTES DE TECHNOLOGIE - 13120 GARDANNE (France)

Pumping suspensions characterized by abrasiveness crystallisation and scale deposit has always met with difficulty because of the behavior of the packing. The alumina industry does not escape this rule. Independent of the high costs of frequent maintenance operations on the packing, the problem of packing flush water arises. This packing needs water or buffer liquid for cooling or protection. Most of this flush water enters the process and has to be eliminated. The removal of this water is an operation high in energy costs. ALUMINIUM PECHINEY has designed and developed a new mechanical seal which is at present manufactured in France by CEFILAC and sold under the trade mark G.P.A. A description of the seal is given : some experimental results collected from the different alumina plants belonging to ALUMINIUM PECHINEY where the use of the G.P.A. tends to be widespread, are given. In conclusion, several suggestions for use and fitting are treated.

Introduction

Numerous pumping circuits handling overflow liquors and slurries are present in an alumina producing plant. In most cases centrifugal pumps are needed to handle these liquors and slurries. The behavior of their packing plays an important role in the reliability of these pumps.

Up till fairly recently, centrifugal pumps were equipped with packing which has two serious disadvantages :

- The poor reliability of this packing necessitates constant supervision, and frequent servicing leads to high maintenance costs. As an example, pumps so equipped at PECHINEY's LA BARASSE plant, require a permanent shift of 2,2 men for 204 installed pumps and the annual cost (labor and supplies) amounts to 1 190 U.S. dollars per pump installed (1 U.S. dollar $\simeq$ 8 FF).

- The need to use a buffer liquid, which is usually water, a large part of which is eliminated in the pumping circuit. This water tends to dilute the fluids that are being transported and must be evaporated if fluid concentration is to remain constant. The water can be evaporated in multi effect evaporators. With a quintuple effect evaporator, heat consumption is approx. 165 thermies per tonne of water evaporated.

If we count on an average water consumption of approx. 0.5 m^3/h per pump, the annual consumption in energy is round about 700 000 thermies, i.e. approx. 14 000 U.S. dollars at the actual cost of the steam thermie (20 U.S. dollars per kth) Servicing costs mentioned previously and the real cost of water must also be added on.

We can therefore calculate that the savings that can be made on evaporation alone are in the neighborhood of 3 500 K U.S. dollars for an average capacity alumina plant (700 000 t/year). This plant requires about 200 pumps to be in operation on the main circuits (except for red mud washing).

The advantages arising from the search for better adapted devices on centrifugal pumps which can limit unwanted water entering the process circuit, while at the same time avoiding high supervision and servicing costs, can therefore be clearly seen.

The mechanical seal device

ALUMINIUM PECHINEY turned towards the use of rotary mechanical seals. These can be described, in comparison with other types of packing, as being devices with low leakage using the rubbing of two solid surfaces that do not buckle, to ensure tightness.

An inexhaustive study of seals offered on the market was first of all undertaken during the last ten years, followed by several tests run in the plant on centrifugal pumps.

If, in general, the devices available on the market were suitable for uncharged liquors, they are less efficient once abrasive suspensions or cristallizing fluids are pumped.

All mechanical seals are designed according to the principle of a rotating ring coming to rest on a stationary ring. Contact is made and maintained between the two surfaces by means of an appropriate mechanism. Generally speaking, this mechanism is made of springs which ensure contact, and a guidage system. The guidage system as well as the springs can be fitted inside or outside the pump volute. It is necessary to use a sliding tightness device axial to the ring housing which moves. Tightness is obtained by means of seal tight bellows, membranes, or O rings.

Now, the major defect in this type of seal is the risk of the contact mechanism (springs) or the sliding device jamming, due to cristallized or settled solids in the stagnant slurry found in the seal, being deposited on the mechanism. The presence of slurry may be inherent in the design of the seal or due to accidental leakage.

ALUMINIUM PECHINEY therefore turned towards researching a seal that would be better suited for pumping cristallizing liquors in a sodium aluminate medium and for pumping red mud and alumina hydrate slurry. This seal which was tested on pumping facilities at GARDANNE is now marketed by the company CEFILAC under the trademark G.P.A. SEAL.

The G.P.A. mechanical seal

Description

This seal (see photo nr 1 and figure nr 2) is essentially made up of three parts :

- a stationary ring made of tungsten carbide inserted in a housing,
- a second ring (also in tungsten carbide) inserted in a housing which is caused to rotate with the pump shaft,
- a spring diaphragm.

The housing is connected via a spring diaphragm to a sleeve which screws onto the pump shaft. The spring diaphragm has multiple functions :

- to ensure initial contact between the rotating ring and the stationary ring,
- to compensate the wearing that may exist between the two contact surfaces of the rings due to rubbing of these surfaces,
- to act as a seal between the rotating housing and the shaft sleeve,
- to position the rotating housing in relation to the pump shaft. This positioning is carried out by flexible regulation of the spring diaphragm and replaces axial sliding of devices that the other types of mechanical seals which were mentioned previously are equipped with.

The seal is, so to speak, on condition that a careful selection of the relative inner/outer diameters of the spring diaphragm and the inner/outer diameters of the rotating ring is made, of the "adjusted" seal type. This means that, whatever the difference between the atmospheric pressure and the pressure inside the volute to the right of the seal, the contact force between the rotating and non rotating surface is theoretically independent of the pressure, and only depends on the initial contact force of the spring diaphragm.

The spring diaphragm is made of a steel washer onto which slits have been arranged radially. This layout enables axial deformation of several millimeters to be obtained easily with minimum stress. The steel washer is coated by means of grinding, with an elastomer (neoprene) which may or may not be coated with a thin sheet of PTFE depending on operating conditions and the type of fluid being handled. The spring diaphragm is fitted in the housing and in a circular groove machined on the shaft sleeve.

Photo 1 - G.P.A. mechanical seal.

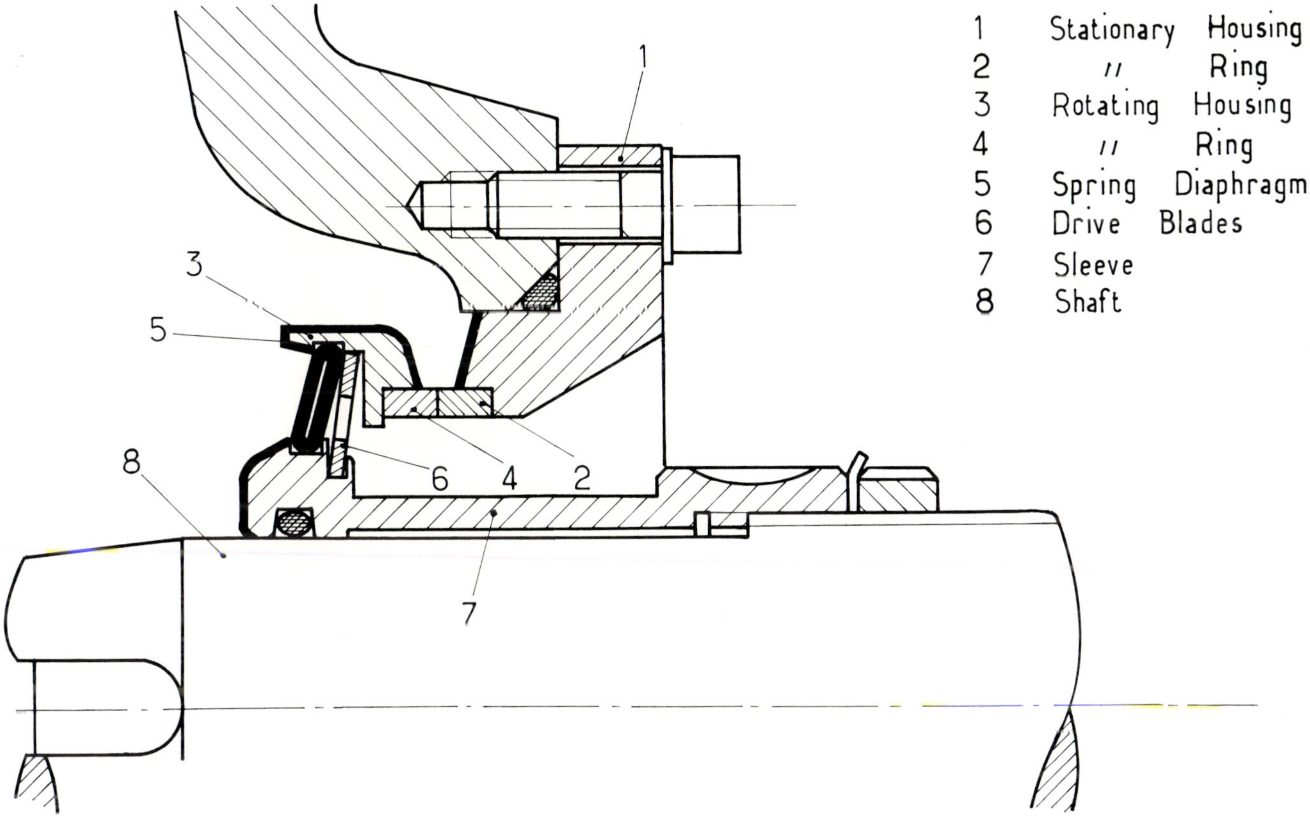

Figure 2 - G.P.A. mechanical seal.

The rotating ring housing is driven by three drive blades positioned at 120° from each other which make possible the transmission of the rotating torque from the shaft sleeve to the ring housing, while at the same time allowing longitudinal movement to be carried out freely. These three drive blades are embedded in holes fitted on both the sleeve and the rotating ring housing.

The seal assembly offers the following advantages :

- simplicity of design,
- spring diaphragm operating by compression and not by compression and bending at the same time, as is the case for other mechanical seal models,
- suppression of axial sliding which is replaced by a flexible regulation movement,
- balancing of seal.

Nevertheless, there are several disadvantages which are set out below :

- In order to obtain correct compensation, the mean diameter of the rings must equal the mean diameter of the spring diaphragm. Now the angle of this spring diaphragm is more or less perpendicular to the pump shaft. Therefore the seal has a diameter larger than most of the other brands on the market.

- Although small, there is a space outside the seal, demarcated between the rotating ring, spring diaphragm and shaft sleeve. This space can be filled with cristallizing liquid or solids which have settled progressively in the case of major leakage on the seal. It is nevertheless possible, as a temporary measure, to inject wash water into the seal. An injection pipe with an inlet in this space can be easily installed.

Dimensions of G.P.A. seal

At present CEFILAC has four types of seal on the market that are capable of equipping rotating machines with a shaft diameter of 40 - 135 mm. There is, a priori, no limit in the dimensions of the seal.

For all four types of seal, the inner and outer diameters of the rings are expressed in millimeters. These values as well as the shaft diameters onto which the seals can be fitted are shown in the following table.

Ring diameter (mm)	Shaft diameter (mm)
72/82	30 to 50
103/113	50 to 78
132/144	80 to 105
160/178	106 to 135

For the same type of seal, the dimensions of the shaft sleeve receiving the spring diaphragms and the drive blades may vary in accordance with the pump shafts.

The seals made by CEFILAC can be adapted for most centrifugal pumps on the market. Without giving an exhaustive list, JEUMONT-SCHNEIDER, K.S.B., MORRIS, SCHABAVER, WARMAN, WORTHINGTON pumps among others, can be cited.

Constituent materials

The constituent materials used in the pumping circuits in ALUMINIUM PECHINEY's plants are a function of the operating conditions and the type of fluid handled. They are essentially of two types :

- Hot fluids (temp. > 100° C) with a high caustic soda concentration :

  . Neoprene diaphragms coated in PTFE.
  . Shaft sleeve and ring housing in stainless steel (AISI 316).

- Low pressure slurries (temp. < 80-90° C) :

  . Uncoated neoprene diaphragms.
  . Shaft sleeve and ring housing in carbon steel.

In both cases the rings are made of tungsten carbide. The sleeve nose, and the surfaces of the rotating housing and stationary housing which are exposed to the abrasiveness of the handled slurries are protected by a coating of chromium oxide which is done by a plasma flame cutter.

Packaging - Fitting and positioning

Packaging. The seals are delivered either "pre-assembled" or in loose parts. "Pre-assembled" means that CEFILAC have made up the mobile assembly which includes rotating ring housing, spring diaphragm, drive blades and shaft sleeve. Only the stationary ring housing is delivered separately.

If the seal is delivered in loose parts, it is recommended to "run in" the rings before assembly. It is also necessary to pre-assemble the seal in the workshop using a simple press device which enables the diaphragm and the drive blades to be installed. The spring diaphragm, once positioned on the mobile assembly will only be energised during assembly of the seal on the pump.

Assembly. The admitted tolerances are listed hereunder :

Stationary ring bearing :

- Perpendicularity in relation to pump shaft : $\pm$ 0.1 mm.
- Eccentricity in relation to pump shaft tolerance : $\pm$ 0.4 mm.

The assembling operation described below corresponds to the most disadvantageous case of a hydraulically regulated pump : it is taken for granted that the volute of the pump is fitted on its frame. Sequence of operations is as follows :

- Installation of stationary ring housing on rear flange.
- Installation of pump mechanical parts on frame.
- Placing of rotating part of pre-assembled seal. Make sure there is a gap of several millimeters between ring faces.
- Installation of turbine and flange-suction-side.
- Positioning of the turbine in relation to suction flange.
- Surface contact by means of slight tightening.
- Filling of pump body with water after the suction and discharge flanges have been sealed.
- Slowly open the seal until a slight leakage appears. This indicates

that the rings are in contact without tightening.

- Shifting of the rotating ring towards the stationary ring so as to obtain tightness by compressing the spring diaphragm in accordance with the recommended tightening. The shaft sleeve is clamped by means of a locknut or pressure screws.
- Static hydraulic test at 6 bars.

Energizing spring diaphragm. Tightening is different according to the type of seal, but the clenching of the spring diaphragm is approx. 2.5-3 mm and corresponds to about 75 % of the maximum deflection.

The applied strength is therefore :

Ring dimensions (mm)	Applied strength (daN)
72/82	56
103/113	60
132/144	85
160/178	140

The contact strength enables tightness to be effected, independent of the pressure present in the pump to the right of the seal. Tests run showed that compensation wasn't complete as contact force looses approx. 12 % of its initial value when pressure in the volute increases from 0-6 bars.

The product P.V. (P = Pressure - V = Speed) of the contact pressure being exerted on the rings by peripheral speed has the following values for a rotation speed of 1000 rpm^{-1}.

Ring dimensions (mm)	P.V.	
	lbs $inch^{-2}$xfoot.sec^{-1}	kg cm^{-2}xm.sec^{-1}
72/82	840	18
103/113	933	20
132/144	1120	24
160/178	1213	26

These values fall within the limits normally admitted for P.V. products concerning surfaces using a tungsten carbide against tungsten carbide contact.

Adaptation of seals - New development perspectives

The seals that are at present on the market only constitute an example of what may be proposed to customers. Research and trials are now underway to find means of adapting these seals to cope with the new preoccupations. Their aim is to improve reliability and make maintenance easier to carry out.

Installation of the seal by fitting it via the suction side of the pump

This adaptation is the answer that the maintenance staff sought in fitting or removing the seal on the jobsite without having to dismantle the mechanical parts of the pump.

The device shown in figure nr 3 has already been implemented on pumps operating at present. The stationary ring housing is fitted on the rear flange via the inside of the pump and kept in position by means of round pellets.

Cartridge seal

The seal is entirely pre-set in the workshop, diaphragm deflection being obtained by means of a clamp. The latter is only removed once the seal has been fitted, the housing is clamped onto the pump flange, and the sleeve is tightened on the shaft.

Figure nr 4 shows assembly of a cartridge seal with it being fitted on the mechanical side of the pump.

Figure nr 5 shows a similar device with assembly on the hydraulic side.

"Reversed" seal

This device, using the assembly of a diaphragm associated with the stationary ring, is feasible in special cases. This type of seal has been successfully tested in ALUMINIUM PECHINEY's plant in GARDANNE. The advantages of this type are set out below :

- Suppression of the drive blades. The gripping of the diaphragm on the stationary ring should suffice to prevent this ring rotating.
- The diaphragm wears less.

Nevertheless several disadvantages must be taken into account :

- The seal is located further away from the pumping zone, hence the risk of slurry accumulating in the gap between the spring diaphragm, the rotating ring housing and the shaft sleeve.
- The "balanced" design of the seal requires that the ratio of the ring mean diameters vis a vis shaft diameter be larger than in the normal design with spring membrane fitted on the rotating ring. This results in the seal being bigger.

Double seal

This device (see figure nr 6) consists of two seals fitted onto a single shaft sleeve with only one component serving as housing for the two stationary rings two diametrically opposed ducts are drilled into the stationary ring housing so that buffer liquid can be injected. The internal seal is adjusted in the same way as the single seal, by positioning the sleeve on the shaft and clamping it with a SKF nut. Tension of the external spring diaphragm is adjusted in two operations :

- Positioning is pre-adjusted by means of a nut screwed onto the shaft sleeve and coming to shoulder the diaphragm via a support washer.

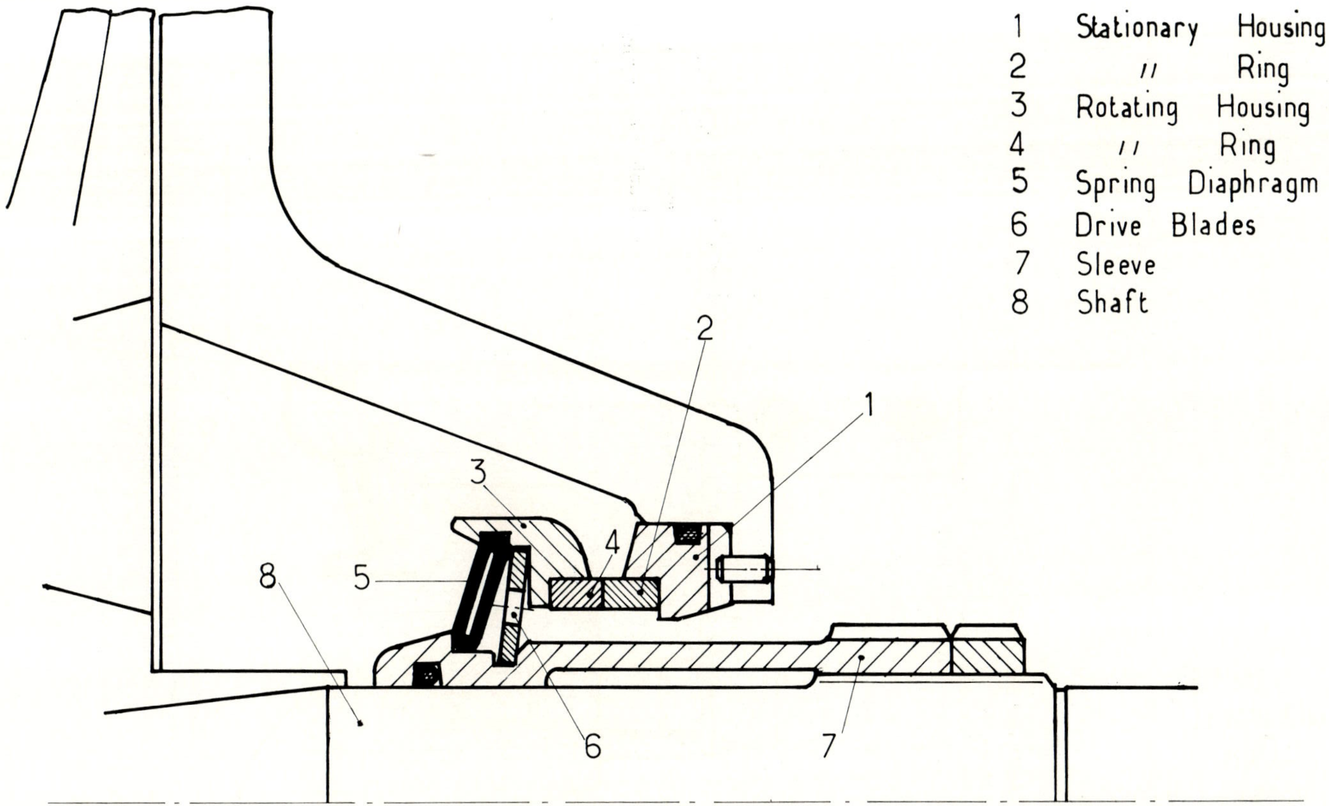

Figure 3 - Fitting of the G.P.A. via the suction side of the pump.

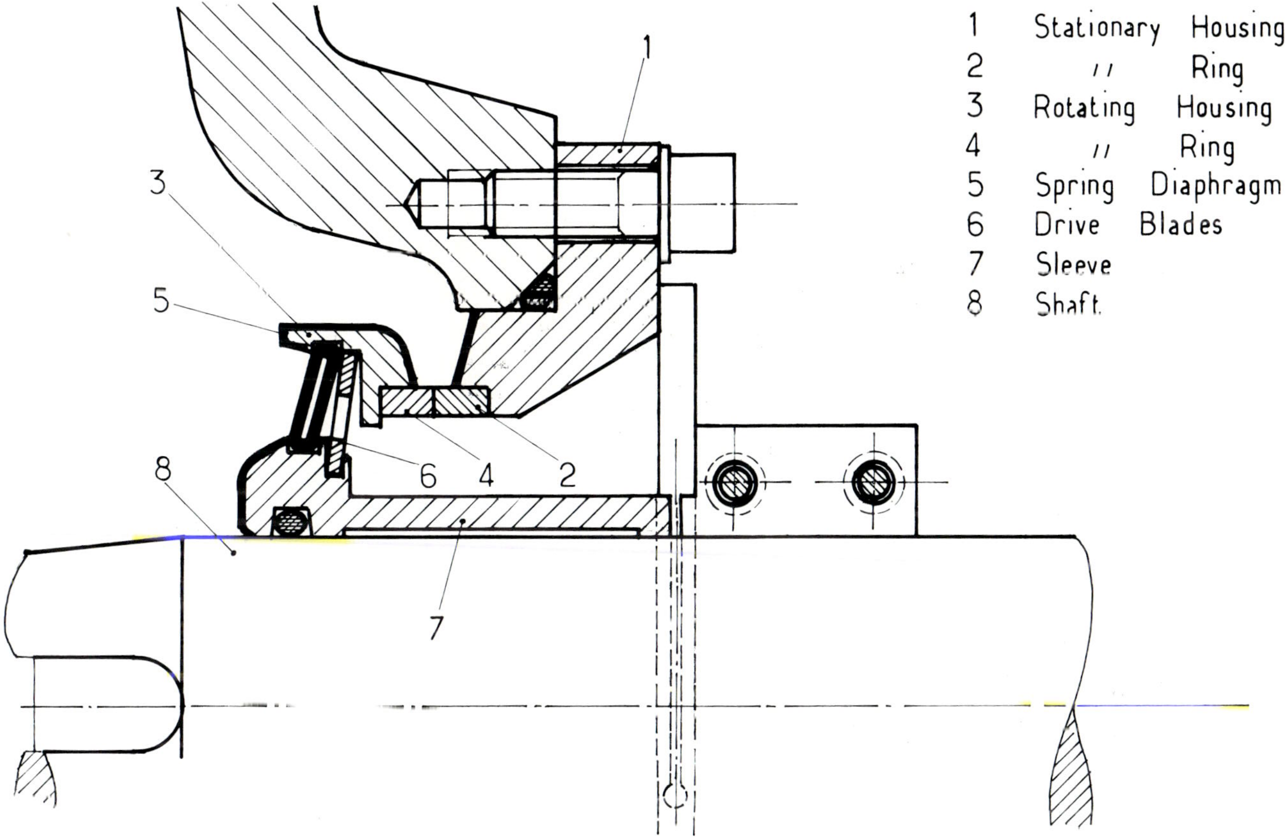

Figure 4 - Cartridge G.P.A. seal. Fitting on the mechanical side of the pump.

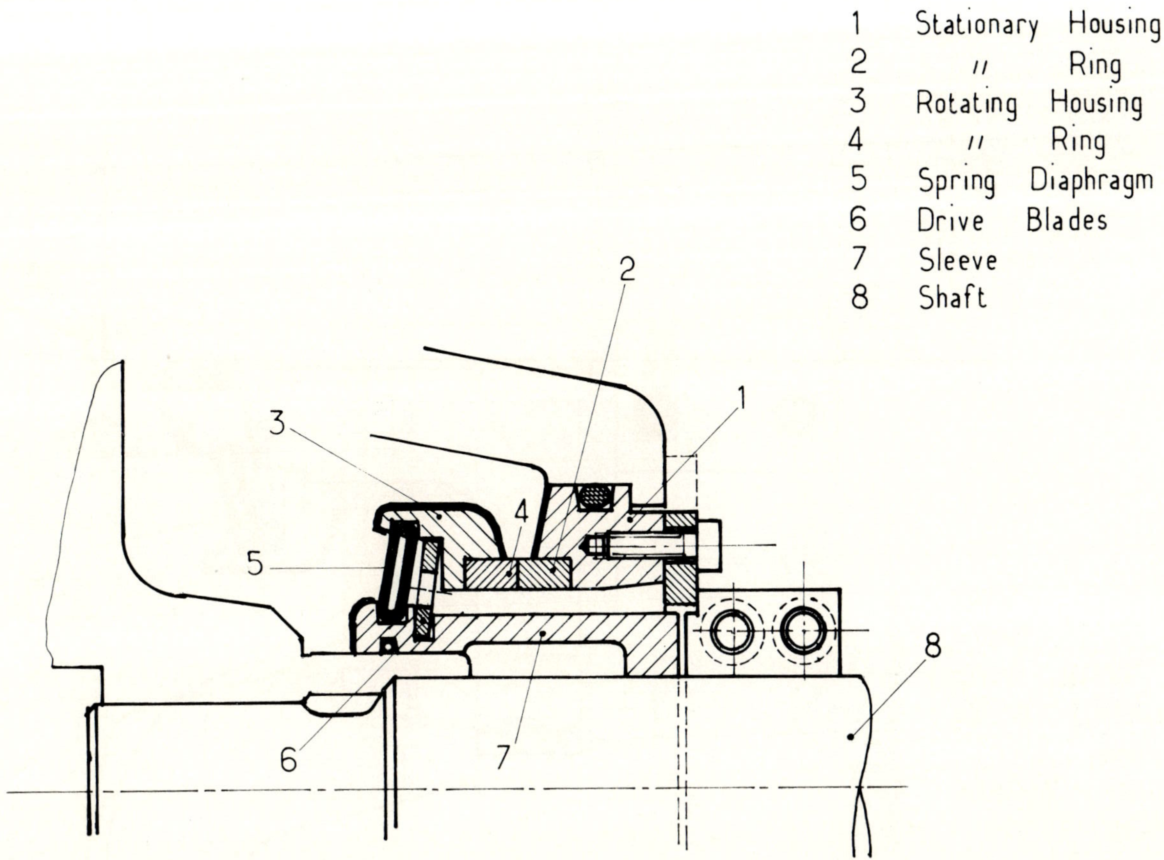

Figure 5 - Cartridge G.P.A. seal. Fitting on the hydraulic side of the pump.

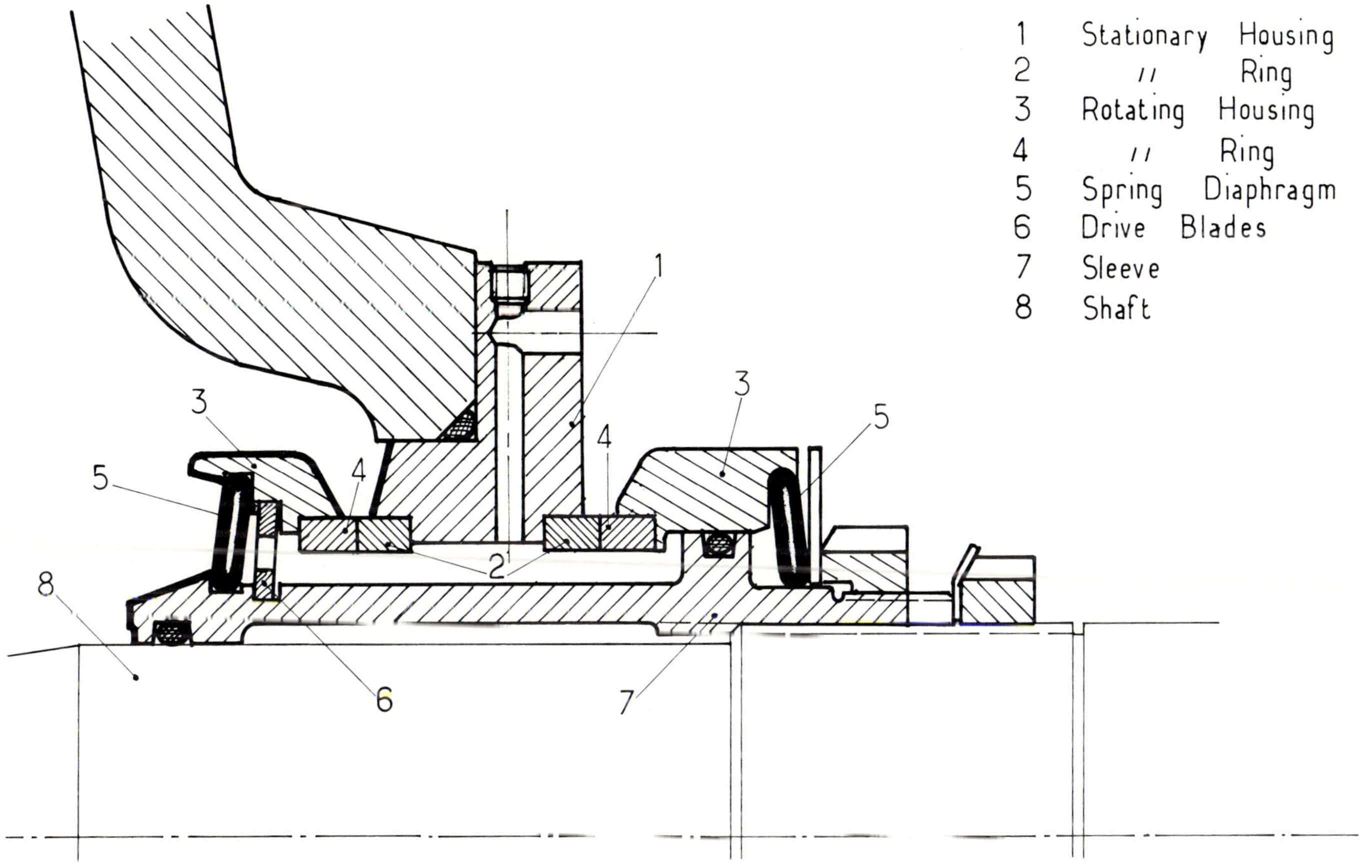

Figure 6 - Double mechanical seal.

- Final adjustment is carried out by shifting the sleeve so that crushing is distributed evenly between the two spring diaphragms.

The double seal corresponds to a "cartridge" type seal design. It is recommended to use this type of seal on centrifugal pumps handling hot liquids which may cause pitting. It can moreover be used in the case of pumping at high suction pressures (> 5-6 bars).

A seal has been undergoing tests for more than seven months on a pump handling red mud slurry at ALUMINIUM PECHINEY's SALINDRES plant. Suction pressure is approx. 10 bars. Packing which was previously fitted on the same pump operating under the same conditions did not last for more than two to three weeks.

Operating results

The following results are given as indications and concern the operating life of seals on pumps in ALUMINIUM PECHINEY's plants at GARDANNE, LA BARASSE and SALINDRES.

At GARDANNE, there are 233 process pumps equipped with mechanical seals out of 295 centrifugal pumps. 195 pumps are operating, 38 are on stand-by.

Servicing was required 246 times in 1982 on mechanical seals. Servicing was not obligatorily motivated by the behavior of the mechanical seal. The latter is in fact serviced every time the pump is repaired, whether or not it is the cause of failure.

Average operating life is, considering only pumps working all the time, 9.5 months. Out of 246 seals, 80 were replaced by new ones and 166 were repaired in the workshop. The spring diaphragms are systematically changed at each servicing operation.

Behavior in different operating sectors

As an indication, mean service life of mechanical seals installed in four of the most important work stations in alumina production are as follows :

GRINDING		GARDANNE	SALINDRES
Slurry pumps after grinding			
	Type	SCHABAVER M 250	SCHABAVER M 250
	RPM	830	780
	Service life (months)	4-5 (*)	20-22

(*) One pump operated without being dismontted for 19.5 months.

EVAPORATION		GARDANNE	SALINDRES
Circulation pumps			
	Type	SCHABAVER M 250	SCHABAVER M 250
	RPM	780	520
	Service life (months)	20-22	16-20(*)
Transfer pumps			
	Type	SCHABAVER M 250	SCHABAVER M 250
	RPM	780	700
	Service life (months)	12-14	10-14(**)

(*) One seal lasted 8 years. (**) One seal lasted 5 years.

SETTLING WASHING		GARDANNE	SALINDRES
Settlers			
- Liquor pumps			
	Type	RATEAU IDB	SCHABAVER M 250
	RPM	1050	⩽ 950
	Service life (months)	6-10	4-6
- Red mud pumps			
	Type	SCHABAVER M 150	SCHABAVER M 100
	RPM	⩽ 750	860
	Service life (months)	6-10	3-5
	Type	KLEIN WD	
	RPM	⩽ 1000	
	Service life (months)	4-8	
Washers			
- Liquor pumps			
	Type	RATEAU IDB	KLEIN P[illegible]R
	RPM	1050	⩽ 1400
	Service life (months)	13-16	6-12
- Red mud pumps			
	Type	SCHABAVER M 250	KLEIN WD
	RPM	⩽ 750	1050
	Service life (months)	6-8	0,5-1,5

PRECIPITATION		GARDANNE	SALINDRES
Seed transfer pumps			
	Type	KLEIN PA 200	
	RPM	< 900	
	Service life (months)	6-10	
	Type	KLEIN NA 250	
	RPM	< 900	
	Service life (months)	6-8	
	Type	JEUMONT SCHNEIDER LC	JEUMONT SCHNEIDER LC
	RPM	< 1270	< 1210
	Service life (months)	10	24-48

Behavior factors for mechanical seals

The preceding results show :

- An average service life of 9.5 months for all the 273 process pumps operating in the two plants.

- Service life reaching 5 years and even 8 years.

- Large differences from one plant to the other for the same type of pump and for the same operating zone (Grinding zone for example).

- Service life for seals equipping pumps in the Evaporating zone on the whole is greater than that for seals equipping other units.

The behavior of mechanical seals is conditioned by 3 groups of factors linked respectively to :

- The seal itself.
- Pump design.
- Design of pumping facility and operating conditions.

For the seals to operate correctly, a lubrication film between the two ring faces in contact must be permantly present. The rings must also be cooled. This assumes that the handled liquor, which comes into contact with the ring housing is continuously renewed. The latter must be dimensioned to ensure that the heat resulting from the rubbing of the rings is eliminated.

A permanent spray system for the G.P.A. seal by means of buffer liquid is useless. Only occasional washing of the seal is recommended, especially in the case of prolonged stoppage before re-startup.

The pump onto which the seal is fitted must be adapted mechanically. This implies an acceptable out of true shaft. The rotating shaft must not have too great a deflection. Each case requires special study and the tolerances to be respected cannot be specified in advance. The pump rotating at low speed is a favorable factor pumps rotating at fewer RPM are a priori more commendable than those rotating at high speed.

It has been noticed poor performances from the seals mounted onto pumps that become badly suited as a result of the changes that time brought upon the facilities onto which the seals are installed. Moreover, the seals suffer more on circuits that are heavily by crust deposit with risks of bad circulation when it is not a question of the ducts being blocked up.

Different factors must be considered :

- Research into the shape of the suction pumps with the aim of limiting head loss to a minimum.

- Checking of available NPSH in relation to NPSH required for all operating conditions.

- Frequency of stoppages and start-up.

- Regulation of the suction tank levels.

How important the role of the design office and the operating conditions can be seen.

HYDRAULIC ACTUATOR FOR ANGLE VALVES

François Lavastre Réjean Saulnier

Alcan Smelters and Chemicals Ltd
Vaudreuil Works, Jonquière, Quebec

A new hydraulic actuator for angle valves has been developed. The system has three main components: a hydraulic unit for delivery of hydraulic fluid, a control panel and a hydraulic motor, with accessories, mounted on the angle valve.

The hydraulic pressure normally used to displace the valve stem is 1100 psi. At the end of each complete stem displacement, a pulsating hydraulic pressure up to 1500 psi is used to break the scale on the valve seat. A zero leakage seal is obtained with a steel/steel contact after a few hydraulic impacts. Time of closure has been reduced to approximated 5 minutes for a 12" x 12" angle valve

Introduction

Most large angle valves installed in Bayer process plant operations are used for on/off services. Scale formation on the valve trim components (disc, seat and stem) and inside the valve body is so severe that it is difficult to achieve leak proof shut off. Extreme physical effort in adverse environmental conditions is sometimes required to operate frozen slurry valves.

The motorization of the more difficult angle valves has been studied over a period of many years by the Alcan organization. Prototype experimental units, using pneumatic or electric actuators have been tried and evaluated. Electric actuators available on the market have been tested successfully in relatively light service, but did not meet our requirements on digester heat exchangers, which we consider a difficult service. Pneumatic actuators were tested, but not developed further after poor initial performance.

The idea of a hydraulic actuator was considered for severe conditions, as on heat exchanger service, where sodalite deposits make it difficult to achieve a leak proof closure.

Statement of Objectives

The main objective was to design a hydraulic actuator to achieve leak proof closure under very difficult conditions.

In order to meet the requirements of the plants, additional features were required:

- In case of hydraulic failure, manual operation is necessary.

- In case of actuator breakdown, it is necessary to be able to replace the damaged actuator without any shutdown of the process.

- The actuator had to be readily adaptable to existing valves, with a minimum of modifications required to present layout.

Description of Equipment

The existing angle-valves have the following main components (see Figure 1):

- body
- yoke
- stem
- seat and disc assembly
- manual operating device

Keeping the original body, yoke and stem, we attempted to optimize seat/disc geometry and to replace the manual operating device by a hydraulic actuator.

Seat-Disc Geometry Optimization

Since the leak proof closure is ensured by the contact between seat and disc, the geometry of faces in contact is very important. The pressure on the seat-disc assembly has a direct relation with the force applied to close the valve. Increasing the seat pressure makes it easier to break the scale and to achieve a tight closure. But the maximum stem forces allowed are limited.

The first limitation to the force applied was given by the mechanical strength of valve components. Calculations, taking into account all safety factors, determined this limit. The second limitation was given by the seating pressure required to break the scale. Laboratory tests determined this limit.

A theoretical relation was defined between the force applied on the valve stem and hydraulic pressure for the actuator. In the laboratory, strain gauges were installed on a valve stem and a practical relation between the hydraulic pressure applied to the valve actuator, mounted on the angle valve, and the stress on the valve stem was established. A noticeable difference was detected between theoretical and practical results (see Figure 2).

Taking into account all the above considerations and data, the possibility of improving seat-disc geometry was investigated. Seating pressure depends upon two parameters: seat joint area and seating angle.

Seating angle for the existing angle valves is around 26°. A reduction in this angle increases efficiency. However, there is a limit to this reduction. As the seating angle is reduced, the valve closure is made easier, but at the same time, the wedge effect is increased, and consequently valve opening is made more difficult. Calculations and tests resulted in the selection of an optimized seating angle.

The force applied to the stem generates a pressure that varies with the contact area between the seat and disc. One limitation on the reduction of contact area was the mechanical properties of the seat and disc material. The contact width between seat and disc was selected to optimize this contact area. (See Figure 3).

Different materials have been tested for seat and disc applications. Some experiments have been carried out with a stellite deposit on the seat and disc. Special steels have been selected for the seat and disc materials.

Actuator Development

The first approach attempted to close the valve by first seating the disc, followed by a rotation of the disc, to break and grind the scale on the seat/disc face. Since the mechanical arrangement required was complex to design, it was rejected. We then attempted to close the valve only by seating. The hydraulic actuator was designed only to translate the valve stem. The hydraulic actuator as now developed (see Figure 1) is compact and the valve needs only minor modifications to fix the actuator to the existing yoke.

The actuator has the following parts:

- electric solenoid
- hydraulic manifold
- hydraulic motor
- coupling
- worm gear

Two hydraulic hoses connect the hydraulic manifold with the hydraulic unit. Electrical wiring connects the control panel with the electric solenoid on the hydraulic actuator.

The electric solenoid receives signals from the control panel for opening or closure of valves. Electric solenoid signals determine the direction of rotation of the hydraulic motor via the hydraulic manifold. Rotation is transmitted from the hydraulic motor to the valve stem by a worm gear transmission.

A control with three push-buttons is used to close, open or stop the valve. A programmable controller is used to control the different steps required during closing or opening.

A standard hydraulic unit is used. It includes oil tank, filter, piping, and a 7.5hp electric motor with coupling driving the hydraulic pump. Some components have been added, such as relief valves. This hydraulic unit delivers hydraulic fluid as required.

A positioner indicator, installed on the hydraulic actuator, includes limit switches to detect the end of travel during closing and opening, and a positioner to stop the valve at a predetermined position. This last device is particularly useful when it is necessary to crack the valve off its seat.

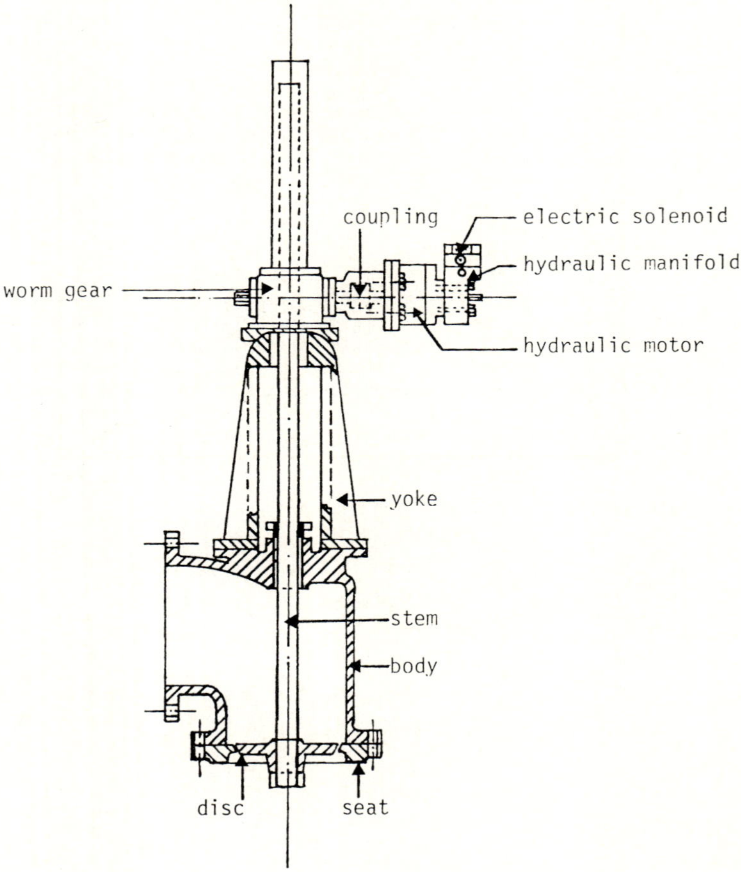

Figure 1: -Hydraulic actuator

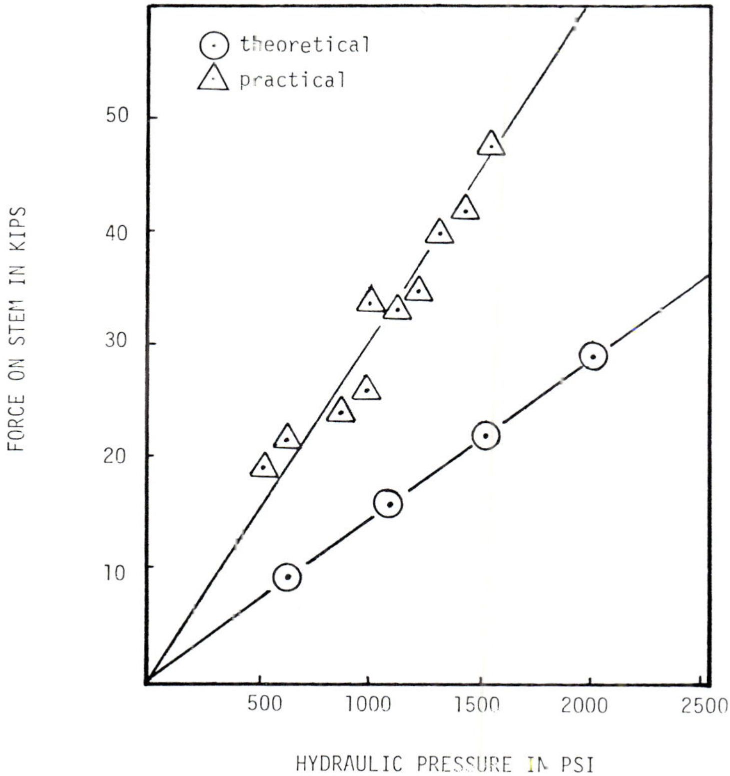

Figure 2: - Force on stem versus hydraulic pressure

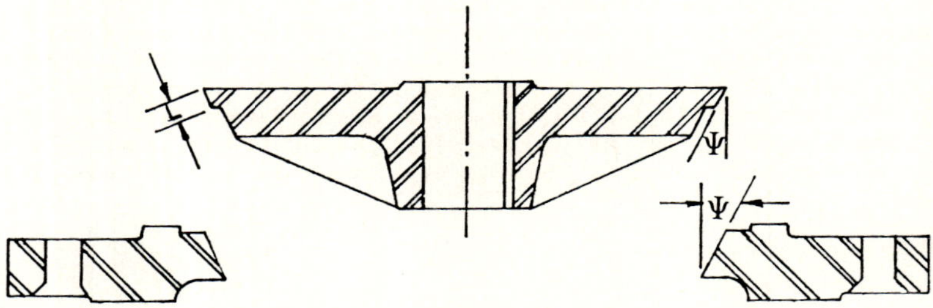

standard seat and disc geometry

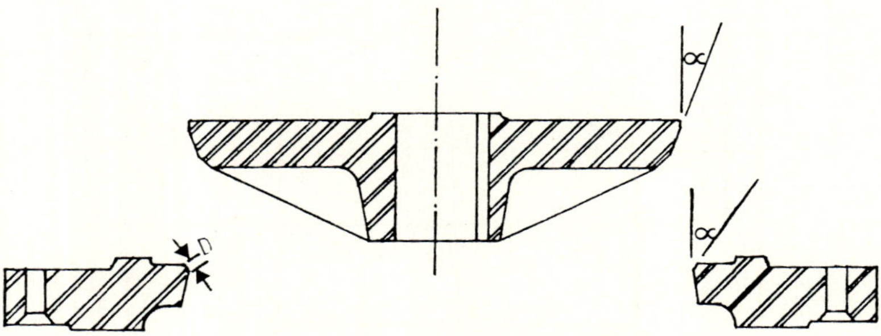

optimized seat and disc geometry

Figure 3: - Seat and disc assembly

Data and Results

A low pressure (1100 psi) has been chosen to control the displacement speed of the stem. This pressure is not sufficient to break the scale. At the end of the stem travel, hydraulic pressure is sharply increased to 1500 psi. The positioner, with the programmable controller, governs a change of the hydraulic pressure from 1100 psi to 1500 psi. This generates a hydraulic impact lasting about 10 milliseconds that breaks the scale (see Figure 4). To wash out the surfaces of the seat disc, the valve is slighly opened and the flow carries away particles of scale. The hydraulic impact is repeated several times. This sequence is performed by the programmable controller. In this way, a metal/metal contact can be obtained and a tight closure achieved in less than five minutes.

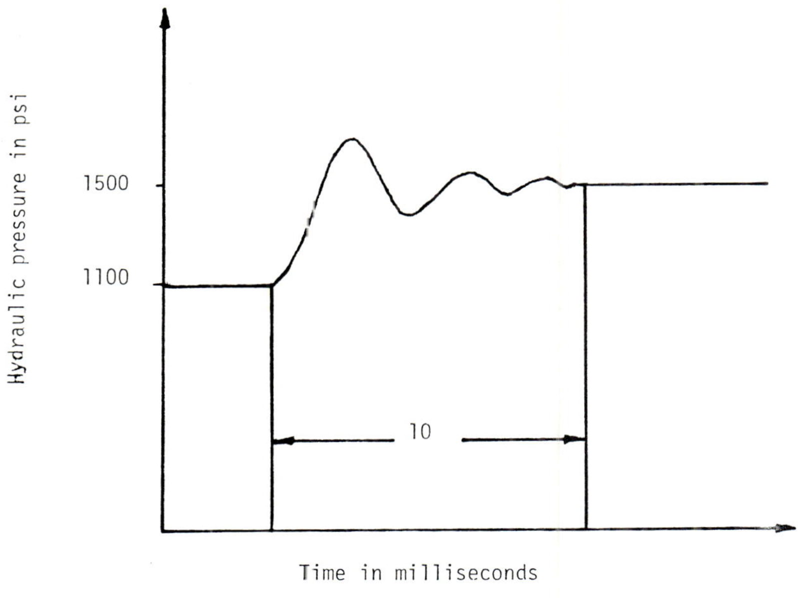

Figure 4: - Hydraulic pressure versus time

One hydraulic actuator has been operational since August 1981. In January 1983, two other actuators were installed in similar services. At the same time a programmable controller was installed to control the different steps of the closing sequence. These actuators are operated about every three weeks.

After more than two years service, a tight closure of the valve is obtained with complete elimination of scale between seat and disc. During annual shutdowns, visual internal inspections have been carried out. The general condition of the valve has not yet required its replacement. Normally, manual valves in this service are replaced every year.

Bench tests, equivalent to 15 years operation have been successfully completed. Since the first installation, no particular maintenance has been required to keep the equipment in operation.

Conclusion

The hydraulic actuator has operated in difficult conditions. This equipment is attractive for existing plants, as it can be installed easily on existing valves. It has a relatively low cost, is compact, and its reliability has been proved. It can be adapted to diverse types and sizes of valves.

DEVELOPMENT OF A DUSTLESS ALUMINA SHIPLOADER

T. Tanabe, S. Asakura, O. Yoshida and T. Kitahara

Nippon Light Metal Company, Ltd.,
Tomakomai Works, Tomakomai, Japan 053

The alumina shiploader, which has been used for loading smelter grade alumina into general bulk cargo vessels at Tomakomai Works, is equipped with special sensors and a dustless hood.

The flexible chute for loading alumina is raised automatically by the sensors in order to eliminate dusting. The special dustless hood attached to the bottom end of the flexible chute restrains the flow of alumina, which minimizes the generation of dust.

Introduction

As the dust which is generated while shiploading alumina and escapes from the hatch causes environmental problems, the development of equipment to restrain alumina dust is greatly needed.

We carried out many shiploading tests to improve the alumina shiploader which had been installed in 1972 and did not have dust-collecting equipment around the chute.

Through many tests the dustless chute of the shiploader was developed. After several improvements, it is now possible to sufficiently eliminate dusting during shiploading.

Outline of Tomakomai Sandy Alumina

The alumina plant started operation in 1972; production capacity is 360,000 metric tons per year.

Export trade of Tomakomai sandy alumina is shown in Table I.

Physical properties of Tomakomai sandy alumina are approximately as shown in Table II.

Table I Export Trade of Tomakomai Sandy Alumina ($x10^3$ mt/y)

1979	1980	1981	1982
96.5	176.0	166.3	256.4

Each year is counted from April to March

Table II Physical Properties of Tomakomai Sandy Alumina

Bulk Density (packed) (g/cc)	BET Specific Surface Area (m^2/g)	α-Al_2O_3 (%)	Angle of Repose (degree)	Size -44 microns (%)
1.1	20-100	0-60	30-34	8

Summary

The alumina, discharged from the alumina silo(25,000 mtx2), is transported by belt conveyors and is taken to the boom conveyor of the shiploader by the travelling tripper. The alumina can then be put into a hatch whenever the shiploader accesses.

Initially, covering a hatch with sheets was the only way to keep alumina dust from the holds, because the alumina was loaded by the telescopic inclination chute which did not have dust-collecting equipment. Since this did not sufficiently prevent dusting and was not easy, it was considered necessary to develop a new method.

After many operations, we found it possible to load automatically without alumina dusting from the holds under the normal weather condition by attaching a special flexible dustless chute to the shiploader.

Besides the dust collector which had already been equipped and was collecting dust at the transfer point of the travelling tripper and the head of the boom conveyor, the new dust collector was attached to collect alumina dust which was generated in the dustless chute.

The boom conveyor and the dustless chute can be folded in the shiploader when not loading.

This alumina shiploader was developed through the cooperation of NLM and IHI (Ishikawajima-Harima Heavy Industries), a world-renowned machinery manufacturer in Japan, which has had extensive experience in designing and constructing shiploaders.

Main specification of the shiploader

Type:	No slewing and chute inclination type		
Material handled:	Sandy alumina		
Bulk carrier:	3,000-30,000 D.W.T.		
Capacity:	Mean 300 t/h (max. 400 t/h)		
Length of boom:	13.6 m (From the center of seaside rail to the center of boom conveyor head chute)		
Travelling distance:	120 m		
Rail gauge:	8 m		
Wheel base:	8 m		
Telescopic inclination chute:		Full extension	11.5 m
		Full retraction	7.0 m
		Angle of inclination	+40 to -40°
Flexible dustless chute:		Full extension	9 m
		Full retraction	3 m

General view: See Figure 1

Schematic flow diagram of dust collecting: See Figure 2

1 Travelling Tripper
2 Bag House
3 Bag House
4 Operating Room
5 Boom Conveyor
6 Telescopic Chute
7 Rotary Joint
8 Dustless Chute

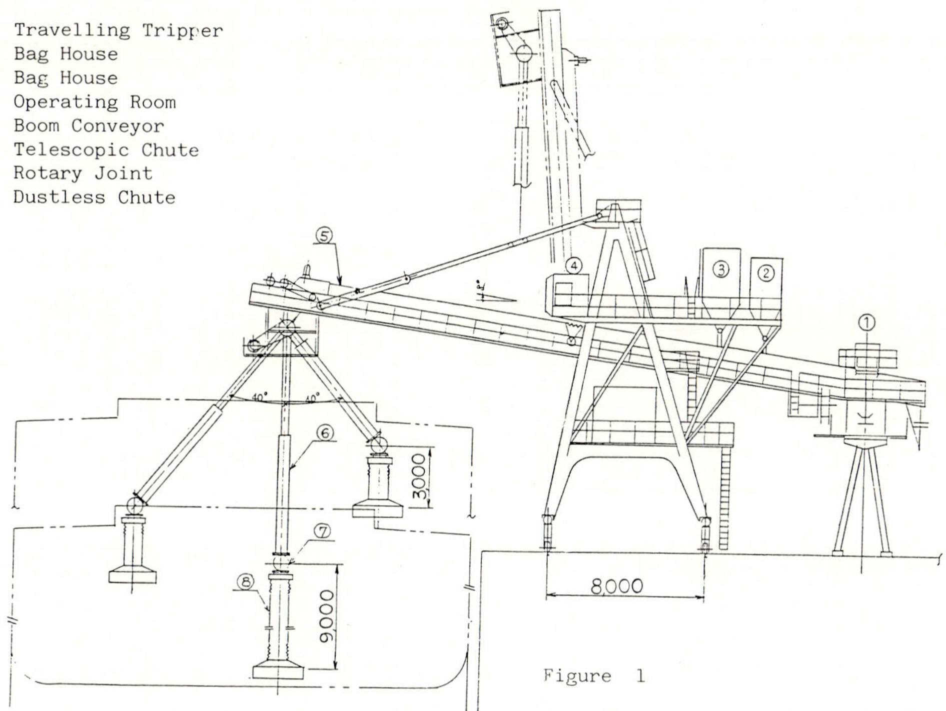

Figure 1

Schematic View of Dustless Shiploader

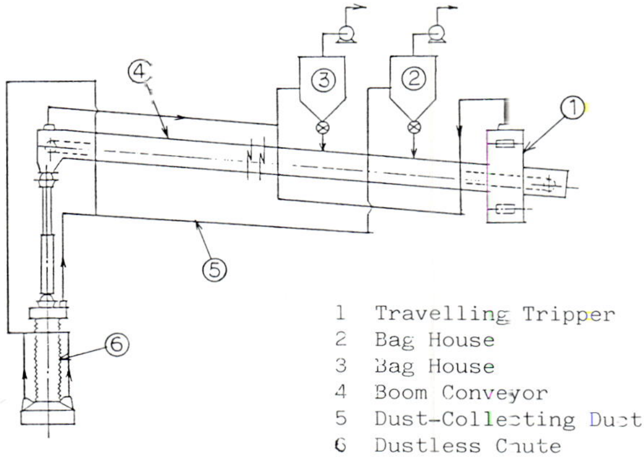

Figure 2 Schematic Flow Diagram of Dust Collecting

Characteristics of Dustless Chute

Structure

This dustless chute is composed of flexible bellows, automatic raising apparatus, internal pipe, rubber skirt, double hood, sensors, dust-collecting duct, etc.

The bottom end of the flexible bellows is covered with rubber skirt and has sensors on the outside for detecting the alumina height.

Wear-resistant rotary joint is used in order to keep the flexible bellows vertical while the chute inclines.

The theory of eliminating dusting; See Figure 3

Alumina, discharged into the telescopic chute from the end of the boom conveyor, falls into the hold and on the way to the hold the impact energy is absorbed by striking the plate of the internal pipe. The dust generated at this time is collected by the dust collector through the duct.

Alumina which falls inside the rubber skirt works as a seal and simultaneously pushes the rubber skirt outward and fluidizes smoothly along the surface of the alumina pile outside the skirt.

Sensors, which hang outside the rubber skirt, perceive that the rubber skirt is pushed outside to a certain angle which means the alumina pile has so grown that the chute should be raised.

As the dustless chute is raised automatically to the desired position, the alumina can be loaded smoothly without any blockage.

The waft dust, generated by fluidizing as alumina is pushed out from the rubber skirt, is collected by double hoods and is recovered by the dust collector.

When to begin raising after sensor perceiving, and how much to travel are freely chosen by the timer.

Figure 3 Dustless Chute Loading a Ship with Alumina

Conclusion

(1) The dustless chute, equipped to the established shiploader, eliminated dusting and has given a good shiploading environment in general bulk cargo.

(2) The shiploader could be operated without lowering loading capacity because of eliminated dusting and it can be used for all types of hatch.

(3) We were able to reduce a number of loading workers due to eliminated dust. And power, required for dust collecting, is small.

(4) As the dustless chute can be attached without drastically modifying a established equipment, a economical effect can be obtained adequately in a comparatively small investment.

NEW COAL BASED ALUMINA PLANT ENERGY CONCEPT

H. Beisswenger
H.W. Schmidt
Lurgi Chemie und Hüttentechnik GMBH
6000 Frankfurt (Main)

F. Kaempf
G. Wargalla
Vereinigte Aluminium Werke AG
4670 Lünen

ABSTRACT

Energy consumption for alumina production has already been reduced to approx. 8 GJ/mt alumina utilizing VAW-BAYER-Technology, incorporating tube digestion of bauxite, red mud filtration and fluid bed calcination. A new concept, the "Combi-Process" can further reduce energy costs by substituting cheap waste coal, for the currently-used oil or gas, to provide the total energy demand. In this concept, the waste coal is gasified and combusted in two process stages, each using Lurgi circulating fluid bed (CFB) reactors. The CFB-gasifier produces medium-BTU-gas to be used as clean fuel for calcination. Residues from the gasifier together with fresh coal are burned in the CFB-combustor, producing process heat for the tube digestors and steam for electric power and process purposes. The coal combustion plant has been in operation since 1982. Installation of the industrial gasifier is scheduled for early 1986.

Introduction

The aluminium technology is a very good example to what extent technical developments in our market-orientated Western World depend on prices of raw materials and energies.

In Europe, prices of energy and bauxite were always well above the comparable prices in the United States. Accordingly, considerable efforts were made to develop heat-saving processes, to invest in heat recuperation equipment and to utilize extraction processes which permit high extraction recoveries (5). In addition, European alumina plants are designed to process bauxite of any origin in order to meet the requirements of the complex supply situation prevailing in Europe.

On the other hand, the North-American alumina producers have built plants with the aim of attaining low production costs by keeping capital cost low and throughput capacity high. Energy cost were not of primary importance. After the explosive price increase this situation has changed considerably. Consumption and cost of energy have now become extremely important to all producers as they account for 30 - 50 % of the price of alumina (1).

In Europe, the efforts to reduce the energy consumed in the production of alumina have caused new and reliable processes. Extraction in the tube digestor and fluidized bed calcination are developments which have resulted in a decrease in heat consumption to approx. 8 GJ/t of alumina. Despite these developments the disadvantages remain that high-price oil, natural gas or high grade coal have to be used as primary energy, unless new processes are developed which permit the use of cheaper energy sources.

The conventional energy source which is abundantly available throughout the world is coal. Today it is also used for the generation of steam and electric energy in alumina plants, although its part only amounts to approx. 15 % aside from gas and oil. The reason for this is that one of the main consumers of heat and energy - the calcining process - necessitates the use of high-grade low-residue fuel such as gas or oil. Another reason is that the costs associated with coal handling kept many alumina producers from installing coal-fired steam generation systems.

Recently, stringent environmental protection requirement for the reduction of sulfur-containing combustion products of the coal are another drawback for operating coal-fired plants in the industrial countries. For example the use of high-sulfur coal of which large quantities are available in the east of the United States becomes much more expensive if the SO_2-emission has to be reduced to the stipulated levels by taking secondary measures. In Germany, the changeover from oil or gas to coal is only economically feasible if the expensive high-grade coal is replaced by cheap waste coal, the byproduct of coal benefication.

As the process technology of alumina production is forced by the market situation to reduce the energy cost by changing over to coal, development of a new process becomes indispensable where

- cheap waste coal with a high ash and low carbon content can be used, and
- the harmful combustion products, such as SO_2, NO_x and others, are discharged to the atmosphere in only the permissible range.

VAW and LURGI have developed a new concept in which the energy supply for an alumina plant is based on coal only with the aim of meeting the above requirements.

The newly developed process - the socalled "Combi-Process" consists of two stages which both basically operate on the circulating fluid bed principle.

In the first process stage coal is partially gasified to produce a combustible gas to be used for the calcination of alumina. The residue of the gasification plant is fed to the combustion plant for process heat and electric power generation.

This paper describes the possibilities of processing in both stages a wide range of fuels - even of inferior quality - by meeting the most stringent environmental requirements.

Application of Circulating Fluid Bed (CFB) - Technology to Coal Combustion and Gasification

The encouraging experience gained with circulating fluid bed (CFB) alumina calciners (3), (6), (7), led to the successful application of this type of reactor for a number of other processes (8), (11). Due to its excellent heat and mass transfer characteristics, and special design features, this technology has also been adopted for combustion and gasification of carbonaceous fuels since the mid-seventies.

Circulating Fluid Bed Combustion

Circulating Fluid Bed - (CFB) - combustors can operate on a wide range of fuels and are suited to generate steam or high temperature process heat.

Figure 1 shows the basic principle of a CFB-combustor for solid fuel.

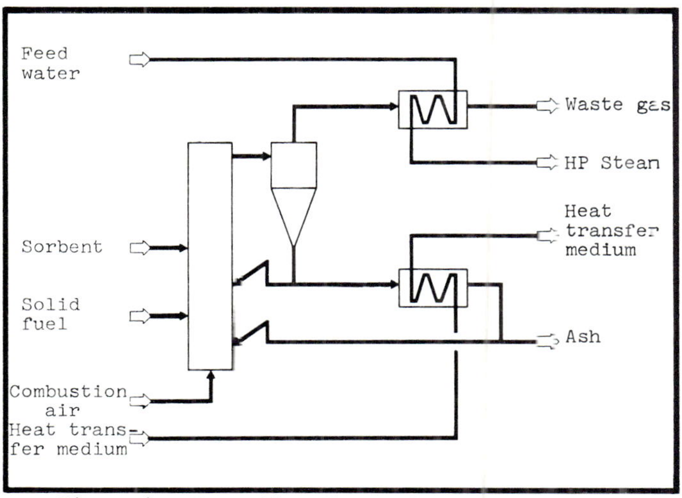

Figure 1: Basic Flow Sheet of C F B - Combustor,

The solid fuel is fed into the base of the CFB-combustor, and is burned at a temperature of about 850°C in a fluid bed.

Combustion air is added in 2 stages, and together with the combustion gases serves as a fluidizing medium for the solids. In the CFB, gas velocity and particle diameter of the solids are matched in such a way that the entire reactor is filled with solids. Part of the solid is continuously entrained from the reactor with the flue gases, captured by a hot gas cyclone, and then recirculated to the reactor to extend retention time. Through this constant recycling of fine-grained solids, the fluid bed in the reactor is stabilized.

Heat removal from the reactor is achieved by the use of an external heat exchanger, which is coupled with the CFB-reactor by means of an ash circuit, ash being fed to the fluid bed heat exchanger at a controlled rate.

In the external heat exchanger, the heat content of the ashes imparts to the heat transfer medium, heated in tube bundles in countercurrent to the ash. The cooled ash is recycled to the CFB-reactor.

Air serves as fluidizing medium for the fluid bed heat exchanger, and is used in the CFB-reactor as secondary combustion air.

The sensible heat of the hot flue gases leaving the cyclone is recovered by passing the gases through a convective bank, and generating high pressure steam. Bed ash is removed from the combustor and its sensible heat is recovered in a fluid bed ash cooler. Fly ashes are continuously removed from the flue gas via a baghouse or an electrostatic precipitator.

This process covers approx. 65 % of the energy demand (process heat and power) of the alumina production. The remaining 35 % of the energy demand can be covered by coal only if the coal is converted by gasification to clean fuel meeting the purity demand of calcination.

Combination of Gasification and Combustion

In the Combi-Process, the CFB-combustor is coupled with a CFB-gasifier. The gasifier is operated ahead of the combustor as shown in Figure 2.

The coal is first charged to the CFB-gasifier, which operates at atmospheric pressure, and a temperature of about 1000°C. By adding steam and oxygen, combustible gases are generated by partial gasification.

Depending on the type of coal and/or the relative gas to steam requirements, the carbon conversion may vary in the range of 50 to 90 %.

Typically, the net calorific value of an oxygen gasified gas to be used for combustion purposes would be about 10.000 KJ/Nm^3. This gas quality is classified as "Medium-BTU-Gas".

Gases leaving the CFB-gasifier are first desulphurized by adding limestone and then cooled in a waste heat boiler. The steam generated is used for gasification or as process steam.

Conventional gas washing processes are used to scrubb the combustible gas to the level of purity required for use as fuel in calcination.

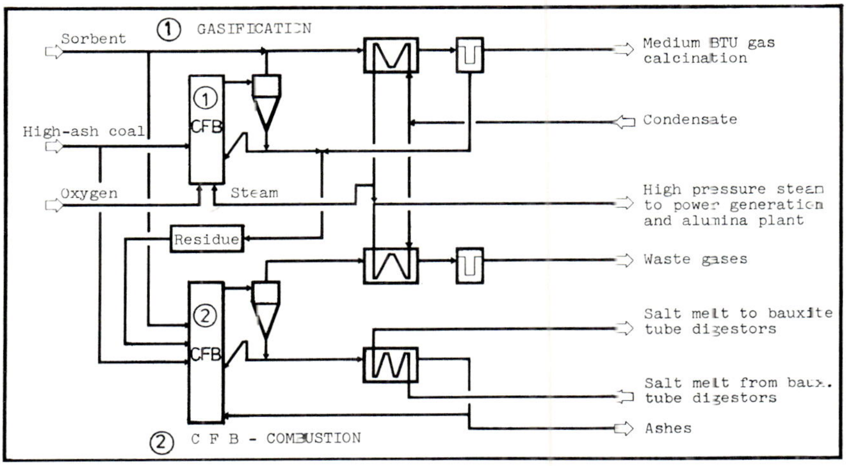

Figure 2: "Combi-Process", C F B - Gasification / Coal Combustion

The solid residue from the CFB-gasifier and the wash liquor from gas cleaning are passed into the CFB-combustion plant, where they are completely burned out. Fresh coal is adidionally fed to the combustor to balance the power, steam and process heat requirements of the overall plant.

Specific advantages of this combined process are:

(1) Complete energy supply of an alumina plant by coal;

(2) Complete exploitation of the coal;

(3) Use of fine-grained coal, containing high levels of ash, sulphur, and other contaminants;

(4) Simple desulphurization of both processes;

(5) Control of combustion at low NO_x and SO_2 emission levels.

Energy Concept VAW Lippewerk Lünen

Circulating Fluid Bed Combustor

The newly developed energy concept was first introduced to the VAW Lippewerk Lünen. This plant is located in the heart of the Ruhr-Area in Germany, where coal mines and large mountains of old tailings are readily available.

The first step in this program was initiated in 1980 by implementation of a CFB-combustor plant. This plant, which has a thermal capacity of about 80 MW, is the largest industrial fluid bed combustor today in the world.

Coal used in this plant shows the characteristics, as specified in Table I:

1.	Type of Coal: Waste Coal		
	Net calorific value	KJ/Kg	13,000
	Surface moisture	%	11
	Volatile constituents	%	15
2.	Chemical Analysis (WF):		
	Ash content:	%	55
	Carbon	%	40
	Sulphur	%	2.5
	Others (incl. F and Cl_2)	%	2.5

TABLE I: Characteristic Data of Raw Coal

As the coal consists of a mixture of old tailings, flotation sludge and coal wash residues, the waste coal contains a relatively high amount of sulphur, chlorides and fluorides.

The CFB-combustion plant is designed to generate:

- high pressure steam for power generation,
- process steam for the alumina plant,
- and high temperature process heat for heating the digestion plant.

The basic flowsheet of the process is shown on Figure 1, e.g. two separate heat transfer media are heated simultaneously by the plant.

The waste heat boiler of the CFB-combustor replaces two of the existing travelling grate boilers and feeds the generated steam directly into high pressure steam (HP-steam) line of the works.

Figure 4: Industrial CFB-Combustor, VAW Lippewerk Germany

The HP-steam is first used in turbo-geneators to generate power, extracted as medium-pressure / low-pressure steam, and then utilized in the alumina plant as process steam.

In the external heat exchanger a salt melt is heated. This salt melt serves as heat transfer medium for three tube digestor units.

Description of CFB-Combustor

The scope of the CFB-combustion plant is shown in Figure 3.

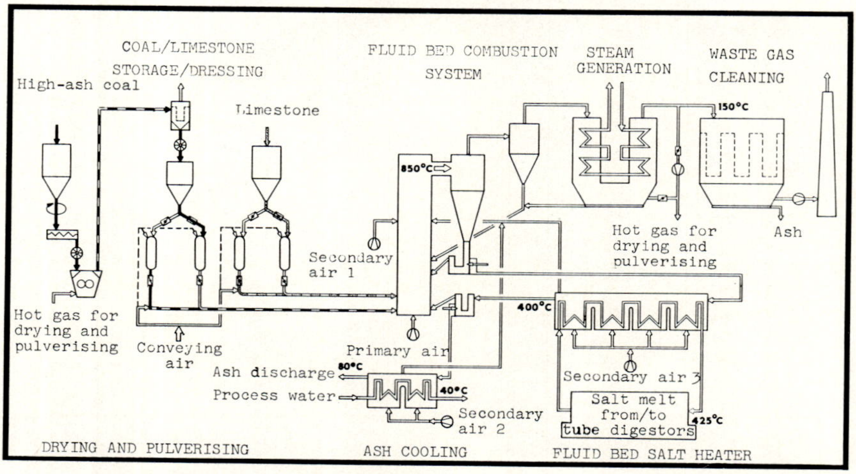

Figure 3: C F B - Combustion Plant, Basic Flow Sheet

The raw coal delivered by rail is unloaded into 2 bunkers and fed at a controlled rate to a drying and crushing plant, where it is crushed to a grain size of below 2 mm. A vertical mill with pneumatic discharge and a classifier on top of it is used to crush the coal to final grain size. Waste gases from the combustion plant serve as a drying medium. The crushed coal and limestone are stored in intermediate silos. Both coal and limestone are fed pneumatically into the combustor by means of two pressurized systems.

The CFB-combustion system comprises of the fluid bed combustor, the recycling cyclone, an additional cyclone, and the solids recirculation system. Coal combustion takes place at a temperature of around 850°C.

The hot ash extracted from the CFB-system, in the external heat exchanger, transfers heat to the salt melt, pumped through tube bundles. The salt melt is heated to 425°C, whilst the ash flowing in countercurrent is cooled to around 400°C.

A two-pass vertical through boiler with natural circulation is used as convective pass. The waste heat boiler can deliver up to 50 t/h steam at 65 bar, 480°C.

For final cleaning of the flue gases, a bag house system is provided downstream of the waste heat boiler.

That part of the ash corresponding to coal ash, free lime and formed calcium sulphate, is finally cooled in a fluid bed cooler. The heat content of the ash is recovered by preheating boiler feed water. Ash disposal does not create any problems by adding to the dry stacked red mud disposal of the alumina refinery (4). Other applications, such as additives in the building or ceramic industry, are considered.

Plant Operation

The facility was commissioned in July 1982, after a 23-months engineering and construction period. Since late 1982, the plant is operated at full capacity of 80 MW thermical. In Figure 4 a photo of the plant is shown.

Process design data, in particular those of coal combustion efficiency and heat transfer, were confirmed (9). The high expectations of the system regarding control at partial loads and response to load changes were fulfilled. The facility can operate flexible between zero and 100 % of salt heating capacity in accordance with plant requirements. Even a sudden interruption in demand by a major heat consumer can be accommodated within a short time by reducing or interrupting the flow of circulated ash.

In Table II typical operating results are presented:

Primary energy input:		280	GJ/h
Coal feed:	approx.	22	t/h
Net calorific value:	approx.	13,000	KJ/kg
Process heat output (salt melt):	approx.	60	%
Process heat output (HP-steam):	approx.	40	%
Combustion temperature:		850	°C
Excess air ratio:		1.15 - 1.20	
Molar ratio Ca/S:		$<$1.5	
Carbon burn-out:		98 - 99 %	
Carbon content in ashes:		0.5 - 1.5 %	
NO_x-emission (as NO_2):		$<$300	mg/Nm³
Retention of sulphur, chlorides and fluorides:		$>$ 90	%

TABLE II: Operating data CFB-Combustion

Energy Structure

Figure 5 shows the energy structure of the alumina plant in 1982 before start-up of the new coal combustion facility.

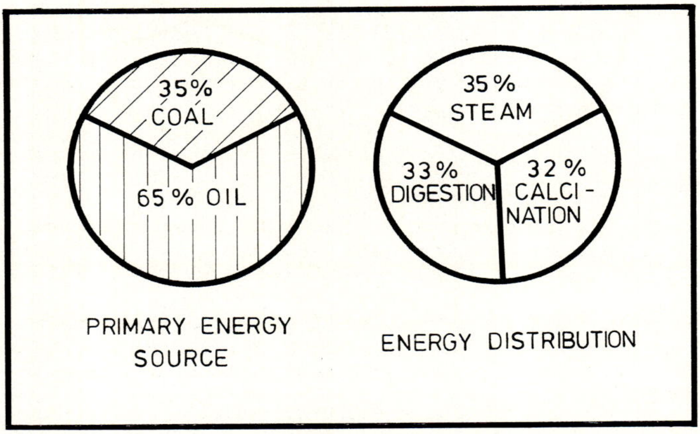

Figure 5: Energy Structure 1982, VAW Lippewerk

At that time the energy demands of the plant areas were as follows:

- Digestion of bauxite in tube digestors requires about 33 % of total heat energy demand. A salt melt is used as high temperature heat transfer medium. It is heated in several units, operating on fuel oil.

- Process steam requirements of several consumers in the alumina plant, such as removal of impurities, red mud separation, liquor cleaning and general heating purposes amounts to 35 % of the total heat energy demand. High pressure steam is generated by 3 old travelling grate boilers, burning top grade coal. By running the high pressure steam over an extraction turbine, power is generated and process steam is extracted from the turbines at the required conditions.

  Calcination of alumina requires about 32 % of the total heat energy demand to be supplied to the calciners as clean fuel in the form of fuel oil.

In 1983, after successful start-up of the CFB-combustion plant, already 49 % of the primary energy demand of the works has been based on the waste coal. Consumption of fuel oil for calcination has remained unchanged, whereas the need for top grade coal could be reduced to 12 % (Figure 6).

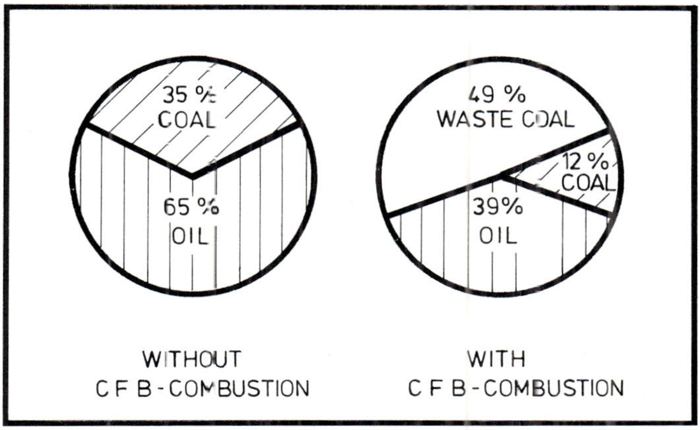

Figure 6: Change of Primary Energy Source, VAW Lippewerk

Future Energy Structure

The implementation of the CFB-combustion plant has already led to a considerable reduction of energy costs for the Lippewerk. In order to further cut costs, it is the intention, to continue the projected energy concept by realization of the CFB-gasification project.

The ultimate objective is, to switch the entire primary energy requirement of the alumina plant over to the waste coal. By implementation of a gasification plant the complete Combi-Process could be realized from 1986 onwards in an industrial scale.

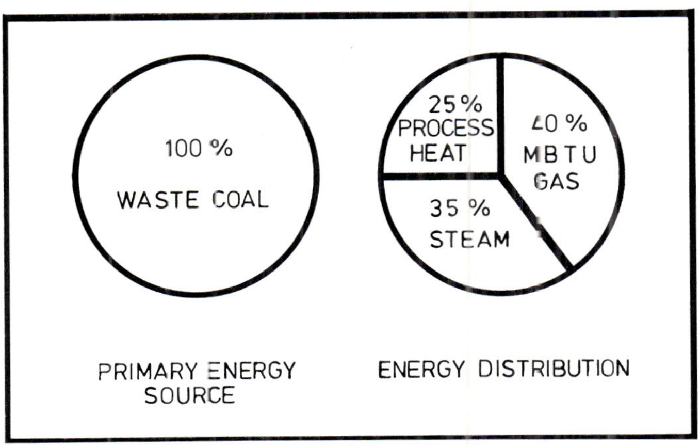

Figure 7: Energy Structure based on 100 % Waste Coal
VAW-Lippewerk

In this case, the energy distribution by both the CFB-combustor and the CFB-gasifier would be as shown in Figure 7:

- about 25 % in the form of high temperature salt melt process heat,
- about 35 % in the form of process steam and the remaining
- 40 % are covered by medium-BTU-gas, generated by the CFB-gasifier.

In mid 1983, gasifying tests with the waste coal, used already for combustion, and a top grade import coal were successfully carried out.

First results are shown in Table III:

	Top Grade Coal	Waste Coal
Gasification Rate:	90 %	70 %
Net calorific value of gases KJ/Nm^3:	10,100	8,800

TABLE III: Pilot Plant Test Results - CFB Gasification

Economic evaluation and process design of the industrial gasification plant are in progress.

The capacity of the plant is designed to produce the medium-BTU-gas required as clean fuel for calcination and for other small gas consumers in the total plant. Excess steam generated by the gasifier will be used as process steam in the alumina plant, thus allowing to completely shut down the remaining old travelling grate boilers.

Commissioning of the industrial CFB-gasifier is scheduled for early 1986.

Other Applications

The new energy concept presented, was tailor-made to meet the demands of the Lünen alumina plant. Special attention was paid to the specific requirements of the VAW-Bayer-Technology and to local preconditions.

The Combi-Process and the two inherent CFB-processes, however, due to their flexible basic principles, can easily be adopted for the energy struture of other alumina refineries.

The CFB-system can also be used for industrial boilers, co-generation plants and utility plants for electric power generation (10).

Table IV shows a summary of other CFB-combustor plants.

Plant	Duisburg Works	Flensburg Work	Ione/USA American Lignite
Type of Coal	Ruhr-Coal	Imported Coal	Lignite
Thermal Capacity MW	208	109	55
Steam Capacity t/h	270	150	66
Steam Conditions bar	145	93	87
°C	535	535	510

TABLE IV: Design Data Lurgi CFB-Boilers

The new CFB-combustion system when used in this context as a steam boiler, meets all requirements of alumina plants, based on different technologies. For typical American designed plants which have a high steam consumption, or in the case of an isolated operated plant with the necessity to generate its own electric power, steam production can be adjusted.

The possibility of using a wider range of fuels, and the very good load response characteristics favour this process for this application.

Summary

A new combustion process for coal using circulating fluid bed technology has been in operation in the Lippewerk Lünen since mid 1982. The encouraging results of the industrial unit during the first year of operation were presented. More than 150,000 t (equiv. 2 Mio GJ) of waste coal were combusted at high efficiency by meeting most stringent environmental requirements.

The energy is converted to process heat, electric power, and process steam for the alumina plant.

Ashes of the combustion plant, at present, are disposed together with the red mud. Utilization for various purposes looks very promising. The completion of the Combi-Process by implementation of the industrial CFB-gasifier is scheduled for 1986.

References

(1) Donaldson;"Energy savings in the Bayer Process" - 110th-AIME-Meeting 1981 - Chicago, TMS-Paper Selection Light Metals

(2) Bielfeldt, K., Winkhaus, G.; "Challenge to Alumina Production Technology in the 80's", ICSOBA 1981 - Hungary

(3) Reh, L.; Schmidt, H.W.; Ernst, J.; Rosenthal, K.; "Experience with the Calcination of Aluminium Trihydrate in a Circulating Fluid Bed"; TMS-Paper Selection, 100th AIME-Meeting - New York 1971

(4) Tusche, K.J.; Heber, W.; "Red Mud Separation by Filtration"; Light Metals 1979, 108th AIME-Meeting - New Orleans 1979

(5) Stahl, W.; Reinsperger, H.; Stockhausen, W.; "Application of Disc Filters in The Bayer Process"; Light Metals 1979, 108th AIME-Meeting - New Orleans 1979

(6) Schmidt, H.W.; Beißwenger, H.; Kaempf, F.; Ernst, H.J.; "Practical Experience with the Operation of LURGI/VAW Fluid Bed Calciners"; Light Metals 1976, 105th AIME-Meeting - Las Vegas 1976

(7) Schmidt, H.W.; Beißwenger, H.; Kaempf, F.; "Flexibility of the Fluid Bed Calciner Process in View of Changing Demands in the Alumina Market"; Journal of Metals, Vol. 32, No. 2, Febr. 1980

(8) Boehm, E.; Reh, L.; Sparwald, V.; "Removal of Impurities in Aluminium Smelter Dry Gas Cleaning using the VAW/LURGI Process"; Light Metals 1976, 105th AIME-Meeting - Las Vegas 1976

(9) Reh, L,; Schmidt, H.W.; Daradimos, G.; Petersen, V.; "Circulating Fluid Bed Combustion, an efficient Technology for Energy Supply and Environmental Protection"; 6th International Conference on Fluidized Bed Combustion - Atlanta 1980

(10) Lund, T.; Moffet, B.; Plass, L.; Anders, R.; "Solid Fuel Combustion in a Circulating Fluid Bed for Co-Generation"; 3rd International Conference on Co-Generation - Houston 1983

(11) Stockhausen, W.; Broedermann, P.; "The Circulating Fluid Bed for endothermic Reactions of fine-grained Materials; TIZ (Tonindustrie-Zeitung) 12/81

ALUMINA AND BAUXITE VI

CHAIRMAN

W.C. Sleppy
Alcoa Laboratories
Alcoa Center, Pennsylvania

ADVANCES IN THE FILTRATION OF ALUMINA TRIHYDRATE

Ronald F. Nunn and Robert M. Talcott

Dorr-Oliver Incorporated
77 Havemeyer Lane
Stamford, CT 06904
USA

In response to changes in Bayer plant operation, an extensive study of the effects of process and operating variables on the filtration - washing of alumina trihydrate was undertaken. The laboratory phase of the study reveals that levels of performance which exceed historical ratings may be possible.

The variables studied include feed slurry characteristics, cake thickness, cake heel effects, vacuum level, vacuum capacity, solids loading, and various wash variables.

Indications are that established design parameters should be revised. Cake moisture is largely controlled by vacuum capacity and solids loading, but significantly shifted by feed liquor concentration. The cake heel left at discharge influences both cake moisture and soda concentration. Exceptionally high cake washing efficiencies are achievable with only two displacement volumes of a single stage wash.

1.0 Introduction

The filtration of alumina trihydrate is an important step in the Bayer process. This study is concerned only with the filtration of hydrate produced by sandy alumina plants.

The three main types of filter used today for this service are the Dorr-Oliver horizontal table filter, the Dorr-Oliver Dorrco rotary vacuum internal drum filter and the horizontal belt filter. The Dorrco filter was superceded by the horizontal table filter some years ago. However, there are still many units of this type of filter in operation.

For some time there has been a trend to the use of fluid bed kilns for trihydrate calcination due to the fact that they require only about two thirds the fuel per ton of alumina that the traditional rotary kiln requires. However, the feed to the fluid bed kiln must be on the order of 2 to 3% coarser than the feed to the rotary kiln in order to compensate for the increased amount of attrition.

2.0 Objective

The objective of this study is to examine the factors affecting the operation of the horizontal table filter on sandy hydrate.

The modern day alumina process demands low moisture and a low soda content in the hydrate feed to the kiln. The kiln is a very inefficient evaporator, so that any reduction in the moisture of the feed will have a significant effect on the cost of fuel for producing alumina. In addition, the benefits of reducing the soda content in the kiln feed are that product purity is improved and soda usage is reduced. The modern day process uses very large fluid bed calciners in some cases, as large as 1600 metric ton/day of alumina. The manufacturers of these calciners typically have to guarantee the fuel usage to the purchaser. Therefore, the performance of the hydrate filter has become much more critical in recent years.

The goal of the study is to upgrade filter performance, in order to provide a cake with less than 10% moisture and less than 0.05% Na_2O using less than 0.4 lb. wash water per lb. of hydrate, while at the same time maximizing the hydrate capacity of the filter.

3.0 Testwork

The testwork was divided into two steps. The first step tested samples of hydrate from Bayer plants in the laboratory. A total of 159 tests were made on three sandy hydrates. The second step will consist of plant tests required to subtantiate the laboratory results.

The factors examined in the testwork were as follows.

3.1 Vacuum Capacity and Level

3.2 Dry Solids Rate, Involving Dry Time Variation

3.3 Feed Slurry Characteristics

3.4 Effect of Leaving Cake Heel on Filter

3.5 Wash Water Rate, Temperature, Method of Application, and Countercurrent Stages

3.6 Cake Thickness

3.7 Filter Cloth Characteristics

4.0 Results

4.1 Vacuum Capacity and Level, and Dry Solids Rate

The solids feed rate and the vacuum capacity are the major factors affecting the moisture of the discharged cake. The ratio of vacuum capacity to solids feed rate

$$\frac{\text{cu. ft./hr./sq.ft.}}{\text{lb. dry cake solids/hr./sq. ft.}} = \text{cu. ft./lb. dry cake}$$

correlates closely with the percent moisture in the cake. In other words the vacuum capacity in terms of cu. ft. per lb. of dry cake when plotted versus cake moisture, yields a single curve relating the key variables of vacuum capacity, solids loading, and cake moisture for a given feed soda concentration. Figure 1 attached shows two such curves. Curve 1 represents results obtained with feed at 160 gpl TTS as Na_2CO_3. This curve is a conservative interpretation, being closer to a maximum percent moisture for well washed cakes. Curve 2 represents data for a feed with 41 gpl TTS as Na_2CO_3. This curve, too, is conservative with most data falling below the line. This curve has been recently reinforced by data on another, similar sample, tested at 120 gpl TTS as Na_2CO_3. These curves can be used to predict plant filtration performance. For example, to obtain a 10% moisture cake in the plant with a feed slurry at 160 gpl TTS Na_2CO_3 would require a vacuum capacity of 2.15 cu. ft. per lb. of dry cake. This relationship can also be shown as in Figure 2. Here the parameters of cubic ft. of vacuum capacity per lb. of discharged dry hydrate are plotted on coordinates of plant vacuum capacity (cfm/sq. ft.) versus plant rate (including 0.8 scale factor) of dry hydrate discharge. Each cf/lb. line is further identified by the cake moisture at various feed liquor concentrations.

With the 160 gpl TTS feed the cake moisture shown on the top (in parentheses) was higher than with the weaker liquors. With the 92-120 gpl TTS feed other variables had a more identifiable effect on moisture and a design maximum and a possible minimum moisture curve were identified. The minimum curve is effectively the same as the curve for 90% of the data with feed at 40 gpl TTS.

The standard vacuum level used was 12-15" Hg vacuum. Some data at 15" Hg vacuum show cake moisture about 0.5% lower.

The gases passing through the filter test cakes to the rotameter were frequently sampled and analyzed for water vapor,

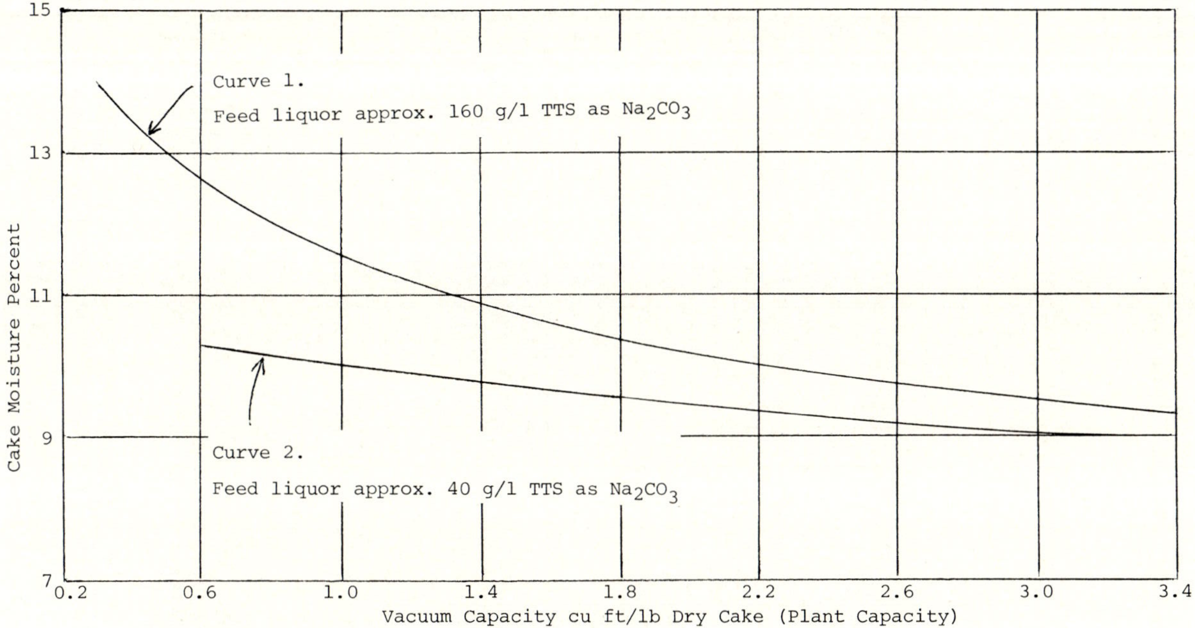

Fig. 1 Cake moisture vs. vacuum capacity per unit dry discharged cake solids (plant design basis for horizontal table filter.)

Limiting Conditions: Total cake thickness - 2.5 - 3.5"; Cake Heel - minimum 0.5"; Cake washed to a soluble soda residual maximum of 0.03% Na_2O based on dry cake.

demonstrating saturation with about equal volumes of vapor and air. Thus test measured gas flows are comparable to plant conditions.

4.2 Feed Slurry Characteristics

The soda content of the feed slurry has a distinct effect on the cake moisture. This can be seen in Figure 1 where Curve 2 representing 40 gpl TTS feed lies considerably below Curve 1 which represents a 160 gpl TTS feed. Figure 2 also shows the effect of variations in the soda content of the feed. Attached to each line are moisture data representing the cake moisture at different feed liquor strengths.

4.3 Effect Of Leaving Cake Heel on Filter

One distinct characteristic of the horizontal table filter is that a cake heel is left on the filter after the bulk of the filter cake has been discharge by the scroll. In contrast, the Dorrco filter and the horizontal belt filter discharge 100% of the filter cakes. This method of operation by the horizontal table filter constitutes a significant benefit with regard to the cake moisture and soda. Analysis of the ½" cake heel shows that, generally, the soda concentration in the heel is ten to twenty times as high as that in the cake discharged from the filter. In addition, cake heels of about ½" had moistures averaging about 4% higher than the discharged cakes. This heel is subsequently reslurried on the horizontal table filter, mixed with the feed slurry, re-filtered and re-washed. The net effect of leaving the heel on the filter is to produce a cake discharge with a lower soda and moisture content than the Dorrco and horizontal belt filters under similar conditions.

4.4 Wash Water Conditions and Cake Thickness

The wash water rate, its temperature, and method of application, all have an important effect on the filtration and washing performance of the filter. Cake washing efficiency on this material is inherently very high. A 99% washing efficiency can be achieved with a 2.5 displacement wash. Thicker cakes, i.e., 2.5"-3.5", wash more efficiently and reliably. The desired residual 0.05% Na_2O based on dry hydrate was obtained by a single stage wash of 0.16 to 0.2 lb. of wash water per lb. of dry hydrate for 3" and 2" thick discharged cakes respectively. Thinner cakes, i.e, 1.0"-1.5", wash less reliably. Washing efficiency data is represented graphically, both for discharged cakes (leaving ½" heel) and for total cakes. This information is represented in Figures 3 through 5 attached. Figures 3 and 4 present the percent solubles remaining (related to an unwashed cake having 100%) versus displacement volumes of wash (where one displacement is identified as the volume of moisture in the cake at discharge - which is 0.11/lb. water/lb. dry hydrate at 10% moisture). The 100% remaining solubles content is approximately 1.45% Na_2O, when feed contains 160 gpl TTS.

Figure 3 represents the data for cakes excluding a ½" cake

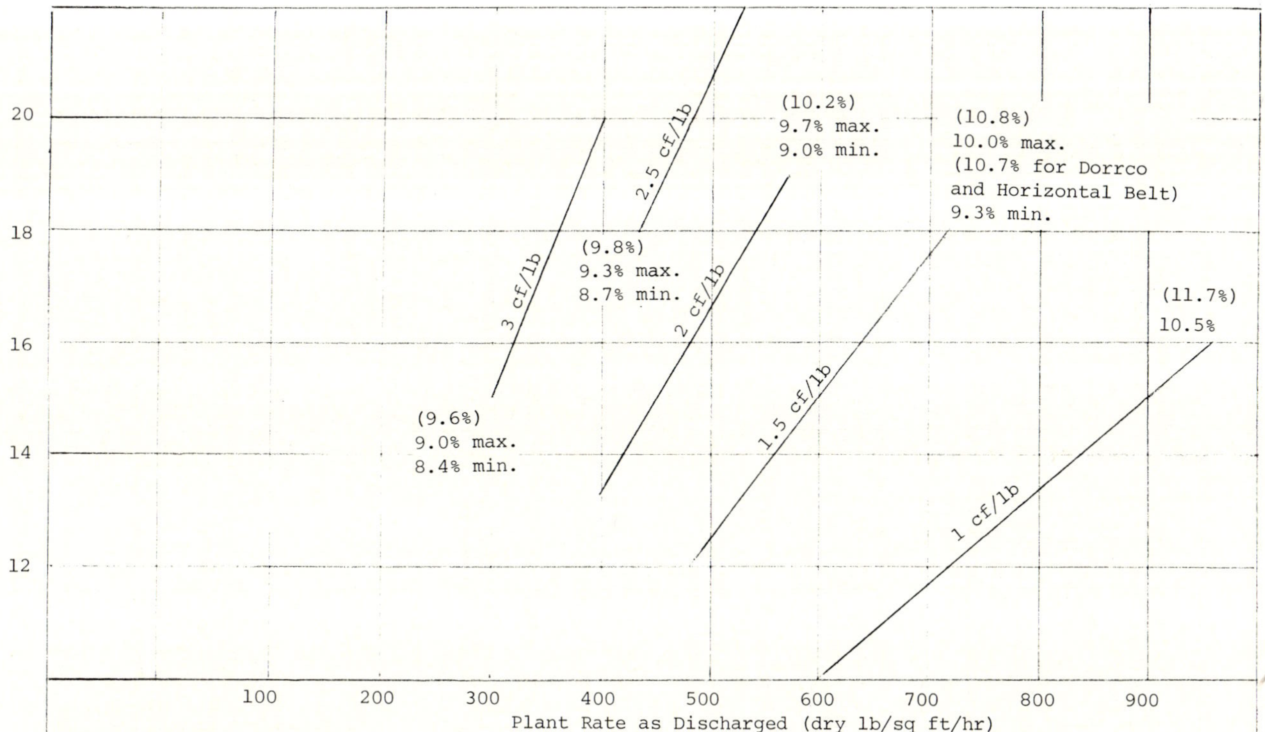

<u>Fig 2 Vacuum Capacity Vs</u>
<u>Plant Solids Rate</u>
Parameters of cu ft/lb and Predicted cake moistures discharged from D-O Horizontal Filter at various feed concentrations.

Cake Moisture Identification	Feed Liquor Strength g/l TTS as Na_2CO_3
(In parentheses)....	160
max..........	90-120
min.........	possible for 90-120 probable for 40

heel - i.e. representative of the Dorr-Oliver horizontal table filter. Figure 4 presents the same data for the total cakes - i.e. representative of Dorrco and horizontal belt filter.

Figure 5 combines the Figures 3-4 data to represent the effect of cake thickness and of leaving the cake heel on the filter. Thicker cakes obviously wash more efficiently. The parameter used here is wash displacements. The 1.5 displacement curve for the horizontal table is close to the two displacement curve for the Dorrco and horizontal belt filters. The two displacement curve for the horizontal table lies between the three and four displacement curves for the Dorrco and horizontal belt filters. The wash water temperature used in most of the tests was 85°C in order to simulate plant operation. Wash water at temperatures up to 95°C yields discharged cakes at moistures averaging 1 full percentage point less for constant solids loadings.

Applying the wash water in two incremental sprays, each spray utilizing half of the total water, was found to be slightly more efficient than one spray only, confirming plant experience. Either spray technique was better than flood washing.

Countercurrent washing showed the expected improved efficiency with two stages but little additional benefit with three stages. This is to be expected when using more than two wash displacements. The CCW curve on Figures 3 and 4 include both two and three stage washing.

4.5 Filter Cloth

Desirable cloths are open enough to permit free flow of air, both on the vacuum and the blowback sections, and also filtrate but tight enough to prevent high solids loss in the filtrate. Satin weaves, due to their high thread count for equivalent air permeability, are the most attractive cloths for this application - twill weaves are next.

Polypropylene is used extensively. The testwork indicated that a 350 cfm air permeability will permit high flow with small loss of fines in the filtrate. A number of different filter cloths were tested. National Filter Media cloth number 224-071-35 yielded the best results.

4.6 Plant Results

After obtaining the foregoing results using leaf tests at bench scale we started a program to substantiate these results at plant scale. At the time of writing, the plant tests are ongoing. Early results indicate that the design and operating parameters established for hydrate filtration at bench scale also hold true on commercial size filters.

5.0 Conclusions

5.1 The tests indicate that leaving a heel of cake on the horizontal table filter significantly reduces the moisture and

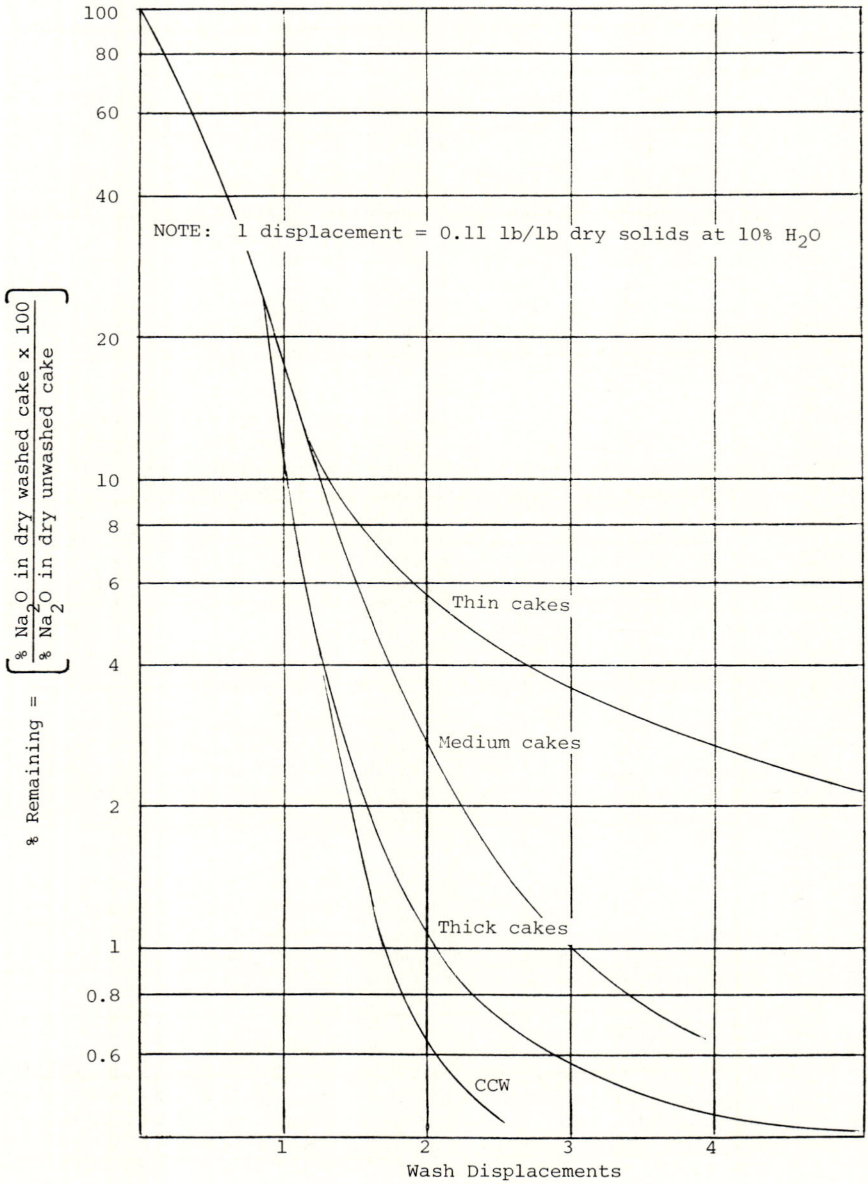

Fig. 3 Washing of discharged cakes - excluding 1/2" heels.

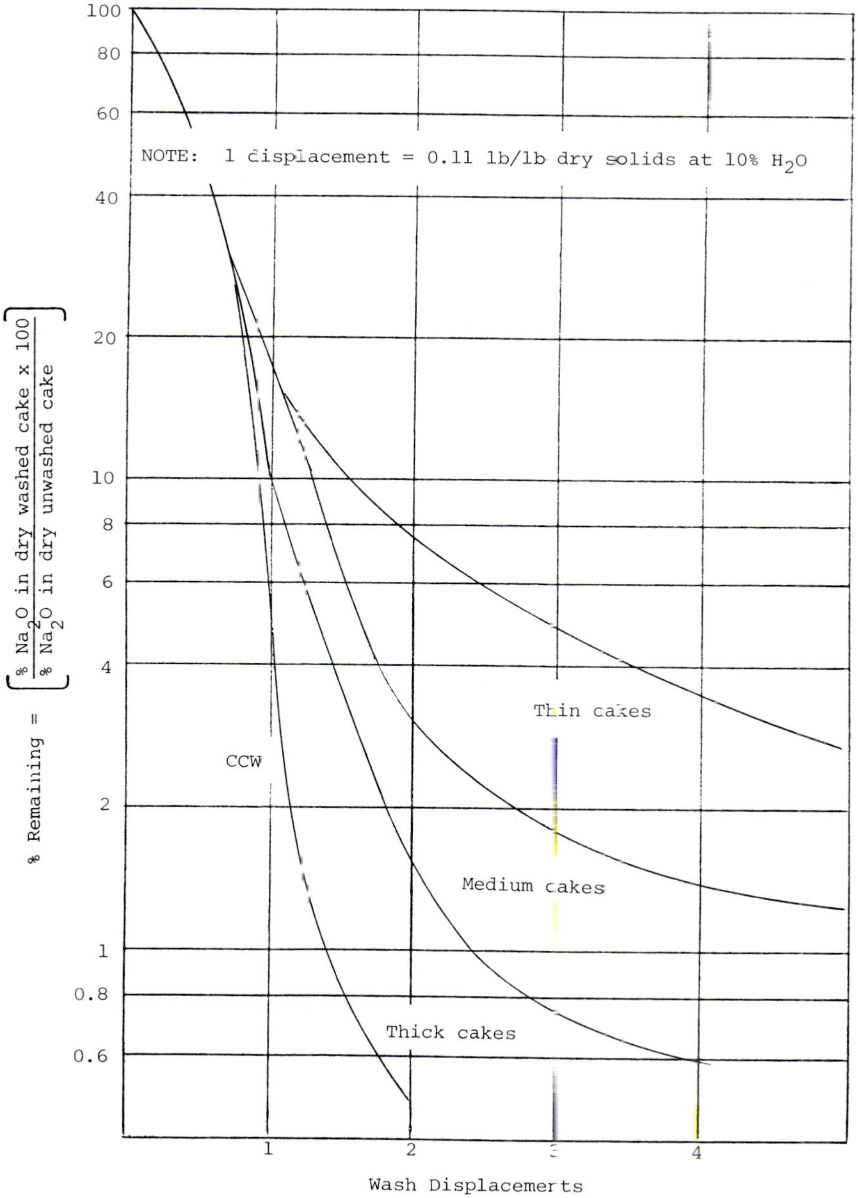

Fig. 4 Washing of total cakes - including 1/2" heels.

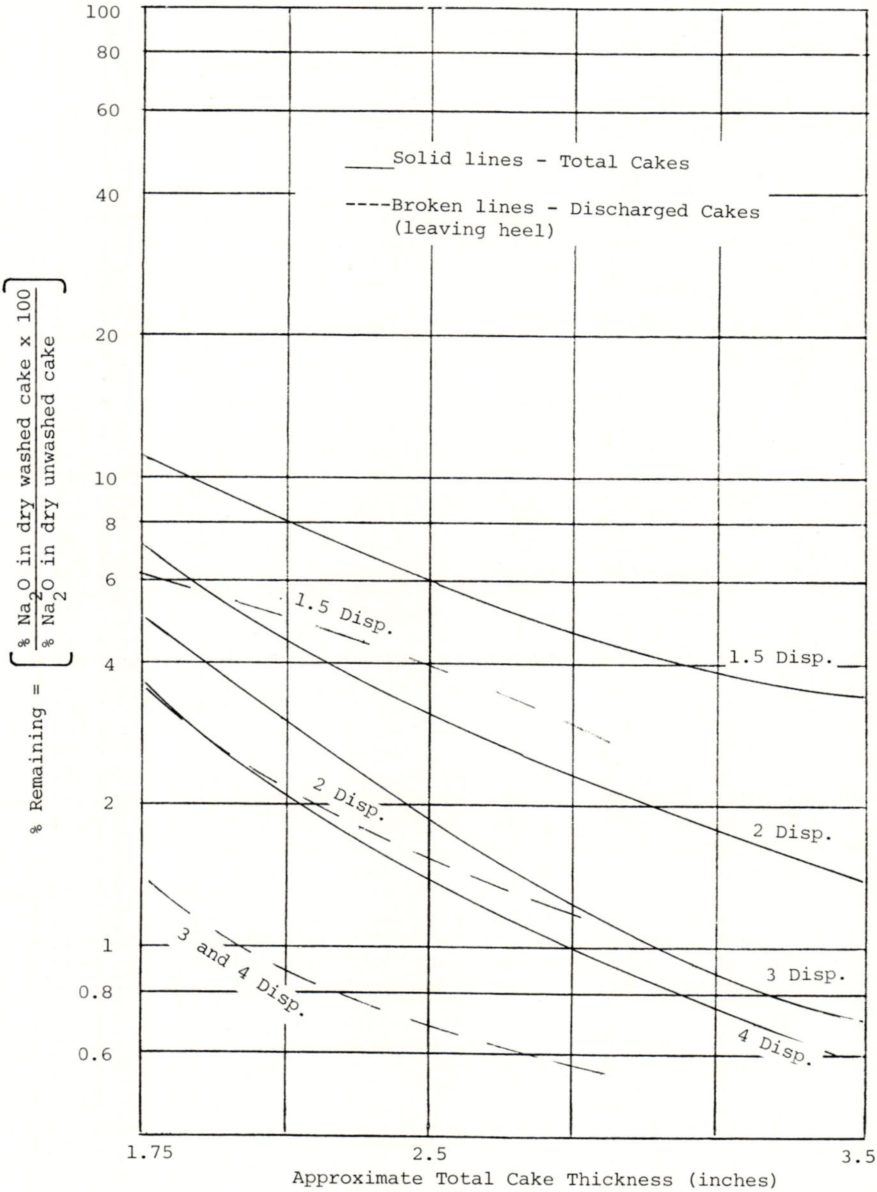

Fig. 5 Cake Thickness Effect on Washing. Parameters are Wash Displacements

soda content of the cake when compared with the Dorrco or horizontal belt filters. Typical moisture and soda values were:

	Wash Water Rate*	% Moisture In Cake	% Na_2O In Cake (D.B.)
Horizontal Belt Filter	0.4	10.7	0.056
Horizontal Table Filter	0.4	10.0	0.022

* lb./lb. dry hydrate

Assuming a calciner fuel cost of $5 per million btu's, the lower cake moisture from the horizontal table filter would result in a fuel savings of about $100,000 per year for a million ton per year alumina plant. Assuming a soda cost of $250 per ton NaOH the lower soda content from the horizontal table filter would result in additional savings of about $170,000 per year. These savings assume the same conditions on both types of filters.

5.2 A survey of commercial horizontal table filters on hydrate filtration service indicates that in general there is considerable room for improvement of the cake moisture, washing efficiency and capacity of these filters, especially those that have been in service for many years.

Modifications have been made to one commercial filter and tests are in progress at the time of writing. Preliminary results indicate that the washing and moisture performance and the filter capacity can be improved.

5.3 The tests indicate that, with feed liquor concentration below about 120 gpl TTS, the horizontal table filter can achieve 10% moisture cake at 0.01-0.05% Na_2O using 0.2-0.4 lbs. wash water/lb. dry hydrate and at filtration rates up to 700 lb./hr./ft^2, when using 18 cfm/ft^2 vacuum capacity (or to 600 lb./hr./ft^2 and 15 cfm/ft^2). The filter area used to calculate this filtration rate is the total cloth area of the filter, which is about one third greater than the "active" area of the filter which is sometimes used to express filtration rates.

This performance meets the stringent requirements of the industry today. The latest fluid bed kiln designs have a 1600 mtpd alumina capacity. The feed to this size kiln would require one 24' diameter Dorr-Oliver horizontal table filter operating under the improved conditions listed above.

UTILIZATION OF HORIZONTAL BELT FILTERS FOR WASHING FINE ALUMINA SEED AND KILN FEED PRODUCT

BY

RICHARD F. CRAWFORD,* DONALD A. DAHLSTROM,** and GERRY MINEAR***

Introduction

In the processing of bauxite to produce alumina (Al_2O_3) for ultimate production of aluminum by electrolysis, the product must have a high purity and a closely controlled size distribution. As the entire alumina flowsheet involves fluids of high caustic concentration, one of the critical measures of product purity is the sodium content. This should be a maximum of 0.04 Wt% soluble Na_2O. In the United States, smelters are based upon sandy alumina. Generally, sandy alumina trihydrate ($Al_2O_3 \cdot 3H_2O$) prior to calcining should have a size distribution in the range of 95% +44 microns and only a very small amount of -20 micron material to meet final alumina product size requirements.

While product quality is mandatory in alumina processing, other factors which significantly influence operating cost, must be considered. First, any water introduced into the flowsheet must be thermally evaporated either in specially designed flash tanks and evaporators or in the product calciners. The alumina industry employs evaporation units which generally remove approximately four pounds of water for every pound of steam. With today's increasing energy costs, it is very desirable to reduce water inputs for filter cake washing in order to reduce evaporation energy consumption.

Secondly, moisture content of product kiln feed should also be minimized. Additional energy must be added over that required for the calcination reaction as the feed moisture increases.

Third, caustic soda must be recovered efficiently for obvious economic reasons. Thus, the final product purity of 0.04 Wt% soluble Na_2O maximum reflects this necessity as well as aluminum reduction plant requirements. It is stressed again, however, that Na_2O recovery should be achieved while minimizing wash water input.

---

* Manager of Process Studies, Kaiser Aluminum & Chemical Corporation, P. O. Box 1031, Baton Rouge, LA 70821

** Sr. Vice President, Research and Development, and *** Product Manager, Eimco Process Equipment Company, P. O. Box 300, Salt Lake City, UT 84110

Finally, organic material in the bauxite and from other inputs converts partially to sodium oxalate in the high pressure/temperature leach and the following processing steps. If the oxalate is not removed, serious processing problems occur. If oxalate concentration is too high, particle size distribution control is adversely affected in the alumina precipitation process. Yield will also decrease and this increases energy consumption and cost. Sodium oxalate concentration on the hydrate seed in the precipitation section should be held to a maximum of 0.1%.

Improving the efficiency of both the oxalate removal and the kiln feed deliquoring and washing steps can reduce process energy consumption and improve product quality.

The oxalate removal step can be achieved by deliquoring the fine alumina trihydrate used as seed for crystal growth in the precipitators, displacing the filter cake liquor with cold water to recover caustic values and then washing the cake with hot water to solubilize the oxalate. This filtrate can then be processed to remove the oxalate. This can also increase the capacity of the precipitators by removing an appreciable quantity of spent liquor.

The deliquoring step prior to the product kiln must achieve both the required purity and minimize wash water input. Additionally, the moisture content must be lowered to a maximum of 10 Wt%.

It is felt both steps can be improved over present practice through the use of horizontal belt filters. The Eimco-Extractor®, which is a horizontal belt filter, is shown schematically in Figure 1. The unit consists of an elastomer drainage belt which slides over a vacuum box. By proper construction, the drainage holes through the rubber belt are confined to the vacuum box opening. Low friction coefficient fluorocarbon plastic slide plates are employed to reduce horsepower drive requirements. A filter media rides on top of the drainage belt on the upper portion of the filter. Synthetic fiber cloths are employed. Because of the design, all filtrates (strong, first wash, second wash, etc.) are isolated by vertical baffles in the vacuum box so they can be separated. Accordingly, countercurrent washing can be practiced. This is very important in reducing water consumption in kiln feed dewatering. This feature also permits the segregated hot and cold water washes for oxalate removal. In addition, the filter cloth is returned to the head end of the filter where it can be washed to prevent blinding. This maintains high filtration rates and minimum moisture content.

Some of the design advantages of horizontal belt filters are:

1. Good feed distribution. Because of the belt principle, feed can be more evenly divided across the width of the filter as compared to a circular disc shape.

2. Efficient multi-stage washing of the cake. An even cake thickness is presented for cake washing and the washing fluids can be isolated by cylindrical rolls or flood dams riding on the cake surface.

3. Variations in feed flow rate and solids concentration can be handled by these units. Cake thickness can be measured and controlled by automatic belt speed variations. Such flexibility is important to maintain product quality.

4. Operator visual analysis of the entire filter operation. The horizontal surface is entirely exposed so that any upset may be quickly observed and the operation can be "fine-tuned".

5. Effective washing and chemical cleaning of the filter media. By use of headers with appropriate spray nozzles, the cloth, which is returned to the head end in a separate path from the drainage belt, can be washed as required. If scaling occurs, the cloth can be washed with appropriate chemicals.

6. Because of the rectangular construction, floor space efficiency is increased. Also, spare filters can be more economically employed to increase availability and improve operating factors.

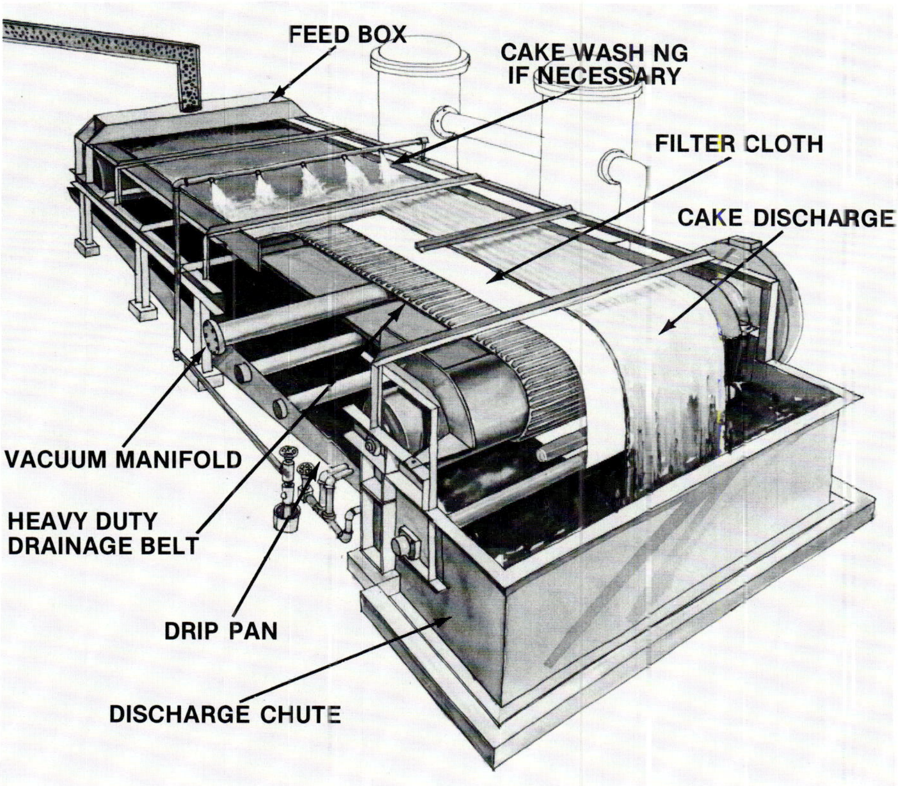

Figure 1
Schematic of an Eimco-Extractor

Application Development

Seed Washing

In precipitation, alumina trihydrate is recovered from the sodium aluminate liquor. A major operating objective in the precipitation process is to obtain the maximum yield consistent with the required particle growth. Since the undersize hydrate is recycled as seed, particle growth must be controlled to obtain a balance between the seed produced and that required to obtain a specified product. This requirement is particularly critical in plants producing coarse alumina.

Contamination of process liquor impairs both yield and particle growth. Sodium oxalate in sufficient concentration in the liquor coprecipitates with alumina hydrate. The solid oxalate tends to concentrate with the fine seed hydrate and can further generate a large amount of very fine hydrate during the precipitation cycle. Removal of oxalate crystals from the seed circuit is sometimes necessary. This can be accomplished by seed heating, seed washing, or selective oxalate crystallization.

Kaiser developed a patented process for oxalate control which employs filtration and washing of fine seed. Facilities utilizing this oxalate removal system have been incorporated into alumina plants worldwide.

The use of the horizontal belt filter was proposed for this application since it had the potential to improve both the installation and operating costs of the seed washing operation.

Kiln Feed

The final unit operation in the Bayer process is calcination, where hydrate is converted to alumina. The function of the hydrate filters is to deliquor the hydrate slurry and recover soluble soda prior to calcination.

The soda content of the filter cake directly affects product quality and the amount of wash water required for soda recovery affects plant evaporation requirements. The moisture content of the filter cake influences calcination fuel requirements.

The use of horizontal belt filters for kiln feed service was proposed as a potential method of improving both installation and operating costs.

Pilot Scale Testing

A development project was carried out at Kaiser Aluminum & Chemical Corporation's Baton Rouge Works in 1970 to evaluate horizontal belt filters in seed washing and kiln feed filtration operations. Pilot scale tests were conducted using an Eimco-Extractor having 0.98 m^2 (10.5 ft^2) of effective filtration area.

A total of 70 pilot scale tests were run. The seed washing studies encompassed 57 tests and the remainder evaluated operation in kiln feed service. Conclusions based on these studies included:

Seed Washing - Oxalate removal could be accomplished effectively on a single-stage horizontal belt filter with a fine seed handling capacity of 1220 - 1464 Kg dry solids/hr-m^2 (250 - 300 lbs/hr-ft^2) of filter area under vacuum.

Kiln Feed - Efficient kiln feed filtration could be accomplished using a horizontal belt filter operating with a hydrate capacity of 5855 Kg dry solids/hr-m^2 (1200 lbs dry solids/hr-ft^2) of filter area under vacuum.

Operation - With comparable feed rate and feed properties, the horizontal belt filter could provide better control and require less wash water than other methods.

Laboratory Scale Testing

Bench scale testing and analysis is commonly performed on many different applications to predict full-scale results. By proper correlation methods (developed from basic theory) filter station requirements can be reliably develped for such important items as discharged cake moisture percent, content of cake with respect to solubles, required wash fluid volume and number of wash stages to achieve desired solubles recovery, and filtration rate per unit area. Furthermore, performance results over a wide range of operating conditions can be predicted.

Many times, it is essential to conduct such experiments in the operating plant. This is the case in alumina processing. Properties of alumina trihydrate slurries can change significantly when samples are shipped to a different location. It is also more convenient, because representative feeds can easily be obtained, difficult analytical procedures are routine, and normal wash fluids with respect to chemical concentration for solubles recovery tests are readily available. Accordingly, bench scale test work has been done in several operating alumina plants on red mud, hydrate seed and product alumina filtration and cake washing, as well as countercurrent sedimentation for separation of red mud and pregnant liquor.

In addition to pilot scale test programs, many bench scale filtration tests have been run to obtain data necessary to size Eimco-Extractors for specific applications. A brief description of the laboratory test procedure which simulates each of the component steps of a commercial Extractor using a test filter apparatus follow:

Equipment - The basic equipment (refer to Figure 2) for bench scale testing of horizontal belt filter application consists of a small drainage grid covered with filter cloth and surrounded by a shim. This filter leaf is supported vertically by a vacuum flask. All system pipes and tubing are of large bore to minimize hydraulic restrictions. A second vacuum flash (not shown in the sketch) is filled with cool water to maintain the vacuum system air at a standard temperature. A gas meter is attached at the discharge side of the vacuum pump to measure airflow though the filter cake.

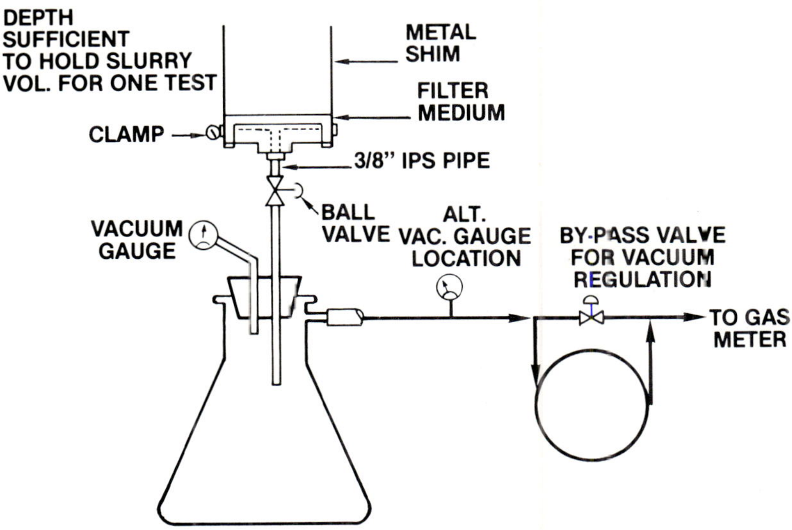

Figure 2
Bench Scale Extractor Test Arrangement

Feeding - A premeasured weight of slurry containing the specified dry cake weight is poured onto the filter leaf surface. Vacuum is immediately applied at the desired level. The time required to form the filter cake (when the last free liquor disappears from the cake surface) is recorded.

Washing - In tests simulating single stage washing, a known volume of solute free wash water is applied after a specified prewash drying period, and the wash time recorded. Countercurrent washing is simulated by using a first wash of known volume and soda concentration, following a drying period, and then applying the final wash using a known volume of solute free water.

Drying - The cake is then allowed to dewater for a predetermined drying period. After releasing the vacuum, the cake is removed from the filter leaf into a container which is sealed immediately and saved for laboratory analyses.

Data from the laboratory filtration tests are reduced, correlated, and usually presented graphically. Correlations developed include:

1. Cake dry solids weight and filter cake thickness relationship
2. Filter cake weight and cake formation time
3. Percent solubles removed and cake wash displacements
4. Cake moisture and dry time correlation factor
5. Air flows and dry times

Scale-up factors between bench and full-scale operation have been evaluated. The factors developed in the laboratory tests have proven reliable for sizing Extractors and vacuum systems.

Laboratory tests conducted in this manner have been utilized to size the Extractors for Kaiser's plants as well as for other services.

Design Considerations

From the previous experimental work and analysis, the Extractors were designed for the two applications. It should be noted that the six advantages of the horizontal belt filter cited in the Introduction were essential to improving the two applications over previous methods. The particular design considerations for the two applications will be considered separately.

Seed Filtration and Oxalate Removal

Feed slurry design conditions were:

Suspended solids concentration - 35 Wt%

Oxalate concentration on dry solids basis - up to 2.5 Wt%

Solids size distribution - 12 - 25% +200 mesh
- 40 - 70% +325 mesh

Feed temperature - 68°C (155°F) max.

Caustic concentration in liquor as NaOH - 12 - 15 Wt%

The first cold water wash should be no more than 0.33 Kg wash/Kg dry solids. It should be noted that this is approximately a 1.0 wash displacement (volume of wash/volume of liquor in the cake prior to wash) in order to maximize the caustic strength in the filtrate. The remaining caustic left in the seed will be recovered from the oxalate filtrate (hot water wash) after the oxalate is removed from the liquor. Thus, the cold wash filtrate will be combined with the strong or mother liquor filtrate with minimum dilution.

Hot water wash should range from 82 - 93°C (180 - 200°F) and 0.85 - 1.0 Kg wash/Kg dry solids, as it is essential to obtain a low concentration of oxalate in the final solids. This is about a four-displacement wash.

The discharged cake should have a maximum moisture of 20 Wt% with no more than 0.05 Wt% of oxalate on a dry solids basis. Another important factor is the suspended solids content of the strong and cold wash filtrates. This liquor will eventually return to digestion, and any solids will be resolubilized. As this consumes additional energy, a filtrate suspended solids content of only 20 mg/liter is desirable.

The flowsheet for the Extractor station is illustrated in Figure 3. It will be noted that the hot water wash is split into two parallel parts to ensure efficient oxalate removal. Also, this filtrate is segregated from the strong and cold wash filtrates by a separation baffle within the vacuum box. As this seals against the drainage belt, a different vacuum from the previous section can be applied if necessary.

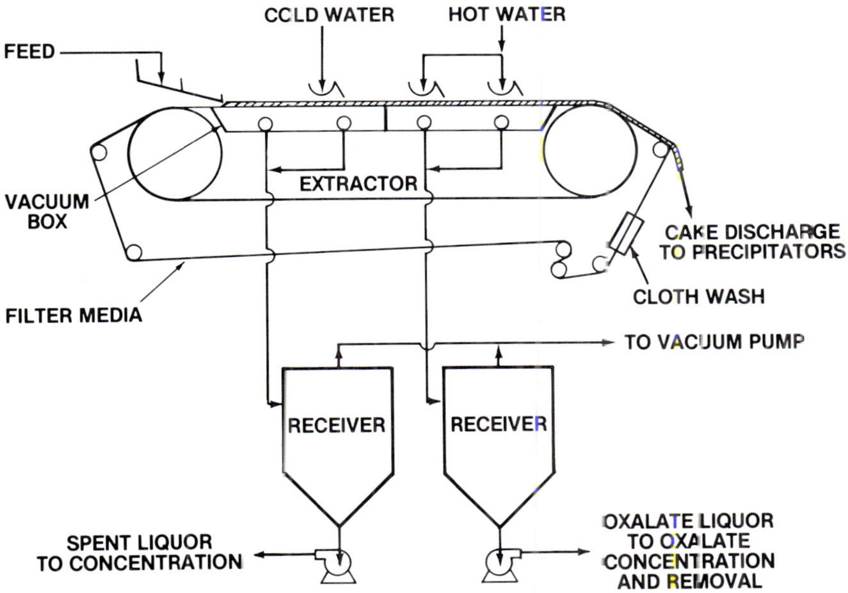

Figure 3
Extractor Flowsheet
Seed Filtration and Oxalate Removal

It is essential that the feed be spread evenly across the width of the Extractor. Thus a special double cascade sawtooth weir is employed to ensure an even cake thickness for most effective cake wash.

Cake wash is practiced with open type weir wash boxes with an elastomer impingement skirt. This ensures an even "gentle" addition of wash water without disturbing the cake.

Side plates on the top of the vacuum box are constructed of a low friction coefficient special fluorocarbon plastic. The slide plates are lubricated by injection of water to further minimize the friction between the traveling drainage belt and the stationary slide plate. The vacuum box is also fitted with protected screw activators, which lower the boxes for easy maintenance. As the caustic liquors are scaling in nature, it is essential to provide for easy and frequent descaling to maximize availability.

The special filter cloth required for good filtrate clarity can be washed continuously if necessary on one or both sides to prevent blinding. High impact spray nozzles are used. This is also true of the elastomer drainage belt.

A drip pan underneath the body of the filter extending the full length of the extractor is also employed to capture any spillage plus the cloth and belt washes.

Belt speed was made variable over a seven : one speed range so that the unit can accommodate available feed slurry. As the feed fluctuates, the speed can be tuned to the feed rate to optimize cake thickness.

Materials of construction in direct contact with the feed, filtrate or cake are stainless steel 316, fiberglass, reinforced polyester, polypropylene plastic (EimcoMet), neoprene covered steel on rolls and pulleys, and Ethylene Propylene Terpolymer drainage belts. These specifications are mainly influenced by the caustic and high temperature.

The Extractors have 25.6 m^2 (275 ft^2) of effective area under vacuum. Design is based on 843.5 metric tons/day (930 short tons/day) of dry solids maximum to each unit.

Product Kiln Dewatering and Washing

As the plant produces sandy alumina, the feed solids are 96 - 99% +325 mesh with 2 - 10% +100 mesh. The feed solids concentration varies between 45 - 47 Wt% with a caustic content in the liquor of approximately 7% as NaOH.

A two stage countercurrent wash is employed to achieve a leachable Na_2O content in the final product of 0.02 - 0.04 Wt%. Because of the high caustic concentration in the feed liquor, a very high recovery of Na_2O must be achieved. The countercurrent wash minimizes input water. Wash water rate to the second wash is between 0.3 and 0.4 Kg wash/Kg dry solids.

Feed temperature is 74 - 85°C (165 - 185°F). Input hot water wash is 85 - 96°C (185 - 205°F). The strong filtrate and first wash filtrate are combined and returned eventually to digestion as spent liquor.

The flowsheet for the Extractor filter station for product kiln feed is similar to Figure 3 except that there are only two wash boxes and the second stage wash filtrate is returned to the first stage wash.

Materials of construction are similar to those employed for seed filtration. This is also true for the design of vacuum boxes, feed and wash boxes, cloth and belt wash systems and filter media. However, flood dams are also included. These are elastomer membranes which ride on the cake and prevent mixing or run-back between feed and first and second washes on the surface of the cake.

Extractors of 94 ft^2 of effective area each under vacuum were employed to process 26.7 - 32.2 MTPH (29.4 - 35.5 STPH) of dry solids and produce a cake moisture of 10 Wt% maximum. However, they are designed to process up to 45.8 MTPH (50.9 STPH) at a discharged cake moisture of 12 Wt% maximum for peak demands.

Plant Experience

Kaiser Aluminum & Chemical Corporation has purchased several Eimco-Extractors for use in the Baton Rouge and Gramercy plants. Applications are detailed in Table I.

TABLE I

EIMCO-EXTRACTOR APPLICATIONS

BATON ROUGE AND GRAMERCY PLANTS

Service	Extractors		Individual Filtration Area		Date Installed
	Number	Model	Ft^2	M^2	
Substrate Alumina (gel)	2	828	212	19.7	1970
Fine Alumina Filtration	1	4624	101	9.4	1972
Kiln Feed	1	4624	101	9.4	1975
Sodium Dawsonite	1	9630	256	23.8	1977
Seed Washing	3	6945	275	25.6	1982
Substrate Alumina (gel)	2	6945	275	25.6	1982
Kiln Feed	7	4624	94	8.8	Delayed

Seed Washing

Three Model 6945 Extractors were installed for use in seed washing service at the Baton Rouge Plant in 1982. Figures 4 and 5 show the feed and discharge portions of one of the Extractors. The distinct color difference between these two points with the white color of the discharged seed is readily apparent The requirements for these filters included:

Capacity - 77,500 pounds of fine seed solids per hour per filter (1340 Kg/hr-m^2 - 275 lbs/hr-ft^2) with a feed solids concentration of at least 35 Wt%.

Filter Cake - Oxalate content of 0.05% or less on a dry basis and a moisture content of no more than 20 Wt%.

Figure 4
Feed End of Extractor
Seed Filtration - Oxalate Removal

Performance of these filters was satisfactory during the period they were in operation. However, production at the Baton Rouge Plant was curtailed before operation was completely optimized.

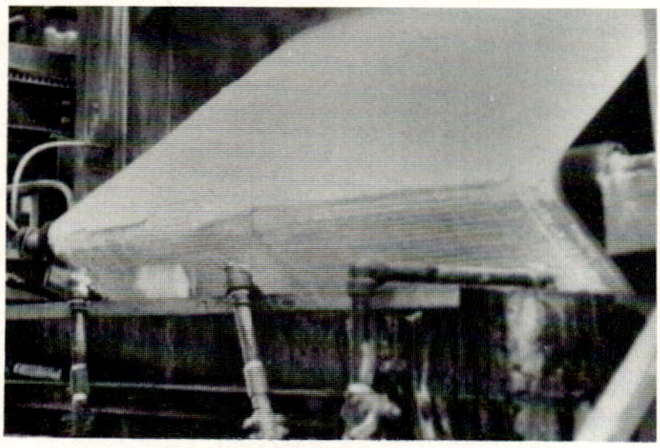

Figure 5
Discharge End of Extractor
Seed Filtration - Oxalate Removal
Note Color Difference from Feed

During start-up, problems were experienced with liquor dripping on the cloth alignment devices and the method of cloth connection. These were solved by design changes.

Problems remaining when production was curtailed included:

Capacity - Above 80% of design capacity had been demonstrated with a feed solids concentration of 30 - 32%. Efforts to increase the feed solids concentration were in progress. It is felt that design conditions would have been achieved.

Figure 6
Calcination Plant Model
Fluid Bed Calciner Plant
Six Extractors in Foreground

Figure 7
Calcination Plant Model
Two Extractor Stations Close-Up

Cloth Alignment - A program to address the problems of maintaining alignment over the entire filter length was being developed when production was curtailed.

The three major operations associated with seed washing are bulk liquor removal, liquor displacement with cold water, and a final hot water wash to remove sodium oxalate. Performance of these filters showed that these three operations can be satisfactorily accomplished in a single unit without repulping between stages.

Kiln Feed

A Model 4624 Extractor was installed for use in kiln feed service at the Gramercy Plant in 1975. Performance of this unit has shown:

Capacity - Up to 5855 Kg dry solids/hr-m^2 (1200 lbs dry solids/hr-ft^2) with cake moistures of about 13% and satisfactory soda control. Lower cake moistures are obtained at lower operating rates.

Maintenance - Lower maintenance costs and better operating factors than other methods for this service.

Wash Water - Lower specific wash water consumption than other methods.

Because of the successful application of the extractor to kiln feed service at Gramercy, Extractors were selected for the new filtration facility connected with the installation of fluid bed calciners at Baton Rouge. Operating data is not available since construction has been delayed due to the production curtailment of the Baton Rouge Plant.

Design of the Baton Rouge fluid bed calcination facility consisted of four calciners and seven Model 4624 Extractors. Figure 6 shows the model constructed for design purposes with six Extractors positioned on an upper floor. Figure 7 is a close-up of two of the Extractor stations. One filter was dedicated to the "special alumina" calciner. The remaining six filters were designed as a multiple filter arrangement with two filters feeding each reduction grade calciner. Each filter was capable of supplying approximately two-thirds of its calciner's hydrate requirement. The requirements for these filters included:

	Normal	Maximum
Solids - lbs/hr-ft^2	625 - 756	1074
- Kg/hr-m^2	3050 - 3690	5240
Free Moisture (% by Wt.)	8 - 10	12
Leachable Na_2O (%)	0.02 - 0.04	0.065
Wash Water Rate (Kg/Kg Dry Cake)	0.3 - 0.4	0.5

Final evaluation of these filters will be made when they have been installed and operated for a sufficient period of time.

THE AMMONOALUNITE PROCESS FOR PRODUCTION OF ALUMINA FROM CLAY

Robert W. Bartlett
Rolf J. Wesely
Anaconda Minerals Company
P.O. Box 27007
Tucson, Arizona 85726

Thomas R. Bolles
ARCO Metals Company
P.O. Box 27007
Tucson, Arizona 85726

The ammonoalunite process begins with two-stage sulfuric acid leaching of calcined kaolinitic clay. The clarified aluminum sulfate liquor is subjected to modest temperature autoclaving in the presence of ammonium ions to precipitate the ammonium analog of alunite, $NH_4Al_3(SO_4)_2(OH)_6$, in a hydrolysis reaction regenerating most of the acid needed in the leach. This acid liquor is recycled while the filtered ammonoalunite is thermally decomposed to alumina. Decomposition gases are scrubbed to recover ammonia and the remaining acid needed for leaching. Experimental results on precipitation parameters, thermal decomposition, energy consumption, and impurity control are given.

Introduction

The shortage of bauxite in the United States has led to several efforts to develop processes for utilizing kaolinitic clays to extract alumina suitable for Hall-Heroult reduction cells. These clays are plentiful in central Georgia, well located with respect to several U.S. reduction plants, and contain alumina concentrations similar to lower grade bauxites.

The various clay processes nearly always begin with calcination to meta-kaolin, which makes the clay more leachable. This is followed by leaching in either sulfuric, hydrochloric, or nitric acid. Precipitation of an intermediate compound usually follows, which is then calcined or otherwise thermally treated to generate alumina.

From leaching considerations alone, a sulfuric acid process is preferred over the other acids because of its lower reagent cost and reduced corrosion. However, sulfuric acid processes have encountered difficulties and high costs associated with recovery of alumina from the pregnant liquor and the regeneration of sulfuric acid for recycle to leaching.

This paper describes a new clay-alumina process based on sulfuric acid leaching. The novelty in the process lies in the precipitation of a new intermediate compound, ammonoalunite, which is subsequently calcined to alumina. The benefits of the process, compared with other sulfuric acid processes, are low energy consumption and immediate regeneration of most of the sulfuric acid during the hydrolysis reaction to produce ammonoalunite. The remaining sulfate and ammonia associated with the precipitate are emitted as gases during its calcination and are scrubbed for recycle to leaching. The process will be described along with its advantages and some of the difficulties with the process in its present, early stage of development.

Process Chemistry

At the outset, it is worthwhile reviewing some aspects of the chemistry involved with precipitation of aluminum compounds from sulfate liquors. Alum, $Al_2(SO_4)_3 \cdot 18H_2O$, can be precipitated from saturated hot solutions of aluminum sulfate on cooling or by evaporation of the water. The yield of alum on cooling is low and the resulting product contains a vast amount of chemically bound water. Consequently, calcination of this product requires large amounts of energy.

Neutralization of aluminum sulfate solutions to precipitate alumina does not permit the regeneration of the sulfate into sulfuric acid for recycle; hence, large amounts of sulfuric acid are wasted and this leads to an uneconomic process.

Heating aluminum sulfate solutions in an autoclave at sufficiently high temperatures can lead to the precipitation of basic aluminum sulfate, $(H_3O)Al_3(OH)_6(SO_4)_2$, but the yield at reasonable temperatures and operating pressures is low, leading to undesirable energy requirements and capital costs for autoclaves. In short, all of these approaches to the recovery of aluminum from aluminum sulfate solutions lead to uneconomic processes in competition with the Bayer Process or even with the best hydrochloric acid approaches to recovery of alumina from kaolin.

A better intermediate product is required--one that can be obtained in high yield from aluminum sulfate solutions with modest energy requirements, and one that contains very little water of hydration to minimize the subsequent energy consumption in calcination to alumina. It is mandatory that the process permit the recovery of sulfuric acid for recycle with minimum sulfate losses.

Hydrothermal processing of aluminum sulfate solutions requires considerable temperature and produces basic aluminum sulfate as already mentioned. If alkali metals are present in solution, the hydrolysis can occur at much lower temperatures to obtain the same yield. In the presence of potassium, the natural mineral alunite, $KAl_3(OH)_6(SO_4)_2$, results, while in the presence of sodium, natroalunite, $NaAl_3(OH)_6(SO_4)_2$, results. These processes occur so easily at modest temperatures that acid weathering of rocks commonly leads to the production of these minerals in nature. On the other hand, the difficult conditions for basic aluminum sulfate, which can be thought of as hydronium or hydrogen alunite, are rarely realized in nature. The use of alkali metals to produce various alunites cannot lead to commercial success because the alkali metal oxide is retained with the solid on calcination and provides gross contamination of the alumina product.

However, if ammonium ion is used, the result of hydrolysis is the ammonium analog of alunite, ammonoalunite, $NH_4Al_3(OH)_6(SO_4)_2$. This compound is not found in nature but it is easily crystallized in the laboratory under temperature conditions much milder than required to produce basic aluminum sulfate. The advantage of ammonoalunite is that the ammonium ion is released as a gas on calcination, which can be recovered by scrubbing and recycled. The ammonium ion is reused and is not a contaminant in the final alumina product.

Process Flowsheet

The ammonoalunite process flowsheet shown in Figure 1 consists of the following steps:

1. Leaching of calcined kaolin clay in the presence of recycled ammonium ion and sulfuric acid. Excess clay is used to neutralize acid and suppress iron solubilization.

2. Solid-liquid separation of the leach residue.

3. Hydrothermal precipitation of ammonoalunite in an autoclave train with concurrent regeneration of sulfuric acid.

4. Solid-liquid separation of ammonoalunite and recycle of the spent liquor to clay leaching.

5. Thermal decomposition of ammonoalunite crystals to gaseous ammonia, water, oxides of sulfur and solid alumina, with scrubbing of the gases by dilute clay-liquor slurry.

6. Purification of the calcined alumina product by a sulfuric acid wash, primarily to remove traces of phosphorus.

The ammonoalunite precipitation occurs with good yield at 150-250°C. Two-thirds of the sulfuric acid consumed in leaching aluminum is recovered in the ammonoalunite precipitation while one-third is recovered from thermal decomposition and scrubbing. Since both sulfuric acid and ammonia are

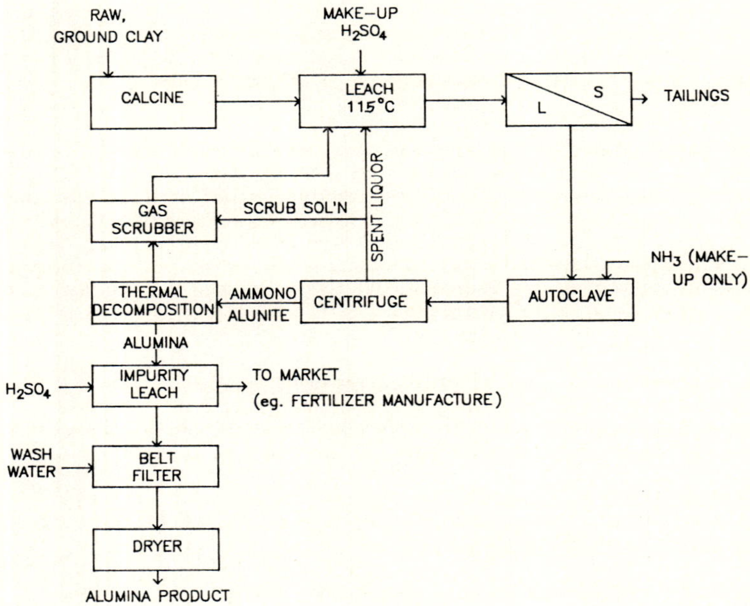

Figure 1 - FLOWSHEET FOR AMMONOALUNITE PROCESS

regenerated and recycled, purchased reagents are required only to make up entrainment losses. Simplified descriptions of the key chemical steps follow:

o Leaching

$$9H_2SO_4 + 3Al_2O_3 \cdot 2SiO_2 \longrightarrow 3Al_2(SO_4)_3 + 9H_2O + 6SiO_2 \quad (1)$$

o Hydrolysis (Ammonoalunite Precipitation and Acid Regeneration)

$$12H_2O + 3Al_2(SO_4)_3 + (NH_4)_2SO_4 \longrightarrow 2NH_4Al_3(OH)_6(SO_4)_2 + 6H_2SO_4 \quad (2)$$

o Ammonoalunite Decomposition

$$2NH_4Al_3(OH)_6(SO_4)_2 \xrightarrow{heat} 3Al_2O_3 + 2NH_3 + 7H_2O + 4SO_3 \quad (3)$$

o Gas Scrubbing and Recovery of Ammonia and SO_3

$$2NH_3 + 4SO_3 + 7H_2O \longrightarrow (NH_4)_2SO_4 + 3H_2SO_4 + 3H_2O \quad (4)$$

The major advantages of the ammonoalunite process are:

1. The ammonoalunite process requires lower autoclave temperatures and gives a much higher yield than the basic aluminum sulfate process.

2. The ammonoalunite, when thermally decomposed, releases ammonia and sulfur oxides as recoverable gases, whereas sodium and potassium alunite, when thermally decomposed, leave the alkali metal sulfate in the alumina product.

3. Ammonoalunite requires less energy to decompose than the various aluminum sulfate solids produced in other sulfuric acid processes. This is because aluminum sulfate solutions when crystallized form various hydrated alum compounds with much bound water such as:

$Al_2(SO_4)_3 \cdot 18H_2O$ (alum)

$NH_4Al(SO_4)_2 \cdot 12H_2O$ (ammonium alum)

$KAl(SO_4)_2 \cdot 12H_2O$ (potassium alum)

Experimental Ammonoalunite Precipitation Yields

Batch precipitation experiments were conducted in a Parr pressure reactor initially using synthetic solutions simulating leach liquor at pH about 2.5 and containing 30 gpl Al_2O_3. These liquors contained dissolved impurity concentrations at the following levels: 0.1 gpl Na_2O, 0.075 gpl K_2O, and 0.3 gpl P_2O_5. Subsequent experiments were conducted with natural calcined clay leach solutions in which iron control was a particularly important factor that will be discussed later.

The study of ammonoalunite precipitation focused on the effect of three variables: ammonia concentration, autoclave temperature, and reaction time. These variables determine the yield that can be obtained at reasonable industrial operating conditions. No excess sulfuric acid was used in the initial experiments because it is necessary to control dissolved iron in the leach by equilibrating the pregnant liquor with an excess of calcined clay.

The ammonia concentration in the autoclave feed solutions was varied relative to the alumina concentration. Molar concentration ratios of zero to three moles of ammonia per mole of dissolved alumina were tested and these results are presented in Figure 2. The purpose in varying ammonia concentration was to see if high molar ratios of NH_3:Al_2O_3 would drive the hydrolysis reaction to the right, increasing the ammonoalunite yield.

The autoclave temperature was examined to determine if reasonable industrial operating temperatures and pressures could be used to precipitate ammonoalunite. Temperatures were varied from 125-250°C. At these temperatures the pressures varied from 35-575 psia. The results of this test work are summarized in Figure 3. Note that potassium is more selectively precipitated than aluminum and that sodium and phosphorus are also precipitated. These are coprecipitated with ammonoalunite. Hydrolysis reaction time was also examined to determine if reasonably sized reactors could be used to precipitate ammonoalunite. The time was varied from one-half to three hours (Figure 4).

Figure 2 shows that as the concentration of ammonia increases, relative to the dissolved alumina concentration in the autoclave feed solution, the percentage of ammonoalunite precipitation increases. An ammonia-to-alumina molar concentration ratio of 3:1 precipitated 90% of alumina as ammonoalunite in this 200°C experiment. At the stoichiometric molar concentration ratio of 0.6:1, only 55% of the alumina contained in the feed precipitates as ammonoalunite at 200°C.

With an increasing ammonia/alumina concentration ratio, the solubility of ammonium alum decreases and ammonium alum may crystallize according to the following reaction:

$$(NH_4)_2SO_4 + Al_2(SO_4)_3 + 24H_2O \longrightarrow 2NH_4Al(SO_4)_2 \cdot 12H_2O \qquad (5)$$

Ammonium alum crystallized from a solution prior to autoclaving would result in an alumina loss. To prevent crystallization of ammonium alum, the solution must be kept hot. Figure 5 shows the solubility of ammonium alum as a function of temperature for solutions with a molar NH_3:Al_2O_3 ratio of 2:1.

Autoclaving temperature is an important variable in ammonoalunite precipitation. The higher the autoclave temperature, the higher the percentage of precipitation. Figure 3 shows that 95% of the alumina contained in the solution can be precipitated at 250°C. Both the temperature and pressure are reasonable for an industrial process. However, it may be more cost-effective to use lower autoclaving temperatures and a lower NH_3:Al_2O_3 ratio and accept a recovery of Al_2O_3 as ammonoalunite significantly lower than 95% of the dissolved Al_2O_3 in the leach liquor. Because the spent liquor is recycled to the leach, all soluble Al_2O_3 not entrained with tailings will eventually be recovered in the precipitation autoclave.

The precipitation yield at various retention times presented in Figure 4 shows that the hydrolysis reaction is very rapid. At 200°C, 81% of the ammonoalunite precipitated within one-half hour. Only 83% had precipitated by the end of three hours. Hence, one-half hour residence time is adequate and further experiments may indicate a shorter residence time.

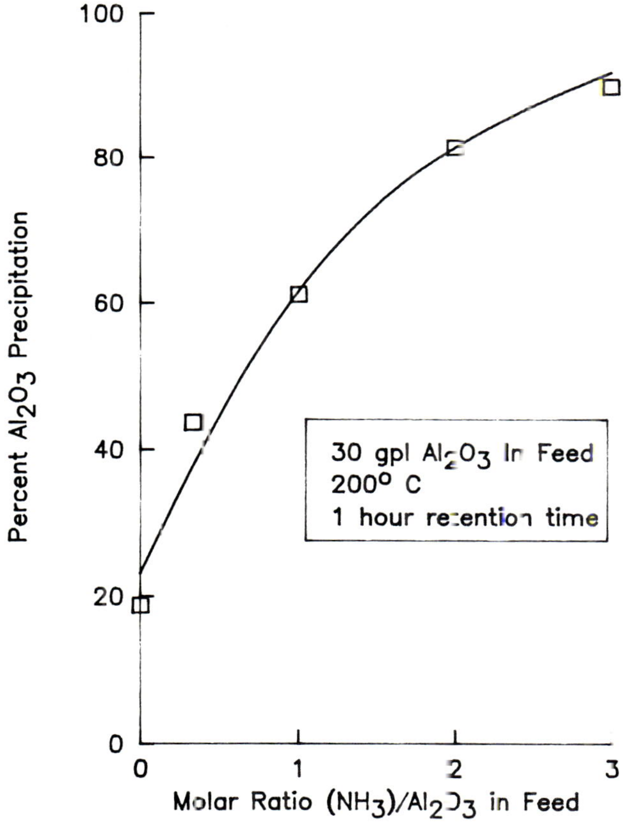

FIGURE 2 – NH_3 CONCENTRATION VS PERCENT PRECIPITATION

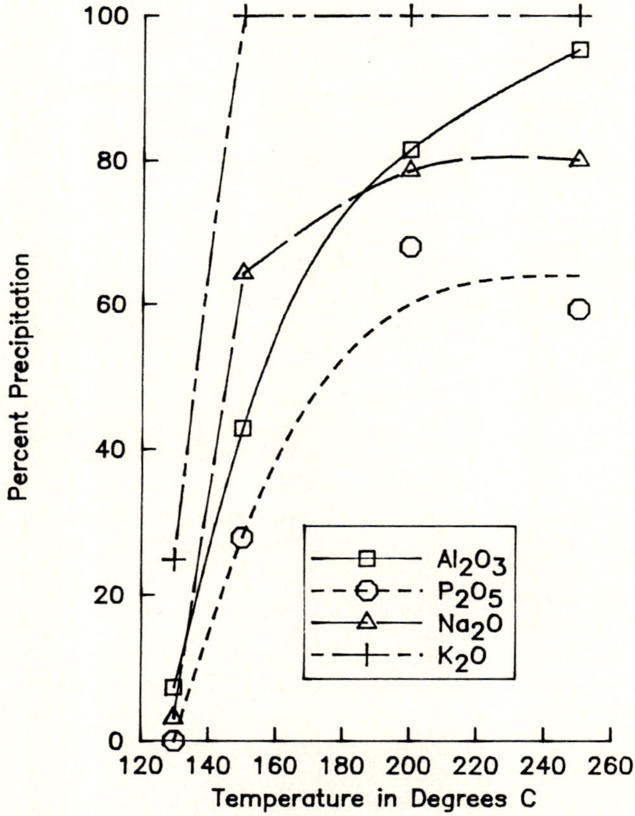

FIGURE 3 — PERCENT PRECIPITATION VS TEMPERATURE

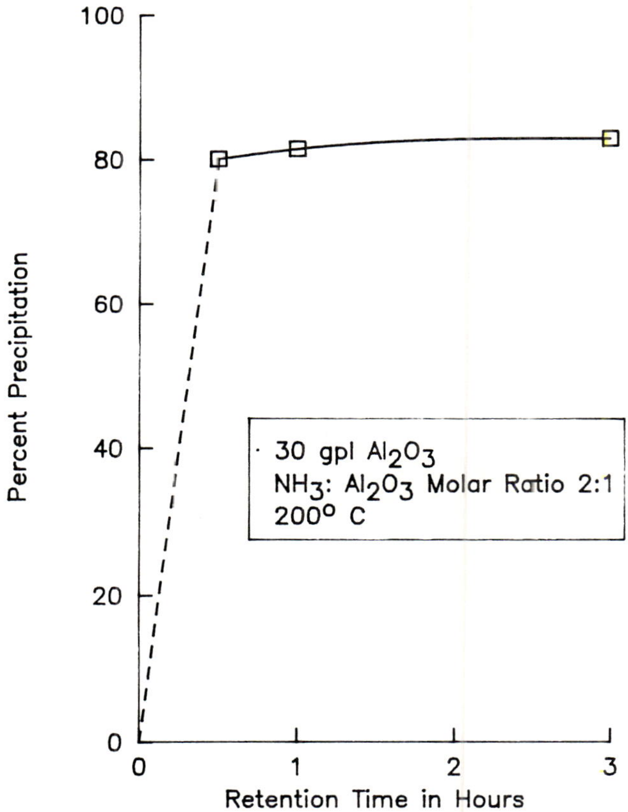

FIGURE 4 — PERCENT PERCIPITATION VS TIME

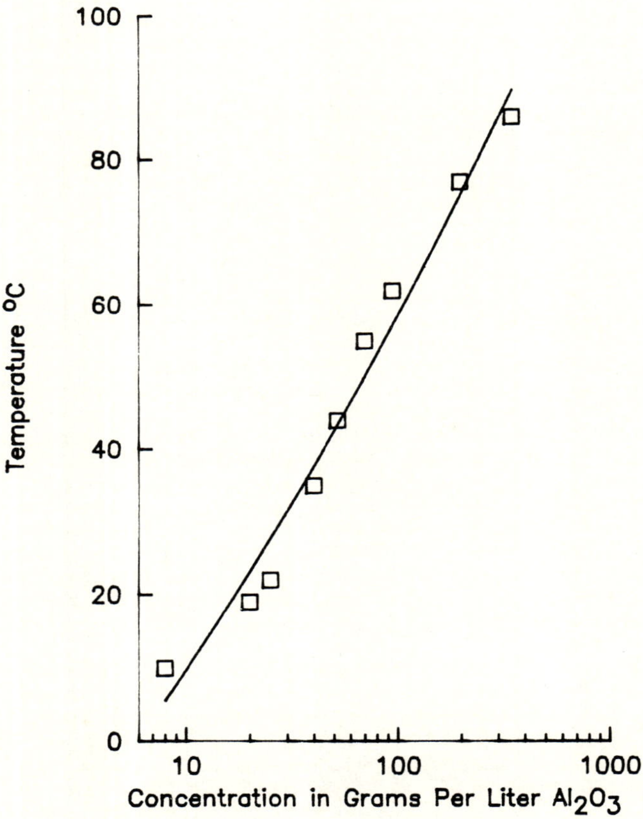

FIGURE 5 – SOLUBILITY OF $NH_4Al(SO_4)_2$ VS TEMPERATURE

Impurity Control

Phosphorus

The impurity concentrations used in making up the synthetic leach liquors used for the autoclave precipitation experiments were derived from an analysis of the feed clay and from extraction efficiencies obtained in prior laboratory work on sulfuric acid leaching of calcinced clay. Current Hall cell specifications (.001% P_2O_5) require removing only 90% of the P_2O_5 in alumina from ammonoalunite. Three general methods of impurity removal were tested in the laboratory. These methods are (1) precipitation of impurities prior to hydrolysis of ammonoalunite, (2) removal of impurities by leaching the ammonoalunite precipitate, and (3) removal of impurities by acid washing the calcined alumina product. Initial work on impurity removal consisted of selectively precipitating potassium, sodium, and phosphorus at lower autoclave temperatures. In the formation of alunites, potassium alunite precipitates first; i.e., at the lowest temperature; then sodium alunite, ammonium alunite, and finally hydrogen alunite precipitate. Seventy-five to 80% of the potassium, 45-50% of the sodium, and 15-20% of the phosphorus could be precipitated at 140°C. But, at this temperature, 25% of the aluminum also precipitates. Because too much alumina was rejected with the impurities, this method of impurity removal was abandoned.

The next step was to look at leaching impurities from the ammonoalunite. Leaching the precipitate with hot water had no effect. A hot acid leach removed 9% of the alumina and only a small portion of the impurities.

Impurity removal from the alumina, after thermal decomposition of ammonoalunite, appeared to be the most promising method of impurity removal. A hot water wash (98°C) of the alumina calcined at 1200°C removed 80% of the P_2O_5 and 40% of the Na_2O and K_2O. When using sulfuric acid at two normal concentration and 98°C to wash the calcined alumina, 2% of the alumina and essentially all of the impurities were removed, meeting specifications. The hot sulfuric acid wash is the recommended purification step at the present stage of process development for control of phosphorus. It should be pointed out that levels of sodium and potassium in the calcined ammonoalunite derived from clay leaching are within specifications for cell grade alumina, and strictly speaking, no treatment to control these impurities is expected to be required for most Georgia kaolinitic clays.

At the present time, the major drawback of the sulfuric acid wash of calcine clay is that for this step to remove phosphorus to cell specification levels requires that the ammonoalunite be calcined at a finishing temperature of 1200°C. At this temperature the calcined product was 100% alpha alumina. Calcination at 1000°C would be needed to provide a product that satisfies the requirements of dry scrubbing of fluoride vapors in Hall-Heroult reduction cells using an alumina dry scrubbing process, which has become prevalent in the industry. When the ammonoalunite is calcined at this lower finishing temperature the subsequent sulfuric acid washing removed 40% of the phosphorus.

Iron

Following the synthetic solution studies, calcined kaolin was leached with sulfuric acid, both with and without ammonium ion being present. This was followed by autoclave precipitation of ammonoalunite. These studies showed that iron was a serious contaminant of the resulting alumina. Later experimental work has shown that most of the iron can be removed if the leach slurry is heated to 115°C in the presence of excess calcined kaolin after leaching in order to neutralize the excess acid and precipitate iron, presumably as ammonium jarosite, $(NH_4)_2Fe_6(OH)_{12}(SO_4)_4$, which is the analog of ammonoalunite, with iron substituting for aluminum. Experimental conditions were obtained where most of the aluminum remained soluble while the majority of the iron is precipitated. Although these experiments have not been optimized in terms of temperature, residence time, ammonium content of the liquor, and amount of excess clay used, the results show that alumina containing iron of less than .2% could be obtained by adding additional clay to the leach liquor and reheating the slurry to 115°C for five hours. Much shorter residence times can be used at higher temperatures if precipitation of some alumina is acceptable.

Ammonoalunite Thermal Decomposition

In order to be a competitive process, the decomposition step must recover ammonia and sulfuric acid. The instability of NH_3 above 450°C dictates that it be recovered at a lower temperature. Industrial calcination of the ammonoalunite should occur in a process in which the material is gradually heated by countercurrent hot gas flow and heat exchange for maximum efficiency. For example, this would occur on passing the material countercurrently through a rotary kiln. A thermogravimetric analysis was run with gradual heating to study the decomposition process. The results are shown in Figure 6.

Forty percent of the weight loss occurred between 300 and 500°C. This initial weight loss corresponds to ammonia and water being driven off. The remaining weight loss occurred between 500 and 1000°C, and corresponds to the emission of sulfur oxides.

Energy and Cost Considerations

Although the process is far from optimized, some preliminary process engineering calculations were made to indicate whether the process had potential for being commercially competitive, if the phosphorus impurity problem could be resolved. This also identified major cost areas. Direct costs are approximately distributed with mining 15-20%, fuel 50%, and all other process costs 30-35%. Fuel usage is very important and concentrated in three major areas: calcination of the clay, heating solutions for autoclave precipitation of the ammonoalunite, and thermal decomposition of the ammonoalunite. Energy requirements for the autoclave depend on the yield of ammonoalunite, $NH_4Al_3(SO_4)_2(OH)_6$, that is obtained in the autoclave, and this in turn depends on the autoclave temperature. Fuel consumption, not including fuel equivalent of electric power used in the process, is about 26 million Btu per ton of alumina product. Although this is a very approximate number at this stage of development, it is less than required for other clay processes.

Conclusions

A laboratory investigation of the ammonoalunite process concept leads to the following conclusions:

1. Ammonoalunite can be precipitated with high conversion efficiency (up to 95% of dissolved aluminum at reasonable process conditions for an

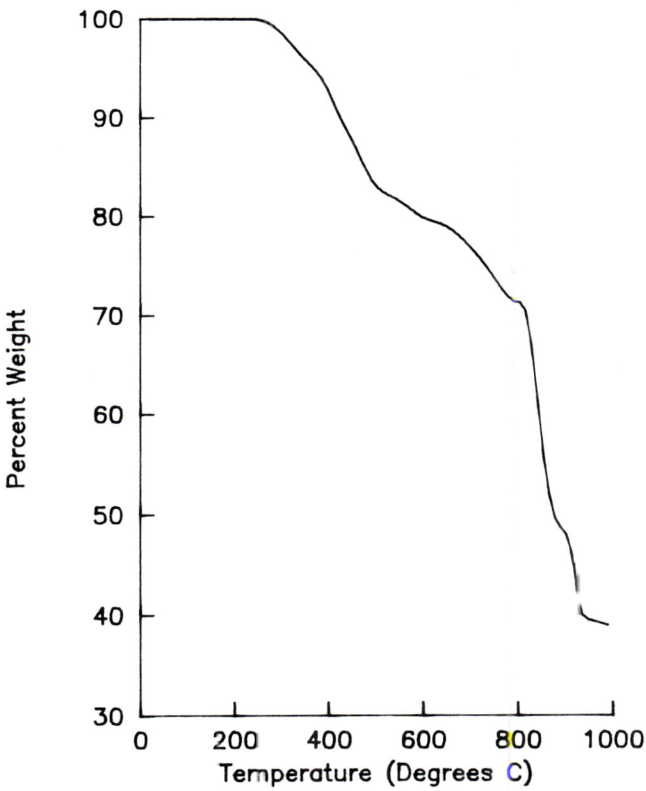

FIGURE 6 — THERMAL GRAVIMETRIC ANALYSIS
DECOMPOSITION OF AMMONOALUNITE

industrial process). The optimum economic conditions may be at a somewhat lower yield and autoclave operating temperature.

2. Iron impurity levels can be adequately controlled by completing the leaching in an excess of calcined clay at a temperature of 115°C.

3. Two-thirds of the sulfuric acid needed is regenerated during the hydrolysis step and is easily recycled back to clay leaching. The remaining acid and ammonia associated with ammonoalunite crystals are recovered by scrubbing the ammonoalunite decomposition gas.

4. Sodium, potassium, and phosphorus can be removed from the product alumina to meet metallurgical specifications for Hall-Heroult cells by using a simple sulfuric acid wash of the calcined ammonoalunite (Al_2O_3). However, for this impurity removal to be sufficiently effective on phosphorus it is necessary that the alumina be calcined at a temperature where its BET surface area is not adequate to meet the specifications of fluoride dry scrubbing.

5. This process requires less energy and lower net reagent consumption when compared with other clay processes, but more energy than a modern Bayer plant with efficient heat recovery.

6. The process is less corrosive than clay processes based on hydrochloric acid or nitric acid.

7. The process does not appear at this stage to have any insurmountable environmental constraints; the slightly acidic silica tailings would have to be neutralized and impounded.

ALUMINA VIA HYDROCHLORIC ACID LEACHING OF HIGH SILICA BAUXITES - PROCESS DEVELOPMENT

D. Elsner, D.H. Jenkins and H.N. Sinha

CSIRO Division of Mineral Chemistry, P.O. Box 124,
Port Melbourne, Victoria, 3207
Australia

A process for the production of cell-grade alumina from high reactive silica bauxite using hydrochloric acid leaching has been developed in the CSIRO Division of Mineral Chemistry. This paper describes the process and presents the flowsheet and the energy requirement of the process. The process is based upon leaching either calcined bauxite at atmospheric pressure or uncalcined bauxite under pressure, followed by crystallization of aluminium chloride hexahydrate using HCl sparging. Unlike other acid processes, effective iron separation is achieved in a single crystallization step. The aluminium chloride crystals are decomposed to alumina in a fluidized-bed and the hydrogen chloride produced is recycled. Preliminary capital and operating costs for an 800,000 tonne/year plant are also given.

Introduction

Currently aluminium is produced from bauxite using a combination of the Bayer and Hall-Heroult Processes. In the Bayer Process, leaching of the bauxite with sodium hydroxide leaves impurities such as iron, titanium and non-reactive silica (e.g., quartz) as an insoluble red mud. Any reactive silica in the ore (i.e., silica combined as clay and other silicates) goes into solution during leaching and is removed from the liquor as sodalite ($2Na_2O \cdot 2Al_2O_3 \cdot 3SiO_2 \cdot 2H_2O$) and represents a loss of both alumina and soda from the process. To reduce the loss of sodium hydroxide, Bayer Plants impose rigid specifications on the amount of reactive silica in the bauxite feed and presently bauxites containing more than 7% reactive silica are considered uneconomic and not readily acceptable as feed (1). Some bauxites need beneficiation by washing and screening to reduce their reactive silica content below 7% prior to shipment to alumina refineries. This beneficiation step leaves large tonnages of bauxite containing more than 10% silica in the tailings. Thus an alternative process to treat high reactive silica bauxites would have the advantage of improving the overall recovery of alumina in the short term and of increasing the life of the mining operation in the long term. Such a process would also be environmentally more acceptable due to a reduced amount of tailings. A hydrochloric acid route to produce alumina was chosen as it produces $AlCl_3 \cdot 6H_2O$ as an intermediate product which can be processed to anhydrous $AlCl_3$ (2). Thus a hydrochloric acid process can produce a product which can be used in either the conventional route to aluminium production or the new Alcoa chloride electrolysis process which is in the development stage.

Our objective was the development of a process on a technical scale which would produce alumina meeting cell-grade specifications using bauxite which is unsuitable as feed to a conventional Bayer Plant due to the high reactive silica content. To this end the following stages were examined on a laboratory and technical scale: (a) calcination, (b) leaching, (c) filtration and (d) crystallization. Regeneration of HCl, hydrolysis of $FeCl_3$/$AlCl_3$ solutions and decomposition of $AlCl_3 \cdot 6H_2O$ crystals to produce alumina was not studied as these steps incorporate known technology.

Process Description

The process as shown in Figure 1 involves calcination of bauxite of composition given in Table I followed by leaching with 30% HCl. Either atmospheric or pressure leaching can be used. However, no calcination step is required when pressure leaching is chosen. In both cases the residue is filtered from the leach liquor, washed and the washings after evaporation are combined with the filtrate.

Gaseous HCl is sparged into the combined filtrate/evaporated washings to crystallize out the aluminium chloride as $AlCl_3 \cdot 6H_2O$. The $AlCl_3 \cdot 6H_2O$ is decomposed in a two-stage decomposition step to produce alumina and hydrogen chloride which is recycled in the process. The solution remaining after crystallization of the aluminium chloride is first stripped to produce a gaseous HCl stream which is used for crystallization. It is further evaporated to produce a 20% HCl solution and a solution containing approximately 45% metal chlorides. This is then hydrolysed to produce more hydrogen chloride which is recycled in the process and metal oxides which are dumped.

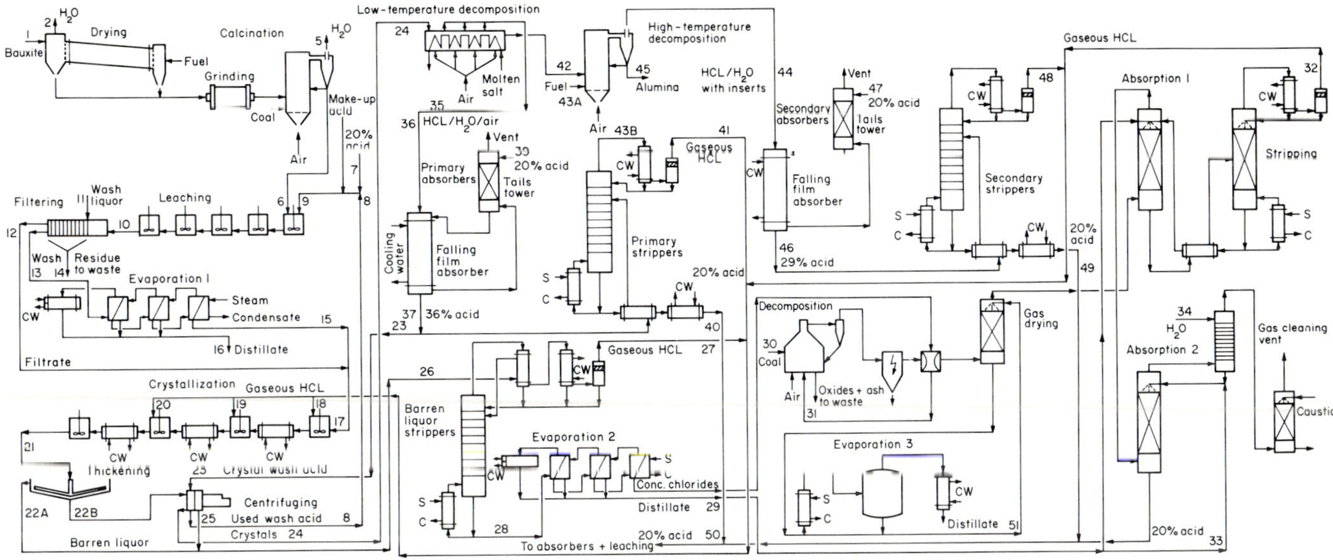

Figure 1 - Process scheme for alumina via HCl leaching of bauxite.

Table I. Chemical composition and mineralogy* of bauxite.

	%
Al_2O_3	55.6
Fe_2O_3	9.5
SiO_2	7.9
TiO_2	2.2
Loss on ignition	24.8
gibbsite	60.9
boehmite	12.2
kaolinite	13.7
quartz	1.5
hematite	9.5
anatase	2.2

*calculated.

Calcination

Calcination of the aluminium-containing mineral whether it be bauxite, clay or coal washery refuse, greatly increases the acid solubility of the alumina. In bauxites, calcination converts the hydrated aluminium minerals diaspore ($Al_2O_3 \cdot H_2O$), boehmite ($Al_2O_3 \cdot H_2O$) and gibbsite ($Al_2O_3 \cdot 3H_2O$) to a form of gamma-alumina and the clay mineral kaolinite to meta- kaolinite. Both the time and temperature of calcination are important if maximum solubility of the aluminium minerals is to be achieved. A temperature between 750 and 850°C with a residence time between 10 and 30 min appears optimum.

X-ray diffraction (XRD) of -1200 +421 μm and of -421 μm samples of bauxite calcined at 600, 700, 750 and 800°C for various times in a muffle furnace was used to determine the conditions required to decompose kaolinite, boehmite and gibbsite. Although the peak in the differential thermal analysis curve corresponding to the temperature of rapid decomposition of kaolinite is 585°C, the results showed that at least 30 min at 700°C are required to produce a kaolinite and boehmite free calcine. Similar results were obtained when calcination times in excess of 15 min were used above 750°C. These results no doubt reflected the influence of particle size, bed depth and the rate of heat transfer to the individual particles. The U.S. Bureau of Mines in their pilot plant study on alumina extraction from clay specify a one hour residence time at a temperature of approximately 750°C when the feed clay is calcined in a fluid-bed calciner (3).

A reliable guide to the reactivity of the bauxite to acid leaching is provided by its surface area. Table II shows the surface areas of two fractions of a calcined and uncalcined bauxite, while Figure 2 shows the effect of the calcination temperature on aluminium extraction.

Table II. Surface areas (m^2/g) of calcined and uncalcined bauxites.

Size of material (μm)	Uncalcined	Calcination temperature (°C)			
		750		800	
		30 (min)	60 (min)	30 (min)	45 (min)
-1200+421	27	125	113	94	95
-421	23	132	97	93	90

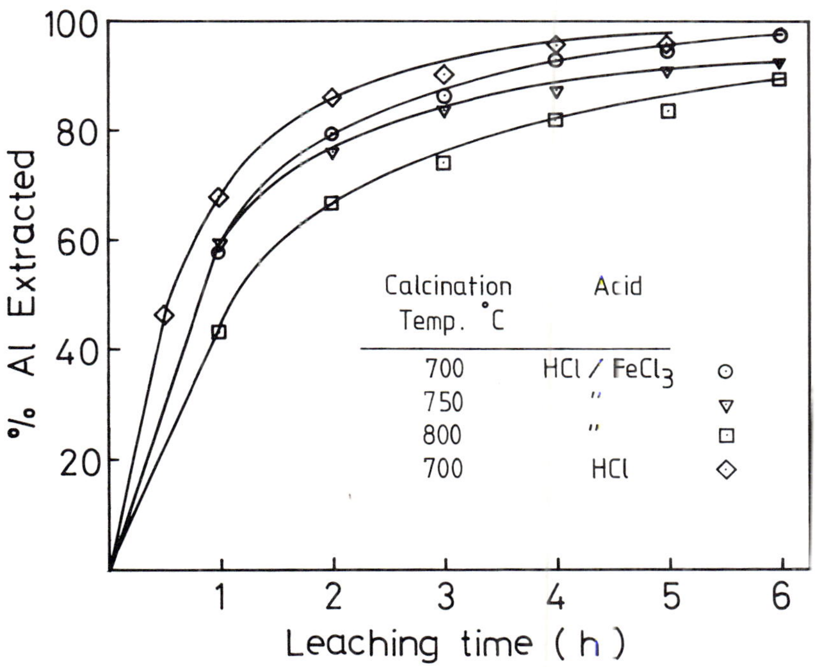

Figure 2 - Aluminium extraction from -1200+421 μm calcine as a function of time: effect of calcination temperature, acid composition (HCl 25%, $HCl/FeCl_3$ 20%-20%), all tests used 20% excess acid above the stoichiometric requirement.

Batch Atmospheric Leaching

Aluminium extraction for both calcined and uncalcined bauxite increased rapidly with time up to about 2 h. As can be seen in Figure 2, between 4 and 6 h are required to achieve extractions in excess of 90%. The sizing of the calcine (-1200 +421 μm, -421 μm), excess acid (3 to 105 g HCl per litre) and the addition of $FeCl_3$ to the feed acid appeared to have little effect on aluminium recovery. Since ferric chloride is one of the reaction products of bauxite leaching with hydrochloric acid it is a suitable additive to increase the chloride content of the leach liquors and thus improve the fuel economy of the acid regeneration process. The effect of agitation on leaching was not studied. Agitation was sufficient to keep the slurry well mixed.

Figure 3 shows the aluminium extraction as a function of time for calcined and uncalcined bauxite (-421 μm) under similar leaching conditions. Approximately 90% of the aluminium can be extracted from uncalcined bauxite in 6 h. This compares with close to 100% for material calcined at 700°C and 87% for calcine produced at 800°C. Iron extractions approached 100% after 6 h for both calcined and uncalcined bauxite.

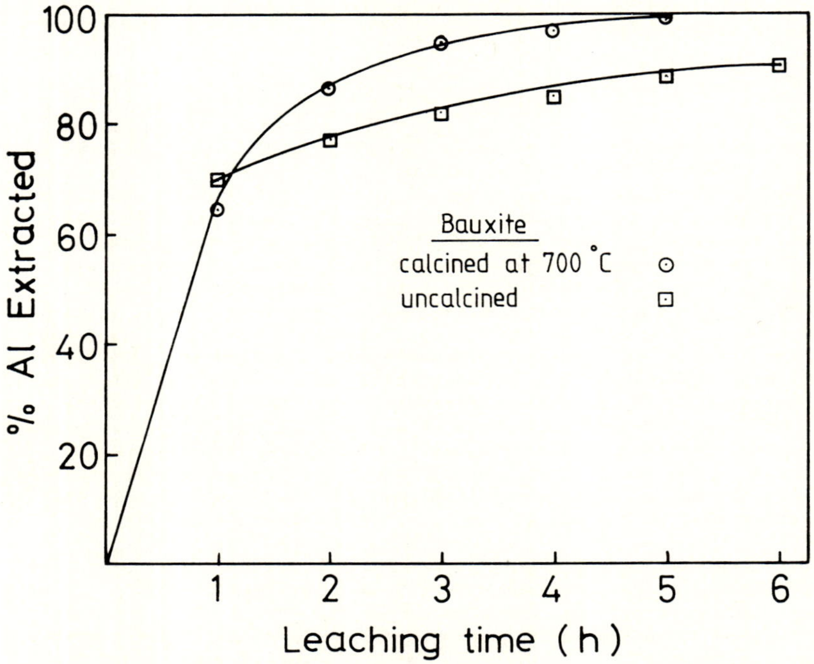

Figure 3 - Aluminium extraction from -421 μm calcined and uncalcined bauxite as a function of time when using 20% HCl and 20% $FeCl_3$ as leachant with 5% excess acid above the stoichiometric requirement.

Batch Pressure Leaching

A series of pressure leach tests was conducted to determine the effect of temperatures above the atmospheric boiling point on the leaching rate of calcine and particularly uncalcined bauxite. These tests were conducted in two 1-gal (US) Pfaudler glass lined steel autoclaves equipped with one baffle and a 3-blade curved-blade turbine operating at 115 rpm. The temperature used in these leach tests were 130 and 150°C. Steam was used as the heating medium for the tests at 130°C, while hot oil was used for the tests at 150°C. Calcine or bauxite was added to the cold acid, the vessel was then closed and heated to the operating temperature. The start of leaching was arbitrarily set at a temperature 5-10 degrees below the nominal operating temperature.

The leaching rate of uncalcined bauxite increased markedly at temperatures above the atmospheric boiling point. A minimum of 6 h is required to achieve aluminium extractions of greater than 90% under atmospheric conditions when the temperature is 110-116°C. However, about 3 h are required at 130°C and less than 2 h at 150°C. These temperatures appear to have no effect on the leaching rate of calcined bauxite since aluminium extractions are comparable at all temperatures 110-150°C for equal residence times, as shown in Table III.

Iron extractions from uncalcined bauxite under pressure conditions were slightly lower than equivalent atmospheric leaches. Leach residues derived from calcine leached at 130°C were pink with numerous larger particles apparently still covered with hematite. Residues from atmospheric leach tests were white.

Continuous Leach Tests

Some continuous leach tests were conducted in the two Pfaudler 1-gal autoclaves. Calcine (-421 μm) and 4 ℓ of leach acid (25% HCl) were charged to both vessels. Eight hundred grams of calcine was added to the first vessel and 900 g to the second. The vessels were then heated to their operating temperature (approximately 100°C) before the calcine and the leach acid were fed continuously into the first vessel via the first vessel's condenser. The calcine feed rate in these tests was varied from 7.1 to 11.1 g/min while the feed acid rate varied from 33 to 47 ml/min. Total operating time varied from 7.5 to 10.5 h to give about 3 displacements. Analysis of process streams showed that steady state conditions were achieved after about 1.5 displacements. Slurry from the first vessel was transferred to the second by peristaltic pump while that from the second vessel was removed to storage by a diaphragm pump. Operation was generally satisfactory, however, a build-up of coarse material was experienced in the first vessel even at the higher flowrate. The solids loading at the end of each test varied from 185 to 194 g/ℓ in the first vessel while that in the second vessel varied between 79 and 115 g/ℓ.

These tests indicated that a minimum of 4 h are required to achieve economically desirable aluminium extractions of greater than 90%. In only one continuous test did the aluminium extraction exceed 90%. Aluminium extractions of 90 to 96% would be expected from batch tests for comparable leaching conditions, i.e., residence times between 2.7 and 4 h and final acidities of more than 20 g/ℓ free HCl. It is possible that inadequate baffling in the two vessels (each uses the thermowell as the baffle) led to short circuiting of the fine material most of which was collected in the overflow from the second vessel.

Thickening of Leach Pulp

The initial work in this area was used to determine whether fines could be satisfactorily flocculated at economically reasonable flocculant addition levels as this would reduce the size of the filtration plant. A considerable number of flocculants were tested including: Separans AP30 and MGL, Superflocs 16, N100, N200, N100S, N2100S, C100 and C110 and Magnafloc R351. Of these only Superfloc N100S and Superfloc N2100S gave good flocculation with subsequent adequate zone settling and a reasonably clear supernatant liquid. Flocculant dosage varied between 132 to 430 g/tonne of residue. Results of settling tests using Superfloc N100S are shown in Figure 4. The settling rate varied between 0.9 and 1.8 cm/min.

Table III. Batch pressure leaching of calcined and uncalcined bauxite. Leachant used: 1. 25% HCl; 2. 20-22% HCl, 18-21% $FeCl_3$. All tests used 5% excess acid over stoichiometric requirements.

Leach test no.	Al extraction (%)	Fe extraction (%)	Bauxite	Sizing bauxite/ calcine (μm)	Leach time (h)	T (°C)	Acidity end of leach (g free HCl/ℓ)	Leachant used
16	94.3	-	uncalcined	-421	2	132	35	2
17	93.6	-	"	-421	2	152	28	"
22	87	92.4	"	-1200+421	2	134	52	1
21	90.4	93.8	"	-1200+421	3	134	43	"
23	95.5	94.8	"	-1200+421	5	134	36	"
25	100	98.2	"	-1200+421	2	152	27	"
31	95.5	98.0	"	-421	2	150	24	"
18	92.9	-	calcined	-421	2	134	43	2
19	84.5	89.2	"	-421	2	134	66	1
20	88.5	-	"	-421	2	134	36	2
24	83.8	90.0	"	-421	2	135	62	1
32	100.8	93.5	"	-421	4	133	45	"
33	101.4	89.2	"	-421	4	149	29	"
26	85.9	95.1	"	-421	2	112	--	"
26	97.3	101.2	"	-421	4	112	26	"

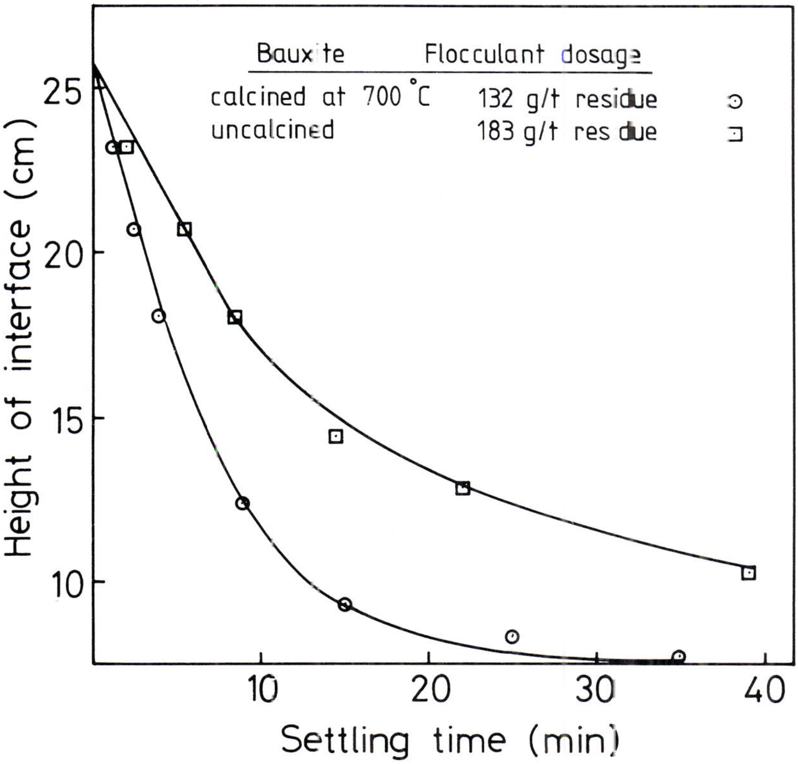

Figure 4 - Interface height as a function of settling time for settling tests on calcined and uncalcined leach residues.

Some remarkable improvement in the settling behaviour was observed when leaching uncalcined bauxite under pressure. Residues from such leach tests settled well without the addition of any flocculant. The addition of flocculant to such slurries did not improve the settling rate.

Filtration of Leach Residue

An extensive series of filtration tests using both vacuum and pressure filtration was conducted using the unthickened slurries from the various leach tests. Most vacuum tests were carried out with a plastic Buchner funnel having a filter area of 50 cm^2. Pressure filtration tests were carried out in a horizontal single-plate pressure filter with a filter area of 255 cm^2. A synthetic twill woven cloth (weight 140 g/m^2) served as the filter medium for both vacuum and pressure tests. The results of filtration tests are shown in Table IV and indicate that the filtration rate rises markedly with increase in pressure. The fast vacuum filtration rates of tests 4, 5, and 6 were obtained with 3-4 mm cakes. Similar pressure filter rates were achieved at cake thicknesses of 10-12 mm. As shown, in Figure 5, most tests exhibit straight line behaviour when the filtration rate is plotted against time on log-log coordinates. However, some data exhibit curvature indicating a change in the specific cake resistance.

Table IV. Filtration of leach residues.

Test	Filtration rate at 10 min (m^3/hm^2)	Pressure (kPa)	SG_{20} of filtrate	Cake thickness (mm)	Sizing of feed calcine (μm)	Slurry solids loading (g/ℓ)
L3	0.12	-63	1.42	2	-1200+421	20
L4	0.24	-63	1.37	3	-1200+421	24
L5	0.24	-63	1.42	4	-421	66
L6	0.22	-63	1.42	3	-1200+421	51
L7	0.16	-63	1.42	4	-1200+421	57
L9	0.31	550	1.40	10	-421	68
L10	0.31	300	1.25	10	-1200+421	70
L11[1]	0.15	550	1.44	8	-1200+421	55
L12	0.24	550	1.43	10	-421	82
L13[2]	0.26	550	1.40	12	-421	82
L14	0.46	550	1.22	12-15	-421	47
L15	0.42	550	1.25	23	-421	74

[1]Tighter woven filtercloth used.

[2]Uncalcined bauxite.

The filtration of thickened leach slurry was also briefly examined but the rates were prohibitively low.

Crystallization

Unlike the U.S. Bureau of Mines process developed in the miniplant program (3) where iron is removed by solvent extraction and in the H+ process where iron is removed as a double salt (iron potassium sulphate) (4), we have achieved effective separation of aluminium from iron in the crystallization step.

In crystallization, the degree of supersaturation is the driving force for both the formation of crystal nuclei and the growth of crystals, and both increase directly with the degree of supersaturation. Where salting out procedures are used to induce supersaturation, local supersaturation is induced at the point of addition and uncontrolled nucleation may result before the additive can be dispersed throughout the solution.

By controlling the HCl content of the crystallizing solution, supersaturation is controlled and coarse $AlCl_3 \cdot 6H_2O$ crystals can be grown. These can attain sizes up to 750 μm. As coarse crystals are easier to filter, contain less adhering mother liquor and are easier to wash than fine crystals, the resulting crystal product exhibits a high degree of purity.

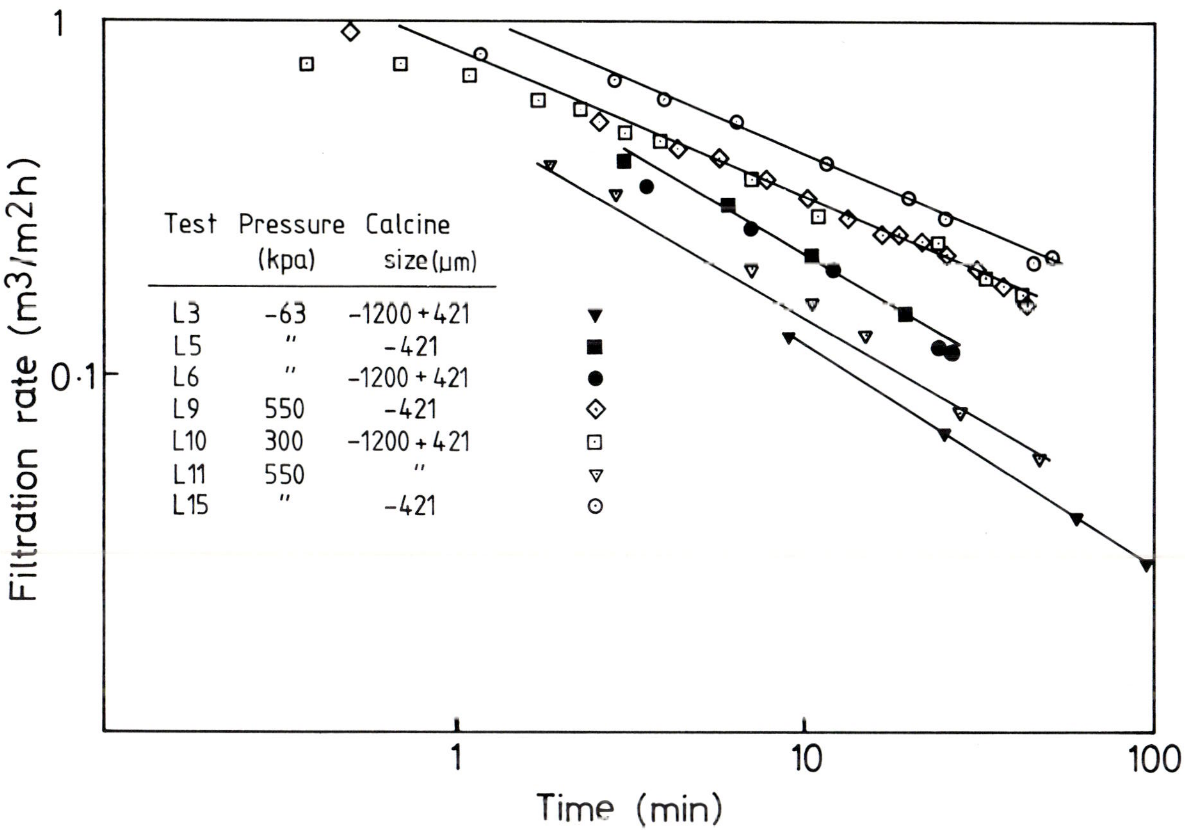

Figure 5 Filtration rate as a function of time for vacuum and pressure filtration tests.

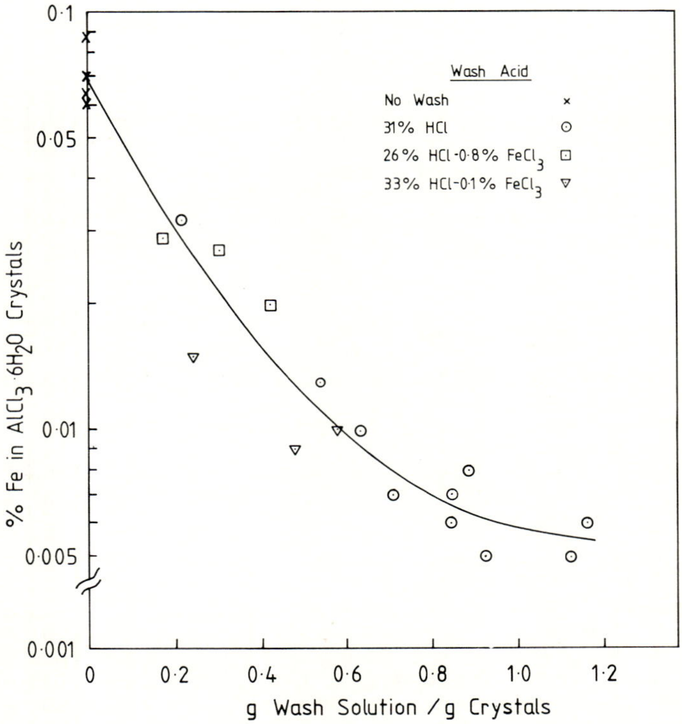

Figure 6 - Iron content of $AlCl_3 \cdot 6H_2O$ crystals as a function of the amount of wash acid used.

Figure 6 shows the effect of washing on the iron content of 300 to 750 µm crystals of $AlCl_3 \cdot 6H_2O$, which were separated from the barren liquor using a laboratory basket centrifuge having an acceleration ratio of 1000. With no washing, the crystals contained 0.07% Fe which reduced to 0.005% Fe when 1.0 - 1.2 g of wash solution per gram of crystals was used. It may be possible to use less wash solution than this on an industrial scale because industrial centrifuges have acceleration ratios as high as 5000.

Decomposition of $AlCl_3 \cdot 6H_2O$ Crystals

The washed aluminium chloride hexahydrate crystals are decomposed in a two-stage fluid bed reactor system as outlined by Marchessaux and Plass (5).

The advantage of this system is that 90 to 95% of the decomposition of the hexahydrate to alumina takes place in a low temperature first stage operating at 200 to 250°C. These temperatures can be attained by indirect heating which reduces the volume of gas to be treated in the absorption stage. The HCl content of the gases is increased and considerable savings in energy consumption are made. The oxide/chloride enters the second stage via a venturi reactor. During a residence time of approximately 30 minutes in this second stage, it is completely decomposed and calcined to alumina at a temperature of about 1000°C.

Regeneration of HCl

(a) Acid Production

Strong acid for crystal washing and for anhydrous HCl is produced from the gases obtained during the thermal decomposition of the hexahydrate. The gases from the low temperature decomposition stage feed one absorption plant which produces 36% hydrochloric acid. The high temperature decomposition stage produces gases with a lower hydrogen chloride content and these are used to produce a 29% hydrochloric acid in another absorption plant.

(b) Anhydrous Hydrogen Chloride

Anhydrous hydrogen chloride for crystallization is produced by stripping the following HCl rich solutions:

1. The 36% acid produced from the low temperature decomposition of the aluminium chloride hexahydrate.
2. The 29% acid produced from the high temperature decomposition.
3. The barren liquor remaining after crystal separation.

(c) Barren Liquor Treatment/Iron Removal

The barren liquor remaining after the bulk of the HCl has been stripped contains about 15% HCl and is first evaporated to produce a weak HCl solution (17.5%) and a concentrated chloride solution. This concentrated chloride solution is further evaporated in a single effect evaporator (not shown on flowsheet) and fed to the decomposition section. Here, evaporation of the remaining water and decomposition of ferric chloride and some aluminium chloride takes place at 850°C in a coal-fired fluid-bed reactor. The Fe_2O_3 and Al_2O_3 formed by the hydrolysis of the solution, plus ash are continuously removed from the reactor. The gases leaving the reactor are dried, absorbed and stripped to produce more gaseous HCl and also acid of azeotropic composition.

Product Alumina

Table V shows some of the physical properties of alumina produced in the laboratory from HCl leaching of bauxite. Also shown are the physical properties of some industrial aluminas (6).

The bulk density, angle of repose and size distribution of the hexahydrate alumina are quite different to the industrial aluminas. However, these properties are dependent on the calcination conditions. Lurgi and Pechiney have investigated the calcination of $AlCl_3 \cdot 6H_2O$ as part of the H+ process and have produced an alumina whose physical properties have resulted in acceptable electrolysis operation and metal quality (5). The analysis of the alumina produced from our tests is shown in Table VI and indicates that the product purity is satisfactory.

Energy Requirements

A heat balance for the process was developed and the net fossil fuel energy requirements for the process has been estimated at 34,600 kJ/kg

Table V. Comparison of physical properties of industrial aluminas with alumina produced from $AlCl_3 \cdot 6H_2O$.

Property	Industrial alumina A	B	C	D	Alumina from $AlCl_3 \cdot 6H_2O$
True density, kg/m^3	3880	3590	3600	3620	3330
Bulk density, stamped, kg/m^3	1240	1090	1220	1190	530
" " , unstamped, kg/m^3	930	890	940	950	420
Spec. surface area, m^2/g	2.4	55	45	42	42.5
L.O.I. 300 - 1000°C, %	0.06	0.37	0.75	0.25	0.14
Angle of repose	38.0	35.0	32.5	32.5	45
Mean particle size, μm	60	74	47	64	165
+74 μm, %	31	50	11	27	83
-44 μm, %	29	10	47	20	12

Table VI. Comparison of impurity levels in alumina from $AlCl_3 \cdot 6H_2O$ with Bayer alumina (%).

Impurity	Alumina from $AlCl_3 \cdot 6H_2O$	Bayer alumina
P_2O_5	0.003	0.001-0.002
MgO	0.002	0.002-0.004
Fe_2O_3	0.023	0.010-0.030
Cl	<0.03	0.01

Al_2O_3 for bauxite leaching and 37,100 kJ/kg Al_2O_3 for calcine leaching. The power consumption has been estimated at 0.23 kWh/kg Al_2O_3 or 2550 kJ/kg Al_2O_3 on a fossil fuel basis for both schemes. At this stage the above figures do not take into account any energy recovery schemes that may be incorporated in the process due to uncertainties surrounding process integration, equipment selection and plant layout.

Preliminary Cost Estimation

Data obtained from technical scale tests was used to develop energy and material balances as well as equipment sizes. The mass balance for the process shown in Figure 1 is contained in the Appendix. Most process plant costs have been derived from cost information submitted by various suppliers either on an equipment only or on a turnkey basis. The factoring method was the basic procedure used to determine total module costs. Factors were selected from Katell (7,8) according to the process conditions and the type of system being considered. Factors proposed by the manufacturers were used to determine the installed cost of turnkey plants. The preliminary capital cost of an 800,000 tonne/year plant using either atmospheric leaching of calcine or pressure leaching of bauxite (in 1981 Australian dollars) was \$790/tonne of alumina. The operating cost for both was estimated at \$145/tonne of alumina.

Reactive Silica Content

A higher silica content in the bauxite would lead to an increased residue output. It has been estimated that the residue tonnage would increase by 1.5 to 2 tonne/h for every percent silica above the base level of 6%. Thus at 12% silica in the bauxite, the residue rate would be about 40 tonne/h compared with 30 tonne/h at 6% silica.

The capital required for a leaching plant treating bauxite containing 12% silica would increase due to additional filters, pumps and possibly larger agitators. The capital cost of wash liquor evaporation would also increase as approximately 30 to 40 tonne/h of extra water would have to be evaporated. The total investment would rise by about 4 to 5%.

The operating costs when treating this high silica bauxite would also rise mainly due to higher chloride losses in the residue requiring additional make-up acid. The cost of this additional acid was estimated at an extra 40%.

These preliminary costs have indicated that when the reactive silica content of the bauxite approaches 10% then the economics favour an acid process.

Conclusion

This laboratory and technical scale study has shown that it is possible to recover alumina from high reactive silica bauxites via a hydrochloric acid route. At present leaching of calcine at atmospheric pressure is the preferred scheme mainly because of the severe corrosive conditions likely to be encountered at conditions above the atmospheric boiling point of leach solutions.

Minimal iron contamination of the product is achieved by producing coarse crystals of aluminium chloride hexahydrate which drain and wash well. These, when calcined, produce alumina meeting cell grade specifications. Iron is removed from the process as Fe_2O_3 which is produced by hydrolysis of the iron containing solution remaining after precipitating aluminium chloride hexahydrate.

The fossil fuel energy requirement for the process was estimated to be 34,600 to 37,100 kJ/kg Al_2O_3 depending upon whether calcine or bauxite leaching was selected.

The capital and operating cost of the proposed process using either calcine or bauxite leaching was estimated at \$635 million (in 1981 Australian dollars) while the operating costs of both alternatives was \$145/tonne of alumina.

This acid process is not a replacement for the Bayer process treating standard bauxite. It is designed for treating bauxites that are unsuitable as feed to the Bayer process due to their high reactive silica content. However, a comparison of the capital and operating costs versus the reactive silica content is instructive and showed that when the reactive silica content approaches 10% then the economics favour an acid process.

Acknowledgement

The authors gratefully acknowledge the financial and other support provided by Comalco Ltd for this project.

References

1. T.G. Pearson, The Chemical Background of the Aluminium Industry, pp. 6-7, Royal Institute of Chemistry, London, 1955.
2. D.H. Jenkins, and H.N. Sinha, "Production of Anhydrous Aluminium Chloride from Aluminium Chloride Hexahydrate", paper presented at 5th International Congress of ICSOBA, Zagreb, Yugoslavia, Sept. 1983.
3. K.B. Bengston, P. Chumberka, R.F. Nunn, A.V. San Jose, G.M. Manarolis, and L.E. Malm, "Alumina Process Feasibility Study and Preliminary Pilot Plant Design. Task 3 Report Preliminary Design of 25 Ton Per Day Pilot Plant. Volume 1. Process Technology and Costs", Bureau of Mines Open File Report 122(1)-80, 1980.
4. Aluminium Pechiney, U.S. Patent 3 862 293 (1975), "Process for Continuous Acid Treatment of Clays and Schists".
5. P. Marchessaux, L. Plass, and L. Reh, "Thermal Decomposition of Aluminium Hexahydrate Chloride ($AlCl_3 \cdot 6H_2O$) for Alumina Production", paper presented at 108th TMS-AIME Annual Meeting, New Orleans, La., Feb. 1979.
6. P. Ravn, and A. Windfeldt, "Dusting Properties of some Industrial Aluminas", paper presented at 104th TMS-AIME Annual Meeting, New York, N.Y., Feb. 1975.
7. S. Kattell, "Factors Utilized in a Preliminary Estimate", American Association of Cost Engineers, Transactions, (1977) pp. 58-59.
8. S. Katell, "Factoring Equipment Installation Costs", American Association of Cost Engineers, Transactions, (1978) p. 51.

APPENDIX. Alumina via HCl leaching of bauxite: Mass balance for 101 tonne/h plant.

Stream No.		1	2	3/4	5	6	7	8	9	10	11	12	13	14	15	16	17	18/19/20	21	22A
Al_2O_3	(t/h)	111.2		111.2		111.2								7.78						
Fe_2O_3	-	19.0		19.0		19.0								0.19						
SiO_2	-	15.8		15.8		15.8								15.8						
TiO_2	-	4.4		4.4		4.4								4.4						
L.O.I.	-	49.6		49.6	49.6															
H_2O	-	27.3	27.3				210.50	340.84	595.3	656.44	84.51	634.52	81.46	27.59	46.35	35.11	680.87	0.27	465.31	179.00
$AlCl_3$	-							1.02	1.02	271.55		261.20	7.76	2.59	7.76		268.96		3.09	1.21
$FeCl_3$	-							1.22	1.22	39.37		37.88	1.12	0.37	1.12		39.00		39.00	15.02
HCl	-						53.2	199.77	255.2	7.40		7.12	0.21	0.07	0.21		7.33	89.26	275.14	105.87
$AlCl_3 \cdot 6H_2O$	-																		481.44	
Fuel	-																			
Air	-																			
Inerts	-																			
Total		227.3	27.3	200.0	49.6	150.4	263.78	542.85	852.74	974.76	84.51	940.72	90.50	58.79	55.44	35.11	996.16	89.53	1263.98	301.10

APPENDIX continued.

Stream No.		22B	23	24	25	26	27	28	29	30	31*	32	33	34	35	36	37	38	39	40
Al_2O_3	(t/h)										1.15									
Fe_2O_3	-										18.43									
SiO_2	-																			
TiO_2	-																			
L.O.I.	-																			
H_2O	-	286.31	340.84	15.86	271.24	450.24	0.54	449.70	403.89			0.06	378.06	11.22		168.19	518.04	177.19	352.96	176.96
$AlCl_3$	-	1.88		0.04	1.78	2.99		2.99												
$FeCl_3$	-	23.98		0.04	22.72	37.74		37.74												
HCl	-	169.27	200.18	9.31	160.36	266.23	179.69	86.55	85.67			19.70	94.51			215.33	304.24	104.07	89.34	44.79
$AlCl_3 \cdot 6H_2O$	-	481.44		479.70																
Fuel	-									10.00										
Air	-														38.22	38.22				
Inerts	-																			
Total		962.88	541.02	504.95	456.10	757.20	180.23	576.98	489.56	10.00	19.58	19.76	472.57	11.22	38.22	421.74	822.28	281.26	442.30	221.75

*Excluding ash.

APPENDIX continued.

Stream No.		41	42	43A	43B	44	45	46	47	48	49	50	51
Al_2O_3	(t/h)		95.93				101.38						
Fe_2O_3	-		0.02				0.02						
SiO_2	-												
TiO_2	-												
L.O.I.	-												
H_2O	-	0.24				13.96		51.90	42.48	0.04	51.86	606.88	34.36
$AlCl_3$	-												
$FeCl_3$	-												
HCl	-	59.27				11.36		22.24	10.75	9.11	13.13	152.43	
$AlCl_3 \cdot 6H_2O$	-		25.77										
Fuel	-			5.36									
Air	-				73.28								
Inerts	-					76.94							
Total		59.51	121.72	5.36	73.28	102.59	101.40	74.14	53.23	9.15	64.99	759.31	34.36

CHAIRMEN

E.L. Cambridge
ARCO Metals Company
Tucson, Arizona

A.T. Tabereaux
Reynolds Metals Company
Sheffield, Alabama

START-UP OF FIRST PRE-BAKE POTLINE IN BRASIL

VALESUL ALUMINIO S.A., SANTA CRUZ, R.J., BRASIL

J. E. Tauscher

Reynolds do Brasil Ltda
Rua da Assembleia, 10-1716
Rio de Janeiro, Brasil
CEP 20011

D. S. Ratcliffe

Valesul Aluminio S.A.
Estr. Aterrado do Leme, 1225
Santa Cruz, R.J., Brasil
CEP 23500

The start-up of the VALESUL smelter at Santa Cruz represented the first phase of the new generation of pre-bake smelters in operation and under construction in Brasil. This paper discusses the planning and programming, the execution and the technical detail used in one of the industries most successful potline start-ups at a new plant. Early performance figures for the potline are also included.

Introduction

VALESUL ALUMINIO S.A. is an aluminum reduction plant with one potline having a nominal capacity of 86000 ton per year. Space exists for two additional potlines and some of the ancillary facilities have provisions to support an expansion of the plant. The company is owned by Shell-Billiton, Compania Vale do Rio Doce (CVRD), and Reynolds International Inc., with management by the two principal partners Shell and CVRD and technical know-how supplied by Reynolds.

VALESUL is located 55 km south of Rio de Janeiro, Brasil near the town of Santa Cruz. Alumina is transported a distance of 30 km by truck from the new deep water port facility on Sepetiba Bay. VALESUL has its own unloading system and two 15000 ton alumina storage tanks at the dock site. An additional 15000 ton storage silo is situated at the plant site.

The cell installed at VALESUL is the Reynolds P-19 design and is being operated at 155 KA. The principle features of the cell are two rows of nine anodes each, center crust breaking and alumina feeding, moveable anode bridge, integral gas ducting system in the superstructure, and computer control. This cell design is in use in three other locations with about 1400 pots installed. A view of one of these pots during start-up is shown in Figure 1, and a view of the potroom in actual operation is shown in Figure 2.

General Plan

The general plant start-up program was developed jointly by Reynolds specialists and the VALESUL staff to coordinate start-up of all the support facilities with the start-up of the potline itself -- REYNOLDS responsible for the technical supervision and training, and VALESUL responsible for all the logistics, manpower, administration, etc. Prior to the plant start-up, it was decided to start 12 cells for use as a training section. In order to avoid advancing the anode plant construction schedule, 4000 baked anodes were purchased and imported into Brasil for use in the training section.

Electrical/Resistor Start-Up Method

Following a pre-determined plan, three pots in Room 1 were started by heating up the cathode and anodes uniformly with a steadily increasing potline amperage. The objective was to achieve uniform anode current distribution and minimize any internal damage from localized overheating, and also to avoid any anode breakage.

(a) Heating Cycle

The rate of increase of amperage established for this resistor start-up was 2000 amperes per hour, a rate that was programmed to reach 130 kA in about 60 hours. During this heating-up period, the ampere loading was maintained uniform on all 18 anodes until the 130 kA holding period was reached where the cryolite melting phase would commence.

The amperage curve is given in Figure 3, and the initial starting load of 50 kA is seen at point A. When the potline was energized and brought to this amperage level, a check was made of all the bus system, the metering, and the anode stem current loading. (Previously, the rectifier - transformer system had been checked at full amperage and had been accepted as

FIGURE 1
View of VALESUL Pot Being Started

FIGURE 2
View of Potroom In Operation

fully satisfactory.) The holding period required to melt cryolite at 130 kA was estimated to be about 16 hours, but was found to be somewhat longer.

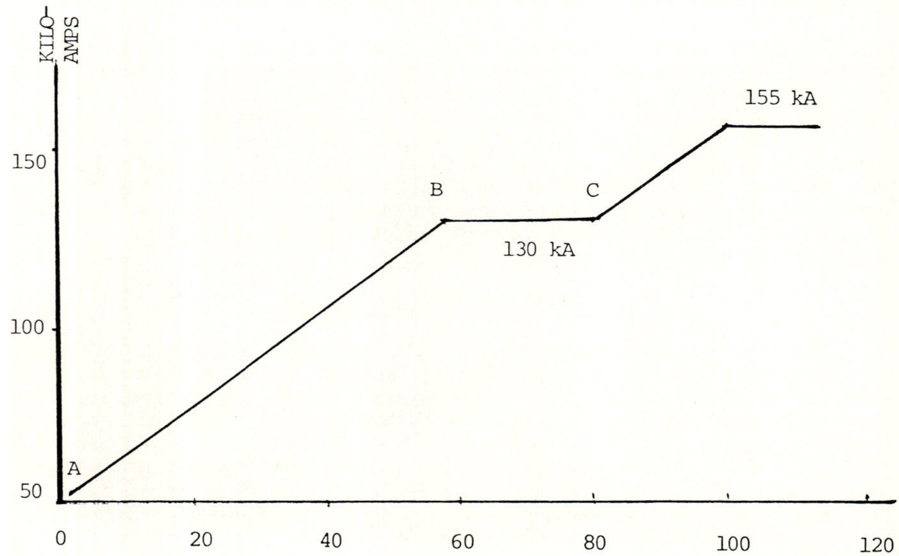

FIGURE 3
Resistor Baking Curve At VALESUL

In preparation for the heating cycle on the first 3 pots, a carbon resistor material was used under the anodes. This consisted of crushed anodes, screened to a size of minus 15mm. The quantity was 2-1/2cm depth under the anodes on two pots, and 5cm under the anodes of the third pot, this extra amount being used to vary the heat input. The pots were covered with insulating boards between the anodes and sidewalls and also through the center area of the pot. This was done to conserve and improve the heat distribution. The method used at VALESUL for maintaining uniform current distribution on all anodes was that of unclamping and reclamping the anode stems. In an earlier AIME paper (1) by Welch, Keniry and Clelland on the smelter start-up un New Zealand, it was reported that this technique was not very successful as many cracks appeared in the cathode. At VALESUL, the experience was quite different.

The amperage distribution was so even that no anodes were broken and no visible cracks appeared on the cathode. In the technique used at VALESUL, the technicians reset the anode clamps on all anodes uniformly with a hand tool every hour, and this was done throughout the heating curve. The clamps on the VALESUL design are the ECL type, modified from those used in earlier smelters of this same design. This reclamping technique was used until the point where the amperage was held constant and the pots were packed with the cryolite mixture.

(b) Operating Cycle

At a predetermined point in the heating cycle, the three pots were packed with a mixture of alumina and bath materials on the sides and ends.

Cryolite was added to a depth of 20cm, then 5cm of alumina was added, then 50 kg soda ash and 50kg magnesite, and then finally more cryolite was added until the sidewalls were packed to the top of the anodes.

As liquid began to form around the anodes and the cell resistance decreased, the anodes were raised off the cathode using the anode bridge motor. The exposed liquid areas were kept covered with the cryolite being fed continually, and cell anode-cathode distance increased by using frequent, small movement of the anode bridge. When all anodes were off the bottom the cell voltage was held at 12-15 volts until liquid depth of 25cm was reached. Cell voltage was gradually lowered to 5.5 volts with pot operation on manual mode, and these three pots were then used to supply liquid bath for the remaining cells which were started using a thermal pre-heat and a hot-bath transfer technique.

Hot Bath Transfer Method

The start-up technique used for all pots after the first 3 consisted of pre-heat by external fuel source, not potline amperage, followed by the transfer of liquid cryolite and metal to the new pot. A propane burner was used under each of the 18 anodes, with the burners being connected to a manifold on each side of the pot. This technique is now normal practice at VALESUL, and the following is the actual standard procedure for the plant:

(1) Check all operating parts of the superstructure and control panel, mechanical and electrical.

(2) Blow out the cathode with compressed air, and free it of any foreign material.

(3) Adjust the bridge to a low enough position to allow bridge to travel up as needed.

(4) Set carbon in place on the cathode bottom and properly clamp and mark with chalk.

(5) Make a mark on the bridge scale to indicate the bottomed-out carbon position.

(6) Raise anodes with the pot jacks to 20cm above the cathode floor.

(7) Set up the baking equipment and proceed to put the pot on bake.

(8) Check the burners continuously during the 48-hour bake time to assure proper heating rate and distribution.

(9) Have sidewall banking material delivered to the pot.

(10) After 48 hours of bake, remove all baking equipment from the pot and take it to the next pot to be put on bake.

(11) Recheck jacking system for proper operation.

(12) Lower anodes to bottom. Loosen clamps and then retighten to assure that all carbon are the same level. Remark with chalk if necessary.

(13) Bank sidewalls with prescribed amounts of material.

(14) Assure that the crucible and lid are ready for bath tapping.

(15) Install pouring trough in proper position at wide end of pot.

(16) Commence tapping bath and check closely to assure that only bath and no metal is tapped. Tap a minimum of 5 tons of bath into a clean crucible.

(17) Pour molten bath into the pot while the cut-in crew stands ready with impact wrenches.

(18) After making sure there is at least 5 to 6cm of anode immersion, notify the rectifier to drop the line load.

(19) With no line current, open the deadheads and insert the insulation boards.

(20) Reapply line load.

(21) Maintain at least 20 volts to warm up the pot. Usually pot voltage will be 20 to 35 volts.

(22) Maintain pot anode effect until carbon are 7cm off the bottom and 25cm of liquid bath is measured.

(23) Do not break or feed pot until the first natural anode effect.

(24) Add 6 tons of molten metal on the shift following cut-in.

(25) Add 6 more tons of metal on the second shift following cut-in.

(26) Add 5cm of alumina on top of carbon and along sides of pot.

(a) Banking

"The prescribed banking procedure mentioned in step 13 is as follows:

(01) Cryolite - 400 Kg, with 200 Kg spread evenly down each side.

(02) Alumina - Approximately 2cm spread evenly on top of this cryolite.

(03) Soda - 100 Kg, with 50 Kg spread along each side.

(04) Magnesite - 100 Kg, with 50 Kg spread along each side.

(5a) Cold crucible bath (if available) - 2600 to 3000 kg with 1300 to 1500 kg spread evenly on each side.

(5b) Cryolite - 8000 kg with 3000 kg on each side and 1000 kg on each end.

(06) If cold bath (5a) is used, cover it with 2000 kg of cryolite, with 1000 kg on each side.

(b) Cut-In

After the molten metal is poured and the line is re-energized, the new pot will come in on an anode effect from 20 to 35 volts. The intensity depends on several things, including bath temperature, anode immersion, and rectifier stability. The liquid bath is immediately chilled by the banking material and begins to shrink. If at least a 20 volt anode effect is not initially obtained, the heat input to the pot is not sufficient to prevent this shrinking. The bath will then shrink away from the carbon, giving a very high intensity anode effect which could shatter carbon and be detrimental to the pot. Light intensity should be maintained by periodically raising the anodes until they are at least 7cm off the bottom. At this time, liquid bath depth should be approximately 25 cm, and the light intensity will gradually diminish. The alumina in the banking material will mix with the liquid bath, and this combined with the warming of the bath will cause the anode effect to extinguish itself. Pot voltage may fall at this time to as low as 4.5 volts. The anodes should be picked up slightly, but normally within 30 minutes pot voltage will climb to between 6 and 7 volts and stabilize.

The first natural anode effect will normally occur within four to five hours and should be allowed to burn for 15 to 20 minutes to warm up the pot and melt the banking material. It can then be extinguished in the normal manner.

Following the first crucible of metal, the voltage is lowered to 5.5 volts and, after the second crucible, to 5 volts. After the second crucible, the sides should be ored with 8 to 10cm of alumina, 3 to 4cm of alumina should be put on top of the anodes, and cover administered into spaces between the carbons until sealed.

When the new pot is no longer used to make bath for other pots, lower the voltage to 4.3 and 631 computer status (2) and 4+0 feed rate. After the next anode effect, voltage is reduced to 4.2 until double carbon sets are completed, at which time voltage is lowered to 4.1 and 621 status (2) and 5+0 feed rate."

Training

At the beginning of the project, a major commitment was made for funding and supporting a comprehensive training program for employees. In general terms, this program can be described in 3 stages that coincided with the construction and start-up period.

The first stage consisted of training for the VALESUL management and supervisory personnel in Reynolds plants and affiliates outside of Brasil. Each trainee was evaluated and an individual training program was developed for each trip. A total of 72 VALESUL employees participated in this stage for a total of 137 man/months.

The second phase involved classroom training using courses provided by Reynolds. There are a series of orientation courses for introduction to the aluminum industry and reduction plant operations, and also a series of technical operations courses on specific production areas. Virtually all VALESUL employees received a minimum of two weeks of classroom training.

The third phase was primarily on-the-job training received from experienced supervisors and Reynolds personnel. This was conducted plantwide as the various facilities were started up.

In order to accomplish the training and start-up, REYNOLDS INTERNATIONAL fielded a team of experts. The team was truly an international group since it had representatives from 10 different countries. All had prior experience with this design in other locations.

The on-line training period is one of the most important times for the operating departments during the start-up of a green field project. In addition to providing hands-on training to new and inexperienced personnel, it allows for an equipment shake-down period. Although the basic cell design was the same as previous Reynolds projects, much of the plant equipment was new, and 70% had to be of Brasilian origin. Thus, VALESUL has a lot of equipment that represents the vendor's first attempt to supply the aluminum industry.

As was mentioned previously, it was decided to start-up and operate 12 pots for 3 months prior to the actual plant start-up. During the operation there was sufficient supervision that training could be done on a one-to-one basis. Emphasis was placed on learning the routine potroom skills such as correct anode setting, tapping, skimming, safety, clean-up, etc. Although some mistakes were made and some equipment failed, major disasters were avoided due to the low workload both on the people and equipment. This low stress environment is necessary for an effective training program.

An important aspect of VALESUL's successful start-up has been the close attention given to anode setting. The anode change crane is a difficult piece of equipment to learn to operate. Therefore, a special course was devised to teach the skills required. The 4 shift foremen took the classroom part of the course first. Then they spent 2 days training with the vendor's commissioning engineer. After that period, they spent several more days practicing with simulated anodes. While the shift foremen were practicing with the cranes, the crane operators were taking the classroom course. The device for measuring anode reference elevation was modified to suit VALESUL requirements, and a special training course was given in this particular operation. The training courses included maintenance as well as production persons, and were given by an experienced training staff.

When the crane operators had finished the classroom course, the shift foremen did the hands-on part of the training. About 50% of the crane operators were trained in the above described manor. The rest of the operators were sent to the classroom course and then placed with an exceptionally skillful crane operator for hands-on experience.

This example is indicative of the comprehensive training carried out throughout the VALESUL plant. It reflects the high degree of coordination that was achieved among the recruiting, personnel, training, and operating departments. This commitment to effective training contributed greatly to the successful plant start-up.

CONCLUSIONS

Commissioning of all VALESUL facilities was successful and the results achieved were excellent, reaching an average current efficiency (in excess of its design criteria) over a 15 month period, operating at an amperage of 155 kA. The main operating statistics (3) are as follows:

		Feb/1982-Apr/1983	May/1983
Average nbr. of pots	(unit)	100.2	216
Amperage	(kA)	155.6	155.3
Voltage per pot	(V)	4.43	4.33
Average production per pot	(kg)	1,128.5	1,144.7
Current efficiency	(%)	90.1	91.5
Power consumption	(kWh DC/kg)	14.65	14.07
Net anode consumption	(kg/kg)	0.479	0.447

Acknowledgements

The authors wish to express their thanks to VALESUL S.A. and to REYNOLDS INTERNATIONAL INC. for their assistance and their permission to publish this work.

References

1) B. J. Welch, J. T. Keniry, and C. W. Clelland, "A Study of Some Aspects of the Influence of Cell Operation on Cathode Life" Light Metals 1982, pp. 299-309.

2) C. A. Wilson, A. T. Tabereaux, "Alumina Control in Center Break Cells", Light Metals 1983 pp. 479-493.

3) Claudio Mazoni Andrade, "Energy and Environmental Conservation at VALESUL Aluminum Smelter", Industry and Environment Magazine, United Nations, September 1983.

DEVELOPMENT OF A HIGH AMPERAGE END-TO-END REDUCTION CELL DESIGN

C.J. Rogers

Alcan International Limited
P.O. Box 8400
Kingston, Ont., Canada, K7L 4Z4

D.W. Macmillan

Alcan International Limited
P.O. Box 1250
Jonquiere, P.Q., Canada, G7S 4K8

W.J. Wright

Alcan Australia Limited
P.O. Box 1,
Kurri Kurri, N.S.W., Australia, 2327

Abstract

In 1979 Alcan initiated a project to construct a 200,000 tonne aluminum smelter on the Queensland coast of Australia. In order to have proven technology and to confirm economics, a development program was initiated on four cells.

This paper describes the design process and details the results of two years of operation of the four 175 kA prototype cells installed within an operating potroom of the Kurri Kurri smelter of Alcan Australia Limited.

Also described are ongoing programs to implement this technology for new construction and to upgrade existing facilities.

Historical Background

During the late 60's and early 70's Alcan built several potlines with end-to-end side feed prebaked anode cells.

In 1979 the company installed its then latest evolution in prebake cell design in a second potline in Kurri Kurri, NSW. This design retained many of the characteristics of the earlier design but alumina was fed to the centre channel between anodes.

Provision was also made in this new Kurri Kurri potline for a facility permitting further development of end-to-end cells. This was to be aimed at reducing substantially the capital costs, as well as cell operating costs.

A design team was formed and using Alcan computer design models, set out to solve the magnetohydrodynamic and thermal problems associated with high amperage end-to-end cells.

The resulting cell design (the Alcan 175E) was successfully tested on four development cells at Kurri Kurri between 1981 and 1982.

The initial intention was to use this technology in a new smelter on the Queensland coast. Regrettably, due to declining economic conditions, work on the Queensland project was stopped in August 1981. Thus the opportunity to apply this newly developed technology was temporarily lost.

Accordingly the attention was then shifted to the application of this technology in the expansion and modernization of existing facilities at Kurri.

A third potline is now under construction at Kurri Kurri and will employ the Alcan 175E technology.

Scope of the Development Program

The scope of the development program was as follows:

(a) The design of a cost effective 175 kA cell.

(b) Construction and operation of four experimental cells.

(c) A trimmer circuit was installed to provide additional amperage.

(d) A second computer control system was provided in parallel with the host line system.

(e) New control logic and operating practices were developed and applied to the test cells; however operations on the cells were performed by the existing potroom labour organization.

(f) A technical group was established to monitor cell operation, to modify and implement control and operating practice changes as required, and to provide specialist training and support to potline operators.

Design Constraints

(a) The test cells were to occupy existing pot stalls in Kurri Line 2.

(b) Existing equipment was to be used to service the cells wherever possible, e.g. for anode changing, ore hopper filling, etc.

(c) The anode current density was established based on economic conditions expected in New South Wales.

(d) Anode geometry changes were limited by the equipment in Kurri Kurri's anode plant.

Cell Process Design

The basic cell design characteristics of the Alcan 175E cell represent a "coming together" of experiences from experimental and production units in Lynemouth, England, Tomakomai, Japan, and Kurri Kurri, Australia.

Relative to the Kurri Kurri Line 2 cell (the most recent construction) the major differences in the Alcan 175E cell are:

(a) The number of anodes is increased by 2 along with the size of each anode for a total increase in carbon area of 13.5%. This maintains similar anode current density.

(b) The cavity size is decreased as a result of a reduction of 39% in the anode-to-side wall distance.

(c) The bottom plate area is decreased by 7% (longer but narrower pot shell).

(d) The busbar arrangement embodies a modified anode riser system.

(e) Alumina feeding is achieved by way of two point breaker-feeders placed on the centreline of the superstructure. The development targets of the experimental cells are indicated in Table 1 and are compared to the existing Kurri Kurri Line 2 operating targets.

Table I. Development Targets Alcan 175E Vs Kurri Kurri Line 2

	Alcan 175E	KK-2 (1983 plan)
Current (kA)	175	158
Voltage	4.3	4.6
Current efficiency (%)	90.5	88.5
kWh/kg	14.1	15.4

Cathode Design. For proper thermal design, experience with the existing Kurri Line 2 cells and experimental cells in England and Japan, has shown that reduction in ASD* and AED* are not only practical, but essential.

The cell lining design itself, which is based on Alcan thermo-electrical mathematical models and experience of experimental and production cells, minimized the tamped mix around the perimeter of the cathode panel. It calls for relatively high heat loss at the mid and upper side wall region of the cell.

To assess the lining design, measurements from installed thermocouples and external heat flux transducers were gathered at regular intervals after start-up. The results of measured data are summarized graphically in Figure 1. Calculated temperature and heat fluxes are presented in Figure 2 and comparison with Figure 1 indicates satisfactory agreement between calculated and observed temperatures and heat fluxes.

Anode Design. The anode configuration and current density was essentially that of the existing Kurri Line 2 cells and consequently the thermoelectrical characteristics were known in advance.

Busbar Arrangement. The cathode busbar arrangement of the Alcan 175E cell was similar to that of the KK-2 cells. Unlike the KK-2 cells, the anode risers are constructed in an "A" frame, providing a self-supporting anode riser assembly, the features of which are clearly visible in Figure 3.

During the design phase of the Alcan 175E cell, a complete analysis of the magnetic field and magnetohydrodynamic implications of the proposed busbar arrangement was conducted.

After start-up, field measurements of the magnetic field intensities in and around the test cells were made. The measurement locations for the magnetic fields in the liquid zones are shown in Figure 4. A comparison of the measured and the calculated fields are presented in Figure 5. Except in the extreme corners of the cells, where the vertical component B_z of the magnetic field is somewhat higher than predicted, the agreement between measured and predicted is quite acceptable.

Alumina Feeding. It was considered important, in achieving the stated development targets, that a sludge-free operation be obtained. The centre bar-breakers of the Line 2 cells were thus replaced by two point breaker-feeders on each cell.

Operating Results

The phase of start-up and early operation was followed by a period of optimization of cell operation. A period of 8 months of operational evaluation then followed. The results summarized in Table II met the stated program objectives.

* ASD - Anode-side Wall Distance, AED - Anode-End Wall Distance

DATE : 30 January 1981

CELL NO : Alcan 175 E

AGE : 130 DAYS

x—x FREEZE RANGE

---- HEAT FLUX RANGE ($W/cm^2 x 10^{-1}$)

•◖ TEMPERATURE MEAN ± RANGE (°C)

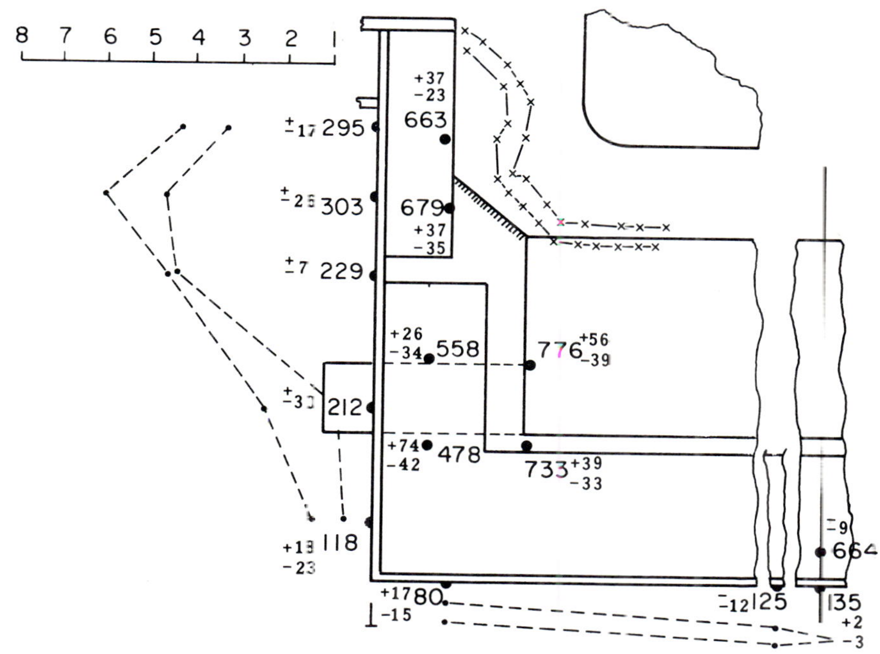

Figure 1 - Partial cell side section with measured heat fluxes, temperatures and freeze at the cell quarter points

• • CALCULATED TEMPERATURES ARE INDICATED AS:
MEASURED TEMPERATURE +/– ERROR IN CALCULATION

⊙ CALCULATED HEAT FLUX ($W/cm^2 \times 10^{-1}$)

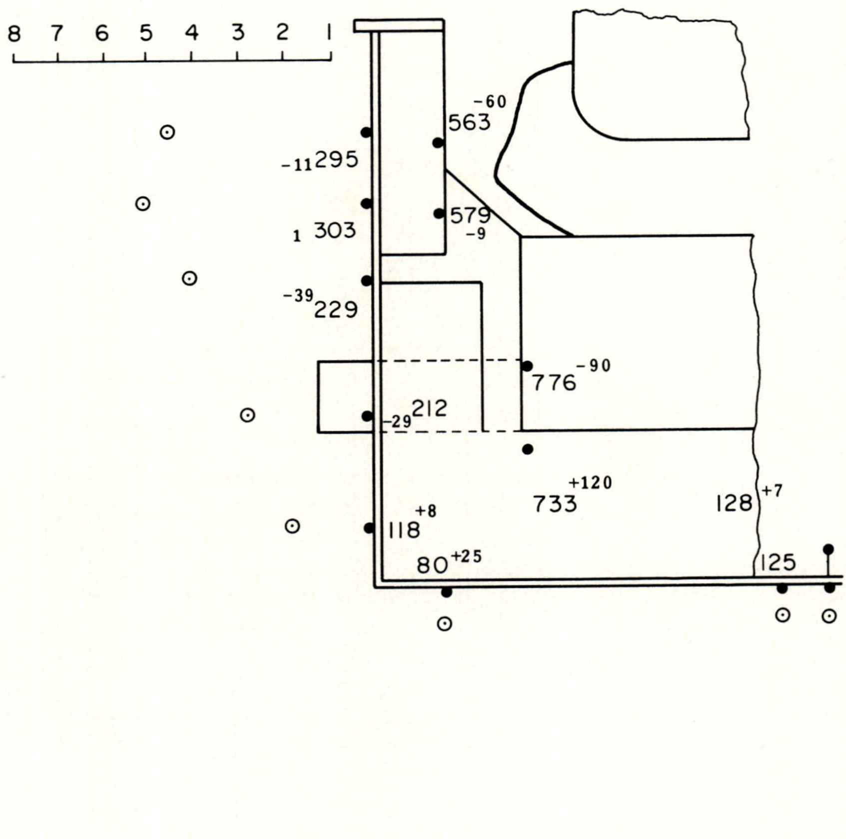

Figure 2 - Summary of thermal 2-D cathode model results

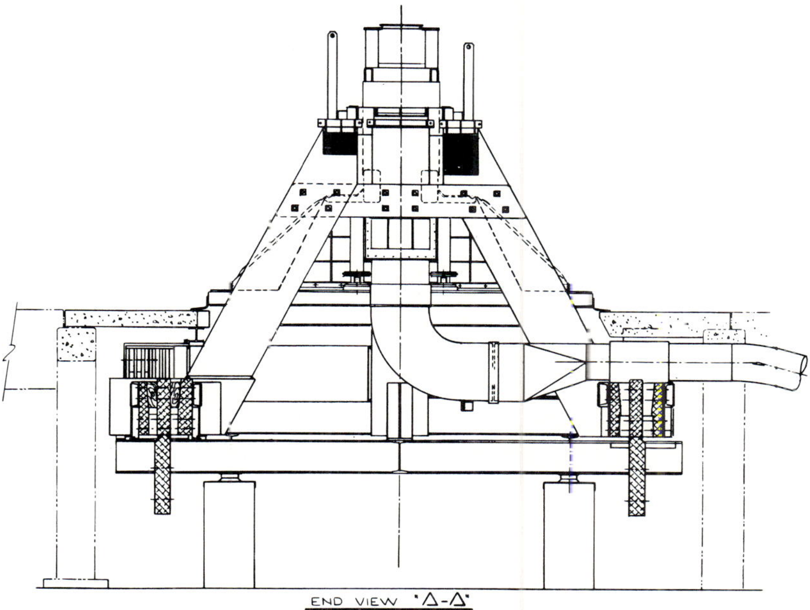

Figure 3 - End view of cell showing "A" frame construction

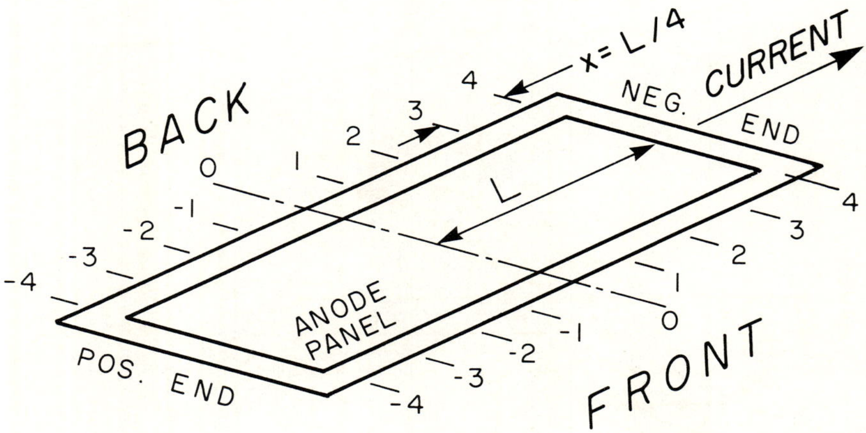

Figure 4 - Measurement locations for magnetic field measurements

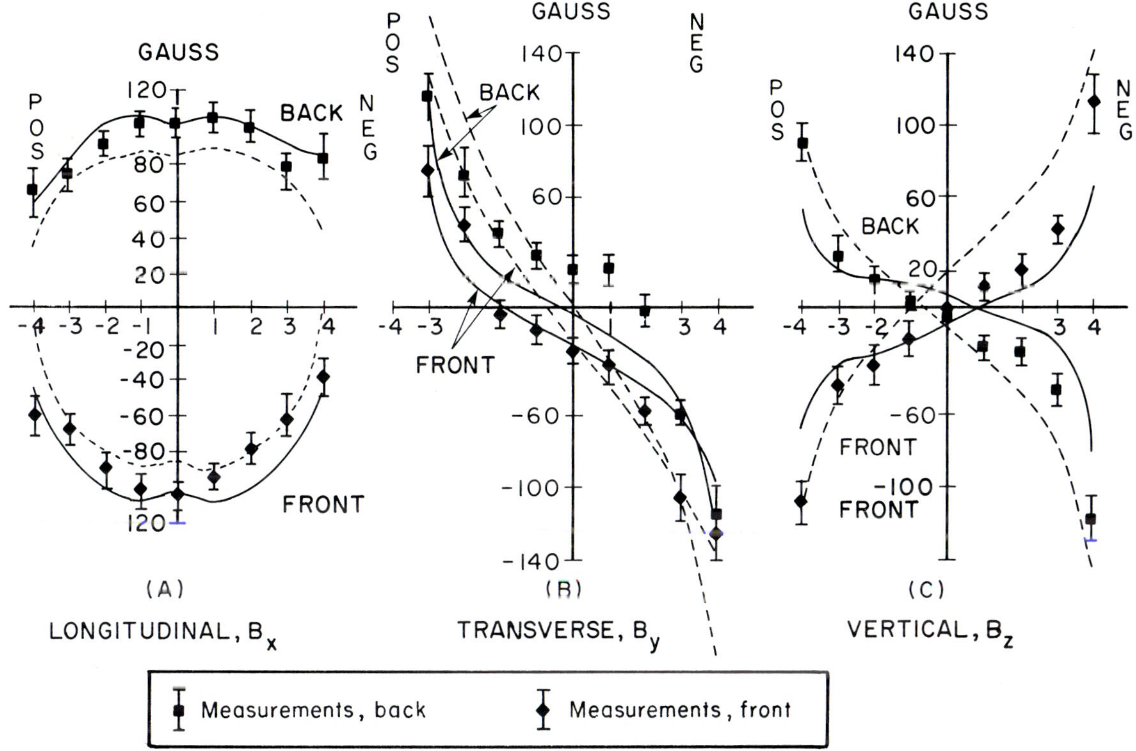

Figure 5 - Comparison of measured magnetic fields in Alcan 175E with calculated values: (A) including shielding effect of steel (solid line) and (B) without steel shielding (dotted line)

Table II. Project Objectives

	Original Development Targets	Actual Performance 8 months Jan.-Aug. 1982
Amperage (kA)	175.0	175.3
Volts (Gross)	4.30	4.33
Bath Ratio	1.20	1.22
C.E. (%)	90.5	90.9
kWh/kg Al	14.1	14.2

Economic Assessment

The major cell design factors contributing to the equivalent annual cost of a potroom are the productivity per unit area of bottom plate (tonne/m^2) and the specific power consumption (kWh/kgAl).

On both counts, Alcan 175E is a major step forward over existing Alcan end-to-end technology. Compared to Kurri Kurri Line 2, power consumption has been reduced by 8.5% and productivity has improved by 19%. This brings Alcan 175E in a similar performance range as competitive side-by-side technologies.

The most recent assessment of Alcan 175E equivalent annual cost against competitive side-by-side technologies suggest that Alcan should endeavour to retrofit the practices of Alcan 175E in existing end-to-end potrooms and plants where possible and economically justifiable.

Technology Implementation

Following completion of the development program, it was decided to employ the results in the expansion program for Kurri Kurri Works. Construction of a third line is now underway making use of the developed technology.

Also within Alcan there are several examples of similar sized cells that could readily be modified to take advantage of the knowledge gained from the Alcan 175E project. One prime example is Line 1 at Kurri Kurri, Australia. This line was completed in 1969 and consists of 120 prebaked anode cells arranged end-to-end in two rows in one building. The cells are hooded and alumina is fed at intervals by breaking of the side crust with a wheel breaker.

Studies have been completed to evaluate the impact of such a retrofit on the operating economics of the existing Kurri Kurri Line 1 cells. A 14% reduction in power consumption and a 15% increase in productivity should be attainable.

Conclusion

A cost effective end-to-end reduction cell technology has been designed and demonstrated and is now being implemented in Alcan Kurri Kurri Works, NSW.

DEVELOPMENT OF THE KAISER ALUMINUM 195 kA CELL

R. D. Zabreznik and E. D. Tarapore

Kaiser Aluminum & Chemical Corporation
Tacoma, Washington 98421
USA

In late 1981, KACC began operation of two experimental reduction cell test sites, one at Tacoma, Washington and another at Chalmette, Louisiana. Altogether, ten pots were installed, six at Tacoma and four at Chalmette. Basic design of all the cells was essentially identical. However, minor variations in bottom insulation and sidewall configuration were installed to evaluate different materials of construction and their performance.

The final cell configuration (Figure I) described in this paper was developed at Tacoma. This cell represents the present state of development.

Figure I - The KACC 195 kA Cell

The Tacoma facility (Figure II) was specifically designed as an experimental center with capability for potline expansion. Along with the six cells, the facility includes a new potroom building, ECL multipurpose crane, sow casting area, injection dry scrubber unit, independent power supply and rodding room. Chalmette's four cells were placed into an existing Soderberg potline using booster rectifiers, a multipurpose crane and other ancillary equipment. Both facilities were manned by operators whose previous experience was limited to Soderberg operation.

Different from Tacoma, Chalmette's four cells operated on lithium modified bath, as is Chalmette's standard practice, and had no basement ventilation. Amperage was varied at both locations to determine the optimal set-point for energy efficient operation.

The variety of cathode linings and two modes of operation were extremely valuable to KACC's large cell program. Using past experience, along with advanced thermal, magnetics and current modelling techniques, KACC was able to develop a high performance cell in a compressed time frame.

Figure II - Aerial View of New Cell Facility at Tacoma

Cell Design

The cell lining and busbar locations were examined using computer models. Selection of prototype design was heavily influenced by the modelling work.

The final cathode design was developed to reduce heat loss while retaining adequate sidewall ledge formation. The computer analysis (1,2) showed good ledge protection for the sidewall and was later confirmed by measurement on the operating cell. Similarly, bottom insulation was chosen to place the critical isotherms in specific areas of the lining. By locating isotherms in areas resistant to disruption by solidifying phases (3), a stable lining configuration is possible.

All ten original test cells had an improved side-riser design. These risers reduce the vertical magnetic field typically found at the ends of end-riser type cells. A magnetohydrodynamic model (4) was used to calculate the cell magnetic field, metal pad topography and metal flow. Actual maximum field values were found to be quite acceptable. In comparison to

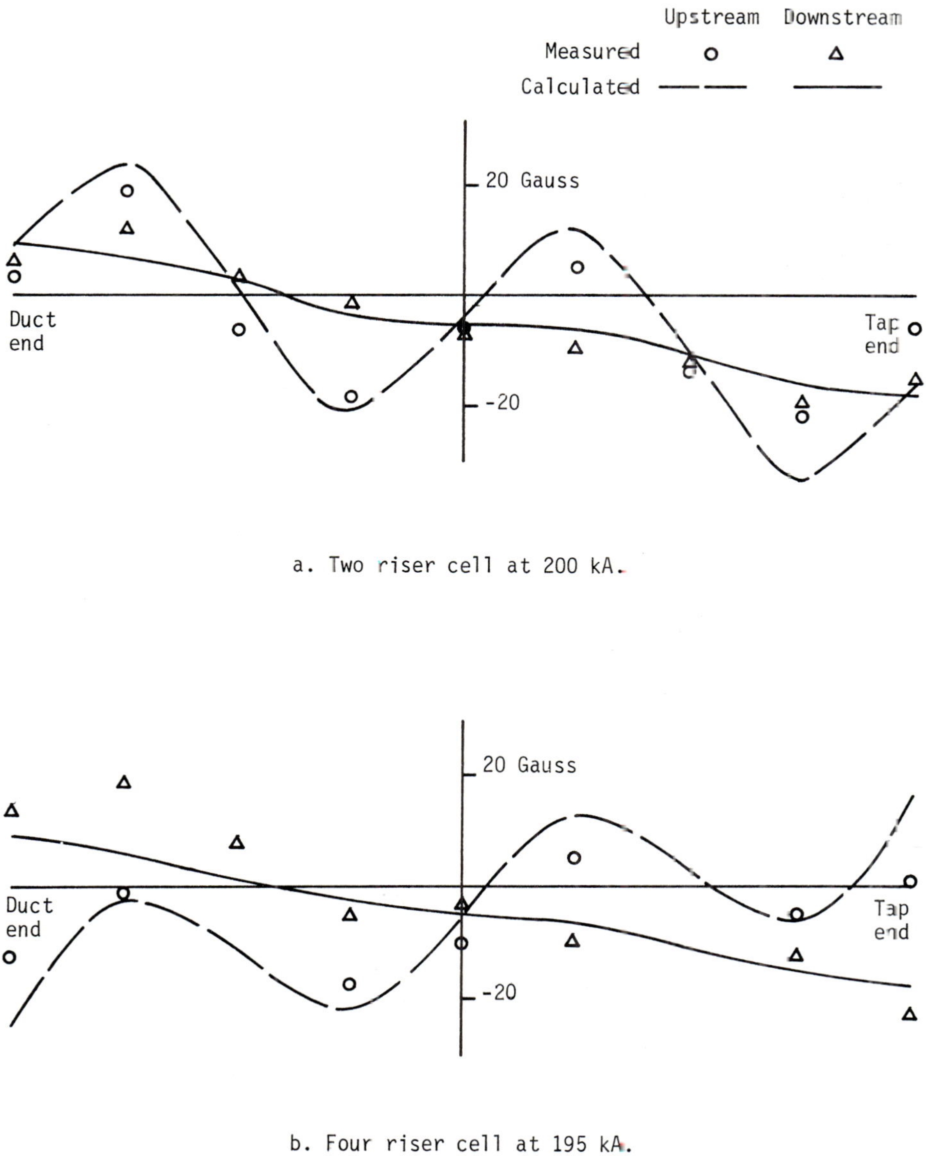

Figure III - Measured and calculated values of vertical magnetic field of cell.

previous end-riser configurations, a relatively flat metal pad was obtained. Also, a special busbar layout was developed for end cells to ensure that their field did not significantly differ from the other cells in the line.

In early operation, there were occasions of metal pad instability with a 30-45 second period. Troubleshooting of the cell indicated non-uniform cathode current distribution with resultant transverse metal pad current flow. This was corrected by a minor cathode busbar modification. An analysis of the hydromagnetic stability of the cell was also conducted. It showed that the negative vertical magnetic field at the tap end of the cell (Figure III) would cause a small metal pad instability to grow until it developed into a periodic wave large enough to noticeably affect cell voltage. End risers were therefore added to the subject cell to reverse the vertical field at the tap end. The modified field (Figure III) effectively damped local instability and did not allow it to propogate.

The anode assembly consists of an aluminum rod and steel yoke with four stubs. The uncoated anode assembly is twenty inches (51 cm) tall. Stub protection collars made from a special aggregate are optionally used to increase anode life.

The cell is also equipped with permanent hardware fixed to the anode buswork allowing continuous determination of anode current distribution through a supervisory computer.

The supervisory computer monitors the cell via a microprocessor control system called CELTROL® (Figure IV). One CELTROL® (5) is used per cell. The unit is programmed to independently control cell voltage, and the feeding of alumina, fluoride and crushed butt bath. It also responds to anode effects, noisy cell conditions, plus controls "anode set" and "tapping" routines.

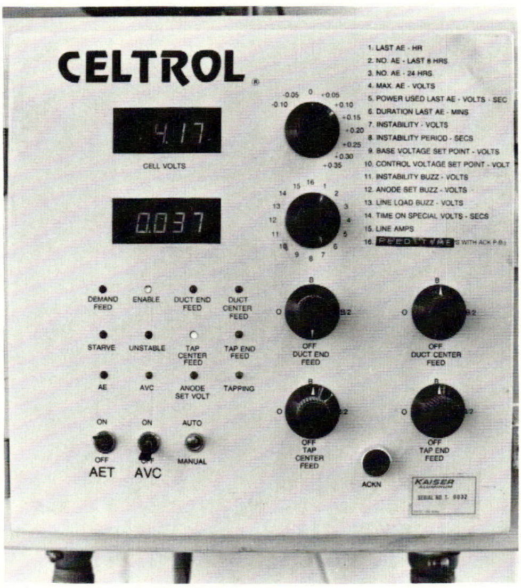

Figure IV - Reduction Cell Control Panel, CELTROL®

Each CELTROL® panel is capable of displaying more than thirty individual pieces of significant cell information for review at the cell by the operator. Included are cell voltage, potline amperage, anode effect frequency, magnitude of the last anode effect, time until next scheduled anode effect, voltage stability index, etc. More analytical information is collected by the supervisory computer and is available to the operator through a variety of routine and special reports.

The cell is supplied with alumina at four points along the center trench. The point feeders are independently controlled, allowing one or more of the feeders to be disabled. The control system automatically compensates by proportionally increasing the feed rate through the remaining feeders. Ore for the superstructure's alumina hoppers is supplied with an automatic air slide system or with the overhead crane.

The hoppers provided within the superstructure for aluminum fluoride and crushed butt bath feed the cell at two of the four alumina feed points. The rate at which these materials are added to the cell is set by the operator via the supervisory computer. Bath ratio and liquid levels can be controlled in this manner.

Process Parameters

During the period reported in this paper, the cell was operated at 195 kA. However, the cell is capable of operating in the 190 to 205 kA range.

Anode change cycle is 22 days. New anodes are referenced based on the actual position of the old butt at the time of change. Freshly set carbons are covered with finely crushed butt bath. Anode cover, being a significant factor in maintaining heat balance, is maintained at a controlled depth of approximately four inches (10 cm).

Chalmette's lithium-modified bath operation, compared to the non-lithium cells at Tacoma, indicated no significant current efficiency advantages. Therefore, standard spar bath chemistry is utilized. Actual bath ratio averaged 1.17 (11% excess aluminum fluoride).

Cell voltage averages 4.16. This voltage includes anode effects and the bus drops associated with one cell.

Tapping of the pot is performed on a 24-hour cycle. The average metal depth is 7.5 inches (19 cm).

Anode setting is a smooth operation requiring a nominal 10 to 15 minutes per change.

Results

The development effort at KACC has resulted in a stable, high productivity cell operating at 195 kA. The four-riser KACC design yields a low vertical magnetic field and an essentially flat and stable metal pad.

The cell established consistent results of:

Current Efficiency	94.7%
DCkWh/lb	5.94 (13.1 DCkWh/kg)
Net Carbon	4.15

Typical results of other cells in the KACC test program were 93-95% current efficiency and 6.0-6.4 DCkWh/lb (13.2-14.1 DCkWh/kg).

Experimental accuracy was assured using routine three-week cyclical copper trace inventory measurements of the metal pad, double weighing of production, and a calibrated current metering system.

Extensive autopsy work on the basic cathode after two years' operation has shown no significant sidewall erosion or cathode heave. Critical isotherms are carefully located by design to create a thermally stable vessel. There is little deterioration of insulating value with age and we project potlife will be 2,000 to 3,000 days.

Fume collection consistently averages over 98%. During anode setting and tapping, the operator actuates a "dual flow" exhaust damper mechanism that increases draft off the cell by approximately 75%

The Next Step

Although these results have fulfilled our initial expectations, improvement is possible with presently known technology. Energy savings are available by reducing voltage drops in both the anode and cathode. Buswork modifications will also contribute. Taken together, the studies completed indicate that a 190 to 210 kA cell, capable of operating at 95% current efficiency and 5.8 DCkWh/lb (12.8 DCkWh/kg) can be constructed.

Acknowledgement

The authors wish to acknowledge the contributions of the many people in KACC who have aided in the development of this cell.

References

1. Y. Arita, N. Urata, H. Ikeuchi, "Estimation of Frozen Bath Shape in an Aluminum Reduction Cell by Computer Simulation," Light Metals 1978, Vol. 1, pp. 59-72.

2. H. Ikeuchi and Y. Arita, "Advanced Heat Design of Aluminum Reduction Cell," Electrochimica Acta, Vol. 25, (1980), pp. 125-129.

3. Private communication, E. O. Strahl, Kaiser Aluminum & Chemical Corporation.

4. E. D. Tarapore, "Magnetic Fields in Aluminum Reduction Cells and their Influence on Metal Pad Circulation," Light Metals 1979, Vol. 1, pp. 541-550.

5. R. A. Mohr, "Aluminum Reduction Plant Distributive Control System," Light Metals 1982, pp. 595-608.

REDUCTION OF POWER CONSUMPTION IN 1940/1950

DESIGNED POTS

O.S. CHAUDHRY

HINDUSTAN ALUMINIUM CORPORATION LIMITED
REDUCTION PLANT, RENUKOOT
MIRZAPUR (UP)
INDIA

In 1940/1950 most of the prebaked pots were designed with the criteria of maximum Kgs/pot day without giving serious consideration to power consumption. The pots laid or rather packed side by side have little scope for changing the cell diamensions and introducing energy efficient technology, without revamping entire pot line and involving major modifications.

The paper describes how HINDALCO has been able to save approximately 1000 KWH/Ton by way of little modifications in aluminium bus rings, anode stubs, anode size etc. and conserving heat by increasing cavity, side & bottom insulation with minor modifications of cradle and pot cell in normal course of operation.

In last 30 years aluminium technology has seen many rapid strides in its technological development. There has been remarkable improvement in energy consumption, size of pots, working condition and pollution control. 1980 generation electrolytic cells are widely operated at less than 13 KWH/Kg of aluminium, with more or less clean atmosphere, having latest electronical gazettes like microprocessors, AET & AVC systems with hardly few persons to man the operating requirements of 50,000/60,000 TPY per line.

Table I. World wide trend of power consumption for aluminium production

Year	Total Production 10^6 T/Year	Cost of Production US $/Kg.	Typical Cell Size	Power Consumption
1900	0.006	0.72	5	28
1930	0.27	0.53	20	25
1940	0.78	0.41	30	24
1950	1.5	0.39	40	20
1960	4.53	0.57	80	17.6
1965	6.58	0.54	100	16.5
1970	10.25	0.63	150	15.5
1975	12.7	0.83	150	14.5
1979	15.23	1.36	175	14.0
1980	17.68	1.38	175	13.5
1981	17.28	1.36	220	13.0
1982	15.08	1.34	220	12.8

These technological developments are being marketed practically by all major aluminium producers in the world, but mostly for new plants. Unfortunately it has not been possible to take full advantage of these technological developments in old plants and specially in those established on 1940/1950 design, without major revamping which involves prohibitive capital expenditure.

This paper deals with various constraints in implementing latest technological conceptions in old plants and how Hindalco has been able to tackle some of these constraints for reducing power consumption in their pot lines.

Technological Developments and Constraints

Before we venture, as to what the constraints are, let us examine in brief the technological developments which led to considerable low power consumption.

The energy required per Kg of aluminium based on Faraday Law of electrolysis can be given :

$$\text{KWH/Kg} = \frac{2.98 \times V}{\text{Eff.}}$$

where V = Average voltage drop per cell
Eff.= Faraday/Current efficiency

Saving of power in aluminium production, seems to be quite simple provided either the voltage can be reduced and/or current efficiency can be improved.

Total voltage (V_{tot}) on pot can be splitted as :

$$V_{tot} = V_{Rev} + V_{Pol} + V_{ERI}$$

V_{Rev} = Reversible decomposition voltage

V_{Pol} = Polarization voltage at anode & cathode

V_{ERI} = Ohmic drops at various places.

These voltages in modern designed & Hindalco cells may be attributed as under:

	Modern Design	Hindalco Existing Pot
V_{Rev}	1.2	1.2
V_{Pol}	0.55	0.56
V_{Cond}	0.15	0.17
V_{Anode}	0.25	0.30
V_{Bath}	1.4	2.0
V_{Lining}	0.4	0.5
$V_{Anode\ Effect}$	0.05	0.05
	4.00	4.78

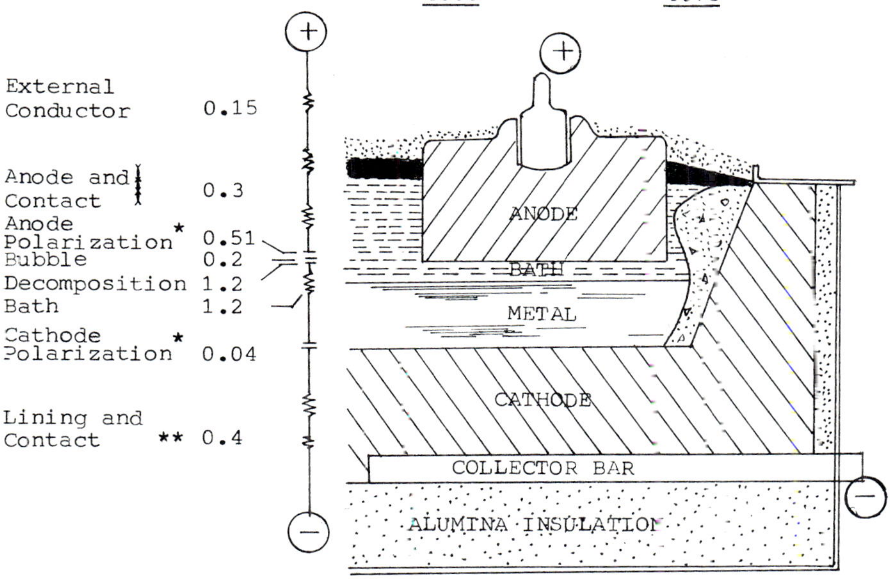

* ANODE AND CATHODE POLARIZATION ARE CURRENT DENSITY DEPENDENT

** CHANGES WITH CELL AGE

Fig. 1 SCHEMATIC REPRESENTATION OF VOLTAGE DROP THROUGH CELL

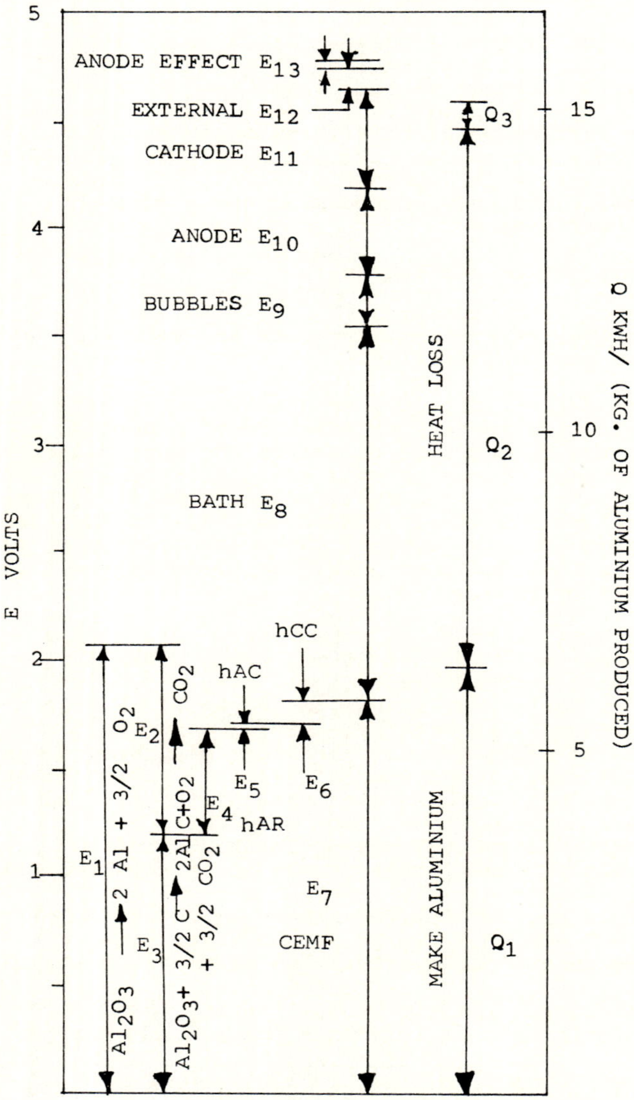

VOLTAGE DROP	IN HINDALCO CELLS	IN NEW DESIGN
E_1	2.1	2.1
E_2	0.9	0.9
E_3	1.2	1.2
E_4	0.48	0.48
E_5	0.03	0.03
E_6	0.05	0.04
E_7	1.76	1.75
E_8	1.75	1.20
E_9	0.25	0.20
E_{10}	0.3	0.25
E_{11}	0.57	0.4
E_{12}	0.15	0.15
E_{13}	0.03	0.05
E_T	4.81	4.00

E_T: AVERAGE VOLTAGE DROP IN THE CELL

hAR : ANODE REACTION POLARIZATION
hAC : ANODE CONCENTRATION POLARIZATION
hCC : CATHODE CONCENTRATION POLARIZATION

Fig. 2 Voltage distribution with respect to energy utilization

While reducing Al_2O_3 with carbon electrode, there is practically little scope to reduce the decomposition voltage once the operating parameters are fixed. The polarization voltage which is mostly a function of current density, Al_2O_3 concentration, operating temperature and surface condition of electrode, can substantially be reduced by using metallic anodes but this will increase the decomposition voltage as gain of exothermic reaction of carbon to CO_2 shall not be available and decomposition voltage shall be 2.1 volts instead of 1.2 volt. Apart from this, metallic anodes have not been found stable at pot operating temperature in presence of AlF_3. The total emphasis consequently has been on reducing ohmic drop and improving current efficiency.

The ohmic drop can be reduced by :

- lowering current density,
- reducing the current carrying path and specially ACD,
- improving the conductivity.

To reduce the current density, which has mostly been done in latest designed pots, apart from extra heavy anode and cathode bus bars, the bigger size of anodes and much bigger pot cells are required. Hindalco pots are operating at approximately 1 amps/sq.cm. against .6 to .55 amps per sq.cm. that of latest design. It simply indicates that for the same current/metal production, the area of the current carrying path and pot shell, in latest designed pots, has to be atleast 1.1/2 times that of Hindalco pots.

The anode cathode distance is a necessary evil and is generally maintained in between 48-50 mm. Even a small reduction of 2-3 mm in this distance shall save approximately .1 volt, but due to horizontal & vertical movement of molten metal on account of heavy magnetic flux any reduction in this distance is liable to short circuit the anode & cathode leading to noisy pots. The minor reduction of 3-4 mm in ACD has been possible in pots designed with more or less flat metal pad with complicated bus bar configuration. Not only this, new thermodynamic conditions created due to lower power input demands heavy insulation and further increase in pot shell size.

The constraints are not limited simply to reduction of current density, ACD and/or increase in conductor size, it involves number of other consequential requirements as well.

- Mechanisation of rodding facilities and handling of massive carbon blocks in baking furnace and potroom.

- Lower heat generation/input in the bath due to reduction in I^2R losses.

- Increased dissipation of heat due to bigger size of anodes, stubs & pot shell.

- Higher current input & higher insulation to compensate lower heat generation and consequential further increase in shell size.

- Heavy magnetic fluxes due to higher amperage, bigger size pots and consequential complicated bus bar arrangement for stable metal pad.

- Changed thermodynamic conditions leading to change in insulation and establishing of new operating parameters.

- Low production per unit area of pot/cell house involving heavy capital expenditure.

Any change either in size of the anode, cathode material and its construction, pot shell size, pot insulation, bus bar configuration, bath chemistry and/or operating parameters including change in ACD, shall lead to thermodynamic and electromagnetic disbalance, off setting the production process, current efficiency and even endangering the life of pots.

As far as current efficiency is concerned, there has been tremendous leap in improving the same. Now pots are operating at as high as 94% current efficiency against 80% that of 1950 pots. The improvement in efficiency has not come free or easy. Bath chemistry and strict process controls are involved. Unfortunately electrolyte that favour Faraday efficiency restrict physical properties of bath that inturn requires much tighter control, demanding complete automization of pot operation with microprocessor/computer based on softwares specially developed for the particular design. Apart from massive expenditure, softwares are not available as off the shelf items. Development of the softwares with respect to particular design again is a time consuming process. Hindalco is installing computer on 10 pots to develop its own softwares.

Accordingly any modification and change in operating parameters requires more or less complete new design for balancing new thermodynamic & electromagnetic conditions and for establishing the operating parameters suiting to these conditions.

So far we have been discussing the various technological developments & contraints. Let us examine, how Hindalco has been able to take advantage of some of these developments.

Hindalco's prebaked pots are based on M/s Kaiser's, USA Mead Plant (Spokane) which was established at fag end of second World War. These pots were designed for maximum Kg/pot day and not for optimum power and raw material consumption. Prebaked pots in Hindalco are laid side by side with practically no space between two adjacent pots and are mounted on cradles on the main working floor as shown in Figure 3.

Since last couple of years, Hindalco has frantically been trying to cut down power consumption and has been evaluating the possibility of increasing anode size, reducing ACD, reducing voltage drop in anode assembly, anode & cathode bus ring and in cathode lining by changing from existing monolithic lining to prebake block lining but the major hurdle, apart from

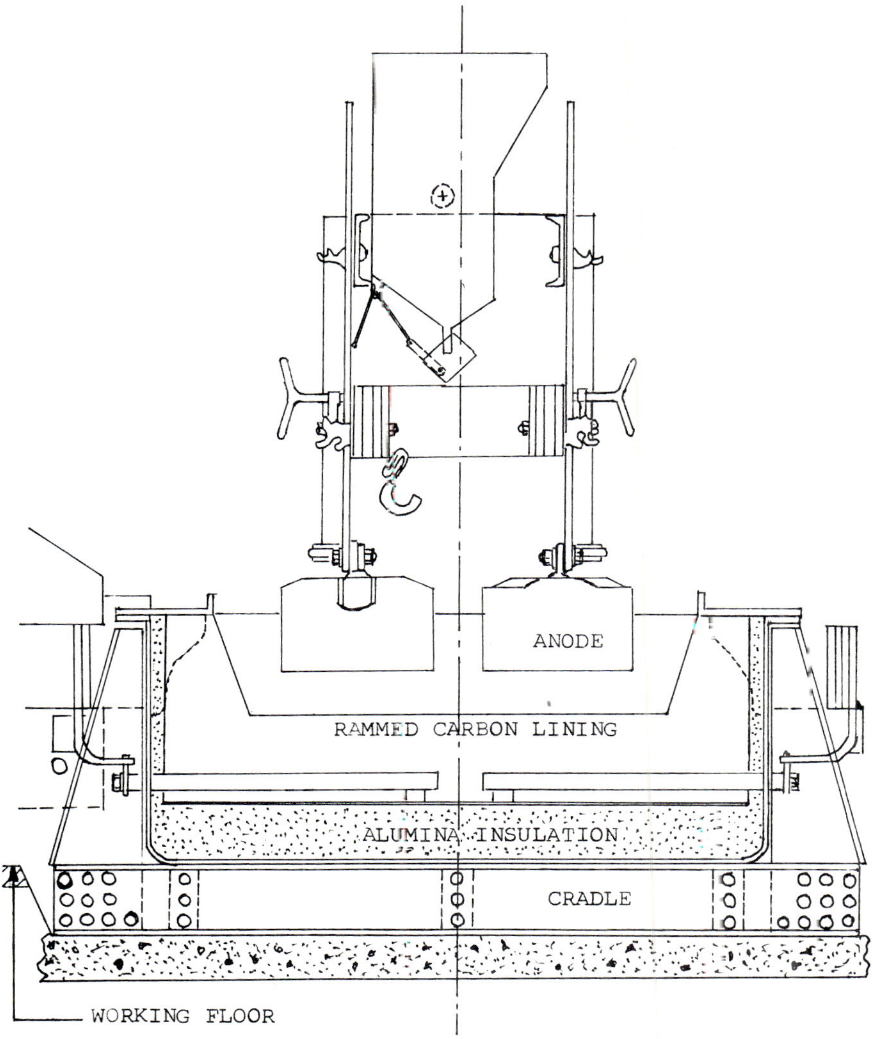

Fig. 3 Existing Hindalco Pot

developing of new operating parameters, has been in changing the pot size to accommodate bigger anodes, block lining and extra insulation for balancing new thermodynamic conditions to compensate lower power input.

Modifications in 1940/50 pots

It was impossible to increase the pot size without reducing the number of pots and involving major modifications of cathode bus ring, anode bus riser, super structure and shutting down a certain portion of the plant completely. The only place available has been to go up or down to increase the pot height to accommodate block lining and higher bottom insulation.

After thoughtful evaluation, the following modifications were carried out in pot shell and cradles (Fig. 4).

- The height of pot shell was increased by 4" to accommodate higher insulation in bottom and prebaked blocks in place of monolithic lining.
- The height of rim plate was increased from 1.1/2" to 4" to retain higher insulation on top of anodes.
- The spherical shape of anode was changed to flat for retaining higher insulation and simultaneously to cut down air burning of carbon.
- Cell cradles which were getting excessively bent were modified for higher bending strength and at the same time bottom beam height was also reduced by 3" to accommodate 4" higher pot shell without any change in anode bus riser and operating practices.
- Pot hoppers were modified to accommodate two/three point feeders per pot.

To cut down the power consumption without disturbing the existing pot line operation, following projects in isolation were taken in number of pots simultaneously to evaluate the gains to assess thermodynamic requirements and to establish new operating practices.

1. Anode-cathode bus area was increased by 16% by adding one extra leaf in situ, special welding technique was developed to carry out aluminium welding on live bus bar. These modifications were carried out on 10 pots.

   The average gain/pot was found to be of 0.022 volts.

2. Anode stub diameter was changed from 4.1/2" to 5.1/2" to cut down anode assembly drops. Special 1000 anodes having bigger dia stub hole were made, the experiment was conducted for three months in number of pots.

   The average gain in voltage per pot was 0.034 volts.

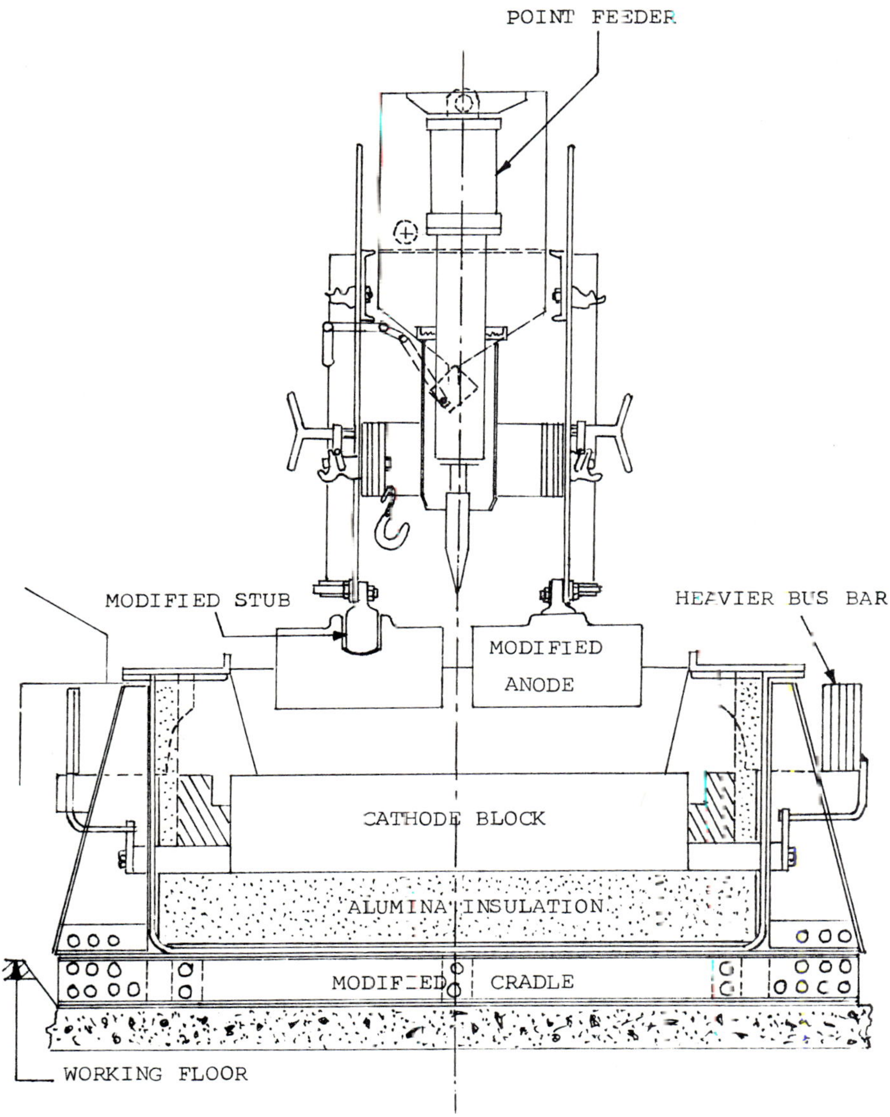

Fig. 4 Hindalco Modified Pot

3. As indicated earlier, it has not been possible to increase the anode size due to limitation of space and limitation in reducing side lining & insulation further. The space available between two adjacent anodes was further squeezed increasing anode size 1/4" all round for existing 22" x 16" size. 3000 anodes were made and used in 10 pots for three months. The increased heat dissipation from top on account of bigger anode was compensated by increasing alumina cover on anodes.

   The net gain in voltage per pot was 0.1 volt.

4. Hindalco has been operating their pots with 4.1/2" metal pad. On investigation of magnetic flux, bath and metal movement (Fig. 5), it was revealed that 4.1/2" metal pad is too low and there are chances of cathode getting bare due to completely lifting of metal near anode bus risers and specially in case of slightly over tapped pots. On carrying out computer simulation it was further revealed that 6" metal pad shall give almost stable metal conditions and with this metal pad the pots can be operated at 45/44 mm ACD without affecting the operation.

(a) 8 Cm. Metal Pad

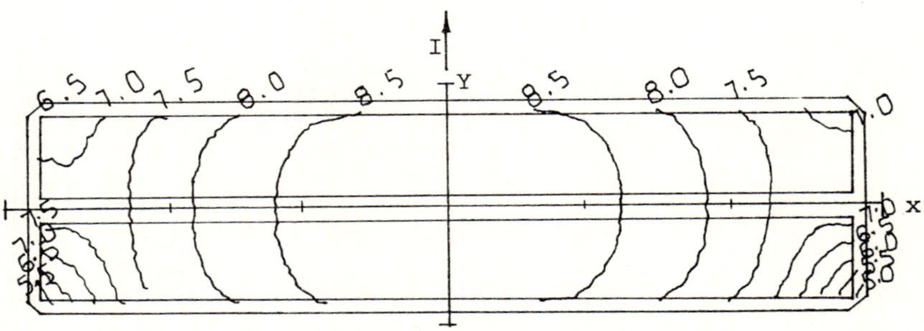

(b) 16 Cm. Metal Pad

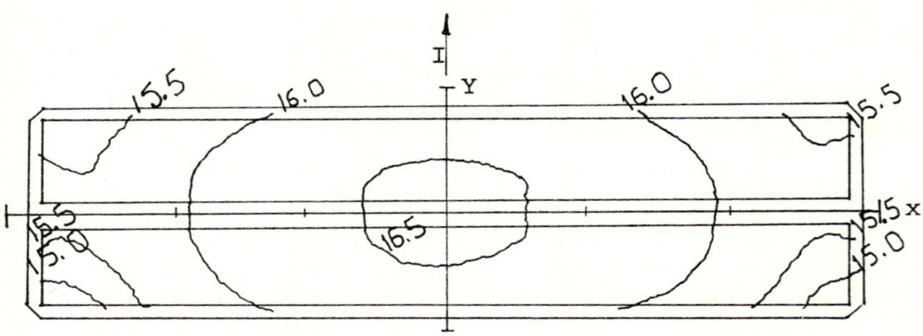

Fig. 5 Metal Contour lines in Cm. with current outwards.

5. After raising the metal pad gradually to 6" and increasing side & bottom insulation, number of pots were taken on 45/44 mm ACD by cutting 0.1 volt on these pots. The behaviour of these pots was carefully monitored for net gain of 0.1 volt.

6. Hindalco is still having monolithic lining and that too based on metallurgical coke. The prebaked block lining was not possible without modifying the pot shell and reconstructing the entire insulation profile. Hindalco average lining drop has been 0.55 volts against 0.4 volts that of prebaked block lining. Three years back, one pot was lined with prebaked blocks to assess the thermodynamic conditions. Based on these datas, 5 pots have been ear marked for prebaked block lining. The necessary thermodynamic conditions and insulation requirement has been simulated on computer. We expect 0.15 to 0.2 volts saving on account of prebaked block lining.

   Expected gains 0.15 volts.

7. Anode shape was changed. Top spherical surface was flattened to hold more alumina. This experiment was conducted in one complete line. There was considerable gain in carbon consumption and at the same time line temperature was observed to be 3°C/4°C higher than that of other lines due to better cover on anodes.

8. Hindalco developed point feeders were installed on three pots and are operating successfully since more than one year now. Apart from achieving more or less continuous feeding of alumina, we have been able to save heat dissipation from centre as the molten bath is only exposed once in 24 hours while killing the anode effect against 18 times that of manual feeding.

<u>Heat balance with low power input</u>

Based on various experimentations, it was assessed that .4/.38 volts can be reduced per pot. The major problem was how to balance the thermodynamic conditions if all these modifications are carried out simultaneously.

The heating KW required with existing profile of insulation was calculated based on:

$$IV = (X \times Eff.+Y)\ I + Q$$

where X = Decomposition + polarization voltage,

Y = Heat required in KW for heating reactants,

Q = Heat loss through convection & radiation in KW

and was found to be approximately 158 KW with above formula as well as on taking actual measurements with flux meter.

Table II. Energy distribution of existing and modified pot in Hindalco

		Existing Pot	Modified Pot
Cell Current	KA	57	57
Cell Voltage	V	4.78	4.43
Cell Temperature	°C	970	970
Energy Input	KW	272.46	252.51
Process Energy	KW	114.39	114.39
Heat Losses	KW	158.06	138.12
Details of Heat Loss (KW)			
1. Through Bottom		18.5	11.1
2. Through Sides		17.25	16.55
3. Through Ends		4.5	4.4
4. Through Deck Plates		29.0	27.0
5. Through Collector Bar		8.0	8.0
6. Through Top Crust and Ore cover		25.0	15.0
7. Through Top of Anodes and Stubs		7.5	7.5
8. Through Exhaust Gases		30.0	30.0
9. External Losses		11.5	11.5
10. Unaccounted Losses		6.81	7.07
		158.06	138.12

New Insulation Profile

To keep the process energy more or less constant, insulation profile was reconstituted to compensate for 20/21 KW less input due to over all reduction in pot voltage by .4/.38 volts. The following changes were made in pot insulation profile.

1. Side & bottom insulation was increased.
2. In the bottom, silicate board was also provided upto the collector bar level.
3. Silicate board was given underneath deck plate to cut down the heat losses through deck plate.
4. Alumina insulation was increased at the top of anodes by raising the rim plate.
5. Side insulation was tapered and redesigned to have bath ledge in molten metal as well.

Hindalco has modified 5 pots and these are under operation since last 6 months. On these pots, we have been able to save 1000 - 1100 units per ton of metal production. Operation of these pots is being watched critically. Mass implementation shall be taken up as soon as critical evaluation is over.

References

1. Brat land, D, Grojotheim K. and Krohm C in Light Metals 1976 Proceeding from 105th AIME Meeting, Las Vegas, Nevada.

2. K. Grojatheim, C. Krohn, M. Malinovsky, K. Matiarovsky, J. Thoustad, Aluminium Electrolysis, Aluminium Verlag Gmbh Dusseldorf.

3. K. Grojatheim & B.J. Welch, Aluminium Smelter Technology Aluminium Verlag Gmbh, Dusseldorf.

4. Haupin W.E. "Cathode Voltage loss in Aluminium Smelting Cells" Light Metal 1, 339 (1975)

5. Stig F. Johansson "Calculation of the heat loss through the bottom of Aluminium Electrolysis Furnaces with consideration of the heat produced in the cathode carbon" Light Metal 1982.

6. O. Fujsima, M. Takasawa, K. Wakaizumi, Y. Arita and H. Ikeuchi "Advance energy saving of Aluminium Reduction Cell" Light Metal 1982.

ENERGY SAVINGS THROUGH IMPROVED CATHODE CONSTRUCTION

Dr. P. Aeschbach
& Mr. H. Friedli
Swiss Aluminium Ltd.
Valais Plants
Steg Smelter
CH-3945 Steg-Gampel

Dr. K. Etzel
Dr. F. Brandmair
SIGRI Elektrographit
D-8901 Meitingen
W. Germany

It is generally recognized that the installation technique and the quality of the cathode blocks decisively influence the service life of the reduction cells. Glueing and sealing techniques - they reduce infiltration into the shell lining - and the use of graphitic block qualities with simultaneous reinforcement of the thermal insulation are, however, also keys to saving energy during the entire service life of a cell. If newly developed graphite cathodes are used, energy savings in the order of magnitude of 0.5 kWh/kg Al can be achieved in the case of an expected potlife of more than 3000 days.

1. Introduction

Efforts have been increased in particular since the last oil crisis, to reduce the electrical power required in aluminium reduction further. The high power prices make measures even with a small savings potential interesting.

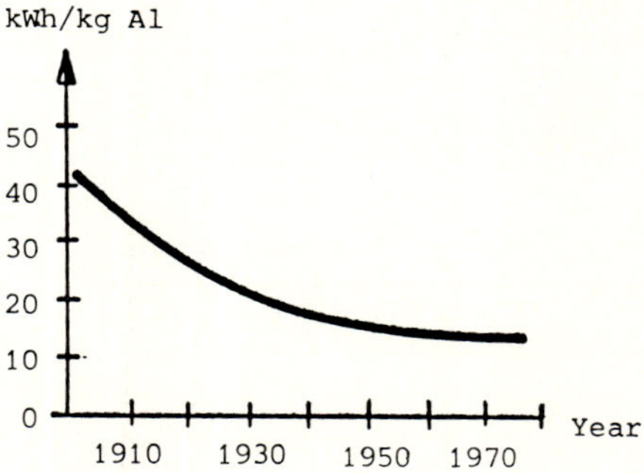

Fig. 1: Development of the spec. el. power consumption (pots with prebaked anodes) acc. to 1)

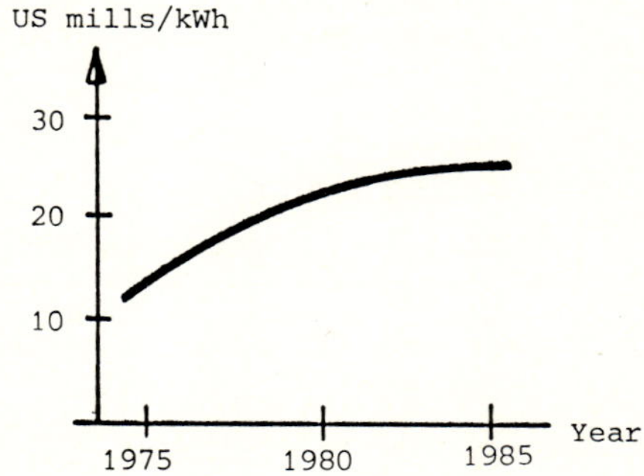

Fig. 2: Development of the power price in the USA
Source: 2)

At the meetings in Chicago[3)] in 1981 and in Dallas [4)] in 1982 as well as in Atlanta[5)] in 1983, we presented our glueing technique in the construction of cathodes and the use of graphitic cathode blocks. The power saving in the test cells mentioned at that time has already been confirmed in more than 150 reduction pots with glued, graphitic cathode blocks and sealed, reinfored insulation. The extremely long service life of these cathodes has also been confirmed.

2. Basics on Power Consumption

If one calculates the power consumption of one of today's reduction pots with prebaked anodes from the electrochemical consumption incl. heating of the raw materials and the heat losses, a total el. power consumption of 13 - 15 kWh/kg Al results:

Table 1

electrochemical consumption (incl. heating of raw materials)		6.4 - 6.5 kWh/kg Al
heat losses		7.0 - 8.5 kWh/kg Al
busbars	1.0 - 1.7 kWh/kg Al	
anodic part	2.8 - 3.2 kWh/kg Al	
cath. part	3.2 - 3.6 kWh/kg Al	
	7.0 - 8.5 kWh/kg Al	
Total el. power consumption		13.4 - 15.0 kWh/kg Al

For the el. power consumption, following formula applies:

$$E \sim \frac{U}{\eta}$$

It follows that a reduction of the power consumption can only be attained by an

- increase in the current efficiency

or a

- reduction of the voltage drop.

2.1 Increase in current efficiency

Factors which influence the current efficiency:

- flux composition
- flux temperature
- magnetic conditions
- mode of operation of the pots.

2.2 Reduction of the Voltage Drops

Factors which influence the voltage drops:

- flux composition
- flux temperature
- cross-sections of busbars
- contact resistances
- el. conductivity of anode block
- el. conductivity of cathode block
- pot mode of operation.

In accordance with the subject of this paper, we shall now concern ourselves with the reduction of the cathode voltage drop.

3. Reduction of the Cathodic Voltage Drop and Realization of Power Saving

It was shown at the AIME Meetings already mentioned that, due to the use of glued, graphitic cathode blocks, the cathode resistance, averaged over the entire cathode life, can be reduced by approx. 1.5 $\mu\Omega$.

It was not possible to make full use of this advantage at first. Conscious of the higher thermal conductivty of the graphitic materials, the first experimental pots were more strongly insulated. Nevertheless bottom crusts formed at times with full utilization of the lower cathode resistance. These crusts were caused, as shown by J. Thonstad, S. Rønning and P. Entner[6]), by an undercooling of the cathode bottom. Also the shell temperature (bottom plate in the middle of the shell) has, compared with the traditional mode of construction, tended to rise.

Model calculations based on the data measured on test pots, showed that the thermal insulation of the cathode shell had to be reinforced again to bring the temperature on the cathode block surface as close as possible to the temperature of the metal. Here it has to be mentioned again that the sealing technique is of vital importance. In order to hinder the thermal insulation being destroyed or to avoid a (preventive) too expensive insulation, infiltrations into the shell lining have to be hindered at all costs.

In order to optimize the contact resistance cathode block - collector bar and the resistance in the collector bar, tests were conducted with all kinds of bar sizes and forms in various attaching techniques:

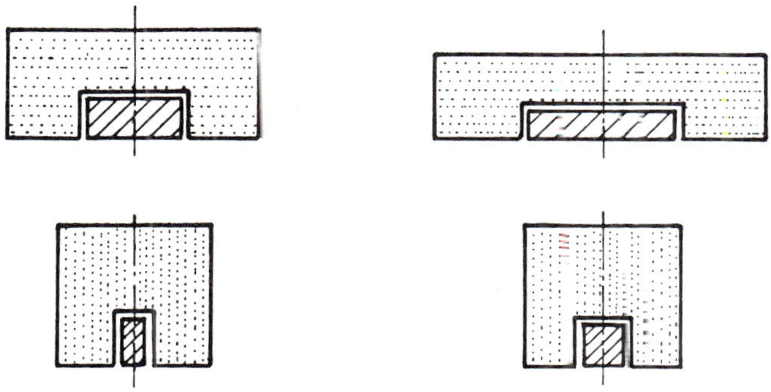

Fig. 3: Cathode Block / Bar Shapes

Here it is clearly illustrated that a solution with a large number of narrow, upright bars represents an optimum. Similar results were reported by Samoilenko and Tsyplakov[7]) in 1965.

4. Results

4.1 Cathode Resistance

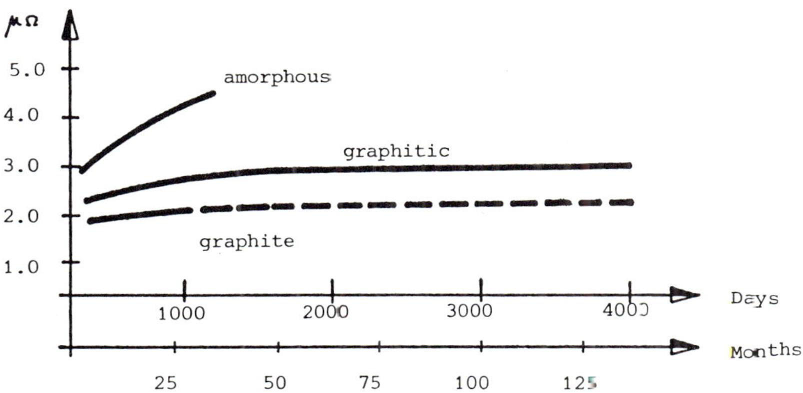

Fig. 4: Cathode resistance in function of the time for different cathode constructions

As can be seen from Fig. 4, a drastic reduction of the cathode resistance is possible by using glued, graphitic cathode blocks. The use of electrographite again brings a reduction of approx. 0.6 µΩ.

4.2 Realization of the Gain in Power

The step by step realization at the Alusuisse Steg plant can be seen from the table below:

Table 2

Year	Av. Cathode Resistance of all Pots (µΩ)	Graphitic[a]), Glued Cath. in Operation Total (%)	Optimiz. cathode bars (%)	Av. of Exp. Potlife (Days)	Av. Pot Resistance[b]) (µΩ)	E kWh/kg Al
1976	4.5	4	-	2100	25.7	./.
1977	4.3	11	1	2400	25.6	
1978	4.2	32	21	2800	25.2	-0.2
1979	3.8	56	45	3000	24.9	
1980	3.5	78	67	3200	24.6	-0.4
1981	3.3	88	77	3500	24.3	
1982	3.2	92	81		24.0	
1983	3.1	95	85		23.9	-0.6
Concrete Goal	2.8		100	3600	23.6	-0.9
Goal Graphite	2.4		100		23.2	-1.3

Remarks re Table 2

a) Various qualities
b) Without anode effects, no essential changes in the other resistances

The statements made 3 years ago at the AIME Meeting regarding a cathode life of more than 2400 days and reduction of the power consumption have not only been confirmed but surpassed.

As a long term goal, using graphite cathodes with optimized bars, a reduction of 1.3 kWh/kg Al is aimed at.

4.3 Cost Consideration

An overall account is not submitted intentionally since the cost factors in the smelters vary considerably. One statement is allowed, however: the additional costs of using graphitic cathode block qualities and a reinforced insulation are more than compensated by the considerably longer cathode life.

The realizable reduction of the spec. el. power consumption of 0.6 kWh/kg Al takes place, therefore, without additional investments.

In Conclusion

Let us once again look at the development of the spec. el. power consumption in aluminium production in the Hall Héroult process.

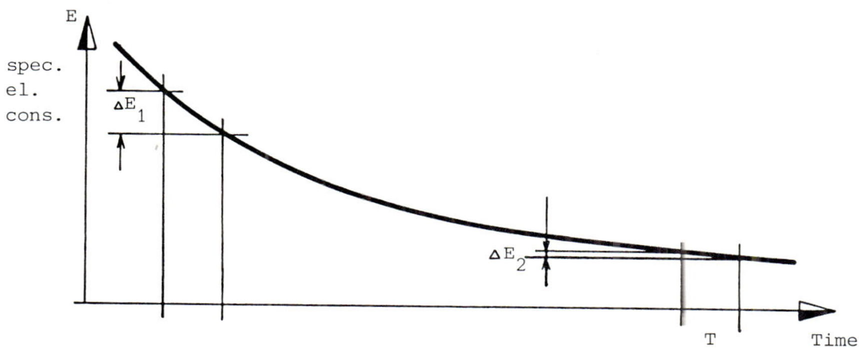

Fig. 5: Development of spec. el. power consumption

In former decades in a period of time T very large power savings - mainly due to better understanding of the process - were attained.

Today we have to concern ourselves with problems of relatively small savings potential and also carry out measures with a modest effect. We have been successful with one step in this direction, with the reduction of the cathodic voltage drop and the realization of the attained gain in power due to the use of glued, graphitic cathode blocks with sealed, reinforced thermal insulation.

References:

1) Pawlek, R., Private Communication
2) Blum, W., Private Communication
3) P. Aeschbach, H. Friedli, "Improved Potlife and Energy Savings by a New Cathode Lining Technique", Light Metals 1981, pp. 389
4) E. Schulz, W. D. Eckel, K. Etzel, P. Aeschbach, H. Friedli, "Applications of a New Carbonaceous Compound Technique for Steel to Carbon Connections in Aluminium Reduction Cells", Light Metals 1982, pp. 661
5) K. Etzel, F. Brandmair, P. Aeschbach, H. Friedli "Glueing of Cathodes and Anodes - A Proven Technology", Light Metals 1983, pp. 885
6) J. Thonstad, S. Rønning, P. Entner, "Formation of Bottom Crusts in Aluminium Pots", Light Metals 1982, pp. 485
7) V. N. Samoilenko, A.M. Tsyplakov, "Improvement in the Carbon Hearth Design of Aluminium Cell", 6 (1965) 6 Soviet Journal of Nonferrous Metals, pp. 50

A NEW APPROACH TO ALUMINUM CATHODE DESIGN AND RELINING

Jostein J. Vadla

Paper provides a critical look at traditional cathode relining procedures and the limitations they impose on cathode design: on the extent of the prebaked portion of the pot bottom, the way collector bars are installed, the bar/block contact area and the cathode voltage drop.

It contends that cathode current collector bar design is commonly deficient from the point of view of minimization of capital and labor expense and cathode voltage drop and proposes a relatively inexpensive means of correcting collector bar design in existing cathodes without expensive redesign of potshells, cradles or flexible connections.

It suggests that the traditional means of relining pots, even in recently built facilities, can be greatly simplified given modern carbon machining and materials.

Introduction

From its inception nearly one hundred years ago, the primary aluminum industry has more or less continuously struggled to reduce energy consumption; a struggle which has received renewed impetus from the producers of energy.

In the eight decades since 1900, consumption of energy by the primary aluminum production has been cut in half from 28 Kwh/kg to about 14.5 Kwh/kg (1). The course of such endeavors with respect to time is such that each increment of progress tends to become more difficult and expensive to attain. The tremendous rise in energy cost in the past decade makes the pursuit of further savings still worthwhile as was ably demonstrated by Todd in his contribution to the Symposium on Aluminum and the Energy Squeeze in Chicago three years ago (2). IPAI reported average world-wide consumption in 1982 at 16,694 Kwh/mt, including such "normal smelter auxilliaries as pollution control equipment, lighting, pumps, motors, etc." This represented a reduction of 0.5% from the previous year. Average North American consumption during the same time was 17,009 Kwh/mt, down 2.8% from 1980 but nearly 2% above world average (3). Clearly, as Todd said in Chicago, there is still much that can be done without the need for great technological breakthrough.

The present paper relates to an area where such savings can be had and which has received relatively little attention in the past. It is an attempt to describe a design for relining cathodes to achieve reduction in capital cost, in labor cost and in the cathode voltage drop (CVD) over the useful life of the cathode. As such, it deals with modifications of cathode blocks, collector bars, materials of rodding, joints between the blocks and around the borders and installation of wall blocks.

Present Relining Practice

The present process of bringing these elements together to make a cathode is typically as follows (disregarding the prior steps of removing the spent lining, reconditioning of the shell, and placement of the thermal insulation):

The carbonaceous cathode blocks are placed slot side up at a rodding station, blocks and bars are pretreated and the void space between bar and block is filled with cast iron or a carbon paste connecting material, the first poured, the last rammed into the slot. This rodding operation invariably requires space, energy and labor and is performed at a location separate from the rest of the process whether this takes place on line or in a separate cathode reline shop.

The now combined block/bar units are next brought to the potshell and placed on top of the heat insulating refractory with or without some bedding mix intervening. The cathode current collector bars ultimately have to extend outside the potshell through portholes and the placement of the combined block/bar, therefore, must employ one of two more or less difficult alternates. Most commonly, each block/bar unit is brought into the potshell aslant, the lower end is extended out its porthole far enough to permit the opposite end to be lowered down inside, after which the unit has to be jockeyed back and centered in the shell. This, of course, imposes a limit on the length of block which can be accommodated in a given shell depending on inside shell width and the required extension of the collector bar. Alternately, cathode blocks can be rodded with bars only slightly longer than their blocks. These units are then set in the shell, short bars with flexible tabs are then inserted through the portholes and butt welded to the bar inside the shell walls. This takes care of the block length limitation but is rather expensive and time-consuming.

Individual blocks are placed 1½" to 2" or more apart (except for experimental variants described in recent AIME (4-9) meetings about which more later) and the spaces between blocks are rammed full of carbon paste either of the traditional hot variety or the recently touted "cold" pastes (10, 11). In either case, the binder portion of these pastes is baked out during the process of bringing the cathode up to operating temperature. The quality of these seams and their effect on pot life are demonstrably subject to variation in paste quality, ramming workmanship and, whether the paste is fuming or stinking or not, the jackhammers involved are not what in today's computer lingo are called "user friendly." There are also questions related to the "cold" paste binder compatibility with blocks (12).

The seams between blocks having been completed to the satisfaction of persons responsible, the next step is the treatment of the space between the cathode block panel and the shell walls. There are many variations on this theme ranging all the way from green, rammed paste to various combinations of refractories, be they oxide, carbon or carbide based but invariably in combination with considerable amounts of ramming paste. The space to be filled with ramming paste is generally rather wide compared to the spaces between blocks and is subject to the same lack of user friendliness referred to a moment ago. To further complicate matters, more or less severe temperature gradients will develop on bake-in and in operation, ranging from about 900° C at the block side to 500° C or less at the shell side. The resulting tendency is for this paste to shrink away from the block ends and permit the access of molten aluminum to collector bars over the ends of the blocks with danger of short life.

There is a general, though not unanimous, consensus that this peripheral seam is probably a major cause of early failure.

The last carbon-related operation in the reline process is the installation of walls which can be of the monolithic, rammed version or a more or less sophisticated block variety which, in the majority of cases, still requires a lot of ramming in front of the blocks, sometimes clear to the top of the blocks. The blocks may be carbon, graphite, silicon carbide or a combination of these.

The common denominator of all these operations is this: They all involve operations and materials that have been generally available for most of the past century and seem out of place in our time of high technology consciousness. Today's blast furnace linings are installed with more sophistication than the much smaller aluminum potlinings and make extensive use of tight machining tolerances and the achievement of a flat starting base in linings as large as 45' diameter hearths. In recent years there have been scattered reports made to the AIME on alternate means of rodding cathode blocks, referred to above, and of cemented joints between blocks as well as cemented or "glued" cathode bars and anode stubs. To be true, this is not exactly avant-garde; cementing of joints in carbon furnace linings dates back into the Fifties and was pioneered by Union Carbide. As a matter of fact, the very first report of cemented joints in aluminum pots is to be found in the collection of papers presented at the very first AIME aluminum oriented meetings, the one in New York in 1962, in a report by Loren Bacon of the then National Carbon Company (13). That work was apparently far ahead of its time.

But even these more recent references, although aware of more modern materials and techniques, fail to take the rather obvious further step: To not only eliminate seams between blocks and use cement rodding instead of paste or cast iron, but to take further advantage of modern means available and eliminate all paste ramming. This would avoid the need for capital and labor expense in providing rodding facilities but also the need for anthracite storage, handling and, last but not least, the backbreaking work of ramming. And instead of doing the relining work in several different locations over several shifts, it can be performed by a small crew in one location in one shift or less. The reline procedure which is proposed here not only allows the elimination of the prerodding of blocks, placing all of the reline operations at shell site, it also provides an opportunity to rectify an almost universal cathode design deficiency--the shape and distribution of the collector bars which in turn affects cathode voltage drop.

Cathode Voltage Drop

CVD is made up of at least four components: The contact molten aluminum/cathode block, the cathode block, the contact cathode block/collector bar and the collector bar. The most sophisticated and widely quoted study of these components is the one made by Haupin (14). Briefly stated, lining drop goes from 350 $\pm$ 50 mv at the start to 90 $\pm$ 20 mv as graphitization proceeds. Carbon to bar contact drop increases from as low as 60 mv at the outset to 350 $\pm$ 50 mv at three years age, becoming the major contributor to the traditional gradual increase in CVD. Loss in the bars remains relatively constant throughout at 80 to 100 mv and the loss at the aluminum to carbon block contact is negligible except in "mucky pots." Haupin depicts a composite curve of CVD vs time starting out at about 500 mv, dropping rapidly during the first half year to about 400 mv, followed by a more gradual increase to slightly more than 500 mv after three years life. Most graphic representations do not show this characteristic, however, but rather a rapid increase from a low point at the start; this relatively rapid rise shifting abruptly after 50 to 100 days operation to a relatively slow rise throughout its remaining life (4, 5).

More recently Stagg (13), lamenting the general lack of attention paid to collector bars, concluded based on metallurgical examination of bars before and after prolonged use: The rise in contact resistance is due to: "1) Removal of iron from the interface zone by diffusion into the steel bars, 2) Deposition of crystalline sodium fluoride deposits at the interface between the cathode and the bar and between the cast iron and the steel within the bars." If item 1 on his list is a problem, it would seem it could be alleviated by carburization of the bar prior to installation.

Hayashi and Mochizuki (16), on the other hand, show how collector bars are caused to grow from direct contact with molten aluminum to the extent that the resulting stress causes block breakage. This would seem to be a problem only in the case of gross infiltration into the region of the bars by molten aluminum in which case dissolution of the bar seems to be a more threatening advent.

Contact Resistance

Electrical contact resistance between two solid surfaces, such as carbon blocks and steel collector bars, depends on a number of variables most important of which are the nature of the contacting materials, the nature of the surfaces contacting and the mechanical lead on the contact. Little can be done about the first except to change from carbon to graphite since graphite-to-metal exhibits lower resistance than carbon-to-metal (17). The second factor can only be affected by ensuring good machining quality. The third factor, the contact load, at first seems difficult to manipulate to one's advantage although of prime importance. The contact resistance of graphite-to-steel, for instance, exhibits a tenfold decrease in value from ambient to 600 psi. To take advantage of this important effect, one must identify the area most subject to load in pot operation and try to maximize that area.

The assembly of blocks, bars and seams becomes essentially a rigid panel at about 500° C in the process of bake-in and has yet another 450° C to go before it achieves operating temperature. This rigid panel about 1½'x12'x35' with steel bars embedded experiences thermal expansion while being tightly constrained by the pot shell and cracles in the lateral plane; in the vertical plane there is little or no constraint to expansion. By the time the bottom is at operating temperature, considerable compressive stress is brought to bear laterally between the side surfaces of the block slots and the side surfaces of the bars. The result is that the only contacting areas carrying appreciable current are the upper side surfaces of the bars/slots.

If one accepts this model it follows that most collector bar designs are deficient because of their shape and distribution in the lining. Most pots will have 5-6" square bars surrounded by cast iron in 6½" - 3" square slots spaced 18 - 22" apart (Fig. 1a). This represents relatively little side surface for the cross-section of bars required to carry the line current. By splitting the bars lengthwise and decreasing the spacing by one half the normal distance the active contact area can immediately be doubled and thus the effect of the contact drop on CVD over the entire life of the pot can be minimized. Figure 1b illustrates how this provision can be achieved in existing potlines without expensive modification to shells and cathode flexible connections. The "tuning fork" bar shown here is made up from two relatively tall and narrow collector bars with a yoke welded on just inside the pot shell to transport the current out the existing portholes to existing flexibles.

Relining

Having revamped the collector bars, the combined rodding - relining can proceed. First, all efforts must be made to make sure the top surface of the insulation is as flat as practically possible by careful screeding. To make sure that it also stays flat during the relining operation, sheets of Masonite or similar material are laid over the insulation and the edges are taped together. Even in the case of very accurately machined blocks proper joint thickness between blocks cannot be realized without a flat surface on which to rest. The overlay is also helpful when accurately locating bar positions. With bars located as close together as 9 to 11" on centers prepositioning bars should be avoided in favor of premarking bar locations by means of appropriate template. In a typical modern prebake, center-break pot in the 140 - 185Ka range, we may assume a cathode peripheral space of 5" to be filled with low density refractory serving as expansion joint between shell and carbon lining to prevent mechanical overloading of either as result of thermal and chemical expansion in actual service. This requirement assumes either odd shaped carbon blocks in the ends of the lining or, probably more practically, separate end blocks without collector bars. The first endblock would be braced in place to the shell with cement applied to the side, the first set of bars placed in position and cement applied to their top surfaces. The first cathode block, also having had cement applied, is then ready to be installed. The

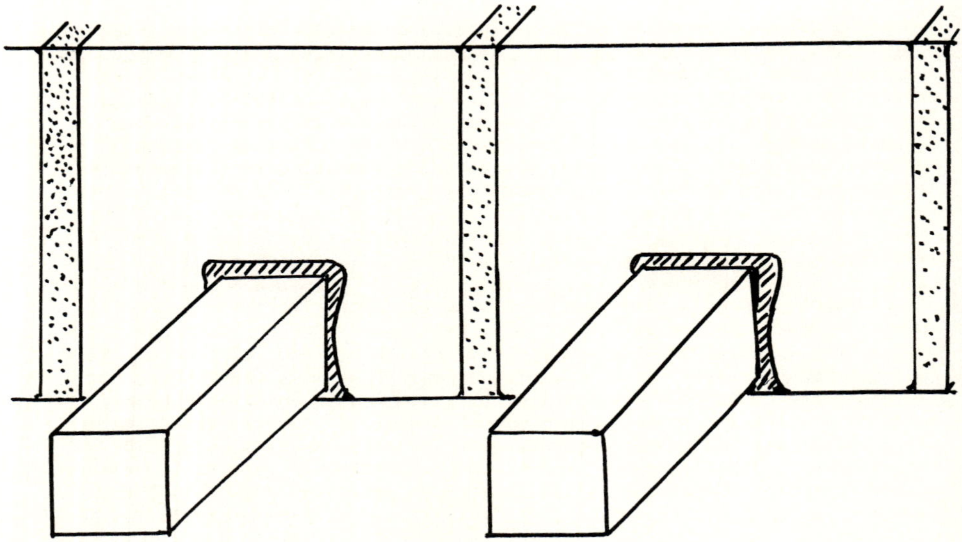

Fig. 1a: Most common type of cast iron rodded bars.

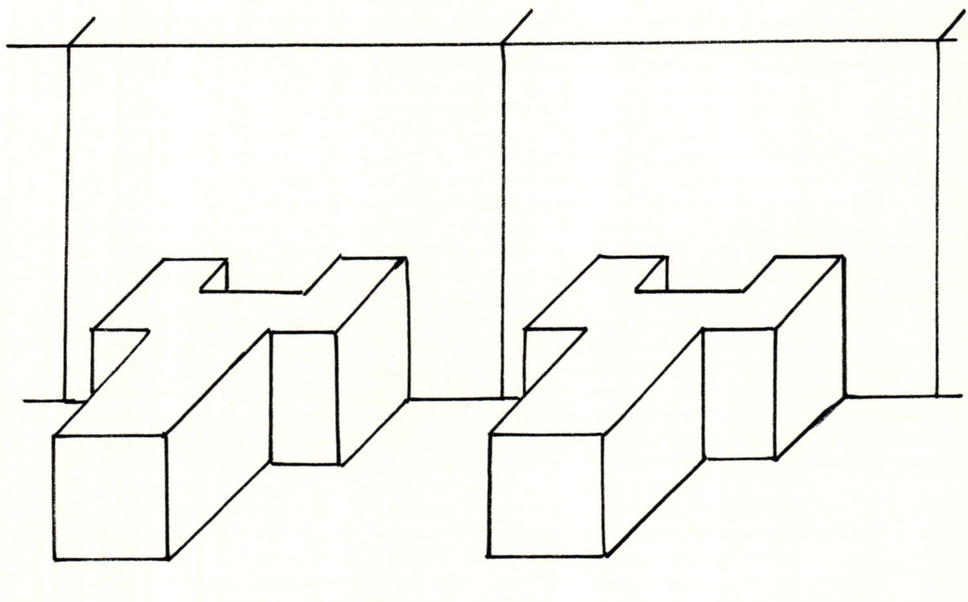

Fig. 1b: Retrofit design doubling active contact area.

key to successful cementing is to prime mating surfaces with plasticizer, to apply a uniform thin layer of cement to mating surfaces, excess cement will needlessly keep the joint open. The incoming block is then lined up in proper position and as soon as the bottom edge of the incoming block is 2-4"below the top of the block already in place two or three manpowers worth of push is applied to the free side to which cement is not yet applied. The block is thus slid into place in contact with the neighbor already there and no complicated jacking is necessary provided the insulation is flat and true and that the blocks have been machined to adequately tight tolerance. The thin layer of cement provides sufficient adherence to keep the block in position.

A second block is immediately set on top of the first. This serves to aid in extruding the bar-cement off the top surfaces and down along the approximately 3mm wide spaces provided along the sides of the bars. It also conveniently puts the block in position for pr ming and application of joint-cement. Meanwhile a new set of bars are placed in its marked off position and the proper amount of cement is applied Joint cement can be plasticized so as to permit spraying onto the block sides or it can be trowelled on. The bar cement must be sufficiently thick not to flow appreciably on a flat surface but still able to be poured or pumped onto the bars.

As soon as the cement application on bars and blocks is completed the top block is picked up, rotated 180° and slid into position in the same manner as the first one, the whole sequence repeated until all cathode blocks are in place, and the last endblock inserted. Joint quality can easily be checked with feeler gauge as the installation proceeds. They should ideally be on the order of 1mm or less. Should any of the joints exhibit appreciably more cement gap, jacking can correct this but should be held off until all the blocks are set. Jacks can then be placed between endblocks and shell wall on both ends, and pressure applied from both ends simultaneously.

The final carbon operation is the placement of wall blocks. Wall block placement to a great extent depends on the function of the wall in the heat balance of the particular pot. If insulation is called for the outside edge of the wall blocks should coincide with the outside edge of the bottom carbon panel with thermosetting joint cement between bottom of wall blocks and top of cathode blocks. If the wall is called upon to aid in dumping heat, the back side of the wall blocks should be flush against the shell wall with some contacting cement or mortar between. In the last case the cemented joint between the bottom of the wall blocks and the top of the bottom blocks will most certainly be ruptured at some point in the operation because of differential movement of the two. This should not be detrimental to continued good operation, however, since the deckplates are holding the wall blocks down and the resulting crack will only be on the order of 1-2 mm in a relatively cool part of the potlining. There will probably be a need to field cut one wall block on each of the pot sides and ends depending on shell tolerance and geometry.

In this manner prerodding of blocks whether by paste or cast iron with attendant capital equipment and labor can be totally eliminated along with paste ramming seams and surrounding space. In the process there should be a bonus in the form of longer pot life since the thickness of the seams can be tightly controlled and the considerable amount of paste of questionable quality has largely been replaced with prebaked material. The relining crews will have to be retrained in the performance of a more exacting, however, more pleasant job.

References

1. K. Grjotheim, B. J. Welch, Aluminium Smelter Technology, p. 3; Aluminium-Verlag GmbH, Dusseldorf, W. Germany, 1980.

2. J. B. Todd, "Energy Reduction in Hall-Heroult Cells with Conventional and Special Electrodes," Journal of Metals, 33 (9) (1981), pp. 42-45.

3. News item in Metal Bulletin dated July 12, 1983.

4. J. J. Vadla, R. E. Milner, "Cemented Collector Bar Assemblies for Aluminum Cell Cathode," United States Patent 4,001,104.

5. P. Aeschbach, H. Friedli, "Improved Potlife and Energy Savings by a New Cathode Lining Technique," Light Metals 1981, AIME, pp. 389-396.

6. J. McSurley, J. Vadla, "Cathode Construction Utilizing Cemented Collector Bars," ibid., pp. 675-681.

7. E. Schulz, W. D. Eckel, K. Etzel, P. R. Aeschbach, H. Friedli, "Application of a New Carbonaceous Compound Technique for Steel to Carbon Connections in Aluminium Reduction Cells," Light Metals 1982, AIME, pp. 661-672.

8. J. J. Vadla, "Current Carbon Refractory Practice in the Aluminum Industry" Paper presented at the American Ceramic Society 35th Pacific Coast Regional Meeting, Seattle, Wa., October 28, 1982.

9. K. Etzel, F. Brandmair, P. Aeschbach, H. Friedli, "Gluing of Cathodes and Anodes--A Proven Technology," Light Metals 1983, AIME, pp. 885-894.

10. D. L. Belitskus, M. B. Dell, G. H. Farrah, "Alcoa's Experience with Seam Mix for Carbon Cathode Blocks at Room Temperature," Light Metals 1979, V. I, AIME, pp. 647-661.

11. I. Letizia, C. Bizzarri, "Use of a Cold Ramming Paste (Seams and Borders) of Electrolytic Cells for Aluminium Production," Light Metals 1981, AIME, pp. 625-635.

12. C. Krohn, M. Sørlie, H. A. Øye, "Penetration of Sodium and Bath Constituents into Cathode Carbon Materials Used in Industrial Cells," Light Metals 1982, AIME, pp. 311-324.

13. L. E. Bacon, "Trends in Cathode Lining Materials, "Extractive Metallurgy of Aluminum" Gary Gerard, Editor, Interscience Publishers, New York, 1962, V. II, pp. 461-483.

14. W. E. Haupin, "Cathode Voltage Loss in Aluminum Smelting Cells," Light Metals 1975, V. I, AIME, pp. 339-349.

15. J. A. Stagg, "Reduction Cell Collector Bars - Interaction with Operating Environment and Influence on Cell Operation," Light Metals 1981, AIME, pp. 397-406.

16. T. Hayashi, F. Mochizuki, "Cathode Carbon Cracking Due to Al-alloying of Collector Bars," ibid., pp. 637-652.

17. National Carbon Company, Industrial Carbon and Graphite Products, Union Carbide Corp., New York, 1957, p. 60.

CHAIRMEN

E.L. Cambridge
ARCO Metals Company
Tucson, Arizona

S. Jacobs
Aluminunm Company of America
Alcoa Center, Pennsylvania

BATH RATIO CONTROL AT THE MEAD WORKS

Richard C. Jeltsch

Kaiser Aluminum & Chemical Corporation
Mead Works
P. O. Box 6217
Spokane, Washington 99207

Bath ratio control was improved from 60% conformance to over 90% conformance to standard over a period of four years. This improvement was achieved simultaneously with a reduction in plant ratio from 1.34 to 1.24. Key factors were: (1) Derivation and computerization of the materials balance formulae which allowed easy calculation of effects on ratio of differing raw materials and changing cathode age. (2) Streamlining of the logistics of sampling, analyzing and making corrective addition. (3) Improvements in analytical techniques. A large portion of the increase in current efficiency of the Mead Plant from 85% in 1979 to 90% in 1983 is attributable to the combination of lower ratio and improved pot operation made possible by good ratio control.

Background

The Mead Works is a World War II vintage smelter designed to operate at 50 KA and over the years upgraded to 74 KA, with an anode current density of 8.6 amps per square inch. There is a KACC fluid bed scrubbing system added in 1974, which recycles fluoride material to the potrooms. At full production six of eight potlines run on this "reacted ore" with two running on "pure" ore. Recovered alumina from pot bottom insulation (alkaline ore) and dross material are used as bath builders and metered on to the belt which delivers ore to holding tanks in the potrooms. During the period 1981-1983 the pots were converted from monolithic to prebake bottom blocks and microprocessors were installed for potroom process control in some potlines. Along with bath chemistry, these improvements also have been significant in lifting plant performance.

The following table shows a comparison of 1978 performance with 1983 with respect to some key bath chemistry variables:

	1978	5/83 YTD
Current Efficiency	85.3	89.6
Ratio Target	1.34	1.24
Ratio Conformance($\pm$ .06)	54%	92%

This paper deals with how this improvement in ratio control was brought about.

Soon after the scrubbers were installed and it became economically and environmentally feasible to run at lower ratios, a plant test was done comparing normal (1.38) ratio with low (1.22) ratio. The study confirmed the gains in current efficiency generally cited in the literature, but found that ratios were harder to control at the lower target, and the decreased alumina solubility of low ratio caused operating problems as evidenced by a doubling of the anode burnoff rate. Further, lower ratio seemed to magnify an already severe silicon carbide brick sidewall erosion problem, so conversion to lower ratio was delayed. When this sidewall problem was solved later, the ratio was lowered to 1.34 but excessive variation disallowed a further reduction.

In 1979 a concerted program of improvement of ratio control was begun. Figure 1 shows the result. 1978 ratio conformance was barely better than 50%; in 1983 it is consistently above 90%. At the same time, the ratio target has been lowered from 1.34 to 1.24 with plans to go lower.

System Mechanics

The old ratio control system consisted of sampling every other week with ratio corrections based on a 6 sample sliding average, weighted toward the most recent samples. There was no correction for age or for differing raw materials input. Control was totally in reaction to events and influences already past.

Initial work showed that individual pot ratios changed more rapidly than originally thought, perhaps due to the relatively low bath volume (less than 3000 lbs) of the Mead pot. Thus, the first step was to increase the sampling frequency to weekly and to discard the six sample sliding average formula in favor of a two-week average weighted 3/4 to

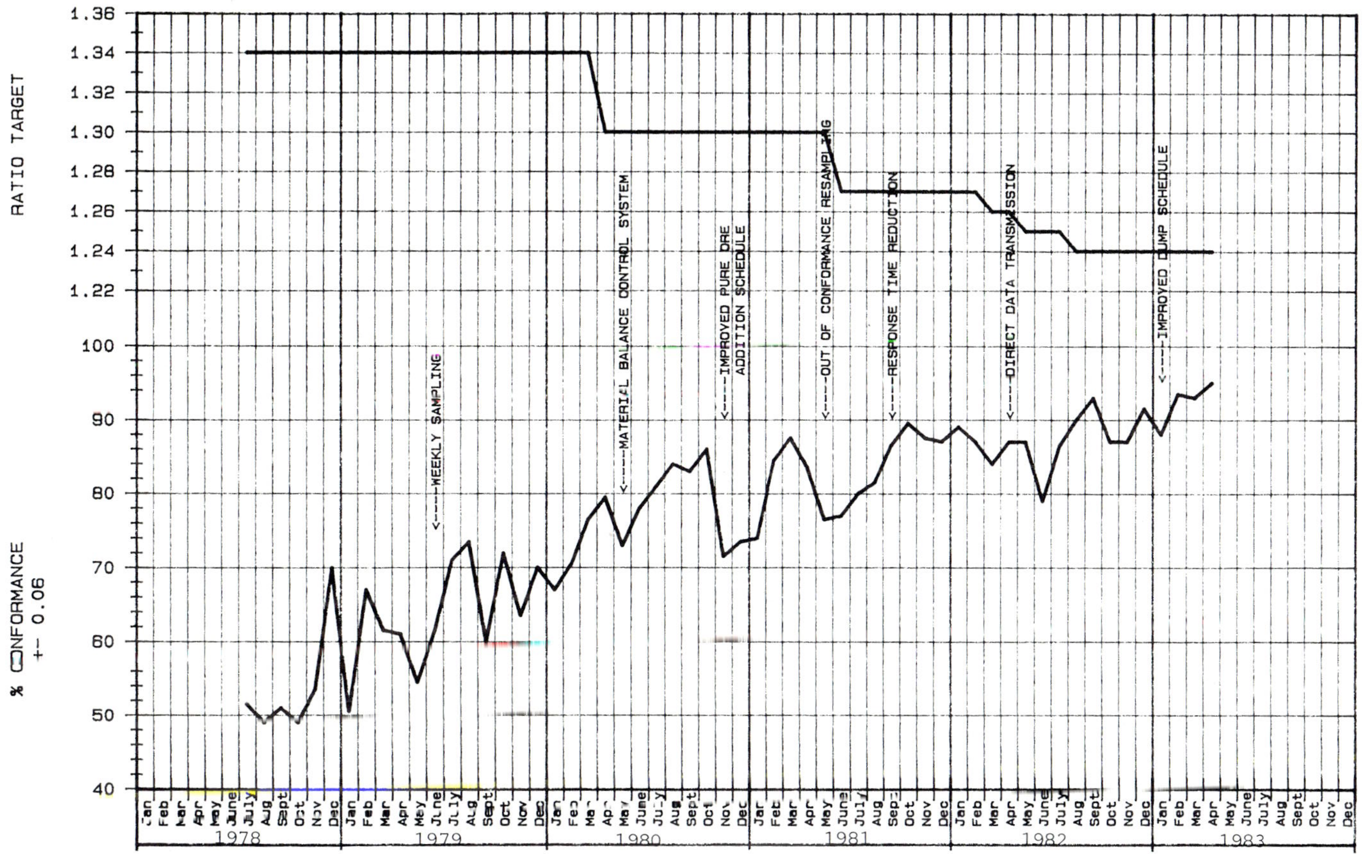

FIGURE 1 - RATIO TARGET AND CONFORMANCE LEVELS
MEAD PLANT 1978 - 1983

the most recent sample result. This change yielded some improvement in control, as expected; but without a real understanding of what caused ratio variations, control was still strictly "back end," and conformance remained in the 60-75% range.

Since weekly sampling gave better control than every other week, it follows that twice a week sampling would be a further improvement. Since this would be costly from a labor standpoint as well as overtaxing the capacity of the analytical machinery, we ran a test where pots which are outside the conformance limits were resampled and then recorrected if found to be still out of conformance. This resulted in considerable improvement in the two pure ore lines which require large amounts of aluminum fluoride but no improvement in the six reacted ore lines which have fluoride recycled from the scrubbers.

The first attempt at "front end control" came in mid-1980 with the derivation of a material balance system which was subsequently computerized. This system, which is basically still in use, corrects for changing sodium uptake as pots age and for differing raw material inputs to the various potlines. The basic variables are called the room adjustment factor (RAF) and age adjustment factor (AAF).

Each pot is assigned to an age group which has an AAF which represents the predicted ratio gain or loss for a week's time, assuming raw material input stays the same. For example, a relatively young pot, say 30 days old with an AAF of -.10, will decline in ratio from 1.25 to 1.15 under conditions where an older pot of 500 days (AAF of 0), would remain at 1.25. Originally there were only four age groups which assumed a sodium uptake curve that was too high. An iterative process of constant updating showed that pot age is a factor in ratio control throughout a pot's life and that older pots manufacture bath due to excess sodium in the lining, thereby requiring greater quantities of fluoride as they age.

The room adjustment factor (RAF) compensates for varying raw material feed to each potroom. Although both rooms of each potline are fed from a common alumina tank, experience has shown that ore tank segregation causes differences between the rooms, and it is necessary to have a separate factor for each room. Combining the RAF with the AAF allows us to anticipate where the next week's ratio will be.

Example:

Pot No.	Age	AAF	RAF	Predicted Change
1	30	-.10	+.05	-.05
2	500	0.0	+.05	+.05

Thus if both pots are on target according to the latest sample, pot 1 would need an addition of soda ash and pot 2 would need an addition of fluoride to keep each of them on target.

The third piece of the system is the material balance formula itself, which, in its simplest application, calculates how much soda ash and fluoride is needed to correct pots 1 and 2 in the example above. In addition, it can be used to predict the effect on ratio of a change in raw materials so that changes in the RAF can be made before the material change affects ratio and subsequently pot operations. For example, if a line is building bath too quickly, the potline supervisor might change his setting of alkaline ore (recycled pot bottoms) from 0.012% to .006% of his ore feed. Since this material is sodium rich, this change has a tendency to lower average bath ratio. By use of the material balance formula, it turns out to have the effect of lowering the ratio from 1.30 to 1.28. Thus, the RAF would be adjusted -.02 and the average ratio would remain the same.

Variability Studies

With the basic system in place, we set out to study ratio variability in the plant to try to pinpoint the causes. As in many systems, the little things (which with the benefit of hindsight should have been obvious) caused some of the biggest problems.

A test in which groups of pots were sampled every day for several weeks showed that the fluoride addition schedule used in the lines using fresh ore had a flaw which caused wide variation in ratio from day-to-day. Although weekly averages showed a steady ratio of 1.31, (target 1.30) daily averages over a two-week period ranged from 1.33 to 1.18. A change in the fluoride dump schedule cut the range of this variation from .15 to .04.

A test of adding fluoride by bag showed no advantage over addition by a crane carried, metered bucket. However, it did point up the necessity for periodic checks of the weight of material dumped from the bucket as well as regular preventive maintenance on the buckets.

An extensive program of sampling in various parts of the scrubbing and material handling system was undertaken. As a result, the dry scrubbing system was confirmed to have a major effect on raw material variability. Other components such as the bath material metering system, which turned out to have a negligible effect, could then be ignored. This study suggested that true "front end" control might be possible, making additions to pots based on analysis of sodium and fluoride in the ore they would be receiving in the next few days. This has not yet been carried to this final step, but the information gained from the study allows us to be much more responsive in adjusting the RAF.

Non-reproducibility of sample results was revealed as a problem. Several problem areas were pinpointed and resolved.

1. Sampling technique was standardized when it turned out that samples taken with hot tongs were more variable than those taken with relatively cool tongs.

2. The effects of impurities were studied. When carbon was found to be a common culprit in sample variability, we started requiring the samplers to break their samples before bringing them to the lab. Any which showed visible carbon were to be discarded and retaken.

3. The sample handling system in the laboratory was complicated and left too many opportunities for samples to become reversed in order or mixed up. This system was streamlined and simplified by the elimination of one grinding step and by the substitution of a paper cup and tray handling system for a cumbersome one which used small envelopes.

4. Seating of the pelletized samples into holders for the x-ray diffractometer turned out to be a major problem. When placed in the holders by an operator using his thumb, the pellet

sometimes would not be straight, causing the plane of the sample to be misaligned with respect to the x-rays striking it. This problem was solved by a simple mechanical pellet loading device.

5. One of the two pellet presses was found to be making pellets with a lip which interfered with the optics in a manner similar to the sample holder problem. Analytical variability was cut in half when the defective press was eliminated.

6. A routine check sample program was set up to monitor sample variability on a week-to-week basis. A duplicate sample from one pot in each room is included with samples from that room. The source of that sample is unknown to laboratory personnel until after the analyses are reported. When this system was first set up, over 15% of the check samples did not match within the tolerance of the x-ray diffractometer. Now this error rate is less than 1%.

Logistical Improvements

With 1135 pots at full production, the logistics of sampling, analysis, calculation of corrections, and addition of fluoride and soda ash are complicated. Two major improvements in this logistic system contributed to better ratio control.

1. Response Time Reduction - By changing work schedules in some departments and emphasizing the importance of the ratio control program in others, the time between sampling and correction was cut from 80-90 hours to 24. This had the obvious benefit of correcting off target pots in a shorter period of time.

2. Direct Data Transmission - Although the actual calculation of additions using the material balance formula was computerized from the beginning, ratio data had to be keypunched into the computer, and additional information had to be transcribed by hand to a sheet that could be read by a crane driver. As well, the room adjustment factor, which is the key to the system, had to be input before the program was run. This meant that any changes to the RAF were based on information that was already a week old. This was alleviated by linking the x-ray computer directly by telephone line to our Modcomp computer used for potroom process control. The data is transmitted directly to the Modcomp, eliminating keypunching. A potroom bath ratio histogram, mean, and proposed RAF as calculated by the computer for each room is accessed on a CRT by the process control engineer. He then decides on the RAF based now on the most up-to-date sample results. The computer then calculates electrolyte additions and prints a sheet for the crane driver, thus eliminating the transcription. The increased flexibility of the computer has allowed for the most recent significant improvement, an improved dump schedule that adds fluoride or soda ash needed for maintenance, as opposed to correction, into the pots at the most advantageous time of the week.

Consistent Pot Operations

A final factor in the Mead Plant's ratio control improvements, which is difficult to quantify, has been improvement of control and workmanship in day-to-day pot operations. Even when ratio conformance was in the range of 60-70%, the lines that were running smoothly, as indicated by stability of temperatures, steady anode effect frequency, and small number of excess anodes changed always had better ratio control. Potroom operators have come to realize that what they do affects ratio. The ongoing conversion to a system of microprocessor pot control has been a key factor. Good ratio control makes for improved pot operation and good pot operation makes for improved ratio control.

Conclusion

The Mead Works is a much more efficient plant today than four years ago. Three key factors were prebake bottomed cathodes, installation of potroom microprocessors and lower ratio operation. The improvements in ratio control described in this paper were a necessary part of the move to lower ratio. We attribute these improvements to careful study of the factors which affect ratio variations and close attention to detail in the conduct of daily operations.

Acknowledgements

The author would like to acknowledge a few of the people who were particularly instrumental in the described improvements at KACC Mead:

Charles A. Brown, Owner, Brown Fluid Systems, Former KACC Mead Development Engineer.
Donald B. Hartvigsen, Technical Superintendent KACC, Mead
Holly N. O'Leary, Development Engineer, KACC, Mead
Michael J. Sawatzky, Chief Chemist, KACC, Mead
Rod D. Zabreznik, Director, Reduction Division Development, KACC, Tacoma, WA.

ROUTINE ANALYSIS OF BATH RATIO: A NEW METHOD

K.Nagy[1], R.Bjørgo[2] and D.Fremstad[1]

[1]SINTEF, The Norwegian Institute of Technology,
N-7034 Trondheim-NTH, Norway

[2]A/S Årdal og Sunndal Verk,
N-5900 Høyanger, Norway

Abstract

A wet chemical method for routine determination of the bath ratio within the aluminium industry was developed and tested. The principle of the method is that the excess of AlF_3 in the sample to be analysed reacts with added NaF to form Na_3AlF_6 without sintering or melting. The remaining excess of NaF after the reaction is measured with a solid-state combination fluoride-selective electrode. The electrode is connected to a computer controlled analytical system. The bath ratios with the associated sample/bath numbers etc. are printed as an analytical report.

The method is fast and eliminates the time consuming sintering procedure which is well known from the literature. The results obtained with this method compare favourably with those obtained by other methods during an international laboratory intercalibration.

Introduction

Control of the NaF/AlF_3 ratio in electrolysis cells is of crucial importance for the economy of aluminium production. Several definitions of this ratio can be found in the literature:

- The weight ratio between NaF and AlF_3 in the bath,
- The mole ratio between NaF and AlF_3 in the bath,
- The excess NaF or AlF_3 in the bath taken as a weight percent of the total melt or of the cryolite content of the melt.

The electrolyte is said to be alkaline when there is an excess of NaF, and acidic when there is an excess of AlF_3. Since modern electrolysis cells are kept on the acidic side, the term "bath acidity" is often used as a synonym for "bath ratio".

Several methods based on different physical or physiochemical parameters have been developed for the determination of bath ratios, (1-4). The most popular method is undoubtedly the bath ratio determination based on X-ray diffractometry. This method depends on the quantitative determination of chiolite and cryolite in acidic baths, or sodium fluoride/crylite eutecticum crystals in alkaline baths.

Among the chemical methods the following deserve to be mentioned:

- Pyro-titration.
  Small tablets containing known amounts of NaF is added to a molten bath sample. Cryolite is then formed from the added NaF and excess AlF_3 in the sample. The neutralization point can be determined by dripping an indicator solution - (phenolphtalein, bromothymole blue) - onto a cooled bead of the titration melt. Titration is continued until the alkaline colour is observed, and the bath ratio is calculated from the NaF consumption. Alkaline baths can be analysed by means of the opposite titration.

- Determination of pH in an aqueous solution of the bath sample.

- Determination based on total cation analysis.

- Sintering with a known amount of added NaF. After sintering and extraction the excess NaF can be determined by a suitable fluoride analysis.

- Dissolution of a finely ground bath sample in a NaOH solution, followed by addition of a known amount of NaF. The remaining excess of NaF after formation of cryolite is determined and the bath ratio is computed. This method is described in a Canadian patent.

The methods above are all afflicted with weaknesses, such as:
- Uncertainty in the presence of foreign substances such as CaF_2, MgF_2, Al_2O_3, etc.
- Insufficient precision,
- Time consuming analytical process,
- Detrimental to the working environment,
- Large instrument investments.

There is therefore still a need for a low-cost method suitable for frequent and accurate bath ratio determinations.

ÅSV, HØyanger Verk was in possession of a powerful laboratory computer, able to handle up to three independent electrode pairs simultaneously. It was used to run the SINTALYZER method for fluoride determinations in emission and immission samples. With a view to exploiting the large capacity of the SINTALYZER instrument also for bath ratio determinations, ÅSV decided - after some introductory experiments - to co-operate with SINTEF in the development of a novel bath ratio method. The results are presented in this paper.

The new method is based on the formation of cryolite from NaF added to a hot water extract of the bath sample and excess AlF_3 present in the sample. Neither melting nor sintering is used and there is no need for acidic alkaline solutions. The excess of NaF is measured by means of an ion-selective fluoride electrode and the bath ratio is computed directly.

In our view, the new method yields results which are at least as accurate as those obtained by other methods; it is less time consuming and can be run at lower cost than any other method.

Choice of Analytical Procedure

An important consideration in the development of the new method was to avoid, - as far as possible, - the known weaknesses of existing bath ratio analyses. The dissolution medium, for instance, should ideally

- dissolve all AlF_3 from the bath samples,
- be a suitable reaction medium for the formation of cryolite from NaF and AlF_3, and
- have the characteristics required for the ion-selective fluoride determination.

The extraction process requires a pH in the range 6.0-7.5. Higher pH-values will lead to increased dissolution of Al_2O_3 (always present in bath samples), and to a concomitant increase in AlF_3 content and acidity values. At pH-values much below 6.0 the analysis will be rendered impossible by the loss of HF during sample boiling.

In order to avoid the formation of complexes such as $AlF_5(OH)^{3-}$ the cryolite formation should take place in the presence of an excess of fluoride. Other species present, -such as chiolite-, must be transformed to cryolite.

Alkaline solutions are not suitable for analysis by means of the fluoride-selective electrode. As is well known, hydroxyl ions interfere strongly with such fluoride determinations.

An unwanted dissociation of AlF_6^{3-} may take place both in strongly acidic and alkaline solutions. This dissocation will, of course, render impossible a correct determination of excess NaF.

Finally, use of the fluoride-selective electrode requires a solution with a reasonably high and constant ionic strength.

It was found in our work that a 1m NaCl solution fulfils the requirements set out above. This solution is neutral, the reaction between NaF and AlF_3 runs stoichiometrically to Na_3AlF_6, the solubility of Al_2O_3 is slight and the ionic strength is high enough for the solution to function as a reference solution for the newly developed SINTEF combination electrode for fluoride and chloride.

Experimental

Sampling and Grinding Procedure

Samples are taken from electrolysis cells by means of a pair of specially constructed tongs which, when closed, forms a hollow sphere with a diameter of 20 mm. The sampling end of the pair of tongs is immersed (while closed) in the bath, opened briefly and then withdrawn. Solidified melt on the outside of the tongs in knocked off before the spherical sample is removed. Samples from a series of ovens are put in labelled boxes.

An automatic Herzog mill is used for sample grinding. After grinding, 98 percent of the sample has a grain size of less than 74 µm (-200 mesh).

Dissolution and Reaction Procedure

Reagents: Water was distilled or de-ionized. The standard dissolving solution contained 58.5g NaCl and 1g F- per litre, standard fluoride solution contained 10g F- per litre and the solution used for dilution contained 58.5g NaCl per litre.

Pilot experiments were carried out using Merck chemicals Na_3AlF_6, NaF, $AlF_3.3H_2O$, Al_2O_3 and CaF_2. Chemicals were pro analysi with the exception of Na_3AlF_6 and Al_2O_3 which were not available in this grade.

Dissolution: Samples of 0.500g are weighed out, transferred to graded 100 ml Erlenmeyer flasks and 50 ml dissolving solution is added. The mixtures are slowly boiled for one hour, allowed to cool, filled to the mark and vigorously shaken. (It is not necessary to use volumetric flasks since the volume accuracy of the Erlenmeyers is better than one percent.)

Aliquots of 5 ml are pipetted into disposable plastic cups, 45 ml of dilution solution is added and the cups are conditioned at 24°C in a thermostatic bath. The solutions are now ready for the determination of excess NaF.

Fluoride Registration Procedure

The NaF-determinations were carried out by means of the fluoride-selective electrode. Analyses were carried out both manually (direct registration and single addition technique) and automatically by means of the computerized SINTALYZER system.

Equipment: Both Orion electrodes and SINTEF solid-state electrodes were used. The two types of electrodes gave identical results both for direct registration and the single addition technique. The SINTEF electrode has the shorter response time and the longer service life.

Potentials were read by means of RADIOMETER's digital voltmeter DVM84. The magnetic stirrer was a METROHM model E649.

Reversible Solid-State Combination Fluoride Electrode. The construction of reversible solid-state ion-selective electrodes has been discussed elsewhere (5-9). In order to reduce noise pickup in the connecting cable and to relax the impedance requirements on the voltmeter, a single-chip operational amplifier within the electrode itself has been integrated. A pure impedance transformation was achieved by a suitable feedback circuit, resulting in a fixed output impedance of about 60 Ohm from the electrode. The voltage stability at the output was about 0.1 mV.

Reversible solid-state contact to the silver salt membrane (AgS/AgCl) was provided by silver-conducting epoxy(8). A fluoride electrode based on a LaF membrane with a newly developed reversible solid membrane contact was also constructed and tested.

Full benefit from this new generation of solid-state potentiometric sensors could not be obtained unless the conventional liquid junction reference electrode was eliminated. However, in a number of practical analytical situations, an all-solid-state combination sensor utilizing two chemically orthogonal membranes could be used. The halfcell potential of either one of the membranes was fixed and used as reference by a suitable choice of buffer solution, with the other membrane retaining its full sensitivity (9). Integral impedance transformation and differential amplification circuitry within the sensor body provided satisfactory noise immunity. The inclusion of a grounded auxiliary platinum electrode gave the necessary potential stability at the measuring site. This solid-state combination ion-selective electrode has no liquid junction potential, which means more potential stability and shorter response time compared with conventional ion-selective electrodes.

Fig. 1 Schematic drawing of the reversible solid-state combination fluoride electrode

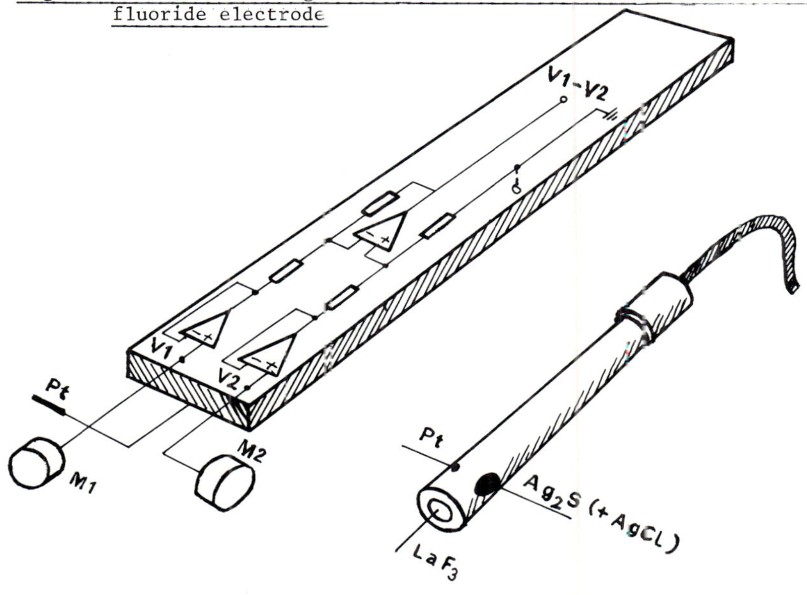

A schematical presentation of the present combination potentiometric sensor is shown in Fig. 1. The membranes were cylindrically shaped with linear dimensions of a few millimeters. The entire construction was sealed and housed inside an epoxy tube, about 10 mm in diameter and 100 mm long. The electrode had external connections, two for carrying the ± 5% V dc voltage from the power supply to the operational amplifiers, and one output signal line. The output signal was displayed on a simple 4½ digit panel meter (Radiometer PHM 84).

Manual Registration Procedure: The plastic cup was put into a small circulation thermostat on the magnetic stirrer and the electrode was immersed in the solution. The electrode potential was read when stabilized, 250 μl standard fluoride solution was added and the potential read again after stabilization. The first stable potential was used for the direct fluoride registration.

At the start of a sample series three standard fluoride solutions were measured. The buffer solution was the same as for the unknown samples and fluoride contents were 0.1, 1.0 and 2.5 mg F^- per beaker. Electrode potentials for these solutions were plotted graphically. The graph was used both to test for linearity of electrode response and to determine the amount of fluoride per beaker by direct registration.

A computational table was constructed to simplify the calculation when using the single addition technique. The table gives bath acidity (defined as w-percent excess of AlF_3 in the sample to be analysed) as a function of potential difference measured at 24°C.

Automatic Fluoride Registration: As mentioned above, automatic bath ratio determinations were made on the computerized SINTALYZER instrument system. A block diagram of the system is shown in fig. 2.

Fig. 2. SINTALYZER system, block diagram

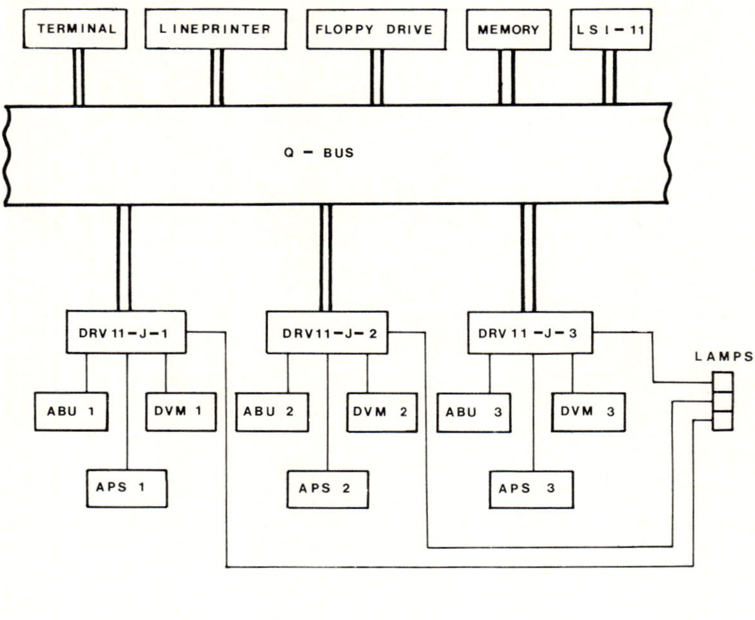

The SINTALYZER method is a computerized version of the manual single addition technique described above. Series of up to 40 samples are analyzed automatically from SINTEF's APS82 sample changer. The program has been written to take into account most variations that can occur in bath ratio determinations while using a minimum of operator input. Thus, samples are identified by means of sample numbers and/or alphanumeric names, sample weights may be fixed for the whole series or can be entered individually (manually or from an automatic balance connected to the SINTALYZER computer), individual correction factors for sample contents of "inert matter" can be used, and the initial addition of NaF during sample pretreatment can be varied.

For control purposes a short sample printout is produced at the end of every analysis. At the end of a sample series, -(which can consist of up to 160 samples)-, an analytical report for the whole series is printed. Facilities are provided for the deletion of spoiled samples or erroneous results from the final report.

For user convenience another type of report, termed a "look-back" report, can also be generated (see fig. 3). A look-back report uses information stored in the computer from earlier runs of samples from the same series of cells and presents these data together with results from the latest series of analyses. This report therefore provides "at a glance" information on the time-development of individual cells

Fig. 3. SINTALYZER look-back report for a series of 20 electrolysis cells

SINTEL II LOOK-BACK REPORT
Generated THU 15-SEP-83 at 17:40:50

Sample No.	Sample Identifier	15-SEP-83	13-SEP-83	09-SEP-83	05-SEP-83
102	A 21	9.62	8.23	12.54	spoiled
103	A 22	10.22	10.79	12.22	9.59
104	A 23	7.89	8.12	11.77	10.46
105	A 24	9.70	10.79	11.62	8.44
106	A 25	7.96	7.89	10.03	9.[illegible]6
107	A 26	7.31	8.23	11.25	11.[illegible]7
108	A 27	11.52	10.79	10.44	12.[illegible]8
109	A 28	9.90	10.77	11.46	12.76
110	A 29	8.12	8.08	10.79	10.[illegible]5
111	A 30	8.94	10.69	11.90	10.44
112	B 21	10.53	11.33	10.91	7.35
113	B 22	8.00	6.94	7.22	7.[illegible]7
114	B 23	11.84	10.44	9.72	8.[illegible]5
115	B 24	5.35	8.18	9.34	7.81
116	B 25	7.48	6.56	8.58	10.48
117	B 26	9.59	10.55	8.78	9.83
118	B 27	7.64	10.57	9.30	10.77
119	B 28	7.81	7.39	7.58	7.73
120	B 29	11.65	11.39	11.58	11.00
Series average:		9.00	9.35	10.41	9.[illegible]5
Standard deviation:		1.72	1.65	1.54	1.[illegible]0

Results and discussion

Results are presented and discussed both for pilot analysis of samples with known composition and for analysis of unknown samples from various aluminium works. The single addition technique was used for most registrations. Many samples were analysed both with the commercially available Orion fluoride electrode and with SINTEF's new solid-state fluoride-selective combination electrode.

Our method is based on the assumption that excess AlF_3 will react quantitatively with added NaF to yield Na_3AlF_6. Our results also indicate that other compounds such as chiolite ($Na_5Al_3F_{14}$) and sodium-calcium-hexafluoaluminate ($NaCaAlF_6$) will react with added NaF to form cryolite. The solubility of AlF_3 in water is relatively high (ca. 6 g/l), and AlF_3 is therefore quantitatively extracted from a finely ground sample during the pretreatment procedure described above. On the other hand, cryolite is only sparingly soluble in water (ca. 0.05 g/l). The solubility of Na_3AlF_6 is of little importance for the analytical result as long as the dissociation of AlF_6^{3-} is very small. (The fluoride-selective electrode registers only free fluoride).

Results from analyses of known samples, (NaF, Na_3AlF_6 and mixtures of NaF, AlF_3 and Na_3AlF_6) are presented in table 1. All samples were subjected to the complete pretreatment procedure described above.

Table 1. Analysis of NaF, Na_3AlF_6 and mixtures of NaF - AlF_3 - Na_3AlF_6

Test no	Added F^-, mg	Na_3AlF_6 mg	AlF_3 mg	F^- reacting to Na_3AlF_6, mg	Remaining F^- after reaction		Excess of AlF_3, w%	
					theoretical	measured	theoretical	measured
1	50					50,42		
2	"					50,82		
3	"					49,80		
4	"					49,50		
5		500				1,06		
6		"				0,89	0	-0,29
7		"				0,92		
8		"				1,10		
9	50		50	33,9	16,1	16,06		
10	"		"	"	"	15,30		
11	10	500	5	3,39	6,61	5,67	1	1,28
12	"	"	"	"	"	5,74	"	1,26
13	50	"	50	33,9	16,1	16,00	10	10,03
14	"	"	"	"	"	15,30	"	10,24
15	100	"	"	"	66,1	65,40	"	10,21
16	"	"	"	"	"	65,65	"	10,13

The analyses of fluoride in pure NaF-solutions show a standard deviation of 0.6 mg F (or 1.2% relative) for four parallel determinations. This uncertainty accords well with our previous experience with such determinations. This standard deviation amounts to 0.18 in the bath acidity units defined above, and is well within the limits set by ÅSV at the outset of this work. (A registration of 3.39 mg F^- corresponds to 1% excess AlF_3). It is worth nothing that a constant standard deviation for the F^- registration will yield a decreasing relative uncertainty for bath ratios with increasing bath acidities.

When 0.5 g cryolite alone is subjected to the complete analytical procedure, 0.99 mg F^- is registered in the final analytical step. This corresponds to 0.29 bath ratio units. In routine analysis a correction can, if necessary, be introduced for this error. No such correction has been applied to the results presented in this paper.

Our results show that AlF_3 reacts quantitatively to Na_3AlF_6 in the presence of an excess of NaF. This statement is valid also when the original sample itself contains Na_3AlF_6.

Table 2. Determination of excess of AlF_3 in synthetical bath samples as a function of added CaF_2 and Al_2O_3.

Test no	Added Na_3AlF_6 mg	Added Al_2O_3 w% of tot. weight	Added CaF_2 w% of tot. weight	Excess of AlF_3 to Na_3AlF_6, w% Theoretical	Measured	
19	500	0	0	1	0,96	0,98
20	"	"	4	"	0,92	±0,078
21	"	"	8	"	1,02	
22	"	5	0	"	1,57	1,50
23	"	"	4	"	1,40	±0,087
24	"	"	8	"	1,52	
25	"	0	0	4	3,81	3,79
26	"	"	4	"	3,87	±0,086
27	"	"	8	"	3,70	
28	"	5	0	"	3,86	4,11
29	"	"	4	"	4,26	±0,218
30	"	"	8	"	4,21	
31	"	0	0	8	7,86	7,84
32	"	"	4	"	7,86	±0,035
33	"	"	8	"	7,80	
34	"	5	0	"	8,07	8,19
35	"	"	4	"	8,16	±0,138
36	"	"	8	"	8,34	
37	"	0	0	12	11,52	11,53
38	"	"	4	"	11,57	±0,036
39	"	"	8	"	11,50	
40	"	5	0	"	11,73	11,85
41	"	"	4	"	11,95	±0,111
42	"	"	8	"	11,86	

In table 2 we present results obtained when varying amounts of Al_2O_3 and CaF_2 are added to Na_3AlF_6. The concentration ratios used correspond to those that can be expected in process samples taken at ÅSV works. Addition of LiF was not included in our pilot experiments. Other experiments, not reported here, show that fluoride from LiF is registered as free F^- along with F^- from NaF. The analytical procedure is identical to the one described in this paper.

Table 2 shows that the presence of CaF_2 introduces no statistically significant analytical errors. This is not surprising, since the solubility of CaF_2 in water is only about 16 mg/l.

Admixture of Al_2O_3 (5% of sample weight), on the other hand, gave significantly increased acidity values in all experiments. On the average, for the four acidity levels examined (1, 4, 8 and 12%), the increase is 0.38 ± 0,096 bath acidity units. A similar effect has been mentioned by G. Kudermann et al. (10). The main reason for this effect seems to be that part of the dissolved Al_2O_3 (Al_2O_3 is partially dissolved during sample pretreatment) reacts to form AlF_3 which will further react with NaF to form Na_3AlF_6. In our opinion these reactions make alkaline dissolution media unsuitable for bath ratio determinations.

If the amount of Al_2O_3 present is known, a numerical correction for this error can be introduced. No such correction has been applied to the results presented in this paper.

In table 3 we present results for unknown samples taken at Granges Aluminium, Sundsvall, Sweden. These samples were analysed at Granges Aluminium by X-ray diffraction and at SINTEF by the present method. The results obtained by the two methods are seen to be in close correspondence. The series standard deviation between methods $(\sqrt{\Sigma(x_1-x_2)^2/11}$ is 0,53 bath ratio units. There is no significant systematic difference between the methods; the difference between the series averages for the two methods is only 0,07 bath acidity units.

Table 3. Determination of excess of AlF_3 in bath samples from Granges Al, Sundsvall, Sweden. Comparison of results from Granges Al and present method

Test no	Determined excess of AlF_3, w%	
	Granges Al	Present method
1	5,11	5,45
2	6,88	7,48
3	7,62	7,90
4	2,00	1,38
5	6,88	6,10
6	6,51	7,01
7	5,45	5,14
8	7,24	6,91
9	7,24	7,26
10	5,45	5,20
11	5,45	5,45

Results from an international bath ratio laboratory intercalibration are presented in table 5. Results obtained with the present method are compared to results obtained at five other laboratories. The X-ray diffraction method was used at three laboratories, one laboratory used the Lewis-acid method, and the last laboratory used an unspecified wet-chemical method. Results obtained with the new method are seen to correspond closely to the average values of the other results.

Table 4. Comparison of results from an international intercalibration for determination of bath ratio (NaF_3/AlF_3)

Lab. and method / No of sample	GrangesAl Sweden (XRD)	El.Kem Lista Norway (Lewis-acid)	ÅSV Sunndalsøra Norway (XRD)	El.Kem Mosjøen Norway (XRD)	Mitsubishi* Japan (Wet-chemical)	Sintef Norway (Present method)
1	1,44	1,46	1,46	1,38	1,40	1,45
2	1,23	1,19	1,22	1,26	1,17	1,23
3	1,24	1,18	1,24	1,18	1,19	1,25
4	1,24	1,16	1,22	1,24	1,20	1,27
5	1,41	1,53	1,46	-	1,44	1,43
6	1,22	1,17	1,20	1,24	1,17	1,23
7	1,23	1,18	1,21	1,23	1,18	1,22
8	1,18	1,16	1,10	1,24	1,17	1,24
9	1,22	1,21	1,25	1,26	1,2	1,27
$\overline{X}$	1,27	1,25	1,27	1,25	1,24	1,29

* The Mitsubishi laboratory spesifies that free Al_2O_3 and CaF_2 is taken care of before calculations

Conclusion

A novel low-cost analytical method for the determination of bath ratios in electrolysis cells is described. The method was tested on synthetic samples of known composition and on unknown samples.

The analyses of synthetic samples showed that the reaction between AlF_3 and NaF is complete and that only free F^- is registered by the fluoride-selective electrode.

The solubility of Na_3AlF_6 influences analytical results only negligibly. Sample contents of up to 8 percent CaF_2 gave no statistically significant analytical errors. A slight increase in acidity values was observed for bath samples with ≥ 5% content of Al_2O_3. This increase is, however, small enough to be accepted in routine process analysis.

Results obtained by the new method has been compared to results obtained by different wet chemical methods and X-ray diffraction analysis in an international laboratory intercalibration. The new method yielded results close to the average results obtained by other laboratories using other methods.

Acknowledgements. The authors want to thank the other participants in the bath ratio intercalibration for their kind permission to use the results in the present paper. We are specially grateful to mr. Lennart Wickstrom of Granges Aluminium for his interest and co-operation.

References

1. Philips' Aluminium Bath Ratio Analysis System

2. Zombori - Major: Timföld és Aluminiumipari elemzések, Müszaki könyvkiado, Budapest 1975

3. H.Neff: Aluminium 53(1977)375

4. S.Bagio, L.G.Olavaria: Aluminium 53(1977)737

5. Nagy K, Fjeldly TA, Johannesen JS (1978) In: Ion-selective electrodes. Elsevier Publ Co, Amsterdam New York, p 49

6. Nagy K, Fjeldly TA (1979) Talanta 26:811

7. Fjeldly TA, Nagy K (1980) J Electrochem Soc 127:1299

8. Fjeldly TA, Nagy K (1981) Ion-selective electrodes 3. Elsevier Publ Co, Amsterdam New York, p 215

9. Fjeldly TA, Nagy K and Stark B: Sensors and Actuators, 3(1982/83)111-118

10. G.Kudermann and K-H, Blaufuss: Erzmetall 33 (1980) Nr.5

ALUMINA DISSOLUTION AND SLUDGE FORMATION

Rudolf Keller

Consultant

R.D. 3

Export, PA 15632

A good dissolution behavior in the Hall-Héroult cell is one of the requirements for an industrially acceptable alumina. Dissolution rate as such is, however, less crucial than the tendency to form sludge. Aluminas that interact with the electrolyte and then disintegrate into coherent agglomerated fragments may have less tendency to form sludge than aluminas that do not react in contact with the bath, although the dissolution rate may actually be smaller.

To form sludge, alumina and bath sink through the metal, overcoming the interfacial tension at the metal-bath interface. It is important how densely alumina accumulates above the metal; the density of the alumina-bath accumulation may in many cases be lower than the density of the metal and sludge will not form. The formation and stability of alumina-bath interaction products and the bath density primarily determine the tendency to form sludge.

Alumina-Bath Interaction Products

Alumina added to bath normally does not sink into the bath particle by particle. Alumina rather forms agglomeration products due to penetration and freezing of bath and rapid phase transformation of alumina into the alpha modification. In the course of this phase transformation, a network of interconnected alumina particles forms, and fragments constituting agglomerated alumina particles are stable even when the intersticial bath is liquid. This effect has been studied, among others, by Gerlach and his coworkers (1, 2). The stability of such alumina-bath interaction products is related to the alpha-alumina content: interaction products of essentially alpha-alumina disintegrate into the original small particles, they "melt apart"; alumina with very low alpha content forms stable aggregates that may only break into large stable fragments of solid alumina and intersticial liquid bath; alumina with an intermediate alpha content disintegrates more easily, into medium-sized fragments. When the alumina particles or fragments sink through the electrolyte, are carried around in the bath by bath motion, or float at the bath surface because of trapped gas, their dissolution rate is determined by fluidynamic conditions and the composition and temperature of the melt. Particles and small fragments generally dissolve much faster than large chunks of alumina-bath interaction products.

Alumina Accumulation at the Metal-Bath Interface

Because of surface tension effects, alumina that sinks to the metal-bath interface does not sink readily through the metal. Thonstad and Liu (3) have calculated that a solid alumina sphere having a diameter of less than about $1\frac{1}{2}$ cm would not overcome the interfacial tension.

Alumina particles, therefore, accumulate at the bath-metal interface. Alumina-bath fragments do likewise, unless they should be large and heavy enough to sink through the metal individually. Such accumulations actually may depress the metal surface locally. The extent of this effect depends on the amount accumulated and the overall density of the accumulation. As Figure 1 illustrates, individual particles, assumed to pack as loose alumina would in air, form denser accumulations than fragments that themselves contain bath.

Sludge Formation

Solid alumina contained in liquid bath below the metal pad constitutes sludge. To form, alumina and bath have to sink below the metal pad. This occurs if the accumulation of alumina is sufficient to overcome the interfacial tension. The density of the accumulation obviously affects the tendency to form sludge.

Figure 2 summarizes an attempt to calculate the critical amount of accumulation necessary to induce sludge formation. Using the relationship given in (3), the critical diameter of a solid alumina sphere was calculated for a bath density of 2.05 g/cm^3, a metal density of 2.30 g/cm^3 and an interfacial tension of 450 mN/m ($g.sec^{-2}$): the sphere of 1.3 cm diameter would contain 4.8 g of alumina. Using the same calculation for a 25 v/o accumulation of Al_2O_3 (as presented in Figure 1), a critical diameter of 3.9 cm results, and the alumina contained in such a sphere would amount to 31 g.

The overall bulk density of an accumulation of particles would be 2.49 g/cm^3. Agglomerated fragments resulting from alumina-bath inter-

Figure 1. Bulk Densities of Alumina Accumulations

	Alumina Particles in Air at 25 C	Alumina Particles in Melt at 950 C	Alumina-Bath Fragments in Melt at 950 C

Density of Al_2O_3	3.98 g/cm^3	3.89 g/cm^3	3.89 g/cm^3
Density of Filler	0	2.02 g/cm^3	2.02 g/cm^3
Volume of Solid	25 v/o	25 v/o	12.5 v/o
Bulk Density	1.00 g/cm^3	2.49 g/cm^3	2.25 g/cm^3

action may each assume this bulk density but the accumulation only achieves a lower overall density because the fragments stack up with bath-containing voids. Assuming a stacking density of 50 v/o for the fragments, an overall bulk density of 2.25 g/cm^3 results. This is less than the metal density and such an accumulation therefore cannot sink through the metal pad!

The bath density may decide whether an accumulation of alumina-bath interaction fragments forms sludge or does not. This is evidenced in Figure 3. With a bath density of 2.12 g/cm^3 rather than 2.05 g/cm^3 (2.09 g/cm^3 and 2.02 g/cm^3, respectively, for the melt saturated with Al_2O_3 within the fragments), the accumulation assumes, according to the model, an overall density of 2.315 g/cm^3. The critical sphere diameter would be 14 cm and at least 2 lbs of alumina would have to accumulate to form sludge.

Often one observes in the laboratory that alumina-bath interaction products hesitate to sink and are obviously specifically not much heavier than the bath. Gas present may increase their buoyancy. This also has an effect on the tendency to form sludge, as indicated in Figure 3.

Conclusions

Sludge formation is crucial for the alumina dissolution behavior. As long as alumina added to the bath does not form sludge, it is, generally, available for dissolution. The dissolution rate of incrementally added alumina may be faster or slower, but this is of secondary importance (it affects, of cousre, the amount of accumulation).

Alumina is wetted by bath but not by metal. At the metal-bath interface it must overcome the interfacial tension to sink through the metal to the bottom of the cell where it may remain as sludge. To form sludge, a sufficient amount has to accumulate; the critical size of the accumulation depends on the overall density of the accumulation involving alumina and intersticial electrolyte. This overall density depends on the packing

Figure 2 Critical Size of Alumina Accumulations

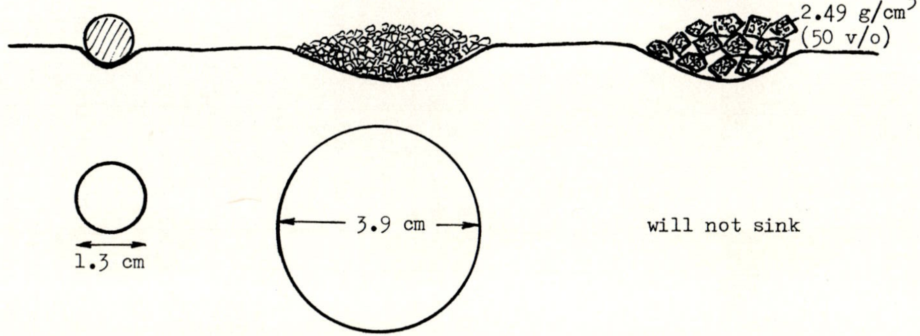

	Solid Al_2O_3 Sphere	Accumulation of alpha-Al_2O_3 Particles	Accumulation of Agglomerated Fragments
Critical Size	1.3 cm	3.9 cm	will not sink
Solid Al_2O_3	100 v/o	25 v/o	12.5 v/o
Overall Bulk Density	3.98 g/cm^3	2.49 g/cm^3	2.25 g/cm^3
Critical Amount of Alumina	4.8 g	31 g	will not sink

Figure 3 Critical Size of Accumulations of Fragments

	Accumulation of Agglomerated Fragments, Bath Density 2.05/2.02 g/cm^3	Accumulation of Agglomerated Fragments, Bath Density 2.12/2.09 g/cm^3	Accumulation of Buoyant Fragments, Fragments 2.20 g/cm^3, Bath density 2.12/2.09 g/cm^3

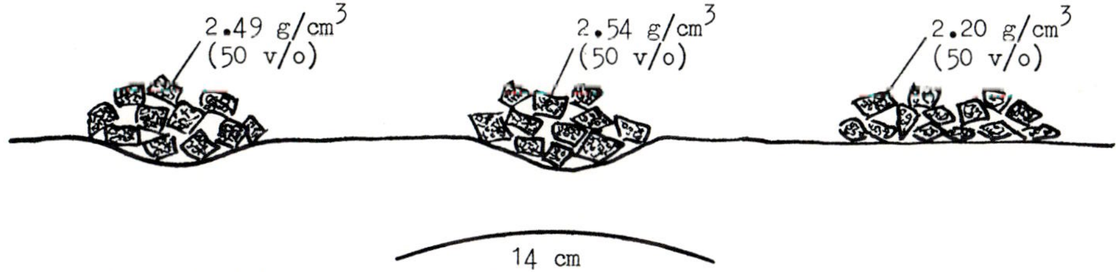

Critical Size	will not sink	14 cm Diameter	will not sink
Solid Al_2O_3	12.5 v/o	12.5 v/o	12.5 v/o
Overall Bulk Density	2.25 g/cm^3	2.315 g/cm^3	2.13 g/cm^3
Critical Amount of Alumina	will not sink	900 g	will not sink

of the alumina, on the density of the bath, and on buoyancy-providing inclusions.

The packing will be relatively dense if individual alumina particles accumulate. This appears to be the case for alumina with a very high alpha content, and thus one may explain the great tendency of such aluminas to form sludge. Aluminas low in alpha content produce fragments of alumina-bath interaction products that stack less dense; a greater amount has to accumulate, or the overall density may actually be so low that sludge cannot form. The fragmentation is expected to depend on the alumina properties and on the mode of addition to the bath (in extreme cases, lumps may form that sink through the metal because they exceed the critical size for the type of accumulation they constitute).

The bath density plays an important role. A low bath density leads to low overall density of accumulations, thus it hinders sludge formation. From intuition, one might rather have guessed that a low bath density would favor sludge formation because alumina would sink faster through the bath to the metal surface.

At this point, we do not know enough to draw definitive conclusions on the relative importance of alumina and bath properties in regard to the tendency of technical aluminas to form sludge; my feeling is that bath properties are generally more important. In selecting bath compositions to arrive at optimum alumina dissolution, density should probably rather be considered than the alumina dissolution rates as determined experimentally under controlled masstransport conditions. Favorable overall alumina dissolution behavior may be expected from electrolytes rich in AlF_3 and LiF.

The calculations presented are based on alumina sinking through a metal surface. Due to bath and metal motion, accumulations may occur rather at the edges of the metal pool. Alumina may sink easier there because one can assume that a liquid bath film exists between metal and cell wall (4, 5). This is not expected to affect the qualitative picture but may have a quantitative effect.

References

1. S. Abd El All, J. Gerlach, U. Hennig, Erzmetall 33, 504 (1980)
2. A. Damoui, J. Gerlach, U. Hennig, Erzmetall 35, 77 (1982)
3. J. Thonstad, Y. Liu, Light Metals 1981, p.303
4. J. Thonstad, P. Johansen, E.W. Kristensen, Light Metals 1980, p.227
5. R.F. Robl, Light Metals 1983, p.449

Acknowledgement

I thank Mr. W.B. Frank, Alcoa Laboratories, for the many stimulating discussions and for reviewing the manuscript.

METAL PAD VELOCITY MEASUREMENTS IN PREBAKE AND SODERBERG REDUCTION CELLS

A. T. Tabereaux and R. B. Hester

Reduction Laboratory
Reynolds Metals Company
P.O. Box 1200
Sheffield, Alabama 35660

Calibration curves have been established for the relationship between measured weight loss per unit length of iron rod immersed and rotated in molten aluminum metal over the practical range of velocities experienced in aluminum reduction cells. The effect of metal temperature, iron content, alloy content of the iron rod, and rod size on the rate of dissolution have been determined. The rate of dissolution of the iron rod can be related directly to an empirical convection mass transfer equation.

Results of measurements for prebake and Soderberg reduction cells suggest basic information and relationships between flow patterns, metal pad velocities, and cell design.

Introduction

The method of measuring metal velocities in aluminum reduction cells developed by Johnson (1) in which iron rods are inserted vertically into the metal pad has proven to be a valuable research technique. The method yields both the magnitude and direction of metal flow at each rod location.

However, the empirically derived relationship used to calculate metal pad velocities in aluminum reduction cells was based on calibrations made in the temperature range 758 to 843°C in metal tapping crucibles, well below those found in reduction cells. The mass transfer relationship has been refined with metal calibration temperatures in the range 950 to 965°C.

Experimental Procedure

Metal calibration experiments were conducted in a gas fired Hausfeld furnace with temperature controller. The furnace has a suspended clay-graphite crucible of the dimensions, 16 inches ID by 26 inches deep. The furnace contained about 200 pounds of aluminum metal at ± 3°C during calibration experiments.

The stirring apparatus shown in Figure 1 consisted of a rotating arm or spindle with two iron rods attached at different radii. The spindle rotated 3/8-inch diameter rods at a specified diameter by a 1/4-hp motor with a variable speed controller. The rotation speed was checked during each calibration run with a digital contact tachometer. To ensure that the metal pool in the crucible was not moving and affecting the calibration results, metal rods were often suspended stationary in the metal pool during the stirring experiments. Results from these stationary rods indicated the metal pool was not being stirred, however more recent data, Figure 13, where rods were rotated individually at a larger radius do indicate stirring of the metal pool was occurring at velocities above 14 cm/sec, Figure 2.

After each set of calibration runs, the iron rods were cleaned of adhering aluminum metal by immersion in a caustic solution for over 24 hours. A uniform section of each rod below the metal immersion line was cut and weighed to determine the weight loss.

The chemistry of the iron rods used for the calibration experiments was determined by spectrographic analysis by an independent metallurgical laboratory.

Equation of Mass Transfer

The correlative equation of mass transfer for 3/8-inch diameter vertical iron rods placed in a stream of molten aluminum established by Johnson has the general form (1):

$$Nu = A + B\ Re^{n}\ Sc^{m} \tag{1}$$

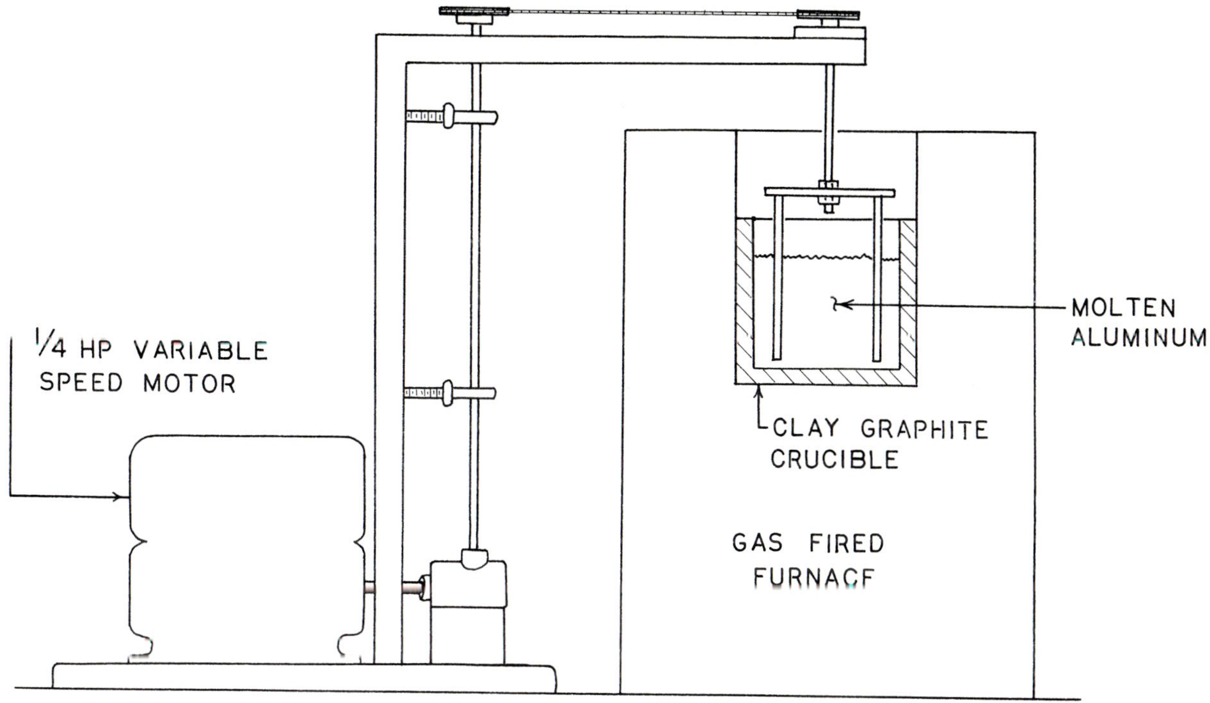

FIGURE I. CALIBRATION APPARATUS FOR METAL VELOCITIES.

where:

$$Nu = \text{Nusselt number for mass transfer} = \frac{k\bar{d}}{D}$$

$$Re = \text{Reynolds number} = \frac{\rho U \bar{d}}{\mu}$$

$$Sc = \text{Schmidt number} = \frac{\mu}{\rho D}$$

$$A^{*}, B, n, m = \text{Constants}$$

and,

k = mass transfer coefficient (cm/sec)

$\bar{d}$ = average rod diameter (cm)
D = diffusivity of iron in aluminum (cm^2/sec)
ρ = density of molten aluminum (g/cm^3)
μ = kinematic viscosity of molten aluminum (g/cm-sec)
U = metal velocity (cm/sec)

Assuming that the dissolution of solid metal is controlled by mass transfer in molten metal, the mass transfer coefficient, k, is defined by the expression:

$$k = \frac{D_o - D_f}{2t \ln \left[1 + \frac{x_o - x_b}{1 - x_o}\right]} \qquad (2)$$

where: D_o = initial rod diameter (cm).
D_f = final rod diameter (cm).
t = time of immersion in metal pad (sec).
x_o = equilibrium mole fraction of iron in molten aluminum at rod surface.
x_b = mole fraction of iron in bulk metal.

The final rod diameter, D_f, is calculated from the final rod weight, W_f:

$$D_f = 2 \sqrt{\frac{W_f}{\rho L}} \qquad (3)$$

where: W_f = final rod weight (gm).
ρ = density of iron rod (gm/cm^3).
L = Length of rod immersion in metal (cm).

The constants B, n, and m in equation (1) were determined by a computer linear regression analysis (SAS General Linear Model Procedure) from calibration experiments. Zero velocity metal flow data were used to determine the constant A. The mass transfer correlation with redefined constants for calibration data measured within the temperature range of 950-965°C is:

$$Nu = 3.57 + 2.8 \times 10^{-6} Re^{1.06} Sc^{2.87} \quad (4)$$

The results of the iron rod measurements used to define the velocity, U, as a function of the mass transfer coefficient, k, are shown in Figures 2 and 13. These are several interesting trends apparent from these results:

- There is a linear relationship between metal velocity and the mass transfer coefficient.
- The slope of the velocity - mass transfer relationship is about 2.6 times higher than reported by Johnson.
- The apparent change in slope above 14 cm/sec in Figure 2 is attributed to stirring of the metal pool. More recent data, Figure 13, where rods were rotated individually at larger radius to minimize stirring, indicate the slope is uniform over the entire range of measurable velocities.
- At velocities above 24 cm/sec, the iron rod immersed in the metal was completely dissolved within the 10-minute test period.

The drag force (coefficient) on the cylinder, pressure resulting from a low pressure region on the rear of the cylinder, is of the same magnitude at Reynolds' numbers in the range of 10^3 to 10^5. At Reynolds' numbers greater than 10^5, the boundary-layer flow becomes completely turbulent with a change in the drag coefficient curve (2). However, the calibration velocities have Reynolds' numbers shown in Table I which are well below that point.

Table I. Metal Velocity-Reynolds' Number Relationship

Velocity cm/sec	Reynolds' Number	Velocity cm/sec	Reynolds' Number
4.0	1041	12.[illegible]	2952
5.0	1273	14.[illegible]	3169
5.9	1477	14.[illegible]	3056
6.9	1704	16.[illegible]	3613
8.0	1957	17.[illegible]	3608
9.0	2115	18.[illegible]	4173
9.7	2204	20.[illegible]	4371
12.0	2771		

Basic Relationships

The effects of several experimental parameters were investigated to determine their effect on the measured metal flow velocity-mass transfer relationship.

Iron Alloy Rods

Increased levels of iron alloy in the rods were studied. The chemical compositions of the various iron rods used in the calibration experiments at 950°C are shown in Table II.

Table II. Iron Rod Chemical Composition

	Pure*	1008	1010	1018	4150
Carbon, %	0.02	0.06	0.11	0.24	0.48
Silicon, %	0.024	0.12	0.06	0.06	0.23
Manganese, %	0.17	0.32	0.39	0.66	0.89
Phosphorus, %	0.002	0.004	0.009	0.003	0.007
Sulphur, %	0.013	0.017	0.026	0.017	0.075
Nickel, %	0.08	0.07	0.07	0.04	0.20
Chromium, %	0.04	0.02	0.04	0.04	0.96
Copper, %	0.06	0.14	0.12	0.05	0.15
Molybdenum, %	NA	0.02	0.02	0.02	0.20

*Commercially pure unalloyed iron.

The results of the calibration tests with various iron alloys are shown in Table III. The presence of any alloy content above that present in the commercially pure iron has a dramatic impact on the dissolution of the iron rod in molten aluminum. Iron rods with increased levels of carbon, silicon, and manganese produced poor results. The dissolution of the alloyed rods was not uniform and exhibited excessive pitting. Only commercially pure unalloyed iron rods with low carbon content, less than 0.02%, can provide the reproducibility necessary for metal flow experiments.

The dissolution of the iron alloy rods in molten aluminum is controlled by the chemical reaction process at iron/alloy layer interface or diffusion process of iron in an alloy layer (3).

Table III. Weight Loss Per Unit Length Iron Alloy Rods

Iron Rod	Velocity cm/sec	Weight Loss, g/cm	Velocity cm/sec	Weight Loss, g/cm	Velocity cm/sec	Weight Loss, g/cm
Pure*	5.1	2.60	15.95	3.86	23.94	5.39
1008	5.1	0.08	15.95	1.38	23.94	0.70
1010	5.1	NA	15.95	2.30	23.94	2.43
1018	5.1	1.39	15.95	2.44	23.94	NA
1020	5.1	1.28	15.95	NA	23.94	2.25
4150	5.1	0	15.95	NA	23.94	0

* Commercially pure unalloyed iron.

Rod Diameter Size

Iron rods of diameter sizes 3/8, 1/2, and 5/8 inches were investigated in calibration experiments to determine the effect of rod diameter size on the mass transfer correlation. The results for 1/2-inch diameter rods are shown in Figure 3. Similarly to the 3/8-inch rods, the 1/2-inch rods also exhibit a linear relationship between metal velocity and mass transfer coefficient and can be calibrated for plant measurements in operating reduction cells. Test results with 5/8-inch rods indicate that they can also be calibrated for metal velocity measurements.

Iron Content

The iron content in the 200 pounds aluminum metal in the calibration furnace increased after each run. The effect of the higher iron levels in the metal upon the mass transfer correlation is shown in Figure 4. There is a 0.02-0.04 x 10^{-3} cm/sec decrease in the mass transfer coefficient per 0.1 w/o increase in the iron content. The iron content in the metal pad of most aluminum reduction cells is in the range of 0.05 to 0.15 w/o.

Rod Immersion Time

An important parameter in both laboratory and plant tests is the rod immersion time in the metal flow. Figure 5 shows the increase in the mass transfer coefficient with respect to the rod immersion time. The mass transfer coefficient increases at a rate of 12 to 19 x 10^{-3} cm/sec per minute immersion time. It is necessary to use the same immersion time in plant test as established during the calibration runs.

Metal Temperature

The results of varying the metal temperature on the mass transfer correlation are shown in Figure 6. The mass transfer coefficient was found to increase at a linear rate of 89 to 127 x 10^{-3} cm/sec per 100 degree centigrade increase in metal temperature during calibration tests. It is evident that the metal temperature has a significant impact on the metal

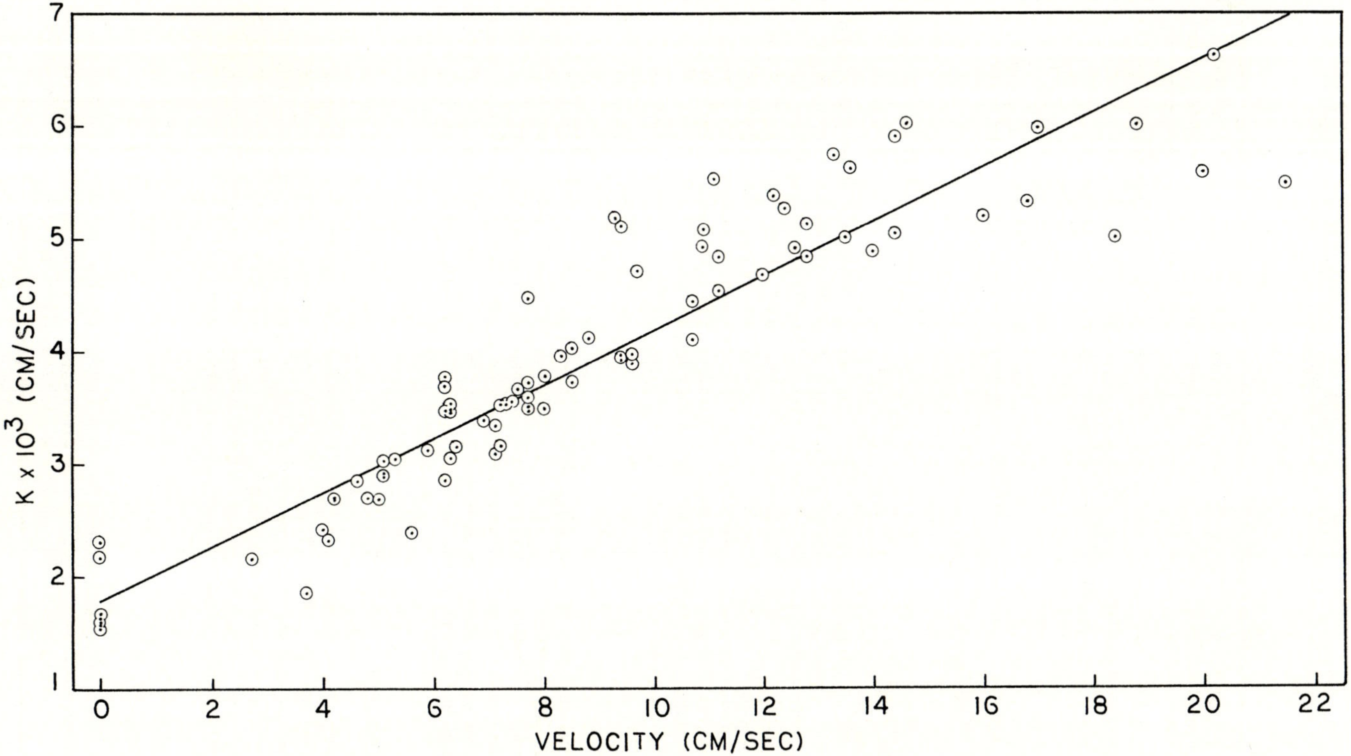

FIGURE 2. EXPERIMENTAL DATA FOR CORRELATION BETWEEN MASS TRANSFER COEFFICIENT AND VELOCITY FOR 3/8 IN. DIA. RODS.

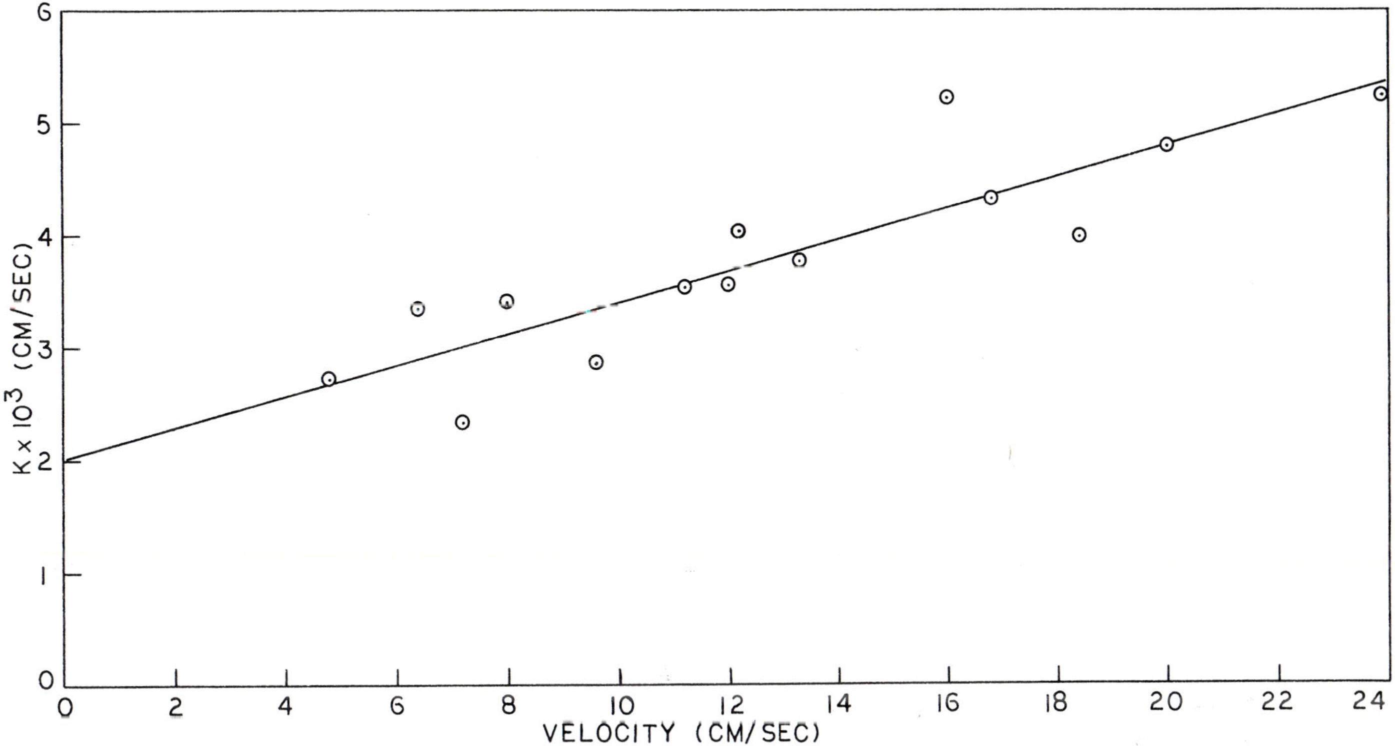

FIGURE 3. EXPERIMENTAL DATA FOR CORRELATION BETWEEN MASS TRANSFER COEFFICIENT AND VELOCITY FOR 1/2 IN. DIA. RODS.

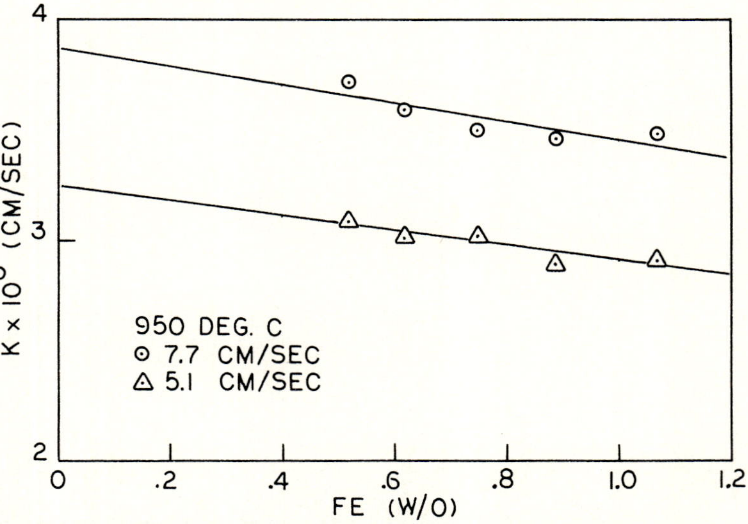

FIG. 4. MASS TRANSFER COEFFICIENT VS. IRON CONTENT IN THE ALUMINUM.

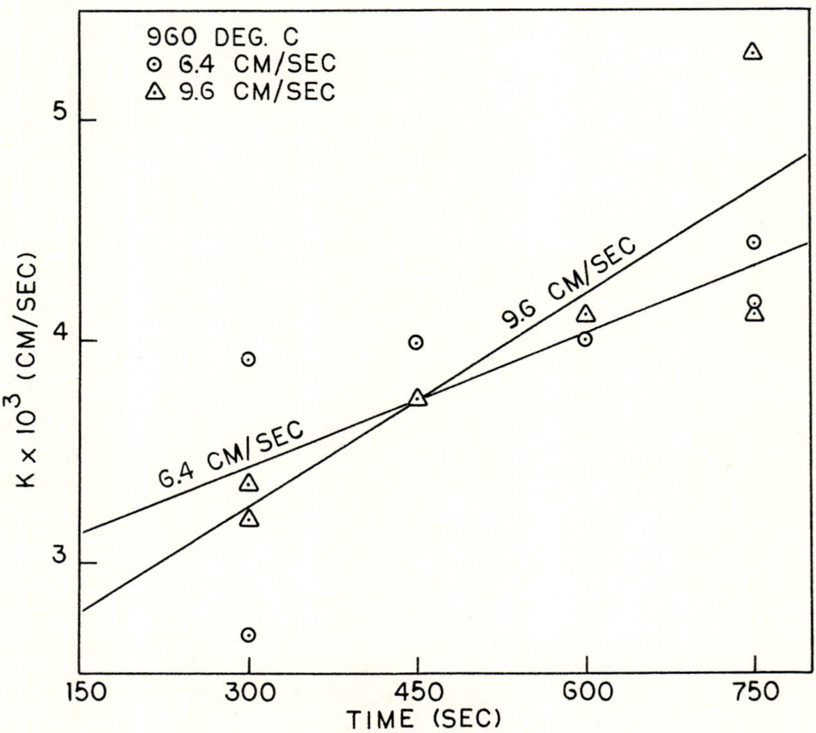

FIG. 5. MASS TRANSFER COEFFICIENT VS. IMMERSION TIME.

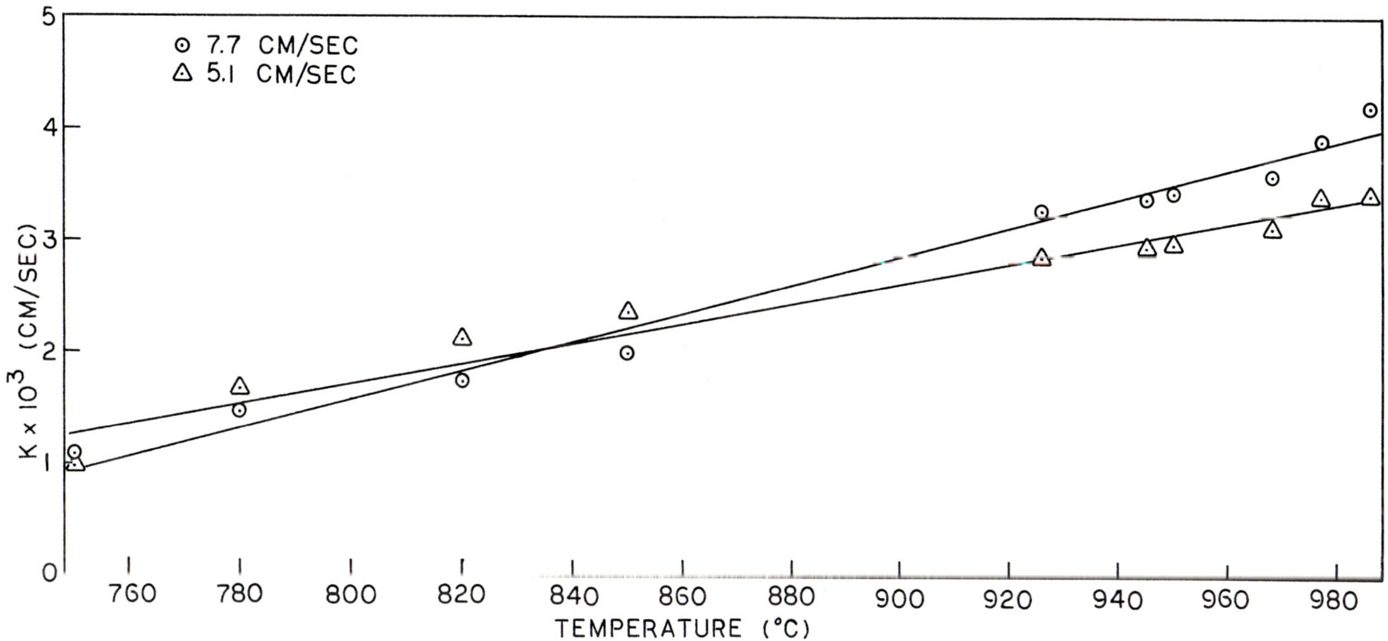

FIGURE 6. MASS TRANSFER COEFFICIENT VS. TEMPERATURE.

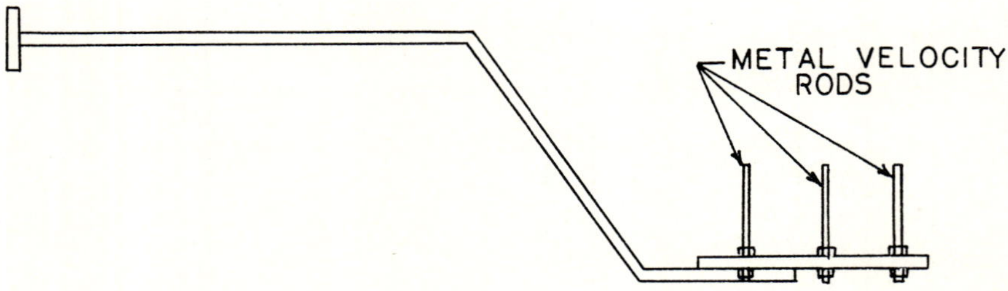

FIGURE 7. SPECIAL METAL ROD HOLDER.

dissolution rate. Metal velocity calibration runs should be conducted at the same temperature range as measured in the reduction cell metal pad.

Reduction Cell Measurements

A number of metal velocity measurements have been made in both Soderberg and prebake reduction cells. The rods were inserted consecutively through small holes in the crust and left in the metal for 10 minutes. The rods were removed in the reverse order. Prebake cell measurements were generally made at locations between the anodes and the cell sidewall and at the ends of the cell. At times measurements were also made through holes drilled in the anodes.

Metal velocity measurements in Soderberg cells were found to be more difficult due to the frozen side ledges and muck near the periphery of the anode. A special rod holder made to measure metal pad velocities simultaneous at three locations under the Soderberg anode is shown in Figure 7. Three rods were inserted into holes in a plate welded to the rod holder and attached with bolts. The rod holder with three rods attached is maneuvered into position under the anode with the plate resting on the cathode cavity floor and the rods pointing upward in the metal pad. Upon removal from the cell metal pad an orientation mark is made on the three rods and they are cut off near the plate with a bolt cutter. It worked especially well in measurements in VS Soderberg cells.

VS Soderberg Cell

Results of metal pad velocity measurements in a 110 kA VS Soderberg cell (end-to-end) are shown in Figure 8. The special Soderberg rod holder was used to obtain measurements at six locations around the anode and three points at each location. The following trends in metal pad circulation pattern are evident:

- One large clockwise metal flow pool.
- Metal velocities were very uniform near the center of the cell.
- The highest and lowest metal velocities occurred near the corners of the cell.

HS Soderberg Cell

Metal pad velocities measured at various locations in a 100 kA HS Soderberg cell (end-to-end) are shown in Figure 9. The following trends in the metal pad circulation pattern are evident:

- Confirmed by metal velocity measurements in other cells, there is one major counterclockwise metal flow pool.
- Anomilies or differing metal flows at some locations were thought to be caused by an excessive accumulation of muck in the cathode cavity of the cell.

End Riser Prebake Cells

150 kA Prebake Cell. Results of metal velocity measurements in a 150 kA prebake cell with end anode riser bus are shown in Figure 10. The following trends in metal pad circulation patterns are evident:

- Two major metal circulation pools. The metal flows from the center of the cell toward the ends of the cell on the downstream side of the cell. The metal circulates around the ends of the cell toward sidewall on the upstream side of the cell.
- The highest metal pad velocities are generally found on the upstream side of the cell as the metal approaches the sidewall after passing around the ends of the cells. The lowest metal pad velocities are generally found in the center of the cell as the two metal pools converge.

76 kA Prebake Cell. Results of metal velocity measurements in a 76 kA prebake cell with end anode riser bus are shown in Figure 11. The following trends in metal pad circulation patterns are evident:

- Four major metal circulation pools. In the center work area of the cell, the metal flows to both ends of the cell. At the ends of the cell, the metal circulates around the ends toward the center of the cell on both the upstream and downstream sides of the cell.

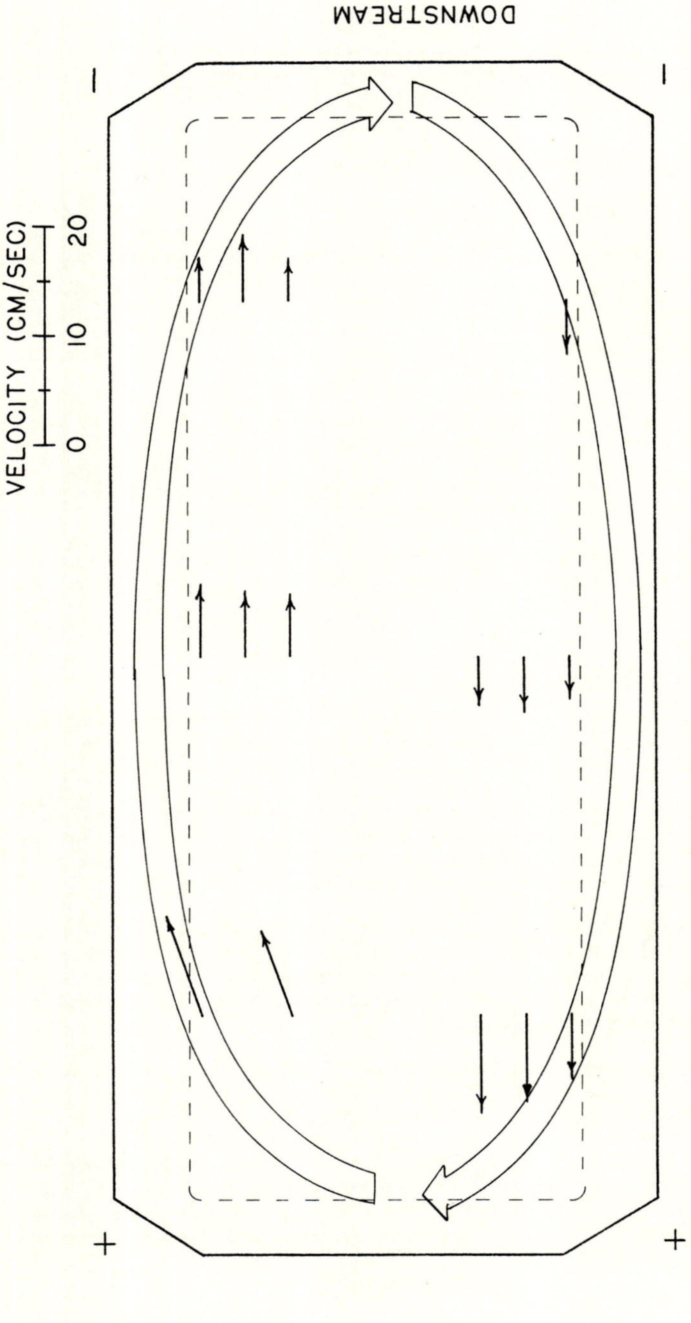

FIGURE 8. METAL FLOW VELOCITIES IN 110 KA V.S. SODERBERG CELL.

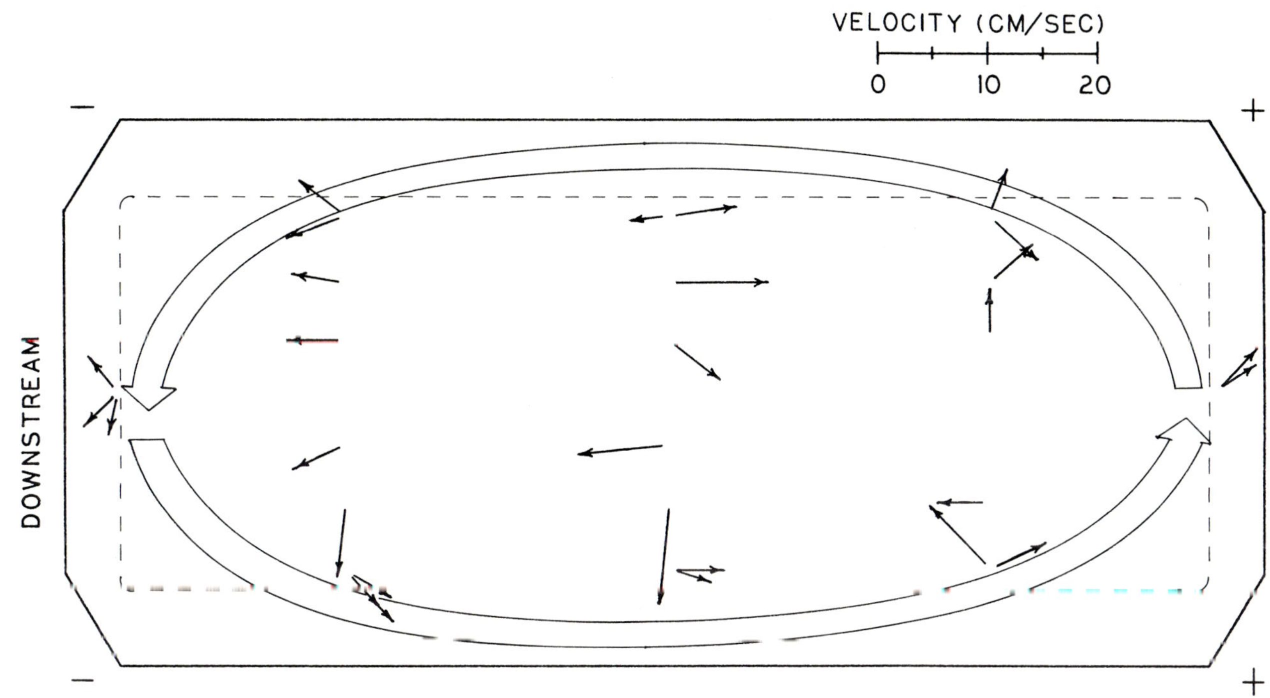

FIGURE 9. METAL FLOW VELOCITIES IN 100 KA H.S. SODERBERG CELL.

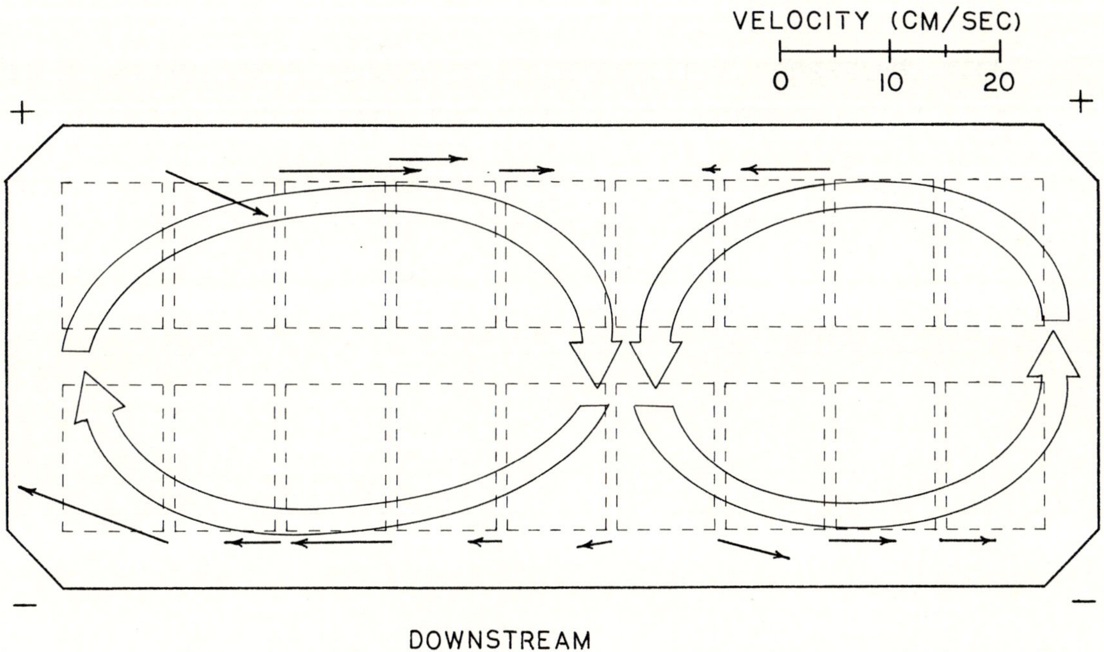

FIGURE 10. METAL FLOW VELOCITIES IN 150 KA END RISER PREBAKE CELL.

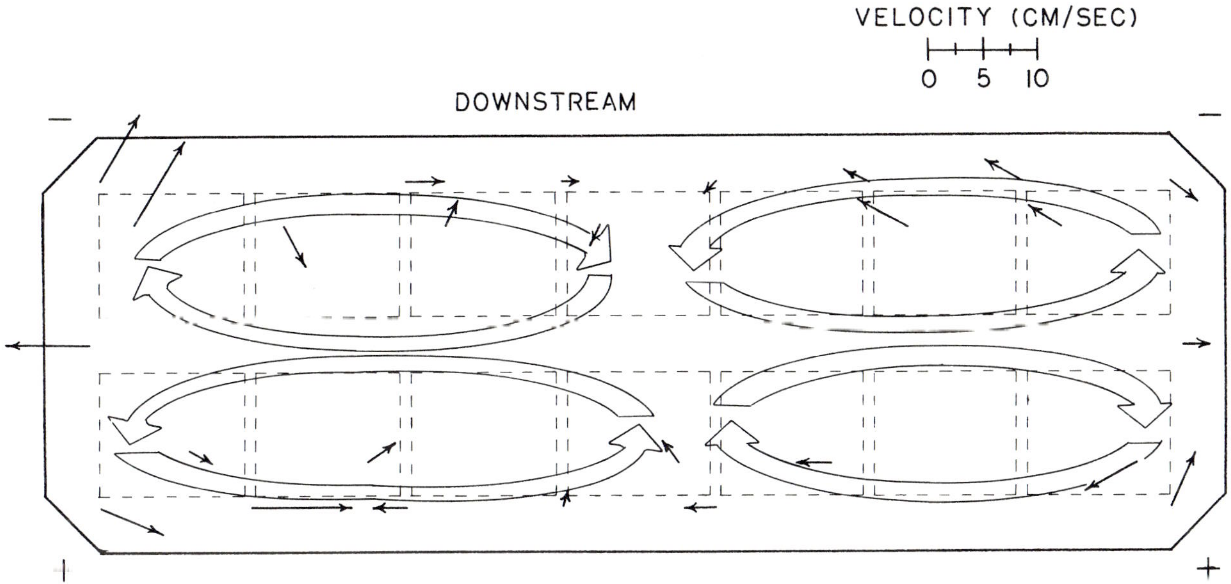

FIGURE II. METAL FLOW VELOCITIES IN 76 KA PREBAKE CELL.

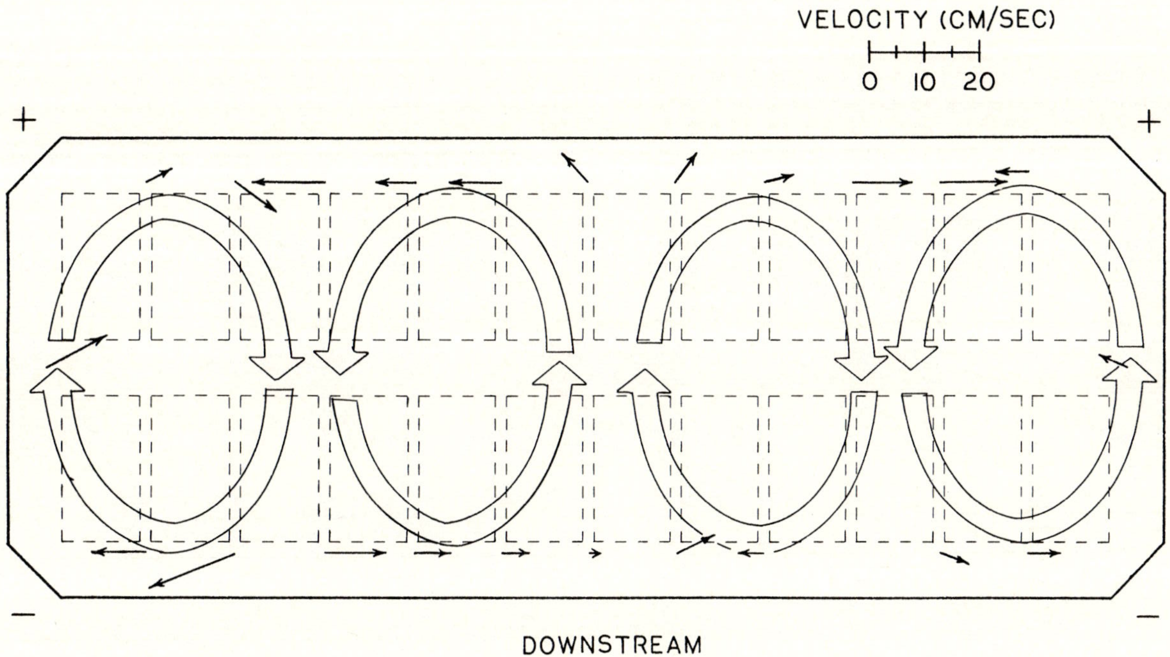

FIGURE 12. METAL FLOW VELOCITIES IN 150 KA QUARTER RISER CELL.

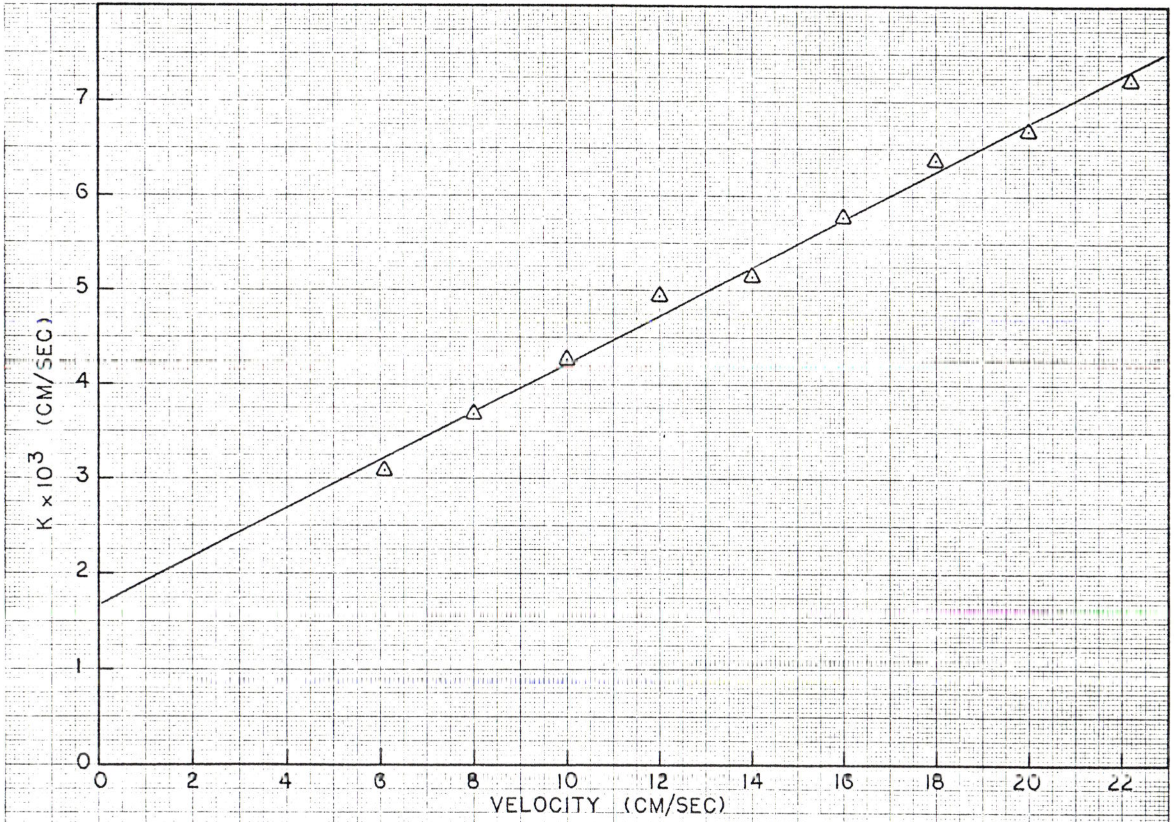

FIGURE 13. **EXPERIMENTAL DATA FOR CORRELATION** BETWEEN **MASS TRANSFER** COEFFICIENT AND **VELOCITY OF** $^3/_8$ **IN. DIA. RODS ROTATED AT A** 5 **IN.** RADIUS.

- The highest metal pad velocities are generally found near the ends of the cell as the metal approaches the sidewall after passing the ends of the cell. The lowest metal velocities are found near the center of the cell as the metal pools converge.

Quarter Riser Prebake Cells

150 kA Prebake Cell. Metal velocity measurements in a 150 kA prebake cell with quarter anode riser bus are shown in Figure 12. The following trends in metal pad circulation patterns are evident:

- Four major metal circulation pools. In the center of the cell, the metal flows across the cell from downstream to upstream side of the cell. On the upstream side of the cell, the metal flows from the center toward the quarter riser where the metal flows across the cell from upstream to downstream sides of the cell. The metal flows around the ends downstream to upstream sides of the cell.

- The highest metal velocities are generally found along the upstream side of the cell and lowest velocities are found on the downstream side of the cell.

Repetitive measurements of metal velocities and flow patterns in each cell have shown that although some variations in metal flow directions and/or velocities occur, the overall average metal flow patterns and velocity trends continue to persist.

From these metal velocity measurements in different reduction cell designs several definite trends are apparent:

- Although there are variations in the direction of metal flow, the overall metal flow circulation pools and/or patterns for each cell are fairly constant.

- Each design of reduction cell had a different and distinctive metal circulation pattern.

- Variation in metal flow directions and/or velocities at some locations can be influenced by cell conditions that can be variable: excessive muck, ledges, or anode changes.

- The highest metal velocities are found near the sidewalls at the ends of the cells where the highest sidewall erosion is generally found.

- The lowest metal pad velocities are found near the center of the cells where the metal pools generally converge.

- Increased metal pad stability may be indicated by a lower average metal velocity and a lower standard deviation of metal velocities.

Conclusions

• It has been demonstrated that the empirically derived mass transfer expression established by Johnson for iron rods in aluminum metal flow is valid for the reduction cell metal pad temperature range 950-965°C, but required that the mass transfer constants be redefined. Other findings from the study included that:

- Only commercially pure iron rods can provide reproducible results necessary for metal velocity measurements.

- Rod with larger diameters, 1/2 and 5/8 inches, as well as 3/8-inch rod can be used for metal velocity measurements.

- Relationships between mass transfer coefficient, iron content, immersion time, and metal temperature were determined.

- Metal velocity measurements in reduction cell metal pads provide important information relating metal circulation patterns and cell bus design.

Acknowledgements

The authors are grateful to Charles Bunt and Michael Grisham for their contribution in conducting the experimental work, and to Beth Barnett for her invaluable assistance with the SAS General Linear Model Procedure.

References

1. A. R. Johnson, "Metal Pad Velocity Measurements In Aluminum Reduction Cells," Light Metals, Vol. 1, 1978, p. 45.

2. J. P. Holman, Heat Transfer, McGraw-Hill, New York, pp. 165-175, 1968.

3. M. Niinomi, Y. Ueda, and M. Sano, "Dissolution of Ferrous Alloys Into Molten Aluminum," Transactions of the Japan Institute of Metals, Vol. 23, No. 12(1982) pp. 780-787.

METAL PAD VELOCITY MEASUREMENTS BY THE IRON ROD METHOD

Brian F. Bradley, Ernest W. Dewing, and John N. Rogers

Alcan International Ltd.
Chalfont Park
Gerrards Cross
Buckinghamshire
England SL9 0QB

Alcan International Ltd.
P.O. Box 8400
Kingston, Ontario
Canada K7M 5T8

Present Address:
Autosense Equipment Ltd.
United Technologies Corp.
Station Approach
Bicester, Oxfordshire
England OX6 7BZ

The rate of dissolution of iron rods in aluminum was determined as a function of velocity at 955°C. The results appeared at first to be in disagreement with earlier work by Johnson (1) at lower temperatures, but it seems that it is the temperature coefficient which is at fault. All the results available have been fitted to

$$v(\text{cm s}^{-1}) = 8.38 \times 10^{-4}\ \delta D \exp(13410/T) - 3.8$$

where δD is the decrease in diameter (cm) of an initially 9.5 mm (3/8 in.) rod in 7 min immersion at temperature T (K). Calculated values of $v > 25$ cm s^{-1} are unreliable.

Measurements in cells at the Invergordon Smelter of what was then the British Aluminium Co. showed that speeds in the upstream half of the cells were generally higher (1.4:1) than in the downstream half, with the highest speeds in the upstream corners. These are the points at which those cells tended to tap out. The flow in the upstream halves generally reversed direction as the cells aged, in association with the growth of a freeze toe on the cathode and a change in the horizontal currents in the metal pad.

Introduction

In 1978 Johnson (1) published a method for determining the rate and direction of metal flow in a reduction cell. It involved putting an iron rod vertically into the cell for a given period of time and observing the dissolution. He reported his calibration results (obtained by moving the rods in a pool of metal) in terms of a mass-transfer coefficient, k, which he defined by

$$k = \delta D/(2t\ \ln[1 + (x_o - x_B)/(1 - x_o)]) \qquad (1)$$

where δD is the reduction of diameter of the rod in time t, x_o is the equilibrium (saturated) mole fraction of Fe in Al at the temperature in question, and x_B is the mole fraction of iron in the bulk of the aluminum in which dissolution is taking place. Johnson also fitted his results to the dimensionless equation

$$Nu = A + B\ Re^n\ Sc^m \qquad (2)$$

where

Nu = Nusselt number for mass transfer = $\frac{k\bar{d}}{D}$

Re = Reynolds number = $\rho\ \frac{U\bar{d}}{\mu}$

Sc = Schmidt number = $\frac{\mu}{\rho D}$

A, B, n, m = constants

k = mass transfer coefficient (cm s^{-1})

$\bar{d}$ = average rod diameter (cm)

D = diffusivity of iron in aluminum (cm^2 s^{-1})

U = metal velocity (cm s^{-1})

ρ = density of molten aluminum (g cm^{-3})

μ = kinematic viscosity of molten aluminum (g cm^{-1} s^{-1})

The values he found for the constants were A = 2.0, B = 0.00853, n = 0.72, and m = 0.36. Although it is not immediately obvious, the resulting equation gives a calibration for all temperatures, since the viscosity, diffusivity, iron solubility, and density are functions only of the temperature. Indeed, none of these quantities is an independent variable; all change at once. Thus the significance of the analysis in terms of dimensionless numbers is not clear, and it only serves to reproduce the experimental temperature coefficient. It should be noted that the equation applies in any case only to rods of 3/8 in. diameter; Johnson remarks that for 5/8 in. rods it does not work, indicating thereby some limitation in the theoretical basis.

We have made new calibration measurements, at a higher temperature than Johnson, have fitted all the results to one equation, and have made measurements in operating cells.

Calibration

Iron rods were obtained from SKF Steel; Table I shows the analysis. Diameter was 3/8 in. (9.52 mm). Calibration was done by mounting two rods in a rotating rig (Figure 1) and immersing them in molten aluminum in a 0.5 ton furnace. It is assumed that the metal remained stationary and was not stirred around by the rods. The background concentration of iron built up (< 0.5%) was low enough not to affect the rate of dissolution

Table I. Analysis of Iron Rods

Element	wt%	Element	wt%
C	0.03	Cr	0.05
Si	0.05	Ni	0.08
Mn	0.18	Cu	0.05
P	0.01	Sn	0.01
S	0.02		

Standard time of immersion was 7 min, and the temperature was 955 ± 10°C. After removal, the rods were cleaned in strong caustic solution for 24 h. Measurements were made as indicated in Figure 2, and Figure 3 shows a plot of reduction in diameter against velocity. Some results from Johnson (1) are shown for comparison.

Figure 1 - Rotating rig for calibrating rod dissolution rate.

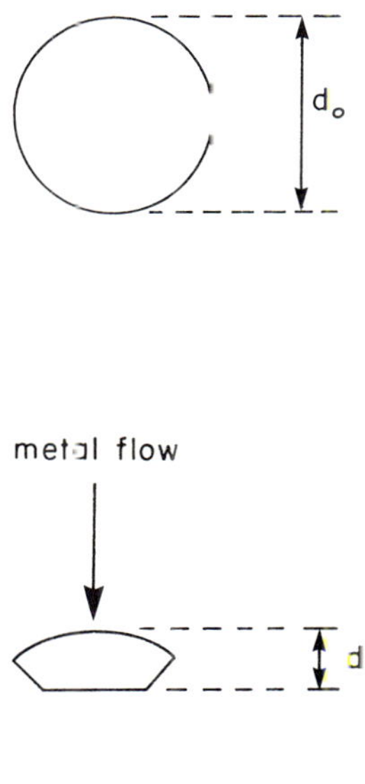

Figure 2 - Cross section of rod before and after immersion (schematic).

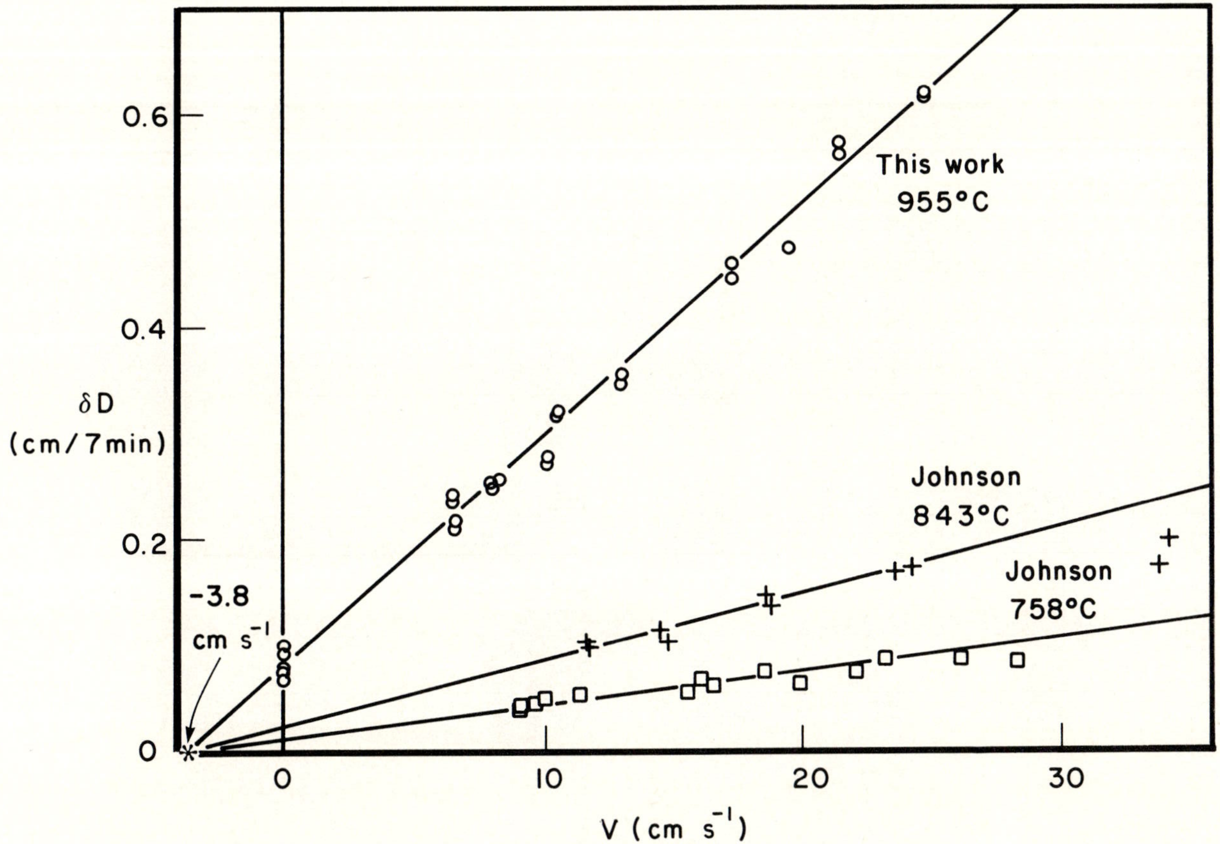

Figure 3 - Reduction of diameter in 7 min vs flowrate.

It is surprising that a simple linear relationship should emerge between the diameter of a rod after dissolution and the speed at which it was rotated through the metal, at least up to 25 cm s^{-1}. The mechanism for iron dissolution in aluminum is complicated, since several Al-Fe phases are formed at the interface and the removal of the dissolved iron from the interface depends on both diffusion and bulk mass transfer rates. In addition, the area and shape of the rod presented to the aluminum changes during the dissolution period. In view of the complexity involved, no theoretical interpretation of the rate of dissolution has been attempted. However, it should be noted that the viscosity of aluminum, and the solubility and diffusivity of iron in it, all follow relations of the type

$$\log x = C/T + D \tag{3}$$

where T is the temperature in kelvin. One might thus anticipate that, for a given velocity, the rate of dissolution would follow a similar law.

The least-squares straight lines representing the results at the various temperatures all have negative intercepts on the velocity axis (i.e. dissolution is still observed at zero velocity). Statistically the intercepts do not differ significantly one from another, and the best value is -3.8 cm s^{-1}. All the lines have been passed through that point, and the slopes are plotted in Figure 4. Equation (3) is seen to work well; the line has been drawn through the point at 955°C, partly because there are more measurements at that temperature than any other, and partly because it is closest to the temperature of prime interest. The results can now be given as

$$v(\text{cm s}^{-1}) = 8.38 \times 10^{-4}\ \delta D \exp(13410/T) - 3.8 \tag{4}$$

where, it must be remembered, T is temperature in kelvin, and δD is the dissolution observed in 7 minutes immersion. If a different immersion time is used, then δD must be scaled pro rata. This equation only works up to v ≃ 25 cm s^{-1}; beyond that Figure 3 suggests that the rate of dissolution becomes almost independent of velocity.

Measurements on Cells

Figure 5 shows a schematic layout of the 130 kA Invergordon cell. It was found that the flow could be characterized by measurements at eight or ten standard positions, as indicated. Because of fume emission regulations, only one side hood could be removed from a cell at one time. It took about an hour to measure the ten points, and the first measurement was generally repeated to check that the flow was unchanged. The check results usually agreed to within 10% in the speed and 10° in the direction.

To insert the rods, holes about 5 cm diameter were broken in the crust. The rods were dropped in rapidly to prevent bath freezing on them on the way down. Each rod had a flat filed on it at the top, and this flat was oriented to be parallel to the side of the cell.

Measurements were made in 20 different cells; in two of them 7 sets of measurements were made over a period of 2 years. Table II gives a summary of the results.

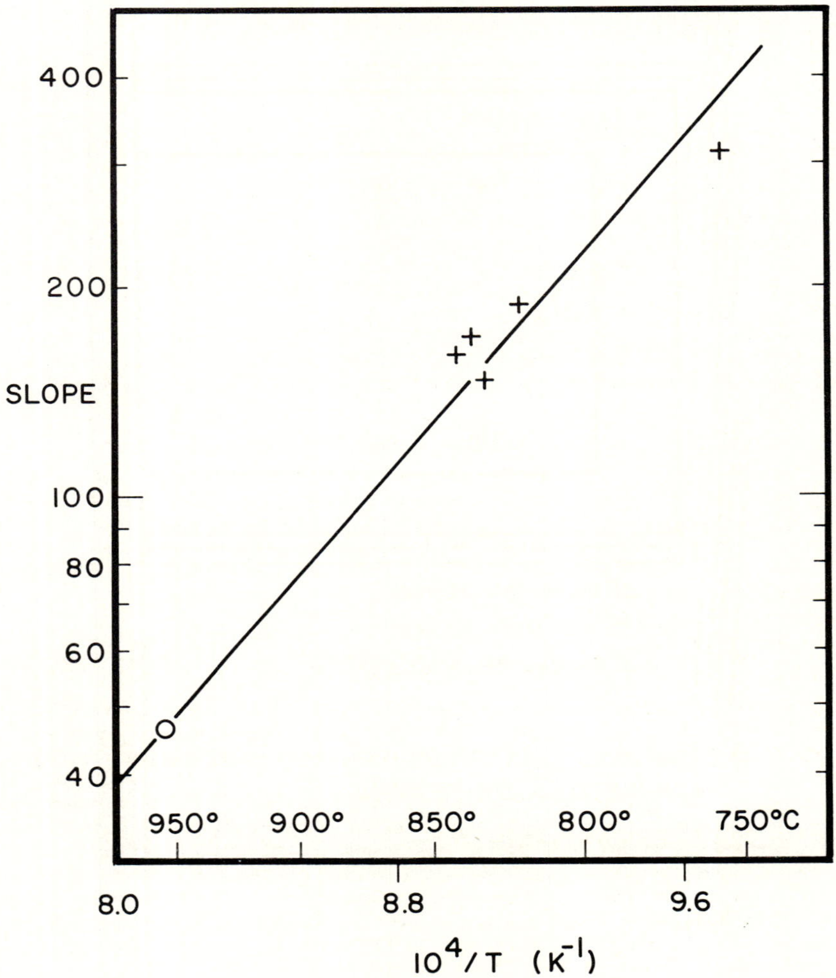

Figure 4 - Slope of lines in Figure 3 as a funcion of temperature.

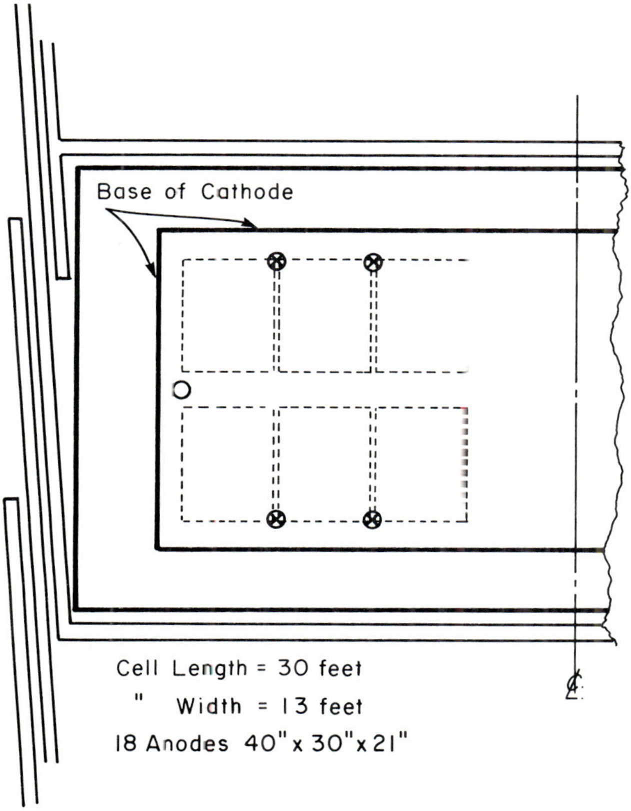

⊗ = STANDARD POINTS FOR IRON ROD IMMERSION FOR METAL FLOW READINGS

O = SUBSIDARY POINTS.

Figure 5 - Schematic layout of Invergordon cell.

Table II. Summary of Velocity Measurements

Age of Cell (days)	Number of Cells	Average Speed (cm/s)	Standard Deviation (cm/s)	Ratio Velocities Upstream/ Downstream	Standard Deviation	Maximum Speed (cm/s)	% Readings over 20 cm/s
<150	10	12.0	2.4	1.6	0.46	27.2	15.8
150-400	10	11.6	2.0	1.4	0.36	22.9	6.2
400-900	19	10.6	1.9	1.3	0.39	25.9	2.6
>900	15	11.4	1.9	1.3	0.55	27.4	4.4
All	54	11.3	2.0	1.4	0.45		

Young cells all had similar flow patterns, but there were wide variations in the flow in older cells. Because of these variations the cells were divided into four arbitrary age groups, and the flows for all cells in a group were combined. The averaged flow patterns are shown in Figures 6-7. There are two important points to be emphasized:

(a) Only for the new cells (< 150 days) can the flow be characterized with any certainty. In all the other age groups there is a wide variation in pattern.

(b) This variation is largest for the cells in the 400-900 days group. As a result, the vectorically average flow gives the impression that the velocities are generally low. This is not the case; in fact the average speed is almost independent of age.

The new cells have a vortex at each end. The metal speeds are higher on the upstream side, and are largest at the corners. A central ridge is generally present on the cathode under the crustbreaker, and this may be high enough to partially divide the metal. There is then a vortex in both upstream corners, the vortices at opposite ends having opposite direction of rotation. The pattern in the downstream half of the cell is less clear, but there is basically again a vortex in each corner. The direction of these downstream vortices varied from cell to cell.

The group of young cells (150-400 days) showed a pattern similar to, but less distinct than, that seen for the first set. The highest speeds were still in the vortices at the upstream corners. No general pattern could be deduced for the flow in the downstream half of the cell. The upstream speeds were still larger than those downstream, but the difference was less marked than in the new cells.

The middle-aged cells (400-900 days) showed greater variations than the two previous groups, with each cell having its own pattern. On averaging the velocities, therefore, no general pattern emerged.

The group of old cells (> 900 days) again showed wide variations, but a clear trend could be seen when the velocities were averaged. Again there was a vortex in each of the upstream corners, but the directions were opposite from those in the new cells. On the downstream side there appeared to be a vortex in one corner, but no pattern could be discerned in

METAL FLOW IN CELLS < 150 DAYS OLD

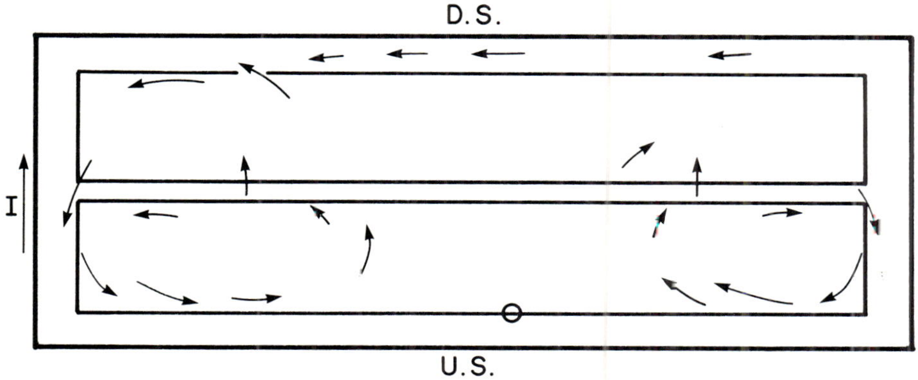

O = NO CONSISTENT FLOW PATTERN - VELOCITIES LOW

METAL FLOW IN CELLS 150-400 DAYS OLD

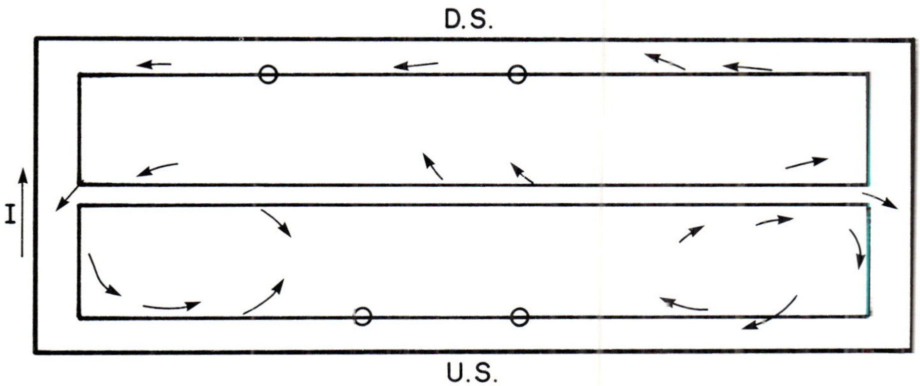

Figure 6 - Flow patterns.

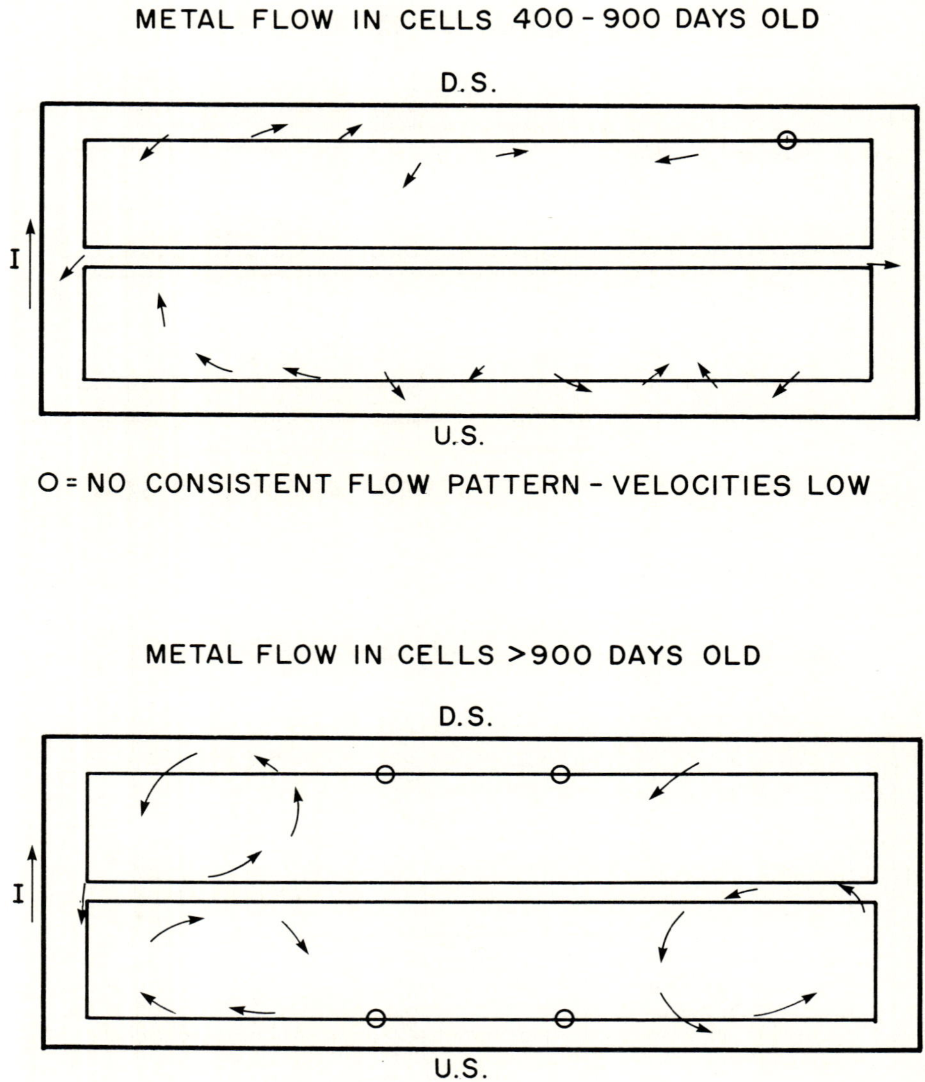

Figure 7 - Flow patterns.

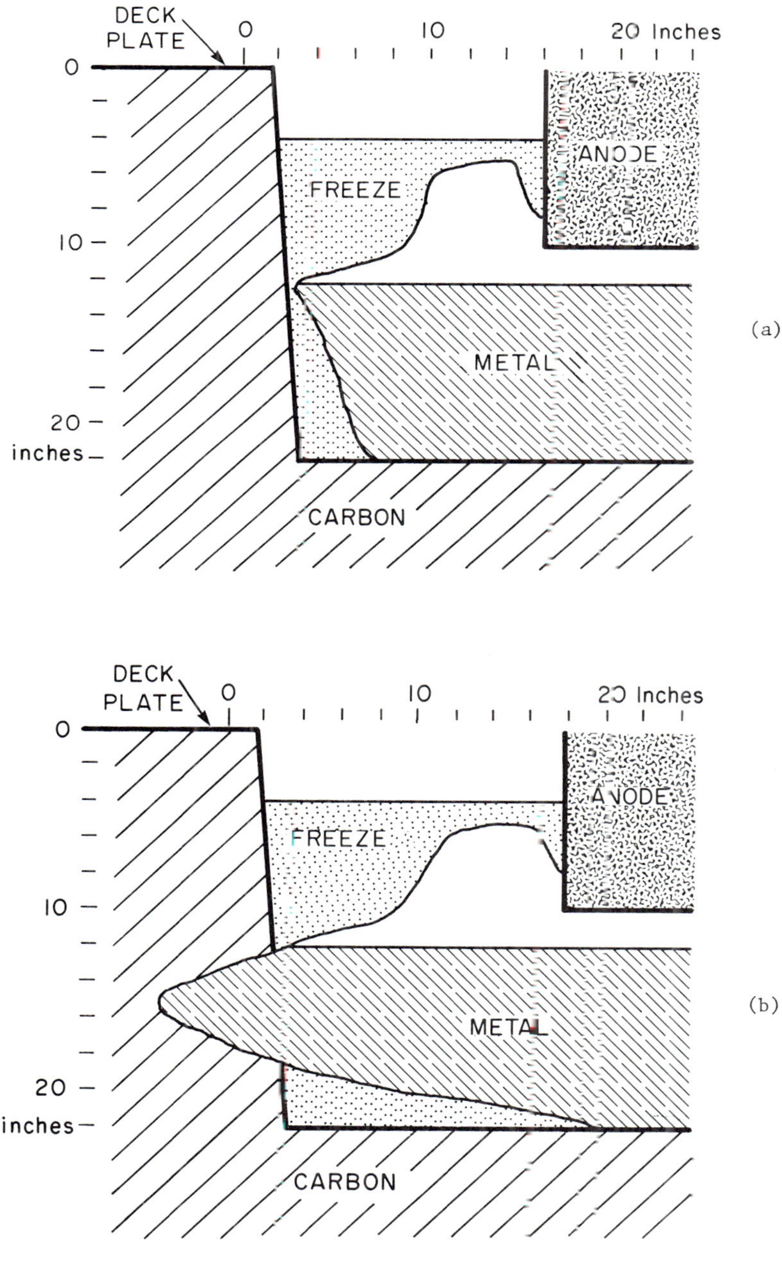

Figure 8 - Freeze profiles (a) <150 days (b) >900 days.

the other. There was again an appreciable difference between the upstream and downstream speeds.

Overall the average metal speed along the anode edge was 11 cm s^{-1}, and the ratio of upstream to downstream velocity was 1.4.

The average trends with age given above were also seen in the two cells followed over 2 years. The first showed the flow characteristic of a new cell when first measured at 62 days, changed to that of a young cell (as seen at 277 days) and to a less well defined pattern (492, 612 and 800 days). Similarly, the other cell initially showed the uncharacterized pattern of a middle-aged cell (527, 735 and 807 days), but as it exceeded 900 days the pattern became established at that expected in an old cell (950, 1031, 1101 and 1283 days).

Discussion

Quite apart from a 'neck' in the iron rod at the level of the metal-bath interface, the velocities derived from measurements of the diameter at the top and the bottom of the metal pad often showed large differences.

The flows measured a week apart in the same cell showed little basic change in pattern, despite larger differences in the anode current distribution. This is not surprising, as localized changes in the forces might produce little change in the overall pattern, which is determined by the forces over a large area.

The positions of greatest change in metal momentum (and hence presumably in sidewall erosion) were at the upstream corners, and in particular at the upstream sides of the endwalls. An examination of the records shows that more cells tapped out at the endwalls than elsewhere, and the age of failure at these points was less than that due to other tap outs. Also more cells tapped out upstream than downstream, and at a younger age.

The shape of the freeze on the sidewalls was measured in about a dozen cells. The measurements showed large differences from cell to cell and, for the older cells, from point to point in the cell. It was also found that the freeze profile could change rapidly with time (e.g. before and after an anode effect). This was particularly true of the freeze toe on the cathode in old cells. Typical results are shown in Figure 8. The major changes with age were the undercutting of the sidewall near the metal-bath interface and the growth of a ledge (or toe) on the cathode which could extend under the anode shadow. It was perhaps the effect of this insulating freeze toe on the distribution of horizontal current in the metal pad which was responsible for the change of circulation pattern with age.

Acknowledgements

Thanks are given in particular to Mr. J.H. Kent of Chalfont Park, and Messrs. G.B. Dallas and J.B. Aitchison of Invergordon for their interest and help. Thanks are also due to Mr. A. Palmer of Chalfont Park who participated in most of the measurements.

References

(1) A.R. Johnson, "Metal Pad Velocity Measurements in Aluminum Reduction Cells", AIME Light Metals, 1978.

CHAIRMAN

E.L. Cambridge
ARCO Metals Company
Tucson, Arizona

TiB_2 COATINGS PREPARED BY PLASMA SPRAYING AND ELECTROPLATING

Douglas W. Townsend and Larry G. Boxall

Martin Marietta Laboratories
Baltimore, Maryland 21227
USA

Laboratory and plant tests have demonstrated that TiB_2 is extremely resistant to the corrosive environment in a Soderberg anode.* TiB_2 coatings were tested on vertical Soderberg studs to prevent sulfide attack in anodes made from high-sulfur petroleum coke. Test studs, plasma-sprayed with TiB_2, were more electrically conductive than conventional studs. Porosity in the TiB_2 coating, however, limited life to 2-3 stud setting cycles. Pore-free TiB_2 coatings were electroplated from a purified molten-fluoride bath on various substrates. A large difference in the coefficients of thermal expansion between iron and electroplated TiB_2 produced stresses which prevent its use on anode studs. Adherent and pore-free TiB_2 coatings also can be applied to other materials, i.e., some few select metals, TiB_2 tile, graphite and aluminum cell cathode block. The electroplated TiB_2 coatings are easily wetted with molten aluminum and are resistant to attack by cryolite and aluminum.

* L.G. Boxall and D.C. Nagle, U.S. Patent 4,354,918 (1982).

Introduction

Iron is the dominant impurity in aluminum metal made in vertical stud Soderberg, VSS, cells. It enters the cell as a minor impurity in raw materials and also arises from dissolution of cathode collector bars, electrolysis and general corrosion of the anode skirts and, most important, from corrosion of anode studs.

Anode stud corrosion can contribute from 0.10 to 0.20 wt% iron to the aluminum. Iron enters the anode as an FeS scale that breaks off each time the stud is pulled (1). The scale is transferred to the bath as the anode is electrolyzed away where it is quantitatively incorporated into the aluminum pad. Stud material is also lost when the studs are cleaned between each setting. As a result of the shortening and rounding, the useful life of the stud may be decreased to 2-4 years.

Sulfiding is pronounced over the bottom 12-18 inches of the stud tip; it is most severe at the tip. Carbonyl sulfide (COS) formed by a reaction between carbon monoxide and sulfur, has been shown to be the principal corrosive agent in Soderberg anodes (1). The corrosive environment is further enhanced by the presence of carbon dioxide and volatile fluorides (2). The corrosion rate of the stud is roughly proportional to the sulfur content of the coke used to make the anode, which usually ranges from 1 to 4wt%. Most of the coke used to make anodes for aluminum reduction is made from residual oils from petroleum refining. In recent years, the sulfur content of the coke has been increasing as the availability of sweet crude oil has decreased relative to the availability of sour crude. Advances in oil refining technology that produce higher yields of more valuable products have tended to reduce the amount of oils available for coking while increasing the sulfur content. This trend is expected to continue.

Improvements in the operation of VSS aluminum cells have resulted in lowering studs several inches closer to the bottom face of the anode before resetting. A decreased voltage drop in the anode of 0.1 to 0.2 volts is achieved with direct savings in reduction energy. However, the studs are placed in a hotter zone, closer to anode gasses, which can increase the corrosion rate of stud tips by 3 to 4 times that previously experienced.

Cell Operation

A sectional view of a typical 105 kA VSS cell is shown in Figure 1. Fifty-two anode studs, deployed in four rows, provide a relatively uniform current distribution in the anode. Studs are typically reset every 14 days when the stud tip approaches within $\sim$ 8 inches of the anode face. The recent industry practice of lowering reset heights has resulted in the stud tips getting hotter and closer to the bath line and coming into contact with more anode gasses. The carbon beneath the studs is formed from stud hole paste that is poured into the hot hole immediately before setting the new stud and is baked very quickly. This carbon typically is more porous than that formed from the bulk of the anode paste that bakes more slowly. It is usually electrolized away faster than the rest of the anode, leaving indentations under each stud. These indentations trap gasses which permeate into the anode through the porous carbon. Furthermore, the anode usually tends to crack vertically, forming a pattern that connects the studs and provides an easy path for gasses to enter the baked part of the anode at the stud tips.

The rate of iron loss from anode studs is almost proportional to the sulfur content in the coke but as little as 20% of the total available sulfur from the coke actually reacts with the iron of the anode studs. This fraction can increase significantly for very permeable anodes. The remaining sulfur electrolyzes at the anode face to produce sulfur oxides which are collected by the anode skirt and sent to the dry and wet scrubber systems.

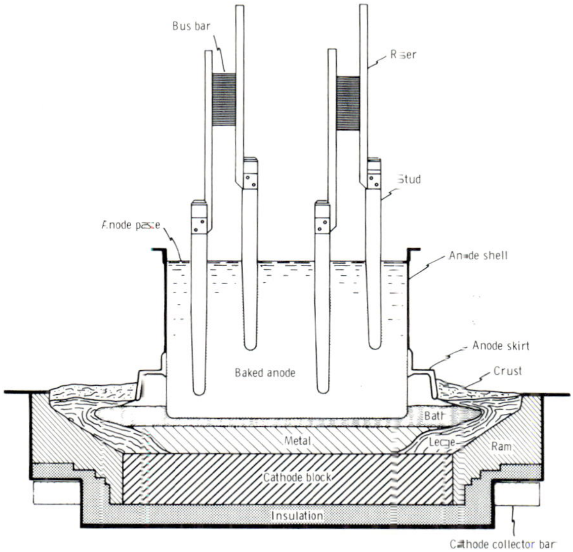

Figure 1. Cross-sectional view through Soderberg reduction cell.

Stud Coating Requirements

Power is carried to a VSS anode by the studs. Over each 2-week setting period, each stud carries an average of 2 kA. Each day, three or four studs are pulled and reset. About five extra studs are either in inventory or in the shot abrading cycle for every cell. A 100 ton/year reduction plant has a total inventory of about 20,000 anode studs, each of which weighs 335 pounds, for a total of 33,500 tons of studs on the plant site. It is therefore imperative that stud bodies be made of only the cheapest possible metal, such as mild steel. However, each stud may have a thin, relatively more expensive coating to prevent corrosion. The following are requirements of a successful stud coating.

- Resistance to attack by the carbon, sulfur, carbon monoxide, carbon dioxide, carbonyl sulfide, and fluoride environment within the anode mass at temperatures up to about 1000°C while still conducting direct current

- Resistance to air oxidation during stud resetting

- Erosion resistance and adherence to the stud metal substrate to prevent coating loss during stud pulling and handling
- High electrical conductivity with low contact resistances to both the stud body and to the anode carbon
- Resistance to thermal shock during the rapid cool-down and heat-up of the stud during resetting operations
- A simple coating procedure together with a long operating coating life to make the application cost effective
- Slow corrosion and erosion loss so as not to significantly degrade metal quality.

Anode Coating Test Procedure

Although numerous materials have been tested as possible substitutes for low-carbon steel for anode studs, only steel alloys containing 2% chromium (3) and aluminum-coated steel (2) are claimed to have corrosion rates lower than low carbon steel. These alloys apparently develop protective aluminum- or chromium-containing oxides when exposed to test gasses, which simulated those produced by the reaction of carbon oxides with sulfur in an anode. Unfortunately, these metals also corrode severely when volatile fluorides are added to the test gas (1).

A test apparatus, Figure 2, was developed to simulate very closely the chemical and electrical environment of a stud tip. Test samples are heated to 900°C in a bed of activated charcoal while carrying an anodic current of 0.23 mA/cm^2. The gaseous atmosphere is 9:1 carbon monoxide (CO):carbonyl sulfide (COS) with volatile fluoride supplied by evaporation from 1 wt% atmolite ($NaAlF_4$) mixed with pitch coke in the bottom heated zone of the stainless steel reactor tube. A standard 4-hour exposure test produces an FeS layer on mild steel with a 0.022-in. loss of metal. The stud loss during a 2-week cycle in the anode can be up to $\sim$ 0.1 inch. Upon removal from the reactor, specimens are water quenched to simulate thermal shock and rapped against a steel anvil to test the coating and protective corrosion product for resistance to mechanical abuse.

Plasma Spraying of Stud Coatings

Titanium diboride, TiB_2, was selected as a possible stud coating material based on preliminary laboratory corrosion tests. Plasma spraying was chosen as a viable method for its application. A very wide range of metals, plastics and ceramics can be applied to many different substrates by this procedure. Commercial equipment is readily available to coat structures as large as anode studs. Studs and test pieces of steel are first cleaned by blasting with 16-mesh alumina grit to produce white metal. Because attempts to plasma spray TiB_2 directly onto steel failed to produce adhesion, a standard bond coat of 309 stainless steel was used to provide adhesion. The stud tips were first preheated to 115°C and sprayed with 0.005 inches of bond coat followed by TiB_2, using a Metco 7M plasma gun with argon-hydrogen and argon-helium gas mixtures, respectively. The stud tips were rotated at 10 rpm during spraying.

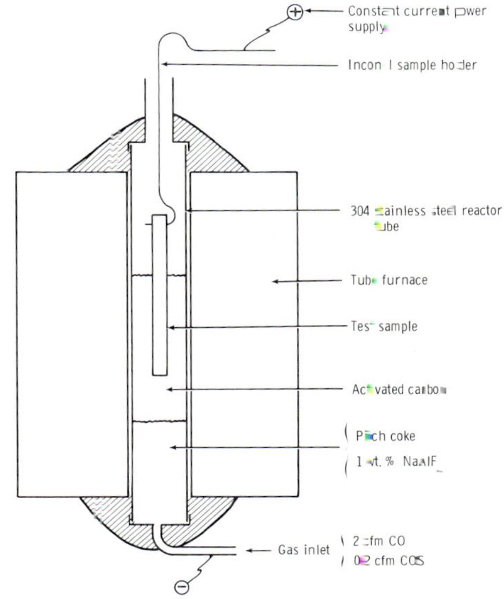

Figure 2. Laboratory test furnace for determining the resistance of candidate stud coatings at 900°C in simulated anode gas under anodic DC current of 0.23 A/cm^2.

The TiB_2 coating was plasma sprayed to a thickness of 0.003-0.010 inches. It is brown, contains about 5 wt% oxygen, and has a hardness of 200-300 kg/mm^2 compared with at least 2500 kg/mm^2 for dense TiB_2, it is electrically conductive as determined by probing the surface with an electrical resistance meter. A cross-section after exposure to the 4-hour laboratory COS test is shown in Figure 3. There was no spalling of the TiB_2 coating and no obvious damage to the base coat during this test.

The TiB_2 coatings, however, were slightly porous and eventually permitted sulfiding gasses to penetrate to the base coat. TiB_2 coatings, plasma sprayed over sintered stainless-steel filter bodies failed to prevent gas flow, even under slight positive pressures. Up to 15% by weight of various fillers, including $MoSi_2$, ZrO_2, aluminum, manganese, and 70:30 manganese:aluminum and 58:42 titanium:manganese alloys, were added to the TiB_2 powder that was plasma sprayed in attempts to reduce the porosity of the coating. Coating porosity was significantly reduced, but not eliminated, by the use of these fillers.

Thirty-one coated stud tips were welded to stud risers and set into anodes for up to two 14-day cycles. DC current measurements showed that coated studs typically carried 3% more current than uncoated studs set at the same height in the same anode position. The coated studs pulled cleanly out of the anodes, free of adhering stud paste. There was no overt sign of sulfidation and only slight oxidation of the hot studs which produced powdery brown to white oxides while air cooling from about 900°C to ambient. This slight oxidation had no effect on electrical conductivity of the TiB_2 coating as determined by probing with an electrical meter.

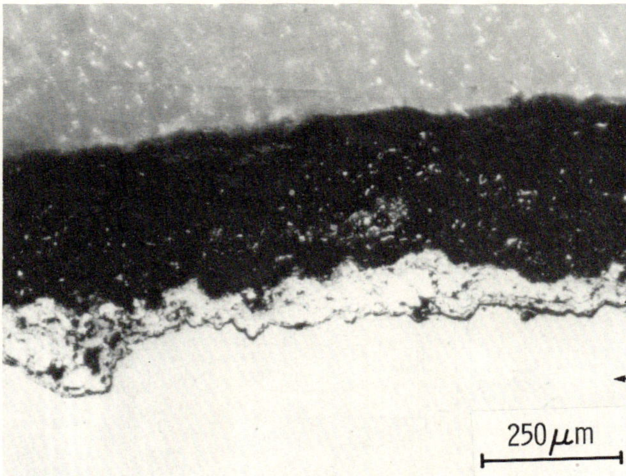

Figure 3. Cross-section of plasma-sprayed TiB_2 over a 309 stainless steel base coat on iron after 4-hour exposure in the laboratory COS tester.

The coatings adhered completely to the hot studs when pulled from the anodes. During cool-down, however, they always spalled off a minimum of about 15% of the surface, as seen in Figure 4, leaving behind the 309 stainless-steel base coat. EDAX analyses by scanning electron microscope, SEM, of sections of exposed 309 stainless-steel base coat showed sulfidation of its nickel, chromium and iron components. A cross section of the sulfided base coat, Figure 5, shows that the outside surface was a mixture of sulfides and oxides, with about 3 wt% sulfur penetrating half-way through the coating.

The plasma-sprayed TiB_2 coating spalls off on cooling because the bond to the base coat is destroyed by sulfidation. Due to a major difference in the coefficient of thermal expansion (CTE), TiB_2 normally shrinks only half as much on cooling after being pulled from the anode as does the iron stud. The sulfided base coat surface is unable to sufficiently restrain the TiB_2 to prevent spalling.

Attack to the base coat in one setting cycle was only about 5% of the normal damage to the iron. However, the cost and effort to clean and recoat the studs between settings made this coating system impractical for plant use. Insufficient tests were run to determine the effect on aluminum metal quality from reduced iron loss from the studs.

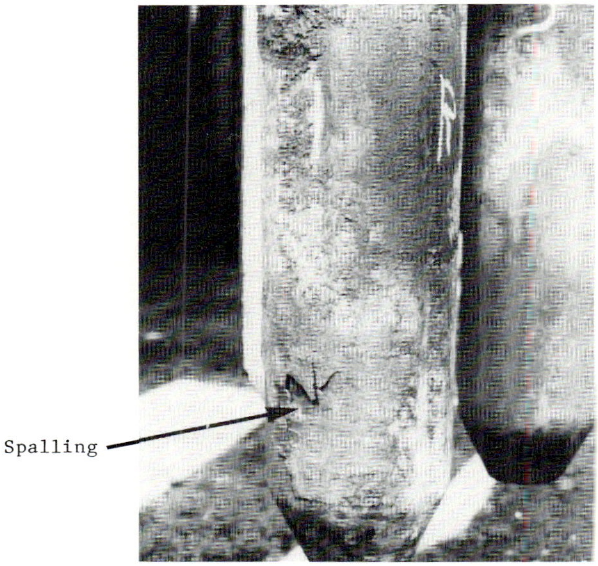

Figure 4. Plasma-sprayed TiB_2 coating on anode studs after a 14-day cycle.

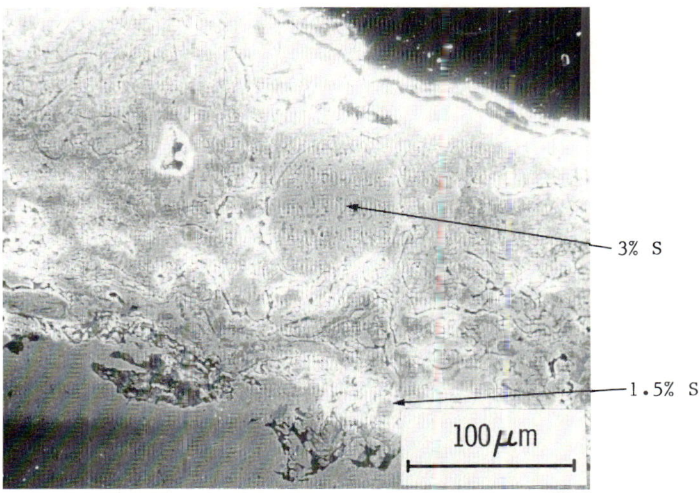

Figure 5. Cross-section of partly sulfided 309 stainless steel plasma-sprayed base coat on an iron anode stud tip after 2 weeks of exposure in a VSS anode.

Electroplated TiB_2

Electroplating was chosen over chemical vapor deposition, ion sputtering and low-pressure plasma spraying as the procedure for applying pore-free coatings of TiB_2 to VSS studs. The pore-free coating would prevent damage to the underlying metal of the studs and thereby prevent spalling. TiB_2 can be electroplated under an inert argon atmosphere from a purified, molten, eutectic mixture of sodium, potassium and lithium fluorides, known as FLINAK, which also contain boron and titanium fluorides (4). The apparatus shown in Figures 6 and 7 was constructed with a capability of electroplating the bottom 12 inches of a stud tip. TiB_2 was electroplated at a rate of up to 1 mil/hour at 700-750°C directly onto iron using a soluble TiB_2 anode.

Electroplated TiB_2 is pore free and exceptionally hard, and has a dull metallic gray to silvery appearance. Both X-ray and EDAX analyses confirmed that the electroplate was TiB_2 and in later tests TiB_2-CrB alloys, when the molten fluoride bath became contaminated with chromium, leached from the stainless steel electroplating apparatus. Electrodeposits were made to adhere to iron, nickel, stainless steel, graphite, Hastelloy, and TiB_2. Figure 8 shows a cross-section of TiB_2 electroplated onto an iron test rod to a thickness of 0.002 inches. This coating protected the iron substrate during the laboratory COS test. Figure 9 shows a SEM micrograph of the surface of an electrodeposit having a characteristic nodular appearance.

Figure 6. Overall view of TiB_2 electroplating apparatus.

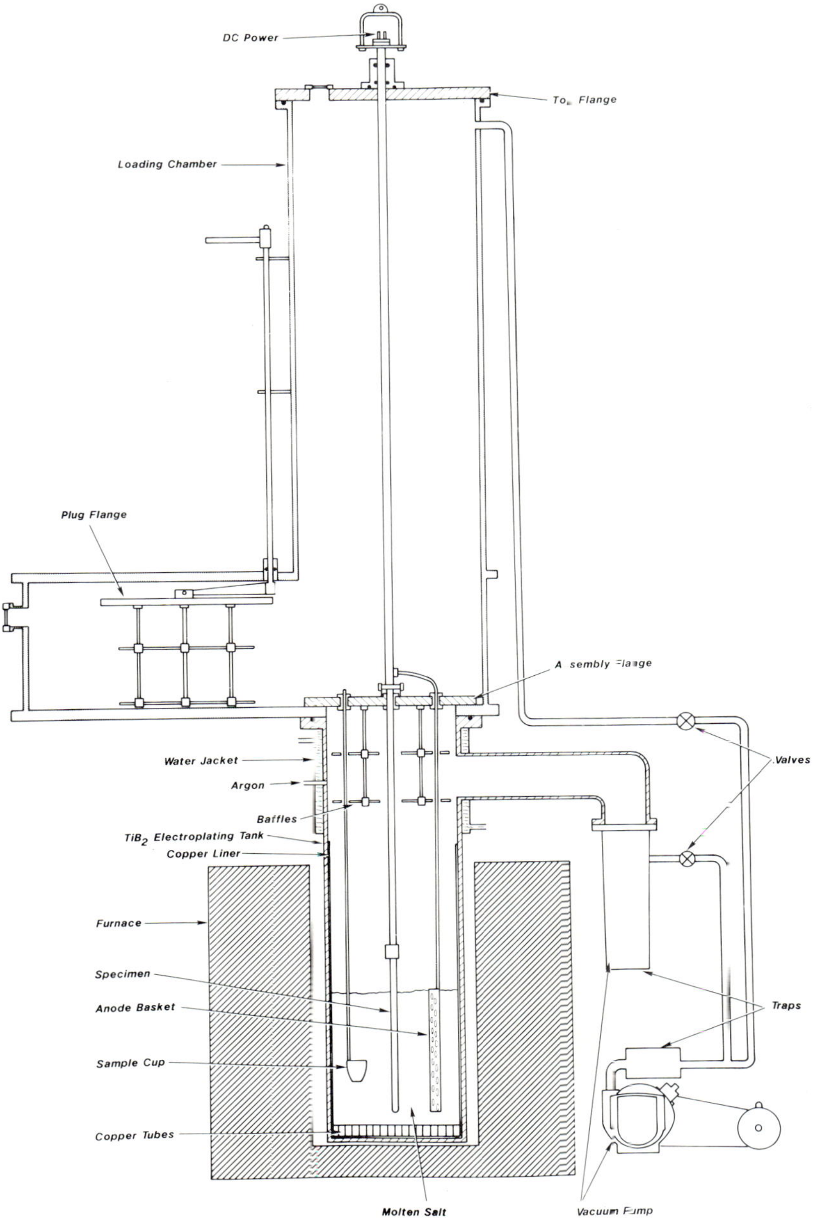

Figure 7. Section view of TiB_2 electroplating apparatus (overall height 8-1/2 feet).

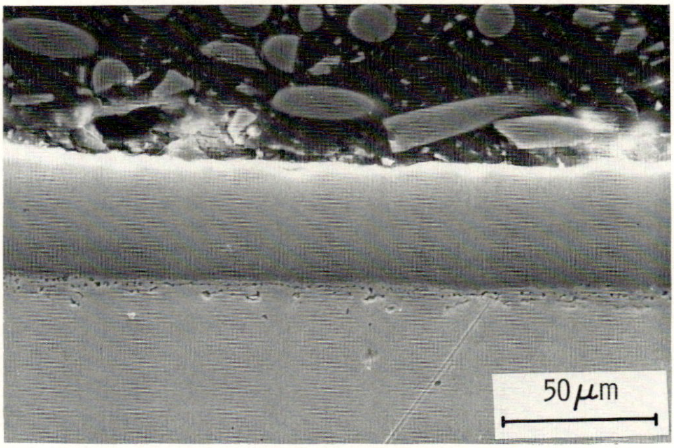

TiB_2

Iron

Figure 8. Cross-section of electroplated TiB_2 on iron.

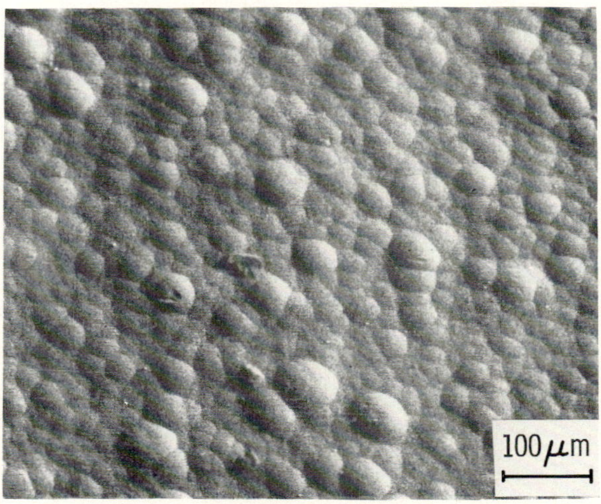

Figure 9. Surface profile of TiB_2 electroplated on iron.

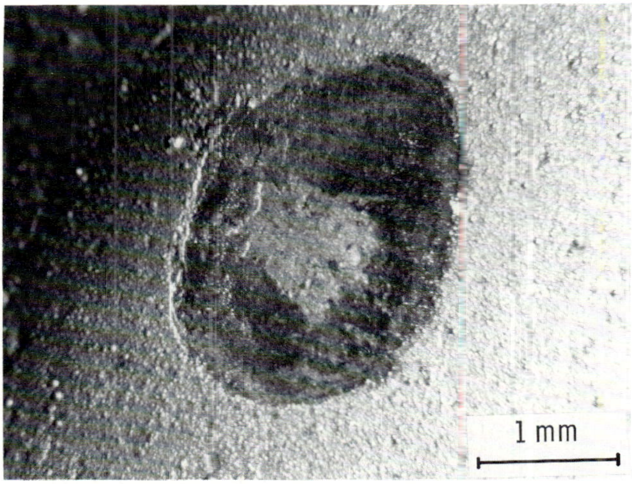

Figure 10. A typical spall of TiB_2 from an iron test rod showing exposed substrate.

However, when TiB_2 and TiB_2-CrB_2 alloy were electroplated onto iron or steel, spall defects (1/10 in. to 1/2 in. in diameter) were found (Fig. 10). This spalling presumably results from the large difference in the CTE's between coating and substrate. On cooling from the electroplating temperature to ambient, additional compressive stresses of over 1,000,000 psi have been estimated to develop in the electrodeposits. The electroplated boride coating typically tapers from full thickness to zero thickness at the spall center. The fracture surface of the spall is conchoidal, a manifestation of the ceramic, rather than metallic, mechanical properties of TiB_2.

TiB_2 electroplates produced at high temperatures (> 740°C) are more prone to spall off iron than those formed at low temperatures. The spalling is reduced significantly when the substrate is 416 stainless steel instead of low-carbon steel (iron). The CTE of 416 stainless steel is 80% that of iron, reducing the compressive stress in the TiB_2 coating to about 625,000 psi. When electroplated onto graphite rods (Fig. 11) and on aluminum reduction-cell cathode block, TiB_2 adheres strongly. However, because TiB_2 has a greater CTE than graphite, the coating shrinks more during cooling and develops tensile cracks (Fig. 12).

Electrodeposits of TiB_2 and TiB_2-CrB_2 appear to withstand tensile stresses of 400,000 psi or compressive stresses of 500,000 psi without spalling or peeling. Accordingly, for mechanical adhesion of TiB_2 and TiB_2-CrB_2 alloys, the CTE of the substrate appears to be restricted to between 5×10^{-6}/°C and 11.5×10^{-6}/°C. Table I shows that TiB_2 is about 10 times stronger in compression than in tension. Thus, if the electroplated TiB_2 is not to crack and is to provide corrosion protection to a substrate, the CTE range must be further restricted to a minimum of about 6.5×10^{-6}/°C to prevent excessive tension in the coating.

Figure 11. A one-inch graphite rod electroplated with 0.010 inches of TiB_2-CrB_2 alloy.

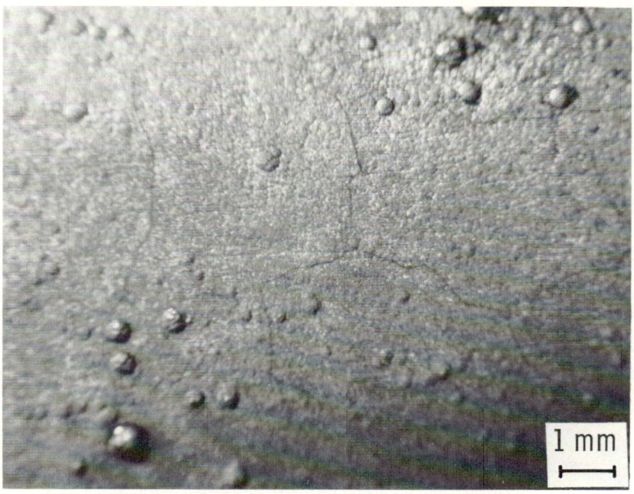

Figure 12. Tensile cracks in the TiB_2-CrB_2 electroplated coating on the graphite rod shown in Figure 11.

TABLE I. Comparison of Physical Properties of Iron and TiB_2

Physical Properties	Material: Iron (1010 Carbon Steel)	TiB_2
Coefficient of thermal expansion at 500°C	$14 \times 10^{-6}/°C$	$8 \times 10^{-6}/°C$
Elastic modulus	30×10^6 psi	80×10^6 psi
Tensile strength	∿ 40 ksi (yield)	18 ksi
Compressive strength	∿ 40 ksi (yield)	> 250 ksi
Electrical resistivity		
20°C	14 μ ohm-cm	10 μ ohm-cm
1000°C	116 μ ohm-cm	60 μ ohm-cm
Hardness		
20°C	240 kg/mm	2500-3700 kg/mm
1000°C	---	100 kg/mm
Thermal conductivity	0.6 W/cm°C	0.8 W/cm°C

The quality of TiB_2 electroplates on various substrates is summarized in Fig. 13 as a function of CTE. Iron-, nickel- and copper-based alloys have a CTE considerably greater than that of TiB_2 so that very high compressive forces develop in the TiB_2 coating on cooling to ambient from either electroplating or VSS anode temperatures. Several of the metals listed in Table II have CTE's compatible with TiB_2. Unfortunately, these are restricted to the refractory metals -- molybdenum, tantalum, columbium, and titanium -- which are so chemically reactive to the plating salt that they require a nickel or copper base coat to prevent reaction with the molten fluoride electroplating bath. Furthermore, all are very expensive metals and could never be cost effective alternatives to carbon steel for VS anode studs.

The compositions of various TiB_2-CrB_2 alloy electrodeposits are shown in Table III. Although the plating bath was made up of chromium-free salts, the electroplating salt acquired chromium primarily by diffusion through the copper liner of the alloy 310 stainless-steel electroplating tank. Chromium impurity accumulated at a rate of a couple of grams per day and accounted for all the CrB_2 in the electrodeposits. The tendency to produce chromium-containing boride alloys increased as the plating temperature was raised and with the time of exposure of the plating salt to the chromium in the stainless steel apparatus. The boride alloys also tend to become harder as the chromium content increases (Table III). A commercial electroplating cell should have sufficient throughput to considerably dilute the chromium concentration in TiB_2. Relatively pure TiB_2 has been electroplated in this laboratory apparatus only below 725°C.

TABLE II. Coefficients of Thermal Expansion of Substrates and Resulting Stress in TiB_2 Electroplates

Substrate	Coefficient of thermal expansion (CTE)	Stress added to TiB_2 coating on substrate cooled to ambient from 750°C (psi)	Cost per pound (dollars)
TiB_2	7.8	0	20.00
CrB_2			
Graphite	4.5	-491,000	8.00
Molybdenum	5.7	-313,000	15.00
Tantalum	6.6	-179,000	170.00
Columbium	7.2	-59,000	34.00
Titanium	9.5	253,000	12.00
Hastelloy B	11.8	595,000	30.00
Stainless Steel 416	12.0	625,000	3.00
Iron	14.1	938,000	0.60

TABLE III. Properties of TiB_2-CrB_2 Electrodeposits

Electroplating temperature (°C)	Hardness (kg/mm^2)	Composition (wt%)			
		TiB_2	CrB_2	FeB_2	S
730	3750	93.6	3.35	1.4	1.7
725	3850	94.2	3.9	-	1.9
740	4100	88.4	11.4	-	0.2
761	3650	47.9	49.4	2.0	0.7
761	4600	79.6	18.8	1.0	0.6
761	4900	38.9	60.2	0.9	-

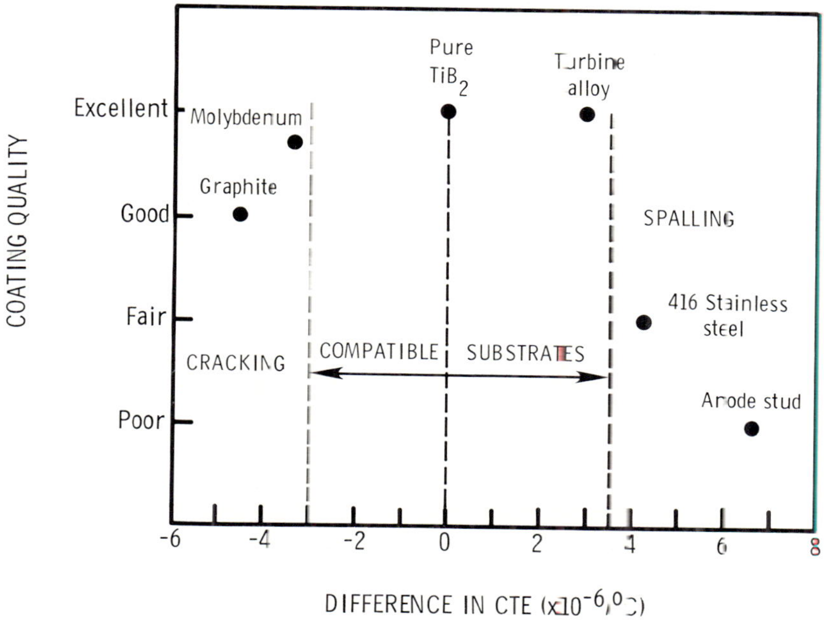

Figure 13. Effect of differences of CTE on the quality of TiB_2 electroplates over various substrates.

TABLE IV. Coating Compositions and Corrosion Losses

Material	Composition (wt %)				Corrosion Loss in COS at 900°C during 4 hours (mils)
	Cr	Ni	Fe	Others	
Iron	--	--	100	------	22
309 SS	23	13	63	1 Si	8.5
IN-657	47	50	--	1.5 Cb	5.0
TiB_2, Chemical Vapor Deposited				100 TiB_2	1.0
TiB_2, Electroplated					
A				100 TiB_2	$\leq$ 1.0
B	3 CrB_2		1 FeB_2	94 TiB_2	$\leq$ 0.5
C	40 CrB_2			48 TiB_2	$\leq$ 0.5

The corrosion resistance of electroplated TiB_2 has been measured in the laboratory COS test. The rates of attack of the TiB_2 and TiB_2-CrB_2 alloys and of metals considered for use as coatings over iron anode studs are shown in Table IV. The corrosion loss of electroplated and CVD-applied TiB_2 and TiB_2-CrB_2 alloys in the laboratory COS is between 0.2 and 0.4 mils ($\sim$1.4% that for iron). Sulfur does not penetrate the TiB_2 coating below the line of corrosion attack. However, the corrosion product of TiB_2 is badly fractured and therefore does not protect against further attack so that attack is linear with time.

A sample of commercial hot-pressed TiB_2 tile was electroplated with TiB_2 and immersed in aluminum metal covered with cryolite at 1000°C for one week to determine resistance to the aluminum reduction cell environment. The electroplated TiB_2 was wetted by molten aluminum and no mechanical damage was observed. A cross-section of the tile electroplated with TiB_2 is shown in Figure 14.

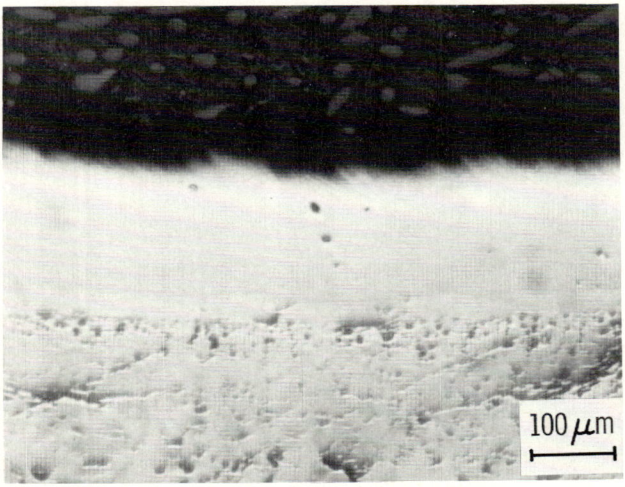

Electroplated TiB_2

Hot pressed TiB_2

Figure 14. Electroplated TiB_2 on hot-pressed TiB_2 tile.

Conclusions

TiB_2 has a corrosion rate about 2-3% that of iron in the simulated Soderberg anode environment. Plasma-sprayed TiB_2 coatings are too soft and porous to prevent long-term sulfidation of underlying metal and spalling of the coating. A continuous layer of full density TiB_2 is unable to withstand the thermal expansion and contraction of a steel anode stud. It therefore spalls and provides inadequate protection against sulfidation.

This work has demonstrated the feasibility of electroplating a pore-free TiB_2 coating on large structures made of select materials. The TiB_2 coating has potential application to protect critical components in sulfidation and molten metal environments.

Acknowledgment

The authors wish to thank Martin Marietta Aluminum for permission to publish this work and for the assistance of H.W. Hayden, K. Ries and M. McClean for plant tests. The assistance of O.C. Keesecker and W. Montague of Martin Marietta Laboratories is also acknowledged.

References

1. J. Brun, K. Dalatun, A. Sterten and R. Tunold, "Corrosion of Anode Contact Spikes and Gas Collecting Skirts in Soderberg Aluminum Cells," Transactions of the Metallurgical Society of AIME, 245 (1969) pp. 671-675.

2. H.G. Johansen, "Iron as a Contaminant in a VS Soderberg Cell," Light Metals 1 (1977) pp. 253-261.

3. M.A. Korobov, "Corrosion of Anode Fasteners in the Electrolysis of Aluminum," Tsvetnye Metally 32 (7) (1959) pp. 52-58.

4. J.D. Kellner, "Process for Electrodepositing Titanium Diboride from Fused Salts," United States Patent 3,880,729, Apr. 29, 1975.

USE OF TiB_2 CATHODE MATERIAL

APPLICATION AND BENEFITS IN CONVENTIONAL VSS CELLS

Larry G. Boxall and Arthur V. Cooke

Martin Marietta Laboratories
Baltimore, Maryland 21227
USA

H. W. Hayden

Martin Marietta Aluminum
The Dalles, Oregon 97058
USA

A low-cost proprietary TiB_2 cathode material* has been developed which circumvents the problems previously experienced in using TiB_2 in aluminum smelting cells, i.e., chemical instability, poor thermal shock and impact resistance, attachment problems, and high cost. During the period 1981-83, the proprietary material was tested as a cathode lining in six Vertical Stud Soderberg (VSS) aluminum reduction cells at the Martin Marietta Aluminum plant, The Dalles, OR. The plant test has demonstrated that the material achieves excellent durability and longevity in commercial aluminum reduction cells, and can be utilized with no modification to conventional cell operating procedures. Current efficiency improvements and energy savings have been realized, and the rate of TiB_2 loss indicates a potential 3-5 year life for the cathode material. Analysis of core samples taken from the experimental cathodes at The Dalles during temporary shutdown confirmed that the cathode lining was functioning as predicted.

* Patents pending

Introduction

Virtually all commercial aluminum is produced in Hall-Heroult design electrolytic cells. Power consumption for the process ranges from about 6 to 8.5 kWh/lb of aluminum produced, while the theoretical thermodynamic requirement is $\sim$ 3.0 kWh/lb (1). The long-term demand for aluminum is anticipated to rise as industry and consumers increase their use of this lightweight metal to reduce energy costs, for example, in transportation. Faced with growing demand and limited supplies of increasingly expensive electrical energy, the aluminum industry has strong incentive to develop a low-energy aluminum reduction cell.

In conventional electrolytic smelting, a significant portion of the "wasted" energy (the difference between theoretical and actual energy requirements) is due to the voltage drop across the resistive cryolite electrolyte bath in the interpolar space between the two electrodes. Reduction of the anode-cathode distance (ACD) in a conventional cell reduces the bath voltage loss, but, due to metal pad instabilities, the cell current efficiency decreases as the reaction products at both electrodes become intermixed. To reduce the ACD without concomitant loss in current efficiency, much work has been done to improve metal pad stability by controlling the cell magnetics. An alternate, more revolutionary, approach being investigated in the industry is the use of a solid aluminum-wetted cathode, such as TiB_2, from which the molten metal is continuously drained. Since in this latter method, the unstable metal pad is replaced with a thin film of aluminum, the ACD can be reduced without a loss of current efficiency.

Extensive industrial research has been carried out since the 1950's on the use of refractory hard materials (RHM) such as TiB_2 for cathode lining applications (2). TiB_2 is very slightly soluble in aluminum, highly conductive compared to cryolite, and most significantly, is wetted by molten aluminum. This property of wettability allows an aluminum film to be electrolytically deposited without the presence of an aluminum pad, because the molten metal skin effectively sticks to the surface instead of agglomerating into separate metal pools. Thus, new cell designs are possible, utilizing, for example, a sloped cathode so that the aluminum film runs off the surface continuously. Hence the ACD can be reduced from the normal 4-6 cm down to 1-3 cm without loss of current efficiency, permitting a 1.0-1.5 V decrease in cell voltage. High current efficiency is maintained because the aluminum film is tightly bound to the RHM surface and therefore has little tendency to intermix with the electrolyte and anode products. Moreover, the interaction between the magnetic field and the thin aluminum film is negligible by comparison with that associated with a full metal pad. Figure 1 illustrates two different cell design concepts in the patent literature for using TiB_2 cathode material in a low-energy cell (LEC).

The two major technical problems arising during previous attempts to make a TiB_2 surface on carbon/graphite blocks, from which the cathode is constructed, are lack of adhesion to the substrate and disintegration of TiB_2 and TiB_2 alloy structures. The generally accepted explanation for disintegration of these structures in aluminum cells is fluoride/molten metal attack and dissolution of oxide impurities segregated on the boundaries between individual TiB_2 grains. The critical component delaying development of a practicable LEC has therefore been the failure to produce a low-cost, durable TiB_2 cathode material.

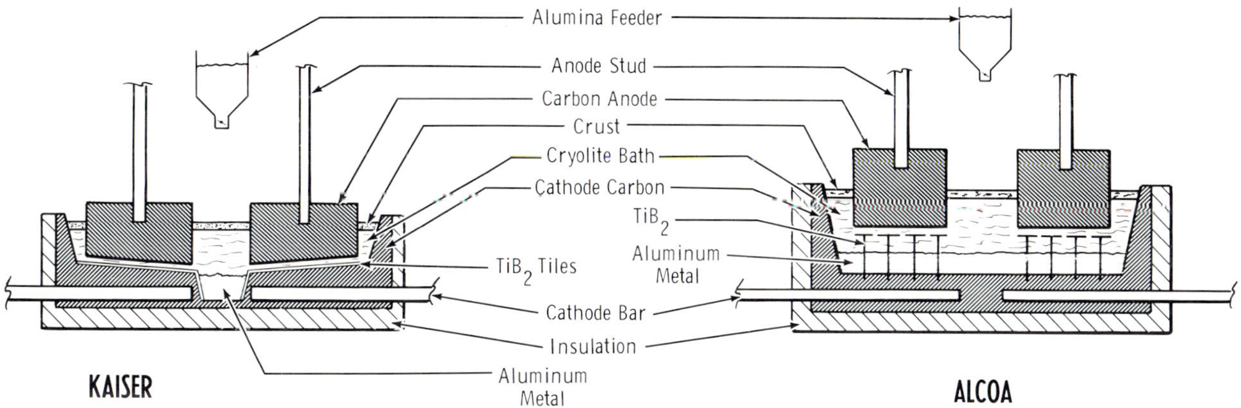

Figure 1. Patented Solid Cathode Reduction Cell Designs.

TiB_2 Cathode Material Requirements

Several papers and patents (2) have documented the requirements for an ideal LEC cathode material. These specifications depend to a certain extent on the particular cell design, but the following properties represent the most important criteria that are commonly stipulated:

1. Wettability by molten aluminum
2. Good electrical conductivity
3. Low solubility and reactivity with molten aluminum and the cryolite-alumina bath
4. Reliable attachment to the underlying cathode substrate
5. Low cost
6. Easy fabrication into any desired shape
7. High resistance to mechanical wear and impact and thermal shock
8. Good resistance to high-temperature oxidation and corrosion by any reactive gases
9. Low thermal conductivity.

The properties of the borides and carbides of the transition metals, particularly TiB_2, make them especially suitable for use as cathodes in aluminum electrolysis cells (2). TiB_2 has a high melting point and hardness, good electrical and thermal conductivity, is wetted by molten aluminum metal, and, in its purest form, is resistant to the aluminum and cryolite bath. However, fabricated TiB_2 parts are brittle, sensitive to thermal shock, not easily attached to a carbon substrate, subject to chemical attack, and expensive. Thus, the early attempts to use hot-pressed TiB_2 tiles were thwarted by mechanical breakage, thermal shock, bath attack along oxide-rich grain boundaries, and attachment problems.

High-purity TiB_2 forms, and composites, made by addition of an oxide stabilizer (3,4) (such as AlN which forms stable oxynitrides), have been used recently (5) to improve the corrosion resistance of the material. Although beneficial, these measures do not address the problem of attaching TiB_2 to carbon over a very wide temperature range. A recent patent issued to Payne in 1978 (6) describes a method for bonding TiB_2 tiles to the carbon cell bottom in-situ during cell startup and operation by forming the adhesive bond as the temperature is gradually raised. This method avoids the detachment problem resulting from the difference in thermal expansion of TiB_2 and carbon during cell startup. However, these differences will cause debonding when the cell is shut down and restarted because the bond is irreversibly formed.

Another patent, assigned to Reynolds in 1982 (5), proposes the use of a tri-layer tile, comprising an intermediate bonding layer between an upper TiB_2/AlN tile and a recessed graphite tile substrate. The bonding layer is present to accommodate thermal expansion mismatch by acting somewhat like a graded glass-to-metal joint. Even in this case, the bond to the carbon will degrade eventually due to molten metal wetting the lower surface of the TiB_2/AlN tile.

None of these developments has led to a long-lived commercial TiB_2 cathode material, as evidenced by recent reports for the joint DOE/Kaiser Aluminum project on developing a drained TiB_2 cathode cell (7). The same series of problems has recurred time after time in unsuccessful attempts to operate a cell utilizing the TiB_2 materials available heretofore, viz. detachment from the carbon substrate, mechanical/thermal shock breakage, disintegration due to chemical attack, or overwhelming material costs.

Material Formulation and Processing

Martin Marietta has developed a low-cost, durable TiB_2 cathode material, consisting principally of a phenolic resin, a furane resin precursor, curing agents, carbonaceous filler, carbonaceous additives, and RHM. This coating composition may be applied wet, directly to a carbon cathode substrate, and then cured and carbonized. The resulting relatively hard, tough surface layer consists of RHM imbedded in a largely non-graphitic and non-graphitizing matrix. The green mix also includes carbon fibers, which act as crack arrestors while the coating material undergoes shrinkage during cure and carbonization. Sufficient RHM is incorporated in the coating to ensure continuous aluminum wetting of the surface. Thus, the resultant composite material is an economic and effective means to obtain the advantages of RHM in cathodes, and eliminates the need for more costly TiB_2 tiles.

A family of coating compositions have been developed to accommodate different carbon substrates, coating thicknesses and aluminum wetting properties. Two typical formulations are given in Table I. All the percentages are by weight prior to mixing.

TABLE I. Typical Cathode Coating Compositions

	Composition 1 (wt%)	Composition 2 (wt%)
TiB_2 powder (-325 mesh)	36.0	37.5
Carbon cement solids	34.2	30.6
Mix liquid	19.7	29.6
Coke particles	10.1	----
1/4 in. carbon fiber	----	2.3

A 1/16-in. thick layer of mix is applied to the prepared surface of a carbon cathode block by trowel, and worked into the block surface. The layer is then built up to 3/8 in. thick, smoothed, and levelled. The coated block is then heated to 100°C at a rate of 25°C per hour, and held at this temperature for 3 hours. Heating is continued at 25°C per hour to 140°C, and the block is held at temperature 16 hours. It is then air cooled to room temperature.

The cured block is carbonized by heating to between 600°C and 1000°C over 24 hours in a protective coke bed to avoid oxidation, then cooled to below 200°C before removal from the coke bed. The integrity of the cured structure and substrate bond remain excellent after carbonization. No changes in cell construction or startup procedure were required for using the blocks coated with composite TiB_2 material.

A variety of cathode block materials, such as Kiowa Carbon, G.R. Stein, and Union Carbide amorphous and CFN grades have been successfully coated. Once carbonized, the TiB_2 coating material has a thermal expansion coefficient almost the same as that of the carbon block, thus eliminating the thermal cycling debonding problems inherent in hot-pressed TiB_2 tiles. Coating formulation changes were required to control the shrinkage of the material during the cure and bake steps, and thus prevent separation and/or damage of the underlying cathode block material.

Material Testing

Testing and scaleup of the cathode coating material involved both passive (or static) and dynamic (electrolysis) exposure tests. The development and evaluation process currently used for assessing materials and processing/formulation modifications comprises a series of sequential tests.

Wettability During Electrolysis

An inverted electrolysis cell, shown in Figure 2, is used for the initial wettability and bath penetration tests. Coatings containing as little as 10 weight percent TiB_2 were partially wet by the molten aluminum metal; at about 20 weight percent the coating was wetted over 90 percent of the surface. A content of 35 to 60 weight percent TiB_2 appeared optimal for achieving maximum wetting at minimum cost. Loss of coating integrity only begins at much higher TiB_2 percentages. Comparison of wettability data from the inverted electrolysis test and from metal pad immersion tests revealed that the coating material is more easily wetted by nascent aluminum generated during electrolysis than by a passive metal pad. This characteristic is a distinct advantage if the material is to be used in a drained cathode cell because these tests effectively show that it is more difficult to achieve aluminum wettability under a deep metal pad in a conventional reduction cell than in a drained cathode LEC.

Durability Under Molten Aluminum and Bath

Static exposure tests are used to assess various different formulations and types of candidate TiB_2 material as cathode linings for aluminum reduction cells. Samples of commercial hot-pressed TiB_2 tiles and proprietary carbon/TiB_2 composite materials were exposed to molten aluminum and molten cryolite bath for 7 to 30 days in externally heated retorts as shown in Figure 3. The samples were subsequently recovered, sectioned, analyzed, and evaluated for durability in the aluminum cell environment. Our significant findings were: 1) the very high-purity TiB_2 tile disintegrated below the aluminum metal line; 2) the stabilized TiB_2 alloy tile was undamaged but had debonded from the carbon cathode substrate; and 3) the various composite coating samples were undamaged below the aluminum metal line, but showed variable exfoliation cracks at the level of the cryolite bath. The exfoliation cracks virtually disappeared when carbon fiber was incorporated in the coating material.

Coatings exposed in retorts for 30 days were compared with core samples taken from experimental VSS cell cathodes lined with carbon/TiB_2 composite that had been operating for between 100 and 300 days. The results correlated very closely, demonstrating the feasibility of evaluating materials for full-scale operation based on short-term static tests. Some care should be used in conducting and evaluating the retort test: 1) the molten liquid levels in the retort must be maintained during the test period, particularly during initial bath penetration of the carbon container; 2) the combined effects of molten aluminum metal and bath on the coating during the brief static retort exposure may prove more aggressive in some respects than prolonged full-scale use under a metal pad; 3) the molten metal in a closed system retort rapidly saturates with TiB_2, whereas, in an operating cell, fresh metal is produced continually; therefore, more TiB_2 will dissolve from the coating on equilibrium solubility

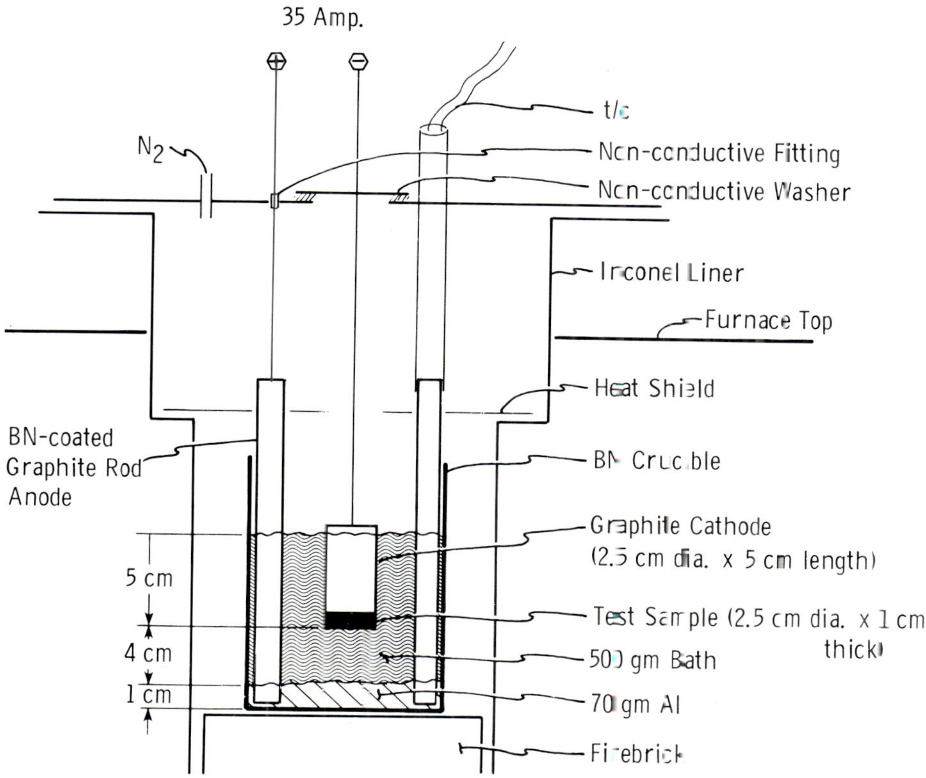

Figure 2. Inverted Electrolysis Test Cell.

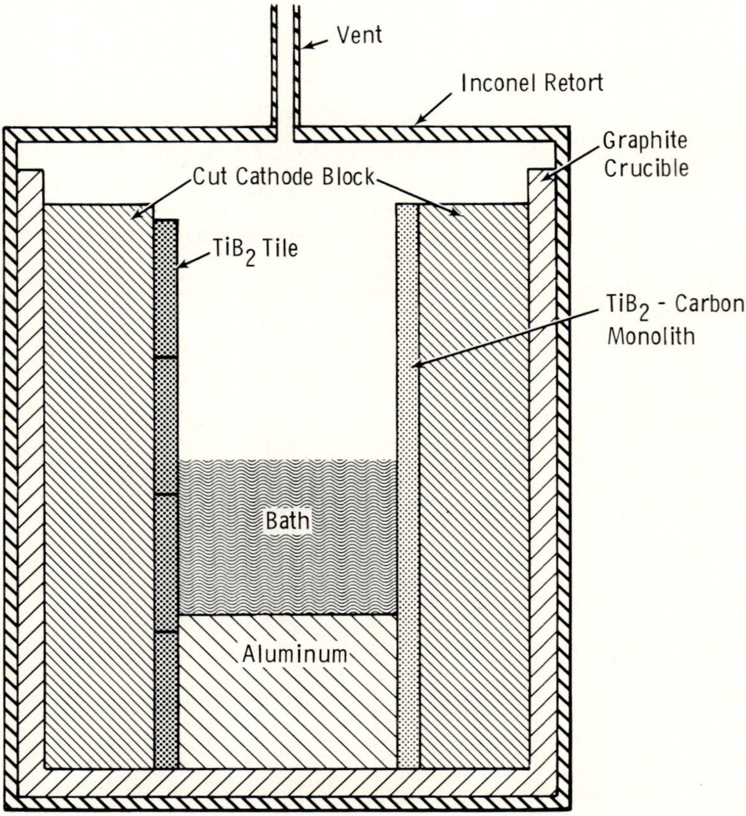

Figure 3. Static Exposure Test Apparatus

grounds in actual cell operation. Tests are in progress in the static retort tester to evaluate the effect of a regular tapping cycle.

Pilot Cell Exposure and Electrolysis Testing

Long-term exposure testing of cathode materials under electrolysis conditions is not feasible in bench-scale test cells. A 1500 A externally heated pilot electrolysis cell is therefore used to evaluate TiB_2 materials under continuous electrolysis for periods up to 3 weeks. The pilot cell (Figure 4) was equipped with automatic temperature control, anode-cathode distance control, alumina feed equipment, and a remote alarm system to permit unattended operation. Use of a replaceable, internally weighted cathode block facilitated the rapid exchange of test cathodes while the cell was operating.

Cathode materials were tested under a 1-in. thick metal pad and in a drained cathode configuration. Two metal wells were cut in the bottom of the cell cavity - one for collecting the aluminum metal (which could be emptied regularly) and the other for exposing test samples to a pad of molten metal. Certain areas of the bottom liner were thus used for metal exposure while the remainder was lined with samples exposed to bath. The operating life of the cell liner was approximately 4 months. At that time, the cell was shut down and the test samples were recovered for analysis.

Results of the pilot cell exposure, in both the static and dynamic configurations, were consistent with those from the retort tests. Figure 6 shows some typical cross sections of cathode samples held under a metal pad during pilot cell operation. In all cases, the coated cathode surface remained covered with a thin film of metal even after the cathode was inverted during removal from the cell. The continuously wetted surface is shown in the enlarged section of the aluminum/coating interface.

The absence of muck on the coated cathode surface in Figure 5 is characteristic of all the coated cathodes tested under electrolysis. By contrast, the uncoated carbon surfaces were consistently found to have a partial covering of muck. We also observed that any aluminum metal remaining in the uncoated 1-in. deep cavity of a metal pad cathode was always separated from the cathode carbon by a thin film of frozen bath material. On a carbon cathode surface, the bath material preferentially wets the surface; on a TiB_2 cathode, molten aluminum expels the bath and wets the cathode instead. Hence, there is little tendency for muck, a bath-rich material, to attach itself to the surface of a TiB_2 cathode; rather, it is easily swept upwards for dissolution in the bath. Plant experience with the TiB_2-coated cathodes confirms that these cells tend to run "cleaner" than cells with conventional carbon linings. It is particularly important for the cathode surface to be muck free in a drained cathode LEC with its reduced anode-cathode distance. The present data indicate that TiB_2-coated cathodes reduce mucking and thereby increase current efficiency in conventional cells. Thus, there will be little tendency for muck to form or settle on a drained cathode surface.

Plant Tests

Six commercial VSS cells with TiB_2-coated cathodes were installed in the Martin Marietta Aluminum smelter at The Dalles, OR, in late 1981 and early 1982. This full-scale plant test of the cathode coating was designed to determine whether:

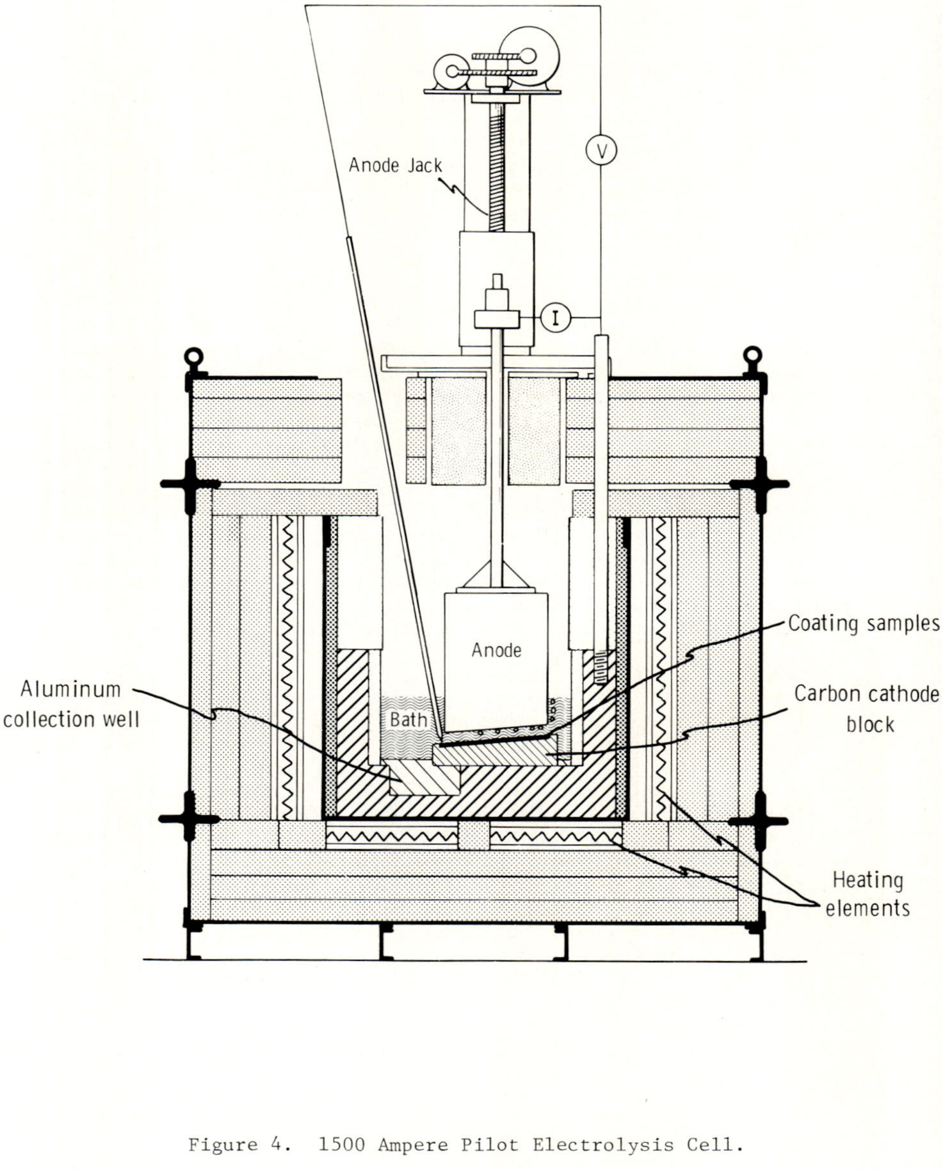

Figure 4. 1500 Ampere Pilot Electrolysis Cell.

Figure 5. Cross section of cathodes after electrolysis tests in aluminum pilot cell.

(a) Conventional carbon cathode
(b) Carbon/TiB_2 coated cathode
(c) Enlargement of interface between cathode coating and frozen aluminum

1. The coating can be applied in a plant environment
2. The coated cells can be started using conventional procedures
3. The coating is durable under operating cell conditions
4. The coating is wettable in commercial use
5. The coating has an acceptable projected life
6. The coating reduces cell mucking
7. The coating has no adverse effects on cell performance
8. The use of the coating in conventional cells results in energy savings or reduced operating costs.

By the end of 1982, it was clear that all of these criteria had been met.

Cathode blocks, coated using the direct spreading techniques described earlier, were used to reline six 105-kA VSS aluminum reduction cells. The only deviation from standard reline and cell startup procedures was to wire brush the excess cathode-ram material from the coated block surface prior to cathode installation in the potroom. No difficulties were encountered in the coke bake startup of the first four test cells or the hot metal startup of the final two.

A production curtailment in June 1982 necessitated the shutdown of a potroom containing three of the test cells. Core samples were taken from two of the coated cathodes before the cells were moved and restarted in another room. Two of the coated test cells were restarted after being out of service for approximately 2 months; another had been out of service for 8 months. No difficulties were experienced in restarting the cells, and performance continued as before the planned shutdown. This demonstrated ability of the cathode coating material to withstand shutdown and restart without sustaining any apparent damage greatly enhances the flexibility and usefulness of the coating material in a plant environment.

The cathode core samples provided an opportunity to examine two different coating materials which had been in service for 310 days and 109 days, respectively. Cross-section photographs and EDS elemental maps across the coating/aluminum and coating/substrate interfaces are given in Figures 6 and 7, respectively. The sections show that a strong bond was maintained between the cathode coating and underlying cathode block even after 310 days of electrolysis in the aluminum cell environment. Moreover, the thin pad of aluminum, left after complete tapdown, can be seen still tenaciously wetting the surface of the coating. The measured decrease in apparent coating thickness was consistent with the coating life estimated from measured Ti loss rate in the metal pad (see below). Extensive scanning electron microscope (SEM) analysis for islands of TiB_2 in the aluminum metal adjacent to the coating did not reveal any significant disintegration or loss of TiB_2 beyond the solubility limit. The measured TiB_2 loss from the coating appears to be almost exclusively due to dissolution and not erosion.

The estimated life for the cathode coating used in The Dalles' test cells is 3-5 years, determined by dividing the total weight of TiB_2 contained in the coating by the average daily loss of TiB_2 in the metal tapped from the cell (based on increase in Ti content beyond the plant average). The effective TiB_2 content in the metal (calculated from the total titanium analysis in the pad) was slightly higher (0.0015%) than that predicted from the published TiB_2 solubility data (8) and varies with

$$\log K_{sp} = 5.22 - \frac{1.62 \times 10^4}{T(k)}$$

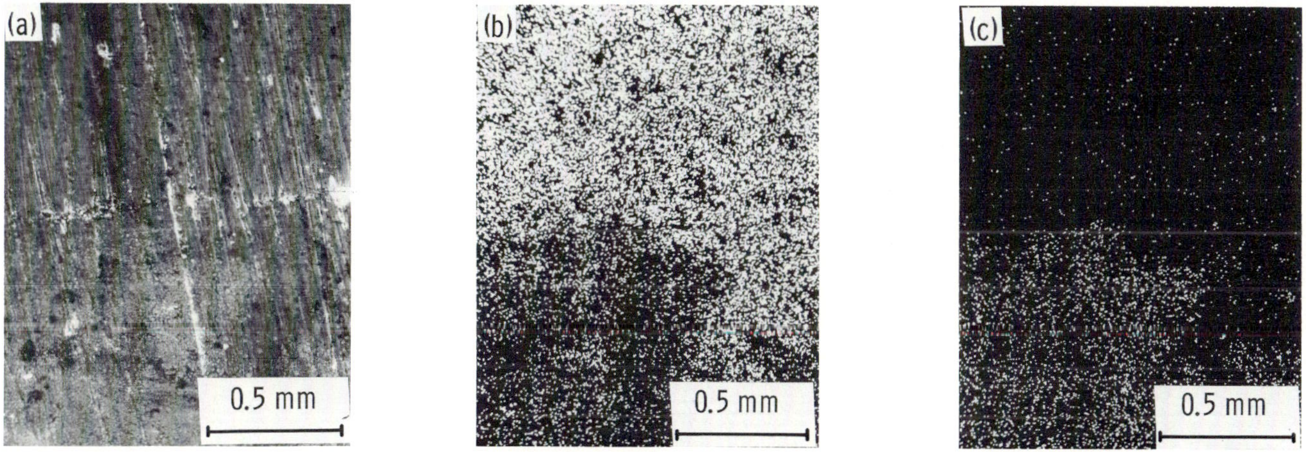

SEM | Aluminum Distribution | Titanium Distribution

Figure 6. SEM photograph and EDS maps across the interface between the aluminum pad (top) and TiB_2 coating material (bottom) in a core sample taken from the coated cathode of a conventional cell after 109 days of operation.

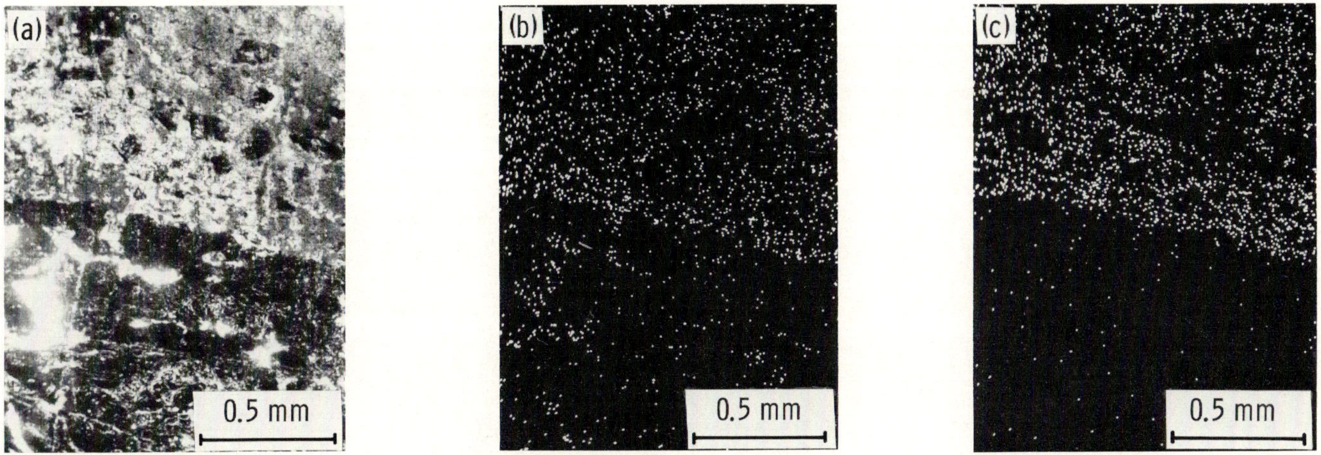

SEM | Aluminum Distribution | Titanium Distribution

Figure 7. SEM photograph and EDS maps across the interface between TiB_2 coating material (top) and cathode block (bottom) in a core sample taken from the coated cathode of a conventional cell after 109 days of operation.

temperature as anticipated. The Ti/B ratio is not stoichiometric in the metal pad since boron is continuously and irreversibly lost to the bath and in the cell fumes. Typical Ti and B levels in the test cell metal pads were 0.0075-0.0110% and 0.0012-0.0018%, respectively, compared to plant average values of 0.0035-0.0060% and 0.0006-0.0010%, respectively. The plant average Ti values were subtracted from the test cell values to obtain the contribution attributable to the coating. The coating life calculation assumes a uniform loss across the entire coated surface. This assumption appears to be valid since there has been no evidence of any lack of uniformity in wear rate even after approximately 700 days operation in the longest running test cell.

Based on the performance up to the end of 1982, we make the following general observations on the differences in performance of the six coated cathode test cells and the rest of the plant:

1. The cathodes tended to run cleaner and recover from upsets more rapidly, particularly with respect to ledge mobility and poor muck adhesion,
2. The coated cathode collector bar current distribution was more uniform,
3. The coated cathode voltage drop was lower
4. The test cell current efficiency was significantly higher,
5. The specific energy consumption of the tests cells in 1982 was lower than the plant average by approximately 0.23 kWh per pound of aluminum produced,
6. The metal grades and cast house operations were not affected by the coated cathodes.
7. The coated cells were easier to operate and were well accepted by the plant personnel.

Conclusions

A low-cost TiB_2 cathode coating with good durability and operational benefits has been demonstrated in conventional VSS aluminum reduction cells. Cells can be retrofitted with the cathode coating and started up without any major disruptions in plant routine. Shutdown and restart of the coated cathodes have no apparent detrimental effect on the TiB_2 coating. Based on the projected coating life of 3-5 years, and the lower processing costs of coating compared to hot pressing, the TiB_2/carbon composite is clearly an economically viable cathode material suitable for use in future drained cathode low-energy cells and present day cells with deep metal pads.

This plant test was originally undertaken primarily to demonstrate a viable TiB_2 cathode material of adequate longevity and durability for drained cathode cells. Secondary results indicate that significant benefits can be gained from the use of a TiB_2/carbon surface in conventional Hall-Heroult cells. These improvements include facilitated cell operation and better current and energy efficiency. Calculations for return on investment indicate that the present payoff period for The Dalles cells appears to be approximately half of the anticipated coating life. The economic benefits will improve substantially if any increase in overall cathode life is realized. An extension in service life of just one year would more than pay for the coating since the expensive and disruptive relining is done less frequently. A complete covering with molten aluminum is expected to protect the bottom from attack by the aggressive bath and muck constituents. Increased longevity is therefore anticipated in the coated cells since the TiB_2 surface is covered exclusively by aluminum for a greater proportion of the time than a conventional carbon surface.

The lack of an economic, durable TiB_2 cathode material has been a major obstacle in past attempts to develop a low-energy cell that can be readily retrofitted into existing smelter facilities. Our new cathode coating material appears to be the key to the development of a viable LEC.

Future

Improved coating formulations and processing techniques are presently being tested in another 14-cell plant test at the Martin Marietta Aluminum Goldendale, WA, smelter as a joint project with the Bonneville Power Administration. The results of this test will be published in the near future. Materials specification tests are continuing to optimize the trade-off between coating quality and material cost.

Acknowledgement

The authors wish to thank Martin Marietta Aluminum Inc., for permission to publish this work, and The Dalles plant management for assistance in setting up and monitoring the full scale tests. Skillful technical work was performed at the plant by K. Ries and M. McClean and at Martin Marietta Laboratories by Dr. D. Townsend, M. Buchta, U. Patel, and W. Montague.

References

1. Grjotheim, K., Thonstad, J., Matiašovský, K., and Malinovský, M., "Energy Efficiency and Bath Composition in Aluminum Electrolysis," Aluminium, 49, 12 (1973) pp. 803-807.

2. Billehang, K., and Oye, H.A., "Inert Cathodes for Aluminum Electrolysis in Hall-Heroult Cells," Aluminium, 56 (1980) pp. 642-648 and pp. 713-717.

3. Long, G., and Foster, L.M., "Aluminum Nitride, A Refractory for Aluminum to 2000°C," Journal of American Ceramic Society, 42, 2 (1959) pp. 53-59.

4. McCauley, J.W., and Corbin, N.D., "Phase Relations and Reaction Sintering of Transparent Cubic Aluminum Oxynitride Spinel (ALON)," Journal of American Ceramic Society 62, 9-10 (1979) pp. 476-479.

5. Kaplan, H.I., et al., "Cathodes for Alumina Reduction Cells," U.S. Patent No. 4,333,813 (1982).

6. Payne, J.R., "Bonding of Refractory Hard Metal," U.S. Patent No. 4,093,524 (1978).

7. Kaiser Aluminum and Chemical Corporation, "Energy Savings Through the Use of an Improved Aluminum Reduction Cell Cathode," DOE Contract DE-AC)#-76CS 40215.

8. Finch, N.J., "The Mutual Solubilities of Titanium and Boron in Pure Aluminum," Met. Trans., 3 (1972) pp. 2709-2711.

CARBON PARTITION BETWEEN CARBOTHERMICALLY REDUCED Al AND Al_2O_3-Al_4C_3 MELTS

Lee M. Ruch, A. F. Saavedra, R. M. Kibby

Reduction Laboratory
Reynolds Metals Company
P.O. Box 1200
Sheffield, Alabama 35660

In direct reduction of Al_2O_3, Al_4C_3 is an important intermediate product and contaminant in the effluent aluminum. Solution models were developed to estimate the partition of carbon between Al-C-O and Al-C melts to be expected in the production of Al by direct carbothermic reduction. Predictions of equilibrum tie lines between the Al-C-O liquids along the Al_2O_3-Al_4C_3 pseudo-binary and Al-C alloys are presented, and compared with experimental data relating the carbon content of carbothermically reduced Al to that of the parent slags. The model is conservative in its prediction of effluent Al_4C_3 levels under actual reduction conditions.

Introduction

Aluminum carbide is an important intermediate product in the carbothermic reduction of alumina. It inevitably becomes incorporated into the metal product to some degree. In Al-C alloys containing more than a few weight percent aluminum carbide, the precipitation of carbide in an inter-locking plate morphology renders the alloys viscous and difficult to purify by conventional fluxing techniques. Since the formation of aluminum carbide from alumina and carbon is a strongly endothermic reaction, purification processes which result in conversion of the carbide back to oxide to effect its removal are likely to consume excessive energy if appreciable amounts of carbide are present in the metal. Therefore, it is important to investigate the extent to which carbon contamination can be avoided under various reduction conditions.

Several reviews of the high temperature chemistry of the Al-C-O system are available (1, 2, 3), and it is widely known that at temperatures sufficiently high to effect reduction, the aluminum product is formed from decomposition of an oxycarbide melt (referred to as "slag"). If this decomposition is conducted in the absence of "reactive carbon" (17), it is reasonable to expect that the partition of carbon between the parent oxycarbide melt and the supernatant metallic melt can be described in terms of an equilibrium distribution of carbides:

$$\underset{\text{(slag)}}{Al_4C_3\ (l)} \rightleftarrows \underset{\text{(metal)}}{Al_4C_3\ (l)} \qquad (1)$$

The description of this equilibrium entails solution modeling of the two phases as outlined below.

Review of Previous Solution Models

In light of the difficulties of making thermodynamic measurements at temperatures of practical interest, especially in the phase field in which liquid Al, and molten oxycarbides (from which Al_4C_3 and Al_4O_4C may precipitate) coexist, solution models assume great importance in carbothermic process development. Cochran's initial treatment (4) and subsequent comprehensive development (5) of solution behavior in the Al-C and Al_2O_3-Al_4C_3 phases formed the basis for more extensive reactor modeling which extended into the Al-Si-C-O system (6). Recently, Sandberg and Motzfeldt (7, 8) have developed a model of solution behavior of Al_2O_3-Al_4C_3 melts based on ideal Temkin mixing of a plausible set of ionic constituents. Although not explicitly stated, Dewing's assessment of activities in the Al_2O_3-Al_4C_3 system may be inferred from his presentation of CO isobars (9).

In formulating solution models which can predict the metal-slag carbon distribution, a number of assumptions are necessary. In models in which the principal precipitating phases (Al_4C_3, Al_2O_3, and Al_4O_4C) are chosen as solution constituents, the thermodynamic properties of crystalline phases, the heats and entropies of melting of these compounds, and a description of their interaction in solution (e.g. ideal, regular solution) are required.

By assuming ideal mixing (of Al, C, and Al_4C_3 in the metal and Al_2O_3, Al_4O_4C, and Al_4C_3 in the slag) and choosing the set of thermodynamic properties shown in Table 1, Cochran derived a model which he used to calculate liquid-vapor equilibria and carbon partition between the two liquid phases. The model shows good agreement with the lower branch of the Al-C liquidus as determined by several experimental studies (10, 11, 12, 13) and with the points of peritectic decomposition of Al_4C_3 and Al_4O_4C in the metal and slag, respectively.

Table I. Assumed Thermodynamic Properties of Al_4O_4C and Al_4C_3

	Al_4C_3		Al_4O_4C	
	Present Work	Cochran(5)	Present Work	Cochran(5)
ΔS_m (eu/mol)	19.10	17.50	20.27	22.50
ΔH_m (kcal/mol)	57.01	56.35	54.01	52.24
T_m (°K)	2985	3220	2664	2332
ΔH_{f298} (kcal/mol)	-49.54	-46.27	-557.82	-553.54

The agreement of the ideal solution model of the slag with the experimental work of Ginsberg and Sparwald (11) along the liquidus curve above the Al_4O_4C peritectic is not as good however (6).

It must be recognized that in view of the inaccessibility of measured values of entropies of melting of Al_4O_4C and Al_4C_3 and uncertainties of extrapolation of other high temperature thermodynamic properties, a number of equally plausible sets of thermodynamic data can be selected to form the basis of solution models which can be fit to the available phase diagram information. The selection task facing the modeler is formidable if internal consistency of the model is to be maintained.

Development of Model

The present model developed from efforts to modify Cochran's ideal solution model to achieve a better fit of the Al_4C_3 liquidus curve above the Al_4O_4C peritectic to the data of Ginsberg and Sparwald (11). Values of thermodynamic properties of solid Al_4C_3 and Al_4O_4C were taken from a critical assessment by Belton (14) and values for other components were taken from JANAF (15, 16). The initial approach was to treat the entropies and heats of melting Al_4C_3 as adjustable parameters (within the limits of believable melting temperatures of the compounds). Since the peritectic points in both metal and slag systems have been determined experimentally and the Al_4C_3 liquidus curves are of greatest interest to the problem of carbon partition, agreement of the calculated curves with above mentioned experimental data was taken as the criteria of success of a given combination of assumed values. We could not find a reasonable set of values which would improve Cochran's fit of the two Al_4C_3 liquidus curves using the ideal solution model; therefore, it appeared that an additional adjustable parameter was required.

This parameter, α, was introduced by adopting a "regular solution" relationship between the activities of the assumed components of the oxycarbide melt as expressed below:

$$\ln\left(\frac{a_i}{N}\right) = \frac{\alpha}{RT}(1-N)^2 \qquad (2)$$

where a_i denotes the activity of the species Al_2O_3 (l), Al_4C_3 (l), and Al_4O_4C (l), R is the gas constant, and N is the mol fraction of component i. Using this parameter and making slight adjustments in the assumed entropies of melting of Al_4C_3 and Al_4O_4C resulted in a significant improvement in the fit of the liquidus lines with experimental data as shown in Figure 1. The best fit was obtained with a value of -6150 cal/mol for α. The liquidus curves in Figure 1 were calculated from the governing solution models (ideal in the Al-C-Al_4C_3 system and "regular" in the Al_2O_3, Al_4C_3, Al_4O_4C system), mass balance equations, and equilibrium constants derived from the appropriate exchange reactions and the assumed thermodynamic properties of the liquid components. Calculations were performed with the RMC program "MODEL3" in APL language.

MODEL3 also contains the function "PERCARB" which calculates the carbide content of solidified metal which was in equilibrium with slag of a given composition and temperature. This calculation invokes the solution model to calculate the activity of Al_4C_3 in the slag which is then set equal to the activity of Al_4C_3 in the metal phase. The solution model is again invoked to arrive at the composition of the metal expressed as mol fraction carbon, N_c. To convert N_c to practical units of weight percent Al_4C_3 in the solidified product, it is assumed that all the carbon present in the melt at high temperature converts to Al_4C_3 during cooling to the Al freezing point (at which the solubility of C in Al is negligible).

Model Results

The results of the calculated carbon partition for temperatures of 2000, 2100, and 2200°C are shown in Figure 2. The abscissa is expressed in terms of mole fraction Al_2O_3 or N*, as defined by Cochran. The curves terminate at the compositions of the Al_4C_3 liquidus in the slag, and they are expected to continue as horizontal lines (not shown), since the carbide activity is independent of N* in the two phase region. The curve shown for 2200°C represents an extrapolation of the model, since it is above the carbide peritectic temperature. The dashed curves in Figure 2 were derived by converting Cochran's values of N_c to w/o Al_4C_3 as described above. The agreement between the two models is reasonable, considering the sensitivity of the activity of Al_4C_3 to the assumed values of properties of liquid Al_4C_3.

Using the model values for activities in the liquid slag phase field, the equilibrium constants for production of Al, CO, and Al_2O from the liquid slag components, and the production of Al_4O_4C and CO by the reaction

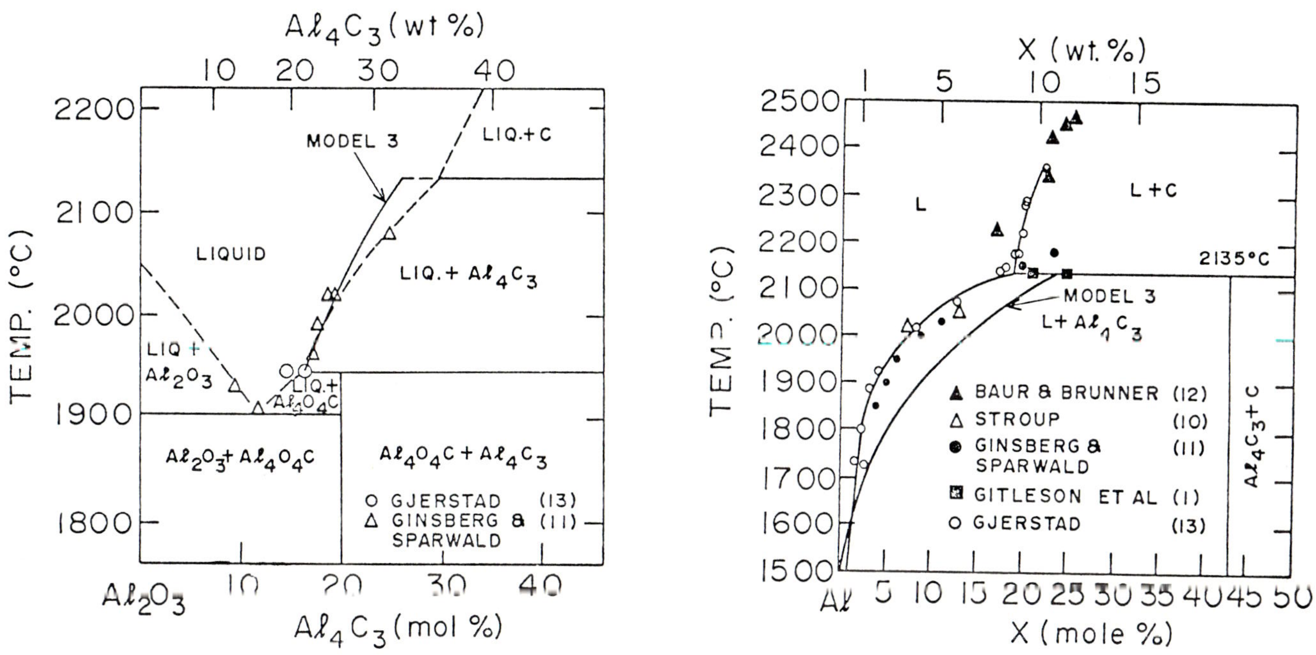

Figure 1.

Phase Diagrams of Al_2O_3-Al_4C_3 Pseudo-binary and Al-C Binary Systems

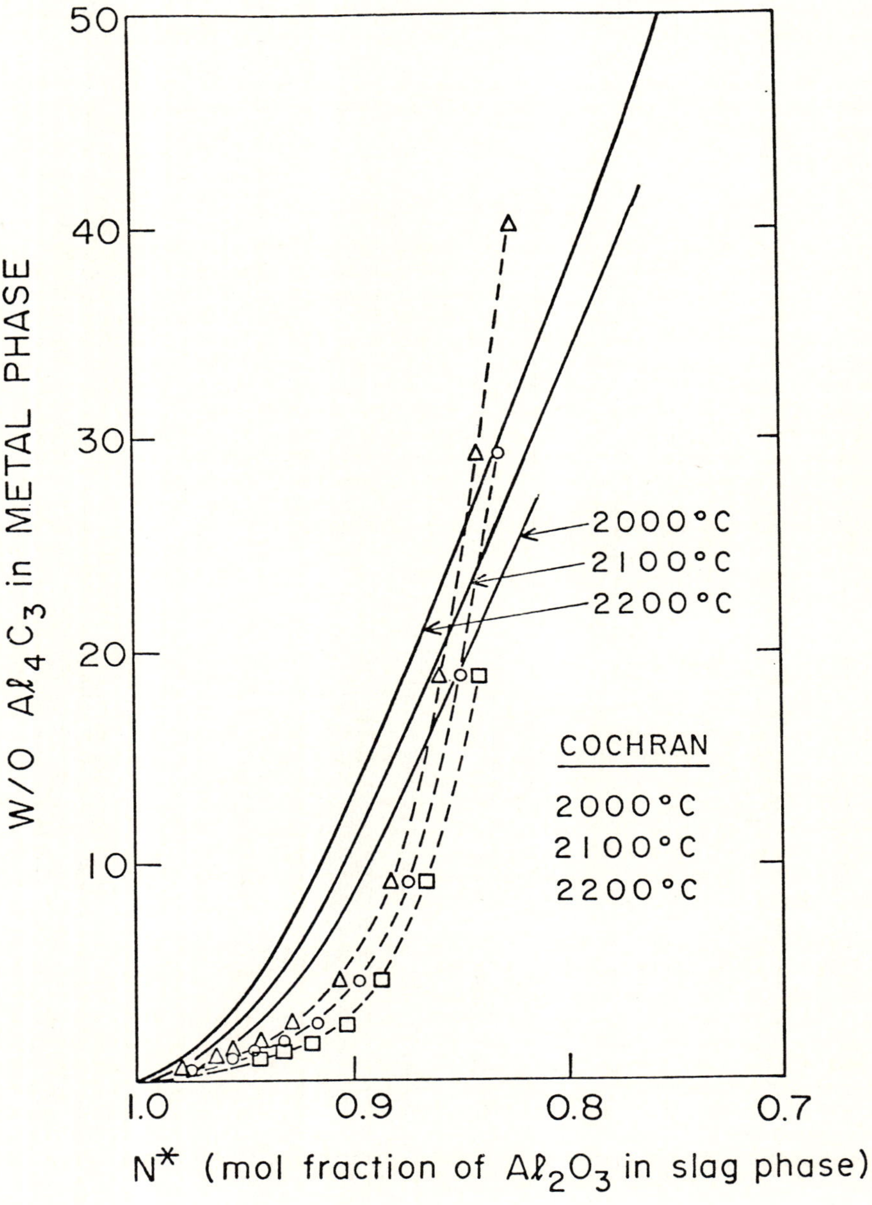

Figure 2.
Predicted Carbide Distribution Between Metal and Slag

of carbon with the slag may be calculated. The good agreement found between the composition dependence of the temperature at which P_{CO} equals one atmosphere for these reactions and the values published by Dewing et al (9) provides additional confirmation of the validity of MODEL3.

The relationship of Figure 2 showing percent aluminum carbide in metal that was in equilibrium with slag only applies in the absence of reactive carbon in the slag.

Experimental Procedures

Achieving equilibrium contact between slag and metal in the absence of carbon is thought to be practically impossible using externally heated laboratory furnaces (i.e., induction or resistance tube furnaces) because of the lack of a crucible material (other than graphite) which can contain metal and slag simultaneously. Internally heated electric furnaces, in which the maximum temperatures are achieved near the center of the crucible, can overcome this problem if they are operated in a manner which allows formation and maintenance of a frozen layer of material next to the carbon wall. Internally heated furnaces are more difficult to control because of the complex interaction of the electric current and the charge materials in various stages of conversion, however. Furthermore, they must be large enough to possess sufficient thermal inertia for stable operation at elevated temperature. Submerged arc furnaces such as those used for smelting of Si and ferro alloys routinely solve the problems of containment, energy delivery, and thermal stability in industrial scale operation.

Experimental Conditions

Experiments were conducted under slag decomposition conditions in which the metal was produced by decomposition of slag in situ. The tests utilized open, submerged arc furnaces lined with carbon and equipped with charge control and tapping systems which allowed continuous operation. Electric current was introduced through graphite electrodes, and charge materials were added directly to the hearth or to charge columns maintained above the melt. A number of furnace and electrode geometries including d.c. electrode to hearth, and multiple top entry electrode (a.c. and d.c.) were tested.

The furnaces were somewhat similar to those described by Bruno (6). Metallurgical alumina and petroleum coke (calcined and/or green) in various forms of agglomeration were intermittently fed to the furnace as power was applied, and the accumulated slag and metal were tapped periodically. Samples were selected from metal and slag which had been tapped simultaneously, and from dip samples frozen to steel bars taken during periods in which distinct pools of metal and slag were present in the furnace. Judging from the rate of evolution of CO, the production of metal took place up until the time of the taps. The presence of frozen skulls at the hearth and sidewalls prevented the carbon lining from dominating the chemical equilibrium between metal and slag.

Furnace operations were conducted so that the slag was substantially depleted of reactive carbon as the time for tapping was approached.

Sampling and Analysis. Samples of metal and slag tapped from the submerged arc furnace used in the slag decomposition experiments were selected from composite ingots of slag and metal. Since the metal phase collected on the surface of the slag pool within the furnace, successful taps consisted of a flow of metal followed by slag. The liquids were tapped into a cast iron ingot mold. Usually it was possible to separate the bulk metal from the slag, although mixing within the chill complicated the analysis. The slag was separated, crushed to minus 1/2-inch fragments, riffled, and sampled for analysis. Analytical samples were stored in desicators, pulverized to -200 mesh, and analyzed for metallic Al, Al_4C_3, Al_2O_3, and total C.

When the taps had sufficient superheat and/or thermal mass, the two liquids had time to separate into discrete layers of slag and metal before freezing in the chill. In these cases the frozen Al-Al_4C_3 ingot could be completely recovered. Ingots of Al-Al_4C_3 alloy could be crushed with some difficulty (due to their ductility) into fragments small enough for riffle sampling. The metal fractured along the faces of yellow Al_4C_3 plates and fragments always appeared yellow. Al_4C_3 hydrolyzes rapidly in contact with ambient moisture, and special care had to be exercised to prevent weathering of analytical samples.

When tapped samples froze rapidly in the chill, quantitative separation of slag and metal was difficult, even though in some runs the chill was lined with insulating refractory. In these cases the ingots were fragmented with a 100-ton hydraulic press and roll crushed. The ductility of the metal particles allowed them to flatten into discrete wafers which were retained on a 12-mesh screen which passed the pulverized slag fines. Analysis of metal recovered by roll-crushing was sometimes suspect, however, due to inclusion of non-metallic material in it, and carbide partition data from metal samples containing more than about 25 w/o non-metallic (excluding Al_4C_3) was discarded. When discrete metal ingots were not recovered, metal in the taps was sampled by drilling samples from pockets of metal which had been trapped in the slag.

When the furnace was tapped prematurely (i.e., before the metal and layers had coalesced), slags frequently contained finely divided particles of metal distributed throughout. It was not possible to separate this metal quantitatively by roll crushing, but composite analysis of slags sometimes showed 1-2 w/o Al as determined by hydrogen evolution during acid hydrolysis.

Metal samples were analyzed for metallic Al and Al_4C_3 by volumetric determination of hydrogen and methane in sealed hydrochloric acid columns. Hydrolysis of Al_4C_3 was catalyzed by additions of $HgCl_2$ to the columns. The samples were also analyzed for total carbon by combustion with O_2 in a standard Leco infrared carbon analyzer. Cu and Sn accelerators were used in the combustion. The residue after boiling in concentrated HCl and firing in air and the total Al content from carbonate fusion and subsequent oxide precipitation were also determined to provide total sample mass balances. Some samples contained minor amounts of Fe and Si which were determined photometrically.

Experimental Results

Figure 3 shows the carbon contents of Al-C alloys as a function of the carbon content of the parent slags. The values of w/o carbon of the slags were plotted directly from the combustion analysis and can be related to the N* of the slag by a simple mass balance if no free carbon is present in the slag (i.e. if all carbon is combined in the form of Al_4C_3 or Al_4O_4C). The corresponding values in the metal were normalized to correct for non-metallic material (typically 10-15 w/o) in the metal phase.

The MODEL3 carbon partition curves at 2000 and 2200°C from Figure 2 are replotted in Figure 3 using the assumption that all carbon in the metal and slag is in the form of Al_4C_3 or Al_4O_4C and Al_4C_3, respectively. Since accurate temperature measurements in the interior of the furnace were not obtained, one might question the validity of direct comparison of the data with the theoretical curves. However, since the metal recovered in these experiments definitely was produced by decomposition of slag at atmospheric pressure, it is reasonable to expect that equilibrium between the metal and parent slag prevailed. This equilibrium is expected to be reflected in the chemical analysis of the quenched products, provided that ambient conditions did not change between the time of decomposition of the slag and the tapping of the products. As noted previously, metal producing conditions (as indicated by gas evolution) usually prevailed at the times of the taps. Furthermore, it is expected that slag decomposition at atmospheric pressure takes place in the approximate temperature range 2050-2150°C (9). Therefore, it is concluded that the comparison is valid.

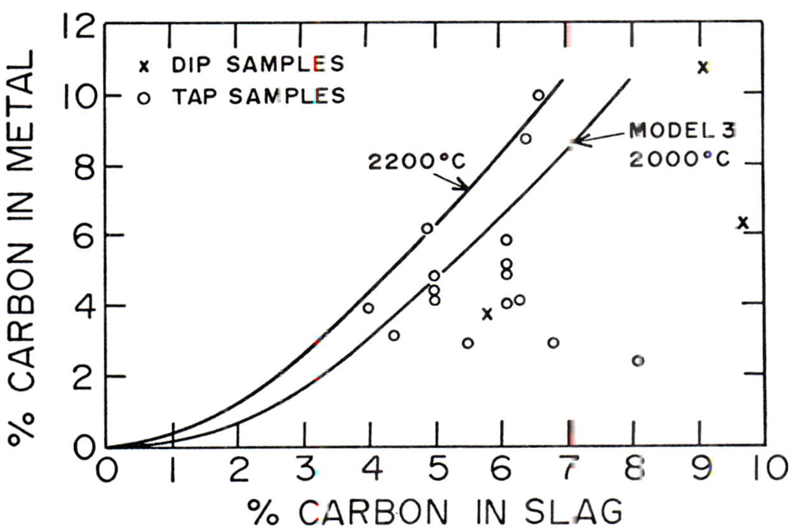

Figure 3.

Experimental Carbon Distribution Between Metal and Slag

The data presented thus indicates that the MODEL3 predictions of carbide content of metal are conservative; i.e. the carbon contents of effluent metal are generally lower than that predicted by MODEL3 for the co-existing slags in the composition range shown. At this point it is not clear whether this difference lies in the thermodynamic assumptions underlying the model or in the assumption that the metal and slag were in fact in equilibrium.

Recommendations for Further Work

1. Further refinements of the model in light of more recent data (8) to explore the possibility of improving the agreement with carbon partition data while preserving agreement with the phase diagrams (assuming that the data represents equilibrium conditions).

2. Exploration of the application of the ionic model recently developed by Sandberg and Motzfeldt (7, 8) to the question of carbon partition.

3. Development of more refined experimental techniques (especially measurement of reaction zone temperatures) for determining equilibrium carbide partition in the appropriate composition and temperature ranges.

4. Definition of the heat and/or mass transfer conditions most likely to account for the apparent deviations from equilibrium on the basis of kinetic limitations.

References

1. G. Gitlesen, O. Herstad, and K. Motzfeldt, "Chemical Investigations Concerning the Carbothermic Reduction of Alumina," pp. 179-196 in Selected Topics in High Temperature Chemistry, T. Førland, K. Grjotheim, K. Motzfeldt, and S. Urnes, ed.s; Unversitetsforlaget, Oslo 1966.

2. K. Motzfeldt and B. Sandberg, "Chemical Investigations Concerning Carbothermic Reduction of Alumina," pp. 411-428 in Light Metals 1979, Warren S. Peterson, ed.; AIME, New York, 1979.

3. K. Motzfeldt, "Method for Carbothemic Production of Aluminum," U.S. Patent 4,314,846, February 9, 1982.

4. C. N. Cochran, "Calculated Equilibrium for Carbothermic Reduction of Alumina," Paper Presented at TMS-AIME Annual Meeting, New York, NY, 1966.

5. C. N. Cochran, "Carbothermic Production of Aluminum,," US Patent 3,971,653, July 27, 1976.

6. M. J. Bruno, "Production of Aluminum-Silicon Alloy and Ferrosilicon and Commercial Purity Aluminum By the Direct Reduction Process," Technical Report, Aluminum Company of America prepared for the Department of Energy, February 1983.

7. B. Sandberg and K. Motzfeldt, "Liquid Aluminum Oxide-Aluminum Carbide Mixtures Considered as Ionic Melts," p. 127 in Proc. Conf. Electrochem. Soc. of Japan, Kyoto, Japan, April 1983.

8. B. Sandberg, "Karbthermisk reduksjon av aluminiumoxsyd," Thesis, Universitetet i Trondheim, Norges tekniske høgskole, Trondheim, Norway, June 1981.

9. E. W. Dewing, et al, "Process for the Production of Aluminum," US Patent 4,099,959, July 11, 1978.

10. P. T. Stroup, "Carbothermic Smelting of Aluminum," pp. 356-372 in Transactions of the Metallurgical Society of AIME, Vol. 230, April 1964.

11. H. Ginsberg and V. Sparwald, "Beitrage zur Aluminiumgewinnung durch Carbothermische Reduktion des Aluminiumoxids unter Besonderer Berucksichtigung des Systems Aluminium-Kohlenstoff" (Contribution on the Production of Al by Carbothermic Reduction of Alumina), Aluminium, 41(1965) pp. 181-193 and 219-230.

12. E. Baur and R. Brunner, "Uber die Schmelzflache im System Aluminium, Aluminiumoxyd, Aluminiumcarbid," Z. Elektrochem 40(1934) pp. 154-158.

13. S. Gjerstad, "Kjemisk-Metallurgiske undersøkelser vedrørende karbotermisk reduksjon av aluminiumoksyd og silisiumoksyd," Thesis, Norges tekniske høgskole, Trondheim, Norway, December 1968.

14. Private Communication, G. R. Belton, University of Strathclyde, Glasgow, December 1974.

15. "JANAF Thermochemical Tables," D. R. Stull and H. Prophet ed.s, US Department of Commerce, National Bureau of Standards, June 1971.

16. M. W. chase et al, "JANAF Thermochemical Tables, 1978 Supplement," J. Phys. Chem. Ref. Data 1(1978) pp. 793-940.

17. R. M. Kibby, "Aluminum Purification System," US Patent 4,216,010, August 5, 1980.

THE PRODUCTION OF RAW Al/SI ALLOYS FROM ITALIAN LEUCITIC MINERALS BY A DIRECT CARBOTHERMIC PROCESS. EXPERIMENTS AND RESULTS ON A PILOT PLANT

M.F.Landi, S.Da Roit, S.Montesi and L.Piras

Alluminio Italia S.p.A
Centro Ricerche.P.Marghera, Via delle Industrie 26,Venezia,Italia
Centro Ricerche. Portovesme,(CA).Italia

The results obtained on a semi pilot scale in the investigation for a direct carbothermic reduction in an arc furnace of poor silico-aluminous minerals, namely Italian leucitic rocks, induced Alluminio Italia to start up a second phase of research the testing on a pilot scale of such results.

To this purpose a pilot plant with a capacity of 1 t/d of raw alloys has been installed by Alluminio Italia at Portovesme.

The plant consists of a section for the set-up of the charge, a section for alloy production, including a triphase arc furnace with graphite electrodes fed by a 1500 KVA transformer, and ancillary equipment.

This paper describes the plant features, the operating conditions adopted and the results obtained.

The program foresees a third phase, that is, a refining process for the raw alloys produced.

Research sponsored by the Commission of the European Communities - CREST - under contract with Soc. Alluminio Italia.

Foreword

This paper follows a previous report presented at the 110th AIME Annual Meeting held in 1981 (1).

The above report illustrated the results obtained on semipilot scale in an investigation aiming at the achievement of raw Al/Si alloys through direct reduction in an arc furnace of a poor aluminous mineral present in fairly good deposits in Italy, that is, the leucitic rocks. It also reported on what bases Alluminio Italia had resolved to carry out such investigation with the financial support CEE (CREST).

This preliminary investigation was followed by the design, construction and installation of a pilot plant by Alluminio Italia at Porto Vesme (Sardinia).

The present paper describes the tests conducted as well as the results obtained on the pilot plant.

Raw Materials

Aluminous Mineral

A leucitic rock was pretreated through grinding and magnetic separation to achieve a percentage lower than 1% in the iron oxide content.

Deferrized leucitic rock - Typical analysis

H_2O	0.96 %
I.L.	1.18 "
SiO_2	53.80 "
Al_2O_3	23.75 "
Fe_2O_3	0.90 "
TiO_2	0.11 "
MgO	0.22 "
MnO	0.12 "
K_2O	15.80 "
Na_2O	1.85 "
CaO	0.62 "
P_2O_5	0.22 "
Other impurities	0.45 "

Particle size analysis

>	3 mm	0.00 %
3 -	2 "	31.70 "
2 -	1 "	42.50 "
1 -	0.54 "	10.80 "
0.54 -	0.34 "	5.80 "
0.34 -	0.09 "	3.00 "
0.09 -	0.06 "	1.70 "
<	0.06 "	4.50 "

Mineralogical composition

- leucite	$K_2O \cdot Al_2O_3 \cdot 4SiO_2$
- orthoclase (sanidine)	$K_2O \cdot Al_2O_3 \cdot 6SiO_2$
- augite	$CaMg(SiO_3)_2 + (Mg,Fe)(Al,Fe)_2 \cdot SiO_6$
- magnetite	Fe_3O_4
- biotite	$(K,H)_2(Mg,Fe)_2(Al,Fe)_2(SiO_4)_3$
- basic plagioclase	$Na_2O \cdot Al_2O_3 \cdot 6SiO_2 + CaO \cdot Al_2O_3 \cdot 2SiO_2$
- apatite	$CaF_2 \cdot 3Ca_3(PO_4)_2$

Raw Material Preparation

Using a technique developed and experienced on the pilot plant, the leucitic rock is ground and added-up with the reducer, ground in its turn, and subsequently converted to pellets having a diameter of ca 15-20 mm. The dried pellets and after charging they reach the reduction zone in the furnace, practically unaltered. Small amounts of clay and starch are also added.

Binder clay composition

I.L.	13.21 %
SiO_2	48.50 "
Al_2O_3	34.60 "
Fe_2O_3	0.97 "
TiO_2	0.80 "
Na_2O	1.20 "
CaO	traces
MgO	"
Other impurities	0.72 "

Reducers. Among the various reducers tested, i.e. lignite, crude and calcined coke, anthracite, lithantrax, this latter was selected. Even though lignite actually gave the best results, it was discarded for various reasons, the main being the low fixed carbon content.

Lithantrax Composition

	H_2O	2.40 %
	Vol. matter	26.34 "
	Fixed carbon	67.99 "
	Ashes,	3.27 "
where	Fe_2O_3	0.26 "
	Al_2O_3	0.62 "
	SiO_2	0.73 "
	S	0.87 "
	Other impurities	0.79 "

As it occurs in similar processes variable amounts of wood chips had to be added. The fixed C content in the chips was taken into account by 80% in its carbon content requested by the reduction process.

The average charge composition, during regular running, was as below:

Charge Composition (Pellets)

Vol. matter + I.L. + H_2O	9.90 %
SiO_2	38.64 "
Al_2O_3	17.73 "
Fe_2O_3	0.71 "
TiO_2	0.11 "
MgO	0.15 "
MnO	0.08 "
K_2O	10.65 "
Na_2O	1.30 "
CaO	0.42 "
Fixed C	19.12 "
Other impurities	1.10 "

Fixed carbon percentage corresponds to approx. 87.5% of the stoichiometric value needed for the reduction process.

Installation of the Pilot Plant

General

The plant design and the direction of the construction works have been carried out by the Alluminio Italia Technical Office.

The total covered area of the Pilot Plant is 36 x 24 m^2. It is designed for a processing capacity of about 5 to 6/t d of mineral and a production of approx. 1 to 2 t/d of raw Al/Si alloy.

Lay-out

As it appears from the lay-out of Figure 1, the plant can be divided into three main sections: the cold area (raw material dressing and dust collector), the hot area (arc furnace and casting facilities), interfaces and ancillary (furnace feeding system, refrigeration system, hydraulic station, fumes collection and treatment).

- The cold section (See also process scheme).

  The leucite is transferred by a bucket conveyor (1) to a hopper (2) and fed to a ball mill (4) by a belt conveyor (3). The ground leucite is transferred by a bucket elevator (5) to the leucite silo (6), the clay used as a binder and the reducer are stored in silos (7) and (8) respectively. Coal of the proper particle size is transported by lorry and launched into silo (8) using compressed air.

  Theree screw conveyors (9) bring the materials to the weighing system (10) and then the proper quantities are mixed together at the mixer (11) and stocked in a hopper ;a screw conveyor brings this mix to the pellet mill (12). The pellets are dried in a conveying screen drier (13) and stocked in containers (14). A dust collecting system with bag filters (16) provides an inside environment almost free from dust.

  The main features of the major components in the cold section are :

  - Ball mill

    Cylinder

- Internal diameter	1.43	m
- length	2.54	"
- rotation speed	26.1	rpm
- power	55	KW
- input particle size	2-3	mm max
- output particle size	0.3	"
- production capacity	3.0	tons/hour

Process Scheme

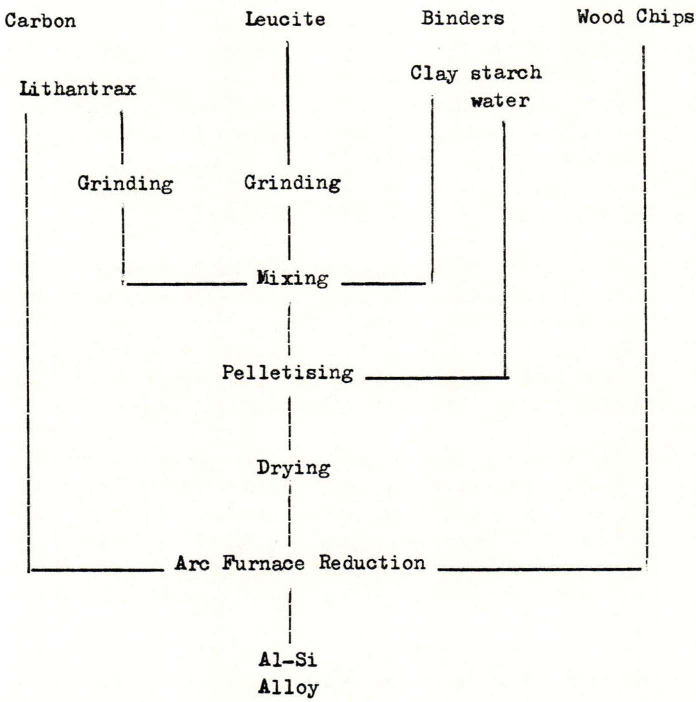

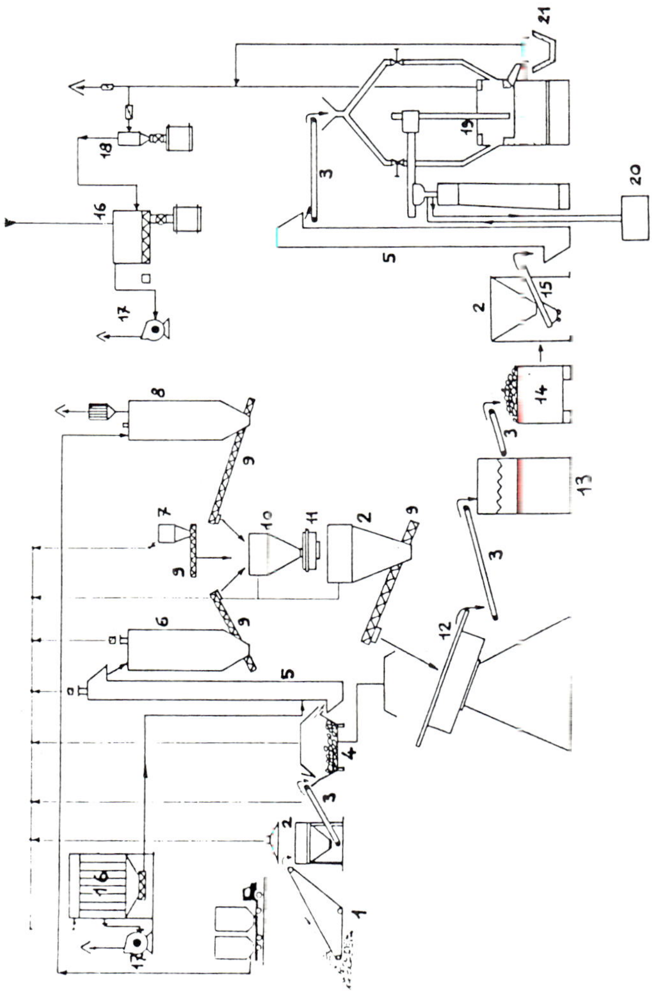

Figure 1 - Pilot plant. Lay-out.

- Silos capacity

  - leucite silo 30 m^3
  - coal silo 30 "
  - clay silo 0.5 "

- The hot section (See Fig. 2)

  The major component of the hot section is a DEMAG three phase electric arc furnace with fixed and hooded tub. Its main features are

  - Mechanical

    - tub diameter 3000 mm
    - " height 2600 "
    - crucible diameter 1850 "
    - " height 1550 "
    - 1 metal tap hole

  - Electrodes

    - type graphite
    - diameter 305 mm
    - travel distance 1600 "
    - electrodes circle diameter 700 to 950 mm

  - Electrical

    - transformer power 1500 KVA
    - voltage, primary winding triphase 6 KV - 50 Hz
    - voltage, secondary winding (22 steps) { 180 to 116 V at constant power; 116 to 40 V at constant current }
    - Electrodes connection : Δ

- Ancillary

  The most significant components are

  - a hydraulic station to move the electrodes
  - a water refrigerating system
  - a system for fumes collection and treatment

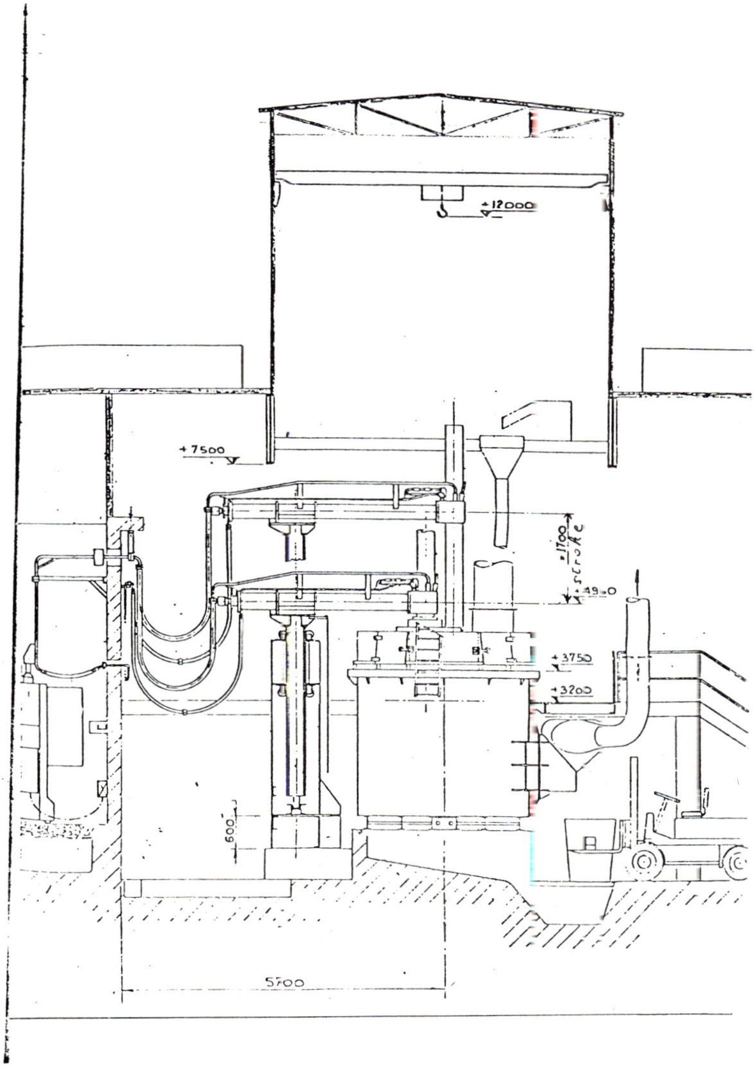

Figure 2 - Pilot plant. Arc furnace.

Tests Carried out and Results

Observations

Due to the high percentage of alkalies present in the charge, particular conditions should be adopted in running the arc furnace to achieve a complete reduction of the charge.

The behaviour of the charge in this reduction process in fact completely differs from that of the minerals so far used to this purpose, such as, for instance, the clays.

The reduction conditions present a very narrow range because of the electrical conductivity of the charge. In order to reach the temperature needed for Al_2O_3 reduction about 2200°C, (as predicted by thermodynamic data and experimentally verified in our tests), the furnace operation has to be resistance-type. The reducing agent must be highly reactive. The charge should be dressed in such a way as to lower its conductivity as much as possible. An adequate porosity is requested to allow a quick removal of the gas CO and the fumes containing alkalies, Al, Si, Al_2O_3, SiO_2 (perhaps some Al_2OC).

The fused metal needs to be cast as often as possible (by an industrial run, in a continuous way).

The following electrical parameters were adopted as steady state conditions

- Current intensity	6.8 to 7.4 KA
- Current density	9.3 to 10 A cm^2
- Voltage electrode to sole	30 to 35 V
- Active power	600 to 680 KW
- Reactive power	400 to 500 KVAR
- Cos φ	0.81 to 0.83

The wood chips were added to the charge at a ratio of about 0.5 to 1, with a 20% excess on the stoichiometric value.

This excess is only apparent because of the consumption of the chips on the surface of the charge in the furnace.

Results

Under these conditions, the last tests gave the results which follow.

They are indicated in comparison with the results obtained at the semipilot plant.

Casting Composition - Tests in the Semipilot Plant

Al	36 to 37	%
Si	57 to 58	"
Fe	3 to 4	"
Ca	1.2 to 1.3	"
Ti	0.3	"
C	0.7	"
Imp.	0.5	"

Casting Composition - Tests in the Pilot Plant

Al	29 to 30	%
Si	64 to 65	"
Fe	4 to 4.5	"
Ca	0.4 to 0.5	"
Ti	0.3	"
C	0.3	"
Imp.	0.3	"

The different composition can be explained as follows. When running in the pilot plant the power density shall be further increased in order to secure a complete alumina reduction. At lower power values, even very little changes in the furnace run can cause an incomplete alumina reduction. Some slag is formed.

Dust on the Filter - Analysis

K_2O	40.66	%
Na_2O	4.96	"
SiO_2	14.14	"
Al_2O_3	6.71	"
Fe_2O_3	0.10	"
CaO	0.97	"
MgO	0.48	"
TiO_2	0.02	"
C	6.5	"
CO_2	23.5	"
S	0.85	"

- Mineralogical composition

K_2CO_3 - Na_2CO_3 - K_2SO_3 - SiO_2 - Al_2O_3 - Si - Al

The ratio between castings and dust agreed with the calculated value, already verified in the semipilot plant, that is, approximately 1 : 1.

The Al and Si losses in the off-gases collected in the dust were 11 and 10%, respectively.

These results are fairly good, but they could certainly be improved carrying out the reduction process at a higher power density.

Weight and Energy Balance

Weight Balance

Fourteen runs have been carried out so far.

The following weight balance was obtained from the last test (10 day running)

Metal Distribution

	Percentage recovered	
	Cast alloy	Dust
Si	90.40	9.60
Al	89.35	10.65
Fe	98.40	1.60
Ti	app.100.00	=
Ca	39.00	61.00
K	=	100.00
Na	=	100.00
Mg	=	100.00

Remarks. The iron content in the casting was much higher that the iron introduced with the pellets.

It comes from the iron tubes used for the spout opening. In the subsequent tests this increase will be eliminated.

In a few castings some unreduced alumina was found as a slag.

Raw Material Specific Consumptions

- leucitic roks	2.60	t/t
- clay	0.17	"
- lithantrax	1.08	"
- wood chips	1.95	"
- graphite electrodes	0.045	"

Energy Consumptions (Raw Alloy Produced 42 Kg/h)

- electrical energy input	664	KWh/h
- total energy input (given as KWh h)	1134	"
- specific electrical consumption	16.195	KWh/Kg
- total energy requested	509	KWh/h

- where theoretical energy for the reduction process | 353 KWh/h
- energy losses (measured) | 498 "

This result no doubt already satisfactory could be further improved by decreasing the Si and Al amount lost by evaporation and collected on the filter.

Potassium Recovery

The potassium, as the sodium, is found in the dusts collected on the filter prevailingly under form of carbonate, with smaller amounts of sulfate. Most of the dusts is then soluble in water. The residue consists of SiO_2, Al_2O_3 and other impurities.

It can be exploited in its turn as secondary product. In this first investigation this aspect has not anyhow been taken into account.

The Corrosion of the Electrodes

The attack of the electrodes just above the charge by the air flow caused serious difficulties.

The temperature of the electrodes rapidly decreases by irradiation starting from the top of the charge; nevertheless it is very high in the first stretch. As a consequence, the graphite can be oxidized. The problem was so far solved covering the electrodes by a 0.3 mm inox steel sheet. A different solution is nevertheless under consideration.

Operating Cost and Capital Cost. Preliminary Evaluation

After this first test phase was completed, we reputed it helpful to carry out an evaluation of an industrial cost of the Al/Si alloy produced by the process under consideration. This evaluation can evidently give merely some rough figures. In fact the specific consumptions could change in an industrial production. Moreover, the industrial plant was supposed to be built in the leucitic deposit area. The ore cost further on indicated is based on this assumption.

Capital Cost

- Data available
  - Raw alloy production: 20,000 t/yr; 60.6 t/day
  - Production days: 330
  - Treatmentment of raw materials
    - grinding: 5 days week 3 shifts
    - granulation and drying: 5 " " 3 "
    - reduction: 7 " " 3 "

- Raw materials
  - deferrized leucite 158 t/day
  - lithantrax 67 "
  - clay 10 "
  - starch 0.2 "
  - wood chips 125 "
  - graphite electrodes 2.75 "
  - total energy 1,030,200 KWh/day

Table I. Capital Cost

Units	N°	t/hr	Unit Cost (1983) ML	Total Cost ML
Mills	3	5	345	1035
Granulators	2	10	115	230
Drying furnace	1	20	300	300
Reduction arc furnace	2	1,34	7200	14400
Fume abatement cooling - softening plant	1		800	800
Silos	4		20	80
				16545
Installations for structures, foundations, utilities				20000
				36545
Contingencies				1455
			Total	38000

To be added
- Financial charges
- Sales charges, headquarters
- Financial charges on the current money

Depreciation 10 years
Incidence on the raw alloy 190 £/Kg

Table II. Operating Cost

Materials	t/t alloy	£/t	t/yr	Total ML yr
Leucitic rock	2.6	25,000*	52,000	1,300
Clay	0.17	50,000	3,400	170
Lithantrax	1.10	170,000	22,000	3,740
Wood chips	2.0	40,000	40,000	1,600
Graphite electrodes	0.045	2,800,000	900	2,520
Fuel oil	0.0005	235,000	100	23.5
	m^3/t	£/m^3	m^3/yr	
Water	-	70	8,000	0.56
	KWh/t	£/KWh	KWh/yr	
Electric power	17,000	28**	340×10^6	9,520
			Total	18,874
Personnel : 1 Responsible + 87 units				2,500
Maintenance: materials				600
Contractors and transports				200
Others				226
			Total	22,400
Incidence on the raw alloy	1120 £/Kg			

* Delivered at the deposit area

** A preferential tariff was estimated, based on the KWh cost which is assigned to the two industrial processes for the aluminium and silicon production.

Charge Composition in the Furnace on Completion of the Test

Once achieved the equilibrium conditions, some tests were intentionally stopped to analyse the composition of the reacting charge.

The elementary analysis and the X ray diffraction exam showed the following compositions :

- Si - Al - βSiC in the area immediately below and on the sides of the electrodes
- Si - Al_2O_3 - βSiC in the intermediate area, at ca 25 cm from the reaction zone
- mullite and leucite towards the upper zone.

No Al_4C_3 or other aluminum oxycarbides were ever found in the cast metal, while SiC and Al_2O_3 were seldom present.

Such alkalies as K_2O or Na_2O are met only in the upper part of the furnace. At roof level their percentage in the charge is appreciably higher than the initial one, no doubt due to the recondensation of the two oxides resulting from the dissociation of the silico-aluminates.

Taking into account the considerable SiO_2 excess over Al_2O_3, and on the basis of a mathematical model and the thermodynamic data available (2)(3), the main reactions are expected to be

1. the reaction giving rise to the Al_4C_3 and Al_4C_4O, aluminum carbides, and SiC silicium carbide
2. the reaction of aluminum carbides with SiC and SiO_2

$$2Al_4C_3 + 3SiO_2 \longrightarrow 8Al + 3Si + 6CO$$

$$Al_4C_4O + 3SiC \longrightarrow 4Al + 3Si + 4CO$$

besides the reduction of the silica access to silicon, considering that the Al:Si molar ratio is approximately 1 : 1,9

The possibility of a similar dissociation of Al_4C_3 to Al and C at 2155°C should also be kept present.

On the other side, according to Cochran's conclusions (2) the reactions taking place (with a theoretical 100% yield under reflux) would seen to lead to the following molar ratio Al : Si = 5.295 : 3, which falls within an intermediate position with respect to the two above reported reactions.

The complexity of the system and the reactions unvolved, anyhow clearly appears. Even small variations in the operating conditions, such as a too fast or too low feeding of the charge, energy variations at the inlet, unexpected coolings, etc, cause therefore an overall change in the formed equilibria, and consequently in the reactions and the final products. The formation of carbides only, as well as a total evaporation of the metals, can thus take place.

Thermodynamic Support Studies

It is not worthwhile lingering over the phase diagrams Si – O – C and Al – O – C (4). They have been in fact the subject of several investigations either under the theoretical aspect or in connection with the industrial development already effected for some industrial processes (Si) or still in the experimental phase (Al).

But only few investigations on the quaternary system Al – Si – O – C are so far available.

Simultaneously to the tests conducted in the pilot plant, a thermodynamic investigation of this complex multiphase multicomponent system was started to improve the approach and test conditions. More precisely the aim of our investigation is the definition of a mathematical model for the identification of the possible reaction products versus the charge composition, the energy supplied and the thermal gradient of the furnace, in a CO pressure versus temperature plot following a similar analysis carried out by Motzfeld for the Al - C - O system (5).

The information collected will be used in their turn in a SOLGAMIX-PV Model or possibly SOLGAMIX-REACTOR.

Moreover, our investigation aimed at defining the optimum distances of electrodes from sole and furnace walls, as well as optimum reaction zones; in other words it had to verify the validity of the present geometry of the pilot furnace in connection with the reaction zone.

Present Program

The result so far obtained are, in our opinion, satisfactory. It should be anyhow emphasized that the aim of the process under investigation is more ambitious. From the raw alloy we expect to obtain a refined alloy along with a Si-rich residual alloy, in its turn capable of being exploited.

The program is therefore now following two pathways

a) - further improvement of the process under investigation by modifying the furnace features;

- use of other national aluminous minerals, such as tuffs;

- Al enrichment of the raw alloy, using aluminous industrial residues, such as the fines of aluminum foundry, recovered refractories, etc., mixed with the leucitic minerals;

b) - investigation and development of an original, economically valid process for the refinement of the raw alloy. To this purpose tests on semipilot scale are still under progress by the Light Metals Institute of Alluminio Italia at Novara. The design of the pilot plant will soon be ready. It shall be located at Portovesme directly downstream the production plant of the raw alloy.

The obtainment of a good separation yield appears to be strictly bound to the impurity content, namely iron, since this element not only substracts aluminum from the alloy, but also adversely affects the separation process. Less adverse effects, though not negligible, are caused by titanium and calcium. The investigation under progress tends to minimize these effects.

References

M.F. Landi, "Experiments of Utilization of Italian Leucitic Rocks for the Al Si Alloys Production by Direct Carbothermic Reduction" paper presented at 110th AIME Annual Meeting, Chicago, Illinois, February 1981.

C.N.Cochran, and N.M. Fitzgerald, "Computer Aided Model of Coke Fired Reduction of Alumina - Silica" paper presented at 109th AIME Annual Meeting, February 1980.

I. Barin, and O. Knacke, "Thermochemical Properties of Inorganic Substances" Springer Verlag, 1973

P.T. Stroup "Carbothermic Smelting of Aluminum", transactions of the Metallurgical Society of AIME, Vol. 230, pp. 356-371, April 1964

K. Motzfeldt, "Carbothermic Reduction of Alumina," paper presented at the ICSOBA-AIM Conference, "New Processes in Aluminium Production," Cagliari, Italy, September 1979.

MICROSCOPIC INVESTIGATION OF CLAY CHLORINATION

E. S. Martin
K. Wefers

ALUMINUM COMPANY OF AMERICA
Alcoa Technical Center
Alcoa Center, PA 15069

Summary

Microscopic and structural investigations of raw, roasted and reacted clays gave the following results. Thermal treatment removes hydroxyl ions from the crystal lattice, leaving a highly disordered aluminum-silicate compound. However, no internal porosity develops, and the surface area of the heated clay does not change significantly. Wetting of the tabular kaolinite crystals by liquid $KAlCl_4$ is a prerequisite for high reaction rates. Molten salt acts as a transport medium and reaction intermediate.

Roasted clay is composed of layers of SiO_2 tetrahedra alternating with layers of Al_2O_3 octahedra (both highly disordered). The two layers are superimposed in such a way that removal of one SiO_2 tetrahedron by chlorination

(1) $SiO_2 + 2COCl_2 \rightarrow SiCl_4 + 2CO_2$

will "uncover" several Al_2O_3 octahedra, leading to a preferential chlorination of Al_2O_3 to form $AlCl_3$.

Because roasting does not change surface area, but reaction rates are controlled by the total surface wet by $KAlCl_4$, it is recommended that high surface area clays be selected as starting material for the chlorination process.

Introduction

Clay minerals having the highest alumina content (38-40% Al_2O_3) belong to the kaolinite group. Dickite, nakrite and kaolinite differ only slightly in their crystal structures. The chemical composition of these three minerals can be described by the formula:

(2) $Al_2O_3 \cdot 2SiO_2 \cdot 2H_2O$ or $Al_2Si_2O_5(OH)_4$.

A structurally disordered, more hydrated form of kaolinite is known as "Fireclay mineral".

To produce anhydrous $AlCl_3$ from clay, the raw material is calcined at temperatures above 650°C to break the hydroxyl bonds and remove OH in the form of water(1). The preroasted clay is then treated at elevated temperatures with chlorine gas and a reducing agent, e.g. CO. Several percent of a "catalyst", such as $KAlCl_4$, are added to increase the reaction rate. Yields in excess of 90% of the available Al_2O_3 have been achieved by this process (1, 2).

Objectives

With the aim of improving process control and economics through an understanding of the reaction mechanisms, starting materials and residues from several process stages were investigated. The following questions were addressed.

1. How does preroasting affect the structure and morphology of the clay particles?

2. What are the sites of reaction between pelletized, calcined clay and gaseous reactants?

   a. Surface of the pellet?

   b. Surface of the primary particle (crystal)?

   c. Intercrystalline pores?

3. What is the effect of the "catalyst"?

X-ray diffraction, transmission electron microscopy, Auger and Ion spectroscopy, and BET surface analysis were used as analytical methods.

Experimental

The material used for the investigations described below was a kaolinite clay. Particle size of the clay was well below 1 μm, as shown in Figure 1. Primary crystals are well developed, pseudohexagonal platelets having a diameter-to-thickness ratio of more than 10. Stacks of platelets form irregular aggregates in which the crystals are arranged parallel to their base plane, but randomly oriented within this plane.

The specific surface area of the raw clay was 22.9 m^2/g. Only the diffraction lines of kaolinite were found in an X-ray diffraction diagram of this sample. Lines of quartz (SiO_2) and anatase (TiO_2) were not discernible due to coincidence with interferences of kaolinite. High intensity and sharp profiles of the kaolinite diffraction lines indicated a high degree of crystalline order, as may have been expected from the crystal habit shown in Figure 1a.

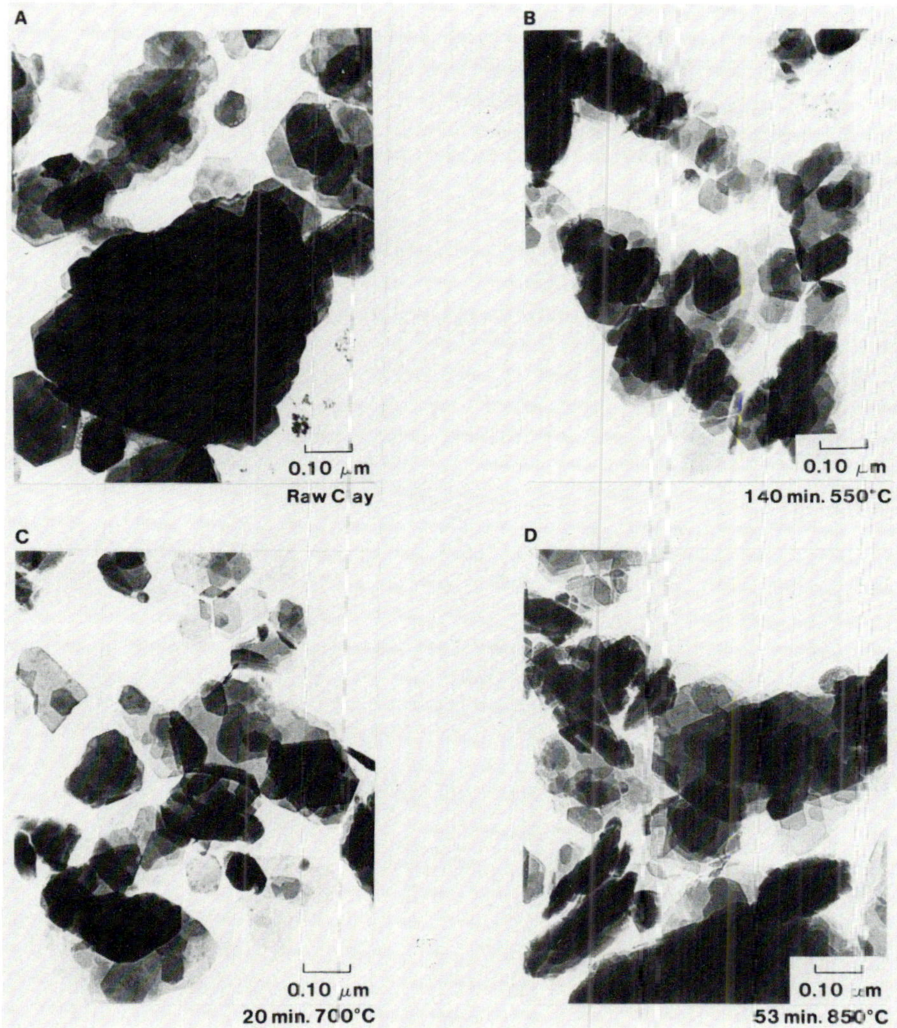

Clay Before and After Various Calcination Treatments
TEM 120,000X
Figure 1

To determine the effect of preheating on crystal habit, structure and surface area, clay samples were dehydroxylated at three different temperatures and two exposure times. A short and a long roast time were chosen for each temperature level:

500°C	35 min,	140 min
700°C	20 min,	80 min
850°C	13 min,	53 min

The roast times were adjusted for temperature by assuming an Arrhenius dependence for the proton transfer process ($E_a \sim 6$ kcal/mole) which controls the rate of dehydroxylation.

Chlorination tests were performed in a fluidized bed reactor. Pelletized, roasted clay was treated at 600°C (in some cases, 700°C) with 0.75 g-mole of $COCl_2$ per hour. Five percent by weight of $KAlCl_4$ were added to the fluidized clay. Reaction times varied between 0.5 and 2 h. In order to separate the effect of $KAlCl_4$ on the reacting clay, three samples were heated in dry nitrogen, while only $KAlCl_4$, but not $COCl_2$, was added. Conditions were: 700°C, 4h, 5% $KAlCl_4$; 700°C, 2h, 5% $KAlCl_4$; 600°C, 2h, 5% $KAlCl_4$.

Experimental Results

Clay Dehydroxylation

Electron microscopic examination of the dehydroxylated clay did not reveal any obvious morphological differences between the raw and the treated material (Figures 1b and 1c). Considering the resolution of the transmission electron microscope, any changes induced by the roasting treatment must be on the order of less than 1 nm (10 Å).

The crystal structure, on the other hand, was strongly affected by thermal exposure. A qualitative evaluation of X-ray diffraction diagrams is listed below:

550°C - 35 min -	Decomposing kaolinite, strong reflections of anatase, TiO_2.
550°C - 140 min -	A few residual kaolinite lines with slightly decreased d-spacing; anatase, TiO_2, and quartz, SiO_2.
700°C - 20 min -	Amorphous clay matrix; anatase, TiO_2; quartz, SiO_2.
700°C - 80 min - 850°C - 13 min 850°C - 53 min	No change compared to 700°C - 20 min for all three temperature/time combinations.

Specific surface areas were not significantly influenced by the roasting conditions. Measured values ranged from 23.8 m^2/g (700°C - 20 min) to 20.6 m^2/g (550°C - 140 min), with no apparent correlation with temperature and time of treatment (Table I).

Clay Chlorination

X-ray diffraction analyses of roasted clay and reaction residues revealed that in all cases X-ray indifferent material constituted the bulk of the samples (Table II). Quartz and anatase found in the roasted clay persisted through the thermal treatment at 700°C in the presence of N_2 and $KAlCl_4$. A few weak lines indicating transition alumina (delta or theta) were found in the spectra of the material treated for two hours at 700 and 600°C, respectively. Sample B yielded a qualitatively similar, but much less intense, diffraction pattern.

TABLE I

CALCINED CLAY SAMPLES

Temperature, °C	Time, min.	Spec Surface Area m^2/g
---	--	22.9
550	35	23.5
550	140	20.6
700	20	23.8
700	80	23.8
850	13	23.2
850	53	22.7

TABLE II

ROASTED AND CHLORINATED CLAY SAMPLES

	DESCRIPTION	CHLORINATION RESULTS* (IF ANY)	
		%Al_2O_3 Recovered As $AlCl_4$	Mole Ratio $AlCl_3/SiCl_4$
(A)	ROASTED CLAY		
(B)	4.0 HR IN N_2 AT 700°C-$KAlCl_4$ PRESENT		
(C)	2.0 HR IN N_2 AT 700°C-$KAlCl_4$ PRESENT		
(D)	2.0 HR IN N_2 AT 600°C-$KAlCl_4$ PRESENT		
(E)	2.0 HR CHLORINATION RESIDUE*	91.3	4.00
(F)	1.5 HR CHLORINATION RESIDUE*	81.4	4.27
(G)	1.0 HR CHLORINATION RESIDUE*	41.4	3.17
(H)	0.5 HR CHLORINATION RESIDUE*	25.6	2.39
(I)	2.08 HR CHLORINATION RESIDUE+	24.2	0.83

*CHLORINATED AT 600°C, 0.75 g-mole $COCl_2$/hr, 2.0 HRS ORIGINAL CHARGE 70.0 gm ROASTED CLAY + 5% $KAlCl_4$

+600°C, 100.0 GRAMS ROASTED CLAY, NO $KAlCl_4$, 0.68 g MOLE $COCl_2$/HR + 0.091 g-MOLE $SiCl_4$/HR

Quartz was the only crystalline phase determined in chlorination residues. A mullite-type structure was indicated by a few weak lines in the pattern of the residue of the 0.5 h chlorination test. Mullite (Al_2SiO_5) normally forms when kaolinite is heated above ∿1000°C. One may speculate that the presence of a molten salt ($KAlCl_4$) had a mineralizing effect on the predominantly amorphous Al_2O_3/SiO_2 composition left after roasting of kaolinite.

Only one, very sharp and intense diffraction line of quartz occurred in the X-ray diagram (Sample F, 1.5 h at 600°C). This phenomenon is generally related to preferred orientation of a specific crystal plane. In this case, it could mean a recrystallization of the SiO_2-network in a plane parallel to the basis of the decomposed kaolinite. Why this reordering was observed in only one sample remains to be explained.

All chlorination residues, as well as the samples heated with $KAlCl_4$ in a nitrogen atmosphere, were investigated by transmission and scanning electron microscopy (TEM and SEM).

It was demonstrated with the micrographs of Figure 1 that roasting by itself does not measurably affect the morphology of the primary kaolinite particles. The shape and size of clay pellets also did not change after the thermal treatment, as could be shown in SEM at low magnification. When four specimens were examined by SEM at high magnifications (2,000 and 10,000X), it was found that calcination and chlorination without $KAlCl_4$ did not change the primary crystals, but that chlorination in the presence of $KAlCl_4$ lead to a significant decrease in particle size (Figures 2a, b, c, d). TEM micrographs confirmed this finding. Compared to the roasted clay (Figure 1), the material heated with $KAlCl_4$ in N_2 showed no obvious morphological change. Some rounding of sharp edges was discernible with Sample I, which had been chlorinated without $KAlCl_4$. Clay chlorinated in the presence of $KAlCl_4$ was severely attacked (Figures 3a, b, c, d).

The peculiar morphology of the particles shown in Figure 3d developed during chlorination, becoming more promounced with longer reaction times, as demonstrated in Figures 4a-d. These micrographs were taken at different magnifications. They illustrate that the primary kaolinite particles do not shrink uniformly with progressing chlorination. Rather, the reactants appear to preferentially attack the inner portions of the platelets, often leaving doughnut-shaped skeletons.

Auger, ISS and SIMS analyses of two chlorination residues (pellets) that had been leached with hot water showed no traces of chlorine. As these analytical techniques require vacua on the order of 10^{-9} torr, any residual, adsorbed chloride may have evaporated during pump down. SIMS (Secondary Ion Mass Spectrometry) indicated a large excess of SiO_2 over Al_2O_3 in the residues (∿9:1) while ISS (Ion Scattering) and Auger Electron Spectroscopy results pointed toward roughly equal quantities of both oxides. These data must be considered quite preliminary, because the methodical problems of surface analyzing porous, rough pellets have yet to be solved.

Discussion of Results

The key to an understanding of the observed reaction behavior lies in the crystal structure of kaolinite.

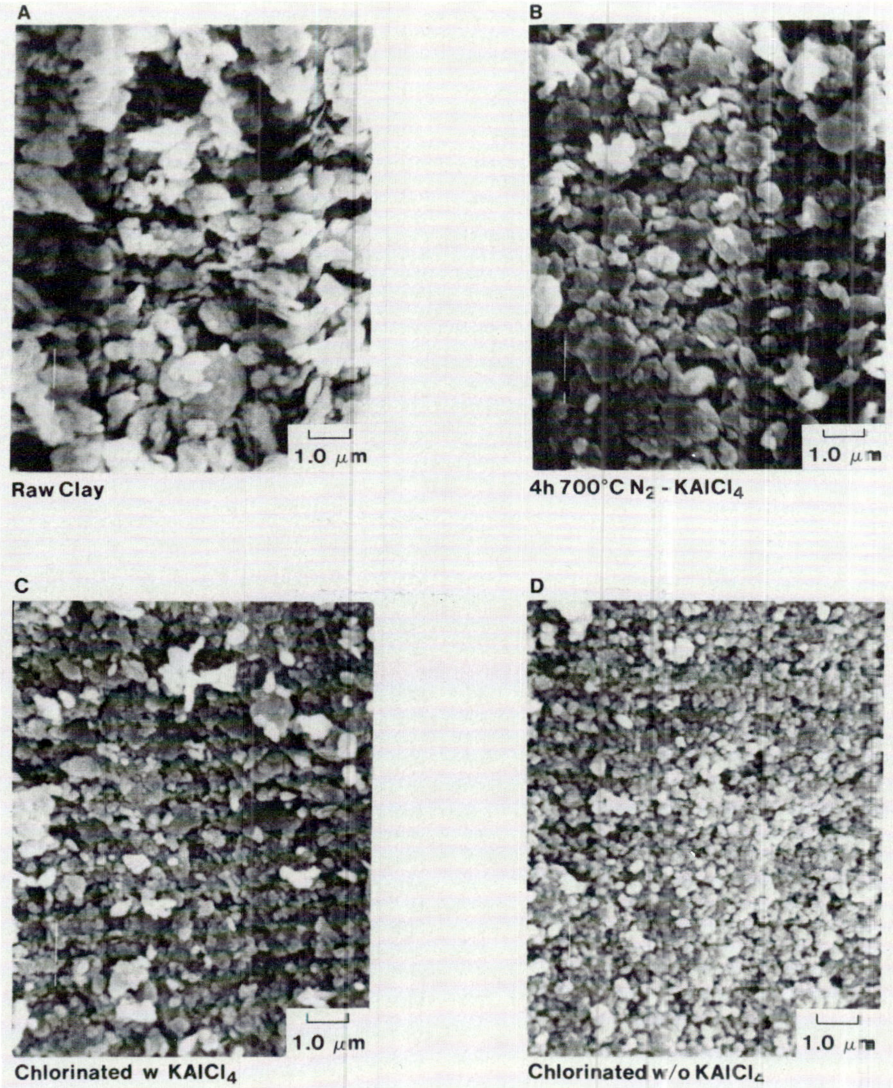

Clay After Various Treatments
SEM 10,000X
Figure 2

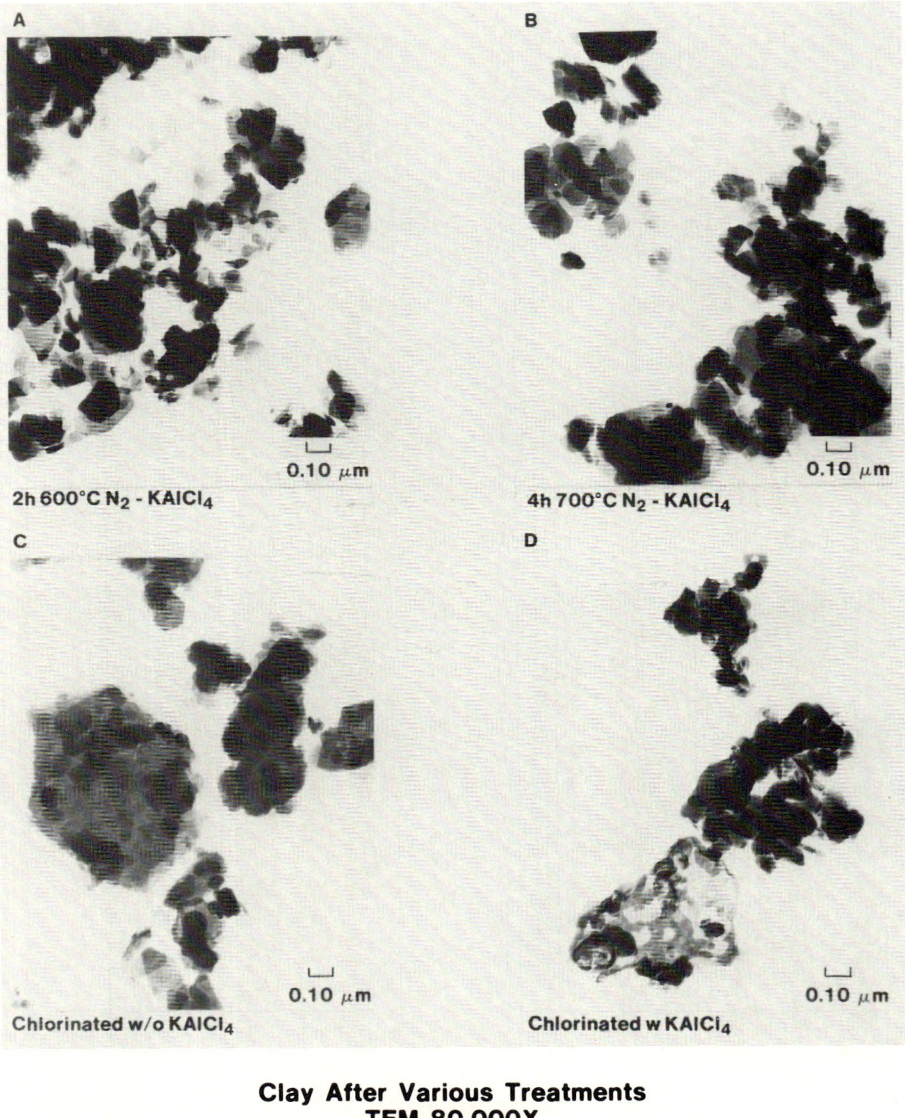

Clay After Various Treatments
TEM 80,000X
Figure 3

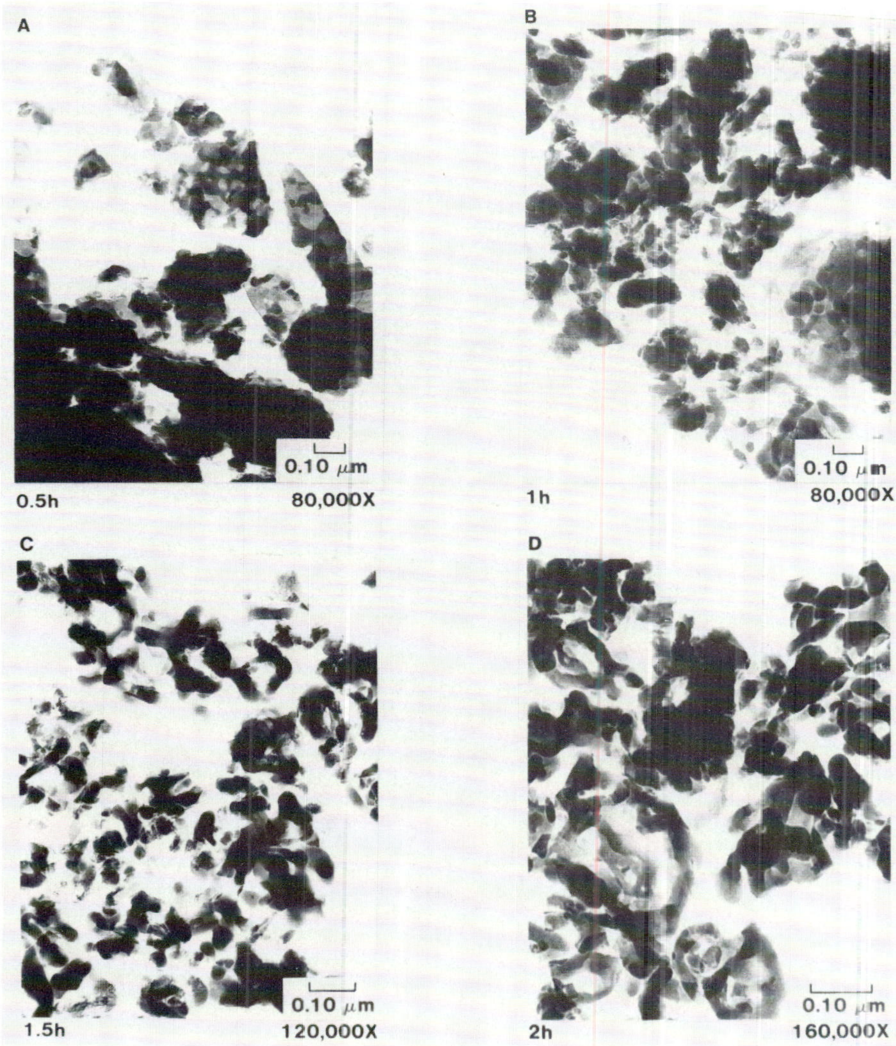

Chlorination Residues after 0.5 - 2 h Treatment, 600°C, $KAlCl_4$, $COCl_2$
Figure 4

A schematic representation of the lattice is given by Figures 5a, b, c, and d. In Figure 5a, the structure is viewed in a direction perpendicular to the short edge of a kaolinite platelet. This projection reveals the pronounced layer character of the structure. Each of the layers consists of a sheet of SiO-tetrahedra which are linked to a sheet of Al (O, OH)-octrahedra through shared oxygen ions.

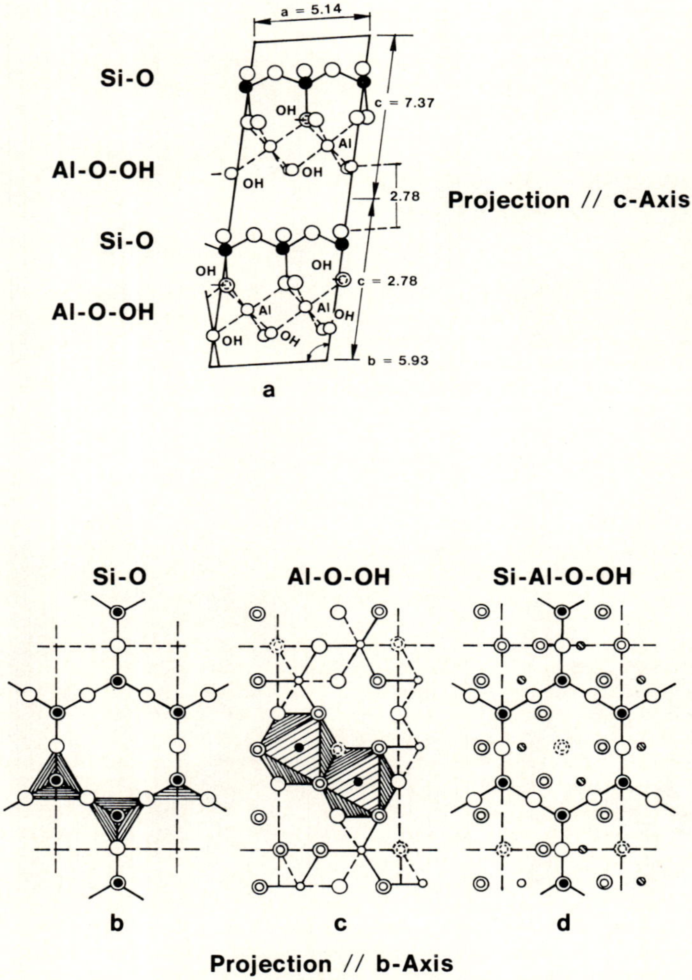

Structure of Kaolinite
Figure 5

The individual elements of these SiO - Al(O,OH) double layers are shown in a projection perpendicular to the base of a kaolinite crystal in Figure 5b (SiO network), 5c (Al(O,OH) and 5d (Super-position of the two networks). The double layers are stacked in an ab-ab-ab sequence. Hydrogen bonds provide the linkage between Al (O,OH) of one double layer and the SiO sheet in the adjacent one. The structure is rather tightly packed, the distance between half the sum of the radii of the O and one OH ion. In other words, there is no "gap" between the double layers. A gap can, on the other hand, be easily seen to exist in the SiO_2 network. Even though the open circles in Figure 5b represent the centers of spheres that touch each other, a channel of roughly the size of one oxygen ion remains in the center of the hexagonal arrangement of tetrahedra. Four hydroxyl ions and two aluminum ions are partially shielded by the overlying oxygen of the SiO_2 network.

Dehydroxylation of kaolinite proceeds via a mobilization of protons (E_a ~5-6 kcal/mole). The reaction sequence is:

$$OH^- \rightarrow O^{--} + H^+$$

$$O^{--} + H^+ + OH^- \rightarrow H_2O + O^{--} \qquad (3)$$

as protons combine with hydroxyl ions to form water, which escapes from the lattice. Due to the sharing of oxygen between the Al(O,OH) and SiO_2 layers, the lattice does not collapse upon complete dehydroxylation, but retains a rudimentary, two-dimensional order ("meta-kaolin")(3). The Al(O,OH) layers are highly disordered after dehydroxylation at temperatures between 600 and 800°C. Above 800°C, reordering takes place and transition alumina forms according to literature. Temperatures over 900 to 1000°C lead to the crystallization of mullite, Al_2SiO_5.

Conclusions

From the morphology of the chlorination residues and the structure of kaolinite, a mechanism for the chlorination reaction may be inferred. Preconditions for the reaction are the presence of the "catalyst" $KAlCl_4$, chlorine, a reducing agent such as CO, and dehydroxylation of kaolinite to the point of a maximum disorder of the Al(O,OH) layers. Maximum disorder implies maximum free energy, consequently, high reactivity. The disordered alumina layers are, however, sandwiched between silica networks. Assuming a layer-by-layer reaction, an SiO_2 layer has to be chlorinated before the underlying alumina is exposed to the reactants. This would result in an $SiCl_4/AlCl_3$ ratio of 1:1, as the structure contains equal numbers of Al and Si ions. The observed, preferential chlorination of aluminum can, therefore, only be explained by a topochemical mechanism which gives chlorine access to the alumina layer through the network of SiO_2-tetrahedra.

The channels left open by the hexagonal arrangement of SiO_2-tetrahedra appear to be the most probable paths for the diffusion of chlorine, as an interplanar "gap" does not exist, even after the removal of the hydrogen bridges. The size of the channel opening is, however, barely large enough to accommodate one chlorine ion (~3.6 Å ϕ), much less an $AlCl_3$ molecule which would have to leave the structure via the same path. The SiO_2 network is, of course, not as ideally preserved as might appear from mere geometric considerations. During dehydroxylation, one oxygen is lost with each two hydrogen ions leaving the lattice. In order to balance the charges in the alumina layer, the aluminum ions will attract oxygen from the SiO_2 network. This leads to a distortion of the silicon-oxygen arrangement, increasing its free energy (and, consequently, its reactivity). The observation that kaolinite samples heated at 700°C and higher were X-ray indifferent (amorphous) supports this assumption. If the SiO_2 layers had retained structural order over extended areas, some X-ray reflections would have occurred.

It is obvious from the structure diagram in Figure 5d that removal of any one tetrahedron from the SiO_2 network will immediately uncover two Al ions and improve access to another three.

The doughnut-shaped dissolution pattern suggests that the base planes of kaolinite platelets were wet by drops of liquid $KAlCl_4$. Concentration of the reaction in the inner portions of the platelets implies that the liquid formed a rather high contact angle with the solid, thus pulling away from the edges. In narrow crevices and gaps between adjacent particles inside the pellet, this wetting pattern is likely to be altered by capillary forces.

The liquid salt acts as a transport medium and a reaction intermediate. First, a silicon dioxide layer is attacked according to the overall reaction:

(4) $$SiO_2\ (KAlCl_4) + 2COCl_2 \rightarrow SiCl_4\ (KAlCl_4) + 2CO_2.$$

Removal of one silicon-oxygen tetrahedron then provides access for the liquid to several aluminum-oxygen sites which will react according to:

(5) $$Al_2O_3\ (KAlCl_4) + 3COCl_2 \rightarrow 2AlCl_3\ (KAlCl_4) + 3CO_2.$$

Two reaction sequences are conceivable. For one, the molten salt may transport dissolved $COCl_2$ to the liquid/solid interface where the reaction occurs. The other is that SiO_2 and Al_2O_3 are dissolved by the flux, forming intermediate oxy-chlorides which are reduced at the liquid/gas interface.

In the absence of $KAlCl_4$, yields are considerably lower, and the $SiCl_4/AlCl_3$ ratio is close to one (4). Obviously, the function of $KAlCl_4$ is more than that of a simple solvent. Whether the selectivity of chlorination in the presence of $KAlCl_4$ can be attributed solely to the layer structure of meta-kaolin, or an intermediate chemical reaction favoring $AlCl_3$, remains to be determined. An understanding of this mechanism should be applicable to processes of $AlCl_3$ formation other than by clay chlorination.

The experimental results reported above show conclusively that the rate of chlorination and the $AlCl_3/SiCl_4$ ratio in the product are strongly affected by the area of contact between clay and the molten salt "catalyst" $KAlCl_4$. An "activated" high surface area kaolinite, analogous to the activated aluminas, can obviously not be produced by thermal treatments. High surface area starting materials can only be obtained by selection of clays with smaller, less perfect primary crystals, such as fire-clay mineral.

The results also suggest that chloride may be removed from chlorination residues by moderate heating under vacuum.

REFERENCES

1. U.S. Patent 4,086,320 (April 25, 1978), E.S. Martin and D.A. Wohleber (to Aluminum Company of America).

2. U.S. Patent 4,096,234 (June 20, 1978), E.S. Martin and D.A. Wohleber (to Aluminum Company of America).

3. F. Freund, Deutschen Keramischen Gessellschaft Brericnte, 44 (1) (1967) pp. 5-13.

4. Private communication, W.B. Frank, Aluminum Company of America. Alcoa Technical Center, May, 1977.

CHAIRMEN

E.L. Cambridge
ARCO Metals Company
Tucson, Arizona

D.J. Donaldson
Kaiser Aluminum and Chemical Corp.
Oakland, California

ALUMINA TRANSPORTATION TO CELLS

Ivan Stankovich, P.E.

Consulting Engineer Corporation
P.O. Box 11182
Piedmont, CA 94611
USA

Until recently the transfer of raw materials (alumina and additives) to the electrolytic cells was via potroom crane and ore bucket, resulting in ore spills, causing internal potroom and external air pollution exposure, extra housekeeping, loss of potlife and higher voltage drop. However, alumina can be transferred from central and intermediate storages to the cell hoppers or cell crust directly, in required portions, automatically, in enclosed dustless conveyors. Various methods of material transportation and automatic control are considered. The new methods of material transportation may simplify and decrease the cost of cell superstructure, as well as the cost of cell maintenance.

PRESENT CONDITIONS

For more than a half century alumina and additives has been transferred from storages to electrolytic cells via buckets either carried by crane (Fig. 1), or by vehicles. Each cell has one or more hoppers which require filling daily.

These operations are causing extensive spillages of raw material in the potrooms (Fig. 2).

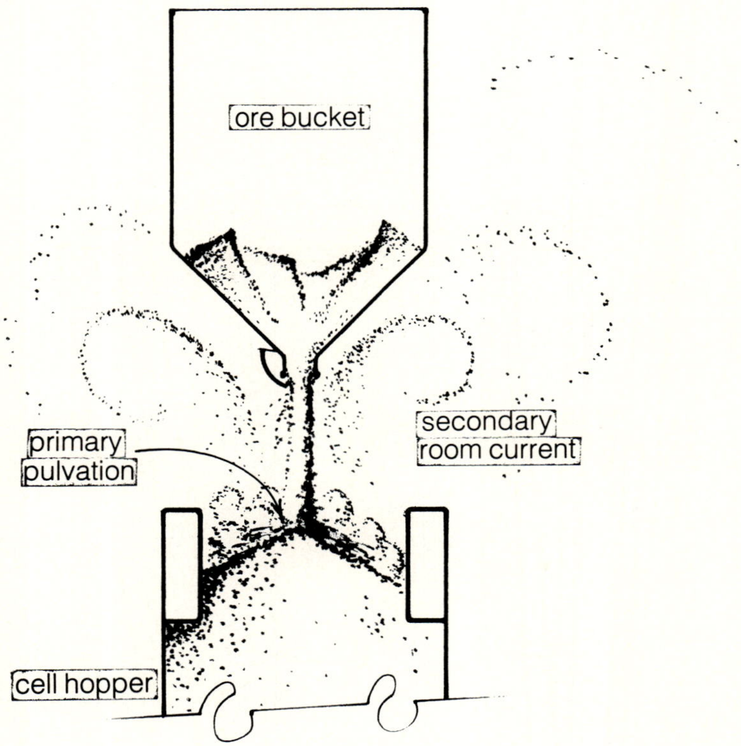

Fig.2 - Disharge of alumina from ore bucket into the cell hopper

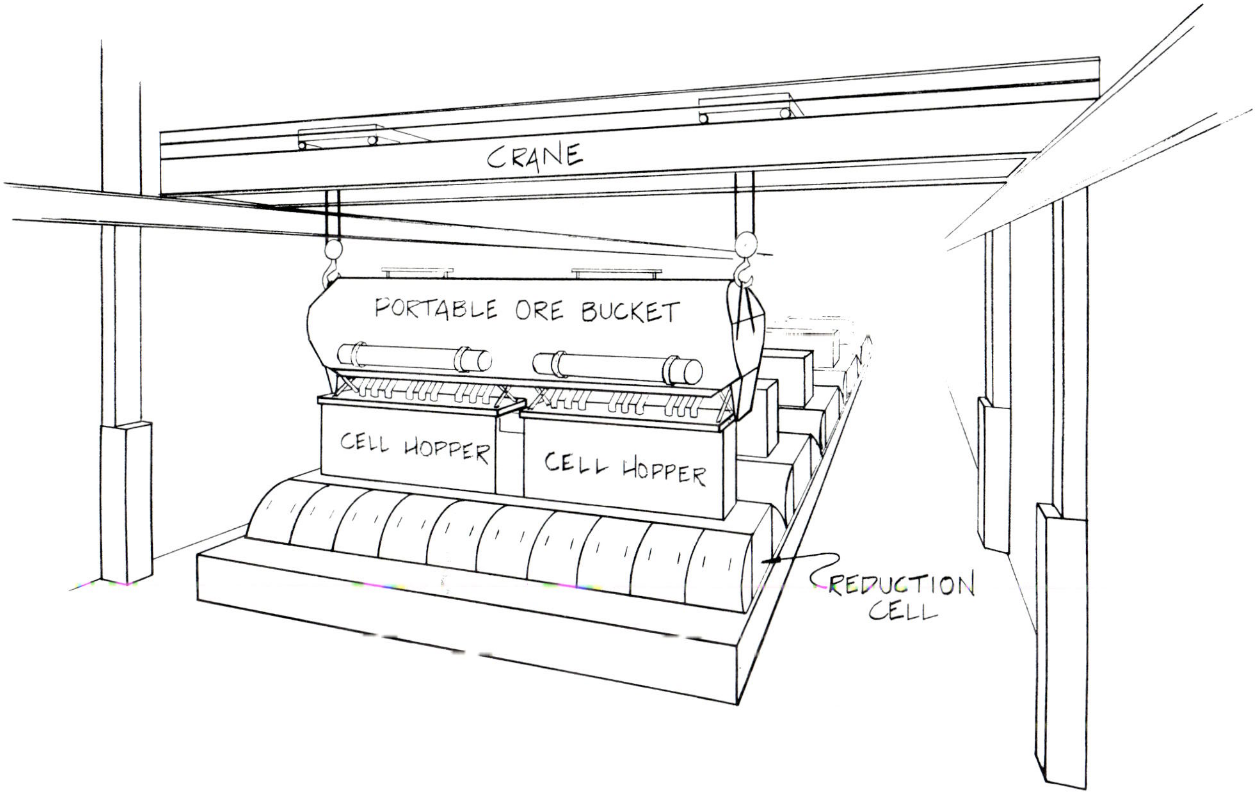

Fig.1 - Portable ore bucket

Investigations and evaluations of potroom dust problems indicate that due to relative high potroom air ventilation velocities (averaging 30 to 40 fpm), practically all sub-44 micron particles exposed to the air currents will be suspended in the potrooms (6). As a result in most instances, this is causing severe worker exposure to dust and atmospheric emissions of particles from the potroom roof monitors, resulting also in extra housekeeping.

Another factor is the use of "enriched" alumina from the dry scrubber installations. It was determined that this type of alumina is in general more free-flowing and contains higher percentages of fines and impurities. The dry scrubbing process using alumina as the adsorbing material (and especially with Soderberg electrolytic cells), turns the white "clean" and nearly-inert alumina into an almost-black and unpleasant material containing fluoride compounds, sulphur dioxide, hydrocarbons and various other compounds.

The measured breathing grain level loading for pot operators in certain exposures is greater in potrooms using "enriched" alumina than in the rest of the plant. Even with the installation of furnaces to process "enriched" alumina before transportation to potlines, the processed alumina is containing fluorides,and therefore, still presenting a potential health hazard.

It was determined that most of the dust in the potroom is generated directly from alumina handling operations; i.e. during filling of the bucket with alumina at the loading station, during discharge of alumina from the bucket into the cell hoppers and (for Soderberg cells) during the anode top cleaning operation.

Attempts to Improve Conditions

During the last decade a number of attempts have been made to improve alumina delivery to electrolytic cells, such as "dustless" buckets, incorporation of "dustless" buckets into the structure of the potroom overhead multipurpose cranes, agglomeration of fine alumina and other.

"Dustless" Alumina Bucket. The most advanced design (Fig. 3) involves the use of baghouse filter with fan and sometimes a small air compressor mounted on the bucket (or on the crane), a retractable hooding system to confine and remove dust as generated, level control to prevent over-filling of cell hoppers, dust-tight bottom dump gates to minimize spilling from bucket during transportation and a sophisticated control system. The development of "dustless" alumina buckets has to be combined with the upgrading of bucket loading stations. In most cases this requires additions of efficient ventilation systems.

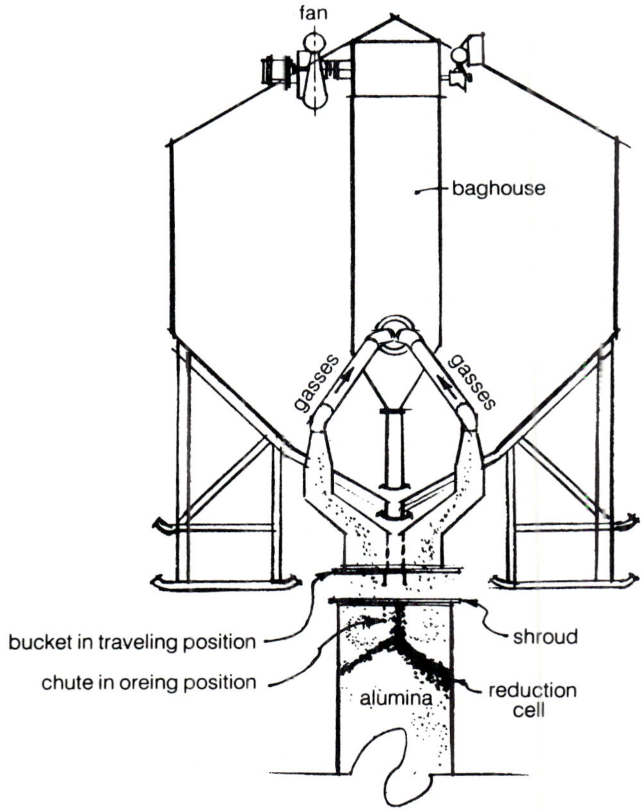

Fig.3 - Advanced alumina bucket during traveling and oreing

Nevertheless, the 10-year experience and survey of many installations has demonstrated that the bucket solution can not satisfy contemporary requirements. The main reason is the inadequacy of the concept itself, requiring operator's skilled personal involvement in hundreds of operations, at each cell, each day, and the relative low reliability of the portable bucket due to the complexity of different mechanisms and therefore resulting also in high maintenance cost.

Bucket Built Into the Crane. An attempt was also made to incorporate an advanced "dustless" bucket, similar as described above into the potroom multipurpose overhead crane (Fig. 4). This concept has practically all the disadvantages of the portable "dustless" bucket, except the control of operations is relatively more convenient. This may reduce some spillages compared to the portable bucket. However, as in the previous case, it will not resolve essentially the hygenic, environmental and housekeeping problems.

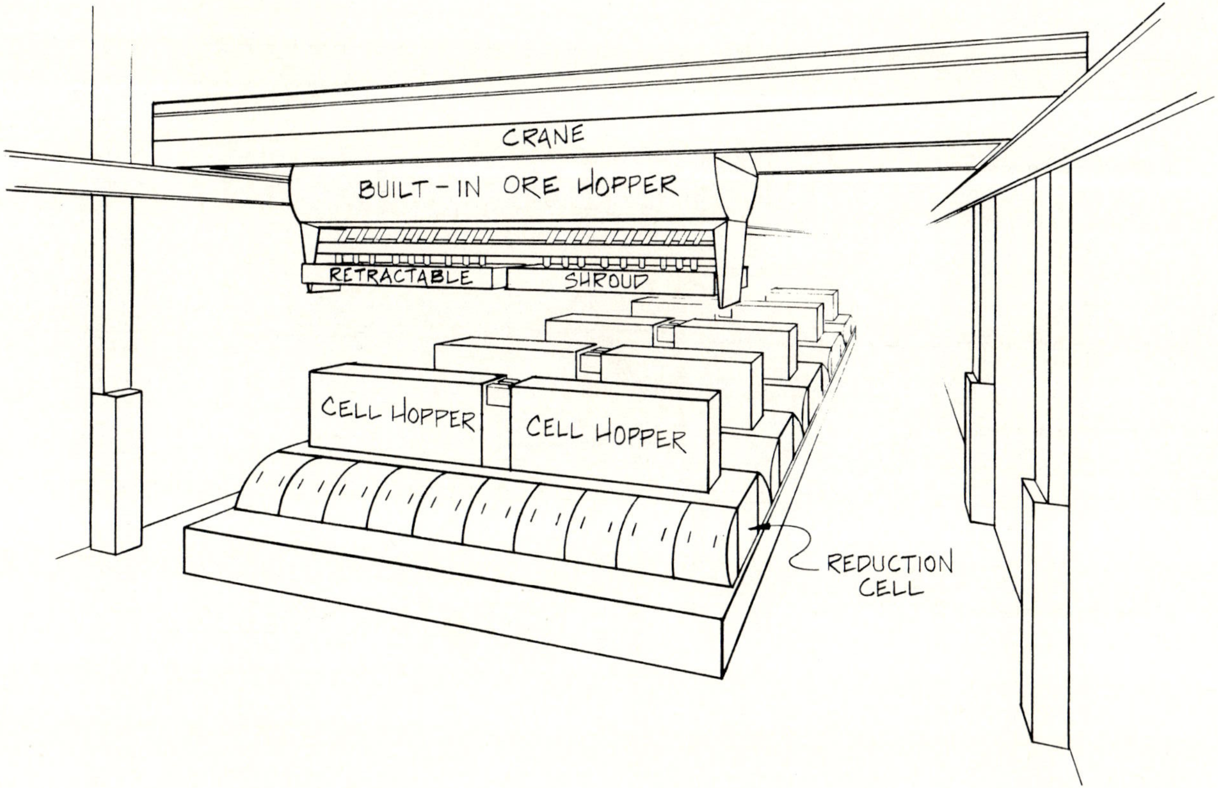

Fig.4 - Ore bucket built into the crane

Agglomeration of Fines. Use of agglomerated "enriched" and processed alumina will reduce dusting, but if not used with an enclosed conveying system, will doubtless result in dusting, as the fragile agglomerated particles are exposed to normal potroom traffic. Once outside the material handling system, the material is very difficult to recover. Ore agglomeration alone will not stop spilling or even reduce it to a minimum.

EMERGING TECHNOLOGY

During the second half of the last decade technology has made futher progress in both: aluminum production and bulk material handling (4).

In the aluminum industry large and "super" electrolytic cells with continuous point feeding, controlled by computers and combined with improved dry scrubbing systems have been introduced.

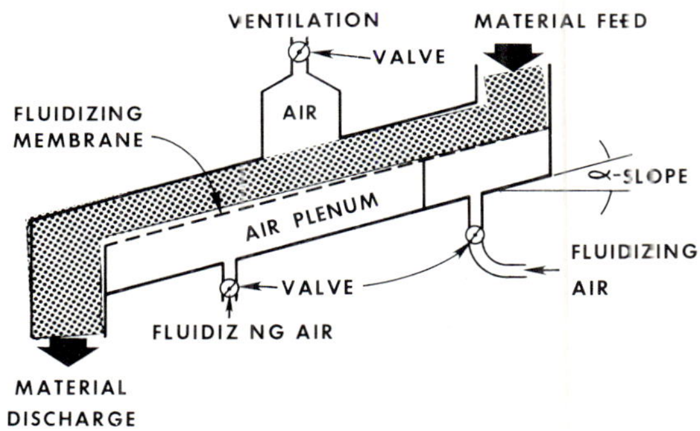

Fig.5 - Choke-fed air-activated gravity conveyor filled with fluidized alumina

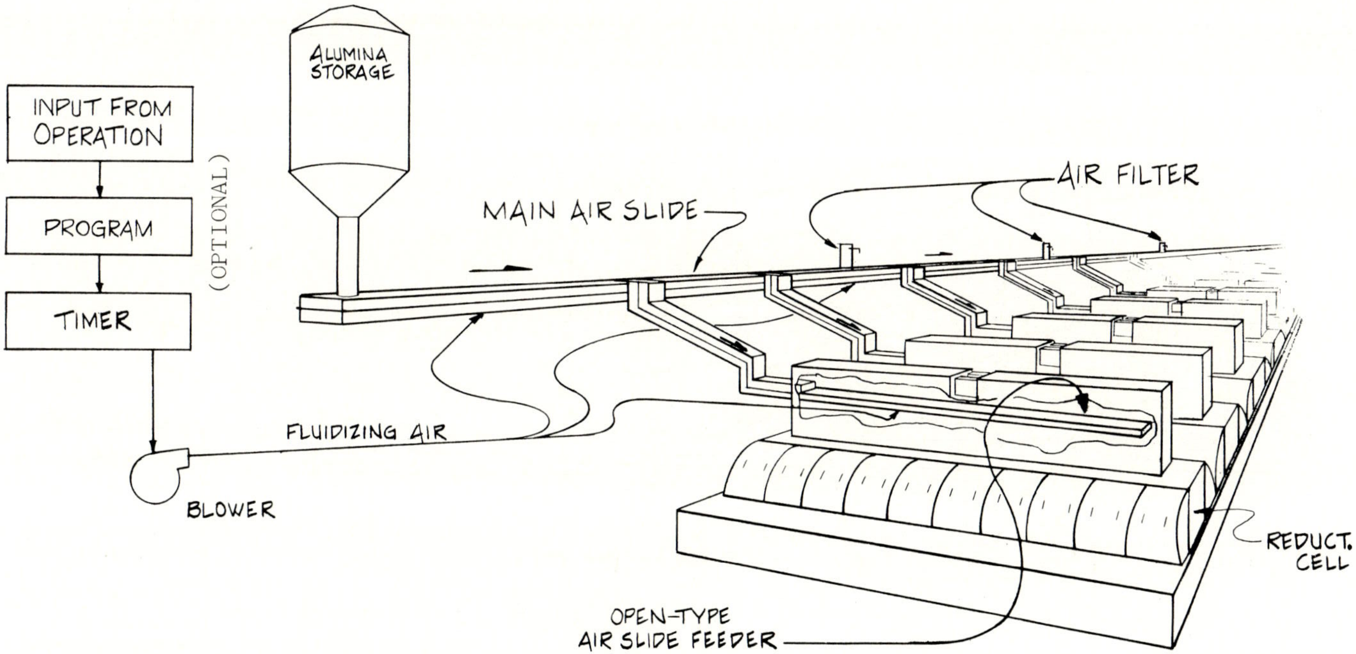

Fig.6 - Alumina transportation to cells - Open-type air slide system

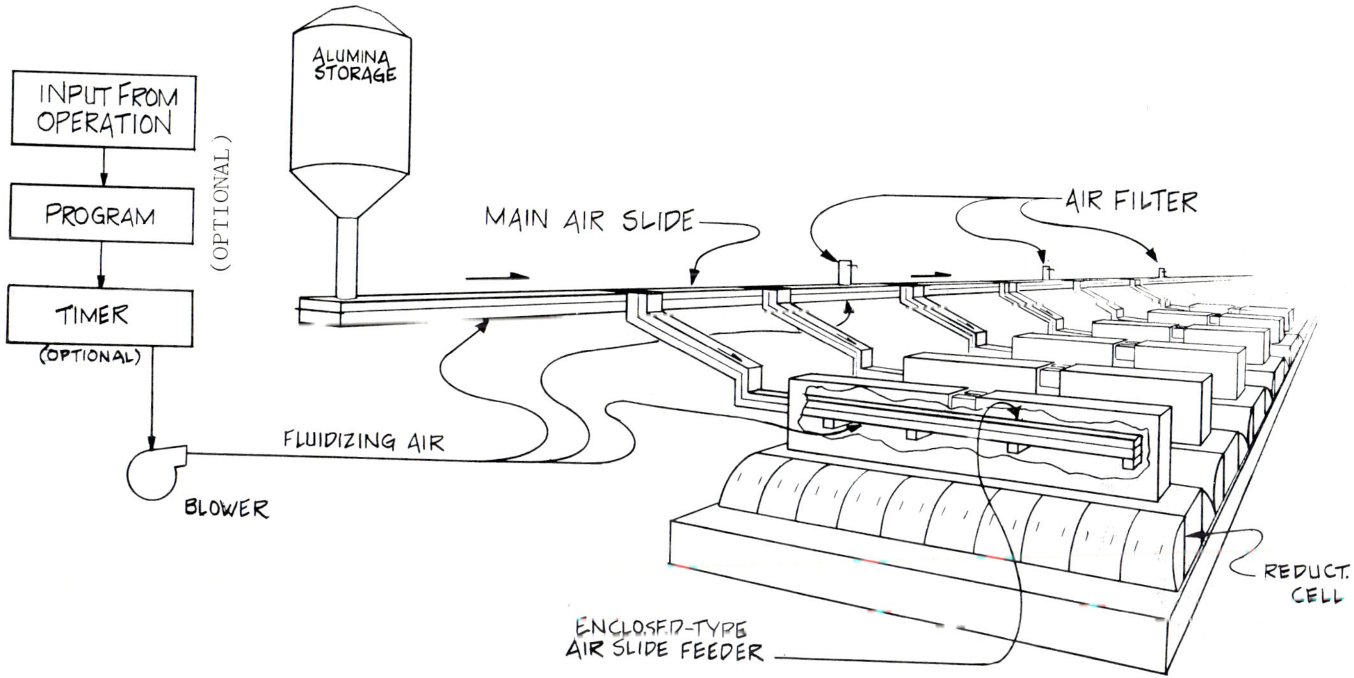

Fig.7 - Alumina transportation to cells - Enclosed-type air slide system

In bulk material handling the fluidization technology (1), reliable system design (3), combined with advanced computer application (2) has experienced a great boost. As a result, choke-fed air-activated gravity conveyors (AAGC) or air slides (Fig. 5), dense-phase pneumatic conveying, cartridge filters, different air-activated material handling devices (4, 5), supported by reliable computer software (2) did emerge.

All these (as often did happen in the history of technology development) has suggested to take a second look on alumina transportation to electrolytic cells. Things, which one time looked unreliable, expensive and doubtful, became much more reliable, economical and real today.

New Approaches

Alumina can be transferred from the storage to any electrolytic cell by automatic, totally enclosed, dustless systems. Such installations, without moving parts, if designed properly (3), can be made reliable (1) and economical. Some examples are given below.

Open-type Air Slide System (Fig. 6). Alumina is transferred from the storage to the electrolytic cells continuously or in accordance with a special operating program by a system, consisting of a main choke-fed air slide with cartridge filters and individual-cell, self-controlling, open-type air slides (7). Special attention is given to the design (including the electrical requirements, heath and erosion resistivity) of the individual cell air slide branches.

If run continuously, all the cell hoppers will be full at all times without the need of any monitoring device or operator's intervention. However, the air rate and the air pressure in the individual-cell air slides shall be tuned accurately.

Enclosed-type Air Slide System (Fig. 7) is similar to the Open-type Air Slide System, except the individual-cell air slides are of the enclosed type. This system, unlike the Open-type, does not require the "fine" tuning of the air rate and the air pressure in the small individual-cell air slides.

Pneumatic Conveyor System (Fig. 8) is similar to the Enclosed-type Air Slide System, except the main air slide is replaced by an automatic dense-phase pneumatic conveyor. If designed properly (2), the dense-phase pneumatic conveyor may be less expensive than the main air slide with supporting steel, access platforms and walkways.

The surge hoppers are kept full at all times without operator's intervention. The alumina feed to individual electrolytic cells is programmed.

Direct Injection System (DIS) consists of penumatic conveying pipes and vessels entirely (Fig. 9). DIS is extremely flexible and capable of delivering desirable (but variable) amounts of alumina to the selected (by operator or computer program) feeding points simultaneously with crust breaking

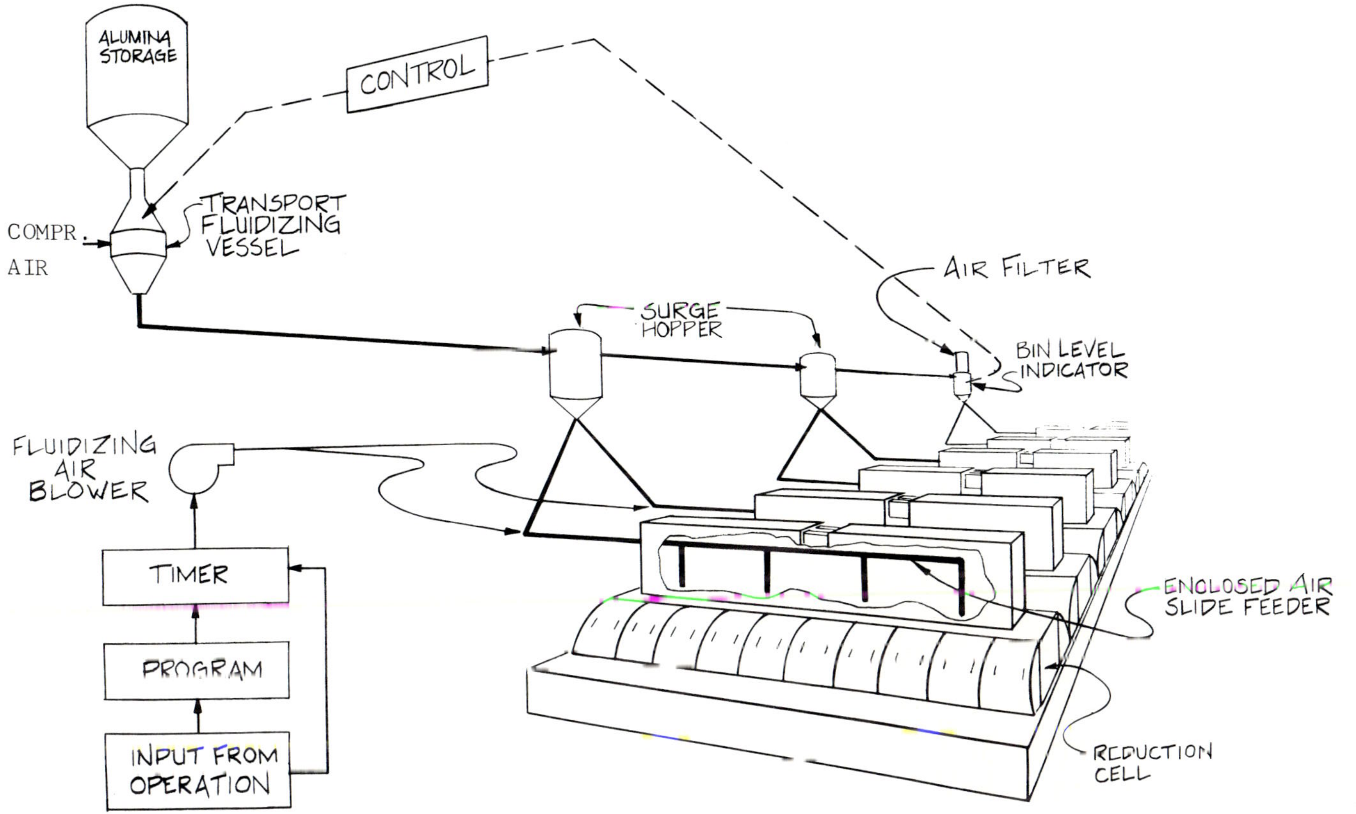

Fig.8 - Alumina transportation to cells - Pneumatic conveyor system

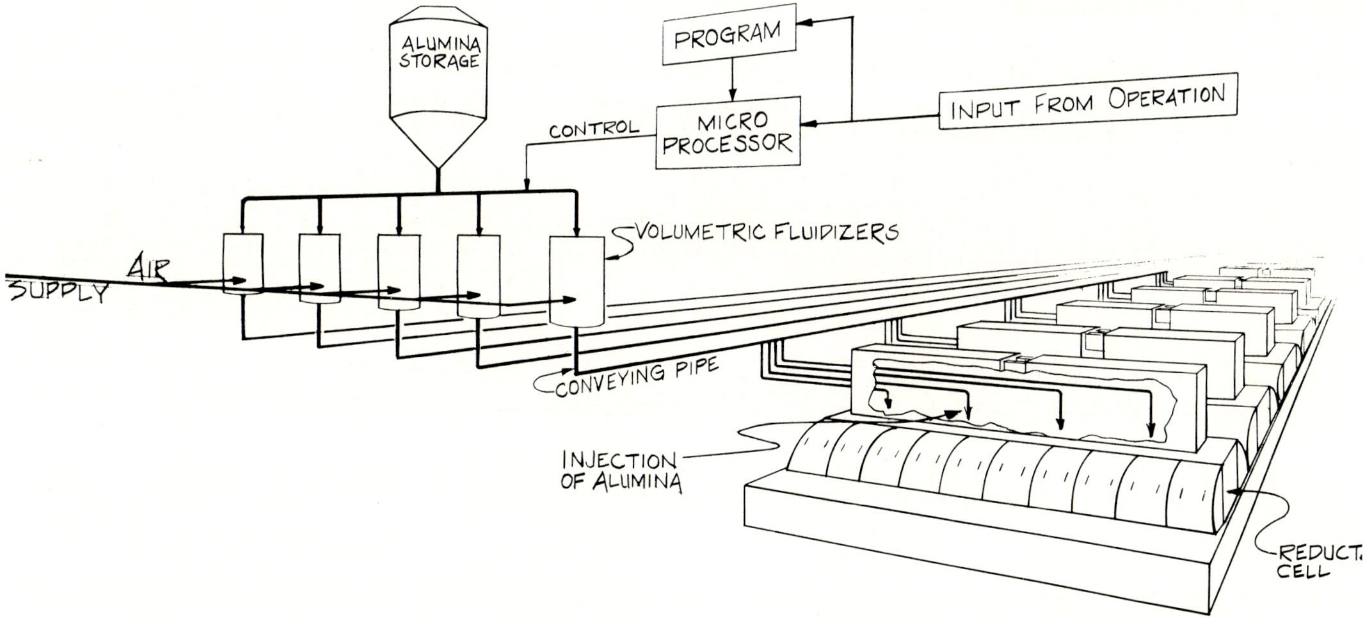

Fig.9 - Alumina transportation to cells - Direct injection system

in the cells. This most advanced and completely automated system, if designed reliably (2, 3) and simultaneously with the electrolytic cell, can affect the design of the cell itself favorably. As an example, it can simplify the cell point feeding system by eliminating the cell alumina volumetric and discharge devices with elaborate mechanisms and individual ore feed control. In turn, this will decrease the volume and the cost of maintenance at hundreds of electrolytic cells.

In addition, the DIS may eliminate or decrease the size of cell ore hoppers. This will decrease the weight and the cost of the cell superstructure. Because of the small quantities of injected alumina, dense mode of alumina conveying and the proximity of alumina discharge to the bath crust, no additional ventilation will be required.

CONCLUSIONS

The enclosed alumina delivery systems in comparison with existing ore buckets have many advantages, allowing to:

- Essentially alleviate the spillage of alumina in the potrooms and therefore improve the hygenic conditions for the operation and reduce OSHA-type problems.
- Substantially reduce the emissions from the roof-top monitors and reduce environmental-type problems from alumina handling.
- Increase potlife due to elimination of spills between cells that over-insulate the cathode.
- Decrease voltage drop due to elimination of spills and contacts between the anode bus and anode rod.
- Lessen the efforts for extra housekeeping in the potrooms and therefore lower the maintenance cost.
- Simplify the multi-purpose crane construction and therefore decrease its cost.
- Increase the crane availability for other potroom work and decrease the workload on potroom crane operators.
- Eliminate the potroom crane filling stations.

In addition, the simultaneous implementation of the most advanced Direct Injection System (Fig. 9), based on dense-phase pneumatic conveying, with the continuous point feeding of alumina could affect the advanced cell design favorably by decreasing the capital and maintenance cost of aluminum production.

REFERENCES

(1) Ivan Stankovich and other, Materials Handling Handbook, Section 27, John Willey & Sons, Inc., New York, N.Y., 1984

(2) Ivan Stankovich, DENSE TM Computer Program for Dense-Phase Pneumatic Conveying, User's Manual, pp. 1-16, West General Associates, Inc., Campbell, CA, 1983

(3) I.D. Stankovich, "A System Approach to Pneumatic Handling", Unit and Bulk Materials Handling, ASME Century 2 Publication, pp. 163-169, New York, N.Y., 1980

(4) I.D. Stankovich "Pneumatic Handling Classification", BHRA, Pneumotransport-4, Vol. 1, pp. F7 81-85, presented at the

International Conference, Carmel, CA, June 26-28, 1978
(5) I.D. Stankovich and K. Woolever, "A Method of Air Lift Design", BHRA Pneumotransport-4, Vol. 1, pp. F4 37-54, presented at the International Conference, Carmel, CA, June 26-28, 1978
(6) I.D. Stankovich, Problems in Control of Particulates in Bulk Material Handling", paper presented at the ASME Winter Annual Meeting, Atlanta, GA, November 27 - December 2, 1977
(7) I.D. Stankovich and other, "Feeding Particulate Material", U.S. Patent 4,016,053, April 5, 1977

CRUSTING BEHAVIOR OF SMELTER ALUMINAS

Douglas W. Townsend and Larry G. Boxall

Martin Marietta Laboratories
Baltimore, Maryland 21227
USA

The mechanism of cell crust formation has been investigated by studying petrographic thin sections of crusts made from different ores. Crusts form when cryolite bath soaks into alumina ore, selectively dissolving and cementing it together. Crusts were made by adding both reacted and unreacted commercial ores to a typical smelter bath. Crusting behavior was correlated with ore type, size fraction, and whether or not the ore was previously reacted with bath fume in a dry scrubber. Reacted ores are wetted faster and form softer crust. Petrographic thin sections of crusts show cemented-together alumina particles that are relatively unattacked by penetrating bath. Crust is formed because γ-alumina dissolves faster in the bath and has a higher solubility limit than α-alumina, causing local supersaturation. γ-alumina continues to dissolve, whereas α-alumina crystallizes to cement ore into crust.

Introduction

Smelter alumina ores with similar physical properties often display very different crusting behavior on reduction cells. In recent years, various aluminum companies have been trading ore so that individual smelters often receive consecutive shipments from different Bayer alumina plants. This practice often results in reduction plant upsets involving cell mucking, escape of excessive amounts of bath fume and difficulty in breaking or maintaining crust.

On a side-break Soderberg cell, an alumina ore that has a relatively low crusting tendency is desirable because it forms a soft crust on a cool area of the cell where there is little agitation. This crust should be easily pushed into the bath by breaker bars and also form quickly before the bath soaks completely through new ore, to prevent loss of fume and heat. Center-break cells and point-fed cells require alumina ore with a relatively high crusting tendency to resist collapsing into the bath in relatively hot and well agitated areas of the cell.

Alumina ores used in western countries generally can be classified as having a sandy texture. Typically 95-98% of the alumina particles will pass a 100-mesh screen, whereas the amount of -325 mesh may vary from less than 10% to more than 35%. The α-alumina content typically ranges from 10% to 30%. Most alumina ore particles are composed of intimate mixtures of both cubic γ-alumina and hexagonal α-alumina crystals as can be seen using an optical microscope and an oil having an index of refraction between that of γ-and α-alumina.

From studies of dissolution rates of individual alumina particles in cryolite bath, Thonstad (1) showed that γ-alumina dissolves faster than α-alumina. The same result was obtained by Gerlack et al. (2) using rotating disks. Although it has been recognized that sandy aluminas tend to form hard crusts, the mechanism of cementing undissolved alumina particles together to form crust has been attributed to networks of α-alumina crystals (3,4,5). These are always found in superficial examinations of the surfaces of ore crusts after cryolite has been leached. The actual cementing together of alumina particles, however, has not been illustrated nor has the process of cementing been described. In this paper, we will further examine the mechanism of crusting of alumina ores.

Alumina Ore Properties

Four alumina ores from different Bayer plants were selected for this study. They can all be classified as sandy aluminas, although they differ from each other considerably in physical appearance. The physical properties of the virgin ores is given in Table I and of reacted ores in Table II. The amounts of chemicals absorbed from the cell fume to make reacted ores in the dry scrubbers are given in Table III.

The physical appearance of the four virgin alumina ores, as viewed by scanning electron microscope (SEM), are shown in Figure 1. Each ore has physical properties typical of sandy alumina, including size distribution, low α-alumina content, high surface area and low angle of repose.

TABLE I. Physical Properties of Virgin Aluminas

Alumina Mesh	A (wt.%)	B (wt.%)	C (wt.%)	D (wt.%)
+100	0.4	5.8	1.2	1.4
100-200	46.5	50.8	53.6	58.8
200-325	36.5	23.3	34.6	29.8
-325	16.6	20.1	10.6	10.0
Sub-screen of -325 mesh diameter μm				
50.8-40.3	22.5	10.8	16.0	15.5
40.3-32.0	42.6	12.0	17.3	23.2
32.0-25.4	22.2	10.4	12.3	16.1
25.4-20.2	7.9	7.7	8.6	11.3
20.2-16.0	2.5	7.5	7.3	9.2
16.0-12.7	1.1	7.8	6.7	7.0
12.7-10.08	0.4	8.3	6.0	5.1
10.08-8.00	0.4	7.7	4.9	3.6
8.00-6.35	0.0	8.6	5.2	3.2
6.35-5.04	0.0	7.1	4.5	2.4
5.04-4.00	0.0	5.0	4.1	1.5
4.00-3.17	0.0	2.9	2.6	0.8
3.17-2.52	0.0	1.4	1.7	0.4
2.52-2.00	0.0	1.2	1.5	0.3
< 2.00	0.0	1.4	1.0	0.3
Specific surface area m^2/g				
+325	67.0	41.2	68.4	65.5
-325	55.6	31.8	38.2	29.7
All fractions	65.1	39.4	65.2	61.9
% α-alumina	20.8	14.7	19.6	20.2
Angle of repose, degrees	28.3	26.8	30.8	29.3
Bulk density (g/cm^3) tamped	1.16	1.25	1.05	1.09
Untamped	1.01	1.06	0.9[illegible]	0.98

Figure 1. Scanning electron micrographs of virgin aluminas: (a) alumina A, (b) alumina B, (c) alumina C, and (d) alumina D.

TABLE II. Physical Properties of Reacted Aluminas

Alumina Mesh	A (wt.%)	B (wt.%)	C (wt.%)	D (wt.%)
+100	0.2	5.9	3.3	0.8
100-200	39.8	57.6	55.4	75.7
200-325	49.5	22.9	36.1	21.8
-325	10.5	13.6	5.2	1.7
Sub-screen of -325 mesh diameter μm				
50.8-40.3	48.8	6.1	17.9	41.5
40.3-32.0	27.5	8.2	24.4	25.5
32.0-25.4	9.0	7.4	13.9	10.8
25.4-20.2	3.6	6.1	8.2	5.5
20.2-16.0	2.0	6.4	6.3	3.8
16.0-12.7	1.5	8.0	5.5	3.1
12.7-10.08	1.3	8.8	4.9	2.6
10.08-8.00	1.4	9.0	4.0	2.0
8.00-6.35	1.2	11.0	4.3	1.7
6.35-5.04	1.1	10.0	3.7	1.4
5.04-4.00	0.9	7.9	2.0	0.8
4.00-3.17	0.6	4.8	1.3	0.4
3.17-2.52	0.3	2.4	0.9	0.1
2.52-2.00	0.3	2.0	0.7	0.2
< 2.00	0.3	1.6	0.6	0.2
Specific surface area m^2/g				
+325 mesh	58.4	37.5	53.2	59.0
-325 mesh	37.6	27.6	33.1	32.4
All fractions	56.2	36.4	52.2	58.4

Experimental

Alumina crusts were made in the laboratory by adding 2-, 4-, and 8-g samples of commercial ores to the surface of a cryolite bath held at 985°C. The basic bath composition was CaF_2 - 3.5%, AlF_3 - 7.5%, and Na_3AlF_6 - 89%. The bath ratio was 1.25. During the course of the experiments, the dissolved Al_2O_3 in the bath ranged from 2% through saturation. About 1600 g of bath was held in a graphite crucible, 8 cm in diameter by 19 cm tall, contained in an Inconel pot. Control thermocouples were kept outside the Inconel pot near the furnace windings. The bath temperature, measured by a thermocouple just above the bath surface, differed by less than 5°C from temperatures measured throughout the bath. There was no tendency to form crust by bath freezing. The top of the furnace was covered with Marinite®; argon flowed over the bath surface between experiments.

TABLE III. Fluoride, Carbon and Sulfur Analyses of Alumina Ores

Alumina	Fluoride (wt.%)	Carbon (wt %)	Sulfur (wt.%)
A Virgin	0.030	---	0.001
A Reacted	0.76	1.27	0.28
B Virgin	0.030	---	0.025
B Reacted	1.07	1.62	0.21
C Virgin	0.005	---	0.010
C Reacted	1.60	0.93	0.21
D Virgin	0.040	---	0.012
D Reacted	1.14	1,99	0.25

A batch of bath salts was warmed to 500°C overnight, melted at 1030°C, and cooled to 985°C, which was from 5-35°C above the melting point of the bath. The minimum freezing point of this bath is 948°C at 8% dissolved Al_2O_3. Crusting tests were made by dropping 2-g samples of unheated alumina from a 1/2-inch tube having a spring-loaded cone stopper from a distance of 5 cm.

Four-gram samples of alumina ore and bath were recovered 10 seconds after wetting for examination of bath penetration before a crust could develop. Crust produced from 8-g samples of alumina ore were recovered three minutes after wetting for studies of the crust-forming process and recrystallization of alumina.

Results

The time for bath to wet the alumina, the nature of the crust, and the times for the alumina to sink and finally to dissolve into both stagnant and stirred bath were observed and recorded in Table IV.

The crust produced by each of the virgin alumina ores, either fell quickly to the crucible bottom in one to four large chunks or broke up into flakes and slowly sank. The following definitions have been used in Table IV to describe the crusting tendency and dissolution behaviors of the ores.

TABLE IV. Alumina Dissolution Properties

Sample Identification	Experiment	Wetting Time (seconds)	Sinking Time (seconds)	Time to Complete Dissolution (seconds)	Crusting and Dissolution Behavior[1]
A Virgin	Unstirred	8	63	375	1a
A Reacted	"	9	FLT[2]	354	2c
B Virgin	"	11	129	237	1b
B Reacted	"	6	FLT	300	2c
C Virgin	"	5	90	283	1b
C Reacted	"	5	FLT	408	2c
D Virgin	"	6	120	353	1b
D Reacted	"	5	FLT	424	2c
A Virgin	Stirred 250 rpm	5	22	161	1a
A Reacted	"	3	10	100	2b
B Virgin	"	5	25	44	2b
B Reacted	"	3	30	49	2c
C Virgin	"	3	9	113	1b
C Reacted	"	5	20	129	2c
D Virgin	"	3	11	126	1c
D Reacted	"	5	15	82	2c
A Virgin -325	Unstirred	4	38	248	1a
A Virgin +325	"	2	200	298	1a
A Reacted -325	"	30	FLT	150	2b
A Reacted +325	"	6	270	504	1b
B Virgin -325	"	4	60	756	2c
B Virgin +325	"	7	90	258	1a
B Reacted -325	"	33	90	536	2b
B Reacted +325	"	3	56	140	2c
C Virgin -325	"	9.5	17	594	2a
C Virgin +325	"	9.5	24	606	1a
C Reacted -325	"	6	139	179	2c
C Reacted +325	"	4	FLT	645	2c
D Virgin -325	"	6	75	402	1a
D Virgin +325	"	5	140	453	1a
D Reacted -325	"	15	FLT	75	2c
D Reacted +325	"	2	FLT	345	2c

1 See text for explanation

2 The aluminum did not sink in the bath; it floated on the bath surface until dissolved

TABLE IV cont'd

Sample Identification	Experiment	Wetting Time (seconds)	Sinking Time (seconds)	Time to Complete Dissolution (seconds)	Crusting and Dissolution Behavior
A Virgin -325	Stirred 225 rpm	9	14	150	1a
A Virgin +325	"	4	13	252	1a
A Reacted -325	"	6	FLT	192	2c
A Reacted +325	"	4	71	120	2b
B Virgin +325	"	4	52	91	2b
B Virgin -325	"	4	14	64	1b
B Reacted +325	"	12	FLT	287	2c
B Reacted -325	"	4	36	97	2c
C Virgin -325	"	3	15	162	1b
C Virgin +325	"	3	35	148	1a
C Reacted -325	"	11	36	155	2c
C Reacted +325	"	3	54	175	1b
D Virgin -325	"	7	20	123	1a
D Virgin +325	"	6	20	244	1a
D Reacted -325	"	3	54	181	2c
D Reacted +325	"	4	41	148	2c

Crusting Tendency

1. Forms a coherent hard crust with mechanical strength, like a white ice

2. Forms a soft crust without mechanical strength, like wet snow,

Dissolution Behavior

a. The crust sank to the bottom of the crucible in one or several large broken pieces and slowly melted away

b. The crust broke up on the surface and sank in a large number of small pieces or flakes, many of which dissolved while sinking

c. The crust largely dissolved away while floating on the surface of the bath with only a few flakes breaking away.

Alumina ore A produced by far the hardest crust, i.e., like a white ice, whereas alumina ore B produced the softest crust, i.e., like a wet snow. Ores C and D produced crusts softer than ores A but harder than B. Compared to virgin ores, reacted ores produce a relatively soft crust due to desorbing gasses. The crust tends to float on the bath because of trapped gas bubbles and carbon particles. Wetting, sinking and dissolving were each accelerated by stirring; however, the basic character of each ore and crust was essentially unchanged.

When each of the ores was divided into a coarse fraction (+325 mesh) and a fine fraction, some changes in crusting behavior were observed. In general, the coarser fraction wetted faster and tended to produce a harder crust than did the fine fraction.

Leaching Tests on Crusts

Crust hardness can be tested by boiling pieces of frozen crust in 3.3% aluminum chloride solution for 10 minutes to remove the cryolite bath. This procedure is derived from a gravimetric analysis of dissolved alumina in cryolite bath. Cryolite bath is completely leached away from the alumina ore and α-alumina platelets.

Crust made from ores A and C remained intact after the bath was leached out, whereas crusts made from ores B and D disintegrated into powder. The soft crusts made from ores B and D apparently contained no extensive large-range network of cemented-together alumina ore particles.

Crust Formation Analysis

Ore crust forms when undissolved alumina particles cement-together into a strong three-dimensional network. When ore is first placed on bath, some freezing occurs as the ore is warmed and alumina dissolves, saturating the intruding bath. To heat alumina ore to bath temperature requires 27.5 kcal/mole for γ-alumina and 26.2 kcal/mole for α-alumina (6). We determined from analysis of photographs of thin sections (Figure 2) that alumina ores A and B each soak up about 35% by volume bath. Saturating this intruded bath with dissolved alumina only requires from 0.3 to 0.6 kcal/mole of ore (7). The heat required from the cell to form crust is essentially only that necessary to heat the ore to the bath temperature. Dissolution and radiation from the surface of the crust causes the crust to remain a few degrees cooler than the bath.

Optical photographs of petrographic thin sections prepared from cryolite-wetted alumina ores A and B along with intruded bath are seen in Figure 2. Conversion of the wetted alumina ores to crust from continued exposure to the bath (3 minutes) are shown in Figure 3. Ore A shows a high degree of particle-to-particle contact in both the wetted ore and hard crust, whereas ore B shows relatively less. Many of the alumina particles actually appear to be free floating in the soft crust. A high degree of continuous physical contact between particles of undissolved alumina is required to form a hard crust. Any process, including agitation, desorption of gas from reacted ore, and additional bath intruding into wetted ore B, clearly disrupts the crust-forming process. During the short time available for crust forming, 12% additional bath intruded the wetted ore and dispersed the ore particles so they did not grow together to form a hard crust. The bath occupies $\sim$ 47% of the volume of the soft crust, whereas only $\sim$ 39% of the volume is occupied by the bath in the hard crust.

Samples of crust that have been air cooled and leached appear to be composed of undissolved alumina ore particles covered by α-alumina platelets, 0.5-µm in diameter, Figures 4 and 5. These thin platelets merely occupy the space filled with the intruded bath. There is no apparent difference in the appearance of these thin platelets that cover the exposed surface of the

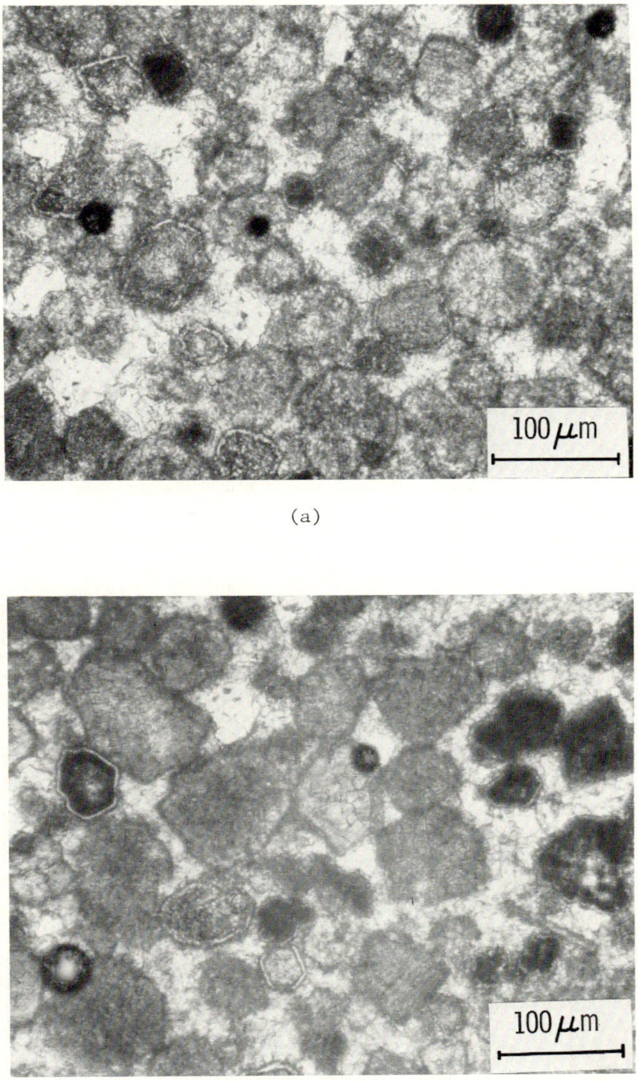

(b)

Figure 2. Petrographic thin sections of wetted cryolite ores: (a) alumina A and (b) alumina B.

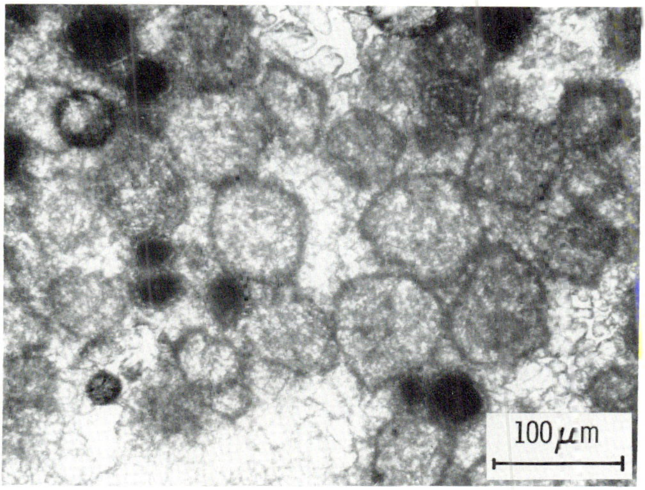

(a)

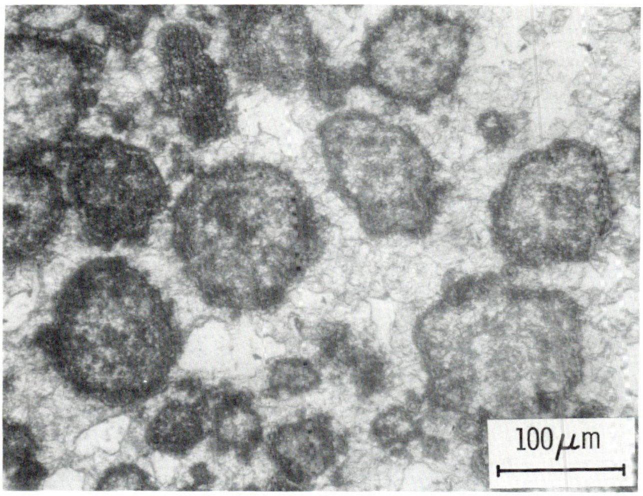

(b)

Figure 3. Petrographic thin sections of crusts: (a) alumina A and (b) alumina B after three minute of wetting by cryolite.

Figure 4. Scanning electron micrograph of leached crust, alumina ore A.

alumina ore particles from hard and soft crust. They merely froze out of the alumina-saturated bath during air cooling and did not contribute to crusting. If these thin α-alumina platelets had formed the crust cement, both crusts should have been equally hard.

It has been postulated by Less (3) that α-alumina platelets constitute the cement that produces ore crusts made from sandy alumina. The rapid growth of α-alumina to cross link or fuse together adjacent grains of alumina ore to form crust has been described by Johnston and Richards (5). The thin platelets shown in Figures 4 and 5, however, are not the ones that cement-together the ore to form crust. The cementing α-alumina platelets can be seen only in Figure 6 where sections have been cut through the crust, happening to intercept the points of contact between alumina ore particles. The crust cement is in fact composed of α-alumina platelets, but each is about three times as thick as the platelets that freeze out of the bath. As seen in an edge view, each of these crust-forming α-alumina platelets in Figure 6 is about 1.5-μm thick. Masses of them form a tight envelope of tangential and radial crystals surrounding the alumina ore particles in a layer about 5-μm thick. These thick crystals in hard crust almost completely fill the space between adjacent alumina ore particles. The difference in thickness between the cementing α-aluminum crystals and those that merely froze when the sample was quickly cooled can be seen in Figure 6.

X-ray analysis of samples of crust taken from bath three minutes after wetting shows that a considerable amount of the γ-alumina originally present in the alumina ore has recrystallized to α-alumina. Alumina, recovered from hard crust made from ore A, was 68% α-alumina compared with only 20.8% in the virgin ore. The soft crust made from ore B had converted from 14.7% to 93% α-alumina during three minutes of exposure to the bath. This shows that the fluoride "mineralizer" in the bath does in fact greatly accelerate recrystallization of the ore during the crust-forming process. The conversion of γ-alumina to α-alumina is thermodynamically favored with a Gibbs free energy of 0.78 kcal/mole (6). In the absence of the fluorides, at 985°C, 10 to 14 hours is required to get these degrees of conversion of γ-alumina to α-alumina (7). The rates of conversion to α-alumina of the ore particles in cryolite bath is rapid but crust forming is an even faster process. It is most significant that γ-alumina can be found in hard crust, because a thermodynamic analysis predicts that γ-alumina should be slightly more soluble in cryolite bath than α-alumina.

(a)

(b)

Figure 5. Scanning electron micrograph of leached crusts after three minutes of wetting by cryolite: (a) hard crust, alumina ore A; and (b) soft crust, alumina ore B.

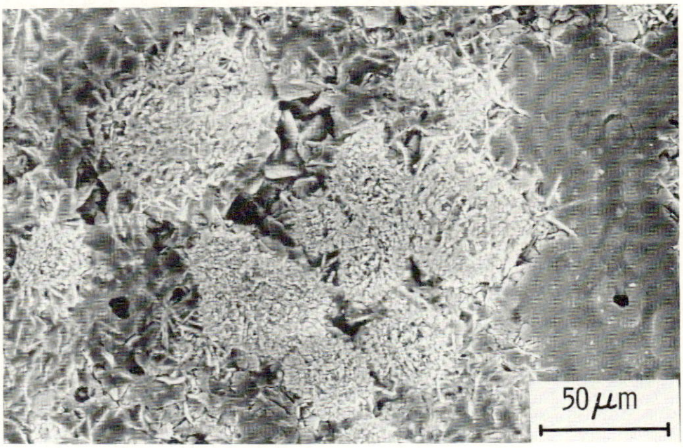

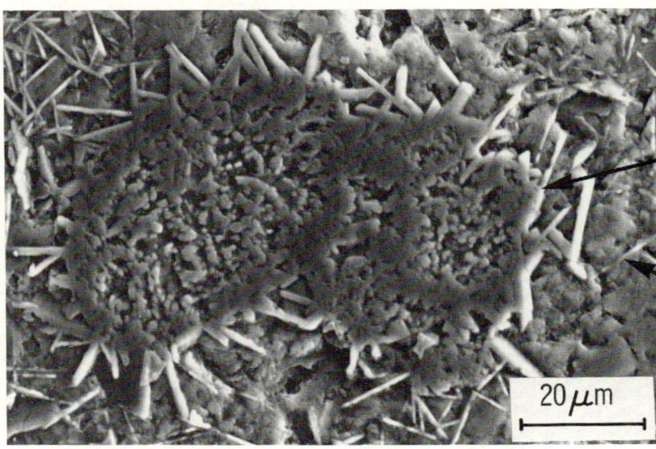

Figure 6. Scanning electron micrographs of cross sections of hard crust after three minutes of wetting by cryolite, alumina ore A.

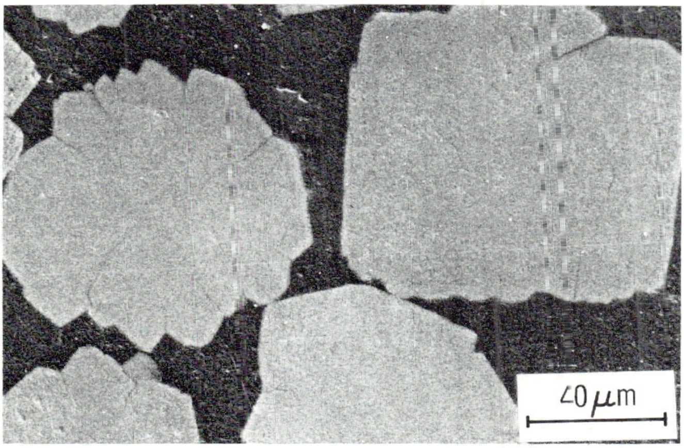

Figure 7. Scanning electron micrograph of cross section of virgin alumina ore A.

Figure 7 shows a cross section of alumina of ore A, which by X-ray analysis contained only 18% α-alumina. The 82% γ-alumina (by difference) has a homogeneous, nearly amorphous structure derived from calcination of aluminum trihydrate crystals (Gibbsite), precipitated in a Bayer plant from sodium hydroxide solutions supersaturated with sodium aluminate. After only three minutes exposure to the bath, the cores of these particles display the characteristic appearance of floury α-alumina of the type usually used in Europe, Figure 6. According to X-ray analysis, the ore has been converted to 68% α-alumina. By difference, the alumina particles taken from the crust still contained 32% γ-alumina so we may conclude that the crust-forming process was continuing when the sample was taken from the bath. This ore forms a very hard crust that simply sinks to the bottom of the crucible in one or a few large chunks and then dissolves into the bath from exposed surfaces without breaking apart.

In contrast, the soft crust formed from ore B contained only 9% γ-alumina after 3 minutes, showing that the crust-forming process was almost completed. Crust made from this ore tends to break up into small flakes and dissolve easily into the bath. This crust is easily broken by stirring. Previous observations (3) that floury high α-alumina has very poor crusting behavior are consistent with our proposed crusting mechanism.

Thus, the most likely mechanism for ore crust formation is the dissolution of γ-alumina into the bath and the simultaneous growth of α-alumina crystals on the same or adjacent particles of alumina ore. It takes several moments for the strong, thick α-alumina platelets to form. The intruded bath transports dissolved alumina by diffusion from less thermodynamically stable alumina particles in the crust. Diffusion distances between dissolving and growing crystals in crusts are very short and the intruded bath quickly becomes slightly supersaturated with respect to α-alumina. The crust-forming process continues for several minutes until the γ-alumina is depleted by dissolution or direct recrystallization within undissolved particles of alumina ore to α-alumina.

Conclusions

We are in general agreement with previously proposed mechanisms for alumina ore crusting. The formation of strong ore crusts does require strong α-alumina platelets to cement-together particles of undissolved ore. Crusting occurs because γ-alumina dissolves and supersaturates the intruded bath, thereby causing the growth of α-alumina platelets between undissolved alumina ore particles. The crusting process must occur within a few minutes because the bath fluorides simultaneously catalyze the recrystallization of undissolved γ-alumina to α-alumina which is a poor crust former. A high degree of particle-to-particle contact in the alumina ore is required for several moments after the aluminum ore is wetted by bath for a strong crust to form. It is still necessary to perform crusting tests to determine the types and relative strengths of crusts formed from different ores. Crust-forming tendencies cannot be correlated with easily measurable physical properties such as relative amounts of γ-alumina and particle size. In general, crust-forming tendencies are reduced by reacting the ore with bath fume in a dry scrubber and by the presence of a high fraction of -325 mesh fines.

Laboratory alumina-crusting tests and optical examination of petrographic thin sections of the formed crusts provide a means to screen ores for their potential impact on established smelter operations. The petrographic thin cross-section photographs provide crust-forming data normally obscured by the alumina-containing bath and can be used to evaluate potential new alumina sources for aluminum smelters.

Acknowledgement

The authors wish to thank Martin Marietta Alumina for permission to publish this work and the acknowledge laboratory assistance of W. Montague of Martin Marietta Laboratories.

References

1. J. Thonstad, F. Nordmo, and J.B. Paulsen, "Dissolution of Alumina in Molten Cryolite," Light Metals, (1971) pp. 213-222.

2. J. Gerlack, U. Hennig, and K. Kern, "The Dissolution of Aluminum Oxide in Cryolite Melts," Light Metals, 1 (1974) pp. 49-61.

3. L.N. Less, "The Crusting Behavior of Smelter Aluminas," Metallurgical Transactions, B, 8B, (1977) pp. 219-225.

4. A.R. Johnson, "Alumina Crusting and Dissolution in Molten Electrolyte," Journal of Metals, 34 (3) (1982) pp. 63-68.

5. T.J. Johnston and N.E. Richards, "Correlation Between Alumina Properties and Crusts," Light Metals, 1 (1983) pp. 623-639.

6. D.R. Stull and H. Prophet, JANAF Thermodynamic Tables, United States Department of Commerce (1971).

7. C.J.P. Steiner, D.P.H. Hasselman, and R.M. Spriggs, "Kinetics of the Gamma-to-Alpha Alumina Phase Transformation," Journal of the American Ceramic Society, 54 (8) (1971) pp. 412-413.

8. O.A. Asbjφrnsen and J.A. Andersen, "Kinetics and Transport Processes in the Dissolution of Aluminum in Cryolite Melts," Light Metals 1 (1977) pp. 137-152.

ALUMINIUM PECHINEY SPECIFICATIONS FOR OPTIMISING THE ALUMINAS USED IN SIDEBREAK AND POINT FEEDING REDUCTION POTS

Y. Bertaud

Aluminium Pechiney
Laboratoire de Recherche des Fabrications
73300 Saint Jean de Maurienne
France

A. Lectard

Aluminium Pechiney
Centre d'Etude de l'alumine
13120 Gardanne
France

The rational operation of reduction pots involves a great many constraints. Some of the major ones, namely automation of feed systems, pollution abatement and improved working conditions, are directly linked to the grade of alumina used. Their respective importance depends, however, on the type of pot in question (Søderberg anode pot, hooded prebaked anode pot using side crustbreaking, hooded prebaked anode pot using point feeding). Aluminium Pechiney has carried out theoretical studies, applied studies, bench scale and industrial tests aimed at optimising the grade of alumina used in each type of pot. A number of our specifications are specially adapted to take into account the specific surface, moisture content, grain size distribution, resistance to powdering and extent to which the alumina crystals are ordered.

The progress made in our know-how and experience enables us to predict what is required of a multipurpose grade of alumina which could be used in all types of pot.

Introduction

For more than two decades following the Second World War, the world of aluminium smelting was, as it were, divided in two:

. North America, with access to cheap energy and trihydrate bauxite, using a sandy coarse alumina,

. while Europe and Japan, handicapped by having to make do with a monohydrate bauxite and expensive energy, found it more advantageous to use floury alumina.

Towards the end of the sixties, changes in the social and economic climates (1) compelled producers to (i) reduce fluorine emission to the atmosphere to next to nothing, (ii) pare capital costs per production tonne, (iii) cut energy consumption and (iv) improve the productivity and the working conditions of their workforces.

These considerations explain the present lack of uniformity in the make-up of Aluminium Pechiney smelters, which variously employ

a) prebake pots which are not completely hooded and which use floury alumina: although efficient (2), these are virtually obsolete and have a very limited future

b) Søderberg pots, equipped for fume collection and wet scrubbing, built in the 1950s

c) hooded prebake pots employing intermittent side crustbreaking and roof-mounted fume treatment facilities, representing a compromise between traditional technology and ecological dictates

d) hooded prebake pots employing automated and virtually continuous point feeding systems, designed to meet present-day social and economic constraints (3, 4).

How best to match the alumina to very different technologies and modes of operation while getting the highest performance poses quite a problem.

Our purpose is to show how an exhaustive analysis of operating constraints can make it possible to optimise the alumina specification for best performance on a particular type of pot. For the sake of simplicity, the discussion is confined to the impact of physical criteria. Much could however be said about the impact of chemical impurities, with particular reference to latter-day technologies entailing the recycling to, and concentration in, the baths of virtually all of the elements present in the starting alumina, some of which have been found to militate against satisfactory performance.

Figure 1 - Pot operating constraints

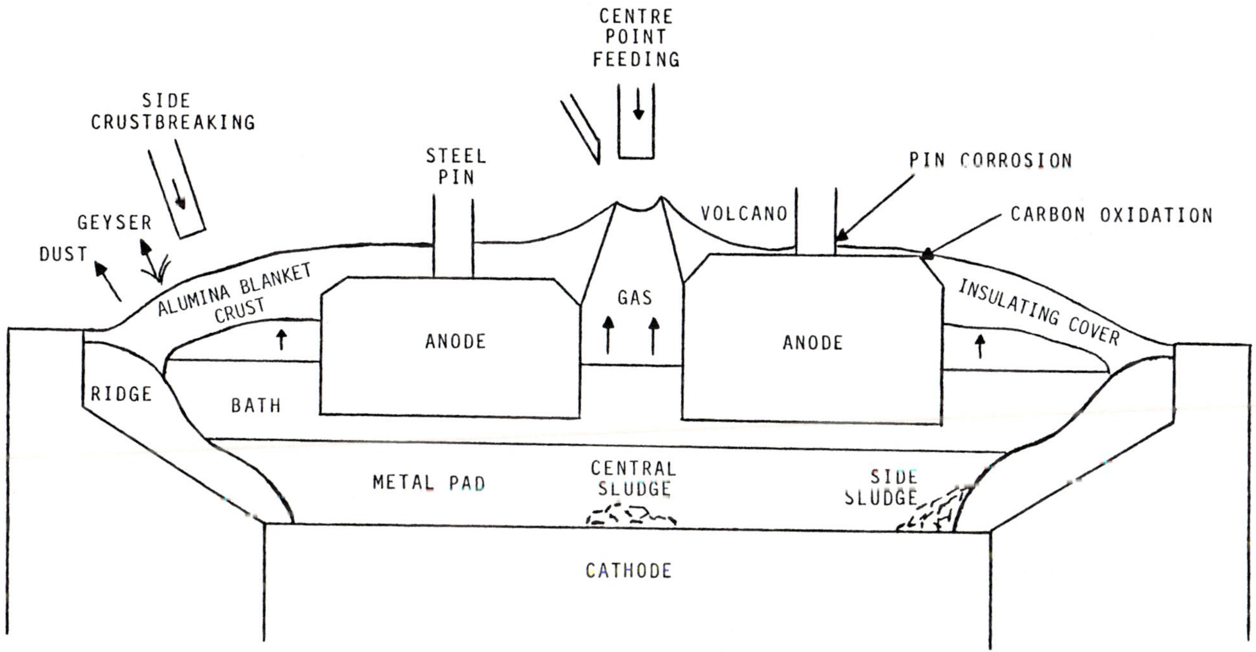

Table 1 - Pot Operating Constraints

Consistent and regular alumina quality is fundamental to satisfactory pot operation.

Acceptable limits on variations in alumina specification will depend on the reduction technology employed.

CONSTRAINTS	OPERATING CRITERIA	IMPORTANCE OF CRITERIA				ALUMINA PROPERTIES CONCERNED
		SÖDERBERG	PREBAKE			
		NO DRY PROCESS TREATMENT	NO HOODS, NO DRY PROCESS TREATMENT	HOODS AND DRY PROCESS TREATMENT	POINT FEEDING DRY PROCESS TREATMENT	
PHYSICO-CHEMICAL	Rapidity of solution	+	+	+	+++	Rate of solution
	Little hydrolysis of bath	++	++	++	++	Moisture content
	Satisfactory protection of anode system (against combustion of carbon and corrosion of steel pins)	0	++	++	+	Impermeability to air and to the bath
	Low heat of solution	0	0	0	++	Crystal structure
	Low overvoltage	++	++	++	++	Lattice water
THERMAL	Effective heat insulation (blanketing)	+	++	++	+	Thickness of layer Thermal conductivity of layer

MECHANICAL	Good stability of alumina blanket	++	++	++	++	Stability of layer
	Formation of a stable crust	++	++	++	++	Wettability of alumina
	Crust easy to break	+	+	+	+	Wettability & reordering of alumina crystal structure
	No formation of volcanos	0	0	0	+++	Behaviour of grains (absence of hydrate)
	Good penetration of alumina into bath	++	++	++	++	Wettability & dispersion in bath
	Good feeding	0	0	0	+++	Free-flowing
ENVIRONMENTAL AND RELATING TO WORKING CONDITIONS	Minimum dust emission (geyser effect and fly-up):					Little susceptibility to entrainment by air or pot fume
	When not carrying out tending operations	+	+++	+	+	
	When carrying out tending operations	+++	+++	+++	+	Ability to outgas
	Minimum emission of fluorine products (at given acidity)	++	++	++	++	
	Good gas-tightness to pot fume	+++	+++	+	+	Formation of gas-tight blanket
	Satisfactory collection/ treatment of HF	+	+	+++	+++	Adsorption
	Minimum retention of SO_2	0	+	++	0/+	Adsorption
PRODUCTION	Maximum reduction current	++	++	++	++	

RANKING ORDER OF CRITERIA: 0 = not important
+ = only slight importance
++ = important (particularly as regards economics)
+++ = fundamental

Table 2 - Main characteristics of aluminas used in Pechiney plants

ALUMINA	AQ	AK	AF	AC	AB	AG1	AG2	AG3
BET specific surface, m^2/g	68	52	69	66	64	52	43	39
Loss on Ignition 300-1000°C, %	1.0	0.4	0.6	0.9	0.7	0.4	0.4	0.3
Total water content, %	4.0	2.5	3.1	3.3	3.2	2.5	1.9	1.8
Hydrate	Yes	ε	No	Yes	Yes	No	No	No
Alpha content, %	29	13	19	< 5	17	25	33	32
Wet sieving, %								
minus 45 μm	3.8	8.2	17.4	6.1	5.6	11	8.7	9.0
minus 30 μm	0.8	6.3	5.7	1.3	2.4	5	2.0	4.2
minus 15 μm	0	5.2	2.4	0	0	0.7	0	0.7
Dry sieving, % minus 325 mesh	9	9	21	8	10	16	14	14
Wet sieving, % after attrition test								
minus 45 μm	17.7	9.2	25.3	12.7	19.7	17.4	21.9	21.1
minus 30 μm	9.3	7.2	9.0	4.7	8.6	8.0	8.9	9.6
minus 15 μm	4.3	6.1	3.9	1.6	2.4	4.0	3.3	2.4
Alcoa attrition index	35	10	20	15	33	25	35	30

Pot operating constraints

Table 1 summarises the main operating constraints impinging directly on the choice of the alumina to be used; these are:

. *thermal - to do with heat of solution*

. *mechanical - to do with free-flowing properties and avoidance of the "volcano" effect*

. *environmental - to do with fume treatment and dust control.*

Their relative importance is therefore seen to vary with the technology employed. For example, centre-worked pots are more sensitive to solution phenomena than side-worked pots, which can even be operated on floury alumina. As against this, dust emission in the potroom when feeding with the hoods open can be something of a problem with sidebreak cells.

The discussion will be confined to the determination of the optimum choice of alumina for use in each of three cases, namely:

i) for side-worked, hooded, prebake pots associated with a dry-process fume treatment plant (FTP)

ii) for centre-worked, hooded, prebake pots employing an automated, virtually continuous, point-feeding system and associated with a dry-process FTP

iii) for Søderberg pots employing only skirt gas collection and roof-mounted wet scrubbing facilities.

Alumina optimisation for side-worked hooded prebake pots with dry-process fume treatment

The basic constraints are:

. *Collection and treatment of hydrogen fluoride*

. *Curtailing fly-up.*

Collection and treatment of hydrogen fluoride

The many variants of the dry treatment process all rely on the adsorption of hydrogen fluoride on alumina. Briefly:

. Current legislation effectively imposes an efficiency of better than 99.99 % for the adsorption of hydrogen fluoride on the alumina in the FTP.

. The gas/alumina contact time in the state-of-the-art systems can be very short, so that the rates of diffusion of the hydrogen fluoride into the pore system and of adsorption both have to be high.

. The alumina has to adsorb and retain the totality of pot fume emission and the pots have to be able to digest the totality of the alumina leaving the FTP.

. Copious emission of fume from side-worked pots occurs only during crust-breaking, when the hoods are open (whence the need for roof-mounted fume collection and treatment).

. Bath acidity is generally low enough to obviate sludge formation arising from the instantaneous introduction of large quantities of alumina (cryolite ratio of approximately 1.25), thus curtailing the production of hydrolysable sublimates.

Table 3 - Characteristics of Dust Emissions in the Potrooms

Grain size, % Laser method CILAS 715 / Alumina / Potlines	AB	AB	AK	AQ
	PA between pots	PA roof wet scrubber inlet	IA roof wet scrubber inlet	IB roof wet scrubber inlet
minus 192 μm	100	100	100	100
minus 128 μm	100	100	100	99
minus 96 μm	98	99	100	98
minus 64 μm	92	95	98	96
minus 48 μm	81	86	97	93
minus 32 μm	54	60	88	73
minus 24 μm	35	40	80	55
minus 16 μm	18	21	64	31
minus 12 μm	13	15	49	18
minus 8 μm	9	11	28	9
minus 6 μm	7	8	18	5.9
minus 4 μm	5.4	6.0	11	4.3
minus 3 μm	4.6	4.9	8.3	3.5
minus 2 μm	3.6	3.7	8.1	3.1
minus 1.5 μm	2.2	2.3	4.6	2.3
minus 1 μm	1.7	1.7	3.9	2
Fluoride content, %		9	3	7
Sodium content, %		3	4.2	2.4
LOI 0 - 1000°C		21		

Under these conditions, the dry process has to be capable of treating an average 10 kg of hydrogen fluoride per tonne of aluminium, and peaks of up to 18 kg/t. These requirements can be satisfied by an alumina with a readily accessible surface area of 35 m^2/g.

Note - Under certain operating conditions, the pots have a tendency to give off more fume, whereupon the natural reaction of managements is to look for an alumina offering a larger surface area. This is the wrong reaction insofar as such an alumina will generally contain more moisture, the effect of which, as a result of hydrolysis of the contents of the bath, is to boost hydrogen fluoride emission very substantially.

Curtailing dust emission

The industrial-scale tests carried out by Aluminium Pechiney at PNL, Intalco and ADG on various batches of alumina (Table 2) have shown that

- the material brought into suspension in the atmosphere is very finely divided (Table 3)

- particle size at the pot is not determined solely by that of the alumina as supplied to the smelter and it is not possible to correlate dust emission solely with initial sieve analysis (Tables 4 and 5).

A closer look at what is happening shows the principal factors to be:

i) the impact of local conditions

ii) the size history of the alumina (i.e. particle size at the pot)

iii) moisture content.

Impact of local conditions. The differences in behaviour of one and the same alumina in two smelters, or even in two adjacent potlines, highlight the influence of ventilation (wind exposure, natural draught, scrubber suction rate, etc.) and of the type of operation (pot tending machine, semi-gantry, negative pressure around the crustbreaker, depth and temperature of the layer of alumina introduced, outgassing by the blanket of alumina, etc.). Local conditions mainly affect the most finely-divided material on the pot.

Alumina size distribution on the pot. The qualification "on the pot" is of fundamental importance, given that particle size will to some extent be affected by the various stages through which the alumina has to pass on the way to the pot, with attrition in airlifts and, particularly, in the dry treatment process, and segregation in storage bins.

To what extent, exactly, can be determined only via a method of analysis which is not itself attritive and which can detect and measure the smallest particle sizes (e.g. wet sieving or the laser analyser).

It can thus be shown that particle size at the pot depends on:

a) the susceptibility of the alumina to attrition, as evaluated by a modified version of the standard attrition index test (5) developed by Aluminium Pechiney in order more closely to simulate actual behaviour in industrial practice (Table 4) (old FTP)

b) the severity of the effects of handling, particularly at the dry treatment stage (Table 5); very considerable progress has been made in this respect (in terms of gas flow and the distribution of alumina in the gas stream) and certain dry processes which are only mildly attritive (e.g. the

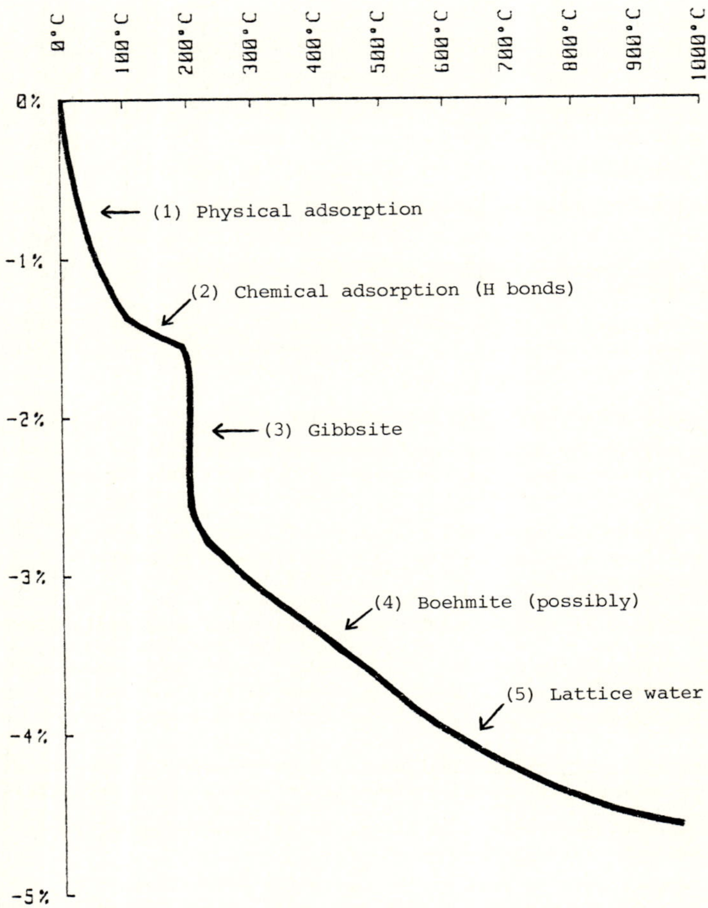

Figure 2 - Constant rate thermogravimetric analysis.
Typical curve for a low-calcined alumina
(according to J. Rouquerol)

Air Industrie process used at Saint Jean) make outstanding resistance to attrition less imperative

c) possible segregation phenomena, a problem which has now been effectively resolved thanks to systems employing rotating feeding systems and multiple discharge points (6) or honeycomb silos.

However, particle size at the pot is not the whole explanation for the variations in dust emissions observed on one and the same potline. Account also has to be taken of moisture content.

Influence of moisture content. There are four "categories" of water to be found in alumina (Figure 2):

i) physically adsorbed water; this is removed at the dry treatment stage

ii) chemisorbed water; this is bonded to the surface and is partially stable up to 300°C

iii) water of constitution (residual hydrates); this comes from the admixture with the alumina of low-calcined dust

iv) lattice water; this is stabilised by defects in the crystal lattice and is an accurate reflection of the degree to which the crystal structure is ordered.

We at Aluminium Pechiney link dust emission to

a) chemisorbed water: the more or less abrupt volatilisation of a substantial quantity of water entraining fines when alumina is introduced into the bath

b) residual hydrate: leading to the formation of spurts or "geysers" of ultrafine material as a result of vapour blanketing and disruption of particles at the bath surface.

In sum, the need is to ban hydrate and curtail the area available for chemisorption.

Conclusion (Side crustbreaking)

Aluminium Pechiney's recommendations to meet the special case of side-worked pots are to employ an alumina with the following characteristics:

. all the grains, and especially the fines, should be thoroughly calcined and hydrate should be totally absent

. the surface area should be such as to ensure the effectiveness of dry treatment of fume (more than 35 m^2/g) while curtailing the potential for chemisorption of water (less than 50 m^2/g)

. the attrition index should be low enough to guarantee a reasonable particle size distribution on reaching the pot - i.e. less than 20 % below 45 µm and less than 10 % below 30 µm - due allowance being made for the degree of attrition inherent to the design of the dry treatment process.

Alumina optimisation for centre-worked hooded pots employing automated and virtually continuous point feeding

The basic constraints are:

. *Collection and treatment of hydrogen fluoride*

Table 4 - Alumina grain size (wet sieving): as supplied, on pots and after attrition test

ALUMINA	AQ			AF			AB			AG1		
WET SIEVING	As supplied	On pots	After attrition test	As supplied	On pots	After attrition test	As supplied	On pots	After attrition test	As supplied	On pots	After attrition test
minus 125 μm	94	99		97		98	86	95	91	84	92	87
minus 100 μm	74	88	83	89	93	92	66	77	77	66	77	76
minus 80 μm	41	65	58	76	81	81	41	56	56	48	59	59
minus 60 μm	14	37	36	44	53	54	17	36	36	27	40	38
minus 45 μm	3.8	18.0	17.7	17.4	24.5	25.3	5.6	18.2	19.7	11.0	18.5	17.4
minus 30 μm	0.8	9.0	9.3	5.7	9.4	9.0	2.4	8.3	8.6	5.0	7.3	8.0
minus 15 μm	0	4.0	1.3	2.4	3.7	3.9	0	3.4	2.4	0.7	3.0	3.2

. *Kinetics of solution*

. *Control of alumina feed*

. *Avoidance of volcano formation*

Collection and treatment of hydrogen fluoride

The pot technology requires holes to be kept permanently open in the crust at the points of operation of feeders.

Economic considerations are such that Aluminium Pechiney seeks to operate at fairly high acidity (cryolite ratio of approximately 1:1 to 1.1 : 1).

Also, the fume treatment system operates at full efficiency even during feeding (which is not the case with side-worked pots). Under these circumstances, the dry treatment process has to adsorb an average 18 kg of hydrogen fluoride per tonne of aluminium, with peaks at 25 kg/t. The adsorption kinetics are such that the alumina has to have an accessible surface area of not less than 45 m^2/g.

Kinetics of solution

This is the major consideration as far as this type of pot is concerned. Should solution be incomplete, then sludge will form under the crustbreaker, the process control system will order the injection of a further quantity of alumina to compensate for the fraction not dissolved and this will in turn contribute to the production of more sludge. The adverse effects are even more pronounced in the case of pots operating with highly acidic baths and, possibly, containing special additives.

The bulk of the Aluminium Pechiney research effort has concentrated on this point and includes: tests carried out on microcells (3), infrared spectroscopic determinations (7), X-ray scattering and electronic microdiffraction analysis (8) and potentiometric determinations on chlorofluorinated baths (9). An understanding of the *mechanism of the process of solution, the degree of organisation of the crystal structure and the part played by water* makes it possible to elicit rules for obtaining a soluble alumina.

Mechanism of solution. The introduction into the reduction bath of a grain of alumina gives rise to two conflicting phenomena, namely:

i) exothermic (10) restructuring of the transition phases (recrystallisation of large crystals in the alpha phase) assisted by the pre-existence of organisable crystallites and the absence of defects to be eliminated (lattice water)

ii) endothermic (10) destruction of the crystal lattice, with preferential attack at sites of defects.

Solution is assisted by the existence of a crystal structure exhibiting a high degree of disorder, a substantial interfacial area available for liquid/solid interaction (mass transfer) and a sufficient reserve of heat in the vicinity of the interface.

Crystal organisation (8). This will be determined by the kiln technology employed in the calcination of the hydrate. Thus:

Table 5 - Comparison of alumina attrition in two different designs of dry system

Alumina / Potlines / Grain size, % Laser method CILAS 715	AB			
	PB - Old Air Industrie dry system		SJF - New Air Industrie dry system	
	As supplied	On pots	As supplied	On pots
minus 192 μm	100	100	100	100
minus 128 μm	90	95	91	93
minus 96 μm	67	80	67	93
minus 64 μm	29	48	27	33
minus 48 μm	14	30	12	17
minus 32 μm	6	14	5	8
minus 24 μm	4	8	3	5
minus 16 μm	2	4	2	3
minus 12 μm	1.6	3.1	1.4	2.3
minus 8 μm	1.2	2.3	1.1	1.8
minus 6 μm	1	2.0	1.0	1.5
minus 4 μm	0.9	1.7	0.9	1.2
minus 3 μm	0.8	1.5	0.7	1.0
minus 2 μm	0.7	1.3	0.7	0.9
minus 1.5 μm	0.7	1.1	0.6	0.8
minus 1 μm	0.4	0.7	0.4	0.5

a) slow decomposition leads to the formation of large crystals of organisable phases (chi, kappa and alpha) and elimination of defects

b) sudden decomposition (involving thermal shock), characterised by the in-situ formation of a boehmite phase, leads to the formation of phases which are less readily organised (gamma and delta) unless they can be held for extended periods at high temperature (theta and alpha).

Part played by lattice water. Residual water is "stabilised" by the presence of lattice defects. It is lost when slow and extended recrystallisation occurs on calcining. It therefore provides a useful picture of the degree of disorder of the crystal structure. The sudden volatilisation of water on contact with the liquid bath brings about disruption of the grains and dispersion of the alumina/bath magma. The technology is such that the resulting emission is completely collected and recycled. It is therefore possible to work with aluminas which are less highly calcined than those recommended for side crustbreaking.

Implications for calcining technology. The considerations just discussed point to the desirability of:

i) subjecting the Bayer hydrate to thermal shock

ii) very short residence time at high temperature for all the constituent particles.

Calcining in the rotary kiln gives a product of mixed composition. A soluble alumina can be obtained only by aiming at high surface area (more than 65 m^2/g) and avoiding over-calcination of fines (dust extraction).

Fluidised bed calcination gives a more uniform product and is carried out at lower temperature. Extended residence times should nevertheless be avoided since they give rise to a theta phase which is highly ordered and therefore not very soluble.

The gas suspension calcination technologies now being developed (11, 12) combine the two desiderata of thermal shock and short residence time and yield a soluble alumina.

It should be emphasized that the correlations observed between crystal organisation, moisture content and specific surface differ to a fair extent as between technologies (3).

Implications for particle size distribution. The process of solution takes place via the transfer of mass across the solid/liquid interface. The greater the area and the reactivity of the interface, the faster the rate of exchange. Hence, for a given quantity by weight, the smaller the particles the greater the surface area and the faster the rate of solution. The presence of large grains has a negative effect. A further requirement is that the interfacial area should be not greatly different from the surface area of the alumina and this will not be the case if the bath solidifies around the particle because of the heat absorbed on solution (10). The need is either for substantial local superheating of the bath, which will be favoured by a disordered crystal structure, or for each grain to be surrounded by a large enough volume of bath; this is obtained if the feed alumina grains are well separated and is assisted by the volatilisation of water, leading to further separation of the individual grains.

Two of these requirements (no coarse grains and sufficient spacing of particles) add up to the need for a narrow distribution of particle size within the range 45-100 μm.

Controlled feeding

The dictates of a virtually continuous feeding system based on the filling and emptying under reproducible conditions of an automatic feeder require a free-flowing alumina (i.e. one with minimum internal friction). The presence of alumina fines and fluoride dust has a strongly negative effect on flow properties. The angle of repose may not be a sufficient criterion and we at Aluminium Pechiney habitually employ a special procedure to determine flow characteristics. Industrial experience shows that the pots can generally accommodate an alumina containing, after attrition at the various process stages, up to 20 % of fines passing the 325 mesh.

Avoidance of volcano formation

Where the point feeding system is employed, the geyser effect mentioned earlier can lead to the formation below the crustbreakers of "volcanos" with walls consisting of alumina ex-hydrate which recrystallises in contact with fluorine fume. In addition to the problems of interference with crustbreaker operation and poor control of alumina feed (with alumina falling onto the outer slopes of the mounds or expelled from the feed hole), collapse or deliberate breaking-up of the mounds leads to the formation of a low-solubility granular magma below the crustbreaker at the centre of the pot.

Although a small quantity of residual water is tolerable, the proportion of hydrate has to be kept low. Aluminas of very high specific surface give rise to the same problems. Experience shows that specific surface should be limited to 80 m^2/g and the proportion of hydrate to less than 0.5 %.

Conclusions (Central crustbreaking)

To sum up, we have seen that the fundamental factors to be considered in the case of centre-worked pots are not to do with working conditions, as in the case of side crustbreaking, but basically to do with the regularity of the introduction and dissolution of the alumina.

A wide range of surface areas is tolerable (45 to 80 m^2/g) but the objective of a degree of disorganisation of the crystal structure means that we prefer a narrower range centered on 70 m^2/g and, where possible, the use of a gas suspension calciner.

Like the too highly-ordered phases, hydrate is to be avoided, or at least kept very low (less than 0.5 %).

A narrow particle size distribution within the range 45-100 μm favours the satisfactory flow of alumina through the feeders and solution in the bath.

On all these grounds, Aluminium Pechiney recommend, for centre-worked pots, the use of alumina of the following specification:

BET surface area	:	65 to 75 m^2/g
LOI 300-1000°C	:	0.6 to 0.9 %
Hydrate	:	< 0.5 %
Fines below 45 μm (-325 mesh)	:	< 20 % at the pot
Retained on 100 μm (+ 150 mesh)	:	< 15 %

Needless to say, the definition of these criteria refers to Aluminium Pechiney pots operated at high levels of performance (current efficiency at least 94 %, kWh/t less than 13 200). Operation at a lower level of performance (current efficiency approximately 92 %, kWh/t approximately 13 500), e.g. using hotter baths of different composition, can accommodate grades of aluminas which differ significantly from the range just specified.

Alumina for Søderberg pots

In contrast to the situation with prebake pots, the gas escapes under the anode skirts and does not percolate through the blanket of alumina, thus curtailing dust emission.

Søderberg potlines are in many cases not equipped with facilities to adsorb hydrogen fluoride on alumina; this reduces the scope for attrition and does away with limitations on surface area. These pots can therefore accommodate any grade of alumina provided total moisture content is low and hydrate is absent (i.e. well-calcined fines). Floury alumina and sandy alumina for side-worked pots are perfectly suitable.

Outlook for a multipurpose alumina

An alumina meant for centre-worked pots cannot satisfactorily be employed on side-worked pots and vice versa. We at Aluminium Pechiney have preferred to optimise each alumina separately, along the lines already discussed. This strategy has been made possible by the flexibility of our production units.

Although the future does hold out hope for effective action, i.e.:

i) separation of production lines

ii) a gradual shift in favour of a single design of pot employing automated centre feeding,

the present position is that aluminas available on economic terms frequently have to be accepted even if they are not perfectly suited to the pots being operated. It is nevertheless possible to arrive at a conpromise enabling the resultant performance penalty to be kept within bounds and a multipurpose alumina to be defined which could be traded everywhere on world markets.

Particle size

Too great a proportion of fines means problems; if over-calcined, they lack solubility; if under-calcined, they fly up.

The conditions for keeping down, or eliminating fines, are:

i) they should be virtually absent in the alumina as supplied

ii) the alumina should have a satisfactory attrition index

iii) the fume treatment technology should involve a minimum of attrition.

A further point to note is that a narrow particle size distribution means better solubility and less occluded air in the alumina blanket.

Calcining and moisture content

These are not unrelated. Correlations exist - but will depend on the technology employed - between BET surface area, crystal organisation and moisture content. While the question of surface area is easily resolved, it is absolutely essential to reach a satisfactory compromise between two mutually conflicting requirements, namely a weakly-ordered crystal structure (to assist solution) and low moisture content (to keep down dust emission). However, gas suspension calcining would seem to constitute an attractive compromise provided a surface area of some 45-55 m^2/g is aimed at.

Control of thermal phenomena

The heat of solution of an alumina in the bath will vary with the crystallographic structure of the alumina, its moisture content, impurities (sodium and calcium oxides) and with bath composition. The range of variation is not great as compared to total enthalpy (32 to 40 kcal/mole). Thermal phenomena with a less obvious impact on centre-worked pots (where superheating is substantial) can be a ractor to be reckoned with on side-worked pots in the vicinity of the ridge (where there is no superheating). This is because the equilibrium obtaining at the solid/liquid interface can be shifted spectacularly by thermal fluctuations which are so small as to be barely perceptible in terms of the total heat balance. This is well known from experience of the effects of a change in alumina or mixing aluminas which differ by virtue of their calcining history or because of segregation occurring at various process stages.

When choosing a multipurpose alumina, therefore, it is important to check not only for compliance with specification but also for identity of calcining history and uniformity from batch to batch.

Conclusions (Multipurpose alumina)

Should the economic climate so dictate, managements prepared to accept a significant performance penalty can employ a multipurpose alumina able, albiet imperfectly, to meet the requirements both of side-worked pots (in terms of working conditions) and centre-worked pots (in terms of solution). The main features of such an alumina will be:

. surface area within the range 45 to 55 m^2/g

. moderate calcination by the gas suspension technique

. very low fines content

. a satisfactory attrition index.

Overall conclusions

We believe we have shown that the most effective approach to obtaining the best possible potline performance is to optimise the alumina specification to meet the specific operating requirements of each type of pot.

The specifications recommended by Aluminium Pechiney are as follows:

Side-worked pots

BET surface area 35 to 50 m^2/g

Proportion of fines at the pot
after attrition < 20 % passing the -325 mesh (45 μm)

Fines well calcined (less than 50 % alpha phase in the fines and no hydrate detectable by X-ray analysis or thermogravimetric analysis).

The attrition index depends on the FTP design.

Automated centre-worked pots

BET surface area 65 to 70 m^2/g

LOI 300 - 1000°C : 0.6 to 0.9 %

Hydrate < 0.5 %

Retained on +150 mesh (100 μm) < 15 %

Passing the -325 mesh (45 μm) < 20 % at the pot.

Strictly speaking, the degree of organisation of the crystal structure, which is fundamental, will be determined by the calcining technology employed. In the case of certain calcining kilns, a slight shift in the limits specified above can be tolerated.

Although dust emission is less of a problem than with side-worked pots, it is essential that the alumina reaching the pot should contain less than 20 % of fluorinated fines of size below 45 μm (to ensure free-flowing properties). Preferably, the objective should be the narrowest possible particle size distribution within the range 45 to 100 μm.

The use of a multipurpose alumina can be contemplated at the cost of a slight loss of performance, very slight in fact if the alumina is obtained by gas suspension calcination set for 45 to 55 m^2/g. This technology is now used by Aluminium Pechiney (12).

Søderberg pots can accommodate grades of alumina designed for use with side-worked prebake pots.

Acknowledgements

The authors wish to thank J. Coursier, P. Tortil and M. Prat of Aluminium Pechiney France, and S. Casuli of Pechiney Nederland, for their very valuable collaboration.

References

(1) M. Keinborg, "L'Elaboration de l'Aluminium de Sainte Claire Deville à nos jours", L'Actualité chimique No. 4, Avril 1983, pp.15-21.

(2) J.P. Dugois, "An example of power saving technology", Light Metals, (1975) pp. 3-9.

(3) E. Barrillon, "The merits and demerits of various types of Bayer aluminas", Erzmetall, Bd. 31 (1978) H 11, pp. 519-522.

(4) M. Keinborg and J.P. Cuny, "Aluminium Pechiney - 180 kA prebake pot", Light Metals (1982) pp. 449-460.

(5) W.L. Forsythe and W.R. Hertwig, "Attrition Characteristics of Fluid Cracking Catalysts", Ind. and Eng. Chem., Vol. 41, pp. 1200-1206.

(6) K.E. Connally, "Particle Segregation in Alumina Silos", Light Metals, (1983) pp. 241-252.

(7) J. Simonato, "Caractérisation par spectrométrie infrarouge à haute temperature de bains fondus alumino-fluorés", Thesis, Orléans University, France, 1979.

(8) Private communication, J.F. Larue and A. Renouprez, Institut de la Catalyse, Villeurbanne, France (to be published).

(9) G. Picard, F. Seon and Y. Bertaud, "A Molten Salt Potentiometric Method for Characterizing Industrial Aluminas", Electrochemica Acta, Vol. 27, No. 3 (1982) pp. 401-409.

(10) Phan-Xuan Dung, "Thermodynamique et Thermocinétique de la Dissolution d'Alumines dans des Mélanges Cryolithiques", Thesis, Provence University, France, 1974.

(11) E. Benny et al., "Application of Gas Suspension Calciner in Relation to Bayer Hydrate Properties", Light Metals (1982) pp. 199-209.

(12) J.P. Henin, "Flash Calcination by Fives Cail Babcock", TMS-AIME 112th Annual Meeting, March 6-10, 1983.

CHAIRMEN

E.L. Cambridge
ARCO Metals Company
Tucson, Arizona

D.W. Macmillan
Alcan International
Jonquiere, Quebec, Canada

ALUMINUM REDUCTION TECHNOLOGY V
Modelling and Analysis

THERMAL ANALYSIS OF SOEDERBERG POTS

T. Ohta and T. Matsushima

Chiba Works, Showa Aluminum Industries K.K.
Ichihara, Chiba, Japan

Taking an example of 100 kA VSS pots at Showa Aluminum Industries' Chiba Works, the authors attempted to obtain the enthalpy required for the electrolytic reaction and have found that the total energy requirement was 6,705 kWh/t-Al or 2.035 v.

It was also found that the total amount of heat emitted from various parts of the electrolytic pot was 168.4 kcal/h or 1.997 v. The total of the reaction energy and amount of the heat emitted was found to conform well to the total amount of energy input.

Also, attempts were made to determine the thermal conductivity of the freeze and the interface heat transfer coefficient between the freeze and bath, on the basis of the thickness of the freeze and the amount of heat emitted from the side wall.

Foreword

Needless to say, improvement of the energy efficiency is extremely important in designing and operating the aluminum smelting pot.

Showa Aluminum Industries (SAL) was one of those companies that made strenuous efforts to improve the energy efficiency from various viewpoints. Some of those efforts proved to be rewarding as have been reported in the past[1), 2)].

In its efforts to achieve such improvements, SAL has been making thermal analyses of the smelting pots for many years. The heat flow meter which has been developed for this purpose has been commercialized and marketed by Showa Denko K.K.

Marked development of the computorized analysis technology has made it easy to estimate the contour of the freeze and the heat emittance distribution on the basis of the pot construction. Accurate measurements are, however, essential for further improvement of the reliability of such estimate.

Energy Required for Reaction

The reactions that are considered to take place within the pot are many but those likely to give non-negligible impacts on the heat balance of the pot are limited.

1) Energy required for reactions

Alumina is supplied at room temperature and the reaction formed substances are taken out of the system at the reaction temperature (T). The energy required for heating and baking the green anode paste is evaluated in paragraph 2) below. The reactions are expressed in the following formula:

Alumina preheating	$Al_2O_3\ (25^oC) \;=\; Al_2O_3\ (T^oC)$	(1)
Electrolytic reaction	$2Al_2O_3\ (25^oC) + 3C\ (T^oC) = 4Al\ (T^oC) + 3CO_2\ (T^oC)$	(2)
Re-oxidation reaction	$2Al\ (T^oC) + 3CO_2\ (T^oC) = Al_2O_3 + 3CO\ (T^oC)$	(3)

When the reaction temperature is 970°C, the energy required for reactions (1) and (2) is calculated under the following formula:

$P_1 = 567$ kWh/kg

$P_2 = \dfrac{5,652}{\eta}$ kWh/kg

$P_3 = -4.384\ (1 - \eta)$ kWh/kg

Where: "η" means the current efficiency and "-", exothermic reaction

2) Reactions related to anode

Of the reactions that occur in relation to the anode other than reaction (2), important ones are baking of green anode paste and oxidation of baked anode with CO_2 and air.

Baking reaction	C (25°C) = C (T°C) CH_x (25°C) = C (T°C) x/2 H_2 (T°C)	(4)
Boudouard reaction	C (T°C) + CO_2 (T°C) = 2CO (T°C)	(5)
Oxidation with air	C (T°C) + O_2 (25°C) + 3.71 N_2 (25°C) = CO_2 (T°C) + 3.71 N_2 (T°C)	(6)

The energy required for reactions (4), (5) and (6) is calculated under the following formula.

P_4 = 0.34 Wc kWh/t-Al
P_5 = 3.91 Cb
P_6 = -5.90 Co

Where Wc : Unit consumption of green paste
Cb : Loss of carbon attributable to the Boudouard reaction
Co : Loss of carbon attributable to the air oxidation

Cb and Co are estimated on the basis of the composition of gas that builds up underneath the crust. Although opinions have been expressed denying the contribution of reaction (4), the gas composition shown in Table 1 comprises the Peason-Waddington formula and shows a tendency of shifting to the CO rich side. About 30 kg of carbon is considered to be consumed by this reaction for each ton of aluminum produced.

Reaction (5) is obtained from the ratio of residual oxigen and nitrogen in the gas. As shown in Table 1, however, it can be neglected in a well controlled Soederberg pot.

Table 1. Typical gas composition

Gas	CO_2	CO	O_2	Inert gas
Composition (%)	67	31	<0.5	bal

3) Reactions related to submaterials

All submaterials are supplied at room temperature and must be heated to the reaction temperature. A part of aluminum fluoride supplied is consumed to neutralize Na_2O in alumina and the balance is considered to be subjected to hydrolysis. Most of cryolite is taken out of the system mechanically. Although hydrolysis of cryolite is considered possible, most of hydrogen fluoride generated is ascribed to aluminum fluoride, in so far as it is estimated on the basis of the hydrogen fluoride concentration of the pot gas.

Neutralization of Na_2O in alumina	$4AlF_3$ (25°C) + $3Na_2O$ (25°C) = $2Na_3AlF_6$ (T°C) + Al_2O_3 (T°C)	(7)
Hydrolysis of aluminum fluoride	$2AlF_3$ (25°C) + $3H_2O$ (25°C) = Al_2O_3 (T°C) + 6HF (T°C)	(8)
Melting of cryolite	Na_3AlF_6 (25°C) = Na_3AlF_6 (T°C liquid)	(9)

The energy required for reactions (7), (8) and (9) is as follows:

P_7 = 7.62 RNa
P_8 = 1.412 (W AlF_2 - 34.32 RNa)
P_9 = 0.479 W cry*

Where RNa = Na_2O concentration (%) of alumina
W AlF2 = Unit consumption of aluminum fluoride
W cry = Unit consumption of cryolite

*At 970°C, cryolite is solid. For the above-calculation, however, it has been considered as super cooled liquid.

4) Total energy required for reactions

The total required energy is calculated as per Table 2 on an assumption that the current efficiency is 90.5%; unit consumption of carbon paste, 520 kg; carbon loss by the Boudourd reaction 30 kg; oxidation loss by air, 0; Na_2O concentration of alumina, 0.40%; unit consumption of aluminum fluoride 35 kg; unit consumption of cryolite 10 kg.

Table 2. Total required energy

Type of Reaction	Required Energy kWh/t	Required Energy Converted to V
(1) Alumina preheating	547	0.163
(2) Electrolytic reaction	6,245	1.897
(3) Re-oxidation reaction	-417	-0.125
(4) Baking of green paste	177	0.054
(5) Boudourd reaction	117	0.036
(6) Air oxidation	0	0
(7) Neutralization of Na_2O	-3	-0.001
(8) Decomposition of AlF_3	30	0.009
(9) Melting of cryolite	5	0.002
TOTAL	6,705	2.035

Measurement of Amount of Heat Emitted from Various Parts of Pot

1) Measurement method

The amount of heat emitted from the surfaces of various parts of the smelting pot was measured with the heat flow meter, SHOWTHERM-HFM, which was originally developed for the thermal analysis of the aluminum smelting pot. It is equipped with sensors having a low heat transmission resistance, which enables the heat flow to be measured without disturbance. Also, the sensors are of such design that they can be easily attached to such parts of pots that are difficult to approach (Fig. 1).

Amount of heat emitted from various parts of the pot was measured by the following method:

(1) Sides and bottom of anode and cathose casings

Measurement was taken with respect to each block, which is separated from another by the reinforcing ribs or the radiation fins. An appropriate number of measuring points was established for each block, and an average amount of heat flow was obtained for each such block.

(2) Spike

Accurate measurement is expected to be obtained only on the flat surface. Measurement was, therefore, taken of spikes positioned at different levels at their top. The amount of heat emitted from each side of the spike was calculated on the basis of the actually measured value at the top of the relevant spike and the surface temperature.

(3) Paste surface

The amount of heat emitted from the top surface of the anode was measured by placing a wire netting over the layer of briquette and attaching the sensor to the netting. Measurements were taken at typical 24 points on the surface of the anode paste, immediately before and after the briquette charging.

(4) Crust and gas ring

The crust surface was segmented appropriately and 10 measuring points were selected. Four measurements were taken during a period from immediately before the crust breaking to immediately before the next crust breaking and the time average was obtained for each measuring points.

(5) Bus bar

32 measuring points were selected for each bus bar and weighted average was obtained with respect to each area represented by the relevant measuring point.

2) Results of measurements

Table 3 shows the heat radiation distribution obtained as a result of the above measurements.

Table 3. Heat emission distribution (unit: kcal/hr)

Location of Measurement		Heat	%
Anode casing	Long end	26.37	
	Short end	4.18	
	Corner	5.08	
		35.63	21.2
Top section of anode	Paste	14.13	
	Spike	7.15	
	Hanger	3.67	
		24.96	14.8
Crust		14.54	
Gas ring		10.55	
		25.09	14.9
Side of cathode casing	Long end	22.72	
	Cathode bar	12.32	
	Short end	7.20	
	Corner	3.02	
		45.26	26.8
Pot bottom		22.50	13.4
Conductors	Anode bus	0.05	
	Cathode flexible bus	2.83	
	Cathode bus	7.48	
	Rising bus	4.60	
		14.96	8.9
		168.40	100.0

The total amount of emitted heat is equivalent to 1.997 V in terms of cell voltage. It was compared with the input power and the heat required for reaction and the heat balance shown in Table 4 was obtained.

Table 4. Heat balance of aluminum smelting pot

Input (v)	Output	(v)
4,040	Reaction heat	2.035
	Heat emission	1.997
	Unknown	0.008
4,040		**4,040**

The unknown factor accounts for 0.2% of the total.

Thickness of Freeze and Amount of Heat Emitted

Ten pots of the same type were selected and measurements were taken of the bath temperature, freeze thickness, surface temperature of the wall on the rear side, amount of heat emitted and ambient temperature, at six measuring points selected in the vicinity of the center section of both long ends of each selected pot. With respect to each measurement item, an average was calculated for each long end.

Fig. 3 shows the relationship between the average thickness of such part of the freeze exposed to the bath "LF" (above the metal-bath interface) and amount of heat emitted from the corresponding section of the external wall of the shell "q".

In this case, the heat flow can be expressed by the following one dimensional equation:

$$q = (Tb - Ta)/(1/hb + LF/\lambda F + LC/\lambda c + LI/\lambda I + 1/ha) \qquad (10)$$

Where Tb = Bath temperature
Ta = Ambient temperature
hb = Solid-liquid thermal comductivity coefficient
ha = Thermal conductivity coefficient of external wall
LF = Thickness of freeze
LC = Thickness of carbon
LI = Thickness of refractory brick
λF = Thermal conductivity of freeze
λC = Thermal conductivity of carbon
λI = Thermal conductivity of refractory brick

The values for LC, λC, λI are known already and the value for "ha" can be obtained from the temperature of and the amount of heat emitted from the wall surface.

Fig. 4 shows the relationship between "(Tb - Ta)/q" and "LF" as obtained by the following formula, which is a variation of formula (10):

$$(Tb - Ta)/q = LF/\lambda F + 1/hb + LC/\lambda C + Lb/\lambda B + 1/hb \qquad (11)$$

The regression formula is (Tb - Ta)/q = 1.43 LF + 0.144, which means λF = 0.70 kcal/m·hr·°C, hb = 156 kcal/m·hr·°C.

Hanpin[6] reports that λF varies in the range of 0.65 - 1.20 kcal/m·hr·°C, depending on the alumina concentration. Also, it is stated in published papers[7), 8)] that the value of "hb" is in the range of 146 - 318 kcal/m^2·hr·°C.

The values of λF and hb obtained on the basis of formula (11) and the regression formula thereof are in agreement with those mentioned above.

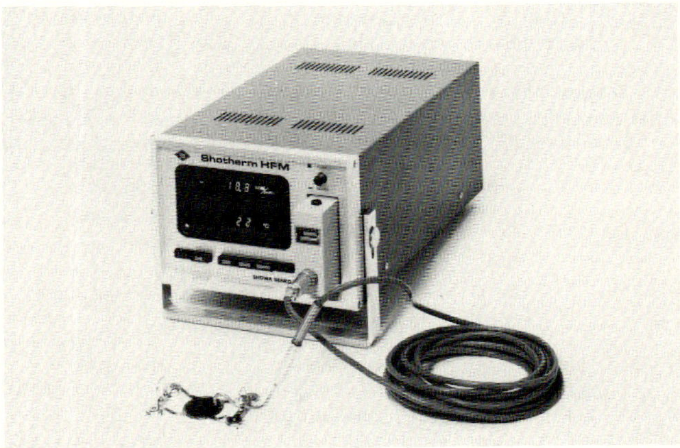

Fig. 1 Heat Flow Meter "SHOTHERM - HFM"

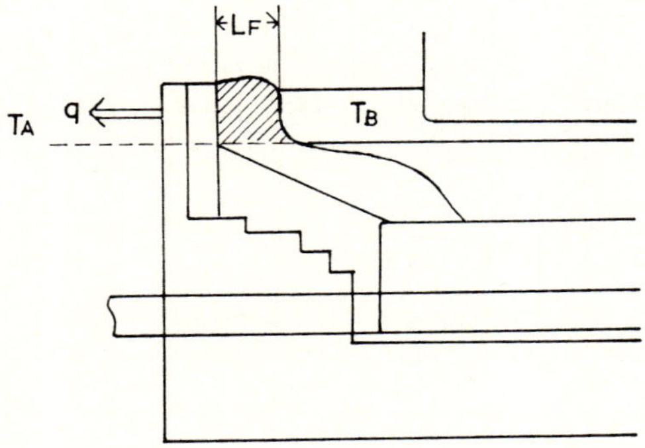

Fig. 2 Cross section of freeze and amount of heat emitted

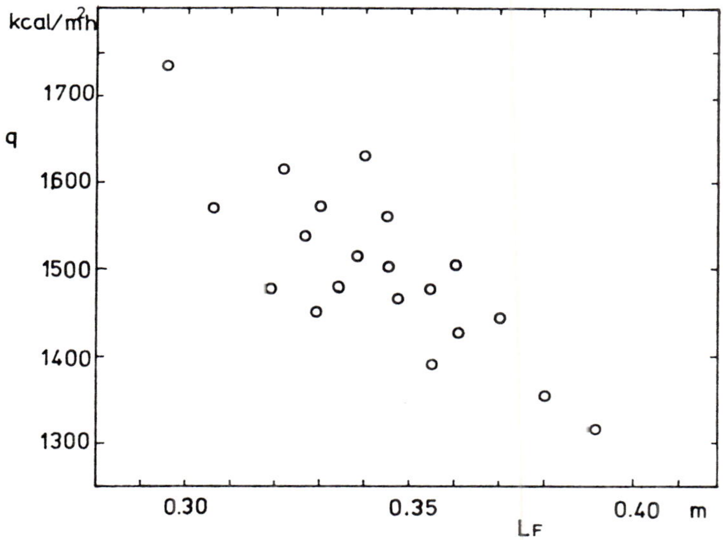

Fig. 3 Relationship between freeze thickness and amount of heat emitted from side wall

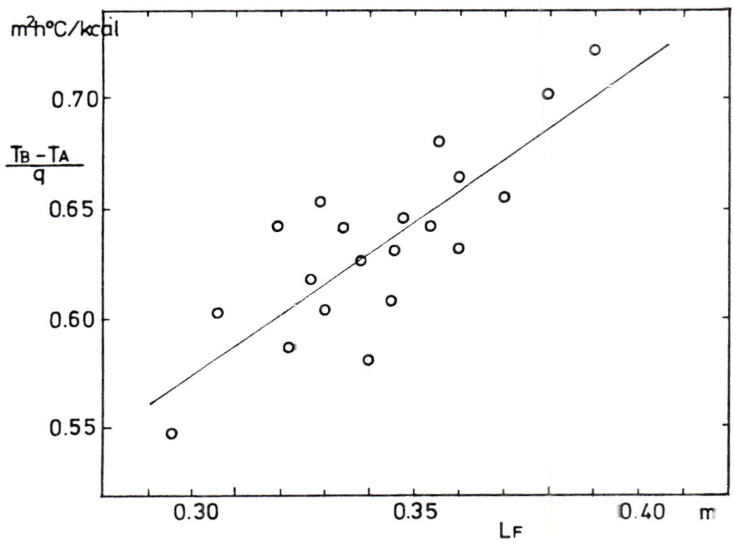

Fig. 4 Relationship between freeze thickness and heat resistance

Conclusion

As a result of thermal analyses of 100 kA VSS pots, the following findings were made:

1) Amount of the energy required for electrolytic reaction, including the energy required for baking of anode paste is equivalent to 1.989 v. The enthalpy variation for all the reactions, including reactions related to submaterials, is equivalent to 2.035 v.

2) The heat emissions from various parts of the smelting pot were measured as follows:

Anode section	about 36%
Crust section	about 15%
Cathode section	about 40%
Bus bar	about 9%

The total amount of heat emitted was equivalent to 1.997 v. When it was added to the amount of heat required for electrolytic reactions, the result was found to be well conforming to the total amount of energy input.

3) The average thermal conductivity of the freeze was found to be 0.70 kcal/m·hr·C and the interface transfer coefficient between solid and liquid, 15.6 kcal/m^2·hr·°C, from the relationship between the thickness of the freeze and the amount of heat emitted from the side wall.

References

1) H. Tsukahara, N. Ono, K. Fujita; Light Metals 1982 Vol I 471 - 482

2) Y. Okada, T. Ohta, K. Hamachi; Light Metals 1983 Vol I 537 - 550

3) J. L. Holm; Doctor Thesis, 1971 Institute of Inorganic Chemistry NTH, Norway

4) K. Grjotheim et al; "Aluminium Electrolysis" Aluminium-Verlag GmbH, Duesseldorf

5) O. Kubaschewski, E. Ll Evans, C. B. Alcock; "Metallurgical Thermochemistry 4th edition, Pergamon Press, 1967

6) W. E. Hanpin; Light metals 1961, 189 - 194

7) K. Arai, K. Yamazaki; Light Metals 1975 Vol I 188

8) Y. Arita, N. Urata, H. Ikeuchi; Light Metals 1978 Vol I 59 - 72

INVESTIGATION OF THE DYNAMIC BEHAVIOR OF ALUMINUM POTS

P. Entner[1], W. Schmidt-Hatting[2], Z. Mitrović[3], U. Gruber[4], D. Rufer[5], B. Fjørnes[6]

1,3,4 Swiss Aluminium Ltd. Reduction Technology Center, CH 3965 Chippis
2 Swiss Aluminium Ltd. Primary Aluminum, CH 8034 Zürich
5 Federal Institute of Technology, CH 8034 Zürich
6 Befag Engineering AG, CH 6300 Zug

Abstract

A perfect process control of aluminum production pots needs relations between input and output values and especially knowledge of all state variables. Because their direct continuous determination is often technically not possible, a numerical model was developed, which describes the dynamic behavior of the system. With this model, the pot behavior is extrapolated to unknown pot states yielding information for pot operation especially with a view to optimization.
The steps for model preparation were:

- creation of the model structure
- identification of the parameter values from experimental data
- analysis of model errors.

The reaction of the pot after changing the alumina feeding is discussed. As a second example, the determination of a stationary pot state is described.

Introduction

In the last twenty years considerable success has been achieved by treating an electrolytic pot for the production of aluminum applying scientific methods. The investigation of the magnetic effects or the calculation of the thermal behavior of the cell have furnished essential information for pot design for instance to decrease the energy consumption. These methods regard the pots as being in a stationary state i.e. no change of current intensity, temperature or alumina consumption with time is allowed.

Looking for methods which treat the dynamic physical-chemical pot behavior, the amount of papers in the available literature is much more modest. Alusuisse in collaboration with the Swiss Federal Institute of Technology, Zürich and Befag Engineering AG, Zug, has developed a model called ALDYM (Alusuisse Dynamic Model).
ALDYM mathematically simulates the reactions of the pots with time due to variations of the process like changes in beam position, alumina feeding, anode effect etc.
The model is used for:

- training process engineers, who can observe by offline calculation the positive or negative results of their way of running the pots.
- optimization of pot operation by extrapolating to unknown pot states without undertaking costly experiments.
- improving the process control by determining process variables which cannot be measured continuously (e.g. Al_2O_3 concentration, electrolyte temperature) but which are necessary for the automatic pot control.

Model Preparation

Structure of the global model:

Studying the stationary state of a cell, the methods of different disciplines are needed: magnetohydrodynamics, electrochemistry, thermodynamics etc. In a similar way the global dynamic model is split up into submodels which treat different aspects of the pots:

- alumina balance model: Al_2O_3 is added to the pot at regular intervals of time. Part of the alumina dissolves or is suspended in the cryolite-electrolyte. Another part may cross the electrolyte and the metal pad to contribute to the formation of bottom sludge, side crust or even bottom crust.
- production model: the electrolytic process produces aluminum at the cathode. Some of this aluminum is lost because of several side reactions especially reacting with anode gas.

- voltage model: the electric voltage measured between the entrance and the exit of the bus bar system is composed of different contributions: voltage drops due to ohmic resistances and electromotive forces due to the electrolytic process in the interpolar gap.

- temperature model: heat is produced and consumed dynamically by different processes: loss of current efficiency, melting or formation of electrolyte crust, heating of alumina. Changes of electrolyte temperature are the result of variations in this heat balance.

In Fig. 1 the structure of ALDYM with its submodels is shown. The submodels themselves are divided into more submodels, for instance for generation and consumption of aluminum (production model) or heat (energy-temperature model) leading to a more complex structure.

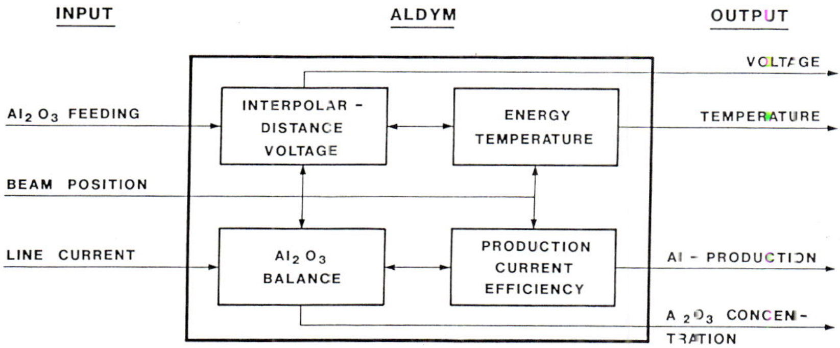

Fig. 1 Schematic structure of ALDYM, the simulation program for dynamic pot behavior. Based on the input values (line current, beam position, alumina feeding rate), the time behavior of pot voltage, electrolyte temperature, aluminum production (current efficiency), Al_2O_3 concentration in the electrolyte are calculated

The submodels are interconnected the output of one submodel being the input of another and vice versa. The temperature of the electrolyte depends for instance, on the interpolar distance, the current efficiency etc. On the other hand, the temperature itself influences the electric conductivity of the electrolyte i.e. the ohmic bath voltage or the reoxidation rate of aluminum.

Mathematical methods

The form of ordinary differential equations was chosen to formulate the time behavior of the pots. The calculation time is shorter compared to partial differential equations thus facilitating the application of ALDYM for the online process control. To solve the differential equations, an adjusted Runge-Kutta procedure for the integration was used.

In the literature, expressions for essential properties or relations are given, for instance the electric conductivity of the cryolite-electrolyte (1) or the maximum solubility of alumina (2) in dependence on bath composition and temperature.

But other values like heat transfer coefficients or solubility rates of aluminum or alumina in the electrolyte depend on hydrodynamic parameters which are special for a given pot type.

The so-called identification process determines the values of these unknown parameters. This is done in such a way that the calculated simulation values fit the experimental results in an optimum way. A program package (NLP) of the Swiss Federal Institute of Technology, Zürich was used (3). NLP contains several algorithms for parameter optimization and for solving nonlinear differential equations. They are selected according to the structure of the problem i.e. the type of the equation system.

Analysis of model errors

The values calculated by the simulation model will always deviate from the measured results. To make sure that this difference results from limited accuracy of the measurements and not from a principle error in the model, three tests were executed:

- the model errors i.e. the differences between measured and calculated values must not be auto-correlated. That means that there is no systematic increase or decrease of the model errors.

- the mean value of the model error is not significantly different from zero. The measured values show no systematic tendency to be gathered on one side of the simulated curve.

- the standard deviation of model errors is equal to or smaller than the standard deviation of the measuring error.

If not all of these conditions are fulfilled by a set of model errors, the model does not represent the real physical behavior of the system. The model has to be modified, which normally means that the structure becomes more complicated.

It may also happen that a too complicated mathematical structure was chosen. If the value of a parameter can be changed within its physically possible range without a significant influence on the simulated values i.e. the model errors, this parameter can be omitted and the structure simplified.

The steps in model preparation are (Fig. 2):

- formulation of the mathematical equations (set-up of model)

- identification of the unknown parameters

* analysis of model errors and eventual revision of the model
* significance check of the parameters and possible simplification of the model

The model is now ready for use. If new experimental values are gained, the prediction by the model and the actual measurements can be compared following the same schedule of Fig. 2 (verification of the model).

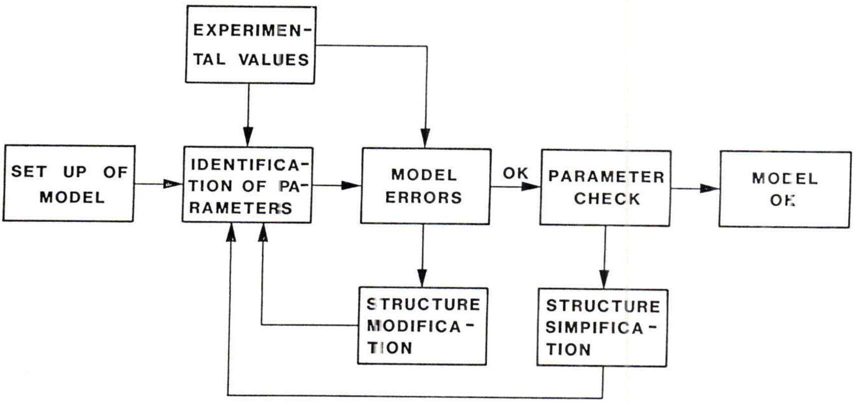

Fig. 2 Steps in model preparation. After the mathematical formulation (model set-up), the unknown parameters are determined from experimental values (identification). The model has to be modified if there are systematic deviations between the measured and calculated values (model errors). It can be simplified if a significant change of parameters shows no influence on the model errors.

Alumina balance

As an example, the formulation of the model for the simplified alumina balance is given. Alumina is added to the pot at regular time intervals by breaking the top crust. A measuring device feeds a volumetrically determined amount of alumina to the pot from an alumina silo (Fig. 3). The Al_2O_3 is first dispersed by the turbulent motion of the bath and then finally dissolved in the liquid electrolyte. Some of the alumina may cross the metal pad and stay there in the form of bottom sludge, a suspension of Al_2O_3 in a saturated solution of Al_2O_3 in cryolite (Lit. 4).

a) ELECTROLYTIC POT　　　　b) Al_2O_3 - BALANCE - MODEL

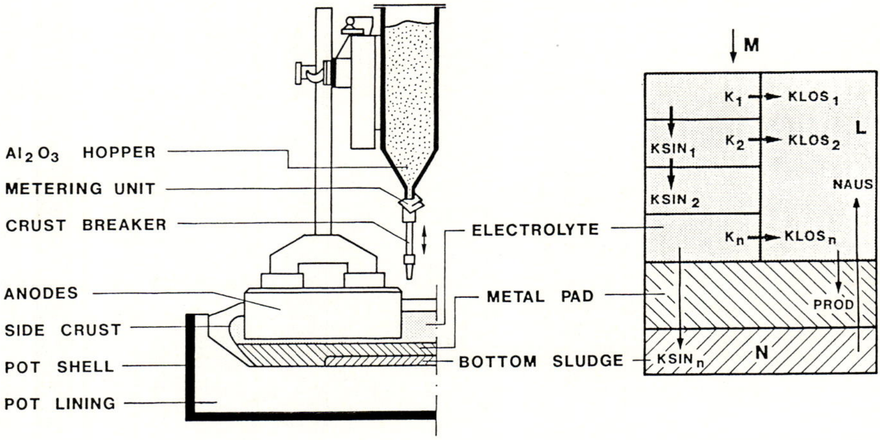

Fig. 3 Half of a cross-section of an electrolytic cell and a schematic representation of the alumina balance model. An amount M of Al_2O_3 is added to the electrolyte.
Part of it ($KLOS_1 \ldots KLOS_n$) is dissolved, part of it ($KSIN_1 \ldots KSIN_n$) crosses the electrolyte to form bottom sludge, from where it dissolves (NAUS). Al_2O_3 is used for the production of aluminum (PROD).
$K_1 \ldots K_n$ and L are the amount of suspended and dissolved Al_2O_3 in the electrolyte and N the Al_2O_3 in the bottom sludge.

To simulate the sedimentation of Al_2O_3, the model divides the electrolyte in layers with different amounts of suspended Al_2O_3 (Fig. 3b). The change of the amount of dispersed alumina (K_1) with time (t) for the first layer is

$$\frac{dK_1}{dt} = M - KSIN_1 - KLOS_1 \qquad (1)$$

where M is the Al_2O_3 added by the feeder, $KSIN_1$ the Al_2O_3 which leaves for the second layer and $KLOS_1$ the Al_2O_3 which is dissolved. One can write for the second layer

$$\frac{dK_2}{dt} = KSIN_1 - KSIN_2 - KLOS_2 \qquad (2)$$

and for the n-th layer

$$\frac{dK_n}{dt} = KSIN_{n-1} - KSIN_n - KLOS_n \qquad (3)$$

The Al_2O_3 dissolved in the electrolyte changes with time according to

$$\frac{dL}{dt} = NAUS + \sum^{n} KLOS_i - PROD \qquad (4)$$

NAUS is the Al_2O_3 dissolved from bottom sludge and PROD the Al_2O_3 used by the aluminum production.

The dissolution terms (KLOS and NAUS) are written

$$KLOS = f(K, \Delta c, T, t_{AE}) \qquad (5)$$

and the suspension terms (KSIN)

$$KSIN = f(K, T, t_{AE}) \qquad (6)$$

K is the amount of dispersed Al_2O_3, Δc the difference between the saturation Al_2O_3 concentration and actual concentration, T the temperature and t_{AE} the time since the last anode effect. Experience has shown that when extinguishing the anode effect, the bath is perturbated, which enhances the dissolution of alumina.

The unknown parameters in the relations were determined by the identification process taking into account that when analysing the Al_2O_3 concentration the sum of suspended and dissolved alumina is determined.

In Fig.4 the Al_2O_3 concentraton with time calculated by ALDYM and determined by chemical analysis is given. The alumina was fed in three ways:

- by stopping the Al_2O_3 addition to wait for an anode effect (NOF)
- by fast feed to quench the anode effect (FFAE)
- by fast feed to reach the Al_2O_3 concentration of the stationary state (FF).

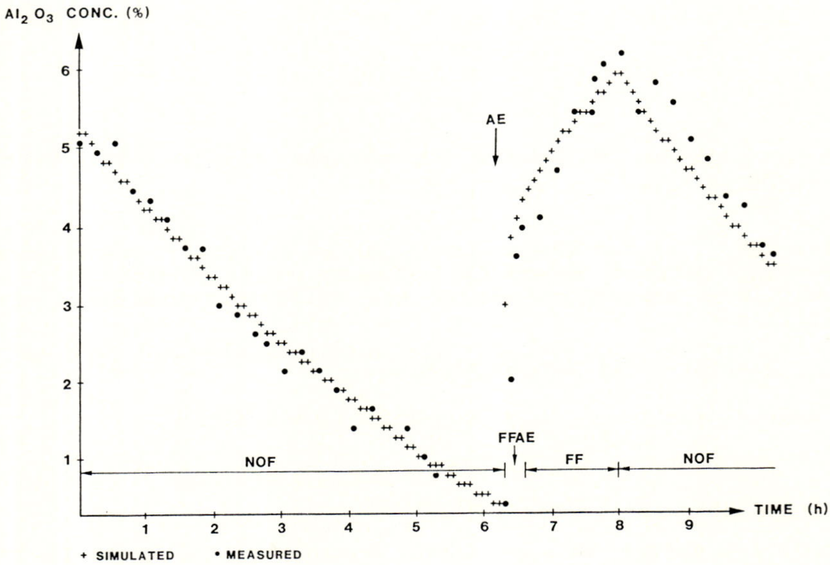

Fig. 4 Verification of the alumina balance model. Al_2O_3 concentration with time calculated by ALDYM (+) and determined analytically from electrolyte samples (•). NOF means no addition of Al_2O_3, FFAE fast feed immediately after the anode effect (AE) and FF fast feed to reach the Al_2O_3 concentration of the stationary pot state.

The agreement of calculated and measured values is better than the error of the analytically determined concentration (∼ 0.3%). Also the other tests of the error analysis were fulfilled.

Determination of stationary states

Running a potline, a stationary state is normally not achieved with the first set of parameters e.g. current intensity, rate of alumina addition etc. Often a rather extensive search by trial and error has to take place to find a pot state which is stable with time.

Similarly it was difficult to operate a simulated pot with ALDYM in a stable way. An algorithm had to be found to determine the parameters of a stationary pot state systematically.

The time behavior of the pot is described by n differential equations

$$\frac{dx_1}{dt} = f_1(x_1 \ldots x_n, H_t, I, M) \quad (7)$$

.
.
.

$$\frac{dx_n}{dt} = f_n(x_1 \ldots x_n, H_t, I, M)$$

where $x_1 \ldots x_n$ are the state variables and H_t, I, M the beam position, current intensity, alumina feeding rate respectively.In a stationary state, the following conditions are fulfilled.

$$f_1(x_1 \ldots x_n, H_t, I, M) = 0 \quad (8)$$

.
.
.

$$f_n(x_1 \ldots x_n, H_t, I, M) = 0$$

yielding a system of n nonlinear equations, the solution of which for the x_i gives the stationary pot parameters. To find the solution, an auxiliary function is formulated.

$$HZ = \sum^{n} g_i f_i^2 \quad (9)$$

where the g_i are weighing factors of the functions f_i according to equ. 8. The solution is determined by NLP formulating an optimization problem

$$HZ = \sum^{n} g_i f_i^2 \rightarrow 0 \quad (10)$$

The value of HZ will reach a minimum value if all f_i are zero.

For the set of values

current intensity: 140 kA
rate of Al_2O_3 addition: 1.31 kg/min
fixed anode beam

the solution of equation 8, that is to say of the optimization problem 10 is

pot voltage: 4.36 V
electrolyte temperature: 960°C
alumina concentration: 5.2%

As a verification these parameters were used with ALDYM, yielding a simulated pot state stable over 24 hours.

Conclusions

ALDYM is able to predict the dynamic behavior of an electrolytic pot with sufficient accuracy. It is thus a suitable tool for investigating the different methods of operating the pots.

By the convenient method of identification and by comparing measured with calculated results, the model parameters are refined yielding a model of higher precision and wider range of extrapolation.

By the mathematically determination of different stationary pot states one can choose the best one according to a given goal. It is obvious that an optimum pot state can automatically be determined by a mathematical algorithm. The first development steps in this direction have already been taken.

Literature

1. G. Choudhary
   "Electrolytic Conductivity for Aluminium Cell Electrolyte between 950-1025°C by Regression Equation," J. Electrochem. Soc. (1973)120(3), 381-383.

2. E.W. Dewing
   "Liquidus Curves for Aluminium Cell Electrolyte," J. Electrochem. Soc. (1970)117(6), 780-781

3. D. Rufer
   "Implementation and Properties of a Method for the Identification of Nonlinear Continuous Time Models," Proceedings of the 7th IFAC World Congress(1978)(Editor: A. Niemi), Pergamon Press.

4. J. Thonstad, P. Johansen, E.W. Kristensen
   "Some Properties of Alumina Sludge," Light Metals(1980), 227-239.

THE CURRENT EFFICIENCY OF A SHORTED ANODE IN A PREBAKE CELL

W. W. Hyland

Aluminum Company of America
Alcoa Technical Center
Alcoa Center, PA 15069

Prebake Hall-Heroult cells experience shorting or contact of one or more anodes to the metal pad for a number of reasons which include pot disturbances such as anode changing, tapping and imbalance. Achievement of maximum operating current efficiency (CE) must address this problem as well as the more common loss mechanisms. A computer model simulation has been developed to identify the magnitude of CE loss from various sizes of metal pad shorts and contacts with the bottom surface of an anode. It was found that most shorts cause a CE loss ranging from 10 to 80% depending on contact cross sectional area. Most shorts are not large enough to cause a total loss of decomposition potential over the total anode surface contacting the bath and therefore the CE loss is generally not 100% for the duration of the short.

Introduction

Electrical shorting of cathodic metal pad to anode occurs in the normal operation of a Hall-Heroult aluminum reduction cell. This shorting occurs more frequently when the anode to cathode space is reduced in an effort to reduce cell voltage and improve process energy efficiency. Shorting occurs more frequently during periods of cell upsets or disturbances. This shorting reveals itself as a reduction of overall cell CE. This work examines the effect of shorting on the CE of the process through construction of a mathematical model of an individual anode. A conventional short is created for purpose of discussion.

Shorting has not been discussed thoroughly in the literature. Other mechanisms of CE loss are covered in detail; for example, in recent reviews by Haupin and Frank (1) and Grjotheim, Krohn, Malinovsky, Matiasovsky and Thonstad (2). A method for detection of shorted or grounded anodes was patented in 1975 by Richards, et al, Reynolds Metals Company (3). Shorting can range from small shorts caused by molten metal welling up between anode bubbles at the crest of a magnetically induced metal wave to complete immersion of the sole of the anode block in the metal pad. In any case, current passes through the anode to metal pad via the short electronically, thus producing a CE loss for that anode over the duration of the short.

Model Development

Background

To determine the CE of a shorted anode, a mathematical model is required. Problems of this nature reduce to a solution of the Laplace Equation in the anode region and bath region, jointed or bounded by nonlinear boundary function expressions for the anodic and cathodic over-potentials. Because of geometry and non-linear boundary functions, an explicit closed form solution is not possible, and numerical methods are required. Klingert, Lynn and Tobias (4) solved a monopolar cell with Taffel polarization in 1964, yielding primary and secondary current distributions for a Leyden Jar problem. Artman, Charap, Hyland and Lembach (5) presented a paper at the 1978 Fall meeting of the Electrochemical Society regarding field mapping of nonuniform electrode structures, including electronic conductive inclusions of bipolar electrodes.

Structure

Solving a shorted anode with a small short is three dimensional. To simplify the problem, a cylindrical coordinate system was chosen such that the system is symmetrical around the vertical axis. By symmetry then, $dV/d\theta$ is identically zero (where V is the scaler potential and θ is the polar angle), and a two-dimensional slice through the structure is used to obtain the potential distribution. Figure 1 is a representation of a cylindrical anode 61 cm in diameter and 43 cm high immersed in bath to a depth of 20 cm. The anode to cathode distance for the purposes of the model was chosen to be 2.0 cm which might be typical for a disturbance wave passing under an anode in a normally operating Hall-Heroult cell. Only one-half of the model needs to be solved by virtue of the symmetry around the centerline axis.

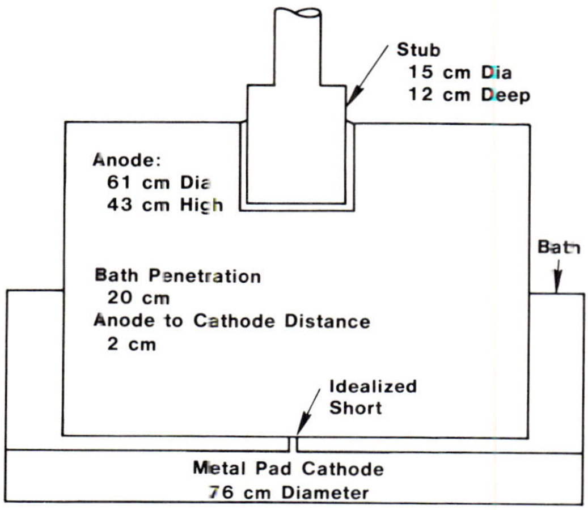

Diagram of Cross Section Through A Cylindrical Anode With Idealized Short
Figure 1

For the resistive medium of the anode and bath, the Laplace Equation in cylindrical coordinates (r, θ, Z) is given as

$$\nabla^2 V = \frac{1}{r}\frac{\partial}{\partial r}\left(r\frac{\partial V}{\partial r}\right) + \frac{1}{r^2}\frac{\partial^2 V}{\partial \theta^2} + \frac{\partial^2 V}{\partial Z^2} \qquad (1)$$

which reduces to

$$\nabla^2 V = \frac{1}{r}\frac{\partial}{\partial r}\left(r\frac{\partial V}{\partial r}\right) + \frac{\partial^2 V}{\partial Z^2} \qquad (2)$$

by virtue of symmetry. Defining a grid coordinate (I, J) corresponding to (r, Z) and using finite difference methods, equation (2) can be rewritten to

$$R_o = V(I+1,J)A + V(I-1,J)B + V(I,J+1)C + V(I,J-1)D - V(I,J)E \qquad (3)$$

Where:

$$A = \frac{1}{h_e+h_w}\left(\frac{1}{h_e} + \frac{1}{2r}\right) \qquad (4)$$

$$B = \frac{1}{h_e+h_w}\left(\frac{1}{h_w} - \frac{1}{2r}\right) \quad (5)$$

$$C = \frac{1}{h_n(h_n+h_s)} \quad (6)$$

$$D = \frac{1}{h_s(h_n+h_s)} \quad (7)$$

$$E = \frac{1}{h_e+h_w}\left(\frac{1}{h_e} + \frac{1}{h_w}\right) + \frac{1}{h_n+h_s}\left(\frac{1}{h_n} + \frac{1}{h_s}\right) \quad (8)$$

$$h_e = r(I+1) - r(I) \quad (9)$$

$$h_w = r(I) - r(I-1) \quad (10)$$

$$h_n = z(J+1) - z(J) \quad (11)$$

$$h_s = z(J) - z(J-1) \quad (12)$$

For the Kth iteration where R_o is a residual, convergence is effected by application of Liebmann's Iteration Method (6) utilizing an overrelaxation factor α such that

$$V(I,J)_{K+1} = V(I,J)_K + \left(\frac{\alpha}{E}\right)R_o \quad (13)$$

Boundary conditions at insulator surfaces are effected by image point and pseudoimage point techniques. A nonlinear exponential grid spacing system is used to enhance solution speed and provide needed resolution in the region of the effected short.

The anode/bath boundary surface is treated by solving for the normal current density on each side of the boundary. Utilizing the anodic overpotential function shown in Figure 2, an iteration is performed flexing the boundary potentials for the anode and bath such that the normal current densities are equal, satisfying a continuity requirement. The anodic over-potential function utilized includes factors for concentration, bubble, reaction overpotential and the equilibrium potential for a bath containing 4% Al_2O_3 at 980°C and a bath ratio of 1.3.

The bath-cathode (metal pad) boundary surface is treated by setting the surface potential equal to the overpotential η_c due to the normal current density at the boundary according to the following function:

$$\eta_c = 0.1076\ i \qquad (0 \le i \le 0.67) \quad (14)$$

$$\eta_c = 0.0753\ \ln\ (i/0.257) \qquad (0.67 < i) \quad (15)$$

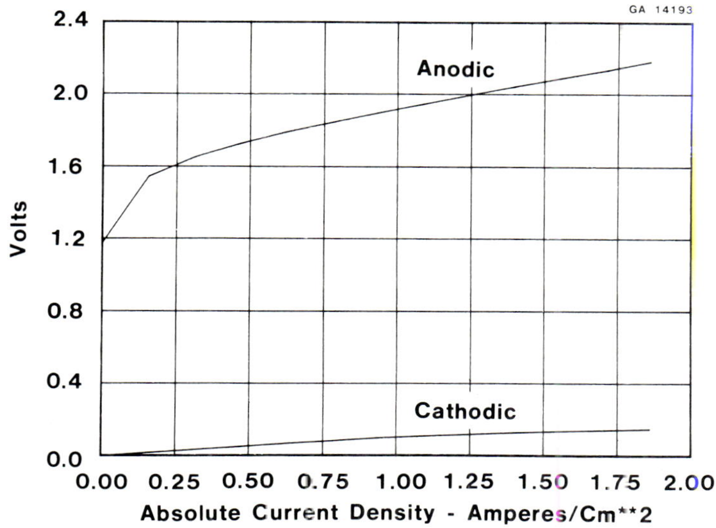

Anodic and Cathodic Overpotential Functions
Figure 2

Equation (15) is determined for the above listed bath from equation (12) in reference 1. This function, equation (15), is negligible at low current densities and therefore a linear behavior is assumed for equation (14).

The solution technique is to begin with an assigned anodic anode rod stub potential and a cathodic metal pad potential defined as 0.0 volt with an arbitrary potential at the bath-anode boundary, solve the anode array, compute a new boundary potential, solve the bath array, compute new boundary potentials and repeat until a convergence criterion is met. The convergence criterion used for this problem is to calculate the total current through the anode and bath by integration around the vertical z-axis for each major iteration such that when the difference between the K and K-1 major iteration is less than some small arbitrary current, iteration halts.

Model Execution

The Unshorted Case

Figure 3 shows the potential map for the unshorted condition of the anode. The anode stub potential is applied to the carbon along the vertical boundary adjacent to the stub side since the radial thermal expansion forces induce much better contact along this boundary surface while the bottom of the stub is assumed not in contact (i.e. an insulator). The potential applied for this case is 3.169 volts with respect to the cathode and the resulting total anode current is 3116 amperes with a bath resistivity of 0.485 ohm-cm and carbon resistivity of 0.006 ohm-cm. The isopotentials are plotted at 0.01 volt intervals for the anode and 0.1 volt intervals for the bath.

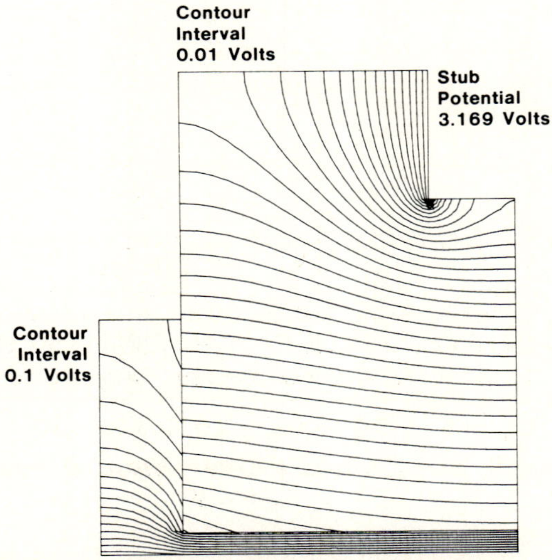

Potential Map for Unshorted Anode and Bath
Figure 3

Current lines (not plotted) are orthogonal to the isopotential lines. The anode to bath anodic overpotential ranges from an average of 1.883 volts at the bottom center of the anode to a minimum of 1.313 volts at the bath top surface along the side of the anode. Maximum current flows at the sharp corner of the anode with an overpotential of 1.899 volts of anodic overpotential. The anode bottom surface area is 2922 cm^2 and the anode average current is 1.066 $amps/cm^2$.

Figure 4 plots the normal cathode current density as a function of radial distance from the centerline of the anode structure. Current density is a maximum 0.8962 $amps/cm^2$ under the center region of the anode and falls off to a minimum of 0.3520 $amps/cm^2$ at the sidewall insulator boundary of the cathode.

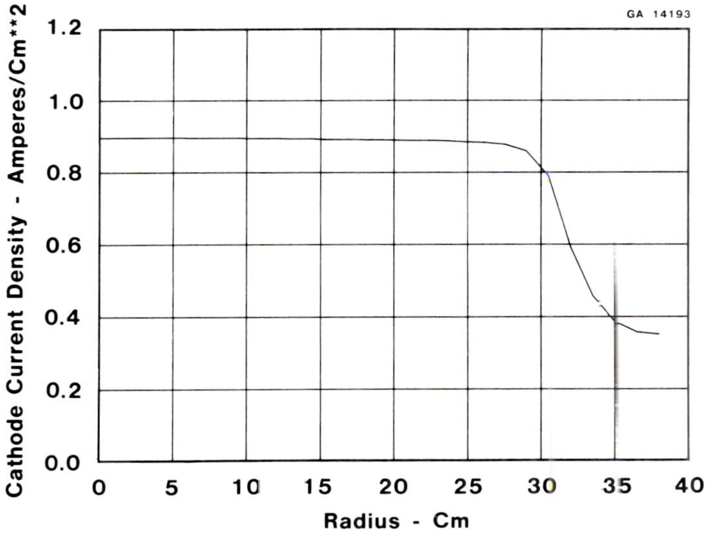

Unshorted Cathodic Current Density
Figure 4

For the purposes of this study, this case will be assumed to be 100% CE, or zero CE loss. Other losses in CE are discussed in the literature. Haupin and Frank (1) discuss recombination, reoxidation of reduced species and electronic conductance.

The Shorted Case

For this case, an idealized cylindrical short or spash of metal is assumed drawn up by hydrodynamic forces making contact at the center of the anode. The short radius is 0.327 cm with a cross sectional area of 0.336 cm^2 and height of 2 cm. Again, 3.169 volts is applied to the anode stub side interface and the cathode is assumed to be at zero potential. The short voltage drop and contact potential drop was assumed to be 0.0615 volts.

The potential map is shown in Figure 5 for this case. Total anode current is 3617 amperes; total bath current is 3122 amperes and the current flowing through the short is 495 amperes yielding a CE for this case of 86.3% CE. Maximum anode to bath overpotential occurs at the anode corner as before at 1.896 volts with a minimum occurring now adjacent to the short of less than total decomposition potential of 1.173 volts assumed for this bath. Since no generation of oxygen occurs to erode the carbon in a small region around the short and electronic conduction taking place through the short preclude erosion in the region of the short, an anode projection should be produced if the condition were allowed to remain for a long time as discussed by Richards, et al. in reference 3.

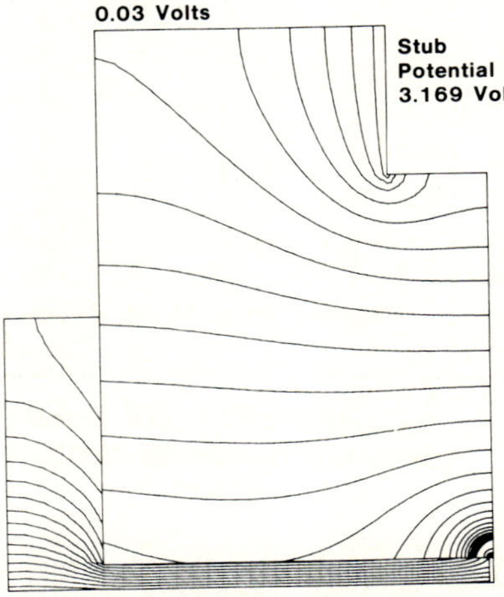

Potential Map for Idealized Metal Pad Short (Contour Interval Above Short, 0.5 Volts)
Figure 5

Figure 6 plots the cathodic current density for the shorted case. The current density at the corner of the short is 0.3476 amp/cm^2 and a minimum of 0.3455 amp/cm^2 at the sidewall insulator boundary of the cathode. Maximum current density occurs at a distance of 19 cm from the centerline of the anode with a magnitude of 0.8566 amp/cm^2.

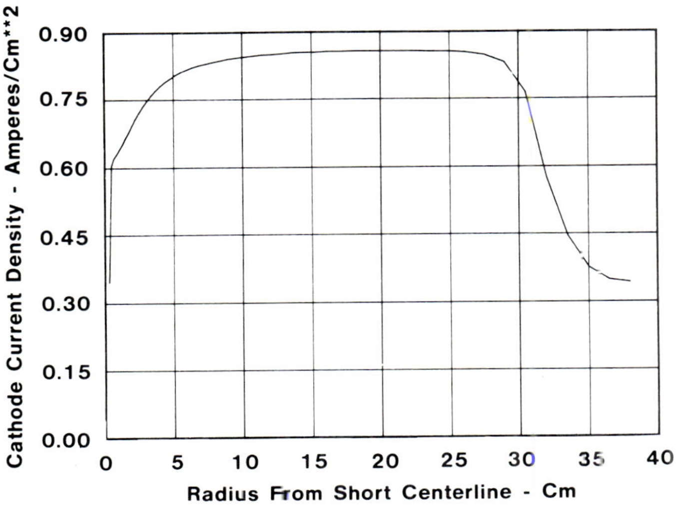

Cathodic Current Density
Shorted Anode to Metal Pad
Figure 6

Table I lists the above case and two other representative cases for the anode shorted to the metal pad.

Table I

Short Radius (cm)	Anode Stub Voltage (volts)	Anode Current (amps)	Bath Current (amps)	CE Loss (%)
0.327	3.169	3617	3122	13.7
0.327	2.370	2158	1654	23.4
1.079	3.169	4781	2827	40.9

Conclusion

The shorting phenomenon in prebake Hall-Heroult cells decreases CE. The model presented indicates the relative magnitude of CE loss for a given anode stub voltage and size of short. For most shorts, the CE loss for an anode ranges between 10 and 80% for the duration of the short. The model allows categorization of the cross sectional area of shorts normally experienced in operation through comparison of model runs with anode voltage and current measurements on candidate pots.

List of Symbols

α	Overrelaxation factor, dimensionless ($1.0 < \alpha < 2.0$)
∇	nabla, directional partial differential operator
$h_{(n,s,e,w)}$	grid spacing dimension, cm
I	grid numbering system, vertical or z direction
i	current density, amperes/cm^2
J	grid numbering system, radial or r direction
K	iteration count
η_c	cathodic overpotential, volts
r	radial coordinate, cm
R_o	residual, volt cm^2
θ	angular coordinate, radians
V	scalar potential volts
z	vertical coordinate, cm

References

(1) W. E. Haupin and W. B. Frank, "Electrometallurgy of Aluminum," Comprehensive Treatise of Electrochemistry, Vol. 2, pp 301-325, Edited by J. O'M. Bockris, B. E. Conway, E. Yeager, and R. E. White, Plenum Publishing Corporation, (1981).

(2) K. Grjotheim, C. Krohn, M. Malinovsky, K. Matiasovsky, and J. Thonstad, "Aluminum Electrolysis Fundamentals of the Hall-Heroult Process," Aluminium-Verlag, Dusseldorf, FRG, 1982.

(3) N. E. Richards, et al, U. S. Patent 3,875,030.

(4) J. A. Klingert, S. Lynn, and C. W. Tobias, "Evaluation of Current Distribution in Electrode Systems by High Speed Digital Computers," Electrochimica Acta, Vol. 9, Pergamon Press Ltd., (1964) pp. 297-311.

(5) J. O. Artman, S. H. Charap, W. W. Hyland, and R. F. Lembaugh, "Electric Potential Mapping for Electrochemical Cells," paper presented at Electrochemical Society Spring Meeting, Seattle, Wash., May 1978.

(6) P. Silvester, Modern Electromagnetic Fields, p 62, Prentice-Hall, Inc., Englewood Cliffs, N.J., 1968.

PHYSICAL MODELLING OF BUBBLE BEHAVIOUR AND GAS RELEASE FROM ALUMINUM REDUCTION CELL ANODES

S. Fortin, M. Gerhardt, A.J. Gesing

Alcan International Limited
Kingston Laboratories
P.O. Box 8400
Kingston, Ontario
Canada
K7L 4Z4

Abstract

A full-scale model of a two dimensional transverse section through a 150 kA prebake aluminum reduction cell was constructed. Gas evolution on the anode face was simulated by passing air through a porous polyethylene plate. Gas bubble behaviour was photographed and analysed. The effects of current density, anode-cathode distance, and anode tilt and electrolyte velocity on gas layer geometry, anode coverage, bubble velocity, and gas release frequency were investigated. The hydrodynamically produced thick bubble front was observed. Under some conditions, up to 2 cm thick bubble fronts were produced.

Introduction

In the electrolysis of aluminum, aluminum oxide dissolved in electrolyte is dissociated, depositing aluminum in a molten metal pool. Oxygen reacts with the carbon anode and is evolved as CO . CO coalesces into bubbles which then travel to, and are released at, the anode edge. Little is known about the behaviour of the gas bubble layer because of the high temperature and opaque operating environment. The following indirect evidence exists.

During the operation of commercial reduction cells, low amplitude voltage instability is observed at about 1-3 Hz. This frequency correlates well with the frequency of gas bubble release at the anode edge. The electrical and gas release behaviour of the anode indicates that molten electrolyte wets the carbon anode in preference to CO . This is in contrast to fluorine which, when evolved during the anode effect, forms a continuous gas film over the anode and stops the electrolysis.

Static surface tension based calculations predict a CO_2 gas bubble layer thickness of approximately 5 mm. This was confirmed by Haupin's (1) measurements of the voltage gradient in the interpolar space in commercial size reduction cells. In these experiments the electrical signal became very noisy as the voltage probe entered the gas bubble layer, approximately 5 mm from the anode face, and lost contact with the electrolyte as the gas bubbles swept past it. Occasional contact between the probe and the gas bubbles was detected as far away as 2 cm from the anode face.

The gas bubble coverage of the anode face in a 60 kA Soderberg cell operating at a current density of 8 kA/m^2 was estimated by carefully measuring the drop in electrolyte level as the current in an operating cell was turned off. From the knowledge of the freeze profile and the electrolyte level change, it was possible to calculate the volume of gas held-up under the operating anode. Assuming a 5 mm gas bubble thickness, this volume implied that 50% of the anode area was covered by gas bubbles.

Geometry of the anode face is known to affect the behaviour of the gas bubble layer. Specifically, the individual anode size, tilt, and immersion are known to be important. Large vertical stud Soderberg anodes are operated at a smaller current density than prebakes. It is suspected that this is due to the larger gas release path for the V.S. anodes. In horizontal stud anodes, a scalloped anode surface is produced as the carbon is consumed around the slightly inclined stud holes. This, and their generally smaller size, aids the gas release and is presumed to allow operation at current densities approaching that of prebakes. A scalloped anode surface is also sometimes produced by gas channelling under the vertical stud Soderberg anodes. This tends to be induced by operation at a very low anode immersion. It is not clear why low immersion should promote gas channelling. At a normal immersion of 15-20 cm, flat anode faces are observed indicating a uniform average current density over the entire anode face and no gas channelling.

The anode burns off parallel to the average cathode surface (the surface of the molten metal). Distortions of the metal surface in the form of tilting or domes are often produced by magnetic forces. These distortions then are transferred to the anode face, producing permanent anode tilt and the evolved gas is preferentially directed to either the center channel or the sidewall space.

Rounding of anode edges is observed. It has been postulated that this is due to some electrolysis taking place on the vertical immersed portion of the anode. The extent of rounding would thus depend on the anode cathode distance and the anode immersion. Rounding decreases the flat area of the anode, increasing the current density on that portion of the anode. The slope of the rounded portion promotes the gas release along the anode edges.

It is the objective of this study to explore the effects of the variation of the cell operation and geometrical parameters on the gas layer behaviour by studying the behaviour of gas bubbles in a two dimensional room temperature water model of the liquid zone of a reduction cell.

Dimensional Similarity Considerations

Dimensional similarity considerations for an invicid, laminar flow system predict that, with 1:1 geometrical scaling, and equality of kinematic viscosity and surface tension, the bubble sizes and bubble velocities should be equal in the model and the CO_2-cryolite system. Effects of turbulence, current flow through the liquid and magnetic fields are not modelled. These cannot be conveniently simulated and for the purpose of this study are a-priori assumed to have only a second-order effect on the bubble behaviour.

A gas bubble which evolves on the anode face by an electrolytic process is electrically insulating, blocks the electric current from reaching the anode, and locally arrests electrolysis. The bubble can grow when gas evolves along the perimeter of the bubble. This tends to favour the formation of many small bubbles. In contrast, the bubbles evolved by gas passing through a porous plate can continue to grow by the addition of gas through the entire anode-bubble contact surface. This favours the growth of large bubbles. Hence, the porous plate model does not adequately simulate initial bubble formation and growth. However, in both cases, bubble growth by coalescence takes place. It has been observed in this study that this method of growth dominates for large bubbles and that the small bubbles forming on the anode surface do not significantly affect the motion of large bubbles.

Construction of a full-scale, room temperature electrolytic anode model was considered and abandoned on the basis of excessive heat generation. A porous plate anode model was adopted to study the geometry and motion of large bubbles.

A water-air system was chosen for the model. The respective liquid properties are given in Table I:

Table I. Liquid Properties

System	Temperature °C	Density Mg/m^3	Kinematic Surface Tension Dynes cm^2/g	Kinematic Viscosity cpoise cm^3/g
Cryolite	950	2.1	60	1.5
Water	25	1.0	70	1.0

This gives a close match for the kinematic surface and viscosity properties. They were judged to be sufficiently close for the evaluation of the effects of changes in geometrical and operating parameters of the reduction cell.

Apparatus

A full-scale model of a two dimensional, transverse slice of the liquid zone of a 150 kA prebake cell was built. A schematic of the model test section is shown in Figure 1. The model consisted of a glass sided tank, 450 cm long, 61 cm high, and 40 cm wide. The anodes were modelled by suspending two aluminum boxes within the tank. Two anodes were suspended within the tank; each was 135 cm long, 66 cm high, and 40 cm wide. Screw mounting of the anodes permitted the adjustment of both the anode elevation and tilt. Gas evolution was simulated by forcing air through a micro-porous high density polyethylene plate which modelled the anode face. Each anode box was subdivided into fifteen air-tight compartments, 27 cm × 13 cm. Air flow in each compartment was individually controlled to ensure uniform gas distribution over the anode face. Along the long edge of the anode slices, a 2 cm wide lip extended below the polyethylene plate to direct the gas along the long axis of the model. The liquid was circulated through the ends of the model and along an external return circuit by a variable speed propeller pump. The bottom edge of the interpolar space was defined by suspending a transparent 9 mm thick Plexiglas sheet parallel to the face of the anodes, modelling a solid cathode surface with little or no metal on it. The bubble behaviour was observed by placing a mirror below the plexiglas plate.

Experimental

The effect of anode-cathode distance, ACD, was investigated by varying the spacing between the plexiglass "cathode" surface and the porous polyethylene "anode" face. Changes in current density were simulated by varying the air flow rate through the porous polyethylene "anode" plate. Since passage of four electrons through the circuit is necessary to release one molecule of CO_2, there is a strict correlation between the current density and gas flow rate of: $10\ kAm^{-2} = 2.71\ Lm^{-2}s^{-1}$. Effects of changes in liquid velocity and anode tilt were also investigated. A convention was set up so that a positive liquid velocity indicates that liquid flow is co-current with the gas bubble motion. Table II gives the range of variation of each parameter. Anode immersion was kept constant at 15 cm.

Table II. Range of Variation of Experimental Parameters

Current Density ($kA\ m^{-2}$)	Liquid Velocity ($cm\ s^{-1}$)	Anode-Cathode Distance (cm)	Anode Tilt (degrees)
4.3	-16.7	3	0
8.7	- 8.0	5	0.43
	- 3.6		1.07
	.0		2.58
	3.6		
	8.0		
	16.7		

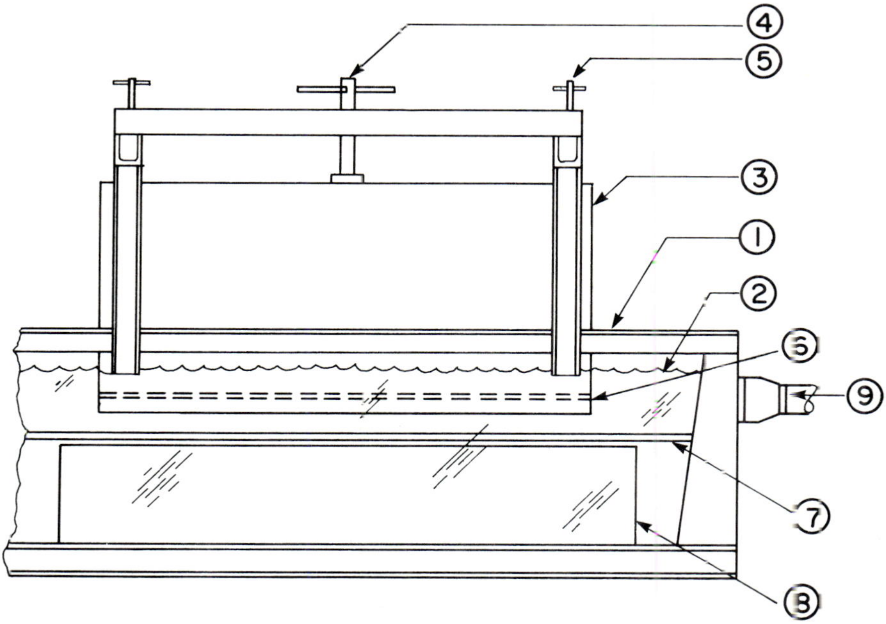

Figure 1 - Schematic of the water model of the aluminum reduction cell: (1) water tank, (2) water level, (3) anode, (4) anode height adjustment, (5) anode tilt adjustment, (6) porous polyethylene, (7) plexiglas plate, (8) mirror, (9) water circulation nozzle.

Measurements were videorecorded as follows. A projection of the bubbles on the anode face was recorded from a mirror inclined at 45° located under the anode. Vertical profiles of the bubbles were recorded just below the anode face with the camera at a nearly horizontal angle. Recordings for each experimental condition were made at two locations - at the centre of the anode and at the anode edge.

Videotapes were then played back in stop-motion and measurements were actually made on the T.V. monitor. The dimensions were scaled to life size by including, in each scene, a square of known dimensions. The time scale was verified by videotaping a stop watch for one minute prior to each experiment.

The measured responses included bubble geometry, gas bubble coverage of anode area, bubble velocity, and frequency of bubble release at the anode edge. Bubble geometry was characterized by the measurement of overall dimensions along and transverse to the direction of bubble motion as well as the height and width of the dynamic bubble front formed at the leading edge of the large bubbles. The model anode was designed such that its long edge extends 2 cm below the anode face. Bubble heights were obtained by videotaping at a nearly horizontal angle and measuring the bubble profiles projected against the extension on the far side of the anode as illustrated in Figure 2.

Average dimensions for five large bubbles are quoted for each experimental condition. Gas bubble velocity was calculated by timing the motion of the bubble between two fixed points. The percent of the anode face covered by gas was calculating by using a 440 square section grid and determining the fraction of the intersection points that were over the gas bubbles. In order to avoid aliasing the data, the average coverage for equally spaced time intervals within one bubble cycle is reported.

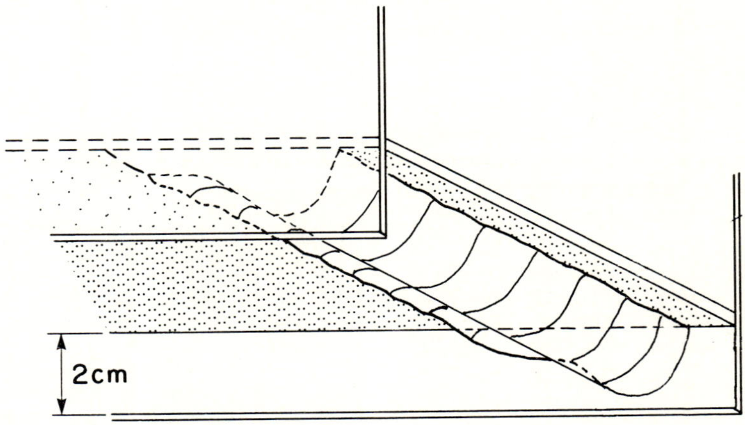

Figure 2 - Gas bubble flowing beneath the porous plate from left to right. Bubble profile, 1, is seen against the extended anode edge, 2.

Results

Bubble Behaviour on a Horizontal Anode Surface

The anode face was judged to be horizontal when, with stationary liquid, no motion of growing bubbles was observed. In this case, small spherical bubbles were generated on the entire anode surface. They grew to a diameter of ~5 mm before expanding horizontally. They coalesced into large individual bubbles until they covered approximately 50-60% of the anode area. At that time they impinged on each other and simultaneously coalesced into a single bubble covering the entire anode face. The edge of the gas sheet almost immediately reached the anode edge and the gas escaped from the anode face with the speed of the air leaving a broken balloon. Clean anode surface was left ready to repeat the process. This process is illustrated in Figure 3.

In this process the time averaged percentage of the anode area covered by gas bubbles was observed to be independent of current density (air flow rate). The gas bubble release frequency increased with the current density. The liquid motion did not alter the overall process. The bubbles were swept slowly towards the downstream edge of the overall anode causing most of the bubble release to take place there.

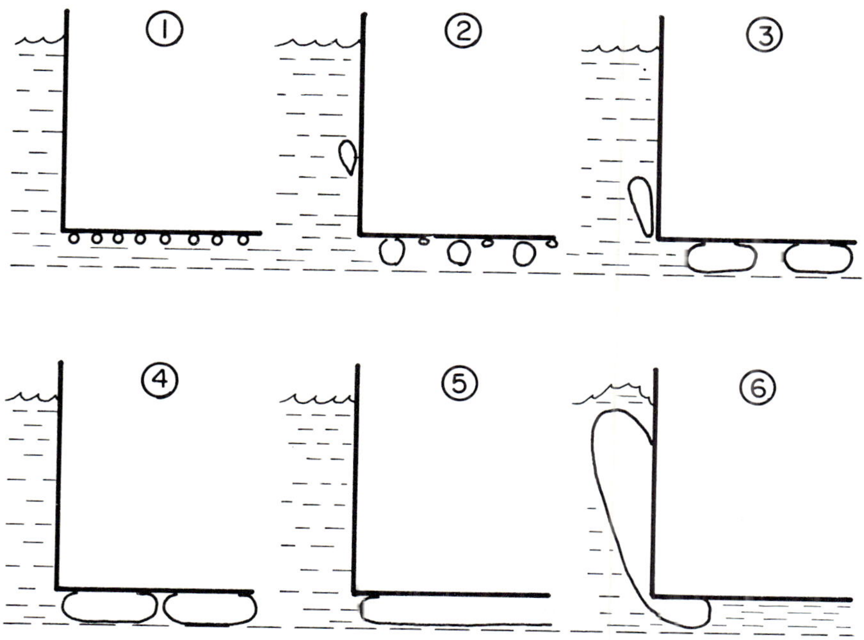

Figure 3 - Steps in the gas release cycle from a horizontal anode face. (1) Formation of bubbles, (2) spherical growth, (3) lateral spread, (4) mutual impingement, (5) coalescence into a single gas sheet, (6) gas release at the anode edge leaving a clean anode surface.

As the coalesced gas layer was released from the edge of the anode all at once, the liquid was sucked into the interpolar gap to replace the gas. The gas bubble rising through the liquid up the anode side created a pumping action which vigorously stirred the anode-sidewall and anode-anode gaps. The result is that the bulk of the disturbance caused by gas bubbles was created at the anode edge and very little underneath the anode. This was particularily evident when a second liquid was used to model the molten pool cathode. Waves were generated on the liquid-liquid interface at the anode edge. They then travelled under the anode.

Bubble Behaviour on the Tilted Anode Surface

Anode tilt drastically affected the behaviour of the bubble layer. A tilt of a fraction of a degree was sufficient to induce significant buoyancy driven motion of the gas bubbles. Examples of typical bubble patterns are shown in Figures 4 to 7.

Small bubbles were created on the entire anode surface. These small bubbles moved slowly. As they began to coalesce they gained velocity. This caused the "sweeping" effect shown in Figure 6 as the large bubbles travelled along the anode face overtaking and swallowing any small bubbles they came into contact with. Bubble growth was rapid and continuous with the bubble gaining in size and speed until it was released at the anode edge. Coalescence of all the bubbles into a continuous layer no longer took place, but a characteristic bubble pattern was developed. This pattern depended on anode tilt, liquid velocity, current density, and location on the anode face.

As the bubble velocity increased, the ratio of transverse to longitudinal dimensions of the bubble increased promoting bubble coalescence in a transverse direction and increasing the separation between successive large bubbles. In many cases this resulted in two dimensional bubbles spanning the entire width of the model section of the anode. The leading edge of the large bubble was preceded by a layer of small bubbles about to coalesce with it. The trailing edge swept the area behind it clean, ready for the formation of new small bubbles.

The shape of a moving bubble was no longer limited by static surface tension balance. The typical bubble profile is shown in Figure 8(A). The thickness of the front portion of the bubble increased in size. As the bubble surface became horizontal the bubble front collapsed, and the trailing portion of the bubble had a thickness similar to those observed at static conditions. In the case of large fast moving bubbles, secondary wavelets were formed along the gas-liquid interface of the trailing portion of the bubble.

Effect of Anode-Cathode Distance. In all cases, a change of ACD from 5 to 3 cm had no effect on the bubble patterns or on any of the measured responses. However, it is interesting to note that the large gas bubbles passing over the Plexiglas sheet, which simulated the cathode, caused enough disturbance to shake the 9 mm thick sheet. This disturbance was more noticeable at the low ACD.

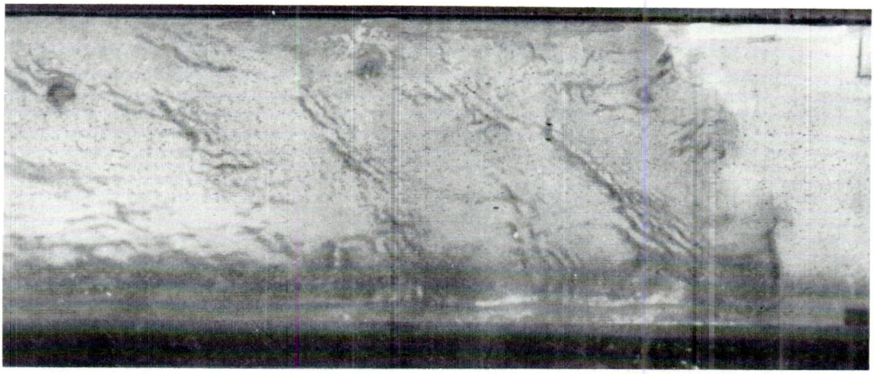

Figure 4 - A bubble covering almost the entire anode face is formed on a nearly horizontal anode with a countercurrent liquid flow.

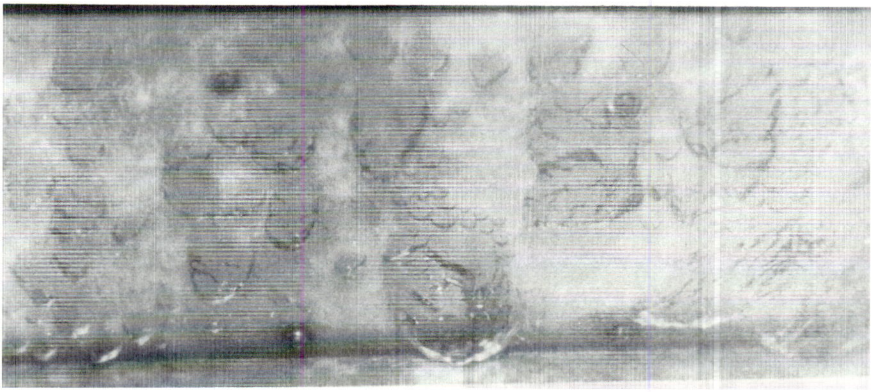

Figure 5 - Gas bubbles growing and coalescing at the center of the anode while travelling from left to right.

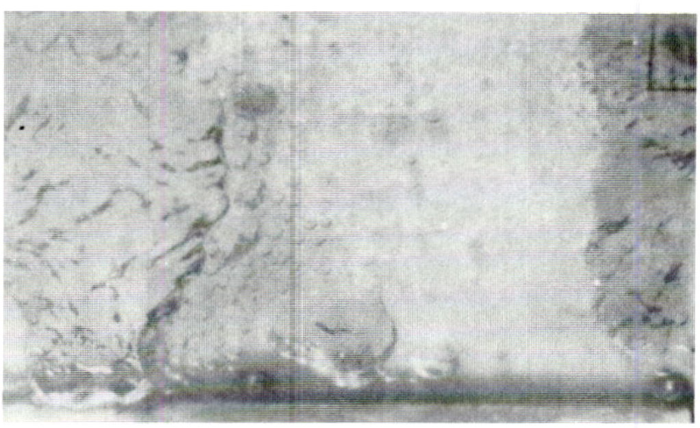

Figure 6 - Sweeping action of the large bubbles as they move from right to left.

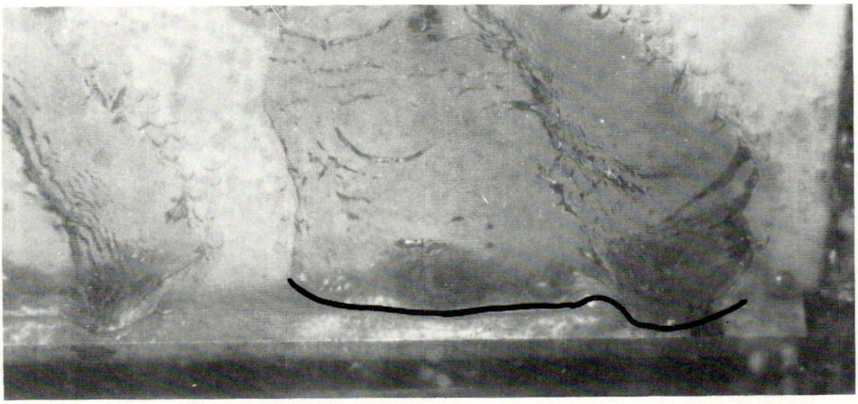

Figure 7 – Large bubble travelling from right to left. A high bubble front if visible and the bubble profile is outlined against a 2 cm wide anode edge extension.

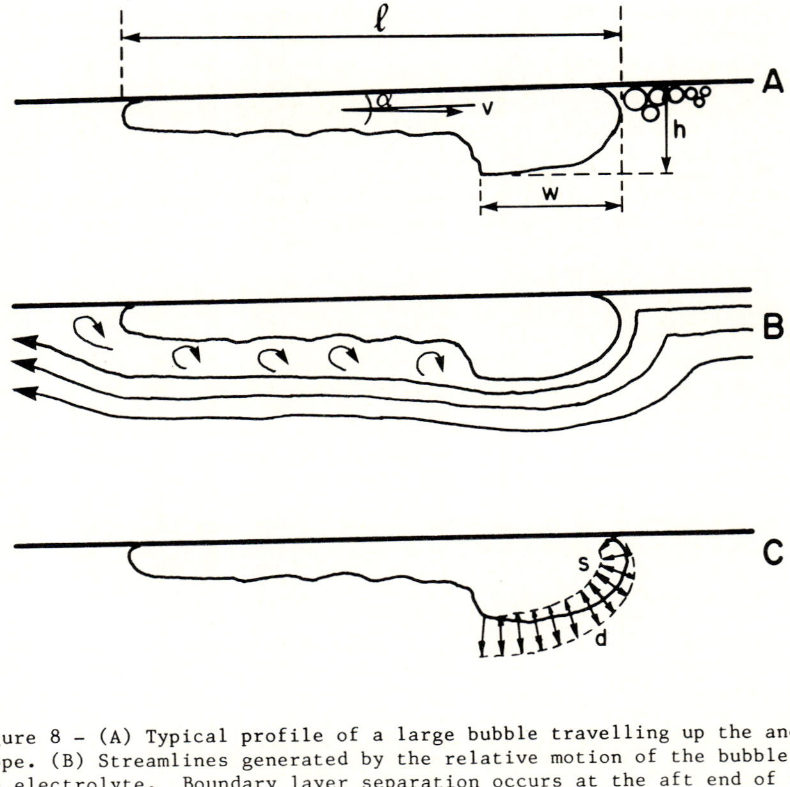

Figure 8 – (A) Typical profile of a large bubble travelling up the anode slope. (B) Streamlines generated by the relative motion of the bubble and the electrolyte. Boundary layer separation occurs at the aft end of the bubble front resulting in the turbulent region downstream. (C) Hydrostatic, s, and hydrodynamic, d, pressure distributions along the bubble front.

Longitudinal Dimension of the Bubbles. This dimension was measured parallel to the direction of bubble motion at the center of the anode. The "length" varied from 2 to 128 cm. It increased with an increase in current density and decreased with increase in liquid velocity or anode tilt as shown in Figure 9.

Transverse Dimensions of the Bubbles. The bubble dimension perpendicular to the direction of motion varied from 4 to 40 cm (the entire width of the anode slice) when measured at the center of the anode. For any bubble not spanning the entire anode slice, this "width" was larger than the corresponding "length" measured parallel to the direction of motion. Figure 10 shows that bubble "width" increased with an increase in current density. At high countercurrent liquid velocities, an increase of anode tilt caused a large decrease in the bubble "width". Tilt had negligible effect on bubble "width" when liquid was flowing in cocurrent direction.

Bubble Profile. The shape of the bubble profile consisted of a thick bubble front and a thin (~5 mm), long trailing portion. Bubble front thickness and width are both defined in Figure 8(A). They were measured for two dimensional bubbles near the edge of the anode at the intersection of the bubble and the extended anode edge. The bubble front thickness varied between 0.8 and 2.4 cm. An increase in current density increased the thickness while an increase in liquid velocity decreased it. Figure 11(A) shows these effects while Figure 11(B) illustrates the effect of anode tilt. With the liquid flowing in a countercurrent direction, an increase in anode tilt caused an increase in front thickness. With a cocurrent flow, an incresed tilt decreased the front thickness slightly.

The measurements of the width of the bubble front illustrated in Figure 12 ranged from 0 to 14.5 cm. An increased front width was recorded with increased current density and anode tilt. An increase in liquid velocity decreased the front width. The effect of the velocity was much more pronounced at a high tilt.

Percent of Anode Area Covered by Gas. In the case of the tilted model anode, the bubble size and the amount of gas passing a given point increased from the low to high edge of the anode. The coverage values were measured at the center of a tilted model anode. The bubble coverage varied from a low of 24% to a high of 90%. The amount of bubble coverage decreased with the increase in liquid velocity and increased with an increase in current density. The anode tilt results depended on liquid velocity and the ACD had no effect.

An increase in current density caused an increase in bubble coverage. The amount of coverage, however, was not proportional to the volume of gas being passed through the anode. Figure 13(A) shows that when current density was doubled, the bubble coverage increased by 10-22%.

The effect of the anode tilt on bubble coverage depended on the direction and magnitude of liquid velocity. Figure 13(B) shows that at a high countercurrent liquid velocity of -16.7 cm/s, an anode with a low tilt of 0.43° had 27% more area covered by gas than an anode with a high tilt of 2.58°. However, at a high cocurrent liquid velocity of 16.7 cm/s there was no significant effect of tilt on coverage.

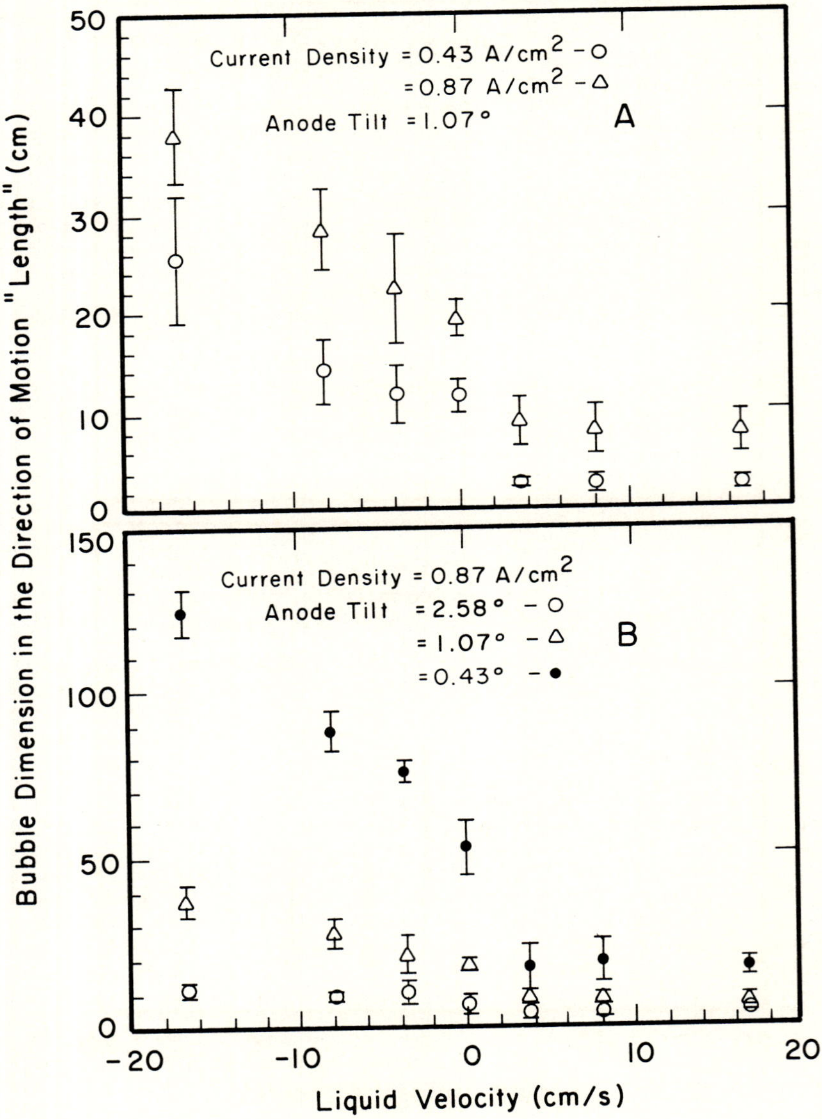

Figure 9 - Effects of liquid velocity, current density, and anode tilt on the gas bubble "length".

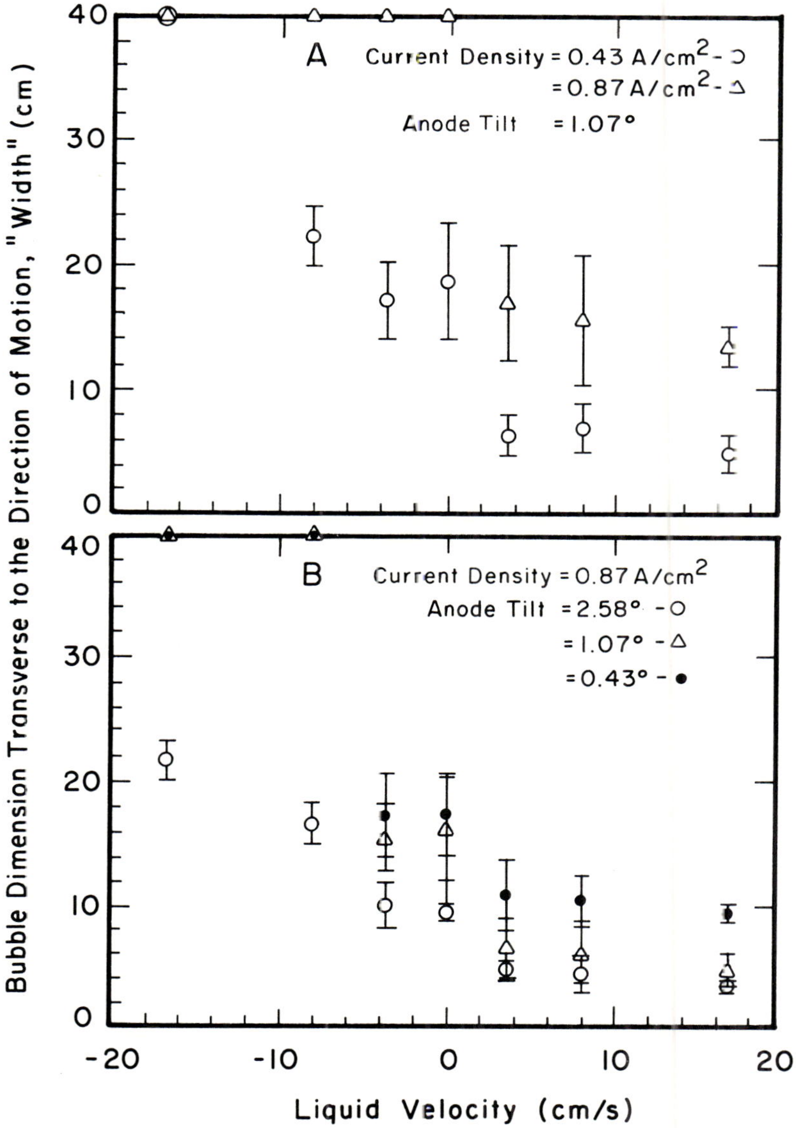

Figure 10 - Effects of liquid velocity, current density, and anode tilt on the gas bubble "width".

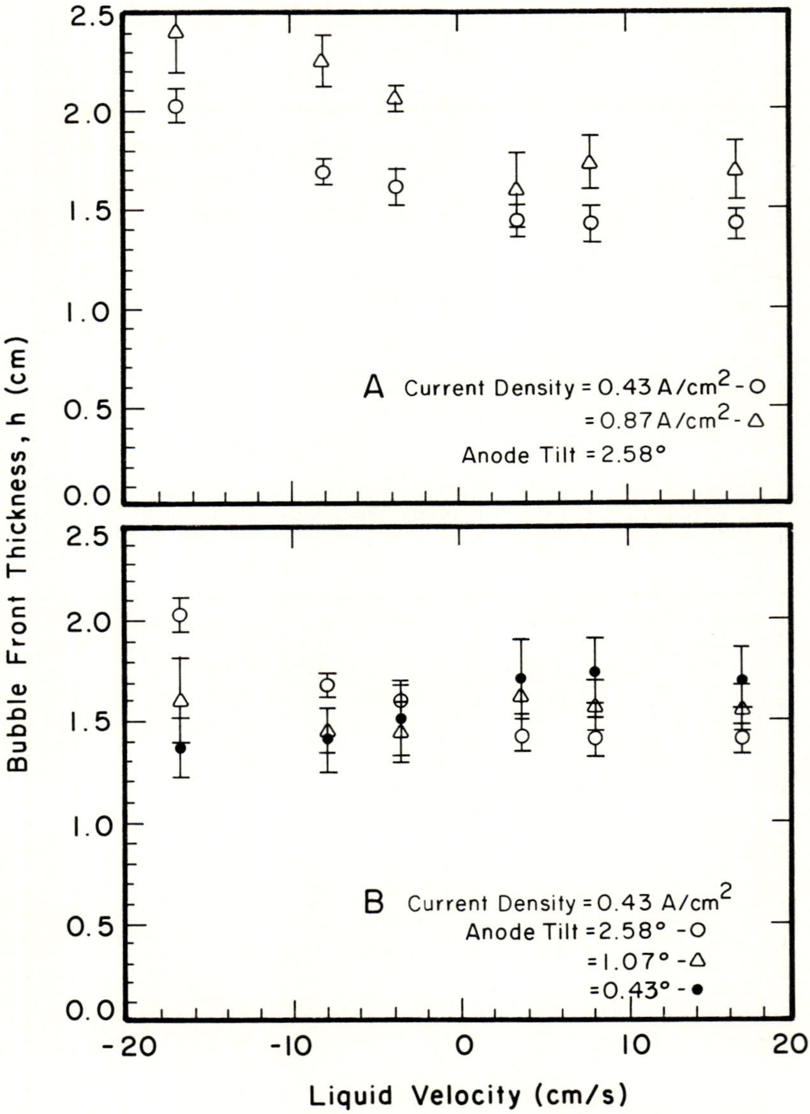

Figure 11 - Effects of liquid velocity, current density, and anode tilt on the thickness of the bubble front.

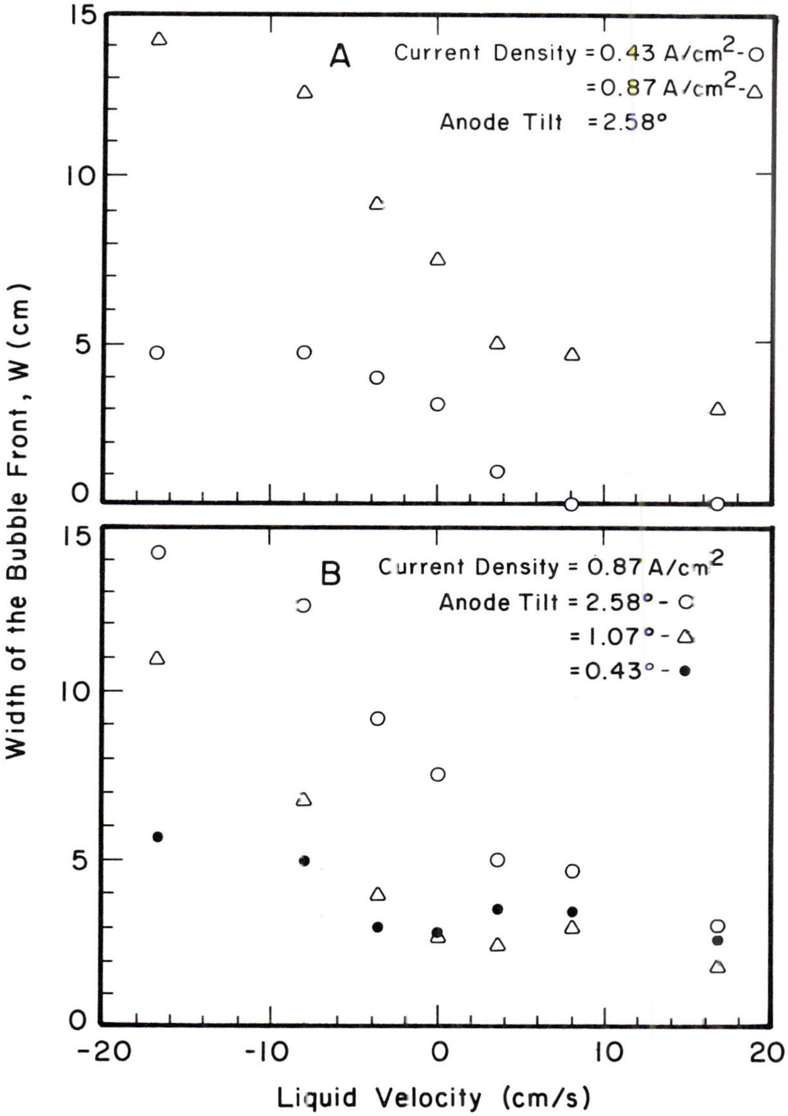

Figure 12 - Effects of liquid velocity, current density, and anode tilt on the width of the gas bubble front.

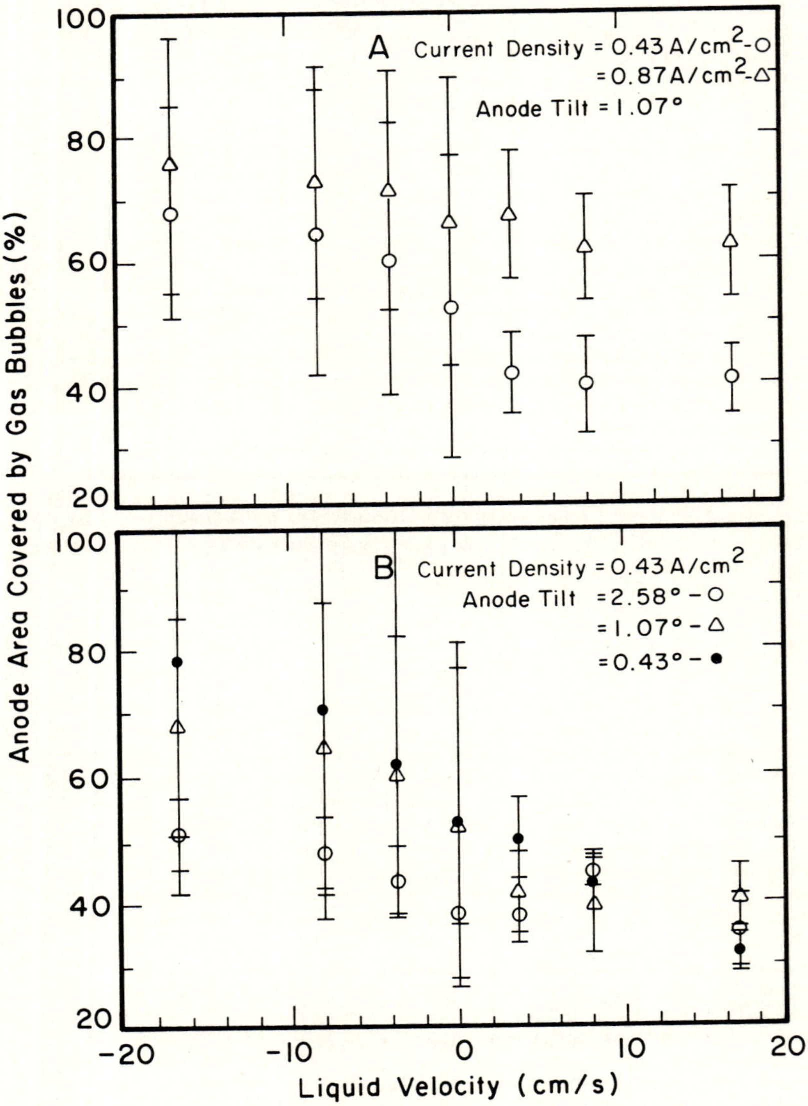

Figure 13 - Effects of liquid velocity, current density, and anode tilt on the percentage of anode area covered by gas bubbles.

Gas Bubble Velocity. With a low anode tilt and the liquid flowing in a countercurrent direction, small bubbles moved down the anode slope. As they grew in size, and they reversed direction and traveled up the anode face. The larger bubbles caught up to and coalesced with the smaller ones. An apparent jump in the velocity of the bubble leading edge occurred when several similarly sized bubbles traveled up the anode together. They grew until at one point they impinged on each other and suddenly coalesced into a single bubble. This registered as a sudden apparent increase in the velocity of the leading edge of the largest bubble. Another anomaly occurred when the leading edge reached the edge of the anode. As the air escaped over the edge, the velocity of the trailing edge of the bubble increased dramatically.

The gas bubble velocities observed at the center of the anode were in the range 8 to 40 cm/s. Bubble velocity increased with an increase in liquid velocity or current density. Both the ACD and anode tilt had only a limited effect. The increase in gas bubble velocity was 10-16 cm/s as the liquid velocity was increased from -16.7 to 16.7 cm/s. The curve has a definite "S" shape with the slope being higher at a slow liquid speed and lower at a high liquid speed (both cocurrent and countercurrent). An increase in current density caused an increase in the gas bubble velocity. Figure 14(A) shows that the bubble velocity is consistently 5 to 13 cm/s faster at a current density of 8.7 kA/m^2 than at 4.3 kA/m^2. Surprisingly, the anode tilt had very little effect on the gas velocity. Figure 14(B) shows that the difference in the bubble velocity on tilts of 0.43° and 2.58° never exceeds 3 cm/s.

Bubble Release Frequency. Bubbles were released from the edge of the anode at frequencies ranging from 0.2 to 3.3 bubbles per second. The frequency increased with an increase in either liquid velocity or anode tilt. The current density and ACD had no effect on the frequency. Figure 15(A) shows that an increase in the liquid velocity from -16.7 to 16.7 cm/s caused an increase in frequency of between 1.4 and 2 s^{-1}.

Conclusions

Table III summarizes the results obtained in this study. The ACD had no effect on gas bubble behaviour. An increase in current density increased the bubble size and thickness of the bubble front as well as gas coverage of the anode face and bubble velocity. Current density had no effect on bubble release frequencies. An increase in electrolyte velocity decreased the bubble size and the gas coverage and increased the bubble velocity and release frequency. An increase in anode tilt decreased the bubble size and gas coverage and increased the bubble release frequency. Tilt had no effect on bubble velocity. There was considerable interaction between the effects of the experimental variables on bubble size and bubble velocity.

Behaviour of the gas layer on a horizontal anode was different than on an inclined anode. On a horizontal anode, the process of bubble nucleation, growth, coalescence and release involved no bubble motion and led to a gas layer thickness of approximately 5 mm. On the inclined surface, gas behaviour was dominated by the motion of large bubbles across the anode surface. Hydrodynamic effects increase the maximum thickness of the bubbles to more than 2 cm. The results of this study are in very good agreement with indirect measurements on the aluminum reduction cells published to date (1).

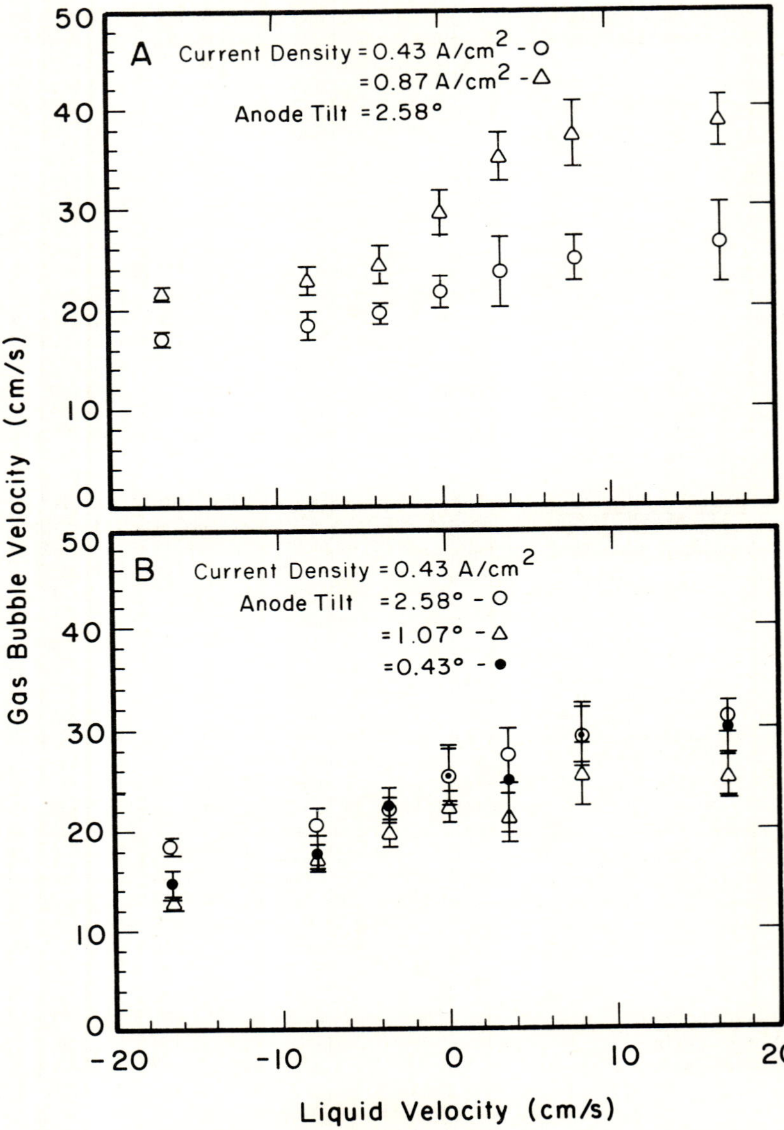

Figure 14 - The effects of liquid velocity, current density, and anode tilt on the gas bubble velocity.

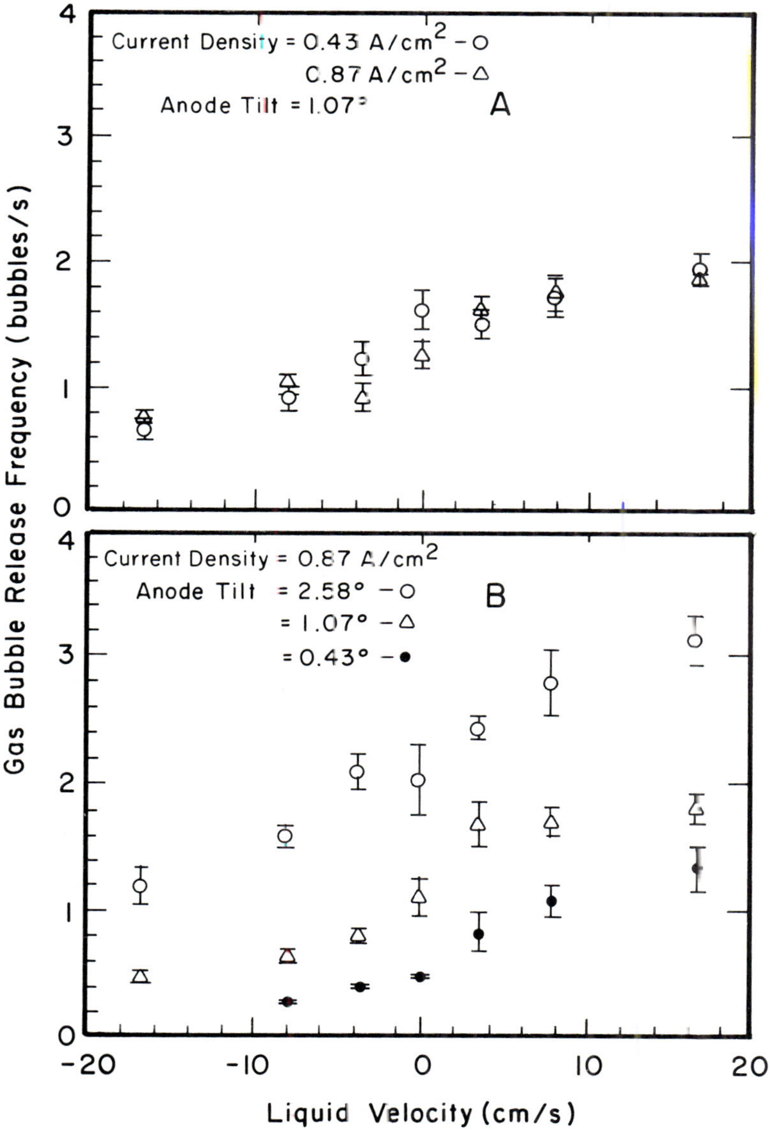

Figure 15 - Effects of liquid velocity, current density, and anode tilt on the gas bubble release frequency.

Table III. Average Slope of the Graph of Response Parameter Against Experimental Variables

Response Parameter \ Experimental Variable	Current Density, i (kA/m^2)	Electrolyte Velocity, v (cm/s)	Anode Tilt, α (degrees)	Range of Values Observed
Longitudinal dimension of the bubble (cm)	2.8 @ v = -16.7 2.2 @ v = 0 1.1 @ v = 16.7	-3.3 @ α = 0.4° -1.0 @ α = 1.1° -0.4 @ α = 2.6°	-52 @ v = -16.7 -22 @ v = 0 -5.6 @ v = 16.7	2-128
Transverse dimension of the bubble (cm)	2.0 @ v > 0	-0.6 @ v > 0	-2.8 @ v > 0	4-40
Bubble front height (cm)	+0.1	+0.01 @ α = 0.4° +0 @ α = 1.1° -0.02 @ α = 2.6°	+0.33 @ v = -16.7 0 @ v = 0 -0.14 @ v = -16.7	0.8-2.4
Bubble front thickness (cm)	~1.0-2.0	-3.3 @ α = 2.6° -0.9 @ α = 0.4°	4.0 @ v = -16.7 0 @ v = 16.7	0-14.5
Anode area covered by gas (%)	+3.6	-0.65	-12.1 @ v = -16.7 -7.4 @ v = 0 0 @ v = 16.7	24-90
Bubble release frequency (s^{-1})	0	+0.04	+0.72	0.3-3.3
Bubble velocity (cm/s)	+1.0 @ v = -16.7 +1.8 @ v = 0 +2.7 @ v = 16.7	+0.4	0	8-46

Acknowledgements

This work is an extension of physical modelling of aluminum reduction cells carried out at Alcan by T. Cambridge and E. Dernede. The measurements of gas coverage in operating the 60 kA Soderberg cells were performed under supervision of O. Sivilotti. M. Schneider and G. Quinn were instrumental in the design and construction of the physical model.

References

(1) W. Haupin, A Scanning Reference Electrode for Voltage Contours in Aluminum Smelting Cells, Journal of Metals. 23 (10) (1971), pp. 46-49.

(2) I.H. Abbott and A.E. VonDoenhoff, Theory of Wing Sections, Dover Publications, New York, New York, 1958.

POTROOM VENTILATION SYSTEM BY SEGREGATED AIR LAYERS

O. Lid   Ø. Haugen

Årdal og Sunndal Verk a.s.
P.O. Box 5177
Maj., Oslo 3
Norway

Introduction

During the seventies, it became more and more obvious that the internal environment in the potrooms was of great importance. Ventilation and gas-cleaning systems played a major part in the efforts of reducing the health-hazards to which workers were exposed.

In this aspect, environmental improvements contributed directly to cutting costs and making it possible to increase the total efficiency in the smelters. Along with this development, the state authorities in Norway set up new and stricter rules for both the internal and external environment, concerning pollution from dust, gases, PAH, operative temperatures and heat stress indexes.

Aluminium producers had to follow up these demands, and present plans to meet the claims in the new laws. When planning new factories, it was a presupposition that the indexvalues had to be complied with before discharge permission was given. Also the limit values for the internal environment have to be observed.

As a result of these new and higher ambitions, ÅSV had to pay more attention to the problems connected with potroom ventilation. In the early seventies, ÅSV had introduced the dry gas scrubber technique in co-operation with Alcan.

When ÅSV in 1977 was planning to build a new smelter at Høyanger in Western Norway, much work was done for pre-studies and tests in a water-model, to determine the best ventilation system for the new potroom. Preliminary investigations and modeltests were carried out in 1977-79, and the new potroom was ready for production in October 1981.

The Høyanger potroom marks a rise in ventilation standard within the ÅSV Group and work is now in progress to give existing potrooms equivalent ventilation systems.

Conventional potroom ventilation

In older potrooms whether with Søderberg pots or prebakes, the ventilation systems have been based more or less on rules of thumb. The main aim has been to achieve a sufficient number of air changes or a reasonable temperature rise to ensure satisfactory ventilation in general. The air distribution in the operator's zone has more or less been neglected as a problem. The system in use can be broadly classified under two categories:

1. Natural ventilation
2. Forced ventilation

Forced ventilation allows a greater degree of control but has, of course, operating costs, which natural ventilation does not have. Natural ventilation has usually been dimensioned to involve larger volumes of air than forced ventilation. For this reason, the latter has often been deemed to result in higher concentrations of pollutants in the working atmosphere, and also higher temperatures. On the other hand, natural ventilation is more susceptible to changes in wind and weather.

To comply with strict pollution control regulations, ventilation air leaving the potroom is, in some cases, passed through roof scrubbers before emission to the outside atmosphere. In these cases, natural ventilation may easily involve excessively large volumes of air with consequent high costs; and forced ventilation has thus been preferred.

In many cases, the air enters the potroom through grilles along the walls, whilst in others the air is drawn in from the cellar and distributed around the pots; or a combination of both systems is used.

Most of the pollutants are drawn off the pots and usually channelled to a scrubber for pot gases. However, some of these pollutants enter the ventilation air as it flows around the pots. Depending upon how the air enters the potroom, and which system is used, pollutants will enter the atmosphere in the operators' zone.

Fig. 1 illustrates a forced ventilation system in which the air enters along the walls. The arrows indicate the flow pattern, i.e. the air enters horizontally, whilst the cooling of the hot air under the roof can result in air flowing downwards along the roof and walls, resulting in a mixing of fresh and polluted air.

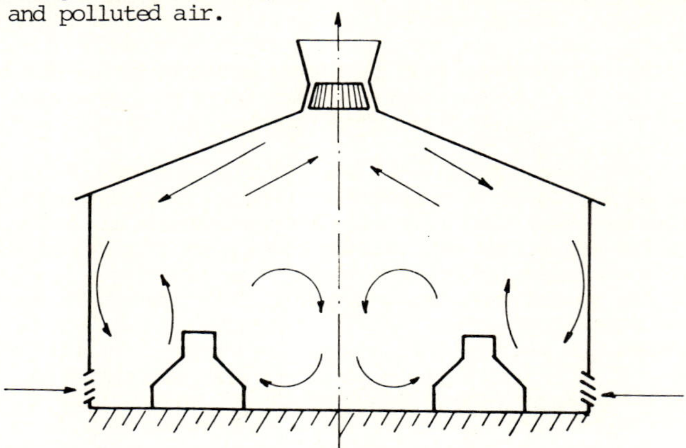

Fig 1 Air entering along the walls

Fig. 2 illustrates a similar forced ventilation system, here the air enters through the cellar. This gives a somewhat better ventilation on account of there being less mixing of fresh and polluted air, but the system still has weaknesses, in the form of down-flowing air, and it is not automatically immune to wind conditions.

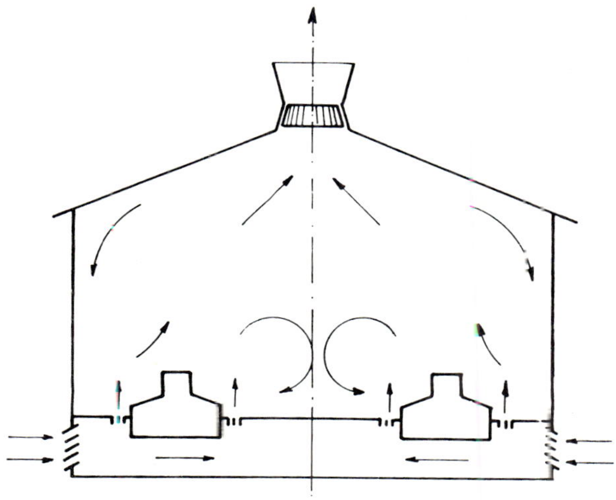

Fig 2 Air entering through the cellar

Not much attention has been paid in the past to the fact that the pots are heat sources which result in rising current of air; and this fact can be utilized, together with an efficient air distribution system to improve conditions in the operators' zone.

In general, it has proved difficult to control air supply and air volume. Doors and windows have been left open, and random openings in walls and roofs have resulted in unintentional air flows, which cannot be measured.

As we will see in the following, it is important that the volume of air be dimensioned on the basis of a study of the potroom in question, and that special attention is paid to the question of air distribution.

Model tests

In the mid-seventies there were carried out studies at the Norwegian Institute of Technology for several Norwegian smelting industries. A model test system was developed that made it possible to try out different solutions for industrial ventilation relatively simply and cheaply.

ÅSV had such investigations made for the new potroom in Høyanger, and much knowledge was acquired, making it possible to design better systems for potrooms.

The ventilation system has been called "ventilation by segregated air layers", because we use the heat transfer from the pots to the surroundings to form a segregation between cooler fresh air in the operators' zone and warmer polluted air higher up.

Such tests are done with water instead of air, which makes it possible to build the models relatively small. Since the physical laws describing flows of air and of water are the same except for the charateristic constants of each medium, an airflow can be simulated by a waterflow.

Potrooms for aluminium are often several hundred meters long. Since the flow pattern in a potroom repeats itself from one pot to another, it is sufficient to build the model of only a few pots.

The Høyanger model was built with four pots. Studies and measurements was made of the two pots in the middle, to ensure that end effects of the model walls did not influence the flow-pattern. The model was of course to confirm geometrically to the actual potroom. Minor details not influencing the flow-pattern can, however, be neglected in the model.

As important as geometrical conformity is the dynamic conformity, which means that the ratio between forces causing the flow pattern must be identical. In a potroom with a high amount of heat dissipated, the ratio between accelerating forces and thermal buoyant forces must be equal in the model and in the actual full scale potroom. (Equal Archimedes' number).

Two dimensional scales must be chosen in the model, - then the other scales are given from the model laws. Naturally one of the chosen scales is the length scale, which is usually 1:15 - 1:20. For the potroom model in Høyanger, the scale was 1:15. The other scale chosen is the emission of convectional heat, which in this model was 1:40. This means that a pot emitting 300 kW is simulated by a modelpot of 7.5 kW.

The potroom is built of transparent plastic while the pots are built of copper plates with heating elements inside. The model is then placed in a water tank, also made of plastic, and connected to water supply and power supply.

To make the flow pattern in the model visible, colour is supplied and mixed with the water. Red colour is used to simulate warm and polluted air, (colour supply in the pots) and green colour visualize the fresh ventilation air. It is then easy to document the model tests with films and photographs.

Segregation of air layers is shown in the model in a very informative way. We can also determine, with a high degree of accuracy, the quantity of ventilation air required to establish a segregation with clean, fresh air in the operators' zone, and how high above floor level the segregation then occurs.

We can also try out different configurations of the air inlet through the floor, and pick out optimal solutions. What we cannot determine with certainty, is how stable the segregation is, since jet impulses are not modelled correctly in an Archimedes'-model. However, experience has told us how large the temperature rise should be to create a stable segregation.

Fig. 3 is an illustration photograph picture.

Fig. 3. Picture showing potroom model in operation.

Ventilation system by segregated air layers

. A typical pre-bake pot in Høyanger emits about 300 kW to the surrounding air. The major part of the heat transfer is by convection, a smaller part by radiation. The air surrounding the pots is heated, which creates an uprising convectional airflow. The volume in this airflow is dependent on the amount of heat transfer by convection from the pots.

$$V \sim E^{1/3} \qquad (1)$$

where V = air volume
E = convective heat transfer

The airflow increases in volume as it rises from the pots and upwards to the ceiling.

When ventilation air is supplied to the potroom in the operators' zone, and with a low velocity, the air needed to increase the volume of the convectional airflow from the pots is taken from the ventilation air. At a certain level above the floor, the air volume in the convectional airlow equals the amount of air supplied from ventilation.

Above this level, the convectional airflow gets its necessary supply of air to expand from resirculated, hot polluted air in the potroom. We then obtain a segregation of air layers at this level, with clean fresh and cool air in the operators' zone, and hot, polluted air above. The main characteristics of the segregation is a sharp temperature rise between the air layers. This is illustrated in fig. 4.

The concentration of pollutions follows the same pattern, since dust, gases etc. leak from the pots and follow the hot convectional airflow. If the pots emit a high amount of pollutive dust, it is even possible to observe the segregation of air layers in the potrooms. However, in the Høyanger potroom it is not possible to see this segregation, since the prebake pots are hooded.

The quantity of air in the air supply systems must be so chosen as to place the segregation between cooler and warmer air approximatly 2.5-3.0 m above floor level. That secures an operators' zone with fresh air and with small amounts of pollution, if the air distribution system is properly maintained.

It is not recommended to increase the air quantity further, since that would decrease the temperature rise that stabilizes the segregation and prevents pollution from reaching the operators' zone.

The connection between air volume V and temperature T is:

$$E = \rho \cdot C_p \cdot V \cdot \quad T \qquad (2)$$

where ρ = specific weight of air kg/m^3
C_p = specific heat capacity of air J/kg K

The right quantity of air for ventilation is not sufficient in itself. It has to be matched by a supply system that introduces the air in the operators' zone through the floor and with a low velocity.

This can be done by making the floor around the pots of perforated concrete. It is recommended that the air velocity through the floor open-

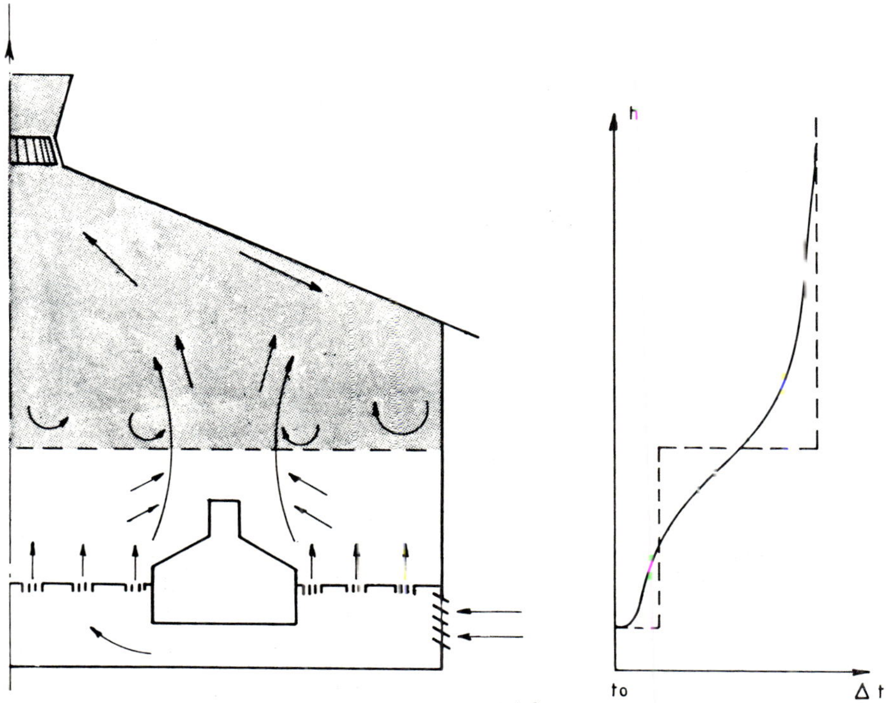

Fig 4 Ventilation system by segregated air layers

ings should be less than 1.5 m/s. This is because higher velocities might break down the segregation higher up.

Also the floor openings must be scattered in such a way that there are no large floor areas without perforated concrete elements.

The temperature rise between the segregated air-layers should not be less than 10 K to obtain stable conditions. Influence from wind- and weather conditions are eliminated by adjustable grids in the air inlet openings into the cellar.

<u>Choise of ventilation system in Høyanger potroom following model tests</u>

The watermodel tests were of great value when deciding on the ventilation system for the Høyanger potroom. A wide range of air volumes were tested out, varying from 2-15 Nm^3/s per pot. The final optimal quantity was chosen to be 11 Nm^3/s per pot, which gave a segregation of air layers appr. 3 m above floor level and a total temperature rise of 20 K from inlet to outlet. A major part of T occurs through the segregation, since preheating from cathode shells is small (2.5 K).

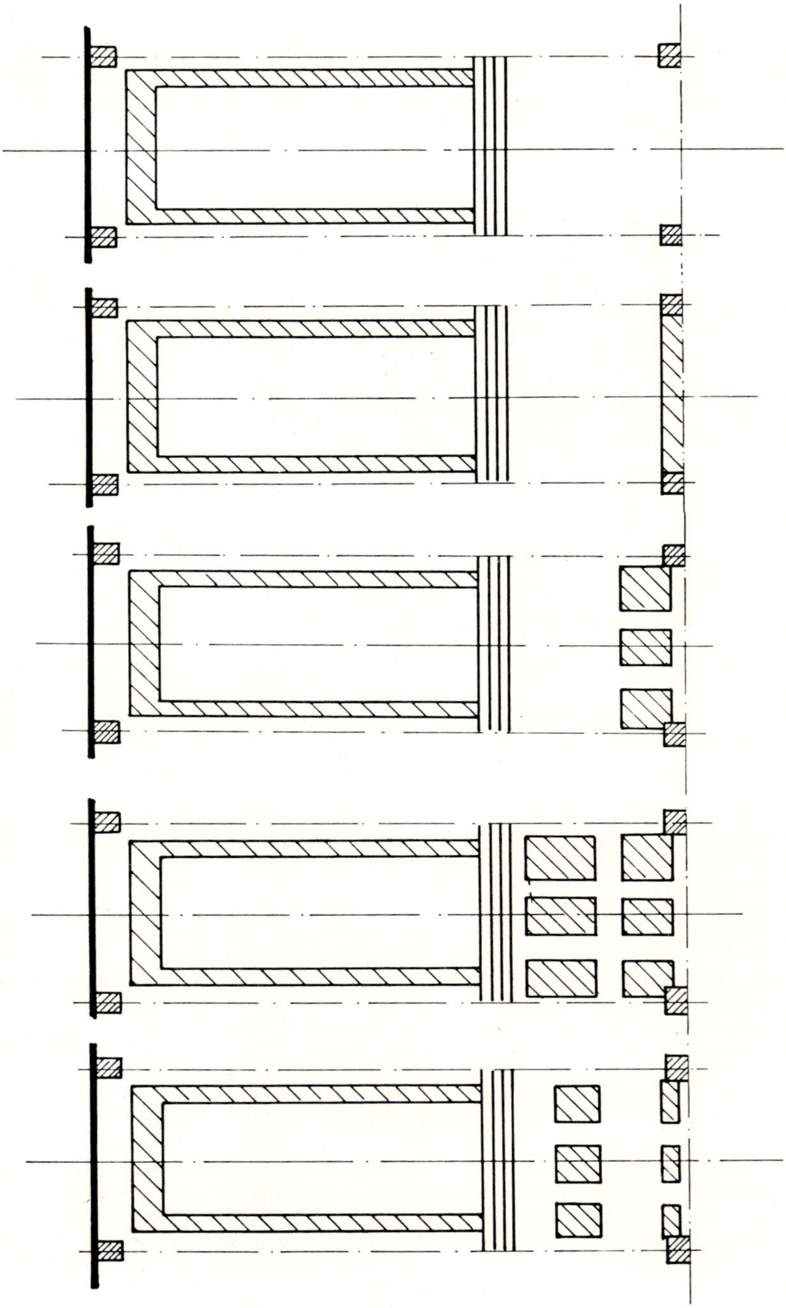

Fig 5 Configurations of floor openings tested in model

Different configurations of floor openings were tried out, - a sample shown in fig. 5.

It was stated in the model test that as large openings as possible were recommended. Practical limitations caused by the building and operation of the pots, made compromises necessary. The chosen configuration of floor openings is shown in Fig. 6

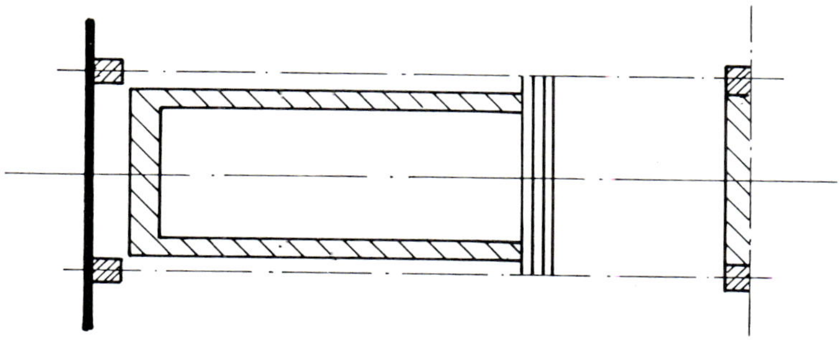

Fig 6 Chosen configuration of floor openings in Høyanger potroom

A major limitation was that the center area of the potroom is situated directly on the ground, which makes floorgrilles in this area impossible. We were however, able to install grilles in the middle, between the concrete pillars, and with air supply to this area with axial fans from the general cellar area (1/3 of total quantity). We have since realized that from a ventilation point of view, there should have been floorgrilles also in the rest of the center area, with a full size cellar for air distribution.

18 axial fans operated through a frequency transformer were installed on the roof, with air-silencers on both sides. Each fan can deliver from 0 to 220 000 m^3/h (61 m^3/s). During the winter season, the amount of air is reduced to obtain comfortable temperature in the potroom.

Effects from wind and weather are reduced by adjustable grids in the cellar walls, regulated according to wind pressure on the wall. In this way we secure that under almost any wind conditions we have the same amount of ventilation air flowing into the cellar on both sides of the potroom. In addition to the ventilation air, gas is drawn off from the pots to a gas-cleaning system. Air quantity for this purpose is taken in together with air for ventilation, and enters the pots through slits in the pothooding.

Experience

Up to now, no complete comparison has been made of the results achieved by segregated ventilation compared with systems we have previously used. This is partly because the starting up of all the pots in potroom A, Høyanger, was delayed on account of the poor market for aluminium in recent years. Another reason is to be found in that conditions are so different in the potrooms the ÅSV Group is now operating.

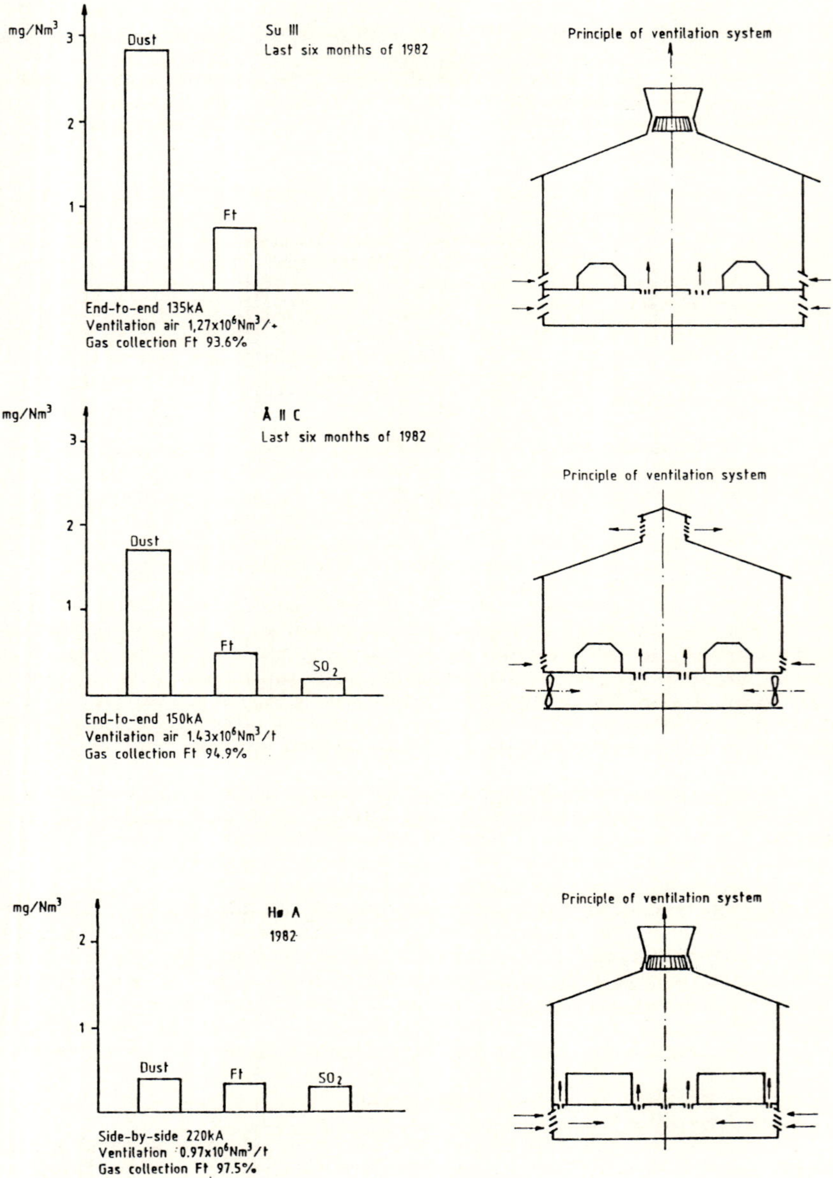

Fig 7 Concentrations of dust, Ft and SO_2 in three different smelters

ÅSV has six potrooms in which the pots are hooded prebakes, and they date back respectively to 1968, 71, 78 and 81. These pots are of somewhat different design, and gas collection efficiency varies on account of different routines for feeding alumina and changing anodes.

Ideally, we would like to have compared two potrooms with the same types of pot and the same collection efficiency for the gas, the one having traditional ventilation whilst the other with so-called ventilation by segregated layers.

In spite of the fact that we have not yet had fully adequate experience, we have endeavoured to collect some typical figures for the working atmosphere in three of the potlines. These figures are taken from the monthly reports on the emission of pollutants and concentrations in the working atmosphere. These measurements are not taken with identical measurement methods. Some of the measurements were taken with stationary equipment, whilst for others, use was made of equipment carried by personnel. The reason for this is that these measurements are primarily used to check developments at the individual smelters, and not for comparisons between the smelters.

Fig. 7 shows the concentration of dust, Ft and SO_2 in graphic form for each smelter. The figures do not refer to the same periods, but are typical average figures. As may be seen from the figures, the ventilation system for Su III and Å II C are essentially the same. In the one case, however, the fans are placed in the roof, whilst in the other the air is drawn through grilles along the bottom of the walls. In potroom A, Høyanger, all the air enters through the cellar.

We have no SO_2 figures for Su III, but for dust and Ft, the concentrations are about the same for Su III and Å II C. Potroom A at Høyanger, however, shows only about one half of the concentrations of Ft and SO_2, and a far lower concentration of dust. In spite of the difference in gas collection efficiency, this shows that a small part of the pollutants which escape from the pots flows into the operators' zone.

The figures for potroom A, Høyanger, refer to 1982, when this potroom was partly in the process of being started up, with the inconveniences this involves.

For comparison, we may mention that the authorities in Norway have laid down the following limits for concentrations in potrooms:

Fluorides	2.5 mg/m^3
Hydrogen fluoride	2 "
Sulphur dioxide	5 "
Dust	5 "
PAH	0.04 "

As may be seen, we are well below these figures, and so far we have only heard positive comments on the working atmosphere.

CHAIRMEN

E.L. Cambridge
ARCO Metals Company
Tucson, Arizona

Kai Grjotheim
University of Oslo
Oslo, Norway

THERMODYNAMICS OF METAL OXIDE REACTIONS IN HALL CELLS

Aaron J. Becker and Don R. Careatti
Aluminum Company of America
Alcoa Technical Center
Alcoa Center, Pennsylvania 15069 USA

Thermodynamic calculations have been used to predict the chemical reactions of silica, titania, and iron oxide with components of the Hall cell. The calculations predict that silica will react with aluminum to produce silicon and alumina, titania will react to form titanium carbide and carbon monoxide, and iron oxide will combine with aluminum to form iron metal and aluminum oxide. In the absence of carbon and aluminum, reaction products depend on aluminum fluoride, cryolite and alumina contents of bath. For example, the calculations suggest that in a Hall bath simulating 1.15 ratio, and aluminum fluoride activity of unity, silica and iron oxide undergo double displacement reactions with AlF_3 to form volatile SiF_4 and FeF_3. At a ratio of 1.5 and a saturated alumina level, no volatile species form, but silica consume cryolite to form nepheline resulting in an increase in bath acidity.

INTERACTIONS OF GASES IN MOLTEN SALTS: CARBON DIOXIDE AND OXYGEN IN CRYOLITE-ALUMINA MELTS

H. Numata

and

J. O'M. Bockris

Department of Chemistry
Texas A&M University
College Station, Texas 77843

ABSTRACT

A radio-tracer method has been developed to measure the solubility of CO_2 in cryolite-related systems at temperatures above 1000°C and also a method to measure the solubility of oxygen.

The high solubility of CO_2 in "cryolite"*, and the highly negative ΔH of solution, suggests complex formation. In the presence of Al_2O_3, the solubility of CO_2 is several times greater than in "cryolite." In contrast, oxygen shows a solubility about a thousandth that of CO_2.

The entropy of CO_2 in "cryolite" indicates a loss of translational degrees of freedom in a manner consistent with complexing. However, in the presence of increased Al_2O_3 (in spite of the increase of solubility) the entropy change upon solution appears to indicate little loss of translational freedom.

It is suggested that CO_2 is complexed in both systems, but more strongly in the "cryolite" system. Complexing here, however, produces the small carbonate (CO_3^{--}) ion, and this compensates the loss of translational freedom arising from the CO_2 complexing with alumina-fluorides.

Measured solubilities for CO_2 were 1.0×10^{-3} moles/liter atm. in "cryolite" at 1000°C, 4.6×10^{-3} moles/liter atm. in $(NaF)_{2,3}\ AlF_3$ + 6.9 mol % Al_2O_3 at 960°C. At 957°C, oxygen showed a solubility of only 3.0×10^{-6} moles/liter atm. in the latter melt.

*This designation refers to a cryolite - 1.5 (weight %) Al_2O_3 system.

INTRODUCTION

The solubility of CO_2 and O_2 in cryolite and in cryolite-alumina have technological importance because of the possible reactions between these entities and aluminum in the Hall-Heroult Process. Thus,

$$2/3\ Al + CO_2 \rightarrow 1/3\ Al_2O_3 + CO$$

Dissolved O_2 may react similarly:

$$2/3\ Al + 1/2\ O_2 \rightarrow 1/3\ Al_2\ O_3$$

Either process would reduce Faradaic efficiency. Because the solubility of these gases is not well known, the situation cannot be quantitatively assessed.

Remarkably, the solubility of oxygen in cryolite seems not to have been measured. As to CO_2, an estimate was made by Forland et al. (1) on the basis of specific gravity determinations, but the first measurements were by Haupin (2) who measured CO_2 solubility in cryolite.

Bratland et al. (3) made also these measurements in cryolite and cryolite-alumina. However, the data of Haupin and Bratland et al. do not agree in systems which are able to be compared. For example, in pure cryolite at 1030°C, Haupin determined the solubility of CO_2 to be 0.10×10^{-6} moles cm^{-3}, but according to the corresponding measurements of Bratland et al. it was $1.17\ 10^{-6}$ moles cm^{-3}.

The theory of the solubility of gases has little predictive power. Thus, if ideality ($\Delta H_{soln}=0$) is assumed, values for CO_2 at 1000°C of around 10^{-9} moles/liter atm result. If the theory of regular solutions (4) is assumed, the values come to around the same value, if the volume of CO_2 used is that given by interatomic radii. The hole theory of molten salts gives realistic results (5) but the predictive power is not better than an order of magnitude and (6) an adjustable parameter has to be used to obtain values agreeing with those of experiment.

More information comes from surveys of the solubilities of gases in molten salts, e.g., that made by Field (7). Here, the solubility at 1000 K (moles/liter atm) varies from 10^{-6} for non-interacting gases through 10^{-4} to occasional highs of 10^{-3} for reacting gases. This is, therefore, the range of orders we may expect for the solubility of CO_2 in cryolite but it is clear that the precise information needed, e.g., to distinguish modes of current-efficiency loss, can only come from accurate experiments.

VARIETIES OF EXPERIMENTAL APPROACH

Haupin absorbed CO_2 in cryolite, using a nickel crucible to withdraw a sample from the melt. After quick freezing, he heated the sample to 500°C, and observed the amount of CO_2 evolved in a mass spectrometer.

Bratland et al. used a volumetric method. The system was evacuated and the volume of gas measured. The system was then filled with CO_2 and, after absorption of CO_2 in the melt had occurred, again evacuated. The difference in volume is due to the dissolution of the CO_2.

Both methods have drawbacks. In Haupin's method, it is likely that not all the CO_2 is desorbed on heating to 500°C. In Bratland's method, the evaporation of the melt during evacuation may cause problems. Further, consideration of the apparatus indicates that a change in volume of about 0.1 cm^3 had to be measured in a total system volume of about 100 cm^3.

In these circumstances, it seemed wise to develop a radio-tracer method. Gaseous CO_2 can be introduced into the melt and equilibrium attained. Then, the gas can be expelled by means of an inert gas and the attendant $C^{14}O_2$ measured by means of a scintillation counter.

In the measurements with O_2, no previous method has been reported. The method chosen depended upon the setting up of a concentration cell between O_2 in equilibrium with the melt and O_2 in equilibrium with Ni-NiO. The electrolyte was ZrO_2 (CaO stabilized)(8).

EXPERIMENTAL

1. General:

The Furnace used was from Lindberg (6Kw) with a temperature range of 1200°C. A section of the furnace, about 7 cm in length, could be maintained at ± 1°C. Arcing took place between the heating element and the stainless steel container of the crucible. But, this was prevented by means of an alumina liner.

The temperature was measured by means of a chromel-alumel thermocouple, using the standard given without independent calibration. The thermocouple, sheathed in inconel, was inserted in the melt for a short time, immediately after each measurement.

The crucible for the CO_2 measurements was made of graphite, A. T. J. grade, and having dimensions 3.56 cm in diameter, and 12.24 cm in length. For the O_2 measurements, the crucible was made out of BN (donated by ALCOA), and had dimensions of 3.05 cm in diameter and 8.89 cm in length.

The materials were cryolite (Fisher Co.), 98.5% pure, the main impurity being Al_2O_3. The moisture was removed by conventional drying. Al_2O_3 was AR grade and not further treated. Oxygen was purified through a molecular sieve. Argon was de-oxygenated while passing through a Ti column at 900°C.

$C^{14}O2$ was produced by syringing H_2SO_4 slowly onto $BaC^{14}C_3$ (New England Nuclear, NEC-009A).

The oxygen sensor was made using ZrO_2 tube (Nippon Kagaku Tokyo Ltd.). NiO and Ni powder of mesh (325) were ground for four hours, making a paste which was pressed into contact with a Ni wire. This electrode was placed against the inside of the ZrO_2 tube; and a Pt wire wound around the outside of the ZrO_2 tubing. Electrical contact was made to the Ni and Pt wires. The potential measured was in the region of -10 to 700 mV.

2. Carbon Dioxide:

At the beginning of a measurement, CO_2 was produced by dropping H_2SO_4 onto $BaC^{14}O_3$. The resulting $Ar-C^{14}O_2$ mixture was stored in the rubber balloon and when a known quantity was needed for the melt, it was measured out with a syringe and introduced into the melt, Fig. 1. A flow meter was used to monitor the amount of Ar being mixed with the CO_2 and this allowed an approximate control to be made upon the relative amounts of Ar and CO_2 being mixed. The precise amount and hence knowledge of the CO_2 in the mixture with Ar was taken by removing a known volume from the sampling system A, and introducing it into the scintillation cocktail.

The CO_2-Ar mixture was fed into the melting system and cycled around the circuit indicated by arrows, Fig. 1. A peristaltic pump (Fischer Co., flow rate 3 to 6500 cm^3/min.) was used to maintain the circulation for about 15 minutes. Two more additions of CO_2-Ar mixtures were necessary to saturate the melt with CO_2 at a given partial pressure (Fig. 2).

The time needed to saturate the melt was about 90 minutes. The saturation was indicated by a constant partial pressure of CO_2 in the gas (as indicated by the sampling system). The circulating system had a volume of 30 cm^3 and the vial of about 5 cm^3.

To compensate for changes of pressure when the sampling system removed some gas, a syringe was used to replace the amount removed until pressure was restored to its former value.

After the condition of equilibrium had been thus established, pure Ar was gently bubbled through the melt for a period of about one hour. The Ar escaping from the melt containing $C^{14}O_2$ was trapped into vials that contained the appropriate scintillation liquid. Measurements continued until the CO_2 was reduced to the background level. At this condition, the procedure was terminated.

One of the difficulties of the present method involves the dead space which is shown in Fig. 1. This space amounts to about 30 cm^3 and the volume of the melt was of the order of 100 cm^3. After the melt had been saturated with CO_2-Ar at a number of partial pressures, Ar was introduced through a tube in the bottom of the container (Fig. 1). The Ar thus introduced was allowed to flow around the system until the dead space was filled with the gas. The time of contact of Ar with the melt was short and soon after filling, the Ar was introduced for the bubbling procedure to force the dissolved CO_2 out of the melt.

A possibility which had to be examined was the exchange of $C^{14}O_2$ with the C^{12} of the crucible. This fact will be discussed in detail later.

3. <u>Oxygen:</u>

The melt was first saturated with oxygen for 90 minutes using the head shown in Fig. 3. Then, Ar gas was flushed over the dead space in the crucible for several minutes. At this point, the head ensemble was exchanged for the second head under N_2 atmosphere contained in the furnace. The second head, thus introduced (main section, Fig. 3) consisted of a bubbling tube and the ZrO_2 oxygen sensor. This was first positioned over the melt. Then, Ar was allowed to flow into the dead space removing any oxygen in the free space above the melt. Thereafter, this head was lowered and screwed tightly into the BN crucible, allowing Ar to bubble directly into the melt. The amount of oxygen being forced out of the melt was followed recording the potential at the ZrO_2 oxygen sensor.

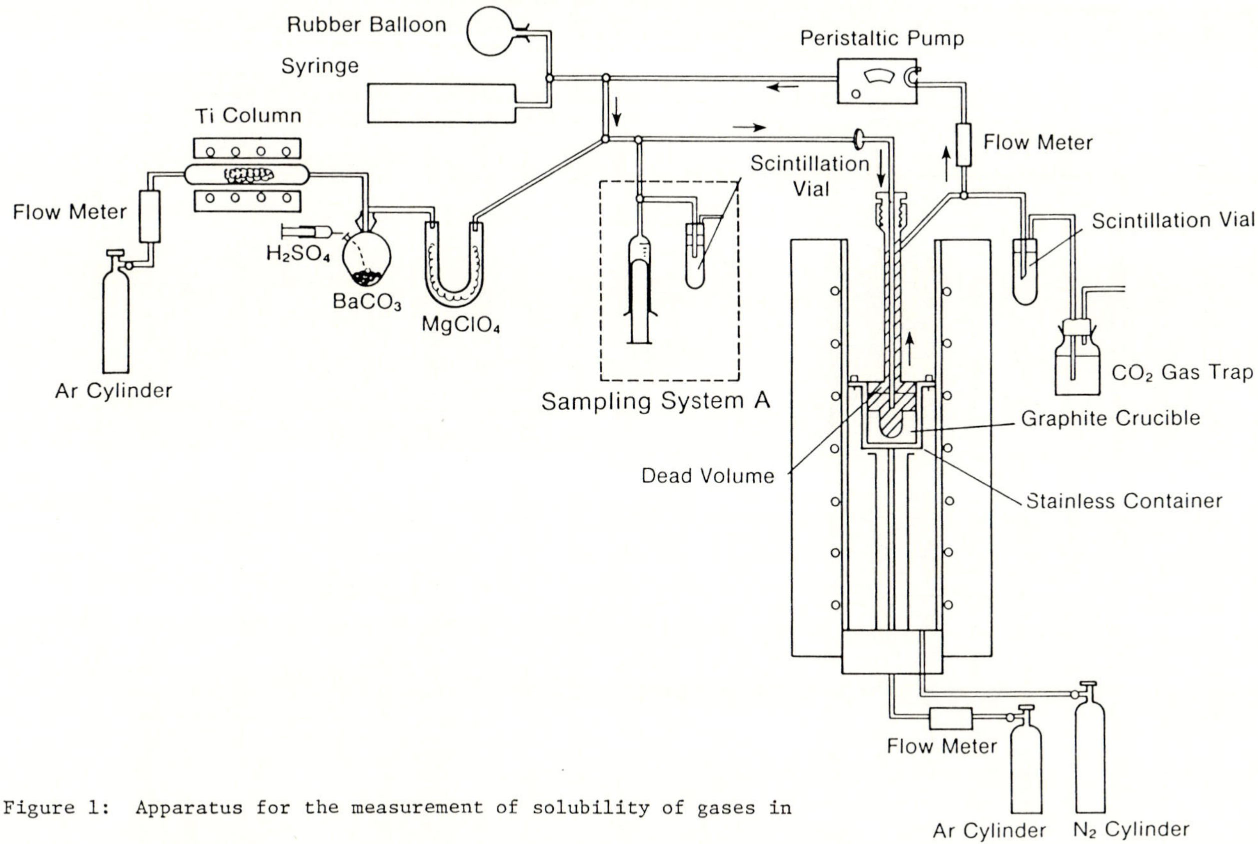

Figure 1: Apparatus for the measurement of solubility of gases in molten salt.

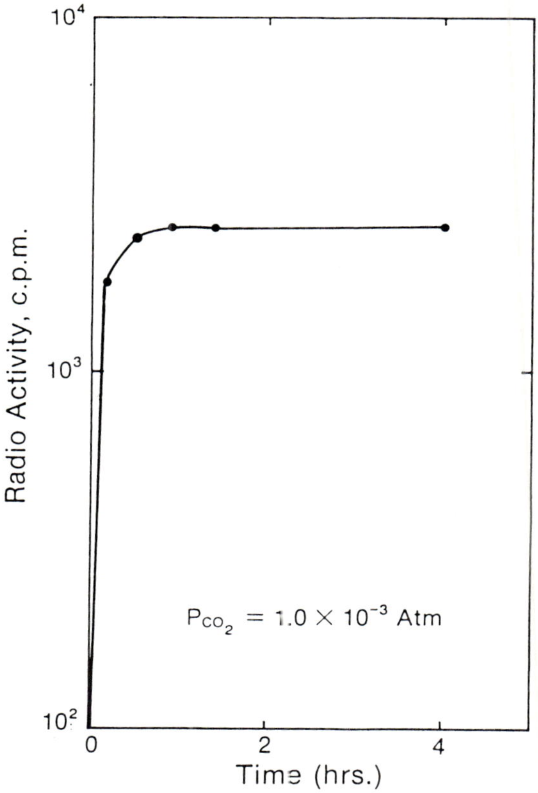

Figure 2: Change of radioactivity of the circulating gas due to the saturation equilibrium between CO_2 gas and melt.

There were two experimental difficulties. After each experiment, the outlet SiO_2 tube, which contained the ZrO_2 sensor broke, due to differences of thermal expansion between SiO_2 and ZrO_2, and had to be cemented to the lid.

Another difficulty (W. E. Haupin)(9), is the contamination of the surface of the ZrO_2 by the volatile "cryolite" melt. This difficulty was reduced by putting an alumina-silica wool trap before the ZrO_2-oxygen sensor (Fig. 3).

The ZrO_2 oxygen sensor is composed of ZrO_2-CaO solid electrolyte and Ni and NiO mixture which is used as a reference, Fig. 4. The potential between the reference Ni-NiO electrode and the test electrode wound outside the ZrO_2 oxygen sensor is:

$$E = \frac{RT}{4F} \ln \frac{P_{O_2}}{P_{O_2}^{1}}$$

Here $P_{O_2}^{1}$ is the reference partial pressure of Ni-NiO determined by thermodynamic data (10) (See Appendix B). P_{O_2} is the oxygen partial pressure at the surface of the test electrode.

The device was tested by using known concentrations of oxygen and Ar in which the P_{O_2} was varied over the range 1.5×10^{-3} to 1.0 atm, obtaining good agreement with the expected values.

RESULTS

1. Test of the Tracer Determination of CO_2 Solubility:

The radio-tracer method described above for the measurement of the solubility of CO_2 was first tested using water at room temperature. A glass cell replaced the crucible described. The rest of the apparatus was as above.

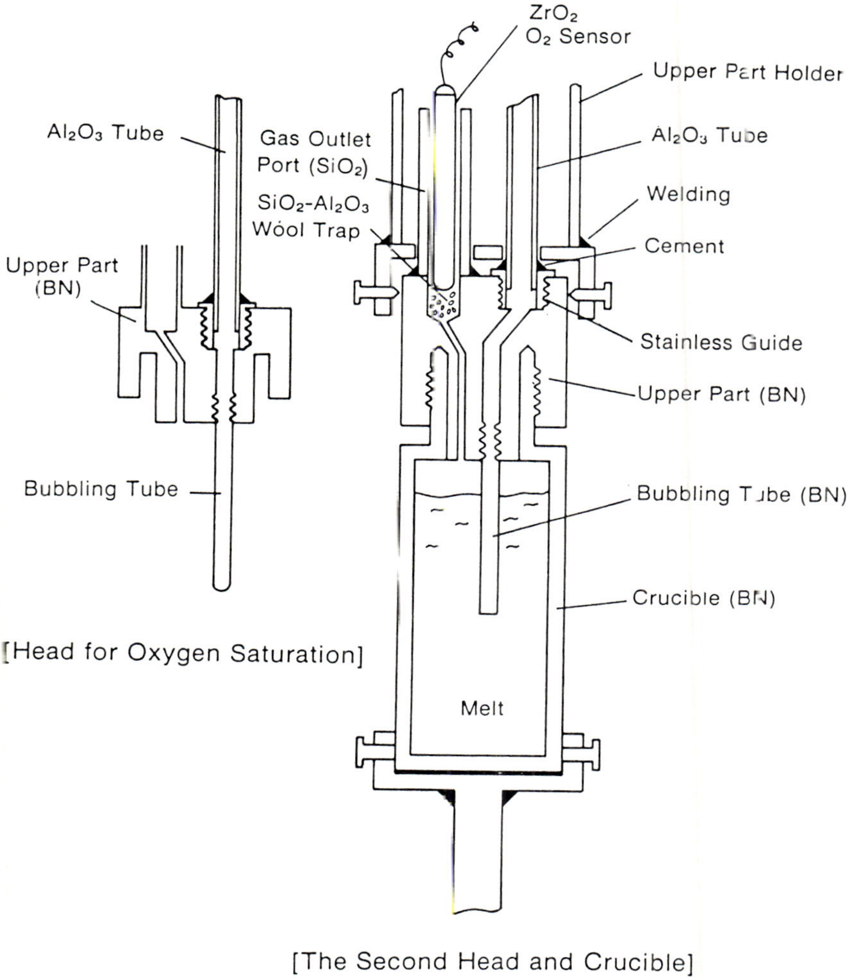

Figure 3: Measurement of oxygen partial pressure in contact with melt.

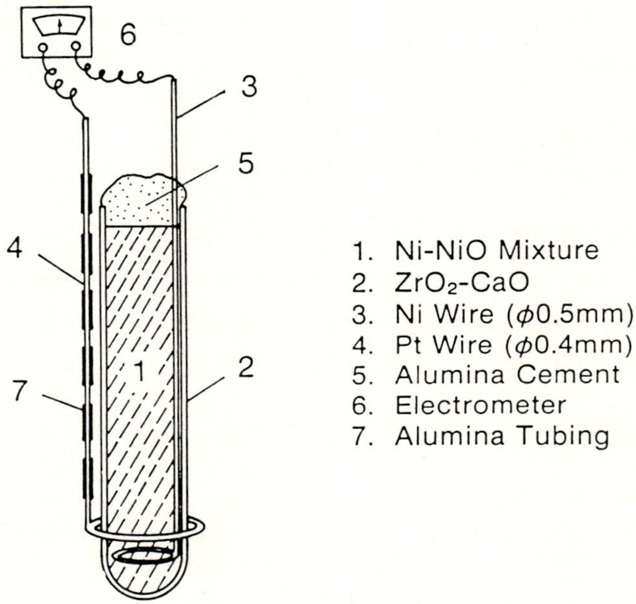

Figure 4: The structure of the ZrO_2 oxygen sensor.

After absorption equilibrium between $C^{14}O_2$ and water is reached, the dissolved $C^{14}O_2$ was forced out using Ar. This is thus absorbed in a scintillation cocktail. The activity of each vial determined in a Liquid Scintillation Counter (Model APL Packard Co.).

The number of counts in the vials decreased exponentially with the bubbling time down to the background level. The total number of counts (J cpm) was obtained by adding the individual activities in every vial. The calculation is shown in Appendix A.

2. CO_2 in Molten Salts:

(a) "Cryolite:"

Results for "cryolite" and "cryolite"-alumina are shown in Table I. In fact, the m.p. of 1000°C (compared with a m.p. of pure cryolite of 1009°C) shows that this "cryolite" contains about 1.5% of Al_2O_3 (see Grotheim et al.)(11,12).

Table . Solubility of CO_2 in Cryolite-Alumina Mixture.

Molten Salt System	Temp. K	Solubility mole/l Atm	Partial Pressure (Atm)
Na_3AlF_6 + 1.5 weight % Al_2O_3	1273	1.06×10^{-3}	9.85×10^{-4}
		0.98×10^{-3}	2.20×10^{-3}
		1.09×10^{-3}	8.68×10^{-4}
	1285	7.73×10^{-4}	1.04×10^{-3}
	1326	5.91×10^{-4}	7.12×10^{-4}
	1348	4.30×10^{-4}	8.74×10^{-4}
$(NaF)_{2.3}$ AlF_3 + 6.9 mole % Al_2O_3*	1253	4.54×10^{-3}	0.774×10^{-3}
	1285	3.98×10^{-3}	0.774×10^{-3}
	1317	3.24×10^{-3}	1.77×10^{-3}
	1366	2.84×10^{-3}	1.75×10^{-3}

*The calculation of mole % Al_2O_3 was made on the basis of $(NaF)_{2.3}$ AlF_3.

The results for "cryolite" are plotted as a function of temperature in Fig. 5. The comparison with Haupin's data is made at 1303 K: where Haupin obtained 1 x 10^{-4} mol/liter atm.

(b) Results in: $(NaF)_{2.3}$ AlF_3 + 6.9 mole % Al_2O_3:

This system was selected for study because it is greatly used in the Hall-Heroult process for Al production. The results are shown in Table I and Fig. 6.

It is possible to compare the results obtained by Bratland et al., in Na_3AlF_6 + 18.6 mole % Al_2O_3 with the present results (Fig. 6) finding a good agreement. The solubility of CO_2 in the Al_2O_3 containing system is 4-8 time higher than that in "cryolite".

(c) Heat of Solution:

This was calculated conventionally and the results are shown in Table II. The calculations of the entropy of solution is discussed below.

3. Oxygen:

The calculation of the moles/liter of O_2 in contact with the sensor during the flushing out of the melt is given in the Appendix B. A typical run pattern is shown in Fig. 8 where the amount of oxygen is shown as a function of time during purging of the molten salt with argon.

Accounting for the 30 cm^3 of the melt, the solubility of oxygen was 3 x 10^{-6} moles/liter atm. at 1230 K in $Na_{2.3}AlF_6$ + 6.9 mole % (4 wt %) Al_2O_3. This compares with 4.6 x 10^{-3} moles/liter atm. for CO_2 under similar conditions.

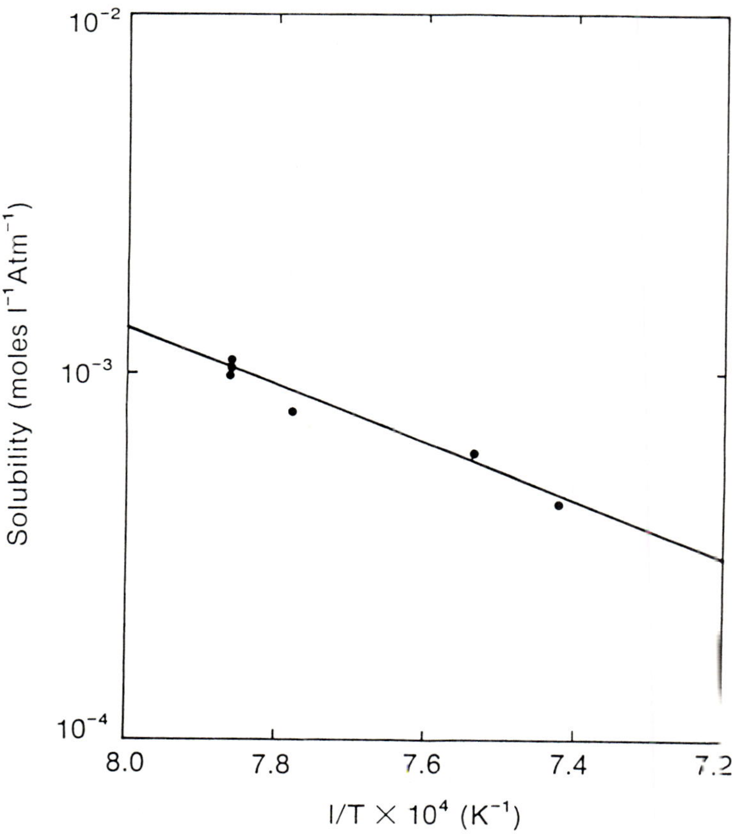

Figure 5: CO_2 solubility in Na_3AlF_6 + 1.5% weight Al_2O_3.

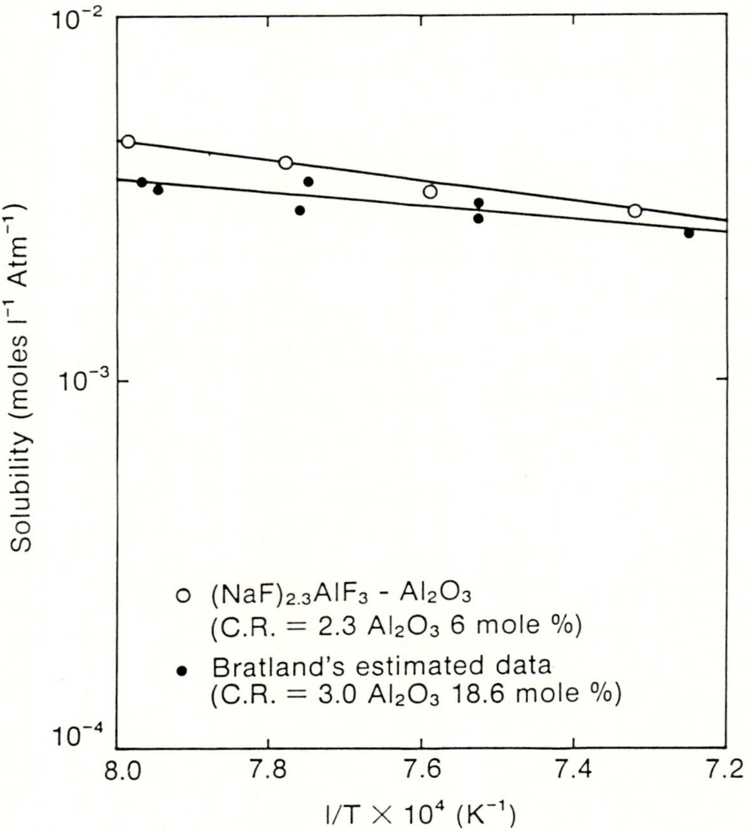

Figure 6: CO_2 solubility in $(NaF)_{2.3}$ AlF_3 + 6 mole % Al_2O_3; Bratlands data (C.R. = 3.0 Al_2O_3, 18.6 mole %), estimated.

TEST OF EXCHANGE REACTION IN SYSTEM

An experiment was carried out to ascertain whether there was any exchange reaction of finite velocity between $C^{14}O_2$ and carbon of the crucible. The apparatus is shown in Fig. 7.

A flow of $C^{14}O_2$ was arranged to circulate around the crucible (Fig. 7) for about 90 minutes, while the crucible was maintained at 1200 K.

After the saturation is reached, Ar was introduced from the tube at the bottom of the container and allowed to flow around the crucible, to flush out $C^{14}O_2$. Thereafter, oxygen was passed in from the bottom tube and allowed to flow around the container. The CO_2 from this experiment was passed into the scintillation system but no radio-active carbon was detected. This proves that no significant exchange reaction such as $C^{14}O_2 + C^{12}$ (graphite) $\rightarrow C^{14}O + C^{12}O$ occurs. Were it to occur significantly, it would increase the apparent solubility of the CO_2.

DISCUSSION

ENTROPY

In considering the solubility of a gas in liquid, one has:

$$\mu^{g}_{CO_2} = \mu^{\ell}_{CO_2}$$

From this, it is obvious that:

$$S^{o,1}_{CO_2} - S^{o,g}_{CO_2} = \frac{\Delta H_s}{T} + R \ln \frac{a^{\ell}_{CO_2}}{a^{g}_{CO_2}}$$

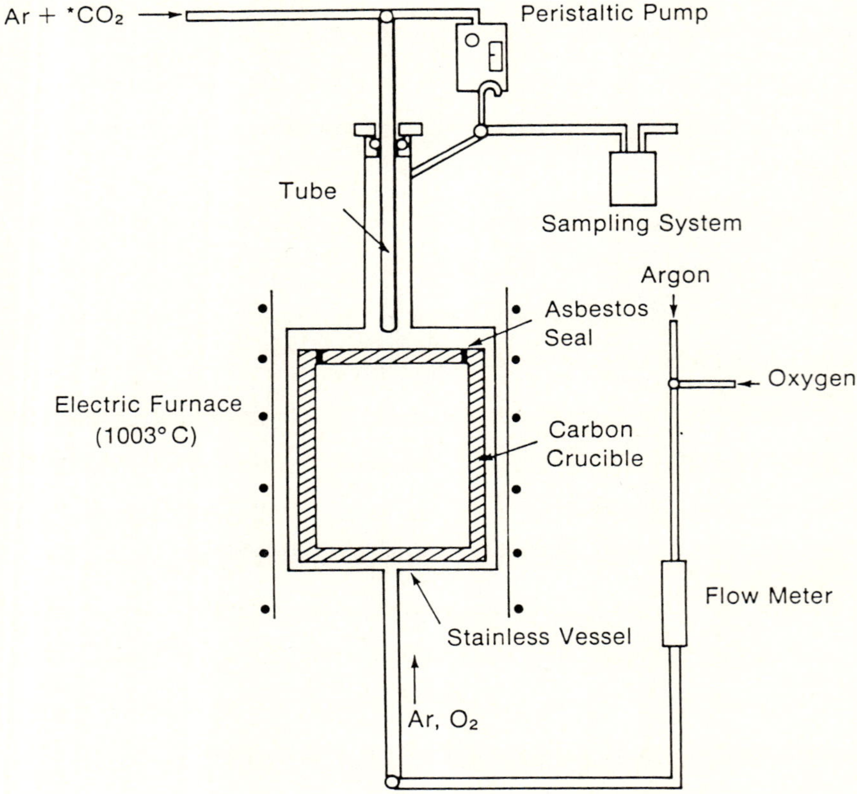

Figure 7: Apparatus for C (carbon) exchange.

According to Grimes (5), the standard entropy of solution of noble gases in molten fluoride mixture is near zero and thus the noble gas must have a full translational entropy when dissolved in molten salts.

In the case of CO_2, we can calculate the entropy in the gas phase by standard statistical mechanical equations (13) obtaining:

$$S_{1273° K} = 42.66 + 15.93 + S \text{ vibr.}$$

Now,

$$(S_{1273° K})^{gas}_{CO_2} = 64.15 \text{ Gibbs mol}^{-1} \quad (14)$$

Hence:

$$S_{vib} = 5.56 \text{ Gibbs. mol}^{-1}$$

Here, CO_2 in the gaseous phase and the CO_2 dissolved in the liquid are in equilibrium. $S^{o,1}_{CO_2}$ and $a^{1}_{CO_2}$ express the entropy of CO_2 gas in the liquid and the activity is unity due to the saturated condition; $S^{o,g}_{CO_2}$ and $a^{g}_{CO_2}$ express the entropy of the CO_2 gas in the gaseous phase mixture with Ar and $a^{1}_{CO_2}$ is the corresponding activity. The solubility measurement is referred to the standard state of $P_{CO_2} = 1$ atm, which is equal to $a^{1}_{CO_2}$. Hence, the entropy change upon solution, $\Delta S^{o}_{CO_2}$, is equal to the heat of solution divided by temperature:

$$\Delta S^{o}_{CO_2} = S^{o,1}_{CO_2} - S^{o,g}_{CO_2} = \Delta H_s/T \quad (4)$$

The standard entropy of solution includes also the entropy of mixing between the gas phase and the liquid, but it is easy to show that, per mole of liquid, the entropy of mixing in the present system is negligible.

Thermodynamically, the solution process of the gas in the liquid is divided into two steps:

$$CO_2\ (P = g) \xrightarrow{I} CO_2\ (P' = SR'T,\ g) \xrightarrow{II} CO_2\ (S,\ soln)$$

where step I involves the isothermal gas expansion from the pressure of the vapor P to P', where P' is the partial pressure where there is equal concentration of the CO_2 in the liquid and the gaseous phase. From the value of the solubility, S, in mol cm^{-3}, P' is calculated as follows:

$$P'V = nRT$$

$M = S$ moles. cm^{-3} $V = 1\ cm^3$ $R' = 82.05\ cm^{-3}$ atm. K^{-1} mol^{-1} thus,

$P' = SR'T$

Step II involves the solution process of the gas into the solvent at equal concentration.

The Gibbs free energy change, ΔG_T, is expressed as the sum of the Gibbs free energy change due to step I, ΔG_I, and to step II, G_{II}. Thus,

$$\Delta G_T = \Delta G_I + \Delta G_{II}$$

$$\Delta G_I = \Delta H_I - T\Delta S_I$$

$$\Delta G_{II} = \Delta H_{II} - T\Delta S_{II}$$

where ΔS_I is expressed as $-R \ln (P'/P_{equil}) = -R \ln (SR'T/P_{equil}) = -R \ln(K_p R'T)$ with $Kp = S/P_{equil}$, ΔS_{II} is the standard entropy of solution, and ΔH_{II} is equal to the heat of solution, ΔH_S. Now:

$$\Delta G_T = (\Delta H_I + \Delta H_{II}) - T(\Delta S_I + \Delta S_{II}) = 0$$

Hence,

$$\Delta H_s - T(-R \ln(KpR'T + \Delta S_{II}) = 0$$

or

$$\Delta S_{II} = \Delta H_s/T + R \ln(KpR'T)$$

In Table II, the entropies of solution in cryolite-alumina mixture (C.R. = 2.3), calculated from the present data, and that of Bratland, exhibit high negative values.

Thus, in cryolite,

$$S^{sol}_{CO_2} = -32.15 \text{ Gibbs. mol}^{-1}$$

Table II. The Standard Entropy and the Enthalpy of the Solution of CO_2 in "Cryolite" and Cryolite-Alumina Melts at 1000°C (1273 K).

Molten Salt	H_s	H_s/T^*	$R\ln(K_pRT^*)$	S_{II}
	K cal/mol	cal/deg mol	cal/deg mol	cal/deg mol
Na_3AlF_6	-35.2	-27.7	-4.41	-32.0
$Na_3AlF_6+AlF_3+Al_2O_3$ (CR**=2.3, Al_2O_3 6.9 mol %)	-13.3	-10.4	-1.72	-12.1
$Na_3AlF_6-Al_2O_3$ (CR**=3.0, Al_2O_3 18.6 mole %)	- 8.5	- 6.5	-2.2	- 8.7

* T = 1273 K
** CR=mole ratio NaF/AlF_3

INTERACTION IN SOLUTION

The value of the solubility of O_2 in $Na_{2.3}\ AlF_6$ + 6.9 mole % Al_2O_3 of $3x10^{-6}$ moles l^{-1} at is in the range expected for non-interacting gases.

Perusal of Table II shows that the largest negative heat of solution is for CO_2 in "cryolite." Thus, the interaction of CO_2 in "cryolite" is more than that in the presence of Al_2O_3.

Correspondingly, for CO_2 in "cryolite," the entropy of solution is highly negative, more so than for the solutions containing alumina and aluminum fluoride. This suggests that complexing is greater in the CO_2-"cryolite" system than in the system CO_2-Al_2O_3-cryolite, or that with added AlF_3. The affinity of Al for oxygen in "cryolite" is therefore greater than that of Al for oxygen in the other melts.

The work of Grimes et al. (5) established that noble gases showed an entropy of solution of around -1.0 Gibbs mol^{-1}. This means that, for noble gases, the molecular movements, and in particular translation (30-100 Gibbs mol^{-1} for many systems) remain essentially unchanged.

The entropies of CO_2 in solution are given in Table II below:

Table III.

System	Value of Entropy $S^{0,1}_{CO_2}$ (Cals. mole -1)
CO_2 in "cryolite" at 1273° K	32.15
CO_2 in "cryolite" + Al_2O_3 up to 1273° K	52.05
CO_2	64.15

Let it be assumed that CO_2 is sufficiently complexed in cryolite so that it no longer has its own degrees of translational freedom in solution. If we subtract the translational entropy, 42.66, Gibbs mol^{-1} from the totally unimpeded value, 64.15, Gibbs mol^{-1}, we get 21.49, and if it is assumed that the vibrational entropy passes from about 5 Gibbs. mol^{-1} in the gas to about 10 Gibbs. mol^{-1} (1273 K), upon solution, one can account for the 32 Gibbs. mol^{-1} observed for the entropy of CO_2 in solution.

A complex which would be consistent with this reasoning (i.e., the total removal of translational freedom) would be:

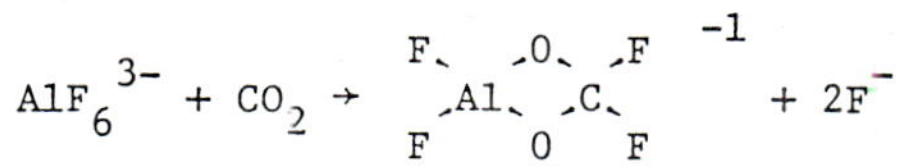

Perusal of Table III and reference of the gas phase value for the CO_2 entropy (64.15 Gibbs. mol^{-1}) shows that less loss of entropy occurs in the Al_2O_3-containing system upon the dissolution of CO_2 than in pure cryolite. The complexing which occurs here, removes less translational entropy than does that in pure cryolite. Nevertheless, something must occur to reduce the entropy by about 9 Gibbs. mol^{-1} from that in the gas phase.

Let it be supposed (5) that the complexing reaction is now:

$$AlOF_x^{1-x} + CO_2 = AlF_x^{3-x} + CO_3^{2-}$$

In this system, the CO_3^{2-} is a relatively small ion. On the basis of Grimes' result for the entropy of noble gases, it should have much translational entropy. On this model the translational entropy would be maintained. One might expect the vibrational entropy in the products to be less than that of the more complexed molecule, say 7 Gibbs. mol^{-1}, and the rotation to be extinguished. The resulting entropy would be around 64.15 - 15.93 Gibbs. mol^{-1} (the rotational entropy) + 2.5 Gibbs. mol^{-1} (the assumed increase in the vibrational entropy) =

50.71 Gibbs. mol^{-1} a reasonable approach to the observed values for CO_2 in Al_2O_3-containing liquids.

The model gives a qualitative and (speculative) quasi-quantitative account of the greater loss of entropy of CO_2 in "cryolite" on the basis that there is a loss of translation in the complexing there which is compensated by means of the translation of the CO_3^- ion, resulting from the preferred complexing of CO_2 with alumino fluoride anions.

SIGNIFICANCE OF RESULTS FOR ALUMINUM TECHNOLOGY

The solubility range determined for CO_2 in all the mixtures was of the same order (10^{-3} moles l^{-1} at^{-1}) as determined earlier by Bratland (3) but in contrast to the results of Haupin (2). Hence, the possible efficiency loss by the reaction of dissolved Al with CO_2 can indeed be estimated as before.

The O_2 solubility in cryolite-alumina melts is much (c. 1000) times less than that of CO_2. Hence, were the aluminum industry to go over to oxygen evolving dimensionally stable anodes, the presence of O_2 in the melt would be unlikely to cause a reduction of efficiency.

ACKNOWLEDGMENT

The authors are grateful to the Aluminum Company of America for financial support. Many thanks, in particular, are due to Mr. Warren E. Haupin for his supportive interaction and for knowledgeable discussion and advice. Also many thanks to Dr. S. L. Lee of Texas A&M University for his help with the scintillation counting.

APPENDIX A

The total measured $C^{14}O_2$ activity in the solution is J c.p.m.

In terms of disintegration per sec. this is $\frac{J}{0.82 \times 60}$ where 0.82 is the efficiency of the scintillation counting employed in this measurement.* 1 Ci corresponds to 3.7×10^{10} (d.p.s.). Hence,

$$C^{14}O_2 \text{ activity} = \frac{\frac{J}{0.82 \times 60}}{3.7 \times 1010} \text{ Ci}$$

The moles of $C^{14}O_2$ displaced from the water are calculated by dividing the above value by the specific activity. The amount of

$$CO_2 \text{ displaced from the water (moles)} = \frac{\frac{J}{0.82 \times 60}}{3.7 \times 10^{10} \times S}$$

The specific activity S is obtained from the New England Nuclear Co. and was 4.45 Ci/mole, the result being 2.00×10^{-2} mole/l with P_{CO_2} = 1 atm.

T. J. Morrison and F. Billett (15) obtained 2.0×10^{-2} mole/l at 25°C.

*This figure results from measurements of the efficiency of counting with ethanolomine carried out by S. I. Lee.

APPENDIX B

The number of moles of oxygen was calculated from the potential difference E of a ZrO_2 sensor. ΔE is defined as:

$$\Delta E = \frac{RT}{4F} \ln \frac{P_{O_2}}{P_{O'_2}}$$

Here $P_{O'_2}$ is given by the equilibrium oxygen partial pressure due to Ni and NiO as follows:

$$Ni + 1/2\ O_2 \rightarrow NiO$$

$\Delta G°_T$ is (9):

$$\Delta G°_T = 56{,}101 - 20.31*(1273)$$

$$= 30{,}079 \text{ cal mol}^{-1}$$

$$\Delta G°_T = -RT\ 2.3 \log \frac{a_{NiO}}{a_{Ni} - P_{O'_2}^{1/2}}$$

Hence,

$$P_{O'_2} = 10^{-10.3269} \text{ atm.}$$

$$E = \frac{RT}{4F}\ 2.3 \log \frac{P_{O_2}}{10^{-10.3269}}$$

Hence,

$$P_{O_2} = 10^{\frac{4F}{2.303\ RT}} - 10.3269$$

The P_{O_2} value thus calculated is that of the O_2 in the Ar-O_2 flow from the melt during purging. The number of moles passing the sensor in a given time interval can be obtained from:

$$n_{O_2} = P_{O_2} \frac{\text{(flow rate)(time interval of observation)}}{1.987 \text{ (absolute temperature)}}$$

A plot of this quantity against time (Fig. 8) shows the period of rise and exhaustion. Its integral gives the total O_2 in the melt.

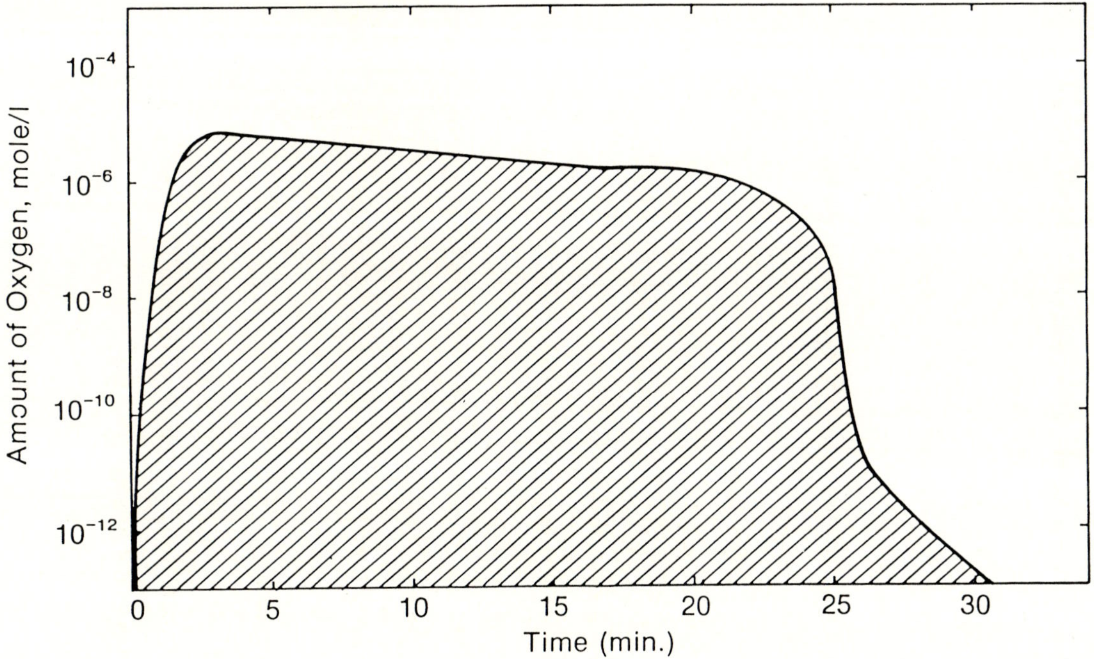

Figure 8: Change of oxygen concentration in gas escaping out of the melt: $(NaF)_{2.3}\ AlF_3$ + 6.9 mole % Al_2O_3 at 1230 K.

REFERENCES

(1) T. Forland, H. Storegraven, S. Urnes: Z. Anorg. Allgem. Chem., 279, 205 (1955).

(2) W. E. Haupin: J. Electrochem Soc., 107, 232 (1960).

(3) D. Bratland, K. Grjotheim, C. Krohn and K. Motzfeldt: J. of Metals, 19, 13 (1967), Oct.

(4) K. Shinoda, "Principles of Solution and Solubility," Marcel Dekker, Inc., New York, 1974.

(5) M. Blander, W. R. Grimes, N. V. Smith, and G. M. Watson: J. Phys. Chem., 63, 1164 (1959).

(6) H. H. Uhlig: J. Phys. Chem., 41, 1215 (1937).

(7) P. Field: "Gas Solubility in Molten Salts" in Adv. Molten Salt Chemistry, Vol. III, ed. by J. Braunstein, G. Mamantov, G. P. Smith, Plenum Press, New York, 1975.

(8) K. Kiukkola and C. Wagner: J. Electrochem. Soc., 104, 379 (1957).

(9) Personal Communication, W. E. Haupin (1982).

(10) O. Kubaschewski and C. B. Alcock: "Metallurgical Thermochemistry" 5th Ed., p. 382, Pergamon Press (1979).

(11) Grotheim et al., "Aluminum Electrolysis," Aluminum Verlag (1982), p. 29.

(12) Holm and Porter, Ibid., p. 28 (ALCOA).

(13) R. Badger, L. Woo, J. Am. Chem. Soc. 54, 3523 (1932).

(14) I. Barin and O. Knacke: "Thermodynamical Properties of Inorganic Substances," Berlin, Springer-Verlag, (1973).

(15) T. J. Morrison and F. Billet: J. Chem. Soc., 3819 (1952).

THE RATE DETERMINING STEP OF METAL LOSS IN CRYOLITE-ALUMINA MELTS

Qiu Zhuxian Fan Liman

The Northeast Institute of Technology,
Shenyang, People's Republic of China

Abstract

The aluminium solubility in cryolite-alumina melts was studied in closed crucibles, and the rate-determining step of metal loss was studied in industrial cells. According to the metal weight loss, the metal solubility is about 0.37%. Samples of liquid electrolyte were taken from industrial cells by means of steel pipes and were analyzed. The metal concentration in these samples was low, less than 0.007% (wt.), and there was no apparent concentration gradient in the bulk of the electrolyte. The diffusion layer was very thin. only 1-3 mm. The rate determining step of aluminium loss in industrial cells may be the dissolution-diffusion combination step.

Introduction

In aluminium electrolysis the dissolution and loss of metal cause a decrease in the current efficiency. To study its rate determining step means to find the way how to increase the current efficiency.

There are several consecutive steps for the dissolution and loss of metal in cryolite-alumina melts. As the cathodic layer is usually very thin, the diffusion step through this layer may be considered to be combined with the dissolution step. Hence there may be three steps:

1. Dissolution-diffusion step, i.e., dissolution of metal at the boundary interphase of metal-electrolyte, and diffusion of the dissolved metal species through the cathodic layer,
2. Transfer of the dissolved metal species through the bulk of the electrolyte,
3. Reoxidation of the dissolved metal species by the anode gases.

The overall reaction may be written as follows:

$$2M(diss.) + CO_2\ (g) = M_2O\ (diss.) + CO\ (g)$$

where M is the dissolved species formed during the dissolution of metal in molten electrolyte. The rate of the overall reaction is determined by the rate of the slowest step (i.e., the rate-determining step). To control this step would result in an effective decrease in metal loss.

Thonstad and Rolseth (3) had analyzed the different steps (totally 4 steps). Here, the changes in metal concentration in the above mentioned 3 steps are listed in Table 1.

Table 1

Changes in metal concentration in the three limiting cases in industrial cells

rate determining	in cathodic region	in the bulk of bath	in anodic region
first step	saturated or not saturated with metal	low concen., without metal gradient	drops down to zero
second step	saturated with metal	with metal gradient	drops down to zero
third step	saturated with metal	high concen., but without gradient	drops down quickly to zero

The dissolution and oxidation of metal are referring to the chemical kinetics, and the diffusion and transfer of the dissolved species in the bulk of electrolyte are dependent on the hydraulic conditions.

In the literature there are various viewpoints presented to the rate determining step, which may be listed briefly in Table 2. Most of the studies were carried out in laboratory cells, only a few of them were made in industrial cells.

It was well explained by Grjotheim et al (1) that the different conclusion concerning the rate determining step may be due to the differences in experimental conditions, and the convection in the melts has been reached in different ways, i.e., by means of a temperature gradient, by mechanical stirring or by gas bubbling. The dissolution of metal is the slow step in vigorously stirred melts, while the mass transfer is the slow step in quiet melts.

In this paper the metal solubility in the cryolite melts and the rate determining steps of metal dissolution and loss in the bath were studied and discussed.

Table 2

Some viewpoints presented to the rate determining step

authours	year	method of experiments	rate determining step
laboratory cells			
Abramov	1953	determine the metal loss in the bath with different depths	second step
Revazyan	1960	determine the metal loss of Cu-Al alloy in the bath	first step
Gjerstad and Welch	1964	determine the metal loss in the bath with different depths, by passing CO_2 over the bath surface	second step
Gerlach	1974	determine the rate of oxidation of metal by passing CO_2 over the bath surface	first step
Thonstad and Rolseth	1981	passing CO_2 over the cryolite melts in contact with metal	first step third step
Qiu and Fan (this paper)	1983	an electroanalytical method to assess the metal concen. in bath	second step
industrial cells			
Arthur	1974	determine the metal concen. at different zones of bath	second step
Thonstad and Rolseth	1976	an electroanalytical method to assess the metal concen. at different zones of bath	third step
Qiu and Fan (this paper)	1983	determine the metal concen. at different zones of bath	first step

Measurement of metal solubility in cryolite-alumina melts

The metal solubility in the cryolite-alumina melts was measured in closed graphite crucibles by weight difference. The electrolyte used was of the following compositions: cryolite (CR=2.7) + 5% Al_2O_3 . The total weight of the bath was 50 grams. The temperature was kept at 1000 C. The weight of metal used was 3.820 ± 0.030 grams. The results obtained were listed in Table 3 and plotted in Figure 1 as curve 2.

Within first three hours, the metal solubility increases with time. Afterwards it tends to be constant, about 0.37% (wt.). The value obtained is about three times as that obtained by Arthur (2) with the gas analysis method, whose data are also plotted in this figure as curve 1. These two curves are nearly parallel.

Table 3
Solubility of metal in the cryolite-alumina melts

time (hrs.)	first set (g)			second set (g)			metal solubility % (wt.)
	w_1	w_2	Δw	w_1	w_2	Δw	
0.25	3.842	3.816	0.026	3.842	3.834	0.008	0.034
0.5	3.836	3.769	0.067	--	--	--	0.134
1	3.850	3.750	0.100	3.850	3.760	0.090	0.190
1.5	3.841	3.710	0.131	3.841	3.692	0.149	0.280
2	3.803	3.586	0.217	3.803	3.623	0.180	0.397
2.5	3.792	3.606	0.186	3.792	3.606	0.186	0.372
3	3.843	3.675	0.168	3.843	3.683	0.160	0.328
4	3.860	3.671	0.189	3.860	3.693	0.167	0.356
4.5	3.795	3.631	0.164	3.795	3.632	0.163	0.327
5	3.830	3.621	0.209	3.830	3.615	0.215	0.424

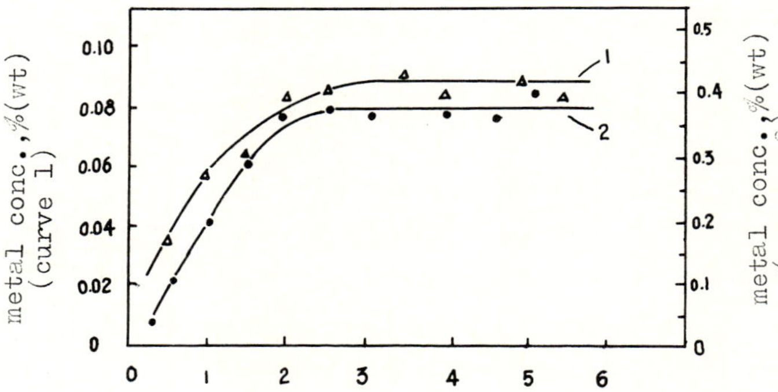

Figure 1. Solubility of aluminium in the cryolite-alumina melts:

Curve 1— pure cryolite, 1030°C, by gas analysis method (Arthur, 1974)

Curve 2— 95% cryolite (CR=2.7) + 5% Al_2O_3, 1000°C, by weight (this paper).

<u>Measurement of the limiting current in laboratory cells</u>

Measured the limiting current in laboratory cells with graphite electrode. The cell apparatus is shown in Figure 2. The ø6mm graphite electrode was capsuled with a boron nitride tube. The graphite crucible is ø50 and H60mm. The bath composition is as follow: cryolite (CR=2.6) + Al_2O_3 8%. The bath weight is 150 grams. The temperature was kept at 1000 C. In order to eliminate the influence of moisture content in the bath, a prelimimary electrolysis at 1 volt was carried out for 20 minutes. After standing for 30 minutes began the measurments of limiting current. The current-voltage curves were recorded with x-y recorder. Some typical curves are shown in Figure 3 and 4.

When the interpolar distance is less than 0.5 cm, the anodic limiting current becomes much lower. The limiting

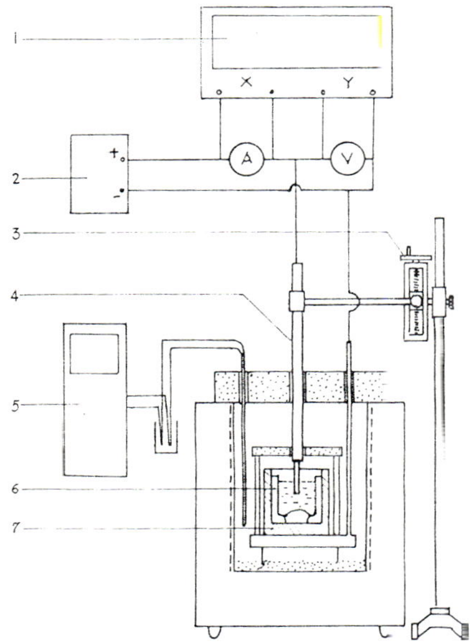

Figure 2. Arrangement for measuring the limiting current. 1-- x-y recorder, 2-- potentiostatic DC source, 3-- handle, 4-- graphite electrode, 5-- temperature controller, 6-- corundum crucible, ø=48mm, 7-- graphite crucible

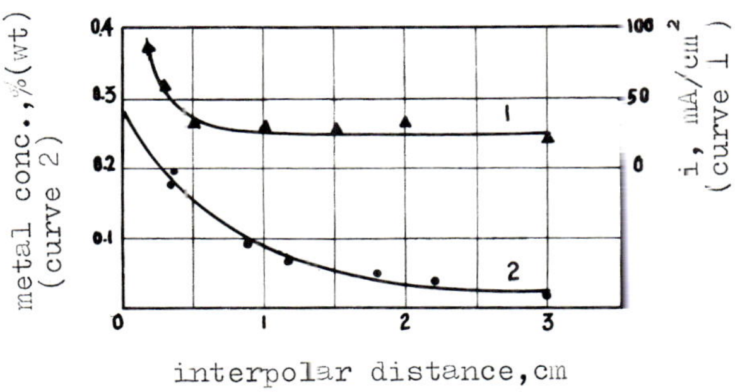

Figure 5. The limiting current densities (curve 1) and metal concentrations (curve 2) at different interpolar distances

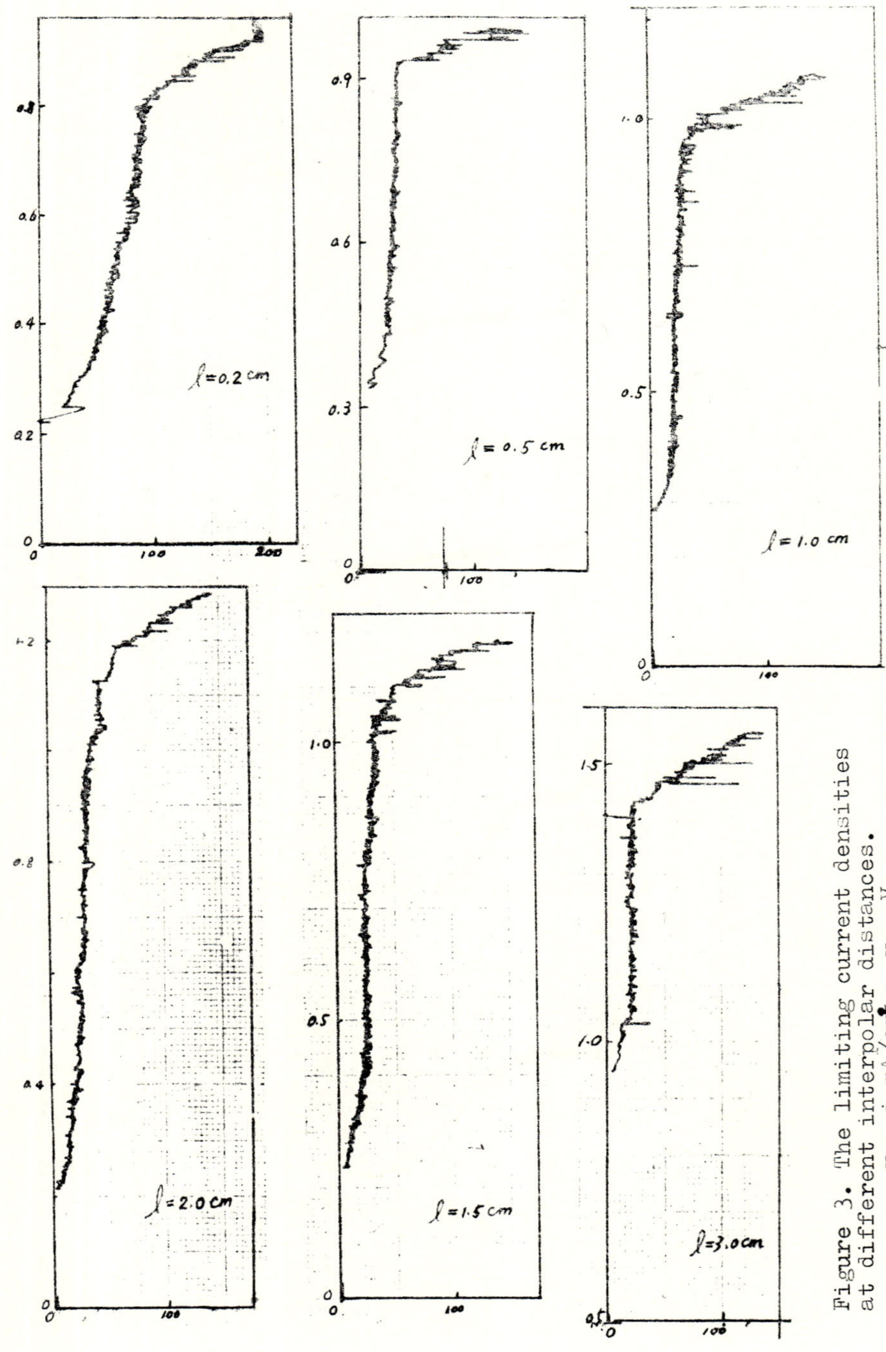

Figure 3. The limiting current densities at different interpolar distances.
x-- i,mA/cm^2, y-- V

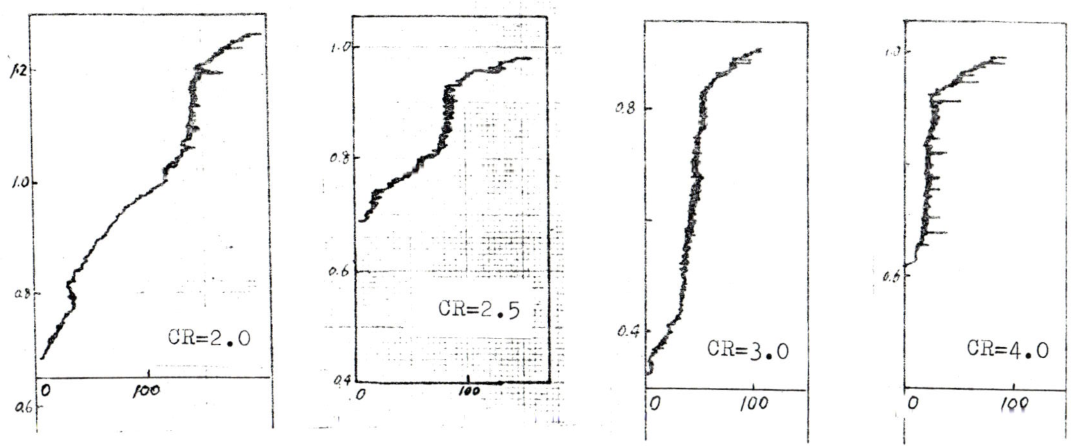

Figure 4. The limiting current densities at different cryolite ratios. x-- i, mA/cm^2, y-- V

current dereases with cryolite ratio. See Table 4 and 5.

Table 4
The limiting current and metal concentration at different interpolar distances

interpolar distance cm	0.2	0.3	0.5	1.0	1.5	2.0	3.0
i, mA/cm^2	85	60	35	30	30	35	25
metal concen. % (wt.)		0.19		0.07		0.05	

Table 5
The limiting current at different cryolite ratios

cryolite ratio	2.0	2.5	3.0	4.0
i, mA/cm^2	140	85	50	25

The bath samples were analyzed for metal content in four layers. The results of analysis are shown in Figure 5.

It is obvious that the higher aluminium concentration corresponds to the higher limiting current.

These phenomena tend to illustrate the mechanism of the dissolution of metal in cryolite-alumina melts: part of the aluminium forms monovlent aluminium ions. Because the limiting current increases with the lower cryolite ratio and with the lower interpolar distance.

At lower cryolite ratio, the monovalent Al^+ ions formed will be more, hence the limiting current would be higher.

When the interpolar distance is shorter, the graphite electrode would contact with more Al^+ ions, hence the limiting current becomes higher.

The above mentioned experimental results tend to show that in the small cells the rate determining step is the second one, because the metal concentration near the metal layer is saturated and there is an appreciable metal gradient in the bulk of bath.

Determination of metal concentration in industrial cells

From 6 industrial cells (1 soderberg cell and 5 prebaked cells) were taken more than 20 "pipes" of electrolyte at different parts.

The sampling equipment is self designed, which is a steel pipe with a diameter of ø 48/41 mm and a length of 500 mm (Figure 6). After sufficient preheating it was inserted into the bath till to the liquid metal layer. Let the molten electrolyte and metal come into the pipe without being stirred. Then a graphite plug was inserted tightly into the lower end of the pipe. After this, the steel pipe together with its contained liquid metal and electrolyte was taken out from the each half section took samples of electrolyte in cell and was cooled down rapidly in air. Then the steel pipe was cut longitudinally into two halves with a saw blade. From each half section took samples of electrolyte in 6 layers, as shown in Figure 7. The lower three layers were in the interpolar region, the orthers were above it. Each sample was analyzed for metal content (by gas analysis), alumina concentration (by chemical analysis), and cryolite ratio (by pyrotitration with NaF).

For the sake contrast, samples of electrolyte were taken from different parts of cell with an iron sampler in the shape of concentric cylinders, thus allowing the bath in the inner cylinder to be cooled down slowly. Then the inner samples were analyzed as mentioned above.

Some typical results of analysis for the "pipe" samples are listed in Table 6.

In the samples from the steel pipes, the metal concentration is very low, less than 0.007% (wt.), and the alumina concentration and cryolite ratio both in the upper and lower layers are without appreciable change. In the samples taken from the inner cylinder, the metal concentration is also very very low, less than 0.007% (wt.). Hence the metal concentration in the both kind of samples are all far from saturation (according to Arthur, the saturation concentration is about 0.01%, according to this paper, it is about 0.37% (wt.).

Table 6
Some typical results of analysis of bath samples

Items	metal % (wt)	alumina concen., %(wt.)				cryolite ratio			
No.of Sample	1-20	3	5	7	20	3	5	7	20
Layer									
1	0.007	3.22	--	--	5.o5	--	--	--	--
2	0.007	2.90	2.40	1.88	4.84	3.01	3.03	2.98	2.50
3	0.007	2.72	2.38	1.67	4.88	3.02	3.00	2.98	2.51
4	0.007	2.60	2.50	1.92	4.62	2.99	3.02	3.00	2.51
5	0.007	2.68	2.92	1.94	4.72	3.03	3.03	2.97	2.50
6	0.007	3.04	2.44	10.98	4.68	3.02	3.03	2.99	2.53

It was found that in the metal-bath interphase there is a brown layer of 1-3mm thickness, which is in parallel with the metal surface. The metal content in it is rather low, still far from the saturation limit, according to some preliminary tests.

In addition, in the lowest layer (the 6th layer) of the bath, small metal globules were found, the max. diameter is about 1-2 mm.

It is concluded that the metal concentration in the bulk of electrolyte in the industrial cells is low and that there is no appreciable gradient of metal concentration. Hence the rate determining step of metal dissolution and loss is the first step. See Figure 8.

This result is in contradiction with that obtained by Arthur (2), who found that there is an appreciable metal concentration in the bulk of electrolyte and its lower layer is saturated with metal. Therefore he concluded that the second step is the rate determining one.

This result is not in full agreement with that obtained by Thonstad and Rolseth (3). They found that there is no appreciable concentration gradient in the bulk of electrolyte, too, but they assumed that its metal concentration is not far from saturation. Therefore they suggested the possibility that the chemical reaction between CO_2 and dissolved metal constitutes the slow step should be considered.

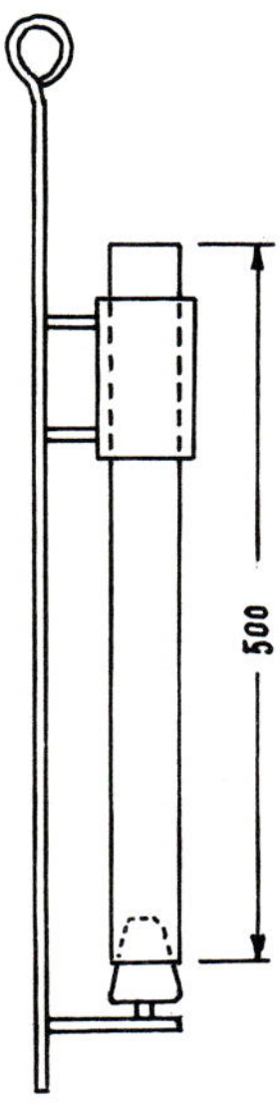

Figure 6. The pipe-type sampler for taking electrolyte from the industrial cells

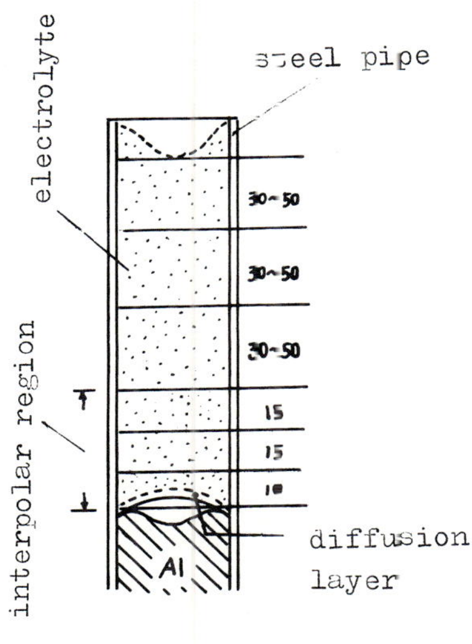

Figure 7. A half section of the "pipe" sample

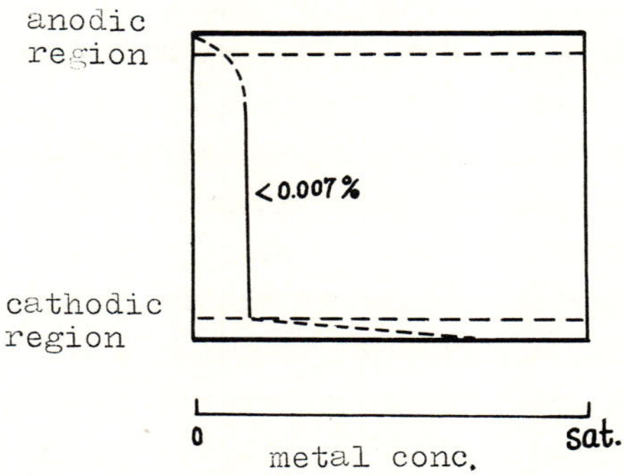

Figure 8. Schematic diagram of the rate determining step of metal loss in industrial cells

Discussion

1. As stated above, the first step is the rate determining step for the dissolution and loss of metal. Why ?

The mass transfer in the cell is rather fast due to various factors, such as the magnetic stirring, gas evolution, temperature gradient etc. And the interpolar distance is rather short. Some gasbubbles even come to the neighborhood of the metal layer. In laboratory closed conditions, it takes nearly 3 hours to achieve the saturation conditions. However in industrial cells this is hardly to be achieved. So the metal concentration in the cathodic region may be not saturated.

In addition the liquid metal layer in the electrolysis cell is under cathodic protection, the rate of its dissolution and loss is about 0.53×10^{-3} g/cm^2 -min. This is according to our determination in 57 industrial cells. This value is much less than that in laboratory cells without passing current through it.

2. In moldern electrolysis cells the current efficiency has

been raised to 90% or more. Why ? Because the industry has adopted some important measures to control the first step. These measures include:

(1) Improve the composition of the electrolyte, so that its surface tension at the boundary of metal is increased,

(2) Decrease the temperature of the electrolysis,

(3) Increase the stability of the metal,

(4) Improve the feeding system, to increase the stability of electrolysis and to prevent from forming much sludge on the cell bottom,

Items (1) and (2) are to decrease the rate of dissolution of metal in the bath, while (3) and (4) to weaken the diffusion of the dissolved metal species through the thin layer near the metal surface. In fact all these measures are very effective, especially the items (1) and (2).

As mentioned above there are three consecutive steps in the process of metal dissolution and loss. When the first step is to be controlled, one must consider the two other steps in the mean time. Because any enhancing of the two other steps would result in an increase in the rate of the first step. For example, more vigorously stirring the bath would cause an enhancing of the rate of mass transfer, in other words, it will increase the rate of dissolution of metal.

At given conditions the rate of all these steps is practically the same, which may be expressed in g/cm^2-min. However, the "ability" of metal dissolution and loss for all these steps is quite different. The rate determining step is of the least ability. Its ability determines the overall rate of dissolution and loss of metal in the bath.

Conclusion

1. From industrial aluminium electrolysis cells bath samples were taken with a pipe-type sampler. The advantage of this type sampler is obvious that the positions of samples are well defined, the bath is not disturbed during sample taking, the operation is more simple and more exact, and a continuous section of solid bath can be obtained.

2. The metal content in the bath samples is low, less than 0.007% (wt.), and there is no appreciable concentration gradient in the bulk of bath. A brown color diffusion layer of 1-3 mm was detected, which is just above the metal layer. Its metal constent is low. Hence the rate determining step for the metal loss in the industrial cells is the first step.

3. In laboratory, the limiting current and the dissolved metal concentration in the bath in contact with a layer of aluminium were studied. It is assumed the second step is the rate determining one. The main reason is that the bath is not so vigorously stirred as in industrial cells.

4. Under given conditions, the three consecutive steps are with the same rate, however the "ability" for metal dissolution and loss is quite different. The rate determining one is of the least "ability".

References

1. K.Grjothem, C.Krohn, M.Malinovsky, K.Matiasovsky, J.Thonstad, Aluminium Electrolysis, 1977.
2. A.M.Arthur, Light Metals 1974, 105-122.
3. J.Thonstad, S.Rolseth, Light Metals 1976, 171-182
4. J.Gerlach, METALL, 1974, No.3, 218-223.

MODEL CALCULATIONS OF THE ELECTRICAL CONDUCTIVITY OF CRYOLITE MELTS

P. Fellner[(1)], K. Grjotheim, and H. Kvande [(2)]

Department of Chemistry, University of Oslo,

Blindern, Oslo 3, Norway.

Three models are presented for calculation of the electrical conductivity of cryolite melts. The first model is based on the assumption that the conductivity is proportional to the number density of the effective electric charges in the melt. The two other models, the series and parallel models, assume that the charge - carrying species do not mutually influence each other. The conductivity then is the sum of the contributions of the pure melts placed in two conductivity cells connected in series and parallel, respectively.

The calculated data show good agreement with experimentally measured values for the electrical conductivity at various bath compositions used industrially. Special consideration is given to the influence of addition of magnesium fluoride to the melt.

Permanent addresses:

(1) Institute of Inorganic Chemistry, Center of Chemical Research, Slovak Academy of Sciences, 842 36 Bratislava, Czechoslovakia.

(2) Årdal og Sunndal Verk a.s., N-5875 Årdalstangen, Norway.

INTRODUCTION

In the industrial production of aluminium it is desirable to have bath compositions with high electrical conductivity, in order to reduce the ohmic voltage drop through the bath. This may contribute to a lower energy consumption by the process. However, in spite of numerous experimental measurements of the electrical conductivity of cryolite melts, the data vary considerably, mainly because of experimental difficulties, as have been discussed by Grjotheim *et al.* (1).

In addition to the need for more accurate experimental work, there is a lack of theoretical models for calculation of conductivity data. It is the purpose of this paper to present three theoretical models for calculation of electrical conductivity data in melts of interest to the aluminium industry. The validity of the models will be tested by comparison of the calculated data with experimentally measured values at various bath compositions.

MODELS

At present in the literature there is only one theoretical model that may be used to calculate the electrical conductivity of molten salt mixtures. Markov and Shumina (2) calculated the equivalent conductivity Λ_{mix} for a binary mixture from the equation

$$\Lambda_{mix} = x_1^2 \cdot \Lambda_1 + 2x_1x_2 \cdot \Lambda_1 + x_2^2 \cdot \Lambda_2 \qquad (1)$$

when $\Lambda_1 < \Lambda_2$. Here, x_1 and x_2 are the mol fractions, and Λ_1 and Λ_2 denote the equivalent conductivities of the pure components in the molten mixture.

In the following, three other theoretical models will be presented and discussed.

The "simpler" model

This model is based on the assumption that the electrical conductivity is proportional to the number density of the effective electric charges in the melt. In cryolite melts with large complex anions the current is carried mainly by cations. Thus, we may assume that the cations move by the influence of an electric field relative to the framework formed by the large anions.

Because the conductivity is proportional to the concentration of electric charges (3), we obtain as a first approximation the following equation for the electrical conductivity κ_{mix} of cryolite melts

$$\kappa_{mix} = A \frac{\Sigma(x_i \cdot z_i)}{V_{mix}} \tag{2}$$

Here, x_i is the mol fraction of component $\underline{i}$ in the melt, z_i is the effective charge of the current-carrying species, V_{mix} is the molar volume of the melt, and A is a proportionality factor that may be determined experimentally. It is expected that A remains constant when the anion framework is not significantly affected by the additions. The numerical value of the factor A may be determined from the conductivity of pure cryolite, extrapolated to 1273 K, since its melting point is 1284 K (1). This calculation gives $A = 93.5\ S\ cm^2\ mol^{-1}$.

Eqn. (2) requires knowledge of the values of z_i and V_{mix}. The latter may be calculated by the equation

$$V_{mix} = \Sigma(x_i \cdot V_i^o) \tag{3}$$

in which the volume parameter V_i^o is calculated from the expression

$$V_i^o = \frac{V_{mix} - x_{Na_3AlF_6} \cdot V^o_{Na_3AlF_6}}{1 - x_{Na_3AlF_6}} \tag{4}$$

The parameter V_i^o does not equal the partial molar volume of component _i_, and it has no exact thermodynamic meaning. However, it permits the calculation of the molar volumes of cryolite melts with sufficient accuracy by using a simple additive rule. Then, eqn. (2) may be written

$$\kappa_{mix} = 93.5 \frac{\Sigma (x_i \cdot z_i)}{\Sigma (x_i \cdot V_i^o)} \tag{5}$$

The series model

The series model is based on the assumption that the charge-carrying species do not mutually influence one another in an ideal mixture. This implies that the electrical conductivity may be calculated as the sum of the contributions of the pure molten components placed in separate conductivity cells connected in series. This model has been described in detail by Fellner (4), and only a brief outline is given here.

Two melts with conductivities κ_1 and κ_2 and volumes V_1 and V_2 are placed in conductivity cells with cross-sectional area A. The resistivities R_1 and R_2 are then proportional to the volumes of the melts

$$R_1 = \frac{V_1}{\kappa_1 \cdot A}, \quad R_2 = \frac{V_2}{\kappa_2 \cdot A} \tag{6}$$

It is implied that the current direction is perpendicular to the conducting walls of the cell.

The conductivity cells are then connected in series, and the total resistivity R_{total} is the sum of R_1 and R_2;

$$R_{total} = \frac{V_1}{\kappa_1 \cdot A} + \frac{V_2}{\kappa_2 \cdot A} = \frac{V_1 + V_2}{\kappa_{mix,s} \cdot A} \tag{7}$$

Here, $\kappa_{mix,s}$ denotes the conductivity of the system consisting of the two cells connected in series, and it is equal to the ideal conductivity of the mixture of the two melts. We may introduce the molar volumes V_1^o and V_2^o and the mol fractions x_1 and x_2 in eqn. (7) to obtain

$$\frac{x_1 \cdot V_1^o}{\kappa_1} + \frac{x_2 \cdot V_2^o}{\kappa_2} = \frac{x_1 \cdot V_1^o + x_2 \cdot V_2^o}{\kappa_{mix,s}} \tag{8}$$

Eqn. (8) may be readily extended for calculation of the conductivity of multi-component mixtures

$$\kappa_{mix,s} = \frac{\Sigma\,(x_i \cdot V_i^o)}{\Sigma\left(\frac{x_i \cdot V_i^o}{\kappa_i}\right)} \tag{9}$$

The series model can be shown to give a good description of systems that exhibit nearly ideal thermodynamic behaviour.

The parallel model

Figure 1 is a schematic illustration of the series and parallel models. The latter model, treated according to the procedure shown above, gives the following expression for the electrical conductivity $\kappa_{mix,p}$

$$\kappa_{mix,p} = \frac{\Sigma\,(x_i \cdot \kappa_i \cdot V_i^o)}{\Sigma\,(x_i \cdot V_i^o)} \tag{10}$$

The parallel model is a boundary model, and the electrical conductivity is always less than that calculated from this model.

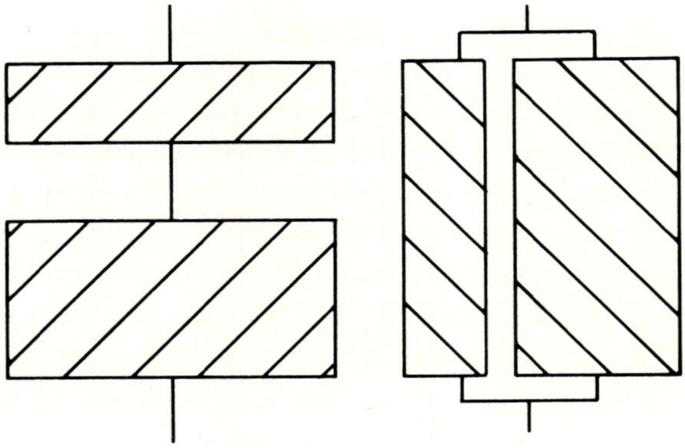

Fig. 1. Schematic illustration of the series and parallel models for calculation of the electrical conductivity of a molten binary mixture with volume fractions of the ratio 1:2.

a) The series model, b) The parallel model.

Table 1. The parameters used for calculation of the electrical conductivity of cryolite melts at 1273 K.

Substance	$V^o/cm^3 mol^{-1}$	$\kappa/S\ cm^{-1}$	z	Remarks
Na_3AlF_6	99.9	2.82	3	
Na_3AlF_6 (un-dissociated)	94.6	3.20		
$NaAlF_4$	77.3	1.70		
Li_3AlF_6	81.8	4.19	3	
NaCl	44.0	4.14	3	$x_{NaCl} < 0.2$
LiF	14.5	9.19	1	only with AlF_3
NaF	21.6	4.95	1	
CaF_2	26.5	1.80	0.45	
AlF_3	50.0		0.5	
MgF_2	22.0		0	
Al_2O_3	60.0		-0.5	

SYSTEMS

In order to test the validity of these three theoretical models, the calculated conductivity data are compared with experimental values from literature dealing with systems of importance to aluminium electrolysis. The necessary data for the molar volume and the electrical conductivity of the pure components are given in Table 1. The values of the effective charge z of the current-carrying species are also included.

$NaF-AlF_3$

This system is relatively complicated because the cryolite anion is partly dissociated according to the reaction:

$$AlF_6^{3-} = AlF_4^- + 2F^- \tag{11}$$

with an equilibrium constant $\underline{K}$ = 0.06 at 1281 K (1). Eqn. (11) is used for the calculation of the mol fractions of these three species. Conductivity data are taken from Table 1.

Figure 2 shows that there is good agreement between the experimental conductivity data and the values calculated using the series model. Because of the structural changes occurring in the melt, the theoretical dependence of conductivity on composition is rather complex. The parts of the plot in regions rich in NaF or AlF_3 are convex to the concentration axis, while the middle part is concave.

However, the experimental data in Figure 2 show considerable scatter, and the fit of the theoretical line may only be interpreted by the series model being able to predict conductivity data in this system with reasonable accuracy. More accurate experimental data are required in order to give direct support to the theory.

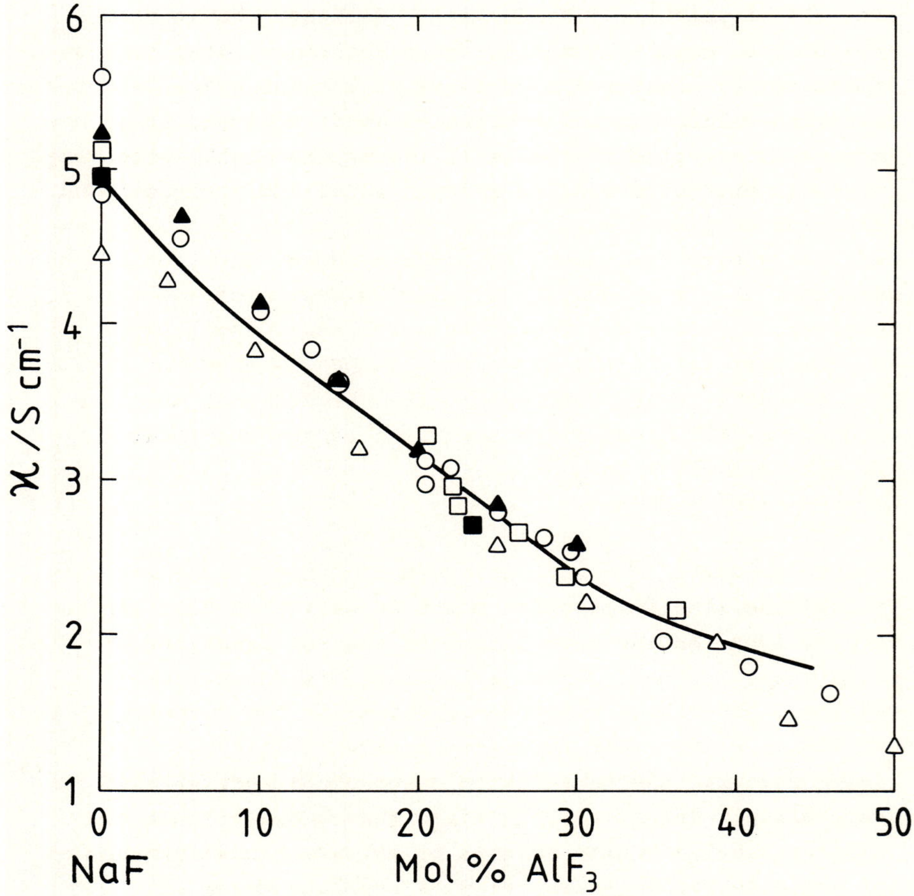

Fig. 2. Electrical conductivity of the system NaF - AlF_3 at 1273 K.

Experimental points: from Grjotheim *et al.* (1).
Solid line: calculated from the series model.

$Na_3AlF_6 - Al_2O_3$

For this system the simpler method was employed, with $z_{Al_2O_3} = -0.5$, as seen in Table 1. One might expect that this value would be zero, since the addition of alumina brings no current-carrying cations into the cryolite melt. However, it is generally accepted that alumina reacts chemically with molten cryolite and forms various anionic aluminium - oxygen - fluorine complexes (1). The concentration of these complexes will vary with the amount of alumina added. Thus, it may be expected that the addition of alumina will decrease the electrical conductivity more than corresponding to $z_{Al_2O_3} = 0$. The best fit to the experimental data is shown in Fig. 3.

$Na_3AlF_6 - Al_2O_3 - CaF_2$

The value of $z_{CaF_2} = 0.45$ is given in Table 1. The mobility of calcium ions in cryolite melts is less than that of the sodium ions, and a value of z less than unity may be expected.

Figure 4 gives data for three sections in the ternary system $Na_3AlF_6 - Al_2O_3 - CaF_2$. It may be shown that the agreement with the experimental data of Belyaev (5) and Yim and Feinleib (6) is satisfactory. This suggests that the simpler model may also be used to calculate reliable values of the electrical conductivity of cryolite melts with additions of CaF_2, for concentration regions where no experimental data are available.

$Na_3AlF_6 - Li_3AlF_6$

Figure 3 shows that good agreement between experimental and calculated conductivities in the cryolite - rich part of this system is obtained, when $z_{Li_3AlF_6} = 3$ (see Table 1). Thus,

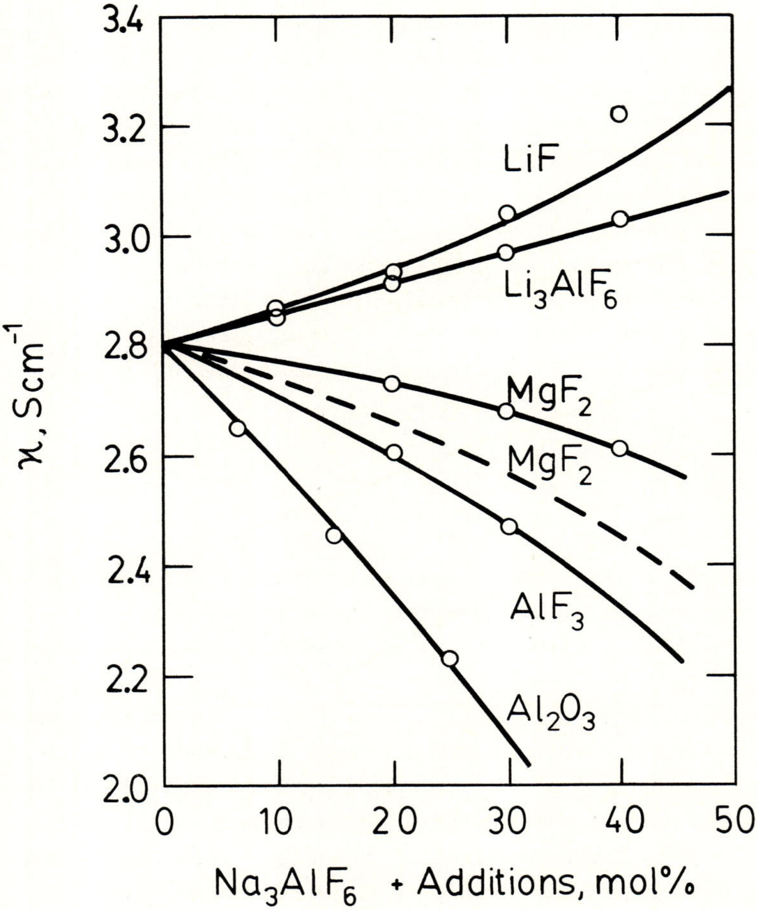

Fig. 3. Comparison of literature data for the electrical conductivity of binary cryolite melts at 1273 K, with data calculated from the simpler model.

Open circles: experimental data, see Grjotheim et al. (1).
Solid line : calculated data.
Dashed line : calculated data, by use of the theoretically estimated parameter $z_{MgF_2} = 0$.

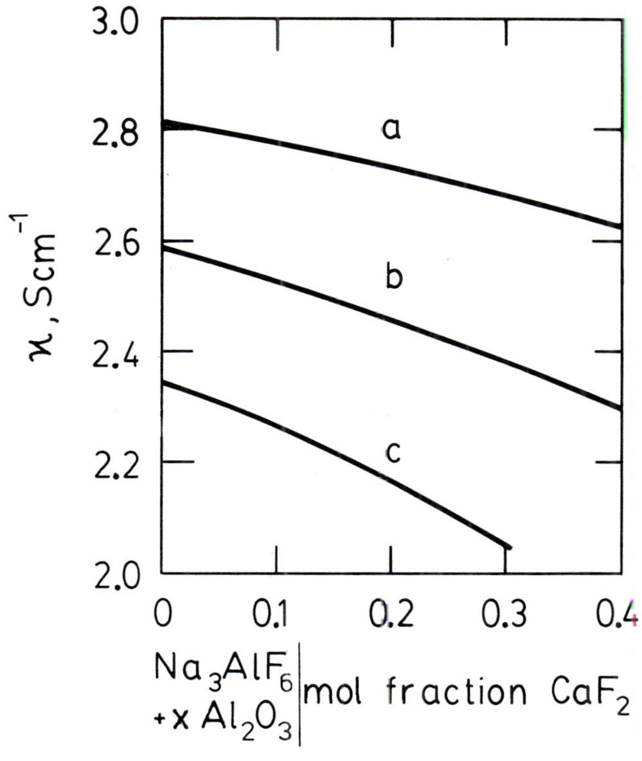

Fig. 4. Electrical conductivity for three sections in the ternary system Na_3AlF_6 - Al_2O_3 - CaF_2 at 1273 K, calculated by the simpler model.

a: x = 0 mol% Al_2O_3.

b: x = 10 mol% Al_2O_3.

c: x = 20 mol% Al_2O_3.

this value of the parameter corresponds to the number of lithium ions formed for each molecule of Li_3AlF_6 dissolved in the melt.

In Figure 5 are plotted conductivity data for the system Na_3AlF_6 - Li_3AlF_6 at 1273 K. The experimental results for this system compare quite well with the series model, and it may be shown that the deviation is less than 2 %. This means that the system behaves practically ideally as regards the transport of electric current through the melt, and it implies that no fundamental changes occur in the structure of the melt when Na_3AlF_6 and Li_3AlF_6 are mixed. The parallel model does not give such good agreement, as shown in Figure 5.

Na_3AlF_6 - LiF

This system is more complex. Figure 3 shows that the simpler model gives results that deviate considerably from the experimental values for molten cryolite containing more than 30 mol% (= 5 wt%) LiF.

Figure 6 gives conductivity data for the system Na_3AlF_6 - LiF. The experimental data fall between the predicted values for the two models, being somewhat nearer to the series model on the Na_3AlF_6 side of the diagram, but nearer to the parallel model on the LiF side. Thus, at high LiF contents the conductivity approaches the behaviour represented by the parallel model.

However, from a practical point of view, the series model is considered quite satisfactory, since the LiF content in the bath does not usually exceed 50 mol% (= 11 wt%).

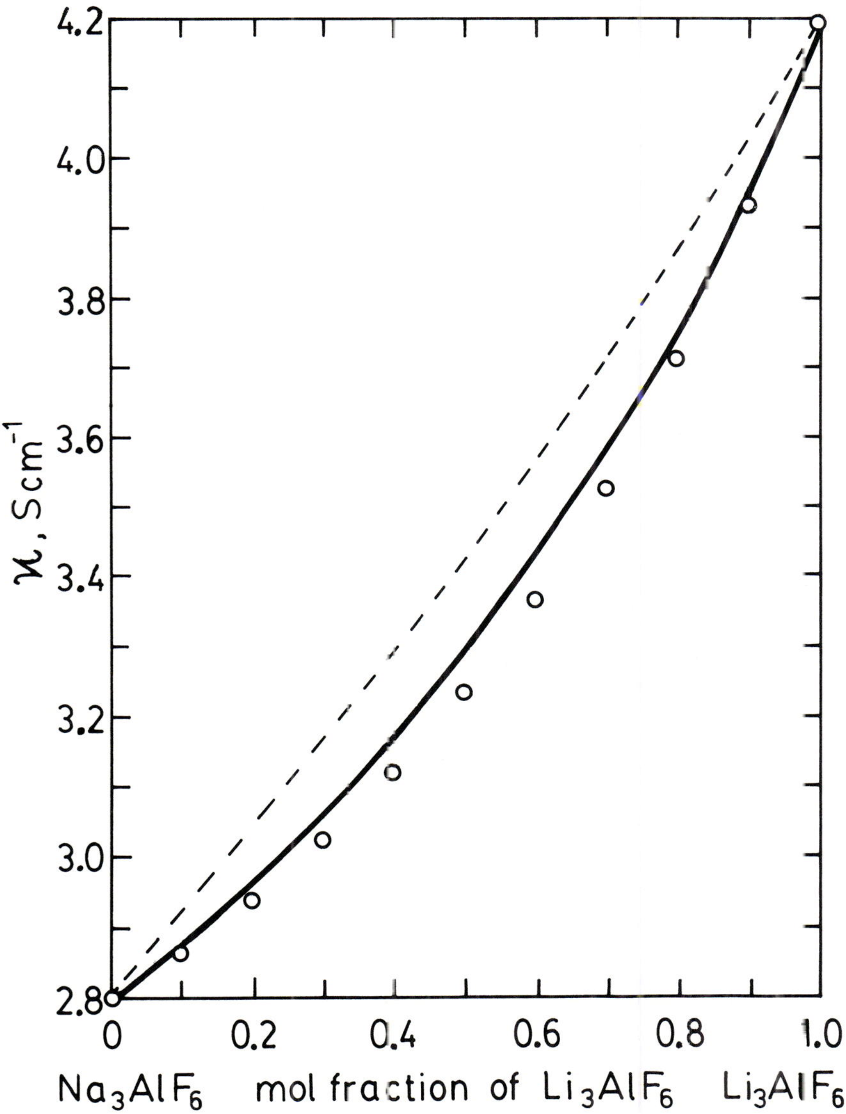

Fig. 5. Electrical conductivity of the system Na_3AlF_6 - Li_3AlF_6 at 1273 K.

Solid line : calculated from the series model.
Dashed line : calculated from the parallel model.
Open circles: experimental data, see ref. (9).

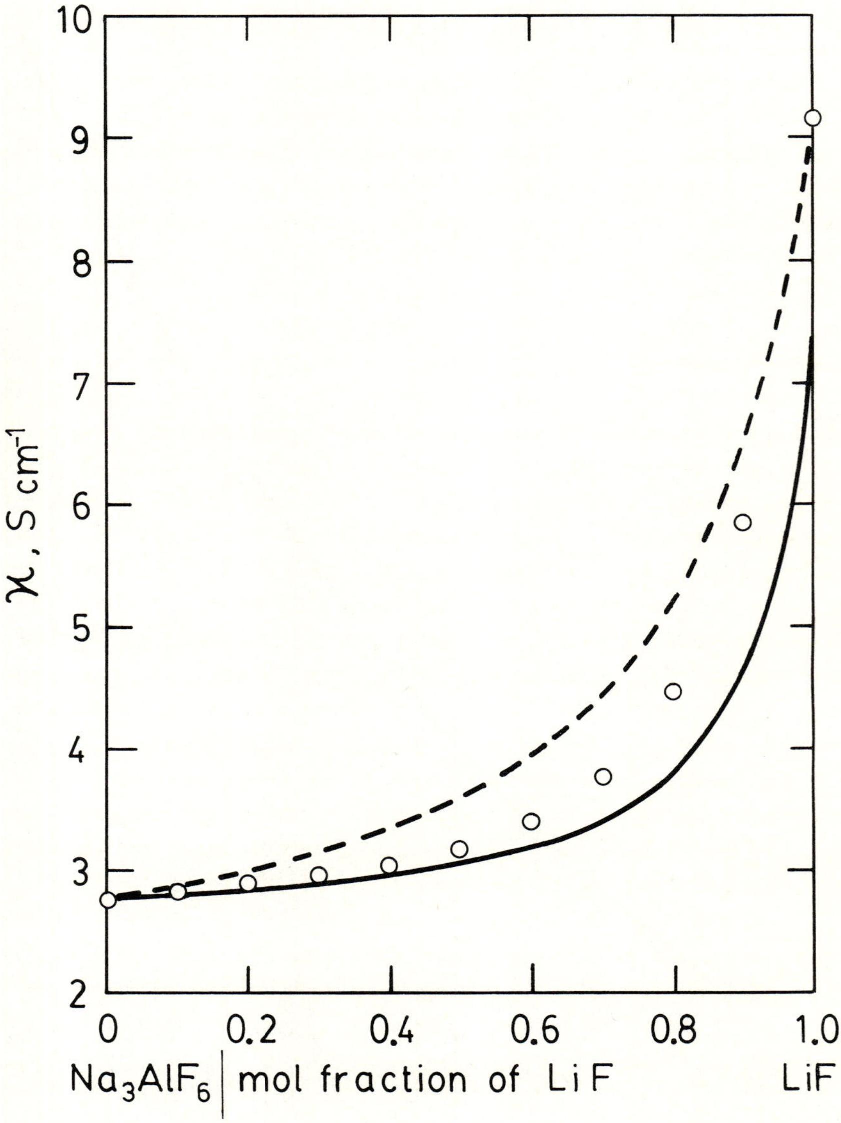

Fig. 6. Electrical conductivity of the system Na_3AlF_6 - LiF at 1273 K.

Solid line : calculated from the series model.
Dashed line : calculated from the parallel model.
Open circles: experimental data, see ref. (9).

$Na_3AlF_6 - Al_2O_3 - LiF$

In this system Al_2O_3 decreases the conductivity, while LiF increases it. Thus, this provides a good test of the suitability of the simpler model to predict conductivity data. The results are given in Figure 7 and show that the model gives data in good agreement with the measurements by Matiašovský et al. (7).

$Na_3AlF_6 - MgF_2$

Data of this system have already been presented in Figure 3. One might expect MgF_2 to behave much like CaF_2 in cryolite melts, and the measurements by Abramov and Kostyukov (8) indicate that MgF_2 lowers the electrical conductivity in approximately the same way as CaF_2. Their data correspond to $z_{MgF_2} = 0.3$, which is close to the value of $z_{CaF_2} = 0.45$. These conductivity values have recently been adopted by Janz and Tomkins (9) in their tables giving molten salt data.

On the other hand, from the extensive measurements by Belyaev (5) it follows that MgF_2 has a much stronger negative effect on the electrical conductivity. Belyaev's data give $z_{MgF_2} = -1.2$. However, the most reasonable data in Figure 3 are obtained when z_{MgF_2} is zero.

Fellner et al. (10) have shown that the reaction of MgF_2 with molten cryolite may be represented by the scheme

$$2\ Na_3AlF_6 + 3\ MgF_2 = 3\ Na_2MgF_4 + 2\ AlF_3 \qquad (12)$$

This implies that MgF_4^{2-} complexes, and not free Mg^{2+} ions, are formed in the melt. According to our model anionic complexes do not contribute to the electrical conductivity. Thus, we need only to consider the influence of reaction (12) on the volume properties of the melt, together with the effect of the AlF_3 formed on the electrical conductivity. These considerations lead to a value of z_{MgF_2} equal to zero.

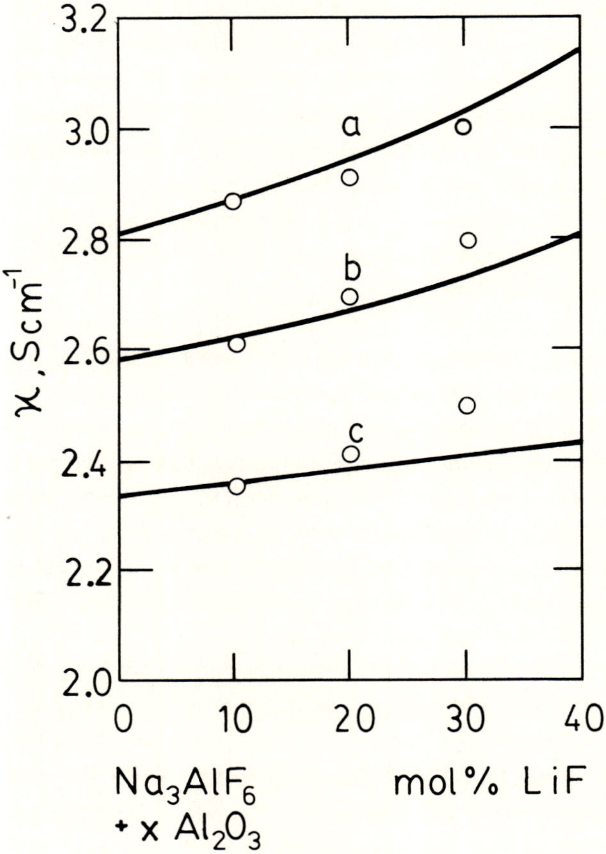

Fig. 7. Electrical conductivity for three sections in the ternary system Na_3AlF_6 - Al_2O_3 - LiF at 1273 K.

Solid line : calculated from the simpler model.
Open circles: experimental data from Matiašovský _et_ _al._ (7).

a: x = 0 mol% Al_2O_3.

b: x = 10 mol% Al_2O_3.

c: x = 20 mol% Al_2O_3.

Na_3AlF_6 - LiF - MgF_2

Grjotheim et al. (11) have recently discussed the influence of the addition of LiF and MgF_2 on the physico-chemical properties of cryolite melts. The advantage of combined addition of LiF and MgF_2 to the bath of the Hall-Héroult process depends greatly on the possibility that the negative effect of MgF_2 on the electrical conductivity can be compensated by LiF. Figure 8 shows the calculated concentration of LiF that is necessary to outweigh this negative influence of MgF_2.

If the experimental data of Belyaev (5) are valid, i. e. $z_{MgF_2} = -1.2$, a large amount of LiF would be needed, while less than 2 wt% LiF would compensate for additions of up to 10 wt% MgF_2 if Abramov and Kostyukov's (8) data are used. The middle curve in Figure 8 is based on $z_{MgF_2} = 0$. From the above discussion these data may be preferred, until new and more accurate experimental data are available for the electrical conductivity of molten cryolite with the addition of MgF_2. Such work is now in progress.

CONCLUDING REMARKS

Because of the practical implications of changes in the electrical conductivity of the bath, Choudhary (12) has developed the following relationship

$$\ln \kappa = 2.0156 - \frac{2068.4}{T} + 0.4349 \left(\frac{NaF}{AlF_3} \text{ wt. ratio} \right) - 0.0207\ (\text{wt\% } Al_2O_3) - 0.005\ (\text{wt\% } CaF_2) - 0.0166\ (\text{wt\% } MgF_2) + 0.0178\ (\text{wt\% LiF}) + 0.0077\ (\text{wt\% } Li_3AlF_6) + 0.0063\ (\text{wt\% NaCl}) \qquad (13)$$

Here, the temperature T is given in Kelvin. Eqn. (13) is claimed to give values of the electrical conductivity with an accuracy better than $\pm$ 4 %. In the present work all conductivity data are calculated at 1273 K. Industrial bath temperatures are lower, and the conductivity data at 1273 K may be recalculated by using the temperature coefficient given by Choudhary in eqn. (13).

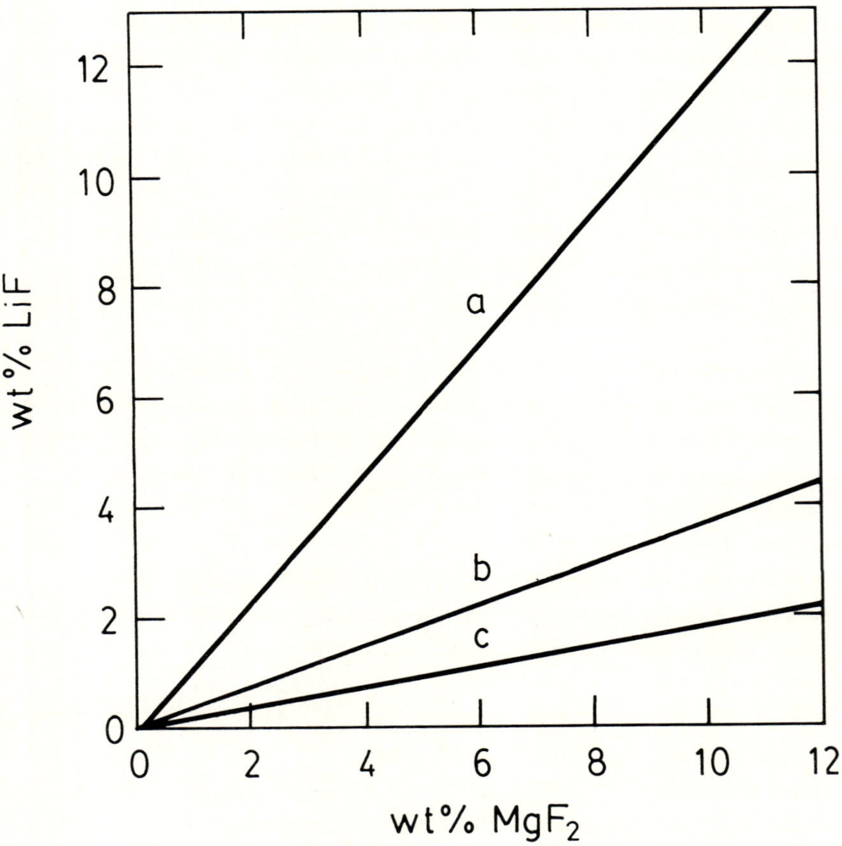

Fig. 8. The calculated concentration of LiF that is necessary to compensate for the negative influence of MgF_2 on the electrical conductivity of a melt with NaF/AlF_3 molar ratio of 2.7, containing 10 mol% Al_2O_3 and 10 mol% CaF_2 at 1273 K.

a: z_{MgF_2} = - 1.2, corresponding to the data by Belyaev (5).

b: z_{MgF_2} = 0 , calculated from the simpler model.

c: z_{MgF_2} = 0.3, corresponding to the data by Abramov and Kostyukov (8).

Acknowledgement

This work has been supported financially by the Royal Norwegian Council for Scientific and Industrial Research (NTNF).

References

1. K. Grjotheim, C. Krohn, M. Malinovský, K. Matiašovský, and J. Thonstad, "Aluminium Electrolysis - Fundamentals of the Hall-Héroult Process", 2nd ed., Aluminium-Verlag, Düsseldorf, 1982.

2. B. F. Markov, and L. A. Shumina, "The Relation Between the Concentration and Electrical Conductivity of Fused Binary Salt Systems", Doklady Akad. Nauk SSSR 110 (1956) 411-413.

3. A. Klemm, Transport Properties of Molten Salts', in "Molten Salt Chemistry", M. Blander, ed., Interscience Publishers, New York, 1964, pp. 537-538.

4. P. Fellner, "The Series and Parallel Models of Electrical Conductivity of Molten Salt Mixtures", Chem. Zvesti, to be published.

5. A. I. Belyaev, "Elektrolit alyuminievykh vann". Metallurgizdat, Moscow, 1961.

6. E. W. Yim, and M. Feinleib, "Electrical Conductivity of Molten Fluorides. II. Conductance of Alkali Fluorides, Cryolite, and Cryolite-Based Melts", J. Electrochem. Soc. 104 (1957) 626-630.

7. K. Matiašovský, V. Daněk, and M. Malinovský, "Effect of LiF and Li_3AlF_6 on the Electrical Conductivity of Cryolite-Alumina Melts", J. Electrochem. Soc. 116 (1969) 1381-1383.

8. G. A. Abramov, and A. A. Kostyukov, see ref. 5, p. 48.

9. G. J. Janz, and R. P. T. Tomkins, "Physical Properties Data Compilations Relevant to Energy Storage, IV. Molten Salts: Data on Additional Single and Multi-Component Salt Systems", U. S. Government Printing Office, Washington, 1981.

10. P. Fellner, K. Grjotheim, and H. Kvande, "Complex Formation in the Systems $NaF-MgF_2$ and $Na_3AlF_6-MgF_2$", to be published.

11. K. Grjotheim, H. Kvande, and K. Matiašovský, "Addition of LiF and MgF_2 to the Bath of the Hall-Héroult Process", in Light Metals 1983, pp. 397-411.

12. G. Choudhary, "Electrical Conductivity for Aluminum Cell Electrolyte between 950 ^{o} - 1025 ^{o}C by Regression Equation", J. Electrochem. Soc. 120 (1973) 381-383.

ON THE ANODE EFFECT IN INDUSTRIAL ALUMINUM CELLS

J. Thonstad[*], F. Nordmo[**],
A. H. Husøy[***], K. Ø. Vee[***] and O. G. Austrheim[***]

[*] Laboratories of Industrial Electrochemistry,
The Norwegian Institute of Technology,
University of Trondheim, N-7034 Trondheim-NTH, Norway

[**] Norsk Hydro, Glomfjord Fabrikker, N-8160 Glomfjord, Norway

[***] A/S Årdal og Sunndal Verk, N-5875 Årdalstangen, Norway

The anode effect (AE) in industrial aluminum cells manifests itself as a sudden voltage rise to a high level. The rate of voltage increase corresponds to double layer charging. The AE voltage depends largely on the wetted anode surface area. For this reason the AE voltage decreases with increasing depth of immersion of the anode. Because the anode surface becomes smoothened during AE, the AE voltage usually increases gradually. AE's normally occur below 50 V, but also a higher voltage range is conceivable from current-voltage characteristics. The highest voltage recorded was 116 V. Abnormally low AE voltages (< 25 V) can be due to an excessively rough anode surface or to shorting to the cathode. AE's are terminated by adding alumina and either lowering the current density by various means or by sparging the anode surface with gas. The effectiveness of some techniques was studied.

Introduction

In industrial aluminum cells the anode effect (AE) usually occurs at a frequency of 0.5 to 3 per day. As the alumina content in the bath gets below a certain level (2-3 wt% Al_2O_3) the cell voltage starts to increase gradually due to increasing anodic polarization (1). If fresh alumina is not supplied to the bath the AE will normally occur within 15-60 minutes. The cell voltage then rises abruptly from the normal level (4-5 V) to the AE voltage (> 15 V).

Laboratory studies of the AE are normally conducted in melts with fixed alumina contents, and the AE is initiated by gradually raising the imposed voltage until the socalled critical current density (ccd) is reached and the AE occurs (1). In industrial cells where the current remains practically constant and the alumina content varies, the normal cd (current density) of $\sim 0.75\ A/cm^2$ should correspond to the ccd for the alumina concentration of the bath at the initiation of the AE. This "critical alumina content" varies within fairly wide limits from 0.5 to 2.2 wt% Al_2O_3; 1-2 wt% Al_2O_3 being the most common range (1). In laboratory cells with graphite anodes this cd ($0.75\ A/cm^2$) corresponds to an alumina content of 0.2-0.4 wt% Al_2O_3 (2). Although alumina contents down to 0.5 wt% Al_2O_3 have been observed in industrial cells (1), the most common values are two to three times higher, as indicated above. Two possible reasons for this discrepancy are,

(a) There may be a concentration gradient with respect to alumina near the anode surface (3).

(b) The anode may be partly covered by gas so that the real cd is higher than the nominal value. A higher cd would correspond to a higher "critical alumina content".

Data on the composition of the anode gas during AE seem to support assumption (a). In industrial cells the anode gas during AE normally contains 14-35% CF_4, the balance being CO and CO_2 (1). Laboratory data (4) indicate that the CF_4 content decreases sharply with increasing alumina content, and the high CF_4 contents found in industrial cells occur only in melts with less than $\sim$ 0.5 wt% Al_2O_3.

Nikitin et al. (5) observed that the gas volume underneath the anode decreased as the AE was approached, and the hydrostatic pressure underneath the anode apparently dropped to zero during AE. Ayushin et al. (6) tried to explain these results by suggesting that the anode during AE is covered by a continuous gas film in contact with the atmosphere rather than by individual gas bubbles.

In laboratory experiments where a gradually increasing potential was applied to graphite anodes, curves of the general shape shown in Fig. 1 were obtained (7). The cd in the e-f range was found to be very dependent on the surface state of the electrode. The cd could be doubled by roughening the surface with a steel brush. With time the electrode became polished and the cd decreased markedly. For a constant current coupling such as in industrial cells this corresponds to increasing AE voltage with time.

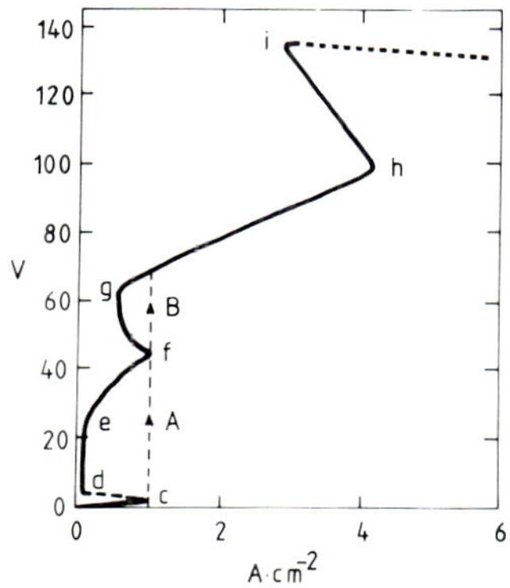

Figure 1. Current-voltage diagram recorded for graphite anode in a laboratory cell by linearly increasing potential. Dotted line indicates possible potential jump in galvanostatic coupling.

If the curve in Fig. 1 is assumed to be representative also for industrial anodes, one might expect to find at least two AE voltage ranges, as indicated on the figure by the symbols A and B. Provided peak f is not surpassed, AE voltages in the range 25-45 should be expected (A). If, however, f is located at a cd less than the operating cd, the AE voltage will jump up to 60 V and higher. (B). The mechanism of charge transfer seems to change along the curve as discussed elsewhere (6,8). Below f the charge seems to be carried largely by numerous sparks originating at protrusions on the surface. Above 60 V a glow discharge appears to take over. Sparks reappear at around 80 V and at 140-110 V an arc discharge occurs (7).

As already mentioned the cd in the e-f range increases with increasing surface roughness of the electrode. Due to the steepness of the curve, the AE voltage may then be strongly dependent on the surface state of the anode. During AE, most of the voltage drop through the cell will be located at the anode surface. The sides of the anode will then carry a larger proportion of the current than they normally do. If the exposed anode surface area is increased by increasing the bath depth, the AE voltage should then decrease. Antipin and Vazhenin (9) found that the AE voltage decreased steadily with increasing bath depth from ~ 80 V at 10 cm to ~ 25 V at 25 cm bath depth.

Once an AE is on certain measures are taken to terminate it. The first step is normally to feed alumina to the bath. Although alumina feeding alone normally does not terminate the AE, it greatly increases the effectiveness of the methods used, and it prevents the AE from reoccurring soon. The various ways of terminating AE's can roughly be divided into three groups

(i) Sparging with air or inserting a fresh wooden pole.

(ii) Lowering or tilting the anode.

(iii) Shunting or short-circuiting the cell.

The first two methods (i and ii) are those normally used. The second method is easily adaptable to automation, and it is therefore favored in modern designs. Items (ii) and (iii) are in principle similar since the cd is lowered in both cases, while (i) is based upon reactivation of the anode surface by vigorous stirring.

In the present work measurements were made on industrial cells and supplemented by laboratory tests to study the initiation of AE, AE voltage, gas pressure underneath the anode, and AE termination.

Experimental and Results

Initiation of AE

The cell voltage prior to AE was not studied systematically. However, it was observed that the slow, gradual increase in cell voltage (1) in some cases was superseded by a much more rapid and irregular rise of 1-2 V during the last 1-2 minutes before the onset of the AE. This latter rise is hardly due to a normal overvoltage increase, but it is rather a prelude to AE, possibly caused by partial blocking of the anode surface. An irregular current distribution at the anode has been observed during this period, as discussed in the following.

In laboratory cells it was shown that once the AE was initiated in a galvanostatic circuit, the voltage rose extremely fast at a rate corresponding to a double layer charging process (7). To study this phenomenon in industrial cells a storage oscilloscope was connected to 140 kA prebake cells. Analysis of the steepest part of the curve depicted in Fig. 2 shows that the voltage rise corresponds to charging a condenser of 350 $\mu F/cm^2$. This is a reasonable value for the double layer capacity of a carbon anode, referred to the geometrical surface area (10).

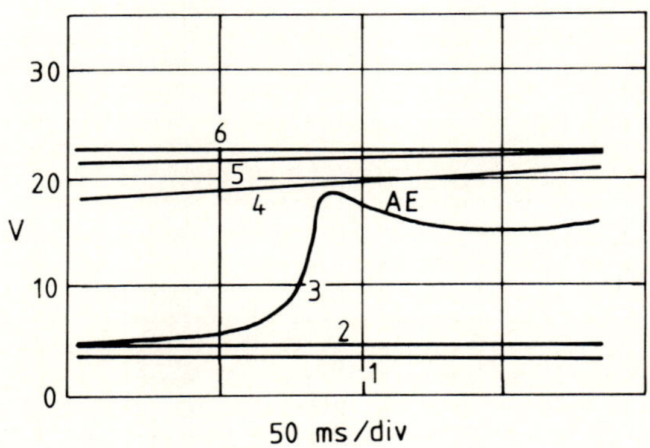

Figure 2. Oscilloscope recording of the initiation of AE on a 140 kA cell with prebaked anodes. Consecutive sweeps are numbered 1-6.

AE Voltage

In the cells under study AE's were terminated by breaking the crust and blowing air underneath the anode with an air lance. It could then take a few minutes before a cell on AE was attended to. During this period the AE voltage would usually increase from say 25 V to 40 V, as shown in Fig. 3. The initial AE voltage could be as low as 10-15 V. There were also cases, particularly for high voltage AE's (> 50 V) where the voltage remained practically constant until the cell was attended to. This is in agreement with the laboratory observation that the anode is polished below 45 V but not at higher voltages (7). As discussed above polishing causes the voltage to rise.

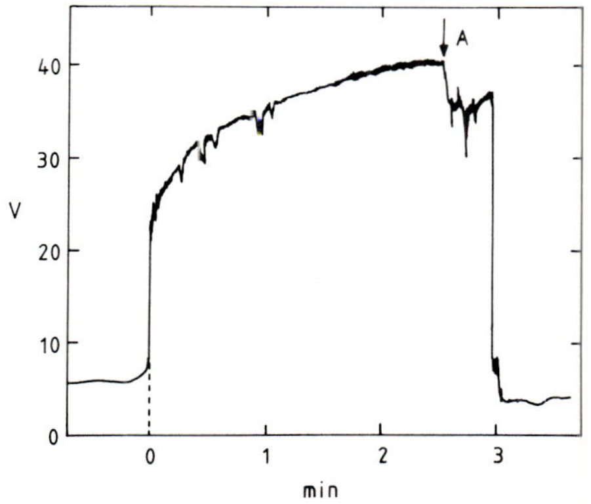

Figure 3. AE voltage versus time for 150 kA cell with prebaked anodes. A - Crust-breaking, followed by air lancing.

An AE normally persisted until the cell was serviced. It might then go on and off a few times before stable normal electrolysis was resumed. However, in some cases the AE tended to go on and off even before the cell was attended to. These AE's were found to occur at low voltages (15-35 V). The time off AE was usually shorter than the time on AE. The frequency of the off periods varied from 3 per min. to 10 per s. Two recordings of such phenomena are shown in Fig. 4.

As mentioned above the AE voltage is expected to depend on the surface condition of the anode, bath depth, type of cell etc. For the 150 kA prebake cells under study it was found that the average AE voltage measured before the crust was broken was 40 V. Out of 133 AE's recorded on different cells 130 fell between 25 and 55 V. One AE occurred at 60 V and one reached 92 V. In the latter case the voltage rose in two steps, first a sharp rise to 45 V drifting slowly to 55, followed by another rapid rise to 85-90 V. After one minute it dropped back to the 40-50 V range again. This behavior can be explained on the basis of Fig. 1 (ranges e-f and g-h).

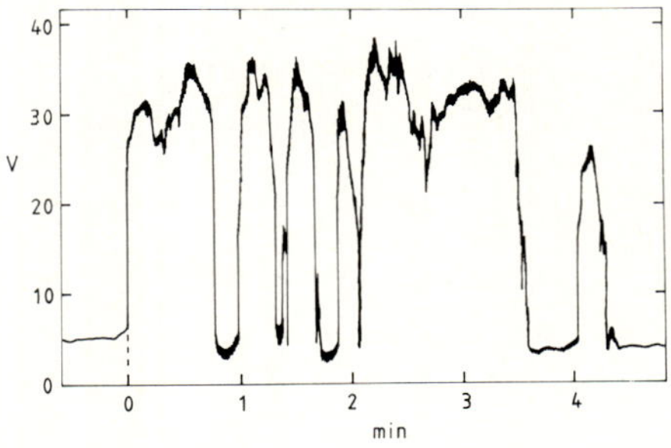

(a)

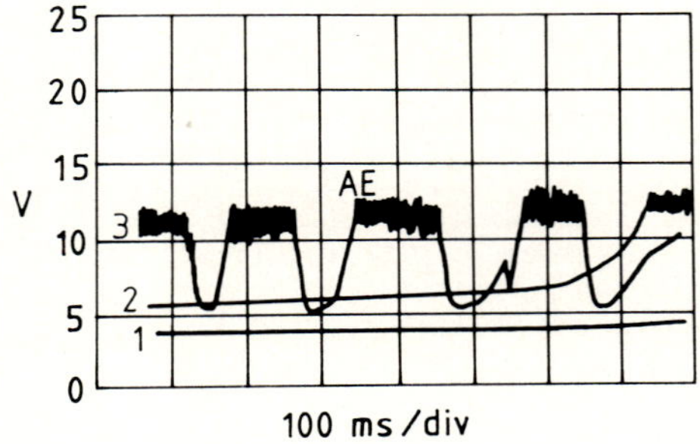

(b)

Figure 4. AE's going on and off. (a) Slow fluctuations. (b) Rapid fluctuations; oscilloscope recording. Consecutive sweeps are numbered 1-3.

High voltage AE's (> 50 V) were found to occur most frequently on VS Soderberg cells operating with low bath depths (10-17 cm). On 84 kA Soderberg cells the average AE voltage was found to be 40 V. Out of more than 300 AE's recorded, about 40% had voltages above 50 V. Twelve AE's were above 100 V, the highest reaching 116 V. In some cases these high voltages were reached in two steps as described above, and the return to normal electrolysis always occurred in steps, as shown in Fig. 5. The voltage rise which in this case occurred in one step, was not clearly recorded due to slow recorder chart speed at this moment.

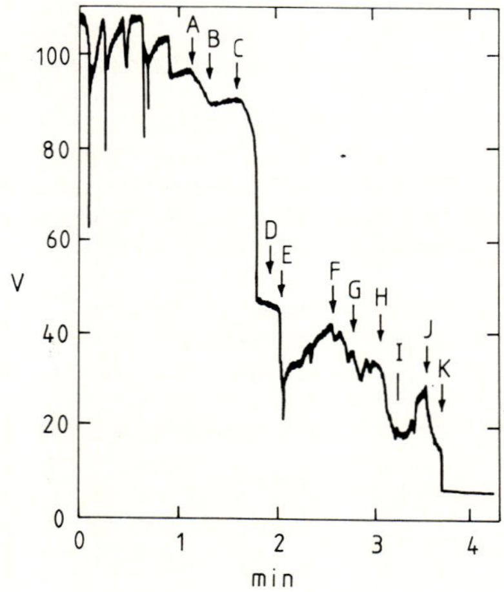

Figure 5. AE voltage versus time trace for 84 kA VS Soderberg cell. Anode lowered at A-B, C-D, F-G, H-I, J-K. Crust-breaking at E. The beginning of the AE was not recorded.

The dependence of AE voltage on bath height was studied in 84 kA VS Soderberg cells running with low bath levels and in 105 kA VS Soderberg and in 150 kA prebake cells with close-fitting, semi-continuous prebaked block anodes using higher bath levels. The results from the latter two types of cells were lumped together because they were very similar. High AE voltages were in both cases associated with low bath levels, as shown in Fig. 6, although this is clearly not the only parameter. In the 84 kA Soderberg cells increasing depth of the metal pad seemed to cause a slight increase in the average AE voltage. The percentage of high voltage AE's (> 50 V) was markedly higher (40-50%) at low and normal bath temperatures, i.e. 945-970°C, than at high temperatures above 970°C (10-20%). The bath ratio (NaF/AlF_3) and the alumina content did not seem to have any significant influence on AE voltage.

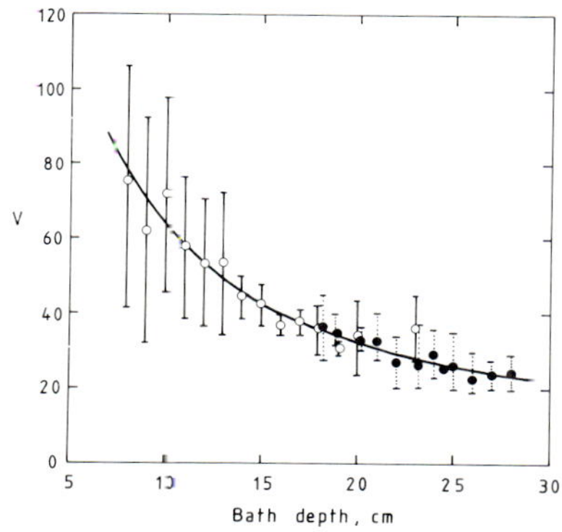

Figure 6. AE voltage versus bath depth. o - 84 kA VS Soderberg cells, (300 measurements), • - 105 kA VS Soderberg and 150 kA prebake cells (50 measurements). Circles represent average values and bars indicate the scatter.

A specific problem in industrial cells is AE's which occur at abnormally low voltages (< 25 V). Such AE's are usually regarded as an indicator of a "sick pot", and corrective measures are usually taken. According to Fig. 1 such low voltages are not to be expected. However, in view of the above treatment three possible reasons for low AE voltages can be suggested,

(a) Large anode surface area due to a rough surface or accumulation of carbon dust.

(b) Large anode surface area due to large bath depth.

(c) Protrusions on the anode (points).

(d) Short-circuiting to the metal.

Low AE voltages may gradually rise to normal values when left on for a relatively long time (5-10 min.). In such cases (a) is probably the reason for the low voltage. If it helps to raise the anode, (d) may be the reason, either due to metal splashing against the anode or points or loose pieces of carbon touching the mtal. Particularly in Soderberg cells protrusions on the anode (points) are known to be one cause of low AE voltage. Such points may short-circuit to the cathode, but even if they do not, protrusions tend to carry very high currents during AE.

In a laboratory cell a graphite anode was furnished with a pencil-shaped protrusion extending 1 cm down from the anode. The surface area of the anode was 11.2 cm^2 and that of the protrusion was 2.4 cm^2. It was insulated from the rest of the anode so that the current through it could be monitored separately when connected in parallel to the anode. During normal electrolysis the protrusion carried about 20 per cent of the total current, while this figure increased to 50-80 per cent during AE. The highest figures corresponded to low AE voltage (25 V). The AE voltage was markedly lower compared to experiments without the protrusion.

The current distribution between individual anodes was investigated on 150 kA cells placed end to end and equipped with 22 prebaked anodes and centerbreak alumina feeding. The currents were obtained by measuring the voltage drop across a section of the anode hangers. The data were compiled by the process computer and written out as averages for 15 minutes and 100 seconds before the AE occurred and during the AE up to the time the cell was serviced. Measurements were taken every 12 seconds. Low AE voltages tended to be associated with cells having non-uniform and fluctuating current distribution prior to as well as during AE. These phenomena were always intensified during AE as shown in Fig. 7. Although most anodes carried roughly (± 20%) average current (~ 7 kA) during AE, some anodes could be far below (0.7 kA) and some far above (10-20 kA) that level. Those carrying high currents (> 9 kA) tended to show strong current fluctuations. The measurements made 15 minutes and 100 seconds before AE showed that whereas the current distribution was fairly uniform at 15 minutes this was not the case 100 seconds before AE. Anodes which carried abnormally high or low currents during AE tended to do so also 100 seconds before AE. Anodes which carried very high currents during AE tended to be on the high side also 15 minutes before AE. Anodes carrying low currents (0.7-3 kA) during AE usually had average currents 15 minutes prior to AE. The pairs of anodes at either end of the cells often carried less than average current during AE. Otherwise there seemed to be no consistent pattern as to which anodes carried abnormal currents. The current fluctuations during AE decreased with increasing anode - cathode distance.

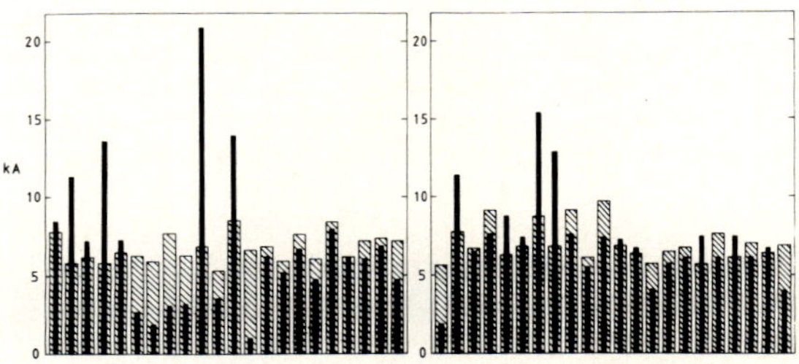

Figure 7. Current distribution on 22 individual prebaked anodes in 150 kA cell before AE (black columns) and during AE (light columns) (measured before the cell was serviced). Each set of two columns (light/black) represents one anode, listed from one end of the cell to the other. The recordings shown were made on the same cell on two consecutive days.

Gas Pressure Underneath the Anode

The gas pressure underneath Soderberg anodes was measured by means of steel pipes which were fed through the anode until they reached the underside. It was ascertained that the tubes were open so that anode gas escaped freely through them before they were connected to a manometer. As expected, the overpressure with respect to atmospheric pressure was roughly equal to the calculated hydrodynamic pressure for the given anode immersion depth. During normal electrolysis the pressure fluctuated 10-15 mm water gauge, probably due to gas bubble movement. These fluctuations are not shown in Fig. 8, which represents a pressure recording. At the onset of AE the pressure did not change, but the fluctuations ceased. When the crust was broken the pressure increased markedly because the broken crust and alumina caused the bath level to rise.

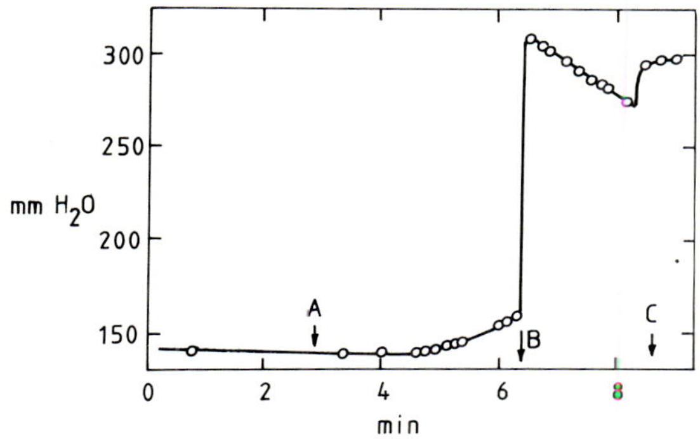

Figure 8. Gas pressure (mm water gauge) underneath a Soderberg anode during AE. A - onset of AE, B - Crustbreaking, C - AE terminated by air lancing.

AE Termination

As outlined in the Introduction AE's can be terminated by several different methods. Some of these techniques were tested in a laboroatory cell, using a one inch graphite rod as anode. The graphite crucible served as cathode. No aluminum was added. The AE was initiated by increasing the applied voltage, and thereafter the power supply was switched to a galvanostatic mode whereby the desired cd could be set.

To terminate the AE air or nitrogen was blown into the melt from a steel tube directed towards the underside of the anode. The AE was easily removed for AE voltages below 50 V, while a fairly strong gas jet was needed at higher voltages. It was immaterial whether air or nitrogen was used. Mechanical stirring of the bath underneath the anode by means of an impeller did not remove the AE except at low voltages (20-30 V, low cds). Vibration applied to the steel rod supporting the anode was equally ineffective.

The anode could be moved smoothly up and down into the melt either manually or by means of a motor. As the anode was lowered, the voltage decreased gradually until the AE suddenly went off. For AE voltages below

50 V the AE went off quickly, while the anode had to be lowered a longer distance for higher starting voltages. The rate of lowering had no apparent effect on the results. When the anode was lowered the exposed anode surface area was increased. On the simplifying assumption that the current distribution is uniform during AE, plots of AE voltages vs. cd could be made. Three such curves are shown in Fig. 9 together with one derived from a similar experiment in an industrial cell.

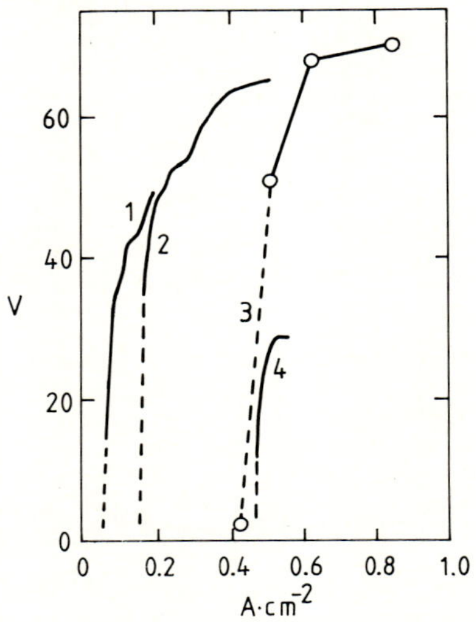

Figure 9. AE voltage versus average cd at constant total current, as the anode is lowered deeper into the melt.
1-3 - Laboratory cell, initial cds: 1 - 0.19 A/cm^2; 2 - 0.51 A/cm^2; 3 - 0.85 A/cm^2 (moved manually); 4 - industrial cell (prebake).

In industrial cells AE termination by blowing air underneath the anode is usually performed manually. In an attempt to automate this process two steel tubes were placed through the anode in 84 kA Soderberg cells, as described above. Pressurized air was applied to these tubes to terminate the AE. It was found that AE termination was greatly facilitated by conducting crust-breaking, preferably on both sides of the cell, prior to air sparging. A typical performance is shown in Fig. 10. The AE normally went off 10-15 s after air was admitted. The method was found to be effective in 77 per cent of all trials. Failures were usually caused by one or both tubes being clogged with frozen bath. Clogged tubes did reopen after some time. Nevertheless, clogging of the tubes seemed to constitute a major shortcoming of this method.

Lowering of the anode is a well established method to terminate AE's in prebake cells. In Soderberg cells it is not so much used because of greater risk of bath overflowing the top of the cell and problems in moving the anode. Tests on the 84 kA Soderberg cells showed that crust-breaking prior to anode lowering greatly facilitated the termination of the AE. In order to be effective the anode had to be lowered 2-2.5 cm. Further

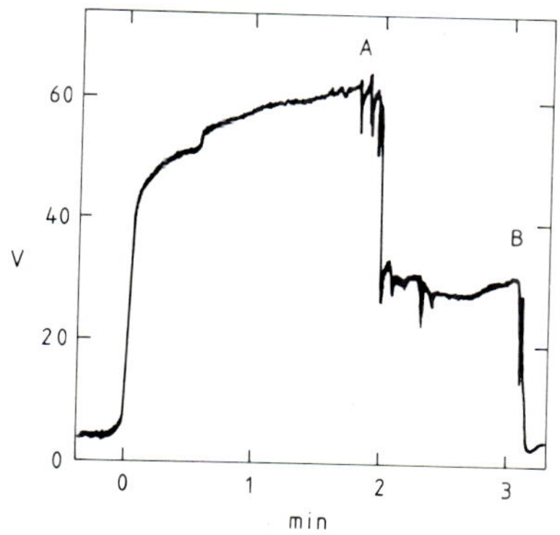

Figure 10. AE termination by blowing air through tubes in Soderberg anode. A - crustbreaking, B - air admitted.

movement tended to cause flooding of bath. As shown in Fig. 11 the voltage did usually drop down to 10-15 V before it finally fell back to normal electrolysis voltage. If the first lowering was not effective, repeated movements up and down were performed, as shown in Fig. 5. It did also help to keep the anode in the low position for some time (~ 30 s). When these techniques were used, the method was found to be 97% effective in terminating AE's. No adverse effects to the cells were experienced.

As mentioned above the risk of flooding of bath can be avoided if the anode is tilted rather than lowered uniformly. Due to mechanical problems this technique could not be tested on the cells under study. Tilting of the anode is being used successfully on potlines with 150 kA and 220 kA prebaked cells (11).

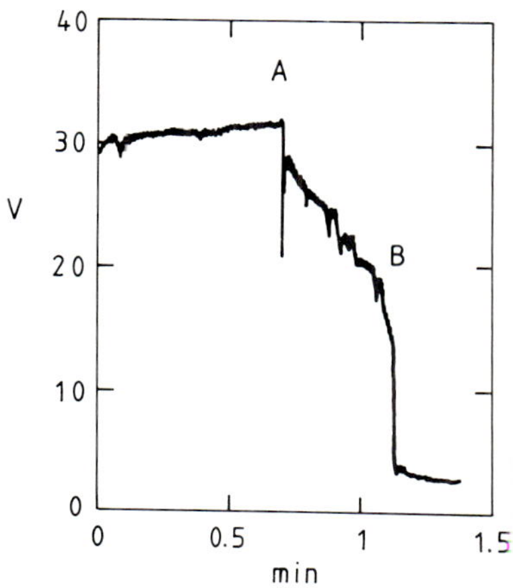

Figure 11. AE terminated by lowering the anode in Soderberg cell. A - crustbreaking and beginning of anode movement, B - AE goes off.

Lowering of Cell Current

If part of the line current is allowed to bypass a cell during AE, an effect similar to that of local short-circuiting might be achieved. To test this idea resistors made of steel bands were placed parallel to Soderberg cells. The resistors were connected when an AE was on and the crust had been broken. It was found to be necessary to have the resistors carry as much as 40 per cent of the total current in order to terminate the AE. A typical curve is shown in Fig. 12.

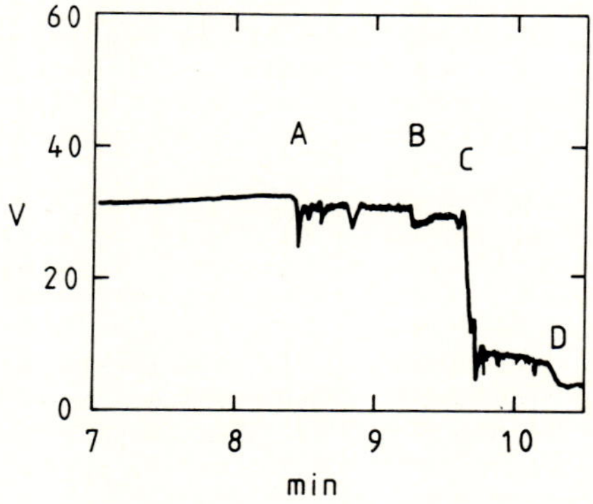

Figure 12. AE terminated by letting 40 per cent of the current bypass an 84 kA VS Soderberg cell. A-B - crustbreaking, C - resistors connected, D - AE off.

Discussion

The curve shown in Fig. 1 undoubtedly represents an oversimplification of the current-voltage relationship on an industrial anode undergoing AE, since surface roughness, uneven current distribution and short-circuiting may interfere. Nevertheless, it is believed that the main features of the curve are valid also for industrial anodes even if the characteristic voltages and the corresponding cds may be shifted somewhat.

As mentioned in the Introduction, the AE voltage range can roughly be divided into three stages, i.e. point sparking below f, glow discharge above g and arcing at i. The voltage during arcing can go down to 110 V (7). Considering the AE voltage range observed in this work, it appears that all these three stages may be encountered in industrial cells. The AE voltage of 116 V appears to be the highest value reported for industrial cells.

The cd in the e-f range in Fig. 1 has been shown to be dependent on the anode surface roughness (7). Due to the steepness of the curve it appears that an anode with a rough surface can sustain AE at a considerably lower voltage than a smooth anode. As the anode surface is smoothened during AE due to preferential attack on protruding particles, the voltage will increase gradually in agreement with Figs. 3 and 1. Samples of anodes taken immediately after AE showed clear signs of the surface being smoothened (12). Some protruding particles may be detached during AE to form carbon dust in the melt. The old notion that the AE tends to clean the anode surface is consistent with this reasoning. The stepwise rise to high AE voltages, first to 40-50 V and then to 60-100 V can also be explained along these lines. As the anode becomes smoothened in the e-f range, the voltage may reach the f peak and then suddenly jump on to the g-h range.

The charging curve shown in Fig. 2 indicates that the anode is insulated by a film of some sort at the onset of the AE, and that a high voltage is needed to allow charge transfer through the film. The nature of this film is not clear, it may be either a solid surface compound or a gas or both (13). For AE's going on and off (Fig. 4) this film must be unstable for some reason. The high frequency case depicted in Fig. 4 b may be related to a pulsating release of gas bubbles, while the low frequency case in Fig. 4 a may be related to intermittent shorting to the metal. The fluctuation and uneven distribution of the current during AE may cause splashing of the metal due to magnetic forces.

For cells in stable operation the AE voltages shows relatively little variation from one AE to another, usually less then ± 5 V. It appears that each type of cell has its own characteristic AE voltage, depending on the operating parameters. Exceptions are the very high and the very low AE voltages. The high AE voltages occur at high cds (low bath depth) and smooth and high quality anodes. If AE cannot be sustained below point f in Fig. 1, the voltage rises to values above 60 V. Such high voltages are believed not to be harmful, except for the high excess energy consumption and overheating of the anode and the cell (4).

A regression analysis showed that the data given in Fig. 6 could be represented by the equation

$$U = \frac{572}{d^{0.96}} \approx \frac{572}{d} \tag{1}$$

where U is the cell voltage and d is the interpolar distance in cm. This result implies that a large fraction of the current passes through the sides of the anode during AE. The anodes of the 84 kA Soderberg cells had the dimensions 576 x 177 cm. Assuming that the cd at the underside is reduced by a factor k compared to the side cd, one can express the average cd by the equation

$$(cd)_{av} = 84000/[576 \cdot 177 \cdot k + 2\,(576 + 177)\,(d - 5)] \tag{2}$$

when the interpolar distance is taken to be 5 cm. Differentiation and combination of the two equations yields an expression for dU/d(cd) which can be set equal to the slope of the curve in Fig. 1 in the e-f range. This slope is ~ 20 V/A cm^{-2}. It can then be shown that for e.g. d = 10 cm the factor k = 0.13. This means that the average cd at the sides will be 3.9 cm^2 as against 0.53 A/cm^2 at the underside, and 36% of the current will pass through the sides of the anode. For d = 10 cm the immersed area of the sides is only 7% of the total anode area.

This calculation can only be regarded as tentative, and it may overestimate the current shift during AE. However, there is some evidence that the current is shifted towards the sides during AE. Measurements of the current distribution through the stubs of 105 kA Soderberg cells showed that during AE the current through the two outer rows of stubs increased by 9% while it decreased accordingly through the two inner rows.

Low AE voltage is commonly regarded as an indication of a "sick pot". As outlined above it may be caused by a number of disorders. If the AE voltage increases with time it is indicative of a rough anode surface. If it increases sharply when raising the anode, short-circuits may be the reason. No simple diagnostic tool seems to be available. Incorrect setting of new anodes is known to have an adverse effect on cell stability. During AE also other factors seem to be at play as appears from Fig. 7. Apart from factors directly related to anode properties, waves at the bath/metal interface may cause local shorting to the anode. Accumulation of alumina sludge tends to aggrevate such conditions.

The gas pressure measurements point in favour of the idea of a gas film being present underneath the anode during AE, rather than discrete gas bubbles. Some workers (5,6) have found the pressure to drop to zero during AE. No indication of such behaviour was found in the present work.

The various ways of terminating AE's all seem to rely on either of two fundamentally different principles. Gas sparging with air or wooden poles appear to clear the anode surface to restore normal electrolysis locally. Calandra et al. (13) showed that the current through the prebaked anode under which an air lance was placed increased sevenfold. The current through the other anodes not directly affected by the sparging then probably decreased sufficiently to make the AE go off. Splashing of metal during sparging contributes to its effectiveness. The other principle is based on various ways of reducing the cd sufficiently to render the AE unstable. At present this method, which is usually conducted by lowering and raising the anode, is preferred for automatic AE termination.

Acknowledgement

The experimental results presented in this work were extracted from a dr.ing. thesis (F. N.) and three graduation theses (K. Ø. V., A. H., O. G. A.) which were partly based on measurements conducted in the aluminum plants of A/S Årdal og Sunndal Verk. Permission to carry out plant studies is gratefully acknowledged, in particular supervision of graduation thesis works by Messrs. H. Holo, M. Leinum, and H. Kvande.

References

1. K. Grjotheim, C. Krohn, M. Malinovsky, K. Matiasovsky, and J. Thonstad, Aluminium Electrolysis, Fundamentals of the Hall-Heroult Process, 2nd ed., Aluminium Verlag, Düsseldorf, 1982.

2. J. Thonstad, F. Nordmo, and J. K. Rødseth, Electrochim. Acta, 19 (1974) pp. 761-769.

3. I. S. Kachanovskaya, N. S. Sirayev, and L. V. Lebedeva, Sov. J. Non Ferrous Met., 13 (10) (1972) pp. 25-27.

4. F. Nordmo and J. Thonstad, 4th Czechoslov. Aluminium Symposium, Banska Bystrica, 1980, pp. 156-165.

5. A. V. Nikitin, V. A. Kryukovskii, and N. S. Nikhalitsin, Sov. J. Non Ferrous Met., 16 (8) (1975) pp. 37-40.

6. B. A. Ayushin, A. I. Begunov, and V. N. Deryagin, Sov. J. Non Ferrous Met., 16 (2) (1975) pp. 27-38.

7. J. Thonstad, F. Nordmo, and K. Vee, Electrochim. Acta, 18 (1973) pp. 27-32.

8. F. Nordmo and J. Thonstad, to be published.

9. L. N. Antipin and S. F. Vazhenin, Elektrokhim. Rasplav. Solei, p. 314, Metallurgizdat, Moscow, 1964.

10. J. Thonstad, Electrochim. Acta, 15 (1970) pp. 1581-95.

11. T. Saksvikrønning, G. Valsvik, and S. J. Hove, Light Metals, (1982), pp. 553-559.

12. J. Thonstad, Travaux ICSOBA (Comité Intern. Etudes Bauxites, Oxydes et Hydroxydes d'Aluminium) No 10, pp. 75-86, 1973.

13. A. J. Calandra, C. E. Castellano, C. M. Ferro, and G. Ccbo, Light Metals, (1982) pp. 345-358.

DETERMINATION OF MELTING TEMPERATURES AND Al_2O_3 SOLUBILITIES

FOR HALL CELL ELECTROLYTE COMPOSITIONS

Steven S. Lee, Kuan-Shaur Lei, Paul Xu and Jesse J. Brown, Jr.

Department of Materials Engineering
Virginia Polytechnic Institute and State University
Blacksburg, VA 24061

Phase equilibria studies have shown that the Na_3AlF_6-AlF_3-CaF_2-Al_2O_3 system contains two ternary compounds, $NaCaAlF_6$ and $NaCaAl_2F_9$, in addition to the well-known binary compounds $CaAlF_5$, Ca_2AlF_7, and $Na_5Al_3F_{14}$. The Hall cell electrolyte composition range lies in the high-cryolite corner of this quaternary system with an invariant point at 56.7% Na_3AlF_6, 30.2% AlF_3, 8.5% CaF_2, 4.6% Al_2O_3 and 685°C. The typical Hall cell composition lies on the boundary between two compatibility tetrahedra; cryolite is always the primary phase to crystallize but, depending on the cryolite content, the secondary crystalline phases can be $NaCaAl_2F_9$, CaF_2 and/or $Na_5Al_3F_{14}$. Twenty compositions were selected systematically throughout the electrolyte composition range. The liquidus temperatures for these compositions were measured to $\pm$ 2°C and the double solubility limit for Al_2O_3 in selected compositions was determined. The liquidus surface over the Hall cell electrolyte composition range is not smooth, but instead contains a deep inflection or swale. The liquidus temperatures over this range can be calculated using the empirical expression $T_{liquidus}$ (°C) = $1009.4 + 4.059\,(CaF_2) - 1.167\,(CaF_2)^2 + 0.968\,(CaF_2)\,(AlF_3) - 0.105\,(CaF_2)(AlF_3)^2 + 0.073\,(CaF_2)^2(AlF_3)^2 + 0.073\,(CaF_2)^2(AlF_3) + 0.002\,(CaF_2)^2(AlF_3)^2 - 4.165\,(AlF_3) - 0.054\,(AlF_3)^2 - 5.33\,(Al_2O_3)$. The double solubility limit for Al_2O_3 was found to vary from 6.5 to 2.0% Al_2O_3 over a temperature range of 960 to 860°C.

This investigation was funded by the Process Metallurgy Division, Alcoa, New Kensington, PA.

Introduction

The industrial production of aluminum involves the electrolysis of a molten bath of alumina and fluoride compounds. Cryolite (Na_3AlF_6) is the major component of the electrolyte and other compounds such as LiF, AlF_3, and CaF_2 are added to modify the electrical conductivity, density and other properties of the bath. Knowledge of the phase relations in the electrolyic cell are important to determine the optimum bath composition and lowest operating temperatures. The objectives of this investigation are to determine accurately the liquidus temperatures of the Hall cell electrolyte compositions and to furthermore determine the solubility of Al_2O_3 in selected compositions throughout this range. In other words, this investigation examines the phase relations in the high-cryolite corner of the CaF_2-AlF_3-Na_3AlF_6-Al_2O_3 system.

A comprehensive review of the phase equilibrium data for cryolite containing systems of importance to the aluminum industry was recently completed (1). This review together with the data on the CaF_2-AlF_3-Na_3AlF_6-Al_2O_3 system by Craig and Brown (2) summarize the present state of knowledge of the phase relations of Hall cell electrolyte compositions.

The phase diagrams for the CaF_2-AlF_3-Na_3AlF_6 and Na_3AlF_6-AlF_3-CaF_2-Al_2O_3 systems are shown in Figures 1 and 2 (2). These diagrams show that the primary phase upon crystallization of Hall cell electrolyte compositions is cryolite solid solution. The secondary crystalline phases are either $NaCaAlF_6$ + $Na_5Al_3F_{14}$ or $NaCaAlF_6$ + CaF_2 depending upon whether the composition is AlF_3 - or CaF_2-rich. The two quaternary invariant points of interest for electrolyte compositions both were reported to occur at low Al_2O_3 contents ($\sim$ 3 mole % Al_2O_3) and about 676°C.

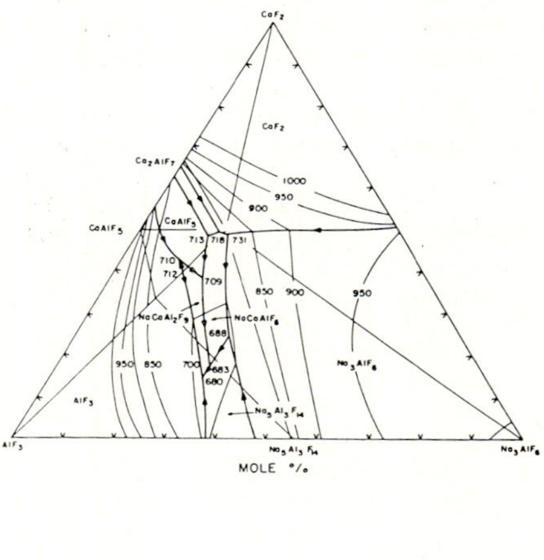

Figure 1. Phase equilibrium diagram for the system CaF_2-AlF_3-Na_3AlF_6.

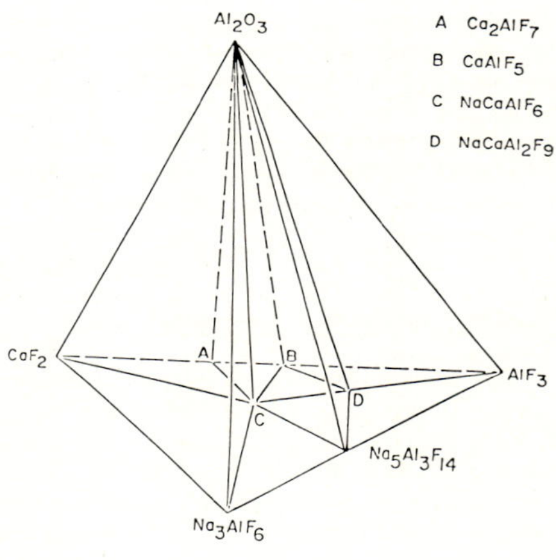

Figure 2. Subsolidus compatibility relations in the quaternary system Na_3AlF_6-AlF_3-CaF_2-Al_2O_3.

Experimental Procedure

Gravimetric weight factors were determined for all chemicals by heating them until no further weight loss occurred. Five- to ten-g batches of CaF_2,[†] handpicked Greenland cryolite (crushed and screened to $\leq$ 200 mesh), AlF_3[¶] and Al_2O_3[¶] were weighed on an analytical balance to $\pm$ 0.1 mg before hand-mixing under acetone. The samples were air-dried and calcined for 15 h at 300°C in fused-silica crucibles. Optical microscopy and X-ray diffraction (XRD) analyses were used to check for the presence of Al_2O_3. Prior to further heat treatment, each sample was encapsulated in a sealed Pt tube (0.125 cm OD by 1.5 cm long). Quench experiments were conducted in a Pt-Pt40Rh-wound vertical tube furnace controlled within $\pm$ 1.0°C. The temperature was measured using a Pt-Pt10Rh thermocouple that was periodically checked against a similar couple which had been standardized against the melting points of Au and Pd.

To prevent volatilization of the fluorides, DTA experiments were performed using samples encapsulated in sealed Pt tubes. The Pt tubes containing the sample and reference material (Al_2O_3) were placed in contact with the Pt-Pt10Rh differential thermocouple. A controller* permitted heating and cooling of the DTA furnace to and from 1400°C at a constant rate which could be varied between 0.5 and 20°C/min. Unless otherwise specified, DTA experiments were conducted at a heating rate of $\sim$ 5°C/min. The DTA was standardized using the α-β quartz inversion (573°C) and the melting points of KI (681°), NaCl (800.8°), and Na_2SO_4 (884.8°C). These references provided standardization points near the temperatures of the invariant reactions observed in the system.

Routine phase identification was performed using an X-ray power diffractometer[o] equipped with a graphite single-crystal monochromator and Ni-filtered CuKα radiation. A 2θ scanning rate of 1°/min was used. Precise measurements were obtained using a scanning rate of 1/4° 2θ/min with an Si disk as an external standard. Optical microscopic examinations were conducted using a petrographic microscope[x] or a universal camera microscope. Refractive indices were measured using Cargille[§] oils and the Becke´ line technique.

Twenty compositions were prepared to systematically study the Hall cell electrolyte compositin range. These compositions are shown in Figure 3.

Experimental Results

Liquidus Temperature Measurements

As is well-known, liquidus temperature measurements in the CaF_2-AlF_3-Na_3AlF_6 system are complicated by supercooling, sluggish compound formation and volatilization. In this study, all compositions were encapsulated in sealed platinum tubes and special care was used to insure the attainment of equilibrium. All quench samples were initially heated to 30°C above the liquidus temperature for 1 hr and gradually cooled to the test temperature.

---

† CaF_2, certified reagent, Fisher Scientific Co., Pittsburgh, PA.
¶ AlF_3, anhydrous, Matheson Coleman and Bell Manufacturing Chemists, Norwood, Ohio.
* Series 60, Leeds & Northrup, North Wales, PA.
o Norelco, North American Phillips Corp., New York, N.Y.
x Model MeF 2, C, Reichert Optische Werke, Vienna, Austria.
§ Cargille Laboratories, Inc., Cedar Grove, N.J.

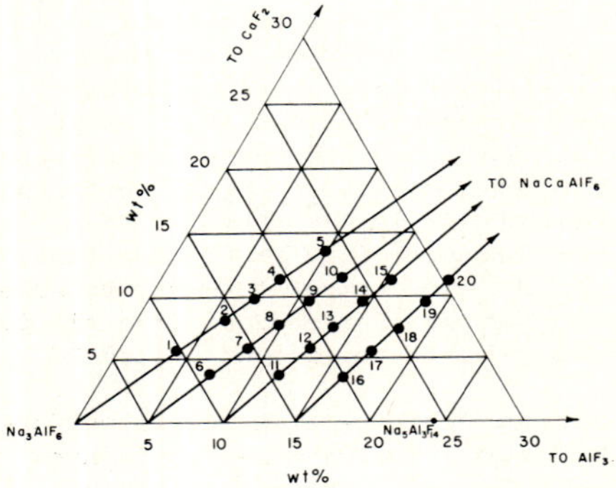

Figure 3. The compositions used to map the liquidus surface of the system Na_3AlF_6-CaF_2-AlF_3.

The results of the liquidus temperature measurements are shown in Table I for the 20 compositions studied (Figure 3) and plotted in Figures 4 through 7. The liquidus surface for the high-cryolite portion of the Na_3AlF_6-CaF_2-AlF_3 system is shown in Figure 8. In all cases the primary crystalline phase was cryolite.

The most noteworthy feature of these results is the deep swale that occurs in the liquidus surface almost directly over the normal compositions used in the Hall cell baths. This feature explains why it has been difficult to predict accurately melting temperature variations in an operating Hall cell as compositional changes occur. It is beyond the scope of this investigation to explain the reasons for this unusual liquidus surface, but it is apparently a reflection of the complex ionic species present in the liquid phase.

In an effort to obtain an empirical relationship between composition and liquidus temperature over the Hall cell compositions, polynomial regression analysis was used to fit the experimental data. It was found that the liquidus temperatures could be expressed as:

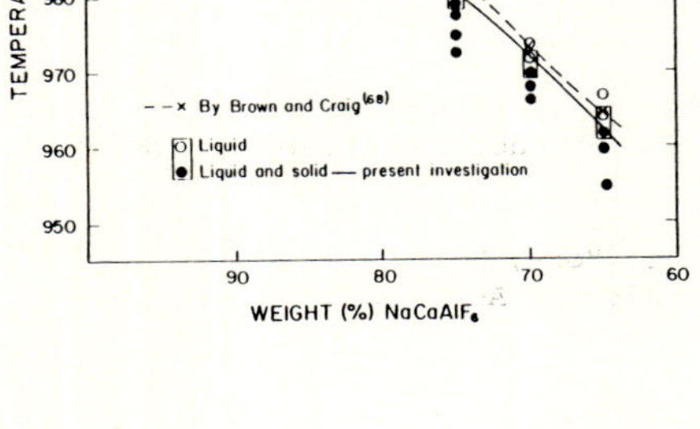

Figure 4. Liquidus curve for a portion of the Na_3AlF_6-$NaCaAlF_6$ join.

TABLE I. THE LIQUIDUS TEMPERATURES OF SELECTED COMPOSITIONS IN THE CRYOLITE-RICH PORTION OF THE SYSTEM Na_3AlF_6-CaF_2-AlF_3.

Composition	Theoretical (Wt.%)			Batch (Wt.%)				Measured Temperatures (°C)	
	Na_3AlF_6	$NaCaAlF_6$	AlF_3	Na_3AlF_6	CaF_2	AlF_3	Al_2O_3	Liquid + Solid	Liquid
1	85.0	15.0	-	90.14	5.74	4.04	0.08	993.4	994.9
2	80.0	20.0	-	86.86	7.63	5.40	0.11	987.1	988.7
3	75.0	25.0	-	83.57	9.57	6.73	0.13	979.9	981.4
4	70.0	30.0	-	80.29	11.48	8.07	0.16	970.0	971.9
5	65.0	35.0	-	77.0	13.39	9.42	0.19	961.7	964.1
6	85.0	10.0	4.5	88.93	3.83	7.10	0.14	991.5	993.5
7	80.75	15.0	4.25	85.89	5.74	8.20	0.17	984.5	986.5
8	76.0	20.0	4.0	82.87	7.63	9.30	0.20	900.5	981.5
9	71.25	25.0	3.75	79.82	9.57	10.40	0.21	969.0	971.0
10	66.5	30.0	3.5	76.79	11.48	11.49	0.23	959.0	961.0
11	81.0	10.0	9.0	84.43	3.83	11.51	0.23	954.5	956.5
12	76.5	15.0	8.5	81.65	5.74	12.37	0.24	945.5	948.5
13	72.0	15.0	8.0	78.86	7.66	13.22	0.26	938.5	940.5
14	67.5	25.0	7.5	76.08	9.57	14.07	0.28	932.5	935.5
15	63.0	30.0	7.0	73.29	11.49	14.93	0.29	927.5	929.5
16	76.5	10.0	13.5	79.93	3.83	15.93	0.31	909	911
17	72.25	15.0	12.75	77.40	5.74	16.53	0.33	896.5	897.5
18	68.0	20.0	12.0	74.86	7.66	17.14	0.34	883	884
19	63.75	25.0	11.25	72.33	9.57	17.75	0.35	874	875
20	59.5	30.0	10.5	69.79	11.49	18.36	0.36	870	872

$$T_{Liq.}\ (^{\circ}C) = 1009.4 + 4.059\ (CaF_2) - 1.167\ (CaF_2)^2 + 0.968\ (CaF_2)(AlF_3) - 0.105\ (CaF_2)(AlF_3)^2 + 0.073\ (CaF_2)^2\ (AlF_3) + 0.002\ (CaF_2)^2 \times (AlF_3) - 4.165\ (AlF_3) - 0.054\ (AlF_3)^2 - 5.33\ (Al_2O_3)$$

for CaF_2 3.8 ∿ 11.25%, AlF_2 5 ∿ 20%, where AlF_3, CaF_2 and Al_2O_3 are the weight percentages of AlF_3, CaF_2 and Al_2O_3 in the cryolite solution, respectively.

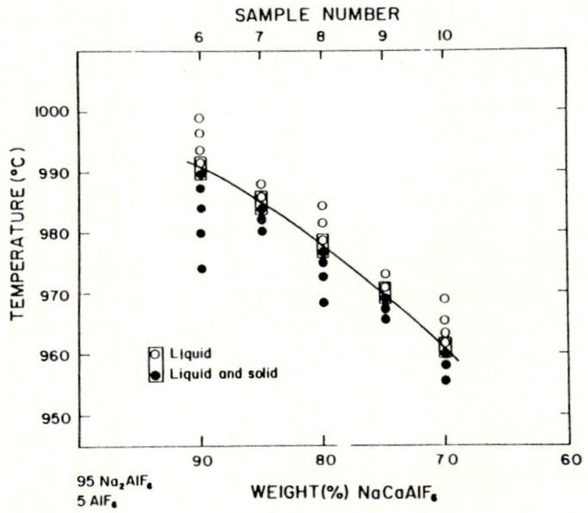

Figure 5. Liquidus curve for a portion of the 95 Na_3AlF_6/5 AlF_3-$NaCaAlF_6$ section.

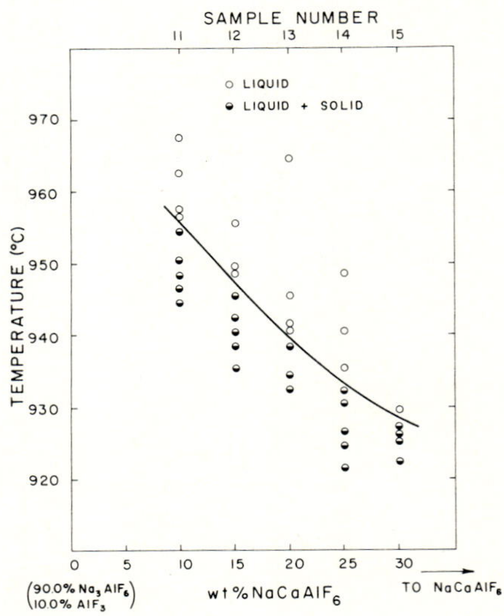

Figure 6. Liquidus temperature for the join 90 Na_3AlF_6/ 10 AlF_3-$NaCaAlF_6$.

From this expression, it is obvious that the effects of CaF_2 and AlF_3 on the cryolite liquidus temperature are not independent. This is in agreement with thermodynamics principles; namely, if thre are two or more solutes in a solution, the activity coefficient of any solution will be affected by the presence of the other solutes. The activity coefficient may be increased or decreased depending on the mutual action of the two solutes. Therefore, the effect of any component on the liquidus temperature will be affected by the presence of other components.

Alumina Solubility Measurements

In an effort to determine the solubility of alumina in Hall cell compositions up to 8 wt% Al_2O_3 was added to selected compositions and the variations in liquidus temperatures were determined. The results of this study are tabulated in Table II and plotted in Figures 9, 10, and 11. The resulting double solubility limits for Al_2O_3 over the composition range studied are summarized in Figures 12, 13, and 14.

For the compositions typically used in the Hall cell (80-86 wt.% Na_3AlF_6, 4-6 wt.% CaF_2, 5-10 wt.% AlF_3 and 3-5 wt.% Al_2O_3), which correspond to compositions 6, 7, and 8, the alumina contents at the double solubility limit are, respectively, 6.6, 5.9 and 5.4 wt.% and the corresponding temperatures are respectively, 958.5, 957.5 and 953.5°C. This means the operating

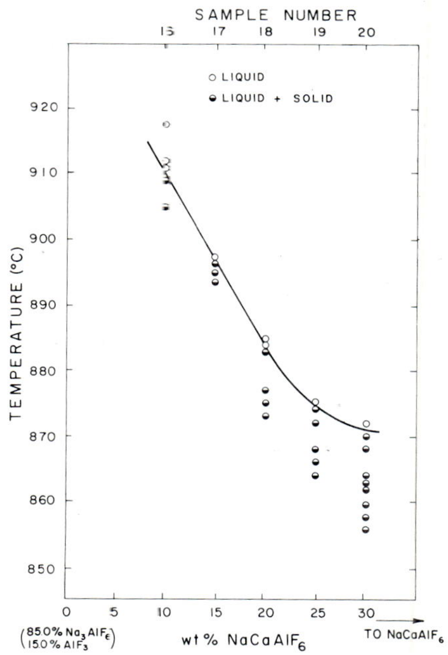

Figure 7. Liquidus temperature for the join 85 Na_3AlF_6/15 AlF_3-$NaCaAlF_6$.

TABLE II. THE ALUMINA SOLUBILITIES OF TEN SELECTED COMPOSITIONS

Composition*	Theoretical (Wt.%)				Batch (Wt.%)				Measured Temperatures (°C)	
	Na_3AlF_6	$NaCaAlF_6$	AlF_3	Al_2O_3	Na_3AlF_6	CaF_2	AlF_3	Al_2O_3	Liquid+Solid	Liquid
6-0	85.0	10	4.5	-	88.93	3.83	7.10	0.14	991.5**	993.5
6-2	83.79	9.8	4.41	2.0	87.15	3.75	7.10	2.0	982.5**	984.5
6-4	82.08	9.6	4.32	4.0	85.38	3.67	6.95	4.0	971.0**	973.0
6-6	80.37	9.4	4.23	6.0	83.59	3.60	6.81	6.0	960.0**	961.5
6-7	79.515	9.3	4.185	7.0	82.70	3.56	6.74	7.0	969.0***	971.0
6-8	78.66	9.2	4.14	8.0	81.82	3.52	6.66	8.0	1000.5***	1002.5
7-0	80.75	15	4.25	-	85.89	5.74	8.20	0.17	984.5**	986.5
7-2	79.135	14.7	4.165	2.0	84.17	5.63	8.20	2.0	976.0**	978.5
7-4	77.52	14.4	4.08	4.0	82.46	5.51	8.03	4.0	966.0**	968.0
7-6	75.905	14.1	3.995	6.0	80.74	5.40	7.86	6.0	959.0***	961.0
7-7	75.098	13.95	3.952	7.0	79.88	5.34	7.78	7.0	988.5***	990.5
7-8	74.29	13.8	3.91	8.0	79.02	5.28	7.70	8.0	1017.5***	1019.5
8-0	76.0	20	4.0	0	82.87	7.63	9.30	0.20	980.5**	981.5
8-2	74.48	19.6	3.92	2	81.20	7.50	9.30	2.0	970.0**	971.5
8-4	72.96	19.2	3.84	4	79.54	7.35	9.11	4.0	962.0**	962.0
8-6	71.44	18.8	3.76	6	77.89	7.19	8.92	6.0	970.0***	972.0
8-7	70.68	18.6	13.72	7	77.06	7.12	8.82	7.0	999.5***	1001.5
9-0	71.25	25	3.75	-	79.82	9.57	10.40	0.21	969.0**	971.0
9-2	69.825	24.5	3.675	2.0	78.23	9.38	10.40	2.0	959.0**	961.0
9-4	68.4	24	3.6	4.0	76.63	9.18	10.19	4.0	950.0**	952.0
9-6	66.975	23.5	3.525	6.0	75.04	8.99	9.97	6.0	968.0***	970.0
9-7	66.263	23	3.487	7.0	74.23	8.90	9.87	7.0	997.5***	999.5
10-0	66.5	30	3.5	-	76.79	11.48	11.50	0.23	959.0**	961.0
10-1	65.835	29.7	3.465	1.0	76.02	11.37	11.61	1.0	954.0**	956.0
10-2	65.17	29.4	3.43	2.0	75.25	11.25	11.50	2.0	949.0**	951.0
10-4	63.84	28.8	3.36	4.0	73.72	11.02	11.26	4.0	940.0***	942.0
10-6	62.41	28.2	3.29	6.0	72.18	10.79	11.03	6.0	999.5***	1001.5

TABLE II. (Cont'd)

Composition*	Theoretical (Wt.%)				Batch (Wt.%)				Measured Temperatures (°C)	
	Na_3AlF_6	$NaCaAlF_6$	AlF_3	Al_2O_3	Na_3AlF_6	CaF_2	AlF_3	Al_2O_3	Liquid+Solid	Liquid
11-0	81.0	10.0	9.0	-	84.43	3.83	11.51	0.23	954.5**	956.5
11-2	79.38	9.8	8.82	2.0	82.74	3.75	11.51	2.0	944.5**	946.5
11-4	77.76	9.6	8.64	4.0	81.06	3.67	11.27	4.0	933.5**	935.5
11-6	76.14	9.4	8.46	6.0	79.36	3.60	11.04	6.0	959.5***	961.5
11-7	75.33	9.3	8.37	7.0	78.52	3.56	10.92	7.0	989.0***	991.0
13-0	72.0	20.0	8.0	-	78.86	7.66	13.22	0.26	938.5**	940.5
13-2	70.56	19.6	7.84	2.0	77.28	7.50	13.22	2.0	928.5**	930.5
13-4	69.12	19.2	7.68	4.0	75.70	7.35	12.95	4.0	917.0**	919.0
13-6	67.68	18.8	7.52	6.0	74.13	7.19	12.68	6.0	942.5***	945.5
13-7	66.96	18.6	7.44	7.0	73.34	7.12	12.54	7.0	974.5***	974.5
15-0	63.0	30.0	7.0	-	73.29	11.49	14.93	0.29	927.5**	929.5
15-1	62.37	29.70	6.93	1.0	72.56	11.36	15.08	1.0	922.0**	924.0
15-2	61.74	29.40	6.86	2.0	71.82	11.25	14.93	2.0	917.0**	919.0
15-4	60.48	28.80	6.72	4.0	70.36	11.02	14.62	4.0	915.0***	917.0
15-6	59.22	28.20	6.58	6.0	68.89	10.79	14.32	6.0	975.0***	977.0
17-0	72.25	15.0	12.75	-	77.39	5.75	16.53	0.33	895.0**	896.5
17-1	71.53	14.85	12.62	1.0	76.62	5.68	16.70	1.0	890.0**	892.0
17-2	70.81	14.70	12.50	2.0	75.84	5.63	16.53	2.0	884.0**	886.0
17-4	69.36	14.40	12.24	4.0	74.30	5.51	16.19	4.0	921.5***	923.5
17-6	67.92	14.10	11.99	6.0	72.74	5.40	15.86	6.0	980.0***	982.0
19-0	63.75	25.0	11.25	-	72.32	9.57	17.75	0.36	874.0**	875.0
19-1	63.11	24.75	11.14	1.0	71.60	9.47	17.93	1.0	867.5**	870.0
19-2	62.475	24.50	11.025	2.0	70.88	9.37	17.75	2.0	861.5**	863.5
19-4	61.20	24.0	10.8	4.0	69.43	9.18	17.39	4.0	912.0***	914.0
19-6	59.925	23.5	10.575	6.0	67.99	8.99	17.02	6.0	972.5***	974.5

*In this item, the first numbers are the composition numbers shown in Fig. 3, and the second numbers are given as the Al_2O_3 contents.

**Cryolite + liquid.

***Corundum + liquid.

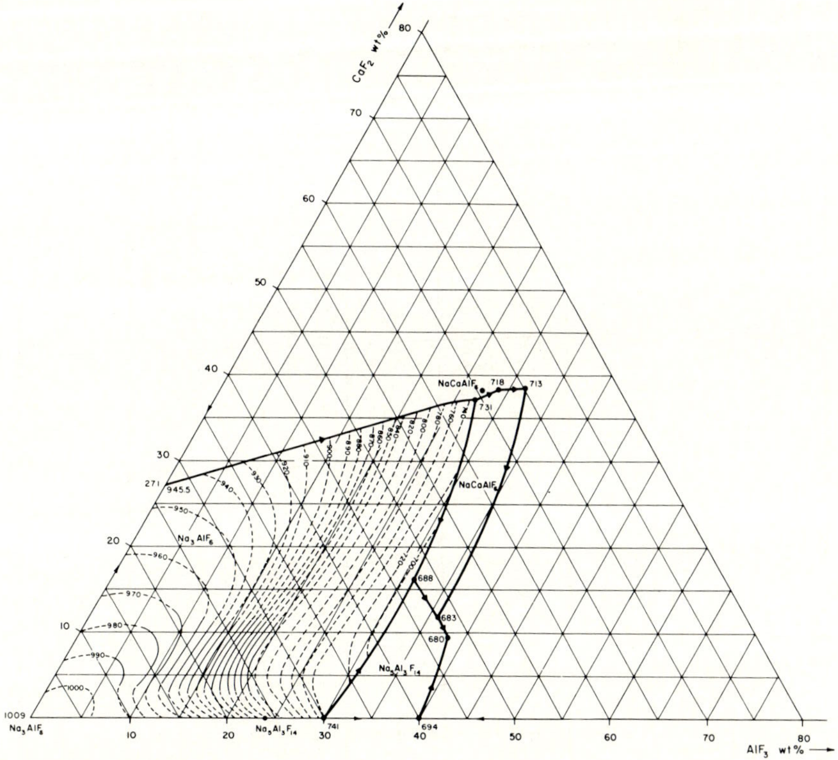

Figure 8. Isotherms for cryolite-rich portion of the ternary system Na_3AlF_6-CaF_2-AlF_3.

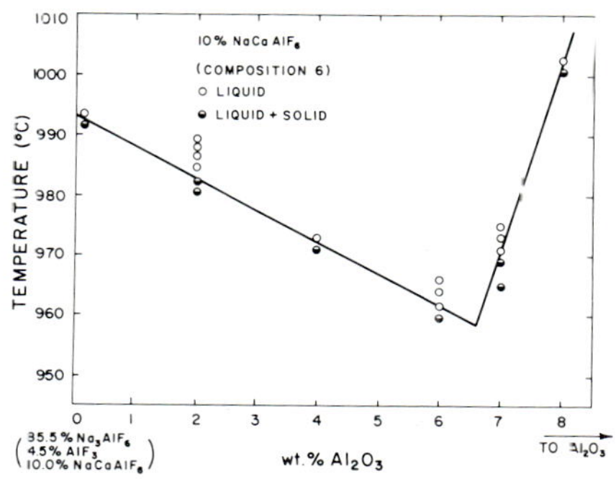

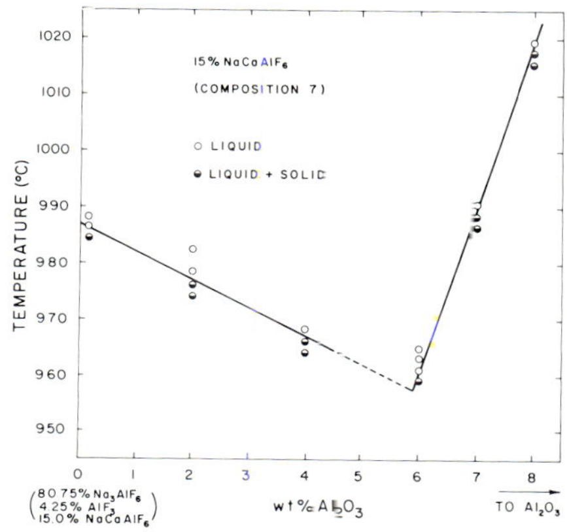

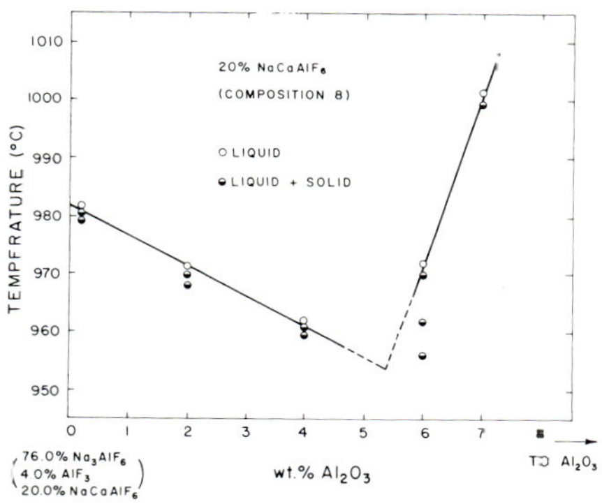

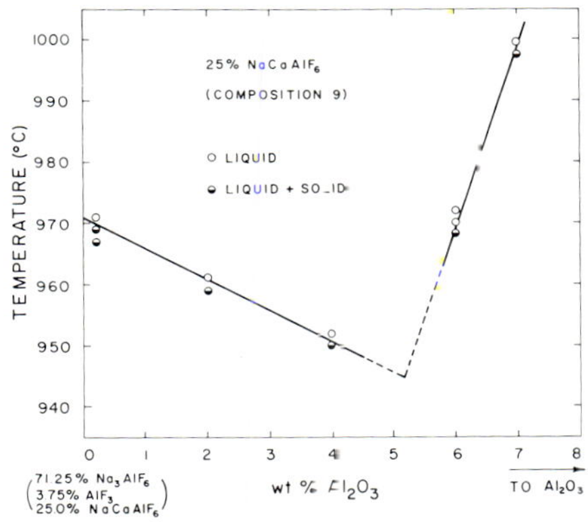

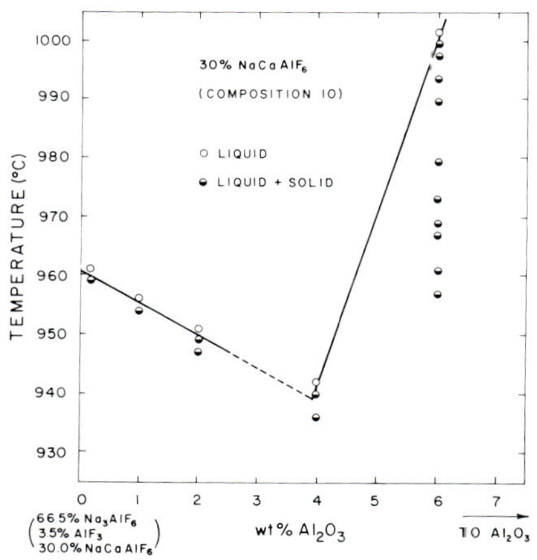

Figure 9. The effect of Al_2O_3 on the liquidus temperatures of the join 95 Na_3AlF_6/ 5 AlF_3-$NaCaAlF_6$.

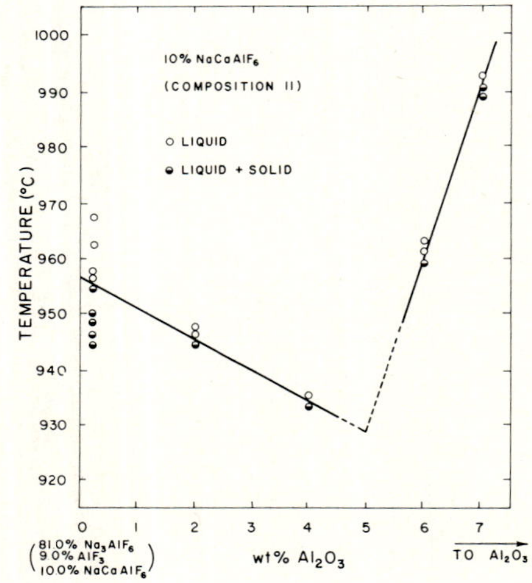

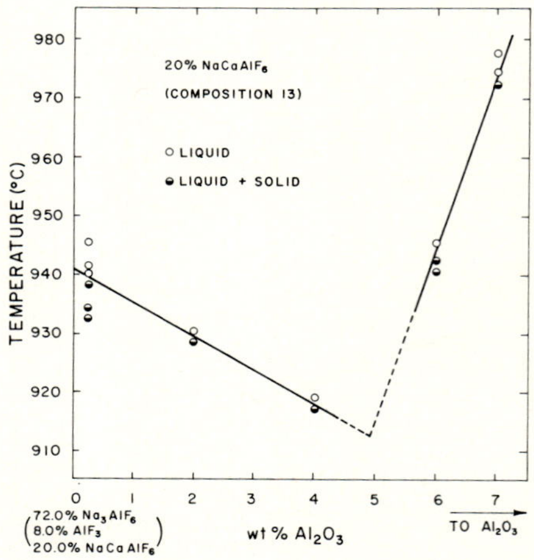

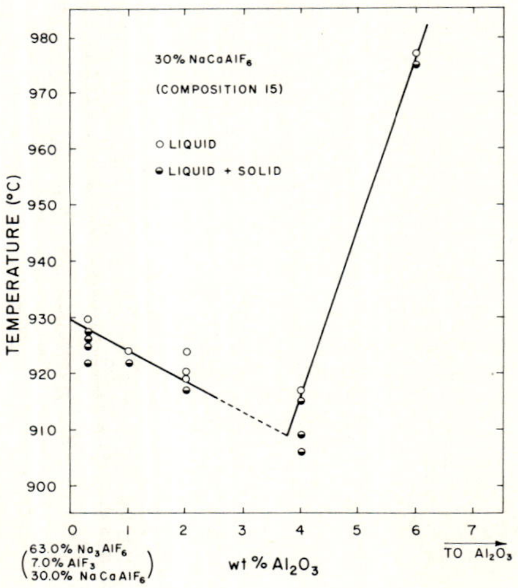

Figure 10. The effect of Al_2O_3 on the liquidus temperatures of the join 90 Na_3AlF_6/ 10 AlF_3-$NaCaAlF_6$.

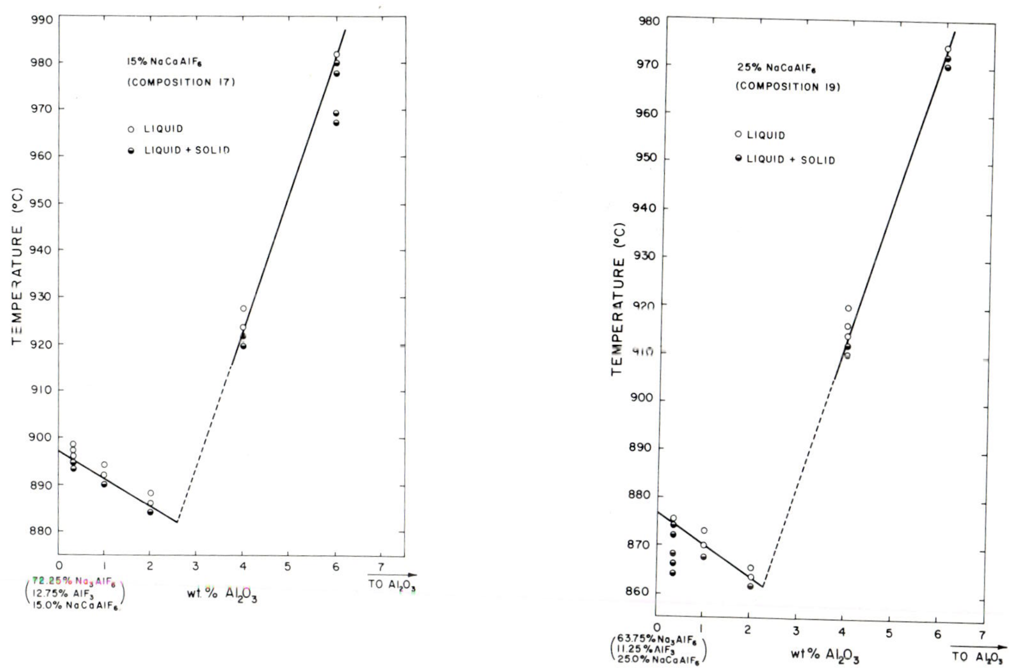

Figure 11. The effect of Al_2O_3 on the liquidus temperatures of the join 85 Na_3AlF_6/15 AlF_3-$NaCaAlF_6$.

temperature for the cell may be as low as 975 ∿ 980°C. For some compositions 11, 12, 13, 15, 16, 17, etc.) having a lower cryolite ratio than that of the typical Hall cell, the melting temperatures are lower. As a result, a lower operating temperature could conceivably be used. However, the alumina contents of the double solubility limit will be less than 5 wt.% and in some cases even less than 3 wt.%.

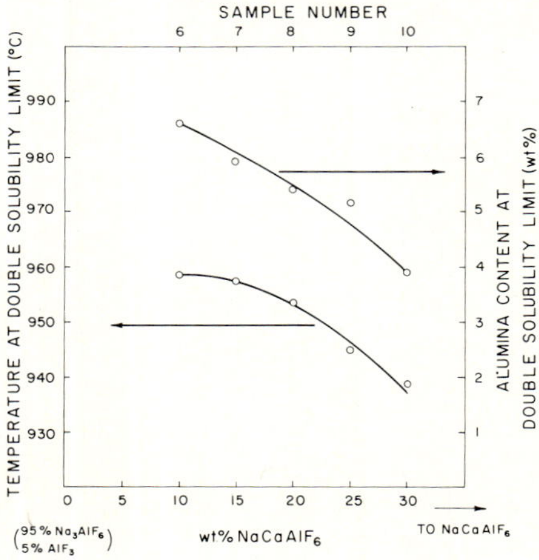

Figure 12. The effect of Al_2O_3 on the temperatures and alumina contents at the double solubility limit of the join 95 Na_3AlF_6/ 5 AlF_3-$NaCaAlF_6$.

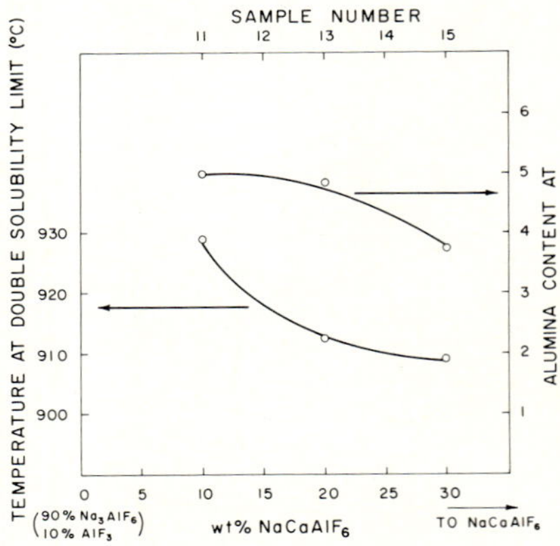

Figure 13. The effect of Al_2O_3 on the temperatures and alumina contents at the double solubility limit of the join 90 Na_3AlF_6/10 AlF_3-$NaCaAlF_6$.

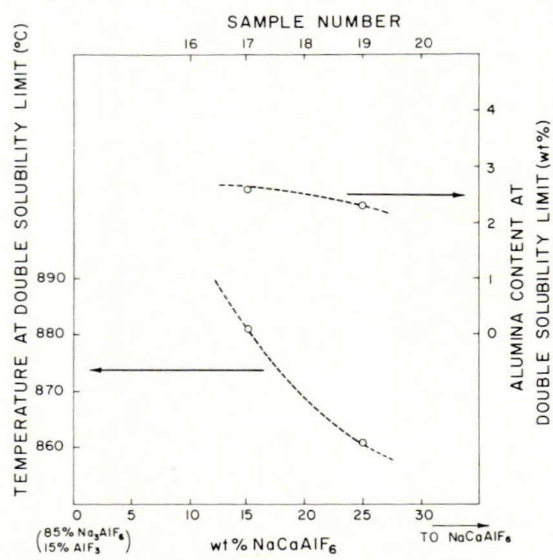

Figure 14. The effect of Al_2O_3 on the temperatures and alumina contents at the double solubility limit of the join 85 Na_3AlF_6/15 AlF_3-$NaCaAlF_6$.

References

D. F. Craig, R. T. Cassidy, and J. J. Brown, pp. 272-345 in Applications of Phase Diagrams in Metallurgy and Ceramics, Natl. Bur. Stand. (U.S.), Spec. Publ. No. 496, Vol. 1, 1978.

D. F. Craig and J. J. Brown, "Phase Equilibria in the System CaF_2-AlF_3-Na_3AlF_6 and Part of the System CaF_2-AlF_3-Na_3AlF_6-Al_2O_3", J. Am. Cer. Soc., 63(5-6) 1980, pp. 254-61.

CHAIRMEN

L.G. Boxall
Martin Marietta Laboratories
Baltimore, Maryland

M.F. Vogt
Kaiser Aluminum and Chemical Corp.
Chalmette, Louisiana

COKER FEEDSTOCK CHARACTERISTICS AND CALCINED COKE PROPERTIES

P.J. Rhedey
Alcan International Limited
Kingston Laboratories
Kingston, Ontario, Canada
K7L 4Z4

S.K. Nadkarni
Alcan International Limited
Arvida Laboratories
Jonquiere, Québec, Canada
G7S 4K8

Abstract

Commercial delayed coker feedstocks and corresponding green cokes were obtained from four refineries. The feedstock materials were characterized by infrared, chromatographic, and solvent extraction techniques in addition to the conventional physical tests and chemical analyses. The green cokes were calcined in a laboratory induction furnace at 1100, 1200, 1300 and 1400°C. The properties of the resulting calcined cokes were related to the feedstock characteristics. The results indicated significant effects of the feedstock materials which should be taken into account during the control of commercial calcination operations.

Introduction

Variations in the composition of refinery coker feedstocks have significant effects on the properties of electrode-grade calcined petroleum coke. In a previous paper the effects on coke structure and bench-scale anode performance were discussed (1). This paper deals with the effects of feedstock on the properties of calcined coke with particular emphasis on those properties which are predominantly indicative of the degree of calcination.

Materials

Commercial coker feedstocks and corresponding green delayed cokes were used from four different refineries. A catalytic cracker slurry oil and a coal-tar pitch were included for comparison. These two materials were coked in the laboratory under a nitrogen atmosphere at 500°C to obtain the corresponding green cokes.

The feedstocks are designated as follows:

Source	Symbols Used in Figures 2 and 3
1 Vacuum Tower Bottom	○
2 Vacuum Tower Bottom	●
3 Vacuum Tower Bottom	△
4 Vacuum Tower Bottom	▲
5 Catalytic Cracker Bottom	□
6 Coal Tar Pitch	■

Experimental

Feedstocks. The feedstocks were characterized by elemental analysis, density (2), viscosity over a range of temperatures using a Brookfield viscometer, coke yield as determined by the Alcan coking value test, asphaltenes content (3), refractive index (4), and infrared spectroscopy, using a NICOLET 7000 series FTIR spectrometer. The infrared parameters reported were obtained by integrating the area under the peak of the spectra over the 670 to 920 cm^{-1} wave number region and multiplying this by the absorbance units. The aromatic/non-aromatic components were determined by silica gel solvent elution technique (5).

Density-refractivity intercept was calculated from refractive index and density (refractivity intercept = refractive index at 20°C $-0.5 \times$ density at 20°C, g/mL) (6). The viscosity-gravity constant (VGC) was derived from Saybolt viscosity at 210°F and specific gravity, 60/60°F (7).

Green Cokes. The green cokes were analysed for volatile matter content, carbon, hydrogen, sulphur, and vibrated bulk density.

Calcined Cokes. To obtain the calcined cokes, green cokes (minus 830 plus 420 micrometers, 20 × 35 Tyler mesh) were calcined in a laboratory induction furnace shown in Figure 1 at 1100, 1200, 1300, and 1400°C. The furnace was purged with nitrogen. The heating rate was 50°C/min and the

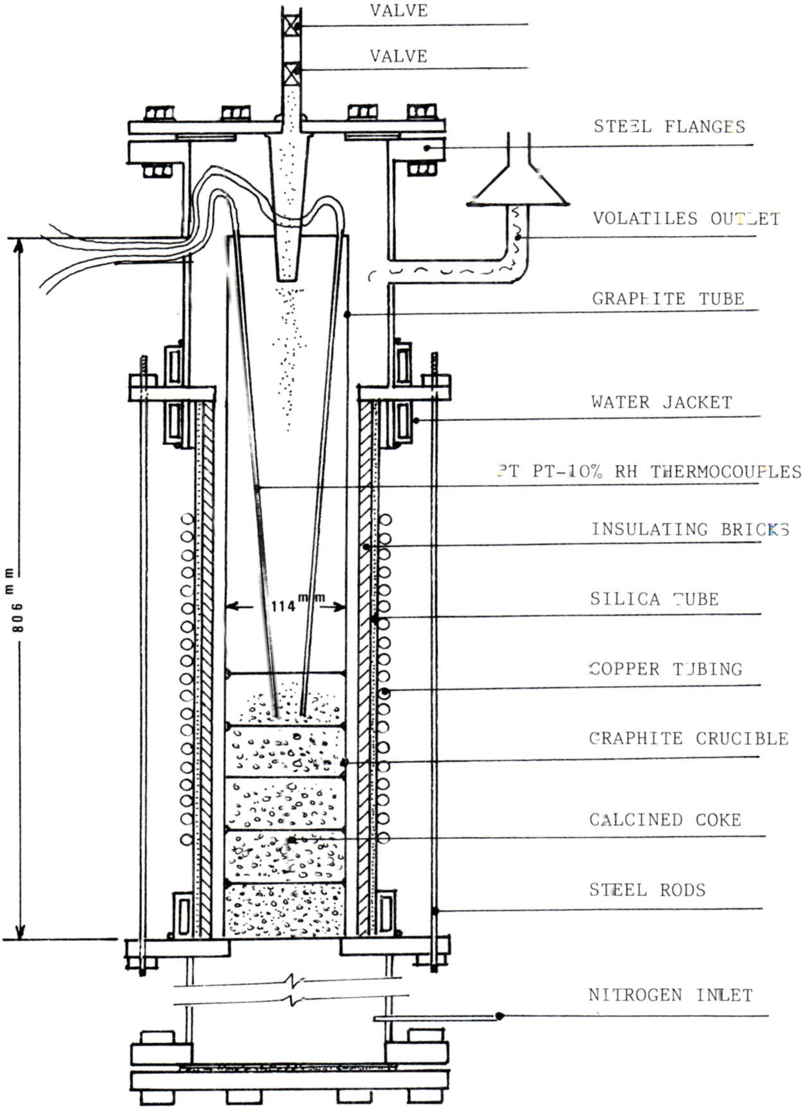

Fig. 1 - Laboratory Induction Furnace Used for Calcination.

holding time at the final temperature was 30 minutes.

One sample of each green coke was also flash calcined (heating rate: approximately 1000°C/min) to 1200°C in a nitrogen atmosphere and then held at this temperature for 30 minutes.

The properties determined on calcined coke were sulphur, mean crystallite thickness by X-ray diffraction method, kerosene density (minus 40 micrometers, 400 Tyler mesh) and vibrated bulk density (minus 830 plus 420 micrometers, 20 × 35 Tyler mesh).

Results and Discussion

The elemental analysis of the feedstocks is given in Table I. Source 3 exhibits relatively high sulphur content. The C/H atomic ratio does not reveal significant differences between feedstocks 1 through 4. For sources 5 and 6 it is in agreement with values expected for slurry oil and coal tar pitch, respectively.

Other properties are given in Table II. Of significance here is the low value of infrared parameter for feedstock 2 indicating the lowest number of low molecular weight benzene ring compounds (less than 5 to 7 rings) for this material.

The results of further characterization of the feedstocks are given in Table III. Non-aromatic and aromatic compounds are reported as determined by chromatographic separation using silica gel absorbent and three selective solvents: n-heptane, benzene, and pyridine. The non-aromatics include paraffins, olefins, and naphthenes which can be removed from silica gel with n-heptane. The aromatic compounds are not removed with n-heptane, but are easily eluted with benzene. The polar compounds which are eluted with pyridine include materials containing oxygen, sulphur, and nitrogen. The not recovered fraction represents volatile, low molecular weight hydrocarbons, as well as pyridine insoluble compounds.

Composition by carbon-type is also given in Table III. These results were obtained by the density-refractivity intercept method which has been shown to be suitable for the characterization of a wide range of viscous petroleum fractions (6).

The calculated viscosity-gravity constant given in Table III is another indicator of hydrocarbon composition. Petroleum hydrocarbons with a VGC near 0.800 are of paraffinic character while those with VGC close to 1.000 are of aromatic nature. It should be noted that the precision of the VGC is very good; repeatability at the 95 percent confidence limit is 0.001 unit.

Inspection of the results in Tables II and III indicates that the aromaticity of vacuum bottom feedstocks increases from source 1 through 4. The significantly higher aromaticity found for the catalytic cracker bottom, source 5, is as expected. The coal-tar pitch shown for comparison exhibits the highest aromaticity based on infrared analysis results and VGC value which is in agreement with published information for coal asphaltenes.

The properties of the green cokes are given in Table IV. Of considerable significance is the low bulk density of the coke from source 2.

Table I. Elemental Analysis of Coker Feedstocks

Source (see "Materials")		1	2	3	4	5	6
Carbon	%	83.8	85.8	84.7	85.8	88.4	93.1
Hydrogen	%	11.5	11.4	10.4	11.2	8.4	4.2
Nitrogen	%	0.43	0.45	0.42	0.31	ND	ND
Oxygen	%	3.61	2.01	2.66	1.82	ND	ND
Sulphur	%	0.66	0.88	1.48	0.87	0.90	0.55
C/H (Atomic Ratio)		0.61	0.63	0.68	0.64	0.38	1.84

ND = Not Determined

Table II. Properties of Coker Feedstocks

Source (see "Materials")		1	2	3	4	5	6
API Gravity		14.4	14.4	10.0	13.0	2.0	ND
Density at 20°C	g/mL	0.970	0.970	1.000	0.979	1.060	1.315
Density at 100°C	g/mL	0.923	0.924	0.964	0.930	1.007	ND
EVT-10 (1)	°C	77	76	105	84	36	165
EVT-1000 (1)	°C	32	33	55	40	14	112
Coke Yield	%	13.1	12.8	18.3	18.2	6.7	58.6
N-Heptane Insoluble	%	7.8	10.1	7.8	13.4	1.0	72.0
Refractive Index		1.544	1.542	1.558	1.553	1.642	ND
Infrared Parameter (2)		0.591	0.474	0.593	0.549	1.547	2.650

(1) Equiviscous temperature at 10 and 1000 poises
(2) Integrated area under peak in 670-920 cm^{-1} wave number region × absorbance units
ND = Not Determined

Table III. Characterization Data for Coker Feedstocks

Source (see "Materials")		1	2	3	4	5	6
Chromatography							
Non-aromatics (oils)	%	88.5	81.2	85.1	70.8	51.0	ND
Aromatics (asphaltenes)	%	8.8	13.1	13.2	18.9	34.0	ND
Oxy (polar) compounds	%	0.9	0.5	1.4	9.7	5.0	ND
Not recovered	%	1.8	5.2	0.3	0.5	10.0	ND
Density - Refractivity Intercept							
Aromatic ring carbons	%	25	24	29	30	66	ND
Naphthene ring carbons	%	18	20	27	14	-	ND
Paraffin chain carbons	%	57	56	44	56	34	ND
Viscosity-Gravity Constant		0.854	0.858	0.862	0.863	1.042	1.550

ND = Not Determined

Table IV. Properties of Green Cokes

Source (see "Materials")		1	2	3	4	5	6
Volatile Matter	%	11.1	10.6	11.2	9.6	5.8	6.1
Carbon	%	92.0	91.2	90.4	91.4	93.4	94.8
Hydrogen	%	3.5	3.6	3.8	3.8	3.6	3.0
Sulphur	%	1.07	1.86	3.82	2.55	1.84	0.34
Bulk Density	g/mL	0.762	0.687	0.732	0.721	ND	ND

ND = Not Determined

This is attributed to the high value of "not recovered" fraction found in feedstock 2 as shown in Table III. It is postulated that the presence of volatile, low molecular weight hydrocarbons is responsible for the low green coke bulk density. These compounds are expected to have a reduced coke yield and create a higher porosity of the solidifying material in the drum during coking.

Table V shows selected properties of the cokes calcined in the laboratory at 1200°C while Figures 2 and 3 show the correlation of kerosene density and mean crystallite thickness with calcination temperature, respectively. It can be seen in Table V that the low bulk density of coke from feedstock 2 prevails after calcination in the laboratory. This behaviour of source 2 coke has also been observed and reconfirmed during calcination trials in a commercial rotary kiln.

Table V. Selected Properties of Calcined Cokes

(Laboratory Calcination at 1200°C)

Source (see "Materials")		1	2	3	4	5	6
Sulphur	%	1.05	2.00	3.68	2.48	1.75	0.32
Mean Crystallite Thickness	Å	26.0	26.8	28.4	28.0	27.0	24.0
Kerosene Density	g/mL	2.05	2.05	2.06	2.08	2.06	1.96
Bulk Density	g/mL	1.006	0.839	0.958	0.993	ND	ND
Bulk Density Upon Flash Heating	g/mL	0.844	0.769	0.820	0.833	ND	ND

ND = Not Determined

Figures 2 and 3 reflect two well-known features which will not be discussed here, namely the very different response of coal tar pitch to carbonization and heat treatment when compared to petroleum fractions and the effect of thermal desulphurization on the real density of high-sulphur petroleum coke. Both aspects have been adequately dealt with in the literature (8,9).

In this paper attention is focused on the usefulness of kerosene density and mean crystallite thickness as measures of degree of heat treatment and control tools for commercial calcination.

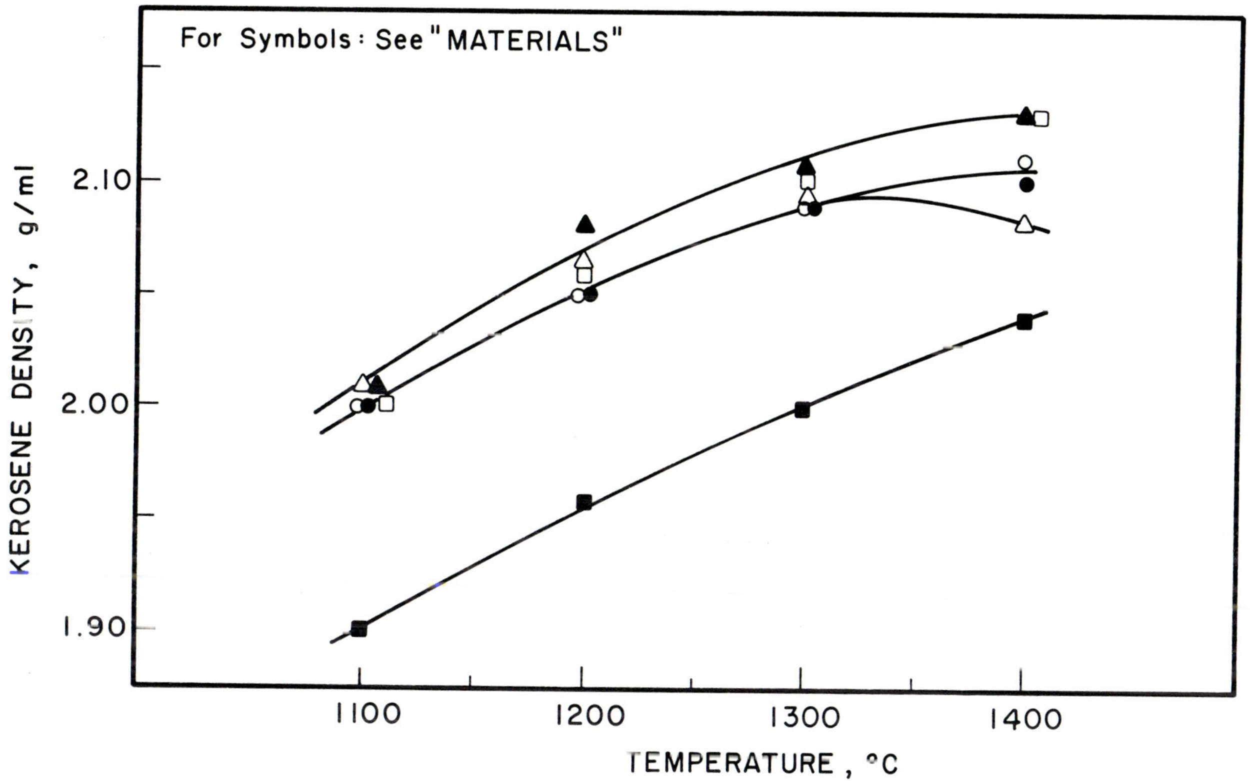

Fig. 2 - Correlation Between Kerosene Density and Temperature of Calcination.

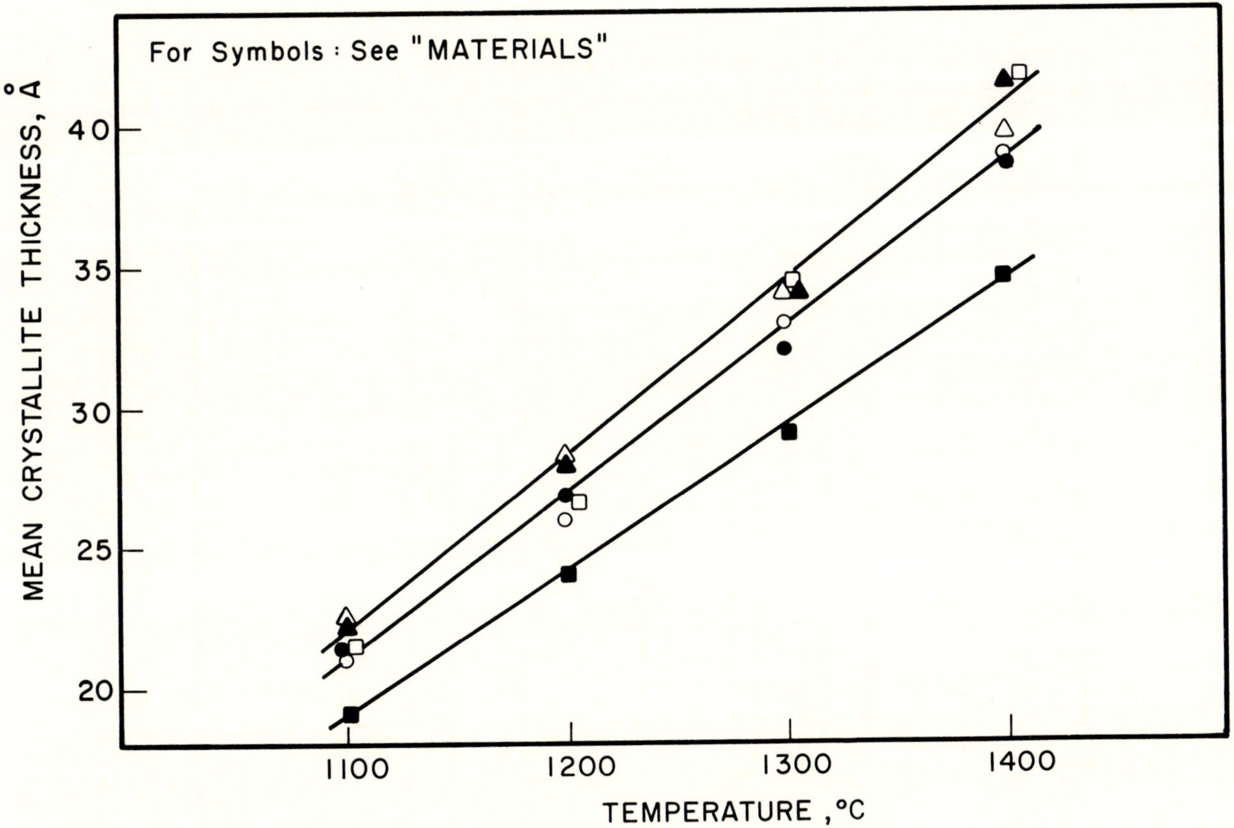

Fig. 3 - Correlation Between Mean Crystallite Thickness and Temperature of Calcination.

The petroleum feedstocks in terms of corresponding calcined cokes are represented by two lines in Figures 2 and 3: cokes 1 and 2 derived from feedstocks of lower aromaticity and cokes 3 to 5 obtained from hydrocarbons of higher aromaticity. This distinction was made on the basis of the aromatic ring carbon concentration as determined by the density - refractivity intercept method (Table III). The lines have been drawn using the least squares method. A statistical evaluation showed that the two lines are not significantly different up to a calcination temperature of 1250°C in the case of kerosene density and up to 1150°C in the case of mean crystallite thickness.

For kerosene density, KD, if one regards the five samples of petroleum based materials as one group, the following equation can be written (with straight line approximation and disregarding desulphurization effects):

$$\text{KD g/mL} = 1.604 + 3.72 \times 10^{-4}\,T,\ ^{\circ}\text{C}$$

Correlation coefficient, r = 0.954
Standard error = 0.0137 g/mL

and the corresponding temperature estimate with 95% confidence limit is within ±70°C. The latter can be reduced to ±60°C if separate correlations are used for samples 1 and 2 and the group of 3 to 5 at temperatures higher than 1250°C.

On the other hand, for mean crystallite thickness, L_c, the equation using all five sample points is:

$$L_c,\ \text{Å} = -48.05 + 6.28 \times 10^{-2}\,T,\ ^{\circ}\text{C}$$

Correlation coefficient, r = 0.989
Standard error = 1.20 Å

and the corresponding temperature estimate with 95% confidence limit is within ±32°C. If two separate correlations are considered, above 1150°C the temperature estimate can be reduced to within ±24°C. The mean crystallite thickness is therefore clearly superior to kerosene density as a calcination control parameter.

The statistical significance of the correlations could have been improved by carrying out calcinations at 1500°C, since however, the trend in KD and L_c is well established this was not felt to be justifiable.

Conclusions

Coker feedstock materials can be characterized by relatively simple tests such as density, refractivity, infrared spectrum, chromatographic separation of structural components, and viscosity-gravity constant. Differences in calcined coke properties can occur depending on the aromaticity of the petroleum feedstocks. Kerosene density can be 0.02 g/mL higher and mean crystallite thickness 1.5 angstroms higher for a catalytic cracker bottom material than for a less aromatic vacuum tower bottom feedstock at temperatures used in industrial calciners. Mean crystallite thickness is a more precise measure of the degree of calcination than kerosene density. There is an indication that the presence of volatile hydrocarbons in the feedstock determined by chromatographic separation as "not recovered" fraction reduces the bulk density of the calcined coke.

Acknowledgements

The authors are grateful to P.J. Ashley and B.R. Pathak of Alcan International's Kingston Laboratories for performing the FTIR studies, and to R.F. Jooste of the same laboratory for carrying out the refractivity measurements.

References

(1) P.J. Rhedey and D. DuTremblay, Refinery Feedstocks, Coke Structures and Aluminum Cell Anodes, Light Metals 1977, AIME, New York, N.Y., Vol. 1, pp. 301-316.

(2) ASTM Method D-287-67 (1977), API Gravity of Crude Petroleum and Petroleum Products (Hydrometer Method).

(3) ASTM Method D-3279-81, Asphaltenes as n-Heptane Insolubles.

(4) ASTM Method D-1218-61 (1977), Refractive Index and Refractive Dispersion of Hydrocarbon Liquids.

(5) L.T. Eby, Properties of High Boiling Petroleum Products, High Aromatics Characterization by Chromatograhy, Analytical Chemistry, Vol. 25, No. 7, 7 July 1953, pp. 1057-1060.

(6) S.S. Kurtz, R.W. King, W.J. Stout, and M.E. Peterkin, Carbon-type Composition of Viscous Fractions of Petroleum: Density-Refractivity Intercept Method, Analytical Chemistry, Vol. 30, No. 7, July 1958, pp. 1224-1236.

(7) ASTM Method D-2501-67 (1977), Calculation of Viscosity-Gravity Constant (VGC) of Petroleum Oils.

(8) H. Tillmanns, The Influence of Quinoline Insoluble Pitch Constituents on the Carbonization Rate and Coke Structure, 15th Biennial Conference on Carbon, Extended Abstracts, Philadelphia, Pennsylvania, 22-23 June 1981, pp. 142-143.

(9) M.H. Wagner, H. Pauls, H. Tillmans, and G. Wilhelmi, Calcination - The Effect of Maximum Heat Treatment Temperature on the Properties of Coke, 16th Biennial Conference on Carbon, Extended Abstracts, San Diego, California, 18-22 July 1983, pp. 582-583.

INFLUENCE OF MAXIMUM TEMPERATURE AND HEAT-SOAK TIMES

ON THE PROPERTIES OF CALCINED COKE

A.F. Bopp, G.B. Groff, and B.H. Howard

Martin Marietta Laboratories
Baltimore, Maryland 21227
USA

Laboratory studies have been performed to determine the effects of maximum temperature and heat-soak times on the properties of calcined petroleum cokes. Mercury porosimetry, BET specific surface area, mean crystallite size, and air reactivity have been measured and correlated with experimental variables. The results of these experiments are being used to gain insight on controlling calcined coke properties by regulating kiln conditions.

Introduction

Coke calcination can be thought to occur in three phases: dehydration, devolatilization, and densification. The high temperatures in a rotary kiln or a rotary hearth furnace complete the carbonization process begun during coking by heating the coke to 1200-1300°C. Devolatilization begins around 500°C when light hydrocarbons are evolved followed by hydrogen around 700°C (1). This devolatilization reduces the atomic H:C ratio from about 0.5 for green cokes to 0.1 for calcined cokes. As the temperature increases, graphitic planes gain mobility through increased kinetic energy and decreased hindrance from side chains, which were lost as volatiles. These crystallites, which are highly disordered in green coke, begin to align to some degree although they are still amorphous compared to graphite. This crystal growth results in densification of the carbon. Temperature basically has no effect on isotropy, ash content, or sulfur except at extremes normally avoided during calcination.

The changes in structure and properties of coke that occur during calcination have been measured in the laboratory and the results published (1-4). These studies focused on a variety of carbons and a full range of temperature, heating rate, and atmospheres. Rhedy's article (3) is considered the landmark work. He observed that for petroleum cokes the heat-up rate through the temperature range 600-900°C was critical to certain coke properties, especially porosity. In other more recent papers (5,6) the importance of a critical temperature region has been reaffirmed.

Studying the changes that occur during commercial calcination is very difficult given the extreme environmental conditions associated with the process. Basically there are two ways to accomplish such a study. One way is to perform a basic parametric experiment by changing calcination conditions, observing the effects on the product and then statistically correlating coke properties with feed properties and calcination conditions. Another way is to interrupt the calcination process at various stages and measure key properties. Despite obvious relevance, no such comprehensive sampling tests have been reported in the literature.

The object of this work was to study the effects of maximum temperature and heat-soak periods on the properties of cokes of interest to Martin Marietta. Two green petroleum cokes from domestic California crudes were heated to calcination temperature which was then held constant for varying times.

In addition to laboratory experiments, special coke samples taken from the Martin Marietta Carbon calciner in Carson, CA, were analyzed to provide insight on coke property changes occurring during commercial calcination. The ultimate goal of this work is to define the key kiln conditions for control of calcined coke properties.

Experimental

-4 + 20 mesh samples of green Cokes A and B were heat treated in a programmable furnace. The samples ($\sim$ 50 g) were placed in ceramic crucibles with drilled lids to allow placement of gas tubes for atmosphere control. Two samples were heated during each run. Samples were kept under an N_2 atmosphere during heating and oxidant gases (air or CO_2) were introduced only during the heat-soak period. The maximum temperatures experienced by the coke samples were 1000°C, 1100°C, 1200°C, and 1300°C. A constant heat-up cycle was used: ambient to 960°C over 45 min then heating to maximum temperature in 60 min. The maximum temperature was maintained for 15, 30, 60, and 90 min, except at 1300°C where only 15 and 30 min heat soak periods were done. For the experiments utilizing a CO_2 purge, the heat soak was maintained for 15 min only. During the heat-soak period, one sample was kept under N_2 and the other was exposed to air for 5 min. For the CO_2 purge, both Cokes A and B were run and exposed to CO_2 simultaneously for the 5-min period. All samples were cooled under N_2. The samples were weighed before and after heat treating to determine mass loss.

Sample temperature was monitored by a chromel/alumel thermocouple placed into the sample bed that recorded the addition of air or CO_2 as either a temperature increase or decrease. However, due to uncertainty of thermocouple placement in the coke bed, these data were used only to ensure qualitatively that the run was completed as programmed.

The -4 +20 mesh calcined material was split and part was retained for crystallite dimension, L_c, measurement. The remainder was screened to -10 +20 for porosity, BET surface area (SSA), and reactivity analyses.

The L_c (ϕ,ϕ,ℓ) was determined on a Phillips APD 3600 using a CuK_α X-ray source and calculated using the Sherrer equation with measured peak widths of the 25.6° 2θ reflection.

The SSA was measured on a Quantachrome Quantasorb as a single-point BET isotherm.

A Micromeritics 9300 coupled to an Apple II Plus Computer collected the porosity data for reduction and storage. Data were taken over the pressure range 0.15 to 165 MPa (2 to 25,000 psi), which corresponds to pore diameters of 94 μ to 0.008μ.

The reactivity of the heat-treated cokes was measured on a Metler TA-1 Thermogravimetric Analyzer. Samples were heated to 550°C at 100°C/min and held at 550°C throughout the test. The sample mass is recorded in real time. Since only a few of these plots show a linear decrease in mass as a function of time, the reactivity was determined as the time required for 2% of the sample to oxidize.

Results

Laboratory Studies

The physical properties and elemental analysis of Cokes A and B are given in Table I. These values are typical for green cokes from LA Basin crudes and compare well except for ash content and metals composition. The results of the evaluation of the heat-treated samples are given in Tables II and III.

TABLE I. Green Coke Property Data

	Coke A	Coke B
Proximate Analysis		
Moisture (%)	0.82	1.03
Volatiles (%)	12.62	11.91
Fixed Carbon (%)	87.16	87.71
Ash (%)	0.22	0.38
Elemental Analysis		
C (%)	91.94	91.58
H (%)	3.89	3.89
N (%)	2.79	2.76
S (%)	1.28	1.45
V (ppm)	354	420
Ni (ppm)	355	307
Fe (ppm)	280	230
Na (ppm)	29	34
Physical Properties		
L_c (A)	8.7	8.8
Macroporosity (mm^3/g)	20.1	17.4
Microporosity (mm^3/g)	28.5	27.1
Total Porosity (mm^3/g)	54.7	50.7
Apparent Density (g/cm^3)	1.276	1.274
Skeletal Density (g/cm^3)	1.332	1.330

The mass loss from both Cokes A and B increased with temperature under both N_2 and N_2/air atmospheres but showed no correlation with heat-soak time. These results are included in Table II. Although under N_2, neither coke lost weight equal to its volatiles content (probably due to differences in coke sizing between the ASTM volatiles test and the heat-treatment experiments reported here), mass loss from Coke A exceeded Coke B by about the difference in analytically determined volatiles. At all temperatures, the samples under N_2/air lost more mass than under N_2, due to loss to oxidation. This conclusion is supported by the fact that this mass loss difference between oxidizing and non-oxidizing atmospheres increased with increasing temperature. Of the experiments utilizing oxidant gases, the CO_2 purge showed the most significant temperature dependence and at the higher temperatures even exceeded the air oxidation loss.

The crystallite size, L_c, increased with heat treatment temperatures and also showed an effect with heat-soak time (see Table II). Figure 1 shows the effects of maximum temperature on L_c at the various heat-soak times under the N_2 atmosphere. From this figure it is obvious that temperature is the controlling factor in crystallite growth, and heat-soak time, while certainly an influence, only has a small effect. The L_c data for the N_2/air tests shows the same functional dependencies and a slightly greater L_c, 0.5 - 0.7A. Although this difference is within experimental error, localized heating from carbon oxidation by O_2 could account for the slight increase in L_c. The L_c's of the samples from the N_2/CO_2 tests and those from the N_2 experiments were basically the same, which supports

Table II (a)

Physical/Chemical Property Data for Coke A

Property	Mass Loss (%)			L_c (Å)			Reactivity (min)			SSA (m^2g)		
Atmosphere*	1†	2†	3†	1	2	3	1	2	3	1	2	3
Temperature (1000°C)												
Time (Min)												
15	10.6	12.3	11.6	15.2	16.2	15.0	11.3	11.0	8.2	1.3	0.8	4.4
30	10.6	12.3	-	15.6	16.4	-	10.4	10.2	-	1.7	0.9	-
60	10.7	13.3	-	15.8	16.8	-	11.8	9.2	-	2.8	1.0	-
90	11.2	12.8	-	16.3	16.8	-	9.6	9.5	-	1.6	2.4	-
Avg.	10.8	12.4	-	-	-	-	10.8	10.0	-	-	-	-
Temperature (1100°C)												
15	11.2	12.4	12.0	18.6	19.1	18.4	11.3	14.3	10.3	1.4	1.0	2.2
30	11.0	12.4	-	18.8	19.8	-	12.6**	12.4**	-	1.2	1.3	-
60	11.2	12.1	-	19.7	20.0	-	12.8	14.6	-	0.6	0.7	-
90	11.0	12.9	-	20.0	20.6	-	15.3	11.5	-	0.4	0.9	-
Avg.	11.1	12.4	-	-	-	-	13.0	13.2	-	-	-	-
Temperature (1200°C)												
15	11.4	13.6	13.4	22.6	24.3	22.8	15.9	19.8	12.4	0.8	0.8	0.8
30	10.4	11.3	-	24.2	24.4	-	17.5	16.6	-	0.5	0.8	-
60	11.6	12.0		24.9	24.9	-	19.8	20.7	-	0.5	0.5	-
90	11.6	12.3	-	25.4	25.6	-	18.1	23.6	-	0.4	0.6	-
Avg.	11.2	12.5	-	-	-	-	17.8	20.2	-	-	-	-
Temperature (1300°C)												
15	13.5	13.7	14.0	27.6	29.0	27.4	10.5	28.9	12.0	1.5	0.5	1.0
30	11.9	13.4	-	28.5	29.6	-	18.9**	25.3**	-	0.7*	0.6	-
Avg.	12.7	13.6	-	-	-	-	16.1	26.5	-	-	-	-

* Atmosphere (1) N_2
(2) N_2 + Air
(3) $N_2 + CO_2$

** Average of two runs

Table II (b)

Physical/Chemical Property Data for Coke B

Property	Mass Loss (%)			L_c (Å)			Reactivity (min)			SSA (m^2g)		
Atmosphere	1	2	3	1	2	3	1	2	3	1	2	3
Temperature (1000°C)												
Time (Min)												
15	10.2	11.8	10.3	15.3	16.7	15.1	11.2	12.6	12.4	1.2	0.8	0.8
30	9.6	11.5	-	15.6	16.7	-	12.8	11.0	-	0.8	1.6	-
60	10.0	11.9	-	16.2	16.8	-	13.0	9.1	-	1.0	1.9	-
90	10.2	11.4	-	16.2	16.5	-	12.3	10.3	-	1.5	2.1	-
Avg.	10.0	11.6	-	-	-	-	12.3	10.8	-	-	-	-
Temperature (1100°C)												
15	10.5	12.3	11.2	19.2	19.7	18.4	13.7	14.6	14.2	0.9	1.1	1.2
30	10.1	12.3	-	19.4	20.1	-	14.2	10.6	-	0.8	1.4	-
60	10.3	12.2	-	20.1	21.1	-	12.1	13.6	-	0.8	1.3	-
90	10.7	12.3	-	20.4	21.0	-	15.2	15.5	-	1.0	1.2	-
Avg.	10.4	12.3	-	-	-	-	13.8	13.6	-	-	-	-
Temperature (1200°C)												
15	10.6	12.6	13.1	23.8	24.5	23.2	21.7	22.8	12.7	0.8	0.9	1.7
30	10.6	13.4	-	24.3	24.5	-	22.8	8.2	-	0.8	1.6	-
60	10.9	13.5	-	24.7	25.2	-	18.3	22.7	-	0.8	1.2	-
90	11.4	14.0	-	25.7	26.4	-	17.9	23.9	-	1.0	1.0	-
Avg.	10.6	13.4	-	-	-	-	20.2	23.1	-	-	-	-
Temperature (1300°C)												
15	10.7	14.6	14.4	28.5	29.6	28.3	29.3	22.4	12.64	0.5	1.4	1.2
30	11.0	12.4	-	29.2	30.7	-	27.4	35.4	-	0.7	0.6	-
Avg.	10.8	13.5	-	-	-	-	28.3	28.9	-	-	-	-

Table III (a)

Porosity/Density Data for Coke A

Property	Macroporosity (mm^3/g)			Microporosity (mm^3g)			Total Porosity (mm^3g)			Apparent Density (g/cm^3)			Skeletal Density (g/cm^3)		
Atmosphere*	1	2	3	1	2	3	1	2	3	1	2	3	1	2	3
Temperature (1000°C)															
Time (Min)															
15	23.2	24.9	23.4	32.5	34.2	36.4	59.6	63.5	64.0	1.760	1.749	1.723	1.880	1.876	1.853
30	26.9	26.8	-	32.7	36.0	-	63.7	67.2	-	1.727	1.754	-	1.845	1.887	-
60	23.8	22.9	-	33.5	35.2	-	61.5	62.4	-	1.749	1.743	-	1.873	1.873	-
90	24.9	27.4	-	35.0	34.5	-	61.2	66.8	-	1.736	1.751	-	1.862	1.880	-
Avg.	24.7	25.5	-	33.4	35.0	-	61.5	65.0	-	1.743	1.749	-	1.865	1.879	-
Temperature (1100°C)															
15	26.0	28.2	25.7	35.5	38.6	39.0	66.1	71.5	64.6	1.759	1.747	1.741	1.892	1.890	1.884
30	29.0	32.6	-	36.6	35.6	-	73.0	72.5	-	1.745	1.753	-	1.878	1.886	-
60	27.7	28.2	-	37.6	36.4	-	70.0	69.4	-	1.763	1.797	-	1.905	1.940	-
90	26.4	28.3	-	36.0	37.6	-	67.1	70.6	-	1.765	1.775	-	1.902	1.919	-
Avg.	27.3	29.3	-	36.4	37.1	-	69.0	71.0	-	1.758	1.768	-	1.894	1.909	-
Temperature (1200°C)															
15	29.3	27.2	31.6	38.3	39.6	39.6	72.8	72.2	76.2	1.758	1.773	1.716	1.903	1.926	1.829
30	35.2	27.6	-	46.1	41.1	-	88.0	74.6	-	1.738	1.742	-	1.913	1.897	-
60	28.0	34.0	-	40.4	44.4	-	74.3	84.4	-	1.758	1.733	-	1.914	1.899	-
90	36.7	26.2	-	45.6	41.3	-	87.7	72.9	-	1.778	1.793	-	1.956	1.957	-
Avg.	32.3	28.8	-	42.6	41.6	-	80.7	76.0	-	1.758	1.760	-	1.922	1.920	-
Temperature (1300°C)															
15	29.8	31.4	31.0	47.6	41.6	50.6	82.8	79.1	87.4	1.700	1.775	1.738	1.966	1.939	1.926
30	28.6	27.4	-	41.4	41.8	-	75.0	75.0	-	1.760	1.783	-	1.920	1.948	-
Avg.	29.2	29.4	-	44.4	41.7	-	78.9	77.0	-	1.770	1.779	-	1.943	1.944	-

* (1) N_2
(2) N_2 + Air
(3) $N_2 + CO_2$

Table III (b)

Porosity/Density Data for Coke B

Property	Macroporosity (mm^3/g)			Microporosity (mm^3/g)			Total Porosity (mm^3/g)			Apparent Density (g/cm^3)			Skeletal Density (g/cm^3)		
Atmosphere	1	2	3	1	2	3	1	2	3	1	2	3	1	2	3
Temperature (1000°C)															
Time (Min)															
15	26.1	26.7	25.2	45.0	45.0	44.8	75.7	76.1	74.4	1.716	1.724	1.717	1.875	1.885	1.869
30	25.0	24.6	-	41.1	42.2	-	70.6	71.2	-	1.736	1.737	-	1.885	1.891	-
60	24.8	25.4	-	41.2	46.1	-	71.0	75.4	-	1.716	1.722	-	1.864	1.888	-
90	25.4	22.8	-	43.2	44.4	-	73.5	75.0	-	1.736	1.756	-	1.894	1.822	-
Avg.	25.3	24.9	-	42.6	44.4	-	72.7	74.4	-	1.726	1.735	-	1.880	1.897	-
Temperature (1100°C)															
15	28.8	29.0	27.6	45.8	45.0	48.0	79.6	78.8	80.2	1.741	1.773	1.733	1.910	1.945	1.908
30	27.3	30.0	-	46.3	47.3	-	78.9	82.2	-	1.772	1.744	-	1.950	1.919	-
60	27.2	29.4	-	45.1	48.0	-	77.3	82.4	-	1.753	1.743	-	1.921	1.920	-
90	28.6	30.3	-	46.8	48.9	-	80.3	84.5	-	1.749	1.743	-	1.923	1.925	-
Avg.	28.0	29.7	-	46.0	47.3	-	79.0	82.0	-	1.754	1.751	-	1.926	1.925	-
Temperature (1200°C)															
15	29.2	28.5	31.8	47.6	50.3	63.4	82.9	84.2	100.6	1.743	1.749	1.736	1.865	1.876	1.971
30	29.8	34.2	-	45.8	58.2	-	80.8	98.5	-	1.775	1.704	-	1.951	1.906	-
60	30.8	32.6	-	50.6	53.4	-	87.1	91.8	-	1.748	1.744	-	1.939	1.944	-
90	29.8	32.4	-	50.6	52.0	-	86.3	91.6	-	1.734	1.718	-	1.922	1.912	-
Avg.	29.9	31.9	-	48.7	53.5	-	84.3	91.5	-	1.750	1.729	-	1.919	1.910	-
Temperature (1300°C)															
15	29.8	31.6	34.6	50.7	58.0	65.5	87.3	93.7	106.6	1.762	1.708	1.725	1.960	1.918	1.968
30	31.1	32.4	-	51.9	52.6	-	89.4	91.4	-	1.756	1.735	-	1.956	1.943	-
Avg.	30.5	32.0	-	51.3	55.3	-	88.4	92.6	-	1.759	1.722	-	1.958	1.930	-

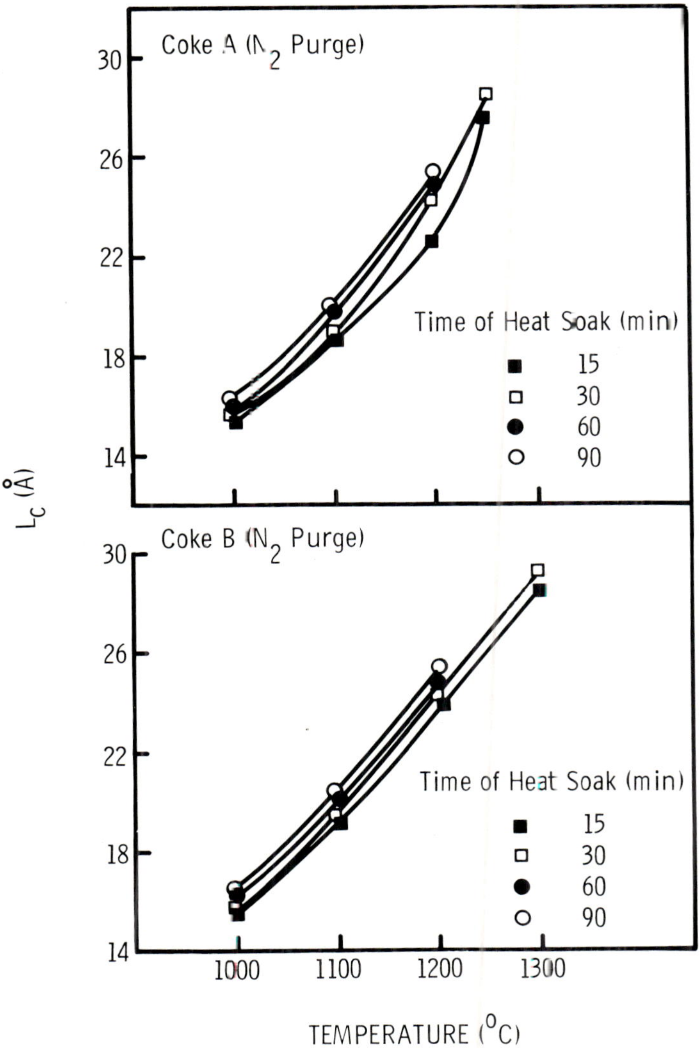

Figure 1. Crystallite size vs. maximum heat-treatment temperature.

particle heating from carbon oxidation as the factor affecting L_c in N_2/air tests.

Like the mass loss data, surface accessible porosity showed a positive correlation with temperature but no relationship to heat-soak periods. There is no apparent difference in macroporosity (pore diameter range of 94μ to 13μ) between Cokes A and B; however, Coke B shows a much greater microporosity (13μ to 0.013μ) than Coke A even though Coke A has higher volatiles. There is also an increase in macro- plus micro-porosity with air addition during the heat soak although this difference may not be experimentally significant. The effect appears to be substantial only in Coke B at 1200°C and 1300°C. Porosity data are listed in Table III. The total measurable porosity at 164 MPa (pore diameters > 0.0083 μ) increased with increasing maximum temperature but likewise showed no differences due to the heat-soak times. The presence of oxidant gases appears to enhance pore development especially at higher temperatures. This effect was observed for both Cokes A and B but was greater in Coke B.

The density of the heat-treated cokes was determined from the Hg intrusion data and included in Table III. The apparent density was determined from Hg intrusion data at 0.1 MPa pressure, whereas the skeletal density was determined from intrusion data at about 164 MPa. Coke A exhibited a higher apparent density but Coke B had the higher skeletal density, reflecting greater microporosity. Both apparent and skeletal densities increased with increasing temperature and showed no clear dependence on heat-soak period. No significant differences in density were observed between the cokes heated in N_2 and the cokes heated in N_2/air. The increase in apparent density with temperature indicates that the effects of increasing L_c and, hence, densification due to crystallite growth, outweigh the effects of increasing porosity which tend to decrease density. No plateau in density at higher temperatures was observed.

As seen in Table II, the SSA measurements showed no correlation with temperature, heat-soak times, or reaction atmosphere. The data showed such relative variation that it was not even possible to average and search for trends, e.g., as for porosity and mass loss.

The air reactivity of the coke samples at 550°C, reported as the time necessary for 2% of the sample to oxidize, is also reported in Table II. This approach was chosen over the more customary initial slope method because not all samples showed a linear mass loss with time. In fact, only about 25% were linear, whereas the rest were quadratic. (The linear mass loss runs were concentrated in the samples exposed to the 1300°C heating cycle). Hence, in evaluating these data, lower time represents higher reactivity. The reactivity data show an inverse relation with maximum temperature and no effect of heat soak time. Further investigation reveals a good correlation with L_c. This correlation is shown in Figure 2 for the N_2 and N_2/air experiments. The reactivity decreases because higher temperatures drive off the reactive hydrogen and alkyl carbons as volatiles. Simultaneously, the growth of crystallite reduces the number of carbon atoms on the edge of the graphitic planes accessible to O_2. Coke B is clearly more reactive than Coke A despite having greater crystallinity. One cause of greater reactivity could be metal impurities. The effects of metal impurities as oxidation catalysts are complex and were not expressly addressed in this study. The reactivity of the air-purged samples is lower than for N_2-purged samples at lower temperatures, but at higher maximum temperatures, the reactivities of the air-purged samples

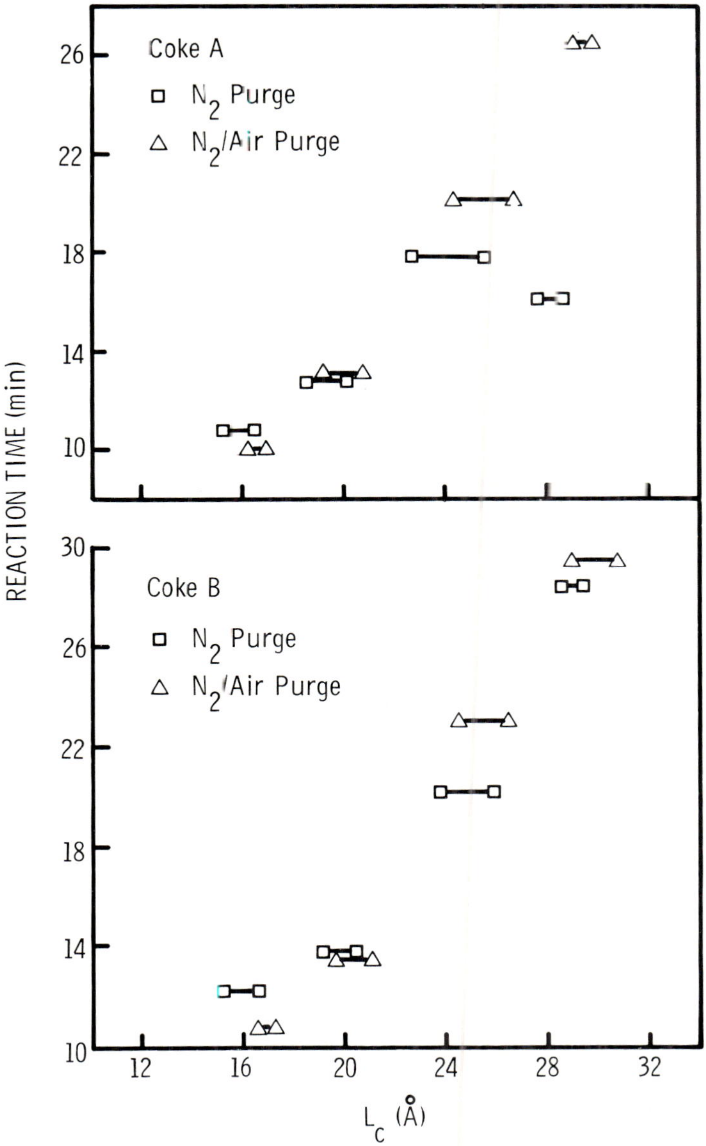

Figure 2. Air reactivity (measured as the time for 2% of the coke to oxidize) of coke samples vs. crystallite size.

exceed the N_2-purged cokes. Hence, the reactivity of cokes exposed to air shows a greater dependence on maximum heat treatment temperature than those exposed to inert gas.

Plant Test

After production at the Martin Marietta Carbon calciner in Carson, CA was terminated, the coke bed was left in place in the kiln, cooled, and sampled every 15 feet up its length. Coke samples at varying stages of calcination could thus be collected for study. Physical and chemical properties are then measured in order to follow changes in the coke during calcination. L_c, which is a good measure of the degree of calcination, and the real density were determined. The L_c results shown in Figure 3 indicate that a rapid change in crystallinity occurs in the kiln hot zone. (In this kiln, tertiary air was added between 50 and 70 feet from the burner to combust volatiles and provide heat for calcination). The real density increases initially, then decreases (probably due to porosity created by the evolution of volatiles, and finally increases during the coke shrinkage/crystal growth phase. (The data point at 216 feet has been disregarded as suspect.) The magnitudes of both L_c and real density are greater than normally observed during calcination. Since the coke bed was allowed to cool through natural convection and conduction, the coke was subjected to a different thermal cycle than production material and was maintained at elevated temperature for a longer period of time ($\sim$ 36 hours). The effects of heat-soak time on coke properties seem to require extended periods to become observable.

Conclusions

In the laboratory experiment reported here, we investigated the effects of heat treatment times and temperatures on key properties of petroleum coke. The three variables studied -- maximum temperature, heat-soak times, and the presence of oxidizing gases -- were varied so as to cover the range of conditions in commercial calcination. Among the laboratory samples, the maximum heat-treatment temperature affected coke properties more than the heat-soak time or the presence of oxidizing gases. These latter two variables had little effect, if any, in these experiments. The effects of temperature, heat-soak times, and calcination atmosphere are summarized below:

- Mass loss, crystallite dimension, porosity, and density increased as a function of maximum heat-treatment temperature, whereas reactivity decreased.

- Heat-soak times affected the crystallite dimension but did not observably affect any other properties.

- Although the effects of oxidant gases were evident in all properties, in some cases they were subtle. The expected effect on mass loss and L_c was observed, but it was only marginal in porosity data. Oxidant gases affected reactivity in a complex manner with maximum temperature.

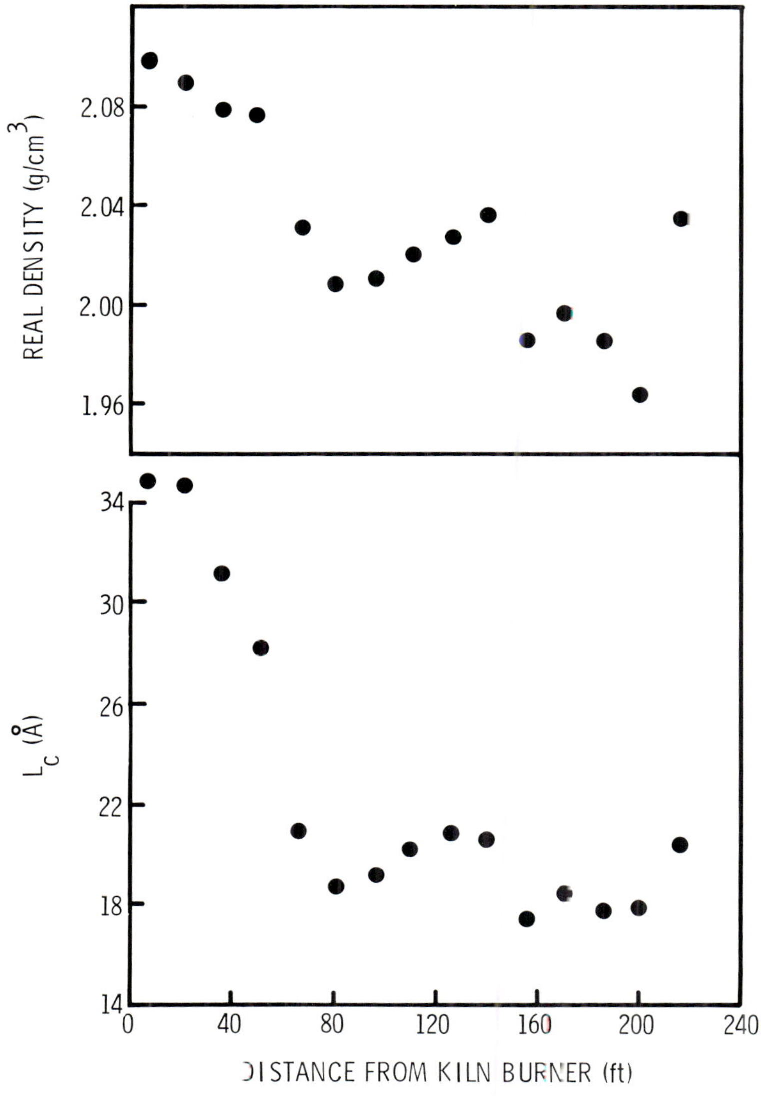

Figure 3. Crystallite dimension and real density from kiln samples.

The BET SSA showed no correlation with any of the variable parameters. This was disappointing since these data would have been particularly valuable in explaining the effects of oxidant gases and the creation of micropores.

The surface-accessible porosity increased with temperature and, to a lesser degree, with the presence of oxidant gases. The porosity in all pore diameter ranges increased. The relation between volatiles content and porosity was not observed, although both cokes had similar volatiles and low heating rates. Apparent and skeletal densities increased with temperature and air exposure, but were lower for the CO_2-purged samples. These data support the earlier conclusion on the predominant effect of calcination temperature.

In addition, coke samples calcined to varying degrees in commercial practice were analyzed to measure the change of key properties during calcination. These samples indicated that much longer heat-soak times are required to show any effects.

References

1. S.S. Jones and R.D. Hildebrandt, Light Metals 1981 (1981) p. 423-440.

2. P.L. Walker, Jr., F. Rusinko, Jr., J.F. Raksyaivsky, and L.M. Liggett, pp. 643-658 in Proceedings Conference Carbon, 3rd, Buffalo, 1957, Pergamon Press, New York, 1959.

3. P. Rhedey, Transactions of the Metallurgical Society of AIME, 239, (1967) 1084.

4. M.P. Whittaker, F.C. Miller, and H.C. Fritz, Industrial Engineering Chemistry Product Research and Development, 9, (1970) 187.

5. R.W. Wallough and F.V. Fair, Carbon, 18, (1980) 147.

6. M. Kakuta, M. Kohriki, and Y. Ohmoto, Tanso, 106, (1981) 102.

DESULFURIZATION OF PETROLEUM COKE

H. H. Brandt and Dr. R. S. Kapner

V.P.-Development, Diamond West Energy Corp.
Salt Lake City, Utah 84103

Professor-Chem. Engrg. The Cooper Union
New York, N.Y. 10003

The incentive to desulfurize petroleum coke is greater today than ever before because of a shrinking supply of low sulfur coke concurrent with a sharply increasing demand. An additional inducement is provided by ever tightening restrictions on SO_x emissions by various environmental agencies. This paper will review some of the past efforts to desulfurize petroleum coke and then describe a new approach developed at The Cooper Union For The Advancement Of Science And Art in New York City.

Free world petroleum coke production has increased dramatically in the past decade. The United States continues to be a dominant factor in the petroleum coke industry, accounting for more than 75% of the total production.

In spite of the phenomenal growth in production, there is real and justified concern about a coke shortage by the aluminum industry. The shortage is in the supply of anode grade coke. This is the coke that is used by the producers of primary aluminum (about one half pound per pound of aluminum). This industry consumes approximately 10 million tons of raw petroleum coke annually, or about 35% of the total free world output. Most of the recently built cokers, as well as many of the older ones, are charged with feed stocks that are high in sulfur and metals contents. The resultant cokes are correspondingly high in sulfur and metals contents. While experience indicates that both of these contaminants can be tolerated by the aluminum producers, they do create serious problems. In the case of metallic impurities, vanadium is the worst offender. It often acts catalytically to increase oxidation of the carbon anode. Furthermore, the metallurgy to remove vanadium from aluminum is very costly. Sulfur presents another problem and the solution to this one is less clearly defined. The problem is air pollution. Two elements in this scenario are on a collision course. (1) Due to shrinking supplies of low sulfur (less than 2.5%) cokes concurrent with increased requirements of the aluminum industry, the average sulfur content of so called anode grade cokes has increased significantly. Ten years ago the average was 2.0% and today's specification is 3.5%. (2) At the same time, SO_x emission limits are being tightened and there are strong indications that we will see a quantum jump in the environmental restrictions on SO_x emissions in the very near future.

Desulfurization of coker feed stocks to produce lower sulfur cokes has proven not to be a solution to the problem. Nor is there cause for optimism in this area. Not only is the technology extremely expensive, it doesn't achieve the expected results. A 50% reduction in feed stock sulfur frequently yields a lesser reduction of the sulfur in the coke.

That leaves us with the challenge to desulfurize the coke itself. There has long been an incentive to remove sulfur from coke since low sulfur coke has always commanded a premium price over its high sulfur counterpart. There have been a dozen or more patents filed on petroleum coke desulfurization technology.

It has been demonstrated in the laboratory that sulfur can be removed from coke thermally, chemically, or thermochemically. Thermal sulfur removal is, as might be expected, a function of time and temperature. A long soak at relatively low temperatures (i.e. several days at 2,000°F) can remove 50% or more of the original sulfur content. This treatment is impractical because of excessive energy costs and its inability to process large tonnages. Higher temperatures and shorter times (i.e. 2,600 to 2,900°F for 45-60 minutes) can also remove 50% or more of the sulfur but create a high degree of porosity and a reduction in real density. Both of these properties are critical in anode making.

Chemical desulfurization techniques have a common drawback. They require pulverizing the coke to expose maximum surface area to the reactant. Thus, the desulfurized coke has no lump content.

Since anodes cannot be made without a coarse fraction, an agglomerating step would be required. This, in turn, would require a coal tar pitch binder and a baking step. The final product becomes a relatively weak lump with prohibitive costs. Some chemical treatments employ alkalies or florides which, in addition to being expensive, may leave residues that are not acceptable.

The accumulated data seems to suggest that a thermochemical process should have the best chance for success. A considerable amount of research effort has been expended in this area. The reported results are frequently inconclusive and occasionally conflicting. For example, one study claims that the sulfur content of petroleum coke actually increase when it is exposed to H_2S at elevated temperatures. Most laboratory results support the contention that H_2S, methane, or other low molecular weight hydrocarbons can induce sulfur removal at elevated temperatures. Many studies indicate the need for pulverizing the coke prior to treatment, others do not. Still others claim that a two stage temperature treatment is beneficial. One thermochemical process reports 90% sulfur removal using a three step procedure. In the first step, the volatiles are removed by treatment with an inert gas at 1,400°F. The temperature is then reduced to 700°F and the coke is partially oxidized, presumably to activate it. The final step consists of re-heating the coke to about 1,700°F and treating it with methane or a similar hydrocarbon.

Some of the inconsistancies in the research are possibly due to inconsistancies in petroleum cokes. It is generally conceded that the coker is "the garbage can" of the refinery and as such it seldom sees an identical charge from day to day. Cokes with virtually identical chemical analyses can exhibit markedly different behaviors when subjected to identical thermal treatments.

To date, we do not have a commercial process for desulfurization. However, there is a new and unique process that promises to succeed where so many previous attempts have failed. This new process has been patented by Diamond West Energy Corp. and the research work, conducted at The Cooper Union in New York City, has been jointly funded by Kennedy Van Saun Corp. and Diamond West.

The process consists of contacting calcined coke with a sulfur bearing gas at elevated temperatures. The concept of treating the calcined coke immediately after it is discharged from the calciner is a key element in minimizing energy costs. The hot coke (about 2,300°F) is fed into a reactor directly from the calciner, where it may be further heated up to 2,800°F and exposed to a sulfur bearing gas for a period of time up to 90 minutes.

The desulfurization chemistry appears to depend on the presence of free sulfur in the contacting gas and works in conjunction with organic sulfur pyrolytically released from the coke. Gases that decompose to form free sulfur or react to form free sulfur are suitable desulfurization agents. These include refinery sour gas, pure and dilute H_2S, mercaptans, mixtures of CO and SO_2 or COS and H_2O.

Laboratory tests on two pound samples of green petroleum cokes with sulfur contents of 4% to 6% have successfully reduced the sulfur levels to as low as 0.7%. The degree of desulfurization can readily be controlled by adjusting the temperature (nominally between 2,500 and 2,800°F), the holding time (between 5 and 90 minutes), and the reactant concentrations or flow rates.

The experimental work has been conducted in horizontal furnace tube reactors containing fixed beds of sized (1/2"x4M) coke electrically heated and controlled at precise temperatures over the entire bed length. The coke is first calcined, following a temperature profile similar to that of commercial calcining. The calcining is immediately followed by desulfurization with pure or dilute sulfur gases metered from cylinders.

The data shown in Table 1 were obtained at reaction temperatures of 2,730°F and are typical of measured desulfurization results at all temperatures between 2,550 and 2,900°F.

Table 1

	THERMAL DESULFURIZATION	WITH ADDED SULFUR GAS
Holding time(min)*	% Desulfurization **	% Desulfurization **
0	0	0
10	15.2	27.5
20	32.1	48.0
30	55.0	66.8
45	66.8	76.8
60	71.2	83.8

Analysis of these and other data suggest that the desulfurization process is a complex sequence of steps tentatively represented as:

$$(\text{-CS-}) \rightleftarrows (\text{-C-}) + S_v \qquad (1)$$

$$S_v + (\text{-CS-}) \longrightarrow CS_2 \qquad (2)$$

$$2S_v + (\text{-C-}) \longrightarrow CS_2 \qquad (3)$$

Reaction (1) constitutes a purely thermal initiation of the desulfurization process described by (1) and (2). Step (1) would appear to be slow, rate controlling and with a high energy. Also, it is very responsive to temperature increases. Hence, the generally observed rapid acceleration thermal desulfurization as the temperature increases from 2,400 to 2,900°F. By adding sulfur from an external source, the desulfurization described in step (2) is no longer limited by the rate at which sulfur is supplied by the coke itself and desulfurization proceeds more rapidly than by a purely thermal process at every temperature. Experimental data show that the desulfurization with added external sulfur, represented by reaction (2) is dependant on temperature and concentration of sulfur in the reactant gas. It has a significantly lower activation energy than reaction (1).

The observation that the activation energy of reaction (1) is greater than that of reaction (2) is confirmed by experimental data which show that the relative rates of desulfurization by thermal processing and by the addition of a sulfur gas decrease as holding temperature is increased. This points out one of the more valuable features of this technology--a high rate of desulfurization can be supported at temperatures where thermal desulfurization is not generally achieved.

Another valuable feature is the preservation of the critical properties real density and bulk density. Real density is a specific gravity measurement of the calcined coke (performed on a sample that has been reduced to minus 200 mesh) and bulk density is a porosity determination (performed on a 35x65 mesh sample). Both of these properties tend to deteriorate to unacceptable levels during thermal desulfurization at high temperatures due to the creation of micro and macro-porosity as sulfur is vaporized and leaves the solid coke matrix. The persistance of high real and bulk density values in cokes desulfurized using this technology is a promising indication that some sort of annealing process occurs during the treatment. Unfortunately, we do not have an explanation for this phenomenon. The data from almost one hundred desulfurization runs strongly supports the claim and is summarized in Table 2.

TABLE 2

COMPARISON OF PHYSICAL PROPERTIES OF DESULFURIZED COKES WITH TYPICAL CALCINED COKE SPECIFICATIONS

PROPERTY	COMMERCIAL SPECIFICATION	SULFUR GAS DESULFURIZED COKE
Real density (g/cc)	2.05-2.11	2.07-2.16
Bulk density (#/cuft)	49 Min.	53-55

Pilot plant work is presently under way at Kennedy Van Saun's facilities in Danville, Pa. We hope to develop hardware that can be retro-fitted to existing calciners.

NEW APPLICATIONS TO UPGRADE COAL TAR PRODUCTS

Dr. Henry Kremer & Mr. Samuel Cukier

Domtar Inc./Organic Chemicals Division

1136 Matheson Blvd.

Mississauga, Ontario, L4W 2V4

Canada

SUMMARY OF WORK

Various commercially available coal tars have been filtered on Pilot Plant Scale to improve quality by reducing Quinoline Insolubles, including ash forming components. The physical properties of filtered tars and coal tar pitches distilled in the laboratory and in plant were determined. The effects of filtration on viscosity were also studied, both in the precursors and in the resultant coal tar pitches.

INTRODUCTION

Demand for premium quality coal tar and coal tar pitch products has increased worldwide in recent years, particularly for the carbon and graphite electrode field. This scenario has initiated renewed research efforts to upgrade the precursor for Electrode Binders and Impregnants. Some researchers have concentrated on improvements to the coal tar pitch, i.e., modifications and/or treatment of the residue after distillation. Others have used new approaches for process developments which benefit the physical and chemical properties of coal tar prior to distillation. Both groups claim technical and economical advantages for justification to their approaches. Time will only show which processes will endure the fluctuating technical and commercial demands of the continously changing market conditions.

Our team has reviewed the state of the art in 1980 and we have set up long term goals for our research. Ever since, we have systematically developed new upgrading processes which have solved persistent long-term problems with coal tar quality.

We reported in 1981 (1). On our success to reduce the water and water soluble inorganic salt (ash forming) content of coal tars. Later, we reported (2, 3), on the composition, structure and physical properties of the Quinoline Insoluble Fraction (QI) of coal tars, namely -resins, inorganic salts and combustible, carbonaceous "dirt" carried over from the coke oven.

Further developments were reported (4,5,6) on a new Domtar® Process which is now being tested exhaustively on large scale Pilot Plant equipment. This process controls the true QI content. It can decrease the QI (and related ash) content to any desired level. It can also increase the QI level without creating undesirable mesophase which is often formed in other types of treatments. Our current report contains data which show the effects of the New Process on viscosity of the tar and the resultant coal tar pitch. It shows also the changes created by reducing the total ash content.

THEORY

The quinoline insoluble fraction of coal tar has chemically different components (2):-

(a) The α-resins (True or Primary QI) are condensed ring aromatic hydrocarbons with a true particle size of less than one micrometer (μm). They appear billiard ball shaped under the electron microscope.

(b) The combustible dirt is mostly coal and coke particles, usually 20-50 μm minimum, sometimes with a size which is visible under optical microscope with low magnification. Their shape is straight or branched needles or granular with amorphous contours.

(c) Inorganic QI's, crystalline, lamellar or amorphous most of the times fused to other (combustible) QI particles. Size is large, mostly above 20 μm.

Many of the various QI particles, primarily due to surface tension, agglomerate to form lumps in the coal tar. These lumps are detrimental to the rheological properties. Often these lumps entrap water and ash forming particles which could not separate out in the designated process equipment, namely in the decanters as liquor or sludge. The lumps also affect the flow and impregnating properties of coal tars and pitches for applications such as coatings, electrodes, etc. The ash forming inorganic components affect the work life and performance of electrodes and as such are detrimental in the coal tar.

The Domtar ® Tar Upgrading Process advocates the use of surfactants for initial treatment of tar (dewatering therefore reducing the water soluble inorganic salt content); deagglomeration of lumps and thereby reducing the total ash content.

The Domtar ® Filtration Process was developed to control natural QI levels and for removal of "dirt". During the scaling up of the process many experiments were conducted and much data was collected. Analysis of the test data revealed changes in tar and pitch viscosity. Data for the latter is still under study and will be reported at another forum. The theory of inorganic QI attachment to combustible particles (to true QI agglomerates and to dirt) was confirmed. Selective filtration of larger particles did reduce the ash content significantly.

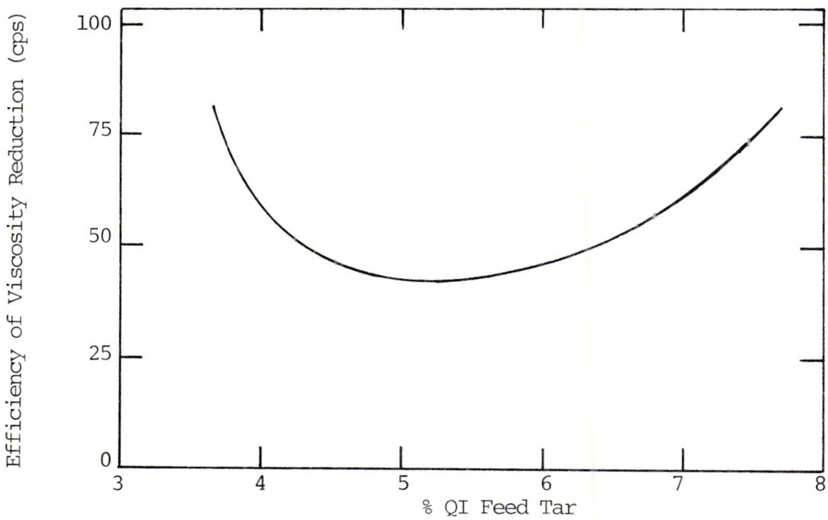

Figure 1: Viscosity vs. QI Content

<u>EXPERIMENTAL</u>

a) <u>Materials used</u> for physical properties, see Table 1&2.

TABLE I - PHYSICAL PROPERTIES OF COAL TARS & PITCHES

FEED TAR		FILTERED TAR				PITCH PRODUCED IN LABORATORY				
Q.I.%	VISCOSITY (cps) @ 70°C	Q.I.%	VISCOSITY (cps) @ 70°C	SP. G. 60°F	OIL FACTOR % @ 100°C PITCH	Q.I.%	X.I.%	C.V.%	Sp °C	VISCOSITY (cps)@ 225°C
3.6	32	0.9	10	1.151	57.5	2.4	22.9	33.1	42	7
		1.1	12	1.165	61.5	2.8	19.3	34.9	41.8	5
		1.6	18	1.164	58.2	4.0	20	38.6	43	5
6.7	46	3.1	20	1.172	60.4	6.3	19.3	48.8	67	10
		2.1	18	1.171	58.2	4.8	22.9	53.5	93	16
6.2	60	1.5	30	1.165	58.3	3.9	20.1	49.2	79	13
5.0	42	2.2	23	1.151	56.1	4.8	23.5	51.6	90	15
4.1	24	0.1	8	1.095	63	0.3	18.8	46	66.5	8
3.4	30	0.7	10	1.139	57.2	1.6	18.1	50.8	80	5
3.6	27	0.3	6	11.125	56.8	0.9	17.6	50.7	76.5	5

TABLE II - PHYSICAL PROPERTIES OF COAL TARS & PITCHES

FEED TAR		FILTERED TAR				PITCH PRODUCED IN PLANT						
Q.I.%	VISCOSITY (cps)@ 70°C	Q.I.%	VISCOSITY (cps)@ 70°C	S.G. @ 70°F	OIL FACTOR % @ 100°C PITCH	Q.I.%	X.I.%	SP.°C	C.V.%	ASH %	VISCOSITY (cps)@225°C	S.G. @60°F
3.6	32	1,4	16	1,134	57.1	2.8	23.05	90.2	52.2	.11	8	1.299
3.6	32	1.6	18	1.155	65.6	4.2	29.3	111.5	58.4	.11	10	1.31
6.7	46	2.1	15	1.14	61.5	6.2	27.8	108.8	57.8	.12	15	1.31
6.7	46	2.1	15	1.14	61.5	5.4	26.2	101.9	53.4	.11	12	1.308
7.7	60	2.7	18	1.172	56.0	6.0	26.2	101.4	54.6	.12	14	1.307
5.0	22	2	7	1.138	64.3	6.4	24.6	102.7	54.5	.15	20	1.307

b) Testing Methods

Softening Point	ASTM D-2319
QI Content	ASTM D-2318
XI Content	ASTM D-2317
Coking Value	British Standard PT-10-79 (Alcan Method)
Ash Content	ASTM D-2415-66-1976
Density	ASTM D-71-72a
Distillation	ASTM D-20
Mesophase	Optical microscope by using polarised light, bright field and interference contrast technique, ASTM Draft Testing Method
High Temp. Viscosity	For Pitches: Brookfield - Thermosel System consists of Standard Synchro-Lectric Viscometer LVF fitted with Spindle SC4-18 For Tars: Brookfield LVF Viscometer fitted with Spindle #1
Oil Factor	ASTM D-20 type distillation until the softening point of the residual pitch is approximately 100°C.

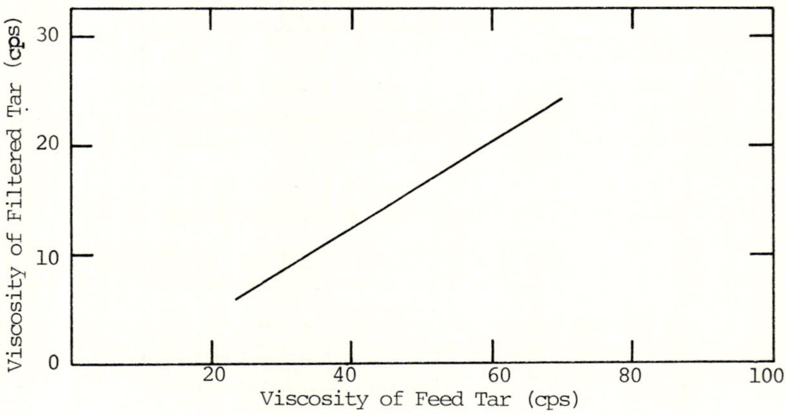

Figure 2: Viscosity Reduction By Filtration of Coal-Tar Blends

PROCEDURE

Coke oven crude tars which were by-products of metallurgical coke making were received from various, mostly Canadian Steel Plants. The crude tars were blended in large storage tanks to provide tars with a designated QI level. The blended tar was heated to 70°C and filtered by means of a proprietary industrial filtration system. The filtered tar was subsequently distilled in the laboratory following a D-20 distillation method, then the resultant pitch was tested for physical properties (see Table 1).

Over 30 production batches were filtered and from time to time, a few batches were collected in a 60,000 liter storage tank and batches were distilled in an 18,000 liter capacity distillation vessel, with indirect gas heating, under atmospheric pressure to produce 100-105°C softening point coal tar pitch. (See Table 2 for Physical Properties)

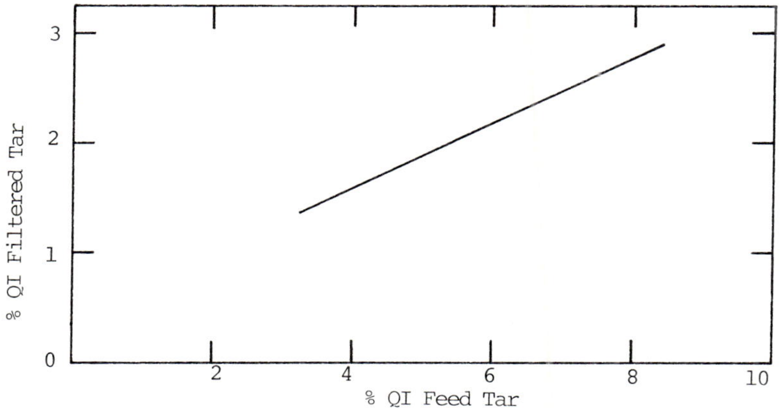

Figure 3: Efficiency of Filtration of Coal Tar Blends

DISCUSSION OF RESULTS

VISCOSITY REDUCTION: In every experiment, the viscosity of the coal tar was reduced by about 50% or more (Table 1&2). We have analyzed the data and have plotted different relationships between variables of which the most meaningful were:

(a) When the efficiency of viscosity reduction is plotted

$$\text{viscosity reduction} = \frac{\text{original viscosity of feed}}{\text{viscosity of filtered tar}}$$

versus the QI content of the feed (Figure 1) it shows a negative bell curve type relationship with a minimum around 4-6% QI level of the feed tar. The significance of these data will be studied later.

(b) The viscosity of the filtered tar has a linear relationship with the viscosity of the feed tar (Figure 2).

QI REDUCTION: The series of experiments reported in Table 1 and 2 were to evaluate process variables and not intended to maximize QI reduction. The rate of QI removal varied between 56-97% of the QI content of the feed prior to filtration. The efficiency of QI removal shows a linear relationship with the QI content of the feed tar (Figure 3).

MESOPHASE: All experiments showed absence of mesophase.

ASH REDUCTION: The ash content of standard coal tar binder pitches varied between 0.25-0.3%. The levels found in the pitches made by the Upgrading Process varied between 0.11-0.12% (Table 2). The reduction was therefore 52-63% with very little variation all through the experiments regardless of other variables. This finding confirms reports published by us and by other researchers (2,7&8) that the majority of inorganic ash forming particles are large or are attached to the larger combustible QI particles.

	COMPOSITE ASH (ppm) STANDARD PITCH	COMPOSITE ASH (ppm) FILTERED PITCH
Na	87	3.5
K	48	1.2
Ca	88	5.5
Zn	73	6.3
Fe	258	100
Si	429	225

FIGURE 4: ASH COMPOSITION

We have observed that in addition to reducing the total ash content, the composition of the ash has also changed drastically. (Figure 4) Most of the key elements as far as electrodes are concerned have been reduced very substantially.

In these series of experiments, no attempt was made to filter quasi 0% QI level. In previously published work (4), we have shown that close to 0% QI level could be achieved and accordingly the ash content also dropped to near 0%.

CONCLUSIONS

- Coke oven tars can be filtered successfully to attain desirable quinoline insoluble levels.
- Significant portions of the ash forming components can be removed by filtration.
- The viscosity of coal tars can be reduced significantly by filtration.
- By varying process conditions, either of the above listed properties can be optimized.
- The proportion of β-resins increases by filtration.
- High quality coal tars can be produced by filtration from low quality (high QI, high ash) coke oven tars. These upgraded coal tars can be used as precursors for the manufacture of high quality coal tar pitches for electrodes as binders and impregnants, precursors for needle coke, for specialty coatings such as tar extenders for epoxies and urethanes, etc.

REFERENCES

1. H.A. Kremer & S. Cukier: Reduction of Ash Content of Electrode Binder Pitches by Dewatering. Proceedings of the Chemical Institute of Canada-Halifax, Nova Scotia (June 3, 1981).

2. H.A. Kremer & S. Cukier: Structures of Insoluble Carbon in Tars and Pitches. Journal of Microscopy, Vol. 132, Dec. 1983.

3. A. Limonchik: Increasing the Oxidation Resistance of Carbonaceous Bodies, Light Metals. Proceedings of AIME, Chicago 1981, p. 459.

4. H.A. Kremer & S. Cukier: Reduction of Quinoline Insoluble (QI) Fraction in Tars. Light Metals 1982, Conference Proceedings of AIME, Dallas 1982, p. 703.

5. H.A. Kremer and S. Cukier: Characteristics of Quinoline Insoluble Particles and their Influence on Coal Tar Pitches. Abstracts 16th Biennial Conference on Carbon, San Diego, July 1983.

6. H.A. Kremer & S. Cukier: Proceedings International Tar Conference, Biarritz, France, June 1983.

7. H.A. Kremer: Recent Developments in Electrode Pitches and Coal Tar Technology Chemistry and Industry. (UK) 18th Sept. 1982, p. 702-713.

8. BART, E.F. et al: Source Factors in Quinoline Insoluble Content of Coal Tar Pitch. Light Metals Proceedings of AIME 1981, p. 479.

CHAIRMEN

L.G. Boxall
Martin Marietta Laboratories
Baltimore, Maryland

V.L. Bullough
Reynolds Metals Company
Sheffield, Alabama

CORRELATION OF RAW MATERIAL PROPERTIES

WITH BINDER CONTENT OF HS SODERBERG PASTE

L. Castonguay, S.K. Nadkarni

Alcan International Limited
Arvida Laboratories
P.O. Box 1250
Jonquière, Québec, Canada
G7S 4K8

The effect of raw material properties on binder requirement of HS Soderberg paste was studied in the laboratory. Green apparent density, initial elongation and elongation after 24 and 72 hours retention time at 170°C were measured for paste produced with various cokes and pitches using different binder contents. Binder requirement was correlated with bulk density of the coke and the softening point and quinoline insoluble content of the pitch. It was observed that the effect of these properties on binder demand is independent of the level of paste fluidity.

Introduction

In the operation of electrolytic cells with Soderberg anodes, the distribution of the paste on anode top affects significantly the electrode performance (1). Control of the paste fluidity generally ensures a good spreading on the anode top.

The flowability of Soderberg paste is closely related to its binder content. However, the raw material properties such as the quinoline insoluble content and the softening point of the pitch and the porosity of the calcined coke influence the fluidity of the paste at a given binder content (1). Therefore, if variation in raw material characteristics occurs, it becomes necessary to vary the binder content of the paste to maintain the fluidity at the target.

This paper describes the correlation between binder requirement and raw material properties to maintain the Soderberg paste fluidity at a given level.

Experimental

Twenty-one test mixes were prepared in the laboratory using 5 different cokes, 5 coal-tar pitches and one petroleum pitch. The analysis of the cokes and pitches is given in Appendices I and II. The aggregate size distribution is given in Appendix III and details of paste preparation are given in Table I.

Table I. Binder Contents Used for Each Coke and Pitch Combination in Soderberg Paste

Pitch No. \ Coke No.	1	2	3	4	5
1				30.0 28.0 26.0	28.0 26.0 24.0
2		34.0 32.5 30.0	34.0 32.0 30.0	34.0 32.0 30.0	32.0 30.0 28.0
3	36.0 35.0 32.0	33.0 32.0 30.0	33.0 30.0 28.5		32.0 30.0 28.5
4		34.0 32.0 30.0		34.0 32.0 31.0	32.0 30.0 28.0
5	36.0 34.0 32.0	33.0 31.0 28.0	36.0 34.0 32.0	34.0 32.0 30.0	32.0 30.0 28.0
6		34.0 32.0 30.0		35.0 32.0 30.0	33.0 32.0 28.0

Coke and pitch were mixed at 160°C for 60 minutes in a sigma blade mixer of 1.0 L capacity. Green apparent density of each mix was measured. Initial elongation of the different mixes was measured on 50 mm long and 25 mm diameter specimens placed on a board with a slope of 7.8° and held at 255°C for 15 minutes. Elongation was calculated as percent increase in length. Elongation was also measured on paste samples held at 170°C for 24 and 72 hours in a similar manner.

Results and Discussion

It has been our experience that to obtain sufficient paste fluidity on anode top, the appropriate amount of binder is such that the initial elongation (IE) would be 60%, elongation after 24 hours at 170°C (VE) would be 25%, elongation after 72 hours at 170°C (SE) would be 15% or the mean elongation (ME) would be 28%, where

$$ME = 1/3 \times \left[0.5\ IE + 1.5\ VE + SE\right] \qquad (1)$$

The mean elongation accounts for thermal stability.

The binder requirement for the various mixes was determined based on the above criteria from the graphs of elongation vs logarithm of the binder content, since it was observed that the elongation varied exponentially with the binder content. These estimates for the different pitch and coke combinations are given in Table II.

Correlations between binder demand % (B) and various raw material properties were attempted by multiple regression analysis. The ones most significantly correlated were quinoline insoluble content of the pitch % (QI), softening point of the pitch °C (SP) and the bulk density of the coke g/mL (BD) determined by Alcan method (2). This correlation is given below.

$$B = A - 31.4\ BD + 0.172\ QI + 0.039\ SP \qquad (2)$$

correlation coefficient = 0.978
standard error = 0.47

where A is a constant.

Part of the standard error is attributed to the errors in measurement of raw material properties.

Since the paste fluidity is affected by the aggregate size distribution and the type of mixer used, the constant A in equation (2) depends on paste manufacturing variables. It was observed that the constant A was a function of the target level of fluidity and did not vary with the raw material properties.

Green apparent density of the various mixes did not vary significantly with the binder content within the range tested, i.e. corresponding to an elongation between 20% to 150% approximately. Average apparent density was calculated for each coke and pitch combination. Results are given in Table III.

A statistically significant correlation was obtained by multiple regression analysis between green apparent density of the paste, g/mL (GD) and softening point of the pitch °C (SP), quinoline insoluble content of the pitch % (QI), bulk density g/mL (BD) and real density g/mL (RD) of the coke. This correlation is given below:

Table II. Binder Requirement for Different Pitch and Coke Combinations Based on Different Fluidity Criteria

Combination		Binder Requirement (wt%) Based on Elongation Measurements			
Coke No.	Pitch No.	Initial Elongation 60%	Elongation after 24 hrs at 170°C 25%	Elongation after 72 hrs at 170°C 15%	Mean Elongation over 72 hrs at 170°C 28%
5	1	25.5	24.9	26.1	25.2
	2	29.7	26.6	25.6	27.0
	3	30.2	28.6	29.1	28.8
	4	31.3	29.4	28.4	29.5
	5	29.8	28.6	27.7	28.4
	6	32.6	29.9	29.9	30.1
4	1	27.8	27.4	27.8	27.5
	2	32.2	29.4	29.4	30.0
	4	33.5	31.4	31.1	31.7
	5	32.2	30.8	31.0	31.0
	6	34.5	32.1	32.0	32.4
3	2	32.8	29.4	29.0	30.0
	3	32.8	32.4	31.7	32.1
	5	33.5	32.2	32.0	32.3
2	2	33.0	30.9	31.7	31.4
	3	32.6	31.7	32.1	31.8
	4	35.6	31.8	32.3	32.7
	5	32.8	32.0	32.5	32.2
	6	34.6	32.6	31.8	32.6
1	3	33.6	33.8	33.6	33.6
	5	35.7	34.5	34.9	34.5

Table III. Average Green Apparent Densities for Different Pitch and Coke Combinations

Combination		Paste Property
Coke No.	Pitch No.	Green Apparent Density g/mL
5	1	1.572
	2	1.614
	3	1.629
	4	1.636
	5	1.631
	6	1.639
4	1	1.531
	2	1.574
	4	1.589
	5	1.591
	6	1.597
3	2	1.576
	3	1.591
	5	1.598
2	2	1.562
	3	1.605
	4	1.590
	5	1.561
	6	1.608
1	3	1.547
	5	1.571

$$GD = A + 0.537\ BD + 0.375\ RD + 0.003\ QI + 0.0004\ SP \tag{3}$$

correlation coefficient = 0.962
standard error = 0.008

where A is a constant.

Conclusion

The binder requirement of the HS Soderberg anode paste is a function of the target fluidity level and the raw material properties and is described by the following relationship:

$$B = A - 31.4\ BD + 0.172\ QI + 0.039\ SP$$

The green apparent density of the paste is independent of the fluidity level within the range of elongation between 20% to 150% approximately. It is a function of the raw material properties as seen from the relationship given below:

$$GD = A + 0.537\ BD + 0.375\ RD + 0.003\ QI + 0.0004\ SP$$

Acknowledgments

The author wishes to acknowledge Mr. D. DuTremblay and Mr. R. LeBlanc who were involved in this study.

References

(1) Bowitz, O., Eftestol, T. and Selvik, R.A., "New Methods for Testing Raw Materials for Anode Carbon Paste", Extractive Metallurgy of Aluminum, Vol, 2, Interscience Publishers, 1962, pp 331-349.

(2) Belitskus, D., "Standardization of a Calcined Coke Bulk Density Test", Light Metals 1982 AIME, Dallas, Texas, pp 673-689.

Appendix I: Properties of Calcined Petroleum Cokes

Coke No.	Real Density g/mL	Bulk Density* g/mL	Porosity** %	Mean Crystallite Thickness Lc Å
1	2.02	0.771	30.0	23.7
2	2.08	0.798	27.0	30.4
3	2.04	0.823	29.5	28.7
4	2.01	0.857	26.0	24.3
5	2.02	0.915	21.0	27.0

* Alcan method (2)
** Measured on -20 +35 Tyler mesh fraction

Appendix II: Analysis of Pitches

Pitch No.	Softening Point °C	Coking Value %	Quinoline Insoluble %	Benzene Insoluble %	Real Density at 20°C g/mL
1*	119.9	44.2	0.6	24.3	1.23
2	123.3	63.0	13.2	31.5	1.36
3	101.3	59.4	16.2	32.9	1.30
4	106.2	63.0	17.5	32.2	1.34
5	106.3	62.6	19.8	36.3	1.37
6	116.2	N.A.	25.1	41.7	1.38

N.A.: Not available
*: Petroleum pitch

Appendix III: Aggregate Size Distribution

Size Tyler Mesh	Cumulative wt%
+2	0
3	5
4	9
10	23
20	34
48	49
200	70
-200	100

TEMPERATURE STABILITY OF SØDERBERG ANODE PITCH

Morten Sørlie

Elkem a/s
R & D Center
P.O.Box 40
N-4620 Vågsbygd, Norway

A number of quality conditions must be met in a good Søderberg binder, both with respect to cell operation and potroom working environment. Among the more important are low evaporation loss of pitch, acceptable rheological properties which are maintained over a wide range of temperatures, and good enough quality of the binder coke to avoid problems in the baking and consumption zones.

Mild heat treatment of several Søderberg pitch binders used in industry revealed large differences in their thermal stabilities and hydrocarbon emissions. Evaporation losses from pitch surfaces may vary by one order of magnitude and changes in viscosity by a factor >1000 after the same mild heat treatment. Variations in pitch viscosity over ten orders of magnitude are well described by a simple empirical equation that utilizes the softening point of the pitches in a reduced temperature expression.

Introduction

Demands for a better environment in the potrooms and new regulations concerning anode emissions have caused several aluminium smelters to modify their operation towards anodes with colder tops and drier paste. However, the use of less binder in the paste may make the anode operation critically sensitive to pitch quality, a problem compounded by the fact that compromises are necessary. A good control with the flow properties of the green paste is important for the whole anode operation. The amount of fine filler particles is crucial to the flow and segregation behaviour of the molten paste [1]. The quinoline insoluble (QI) fraction of the binder pitch may act as solid particles in this respect, requiring this component to be considered along with the finest coke filler fractions.

Apart from small amounts of coal dust and ash carry-overs, the QI in coal-tar pitches usually contains two different kinds of particles. One is "natural" QI (also called "primary" QI), made up of spherical particles formed during severe vapour phase cracking (1000 °C and more) of tar vapours in the coke oven [2]. These solid pyrolysis particles are small, usually less than 1 μm in diameter, and have an atomic carbon to hydrogen (C/H) ratio between 3.5 and 4.0 [3]. The second type of particles often found in anode pitches are mesophase spheres, comprising the "secondary" QI or "thermal" QI. The mesophase is an anisotropic liquid crystal phase which formes and separates from the isotropic pitch during heat treatment. The frequently observed spherical shape is caused by surface tension phenomena. Average mesophase diameters are usually much larger than primary QI diameters and values ranging from optically observable (approximately 1 μm) to >100 μm are common. Since the mesophase is a liquid phase, its C/H ratio is considerably less than that of primary QI; a ratio of about 2 - 2.5 has been indicated.

The mesophase content in coal-tar pitch has been the subject of much discussion in recent years. It has been claimed [4] that the mesophase acts as solid particles in the green paste stage and therefore does not contribute to the properties of the binder. Another report [5] maintains the conflicting view that the mesophase spherolites are destroyed upon mixing and adhere to the filler particles, thus decreasing the wettability of coke particles by the pitch. The mesophase may form a very viscous envelope around the filler particles and this envelope may cause an increase in the viscosity of the green paste. A high paste viscosity is believed to be undesirable [2], as more pitch is needed to correct this deficiency.

Søderberg anode pitches

The three pitches used in the present laboratory investigation are industrially used coal-tar based binders for the vertical stub Søderberg anodes. Table 1 contains some analysis data for the pitches, lettered A, B, and C. Flowability characteristics of the pitches in mixtures with fine coke were all typical of the Søderberg-type binders.

Pitches A and B are extremes with respect to mesophase content (Figure 1). Pitch A contains no observable mesophase (optical microscopy, >1 μm diameter), whereas more than 50 % of the QI material in pitch B is optically quantifiable mesophase spheres up to about 40 μm in diameter. Pitch C is a more "typical" Søderberg anode binder, whose mesophase content constitutes only a minor part of the total QI fraction.

Table 1. Properties of pitch binders studied.

Pitch		A	B	C
Softening point (Mettler)	°C	120.7	113.4	105.6
Density, 20 °C	kg/m^3	1326	1315	1317
QI	%	10.4	11.9	11.9
TI	%	32.0	32.9	34.0
β-fraction	%	21.6	21.0	22.1
Coking value (Alcan)	%	59.2	57.5	56.8
Mesophase	%	0.0	6.0	3.5
$(C/H)_{QI}$		3,89	2.59	2.98

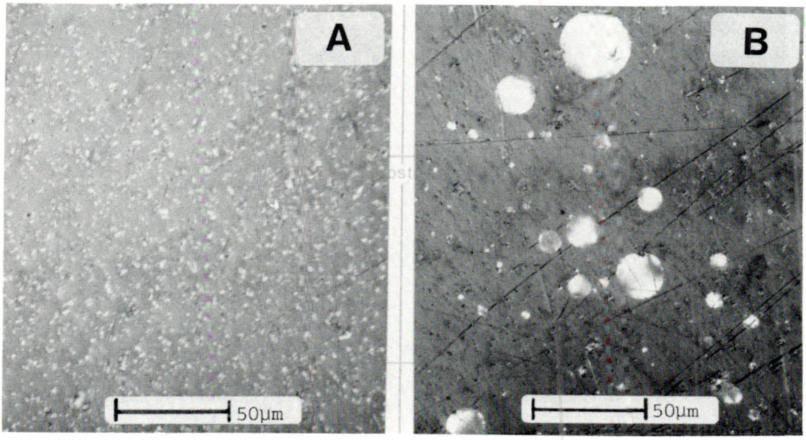

Figure 1 - Microphotographs of pitch A and pitch B show their difference in QI matter. Pitch C has a visual QI appearance between these extremes.

The mesophase content reflects partly the thermal treatment which the binder has been subjected to during the tar distillation process. In order to achieve an acceptable QI content, pitch B has apparently been subjected to the most severe heat treatment resulting in 6.0 % optically identifiable mesophase; pitch C has received a somewhat less severe treatment (3.5 % mesophase), whereas pitch A has been given a mild heat treatment so that no mesophase can be optically determined. The atomic C/H ratio of the QI fraction $(C/H)_{QI}$, verifies this, and is 2.59, 2.98, and 3.89 for pitches B, C, and A, respectively. The β-fraction which is the difference between TI (toluene insoluble) and QI fractions and often believed to indicate carbonization and mesophase forming properties of pitches, differs with only about 1 %.

Heat treatment is also used by tar distillers to increase the softening point of the binder. An alternative way to achieve this is to concentrate the heaviest fractions by a vacuum treatment of the straight-run pitch. This may be the reason why pitch A simultaneously has a high softening point and zero mesophase content. The removal of low boiling compounds may also contribute to the higher coke yield and density of pitch A (Table 1).

Figure 2 - Pitch weight loss from evaporation at constant temperature 200 °C. Open symbols: pitch surface exposed to surrounding air (open system). Closed symbols: Aluminium foil tightly fitted around crucible top to protect pitch surface (semi-closed system)
o, ●: pitch A; △, ▲: pitch B; □, ■: pitch C.

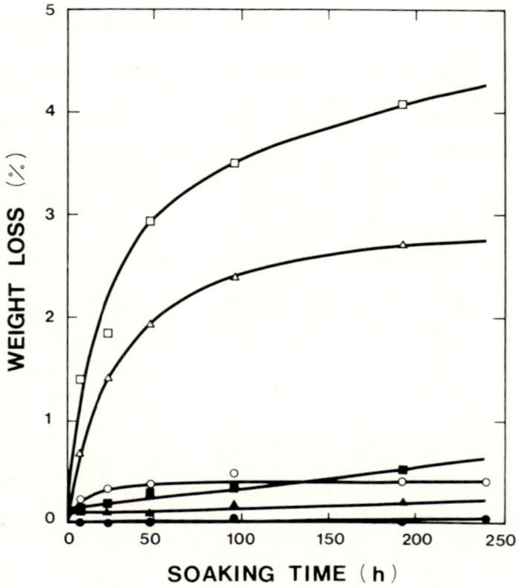

Evaporation loss from pitch surfaces

The weight loss during mild heat treatment of pitches A, B, and C showed large individual differences. In each experiment a container with 60 g pitch was used with an exposed pitch surface area of 18 cm^2. Figure 2 shows weight loss curves for the three anode pitches held at 200 °C for various lengths of time. Open symbols represent experiments where the molten pitch surface was exposed to the surrounding air. Closed symbols represent a "semi"-closed system where an aluminium foil was tightly fitted around the upper rim of the container. The foil allowed for vapour pressure equilibration with the surrounding atmosphere but also for diffusion of minor amounts of pitch vapour components out of the system. In contrast to the former case, no pitch vapour transport due to air convection above the pitch surface, could take place.

As shown in Figure 2, the evaporation loss of pitch C is 10 times larger than pitch A both in a system exposed to the surrounding atmosphere and in a semi-closed one. Evaporation from the exposed surface of pitch A reaches a constant value of 0.4 % after a period of about 50 h at 200 °C, while weight losses in pitches B and C continue for the duration of the experiment. Due to the limited pitch reservoir, the evaporation rate is continuously reduced as the melts are depleted of their low boiling compounds. Keeping in mind that the measurements were performed with small pitch volumes, the calculated pitch losses are 130, 900, and 1400 g/m^2 pitch surface during a 200 h period for pitches A, B, and C, respectively. In the same order the pitch evaporation rates are 0, 4.8, and 6.4 g/m^2h, and 0, 0.4, and 1.5 g/m^2h, at 50 and 200 h, respectively.

Weight loss curves for pitches B and C in the semi-closed system appear to intercept the y-axis above zero. This value corresponds to the moisture content in these two pitches which were flash evaporated at the beginning of each experiment. Pitch A does not contain water in detectable amounts. The major part of the evaporation loss is thus caused by loss of organic pitch components. Identification of pitch vapour components has not yet been performed, but all three pitches are highly aromatic coal-tar pitches; consequently low pitch evaporation losses, like those found for pitch A, can contribute to improved working conditions in the Søderberg potrooms.

Changes in pitch properties during mild heat treatment

The observations described above are probably functions of process variables in the tar distillery and also reflect the quality of the coal-tar precursor. Pitch properties change somewhat during the isothermal heat treatment at 200 °C. As one might expect, the pitch softening point increases (Figure 3). The increase is largest for pitch C which has the lowest softening point to start with (Table 1). An increase of about 15°C is observed in this pitch over a period of 200 h, while the softening point of pitch A only increases about 3°C under the same conditions.

No significant softening point differences or trends are, however, observed for pitches which have been exposed to the surrounding atmosphere during the isothermal heat treatment and pitches which have been heat-treated in the semi-closed system described above, although the difference in loss of volatiles is as much as 3.5 wt % in the case of pitch C. Prolonged soaking times eventually give pitch B a higher softening point than pitch A (Figure 3).

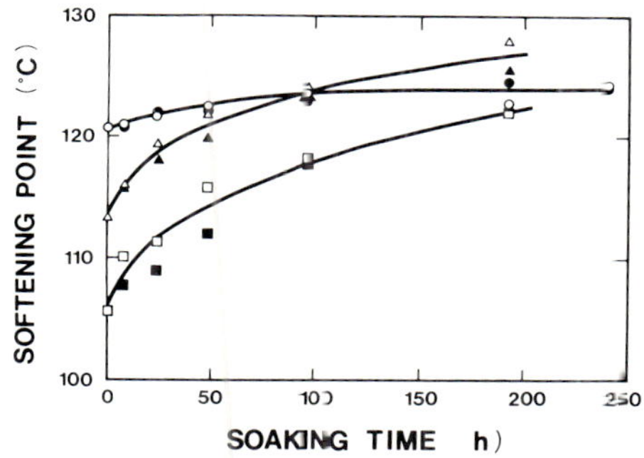

Figure 3 - Changes in pitch softening point (Mettler) during isothermal heat treatment at 200 °C. See Figure 2 for symbol explanation.

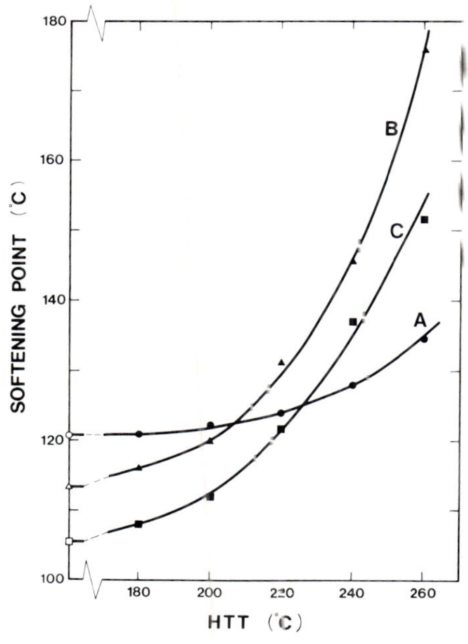

Figure 4 - Pitch softening point (Mettler) as a function of heat treatment temperature (HTT) in semi-closed system. Open symbols are softening points of untreated pitches (starting values).

Changes in QI content at 200 °C is almost zero for all three pitches, while the total content of TI material increases about 1.5 - 2 %. There is no significant discrepancy between TI values measured in pitch contained in an open crucible, or a semi-closed one, in accordance with the observations for the softening point.

The temperature stabilities of the anode pitches, as well as their relative differences, are better illustrated in Figure 4 where the softening point is plotted as a function of heat treatment temperature (HTT) at constant soaking time 48 h. These experimental values are from the semi-closed system only. Pitch A, which at the start has the highest softening point, is exceeded by pitch B at a HTT less than 210°C and by pitch C at HTT less than 230°C. Total changes in softening points at HTT 260°C are 14, 62, and 46°C for pitch A, B, and C, respectively. The pitch evaporation losses are 0.30, 0.84 and 1.71 %, in the same order as above.

Changes in both QI and TI take place rapidly above 200 °C. In the semi-closed system, a soaking time of 48 h at 260 °C increases the QI content in the pitch to 39.3, 50.4, and 46.6 % for pitches A, B and C, respectively, taking into account the weight loss due to evaporation of low-boiling pitch components.

Obviously, the above changes in pitch properties are not primarily caused by the removal of "solvent" compounds of low boiling points and low molecular weights. It is known [6-8] that the changes in pitch properties will not take place at such low temperatures if oxygen is excluded. This was verified with pitches A and B which were sealed into quartz glass ampoules under an argon atmosphere and kept at 240 °C for 48 h. No significant change in pitch softening points could be measured. If, however, oxygen is present, changes in pitch properties have been reported [6] to start above 200 °C.

The observed changes in softening point, QI, and TI may therefore be the result of reactions with atmospheric oxygen. As previously discussed, no significant differences in pitch properties could be observed in experiments where the surrounding air had free access to the pitch surface and in cases where the supply of oxygen was restricted (semi-closed system). It seems therefore likely to assume that oxygen may initiate pitch polymerization or condensation reactions without being consumed in significant quantities. Published studies [9] of mild air oxidation of pitch, at comparable temperatures, show that this treatment leads to increased average molecular weight without oxygen being chemically bonded to any significant extent in the product. The reported results were consistent with an oxidative mechanism involving dehydrogenative condensation of high molecular weight aromatic components (ArH) in the pitch:

$$2ArH + \tfrac{1}{2}O_2 = Ar\text{-}Ar + H_2O \qquad (1)$$

The organic reaction products are larger and higher in molecular weights and, are thus likely to increase softening points and QI and TI fractions of the pitch.

It is not realistic to assume that thermal pitch reactions which take place above the baking zone in dry top Søderberg anodes proceed without any access to oxygen. Air is always present in pores in the paste and will follow the flow of paste down towards the baking zone. These pores can always be seen when samples from the upper part of the anode are examined. The access of small amounts of air will therefore give a more realistic environment for studying the low temperature thermal reactions which take place well above the baking zone, than performing similar experiments under an inert atmosphere.

The observed increase in QI content with HTT is not followed by a simultaneous increase in observable mesophase. In the case of pitch A, no mesophase is observed by optical microscopy after heat treatment at 200°C, although the QI content increases from 10.4 to 14.7 %. In pitches B and C an increase in mesophase content is observed, but not nearly enough to account for the increase in QI fraction. Figure 5 shows the deviation between QI analysis and optical mesophase content,

$$\Delta QI = QI_{total} - \text{Mesophase}, \tag{2}$$

plotted as a function of HTT. The loss of volatile compounds has been taken into account. It is shown that the rate of deviation increases most rapidly for the most reactive, or thermally least stable, pitch B.

It should be possible to spot any growth of existing optical mesophase upon heat treatment by a microscopic point counting technique. This is confirmed to some extent with pitches B and C, but as Figure 5 shows, the optically quantifiable mesophase cannot account for the total increase in QI. Excluding any new quinoline insoluble phases it seems likely that the difference consists of mesophase particles too small to be identified in the optical microscope (< 1 μm). A mild heat treatment may thus not only result in growth of existing mesophase spherolites, but also in the nucleation of new ones. Although discussions involving mesophase spherolites most often are concerned with particle sizes that can be identified by optical microscopy, it has been shown [10] that in the early stages of formation the particles can only be observed by transmission electron microscopy techniques.

The quinoline extraction and mesophase point counting techniques have their limitations. The room temperature microscopic techniques are only strictly valid for measuring the observable anisotropic phase present at the solidification temperature of the pitch; extrapolation of these results to molten pitch temperatures should be done with great care [11]. The ani-

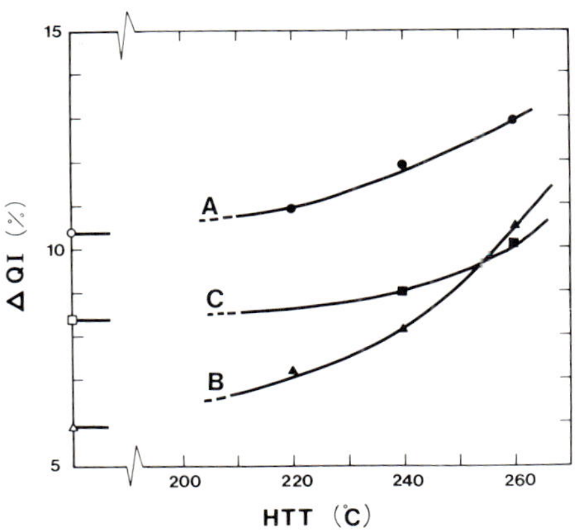

Figure 5 - The difference between QI content and optically observable mesophase in original and heat treated pitches (semi-closed system).

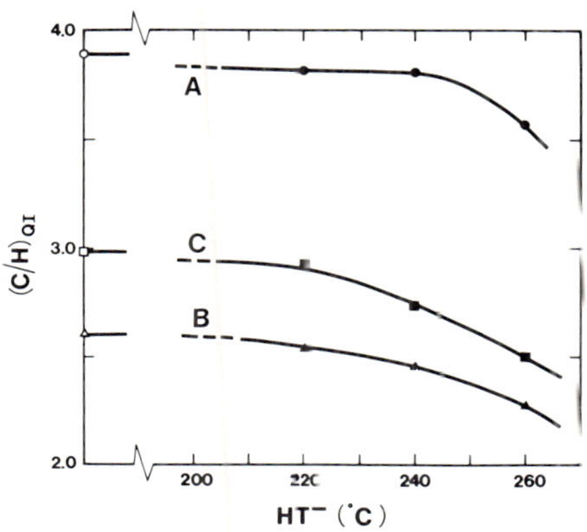

Figure 6 - Atomic C/H ratio of QI fraction of original and heat treated pitches (semi-closed system).

sotropic mesophase has no uniform molecular composition, but contains components over a wide range of molecular weights. It has been found [4] that for certain pitches the mesophase content can partly be dissolved in quinoline, thus making the QI content an underestimate of the total particle content in the pitch. In terms of thermodynamic considerations it will be expected that the smallest mesophase particles also will be those most easily dissolved in the quinoline solvent.

Changes in atomic C/H ratios of the quinoline insolubles as a function of HTT (Figure 6) also illustrate the increased content of secondary QI material. The positioning of the curves relative to the y-axis serves as an indication of their content of mesophase in the QI fraction.

Viscosity measurements

The pitch viscosities, before and after heat treatment, were measured in the temperature range 60.5-203°C. These measurements, which covered viscosities from far below the softening point to the liquid low viscosity range, were all performed on a viscometer built at Elkem a/s, R & D Center [12]. The instrument is a parallel plate compression viscometer on-line interfaced to a microcomputer for automatic data aquisition and evaluation of absolute viscosities. Experimentally, the plate separation is measured at a given temperature as a function of time [13].

The measuring range of the instrument covers viscosities differing by 10 orders of magnitude, from as low as 0.1 Pas to as high as 1,000,000,000 Pas (Figure 7). Coal-tar pitches behave more like super-cooled liquids than solids below their softening temperatures, and viscosities may thus be determined at temperatures well below their normal fluid range.

Arrhenius plots of logarithmic viscosity vs. 1/T (Kelvin) for untreated pitches A, B, and C, as well as plots of the same pitches heat treated at 260°C (48 h), are shown in Figure 7. The untreated pitch A was most thoroughly examined and shows a near constant activation energy for viscous

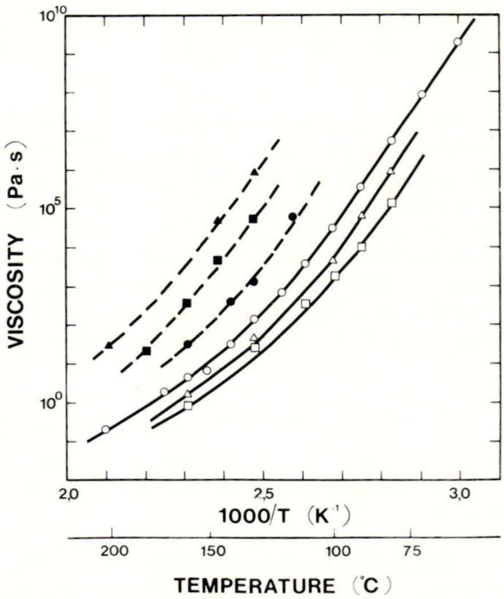

Figure 7 - Viscosites of untreated binder pitches (open symbols) and the same pitches heattreated at 260°C for 48 h (filled symbols).
o,●: pitch A; △,▲ : pitch B; □,■: pitch C.

Figure 8 - Plot of binder pitch viscosities at 150 °C vs. HTT. Values on ordinate corresponds to untreated pitches.

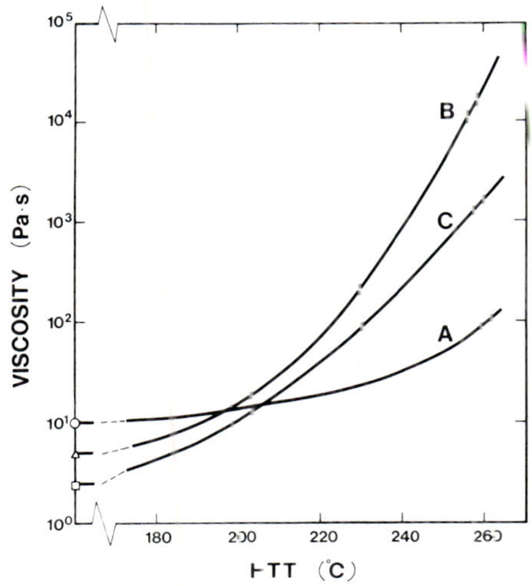

Table 2. Parameters in 3rd order polynomial fit of log(viscosity) [Pas] vs. T/T_s (equation (3)).

A(0)	A(1)	A(2)	A(3)	Std.error
292.22	-729.96	617.60	-177.22	0.23

flow at temperatures below its softening point and a small monotonous decrease in the activation energy at temperatures above. It can also be seen that the activation energy for viscous flow of pitches B and C does not differ much from what is observed in pitch A as the curves have a near parallel displacement determined by the difference in their softening points. The same is also valid for the heat treated pitches shown in Figure 7.

Plots of log(viscosities) vs. HTT yield curves which qualitatively are very similar to those presented in Figure 4. This is shown in Figure 8 with viscosity values interpolated at 150°C. It might thus be possible to calculate the viscosities of any examined pitch at any temperature provided that its softening point is known. This is demonstrated in Figure 9 where all experimentally determined viscosities have been plotted against the reduced temperature T/T_s, where T is the temperature at which viscosity was measured, and T_s is the softening point (Mettler) of the pitch in degrees Kelvin. A 3rd order polynomial may satisfactorily be fitted to these values:

$$\log(\text{viscosity}) = A(0) + A(1)\frac{T}{T_s} + A(2)(\frac{T}{T_s})^2 + A(3)(\frac{T}{T_s})^3 . \qquad (3)$$

The parameters for calculating pitch viscosities (in Pas) are given in Table 2.

The softening point (Mettler) viscosity calculated from Equation (3) is 435 Pas. The Mettler method gives a somewhat higher softening point than the often quoted Ring and Ball (R & B) method. A careful examination of both methods with several coal-tar pitches gives the following relation between the softening points (t_s) in °C

$$t_s(\text{Mettler}) = t_s(\text{R \& B}) + 4.7 \qquad (4)$$

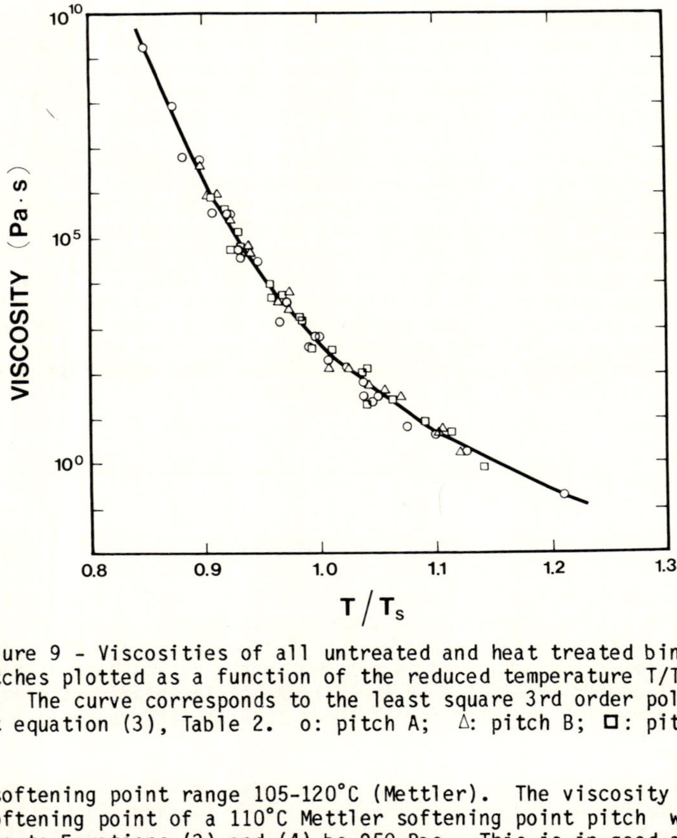

Figure 9 - Viscosities of all untreated and heat treated binder pitches plotted as a function of the reduced temperature T/T_S (in K). The curve corresponds to the least square 3rd order polynomial fit equation (3), Table 2. o: pitch A; Δ: pitch B; □: pitch C.

in the softening point range 105-120°C (Mettler). The viscosity at the R & B softening point of a 110°C Mettler softening point pitch will according to Equations (3) and (4) be 950 Pas. This is in good agreement with the expected value of about 800 Pas [14].

Carbonization behaviour of heat treated pitches

Pitch A yields a near constant coking value (corrected for pitch volatilization during heat treatment) independent of soaking time (200 °C), while pitches B and C show an increase in coke residue under similar conditions, yet without reaching the value found in pitch A (Table 1) for the duration of the experiment. Contrary to what was found for changes in softening point, QI, and TI upon heat treatment time, the coking values from open and semi-closed systems differ appreciably for pitches B and C, yielding lower increases in coke residues in the open system. This may either be caused by the rapid evaporation loss of low boiling reactive pitch components, which otherwise might increase pitch polymerization rates at elevated temperatures, or by higher oxygen concentrations which in these melts result in formation of more CO_2 during coking.

Coking values at different HTT follow qualitatively a similar pattern as the softening points (Figure 4), QI, and TI. Pitch A, which originally has the highest coking value is at HTT 220 - 260 °C surpassed by pitches B and C which show a much more rapid increase in carbonization rate at elevated temperatures (180 - 260 °C).

The final carbon residue of the coked binder pitch in the baked Søderberg anode is, however, what is most important with respect to anode quality and operation. Possible anode reactivity and dusting problems are believed mainly to be determined by the binder coke quality [15], which again is closely related to the coking behaviour of the pitch.

The apparent lower coking values obtained with pitch A at HTT's >220 °C do not imply that it will give a lower amount of fixed carbon in the baked anode, only that the carbonization reactions are shifted to higher temperatures compared to pitches B and C. This is seen in Figures 10 and 11 where the coking values, corrected for pitch volatilization losses, are plotted as a function of softening point (Figure 10) and TI (Figure 11). Both the softening point and the TI fraction may serve as a measure for the extent of carbonization or polymerization in the pitch prior to the standardized coking procedure. An increased yield of binder coke will tend to lower the microporosity of the baked anode and may thus reduce the dusting problem [15].

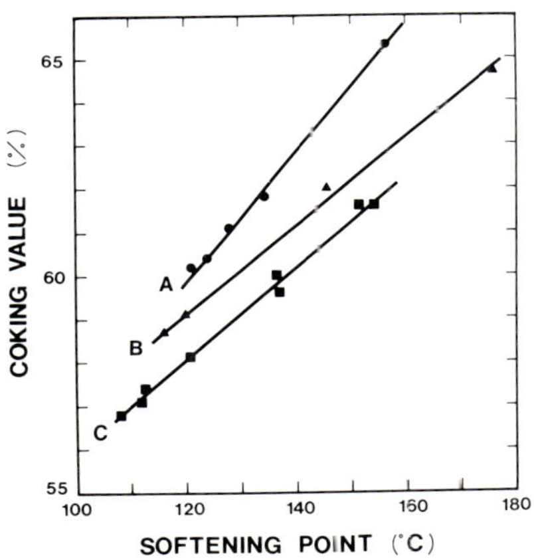

Figure 10 - Coking value (Alcan) of heat treated binder pitches vs. their softening points (Mettler).

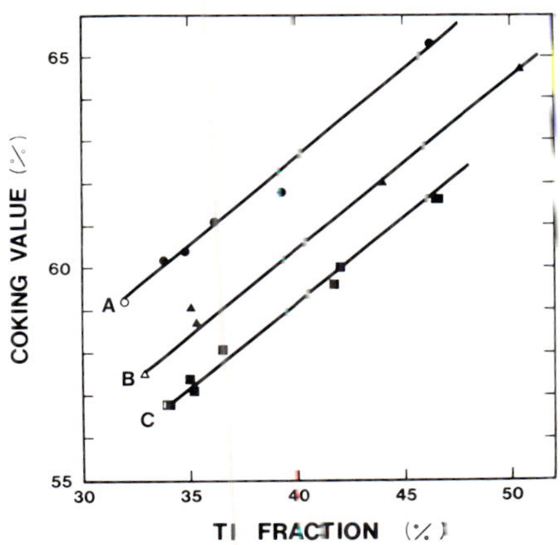

Figure 11 - Coking value (Alcan) of untreated (open symbols) and heat treated (filled symbols) pitches vs. TI fraction.

Concluding remarks

As the preceding discussion shows, thermal reactivity and carbonization behaviour can vary considerably between typical Søderberg binders. The pitches pass through polymerization and mesophase forming stages during the carbonization process and it is reasonable to believe that the rate at which these reactions take place place may severely influence their properties as binders both in the viscous and the baked part of the anode. Especially, the type of QI material may play an important part in pitch binder performance in the anode. It seems that a large content of primary QI retards the mesophase forming and coalescing reactions, thus shifting the rapid increases in viscosity to higher temperatures. It has been claimed [16] that binders which undergo minimal changes in viscosities, when subjected to heat treatment, are required for good anode operation. Tests for viscosity stability may thus be helpful in defining more accurately the quality of a given pitch. As binder quality and properties are related to the quality of the pitch precursor, as well as to process variables in the tar distillery, a more complete knowledge is needed to better understand and specify optimum binder properties for the Søderberg anode.

Acknowledgements

This work is part of a program jointly supported by Elkem a.s, Mosal Aluminium and the Royal Norwegian Council for Scientific and Industrial Research (NTNF). The author wishes to thank Mr. H.Gran for performing several experiments and Mr. J.Lango for doing most of the pitch analyses.

References

1. O. Bowitz and O. Sandberg, "Søderberg anode carbon in cells for electrolytic production of aluminium", Trans. Met. Soc. AIME 224 (1962) 53.

2. G.R. Romovacek, "The influence of quinoline insolubles on the performance of industrial pitches", Prepr. 3rd Int. Carbon Conf. (1980) 303.

3. H.R. Sulkowski and J.W. Bluhm, "Examination of quinoline insolubles in tars and pitches by scanning electron microscope and energy dispersive X-ray", Light Met., Proc. Sess. AIME Annu. Meet., 107th (1) (1978) 289.

4. D.R. Ball, "Estimation of secondary QI", Prepr. 3rd Int. Carbon Conf., (1980) 299.

5. D. Auguie, M. Oberlin, A. Oberlin, and P. Hyvernat, "Formation of thin mesophase layers at the interface between filler and binder in prebaked anodes. Effect of mixing on mesophase", Carbon 19 (1981) 277.

6. T. Watanabe, "Studies of pitch for Søderberg anodes", in Extractive Metallurgy of Aluminum, Vol. 2, ed. G.Gerard, Interscience Publishers, New York, 1963, p. 351.

7. T. Imamura, M. Nakamizo, and H. Honda, "Formation of carbonaceous mesophase at lower temperature", Carbon 16 (1978) 487.

8. C.R. Mason and N.R. Turner, "Coal tar pitch binders for carbon and graphite electrode manufacture", paper presented at International Tar Conference, Eastbourne, 1980, p. 28.

9. J.B. Barr and I.C. Lewis, "Chemical changes during the mild air oxidation of pitch", Carbon 16 (1978) 439.

10. J.J. Friel, S. Mehta, G.D. Mitchell, and J.M. Karpinski, "Direct observation of the mesophase in coal", Fuel 59 (1980) 610.

11. S. Chwastiak, R.T. Lewis, and J.D. Ruggiero, "Quantitative determination of mesophase content in pitch", Carbon 19 (1981) 357.

12. K.Tørklep, Elkem a/s, R & D Center, to be published.

13. G.J. Dienes and H.F. Klemm, "Theory and application of the parallel plate plastometer", J. Appl. Phys. 17 (1946) 458.

14. British Carbonization Research Association, "The physical properties of coal tar products", Special publication 21, Chesterfield, 1978, pp. 33.

15. O.Bowitz, O.C.Bøckman, J.Jahr, and O.Sandberg, "Surface disintegration of carbon electrodes", Ind. Carbon Graphite (1958) 373.

16. H.W. Nelson, "Viscosity stability of binders for Søderberg electrodes", in Extractive Metallurgy of Aluminum, Vol. 2, ed. G. Gerard, Interscience Publishers, New York, 1963, p. 321.

EVALUATION OF VOLUMETRIC ANODE MIX FORMULATION

David Belitskus

Alcoa Laboratories
Alcoa Center, Pennsylvania 15069
USA

Volumetric anode mix formulation was examined as a potential means of maintaining a high bulk density aggregate sizing and of maintaining an optimum binder pitch level. Effects of coke and sizing changes on aggregate bulk density and effects of coke, sizing, and binder level changes on bench scale anode properties were compared from weight and volume standpoints. Volumetric blending of coke fractions was found to reduce deviation from a maximum bulk density aggregate only slightly. Binder pitch addition based on coke aggregate volume significantly decreased, but did not eliminate, deviations from the optimum binder level.

Introduction

For prebaked anodes for aluminum smelting, use of an aggregate sizing giving a high bulk density is advantageous (1). An empirical method for producing the densest sizings for any coke has been described (2). Typical sizings used commercially have been reported frequently. In all cases, sizing is given in terms of weights of fractions. It has been shown, however, that vibrated bulk densities of different lots of calcined petroleum coke vary appreciably with particle size (Figure 1). This means that the use of a constant sizing by weight can result in quite different ratios of coarse, intermediate, and fine particles for different cokes, i.e., quite different sizings by volume. One object of the work described herein was to determine whether the differences in maximum density sizings from one coke to another when considered on a weight basis would be reduced or eliminated when considered on a volume basis.

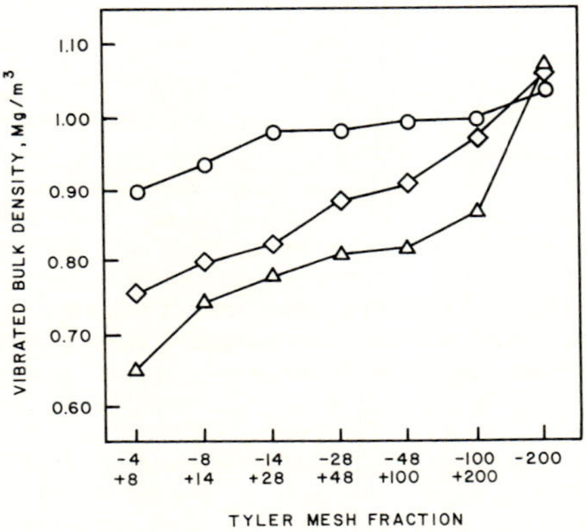

Figure 1 - Vibrated bulk density as a function of particle size for three calcined coke samples.

Use of an optimum binder pitch level is essential for the production of consistantly high quality anodes. A second object of this work was to determine whether there is any advantage to adding binder according to coke aggregate volume rather than by weight.

Consideration of Volumetric Aggregate Sizing

Previously reported data on empirical determinations of maximum bulk density sizings for 12 calcined cokes (3) were convenient for determining whether volumetric blending of fractions could result in an invariant maximum density sizing. The sizings investigated in the referenced work were limited to those that could be produced from typical commercial coarse, intermediate, and fine fractions, although they were reported in terms of more closely sized fractions such as -4+8 mesh, -8+14 mesh, etc. Bulk densities were carried out on the more closely sized fractions also.

Scatter in the data from the bulk density maximization work described above was large, so it was found useful to treat the results in terms of three groups of four cokes (high, medium, and low bulk density cokes) and to average the compositions of fractions and bulk densities of fractions of the four cokes of each group. Cumulative weight percentages and -28+48 mesh coke bulk densities are given in Figure 2. It can be seen that as coke bulk density increases, the maximum bulk density sizing becomes more coarse, on a weight basis.

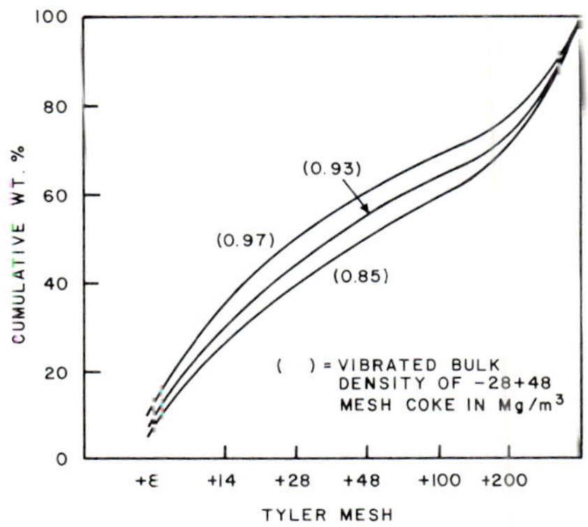

Figure 2 - Average maximum bulk density sizings for three groups of calcined cokes having high, medium, and low densities.

Table I gives the average weight percentages of fractions for the maximum bulk density sizing for each of the three groups of cokes. Average vibrated bulk densities of fractions for the coke groups are given in Table II. From the data in Tables I and II, volumes per 100 Mg of aggregate were calculated (Table III). Finally, Table IV gives the volume percentages of fractions, calculated from the data in Table III.

Table I. Average Weight Percentages of Fractions for Maximum Bulk Density Sizings of Three Groups of Calcined Cokes (Ref. 3)

	Weight %						
Coke Type	-4+8	-8+14	-14+28	-28+48	-48+100	-100+200	-200
High Bulk Density	10.3	21.8	15.9	9.7	10.0	10.0	22.3
Medium Bulk Density	9.4	18.7	15.1	10.1	8.3	13.8	24.6
Low Bulk Density	6.0	17.2	15.3	9.1	11.2	14.6	26.6
Spread:	4.3	4.6	0.8	1.0	2.9	4.6	4.3

Table II. Average Vibrated Bulk Densities of Fractions of Three Groups of Calcined Cokes (Ref. 3)

Coke Type	Vibrated Bulk Density, Mg/m^3						
	-4+8	-8+14	-14+28	-28+48	-48+100	-100+200	-200
High Bulk Density	0.87	0.91	0.95	0.97	0.99	0.99	1.06
Medium Bulk Density	0.83	0.88	0.90	0.93	0.92	0.97	1.07
Low Bulk Density	0.78	0.79	0.81	0.85	0.90	0.94	1.07

Table III. Volumes per 100 Mg of Aggregate Calculated from Weight Percentages (Table I) and Vibrated Bulk Densities (Table II)

Coke Type	Calculated Volume per 100 Mg of Aggregate, m^3							
	-4+8	-8+14	-14+28	-28+48	-48+100	-100+200	-200	Total
High Bulk Density	11.8	24.0	16.7	10.0	10.1	10.1	21.0	103.7
Medium Bulk Density	11.3	21.3	16.8	8.9	9.0	14.2	23.0	104.5
Low Bulk Density	7.7	21.8	18.9	10.7	12.4	15.5	24.9	111.9

Table IV. Volume Percentages of Fractions, Calculated from Data in Table III

Coke Type	Volume %						
	-4+8	-8+14	-14+28	-28+48	-48+100	-100+200	-200
High Bulk Density	11.4	23.1	16.1	9.6	9.7	9.8	20.3
Medium Bulk Density	10.8	20.4	16.1	8.5	8.6	13.6	22.0
Low Bulk Density	6.9	19.5	16.9	9.6	11.1	13.8	22.2
Spread:	4.5	3.6	0.8	1.1	2.5	4.0	1.9

From Table IV, it is obvious that maximum bulk density sizing is not invariant on a volume basis. By comparison with Table I, it can be seen that overall variation is only slightly reduced on a volume basis, although variation in the -200 mesh fraction is reduced considerably. Since maximum bulk density sizing is not invariant on a volume basis, the change in maximum density sizing with intrinsic coke bulk density involves more than merely keeping equal volumes of fractions, i.e., equal amounts of particles of each size category. Low bulk density cokes require a decrease in the number of coarse coke particles in the aggregate. Maximum density sizing is obviously a compromise between minimization of interparticle void space and minimization of the amount of very low density coarse particles.

Consideration of Volumetric Pitch Addition

In the work upon which the above calculations were made, not enough binder pitch levels were used in anode fabrication to accurately determine the optimum level for each coke. Therefore, data from other previously reported work (4) were used. In that work, two series of experimental anodes were fabricated. One series involved a single coke, four aggregate sizings (a maximum density sizing and three sizings that differed greatly from that sizing), and five or six binder pitch levels for each sizing. The second series involved three cokes varying significantly in vibrated bulk density, three aggregate sizings (all close to a maximum bulk density sizing), and four or five binder levels for each sizing.

Maximum baked apparent density was a primary criterion for estimating optimum binder content. However, since optimum binder content is not always clear from baked apparent density values alone, electrical resistivities were determined to supplement the baked apparent densities. Resistivity at room temperature was calculated from the voltage drop along a 25 mm length of a 150 mm long specimen carrying 6A of current.

The sizings used for the first series are given in the original report on these anodes. Aggregate vibrated bulk densities are given in Table V. Baked apparent densities and electrical resistivities are given as functions of binder content in Figure 3. From these data, optimum binder levels on a weight basis were estimated. Based on the bulk densities given in Table V and an assumed binder pitch density of 1.30 Mg/m^3, optimum levels on a volume basis were calculated. Both results are presented in Table VI. It is obvious that, for these anodes, optimum pitch content was essentially invariant when considered on a volume basis.

Table V. Vibrated Bulk Densities of Aggregate Blends (Ref. 4)

Sizing Designation	Vibrated Bulk Density, Mg/m^3
a	1.26
b	1.17
c	1.11
d	0.98

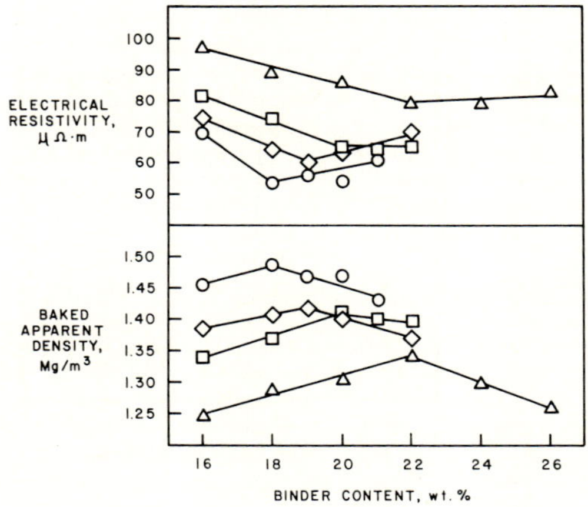

Figure 3 - Effects of aggregate sizing and binder level on anode properties. (○ , sizing a; □ , sizing b; ◇ , sizing c; △ , sizing d. See Reference 4 for sizing details.)

Table VI. Optimum Binder Contents on a Weight and Volume Basis for Anodes Produced Using Aggregate Sizings "a-d" (Ref. 4)

Sizing Designation	Optimum Binder Content Wt %*	Optimum Binder Content Vol %
a	18	17.6
b	19	17.4
c	20	17.6
d	22	17.5

* Estimated from Figure 3

The original report (4) gives the aggregate sizings used for the second series of anodes. Aggregate vibrated bulk densities are given in Table VII. Figures 4-6 give baked apparent densities and electrical resistivities as functions of binder content. Table VIII presents estimated optimum binder contents on a weight basis as well as calculated optimum levels on a volume basis.

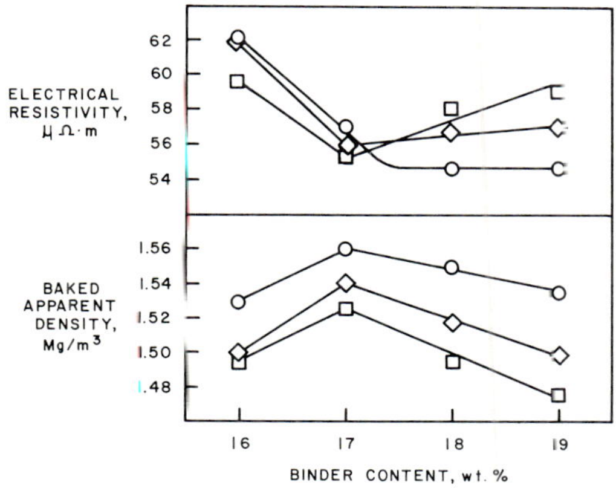

Figure 4 - Effects of aggregate sizing and binder level on anode properties for a high bulk density coke. (○ , sizing 1; ◇ , sizing 2; □ , sizing 3. See Reference 4 for sizing details.)

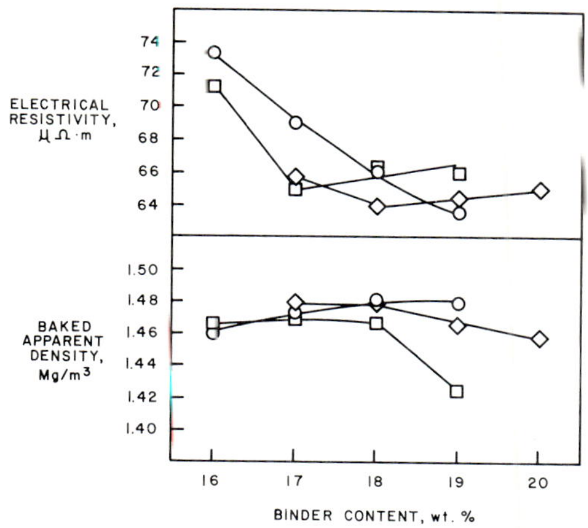

Figure 5 - Effects of aggregate sizing and binder level on anode properties for a medium bulk density coke. (○ , sizing 1; ◇ , sizing 2; □ , sizing 3. See Reference 4 for sizing details.)

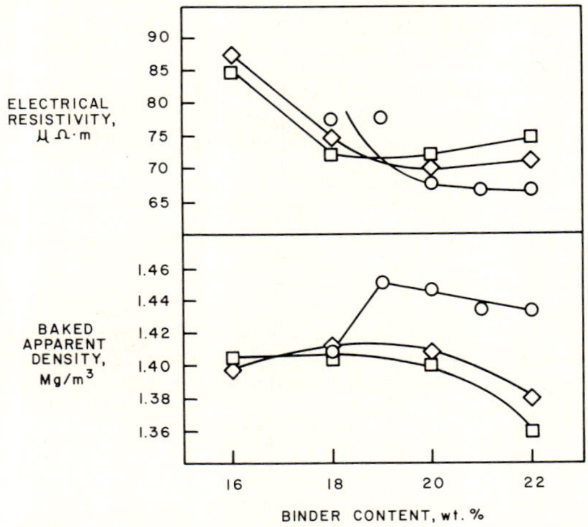

Figure 6 - Effects of aggregate sizing and binder level on anode properties for a low bulk density coke. (○ , sizing 1; ◇ , sizing 2; □ , sizing 3. See Reference 4 for sizing details.)

Table VII. Vibrated Bulk Densities of Aggregates from Three Cokes and Sizings "1-3" (Ref. 4)

Sizing	Coke Bulk Density Classification	Vibrated Bulk Density, Mg/m^3
1	High	1.33
1	Medium	1.25
1	Low	1.15
2	High	1.33
2	Medium	1.25
2	Low	1.14
3	High	1.30
3	Medium	1.25
3	Low	1.13

Table VIII. Optimum Binder Contents on a Weight and Volume Basis for Anodes Produced using Aggregate Sizings "1-3"

Coke Bulk Density Classification	Optimum Anode Binder Content					
	Sizing 1		Sizing 2		Sizing 3	
	Wt %*	Vol %	Wt %*	Vol %	Wt %*	Vol %
High	17.5	17.8	17	17.3	17	17.0
Medium	19	18.4	18	17.5	17	16.5
Low	20	18.1	19	17.1	19	17.0

* Estimated from Figures 4-6

It is apparent from Table VIII that with a given sizing (in terms of weight) but with a changing aggregate bulk density due to a coke change, optimum pitch content based on coke volume was much less variant than when considered on a weight basis. For example, for sizing "1" optimum binder level varied from 17.5 to 20% on a weight basis, but only from 17.8 to 18.4% on a volume basis.

Since for each coke all three sizings resulted in very similar vibrated bulk densities, relative differences in optimum binder content with changing sizing on either a weight or a volume basis were about the same. It is obvious that optimum binder content does not necessarily correlate with total porosity of an aggregate, i.e., with the sum of inter- and intraparticle porosity. While an increase in interparticle porosity is probably directly associated with an increase in binder requirement, the effect of an increase in intraparticle porosity may be complex. Change in binder requirement probably depends upon change in the ratio of total particle porosity to surface-accessible particle porosity and upon change in the average size distribution of surface-accessible pores.

Summary and Conclusions

Blending of coke fractions by constant volume has little advantage over blending by weight in maintaining a high bulk density aggregate sizing. Trial-and-error bulk density determinations or predictive equations based on an understanding of factors affecting bulk density are more suitable for this purpose.

Binder pitch additions based on coke aggregate volume rather than weight can significantly reduce, but not eliminate, deviations from the optimum level. Deviation is not eliminated because total aggregate porosity does not necessarily correlate with pitch requirement. An understanding of the nature of the porosity (such as interparticle versus intraparticle porosity, surface-accessible particle porosity versus total intraparticle porosity, and pore size distribution) is required for maintaining an optimum binder content under any changing conditions.

References

1. D. Belitskus, "Optimization of Raw Materials and Formulations for Hall-Heroult Cell Electrodes," Third Yugoslav Symposium on Aluminum, (Preprints) Edited by Prof. Andrej Paulin, University of Ljubljana, pp. 127-143 (1978).

2. R.W. Peterson, "Aggregate Optimization Using a Y-Blender," Light Metals 1982, The Metallurgical Society of AIME, Warrendale, Pennsylvania, pp. 691-702 (1982).

3. D. Belitskus, "Evaluating Calcined Coke for Aluminum Smelting by Bulk Density," Aluminium, v. 51, pp. 171-174 (1975).

4. D. Belitskus, "Effects of Coke and Formulation Variables on Fracture of Bench Scale Prebaked Anodes for Aluminum Smelting," Metallurgical Transactions B, v. 9B, pp. 705-710 (1978).

Anode Consumption Test Cell

P. P. Russell, K. R. Weisbrod and S. S. Jones

Primary Metals Research
ARCO Metals Company
Tucson, Arizona 85726
USA

A laboratory anode consumption cell with four anodes was constructed to evaluate anode performance free of air oxidation effects. One standard anode and three test anodes were typically used to effectively evaluate changes in coke and pitch source, granulometry, and pitching levels.

The average anode consumption rate of prebaked plant anode cores was 111 ± 1.5 percent of theoretical. Laboratory-made Soderberg-type anodes showed a lower performance of 125% carbon consumption. Under nonideal conditions, such as during excessive dusting of the anodes, the standard anode may be used to normalize data to standard conditions.

Introduction

In order to optimize anodes for aluminum production, a laboratory test cell was constructed. Excess carbon consumption during electrolysis in a commercial reduction cell typically arises from airburn, carbon dioxide oxidation, and carbon dust formation because of the preferential attack of the binder coke. While carbon dioxide and air oxidation may be semiquantitatively evaluated by standard tests already in the laboratory, an accurate test indicating carbon consumption during electrolysis under the same conditions was not readily available.

Most early anode consumption cell designs, such as the cell described by Hollingshead and Braunwarth and others,(1-3) made use of a single anode to evaluate the effect of cell operating conditions and different anode compositions upon anode consumption. During and between runs, cell temperature, bath ratio, and anode current densities may change. Often cell variations would not allow the investigator to effectively evaluate changes in coke or pitch source, granulometry, or optimum pitching level. Design of a four anode cell, with one standard anode for reference, was undertaken to obtain this objective.

Experimental Apparatus

Cell Design Criteria

In order to obtain laboratory results from a cell of manageable size which represents conditions expected in a commercial cell, the following design guidelines were followed.

(a) The overall size of the cell must be sufficiently small to be operated manually with a minimum of special hardware.

(b) The anode dimensions should easily accommodate the larger aggregate in prebaked anodes.

(c) A minimum distance of two anode diameters between anodes was desirable to reduce electrode potential interactions.

(d) An electrically insulating cathode sidelining was needed to reduce the radial current flux, shorting of the cell if excessive dusting occurs, and air oxidation of the graphite crucible.

(e) A semicontinuous alumina feed system was desired to prevent anode effects and consumption of alumina insulators.

(f) Independent control of DC power to each anode was required along with current and voltage limiters for safety.

Cell Arrangement

An overall view of the experimental system is given in Figure 1. An instrument rack containing the four DC power supplies and other instrumentation is shown along with the cell and auxiliary equipment. Details of the cell are presented in Figures 2 and 3.

Figure 1 - The Overall Experimental Arrangement

Figure 2 - Anode Arrangement after Electrolysis

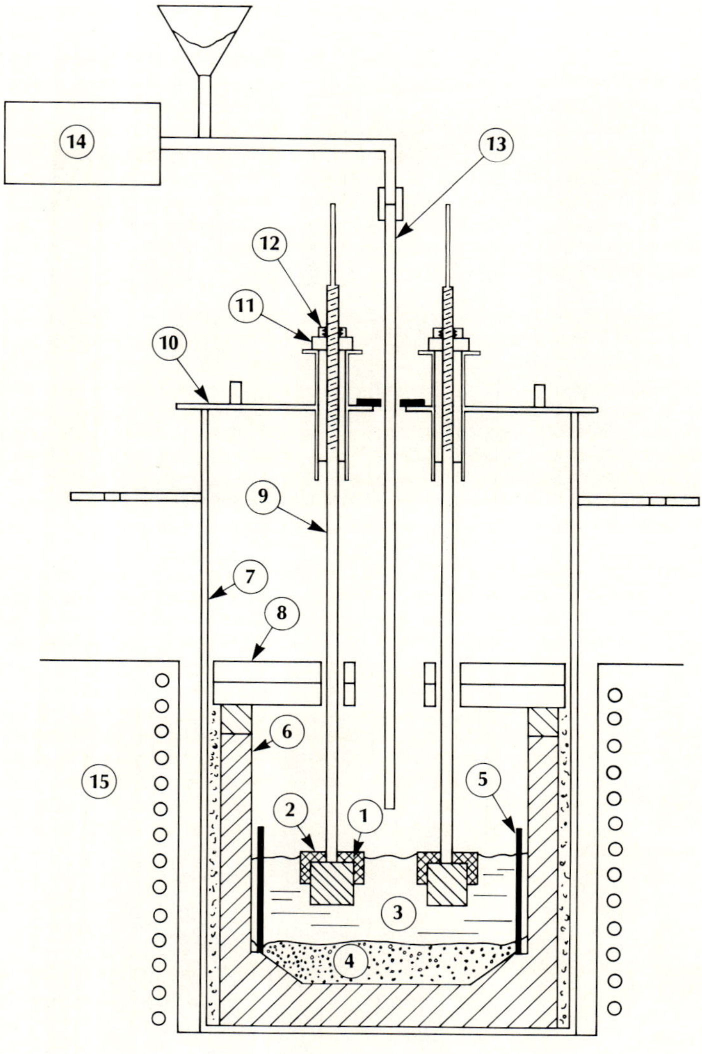

1. One of four carbon anodes; 2. Boron nitride cap; 3. Electrolyte; 4. Aluminum pad; 5. Alumina liner; 6. Graphie crucible; 7. Inconel 601 shell; 8. Insulating boards; 9. 316 stainless steel support rods; 10. Lid; 11. Ceramic electrical insulators; 12. Anode height adjustment nut; 13. Alumina feeder tube; 14. Alumina screw feeder; 15. Resistive furnace.

Figure 3 - Arrangement of the Electrolytic Cell

Anode Configuration. Four anodes 50 mm in diameter by 50 mm high were placed in a 140 mm center to center square pattern and suspended in the bath. Electrical contact was made by threading the anode upon a 316 stainless steel rod with a protective Al_2O_3 sleeve to reduce corrosion. Boron nitride caps, which left the bottom and lower 25 mm of the sides of the anodes exposed, were used to prevent air burn of the anodes. Proper anode alignment was provided by stainless tubes welded perpendicularly to the lid which contained ceramic inserts between the tube and the stainless support rods. Anode height was adjusted with threads on the stainless rod and a nut which rested upon the ceramic insert. Figure 2 illustrates the relative orientation of the four anodes as they were withdrawn from the cell.

The Cathode and Cell. Electrolyte and aluminum bath depths of 100 mm and 50 mm, respectively, were contained by a graphite crucible with a 330 mm ID. A 305 mm ID Al_2O_3 liner insulated the graphite crucible walls. Electrical contact for the cathodic graphite crucible was made through the Inconel 601 containment shell. Fluid coke packed between the sides of the crucible and shell provided additional electrical contact and room for expansion. Heat losses were reduced by two layers of 25 mm thick alumina fiberboard insulation which rested upon the top of the graphite crucible. A timed screw feeder operated intermittently to maintain the alumina concentration within the desired range.

The Electrical Arrangement. Cell temperature was controlled by varying the power to the resistive furnace with temperature sensing of the Inconel shell. Typically, the bath temperature was 15°-20°C less than the shell temperature while the cell was operated at 50 amperes per anode.

Power to the four anodes was supplied by four independent DC power supplies with 10V-80 ampere capability. Figure 4 outlines the general layout. Either constant voltage or constant current capabilities were supplied. The current and voltage of one anode at a time were monitored on two digital indicators--the voltage was recorded continuously on a multipoint recorder. An amp-hour meter indicated total coulombs passed for each anode.

Procedure

The anode consumption cell has been operated near plant conditions. Typical control values included:

(a)	Bath Ratio	1.1-1.3
(b)	CaF_2 Concentration	3-7%
(c)	Temperature	935°-980°C
(d)	ACD	4.4 mm

Typically, one standard anode was included for direct comparison with three test anodes.

Cell operation was begun by adding sufficient aluminum and bath to obtain the required levels. Anodes were brought to operating temperature by preheating the cell 10° to 15°C higher than the desired temperature.

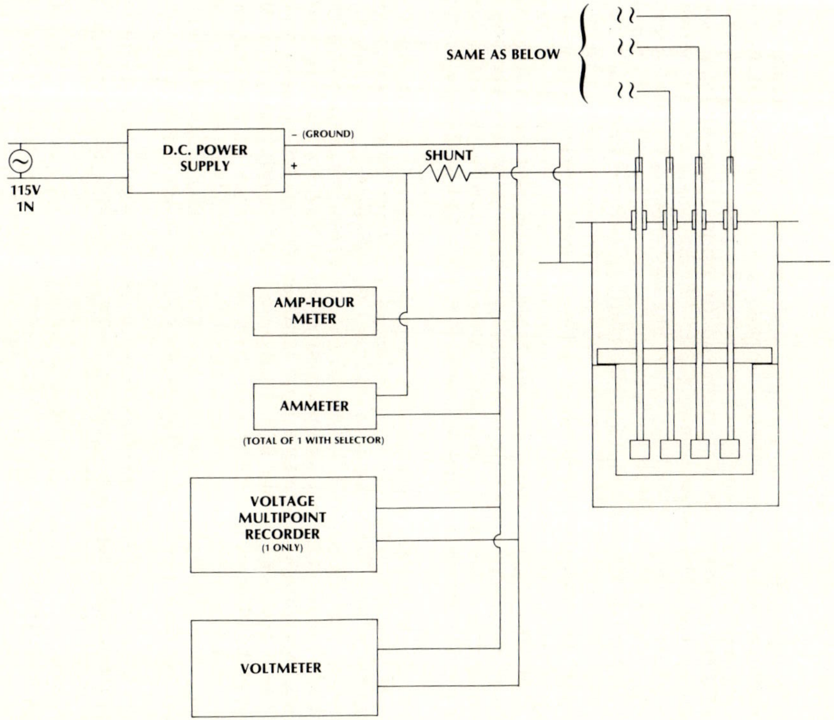

Figure 4 - Electrical Schematic

The applied current densities were typical of industrial practice. Initial runs indicated that the bottom and exposed sides of the anodes were consumed at essentially the same rate. Hollingshead and Braunwarth observed similar results. The total exposed area was, therefore, used to calculate current densities. Initial densities at 50 amperes per anode were 0.82 A/cm². After 200 ampere-hours, about 20 wt % of the anode was consumed to give a final current density of 1.05-1.24 A/cm².

After completion of an experiment, the anodes were removed and immediately placed in a nitrogen purged vessel to cool. Because of this procedure as well as the absence of carbon loss from under the caps during electrolysis, no correction for air burn was applied to the results.

Anode carbon consumption was determined after the anodes were removed from the rods and weighed, crushed, and ashed for 24 hours at 650°C. Consumption was expressed as a percentage corresponding to formation of CO_2 at 100% current efficiency (0.1118 gm/amp-hr), corrected for the ash content of the original carbon.

Typical Results and Discussion

In order to reduce interactions among anodes from potential distribution effects, the anodes were separated by a minimum of two diameters. If the anodes were not uniformly consumed around their circumference, interactions among the anodes would be indicated. Initial measurements of the anodes, however, indicated uniform consumption. Therefore, anode interaction effects were expected to be minimal.

Cores from prebaked plant anodes were adopted as the standard test anode. From Table I, the typical performance of the prebaked anodes may be compared with laboratory-made Soderberg anodes. For some of the initial runs, air circulation within the cell was not sufficient to burn the coke dust floating on the electrolyte. As a result, the percent carbon consumption was unusually low, indicating current leakage to the cathode. For runs where this appeared to occur, normalization of the consumption data, based upon the expected performance of the standard, yielded surprisingly good agreement with other Soderberg data. From this example, the multiple anode reduction cell appears to effectively evaluate the relative performance of anodes, even under less than ideal conditions.

Table I. Typical Performance of Soderberg and Prebaked Anodes

Anode Type	Temperature Range (°C)	Anode Consumption (%)	Standard Deviation (%)
Prebaked Plant Anode Cores	940-950	111.1	1.5
Soderberg Laboratory Anodes (no apparent dusting)	972	125.6	3.0
Normalized Soderberg Laboratory Anodes (dusting present)	965-975	124.9	2.6

The increased carbon consumption of Soderberg anodes compared to prebaked anodes is also apparent from their greater roughness after consumption as shown in Figures 5 and 6. Because of the lower bake temperature of the binder pitch coke, it was preferentially attacked leaving the aggregate exposed. Thus, higher carbon consumptions were observed.

Figure 5 - A Prebaked Anode after Electrolysis

Figure 6 - A Soderberg Anode after Electrolysis

Conclusions

The degree of anode carbon loss from dusting during electrolysis in a commercial cell may be predicted from laboratory cell results. Corrections available from other tests for air burn may be applied to adequately predict overall carbon consumption rates.

This four anode test cell can effectively evaluate the relative performance of anode materials since standard and test anodes are exposed to nearly identical electrolysis conditions.

References

1. E. A. Hollinghead and V. A. Braunwarth, "Laboratory Investigation of Carbon Anode Consumption in the Electrolytic Production of Aluminium," pp. 31-50 in Extractive Metallurgy of Aluminium, Vol. 2, Interscience Publishers, New York, 1963.

2. V. A. Cammarota Jr. and D. Schlain "Anode Materials for Aluminium Reduction - Evolution of Lignite as a Carbon Source," U.S. Bureau of Mines Report 7370, April 1970.

3. N. F. Siviour, J. W. Graham, and G. V. Cullen "Performance of Electrodes containing Brown Coal Char in Alumina Reduction," Proceedings of Australasian Institute of Mining and Metallurgy, No. 237, March 1971, pp. 55-62.

INTERRELATION BETWEEN COMPOSITION AND PROPERTIES OF COAL TAR PITCHES AND PROPERTIES OF ANODE

by Michael L.Itskov, Rem V.Svoboda, Vera I.Denisenko, Edward A. Kolodin, German F.Vedernokov

VAMI, Leningrad

ABSTRACT

The laboratory studies of coal tar pitches which serve as a binder in production of an anode paste with a different softening temperature have shown that from a point of view of characteristics of an aluminium cell anode the pitch with a softening temperature of 110±3°C has the optimum properties.

The semi-commercial tests of an anode paste manufactured with the use of the above pitch have confirmed its advantages in the major operating characteristics compared with a paste produced with the use of a medium-temperature pitch.

INTRODUCTION

A major amount of an electrolitic aluminium in the USSR is produced in the Soderberg cells. This fact, along with an increase of amperage of an individual cell place stringent requirements upon quality of the Soderberg anodes. The anode's properties influence not only a carbon consumption in the electrolitic aluminium production, but also the performance figures and a level of the electrolysis technology. The anode operating characteristics depend on a design of the anode assembly and a procedure used, and on a quality of the anode paste which to a qreat extent is determined by properties of a binding pitch used. Its complex individual composition, phase and thermal transformations upon heating determine the physical-chemical interaction of components of the anode paste, the conditions of formation of a carbon structure, the mechanical and electric properties of the anodes and their consumption rates.

At present a medium-temperature pitch is mainly used as a binder in production of an anode paste for the Sodeberg cells. It has a softening temperature 67-73°C determined by a "ring-and-rod" method adopted in the USSR.

In the last few years some research was done on studying a pitch with a higher decomposition temperature, 85-90°C. It has been shown (Lasukova et al., 1975, Sverdlin et al., 1976) that this type of binder is characterised by a higher density, an increased content of the high-condensed organic compounds accompanied by an increase in the coking yield, by a lower liberation of volatiles during the anode paste baking, a reduction of its transportation porosity, an electric resistance and a reactivity.

A revealed tendency of improvement of operating characteristics of the anode paste as a softening temperature of the binder increases makes it practical to enlarge a range of softening temperatures of the pitches being stadied and to select a pitch ensuring the optimum properties of a baked material.

LABORATORY STUDIES

During the laboratory investigations the following was studied: a group composition and aromaticity indices of the pitches, their pycnometric density, a C/H ratio, a volatiles' content and a coking yield. The aromaticity indices were determined by an infrared spectroscopy as a ratio between an optical density of the out-of-plane deformation vibrations of CH-aromatic groups (D-750) and an optical density of the asymmetric stretching vibrations of CH_2-aliphatic groups (D-2930), and as a ratio between an optical density of the stretching vibrations of CH-aromatic (D-3050) and CH_2-aliphatic (D-2930) groups. The specimens of anode pastes which were thermally treated up to 1,000°C, were tested for a mechanical strength, an electric resistance, a chemical and electrochemical destructability in a stream of CO_2 and in a stream of NaOH dissolved in water, a back e.m.f. value and an amount of C separated as a result of a pyrolitic decomposition of the hydrocarbon gases during the binder's coking.

The methods used for determining a softening temperature ("ring-and-rod"), the aromaticity indices and a C/H ratio, a group composition and an ash content of the binding materials, a compressive strength, a resistivity, a reactivity in CO_2 of the baked anode paste specimens, during the laboratory tests are universally adopted in studying the properties of carbon materials.

For determining a coking yield a unit was developed schematically shown in Fig. 1.

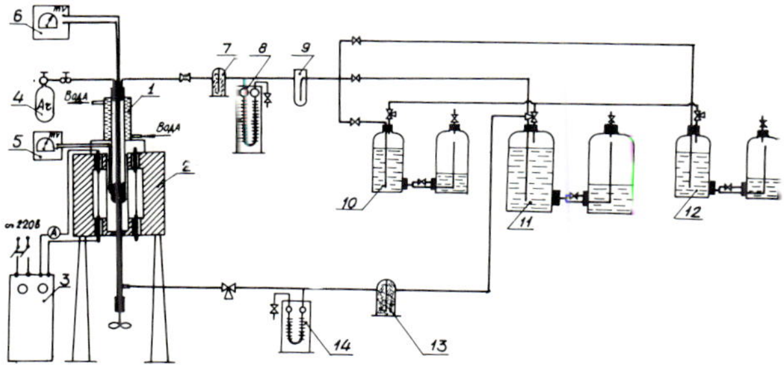

Fig. 1. Unit for Determining Total Coking Yield (Shown Schematically)

1. Retort 2. Silit Furnace 3. Variator 4. Cylinder with Inert Gas 5. Millivoltmeter 6. Millivoltmeter 7. Tischenko Bottle with Coke Packing 8. Manometer 9. Back Container 10. Gas Collector 11. Gas Collector 12. Gas Collector 13. Tischenko Bottle with $CACl_2$ 14. Manometer.

The main difference of the method proposed from those used in a research practice earlier consists in condensing the tar compounds upon water cooling of a top section of the retort and their return into a coking zone and through a consequtive circulation of the coking gases through the spesimen at 600, 700, 800, 900 and 1,000°C. The use of this method gives a possibility of determining a total coking yield for the anode pastes taking into account the cracking and pyrolisis processes which take place during a formation of the commercial anode. For comparison the tests were carried out without circulating the coking gases. The same apparatus was used for determining a coking yield for the pitch.

An electrochemical oxidability and dusting of the baked samples of anode pastes was determined in an alkaline solution. A cell proposed by Watanabe forms a basis of the method (Watanabe, 1963). A difference of this method consists in the choice of dimensions of a baked specimen of the anode paste being tested, in a size of the cell, a quantity of the electrolyte used as well as in experimental conditions, specifically, a stabilization of an electrolysis temperature through the artificial cooling.

Fig. 2 gives schematically a cell used for determining an electrochemical reactivity of the baked anodes.

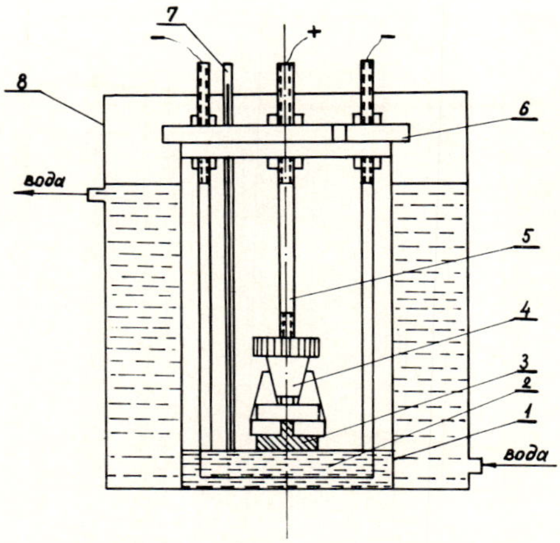

Fig. 2. Cell for Determining in Alkaline Water Solution of an Electrochemical Reactivity of Baked Anode Pastes
1, Sleeve; 2. Cathode; 3. Specimen; 4. Screw Clamp; 5. Current-Carrying Rod; 6. Cell's Cover; 7. Thermometer; 8. Cooling System.

The cell consists of a glass sleeve (1) and the electodes inserted into it. A specimen (2) serves as an anode and a 20 mm high steel ring serves as a cathode (Watanabe, 1963). A 4N solution of NaOH serves as an electrolyte. An electrolysis is carried out for 5 hours at a current density 0.7 A/cm^2. From the data obtained an electrochemical oxidability, , a dustability and a total destructability are calculated. A relative error of the method is 2.5%.

A back e.m.f. was measured by an impulse technique given in detail in (Vetyukov, Starodubrovsky, Pryakhin, 1977).

An electrolysis was carried out at a current density 0.7 A/cm^2. For measuring a back e.m.f. the specimens were made of baked anode paste, having a working section dia. 25 mm. The specimens were used as anodes in a laboratory electrolysis of the cryolite-alumina melts.

Under laboratory conditions the physico-chemical properties were studied of 11 specimens of pitch with a softening temperature from 70 to 147°C and of the anode pastes obtained with their use.

As it follows from the experimental data (Fig. 3), an increase of the pitch softening temperature is accompanied by an increase in content of the compounds insoluable in toluene and quinoline (α - and α_1 - fractions) at the expense of a reduction in a quantity of γ -fraction soluable in isooctane. Here attention is drawn to the fact of an insignificant change in amount of β -and γ -fractions for the softening

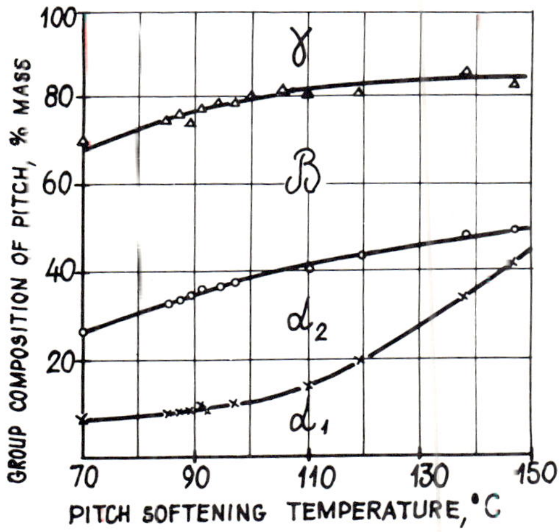

Fig. 3. Relationship Between a Group Composition of the Pitches and a Softening Temperature.

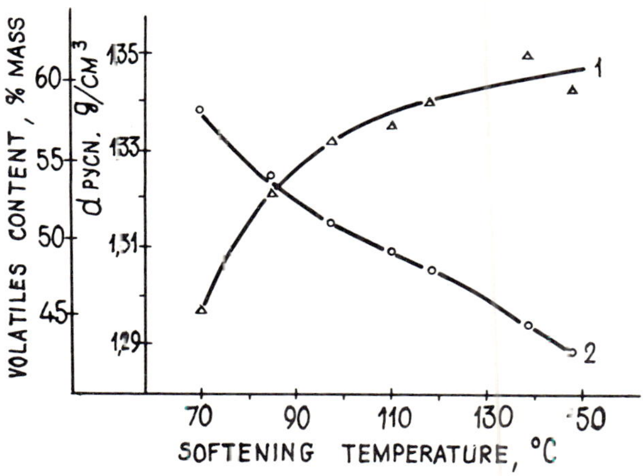

Fig. 4. Variation of a Pycnometric Density (1) and of a Volatiles' Content (2) in the Pitch with a Softening Temperature.

temperatures over 110°C. This permits to assume not only a consequtive transformation of the pitch components according to a widely known scheme $\gamma \rightarrow \beta \rightarrow \alpha_2 \rightarrow \alpha_1$ but it confirms a possibility of the concurrent reactions of destruction and synthesis of the pitch components with decomposition of individual micelles of α_2-and β -fractions and formation of α_1 -, β - and γ -groups. An amount of α_2 - and β -fractions ensuring a high coking yield and preserving the satisfactory rheological properties, and which are considered, in the light of the present-day ideas, the most valuable components of the pitch (Hüttinger, 1970; Gasienko et al., 1977) increases considerably with an increase of the softening temperature from 70 to 110°C. At the same time with an increase of the pitch softening temperature a pycnometric density of the binder increases and a volatiles' content of the binder decreases (Fig. 4).

When the coking of an anode paste takes place in an anode of the aluminium cell, the processes of thermal transformations pratically go to completion and the maximum yield of carbon from the binder is achieved. The tar products resulting from a thermal destruction of the pitch undergo a deep cracking which permits to considerably increase a coking yield.

Besides, an increase in the share of coke from the tar compounds increases a quality of the anode because the coking of these products takes place on a porous surface thus decreassing a volume of the voids, eliminating microcracks and microdefects. In this connection it is evident that an advantage will be with the pitch having the largest coking yield. A significant increase of the coking yield from the pitch is observed in the softening temperature range 85-110°C, reaching practically maximum values at 110°C (Fig. 5). It is explained by a sharp increase of the pitch aromaticity indices which is connected with a tighter packing of the structure which results from a buildup of the high-condensed aromatic compounds without their noticeable decomposition.

At the same time an increase of the aromaticity indices ensuring an increase of the coking yield results in a rise of viscosity and of a surface tension (Fig. 6).

The baked specimens of an anode paste for the laboratory studies were prepared from calcined coke whose properties are given in Table I, and a size distribution of the coke mix is given in Table 2.

TABLE I. Properties of the Cokes Used in Preparation of Anode Pastes for Laboratory Studies

Description	Unit of Measure	Value
Pycnometric density	g/cm^3	2.02
Resistivity of powder	$\Omega \cdot mm^2/m$	616.0
Sulphur content	%, mass	0.35
Ash content	%, mass	0.36

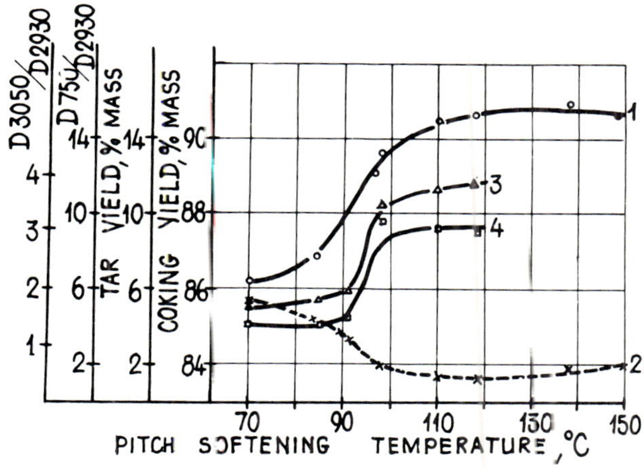

Fig. 5.Relationship Between the Pitch Properties and a Softening Temperature

1 - coking yield; 2 - tar yield; 3 - a ratio of an optical of the out-of-plane deformation vibrations of CH_{ar} - groups (D 750) to an optical density of the asymmetric stretching vibrations of CH_2 - aliphatic groups (D 2930); 4 - a ratio of optical densities of the stretching vibrations of CH - (D 3050) and CH_2 - (D 2930) groups.

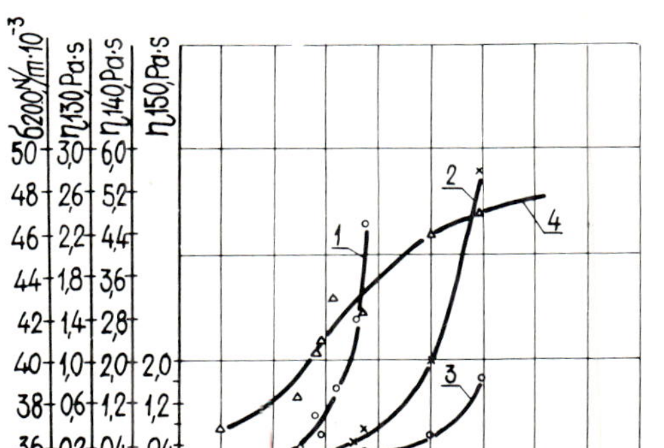

Fig. 6. Relationship Between the Rheological Properties of Pitches and a Softening Temperature

1. Viscosity at 130°C; 2. Viscosity at 140°C
3. Viscosity at 150°C; 4. Surface tension at 200°C.

TABLE 2. A Size Distribution of the Pitch Mix Used for Preparation of the Anode Pastes from Pitches Having a Different Softening Temperature

Description	Grain size						
	+6	+4 -6	+2 -4	+1 -2	+0.16 -1.0	+0.08 -0.16	-0.08
Fraction content in a composition of the coke mix % mass	1	12	13	13	21	10	30

When studying the processes of a thermal transformation of the anode paste according to the technique given above under the conditions which exist during a formation of the Soderberg anode the results were obtained showing an increase of the coking yield from 93.9 to 96.6% (mass) when a softening temperature of the binder used increases from 70 to 147°C (Fig. 7). Most evidently this increase shows up in the range from 70 to 110°C which is connected with a noticeable rise of the aromaticity indices and of the coking yield.

A pronounced effect on properties of the baked anode paste specimens has a quantity of the pyrolitic carbon liberated from the coking

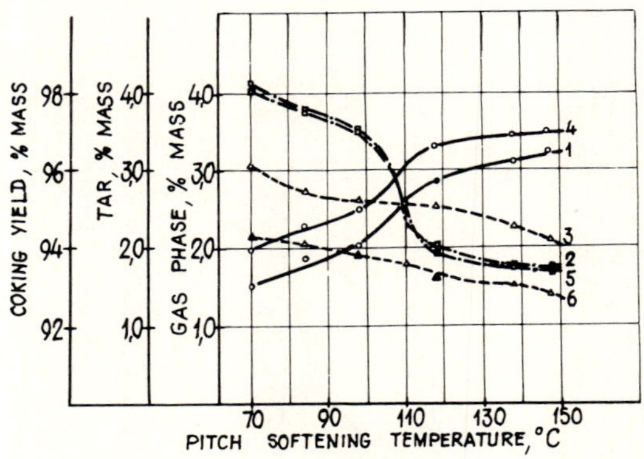

Fig. 7. Variation of a Mass Change of Coking Components of the Anode Pastes with a Pitch Softening Temperature.

- without circulation 1. Coking carbon residue; 2. Tar compounds; 3. Gas phase

- after circulation at 1,000°C 1. Coking carbon residue; 2. Tar compounds; 3. Gas phase

gases on a porous carbon backing in a process of thermal transformations of the binder during a formation of the Soderberg anode, thus determining to a considerable degree a value of the chemical and electrochemical oxidability, dusting ability and of the total destructability of the anode.

A pyrolitic carbon liberating in voids of the laboratory anode at 0.2% of the initial specimen mass decreases its porosity by 2% and reactivity (in the electrolytical production of aluminium) by 12%, and an increase of the pyrocarbon quantity up to 1% decreases a porosity by 9% and a reactivity by 21%.

An anode paste coking process is accompanied by a liberation of a considerable amount of the gaseous products where the major components are methane and hydrogen. Besides, the coking gases also contain some amounts of CO_2, CO, ethane, propane, as well as nitrogen of the air adsorbed by a surface of the anode paste.

With an increase of a pitch softening temperature a share of the hydrocarbon gases in a total volume of the coking gases decreases, and that of hydrogen - increases (Fig. 8). A study of the processes of a thermal decomposition of the coking gases has shown that the maximum quantity of pyrocarbon from the hydrocarbon gases is formed upon coking the anode pastes prepared from pitches with a softening temperature 70-110°C and is 0.9 to 1.0% of a mass of the initial specimen (Fig. 9). A decrease in the amount of pyrocarbon with a further rise of the softening temperature is connected with a consolidation of the binder and an increase in content of the high-condensed aromatic compounds. During their thermal decomposition both a total volume of the coking gases and a content of methane in them decrease.

An improvement in the binder's quantity with an increase of the softening temperature from 70 to 110°C is accompanied by a minimum of porosity, electric resistance and by a maximum of strength of the anode pastes' baked specimens (Fig. 10 and 11).

An increase of porosity, electric resistance, reactivity and a reduction of the mechanical strength at the softening temperatures over 110°C is caused by worsening of the pitches' wetting properties and, in particular, by an increase of the surface tension with a noticeable increase of the α_1-fraction content.

An important parameter which in many respects determines the power parameters for electrolysis of the cryolite-alumina melts is a back e.m.f. value which at the same time to a considerable degree characterizes a reactivity of the anode.

It is seen from Fig. 12 that an increase in the softening temperature of the pitch used from 70 to 110°C results in a rise of the back e.m.f. which is caused by a decrease of reactivity of the anode paste with an increase in the pitch's consolidation degree and of the coking

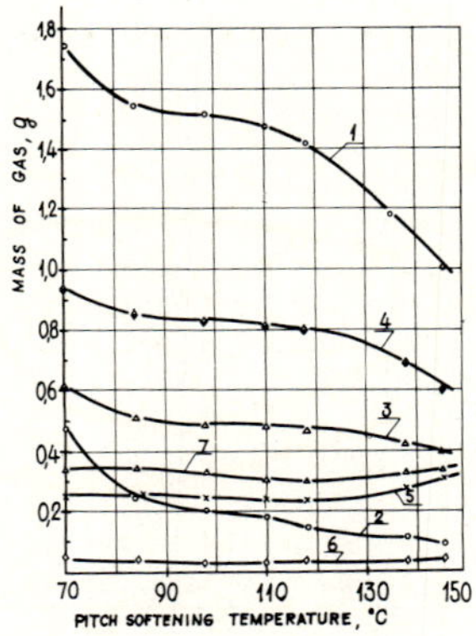

Fig. 8. Variation of a Mass Change of the Gas Phase Components in Coking the Anode Pastes with a Pitch Softening Temperature

- without circulation
1 - sum of hydrocarbon gases ($CH_4+C_mH_n$); 3 - hydrogen; 5- CO_2

- after circulation at 1.000°C
2 - sum of hydrocarbon gases ($CH_4+C_mH_n$); 4 - hydrogen; 6 - CO_2
7 - CO.

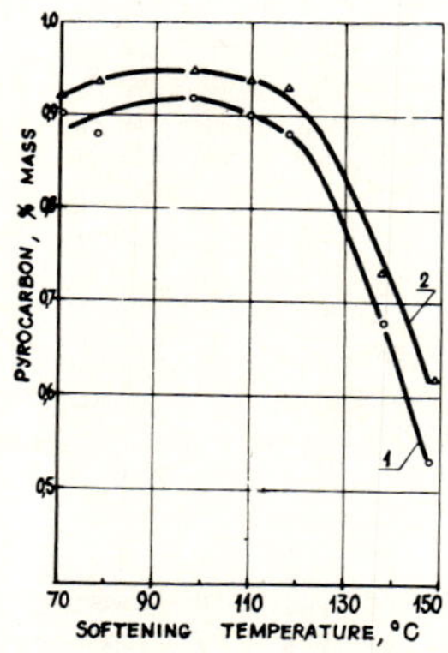

Fig. 9. Variation of a Weight Gain of the Coking Carbon Yield of the Anode Pastes as a Result of Circulation of the Coking Gases with a Pitch Softening Temperature

1. Experimental data
2. Calculated data

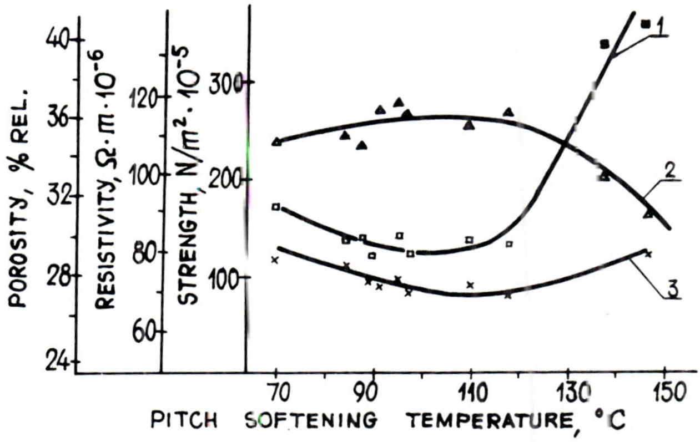

Fig. 10. Variation of a Change of Physical Properties of the Baked Anode Pastes with a Pitch Softening Temperature

1. Porosity
2. Mechanical strength
3. Electric resistivity

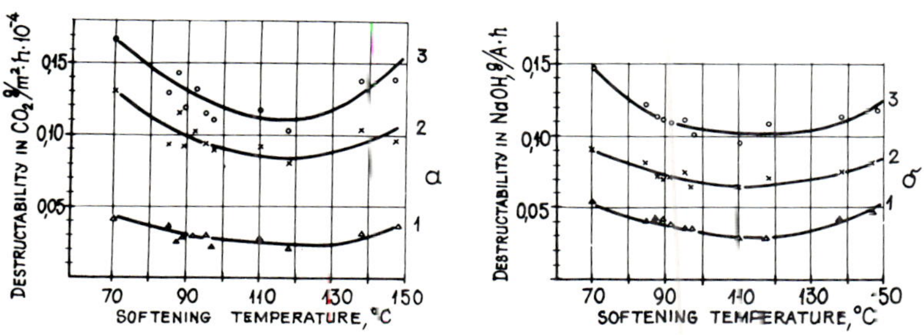

Fig. 11. Variation of a Change of Reactivity Parameters of the Baked Anode Pastes with a Softening Temperature

"a" - in CO_2 flow, 1 - oxidability, 2 - dusting, 3 - total destructability

"b" - an electrochemical reactivity in an alkaline water solution: 1 - oxidability, 2 - dusting, 3 - total destructability.

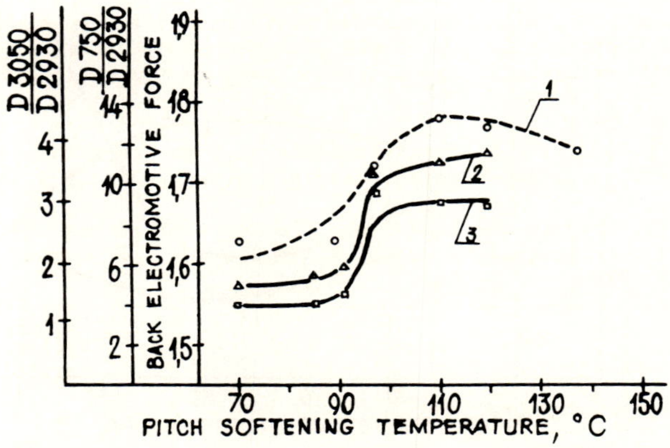

Fig. 12. Variation of a Change of the Back E.M.F. and of the Pitches' Aromaticity with a Softening Temperature
1 . Back E.M.F.
2. Ratio of an optical density of the out-of-plane deformation vibrations of CH - aromatic groups (D 750) to an optical density of the asymmetric stretching vibrations of CH_2 - aliphatic groups (D 2930)
3. Ratio of optical densities of the stretching vibrations of CH (D 3050) and of CH_2 (D 2930) groups.

yield. A reduction of the back e.m.f. value with a further rise of the binder's softening temperature is caused by a considerable increase of reactivity of the baked anode pastes due to an increase of their porosity.

As a whole, the data obtained under the laboratory conditions demonstrate an improvement in properties of the binding materials with an increase of their softening temperature from 70 to 110°C. An anode paste prepared from a pitch with the softening temperature 110°C has the optimum operating parameters.

SEMI-COMMERCIAL TESTS

The laboratory testing results were checked during the semi-commercial tests of an anode paste.

When preparing an anode paste for the semi-commercial tests with the use of a high-temperature pitch the mixing of components of the anode paste (coke and pitch) in the solid state was done according to a technique developed by Yanko E.A. et al. An anode paste with the use of a medium-temperature pitch was done according to a generally used technology when the pitch was fed into a mixing machine in a melted state. A coke mix with the maximum grain size 6 mm, consisting of 80-85% of a pitch coke and 15-20% of a petroleum coke, was used as a filler for preparing the anode pastes used in the semi-commercial tests.

Properties of the binding materials used in preparing an anode paste for the semi-commercial tests are given in Table 3.

TABLE 3. Physico-Chemical Properties of the Pitches Used for Preparing the Anode Pastes for the Semi-Commercial Tests

Description	Unit of measure	Pitch characteristics	
		High-temperature pitch	Medium-temperature pitch
Softening temperature	°C	110	65
α-fraction content	% mass	40.6	27.1
α_1-fraction content	% mass	16.7	12.8
Volatiles' yield	% mass	48.5	62.6
Pycnometric dendity	% mass	1.337	1.259
Ash content	% mass	0.30	0.17

The semi-commercial tests were carried out in a 6 kA cell. Its main performance figures when using the anode pastes from the medium- and high-temperature pitches are given in Table 4.

TABLE 4. Electrolysis Parameters when Using the Anode Pastes from Pitches with Varying Softening Temperature

Description	Unit of measure	Values	
		Anode paste from pitch with T_S = 110°C	Anode paste from pitch with T_S = 65°C
Current	A	6,130	6,000
Aluminium produced	kg	2,396	1,302
Current efficiency	% rel	87.2	88.0
Bath temperature	°C	960	950
Specific consumption of anode paste per 1 t of Al	kg	509	525
Specific consumption of anode paste	g/A.hr	0.149	0.154
Carbon content in bath skim, per 1 t of Al	kg	9.5	23.3 contd

TABLE 4

Description	Unit of measure	Values	
		Anode paste from pitch with T_s = 110°C	Anode paste from pitch with T_s = 65°C
Specific carbon content in bath skim	g/A.hr	0.0031	0.0077
Anode consumption rate	cm/day	2.2	2.3
Voltage drop in anode	V	0.38	0.40

CONCLUSION

An analysis of results of the laboratory studies and semi-commercial tests permits to draw a conclusion that the use of an anode paste from a high-temperature pitch results, as has been shown above, in a higher mechanical strength, a lower electric resistance and a lower reactivity. It ensured a decrease in the following: an anode consumption by 16 kg/t Al, an anode consumption rate by 4.5%, a bath skim output - by more than 50%, an anode voltage drop - by 20 mV. Besides, during the semi-commercial tests a reduction in the gas evolution from an anode surface is noted due to a decrease in the volatiles' content in a high-temperature pitch and an increase in viscosity of the anode's liquid phase. This permits to improve the working conditions and to reduce a cancer hazard in the electrolysis shops.

But the introduction of an anode paste from a high-temperature pitch, having the optimum properties (T_s = 110 ± 3°C) requries a development of the new equipment for producing the anode paste from a dry mix for charging into anodes in a pelletized form.

REFERENCES

Lasukova, L.P., et al.,1975, "Criteria characterizing a high-quality anode binder", Tsvetnie Metalli, No. 8, pp. 41-44.

Sverdlin, V.A., Diblin, B.S., Vetyukov, M.M., 1976, "On a porous structure of the carbon anode", Tsvetnie Metalli, No. 11, pp. 37-41.

Watanabe T., 1963, "Studies on pitch for Soderberg anodes", Extractive Metallurgy of Aluminium, vol. 2, Aluminium, pp. 351-373.

Vetyukov, M.M., Starodubrovsky, S.S., Pryakhin G.V., 1977, "A study of the non-stationary electrolysis conditions of the cryolite-alumina melts by an impulse technique", Tsvetnie Metalli, No. 6, pp. 28-29.

Hüttinger K.J., 1970, "Eine kinetische Analyse der Steinkohlenteerpechpyrolyse", Bitumen, Teere, Asphalte, Peche und verwandte Stoff, No. 12, s. 528-531.

Gasieńko, S., Swietlik, U., Gastrzebski, J., 1977, "Wlasnosci i struktura skladnikow grupowych paku weglowego", Koks-Smola-Gas, r.22, No. 11, s. 283-287.

Domrachev, B.P., and Korobov, M.A., 1972, "On consolidation of the baked anode paste with a pyrolitic carbon for reducing an anode consumption", Tsvetnie Metalli, No. 7, pp. 38-41.

CHAIRMEN

L.G. Boxall
Martin Marietta Laboratories
Baltimore, Maryland

A.F. Bopp
Martin Marietta Laboratories
Baltimore, Maryland

CARBON TECHNOLOGY III

FACTORS IN THE DESIGN OF REDUCTION CELL ANODES

D. G. Brooks and V. L. Bullough

Reduction Laboratory
Reynolds Metals Company
P. O. Box 1200
Sheffield, AL 35660 USA

Introduction

As the aluminum industry tends toward the large, multi-stubbed anodes for low current density operation, the design factors affecting the mechanical and thermal stresses acting on an in-service anode become increasingly important. The design should be such that the resulting stresses are well removed from the mechanical limits of the carbon while, simultaneously, providing for minimum stub-to-carbon contact drops.

Anode electrical connections are secured by placing the steel stub(s) in the anode stub hole(s) and pouring molten cast iron in the annular spacing. During cell operation, the anode carbon can be mechanically stressed by the strain imposed through the stub hole by the thermal expansion of the stub and cast iron. Thermal stresses occur when an anode block, at ambient temperature, is set in the reduction cell with bath temperatures on the order of 950°C. When these stresses exceed the mechanical limits of the carbon, anode failure or cracking can occur.

Experiments were performed to: (1) Verify the reliability of carbon strain calculations through a stub hole brought on by the expansion of the steel stub and cast iron; (2) Determine the strain or "tightness of fit" necessary to minimize the stub-to-carbon contact resistance; and (3) Determine, at least qualitatively, the degree of thermal stress exerted on an operating anode.

Calculation of Stub Hole Strain

A simple series of calculations was used to determine the degree of carbon strain exerted across a stub hole based upon the elementary principles of thermal expansion. The calculations assume that strain can be estimated by the following relationship

$$\varepsilon = (E_n - d)/\ell_o \ , \ E_n \geq d \tag{1}$$

where

ε = Strain across stub hole; inches/inch

E_n = Net expansion within stub hole including effects of stub, cast iron and carbon; inches

d = Air gap between cast iron and carbon surfaces at ambient temperature; inches

ℓ_o = Stub hole diameter with free expansion, inches

The air gap arises because of the cast iron and stub shrinkage after casting.

The expansions and air gap can be calculated from the coefficients of thermal expansion (CTE), cast iron shrinkage data and casting and operation temperatures.

Procedure

Deflections in and around a stub hole were determined experimentally and analyzed in much the same fashion as that reported by Peterson(1). A double-stubbed prebake anode measuring 31" x 52" and a steel yoke were prepared for testing. Provision was made for measuring deflections by threading 3/4" carbon steel rods at selected positions in the top of the block. Slat steel was welded to the rods to provide convenient measuring points. Deflections were measured with dial calipers (0.001"). Thermocouples were installed at strategic locations within the block and yoke for obtaining temperature gradients and distribution. Inconel tubing was placed within the stub holes prior to casting to provide thermocouple wells for measuring stub hole temperatures. See Figure 1 for a schematic representation.

Measurements of deflection were performed while using a baking pit of an open top, ring carbon baking furnace as a heating source. Here, the vertical temperature gradients could simulate, in some way, those gradients that would be expected in a reduction cell. Temperatures were recorded with a data logger.

Deflections were measured and recorded manually at regular time intervals. The temperature distribution across the slat steel was also recorded to provide for expansion corrections, which was negligible in most cases. Deflections were measured at three vertical points and extrapolated to the carbon surface. The experiment continued for approximately 72 hours.

Upon completion of this experiment, the assembly was pulled from the baking pit, cooled and prepared for additional testing while in service in a reduction cell. The thermocouples were left intact and the steel works

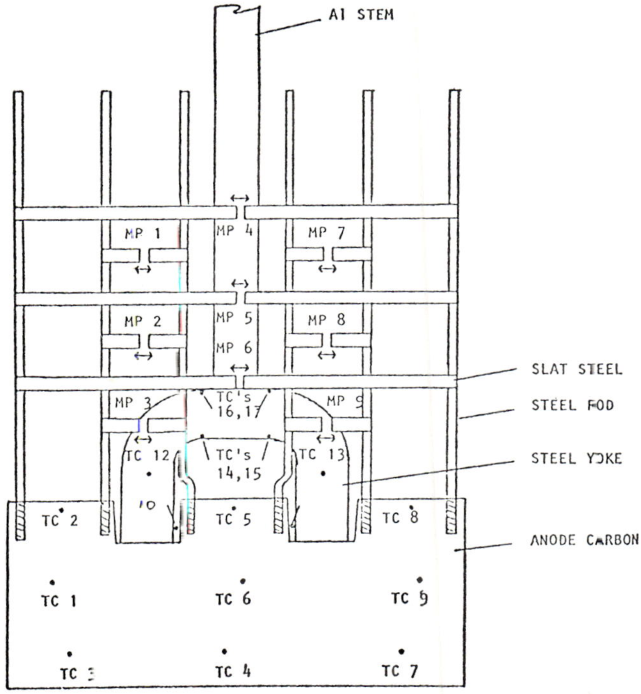

Figure 1 - Experimental Set-Up for Measuring Temperatures And Expansions of Anode/Yoke Assembly.

used in measuring deflections were removed, with the exception of about 8" of steel rodding on each side of each stub hole. These points were used as voltage probes for measuring stem-to-carbon voltage drops.

The prepared assembly was then set in a reduction cell operating at 152 kA as a corner anode according to the normal setting pattern and procedures. Both temperature and voltage data were recorded with a data logger. Refer to Figure 2. Data were obtained for the first 45 hours of service life.

Results

Deflection Experiment Results

The deflection experiments were carried out to determine the amount of strain being imposed on the carbon from the thermal expansion of the yoke steel, and to provide an experimental point to verify the reliability of calculations. An attempt was also made to measure experimentally the bowing of an anode as reported by other investigators(_). The bowing action is believed to result in an increased interference fit (strain) through the stub holes.

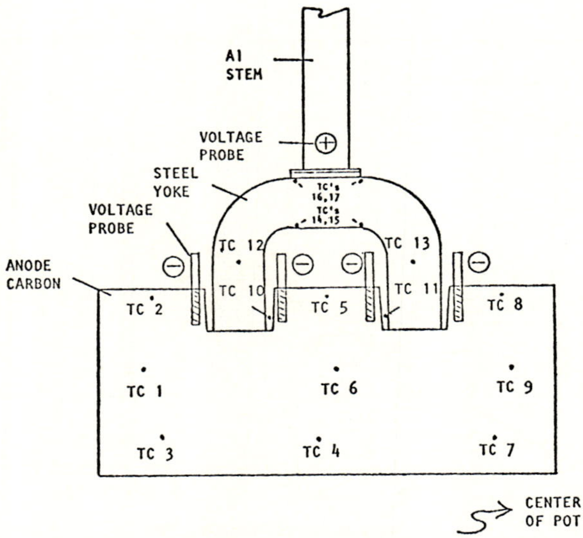

Figure 2 - Experimental Set-Up for Determining Thermal Gradients and Contact Voltge Drops.

Substantial scatter was found in the deflection measurements, as expected. A large number of readings were obtained to partly compensate for this scatter.

Block expansion versus mean block temperature is demonstrated in Figure 3. The scatter in the experimental points is apparent; however, the experimental data are in good agreement with the predicted expansion as determined from the coefficient of thermal expansion (CTE). The mean block temperature was obtained by two-dimensional integration over the height and length of the block. CTE values were determined experimentally with a dilatometer furnace.

If significant bowing of the anode occurs, it would be most apparent in the block deflection measurement of Figure 3, resulting in the experimental deflection being less than the predicted expansion. The bowing effect was not observed in this case, although a slight effect may not have been detectable because of the data scatter.

Expansion effects across a stub hole are shown in Figures 4 and 5. When the temperature first begins to rise, the expansion is expected to follow that of carbon alone. This is because of the air gap that is initially present in an anode/yoke assembly that is adjoined with cast iron. The air gap provides a buffer for free expansion of the steel stub and cast iron, and throughout this expansion the carbon is not stressed.

The thermal expansion coefficient of steel is greater than that of carbon, causing the connection to eventually tighten. Upon tightening, the stub hole would be expected to expand at the rate of the yoke steel, cast iron and residual carbon between reference points, all combined.

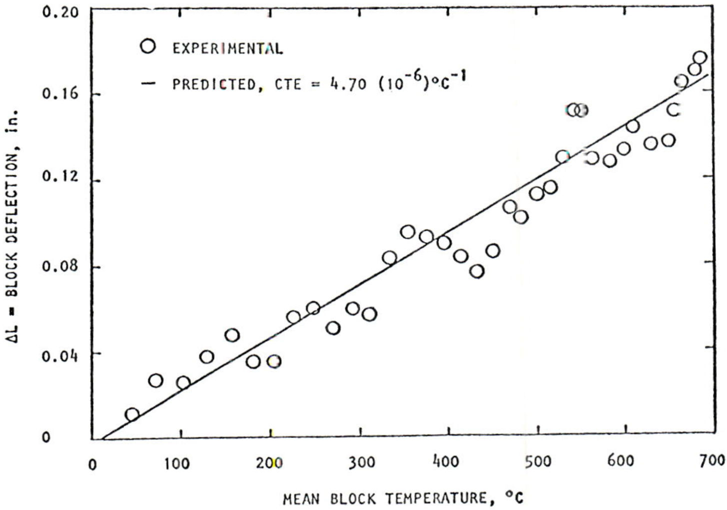

Figure 3 - Linear Block Deflection as Function of Mean Block Temperature.

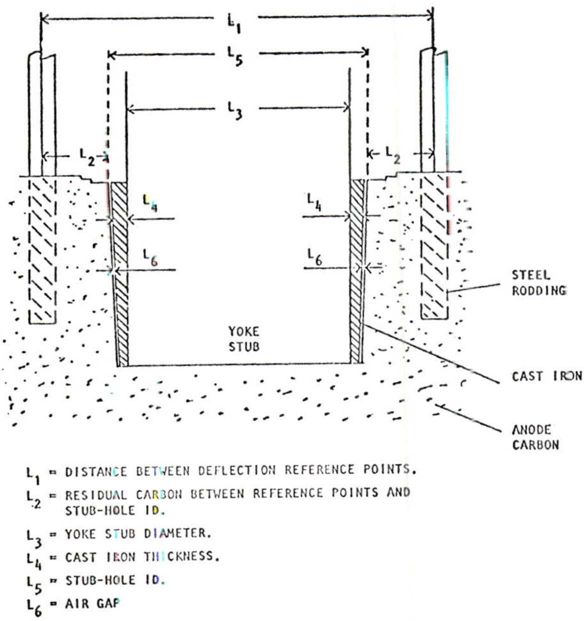

Figure 4 - Schematic Representation of Anode/Stub Electrical Connection.

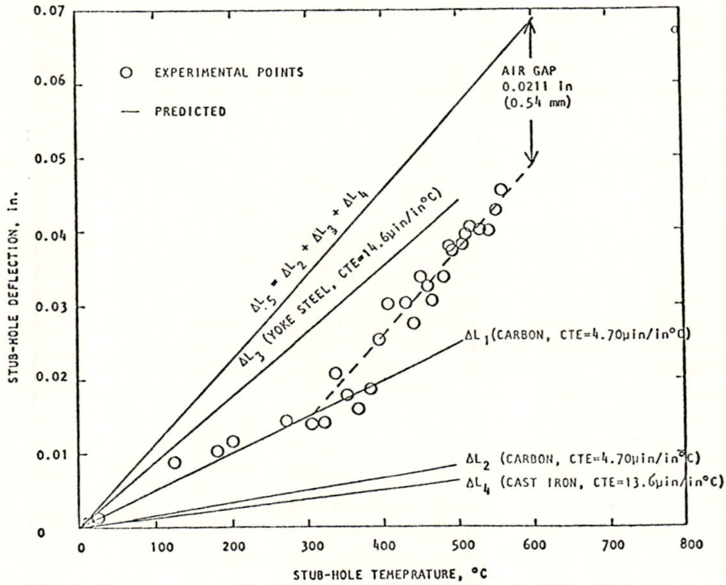

Figure 5 - Experimental and Predicted Expansions Within Stub Hole. Refer to Figure 4 for Schematic Representation.

The predicted effects, along with experimental measurements, are shown in Figure 5. Initially, before the stub/carbon connection tightens, the stub hole is expected to expand at a rate designated by ΔL_1, which is the expansion of anode carbon between reference points. The experimental points follow this expansion until temperatures reach about 300-350°C.

Predicted expansions of the residual carbon, steel stub and cast iron are given by ΔL_2, ΔL_3 and ΔL_4, respectively. The additive effect is also given, ΔL_5. Above 300-350°C, the experimental data show expansion at very nearly the same rate as ΔL_5 which is indicative that the stub/carbon connection has tightened and the carbon is being stressed.

Had there been no air gap initially present, the measured expansions would have followed the line given by ΔL_5 from the beginning. Therefore, a measure of the air gap can be obtained from the difference of the intercepts between the line given by ΔL_5 and the parallel experimental expansion. Forcing the rate of expansion (slope) of the experimental deflection to be parallel with ΔL_5, and performing a least-squares fit on the intercept alone, yields an air gap of 0.0211" (0.54 mm), which is in excellent agreement to that predicted from casting temperatures, anode design and cast iron shrinkage data.

These results demonstrate that strain loads through a stub hole can be estimated with a high degree of reliability from simple calculations based on the elementary principles of thermal expansion together with casting and operational temperature data.

Cell Operation Results

The information obtained on an in-service anode is useful in determining the temperatures at which the stub/carbon connection resistance is minimized. This temperature is representative of some strain load on the carbon and can be used for sizing an optimum, or near optimum, stub hole diameter to achieve minimum voltage drops at an acceptable carbon stress.

The operating thermal gradients can also be used to analyze, at least qualitatively, the thermal stresses imposed on an anode.

Figure 6 represents the decrease in contact voltage drop with increasing stub hole temperature as the anode service life progresses. Initially, the voltage drop increases rapidly as the anode begins to draw current. This is followed by a moderate rate of decrease in resistance as the yoke steel and connections are heated. After about 24 hours, the voltage drop becomes relatively constant, while stub temperatures continue to rise slowly. This is indicative that the decrease in contact resistance brought about by carbon strain has been fully realized; further stress through the stub hole is of little or no benefit and increases the risk of breakage.

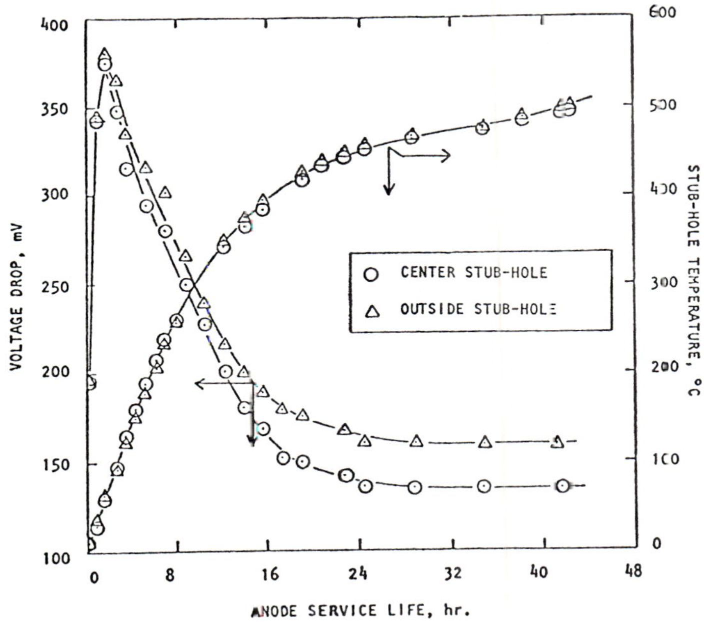

Figure 6 - Contact Voltage Drop and Temperture Response Versus Anode Service Life.

The service life of a low density anode is on the order of 20 days, during which time the carbon is consumed and the stub/carbon connections continue to be lowered closer to the bath. Thus, stub temperatures will continue to increase, possibly to as much as 700°C resulting in considerably more stress on the carbon. The stub holes should be appropriately sized so that the mechanical stresses exerted on the carbon result in minimum contact voltage drops, yet are sufficiently removed from the stress limit of the carbon.

The analysis of thermal stress requires a knowledge of the temperature gradients developing through the stressed body. These gradients have been measured experimentally and are depicted in Figures 7 and 8. An important point to note is that the overall gradient is not important, but rather the temperature profile through the block.

A rigorous analysis of thermal stresses in a reduction cell anode is very complicated. To simplify the analysis, the theory of thermal stresses in beams is used. While this is an oversimplification of the actual system, it can be used to analyze, at least qualitatively, for the relative magnitudes of thermal stress. In such a case the distribution of thermal stress is given by(2)

$$\sigma_T = \frac{-\alpha TE}{1-\nu} + \frac{E\alpha}{2b(1-\nu)}\int_{-b}^{b} T\,dy + \frac{3E\alpha y}{2b^3(1-\nu)}\int_{-b}^{b} Ty\,dy \qquad (2)$$

where

σ_T = Thermal stress
α = Coefficient of thermal expansion
E = Modulus of elasticity
ν = Poisson's ratio
2b = Anode thickness (height)
T = Temperature
y = Depth coordinate through anode (center @ y = o)

Equation 2 can be expressed in dimensionless form as

$$\sigma^* = \frac{\sigma(1-\nu)}{E\alpha To} = -T^* + 1/2\int_{-1}^{1} T^*dy^* + \frac{3y^*}{2}\int_{-1}^{1} T^*y^*dy^* \qquad (3)$$

where

To = Initial temperature
T* = T/To
y* = y/b
σ^* = Dimensionless thermal stress

Equation 3 is a measure of the relative thermal stress as a function of position (depth) through the anode block. This equation has been used in conjunction with the thermal gradients to obtain relative values of thermal stress, and the results are given in Figure 9, where maximum dimensionless thermal stress (tensile) is plotted as a function of service

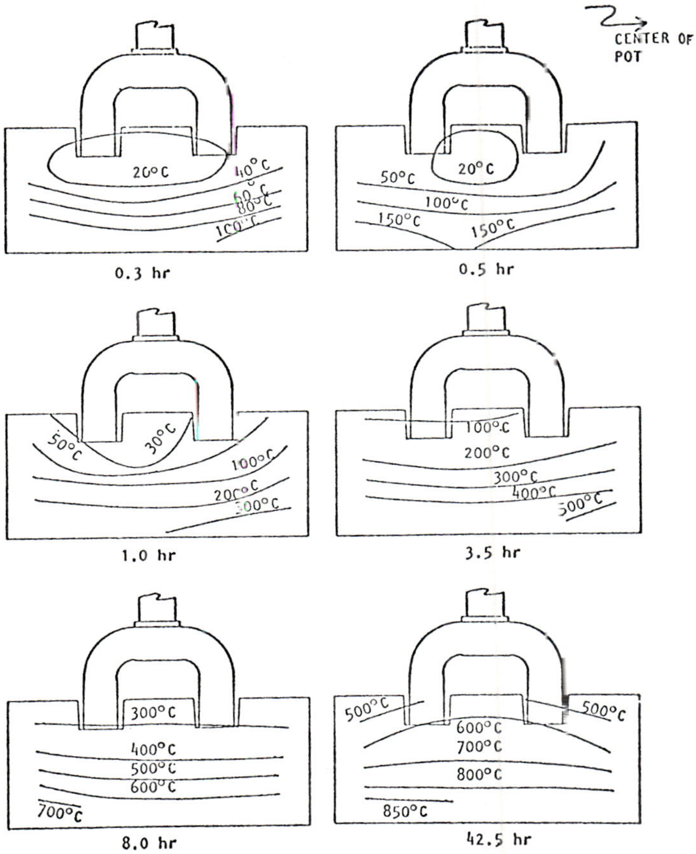

Figure 7 - Temperature Isotherms in an Operating Anode As a Function of Service Life.

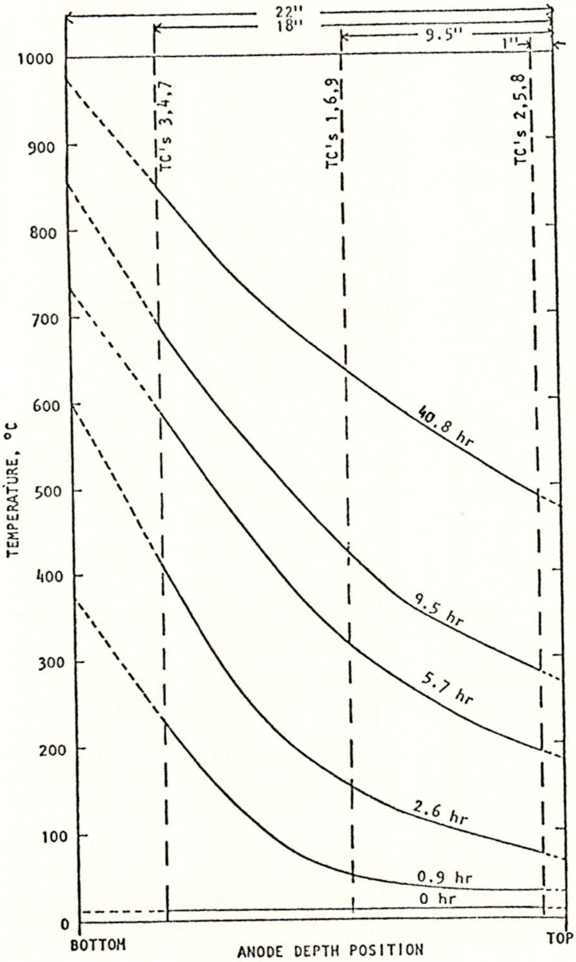

Figure 8 - Temperature/Depth Profiles in an Operating Anode as a Function of Anode Service Life.

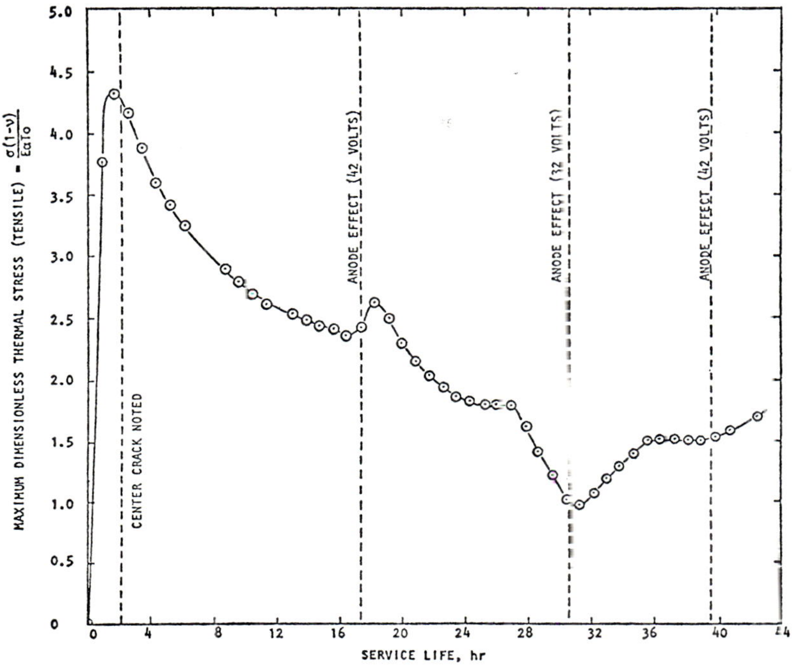

Figure 9 - Relative Magnitudes of Thermal Stress Exerted on an Operating Anode as a Function of Anode Service Life.

life. These results demonstrate that the most severe thermal stress occurs within the first few hours of operation. This is the thermal shock effect, and is expected when a cold anode is first set in hot bath.

It is interesting to note that the anode used in this test cracked between stubs shortly after set. This crack was first noted at 2 hours 8 minutes after set, but could have occurred a short time before. This is a period of high thermal stress, and suggests that thermal stress is a major factor resulting in early anode breakage, particularly the center cracks between stubs.

The results of Figure 9 also hint that anode effects can increase thermal stress, although it is not conclusive that this is so. In every case, the thermal stress tends to rise following an anode effect, but the magnitude does not appear to be significant when compared with the initial thermal shock. Also, the continual rise in thermal stress over a period of 6 hours, as seen following the second anode effect, is difficult to rationalize since an anode effect lasts for only a brief period of time, and the heat generated is expected to dissipate quickly.

An anode that does not crack early in service is not likely to fail later from thermal stress alone. However, some degree of thermal stress is always present and could be a contributing factor when coupled with the mechanical stresses that may occur later in the service life as the yoke and the stubs continue to increase in temperature

Discussion

Mechanical Effects

The decrease in contact drop with anode service life, as shown in Figure 6, could be attributed partly to a temperature effect. Laboratory results demonstrate that both temperature and contact pressure affect contact resistance. However, as demonstrated in Figure 10, the effect of increasing either temperature or contact pressure is minimal once the contact pressure becomes sufficient; a moderate pressure of 150 psi in this case. Thus, the minimum voltage drops attained in Figure 6 is believed to result from obtaining a sufficient contact pressure, i.e., a sufficient carbon strain.

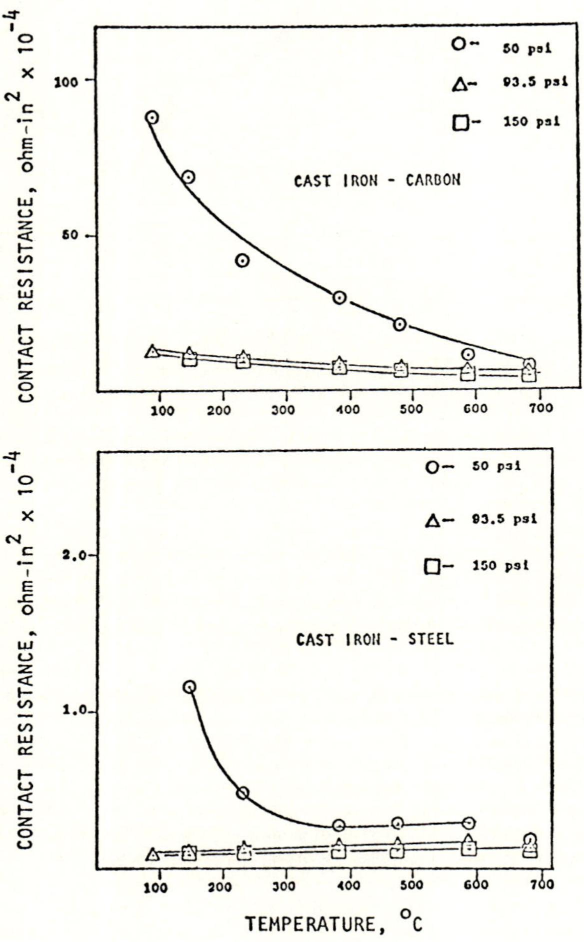

Figure 10 - Contact Resistance as a Function of Temperature and Pressure.

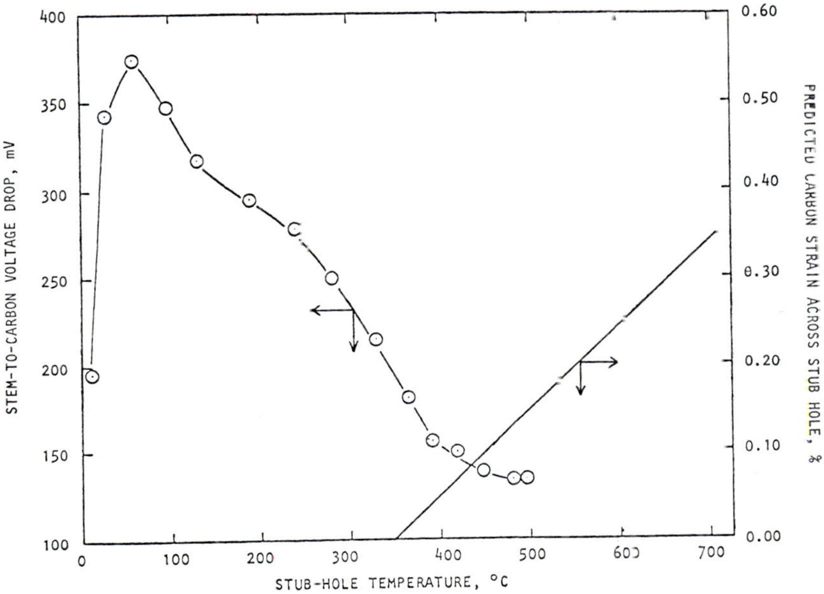

Figure 11 - Voltage Drop and Strain Load Through Stub Hole As a Function of Stub-Hole Temperature.

It is demonstrated in Figure 6 that, for the particular design in this test, stub hole temperatures of 450°C and above yield minimum contact drops. These temperatures also correspond to certain strain loads through the stub hole. Both effects are presented in Figure 11. The contact resistances are minimum at strains of greater than 0.10%

How much carbon strain is desired is a compromise. It is possible to design for a carbon strain that is just sufficient to minimize contact voltage drops; but this leaves little room for error or minor modifications in rodding or potroom practices, and risks the possibility that the resulting contact drops will be greater than desired. It is better to design for a moderate carbon strain such that minor changes in operations would always provide for minimum contact resistance while remaining below the mechanical limits of the carbon.

Based on this study, it is believed that a carbon strain of 0.20% at 600°C would result in a near optimum design. A strain of 0.20% is well removed from the strain limit of most carbon bodies, so that temperatures above 600°C (up to about 700°C) would not result in breakage. Furthermore, a strain moderately below 0.20% would still provide for low contact resistance; thus, obtaining low voltage drops early in the anode service life.

Using these design criteria, the desired stub hole diameter can be calculated. The important variables are stub diameter, CTE of the stub and CTE of the anode carbon. A set of design charts has been prepared and is given by Figure 12. These charts cover the range of variables normally encountered. Thus, if the above variables are known, a near optimum stub hole diameter can be estimated.

It should be pointed out that these charts are valid only for the anode/stub assemblies adjoined with cast iron. Also, the amount of preheat on the stubs prior to casting is considered mild, i.e., sufficient for removing moisture.

These design charts should provide a logical starting point when designing a new anode/stub assembly, either as a wholly new cell design or as a retrofit of existing cells.

Thermal Effects

Practical means of reducing thermal stress are not readily apparent. Invariably, a cold anode is going to be thermally stressed when placed in hot bath. The degree of thermal stress can be influenced by the mechanical and thermal properties of the carbon.

The effect of carbon properties has been discussed previously(3). The thermal stress resistance of carbon can be enhanced by increasing the parameter:

$$TSR = \frac{FS \times k}{E \times \alpha} \tag{4}$$

where

TSR = Thermal stress resistance
FS = Flexural strength
k = Thermal conductivity
E = Modulus of elasticity
α = Coefficient of thermal expansion

Therefore, a change in carbon processing to increase FS, k and/or decrease E, α will result in a more stress resistant anode.

Two different anode types have been tested in a reduction cell operating at 152 kA. The properties of these anode regarding thermal stress resistance are given in Table I.

Table I. Anode Carbon Properties Pertaining to Thermal Stress

Anode	FS, psi	k, W/m°C	$E \times 10^5$ psi	$\alpha \times 10^{-6}$ μin/in°C	$TSR \times 10^3$ W/m
Type A	1090	5.06	3.31	4.56	3.65
Type B	1578	5.70	3.16	4.70	6.02

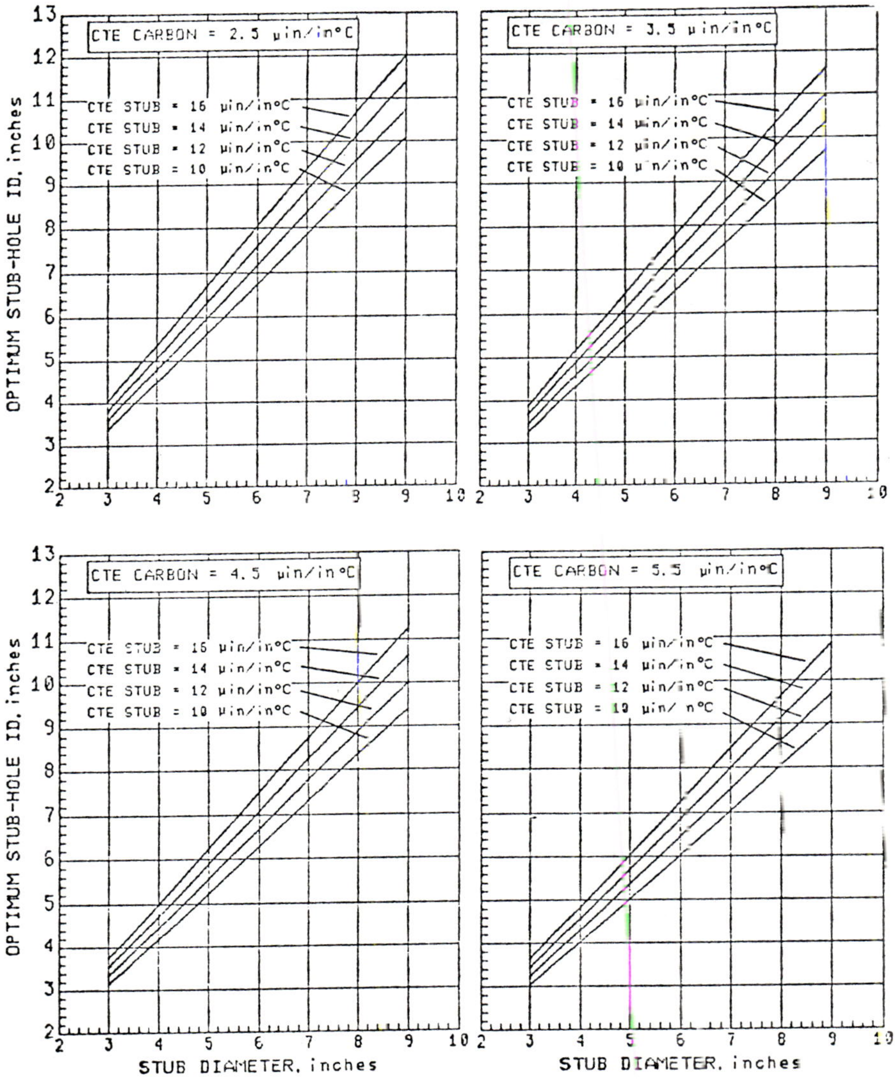

Figure 12 - Design Charts for Sizing Optimum Stub-Hole ID With Cast Iron Connections.

The results from cell operations indicate that the Type B anode is a superior anode to Type A regarding resistance to cracking. Whether or not this is directly attributal to improved thermal stress resistance is not known in that the increased flexural strength would also result in improved resistance to mechanical stress.

Conclusion

Minimum stem-to-carbon contact resistance in an operating anode are obtained at a moderate strain load of greater than or equal to 0.10%. Increased strain offers little benefit while increasing the risk of anode breakage. Design and operational data provide a basis for estimating the strain imposed across a stub hole, allowing for the sizing of a near optimum stub hole diameter.

Thermal stresses are most critical within the first few hours of anode service, but some degree of thermal stress is always present.

Acknowledgement

The authors would like to acknowledge the contribution of C. J. McMinn of the Reynolds Metals Company Reduction Laboratory regarding the information of metal/carbon electrical resistances.

References

1. Peterson, R. W., "Studies of Stub to Carbon Voltage," Light Metals 1978, p. 367.

2. Boley, B. A. and J. H. Weiner, Theory of Thermal Stresses, John Wiley & Sons, Inc., 1960.

3. Brown, J. A. and P. J. Rhedey, "Characterization of Prebaked Anode Carbon by Mechanical and Thermal Properties," Light Metals 1975, p. 263.

TEMPERATURE AND STRESSES IN CARBON ANODES

IN AN ALUMINUM REDUCTION CELL

Tu-Lung Weng
Kaiser Aluminum and Chemical Corporation
Center for Technology
Pleasanton, CA

Michael B. Hsu
MARC Analysis Research Corporation
Palo Alto, CA

Temperatures and stresses in a carbon anode used in the alumina reduction process have been determined using two- and three-dimensional finite element models. In this work, all material properties except the coefficient of thermal expansion are assumed to be temperature independent.

The anode temperature distribution is essentially a one-dimensional pattern with minor deviations near the edges and corners. The temperature rise in a newly-placed anode is very rapid - mainly caused by heat conduction through the bottom of the anode from the molten bath. Large stresses are developed in the lower region of the anode shortly after anode change and at the top surface of the anode between the stub holes in the later period in the anode life.

Introduction

Carbon anodes are consumed in the aluminum reduction process. As each carbon anode is consumed it is replaced with a new one. The new anode is suddenly exposed to a severe thermal shock as it is submerged in the 950°C molten bath. The thermal shock induces thermal stress that often initiates cracks in the carbon anode. As time passes, the temperature of the carbon anode rises through heat conduction from the molten bath, heat convection from the hot surroundings and the heat generation from the current flow. Thermal stresses generated from this temperature rise may further alter the integrity of the carbon anode.

In this investigation, the temperatures and stresses in carbon anodes of two different sizes: anode A - 60 in. x 42.5 in. x 20 in. and anode B - 30 in. x 42.5 in. x 20 in. are determined using a finite element program*. Also, the locations and time history of critical thermal stresses for potential initiation of cracks in the carbon anodes are quantified. Two- and three-dimensional models are used; the two-dimensional model is used in a preliminary study of the important parameters in the problem, but the three-dimensional model is used in the primary analyses of the problem.

Finite Element Models

Due to the condition of geometrical symmetry, it is necessary to model only one-quarter of anode A and one-half of anode B for finite element analysis. Since one-quarter of anode A and one-half of anode B are identical in dimension, the same finite element meshes consisting of 64 8-node brick elements and 3 truss elements (one embedded) as shown in Figure 1, can be used. The brick elements are used for modeling the carbon anode and the truss elements are chosen for the steel yoke and stub. Both elements assume a linear variation of temperature (or displacement) within the element and have one nodal degree-of-freedom

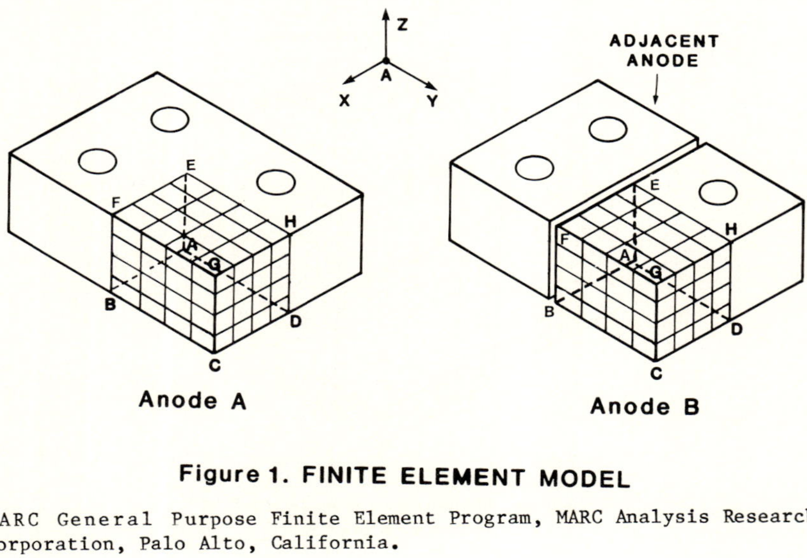

Figure 1. FINITE ELEMENT MODEL

* MARC General Purpose Finite Element Program, MARC Analysis Research Corporation, Palo Alto, California.

(temperature) for thermal analysis or three degrees-of-freedom (displacements) at each nodal point for stress analyses.

Figure 1A shows the relative positions of the finite element model in anode A. The surfaces ABFE and ADHE of the model coincide with the planes of the geometrical symmetry of the anodes. The anode is supported with a cross-legged steel yoke with four stubs; thus, one of the steel legs is included in the model. Figure 1B shows anode B and its adjacent anode. The same cross-legged steel yoke with four stubs is used to support two type B anodes. As shown, only one-half of anode B and one of the four legs of the steel yoke are needed for modeling. The thermal analysis treats the problem as a case of transient state with time-dependent boundary conditions and the stress analysis performed at selected time steps for critical thermal stresses treats all the materials as linear elastic and isotropic.

In this study, current flow in the anode and the reduction of anode size by electrochemical and chemical oxidation are neglected. However, heat generated from the current flow is taken into account indirectly by specifying the temperaures of some nodal points using measured values.

Initial Conditions and Boundary Conditions

Thermal Analysis

For thermal analysis, the same temperature boundary conditions can be used for anode A and anode B. Figure 2 depicts the boundary conditions for thermal analysis, consisting of convective boundary through free surface and the crust, insulated boundary at any plane of symmetry and any surface in contact with an adjacent anode.

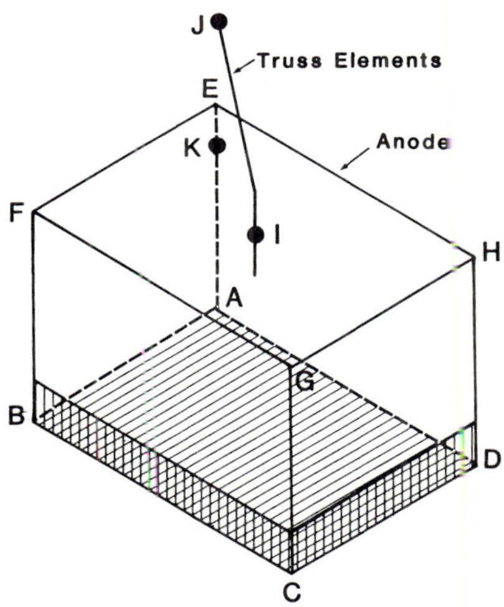

Figure 2. BOUNDARY CONDITIONS FOR THERMAL ANALYSIS

Surfaces ADHE and ABFE of the anodes are treated as an insulated boundary. For anode A, these insulated surfaces coincide with the planes of symmetry. For anode B, surface ADHE is the plane of symmetry and surface ABFE is in contact with the adjacent anode. Temperature distributions in anode B and its adjacent anode are almost alike and, hence, heat flow between the two anodes across the surface ABFE is negligible. Because of this situation, the surface ABFE of anode B can be treated as an insulated boundary.

When the lower portion of a new anode (initially at 25°C) is plunged into the 950°C molten bath, a thin layer of solidified bath material or crust forms immediately on the bottom of the anode and the lower portion of the two free surfaces CDHG and CBFG of the anodes. In this work, it is assumed that the 5 in. bottom portion of the two free surfaces is covered with crust and that the thickness of this crust is uniform at 0.12 in. Moreover, for computational simplification, it is also assumed that the crust does not melt away as the temperature of the anode bottom increases and, thus, the value of the film coefficient of the crust (thermal conductivity/crust thickness) remains constant throughout the anode's life. Heat transfer through the crust improves as the crust melts away; therefore, this assumption results in calculating slightly lower anode temperatures.

After anode change, it is assumed that the new anode is left uncovered for 1.5 hours before being covered with 50°C ore material. During the initial 1.5 hours, the top surface of the new anode is open to a 120°C environment, the upper portion of the new anode surface CDHG is exposed to 300°C cell sidewall, and the upper portion of the new anode surface CBFG is exposed to the 800°C adjacent anode. Heat transfer between the surroundings and these surface areas of the anode are assumed to be by convection. At 1.5 hours, the anode is covered with 50°C ore material. In order to simulate this change of the surroundings, the values of the film coefficient for these surfaces are allowed to increase stepwise at time 1.5 hours. Boundary conditions and the values of the film coefficients before and after 1.5 hours for the thermal analysis are summarized in Table I.

TABLE I. THERMAL BOUNDARY CONDITIONS AND FILM COEFFICIENTS

LOCATION	BOUNDARY CONDITIONS	FILM COEFFICIENT ($cal\text{-}cm/hr\text{-}cm^2\text{-}°C$) TIME < 1.5 HR.	TIME > 1.5 HR.
Top, EFGH	Convection	0.057	0.405
Bottom, ABCD	Convection	8.400	8.400
Face, CDHG	Convection	0.021	0.147
Face, CBFG	Convection	0.115	0.810
Face, ADHE & ABFE	Insulation	--	--
Point, I, J, & K	Specified Temperatures	--	--

In addition, the temperatures at nodal points I, J and K in Figure 2 were measured and used as time-dependent temperature input for these points in thermal analysis.

Stress Analysis

The fixed condition (zero nodal displacement) is prescribed at the upper end of the steel yoke, J. The planes of symmetry are made to remain in the original planes. In anode B, surface ABFE contacting the adjacent anode is allowed to distort, and linear springs are attached to this surface to determine the forces acting on the surface and the movements of the surface relative to the adjacent anode. The signs of the spring forces indicate whether or not the two anodes are in contact or separation.

For the sake of simplification in the modeling, the truss elements are used to represent the steel yoke and stub (including cast iron filling) and the gap between the cast iron filling and the carbon anode stub hole is neglected.

Material Properties

Calculations are performed using room temperature physical properties with the exception of the coefficient of thermal expansion. Room temperature physical properties of the anode carbon and steel are given in Table II.

TABLE II. PHYSICAL PROPERTIES OF ANODE CARBON AND STEEL AT ROOM TEMPERATURE (25°C)

PROPERTY	CARBON	STEEL
Specific Heat (cal/g °C)	0.167	0.097
Thermal conductivity (cal-cm/sec-cm^2-°C)	0.013	0.130
Coefficient of Thermal Expansion (10^{-6}/°C)	1.50	2.00
Density (g/cm^3)	1.55	7.06
Young's Modulus (10^6 psi)	1.20	29.00
Poisson's Ratio	0.15	0.30

Temperature dependent coefficients of thermal expansion (CTE) in 10^{-6}/°C for anode carbon and steel are expressed by a quadratic equation for a lower temperature range and by a constant for a higher temperature range as follows:

Anode Carbon:

$$CTE = 1.5 \times 1.23 \times 10^{-2}(T-25) - 0.094 \times 10^{-4}(T-25)^2 \text{ for } 25°C<T<680°C$$

$$CTE = 5.523 \text{ for } T \geq 680°C$$

Steel:

$$CTE = 2.0 + 4.591 \times 10^{-2}(T-25) - 0.446 \times 10^{-4}(T-25)^2 \text{ for } 25°C<T<540°C$$

$$CTE = 13.815 \text{ for } T \geq 540°C$$

Results and Discussion

Temperatures

Surface temperatures on both anodes A and B are shown in Figures 3 and 4. Surface temperatures at the bottom and the insulated surfaces, referred to as farside surfaces, at 1.5 hours and 240 hours are shown, respectively, in Figures 3A and 4A, whereas surface temperatures at the top and the free surfaces, referred to as nearside surfaces, at 1.5 hours and 240 hours are shown, respectively, in Figures 3B and 4B.

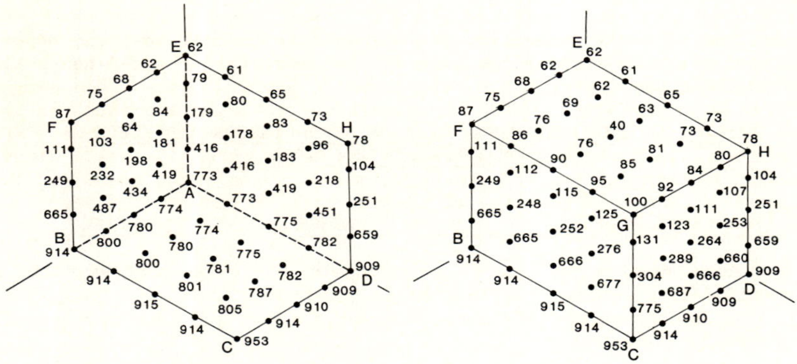

(A) FARSIDE SURFACES

(B) NEARSIDE SURFACES

Figure 3. ANODE SURFACE TEMPERATURES AT 1.5 HOURS

Temperature variations on the bottom and the top of the anode are very small at 1.5 hours and become even smaller at 240 hours. Temperature gradient through-the-height of the anode is very large at 1.5 hours and relatively small at 240 hours. Temperature gradient through-the-anode height (from point C to point G in Figure 3B) at 1.5 hours is about 850°C. The anode temperature distribution is essentially a one-dimensional pattern in the horizontal plane with minor deviation near the edges and corners. This temperature distribution appears to be influenced heavily by heat conduction through the bottom of the anode from the molten bath.

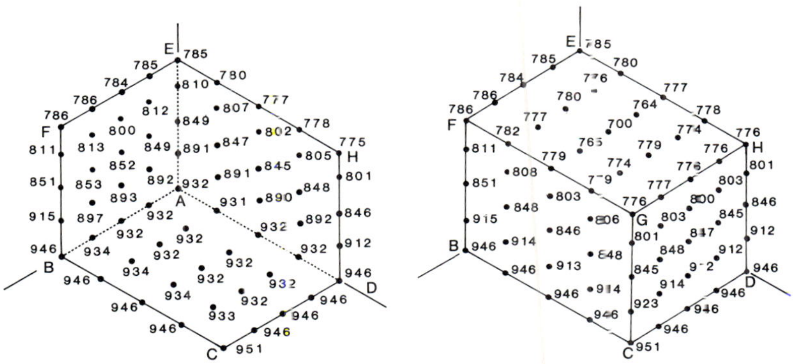

(A) FARSIDE SURFACES

(B) NEARSIDE SURFACES

Figure 4. ANODE SURFACE TEMPERATURES AT 240 HOURS

The bottom surface temperatures obtained by thermal analysis appear to be on the low side of the expected temperatures. By 240 hours the bottom surface temperatures are very close to 940°C. Since a constant convection boundary condition was assumed to exist along the entire bottom surface in the thermal analysis, adjustments to the film coefficient and the bath temperature may be required to obtain higher anode bottom temperatures. These adjustments can only be made on the basis of better understanding of the interaction between the anode and the molten bath.

An examination of nodal temperature histories suggests that a steady-state temperature distribution is reached in about 168 hours (7-days). The temperature differences between the 7th and 10th day are all within 10°C.

The current model predicts a continuous increase of surface temperatures even after cold ore-cover is added to the surfaces at 1.5 hours. Because of the low thermal conductivities of air and ore-cover, the temperature distribution of the anode is found to be insensitive to heat transfer on surfaces exposed to air and ore-cover material.

The thermal gradients in the anode will alter if additional conditions such as temperature dependent thermal properties, non-uniform crust thickness at the bottom of the anode and current flow are considered.

Thermal Stresses

Thermal stress distributions in the anodes at 1.5 hours and 240 hours after anode change are determined. Maximum stresses in anode A and anode B at these timesteps are given in Table 3.

TABLE III. MAXIMUM STRESSES IN ANODES A AND B

	Maximum Stresses, psi.			
	1.5 Hours		240 Hours	
	Anode A	Anode B	Anode A	Anode B
Shearing Stress	570	450	720	130
Principal Stress	670	650	1470	140
Tensile Stress in x-direction	610	650	960	80
Tensile Stress in y-direction	670	530	1360	70
Tensile Stress in z-direction	250	280	100	50

The maximum shearing stress is equal to one-half of the difference between the largest and smallest principal stresses. Maximum principal stress is the largest of the three principal stresses computed from the three normal stresses and the three shearing stresses. Maximum tensile stresses are the largest normal stresses acting in the directions of the coordinate axes, x, y and z.

At 1.5 hours high shearing stresses develop in the lower portion of both anode A and anode B. The maximum shearing stresses in anodes A and B are 570 psi. and 450 psi., respectively, and are found in one of the surface elements located in the bottom layer of the anodes. The values of the shearing stress in the anodes are somewhat below the shearing strength of anode carbon, which is estimated to be about 600 psi. However, the large concentration of relatively high shearing stress may contribute to occassional chunking of lower corners and lower portions of the anode shortly after anode change.

Maximum principal stresses in anode A and B are, respectively, 670 psi. and 650 psi. These stresses are lower than the tensile strength of the anode carbon of 800 psi. These stresses are found in one of the internal elements.

At 240 hours after anode change, high tensile stresses develop in the horizontal direction in the top portion of anode A, but not in anode B. A maximum principal stress of 1470 psi. is found in a top surface element in anode A. Maximum tensile stress of 1360 psi. in the y direction in anode A may explain the frequent vertical crack in the plane of symmetry ABFE late in the life of the anode. A high shearing stress of 720 psi. is found in one of the internal elements, but does not seem to have a serious consequence. Stresses in anode B are generally very small.

At late time (240 hours) high tensile stresses are found near the steel stub and are probably caused by the idealized modeling of perfect contact between the steel stub and carbon anode. The need for detailed modeling of this region in future studies is evident.

It is apparent that the critical thermal stress occurring early on is primarily caused by the large thermal gradient in the anodes and, later on is caused by the differential thermal expansion between the steel yoke and the carbon anode.

Deformation of Anodes

In addition to the thermal stresses, deformation of the anodes can also be determined from finite element stress analysis. In Figures 5A and 5B, the deformed one-quarter section of anode A and the deformed one-half section of anode B at 1.5 hours after anode change are shown, respectively.

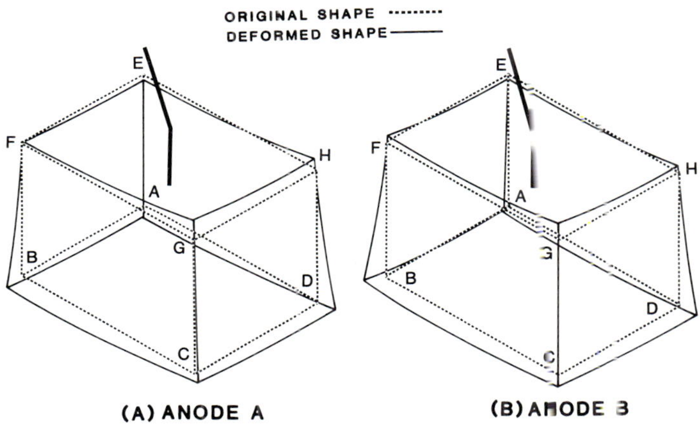

Figure 5. DEFORMED ANODES AT 1.5 HOURS

In these figures, the original shape of the anode is depicted by dashed lines and the deformed shape of the anode is depicted by solid lines. Both anodes wrap upward due to the severe thermal gradient in the lower portion of the anodes, while the planes of symmetry remain in the same planes. Surface ABFE in anode B is a free surface adjacent to the next anode and the lower portion of this surface deforms outward to butt against the adjacent anode, especially at point B, as seen in Figure 5B.

In Figures 6A and 6B, deformed anode A and anode B at 240 hours after anode change are shown, respectively. At this time, the anodes are slightly enlarged but return almost to their original shape. At 240 hours, face ABFE of anode B moves slightly away from the adjacent anode, as shown in Figure 6B. The effect of butting between adjacent anodes B at early time is not investigated. Additional stresses produced from butting against the adjacent anode depend strongly on the value of the rigidity of the linear spring elements chosen to simulate the external constraints on surface ABFE. However, the present investigation provides a fair idea on how surface ABFE moves relative to the adjacent anode.

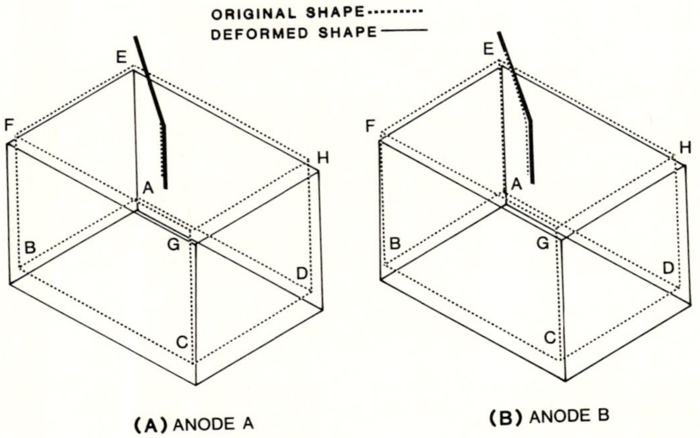

Figure 6. DEFORMED ANODES AT 240 HOURS

Summary and Conclusions

In this study a transient thermal analysis and a number of elastic thermal stress analyses are performed on two anodes. The anodes are: anode A-60 in. x 42.5 in. x 20 in. and anode B-30 in. x 42.5 in. x 20 in. The transient thermal analysis is carried out for 240 hours, and temperatures at 1.5 hours and 240 hours are used for thermal stress calculations. The MARC general purpose nonlinear finite element program is used for both the thermal and stress analyses.

It is observed that the heat transfer in the anode appears to be heavily influenced by heat conduction from the molten bath through the bottom of the anode. Surface temperatures of the anode at 1.5 hours are found to be reasonably close to measured temperatures. The bottom surface temperatures appear to be on the low side of the crust melting temperature (940°C); however, the bottom surface temperatures at 240 hours closely approach 940°C. It is also been observed that a steady-state temperature distribution is reached in about 168 hours (7-days). The current model predicts a monotonical increase of surface temperatures even after ore-cover is added to the surfaces at 1.5 hours.

The results of the stress analyses indicate that the critical thermal stress occurring at the early time (1.5 hours) is primarily caused by the large thermal gradient in the anodes, and at the late time (240 hours) by the differential thermal expansion between the steel yoke and the carbon anode.

Maximum shearing stresses in the lower region of both anodes at early time are found to be slightly below the critical shear strength of carbon; however, the concentration of high shearing stresses clearly poses potential problems for the lower part and corners of the anodes.

The normal stresses may cause failure in the top region of anode A at late time (240 hours). A maximum principal stress of 1470 psi. in anode A can initiate cracks at the top of anode A. No high shearing and normal stresses are found in anode B at this late time.

At late time (240 hours) high tensile stresses are found near the steel stub in both anodes. These high stresses are local and caused by the idealized modeling of perfect contact between steel stub and the carbon anode.

The finite element models developed in the current study can be adequately used for the prediction of the global thermal-mechanical responses of the anodes. In future investigations, modifications can easily be made to the models to account for various local behaviors.

Acknowledgement

The authors wish to thank Kaiser Aluminum and Chemical Corporation for giving permission to publish this paper.

A REFEREE METHOD FOR EVALUATING ANODE COKE DENSITY

V. L. Bullough, A. L. George and E. R. Russell

Reduction Laboratory
Reynolds Metals Company
P. O. Box 1200
Sheffield, AL 35660, USA

Real density or specific gravity is one of the indices used by the producers and users of calcined petroleum coke to measure the degree of calcination. The normal pycnometric method used indicates an average value without providing any information about the uniformity of the calcined product. A density separation method has been developed based on the use of mixtures of organic liquids of known specific gravity. This method permits the isolation and characterization of fractions of a calcined petroleum coke.

Introduction

In the electrolytic production of aluminum by Hall-Heroult process, the major component for anode manufacture is calcined petroleum coke. This coke is selected primarily because of its high purity and relatively low cost as a by-product of petroleum refining. The green coke as produced, however, requires additional heat treatment to produce a material that is dimensionally stable and electrically conductive. Calcination is generally accomplished in a rotary kiln at finishing temperatures above 1200°C. The quality and uniformity of the calcined product is effected by bed depth, heating rate, retention time and finishing temperature.

Generally, the purchase specifications for calcined petroleum coke include specifications for its electrical conductivity and real density. The usual procedure for the real density determination is the pycnometric measurement of a fine coke powder (-74 microns) in kerosene, toluene or water containing an aerosol. Alternately, the crystallite thickness may be calculated from an X-ray diffraction pattern, and this value related to real density through correlation. All of these methods give only an averaged value and do not indicate the uniformity of the calcined product. Often, to assure adequate uniformity and protect against undercalcination, a real density is specified that is higher than optimum.

We have developed a procedure to measure the calcining uniformity based on a principle published by Locke and Austin(1) establishing the amount of coke powder that floats in mixtures of organic liquids of different specific gravities. Locke and Austin demonstrated the ability to fractionate synthetic carbons by gradually increasing the density of aromatic hydrocarbons by the addition of bromoform.

Method Development

Our initial development involved placing portions of calcined coke powder (-100+200 mesh fraction) in graduated cyclinders and dispersing the coke in mixtures of benzene and bromoform of different specific gravities. The percentage of float and sink were estimated from the volumes of coke at the bottom and top of the graduated cylinders as shown in Figure 1. This method could reliably predict the distribution of coke densities, but, it was difficult to read the meniscus and the coke fines had a tendency to cling to the cylinder above the liquid level. We found it useful, as did Locke and Austin, to remove -200 mesh fraction of the material from our sample to reduce the time necessary for the sample to come to equilibrium.

To separate different density fractions from a weighed coke sample, an apparatus was developed to allow the removal, filtration and weighing of the float fraction. This was accomplished by constructing tubes with a stopcock in the middle that prevented back-mixing during separation of the two fractions. This apparatus is shown in Figure 2. We had suspected that difficulties in reading the meniscus of the float fraction in low specific gravity mixtures could produce an error. This was demonstrated in Figure 3 where the same commercially calcined coke sample was fractionated both volumetrically and gravimetrically. It can be seen that an error of 3% to 4% could easily occur in estimating volumetrically the amount of undercalcined material. It can also be seen that the real density interpolated by the two methods is the same.

Results are usually plotted on probability coordinates where Gaussian distribution would be indicated by a straight line(2). If a linear relationship exists, calcination uniformity can be estimated from the line

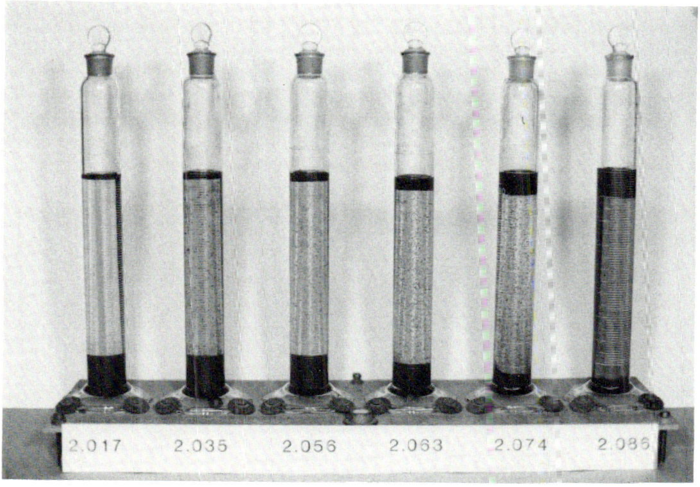

Figure 1 - Volumetric Estimation of Coke Density Gradient.

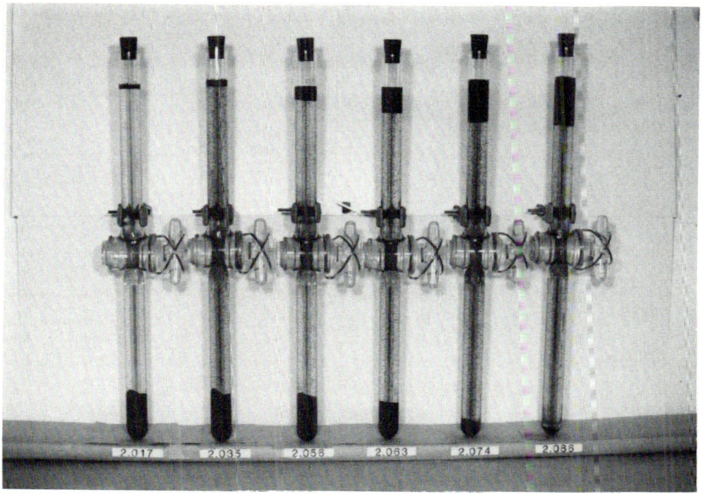

Figure 2 - Gravimetric Determination of Coke Density Gradient.

slope, and the standard deviation (σ) can be interpolated from the intercepts at the 16 and 84 percentile, the values of the mean $\pm$ 1 σ. The $\pm$ 2σ population can be determined by the 2.5 and 97.5 percentile. Figure 4 indicates the Gaussian behavior of the same coke calcined to different finishing temperatures. Interpolation of the mean or 50 percentile from the graph should give the same real density value as the value obtained by the standard pycnometric method. The curvature at the higher specific gravity end of each curve may be due to analytical difficulty in recovering and measuring a small amount of coke. Table I shows a comparison of real density values obtained by the pycnometric method and values interpolated from the graph. Figure 4 shows the family of almost parallel curves are linear over nearly 90% of the distribution. Table I shows a slightly poorer reproducibility for Sample No. 224 which was calcined to a higher finishing temperature.

Table I - Comparison of Real Densities Determined by Water Pycnometer with Real Densities Interpolated by Float Sink Separation in Benzene-Bromoform Mixtures

Sample No.	Pycnometric Density, g/cc	Float-Sink Density, g/cc	Float-Sink σ	Float-Sink $\pm$ 1 σ
224	2.04	2.04	0.024	0.048
225	1.93	1.93	0.022	0.045
226	1.88	1.88	0.022	0.043
227	1.82	1.80	0.023	0.046

The Effect of Particle Size

A study of the density gradient of different particle sizes was conducted on a commercially calcined coke. The results are shown in Figure 5. As expected, this study indicated that the larger particle sizes have pore structures excluding the liquid phase at the separation time that we studied (overnight). For estimating coke density distribution, the -100+200 mesh fraction seems to be the same as the -200 mesh material.

The Effect of Separation Time

Normally, our practice is to allow a series of samples to stand overnight before separation of the float fractions. By this time, the samples at the extremes of density have separated very clearly. However, for the samples near the median, there is always some material that remains suspended having a density nearly identical to the specific gravity of the liquid. An experiment was conducted to examine the effect of incomplete separation. In one set of samples, separation of the float fraction was made after two hours and compared with the float fraction of the same sample allowed to stand overnight. The results shown in Figure 6 indicate that the weight of suspended coke is minor and does not adversely affect the float/sink separation. Two hours appears to be an adequate time for sample equilibrium.

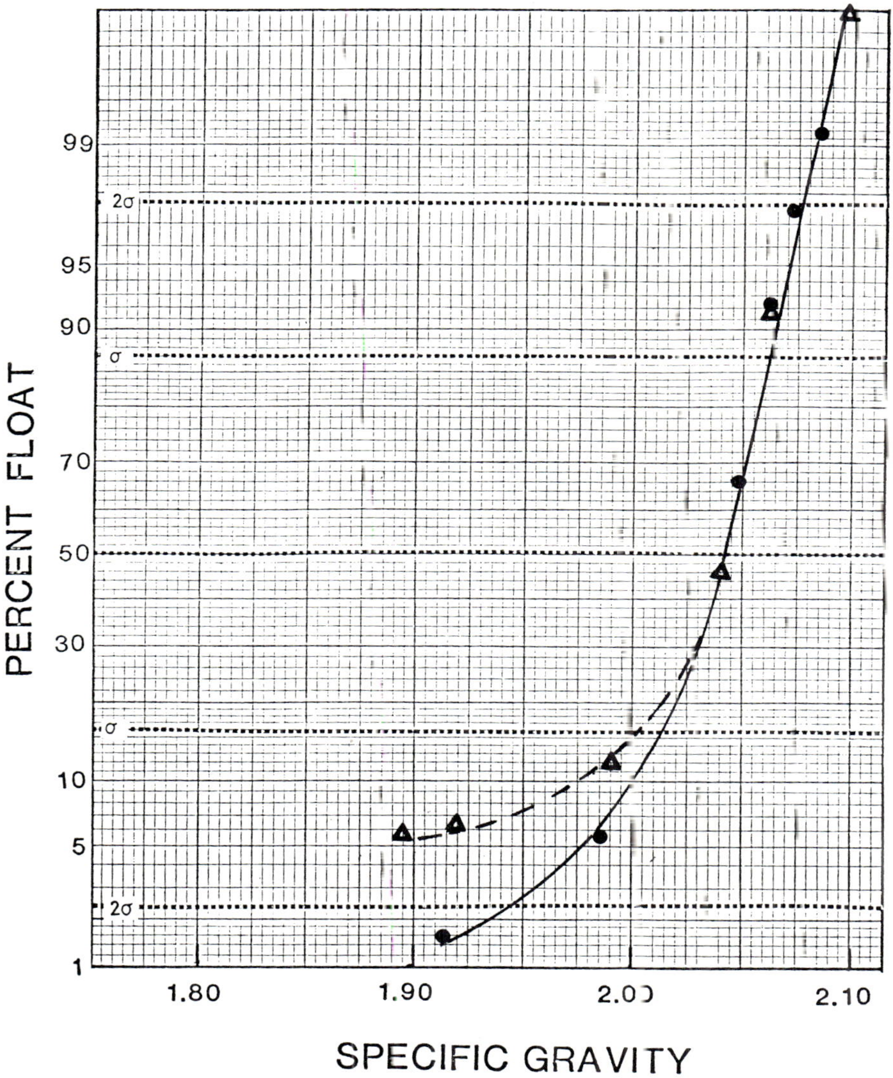

Figure 3 - The Determination of Density Gradient on the Same Sample Using Volumetric ▲ and Gravimetric ● Measuring Methods.

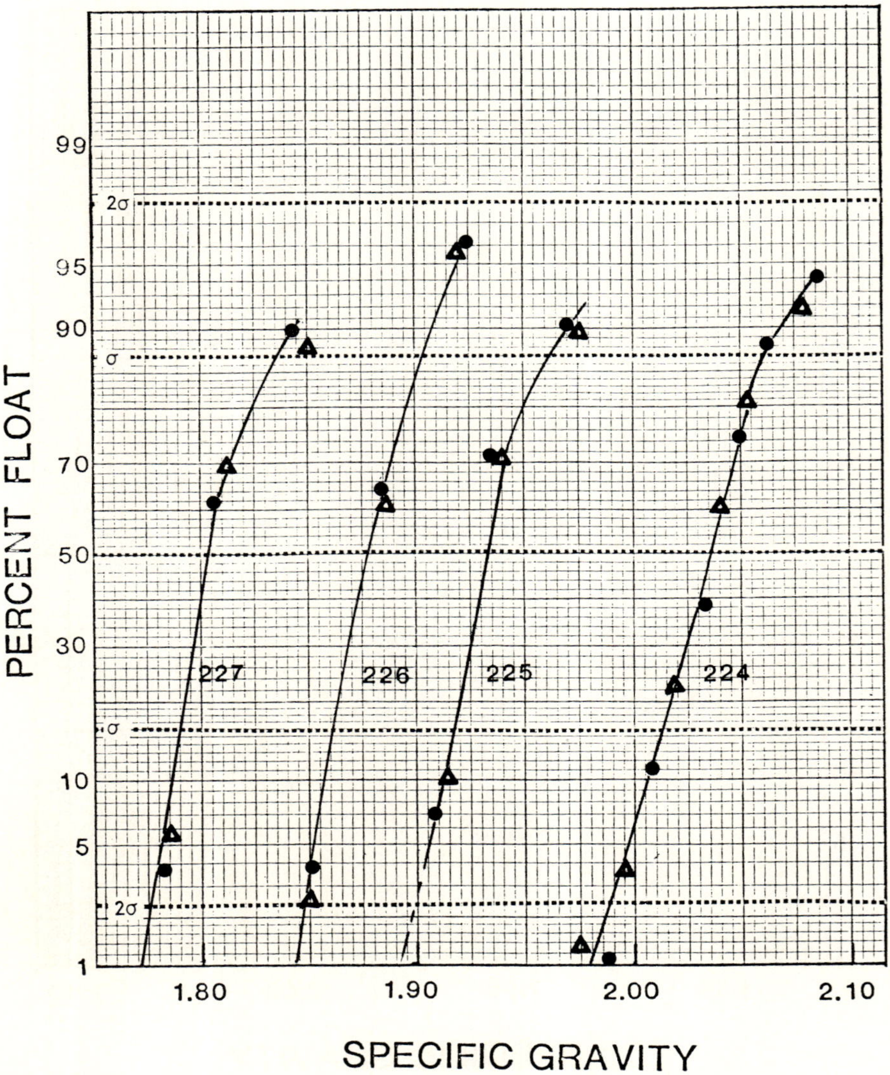

Figure 4 - Measurement of Density Gradient in the Same Green Coke Sample Calcined to Different Temperatures. (▲, ● Duplicate Samples).

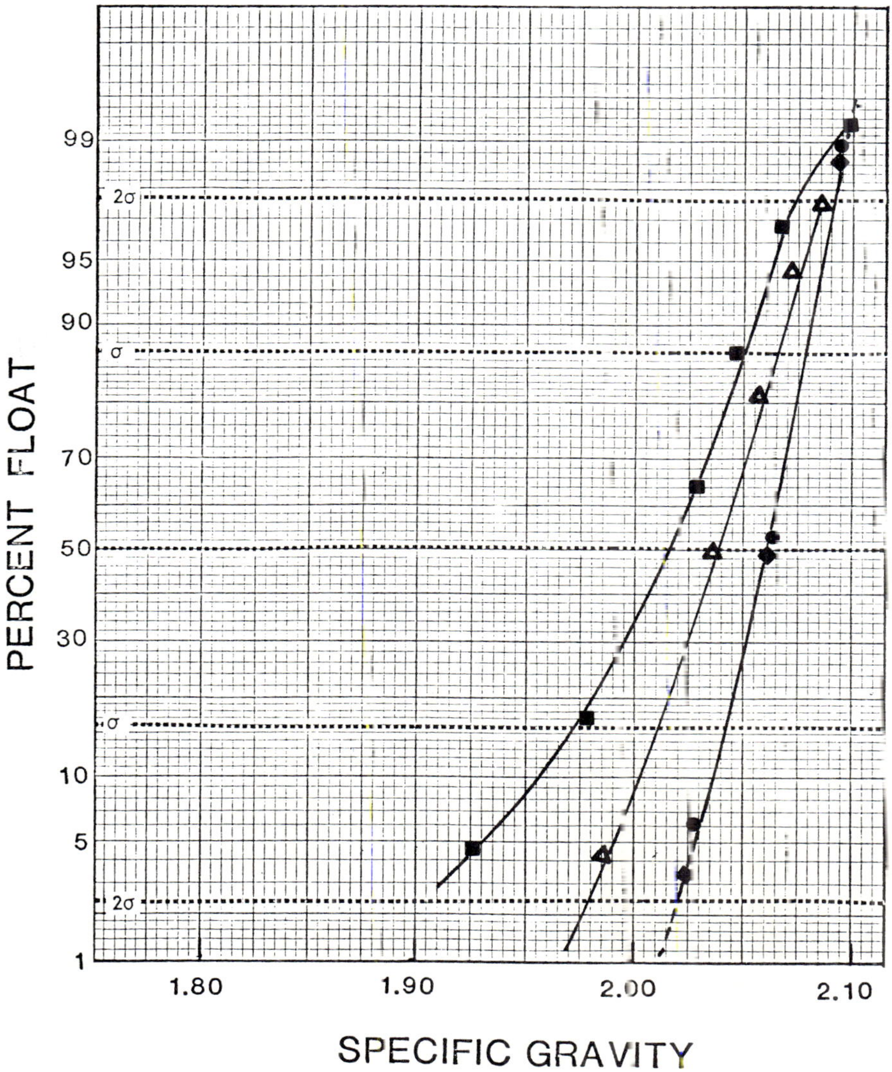

Figure 5 - The Effect of Particle Size on the Determination of Density Gradient. (■ 35/65 mesh, ▲ 48/100 mesh, ● 100/200 mesh and ◆ -200 mesh).

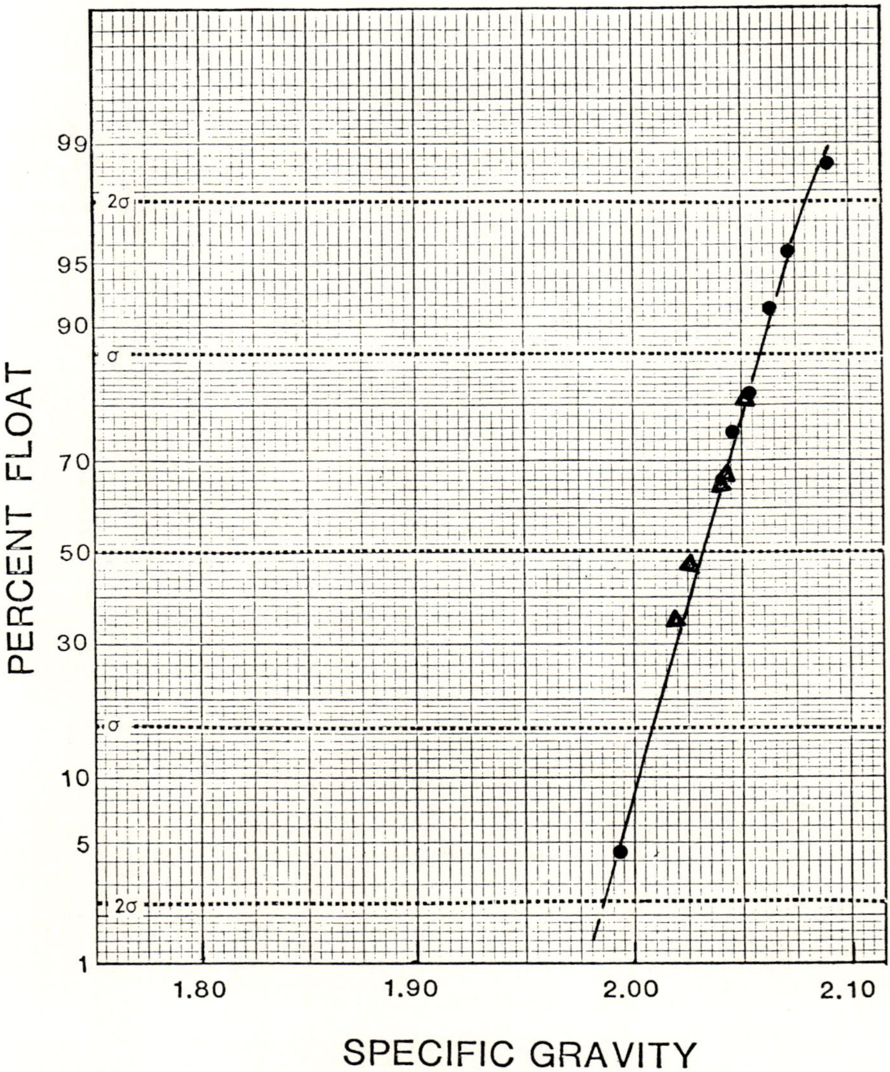

Figure 6 - The Effect of Separation Time on the Gravimetric Determination of Density Gradient. (▲ 2 Hours, ● Overnight).

Reproducibility

A sample of the same coke was riffled, split into four fractions and the density gradient of each fraction determined. The close fit of data illustrated in Figure 7 for these determinations demonstrates excellent reproducibility.

Effects of Solvent

To reduce worker exposure to benzene vapors, monochlorobenzene and toluene were substituted for benzene in the specific gravity mixtures. The substitution of chlorobenzene for benzene seems to have introduced a determinate error of about 0.02 g/cc in real density. This may be due either to poorer wetting of the coke particle or to an exclusion of the larger molecule from some of its finer pores. We have not resolved the cause of this difference in coke real density values, but it would lead us to question whether an organic liquid can be substituted randomly in the pycnometric determination of the real density of calcined petroleum coke. These results are shown in Figure 8.

Detection of Puffing

Rhedey(3) and others have shown that while the density of calcined coke normally increases with increasing calcination temperature, in some cases a decrease in density takes place that is attributed to the "puffing" that takes place during desulfurization. A comparison of a coke thermally desulfurized in a pilot plant at temperatures above 1500°C with the same coke calcined to approximately 1300°C in a rotary kiln is shown in Figure 9. It can be seen that the density distribution has shifted radically to a lower value in the desulfurized coke. An examination of the mercury intrusion porosity spectrum shows that this change is associated with pores smaller than 0.2 micron. The density gradient indicates Gaussian distribution in the desulfurized coke, which is an indication of uniform heat treatment. This might be associated with the operating conditions of the pilot plant.

The Effect of Calcining Temperature

Samples were taken during a kiln start-up to measure the calcining uniformity at different temperatures. The results of this study are shown in Figure 10 where it can be seen, with the exception of tailing at the lower real density, that coke density can be shifted and controlled within reasonable limits for a given feedstock. It is interesting that in this study the tailing was at the low density end of the distribution -- the opposite of the effect observed in Figure 4.

Commerically Available Cokes

The density gradient on several commercially available cokes were measured in bromoform/chlorobenzene mixtures. The results of this study are shown in Figure 11. These cokes represent samples from different geographical regions that were calcined in different types of calcining equipment. All are "sponge" delayed cokes that are offered in the commerical market as electrode-grade cokes. To our knowledge all are used successfully in electrode manufacture of both Soderberg and prebake type electrodes. The linearity and similarity of line slopes reflected in Figure 11 would indicate that delayed sponge petroleum coke has a Gaussian distribution over 95% of its range. We have not investigated if the

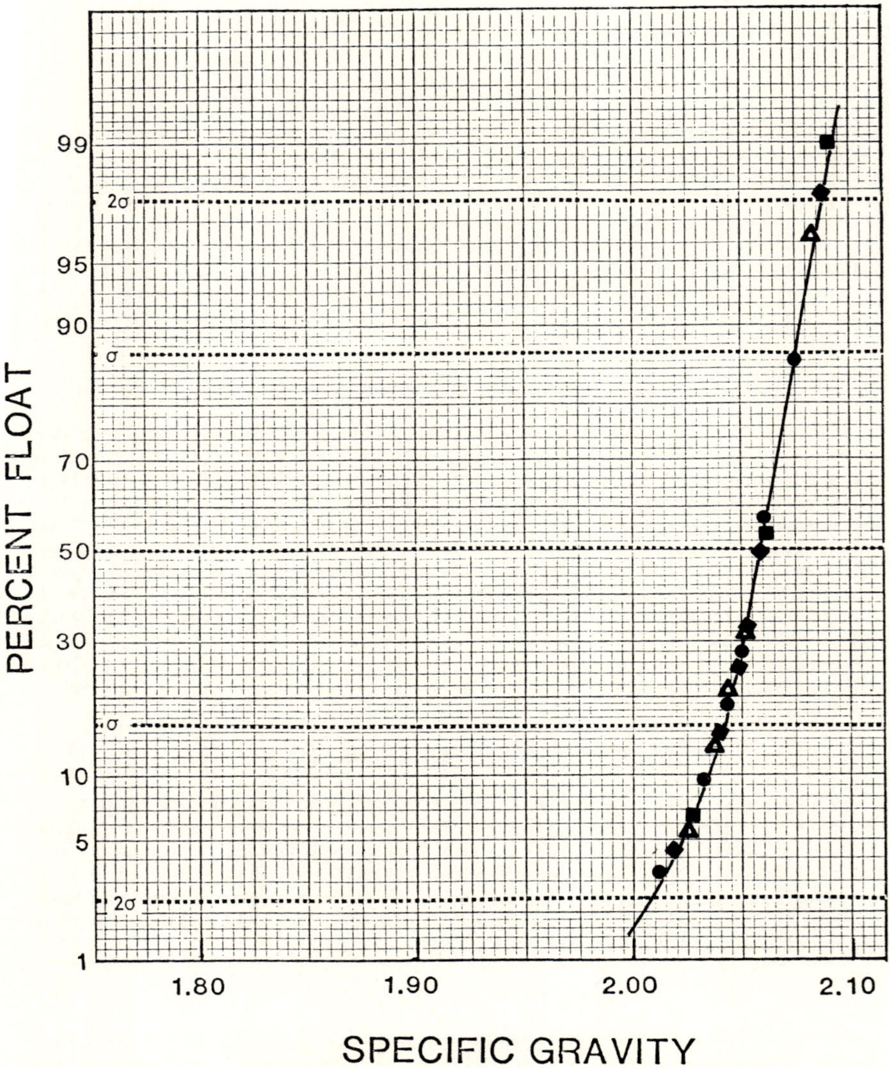

Figure 7 - Reproducibility of Density Gradient as Illustrated By Replicate Determination of the Same Sample.

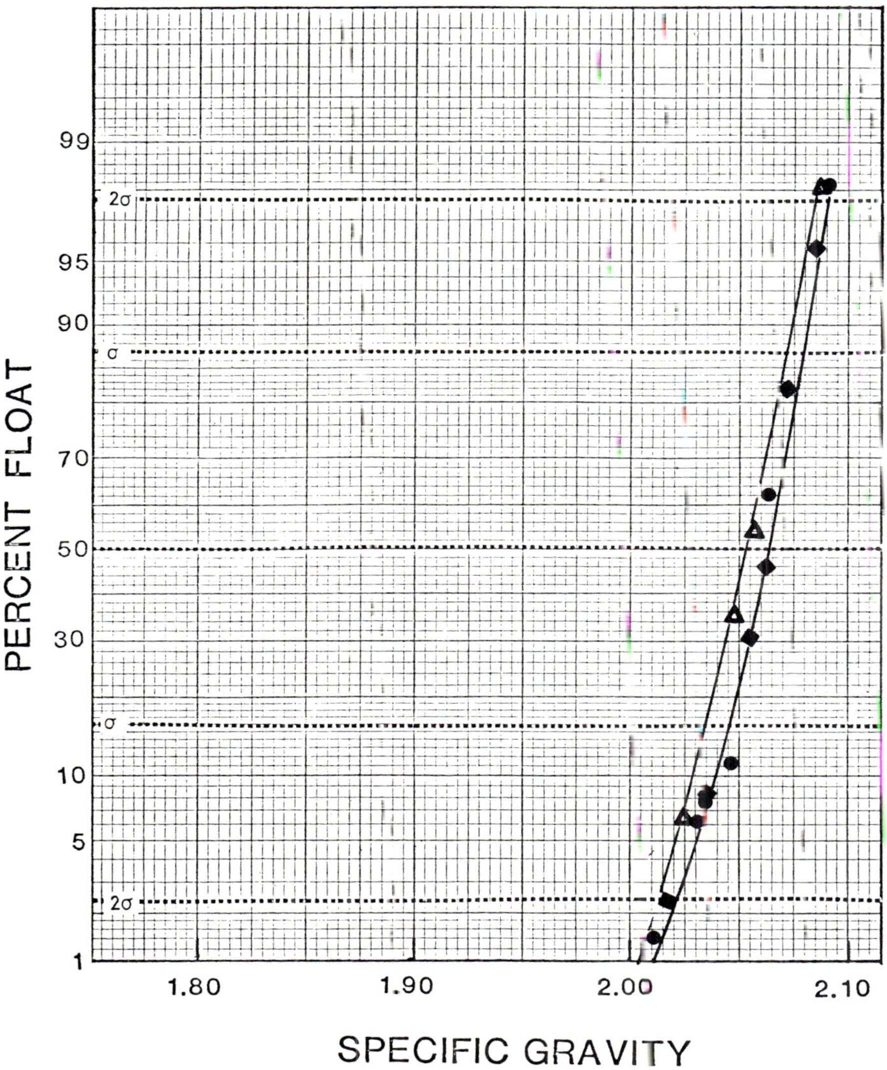

Figure 8 - The Effect of Different Solvent Combinations on the Measurement of Density Gradient. (● Bromoform-Benzene, ◆ Bromoform Toluene, △ Bromoform-m-Chlorobenzene).

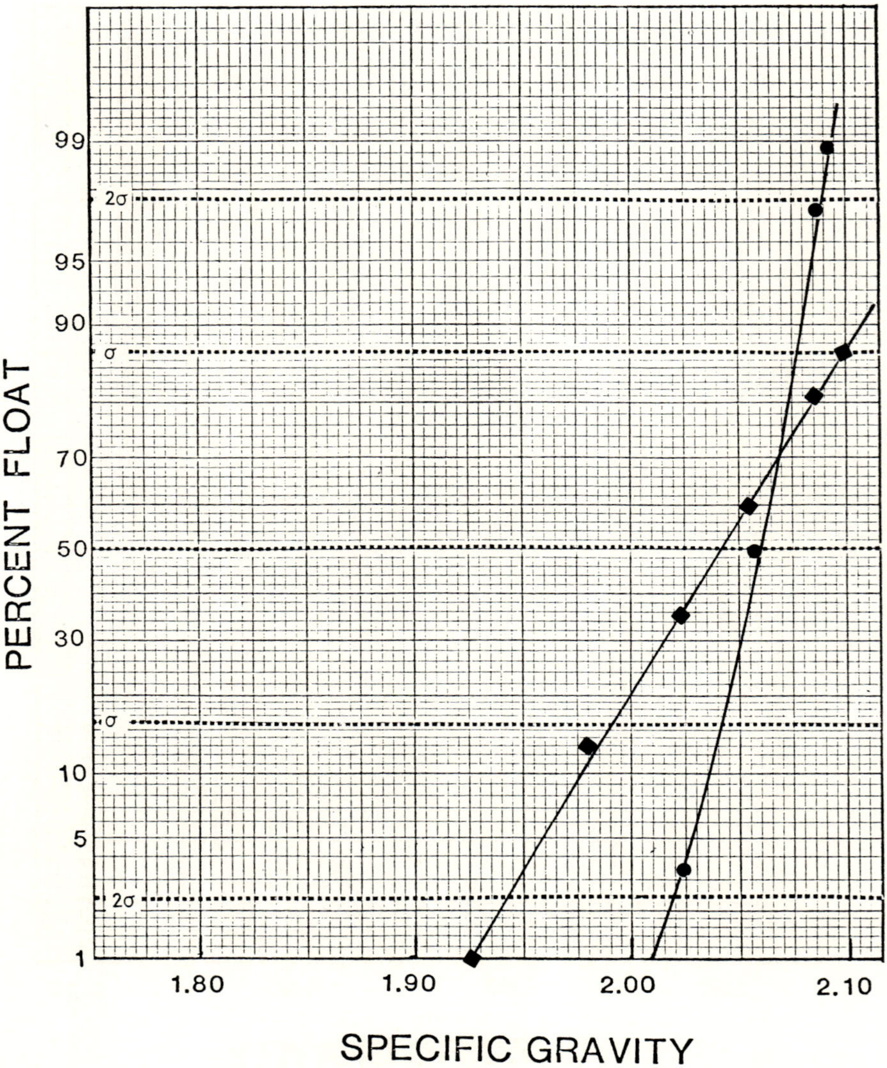

Figure 9 - A Comparison of the Density Gradient in a Coke Thermally Desulfurized ◆ With the Same Coke Normally Calcined ● .

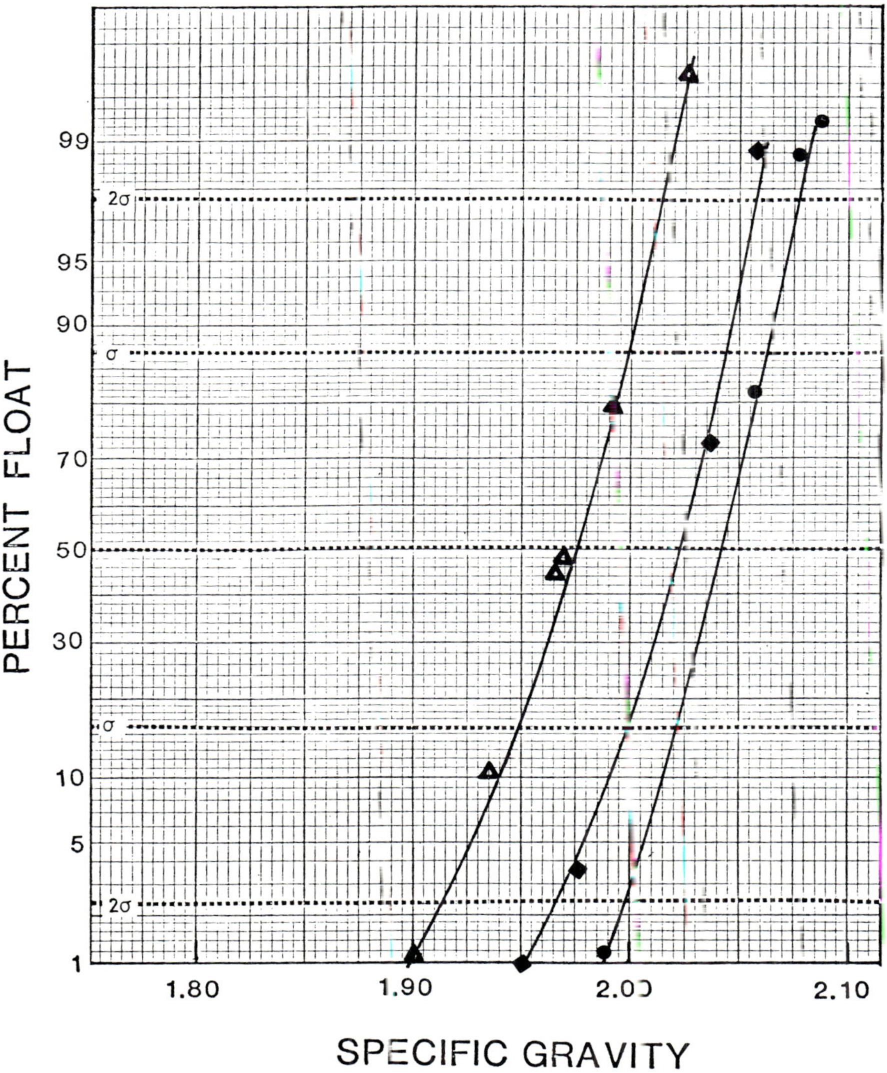

Figure 10 - The Effect of Calcining Temperature on the Density Gradient in a Commercially Calcined Coke
(▲ 1040°C, ◆1150°C, ●1320°C).

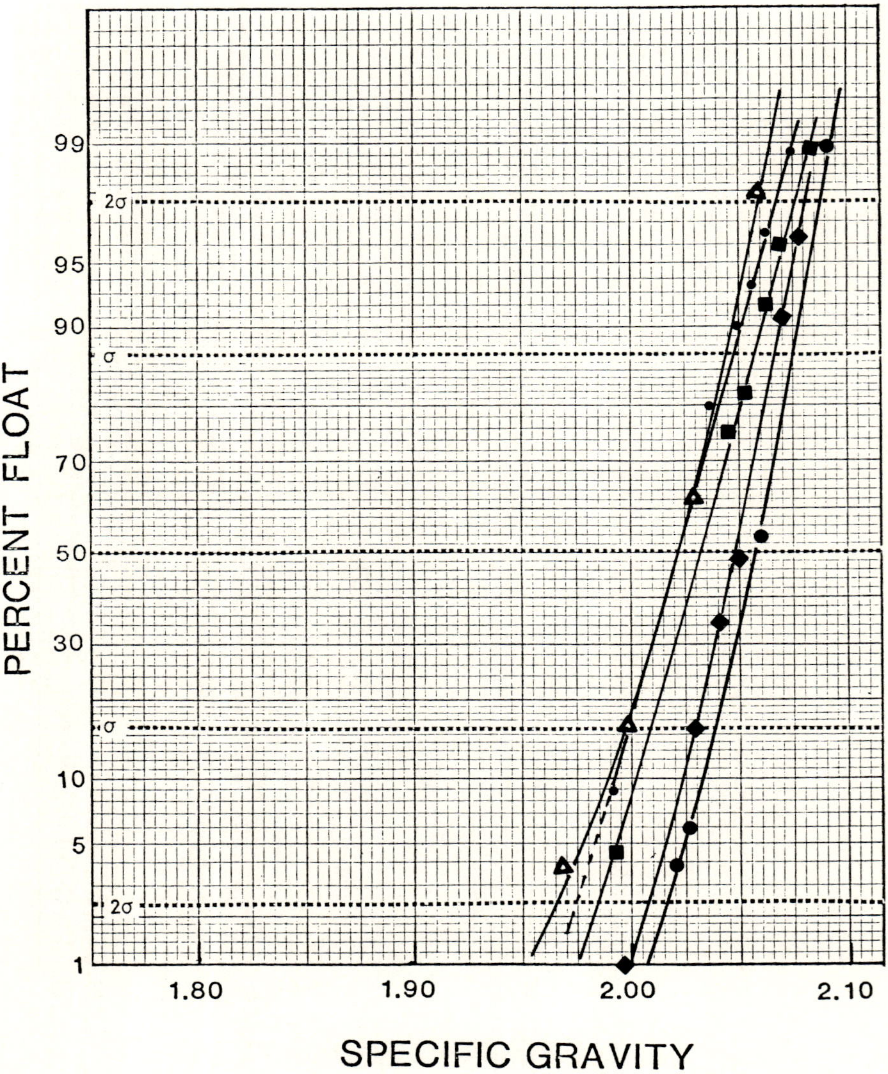

Figure 11 - The Density Gradient of Several Different Commercially Calcined Cokes.

offsets in real density are a result of different calcining temperatures or of variables associated with the feedstocks.

Summary and Conclusions

The particle density distribution in the calcined petroleum cokes studied appears to be Gaussian over nearly 95% ($\pm$ 2 σ) of the population. While we have not established if tailing at both the high and low end of the distribution is real or an analytical error, the tailing reproduces in duplicate samples.

Determinate errors of up to 0.02 g/cc in real density can be introduced by an in appropriate selection of solvent pairs. Our work indicates that benzene bromoform or toluene bromoform most nearly approach the mean as normally reported by pycnometric procedures for real density determination. The amount of coke suspended after two hours settling time appears unimportant in the determination of real density.

Application of density gradient separation provides a tool to evaluate the quality of coke calcining and a means to optimize kiln performance.

References

1. L. D. Loch and A. E. Austin, "Fines Pore Structure -- Crystallite Size Relationships in Carbon," Proceedings of the First and Second Conferences on Carbon, pp. 65-73; Waverly Press, Baltimore, Maryland, 1956.

2. John B. Kennedy and Adam M. Neville, Basic Statistical Methods for Engineers and Scientists, 2nd ed., p. 148; Harper and Row, New York, N.Y., 1976.

3. Paul Rhedey, "Structural Changes in Petroleum Coke During Calcination," Transactions of the Metallurgical Society of AIME, 239, July (1967) pp. 1084-1091.

EFFECT OF BUTT PARTICLE SIZE AND BUTT FRACTION ON PREBAKE ANODE PROPERTIES

Tu-Lung Weng
Center for Technology, Pleasanton, CA

Victor M. Vera
Mead Works, Spokane, WA

Kaiser Aluminum & Chemical Corp.

The effects of butt particle size and butt fraction on anode properties have been investigated using carbon anodes containing 25% and 30% of 1-1/2-in. and 3/4-in. maximum particle size butts. These butt particles are produced, respectively, by the existing and a newly installed butt crushing and screening system at Mead plant. The study was conducted using laboratory-scale and plant-size anodes. Quality of green and baked anodes was inspected utilizing ultrasonic pulse velocity techniques. The properties of baked anodes were evaluated using core samples.

It was found that anode quality improved as the maximum butt particle size was reduced from 1-1/2 in. to 3/4 in. The improvement of anode quality is characterized by higher ultrasonic velocity, greater strength and lower electrical resistivity. This decrease in resistivity is estimated to result in a 25 to 30% power saving across the anode.

INTRODUCTION

During the last few years, high electrical resistivity of anodes at Mead plant has been noted. Higher electrical resistivity of anodes results in higher voltage requirement and, subsequently, higher power consumption per pound of aluminum. In an effort to determine causes for the higher electrical resistivity of Mead anodes, a series of experiments has been conducted.

A preliminary study at Permanente carbon laboratory* showed that high electrical resistivity anode problem is partly due to poor internal structure characterized by chicken wire and horizontal cracks, and that the quality of anodes improves with smaller butt particle size.

Mead also conducted experiments using anodes made with 25% and 30% butt fractions having maximum butt particle sizes of 3/4 in. and 1-1/2 in. In all cases, the anodes made with the 3/4-in. maximum butt particle size were better than those made with the 1-1/2-in. maximum butt particle size by exhibiting (1) lower electrical resistivity, (2) higher ultrasonic pulse velocity, and (3) fewer internal cracks.

Currently, Mead is in the process of installing a butt crushing and screening system to make a maximum particle size of 3/4 in. In order to determine the quality of anodes fabricated from the smaller butt particles produced with the newly installed butt crushing and screening system, properties of anodes containing 25% and 30% of 1-1/2-in. and 3/4-in. maximum butt particles have been measured using laboratory-scale anodes made at Permanente and full-size anodes produced at Mead.

LABORATORY-SCALE ANODES

Laboratory-scale anodes, 4.5-in. dia. x about 6.0-in. length, were fabricated with green paste taken from lots used to fabricate full-size anodes. The green paste was heated to 150°C in an oven and pressed in a preheated mold under a pressure of 5000 psi. The green anodes were subsequently baked to 1150°C according to the following schedule: 10°C/hr from room temperature to 650°C, 50°C/hr from 650°C to 1150°C and soak for 2 hr at 1150°C. The baked anodes were allowed to cool in the furnace.

Density and ultrasonic velocity were measured on the green and baked laboratory-scale anodes. Additional baked anode properties such as electrical resistivity and oxidation resistance were also determined on core samples. Property data on the laboratory-scale anodes are summarized in Table 1. When maximum butt size was decreased from 1-1/2 in. to 3/4 in., the improvement in anode quality was characterized by higher anode density, higher ultrasonic velocities, lower electrical resistivity, and lower airburn resistance index. Since the laboratory-scale anodes were made under closely controlled conditions, the corresponding property values can be regarded as the best property values obtainable from the plant aggregates.

FULL-SIZE ANODES

Full-size prebaked anodes were made at Mead from a mixture of (1) Santa Maria delayed coke, (2) Mead-Exxon fluid coke, (3) USS Steel Ironton coal-tar pitch, and (4) crushed butts.

---

*Now located at the Center for Technology, Pleasanton, CA

Table 1

Effect of Maximum Butt Size on Properties of Laboratory Scale Anodes Containing 25% Butt Fraction

Butt Maximum Particle Size in.	Anode	No. of Anodes	Density g/cc	UV 10^6 cm/sec V	UV 10^6 cm/sec H	ER 10^{-3} Ω-cm V	AB mg/cm²-hr 550°C	AB mg/cm²-hr 650°C
1-1/2	Green	5	1.592	.264	.270	-	-	-
1-1/2	Baked	5	1.544	.240	.252	6.5	84	253
3/4	Green	3	1.577	.259	.272	-	-	-
3/4	Baked	3	1.552	.255	.262	5.5	79	250

Optimum green anode density was obtained by using an optimum dry coke aggregate composition determined by vibrated bulk density (VBD) and the pitch with an appropriate QI value (1). Ball mill input and output products as well as the fines products were monitored very closely, since all the dust collector catch blends with the fines. At Mead, 75% fluid coke and 25% Santa Maria delay coke are used to fabricate anodes.

For this investigation, more than eight hundred experimental anodes were made using the mentioned procedures, with 25% and 30% butt fraction using a minus 3/4-in. + pan butt particle size in the aggregate. Control groups of the same butt percentages but with a minus 1-1/2-in. + pan butt particle size were also made.

ANODE PROPERTY EVALUATION AND DISCUSSION

Density and ultrasonic pulse velocity of full-size green and baked anodes were measured. Anode density was determined via water submersion method. Ultrasonic pulse velocity through the anodes was determined from the ultrasonic transit time measured using a James "V"-meter and a pair of 150 KHZ transducers (2). As illustrated in Figure 1, a total of twenty-five measurements of ultrasonic transit time were made on each anode: four measurements in the vertical direction, twelve measurements in the width direction, and nine measurements in the length direction. Core samples were also taken from baked anodes to obtain comprehensive property data.

Ultrasonic velocities of the anode are calculated simply by dividing the measured transit time by the distance between the transducers. Frequency distributions of ultrasonic velocity in the vertical, length and width directions for anodes with 25% 1-1/2-in. butt particles and 25% 3/4-in. butt par-

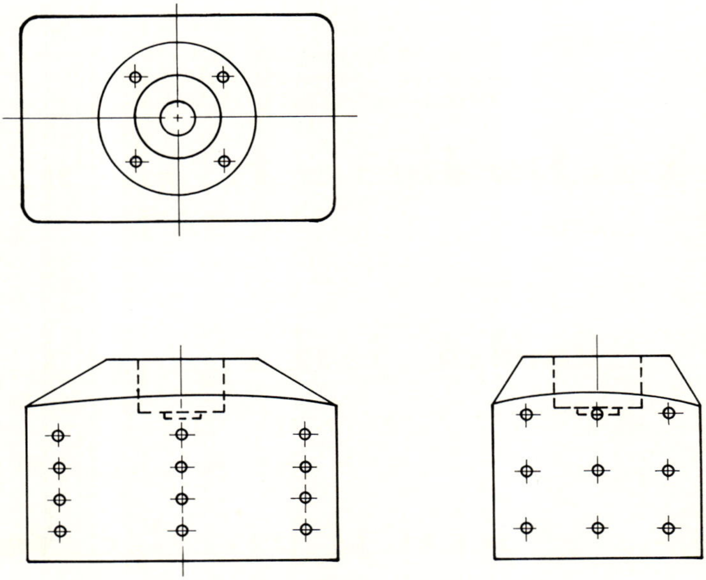

Figure 1. LOCATION OF TRANSDUCERS FOR ULTRASONIC PULSE VELOCITY TEST ON MEAD ANODES

ticles are shown, respectively, in Figure 2 and Figure 3. An anode with good quality contains fewer internal cracks and, hence, is characterized by a high ultrasonic velocity. A group of anodes having uniform quality exhibits a small range of ultrasonic velocity in the frequency distribution plot. Comparison of Figure 2 and Figure 3 discloses that the reduction of butt particle size from 1-1/2 in. to 3/4 in. resulted in a significant increase of anode ultrasonic velocity and a marked decrease of the range of ultrasonic velocity frequency distribution, indicating substantial improvement in overall anode quality.

Density and ultrasonic velocity of the green and baked anodes are given, respectively, in Tables 2A and 3B. While the density of green anodes with 1-1/2-in. butts appeared to be slightly higher than that of green anodes with 3/4-in. butts, the ultrasonic velocity appeared to be unaffected. However, the ultrasonic velocity in the baked anodes increased by 60 to 100% with reducing the butt particle size from 1-1/2 in. to 3/4 in.

As the butt fraction increased from 25% to 30%, the quality of baked anodes containing 1-1/2-in butt particles deteriorated slightly, but the quality of baked anodes containing 3/4-in. butt particles improved. The use of 3/4-in. butt particles not only improved anode quality, but also allowed a larger butt fraction to be used without sacrificing the anode quality.

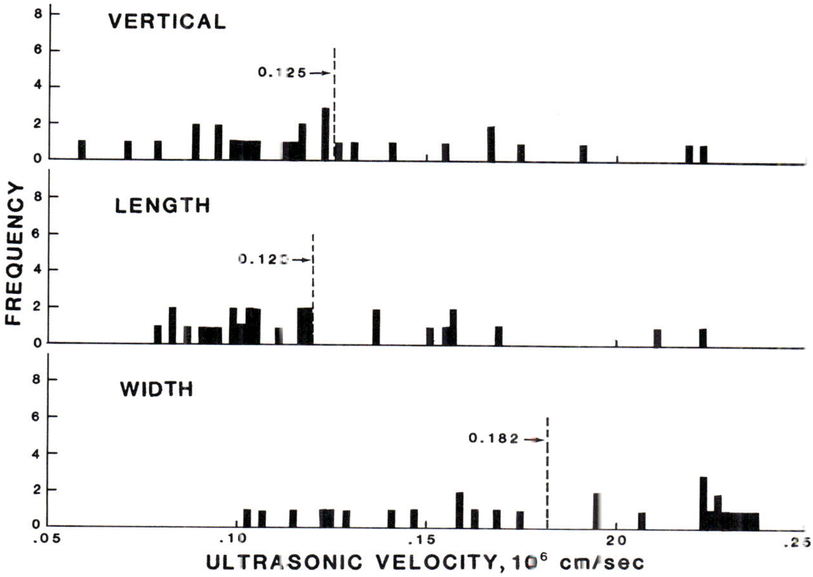

Figure 2. FREQUENCY DISTRIBUTION OF ULTRASONIC VELOCITIES IN ANODES WITH 25%, 1-1/2 IN. MAXIMUM BUTT PARTICLES.

Core samples were taken from the vertical and horizontal directions of the baked anodes. Anode properties determined on these core samples are summarized in Table 3. It is noted that ultrasonic velocity data for core samples from anodes fabricated with 25% 1-1/2-in. butt particles were considerably higher than those measured in full-size anodes and, also, that ultrasonic velocities of these core samples were as high as those of core samples taken from the anodes fabricated with 3/4-in. butt particles.

The discrepancy between ultrasonic velocities of core samples and full-size anodes containing 1-1/2-in. butt particles can be explained. Photographs of the cross sectional view of an anode containing 30% 1-1/2-in. butt particles and an anode containing 25% 3/4-in. butt particles are shown in Figure 4. It is evident that the quality of the anode fabricated with 3/4-in. butt particles was much better than that of the anode fabricated with 1-1/2-in. butt particles. The anode with 1-1/2-in. butt particles contained many severe internal cracks. Ultrasonic velocities measured on full-size anodes substantiated this observation. Because of the severe internal cracks in the anodes with 1-1/2-in. butt particles, core samples taken from these anodes usually broke into several sections along cracks. Since the ultrasonic velocity tests were conducted on short core sections free from the severe cracks, the values obtained were superficially high for the core samples. For the same reason, the values of electrical resistivity and coef-

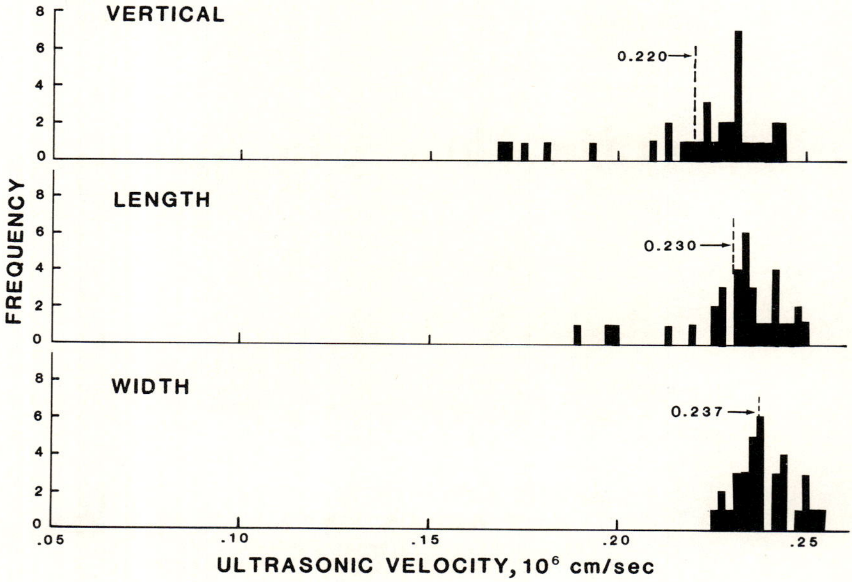

Figure 3. FREQUENCY DISTRIBUTION OF ULTRASONIC VELOCITIES IN ANODES WITH 25%, 3/4 IN. MAXIMUM BUTT PARTICLES.

ficient of thermal expansion reported for the core samples were lower than those expected for baked anodes containing 1-1/2-in. butt particles.

The value of electrical resistivity for anodes containing 1-1/2-in. butt particles has been estimated using an electrical resistivity versus ultrasonic velocity relationship observed for other types of carbonaceous materials and the ultrasonic velocity measured from the full-size anodes. The electrical resistivity of the anodes in the horizontal direction is estimated to be 8.9×10^{-3} ohm-cm for 25% butt fraction, and 9.1×10^{-3} ohm-cm for 30% butt fraction. The electrical resistivity of the anodes in the vertical direction would be approximately 9% greater than the values estimated for the horizontal direction. With the estimated values, we determined that the electrical resistivity of the anode could be lowered by as much as 25 to 30% by reducing the butt particle size from 1-1/2 in. to 3/4 in. Potential power saving which could be obtained from the lowering of the electrical resistivity is estimated to be 0.048 volt/pot for Mead plant.

(A) BUTT SIZE: 1-1/2 in.
BUTT FRACTION: 30%
Many Severe Internal Cracks

(B) BUTT SIZE: 3/4 in.
BUTT FRACTION: 25%
No Severe Internal Cracks

Figure 4. INTERNAL STRUCTURE OF BAKED ANODES

Table 2

Full-Size Green and Baked Anode Properties

(A) Green Anodes

Butt Size, in.		1.50	1.50	0.75	0.75
Butt Fraction, %		25	30	25	30
Density, g/cc		1.598	1.624	1.581	1.589
UV, 10^6 cm/sec	V	0.246	0.250	0.247	0.252
	H*	0.259	0.268	0.257	0.259

(B) Baked Anodes

Butt Size, in.		1.50	1.50	0.75	0.75
Butt Fraction, %		25	30	25	30
Density, g/cc		N/A	N/A	1.570	1.570
UV, 10^6 cm/sec	V	0.125	0.114	0.220	0.234
	H*	0.151	0.146	0.234	0.240

*Average of width and length directions

Table 3

Full-Size Baked Anode Core Properties

Butt Size, in.		1.50	0.75	0.75
Butt Fraction, %		25	25	30
Density, g/cc		1.531	1.504	1.515
UV, 10^6 cm/sec	V	0.236	0.228	0.243
	H	0.237	0.241	0.244
ER, 10^{-3} Ω-cm	V	6.72	6.43	6.00
	H	6.33	6.03	5.98
CTE, 10^{-6} cm/cm (25-100°C)	V	4.34	3.42	3.77
	H	3.75	3.62	3.72
Flexural Strength, psi	V	740	980	1020
Comp. Strength, psi	V	7450	6960	7930
AB, mg/cm^2-hr	550°C	80	90	90
	650°C	250	250	250

ACKNOWLEDGMENT

Many persons from both Permanente laboratory and Mead plant contributed much invaluable assistance and cooperation to this work. However, we would like to express our special thanks to W. H. Goodnow, R. D. Zabreznik, D. E. Hartvigsen and J. E. Daniel for their helpful advice and support, and R. Mora for his assistance in various phases of this work.

NOMENCLATURE

UV	Ultrasonic Velocity
CTE	Coefficient of Thermal Expansion
ER	Electrical Resistivity
AB	Air Reactivity (Airburn Rate)
V, H	Vertical and Horizontal Directions

REFERENCES

1. S. S. Jones, R. D. Hildebrandt, and M. C. Hedlund, "Variation of Anode Performance with Coke Quality," Paper No. A77-97, presented at the 106th AIME Annual Meeting, Atlanta, GA, 1977.

2. T. Weng, "Quality Evaluation of Carbonaceous Materials by Ultrasonic Techniques," Light Metals 1981, AIME, pp. 517-528

REMOVAL OF SOLID BATH FROM SPENT ANODES

J. C. Broad and B. D. Farmer

New Zealand Aluminium Smelters Limited
Tiwai Point
Private Bag, Invercargill,
NEW ZEALAND

The removal of frozen electrolyte from the used anode butt has long caused problems in the industry - from high fluoride exposure and heat stress for potroom operators, to effects on new anode quality and refractory damage in anode baking.

To minimise these problems New Zealand Aluminium Smelters Limited has developed a system to vibrate spent anodes clean. The development of these vibrating machines through prototype stages to the design and construction of a plant to handle anode cleaning for a 244 000 tonne per annum smelter is described.

This plant is now installed and operating at two Comalco managed smelters.

Introduction

New Zealand Aluminium Smelters (NZAS) operates a 244 000 tpa smelter located at Tiwai Point, Invercargill, New Zealand. Spent anodes from the 155 kA pots weigh 220kg and have 70kg of frozen bath and alumina adhering to them.

This bath was manually removed by pot operators using crowbars for coarse cleaning and wire brushes for fine cleaning. This operation was done on the floor beside the pot while the anodes were hot. The bath cleanings were shovelled back into the pot.

The cleaning operation was the most physically demanding and unpleasant aspect of the pot operator's duties and led to high labour turnover and involved constant supervisory effort. In the mid-seventies fine cleaning and shovelling bath back into the pot were removed from pot operators duties. Fine cleaning was undertaken at a chipping and air blast station in the Rodding Room. Bath was processed in a grinding plant and added back to the pot with the alumina.

In the late seventies NZAS resolved to design an anode cleaning system. A design group involving Engineering and Production personnel listed possible anode cleaning techniques including: chemical, flail, pneumatic hammer, pusher and ripper, rotating cutter, ultrasonic, vibration, robots, air blast and water jet cleaning methods.

Design Phase

Most of these ideas were examined in detail and some were given trials. A series of experiments on chemical cleaning were made but the cleaning efficiency was low and the bath could not be reclaimed economically.

Pneumatic hammers were used to clean anodes with limited success. This process was slow, noisy, not well suited to automation and a high maintenance method.

Vibration Cleaning

Out of the remaining ideas the most promising appeared to be vibration cleaning.

Trials were held in 1978 using an anode mounted on a vibrating frame (figure 1) using equipment from a dross recovery plant. These trials confirmed that the vibrating process could clean anodes but that the prototype's geometry and the vibrational amplitude were far from optimum values.

A second prototype cleaning machine (figure 2) was built in 1979 consisting of an anode support frame mounted on four coil springs and two vibrating motors close to the centre of gravity of the anode and machine. Early tests on the second prototype indicated that the vibration cleaning system was effective and mechanically simple and quiet. This was the process for which we had been looking.

Figure 1
First Prototype

Figure 2
Second Prototype

Figure 3
Third Prototype

Figure 4
Production Unit

Anode Rod Endurance Tests

Endurance trials on anode rod assemblies were carried out using dummy weights welded to the anode yoke assembly. Tests using spent butts were not feasible because the carbon was dislodged from the yoke after approximately 20 minutes of continuous vibration. The assemblies tested fatigue fractured through the aluminium weld joining the aluminium rod to the transition joint in times varying from 2.25 hours to 12.5 hours. These test times were equivalent to an operating life of between 3.7 years and 20.5 years for an anode assembly, based on an anticipated vibration cleaning duration of two minutes. The time of vibration was determined from cleaning tests on cold anode assemblies.

Since clamps on the aluminium rod alone were not sufficient to prevent assembly fatigue, a yoke support was designed to securely clamp the yoke just below the transition joint. Endurance tests then gave an assembly a minimum operating life of 36.9 years. These fatigue tests were stopped after 24 hours of continuous vibration with no signs of fracture apparent, indicating that vibration cleaning would have no effect on the working life of the anode assembly.

The vibration time on the production version has been reduced from two minutes to 30 seconds giving an assembly operating life of 148 years minimum.

Endurance tests were undertaken on the copper rod anode assemblies from the Comalco Bell Bay smelter with no fatigue problems detected.

Production Prototype Machines

Following the success of the second prototype a further much longer series of trials were undertaken to determine optimum cooling times, cleaning efficiencies and develop the machine to the stage where production versions could be built to clean 480 anodes per day.

In early 1980 work began on the third prototype machine (figure 3) designed to simulate as closely as possible the final on line installation. The new machine had hydraulic powered rod and yoke clamps to replace the lever operated clamps; larger rotary vibrating motors; a powered overhead conveyor for feeding anode assemblies into the cleaning machine; a raise and lower system using coil springs with hydraulic hold down clamps to steady the machine when the anodes were fed into the machine.

Temperature for Cleaning

Trial cleaning of hundreds of anodes were made to determine the optimum temperature to clean anodes. The results of these tests confirmed anodes were easier to clean with decreasing temperature. Sixteen hours of natural draft cooling (figure 5) and a bath temperature of less than $80^{\circ}C$ is optimum for vibration cleaning.

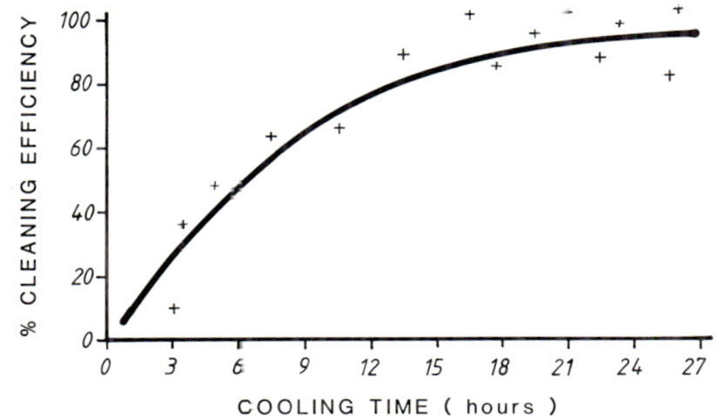

ANODE CLEANING EFFICIENCY Versus COOLING TIME

Figure 5

Sixteen hours cooling time, was used for the design of the present full scale cleaning system. Trials made in 1978 also examined the possibility of cooling anodes on the floor, and force cooling anodes with fans. Floor cooling was rejected because of the problems with cleaning up and the double handling involved. Force cooling was abandoned because of airborne dust, and the high capital cost of the large dust extraction system that would have been required.

Machine Improvements

During these trials approximately 50% of the anodes cleaned had large loose pieces of bath sitting between the front and rear stubs which could not be vibrated off. A trial was made with only one of the two vibrating motors running in an attempt to set up a sideways vibration to dislodge the bath. This was unsuccessful. Also a small amount of bath material tended to bridge between the front or the rear pair of stubs. To remove this bath and reduce the overall cleaning cycle time a hydraulic pusher was designed.

The pusher was operated 20 seconds after commencement of the vibration cycle. The vibration cleaning time at the end of this development had been reduced from 120 seconds to 90 seconds with the pusher operating. The overall machine cycle time was then two minutes and 55 seconds, including loading and unloading.

The coil suspension springs were replaced with rubber air bellows which resulted in a quieter system with much improved vibrational isolation. This change also eliminated the need for the hydraulic table hold down clamps, as the air bellows could also be used as the raising and lowering device.

A D.C. voltage injection system for braking the rotary vibrating motors was also installed to eliminate the alarming elephantine lurching of the machine during the deacceleration phase through the critical frequency range of two to three cycles/second.

As the final part of the development process the frame design for the NZAS production machine (figure 4 and figure 6) was optimized for minimum stress concentration using a specialist computer programme based on finite element theory. This also led to optimization of the machine weight distribution. As a result the final version of the machine is much more compact than the final prototype, and this should result in a longer service life.

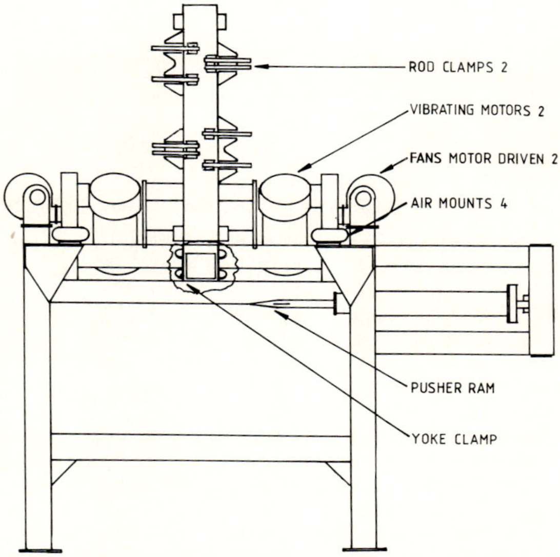

Figure 6 - NZAS production vibration cleaning machine

Production Units

Two types of vibration cleaning machines have been built and are now in regular use.

The anode cleaning system at NZAS uses a fixed machine with the anodes fed into the machine by overhead traversing devices.

The anode cleaning system at Boyne Smelters Limited (BSL) was built before the NZAS system and uses a vibration cleaning machine similar to the last NZAS prototype version. The prototype was built at Sumitomo's Toyo Smelter in Japan to our specifications so that the much larger anodes for the new Comalco managed smelter at Boyne Island in Queensland, Australia could be test cleaned. These tests were successful.

The machines operate as follows: the anode is clamped firmly to the top frame of the machine by three pairs of hydraulic clamps. Two pairs of clamps secure the aluminium rod and one pair of clamps holds the yoke just below the transition joint. Two large rotary vibrating motors are mounted on either side of the anode at the centre of gravity of the combined top frame and anode assembly. The top frame of the machine is mounted on four rubber air bellows which isolate the machine vibrations from the lower support frame. These bellows are also used to raise and lower the top frame. The bellows are inflated after the anode is clamped onto the machine to take tension off the chain linking the anode to the overhead conveyor prior to vibration.

After 15 seconds of vibration (figure 7) the pusher moves across to the centre of the anode. If bath material is still left on the butt the point on the pusher head acts as a stress concentrator rapidly initiating cracking.

Figure 7 - Vibration

Figure 8 - Cleaned Anode

Once at the centre of the butt the pusher kicks backwards between the rear stubs and then forwards between the front stubs. This wiping action is designed to clear any hard bath bridging between the stubs. The pusher then centralizes again and continues to push to the far side of the butt before retracting completely. This action removes any bath too big to vibrate through the gap between front stubs. The effect of the cleaning system can be seen in the reduction in ash in baked anodes shown below (figure 9):

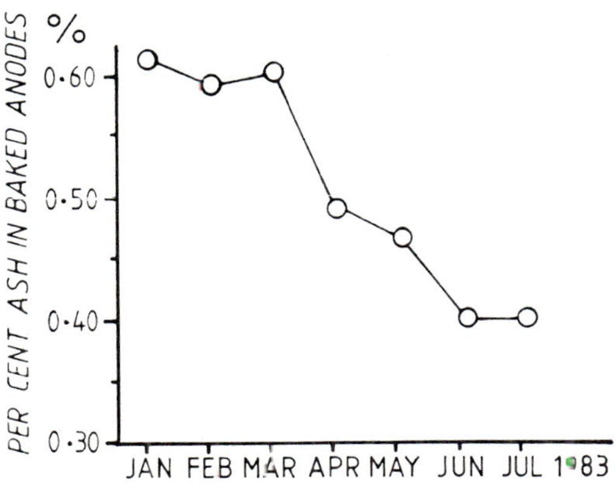

Figure 9 - Ash in Baked Anodes 3 month moving average less raw material contribution.

Plant Layout

New Zealand Aluminium Smelters Limited

The NZAS cleaning plant (figure 10) is highly automated requiring only two men to operate it. Uncleaned anodes are loaded onto an overhead conveyor which runs across one end of the potlines.The length of this conveyor ensures that all anodes have a minimum cooling time of 16 hours. Bath material dislodged from the anodes during the cooling process falls onto sloping concrete walls and into a central aisle where it can be picked up by a front end loader. About five tonne of bath material per potline per day is collected here.

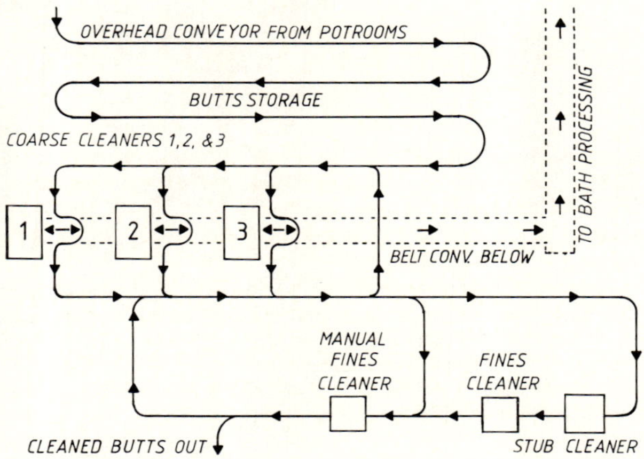

Figure 10 - Layout of NZAS Anode Cleaning Plant

The anodes leave the cooling section and travel by overhead conveyor to the coarse cleaning building. The anodes are first transferred by overhead traversing devices into the vibration cleaning machines. NZAS have three coarse cleaning machines, two of which are designed to handle all the anodes. The other machine is kept on standby.

Bath material vibrated from the anodes falls through a grizzly screen designed to catch any carbon shaken off the yoke onto a conveyor belt. Any carbon pieces are fed down a bypass chute into a portable bin. The bath material is fed by belt conveyors into a storage bin at the bath plant where it is processed by crushing for redistribution to the potlines.

The coarse cleaned anodes are then conveyed from the vibration cleaning machines through a shot cleaner where the remaining bath is removed. The anodes are then conveyed back into the rodding room.

Boyne Smelters Limited

The BSL anode cleaning system is also highly automated. Anode cooling relies on the storage of anodes in pairs in pallets and therefore the overhead conveyor at Boyne is substantially shorter than the NZAS chain. From the cooling storage area the anodes are loaded onto the conveyor and taken to the two coarse cleaning machines.

The anode remains stationary while the vibration cleaning machine moves forward on rails and clamps the rod and yoke. The machine then raises the anode and commences the vibration cycle.

The bath material vibrated off the anodes is fed through a chute onto a grizzly screen and down onto a conveyor belt which feeds the material into a receiving hopper at the Bath Processing Plant.

After leaving the coarse cleaning machines the anodes are fine cleaned in a shot cleaner before continuing the normal circuit.

Design Options

The NZAS and BSL coarse cleaning systems are high capital systems designed for low manning levels in the potrooms and to be operated with the minimum manning required for safety and good housekeeping. Since these plants have been built and operated experience has shown many different layouts are possible. The system of anode cooling is variable.

Bath material could be dropped directly into large truckable skips, eliminating the belt conveying system. This would be justified if the bath crushing plant is either a long distance from the coarse cleaning plant or if more than one conveyor is required.

The choice between the fixed and moveable vibration cleaning machine configurations depends on the site layout and economic factors. A cleaning machine on rails may be more suitable for incorporation into an existing overhead chain system as it eliminates costly chain conveyor modifications.

Developments in Anode Cleaning

Anodes leaving the vibration cleaning process still have some bath on the steel stubs which is presently being removed manually and by the shot blasting process. NZAS have developed an automatic stub cleaning machine to remove this bath and are building a production scale prototype at present. NZAS are also building a manual fine cleaning station to be used when the shot blast system is out for extended maintenance.

A recent study of the pusher operation has resulted in a more efficient pusher head configuration. Studies currently in progress on machine cycle time should lead to reductions in the overall cycle time. At present NZAS are operating the coarse cleaning machines on an average cycle time (figure 11) of two minutes 24 seconds (25 butts/hr).

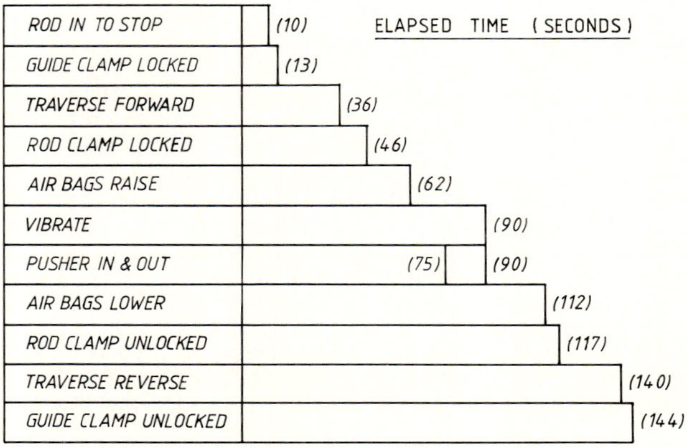

Figure 11 - NZAS Vibration Cleaning Cycle

Conclusion

NZAS has developed and built anode cleaning machines which are effective and reliable. Production units have been installed and are operating in two plants.

Exposure of potroom operators to fluorides, heat and dust has been reduced and accidents associated with manual cleaning eliminated.

Potroom labour turnover has been reduced.

Ash level in Carbon Anodes have been reduced by over 30%.

CHAIRMEN

L.G. Boxall
Martin Marietta Laboratories
Baltimore, Maryland

P.J. Rhedey
Alcan International Limited
Kingston, Ontario, Canada

NORANDA ANODE BAKING FURNACE - DESIGN AND PERFORMANCE

J. Auton, W.F. Crowell, R.R. Ebersole, W.W. Ellison, J. Kaiser

Noranda Aluminum,
P.O. Box 70, New Madrid, Missouri, U.S.A., 63869.

J.G. Peacey and E.J. Chin

Noranda Research Centre,
240 Hymus Blvd., Pointe Claire, Québec, Canada, H9R 1G5.

Abstract

Noranda Aluminum's third anode baking furnace with an initial production capacity of 55,000 tpy of baked anodes was started up in May 1983. The furnace was designed by Noranda Aluminum to satisfy the following requirements:

- low fuel consumption and operating costs
- high productivity and low capital cost

The key design features of the furnace and the firing practice employed are described. Also the operating performance of the furnace is compared with Noranda Aluminum's other baking furnaces.

Introduction

In mid-1980, it was decided to build a third potline and anode baking furnace at Noranda Aluminum's, New Madrid, Missouri smelter. The main requirements for the new anode baking furnace were as follows:

(a) A baked anode production capacity of 55,000 tpy, about 20% higher than each of the two existing bake furnaces.

(b) The furnace had to fit within the limited available space, which was constrained on three sides by the existing bake furnaces and anode storage buildings.

(c) Elimination of the soil settlement problem associated with the two existing in-ground baking furnaces.

(d) Low furnace fuel requirements, flue maintenance and capital costs.

(e) Dry exhaust gas cleaning system.

After thoroughly evaluating the available bake furnace technology, it was decided, late in 1980, to build a furnace using Noranda Aluminum's own experience and know-how. This decision was based on the following:

(a) Development work then in progress on the two existing furnaces indicated that fuel consumption potentially could be reduced to much lower levels, comparable to other available bake furnace designs, by sealing the furnace better and using more-efficient firing practices.

(b) An open-pit furnace of Noranda Aluminum's own design was estimated to have a much lower capital cost than the other bake furnace designs.

(c) An open-pit baking furnace using Noranda's firing practice and Kaiser-type dry scrubbers was familiar technology for Noranda personnel.

Some of the important design features used to meet the key bake furnace requirements are summarized in Table I.

Table I
Design Features to Meet Bake Furnace Requirements

Requirement	Design Feature
Low fuel consumption	- tight furnace construction - reduced weight brick & packing coke/unit anode carbon - low pressure drop across flue - lower heat loss - smaller furnace surface area, better insulation - efficient firing practice
Long flue life	- shorter flue
Good temperature distribution	- flue design for good gas flow distribution
High productivity	- narrower pit and flue
Low capital cost	- simple furnace design - above ground construction - low exhaust volume
Soil settlement	- above ground construction - forced air-cooled bottom

Furnace Sizing

The physical sizing of the furnace was determined, based on:

- individual anode size,
- potroom requirements,
- anticipated firing practice and baking times,
- constraints on maximum furnace length

The key design parameters for the new baking furnace are compared with those for the existing baking furnaces in Table II. The new bake furnace has 46 sections with 7 pits per section compared with 36 sections, each with 6 pits in the existing furnace. An additional pit was added to each section to give about the same weight of baked anodes per section. The pits in the new furnace are about 20% shorter than those in the existing furnace to improve flue life. The pits and flues are also narrower to improve heat transfer to the anodes and permit faster firing and cooling rates. The relative heat capacity per unit of baked anode was reduced by more than 10%. Heat losses from the furnace were also significantly reduced.

Table II
Bake Furnace Design Comparison

Parameter	New	Existing
Anode size, in.	28x28x20.7	31.5x22.5x20.7
Baked Anode production, tpy	55,000	45,000
No. sections	46	36
No. pits	7	6
Pit length, in.	158	196
depth, in.	164	150
width, in.	25	28
Flue width, in.	9	11
Anode packing arrangement	5x5 (25)	8x4 (32)
Weight baked carbon, ton/section	72.5	72
Relative heat capacity/unit baked carbon	0.88	1.0
Firing practice (1980 design)		
No. fires/furnace	3	2
No. sections in fire group	5-6	7
No. cooling sections	6-5	6
Packing/unpacking sections	4	5
Firing time/section, h	24-28	18-22

The new baking furnace was designed to be able to operate with three 5 or 6 section fires but the potroom baked anode requirements are currently being satisfied using only two 6 section fires.

Flue Design

Once the basic dimensions of the new bake furnace had been established, a considerable amount of modelling work was undertaken to design a flue baffle arrangement to give the required combination of good gas flow and hence temperature, distribution and low pressure drop. Water flow modelling tests were carried out on models (1:12 scale) of several flue baffle arrangements, using black ABS particles as tracers. Potassium permanganate solution was also injected to simulate the flow of hot combustion gases from the burners. A typical photograph from a water modelling test on one of the flue designs is shown in Figure 1.

↓ FLOW

Figure 1 - Photograph of Flue Modelling Test

The estimated flue gas flow distribution in the new furnace is compared with that in the existing baking furnaces in Table III.

Table III
Flue Gas Flow Distribution Comparison

Furnace	Sectional Flue Gas Flow Distribution, %		
	Top	Middle	Bottom
Existing	15	65	20
New	15-20	55	25-30

The new flue design gives a greater gas flow to the bottom section of the furnace, which has always been the coldest section. Measurements also showed that the pressure drop across the new flue was 15% lower.

Furnace Construction

An above-ground furnace was chosen primarily because of groundwater problems due to the close proximity of the Mississippi River. Also excessive heat loss from the existing in-ground furnaces had caused severe differential soil settlement. To prevent this in the new furnace, Rust International designed a special forced-air cooled concrete bottom slab.

Furnace Performance

The new baking furnace was started up in late May 1983 and has proved so far to be a very easy and efficient furnace to operate. To-date the main firing practice employed has been a six-section fire similar to that described in an earlier paper([1]). This fire, Figure 2, is composed of four post-bake sections, two burner sections and four preheat sections. The last preheat section and the section ahead of the exhaust manifold are both covered with a fabric cover to reduce air inleakage. The furnace draft is controlled on the PH4 section at about 0.3-in. W.G. on the inside flues.

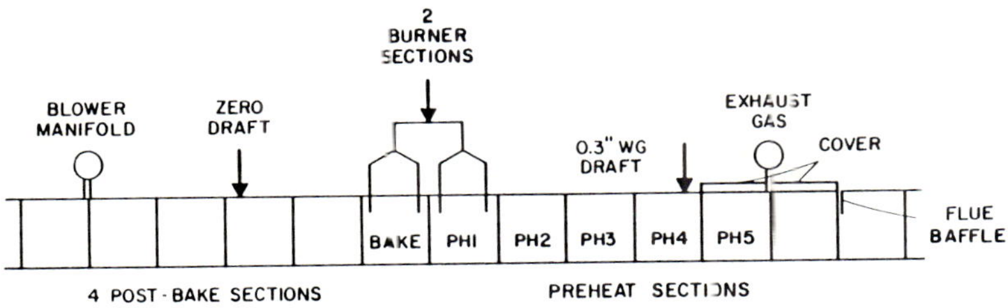

Figure 2 - Present fire arrangement with furnace covers and blower manifold.

The monthly average fuel consumption for the furnace is given below:

June 1983	2.7 million Btu/ton baked anode
July 1983	2.2 million Btu/ton baked anode
August 1983	2.0 million Btu/ton baked anode
September 1983	1.9 million Btu/ton baked anode

After the bake-out in June, the fuel consumption has been averaging about 2.0 million Btu/ton baked anode despite having to idle on 4 section fires for several weeks because of high baked anode inventories. This fuel consumption is significantly lower than the average fuel consumption on the two existing furnaces of about 2.8 million Btu/ton baked anode, using similar firing practices.

Conclusions

Noranda Aluminum has designed and built a modern, above-ground, open-pit anode baking furnace. Based on the performance to-date, using energy-efficient firing practices, this type of furnace is probably the most cost-effective way of baking anodes.

Acknowledgements

The authors wish to acknowledge the important contributions made by the many Carbon Operations and Engineering Department personnel at Noranda Aluminum, Inc. involved in the construction, start-up and operation of the new anode baking furnace. Appreciation is also expressed to Rust International, Birmingham, Alabama, which performed the detailed design and engineering and to Noranda Mines Limited for permission to publish this paper.

Reference

1. Peacey, J.G., Auton, J., and Kaiser, J., "Reduced Fuel Consumption in Noranda Aluminum's Open-Pit Baking Furnaces", Light Metals 1983, (ed. by E.M. Adkins), AIME, New York, 1983, p.777-789.

MATHEMATICAL SIMULATION OF A HORIZONTAL FLUE RING FURNACE

R.T. Bui, E. Dernedde*, A. Charette, T. Bourgeois

Université du Québec, Chicoutimi, P.Q., Canada G7H 2B1
*Alcan International Ltd., P.O.B. 1250, Arvida, Canada, G7S 4K8

A mathematical model is developed in which the horizontal flue ring furnace is treated as a continuous counterflow heat exchanger. The model integrates the equations of heat transfer between flue gas and solids along the entire length of the furnace, and the heat conduction equations within the solids (bricks, coke and anode). Infiltration and volatiles are treated by appropriate sub-models. Simulation runs are made which succeed in reproducing the operating conditions of the Arvida ring furnaces.

Introduction

A mathematical model for the horizontal flue ring furnace used in the baking of carbon anodes at the Arvida Works of Alcan Smelters and Chemicals Ltd., Arvida, Canada, has been developed for the purpose of analyzing the behavior and performance of the furnace in response to various parameter changes. Published literature on the modeling of the overall process in the horizontal flue ring furnace is scarce [1]. The purpose of this work is to develop a computer model not too cumbersome yet capable of taking into account the numerous aspects of the baking process, the most important of which being the air infiltration, the draft control, the release of volatiles and their burning.

The model

The model is based on the assumption that the furnace is a continuous counterflow heat and mass exchanger, where the gas flows in flues from right to left (Fig. 1), and the solids " flow " in pits from left to right. The gas flows at changing velocity determined by its mass, pressure and temperature. The solids " flow " at constant velocity equal to the length of one pit divided by fire-step time. This calls for a balance equation for the gas to wall exchange.

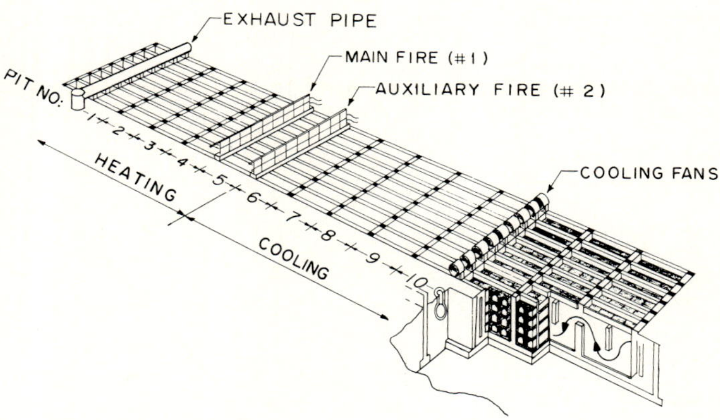

Fig. 1 A three-dimensional view of the ring furnace

This balance equation is written for a control volume extending from the top of the uppermost anode to the bottom of the lowermost anode, and from the centerline of a flue to the centerline of the adjacent pit, and of length δx along the furnace. This yields:

$$\dot{m}_g C_{pg} \frac{\partial T_g}{\partial x} + \alpha \frac{\partial \dot{m}_g}{\partial x} C_{pg} (T_g - 300) - 2h_T R(T_g - T_w) + \dot{m}_f H_f + \alpha\beta\dot{m}_v H_v = 0 \qquad (1)$$

$\dot{m}_g$ = gas mass flow in flue [kg/s]

C_{pg} = specific heat of gas [J/kg.K]

T_g = gas temperature [K]

300 = ambient air temperature [K]

h_T = total heat transfer coefficient between gas and wall (convection and radiation) [W/m².K]

R = depth of pit [m]

T_w = wall temperature [K]

$\dot{m}_f$ = fuel injection per unit furnace length [kg/m.s]

$\dot{m}_v$ = volatiles (CH_4, H_2, tar separately) release rate per unit furnace length [kg/m.s]

H_f,H_v = heat of reaction of fuel and of volatiles [J/kg]

α = binary variable, 1 for heating zone, 0 for cooling zone

β = percent of volatile release that combusts.

The heat thus exchanged from the gas to the brick wall (in the heating section, or conversely in the cooling section), is then transferred through the three solids (brick, coke, anode) by conduction. This calls for a set of equations describing the two-dimensional heat conduction at different nodes inside and on the face and interface of the solids. These equations are of the type:

$$T_o^{t+\Delta t} = T_o^t + \frac{\Delta t}{\rho C_p} \cdot \left[\frac{K_1}{\Delta z^2} (T_1^t - T_o^t) + \frac{K_3}{\Delta z^2} (T_3^t - T_o^t) + \frac{K_2}{\Delta y^2} (T_2^t - T_o^t) + \frac{K_4}{\Delta y^2} (T_4^t - T_o^t) \right] \quad (2)$$

written for an internal node as represented in Fig. 2. The conductivities are taken as arithmetic means of the two values at the two nodes concerned.

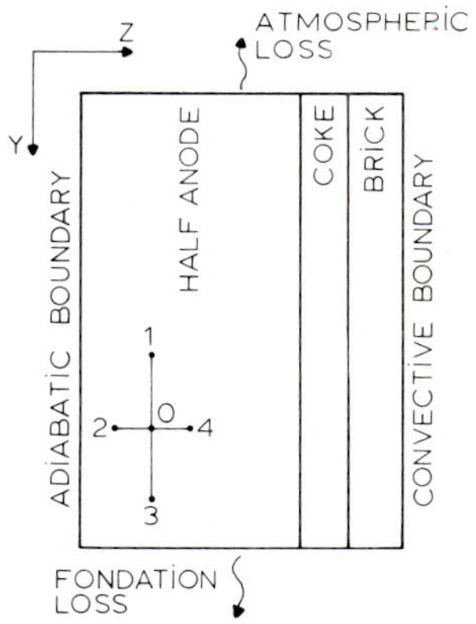

Fig. 2 The control volume

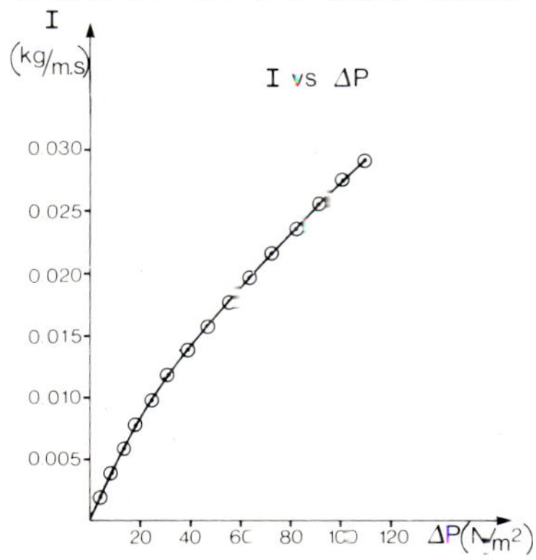

Fig. 3 Infiltration vs draft

For nodes on the convective boundary (brick wall), the K_4 term is replaced by a convection term. For nodes on the adiabatic boundary (anode centerline), the K_2 term simply vanishes. For nodes along the top boundary, the K_1 term is replaced by a term representing the heat loss to the atmosphere. For nodes along the bottom boundary, the K_3 term is replaced by a term representing the heat loss to the foundation. These loss terms are supplied by a separate calculation of losses performed on the furnace, then introduced into this model as a curve in order to avoid making the model unduly cumbersome.

Equation (1) together with the equations (2) constitute the main body of the model. Using appropriate boundary conditions, its solution yields the temperature of the flue gas and of the different solids at different locations, along the entire length of the furnace.

The solution is performed by Euler explicit method using finite differences. The criterion for choosing the solution method is speed of execution.

Infiltration and draft

Infiltration and draft are related. To obtain this relation, we proceed as follows. Starting with a set of experimental data on gas mass flow profile along the furnace $\dot{m}_g(x)$, and a set of experimental data on draft profile along the furnace $\Delta P(x)$ we differentiate to obtain the infiltration profile:

$$I(x) = -\frac{d\dot{m}_g(x)}{dx}$$

then plot I(x) versus $\Delta P(x)$ to have the experimental curve:

$$I = f(\Delta P) \tag{3}$$

which is then curve-fitted with a polynomial (Fig. 3). Thus knowledge of draft ΔP will give infiltration I. In order to determine ΔP, we write a momentum balance for the gas in a slice Δx of the flue, then use relation (3) and this momentum balance alternatively to calculate ΔP for each slice of the furnace in an iterative manner, all the while insuring that ΔP just behind the fires (pit No. 5) must pass through zero as required for good operation of the furnace.

The volatiles

As for volatiles, two problems must be tackled: volatile release from the anodes, and volatile burning. An approach to the former is presented here, while the latter requires trial-and-error to determine the percentage of volatiles actually combusted.

Laboratory analysis provides us with the percentage volatile release for each species of volatiles (methane, hydrogen and tar) in kg volatile per kg anode as function of anode temperature T_a. Multiplied by the mass of anodes per meter of furnace length, this yields the volatile release in kg volatile per meter of furnace. Such a curve $m_v(T_a)$ when differentiated gives $\partial m_v/\partial T_a$. On the other hand, starting from a typical anode temperature profile $T_a(t)$, we differentiate to obtain $\partial T_a/\partial t$, and a multiplication yields

$$\dot{m}_v = \frac{\partial m_v}{\partial T_a} \times \frac{\partial T_a}{\partial t} = \frac{\partial m_v}{\partial t} \qquad \left[\frac{\text{kg vol}}{\text{m.h}}\right]$$

The same operation is performed three times, for methane, hydrogen and tar respectively. Fig.4 presents the three resulting curves giving the release rate of each of three species of volatiles as function of anode temperature.

As for volatile burning, we assume that combustion occurs in the flue and not in the solids, and we adjust the parameter β so as to obtain the

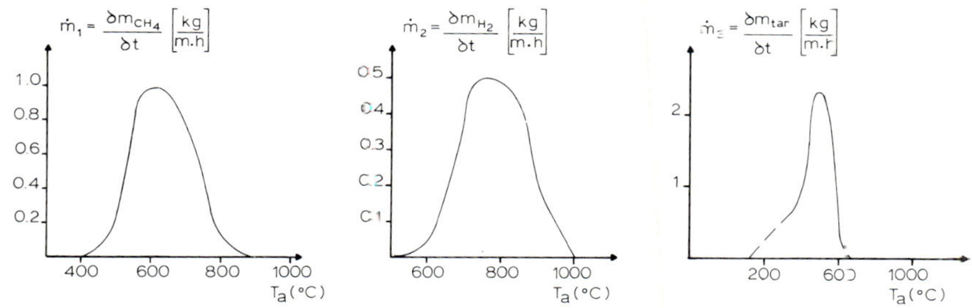

Fig. 4 Release rate of methane, hydrogen and tar as function of anode temperature

right exhaust gas temperature which is $T_g(0) = 740$ K. At each point along the furnace, it is also verified to see if gas temperature is high enough to insure volatile combustion, knowing the ignition temperatures to be 848 K for hydrogen, 903 K for methane and 695 K for tar, the latter being an average. It turns out that this condition is always satisfied, gas temperatures at the points concerned being consistently higher than the ignition temperatures mentioned. As a result, it is found that volatile combustion is spread from pit 1 to pit 4, and that $\beta = 0.78$, meaning that in total, 78 percent of the volatiles released actually combust.

Procedure for computation

First and foremost, we must bear in mind that the thermal properties of the gas as well as of the three solids are variable. Thus in equations (2), specific heats C_p and conductivities K vary from one solid to another and also within the same solid as temperatures vary. In equation (1), h_T, which is the sum of convection heat transfer coefficient and radiation heat transfer coefficient, $h_T = h_c + h_r$, varies with gas flow, gas composition and gas temperature. Also in equation (1), C_{pg} varies with gas temperature and composition.

The above is to say that some preliminary calculation needs to be done before we can enter the model itself. We proceed as follows:

1- The infiltration sub-model is solved first. Starting from an assumed profile of gas temperature and a known " target " value (gas flow at entrance of cooling zone, or at burners, or at exhaust), we obtain the gas flow profile $\dot{m}_g(x)$ and the draft profile $\Delta P(x)$ along the furnace.

2- Then, starting from $\dot{m}_g(x)$ thus obtained, calculate h_c using the Dittus-Boelter correlation. Starting from an assumed value of $\dot{m}_f(x)$, calculate h_r. Add the two to obtain h_T. Also, C_{pg} is calculated based on the concentration of CO_2, H_2O, O_2 and N_2 at each slice along the furnace.

3- Input the values of h_T and C_{pg} thus found into the model itself in order to solve equations (1) and (2) from one end of the furnace to the other and obtain a new gas temperature profile $T_g(x)$ and solid temperature profiles $T_w(x)$, $T_c(x)$, $T_a(x)$ where subscripts w, c, a are for wall, coke and anode respectively. Since we start from one end and move toward the other, this part of the computation is referred to as " shooting " . Although it is not strictly necessary to do so, we used to start from left, using the initial solid temperature 50°C and the exhaust gas temperature 467°C as boundary values and in doing the " shooting " , we aim at the known values of maximum gas temperature and maximum baking temperature. Air inlet temperature at the right end of furnace is also a useful control target. Shooting is done by adjusting the fuel flow $\dot{m}_f(x)$ so as to reach the known targets. Also, the value of β is adjusted the same way to find out the percentage of volatiles burned.

4- Finally, compare the new $T_g(x)$ profile found in step 3 with the assumed $T_g(x)$ profile used in step 1. If they differ, go back to step 1. Otherwise, computation ends. Usually, no more than two such iterations are needed.

The final product of this exercise is a temperature profile along x for the gas and for each solid, and also a two-dimensional isothermal map at each section along the furnace.

Results

Fig. 5 presents the temperature profiles along the furnace for the wall and for the anodes, calculated and experimental. Temperature profiles for the gas, the brickwork, the coke are also available from the model but are not reproduced here, to avoid overloading the figure. Anode temperatures are taken at the geometrical centre of the pit.

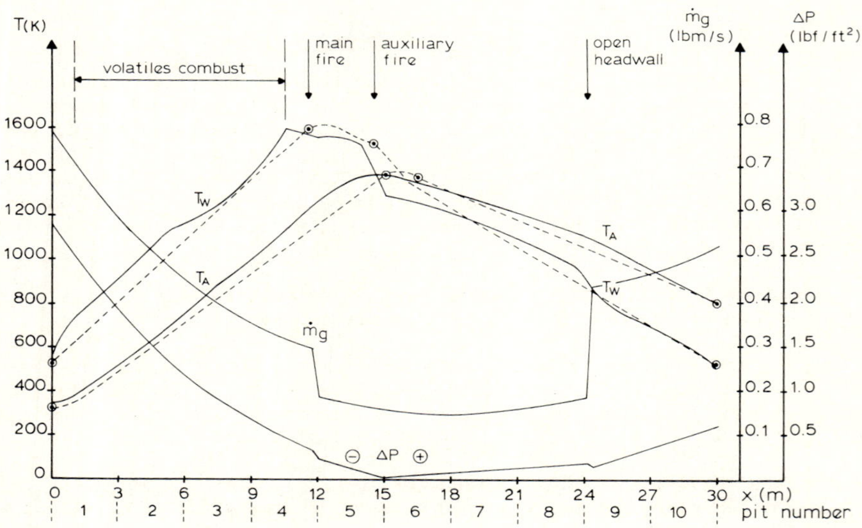

Fig. 5 Profiles for wall and anode temperatures along the furnace, calculated versus experimental. Experimental data come from several sets of measurments, important data points are encircled.

It can be seen that the calculated profiles compare reasonably well with the experimental ones. Note that the experimental data used for comparison in Fig. 5 come from more than one set of measurements but they are all consistent with one another.

Several simulations are now made, some of them are of a " theoretical " nature such as lengthening or shortening the heating and the cooling zones, while others reproduce the physical experimentations which have actually been performed on the Arvida furnaces in the recent past. The latter series includes simulation of the changing of baking temperature, the changing of cooling air flow, the reduction of infiltration by use of pit covers, the changing of fire-cycle time, and the injection of water into the flues. In all these tests, the calculated results agree with the measurements from fairly well to very well.

The last test, concerning water spray cooling, is presented here partly because the results of the experimental investigation by Holdner and Proulx are available to the readers [2]. We inject the same flow of water as used in the investigation i.e. 38 liters per hour per flue. The anode temperature profiles found are presented on Fig. 6 together with the experimental results by Holdner and Proulx. It can be seen that they are in very good agreement. Other elements of comparison (fuel flow, heat transfer coefficient, specific heat of gas) also agree although not mentioned on the figure. Notice that the simulation shows the changes in anode cooling rate the same way they have been measured experimentally. In particular, the anode temperature difference between air cooling and water spray cooling, experimentally found to be 140°, is given as 136° by the simulation.

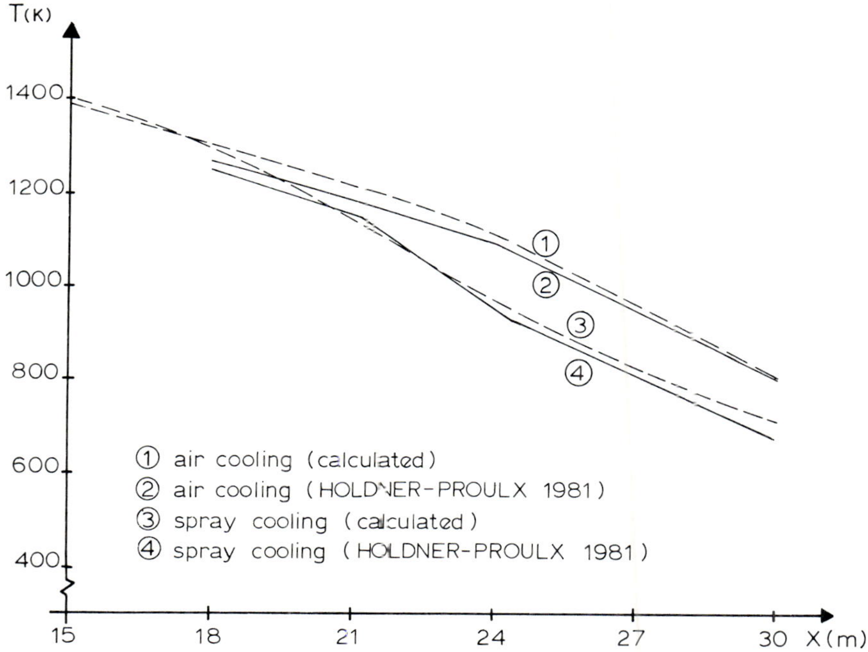

Fig. 6 Profiles for anode temperatures in the cooling zone, with air cooling and water spray cooling, calculated versus experimental. Experimental data comes from Holdner and Proulx's investigation.

Conclusion

It is believed that the model proposed is simple enough to be useful yet elaborate enough to be representative of the process, provided the necessary data be available so that appropriate calibration can be made. It constitutes a good tool for process evaluation as well as for process improvement.

Acknowledgements

The authors express their appreciation for the numerous constributions from Messrs. André Proulx and Ray Peterson of Alcan Smelters and Chemicals Ltd.,Arvida, and Mr. Serge Lavoie of Alcan International Ltd.,Arvida. Special acknowledgement is due to Mr. Donald N. Holdner of Alcan Products Ltd., Arvida, who was instrumental in starting this project and contributed to it in many ways. Financial support comes from NSERC of Canada, Alcan and the Fondation de l'UQAC. The authors are grateful to Alcan International Ltd., Arvida, for authorizing the publication of this work.

References

[1] Keller F. and Disselhorst J.H.M. " Modern Anode Bake Furnace Developments " . Lightmetals 1981, pp. 611-621.

[2] Holdner D.N. and Proulx A.L. " Water Addition to the Cooling Sections of Horizontal Flue Ring Furnaces " . Lightmetals 1981, pp. 597-609.

OPERATIONAL EXPERIENCE WITH COLD RAMMING PASTE RHEINFELDEN 1AK

A.J. Franke L. Förster A. Sturm

Aluminium-Hütte Rheinfelden GmbH
Friedrichstrasse 80
7888 Rheinfelden
W-Germany

Workplace measurements during potlining with cold ramming paste RHEINFELDEN 1AK show concentrations of

- polyaromatic hydrocarbons below 40 $\mu g/m^3$

- benzo(a)pyrene below 5 $\mu g/m^3$

The present paper outlines how the quality of seams and joints influences potlife. The procedures of tamping, preheating and first weeks' operation of the pot are evaluated. Conclusions are drawn about what to do and what to avoid in order to obtain a good potlife. Results from the Rheinfelden smelter are communicated.

1. Protection of workers' health

For decades hot ramming paste has been applied in potlining with good operational success.There is no evidence that occupational diseases have resulted from this practice.

The binders in hot pastes are usually based on coal tar pitch. Processed at temperatures between 100 and 160 °C (210 to 320 °F), the different types of hot paste emit Polyaromatic Hydrocarbons (PAH). Some of these PAH compounds like benzo(a)pyrene (BaP) are considered to increase the risk of cancer if a certain concentration is exceeded. Others are harmless. It is generally accepted that no risk is involved if, during the working shift, the atmosphere contains less than 40 $\mu g/m^3$ Polyaromatic Hydrocarbons resp. less than 5 $\mu g/m^3$ Benzo(a)pyrene. This means that by keeping the PAH below 40 resp. the BaP below 5 $\mu g/m^3$, the worker is on the safe side. It does not imply that beyond these values, health is in jeopardy.

There are two steps in qualifying a ramming paste as to how much PAH it emits. In a laboratory experiment, the paste is kept at processing temperature in a flask with a constant air stream passing through. The concentration of PAH in the air stream is measured. This experiment is valuable in comparing the PAH-emission of different binders. It is used in the developing stage of the paste.

The second step is to measure the PAH exposure of the worker when using the paste in potlining. All figures given in this paper refer to this exposure test. The procedure of sampling and analysis are described by STEINEGGER and GLAUS (1). The worker carries a battery driven sampler during his shift. The analysis of the pitch volatiles sampled indicates to what PAH-concentrations he has been exposed. Table 1) exhibits the exposure of workers during the tamping of joints and seams of a pot relined in the potroom in the usual manner. No special suction equipment was used.

Table 1
Exposure of pot repair workers during tamping with hot resp. cold paste

Compound	Commercial Hot Paste processed at 145°C(290°F)		Cold Paste RHEINFELDEN 1AK processed at 20°C (70°F)	
	joints($\mu g/m^3$)	seams($\mu g/m^3$)	joints($\mu g/m^3$)	seams($\mu g/m^3$)
Phenanthrene	15	26	3.5	5.3
Anthracene	1.1	3.0	0.3	0.4
Fluoranthene	16	36	3.3	5.5
Pyrene	9.2	22	1.6	4.2
Benzo(a)anthracene	5.8	12	0.9	1.4
Chrysene	4.9	10	0.9	1.6
Benzo(b)fluoranthene	1.9	4.4	0.3	0.6
Benzo(k)fluoranthene	1.3	2.4	0.1	0.2
Benzo(e)pyrene	2.4	5.2	0.2	0.5
Benzo(a)pyrene	1.8	4.8	0.2	0.3
Benz(g,h,i)perylene	0.4	0.6	0.1	0.1
Total PAH	59.8	126.4	11.4	20.1

Table 1) shows that PAH exposure below 40 μg/m³ and BaP exposure below 5 μg/m³ is possible when using the cold paste. Using hot paste, these limits can be exceeded.

2. General aspects of potlife

In most aluminium smelters, individual pots attain a total potlife considerably above average. This happens in potlines where all pots are of the same type. Investigating what distinguishes these long-life pots from the rest, normally little is found: cell construction, design of lining and lining materials are in many cases the same for all of them. Thus, it has to be concluded that the pot per se, its design and its construction materials, can be good for a potlife of e.g. 10 years. However, it depends on lining, start-up and operational procedures whether this ultimate is attained.

One type of pot failure is leakage through the sidewall. Its main cause are strong metal movements which erode the side lining. These movements are induced by electro magnetic forces. The maximum erosion occurs at the level of the metal to bath boundary. If a pot by its very design is prone to sidewall leakage, then its total potlife will be much shorter than what could be expected from the quality of the potlining materials. Within certain limits, adaption of operational parameters like increasing the metal height or decreasing the cryolite ratio helps prevent erosion. The best remedy is to improve the magnetic balance of the pot. Details are given by BLANC and ENTNER (2). The magnetic calculations must not be applied isolatedly in pot design. They are part of a general magnetic, energetic and mechanical model. BOSSHARD, KNAISCH, SCHMIDT-HATTING and BLANC (3) describe how this mathematical model has been successfully applied when designing the new 180 kA pots in RHEINFELDEN.

The other and more common failure is the early deterioration of the cathode carbon. It shows up as iron contamination of the metal and/or as excessive upheaval of the bottom. Both symptoms require the existence of cracks in the cathode carbon. Avoiding crack formation in early pot age is a prerequisite to getting the pot to operate the whole possible lifespan of e.g. 10 years.

Cracks result from stresses. The first major stresses in the cathode carbon occur when preheating the pot for start-up. The theory of thermal stresses in elastic bodies is outlined by TIMOSHENKO and GOODIER (4). No stresses exist in an unrestrained bottom block with different temperatures on upper and lower side as long as the temperature drop is the same inch by inch inside the block. If, however, the temperature varies nonlinearly inside the block, tensile stresses arise. In case of a large change of the temperature gradient over a short distance, disrupture of the carbon may result.

To prevent this in practice, slow preheating is applied to the pot with an uniform temperature increase avoiding sudden heat waves passing through the carbon. Hot spots are dangerous.

The next stress situation appears after start-up when sodium is deposited electrolytically and diffuses into the carbon. SØRLIE and ØYE (5) give a comprehensive review of the reactions in cathode carbon and their impact on potlife.

Sodium diffused into the cathode carbon causes swelling. The least deformation is experienced by graphite, followed by calcined anthracite, then metallurgical coke and petrol coke. In most smelters, calcined anthracite is the standard material for potlining.

The sodium expansion may be handled mathematically like the thermal expansion: little harm is done to the unrestrained block if the sodium concentration varies linearly from the top to the bottom. Sudden changes of the sodium concentration inside the block cause severe stresses and may lead to disrupture.

As sodium deposition during aluminium electrolysis is inevitable, the best solution is to reduce the deposition rate to a minimum. The carbon has then more time to dissipate the sodium and the concentration gradient is more even.

Under conditions of electrolysis, sodium is generated by the reaction

$$Al + 3\ NaF \quad = \quad 3\ Na \ + AlF3$$

From this equation the influence of the bath ratio can be seen: The higher the AlF3-excess (i.e. the lower the cryolite ratio), the less sodium is generated. WILKENING and BUSSE (6) show that the Na-generation increases with the cathodic current density and so does the swelling. Sodium deposition augments with temperature. However, this is counteracted by the fact that sodium penetration into the carbon cathode decreases with increasing temperature.

Operating the pot from start-up with an acid bath (e.g. 1.3 CR, resp. 6 % AlF3-excess) is a way to prevent early degradation of the cathode carbon by sudden sodium swelling. At the same time it has to be ensured that the current distribution in the whole cathode is as even as possible. No parts of the bottom within the confines of the anode projection should be sealed by precipitated bath constituents. Overfeeding with alumina has to be carefully avoided in the first few weeks after start-up. For the first six weeks, the anode effect frequency should be kept as high as two a day.

The use of graphitic blocks for potlining as described by AESCHBACH and FRIEDLI (7) reduces the risk of swelling-induced cathode deterioration because graphite by itself is more resistant to sodium attack.

From potline experience we know that the big cracks in the cathode leading to iron contamination and termination of cell life are not something that shows up in the first months after start-up. It normally takes at least three years. These big crevices are not the immediate consequence of preheating or sodium swelling. They usually are caused by heaving of the bottom as a result of large crystallized deposits inside the lining. Their thickness is of the order of several inches as may be seen when digging out such a pot. Chemical analysis shows that NaF is the main compound. Yellow films of Al4C3 can be seen in abundance. They are formed by the reaction

$$4\ Na_3AlF_6\ (liq.) + 12\ Na\ (in\ carbon) + 3\ C = Al_4C_3\ (sol.) + 24\ NaF\ (liq.)$$

The NaF solidifies to columnar crystals which can heave the carbon.

After about 3 months of cell operation, sodium has diffused all over the cathode carbon. It depends on the availability of liquid cryolite whether the NaF-forming reaction takes place. Under conditions of electrolysis, sodium impregnated carbon is quickly wetted by liquid bath. The more cracks are present inside the carbon, the easier the electrolyte penetrates and reacts. This explains why it is so important to avoid crack formation during preheating and during the first few weeks after start-up of the pot.

3. The role of joints and seams

In most smelters the standard potlining is a refractory filling up to the collector bar level, then calcined carbon blocks interconnected by rammed paste or glue and side walls lined with calcined blocks. The gap between side- and bottom blocks is filled with rammed paste. More details can be seen from Fig. 1.

It is not the task of the joints to weld the blocks together. The joint paste has to fill the space between the blocks and has to prevent metal and bath from penetrating the lining. Once buckling and warping of the cathode sets in, the pot is due for termination irrespective of how good the blocks stick together: If the joints do not break, then the blocks must.

The situation is similar for the seams. Again, the main task is to prevent metal and bath from penetrating the lining. In addition, the seam is subject to a certain metal/bath movement and has to be hard enough to withstand abrasion as well as the sideblock itself.

In both cases an excellent compaction of the paste after tamping is of paramount importance. This requires a well formulated paste and a very tight and uniform ramming.

RHEINFELDEN

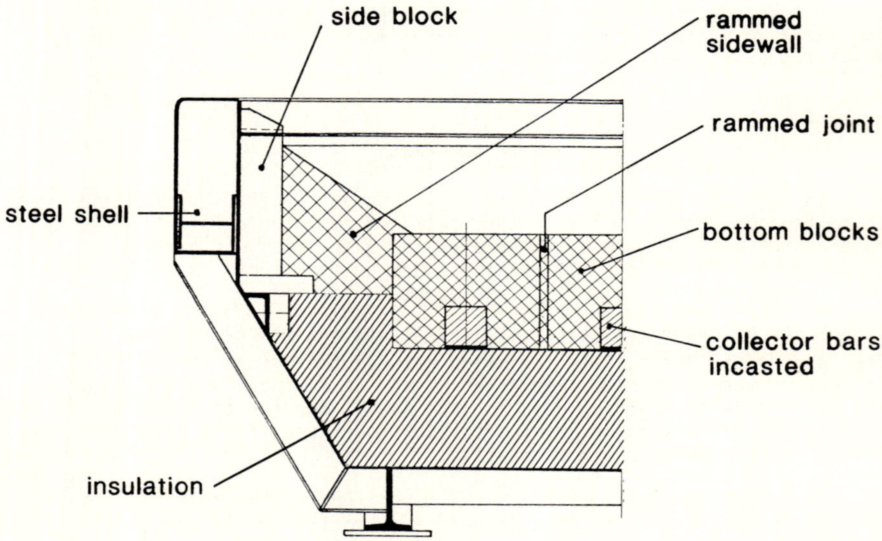

Fig. 1 Section of an 100 kA-pot in RHEINFELDEN

4. Quality requirements for the ramming paste

Complying with today's hygiene standards, as outlined in paragraph 1, is one requirement. However, this is not enough.

With reference to paragraphs 2 und 3, the paste has to meet the following requirements:

- Minimum volume change during baking, no swelling.

- Minimum porosity after baking.

- Good abrasion resistance.

The paste is composed of dry aggregate and pitch. The aggregate has to be sized for maximum packing density. The amount of pitch has to be minimized, just sufficient to cover all particles of the aggregate. During baking about 50 - 60 % of the pitch evaporate and the remaining 40 - 50 % form so-called binder-coke bridges between the aggregate particles.

The porosity of seams and joints is important for the cryolite penetration. In the absence of cracks, this is how electrolyte gets into the lining. The lower the porosity, the longer it takes. Highly compact seams and joints of low porosity are a prerequisite for attaining a good potlife.

Selecting the right aggregate material means optimizing between sodium and abrasion resistance. Petroleum and metallurgical coke have a poor sodium resistance, graphite is poor in abrasion. There remains anthracite. The question is gas- or electrically calcined material. The latter may be superior in terms of average sodium resistivity, but usually its quality it is rather inhomogeneous. Evaluating all aspects, we consider a good gas-calcined anthracite to be the best choice.

Cold Ramming Paste RHEINFELDEN 1AK is based on gas calcined anthracite from a West German mine. In this anthracite the volatiles are below 0.5 %, the real density as measured in xylene is 1.80 ± 0.05 g/cm³, the ash is 3.5 ± 0.5 %, the sulphur content is 0.65 ± 0.05 %, the iron 0.20 ± 0.07 % and the silicon content is 0.7 ± 0.1 %. The residual hydrogen is 0.13 ± 0.03 %. The calcining temperature is 1250 to 1290 °C. The anthracite grain after calcination is hard, its resistance to wear and tear is high.

The Cold Ramming Paste RHEINFELDEN 1AK has the following properties after baking:

- Shrinkage below 0.2 %.

- Total porosity 15 to 17 %.

- Bulk density 1.45 g/cm³.

- Abrasion resistant, due to hard anthracite.

The formulation of a paste originates from the desired properties of the seams and joints. For cold paste this can be done straight, as there is no further processing step between production and ramming of the paste. Hot mix, on the other hand, needs some excess pitch to compensate for evaporation losses during heating up. A seam or joint tamped with hot paste, therefore, is likely to enter the baking process with more pitch than needed for aggregate wetting. The result after baking is additional porosity. This explains why cold mix usually yields a better compaction than hot mix.

The tamping of seams or joints with cold paste RHEINFELDEN 1AK is done at room temperature out of the original package. This guarantees that the seam or joint enters baking in optimum composition. Good tight tamping, of course, is necessary.

5. Potline Performance

The 100 kA sidework-pots in RHEINFELDEN exhibit good potlife year for year (see Fig. 2). In 1982 the average service time of the pots shut down for repair was 2648 days i.e. more than 7 years.

Besides good craftsmanship in lining the pot, we consider slow preheating, slow sodium swelling, compact seams and joints free from cracks to be crucial in attaining this potlife. The blocks used in RHEINFELDEN are anthracite based in good commercial quality.

Different smelters have different baking and start-up techniques. The characteristics of pot design have to be considered. However, the scientific principles underlying the succesful techniques do not vary from potline to potline.

Two men are needed for ramming joints and seams when using cold paste RHEINFELDEN 1AK. A third one feeds the mix into the tamping area. Breaks are permissible, even over night, provided the seam or joint is kept clean and free from dust. Good and uniform compaction is the main task. The ramming schedule can be fitted to the workforce available. Considerable productivity improvements are possible as compared to the use of hot paste.

The first major test for the cold ramming paste RHEINFELDEN 1AK came in 1980/81 when 52 new 100 kA-pots were constructed in the rebuilt potrooms 12/13. By end of August 1983, these pots had an average age of 786 days. Not one has failed. The metal quality is better than 99.90 % in each pot (see Fig. 3). This of course requires the use of pure raw materials. The alumina is from Gove, Australia and has typically 0.006 % Fe_2O_3 and 0.017 % SiO_2. The anodes have 0.025 % Fe and 0.007 % Si. The anode butts amount to 20 % of the new anodes. The metal is needed in this grade because of the special product mix in the RHEINFELDEN casthouse. However, these figures show that after more than 786 days all 52 cathodes are still free of any damage. We expect a potlife in the order of 3000 days.

RHEINFELDEN

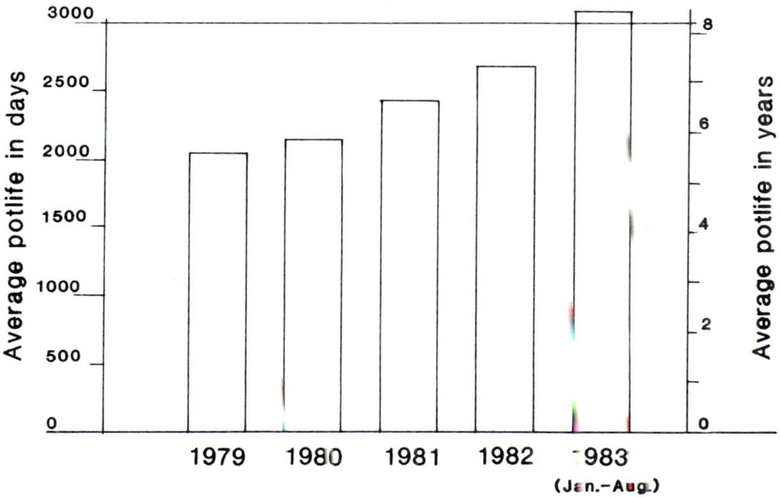

Fig.2

Average life of pots stopped in RHEINFELDEN (100kA,40 000 tpy production)

RHEINFELDEN

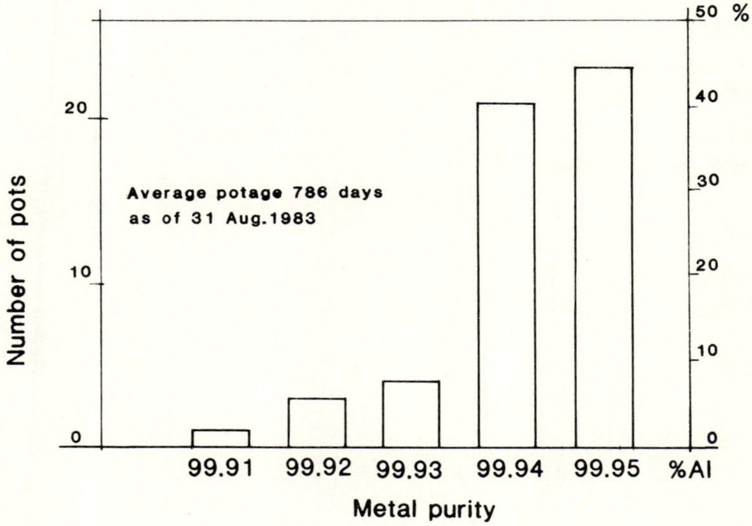

Fig.3 Metal quality of potlines 12/13(100kA) in RHEINFELDEN as of August 1983

52 pots started up in three steps from
6 Nov.1980 through 13 Nov.1981

The average energy consumption of these 52 pots from January through August 1983 was 13.40 kWh/kg Al. This too, shows that the cathodes must be still in perfect condition.

With reference to Fig. 2, it should be noted that the pots terminated so far still had been rammed with hot mix. The cold ramming paste RHEINFELDEN 1AK was developed during the late 70ies and was introduced as standard potlining material in 1980. Today it is the only paste used in RHEINFELDEN for potlining.

6. Conclusions

- Using cold ramming paste RHEINFELDEN 1AK in potlining, the regulations concerning PAH exposure are met.
- Avoiding crack formation during preheating, start-up and early pot age is a prerequisite for good potlife.
- The seams and joints must be well compacted to prevent electrolyte penetration.
- Cold Ramming paste RHEINFELDEN 1AK fulfills the requirements necessary for a good potlife.

(1) A. Steinegger, R. Glaus; "The Determination of Polycyclic Aromatic Hydrocarbons at the Workplace in Carbon Plant and Reduction Plant"; pp 977 - 989 in Light Metals, 1981.

(2) J. M. Blanc, P. Entner; "Application of Computer Calculations to Improve Electromagnetic Behaviour of Pots"; pp 285 - 295 in Light Metals, 1980.

(3) E. Bosshard, O. Knaisch, W. Schmidt-Hatting, J. M. Blanc; "EPT 18: The new 180 kA-pot of ALUSUISSE"; pp 595 - 605 in Light Metals, 1983.

(4) S. Timoshenko and J. N. Goodier "Theory of Elasticity", Chapter 14, 2nd edition, Mc Graw-Hill, New York 1951.

(5) M. Sørlie, H. A. Øye; "A Survey on Deterioration of Carbon Linings in Aluminium Reduction Cells"; pp 635 - 642, Metall, June 1982.

(6) S. Wilkening, G. Busse; "Evaluation and Production of Carbon Cathode Blocks"; pp 653 - 674 in Light Metals, 1981.

(7) P. Aeschbach, H. Friedli; "Improved Potlife and Energy Savings by a new Cathode Lining Technique"; pp 389 - 396, Light Metals, 1981.

NEW DEVELOPMENT IN POT WALL LININGS

Dr. K. Etzel
Dr. F. Brandmair
SIGRI Elektrographit
D-8901 Meitingen
W. Germany

Dr. P. Feschbach
Mr. H. Friedli
Swiss Aluminium Ltd.
Valais Plants
Steg Smelter
CH-3945 Steg-Gampel

Due to improvements in the quality of amorphous-type cathode blocks and the development of new graphitic qualities and as a result of further developments in installation technique (e.g. glueing, sealing), the service life of the cathodes has generally increased in the last few years. Now the pot wall lining is more often the weakpoint than was the case before and leads to a failure of the cell. This trend is reinforced by the increased application of centerbreaking/pointfeeding respectively.

Based on ideas regarding the mechanism of the destruction of the pot wall lining, improvements are introduced both as regards the design and choice of material for the structure of the pot wall lining as well as new repair methods, which can be applied without shutting down the pots.

The known typical damage that occurs on the edges of the bath, which have not been covered with a natural sidewall of frozen bath for some time, is the erosion especially at the height of the metal level. In extreme cases the lining is eroded as far as the steel plate of the shell, and finally the metal breaks through the wall and flows out of the pot.

The damage to the sidewall can be caused by abrasion and erosion (as a result of electromagnetically caused metal and bath movements and convection) and by corrosion (cf. the tests conducted by Dewing[1]) and Grjotheim, Naeumann and Øye[2]).

This corrosion problem can be controlled by aiming at the following basic modes of procedure:

a) lining the edge of the pot with a material that absolutely withstands the corroding attack of melt and metal.

b) lining the edge of the pot in a manner as to force the formation of a protective natural sidewall of frozen electrolyte.

A solution with a corrosion-resistant lining would have the very great advantages that the lateral heat losses could be considerably reduced by means of good insulation and the specific energy consumption per unit of weight of metal also improved.

If the formation of the natural sidewall ledge is to be forced, then the passage of heat through the edge lining has to be increased to such an extent or the insulating effect reduced to such an extent that the melt tends to have a temperature below the liquidus point. This can be achieved by reducing the thicknes of the lining or by the choice of better heat-conducting lining materials or by a combination of both measures. Care has to be given in any event to a good contact of the lining with the steel plate of the shell, for any air gap that forms can quite considerably change the originally calculated relations. This solution quite consciously tolerates heat losses which can influence the energy balance.

A compromise finally is

c) to design the edge of the bath in such a way that as a rule a protective natural sidewall crust is formed, but the side facing the bath is resistant to corrosion by melt and metal to such an extent that longer periods, during which the natural sidewall ledge is melted, are survived without essential damage.

We are presently successfully solving the problem in an optimum manner in accordance with alternative a) by providing a corrosion-resistant lining material.

Available are the means for a solution according to b). Materials that would allow a compromise according to c) are being tested and show good results to date.

In Table I the properties of different lining materials are compiled and characterized in more detail.

The classic lining materials out of more or less pure carbon can be subdivided into the groups

carbon ("amorphous" carbon)
semi-graphite ("graphitic" materials) and
electrographite (graphitized carbon),

cf. AIME 1983 (3). Based on amorphous carbon, the heat conductivity and the resistance to electrochemical corrosion increase when progressing in the direction of electrographite.

With the efficient heat conducting materials made of semi-graphite and electrographite, the solution according to b) can be realized with a corresponding design of the wall thickness. The relative resistance to cryolite melt also makes it possible to survive shorter term periods in which the natural sidewall has melted.

The material type Bos-CN (see Table) has a considerably higher resistance to molten cryolite than the materials consisting of carbon of different forms listed in the first three columns of the table. It allows a solution according to c) with proper dimensioning of the wall thickness and the heat passage values. If pots lined in this manner are not run excessively long under extreme conditions, in particular as regards the temperature, then long term troublefree operation should be possible.

It has already been emphasized that for edge linings according to the alternatives described under b) and c), a good contact to the shell plate has to be created and maintained, since a thin gap already considerably reduces the passage of heat. Therefore, a process must be interesting that makes it possible to eliminate a gap which has formed during pot operation.

In the sector of the lining of blast furnaces, very positive experience has been made for quite some time now with liquid repair pastes, so-called Mux pastes, which are insered in hollow spaces with a pump, where they very quickly solidify under the effect of the temperature.

Table: LINING MATERIALS FOR THE EDGE OF THE POT

Properties	Carbon (amorph)	Semi-graph-ite	Graphite (graph-itized)	Burn-off and Corr-osion Resistant Varieties	
				7 BOS-CN	5 BOS-CN
App. den-sity g/cm^3	1.54	1.70	1.62	2.25	2.05
Real den-sity g/cm^3	1.90	1.92	1.95	2.80	2.47
Total open porosity	16	15	21	18	15
Spec. el. resis-tivity $\Omega\mu m$	40	15	12	900	130
Bending strength N/mm^2	7	14	10	20	20
Pressure strength N/mm^2	27	30	29	60	70
Heat con-ductiv-ity W/mK	8	45	116	9	16
Alkal-ine resist. ASTM C 454 1st test 2nd test	 2 -	 1 2	 1 1	 1 1	 1 1
Oxid. resist. Burn-off test	0-1	1-2	3-4	10	9

The first attempts to install such systems in reduction pots are underway.

References:

1) E. W. Dewing, "The Solubility of Aluminium Carbide in Cryolite Melts"
Trans. Met. Soc. AIME 245 (1969) 2181

2) K. Grjotheim, R. Naeumann, H.A. Øye
"Formation of Aluminium Carbide in the Presence of Cryolite Melts"
Light Metals, Volume I, 1977, pp.233

3) K. Etzel, F. Brandmair, P. Aeschbach, H. Friedli
"Glueing of Cathodes and Anodes - A Proven Technology",
Light Metals 1983, pp. 885

CHEMICAL RESISTANCE OF CATHODE CARBON

MATERIALS DURING ELECTROLYSIS

Morten Sørlie* and Harald A. Øye

Institutt for uorganisk kjemi
Norges tekniske høgskole
Universitetet i Trondheim
N-7034 Trondheim-NTH, Norway.

Commercial cathode materials for use in Hall-Heroult electrolysis are characterized physicochemically and the properties are correlated with deterioration resistance *versus* sodium vapour and interaction with melt-liquid aluminium. Resistant materials were characterized by the (002) X-ray peak, the crushing strength and the porosity. An unsuccessful attempt to graphitize anthracite by interaction with sodium vapour is described. Finally, penetration velocities and profiles are given and discussed.

* Permanent address:
Elkem a.s., R & D Center
P.O. Box 40
N-4620 VÅGSBYGD, Norway.

Introduction

This work is part of a study of the deterioration mechanism of cathode carbon during aluminium electrolysis. A literature survey has been given in Metall [1] while the first account on our experimental work was presented at the 1982 AIME Annual Meeting [2]. A more extensive account of the relationship between physicochemical properties and resistance towards sodium attack is recently printed in Metall [3].

Failures associated with the cathode carbon lining are the most common causes for discontinued production in aluminium reduction cells. During electrolysis sodium and electrolyte are continuously being absorbed by the carbon lining, creating problems of swelling, cracking, arching, heaving and crumbling of the cathode.

The aim of the present study is to find correlations between physicochemical properties of the cathode carbon and its resistance towards deterioration by sodium attack and penetration by sodium and bath. Some experiments to study a possible sodium induced graphitization are also described.

Characterization of laboratory cathodes and their resistance towards sodium

The cathodes employed were either fabricated in the laboratory from green cold type tamping mix [2,4] or made from cores drilled out of prebaked cathode blocks. The finished cathode had a diameter of 25 mm and a length of 60-75 mm. In order to cover a broad range of industrially important carbon qualities we have included four different cold type tamping pastes and four different prebaked cathode blocks. Physicochemical properties are given in Table 1. The numbering of the materials is the same as in ref. [3]. All are commercially available and currently used in the aluminium smelting industry. The resistance of the various materials towards cracking and exfoliation due to sodium attack are also listed in Table 1.

Two of the materials (*M-V, B-VIII*) may be regarded as highly graphitic and can be classified as being strongly resistant to chemical deterioration from sodium attack. A partly graphitized anthracite (*M-I* and *B-III*) has a far better sodium resistivity than a partly graphitized calcined coke (*M-IV*) [3] which is very susceptible to sodium cracking and exfoliation. The amorphous anthracitic material *M-II* also shows a poor resistance towards sodium. The block materials *B-II* and *B-IV* have intermediate properties.

There is no single material parameter which can determine the suitability of material in the aluminium smelter cathode [3]. With a few exceptions, strength appears to be inversely related to the graphite content and it is of interest to observe that the material which showed the poorest sodium deterioration resistance (*M-II*) originally had a particularly high crushing strength (Table 1). The inferior materials *M-II, B-II* and *B-IV* did also have a low porosity.

Usual X-ray characteristica could not readily be employed for characterization of the chemical resistivity of anthracitic cathode carbon materials [3]. Neither graphitization index nor mean stacking height of layer planes (L_c) appeared to be applicable for evaluation of deterioration resistivity with respect to cathode carbons. Neither did the asymmetry of the most important diffraction peak give any clear indication of the graphite content. However, a X-ray diffraction intensity ratio between a natural graphite standard and the tested material gave a correlation between the anthracitic cathode carbon materials and the observed deterioration resistances [3]. The inferior materials, seam mix *M-II* and carbon blocks *B-II* and

Table 1. Physico-chemical properties and sodium resistance of cathode carbons.

	Cold type tamping paste				Prebaked cathode blocks			
	M-I	*M-II*	*M-IV*	*M-V*	*B-II*	*B-III*	*B-IV*[a)]	*B-VIII*
Carbon filler material	Anthracite, partly graphitized	Anthracite, calcined but highly amorphous with respect to X-ray diffraction	Soft and partly graphitized calcined coke filler	Highly graphitic	Anthracite, partly graphitized			Highly graphitic
PHYSICOCHEMICAL PROPERTIES								
Ash content (%)	1.9	4.4	0.41	0.32	5.2	3.7	3.6	0.03
Porosity (%)	18.6	14.4	25.7	33.4	9.9	18.6	11.1	21.1
Crushing strength (N/mm^2)	15	60	37.5	8.5	28	21	52	25
d_{002} (Å)	3.3684	b)	3.3797	3.3684	3.3797	3.3797	3.3671	3.3671
Graphitization index	0.83	b)	0.70	0.83	0.70	0.70	0.85	0.85
L_c (Å)	177	b)	107	62	83	100	180	63
(002) intensity ratio	0.08	0	0.17	0.46	0.02	0.06	0.02	0.40
RESISTANCE TOWARDS SODIUM ATTACK								
Crucible test, 1000°C — CR ratio ∞ (NaF)	*	***	***	-	**	**	**	-
4.0	*	**	**	-	*		-	-
3.2	-	**	*					
3.0 (Na_3AlF_6)	-	*	*		*		*	
2.5	-	*	*					
1.5	-	-	-					
Sodium vapour test, 883°C	*	***	***	-	**/***	*	**/***	-

a) Sidewall blocks; b) Can not be determined; ***: Extreme cracking and exfoliation; **: Heavy to moderate cracking; *: Hairline cracks, little damage; -: No effect.

B-IV, have (002) intensity ratios less than or equal to 0.02 while acceptable materials have ratios larger than or equal to 0.06. The seam mix of calcined coke origin (*M-IV*) again does not fit into the pattern and will probably remain susceptible to chemical deterioration unless it is more fully graphitized.

In Situ graphitization of aluminium smelter cathodes

It has been known that aluminium smelter linings graphitizes with age [5,6] and that after a period of about 1 year, the anthracite filler is nearly completely converted to graphite [7]. Dell [5] found that the graphitization of the lining was equivalent to that of anthracite heated to 2500°C.

The operating temperature of a Hall cell is about 970°C and there is no evidence which supports that the lining or part of it ever comes anywhere near thermal graphitizing temperatures. The ordering process which takes place at temperatures 850-1000°C must be catalyzed and the intercalation of sodium into the carbon structure is believed to play a crucial part as it modifies the carbon structure and enlarges the distance between the graphite layers adjacent to the intercalation sites [1].

Short time exposure of anthracite to sodium vapour was not found to lead to graphitization [8], but Rapoport [9] reported structural ordering in anthracite to take place during laboratory electrolysis experiments where samples were cathodically polarized for up to 95 h with temperature fluctuations of 100°C. The reason for this was believed to be the repeated formation and decomposition of sodium intercalation compounds caused by the temperature cycling.

The exposure of anthracite to sodium vapour at 833°C (the b.p. of sodium) was shown to cause extensive damage due to swelling and exfoliation [2]. Experiments were therefore initiated to test if long term exposure to sodium had any influence on the structural ordering of anthracite. Samples of calcined amorphous anthracite were together with sodium metal sealed under argon atmosphere in heavy wall cylindrical steel vessels (by welding). The steel cylinder was placed in an alumina crucible and the assembly was finally sealed in quartz ampoules under vacuum. The samples were kept in a furnace where the temperature was allowed to cycle 15°C around 883°C, the boiling point of sodium metal. Samples were removed at regular intervals, cut open and the carbon content analyzed by X-ray diffraction after removal of its sodium content by acid leaching. Over a period of 10 months no structural changes detectable by X-ray diffraction analysis were found.

These results seem to exclude that the low temperature graphitization of aluminium smelter linings is caused solely by the mechanism of cyclic formation and decomposition of sodium intercalation compounds. These results lend support to the mechanism suggested by Waddington [6]. He observed that graphitization only seemed to occur in the parts of the cathode which had been penetrated by both sodium metal and molten electrolyte and not in sections where only sodium metal was present. Based on this he proposed that the carbon first was converted into a lamellar compound by the penetrating sodium vapour and then later, when liquid bath reached the compound, it decomposed with the formation of graphitic material [6]. The carbide forming reaction may well be the decomposition step.

$$4Na_3AlF_6 + 12Na + 3C = Al_4C_3 + 24NaF \qquad (1)$$

Penetration of sodium and bath during reversed polarity electrolysis

Electrolysis were carried out at 1000°C and the parameters varied were electrolyte cryolite ratio (CR)*, current density (CD) and duration of electrolysis. The experiments were performed in the reversed polarity apparatus which has been described previously [2].

After removal of excess electrolyte or cathodic deposits [4] from the cathode surface, the cathodes were cut into 3-5 mm thin discs. One surface of these discs was then analyzed for sodium, aluminium and fluorine by means of X-ray fluorescence spectroscopy. It was important to perform these analyses as quickly as possible after cutting the cathode, as sodium metal which was present in the carbon lattice readily oxidized near the surface. Due to its hygroscopy and subsequent volume expansion it could else accumulate on the surface as sodium hydroxide.

The X-ray fluorescence analyses gave concentration profiles like those shown in Figures 1 and 2. Cryolite (Na_3AlF_6) was used as an analytical standard and the values given are net concentrations. Any sodium and aluminium which might have been present as mineral impurities (ash) were subtracted from the results by using a dummy sample made of the same unelectrolyzed carbon material. Due to its low atomic number the fluorine X-ray analysis is believed to give the least accurate values.

The concentration profiles shown in Figures 1 and 2 correspond qualitatively to those earlier presented by Dell [5] which show that sodium usually was the diffusing atom with the highest concentration in carbon. The sodium concentration profile consists of two superimposed sodium waves, firstly the metallic sodium which migrates through the carbon lattice by means of an intercalation mechanism [2], then secondly follows the diffusion of molten bath by capillary attraction through the open porous network of the carbon specimen. The migration of this second wave is probably made possible by changes in bath/carbon wetting conditions caused by changes in surface properties of the sodium-carbon compound. The form of the concentration profiles was explained by Dewing [10] to be caused by the deposition of aluminium carbide on the pore walls by reaction between migrating cryolite, carbon and intercalated sodium metal (Equation (1)). The solid aluminium carbide does not migrate further into the carbon but is deposited near the surface, while the liquid electrolyte with increased excess of sodium fluoride continues to migrate through the electrode in the wake of the sodium metal front. Figure 3 shows photographs from an electron microprobe analysis of a cylindrical carbon electrode (cathode) cut perpendicular to its axis. The metallic sodium wave has already passed through this plane and the molten electrolyte, now completely depleted of aluminium fluoride, has just started to penetrate. A vein of nearly pure sodium fluoride formed according to Equation (1) can be seen at the upper right hand side corner (BSE, Na, and F images).

The electrode in Figure 3 is the *M-I* type made with a partly graphitized anthracitic filler. From the sodium X-ray emissions (which density is approximately proportional to the sodium concentration) it can be seen that the sodium concentration is larger near the grain boundaries of the filler grains than within the particles. This is contrary to what was found in cases where the carbon aggregate was made of a low calcined amorphous anthracite [2]. In that case it was shown that sodium had an equally high prefe-

---

* CR = mol NaF/mol AlF_3. Acidic compositions at CR < 3.0; basic at CR > 3.0.

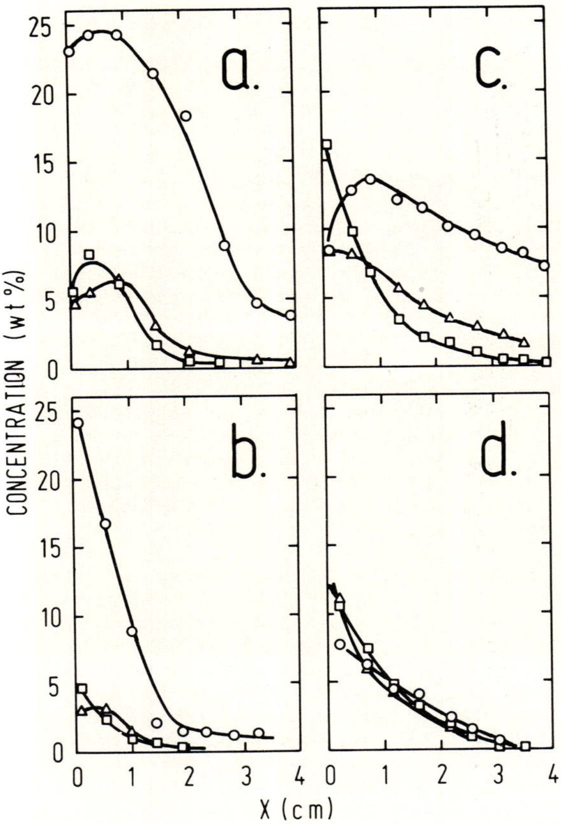

Figure 1. Concentration profiles after 4 h electrolysis at 1000°C using seam mixes *M-II* and *M-V* as cathode materials.
o: Na; □: Al; Δ: F.

a. *M-II*
CD = 0.43 A/cm^2
CR = 4.0

b. *M-II*
CD = 0.43 A/cm^2
CR = 1.5

c. *M-V*
CD = 0.46 A/cm^2
CR = 4.0

d. *M-V*
CD = 0.46 A/cm^2
CR = 1.5

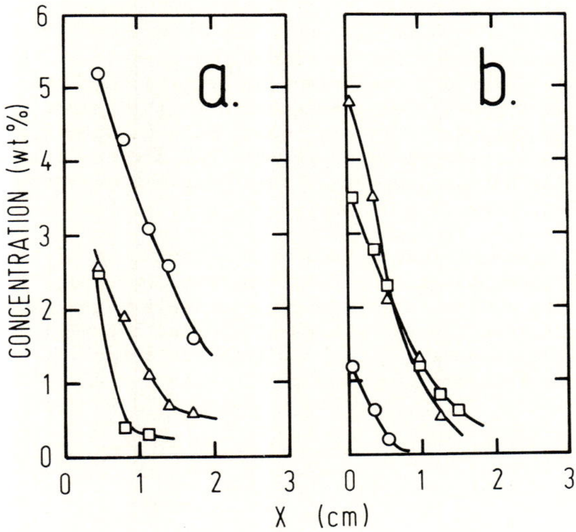

Figure 2. Concentration profiles in *B-VIII* prebaked cathode block after electrolysis at 1000°C with CD = 0.15 A/cm^2.
o: Na; □: Al; Δ: F.

a. CR = 4.0, 4 h

b. CR = 1.5, 2 h

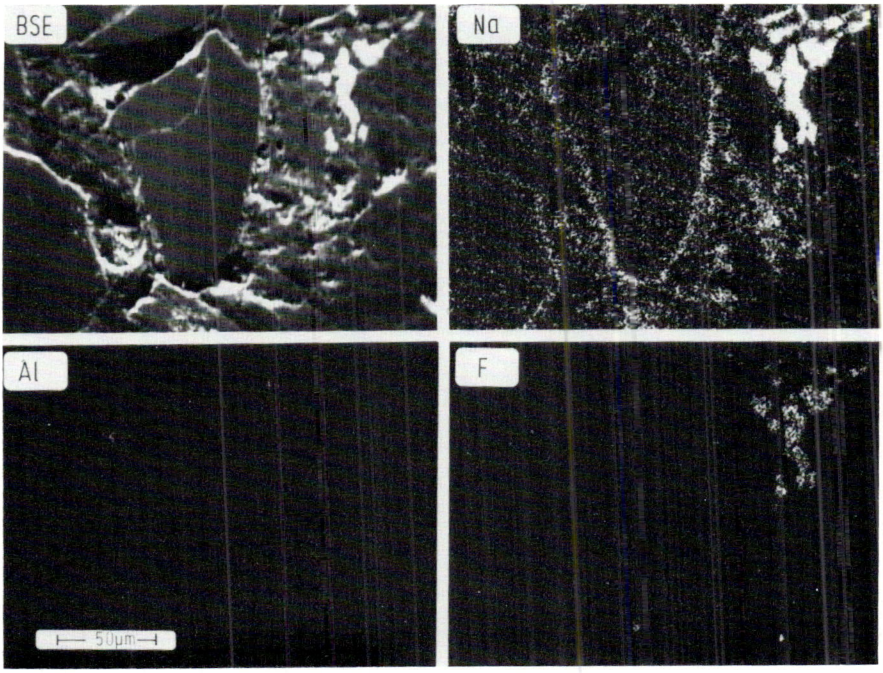

Figure 3. Qualitative microprobe analysis of *M-I* carbon electrode after electrolysis. The photographs shown are backscatter electron (BSE) together with sodium, aluminium, and fluorine X-ray emission images.

rence for the non-porous anthracite grains as for the highly porous binder coke. In the partly graphitized *M-I* filler (Figure 3) the better ordering of the lattice has made the anthracite less susceptible to sodium penetration than the binder coke.

The concentration profiles in Figures 1 and 2 tell about the ease with which sodium metal and electrolyte penetrate into various cathode carbon materials. Sodium penetrates anthracite easily (Figures 1a,b) and the melt follows slowly in the wake of a high concentration of sodium metal. Graphite is much more resistant towards sodium metal intercalation (Figures 1c,d) and the sodium metal wave is not as pronounced here and the bath components follow more closely, but due to its higher open porosity (Table 1) more melt can accumulate. In Figure 1d (graphite *M-V*, acidic melt) it is not possible to distinguish the sodium metal front from that of the migrating bath. This is further emphasized in Figure 2 (prebaked graphitic cathode block *B-VIII*) where the melt penetration closely follows the sodium metal front. In the acidic melt the concentration of fluorine has even become higher than the total sodium concentration and reflects roughly the melt composition near the electrolyte surface.

Sodium metal and electrolyte diffusion in cathode carbons

Previous work [2] indicated that diffusion was the mechanism by which sodium metal and melt migrated through the lattice. By considering the cathode as an infinite solid and assuming a constant activity of sodium metal at the electrolyte/carbon interface during electrolysis (or an infinite reservoir of diffusing species), the square of the diffusion length along the axis of the electrode, measured at a constant concentration of the diffusing species, is a linear function of the diffusion time, t [11]. The solution to Fick's law

$$\frac{\partial c}{\partial t} = D \frac{\partial^2 c}{\partial x^2} \quad (2)$$

under the above circumstances is

$$\frac{c_i - c_x}{c_i - c_0} = \frac{2}{\pi^{\frac{1}{2}}} \int_0^k e^{-y^2} dy \quad (3)$$

where

$$k = \frac{x}{2(Dt)^{\frac{1}{2}}} \quad (4)$$

c_i = concentration of diffusing species at the interface x = 0 which during electrolysis is assumed to be constant; c_x = concentration at diffusion depth x at time t; c_0 = concentration (constant for all x) at the beginning of diffusion. By measuring x as a function of t at constant c (c_x = cut-off concentration) the left hand side of Equation (3) is constant. This means that also the right hand side of Equation (2) and thus k is constant. A plot of x^2 *versus* t should then yield a straight line for all diffusion controlled processes.

The quick method by cutting the laboratory cathode along its axis and measuring the penetration depth of sodium metal by pressing the exposed surface towards a paper wetted by phenolphthalein acid/base indicator [2] proved to be a quite accurate way of determining the diffusion depth of Na metal in anthracitic materials (at a constant cut-off concentration). Remeasurements by means of X-ray fluorescence techniques of the same cathode materials subjected to the same experimental conditions gave nearly identical curves at a cut-off concentration of 3 wt% Na. All further discussion is referred to this cut-off concentration.

Figures 4-6 show the penetration of sodium in the amorphous low calcined anthracite (*M-II*) as a function of current density and cryolite ratio. Figure 4 shows that the penetration follows Equation (3) and that the penetration velocity is highest at the highest current density.

During electrolysis sodium is formed at the electrode either by a primary charge transfer reaction or from the equilibrium

$$3NaF + Al = AlF_3 + 3Na \quad (5)$$

Equation (5) is shifted towards the right hand side in basic melts, thus increasing the interfacial concentration of sodium, c_i, in Equation (2). More sodium penetrates the electrode in a given time thus moving the constant cut-off concentration further into the carbon material. This is illustrated in Figure 5 where the sodium penetration depth into amorphous anthracite (*M-II*) increases markedly as the cryolite ratio increases. The change is most pronounced in the acidic region. The movement of the sodium front

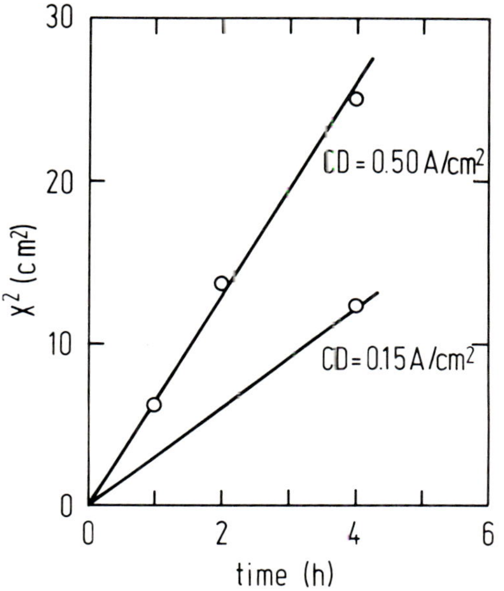

Figure 4. Square of sodium metal penetration depth (3 wt% cut-off) plotted *vs.* electrolysis time at 1000°C with CR = 4.0 using amorphous anthracitic seam mix *M-II* as cathode.

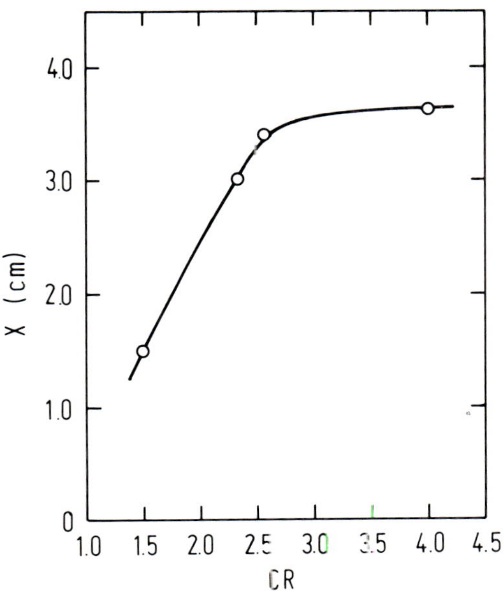

Figure 5. Penetration depth of sodium in *M-II* electrode (3 wt% cut-off) during electrolysis for 4 h at 1000°C, CD = 0.15 A/cm^2 in cryolite electrolytes of varying acidity

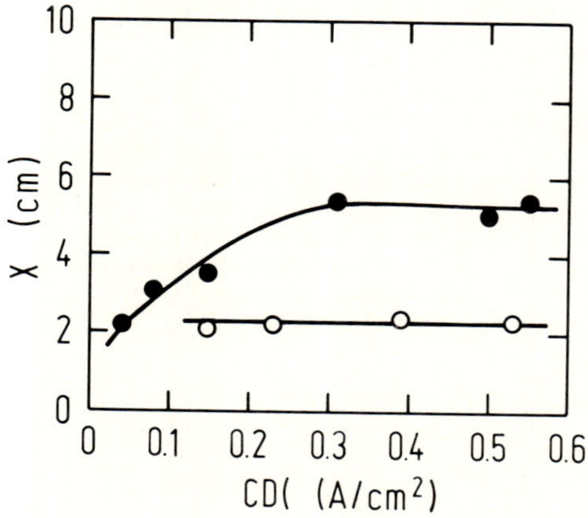

<u>Figure 6</u>. Sodium penetration (3 wt% cut-off) in *M-II* during 4 h electrolysis at 1000°C at various CD's.

●: CR = 4.0
o: CR = 1.5

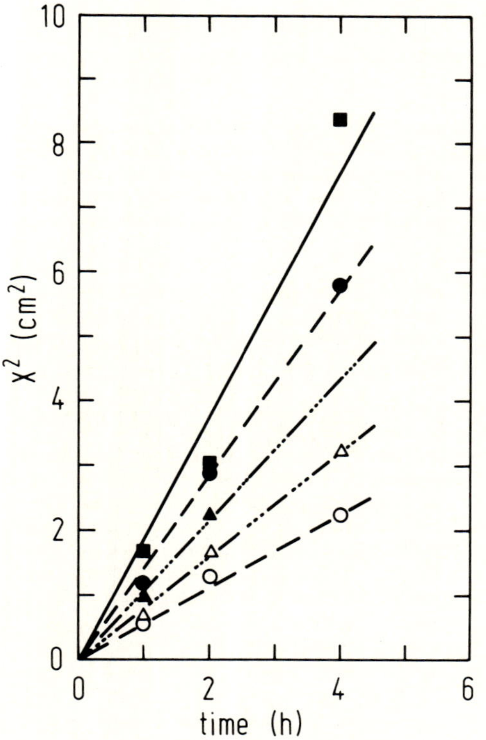

<u>Figure 7</u>. Square of sodium penetration depth (3 wt% cut-off) in three graphitic or partly graphitic seam mix cathodes plotted *vs*. electrolysis time at 1000°C with CD = 0.15 A/cm^2.
Open symbols: CR = 1.5.
Filled symbols: CR = 4.0.

o,● --------: *M-I*
■ ————: *M-IV*
Δ,▲ —·· —: *M-V*

further into the carbon is caused by the increased sodium equilibrium concentration at the interface. At basic compositions the curve seems to flatten out. Due to a high sodium activity in the basic melt, combined with a rapid removal of sodium by intercalation into the carbon, the presence of liquid aluminium can no longer be maintained at the electrode surface at low CD's.

This is also apparent when the current density is varied (Figure 6). It is evident that a constant sodium activity (*i.e.* presence of aluminium metal) at the electrode/bath interface is reached at CD = 0.3 A/cm^2 in the basic electrolyte CR = 4.0 while aluminium metal appears to be present below CD = 0.15 A/cm^2 in the acidic CR = 1.5.

Penetration time to a depth of 30 cm into various carbon materials is calculated in Table 2. The highest sodium penetration rate is

Table 2. Time calculated for sodium (3 wt% cut-off concentration) to penetrate a 30 cm high carbon slab at 1000°C.

CD	CR	*M-I*	*M-II*	*M-IV*	*M-V*	*B-III*	*B-VIII*
(A/cm^2)				Time (days)			
0.15	1.5	67	67		47		>200
	4.0	26	12	20	35	27	109
0.50	4.0		6				

found in amorphous low calcined anthracite (*M-II*) while the slowest migration is observed in the prebaked graphitic cathode block (*B-VIII*).

The penetration of sodium into some of the partly graphitized cathode blocks is shown in Figure 7. The penetration velocity follows Equation (3) but the penetration mechanism is more complex. These materials are more porous than the amorphous anthracite (Table 2) and the total movement of sodium (metallic or ionic) is due to an interplay of intercalation and movement of melt in pores.

In basic melts (high activity of metallic sodium) it is seen from Figure 7 that the penetration velocity is lowest in the highly graphitic material *M-V*. The penetration is higher for the partly graphitized material *M-I* and still higher for the material made from calcined coke (*M-IV*).

In acidic melts the penetration velocity of sodium is as expected lower for all materials but the order is reversed, *M-V* > *M-I*. This latter order is, however, the order of porosity and the reversed sequence is probably due to that the measured Na cut-off (3 wt%) is a cut-off for melt movement, the melt moving more easily in the more porous material. As already mentioned this is also indicated in Figure 1 where the melt is trailing behind metallic sodium in the amorphous material (Figures 1a,b) and for basic melts in the graphitic material (Figure 1c), while sodium, aluminium and fluoride are moving close together in graphitic carbon when the melt is acidic (Figure 1d).

Acknowledgement

We wish to thank Bente M. Faaness for experimental assistance and the *Royal Norwegian Council for Scientific and Industrial Research (NTNF)* and the *Norwegian aluminium industry* for financial support.

References

1. M. Sørlie and H.A. Øye, Metall 36 (1982) 635.
2. C. Krohn, M. Sørlie, and H.A. Øye, Light Met., Proc. Sess. AIME Annu. Meet., 111th (1982) 311.
3. M. Sørlie and H.A. Øye, Metall 37 (1983.
4. O. Herstad, C. Krohn, M. Sørlie, and H.A. Øye, Aluminium 59 (1983) 200.
5. M.B. Dell, in Extractive Metallurgy of Aluminium, Vol. 2, ed. by G. Gerard, Interscience Publishers, New York, 1963, p. 403.
6. J. Waddington, in Extractive Metallurgy of Aluminium, Vol. 2, ed. by G. Gerard, Interscience Publishers, New York, 1963, p. 435.
7. W.E. Haupin, Aluminium 52 (1976) 446.
8. J.G. Hooley and P.T. Hough, Ext. Abstr. Program - Bienn. Conf. Carbon, 13th (1977) 454; 14th (1979) 286.
9. M.B. Rapoport, I.M. Maltseva, and M.L. Bluyshtein, Dokl. Chem. Technol. (English translation) 177 (1967) 180.
10. E.W. Dewing, Trans. Met. Soc. AIME 227 (1963) 1328.
11. A. Tiselius, Z. Phys. Chem. 169A (1934) 425.

ELECTROCHEMICALLY INDUCED CRYSTALLOGRAPHIC CHANGES IN GRAPHITES IN MOLTEN SALTS

Arshad Hafeez

Alloy Technology Division
Alcoa Technical Center
Alcoa Center, Pennsylvania 15069
USA

DC polarization studies of graphite electrodes in $NaCl-LiCl-AlCl_3$ molten electrolytes at 725°C suggested the presence of up to four cathodic reactions. The data showed that irreversible changes had taken place on the graphite surface at each successive reduction reaction. Increase in crystallite size (Lc) was associated with electrodes cracking and swelling when they were held at certain potentials. They were most severely affected at the more electro-negative potentials. Deterioration was also found to increase with temperature.

These findings suggest that crystallographic changes in graphites are related to an electrochemical effect; therefore, graphites can be evaluated by measuring the crystallite size (Lc) after graphites are polarized electrochemically.

Introduction

Graphites are made up of stacks of parallel atomic layers separated by a distance d_{002} (typically 3.35 A°) in the C-direction. The term "intercalation" (1) is used when reactive species are inserted between graphitic parallel layers which have not been moved aside prior to the reaction. Previous studies (2-3) showed that graphites cracked and swelled due to formation of Al_4C_3 and/or Na or some other intercalation during Hall-Heroult cell operation. Techniques to understand mechanisms of graphite by Na penetration has been conducted (4-5). At different cryolite ratios, resistance to Na penetration during electrolysis could be related to the type of graphite (4). After measuring Na penetration profiles into vitreous carbon cathode, Welch, et al (5) concluded that saturation of Na content occurs prior to failure of the cathode. Therefore, as the Na uptake in the graphite is complete (saturated) the d_{002} parameter (1) and the crystallite height (Lc) increases while the width (La) in the a-direction (Figure 1) remains the same.

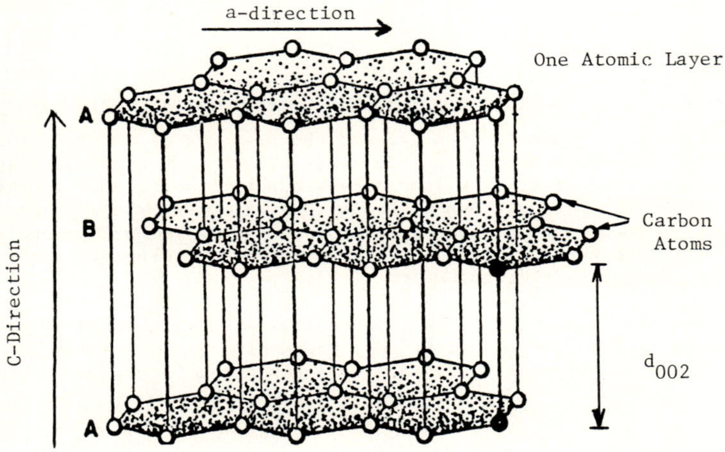

Figure 1 - Sketch of hexagonal graphite (6) showing the d_{002} in the c-direction and crystallite a-direction. The crystallite height Lc (not shown) is comprised of several d_{002} atomic layers making up one crystal.

Therefore, graphite cathodes deteriorate when they are polarized due to either an electrochemical reaction or a cathodic reaction product. These observations and hypothesis could be tested by cathodic polarization of graphite. The cathodic deposits or reactions at different potentials can then be correlated with their effects an graphite surfaces and crystallographic changes within graphite cathodes.

Experimental

Graphites A, B, and C from several graphite companies were investigated (physical properties are given in Table I). Electrochemical studies for the above electrodes (18" long and 0.25" O.D.) were made in a molten NaCl-LiCl-$AlCl_3$ electrolyte. The NaCl and LiCl salts were obtained from Fisher Scientific (reagent grade) and the $AlCl_3$ from within Alcoa.

TABLE I

Physical Properties of As-Received Graphites

Graphite Grade	A	B	C
Bulk Density g/cc	1.57	1.77	1.76
Max. Grain inches	0.060	0.260	0.065
C. T. E. in/in°C x 10^{-6}	12.5 (WG) 27.5 (AG)	20	2.8 (WG) 4.2 (AG)
Tensile Strength psi	727 (WG) 640 (AG)	1100	2800
El. Conductivity BTu/ft^2/ft/hr°F	90 (WG) 62 (AG)	80	--
El. Resistivity ohm - in x 10^5	31.6 (WG) 47.6 (AG)	34	6.3 24 x 10^{-5}
Sample Diameter inches	6-9	8	--
Ash %	0.12	0.3	0.5
Lc A°	400	400	400-430

WG = Grain

AG = Against Grain

Because of the hygroscopic nature of NaCl, LiCl, and $AlCl_3$, special precautions were necessary to obtain an anhydrous oxide-free electrolyte. The electrochemical cell shown in Figure 2 was loaded with salts and assembled in a dry box where a continually purified argon atmosphere was maintained. The electrolyte was contained in a graphite crucible which was also used as the auxiliary electrode. The working electrode or graphite (to be investigated), a Ag-AgCl vycor membrane reference electrode, and a thermocouple inside a quartz tube were immersed in electrolyte. The arrangements of the electronic measuring instruments to obtain polarization data is given in Figure 3. The Cell was heated by a Lindberg crucible furnace. An argon gas atmosphere was continuously maintained throughout each experimental run.

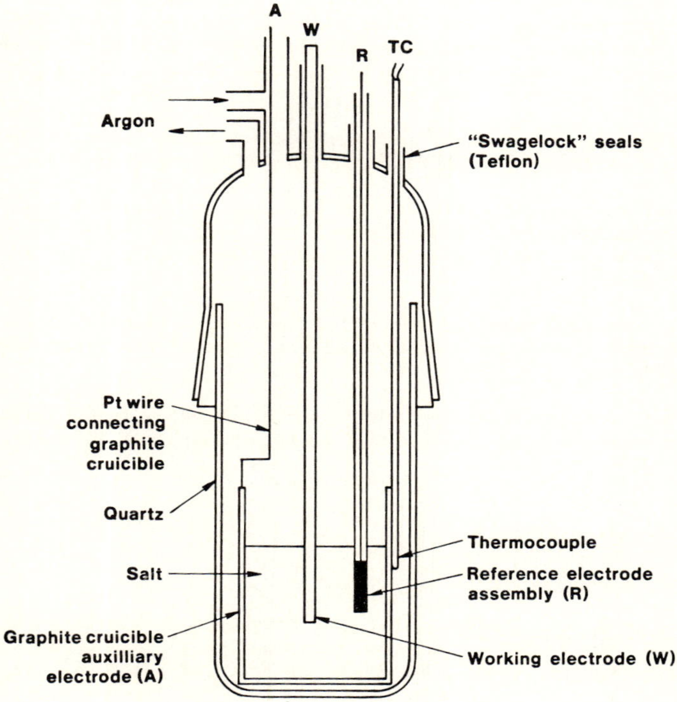

Figure 2 - Electrochemical Cell

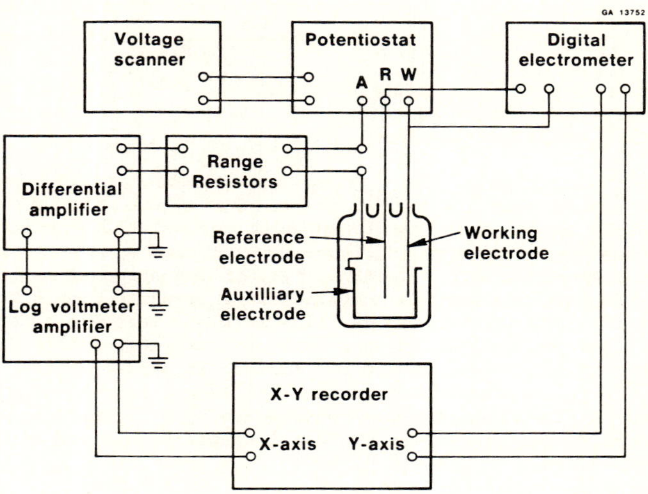

Figure 3 - Electronic equipment and connections to the electrochemical cell

The studies were divided in three parts to demonstrate the relationship and important reactions taking place on graphites during electrochemical polarization. Graphites were analyzed for reaction products and crystallographic changes using electron microprobe, atomic absorption, spectrographic analysis, and X-ray diffraction techniques.

Polarization data was obtained to study reaction products and temperature effects on graphites using the following techniques:

(a) A series of graphite samples was prepared for crystallographic and x-ray analysis by cathodically polarizing to -0.4V, -0.8V, -1.1V, 02.0V, and -2.5V, respectively, half cell potentials for 3 hours at 725°C.

(b) Slow scan (500 mv/min.) cyclic voltammetry of electrodes was used to determine irreversible changes within range of successive cathodic reactions.

(c) Cathodic polarization (potentiodynamic) curves were obtained for electrodes immersed in electrolyte temperatures of 700, 725, 750, and 800°C.

Results and Discussion

Part 1

Electrochemical Studies. The polarization data suggests the presence of three or four cathodic reduction reactions on the A, B, and C graphites (as shown in Figures 4a-d). The sequence of reactions under study are:

(i) $Fe^{(2+ \text{ or } 3+)} + (2 \text{ or } 3)\, e^- \longrightarrow Fe$ (1.35V)

(ii) $Al^{(3+)} + 3e^- \longrightarrow Al$ (1.84V)

(iii) $Na^{(1+)} + 1e^- \longrightarrow Na$ (3.25V)

(iv) $Li^{(1+)} + 1e^- \longrightarrow Li$ (3.29V)

Decomposition or anode-cathode voltages at 750°C given in parentheses may explain the sequence of reactions (ii), (iii), and (iv) in a $NaCl-LiCl-AlCl_3$ molten electrolyte.

In this section data, observations and evidence supporting each reaction are presented. The half cell potential represents the point on the curve being discussed; surface appearance is the observed graphite surface after each experiment in series (a); and Lc (A°) is the crystallite size measured in C direction (6). Because of small changes observed in d_{002} values, comparisons of graphites with Lc is quite satisfactory (7).

Reaction (i)

$Fe^{(2+ \text{ or } 3+)} + (2 \text{ or } 3)e^- \longrightarrow Fe$

Half Cell Potential: -0.8V

Surface Appearance: No attack or deposit (at -0.4V)
Rust film (at -0.8V)

Lc Increase: Not significant

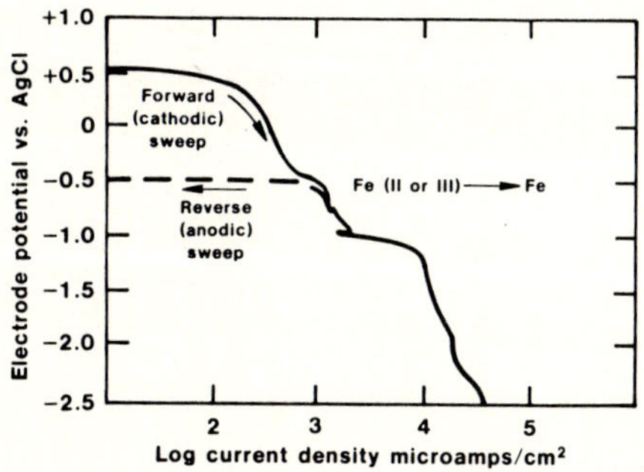

Figure 4a

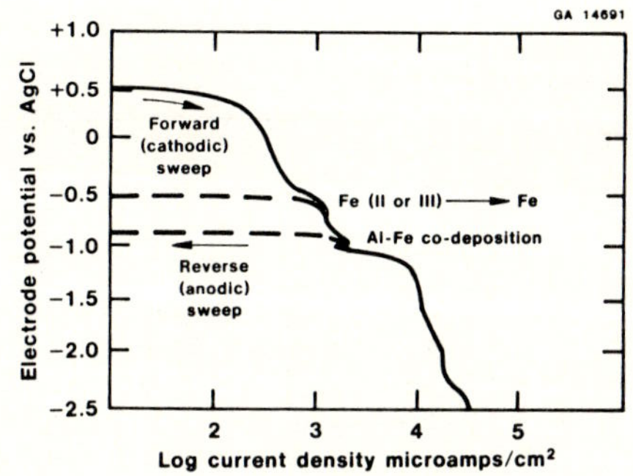

Figure 4b

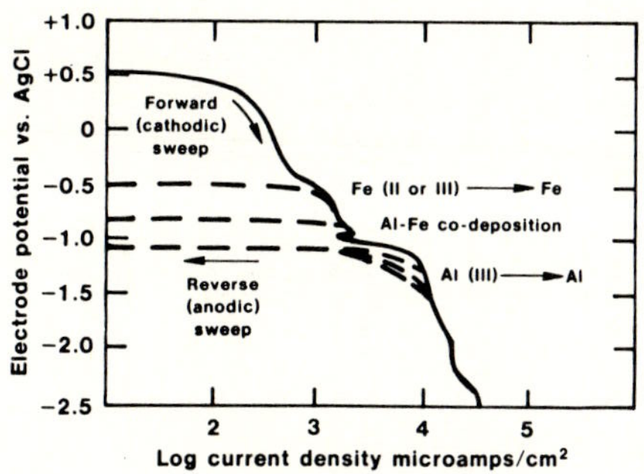

Figure 4c

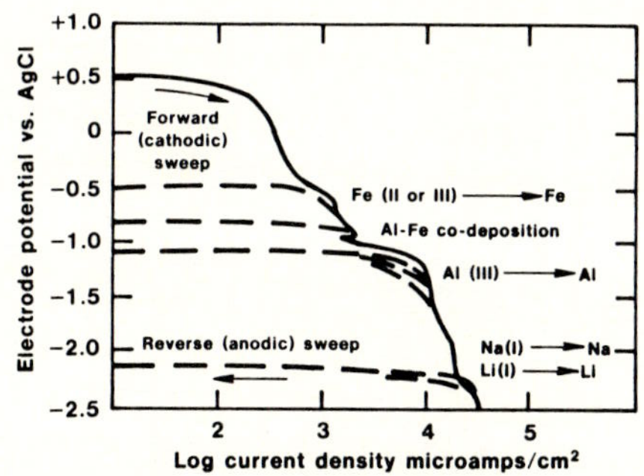

Figure 4d

Figure 4 - Typical cathodic and anodic polarization data showing various reactions at the cathode surface.

Evidence Supporting Reaction (i):

(a) A "Rust" film was observed on the graphite surface held at -0.8V but was not observed at -1.3 and -2.5V.

(b) Scrapings taken from electrodes (held at -0.8V, -1.3V, and -2.5V) show the insoluble Fe content (in H_2O and HCl) is an order of magnitude higher than for electrodes held at -0.8V than -1.3V and -2.5V (see Table II), indicating deposition of Fe on the graphite surface at the half cell potential of -0.8V.

(c) Electrode surface characteristics change from graphite to Fe surface characteristics at the reaction (i) half cell potential. Figure 4a shows the reverse sweep (anodic) potentials are more cathodic than their initial values and arrest potential.

Comments

The likely source of iron are iron compounds used in graphite production, during the mixing stage (135 to 160°C), FeO is used as an "anti-puffing agent" to produce denser graphites. At the baking stage (815°C), FeO combines with S gas to form FeS. In practice, FeS in the graphites should be removed by volatilization during the graphitization stage (2500°C) but it appears that Fe and Fe compounds concentrate non-uniformly near the graphite surface.

TABLE II

ATOMIC ABSORBTION ANALYSIS OF DISTILLED WATER IN WHICH A AND B GRAPHITES WERE BOILED

Potentials Held	Graphite Grade	Na mg/l	Li mg/l	Al mg/l	Fe mg/l	Fe mg/l
-0.8V	A	62.8	13.5	0.10	1.87	2.44
-1.3V	A	136.0	35.0	15.7	0.62	0.74
-2.3V	A	4254	843.0	18.8	0.10	0.17
-0.8V	B	339	59.6	0.65	1.72	2.03
-1.3V	B	147	35.2	5.70	0.22	0.40
-2.3V	B	593	196	28.20	0.10	0.10

*HCl Soluble

Reaction (ia)

Co-deposition of Al and Fe to form intermetallics.

Half cell potential:	-1.0V
Surface appearance:	No change
Lc increase:	Not significant

Evidence Supporting Reaction (ia)

(1) The absence of rust film on the electrode surfaces (held at potentials more electro-negative than -0.8V) indicates that the Fe film was removed by deposition of Al.

(2) "Break" observed (Figure 4b) between reactions (i) and (ii) indicates co-deposition of Fe and Al (8).

(3) Surface characteristics change at the reaction (ia) potentials (-1.0V), since potentials on the reverse (anodic) scan were more cathodic than the reaction (i) potentials (see Figure 4b). This indicates that the surface no longer has the characteristics exhibited by reaction (i) deposits.

Comments

The formation of Al-Fe intermetallics (probably Al_3Fe) during electrolysis may be due to diffusion of Fe to Al or vice versa. The contribution to graphite deterioration may be caused by Al_3Fe particles setting up internal stresses (9) in graphites which weakens the surface.

Reaction (ii)

$$Al\ (3+) + 3e^- \longrightarrow Al$$

Half Cell Potential:	-1.3V		
Surface Appearance:	Some micro cracks		
Lc Increase (from -0.8V):	A	420 → 600A°	
	B	500 → 510A°	
	C	490 → 530A°	

Evidence Supporting Reaction (ii)

(1) Al droplets on cathodic surface were confirmed by x-ray diffraction and microprobe.

(2) When scan was reversed from -1.3V, potentials are more electro-negative than the reverse potentials from -0.8 and -1.0V corresponding to reactions (i) and (ia), respectively. The return potential of -1.1V indicates that the graphite surface has a film of Al, and thus the cathode is operating like an Al surface (see Figure 4C).

Comments

A comparison of increase in Lc values from as-received and graphite held at -1.3V shows that for the Lc increase (of 200A°) is double the increase observed for the B and C graphites. In continuous long term cell operations, the extent of deterioration for graphite A was substantially higher than graphites B and C.

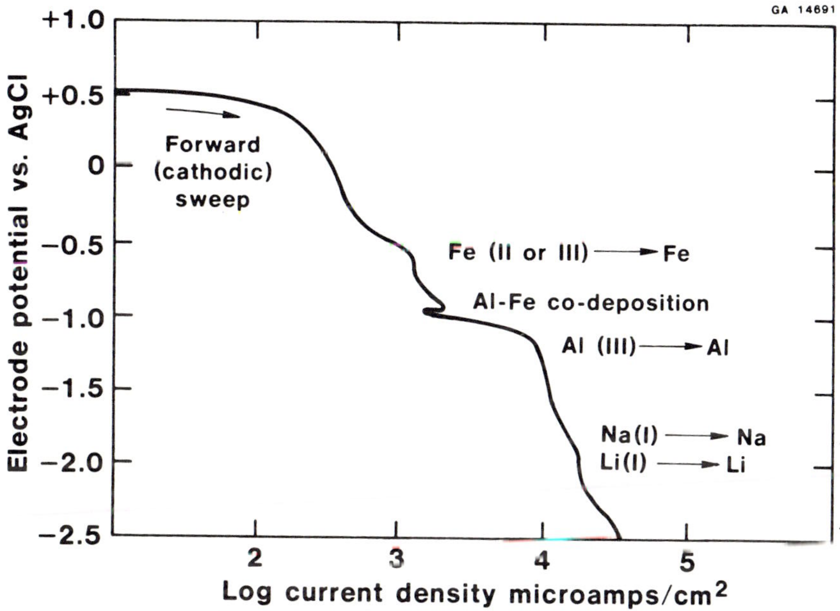

TABLE III Lc(A°) FOR GRAPHITES HELD AT VARIOUS POTENTIALS AS SHOWN ON THE TYPICAL CATHODIC POLARIZATION CURVE

	A	B	C
As-Received	400	400	430
-0.4V	430	460	470
-0.8V	420	500	490
-1.3V	600	510	530
-2.0V	600	460	510
-2.5V	660	750	690

Reaction (iii) and Reaction (iv)

(iii) Na(+) + 1e⁻ ⟶ Na

(iv) Li(+) + 1e⁻ ⟶ Li

Half Cell Potential:	-2.5V	
Surface Appearance:	Severe cracking and swelling	
Lc Increase (from As-Received):	A	400 ⟶ 660A°
	B	400 ⟶ 750A°
	C	430 ⟶ 690A°

Evidence Supporting (iii) and (iv)

(1) Cracking and swelling of electrodes by Na and/or Li deposites are typical characteristics of Na and/or Li penetration into graphites.

(2) The increase in Lc by 250A° (from as-received graphites) for the B graphite is considerably larger than increases at the Fe and Al deposition potentials.

(3) Reverse (anodic) scan potentials are more electro-negative (see Figure 4d) after reaching -1.6V than the earlier reverse potentials. At potentials more electro-negative than -1.6V, the graphite surfaces has changed from an aluminum surface to either Na or Li a co-deposit of both.

Comments

No conclusive evidence has been made for which one of the two reduction reactions takes place first or whether Li and Na co-deposit. Therefore, a technique to analyze Na or Li metal should be developed.

Summary

Graphite performance characteristics in molten salts can be evaluated using electrochemical polarization of electrodes followed by a measure of crystallite size (Lc). The Lc data of graphites held at increasing cathodic potentials (given in Table III) show two major major points of interest: (1) The Lc increases marketly for A graphites at the $AlCl_3$ reduction potential (-1.3V) as compared to -0.8V while the better performing graphites (B and C) do not, and (2) A large increase of Lc for B and C graphites does occur at -2.5V, implying that reaction(s) other than $AlCl_3$ reduction take place.

Effect of Temperature:

Potentiodynamic data of graphites polarized at various temperatures showed no significant difference in the typical cathodic polarization curves. Graphites were unaffected when polarized at 700 and 725°C. At temperatures of 750°C, cracks on the graphite surfaces were observed while at 800°C increased cracking and significant swelling occurred after 3 hours exposure.

Conclusions

(1) A promising method to evaluate the performance of graphites in molten salts has been developed.

(2) The extent of graphite swelling and cracking is related to at least three, possibly four cathodic reactions.

(3) Increased cracking and swelling of graphites occur at temperatures above 750°C in a $NaCl-LiCl-AlCl_3$ electrolyte.

(4) Data from cyclic voltammetric scans shows that the graphite surface is changed irreversibly during cathodic polarization at successive reduction potentials.

References

1. Philip L. Walker, Jr. and Peter A. Thrower, Chemistry and Physics of Carbon, Volume d12, pp. 1-37; Marcl Dekker, Inc., New York, NY, 1975.

2. M. B. Dell, Extractive Metallurgy of Aluminum, Volume 2, p. 403, Interscience, New York, NY, 1963.

3. E. W. Dewing, Trans. Met. Soc. AIME 227 (1963) 1328.

4. C. Krohn, M. Sorlieg, and H. A. Oye, "Penetration of Sodium and Bath Constituents into Cathode Carbon Materials Used in Industrial Cells," Light Metals, AIME Annual Meeting 111th (1982), pp. 311-324.

5. G. J. Houston, B. J. Welch, and D. J. Young, "Degradation Processes in Carbon Cathodes Used in Aluminum Smelting," CHEMECA: Process Ind. (1980), pp. 124-128.

6. S. K. Das, PhD. Thesis, The University of Michigan, Ann Arbour (1974).

7. M. A. Short and P. L. Walker, Jr., Carbon, 1, 1 (1963).

8. Private Communication, D. D. MacDonald, Ohio State University, May (1981).

9. T. Hayashi and F. Mochizuki, "Cathode Carbon Cracking Due to Al-Alloy of Collector Bars," Light Metals, Carbon Technology Session AIME Annual Meeting 110th (1981), 637.

CHAIRMEN

W.S. Peterson
Spokane, Washington

G.E. Spangler
Reynolds Metals Company
Richmond, Virginia

PIPELINE FOR TRANSFERRING MOLTEN ALUMINUM ALLOYS

O.G. Sivilotti and J. Sulzer

Alcan International Limited
Kingston Laboratories
P.O. Box 8400
Kingston, Ontario
Canada
K7L 4Z4

A.F. Tregenza

Alcan Canada Products Limted
P.O. Box 2000
Kingston, Ontario
Canada
K7L 4Z4

Abstract

This paper describes the development of an electrically heated pipeline of modular construction to transfer molten metal at rates up to 450 $kg.min.^{-1}$ over a distance of about 35 meters, from tilting furnaces to a pilot plant casting installation. Included is a discussion of the requirements for the line, an outline of the design concepts, and a description of the prototype testing in the laboratory of the first modules constructed, including results on thermal losses upon initial filling as well as on steady state flow of metal, metal drainage behaviour, and metal skull buildup. Finally, the installation of the complete pipeline in the pilot line is described together with the operational performance in that plant.

Introduction

In the aluminum industry molten metal is commonly transferred from a melting or holding furnace to a casting facility by means of an open trough system, either intermittently or continuously. These have the advantage of easy skull removal after each casting run, however, such troughs tend to dissipate heat so that there can be a considerable drop in temperature of the metal between entry and exit of the system. This limits the distance the metal may be transferred, dependent also on the trough materials and on the dwell time of the metal in the system. Moreover, there is some problem in starting up the system, because the first metal flowing into the system loses some heat to the trough lining depending of course on how much preheating was possible. Superheating the metal in the furnace is a way of offsetting these losses, however there is a practical upper limit to the metal temperature. The higher the metal temperature, the greater the melt oxidation losses and the higher the metal gas contents. Also, as the trough materials heat up at the beginning of casting, the temperature losses gradually diminish, hence demanding diminishing superheat in the metal. This poses a production control problem in that it necessitates controlling rates of temperature reduction in the large volume of metal in the furnace.

These matters become concerns only in transferring molten metal over relatively large distances, where the temperature drop of the metal in transit is in the order of 30 to 50°C, or greater. Such losses occur with trough systems exceeding 9 meters and metal flow rates less than 450 $kg.min.^{-1}$. Covers may be provided to limit thermal losses from the metal however these offer no solution for the initial large temperature losses during filling of the trough systems. Furthermore, such covers provide no protection against metal surface oxidation, and hence skull formation, which becomes significant for long trough lengths.

Alcan's Special Requirements

A prototype continuous casting unit was to be placed into an existing Alcan production line. The only suitable location for this unit placed it 45 meters from the furnaces that were to supply the molten metal for casting. This meant tapping into an existing troughing system 8.5 meters from the furnaces which left 35 meters more for any new system to bridge, and obviously, it called for a system with very low heat losses in order to operate the furnaces at temperatures no higher than 750°C. Since metal casting temperatures of up to 710°C were part of the requirements for the prototype casting unit, it left a maximum permissible temperature loss of only 40°C for the entire transfer system. This includes metal temperature losses in the tundish as well as losses in any in-line degassing and filtration units. Furthermore, this temperature loss was to apply to metal transfer rates of 100 to 450 $kg.min.^{-1}$.

A survey of the existing conventional troughing sytem produced the temperature loss curve shown in Figure 1, under steady state conditions, over the range of flow rates of 500 to 3000 $kg.min.^{-1}$. Extrapolation of this data to the lowest transfer rate required, namely 100 $kg.min.^{-1}$, suggests a loss of more than 35°C for troughing over 30 meters in length. Consequently a loss of 10°C could be expected in 8.5 meters of the conventional troughing to be used for the prototype casting installation, plus another 5°C loss in an in-line degassing unit, and another 5 to 10°C loss in an in-line filter and tundish to the casting unit. This leaves a maximum permissible temperature loss of only 15°C for the remaining 35 meter transfer length to be installed.

A highly insulative pipeline was considered an attractive possibility for a number of reasons. It would envelope the metal completely thereby preventing radiation losses; it would present minimal surface area to air and therefore it would be least prone to surface oxidation over prolonged casting periods; and, if necessary it could be fitted easily with external electric resistance heaters for efficient preheating and to prevent excessive heat losses during casting, should this be found necessary. All these considerations are relative to conventional open troughs. Fibrous aluminum silicate tubing of suitable dimensions were readily available from various refractory companies and calculated thermal losses, as summarized in Table I, indicated that the metal temperature losses, at least at steady operating states, were within the thermal requirements, as described earlier.

The calculations pertain to a supposed pipeline 35 meters in total length, comprizing segments of 100 mm internal diameter and 180 mm external diameter, delivering 700°C molten aluminum at 100 to 450 $kg.mm^{-1}$. The thermal conductivity of the pipe material was estimated at 0.11 $W.m^{-1}\,°C^{-1}$. The conductivity of the molten metal was taken as 0.225 $cal.cm^{-2}.sec^{-1}$. The heat loss at steady state through the pipe is given by equation 1.

$$Q = \frac{(T_i - T_o)}{\dfrac{\ln\left(\frac{R_o}{R_i}\right)}{2\cdot\pi\cdot C\cdot \ell}} \qquad (1)$$

where Q - heat loss (W)

C - thermal conductivity of the material

T_i - inside pipe wall temperature (700°C)

T_o - outside pipe wall temperature (124°C, from Carborundum Product Specifications)

R_i - inner pipe radius (0.050 m)

R_o - outer pipe radius (0.090 m)

ℓ - pipe length (35 m)

The temperature loss from the metal at various metal flow rates is given by equation (2).

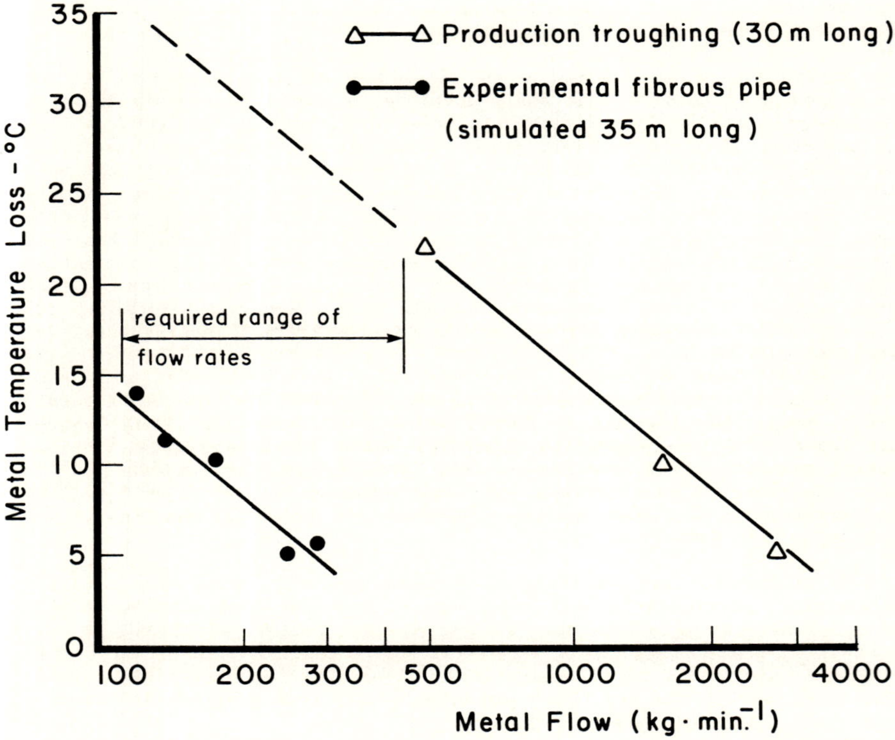

Figure 1 - Comparison of metal temperature losses in production troughing and the experimental fibrous pipe

$$\Delta T = \frac{K.Q}{F.H_{sp}} \quad (2)$$

where Q - is the heat loss (W)

ΔT - is the temperature loss (°C)

F - metal flow rate (kg.min.$^{-1}$)

H_{sp} - specific heat of the metal (cal.g^{-1}.°C^{-1})

K - constant (0.01434)

Table I. Calculated Metal Temperature Losses for a Fibrous Refractory Pipeline

Metal Flow Rate ($kg.min.^{-1}$)	Temperature Loss (°C)
115	13.1
225	6.7
350	4.3
450	3.4

Some other important aspects had to be addressed, namely, the system would have to be relatively maintenance free, demanding infrequent entry into the pipeline for metal skull removal, liner replacement, heater replacement (if heaters are necessary), or any other disruptive maintenance. Also, the line was to supply metal for several shifts per week interlaced into the production schedule, which involved rather lengthy casting periods interspaced with rather lengthy shut down periods. The metal transfer system therefore had to be brought on-line with relative ease after each shut down period. In addition, it was to be capable of handling a wide variety of alloys.

No information was available on skulling tendency, or erosion rate of such a pipe, particularly with regard to a line supplying aluminum base, magnesium containing alloys over rather extended periods of time. Consequently, no predictions could be made on the potential life or frequency of maintenance likely to be encountered, or on transient heat losses when metal first enters the pipeline containing skulls from previous casts.

Feasibility tests were therefore undertaken in the laboratory to provide some information on these aspects of such a pipeline.

Laboratory Feasibility Tests

An experimental pipe segment was assembled as shown in Figure 2. It comprised two tube segments, each 1.22 meters long 100 mm internal diameter, 178 mm external diameter, suitably cut to fit the configurations shown schematically in Figure 2. The tube segments were fitted into a thick-walled aluminum alloy casing for structural support, leaving a 6 mm air space between the fibrous tubes and the casing, as insulation. Before assembly the inner walls of the tubes were sprayed with hardener/release agent. This treatment was considered essential to resist abrasion during skull removal, and to resist wetting of the aluminum alloy to the fibrous silicate tube walls. The segments were then heated to 260°C for drying, and then they were cut and assembled into the casing. Vertical end members were provided, together with refractory lined basins, ensuring that the experimental pipe segment was full of molten metal throughout each test. Constant metal head in the up-stream basin was achieved with a conventional refractory float and dip tube arrangement. Predetermined

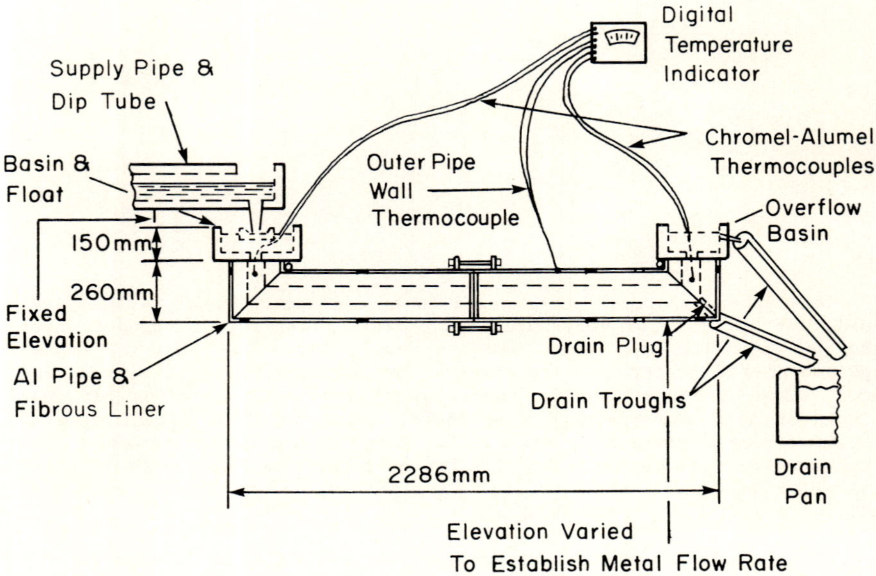

Figure 2 - Schematic layout of the experimental fibrous pipe segment and metal pouring arrangement

metal flow rates were established and maintained by appropriately raising or lowering the exit end of the experimental pipe segment, that is, by adjusting the elevation of the overflow weir. Metal was supplied from a 4,500 kg tilting furnace to the experimental segment via another pipe segment as shown in Figure 3.

The choice of alloy was based on the concern about skull formation and removal. A relatively high magnesium containing alloy, AA-5454, was selected because it represents an alloy series which is relatively aggressive to silicate refractories, is prone to oxidation, and it has a fairly long freezing range which would influence the ease of draining from a non-heated pipe segment. All tests were of 1 1/2 to 2 1/2 hour duration, depending on the metal flow rate under investigation. Melt weights were of the order of 900 to 1300 kg each, and all melts were fluxed with chlorine gas for 10 minutes as part of the batch preparation.

The differential temperature between upstream and downstream locations was determined by measuring the absolute temperatures at each location using chromel/alumel thermocouples. They were calibrated by placing both in the upstream metal basin at the beginning of each test, and any discrepancy was noted and the necessary corrections were made to the difference in temperatures measured.

Figure 3 - Photograph of the metal supply and experimental pipe layout

Experimental Procedure

A series of 10 tests was conducted in which the metal temperature losses were measured until steady state was achieved. In addition, the external refractory pipe wall temperature was measured and recorded. Temperature losses were measured for various metal flow rates in the range of 4.5 kg.min.$^{-1}$ to 13.5 kg.min.$^{-1}$, these being scaled down to represent similar metal dwell times in a pipeline of 35 meters in length, with metal flow rates ranging from 115 kg.min.$^{-1}$ to 350 kg min.$^{-1}$. Only one predetermined flow rate was maintained during each test. The system was filled at much higher flow rates, namely 25 to 45 kg.min.$^{-1}$ to avoid premature freezing of the metal onto the cold inner pipe walls and the cold metal skulls left in the pipe from previous days' tests. These skulls were removed only temporarily between tests in order to examine the condition of the inner pipe walls.

The flow rate was measured by timing with a stop watch the filling of a 4.5 kg ladle from the overflow weir. Adjustments were made when necessary to the pipe slope. At the end of each test the supply of molten metal was stopped by tilting the furnace back, then the horizontal pipe segment was tilted by lowering the metal exit end by 68 mm, a slope of about 3%. Then the drain plug was knocked out. With a metal depth of only about 50 mm left in the pipe the 45 degree joint was opened in order to speed up drainage of the remainder; it was not safe to attempt this with large quantities of metal still in the pipe.

Of major interest in this work was the nature of the skull formation, i.e. the extent of skull buildup and its effect on metal temperature when left in place following each test, and whether it adhered to the pipe wall and consequently damaged the inner pipe surface during removal. Therefore, after cooling down from each day's test run, the skulls were pulled to see if they stuck to the inner wall of the pipe. The inner wall was examined for wear and softness, and then the skulls were returned to the pipe prior to the next test. After the total of 10 repeat test runs the accumulated skulls were weighed. Those from the upper delivery pipe were removed, weighed, and discarded after each run.

Experimental Results. The differential steady state temperature losses for the various simulated metal flow rates are shown in Figure 1, for comparison with the results from conventional troughing. The measured steady state losses were in excellent agreement with the calculated losses given in Table I, and certainly met the requirements for the proposed Alcan installation. However, the transient losses, as seen in Figure 4, did not meet the requirements. Temperature losses of over 20°C occurred in the first 30 minutes of each test with intermediate and low flow rates, and equilibrium conditions were reached after only one hour into the test. Indeed metal temperature losses of 35 to 50°C were experienced on first filling of the experimental pipe segment, which would undoubtedly result in freeze-off of metal in the feeder to the continuous casting machine. Heat flow conditions through the pipeline were the same for each test run, also reaching equilibrium gradients only after one hour into the test run. The outer surface temperature was much higher than expected from the data sheets, however this could have been due to the fact that the pipe was enclosed in a casing.

The pipe segments stood up remarkably well during the 10 day test series. The final skulls pulled away freely from the inner hardened surface without evidence of surface abrasion or mechanical interlocking, nor was there any evidence of softening of the surface. Damage had occurred only around the drain plug. A steel tool had to be used to knock the plug free as it had been cemented into place.

The skulls were rather heavy metallic skulls containing much metal among layers of oxide. The skulls weighed 1 kg in comparison with only 0.045 to 0.063 kg weight of each skull removed from the upper supply pipe. This showed that the accumulated skulls had not fused with incoming metal and as a result the oxide layers may have hindered the final drainage from the pipe.

It was concluded that an unheated, insulated pipe fell short of the requirements, mainly because of high initial temperature losses and also due to excessively heavy skull accumulations. Therefore, it was decided to surround the pipe with heaters.

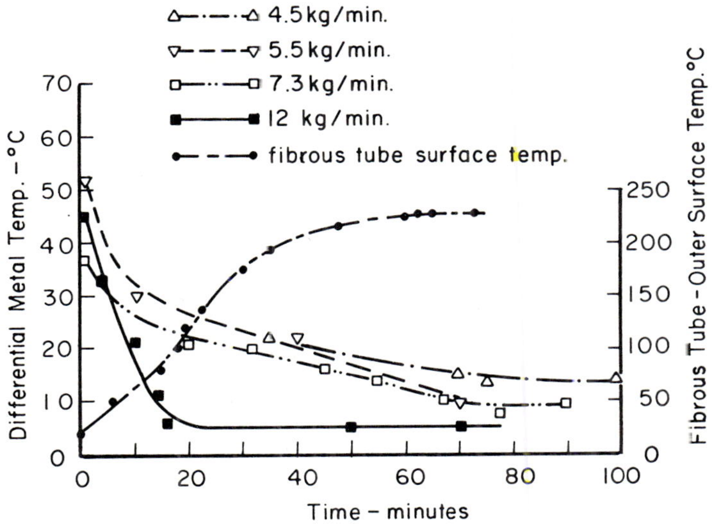

Figure 4 - Examples of metal temperature losses with time at various metal flow rates and time to establish constant thermal gradient through the pipe.

Heated Pipeline Design (1)

The pipeline for the production plant consists of 30 individual fibrous segments, each 1200 mm in length, with inside and outside diameters of 127 mm and 178 mm, respectively, spanning the 35 meters from the furnace room to the continuous casting machine, part of which is shown in Figure 5. As installed each segment is surrounded by electrical heating elements, 2 kW each, encased in 0.7 mm thick stainless steel sheet running almost the full length of the segment. This sheathing ensures uniform heat dissipation from the electrical elements to the fibrous segments, termed liners. A 25 mm thick outer layer of insulation fills the space between the heaters and an outer thick-walled steel pipe and water jacket, as seen schematically in Figure 6. The insulation is constituted of mats filled with a fine silica-type material. The steel pipe encases and supports the whole assembly.

The liner is held in place concentric with the steel casing by means of a tapered, hardened annular ring, called a plug, of fibrous silicate material as is also shown schematically in Figure 6. The smaller end face of each plug abuts the end of the thermal insulation and seals the heater cavity against any metal leakage between pipe segments. The larger end face is flush with the steel end flange with which the segments are bolted

Figure 5 - Photograph of the up-stream half of the molten metal pipeline, as installed in Alcan's production plant.

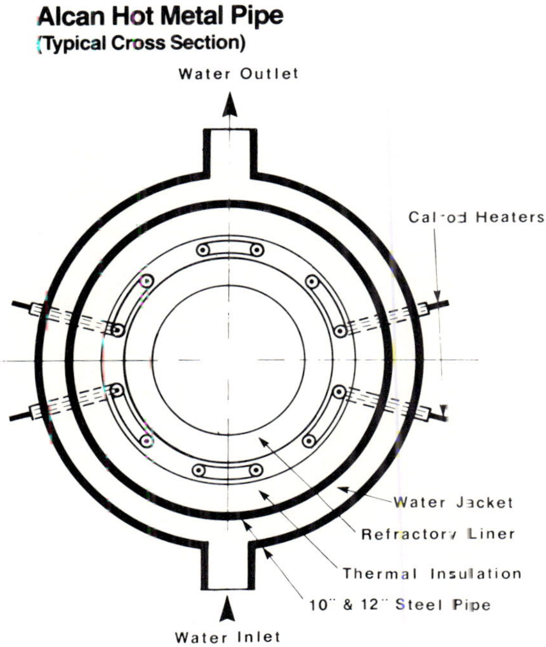

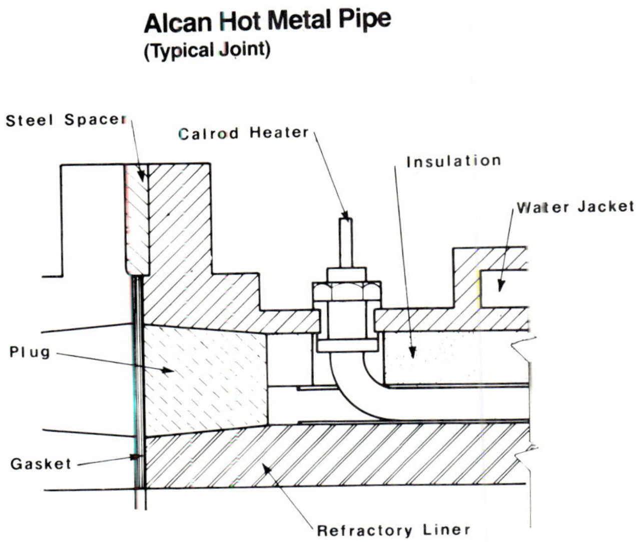

Figure 6 - Schematic drawings of the design of a pipeline segment, showing heater placement and end construction.

to one another. A fibrous paper gasket is placed between segments as a metal seal, however the bolting stress is carried by a larger steel ring of a thickness that will allow only 50% gasket compression and then the bolting compression stress is transmitted through the steel spacer ring only, thus a strong positive non-leaking joint is achieved.

All segments are of identical design and hence interchangeable anywhere in the pipeline. Each has its own water-in and water-out connection, its own electrical cable and individual thermocouple temperature sensor located within the heater cavity. Thermocouple leads are brought to banks of proportional controllers, as seen in Figure 7, also one controller per pipe segment. On-off lights on the controllers indicate whether or not the heaters are operating. Flexible water hoses connect the water jackets with one main supply, pressurized by a single pump creating a flow rate of 25 to 35 $\ell.min.^{-1}$ through each pipe segment; four segments are connected in series following which the water is fed to open drains on a collector pipe. Should a metal leakage occur which would overheat the water in the cooling jacket, steam would form and would be visible, rising from the open drain. The cooling water temperature increases by only about 2 to 3°C per jacket, maintaining the outer jacket temperature virtually at room temperature. This is compatible with the construction of each segment. The fibrous liner expands by only 0.7 to 0.8 mm over a temperature range of 700°C, which is just sufficient to maintain adequate gasket pressure between segments. For this reason the outer steel casing is allowed to expand very little.

Heaters were selected to provide enough power to preheat the liners to metal temperature, in the order of 705°C, in just two hours, in order to bring the line into service in a short time. As this is by far in excess of maintaining the line at temperature, proportioning controllers are installed. Calculated and measured line requirements are 1.5 kW per segment during steady state conditions. Since the liners are preheated to metal temperature it is an adiabatic system - the metal neither gains nor loses temperature throughout the line segments and therefore the control over metal temperature is exercized as with any normal short transfer system, namely by controlling metal temperature in the casting furnace.

The pipeline has several elbows where a change in direction was required for placement along suitable paths in the plant. It was decided not to heat these because of their relatively short lengths, and because of the desire to have only a single heated segment for the simplicity of stocking standard pipe segments as spare parts. The elbows are fitted with thick fibrous hardened liners of inside and outside diameters of 127 mm and 250 mm, respectively, therefore filling the space otherwise occupied by the heaters and the insulation in the 1200 mm long segments. There are only four elbows in the line which reduces the metal temperature by less than 10°C. Special transition pieces with the existing open troughing, and in-line filter and degasser are of this same construction.

The pipeline is subdivided into sections of four 1200 mm segments each, for draining purposes. At each juncture of these sections a hoist is provided to lift the ends of two 4-segment units simultaneously hence draining about 10 meters of line at once. An 8 degree tilt was found best. The heaters are left on of course during draining to keep the metal in the liners above the melting point while it flows to the exits. Only the last vestiges of metal left in the line need to be drained, because the transfer system is essentially level-pour, and connected to tilting furnaces that permit return flow of metal into these, as they are tilted

Figure 7 - Photograph of a controller bank and an open elbow joint in the pipeline.

back to their holding positions. Therefore as soon as they are activated to tilt back, metal flow to the caster ceases and shortly thereafter return flow commences. The pipe joints are broken only when the metal level in the line has dropped to no more than about 30 mm in depth. This metal is drained into cast iron pans of about 1500 kg capacity, for eventual return to the furnace after the metal has solidified.

Prototype Testing

Before the line was built two prototype 1.2 m long segments were constructed and tested in the laboratory in a similar manner to that described earlier for the fibrous pipe feasibility tests. Much of the gasketting design work and liner manufacturing, coating and impregnation practices were developed during this stage. The first trials were done with two 1 kW heaters built into 1200 mm segments. This was sufficient power during steady state operation however it was inadequate for practical purposes because it took 7 hours for the liner inner wall temperature to reach 660°C. The long preheat time was of no consequence for the laboratory test work and therefore no change was made at this stage; it was considered impractical for the production line. Later, the first few production segments were assembled with the two 2 kW heaters and a few test runs were made to verify that no problems would be encountered by their use.

The first test with two prototypes were carried out without molten metal. Instead the second, or downstream segment, was fitted with four thermocouples held against the inner wall face by a cross-shaped member of refractory. This bank of thermocouples was placed 660 mm from the exit end; monitored were top, bottom and the two sides. The results are plotted in Figure 8, together with the results of the first production segments of the heated pipeline. The air space temperature between heater element rods was measured by a theromocouple built in especially for these tests. It was located at the top of the segments, 200 mm from the downstream end of the pipe segment. Water was supplied at 23 $\ell.min^{-1}$ per pipe segment, at a temperature of 16 to 19°C. There was a rise of only 1 to 2°C across the segment which verified the excellence of the thermal insulation used around the heaters.

The results demonstrate the relative uniformity of the temperature rise among the four test locations on the pipe liner inner walls. The air space temperatures were nearly identical with the controller temperatures and hence were not plotted in this figure. The data illustrates that the stainless steel sheets enveloping the heater element rods distribute the heat well. There is some lag between the inner wall temperature rise and that of the heater air space, as expected, however by the time the inner wall reaches metal temperature, about 660°C, the lag is less than 50°C. These first production units were tested at only 203 V, whereas in the production plant they were to be supplied from a 240 V line. It was therefore safely concluded that the segments would have, and did have, a preheat time of only about 2 hours, which was the design goal.

Test results with molten metal were highly satisfactory. Thermal losses were less than 5°C and in fact when the entry temperature was lower than the controller settings, the metal actually came out hotter than supplied to the segments. Furthermore metal skulls were much thinner than without heaters surrounding the fibrous liners.

Commissioning Procedures and Results

After the pipeline had been installed, the cooling water was turned on to a total of twenty 1.2 meter long segments of the 27 segments in the system. With the main supply valve fully opened, all individual valves to the 5 meter long groupings were calibrated to pass 35 $\ell.min,^{-1}$ and then the valve handles were removed to maintain the settings. Calibration was done with a 35 ℓ marked pail and a stop watch. Next, the inlet water temperature gauge at the main valve was checked with a glass thermometer and also at each 5 meter pipe grouping; the outlet temperatures were measured with this glass thermometer at 15 min intervals during the preheat treatments.

The first preheat was carried out on only half of the pipeline. The controllers were first set at 400°C and the heating rate was checked every 15 min with these same indicating controllers. This temperature was reached by all sections in about one hour and after holding for 45 min at this temperature all controllers were raised to a set point of 600°C. This temperature was reached in only 45 min and after another hold of

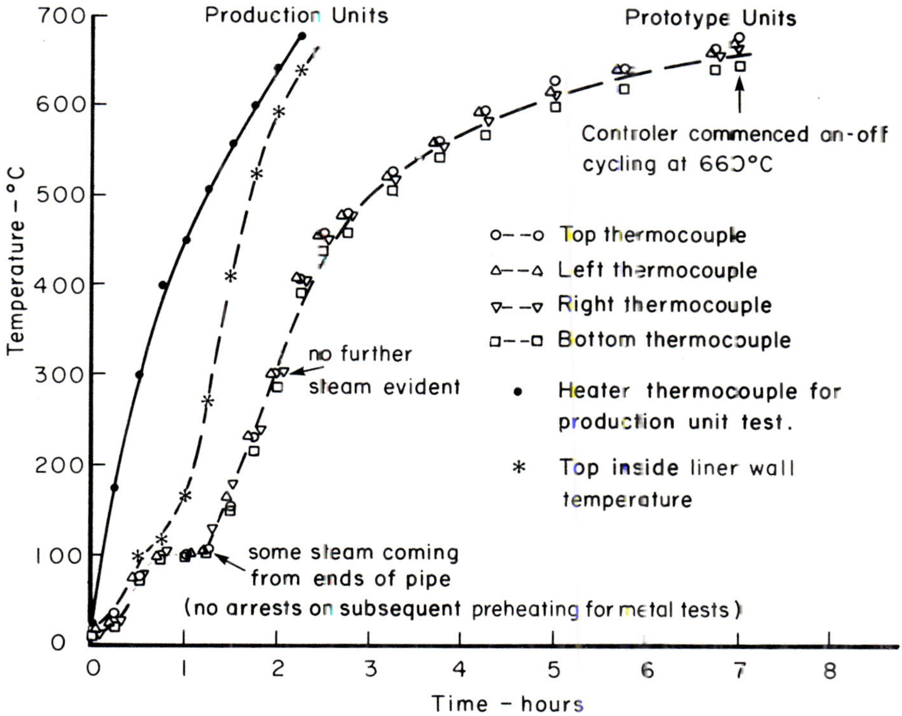

Figure 8 - Dry test results on first heating of prototype and of first production pipe segments

45 min the set point was again raised to 710°C which was reached in another 45 min and then the heaters were shut off and the line was allowed to cool with the cooling water left on. This took in the order of 5 to 6 hours. Next, all liners and pipe joints were examined with a stainless steel mirror for stress cracks before proceeding to the next preheat. No problems were encountered.

A second preheat included a set point at 400°C, and another at 710°C giving a total of only 2 hours for the preheat, as was planned for the line. Then the pipeline was held at temperature (710°C) for another 2 hours prior to shutdown. Also, water temperature measurements were made every 15 min. The general shape of temperature increases are shown in the attached Figure 9 plotted from one representative indicating controller. The line was checked after one year of use and the heating characteristics had remained the same. The water temperature increased 6 to 7°C per 4 section grouping during steady state with heaters at 710°C, and there was very little variation among sections. The line was inspected again and all was in order.

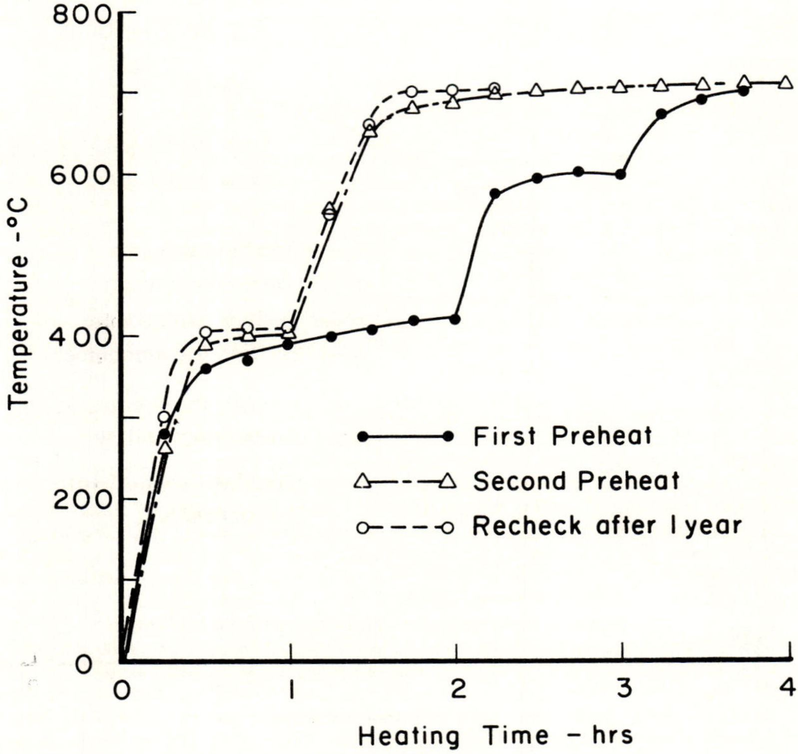

Figure 9 - Rates of temperature climb on heating the line.

Following this, the line was prepared to receive molten metal for the first time. It was planned to put metal into only the first 12 units. A riser section was installed at the end of these units to permit raising the metal head to about 254 mm in the up-stream trough in order to permit running the line full. Thermocouple readings were taken at the pipe entry and at the overflow section to determine the metal temperature losses from the first metal coming through the line. Pouring time was about 20 min, that is, the time to fill about three drain pans of about 1350 kg each. Correcting for the time taken by the metal to reach the exit, the temperature losses were only about 4-5°C for flow rates ranging from 100 kg.min.$^{-1}$ at the beginning to about 200 kg.min.$^{-1}$ for the bulk of the pour. This demonstrated that a starting temperature loss of only 10°C would prevail over the complete line at the beginning of each cast, and probably less during the cast. The temperature losses were obviously occurring in the non-heated pipe joints.

After the line had cooled to room temperature, the thin metal skulls were removed, and the interior of the line was inspected again. No damage had occurred. Following these tests, the remainder of the line was preheated in the same manner as the first 12-unit section, and all was ready for operation.

The pipeline was in service for a period of about 3 years, which was as long as it was required for the prototype continuous casting unit. During this time much was learned about operating such a heated closed pipeline. Twice liner segments overheated accidentally which caused leakage of metal into the heater cavities. These shorted electrically and the liners simply could not be preheated. This caused no problems to personnel because the whole line was electrically grounded. One cold segment in the entire line produced hardly a reduction in metal temperature because whatever temperature was lost in this one segment, was regained in subsequent segments, because the failed units were mid-way along the pipeline. They were replaced by spare segments at the earliest opportunity.

The leakage had been caused by a crushed inner liner. The controller had failed in the power-on position and consequently the heaters increased the temperature around the outer surface of the liner to nearly 1000°C. These rod heaters had a specified maximum temperature of only 820°C yet they had not failed at this temperature. This overheating caused undue liner thermal expansion, however since the gaskets were not designed to accommodate more than about 1 mm expansion, the hot liner was crushed by adjacent segments. Following this experience the controllers were modified so that similar failures would result in power-off.

Skull formation and liner life were excellent for non-magnesium containing alloys, such as, AA-1100 and AA-3003. The skulls were so thin that they could be blown from the cold line with compressed air. These alloys drained well from the line because of their short freezing range. Longer freezing range alloys, such as AA-3004 type, left more skull weight in the units and for these the skulls had to be removed manually, and hence liner abrasion was more prevalent than with the short freezing range alloys. Skulls were normally removed after 5 to 8 casts. There was no detrimental effect on metal quality; the skulls remained in the line rather than being flushed through during line filling.

One long freezing range alloy was transferred through the line, namely AA-5182 alloy. The high magnesium content of this alloy made it very prone to wetting despite the treatment given the liners; the treatment was not sufficiently protective towards oxidation reaction with the silicates of the liner. Also, the long freezing range caused sluggish metal flow during tilt drainage. Ingress of cold air into the line during tilting reduced the metal temperature below the liquidus temperature of the alloy, and consequently metal flow was impeded. This occurred only at the non-heated elbows, which were of course suitable drainage points in the line.

Liner reaction with the molten alloy is by far the most common cause of liner gradual failure, mainly because it delaminates the liner, a bit at a time, each time the skulls are removed from the line. Work was initiated, and was successful in developing better coating and impregnation compounds, especially to resist attack from high magnesium bearing alloys.

The pipeline served very well the purposes of the prototype casting development work which could not have been carried out without the heated pipeline. Excess metal temperature losses in the transfer system were never a problem, nor was metal head or flow control in spite of the long transfer distance. Furnace practices were not changed for any alloy to accommodate the long molten metal transfer line.

References

(1) Olivo G. Sivilotti, J. Sulzer, and Alan F. Tregenza, "Means and Method for Containing Flowing or Standing Molten Metal", US Patent No. 4,345,743, issued 24 August 1982.

AUTOMATION OF SLAB CASTING

J. MORICEAU

Research Center of Aluminium Péchiney
b.p 24, 38340 Voreppe (France)

and

P. HEUILLARD

Aluminium Péchiney Noguères
b.p 11 Pardies, 64150 Noguères (France)

An automatic casting pilot was developed in the Research Center in Voreppe and after a first implementation by Cégédur Péchiney In Issoire cast house, is now operating in Aluminium Péchiney cast house in Noguères.

By means of a micro computer associated with a programmable logical controller, many parameters and functions are controlled, monitored or supervised : furnace and degassing box temperature, furnace tilting, melt level in troughs and moulds, water flow rate (pulsated during starting), casting speed and preset ingot length, refinig wire feeding, degassing box, fully automatic start and end of drop.

Melt level in each mould is measured by a sensor specially developped, with a movable electrode without contact with the metal. This sensor uses electrical capacity measurement between the melt and the electrode. The liquid flow rate is regulated, by a motor driven plug on the spout.

Industrial operation has showed improvements of casting reliability and ingot characteristics, more safety especially during starting and less casting rejects.

Introduction

It has been more and more evident, for many years, that the automatic control of d.c. casting was a right way to improve the reliability and safety of casting operation and quality of ingots.

The automated casting developped in Voreppe Research Center may appear quite similar to other systems described in preceding papers, but actually it includes original particularities, as follows :

- the melt level measurement in moulds is performed using a new sensor with no contact with molten surface
- the program and process are extended to control all operations connected with casting (furnace, degassing box, refining rod..)
- the program includes sophisticated software to help the person who creates or brings up to date casting instructions, set points or supervision data
- the safety is particularly elaborated to allow safe emergency stopping in all critical conditions, even in case of computer failure, without any human supervision

At the end of the first development in the Research Center, the system was first tested in the cast house of a fabricating plant (Cegedur Péchiney in Issoire), on a unit casting 10 ingots of hard alloys.

After that, an improved system was set up in the cast house of a reduction plant (Aluminium Péchiney in Noguères). The equipment is there a combination of the original process entirely based on a mini computer and the current technology for measure and regulation in the plant.

The system began to work in Noguères at the beginning of 1983. It will be installed in the near future in all new fabricating and reduction plant cast houses of the Group.

Melt level measurement

Melt level in the mould is the only casting parameter requiring an unconventional sensor.

In many known automatic control systems for d.c. casting, the measure is provided by means of a float bearing a rod moving through a linear motion transducer. That solution generaly ensures accurate measures in steady state conditions. However the method involves some difficulties :

- disturbing the oxide skin at the surface of the melt
- possible measure deviation in case of binding of the rod fixed above the float
- limited range making it difficult to measure the melt level from the very beginning of stool cup feeding

One may notice that these problems are more critical in the case of electromagnetic casting.

To avoid these defects we experimented and developed a new sensor working with no contact with the melt.

Basically, the sensor uses the measurement of an electrical capacity between the surface of the melt and a movable electrode. A servo-mechanism maintains this electrode at a constant distance above the surface of the melt, then the position of the electrode is given by a linear transducer.

The body of the sensor and the mechanical parts were designed to endure industrial conditions (heat and atmosphere) in a cast house, above molten metal.

The sensor measures the level with a very good accuracy throughout the 6-inch travel. There is an internal auto-test circuit with a digital output to check any failure of the sensor.

The sensors are held on the side of the trough. If the trough does not remain perfectly straight and horizontal, a zero reset operation is necessary during preliminary checking phase before start of casting. This is done by reading the level of a calibrated sheet put on the top of each mould under the sensor.

Melt level control

In order to control the melt flow rate through the dip tube, a motor driven plug is placed above each dip tube in the trough.

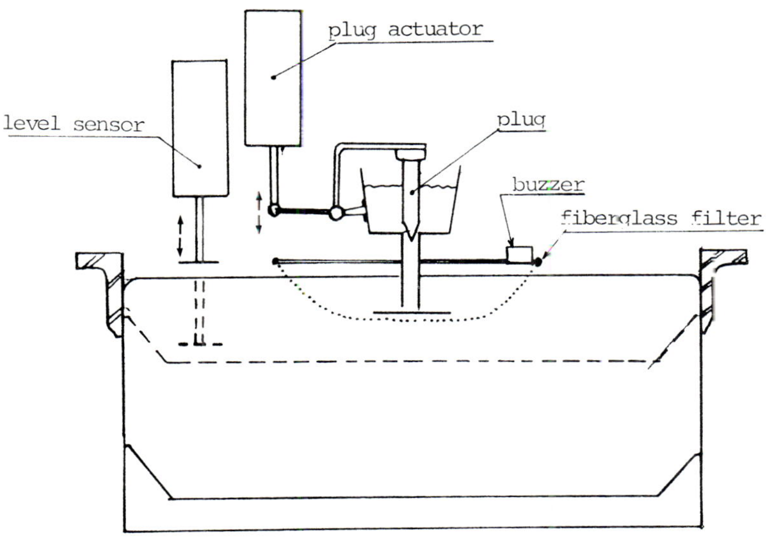

Figure 1 - Melt level regulation (schematic). In dashed line position before starting

Figure 2 - General view of the system installed in Aluminium Péchiney Noguères Cast house. The 3 ingots are 24-inch thick and 44-inch large. The sensors may be moved on the holder to cast ingots up to 86-inch large.

The plug can be moved upwards or downwards by a linear actuator placed on the side of the trough.

There are two ways to control the flow rate :

- partial opening of the dip tube by moving the plug up to a constant height, computed from the flow rate set point using hydraulic laws (flow rate through a hole at the bottom of a vessel)
- actuating the plug by a regulator, according to the difference between measured level and set point at the same time

During the very first moments of stool cup feeding. the first method allows a monitoring of the flow rate until the layer of molten metal reaches the area at the vertical of the sensor. Then the regulators ensure that the levels in all the moulds fit with the set point. The set point is generally variable as a function of time.

System operation

Basically, the management of the casting operations is done by a micro computer which performs many main functions, considered as a whole :

- data storage, management and processing
- control of sequences and setting of parameters
- supervision and checking of working parameters and general behaviour
- continuous display : measures, prompts, warnings
- delayed printing of a casting report

There is an extensive series of elementary control functions :

- furnace temperature
- furnace tilting to get a constant level in the trough
- cooling water flow-rate and during starting, pulsated flow
- speed and length
- melt level in each mould
- trough lifting after casting
- refiner rod feeder
- AlPur or SNIF degassing box including gas flow rate control if adequate valves exist
- wipers operation if any
- for electromagnetic casting AC, intensity in inductors

Before casting, the operator has to input only few data : alloy number, size, number and length of ingots.

During a short checking phase, a few minutes before casting, the system checks the correct value of the main parameters (temperature, water flow, stool cups height), the good functionning and positionning of melt level sensors and plugs and also the state of the "watch-dog".

If start is authorized, the operator may push a key and all the sequences are performed without any human operation until complete stop of the casting. The degasing treatment in the ladle is started, at the very moment the refiner rod begins unrolling and the furnace is tilted up to fill the trough. When the level is correct, all the plugs raise to feed the stool cups and moulds. Buzzers fixed on the holders of fiberglass filters are actuated to avoid molten metal accumulating. As soon as each melt level sensor is able to detect the first molten layer, the levels are controlled by regulators according to the variable set points. The casting machine is started and the parameters move towards steady state conditions.

A few moments before the preset length, the furnace is tilted down, each plug is lowered at the right moment when the actual length of each ingot is correct. Then the trough is lifted up, the plugs are raised, the degassing box and rod feeder are stopped as well as the casting machine. After a computed time, the ingots are moved down below the moulds and the water flow is interrupted.

The casting report is printed in a synthetic form giving the averages and variation ranges of casting and treating parameters, and summarizing if needed all the anomalies noticed during the drop.

Safety function

As a guarantee for ingots quality and casting safety, emergency stop automatically occurs if some parameters remain out of a critical safe range later than a recorded delay.

During startup phase, emergency stop may happen in case of excessive deviation from level set points or in critical situations like hang up ingot or bleeding out which may be detected thanks to the computer.

In case of failure of the microcomputer or of the controller, it is also necessary to execute automatic emergency stop.

That is why a special microprocessor the "watch dog", outside the computer, controls emergency stop :

- tilting down the furnace
- plugs lowering and trough lifting up
- casting machine, treatment in ladle and rod feeding stopping

Before each casting, the computer and "watch dog" microprocessor mutually test their good working. During casting, the "watch dog" must receive regular pulses from the computer, otherwise it stops the process.

In case of failure of the mechanism actuating the plug, an electromagnetic bolt is released allowing the plugs to fall down and close the dip tubes.

Description of equipement

The system installed in Noguères Cast house is designed with two levels.

On the first level the process is connected with a programmable logical controller, external regulators and adequate analog I/O interfaces.

The second level is based on a DEC 11/23 microcomputer with a cassette tape mass memory, connected with the first level and with peripherals placed near the casting pit : control desk, coloured video screen and line printer.

A software was developed to greatly simplify the operations, so that the operator does not know that a computer is present : the communications only require a simple numerical keyboard, to input data or answer questions displayed on the video screen.

Data management

The system requires a number of data : fixed or variable set points for regulation of casting parameters, delays and limits of deviation for checking, physical or technical characteristics of equipment. The data are recorded in files in the mass memory of the computer : one file for common data and another with one page for each product (alloy and size).

The files are also used to store all the informations and instructions about casting equipment and tools : reference and main characteristics of moulds, spouts, filters... So, all the instructions and specifications needed for melt treatment and casting of a given product are stored and may be displayed and printed after identification of the alloy and size.

A program was written to create or bring up to date the pages of the file. That program includes a part of Cast house knowledge.

For some parameters and to help the operator, the program suggests a value which is calculated from physical or metallurgical laws.

All input data are checked before being recorded and incorrect values are not accepted. The program verifies that each value lies between calculated boundaries which can be physical or geometrical characteristics of the equipments, maximum allowed deviation from suggested values or limits calculated from other parameters. For instance the melt level set point in the mould is compared with the position of the solidification front in the mould to optimize the cortical zone and avoid surface defects (this involving the height of the mould and casting rate). Or in another example the asked melt flow rate through the dip tube is compared to theoretical flow rate given by Bernouilli's law according to the diameter of the dip tube.

If some particular parameters are modified, the program also suggests a change in the other parameters which are physically connected.

The detailed inspection of the input data reduces to a very minimum the possibility to get wrong practices or instructions impossible to execute.

The know-how enclosed in such a program is a complementary but indivisible part of our automatic casting system.

Advantages of automated casting

Most of the advantages of automatic control of ingot casting have been already reported :

- less rejects for false starts, hang up ingots, bleeding out, or butt cracks
- better and more consistent quality : surface quality, butt geometry, structure (thinner shell or chill zone)
- more safety to start large ingots casting, because no human presence near the pit
- less hours/ton since one operator only can start or stop a drop with many ingots
- quality and scrap independant on operator's skill
- easier development of new practices

In Noguères Aluminium Péchiney Cast house, it was observed that rejects on a casting unit producing 24-inch thick ingots of alloys for can stock was decreased by 50 % since automatic casting was used.

Conclusions

Fully automated d.c. casting proved to give ingots with best and more consistent quality, geometry and structure, with less rejects or scrapped butts and safer startups.

TWIN ROLL CASTING OF ALUMINIUM:

THE OCCURRENCE OF STRUCTURE INHOMOGENEITIES AND DEFECTS IN AS CAST STRIP

H. Westengen
Årdal og Sunndal Verk a.s. Metallurgical Research Laboratories
N-6600 Sunndalsøra, Norway

K. Nes
Årdal og Sunndal Verk a.s.
N-5875 Øvre Årdal, Norway

Continuous strip casting provides an attractive shortcut from the liquid metal to final gauge sheet. Twin roll casters are by far the most widespread machines in commercial operation, mainly due to their flexibility and relatively low investment costs. Strip produced by these casters is presently appreciated as a well suited alternative to conventional material for a number of applications, and in some instances improved properties are obtained (1). Twin roll casters have been in commercial operation for nearly three decades, however, the literature on the subject is rather incomplete. Most of the available information has been published by the machine producers, hence attention is mainly paid to the positive aspects of the methods. However, some interesting papers have recently been published. A general description of casting and annealing structures is given by Nes and Slevolden (2), while Jin et. al. (3) consider the formation of centreline segregates during casting. Althoff (1) summarizes some aspects of mechanical properties, and a couple of papers desribe new alloys which are especially suitable for strip casting (4,5). A contribution to the understanding of the influence of casting parameters on the solidification zone is presented by Iricibar and Jin (6).

In the present paper a survey of some typical structural inhomogeneities and casting defects occuring during twin roll casting will be given. While most of the defects can be eliminated through careful control of casting conditions and operational practice, it will be evident that in order to become fully competitive with conventional materials for critical applications, further research to upgrade the twin roll casting process will be necessary.

The Twin Roll Casting Process

General aspects

Although differing in some important details, the main principles of machine design and operation are common to the commercially available twin roll casters (7,8,9). Liquid metal from the caster furnace is fed through a launder system to the headbox in front of the casting nozzle, which distributes the metal to be solidified in the gap of the water cooled rolls over the desired strip width. Grain refining elements are added in the casting furnace and/or in the launder. An in-line melt cleaning unit may be installed in the launder system to remove nonmetallic inclusions and gases. Recently built equipment usually includes a coiler with a powerful drive to secure tight coiling and reduce problems caused by sticking of the strip to the roll surfaces.

Solidification front

A schematic drawing of the rolls cavity showing the solidifying strip is shown in fig. 1. Some important parameters are defined in the figure.

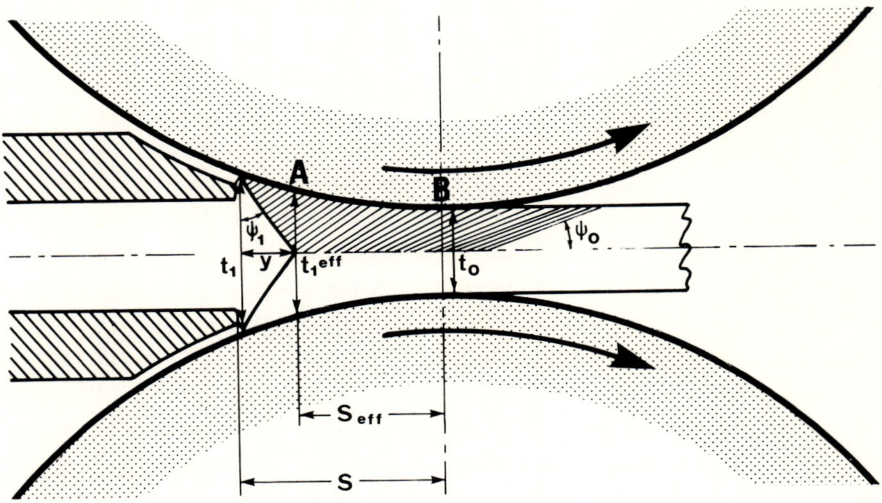

Figure 1 - Schematic drawing showing the rolls cavity, nozzle and solidifying strip.

The typical grain structure of as cast strip, as seen in a longitudinal cross section, fig. 2, contains some information regarding the solidification front. It is assumed that the grains grow normal to the solidification front, and that deformation is essentally confined to the region between A and B in fig. 1. From geometrical considerations the following relationships apply:

$$y = (t_1/2) \tan \Psi_1 \quad (1)$$

$$S_{eff} \sim S-y \quad (2)$$

$$t_1^{eff} = S^2_{eff}/R + t_0 \quad (3)$$

$$\tan \Psi_1 = (t_1^{eff}/t_0)^2 \tan \Psi_0 \quad (4)$$

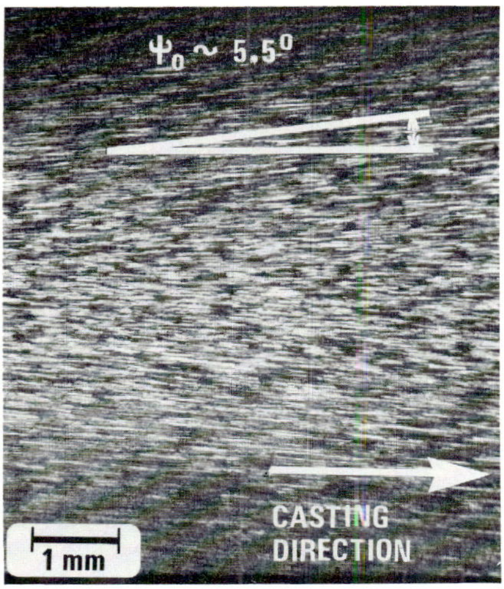

Figure 2 - Typical grain structure of properly grain refined strip. Casting conditions: tip setback S=70 mm, casting speed . 95 m/min., liquid metal temperature in the headbox 685-688°C.

These equations can easily be solved by an iteration procedure allowing the sump length y to be evaluated. For the particular example in fig. 2, Ψ_0 is measured to be approximately 5.5 deg corresponding to a sump length y = 4.8 mm.

The influence of roll velocity or casting temperature on the sump length can be deduced from measurements of the forward slip, as demonstrated by Iricibar and Jin (6). As an example, they estimated an increase in the sump length of magnitude 1.6 mm when the roll velocity was increased from 780 to 1260 mm/min., other parameters kept constant.

Deformation in the rolling gap

The reduction in cross section is given approximately by the expression (ref. fig. 1.):

$$\% \text{ red.} = (t_1^{eff} - t_0)/t_1^{eff} \qquad (5)$$

For the particular example discussed above this amounts to approximately 58 %. The metal leaves the rolls with a temperature typically around 300°C. Deformation thus takes place in the temperature range between the solidus temperature and 300°C. The fairly low exit temperature prevents recrystallization leaving the strip in a deformed condition, fig. 3. Fig. 4a shows the hardness profile across the sheet thickness. When compared with the work hardening curve fig 4b, it follows that the deformation introduced during casting corresponds to 10-25 % reduction in area by cold rolling.

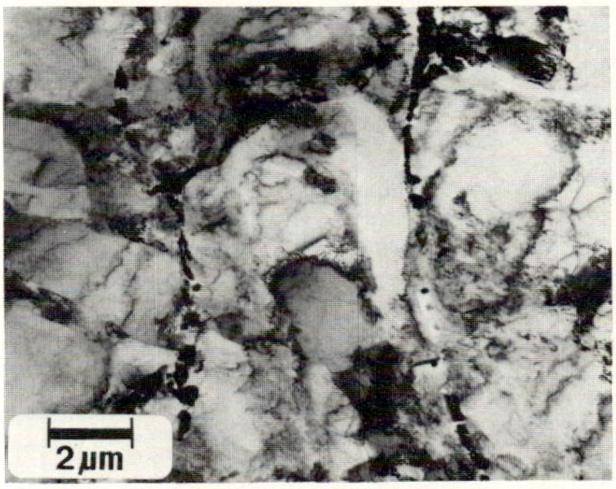

Figure 3 - TEM micrograph showing the deformation structure of as cast strip.

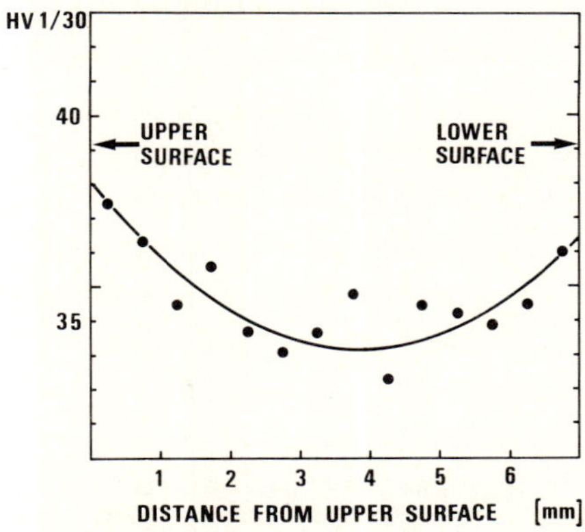

a)

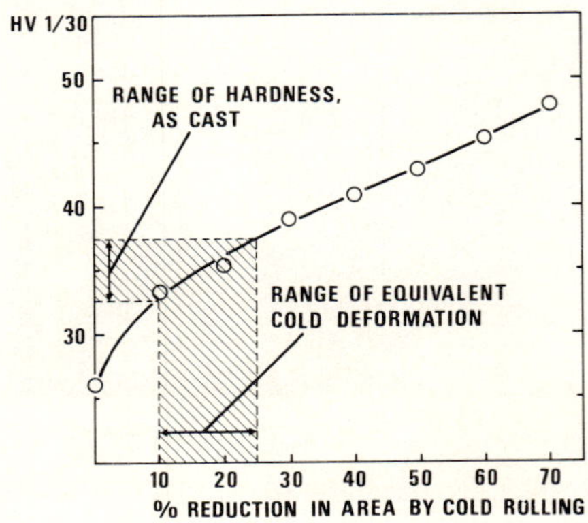

b)

Figure 4 - a: Variation in hardness across the strip thickness, for as cast strip. Al-99.5 wt%. b: Work hardening of the same material during cold rolling after soft annealing.

Solidification microstructure

Solidification occur at high cooling rates, favouring a cellular growth pattern, fig. 5a. The variation in the intercellular spacing across the strip thickness is shown in fig 5b.

By using a formula orginally deduced from measurements of secondary dendrite spacing d and solidification rate v (10):

$$d \cdot v^{0.33} = 33.4 \quad (\mu m (K/s)^{0.33}) \qquad (6)$$

the solidification rate is estimated to be in the range 10^2 K/s to 10^3 K/s. These values are typically 2 orders of magnitude larger than those of conventional DC-casting (10). The high solidification rate has important consequences for the as cast microstructure:

Size distribution of primary constituents. The small cell size, fig. 5b, directly influences the size of the primary constituents. Compared with conventionally DC-cast and hot rolled material the size of the inter-metallic particles are typically reduced by a factor of approximately 5. This is demonstrated in fig. 6.

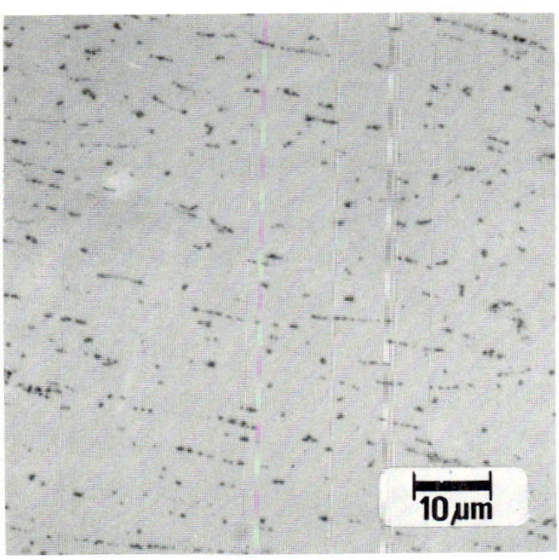

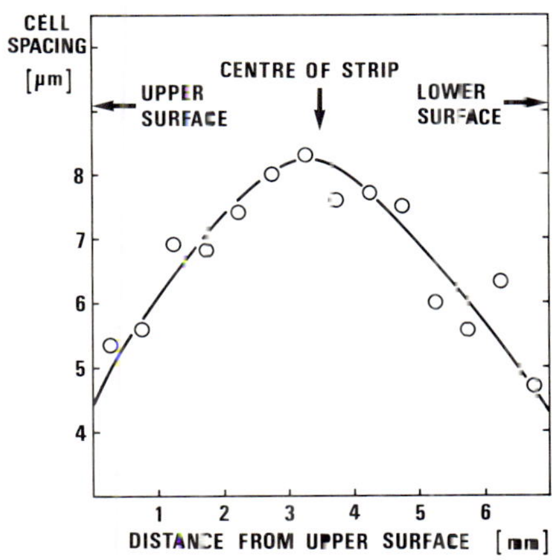

a) b)

Figure 5 - a: Typical solidification cell structure as observed by light microscopy, longitudinal cross section, .5 mm from the surface. b: Variation in cell spacing across the strip thickness. Alloy Al-.5 wt%Fe-.2 wt%Si.

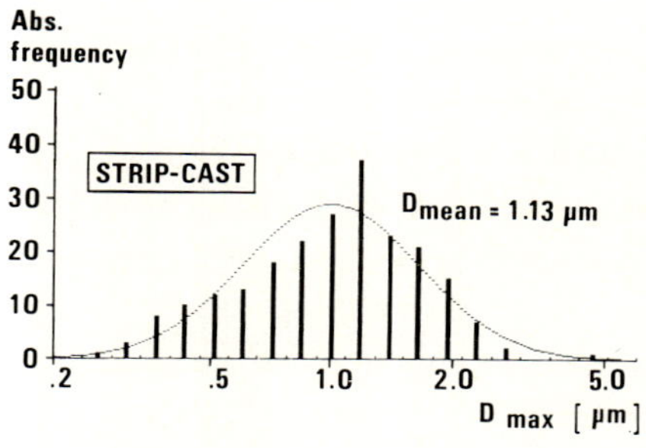

a)

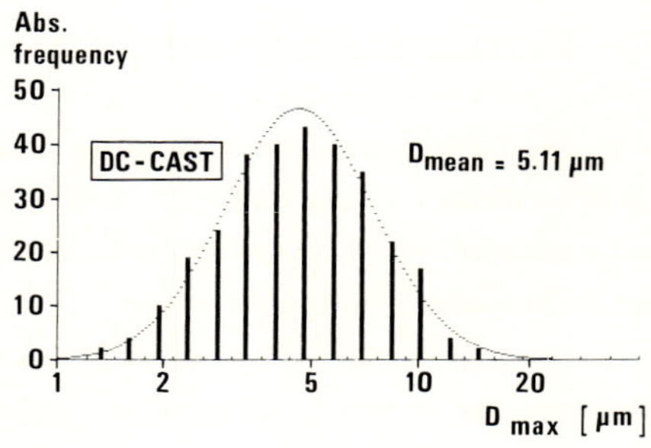

b)

Figure 6 - Size distributions of primary constituents. Alloy AA 3004. a: Strip cast, homogenized 24h/550°C, cold rolled to 2.5 mm and softannealed. b: DC-cast, homogenized, hot and warm rolled to 2.5 mm followed by softannealing. (The guoted mean values are the arithmetic means).

The reduced particle size has some important positive and negative consequences:

The newly developed alloys AA 8006 and AA 8007 utilize a fine dispersion of intermetallic particles to stabilize a grain size in the range 1-5µm. These socalled low-eutectic alloys, show enhanced strength without a severe loss of formability, due to grain boundary hardening. To obtain the necessary dispersion level, a high cooling rate must be applied during casting, favouring the twin roll casting method (4).

Surface galling is frequently observed when cans are produced from strip cast material by the drawing and wall-ironing method. This phenomenon is believed to be caused by the small size of intermetallic constituents, however, other factors related to the as cast surface probably also have an effect.

Types of primary constituents. The high solidification rate favours the formation of metastable compounds during casting. Similar phenomena are observed in DC-cast ingots where the variation in solidification rate across the ingot thickness causes the formation of a number of different phases, giving rise to the socalled "fir-tree" zones (11). The tetragonal phase Al_mFe which forms at cooling rates exceeding 10 K/s (10) is a dominating phase, together with the bcc α-AlFeSi, the formation of which is also reported to be favoured by high cooling rates (12).

Elements in solid solution. As-cast strip shows considerable supersaturation of elements like Mn, Fe, Cr, Zr etc. when these are present in the material. As a particular example, the electrical resistivity of an as-cast AA 3004 alloy, containing 1.0 wt% Mg, 1.0 wt% Mn, .13 wt% Si and .35 wt% Fe was measured to be 6.54 μΩ cm^{-1}. The amount of Mn + Fe in solid solution is approximately given by:

$$\text{wt\% (Mn + Fe) in solid solution} \sim (\rho - \rho_{Al\ 99.99} - 0.5\ C_{Mg})/3.6 \qquad (7)$$

where ρ is the resistivity of the alloy, $\rho_{Al\ 99.99}$ is the resistivity of Al (99.99 wt%) and C_{Mg} is the concentration of Mg. The factors 0.5 and 3.6 are taken from ref. (13). They represent the contribution from Mg and Mn in solid solution to the resistivity in units μΩ $(cm\ wt\%)^{-1}$. From the measured electrical resistivity of as-cast AA 3004 it follows that 0.92 wt% Mn + Fe is in solid solution.

Due to precipitation of Mn and Fe during subsequent anneals, grain size problems are encountered (2). Possible means to avoid these problems include homogenizing and flash-annealing. A positive aspect of the high supersaturation is that recrystallization is retarded, giving possibilities for the production of temper-annealed sheet with attractive mechanical properties (1).

Structure Inhomogeneities and Casting Defects

Centreline segregates

Centreline segregates are frequently observed, especially for materials with higher alloy contents, fig. 7. In a recent report by Jin et.al. (3) these inhomogeneities are attributed to pressure-induced channel segregation. Under the action of the pressure exerted by the rolls, liquid metal enriched in alloying elements is squeezed from colder to hotter regions forming channels. When this liquid solidifies, typical eutectic colonies are formed. Locally the composition may even become hypereutectic, causing large preeutectic particles to form, fig. 7. According to Jin et.al. (3), the volume fraction of centreline segregates increases with alloying additions and casting speed. These inhomogeneities seem to be inherent to the roll casting process. No harmful effects have been observed on the final properties of the sheet.

Grain structure

It is a general experience that a fairly large amount of grain refining additions must be applied to secure a homogeneous, fine-grained structure. Figs. 2 and 8 show various grain structures including a properly grain refined structure, fig. 2, feather crystal formation caused by too little grain refiner, fig. 8a, and coarse grains formed when no grain refiners are added, fig. 8b.

Practical experience indicates that a combination of Ti-containing hardener added in the casting furnace and Al5Ti1B-rod added in the launder provide the best reliability during long term continuous running of the machines.

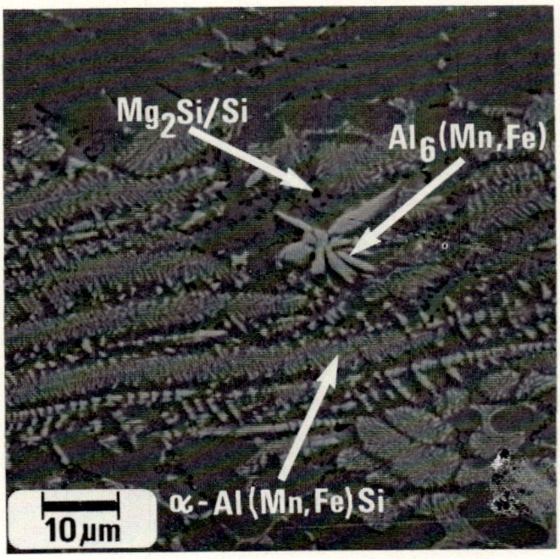

Figure 7 - Centreline segregates as observed by scanning electron microscopy. Alloy AA 3004.

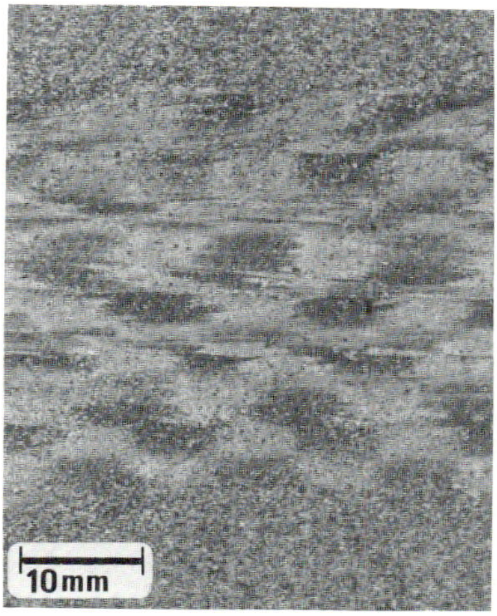

a)

b)

Figure 8 - Unfavourable grain structures. a: Feather crystal formation. b: Coarse grain structure.

Level lines

A surface inhomogeneity termed level lines in analogy with DC-casting is shown in fig. 9. As discussed by Nes (2) this inhomogeneity, which is visible on the as cast sheet, is connected to a subsurface variation in the cell spacing, indicating that the solidification rate close to the surface varies in a periodic way. This is attributed to a cyclic breaking up of the oxide film enveloping the liquid metal when it leaves the tip of the nozzle.

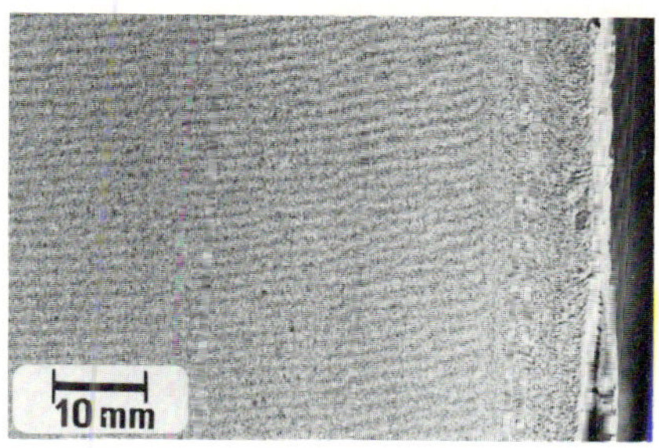

Figure 9 - Typical level lines. The Sample is etched to emphasize the effect.

Surface streaking

Streaks on the surface are often observed during casting. These may or may not be an indication of subsurface defects.

Streaks with underlying voids. A serious casting defect is the formation of more or less continuous stringers of voids, which can extend for hundreds of meters along the strip. Fig. 10a shows an example as seen in a transverse cross section. The occurrence of these defects is more frequent in Mg-containing alloys than in pure aluminium or Al-Mn alloys. The defects are usually accompanied by surface streaking of the as cast strip, fig. 10b.

A simple bend test provides some information on the seriousness of the defects, fig. 10c. It follows from fig. 10a that the grain structure in the vicinity of the voids is distorted, indicating that some restriction to metal flow causes the defect formation.

There are two alternative explanations for the formation of the restrictions:

- Local build up of impurities and clogging of the nozzle.
- Formation of gas bubbles (H_2) at the outlet of the nozzle.

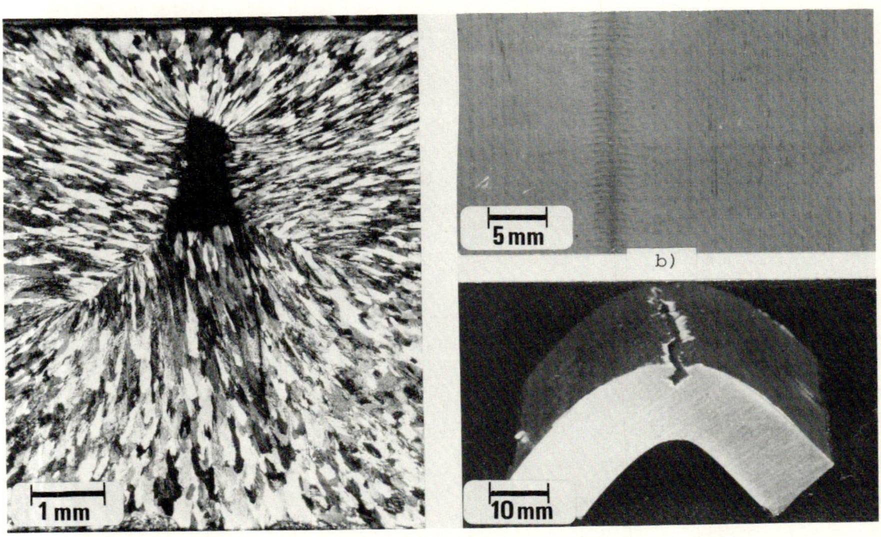

a) c)

Figure 10 - Surface streaking with underlying voids. a: Transverse cross section showing the grain structure and void. b: Appearance of the as cast surface. c: Cracking after bend testing. Alloy Al - 0.6 wt%Mg - 1.0 wt%Mn.

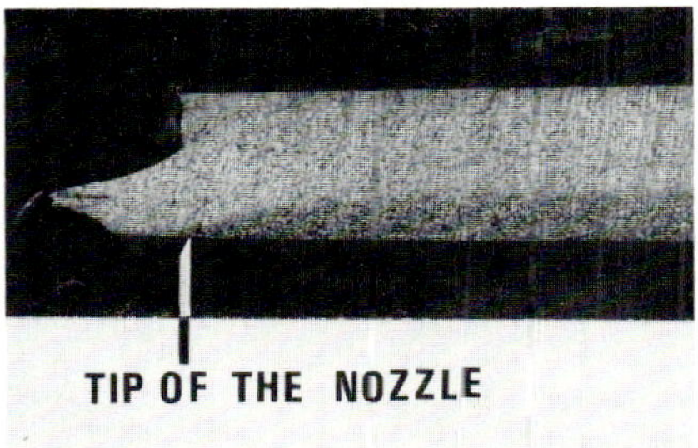

a)

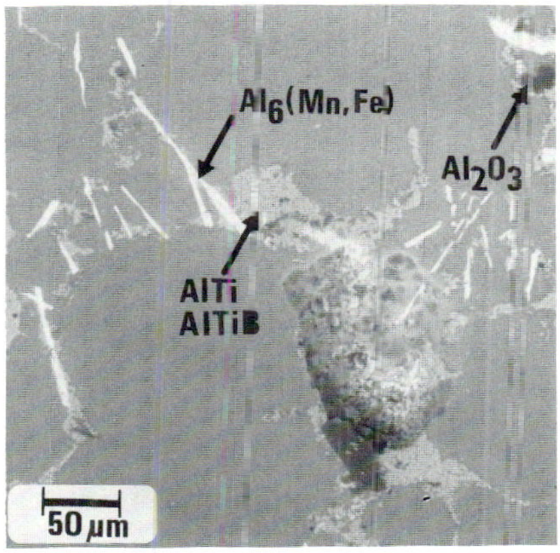

b)

Figure 11 - a: Longitudinal cross section of the material inside the casting nozzle. Etched to reveal the segregates at the bottom of the nozzle. b: Micrograph of the botton layer showing the various constituents which are present. Alloy Al - 0.6 wt%Mg - 1.0 wt%Mn.

Investigations of the casting nozzle after finishing a casting cycle have revealed build-ups of a slurry at the bottom of the nozzle, fig. 11. Detailed examinations in the scanning electron microscope show agglomerates of Ti and B-containing particles as well as oxides, fig. 11b.

It is also found that Mg-containing melts attack the styrite material in the nozzle leading to erosion and formation of spinels. The more or less stationary bottom layer in the nozzle can become several mm thick and possibly cause local clogging.

Measurements of the H_2-content in materials containing void defects generally show high values. As an example, the material containing the defect in fig. 10, was found to contain 0.29 ppm H_2. The low metal throughput necessitates keeping a high temperature in the casting furnace, thus increasing the H_2-content, especially for Mg containing melts. The need for improved launder systems and efficient degassing units is evident.

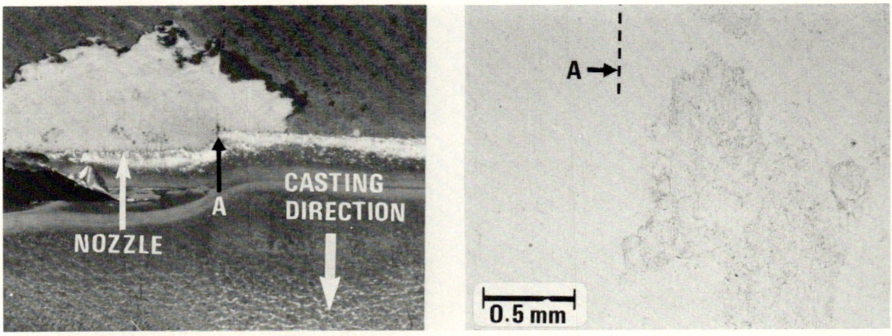

a) b)

Figure 12 - a: Close up view of the remnants of the tip of the nozzle showing displacement of two adjacent nozzle segments. b: Segregation of Fe/Si caused by the distorted metal flow.

Surface streaks caused by segregates. Any irregularities in the way metal is fed to the solidification zone can cause segregation. Fig. 12a shows a close-up view of the remnants of the nozzle tip after a casting cycle. It is evident that the two adjacent nozzle segments have been displaced, see position A. The resulting distortion of the solidification zone causes surface streaking on the strip. The micrograph in fig. 12b shows massive segregation associated with the streaking.

Unharmful streaks. Finally, to add to the complexity, surface streaks may occur, which are not connected to any internal defects. These disappear during cold rolling and is caused by local variations in the surface condition of the casting rolls, deposits of Al or oxides etc.

Surface defects caused by damaged roll surfaces.

Irregularities in the surface of the casting rolls will be transmitted to the strip surface. The most common damage to the rolls is caused by thermal fatigue. After a certain operation period, the surface of the rolls cracks and protrusions are formed along the edge of the cracks. Accordingly, intrusions are formed in the strip surface which eventually becomes so serious that the rolls have to be reground. Fig. 13a shows the nearly perfect satin matte surface obtainable with newly ground rolls. This can be compared with fig. 13b showing the surface prior to changing the rolls (after casting 450 metric tonnes for this particular example).

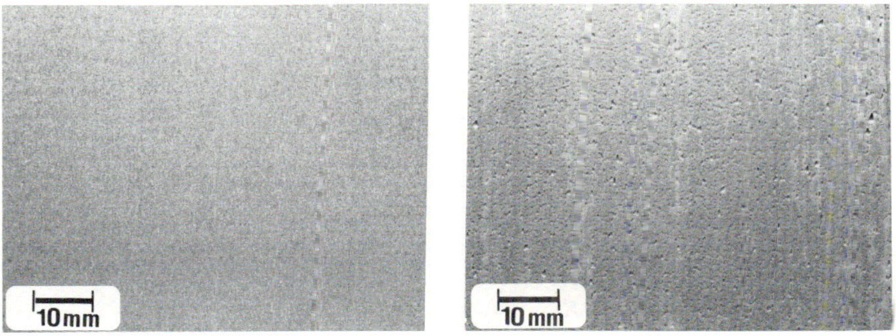

a) b)

Figure 13 - Typical examples of good and bad strip surface caused by changes in the surface quality of the rolls. a: New rolls. b: After casting 450 metric tonnes.

Rolled-in particles from the nozzle material

In the SCAL Jumbo 3C Caster, the nozzle is fitted by running the rolls in contact with the nozzle tip prior to casting. The clearance between the

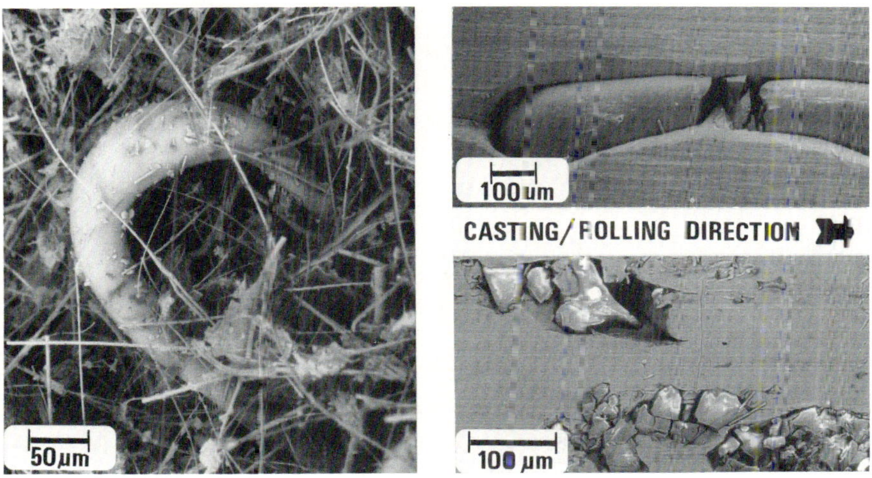

a) c)

Figure 14 - SEM-micrographs, a: Nozzle material, note the coarse particle embedded in the fibrous matrix. b: Particle from the nozzle rolled into the surface of as cast strip. c: Defects in 70 µm foil caused by the nozzle particles.

nozzle and the rolls during casting is very small and under unfavourable circumstances direct contact can occur. The nozzle material consists of Al_2O_3-SiO_2 fibers with an added binder, however, a coarse fraction of particles is also present, fig. 14a.

When the rolls are in direct contact with the nozzle, particles can be deposited on the roll surfaces and further transmitted to the as cast strip where they are rolled in during the deformation, fig. 14b. During further rolling these particles will cause serious defects in the sheet, fig. 14c. Careful adjustment of the nozzle minimizes the number of these defects, however, improvements in the nozzle material is imperative for complete elimination of the problems.

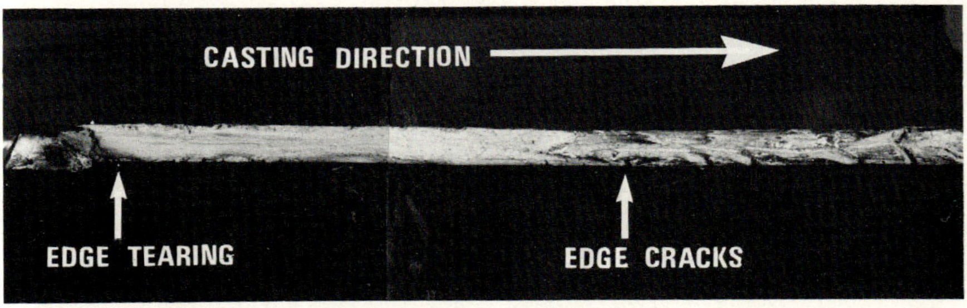

a)

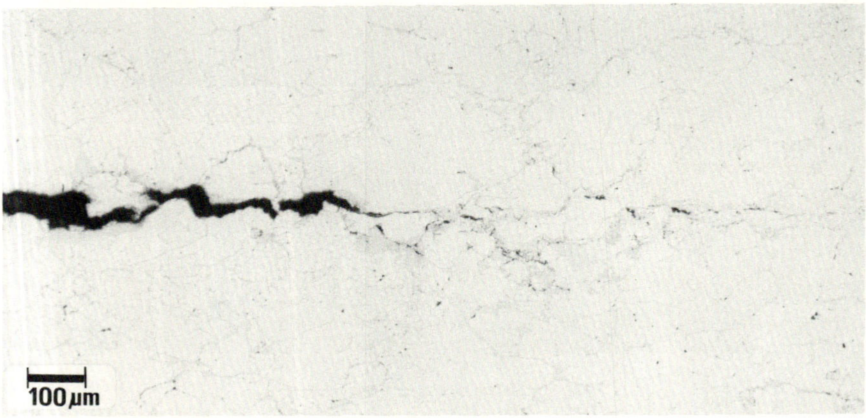

b)

Figure 15 - a: Typical edge cracks which develop into an edge tearing. b: Segregation pattern in the vicinity of an edge crack.

Edge cracking and tearing

Edge cracking is a common feature of twin roll cast material. The cracks can be seen in fig. 15a and more detailed examination reveals a segregation pattern in the vicinity of the cracks, indicating that they are formed when the metal is partly solidified, fig. 15b. Alloys with a large solidification interval are more prone to edge cracking than low alloyed materials. Furthermore, increasing the tip setback favours crack formation due to the heavier deformation. No simple method to eliminate edge cracks is known, except, of course, on-line edge trimming.

Moderate edge cracking can be taken account of by proper dimensioning of the strip width. However, the cracks occasionally develop into edge tearings, fig. 15a. These can extend several cm into the strip and cause dimensional problems during further fabricating.

Sticking

An upper limit to the casting speed is obviously determined by the rate of heat extraction from the melt through the roll walls. Solidification becomes unstable when the metal does not completely solidify prior to leaving the rolls. However, in practice, other speed limiting phenomena occur, in particular that of adhesion or sticking of the strip to the roll surfaces. The sticking tendency increases with roll velocity and temperature of the liquid metal. Furthermore, alloys with a wide solidification range are more susceptible to sticking than others. The sticking problems can be reduced by using efficient lubricants (MgO, graphite) and applying tension to the strip. According to Bercovici (14), macroscopic sticking, which can cause serious strip damage is preceeded by microsticking which can be detected by comparing the driving forces on each of the rolls. This phenomenon has been used in the design of an automated control system to optimize productivity (14), however, such systems are not in common use.

Concluding Remarks

The present summary of structure inhomogeneities and casting defects is far from complete. It does, however, indicate the variety of phenomena that needs attention. To the metallurgist, the continuous strip casting process presents new and interesting tasks. Furthermore, problems related to the equipment emphasize the need for further developments by the machine designers and production engineers. The most acute problems are related to the feeding of liquid metal to the solidification zone. Improvement of the nozzle material is imperative. With regard to the economy of the process, improved life-time of the rolls is important. A basic understanding of edge cracking and sticking could substantially contribute to a reduction of the scrap percentage and increase productivity of the process.

Acknowledgements

The authors are grateful to Årdal og Sunndal Verk a.s. for permission to publish this paper, as well as to the staff of the Research Laboratory for their assistance with experimental work.

References

1. J. Althoff, "Einfluss der Wärmebehandlung auf die mechanischen Eigenschaften von bandgegossenem Aluminiumhalbzung", Aluminium 56 (3) (1980) pp. 199-202.

2. E. Nes and S. Slevolden, "Casting and Annealing Structures in Strip Cast Aluminium Alloys", Aluminium 55 (5) (1979) pp. 319-324.

3. I. Jin, L.R. Morris and J.D. Hunt, "Centreline Segregation in Twin Roll Cast Aluminium Alloy Slab", pp. 873-888 in Light Metals 1982, J.E. Andersen, ed.; AIME, New York, N.Y. 1981.

4. L.R. Morris, "Structure of Continuously Cast Eutectics Used in the Manufacture of Fine-Grained Aluminium Sheet", pp. 370-376 in Solidification of Metals, The Iron and Steel Institute, London, 1967.

5. E. Nes and S. Slevolden, "Mechanical Properties of New Strip Cast AlMnZr Alloys", Aluminium 55 (5) (1979) pp. 398-402.

6. R. Iricibar and I. Jin, "Forward Slip During Roll Casting of Aluminium Strip", in Pacific Northwest Metals and Minerals Conference, Spokane, Washington, 1982.

7. C.J. Moser, "Continuous Casting - Hunter Technology", pp. 837-849 in Light Metals 1982, J.E. Andersen, ed.; AIME, New York, N.Y., 1981.

8. J.M. Chateau and P. Crouzet, "Technological Break-through in Continuous Strip Casting: The Jumbo 3C", in Light Metals 1978, J.J. Miller ed.; AIME, New York, N.Y., 1978.

9. A. Odok and G. Thym, "Moderne Kontinuerlich arbeitende Bandgiessverfahren für die Herstellung von Aluminium Breitband", Aluminium, 50 (6) (1974) pp. 411-414.

10. I. Miki, H. Kosuge and K. Nagahama, "Supersaturation and Decomposition of Al-Fe Alloys during Solidification", Journal of Japan Inst. Light Metals, 25 (7) (1975) pp. 1-9.

11. H. Westengen, "Formation of Intermetallic Compounds During DC Casting of a Commercial Purity AlFeSi Alloy", Zeitschrift für Metallkunde 73 (6) (1982), pp. 360-363.

12. M. Ryvola, L.R. Morris and G. Marchand, "Nucleation of Intermetallic Phases on the Cast Surface of Al-Mn Alloys", pp. 23-29 in Microstructural Science, Vol. 6, J.E. Bennett, ed.; Elsevier North-Holland, 1978.

13. D. Altenpohl, Aluminium und Aluminiumlegierungen, p. 526; Springer Verlag, Berlin, 1965.

14. S. Bercovici, French Patent no. 8101185, 1981.

ROLLING ASPECTS OF TWIN ROLL CASTING

R. Iricibar and I. Jin

Alcan International Limited
Kingston Laboratories
Kingston, Ontario, Canada

In the twin roll casting process the metal is deformed by rolling even before it has solidified completely. Hot metal and rough rolls all contribute to create an extreme form of hot rolling on the exit side of the roll gap. Under these conditions high friction develops at the roll-slab interface. This high friction generates a steep pressure distribution along the arc of contact where pressures developed are several times higher than the flow stress of the metal being cast. High deformation gradients towards the centre of the slab affect the shape of the sump towards the entry side of the gap.

This paper addresses some of the aspects associated with the rolling side of the roll caster. Features like the pressure distribution, the extrusion and sticking phenomena, and the effect of the front tension are examined from a fundamental standpoint and some experimental observations discussed.

Introduction

The twin roll caster produces a coilable reroll directly from the liquid metal. The recognized advantage of this process is that (when applicable) it by-passes the ingot casting and re-heating stages and, in particular, the need for roughing passes and therefore makes hot-mills unnecessary.

It is true that roll casting does away with hot mills, what is not often recognized is that it does not eliminate hot rolling. In this paper we focus on the rolling side of twin roll casting. After defining the region and the conditions under which the metal is being rolled, we look into some of the consequences of this rolling on the operation of the caster.

The Rolling Region

In roll casting, hot metal from a tundish is poured through a feeding nose into the convergent cavity formed by the walls of two rotating rolls (Figure 1). Heat is extracted from the melt into the roll shells which are water cooled from the inside. As solidification is completed within 10 or 20 mm from the nose tip (the 'casting' zone) the solid metal and the roll shells remain in contact over an arc of length 30 to 80 mm long (the 'rolling' zone) depending on the type of caster and operating conditions. In the rolling zone, the strip is reduced by 15 to 50% depending on the length of the 'rolling' zone and the roll diameter.

As implied by the brief description of the process given above, the solid (or 'mushy') metal enters the rolling region at a temperature close to the melting point of the particular alloy being cast. At the exit point the temperature of the slab is usually between 250 and 350°C.

In addition, although the shells are usually finished to an r.m.s. of about 2 μm (80 microinches), after a short time of operation they become very rough. This is initially due to the formation of a parting layer (10-20 μm thick) made up of the fines sprayed on the shells to avoid sticking as well as of iron and aluminum oxides.

We thus have identified a rolling condition in which very hot material is rolled with rough rolls. This in turn implies that the metal being rolled will yield at low stresses, show very little work hardening and that high friction will characterize the interface between the slab and the roll. This conclusion is basic to most of the discussion presented below.

Rolling and Caster Productivity

The rolling person is interested in understanding the rolling process to be able to predict whether separating forces and torques to be delivered by the mill (for a given alloy and practice) are within the limitations of the equipment. In the case of the roll casters, neither the separating forces nor the required torque are limiting factors by themselves.

In these machines the productivity is limited by slab defects ('heat-lines', edge cracking) or by sticking of the strip to rolls. In turn the operating conditions which avoid most of these defects (a large tip set-back for instance) also require large separating forces and torques. The drive for higher productivities has therefore brought about larger casters designed to withstand loads and torques which might be unnecessary if the slab defects could otherwise be controlled.

It is clear that defects like edge cracking and sticking must be rolling defects. The issue is not as obvious in the case of heat lines, although the fact that this defect does not appear in the absence of rolling indicates that it maybe due to a combined effect of both the casting and the rolling.

In this context the understanding of the rolling process is crucial and in particular of the distribution of pressure along the arc of contact.

The Distribution of Pressure Along the Arc of Contact

Under the rolling conditions that we have identified (high temperature and rough rolls) friction effects will be dominant and we expect the pressure profile to be a very steep hill. This may be seen by considering Figure 2 where we have plotted two pressure distributions determined experimentally (1) for the rolling of aluminum with smooth and rough rolls. Notice that the steepness of the friction hill increases with increasing friction and also that the maximum pressure is several times greater than the material flow stress as given by the base of the hill. This feature indicates that the detailed shape of the friction hill may be expected to be fairly independent of minor changes in material properties.

Against the background of the above discussion we assume in what follows that main features of the pressure distribution may be described by assuming that the sides of the hill are straight. This assumption can be seen to be justified by reference to the 'rough rolls' curve of Figure (2). Alternatively, the simpler case of slab compression between parallel plates with high friction may be considered for which the 'straight' sides assumption holds exactly for the extreme case of sticking friction.

Using this assumption we have computed, using simple geometrical considerations, an approximate pressure profile for any given operating condition from a knowledge of the separating force, the torque delivered by the rolls and an estimate of the material properties at the entry and exit points. Figures 3 and 4 show the results of such computations applied to a cast of commercially pure Al in a Hunter Super Caster and in a Jumbo 3C Supercaster respectively. In these Figures, the base line a-a' represents, in a schematic fashion, the yield stress (in restricted, plane strain flow) of the underlaying material.

The general features of the curves are consistent with the assumptions used to construct it. Notice in particular that friction contributions have raised the peak pressure to a value an order of magnitude greater than the flow stress of the metal being rolled.

In considering these curves the reader should note that close to the sump line (entry) and exit points the 'straight-sides' approximation should not be expected to be a good approximation to the actual pressure distribution. This is due to the peculiar deformation zone geometry at the edges and also to friction conditions different from those existing along most of the arc of contact.

An inkling of the real shape of the pressure distribution close to the exit may be obtained by analogy with the case of compression between two parallel plates. This we have done in Figure 5 assuming that either dry friction (Figure 5-b) or sticking (Figure 5-c) occurs in the vicinity of the

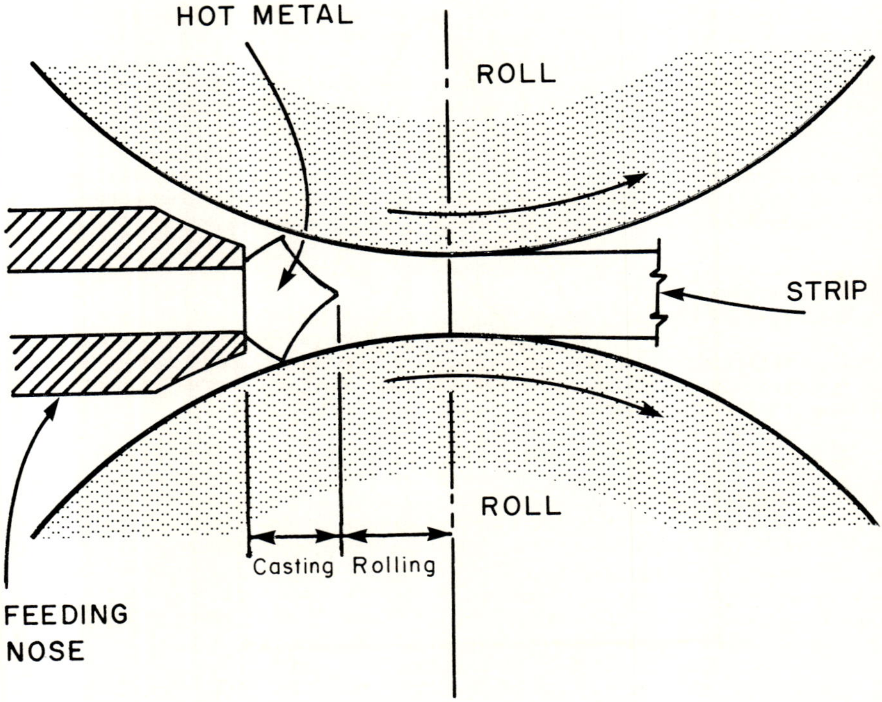

FIGURE 1

SCHEMATIC DRAWING OF THE ROLL CAVITY
SHOWING THE 'CASTING' AND 'ROLLING' ZONES

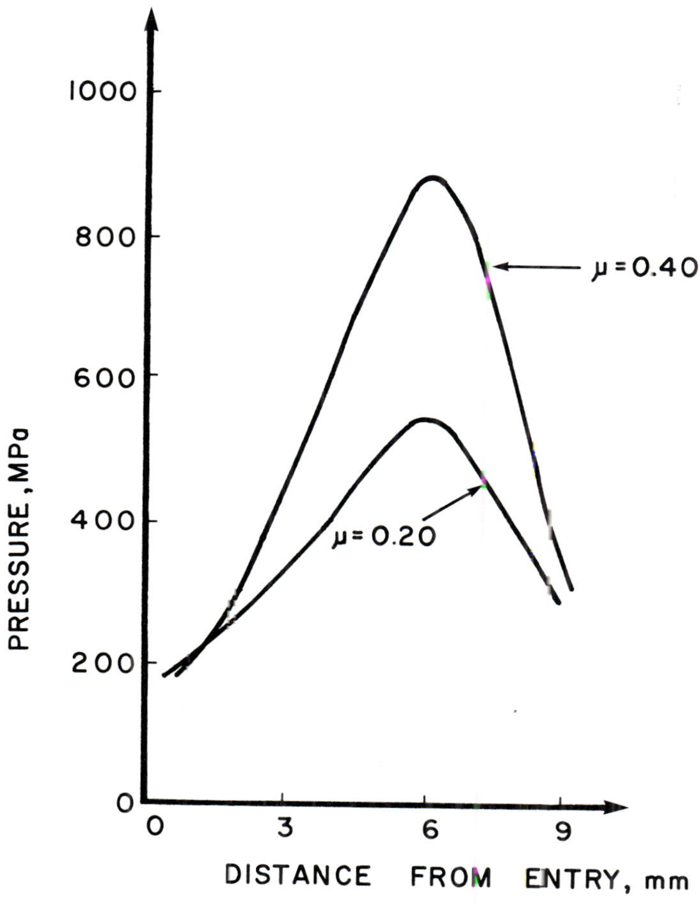

FIGURE 2

EXPERIMENTAL PRESSURE DISTRIBUTION FOR THE ROLLING OF ALUMINUM WITH ROUGH (μ = 0.40) AND SMOOTH (μ = 0.20) ROLLS[1]

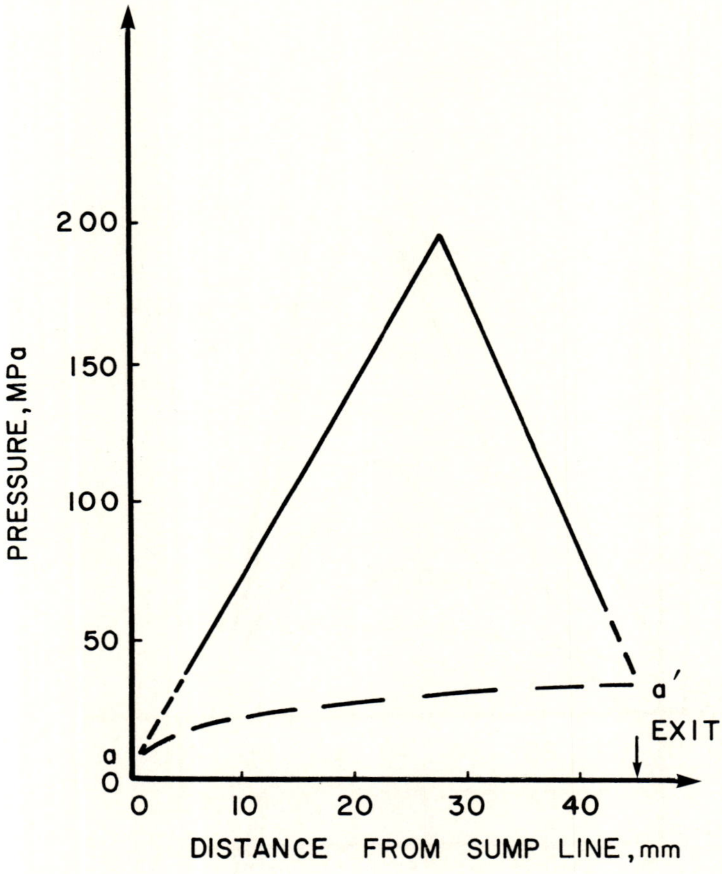

FIGURE 3

PRESSURE DISTRIBUTION CALCULATED USING THE 'STRAIGHT SIDE' APPROXIMATION FOR A HUNTER SUPER CASTER

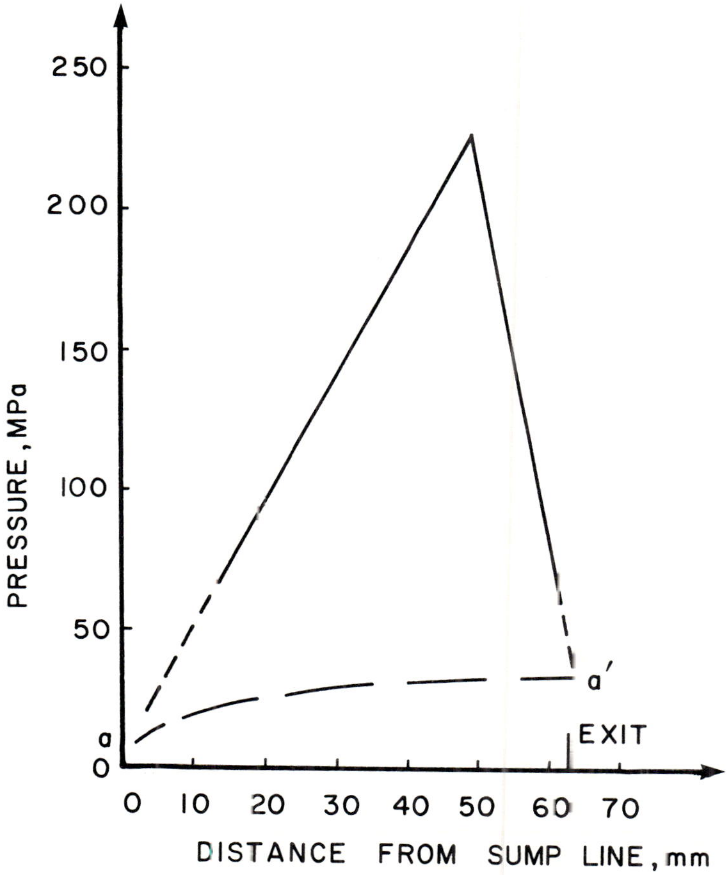

FIGURE 4

PRESSURE DISTRIBUTION CALCULATED USING THE 'STRAIGHT SIDE' APPROXIMATION FOR SCAL JUMBO 3C

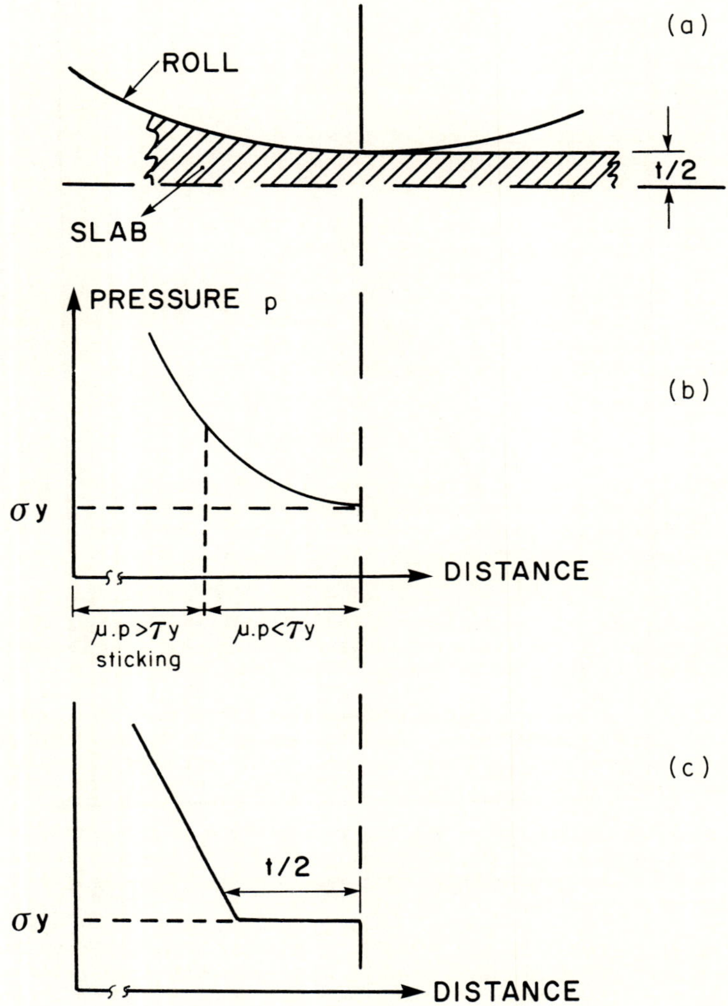

FIGURE 5

PRESSURE DISTRIBUTION CLOSE TO THE EXIT POINT OF THE ARC OF CONTACT, (b) FOR DRY FRICTION, AND (c) FOR STICKING FRICTION.

μ = friction coefficient
σy = yield stress in compression
τy = yield stress in shear of the slab

exit point. The feature to note from these graphs is that we should expect the pressure curve to flatten significantly as points closer to the exit point are considered.

In addition, the model is very sensitive to the specification of the entry point and this in turn requires knowing the sump depth for each operating condition. Fortunately, experimental determinations of this parameter showed us that the interaction between the rolling and casting regions lead to changes in sump depth with casting rate (within normal production ranges) much milder than would be expected from exclusively thermal considerations.

We have thus found using the 'straight sides' approximation a very steep friction hill and that friction contributions have raised the peak pressure to a value several times higher than the yield stress of the underlying metal. We therefore expect the pressure curve itself to be fairly independent of minor changes in material properties. With this characteristic of the friction hill and the limitations of the model in mind we now turn to its application to the understanding of the 'sticking' problem.

A Model For Sticking

'Sticking' is a condition that tends to develop at high casting rates and high casting temperatures although the distinction is not clear because these two variables are usually correlated in normal operating practice: increasing the casting rate increases the incoming hot metal temperature.

When 'sticking' occurs the slab adheres to either or both of the rolls somewhere along the arc of contact. Although local sticking has been sometimes found to occur, the condition usually develops and ceases suddenly all across the width of the slab as Figure 6 shows very clearly.

The usual cure for a sticking condition is to decrease the casting speed, which limits productivity. The alternatives followed by the industry have been the use of parting layers sprayed on the rolls or the shift to horizontal casters which allow a front pull to be exerted on the slab.

The significance of the effect of front tension on the occurrence of sticking is that it is not one of simply pulling the slab away from the roll as might be intuitively expected. Sticking is effectively retarded by the application of a front pull.

The characteristics of the pressure distribution that we have uncovered and the way they are affected by changes in the operating variables may be used to explain all of these observations. For this we need to define what we understand by the sticking region and discuss the effect of casting rate and front tension on the pressure hill.

The effect of front tension on the shape of the friction hill is well known from rolling theory. Applying a front tension to the slab will lower the base value of pressure at the exit by a similar amount. The effect on the peak height and position may be calculated by means of the 'straight side' model modified to account for this effect of front tension. For the effects of this discussion, however, it suffices to note that the net effect of the pull will be to displace the exit side of the friction hill, as shown by the dotted line in Figure 7 for a front tension of roughly 7 MPa. (1 ksi).

Turning now to the extent of sticking let the line b-b' in the same Figure 7, be the locus of the yield stress in shear divided by the friction

FIGURE 6

A ROLL CASTER IN OPERATION UNDER STICKING CONDITION
SEEN FROM THE EXIT SIDE.

Notice the marks left on both roll and slab wherever sticking occurred.

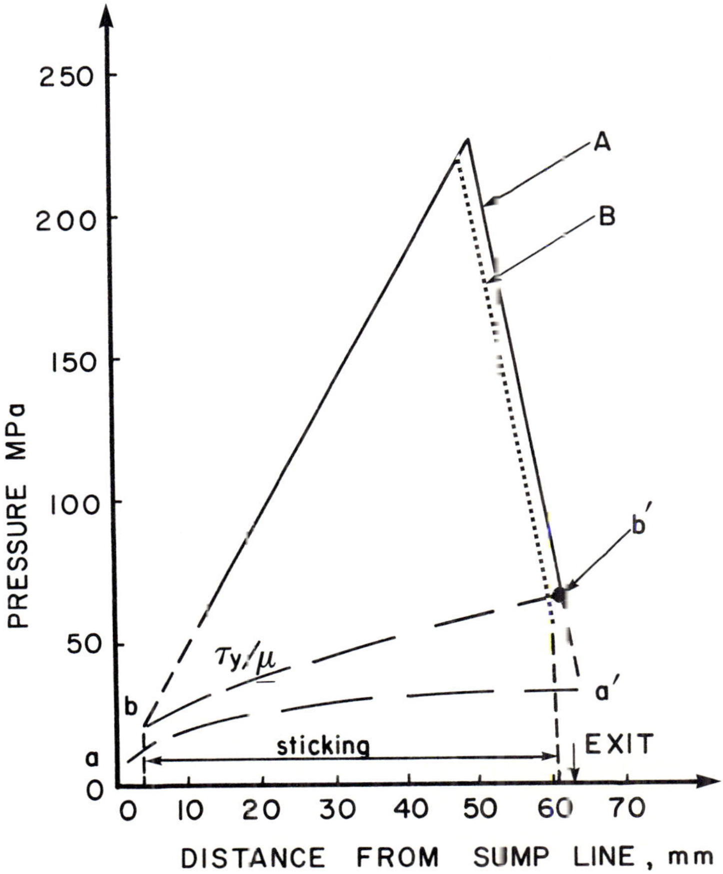

FIGURE 7

CALCULATED PRESSURE DISTRIBUTION SHOWING THAT STICKING WILL OCCUR WHEREVER THE LOCAL YIELD STRESS IN THE SHEAR IS SMALLER THAN THE PRODUCT OF THE FRICTION COEFFICIENT TIMES THE LOCAL PRESSURE.

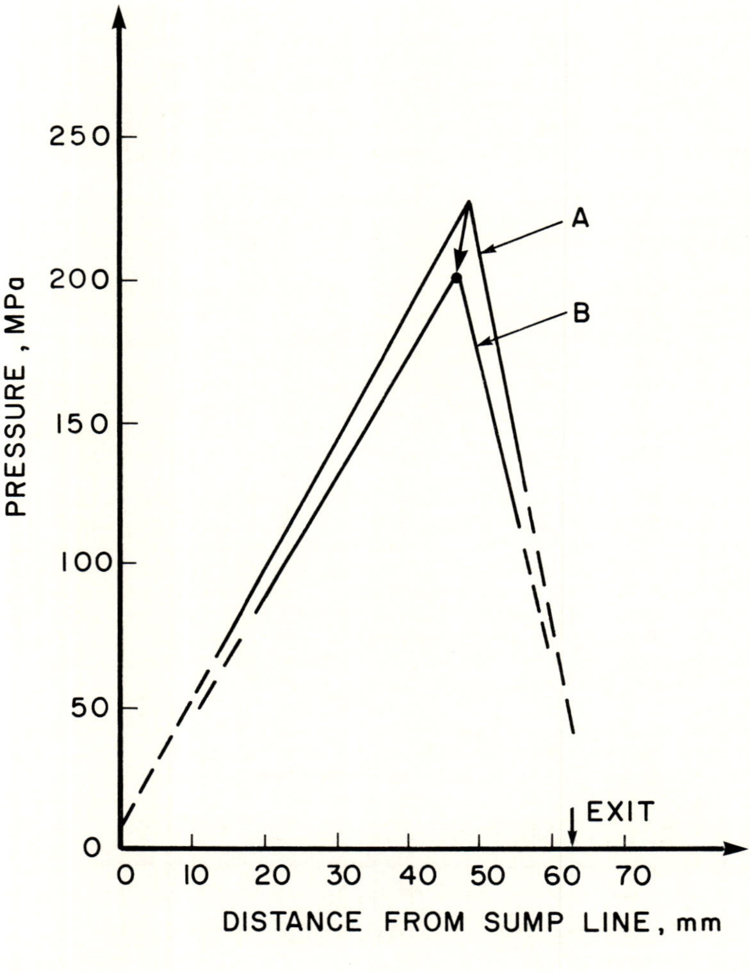

FIGURE 8

CALCULATED PRESSURE DISTRIBUTION SHOWING
THE EFFECT OF INCREASING CASTING RATE
SCAL JUMBO 3C

coefficient for the casting conditions in use. The slab will therefore stick to the rolls at all points within the region b-b', which we call the 'sticking region'. The line b-b', which was arbitrarily chosen at the level shown for the sake of clarity, implies the choice of a friction coefficient of approximately 0.2. Notice that even for such a low value of friction the height of the pressure hill is such that sticking should occur along most of the arc of contact. In fact, the point b' is only a few mm away from the exit point and, as discussed above, either Figure 5-b or 5-c would be a better representation of the actual pressure distribution within this region.

This being the case, even minor changes in any of the parameters (yield stress, friction coefficient) defining the curve b-b' will affect very significantly the position of point b'.

Assume now that due to a local fluctuation of the sump depth, or of the heat extraction rate through the shell, etc. the sticking region at some position across the width of the slab moves ahead of the average position of the forward edge. At this point the slab will stick to the roll and will therefore exert a drag on the rest of the material which, being close to the exit, moves faster than the roll surface due to the extrusion effect. This drag will be felt like a front pull by the portion of slab associated with the instability, while the rest of the strip will feel a front push.

A front pull on the slab will cause a displacement of the exit side of the hill of the kind shown in Figure 7. Such a shift in pressure will cause this protrusion of the sticking zone to recede to the average position of the zone edge. The average position of the edge of the sticking zone will be affected by this drag in an opposite sense: i.e. it will show a (weaker) tendency to move towards the exit. The net effect will tend to dampen the formation of any protrusions (or intrusions) of the zone edge. The present model of sticking thus requires the edges of the sticking region to be straight lines across the width of the rolls.

If the casting rate is now assumed to increase, using the 'straight sides' approximation the effect of this change can be also estimated from the knowledge of separating force and torque delivered. As an instance, we have reproduced in Figure 8 the pressure profile of Figure 4 as well as the one resulting from an increase in casting speed from the initial .0126 m/sec to .0160 m/sec. Notice that the model suggests that the peak pressure will drop and predicts a slight displacement of the peak position. None of these shifts affects the essential features of the pressure distribution discussed above.

Following the increase in casting rate we may also expect that, if it changes at all, the sump depth should increase mildly therefore decreasing the length of the actual arc of contact. As may be seen from Figures 9 and 10 the effect of increasing the casting rate and the ensuing shortening of the arc of contact should also increase the overall temperature of the solid slab being rolled. This temperature increase will lower the yield stress of the metal and increase the relative length of the sticking zone bringing the edge of the sticking region very close to the exit point.

With these elements at hand we can now describe the mechanism of sticking as follows: during casting operation at moderate speeds the sticking region extends over most of the arc of contact separated by straight boundaries from the non-sticking regions close to the entry and exit planes. As the speed is increased, the shape of the pressure hill changes only slightly while the front edge of the sticking zone moves towards the exit due to material softening with rising metal temperature. When the leading edge of

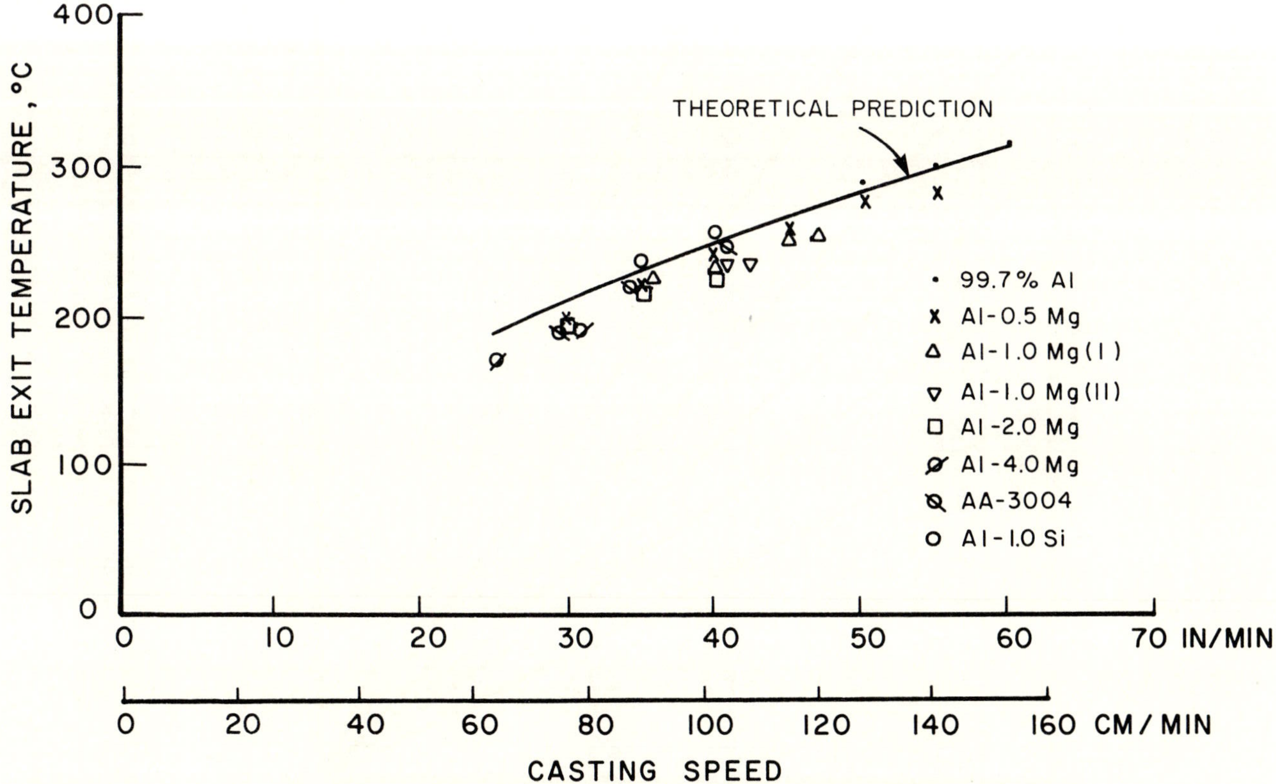

FIGURE 9

SLAB EXIT TEMPERATURE VS. CASTING SPEED (6.5 mm SLAB) (63 mm TIP SETBACK)

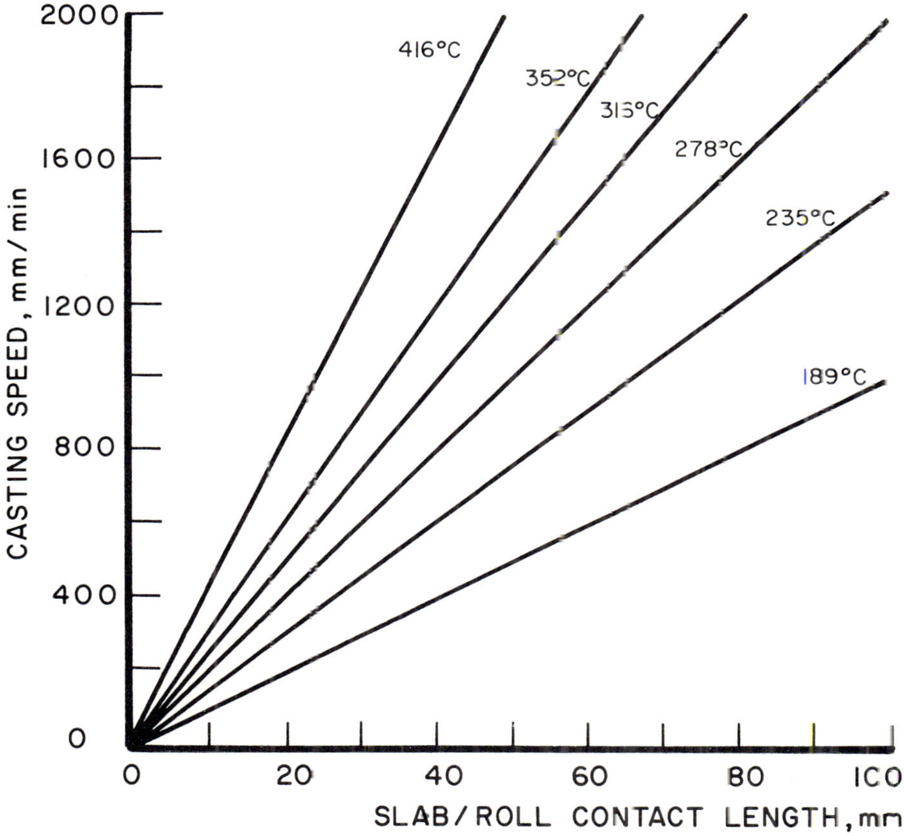

FIGURE 10

SLAB EXIT TEMPERATURE AS A FUNCTION OF CASTING SPEED AND CONTACT LENGTH

(Heat Transfer Coefficient = 0.4 cal/cm^{2}s ^{o}C)
(Slab Thickness = 6.5 mm)

the sticking edge reaches the exit 'sticking' (in the operating sense) occurs. As front tension is applied, the exit side of the pressure hill moves inwards and the pressure near the exit drops. Following the drop in pressure the forward edge of the sticking region recedes back into the roll gap and the 'sticking' condition is thus cured, allowing the caster to operate at somewhat higher speeds than without coiler tension.

Reference

(1) By Siebel, E., & Lueg, W., as given by Orowan, E. in "The Calculation of Roll Pressure in Hot & Cold Rolling", Proc. of the Institute of Mechanical Engineers, V150, 1943, p. 141.

NEW CASTING METHOD FOR IMPROVING BILLET QUALITY

John P. Faunce(1), Frank E. Wagstaff (2), Howard Shaw (3),

(1) Martin Marietta Laboratories
1450 South Rolling Road
Baltimore, Maryland 21227
USA

Extrusion billet cast by the direct chill (DC) casting process frequently has inconsistent quality. This can adversely affect extrudability of the billet, for example, by causing a poor as-extruded surface. A recent development, the Wagstaff AIR-SLIP™ casting mold, incorporating several proprietary features, overcomes the problem of inconsistent quality and produces billet to exceptional metallurgical standards. Samples of 6xxx alloy aluminum billet cast by the AIR-SLIP™ process have had ≤ 0.005" peripheral segregation and an extremely smooth as-cast surface.

Data obtained from over 10 x 10^6 lb. of production of various alloys of extrusion billet are used to demonstrate how the AIR-SLIP™ casting process has significantly improved both billet quality and overall cast shop productivity.

(2) Wagstaff Engineering, Inc.
3910 N. Flora Road
Spokane, Washington 99216

(3) Martin Marietta Aluminum
P.O. Box 711
The Dalles, OR 97058

Introduction

All aluminum alloy extrusion billet is cast by some variant of the basic direct chill (DC) casting process, a process for billet casting which has been described at length in the literature (1,2,3). Numerous authors, for example Bergmann (4) and Collins (5), have described the problems in using the D.C. process to cast billet which has both: 1) a uniform as-cast surface free of defects such as tears and cold folds; and 2) little or no segregation near the periphery of the billet -- often referred to as liquation.

Defects remaining from the casting process are known to affect the characteristics of subsequent extrusion (6,7). Thus, research has been directed at ways to eliminate these defects and produce billet which will improve extrudability, for example, by permitting higher extrusion rates and/or better mechanical properties in the finished extrusions.

To overcome one or more of the problems associated with the basic D.C. casting process, many different casting systems and methods have been developed. These include, but are not limited to, electromagnetic casting (8,9), the variable chill depth mold system (10), level feed casting (11), and others.

A recent development, the Wagstaff AIR-SLIP™ casting mold, incorporating several proprietary features, overcomes the problems of inconsistent quality and produces exceptional billet. Additionally, the process can significantly improve overall cast shop productivity.

This paper discusses the equipment and procedures used to cast extrusion billet with the Wagstaff AIR-SLIP™ casting mold. It also presents quality, recovery, and productivity data based on casting of over 10×10^6 lbs. of extrusion billet using the AIR-SLIP™ technology.

Background

The cast shop at Martin Marietta Aluminum's plant at The Dalles, OR, casts 5 1/8" - 12" diameter extrusion billet. Production is predominantly 6xxx alloys (e.g., 6060, 6063, 6061, etc.), although other alloys such as 1435 and 3102 are also cast. Typical production rates are 12,000,000 lb./mo. Since the majority of this metal comes from adjacent potrooms, or reduction cells, cast shop processes and practices are designed to minimize any delays that would adversely affect the potrooms' tapping schedule. Furthermore, the quantity of scrap billet must also be minimal since the cast shop furnaces are designed to be charged with molten metal rather than solid mill scrap.

Prior to 1983, The Dalles used two types of tooling to cast extrusion billet. The majority of the billet (i.e., 5 1/8" diameter up to and including 8" diameter) was cast using a type of hot top mold. The remaining sizes (over 8" diameter through 12" diameter) were cast with a conventional type mold employing a downspout and float to control the molten metal head in the mold. Although this arrangement produced acceptable billet, there were several operational and quality problems that included:

- As-cast surface quality, particularly on the smaller diameter billet, namely 5 1/8" - 8" diameters, tended to be inconsistent both within a given drop and from drop to drop.

- Production rate (i.e., numbers of ingots/drop) was restricted due to mold design.
- Billet recovery was adversely affected by the inconsistencies in as-cast surface quality.
- Mold lubricant usage was high.

Figure 1 shows the typical as-cast surface of 6" diameter 6063 and 6061 alloy billet cast at The Dalles with a conventional hot top mold.

Consequently, in 1982, a project was begun to eliminate these problems when casting extrusion billet. The goals of this project were:

- Improve uniformity of as-cast surface quality
- Increase the number of ingots cast per drop
- Improve billet recovery
- Improve internal billet quality.
- Reduce consumption of casting lubricant.

Many of the newer casting methods mentioned previously were felt to offer advantages over existing Dalles' tooling. However, all had at least one or more overriding disadvantages (e.g., high initial cost) which prevented their consideration.

A recent development, the Wagstaff AIR-SLIP™ casting mold was determined to have all of the features necessary to produce the desired billet quality at high production rates. Furthermore, adoption of the Wagstaff AIR-SLIP™ casting mold was determined to entail fewer process and practice changes than for other types of tooling. Hence, casting trials employing this technology were begun at The Dalles in October 1982.

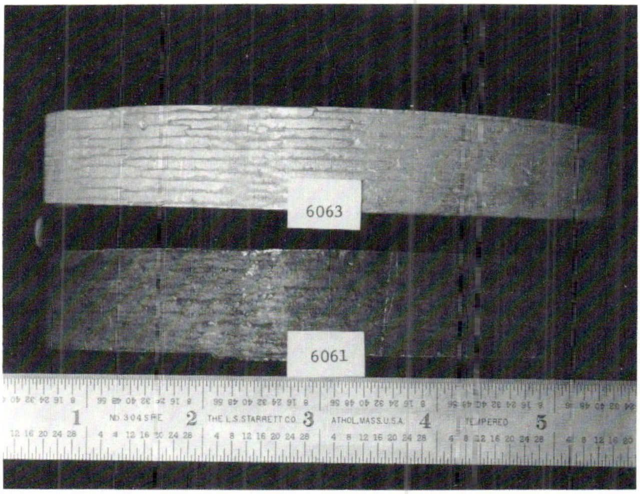

Figure 1 - Typical as-cast surface of 6" diameter 6063 and 6061 alloy cast prior to 1983 using The Dalles' hot top mold.

Description of the AIR-SLIP™ Mold

The Wagstaff AIR-SLIP™ casting mold (hereafter designated as WASM) was developed specifically for high-volume production of extrusion billet of exceptional metallurgical quality. It represents a state-of-the-art development in hot top billet casting and is similar in outward physical appearance to the Wagstaff MAXI-CAST™ mold (12,13).

The WASM, shown in Figure 2, consists of: an aluminum alloy mold body; a graphite insert that mates to the mold body; an aluminum alloy jacket surrounding the mold body to provide both a means of guiding the bottom block or stool cap into the mold, as well as forming part of the chamber for mold cooling water; and a ceramic orifice plate, or transition plate, to connect the mold to the thimble which joins the mold and the pouring pan or distribution trough. This design permits a reduction in the distance between mold centerlines, resulting in a very high mold density for a given casting unit, as illustrated in Figure 3. Table I gives the number of ingots cast with WASM molds for each size of billet produced at The Dalles and compares this to the number cast with conventional molds. It should be noted that the increase in the number of molds for a given size made possible by the WASM technology was achieved with only minor changes to the existing casting unit.

Mold length of the WASM is very short. This, coupled with a high volume of cooling water, provides a rapid solidification rate. In fact, with the WASM, most of the heat transfer during solidification takes place by direct water contact on the ingot shell after it emerges from the mold rather than through the mold wall. This rapid solidification results in a very uniform, fine-grained microstructure.

Figure 2 - Overall view of a Wagstaff AIR-SLIP™ casting mold for 6" diameter extrusion billet.

Figure 3 - View of Wagstaff AIR-SLIP™ casting unit for 5" diameter billet showing the high mold density.

TABLE I

Effect of Mold Design on Number of Ingots Cast Per Drop

Ingot Diameter, Inches	Number of Ingots Cast Per Drop with Conventional Mold	Number of Ingots Cast Per Drop with Wagstaff AIR-SLIP™ Mold
5 1/8	48	60
6	28	60
7	28	54
8	28	48
9	18	30

To improve the as-cast surface, the WASM relies on reducing the length and duration of contact between the solidifying ingot shell and the mold wall. This is accomplished by having both a very short mold (as described above) and a proprietary method for applying a mold lubricant and a gas. Precise control of this lubricant/gas mixture not only improves the as-cast surface but significantly reduces lubricant consumption.

A further feature of the WASM mold is a means of self-centering the bottom block or stool cap. This greatly reduces set-up time prior to casting and eliminates damage and wear to the mold bore.

Development of WASM Casting Practices

At the time of writing, not all sizes of WASM had been placed in use at The Dalles. Therefore, this section will deal only with development of practices for the 6" diameter size WASM mold since this was the size first put into production.

The 6" diameter WASM unit -- with 60 ingots/drop -- was installed in May 1983 and subsequently has been used to cast 1235, 6060, 6063, 6061, and 6463 alloys. No changes were made to the casting pit or casting pit controls to accommodate the WASM other than to install a larger diameter inlet pipe for the mold cooling water due to the increased water flow rates (it should be noted that part of this increase was required in going from 28 ingots/drop with the old tooling to 60 ingots/drop with the WASM). Water flow rates up to 2,500 gpm are now obtainable.

There is very little difference in the operation of the WASM compared to The Dalles' hot top unit. In fact, casting with the WASM is actually easier, since the mold/bottom block alignment is simplified by the self-centering bottom block design. Due to the short length of the WASM, the depth of entry into the mold by the bottom block is critical. However, this is controlled by use of a special gage inserted into the thimble. A depth indicator on this gage shows the operator precisely how far into the mold the bottom block has traveled. Other preparation at the start of the drop (e.g., drying bottom blocks, etc.) is the same as with conventional molds.

Each WASM mold has an individual control to adjust mold lubricant and gas flow. Generally, this is done prior to casting and, once set, does not require resetting between drops unless an alloy change occurs. In some instances, lubricant flow rates have to be set slightly higher when a mold is first placed in service, but can be reduced after 2-5 drops have been cast through the mold. Visual inspection of the molds between drops is recommended to ensure that lubricant flow is uniform. Typical flow rates are quite low, for example, a total of approximately 10-32 oz. of lubricant for a 35,000 lb. drop of 6" diameter billet, depending on the alloy being cast.

Experience has shown that the flow rate for mold cooling water should be reduced at the start of the cast to ensure that no water accidentally splashes into the mold prior to the molten metal entering the mold. After the mold is filled with molten metal, the cooling water is ramped up (either manually or automatically) to the desired value.

It is very important when casting with WASM to maintain a constant molten metal level in the distribution trough/pouring pan throughout the entire drop. Additionally, this level should be kept somewhat lower than when casting with conventional molds. It is not necessary to wait until the level in the pouring pan is at the desired depth before starting the platten. In fact, best results have been obtained by starting the platten soon after the thimbles are full and only a shallow layer of molten metal is in the pouring pan.

Actual casting conditions for WASM and conventional molds are quite similar. Typically, it is possible to operate with lower metal temperatures with the WASM, in large part due to the advanced refractories -- that limit heat loss -- used in the pouring pan and mold. Drop rates up to 25% faster than those with an equivalent size conventional mold have been achieved with the WASM with no sacrifice in as-cast surface quality.

However, drop rate is dependent on many factors (e.g., temperature, lubricant flow rate, etc.) and therefore this type of increase is not achieved for every alloy. Additionally, as drop rate increases, the likelihood of ingot cracking increases, particularly for certain alloys. Consequently, for 6" diameter 6063 alloy, drop rates of 6-7 ipm are typically used, even though higher rates may be possible.

At the completion of a drop, the procedures for "plugging off" the WASM are only slightly different than for a conventional hot top type mold. As with a conventional mold, the metal level in the pouring pan is allowed to drop. As this level drops, the drop rate is reduced while, simultaneously, the mold water is reduced to about 50% of the value used for casting. Under these conditions, the ingot will emerge from the mold fully solidified. At no time have there been any problems with the WASM due to accidental draining of the mold before all the metal had solidified.

Since the as-cast ingot surface produced with the WASM is typically much smoother and brighter than those with conventional molds, minor handling marks become very evident. Thus, care should be taken in removing the ingot from the casting pit. The extent to which these handling marks are a concern will vary from one cast shop to another. In our experience, handling marks have not been a cause for scrap at either the casting pit or billet saw, even though The Dalles' handling practices and equipment were not specially modified to accommodate WASM.

Casting Results

The WASM produces extrusion billet superior to The Dalles' previous casting tooling with respect to productivity, quality, and recovery. Table II summarizes productivity comparisons for 6" diameter 6063 alloy cast with the WASM versus conventional hot top molds. As can be seen, use of the WASM resulted in a significant improvement. Additionally, and perhaps even more important, is the fact that the WASM produces a billet of superior metallurgical quality. Figures 4-9 show typical microstructures of 6xxx alloy extrusion billet cast using WASM and conventional molds. Use of the WASM resulted in:

- Elimination of peripheral cold folds or cold shuts, as well as virtually all other types of surface defects (e.g., tears).
- Reduction in the segregated area at the as-cast surface of the billet from ≤ 0.250" to ≤ 0.005".
- A finer subsurface microstructure with constituents of more uniform size and distribution.

Improvements in quality such as those mentioned above are desirable for improved extrudability since defects remaining from the casting process are known to cause problems during subsequent extrusion (e.g., affect as-extruded surface quality). The improved metallurgical quality imparted by the WASM results from an increased solidification rate, due in large part to the very short mold length combined with several proprietary features.

Additionally, the WASM has improved cast shop productivity. A maximum of 6 drops (i.e., 168 ingots) was possible in an 8-hour shift using conventional 6" diameter tooling. However, 4.5 drops (i.e., 270 ingots) per 8-hour shift are now obtained using the WASM mold. Thus, productivity has increased by 61% through use of the WASM mold. Furthermore, overall recovery has increased due to reduced surface scrap, reduced head and butt

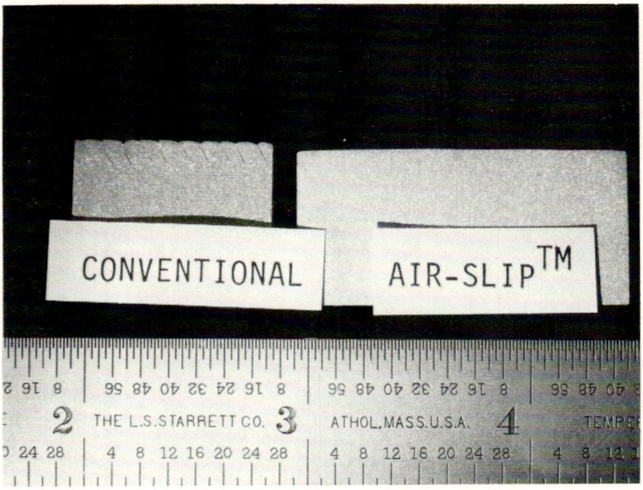

Figure 4 - Macrostructure of 6" diameter 6063 alloy billet cast with a conventional (left) and WASM (right) mold.

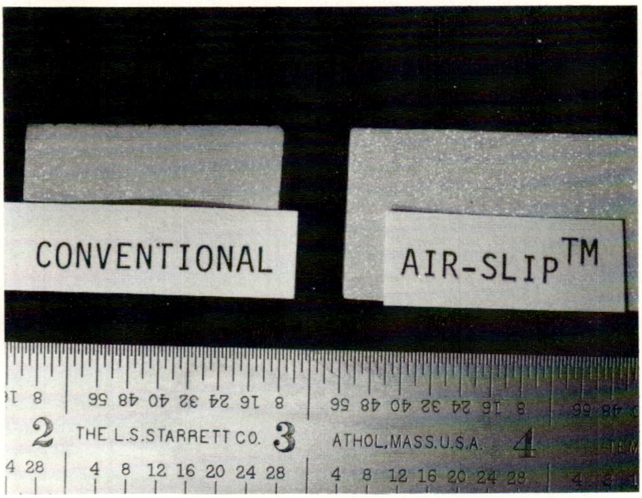

Figure 5 - Macrostructure of 6" diameter 6061 alloy billet cast with a conventional (left) and WASM (right) mold.

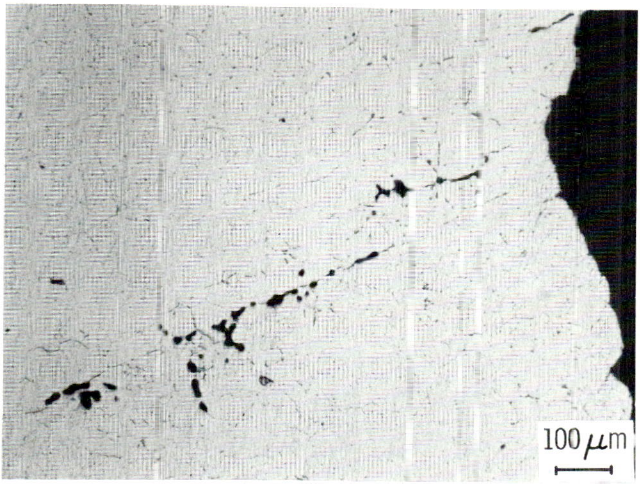

(a)

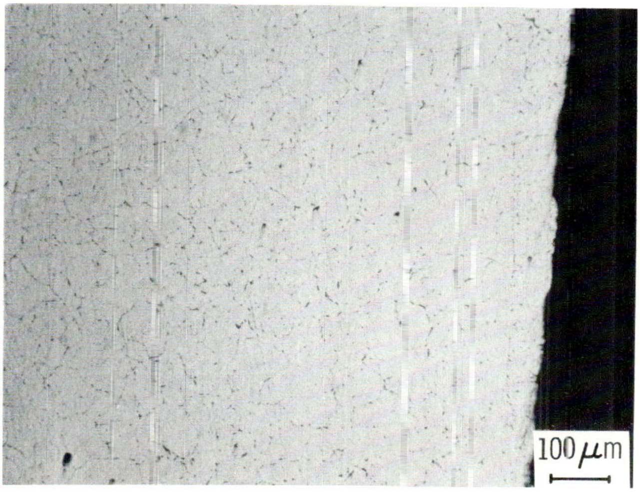

(b)

HF etch

Figure 6 - Typical microstructure at as-cast surface of 6" diameter 6063 alloy cast using: (a) conventional mold; and (b) WASM mold.

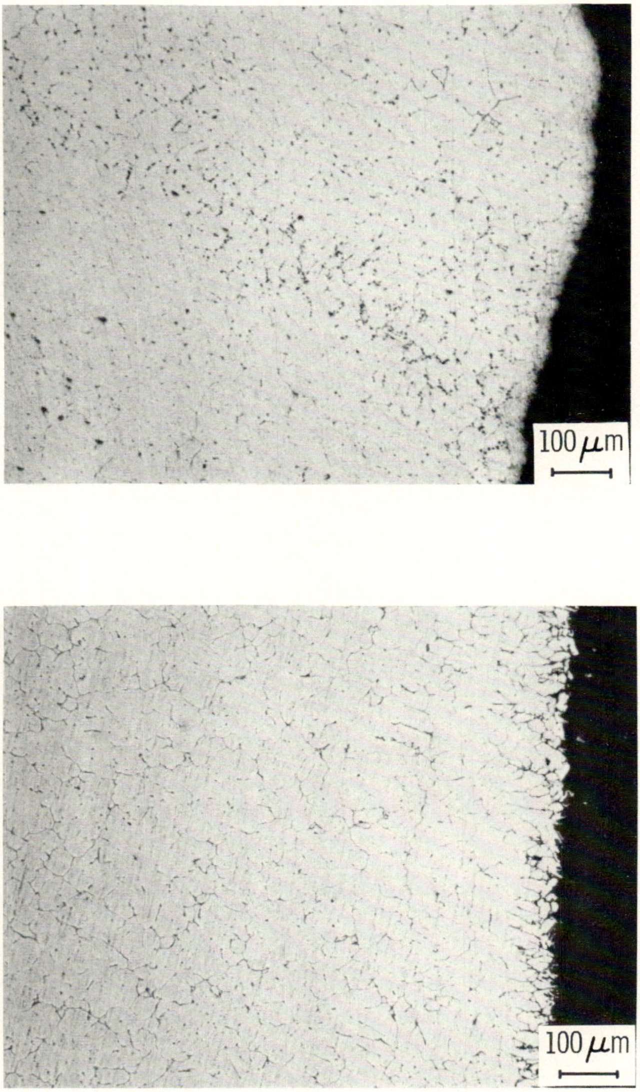

(a)

(b)

HF etch

Figure 7 - Typical microstructure at as-cast surface of 6" diameter 6061 alloy cast using: (a) conventional mold; and (b) WASM mold.

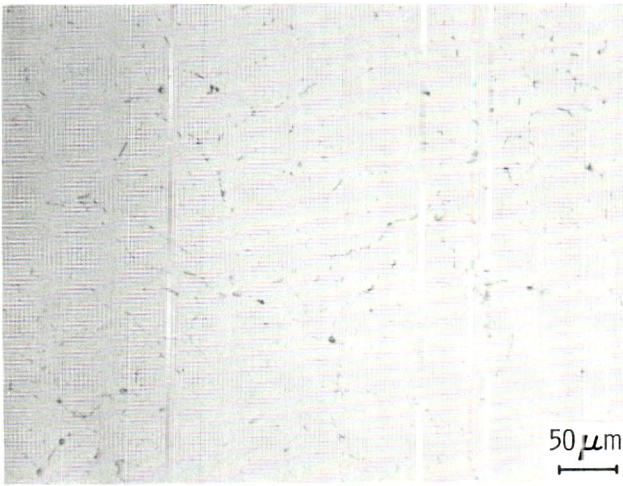

(a)

(b)

HF etch

Figure 8 - Typical microstructure at mid-radius of cross-section of 6" diameter 6063 alloy cast using: (a) conventional mold; and (b) WASM mold.

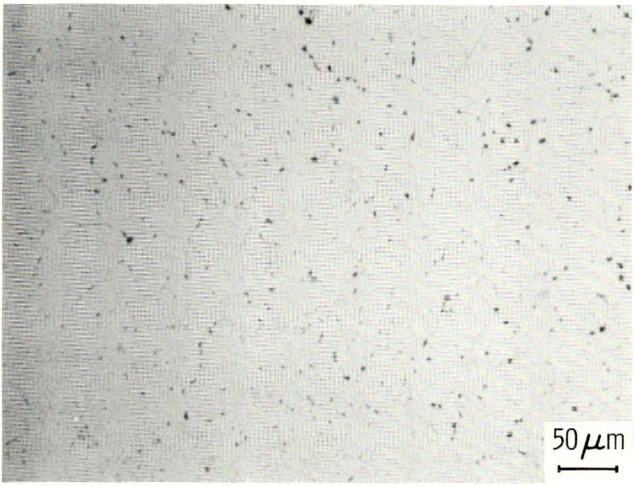

(a)

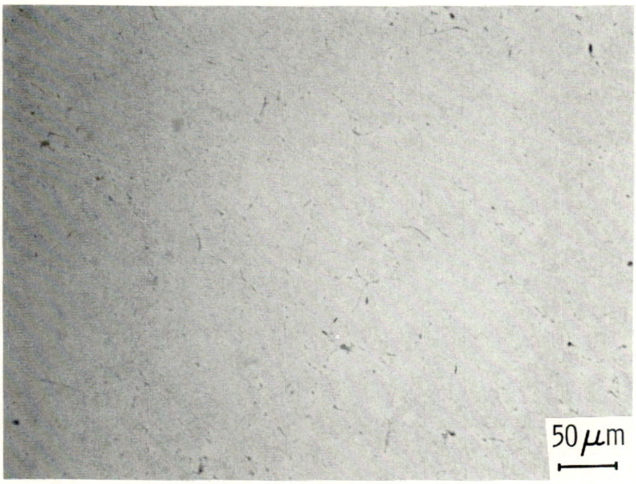

(b)

HF etch

Figure 9 - Typical microstructure at mid-radius of cross-section of 6" diameter 6061 alloy cast using: (a) conventional mold; and (b) WASM mold.

crops, and straighter ingots. Actual recent recovery has not been calculated as of this writing since the figure would have been biased by scrap incurred during training of the operating crews and normal "de-bugging" of the equipment. For example, head and butt crops are currently less than those incurred when the WASM molds were first used even though initially these crops were less than those for an equivalent size conventional mold.

Table II

Comparison of Productivity of WASM Versus Conventional Mold(1)

	WASM(2)	Conventional Mold
Production rate (lbs./hr.)	59,935	28,479
Lbs./Man Hr.	19,978	9,493
Crew Size (men/unit)	3	3
Average Lubricant use	18 oz.	192 oz.

Notes:

1) Based on data for 6" diameter 6063 alloy.
2) Approximately 6,000,000 lbs. cast weight

Summary

The Wagstaff AIR-SLIP™ casting mold has been used by Martin Marietta Aluminum for approximately 8 months to cast 6xxx alloy extrusion billet. This process was relatively simple to install and operate by virtue of being compatible with existing processes and practices. The WASM has resulted in a significant improvement in billet quality while simultaneously improving overall cast shop productivity and recovery.

Acknowledgement

The authors wish to thank the managements of Martin Marietta Aluminum and Wagstaff Engineering, Inc., for granting permission to publish this paper. Additionally, we wish to thank our co-workers who helped with its preparation.

Special thanks are given to Mr. Richard Collins, Wagstaff Engineering, and Mr. Paul Goodall, Martin Marietta Aluminum, who were largely responsible for the implementation of this technology at The Dalles plant; Mr. James Watts, Martin Marietta Laboratories, who performed the microscopy; and Mr. James Shaver, Martin Marietta Aluminum, who collected and organized data on the operation of the AIR-SLIP™ casting unit.

References

1. Baker, C., and Subramanian, V., "DC and Continuous Casting," Aluminum Transformation Technology and Applications, pp. 335-362, American Society for Metals, 1980.

2. Emley, E.F., "Continuous Casting of Aluminum," International Metals Review, (1976) pp. 75-115.

3. Davis, D., "Aluminum Billet Casting," Light Metal Age, 37, (1,2) (1979) pp. 26-31.

4. Bergmann, W.J., "Aluminum Ingot Surface Improvement by Advanced Continuous Casting Technology," Aluminium, 51, (5) (1975) pp. 336-339.

5. Collins, D.L.W., "A New Explanation of the Surface Structure of D.C. Ingots," Metallurgia, 34, (10) (1967) pp. 137-144.

6. Franz, E.C., "Metallurgical Factors Affecting Finishing Characteristics of 6063 Alloy Extrusions," Light Metal Age, 40, (7,8) (1982) pp. 6-10.

7. Scharf, G., and Lossack, E., "The Influence of Aluminum Alloy Casting Structures on Homogenization and Extrudability," Proceedings of the Second International Aluminum Extrusion Technology Seminar, pp. 311-320, Aluminum Association, 1977.

8. Dunn, E.M., "Metallurgical Structure of Electromagnetically Cast Extrusion Billet," Light Metals 1979, AIME, pp. 671-682.

9. Gelseleo, Z.N., J. Met., 23 (1971) pp. 38-39.

10. Wilkens, R., F., T., "The Variable Chill Depth Mould System," Light Metals 1983, AIME, pp. 907-920.

11. Anderson, R.V., and Harris, J.R., "Level Feed Billet Casting at Kaiser Chalmette," Light Metals 1981, AIME, pp. 827-843.

12. "Hot Top Caster Doubles Extrusion Billet Production," Modern Metals, 38, (12) (1983) pp. 34-35.

13. Wagstaff, F.E., and Keeler R., "Wagstaff Maxi-Cast™ Billet Casting System at Reynolds Massena," Proceedings of the Pacific Northwest Metals & Minerals Conference, 1982.

MOULD TEMPERATURES DURING DC-CASTING

OF 8 INCH DIAMETER EXTRUSION INGOTS IN ALLOY 6063

E.K. Jensen
Mosal Aluminium, Lista Aluminiumverk
N-4551 FARSUND, Norway

Some experimental casting data are presented as well as analysing of the results by means of a two-dimensional heat transfer mathematical model. Several details concerning the heat transfer within the mould regions are still not fully clarified. Measurements and calculations indicate that realistic maximum mould temperatures during normal DC casting are within a band ranging from about 50°C to about 100°C, depending upon casting conditions/mould design. Heat is transferred to the mould for a large part in a short "contact" zone of about 2-4 mm or possibly a somewhat longer high intensity heat transfer zone of about 5-10 mm. Maximum heat intensities indicated are about 2-10 MW/m2, max. heat transfer coefficients about 5-10 kW/m2 °C, oil film thickness about 5-10 µm. The question is raised whether such a high heat flow can be transferred through the oil film by conduction alone or whether additional factors are operating.

Background

It is well known that the quality of the DC-cast ingot surface and nearest subsurface-structure zone is determined by the solidification and initial shell formation within the mould.

A better control with surface quality requires a better understanding of the initial solidification phases and heat transfer within the mould region. A good physical model for the functioning of the lubricating oil film is thus also important. Prevailing mould temperatures reflects the actual heat transfer conditions, the role of the oil film and the thermal conditions which the oil film has to withstand. These conditions should be major factors to consider in trying to optimize and improve lubricating oils.

Results reported in the litterature show a large spread in measured mould temperatures, however, e.g. from the order of less than about 100°C, Fortina et al (3) to 450°C, Baumgarten et al (4).

This paper tries to analyse experimental data from Rönigk (1) and this author (5), and to clarify probable limiting conditions for possible heat transfer and mould temperatures. The author is grateful for stimulating discussions within an international DC-casting group of the German society "Deutsche Gesellschaft für Metallkunde", especially Mr. Rönigk.

The overall heat balances

Fig. 1 and table 1 summarize measured heat transfer. In table 1 is also shown calculated heat transfer. The differences between calculations and measurements are quite negligible and easily accounted for in terms of possible measuring deviations and losses due to vapourization of water etc.

Though main interest is focused on the mould, for the sake of order also calculated isoterms in the ingot are shown, fig. 2, to document overall agreement with the casting experiments.

Cooling conditions in the calculations have been chosen so as to reproduce the observed depth of the liquid pool in the ingot center = 80 mm. Also the calculated secondary dendrite cell size of about 23 µm at 20 mm distance from the surface agrees favourably with the measured size of 22 µm, fig. 3 (the constants in the relation between cell sizes and solidification times have been determined earlier in separate directional solidification experiments).

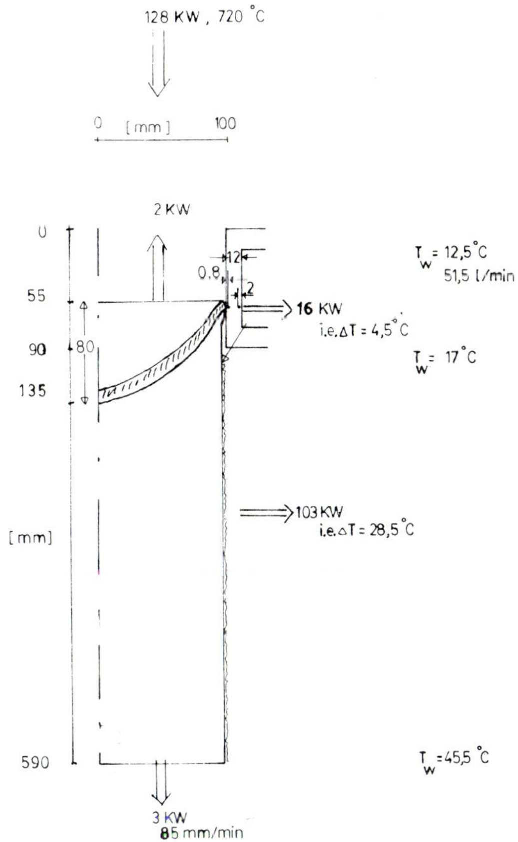

Fig. 1. DC casting of 195 mm diameter extrusion ingot.
Values as measured by Rönigk (1).

	Measured by Rönigk (1)	Calculated case 645502
Heat loss, top surface of liquid Al	2 kW	2 kW
Primary cooling mould	16 kW (4,5°C)	16 kW
Secondary cooling below mould	103 kW (28,7°C)	106 kW (29,5°C)
Bottom of ingot	3 kW	3,7 kW
	124 kW	127,7 kW
Heat content of liquid Al entering the mould	128 kW	127 kW
Q	4 kW (1,1 °C)	0,7 kW

Table 1

194,5 mm Ø ingot, 85 mm/min casting rate, 720°C casting temperature, 51,5 l/min cooling water.

(The temperature increase of cooling water corresponding to the amount of heat transferred is indicated in paranthesis).

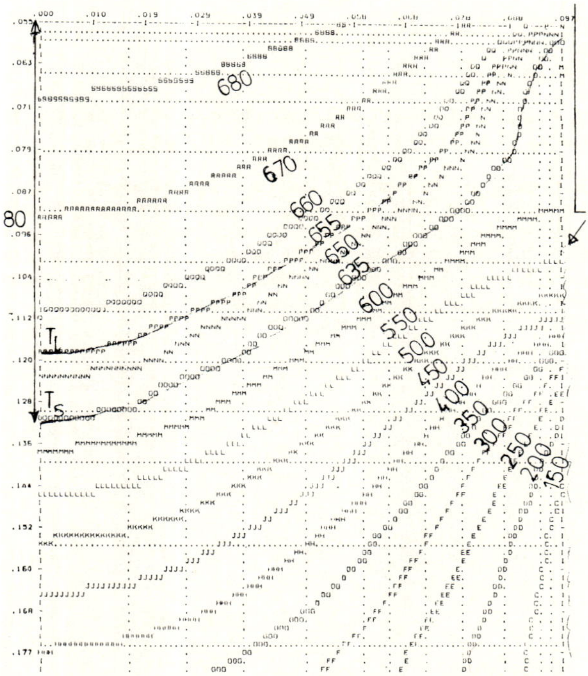

Fig. 2.
Calculated isotherms including liquid pool depth during DC floater casting of 8" diameter extrusion ingots.

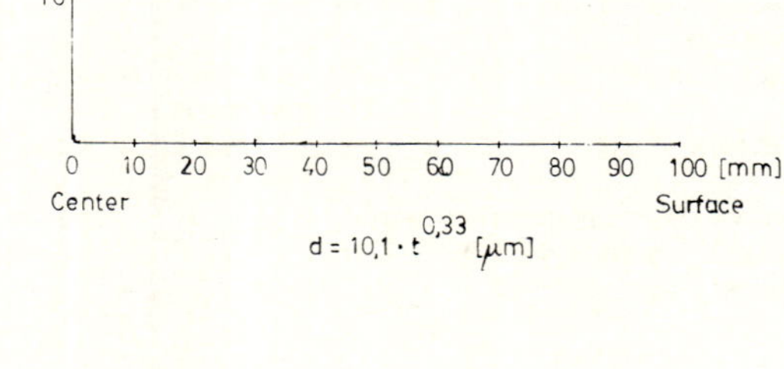

Fig. 3.
Measured ⊗ (1) and calculated secondary dendrite arm spacing as a function of distance from ingot center.

Measured mould temperatures

DC-casting by using floater

Some results reported in the litterature (3) as well as measured by B. Rönigk (1) are illustrated in fig. 4.

A calculated temperature profile is also shown. All temperatures are well below 100 degrees. Maximum temperature measured seem to occur about 10 mm below the metal level in the mould. (Though the experiments (1) and (3) might not be directly comparable in every detail, they serve to document a probable normal temperature range in the mould during casting).

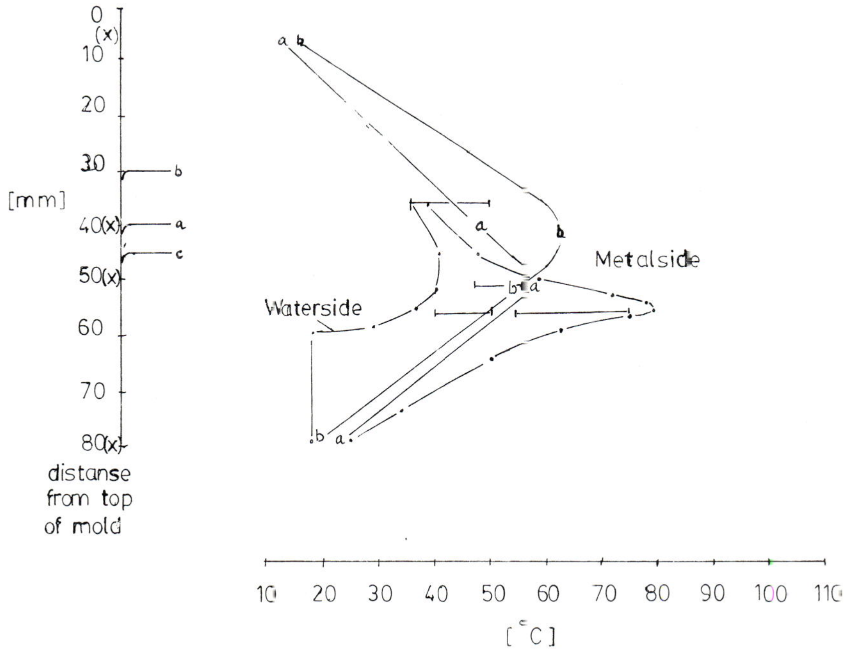

Fig. 4

Temperatures measured and calculated in the mould during casting of about 8" dia ingots AlMgSi 0,5 (AA 6063).

Metal level (head) in mould

a)	40 mm)	Ref. Fortuna et al (3)	180 mm dia mould
b)	50 mm)	"	10 cm/min.
c)	35 mm ├──┤	Ref. Rönigk (1)	194,5 mm dia mould 8,5 cm/min
(x)		Thermocouples in mould wall	
	35 mm •──•	Calculated case 644428 (re. also fig. 10 b)	

Casting trials with level pouring moulds

A series of casting trials have been carried out by this author and coworkers at one of the Mosal plants in Norway (5). The experiments were carried out with a normal production mould of the level pouring (hot top) type with 0,5 mm dia thermocouples mounted along the inner mould surface. The thermocouples were mounted from the water cooled side of the moulds to a depth of about 0,5 mm from the liquid metal side, at 2 mm intervals in the casting direction, covering from 0 to 20 mm of the mould immediately below the insulated upper part of the mould (hot top). Results are shown in fig. 5.

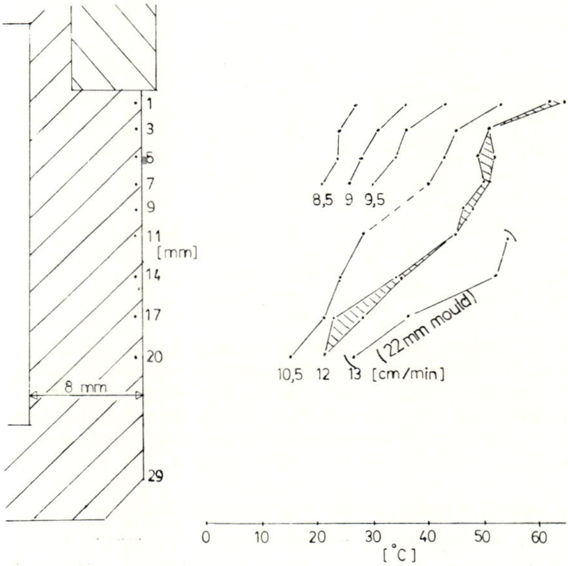

Fig. 5
Measured temperature profiles in mould wall 0,5 - 1 mm from liquid Al/solidified shell during level pouring casting of 8" dia extrusion ingot AlMgSi 0,5 (AA 6063), 100 l water/min.

We see that the maximum mould temperatures are measured immediately below the mould insulation, decreasing rapidly the first 10-20 mm below the insulation.

Max. temperatures range from about 30 to about 70°C, depending on casting rate, which was varied from 3,5 to 13 cm/min. As indicated by a band, reproducibility of results were within 5-10°C.

Thus, also with level pouring moulds, mould temperatures below about 100°C are indicated.

<u>Heat transfer from ingot to the mould</u>

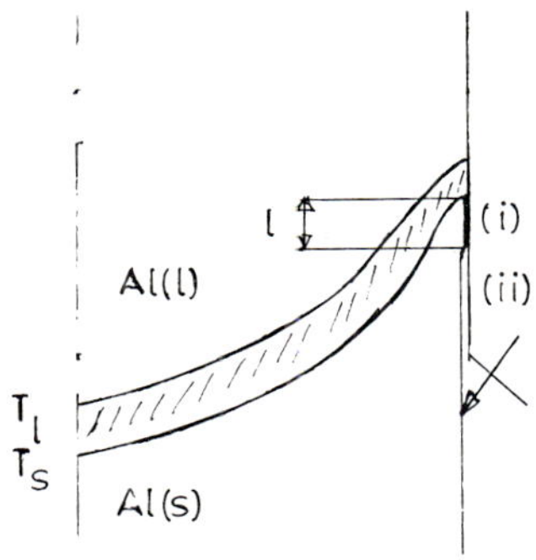

Fig. 6

During the initial high heat transfer zone of length l, re. fig. 6, fairly direct contact (i) between the solidifying shell and the mould is made, only moderated to some degree by the presence of the thin oil film. Below this zone the "air" gap (ii) develops and relatively low heat transfer prevails until direct water cooling occur below the mould.

For the varying conditions, the simple formula

(I) $q = \alpha \cdot \Delta T$ [W/m2]

can be used to describe the various heat transfer intensities q. Heat intensities vary, as a consequence of varying heat transfer coefficients α and temperature drops ΔT across the discontinuity between the solidifying shell and the mould wall.

In all the following calculations, a total amount of heat transferred to the mould of 16 kW has been assumed to simulate the measurements (1).

Heat transfer of a given amount of heat might be discussed in terms of various contact lengths l. A short contact length would thus correspond to a high α -value and vice versa.

Total heat transferred to the mould over the contact length l equals

(II) $Q = \alpha \cdot \Delta T \cdot \pi \cdot D \cdot l \quad [W]$

From (I) and (II) we get

(III) $\Delta T = \frac{q}{\alpha} = \frac{1}{\alpha} \cdot \frac{Q}{\pi D} \cdot \frac{1}{l}$

(IV) and $l = \frac{Q}{\pi D} \cdot \frac{1}{q}$

Mould temperatures and total heat transferred have been measured by (1), also a maximum value for $l \sim 4$ mm has been observed. α -values are, however, not known "a priori" and various α -values could be tried out in the calculations to get a good fit with measured mould temperatures. α -values can also be calculated backwards when known heat intensities q are given as input in the model calculations.

This is illustrated in fig. 7.

Lines of different α -values passes through origo in fig. 7 where temperature differences Δ T are shown as a function of q. The broken line refers to a number of calculated cases for a constant heat of 15-16 kW transferred to the mould as measured by (1). Corresponding l-values are also shown on the abscissa axis. (These l-values are strictly correct only for a constant value of heat $Q \sim 16$ kW transferred to the mould in the contact zone). Calculated Δ T and l decreases as heat intensities and mould temperatures increases.
The slope of the curve is steadily decreasing, however.

Fig. 8 illustrates directly the calculated average shell surface temperatures and average mould temperatures in the contact zone. Mould temperatures for realistic contact lengths
seem to be well below 150°C.

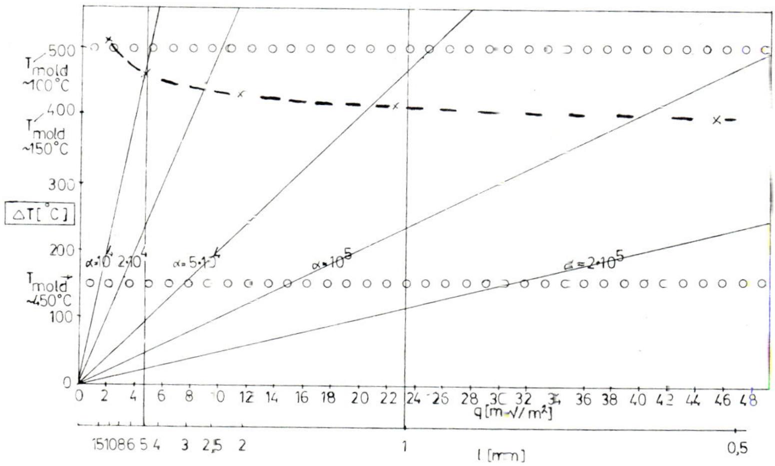

Fig. 7. Calculated temperature drop ΔT across solidified shell - mould wall discontinuity as function of heat intensity q/contact length l in the high heat transfer contact zone.

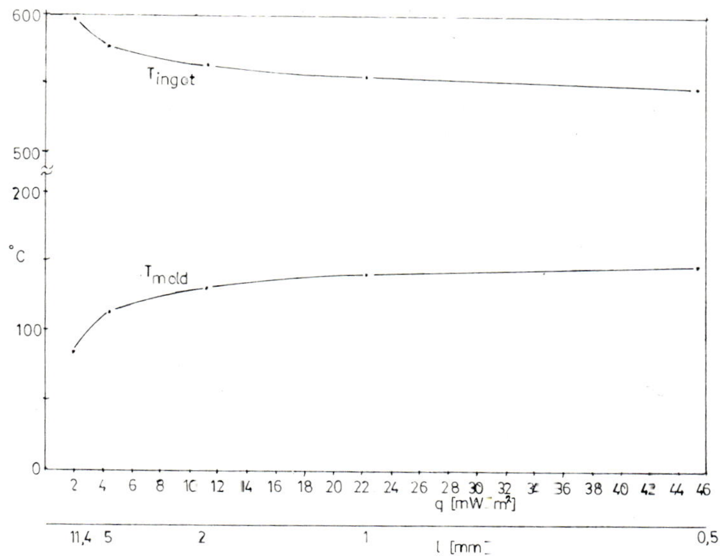

Fig. 8. Calculated average surface temperature of solidifying shell as well as average mould surface temperature in high heat transfer contact zone in the mould.

Based on these calculations, it is very difficult to visualize mould temperatures as high as 400-450°C(4),(lower 00-line, fig. 7) at least under normal conditions. Even if an l-value as low as 0,5 mm is assumed, corresponding to α = 125 kW/m2 °C and q = 47 MW/m2, the temperature drop Δ T decreases only to about 400°C. Assuming an ingot temperature of about 550-600°C, this would indicate a mould temperature of only max. 150-200°C. (See also fig. 11 c).

It was shown earlier that even when taking a variation in casting speed into account, this conclusion is not noticeably altered (re. fig. 5).

Thus in order to try to possibly explain mould temperatures of 400-450°C, one has to assume Δ T equal to about 150°C, and thus extremely small and very unlikely contact lengths, like-wise unlikely high q values.

Assuming l = 4 mm (re:the other end of the abscissa) i.e. α = 14 kW/m2 °C and q = 6 MW/m2, ΔT = 450°C and mould temperature of max. 100-150°C are indicated (upper 00 line, fig. 7).

Even q = 6 MW/m2 for the 4 mm contact length is a very high q-value. It is perhaps questionable if such a large heat intensity can be conducted through the (durable?) oil film without the film immediately vanishing. And yet the oil film is normally seen to last regularly e.g. for a drop of up to about 7 m casting length or up to about 1 h casting time. It can thus be questioned whether a physical model/understanding of the oil film functioning only as a pure "separating film medium" is correct. Perhaps a physical model adopting "tribological" features might be more fruitful. Such an alternative model could perhaps include (numerous) small points of direct contact between the solidifying shell and mould wall in the high intensity heat transfer zone, bridging the "oil gap" for a short time locally.

We have seen that max. mould temperatures measured by Rönigk (1) and Fortina & Gatto (2) are 65-75°C under normal conditions. This would correspond to a heat intensity of less than 2 MW/m2 and an apparent contact length of about 15 mm in the diagrams, fig. 7, 8. This length is clearly longer than the about 4 mm indicated by observations on the mould surface according to (1). This could perhaps be taken as a sign that the effective length of a fairly high heat transfer zone is somewhat longer than the actual "contact" zone.

Other factors influencing the more detailed calculation of mould temperatures are discussed in the following paragraphs.

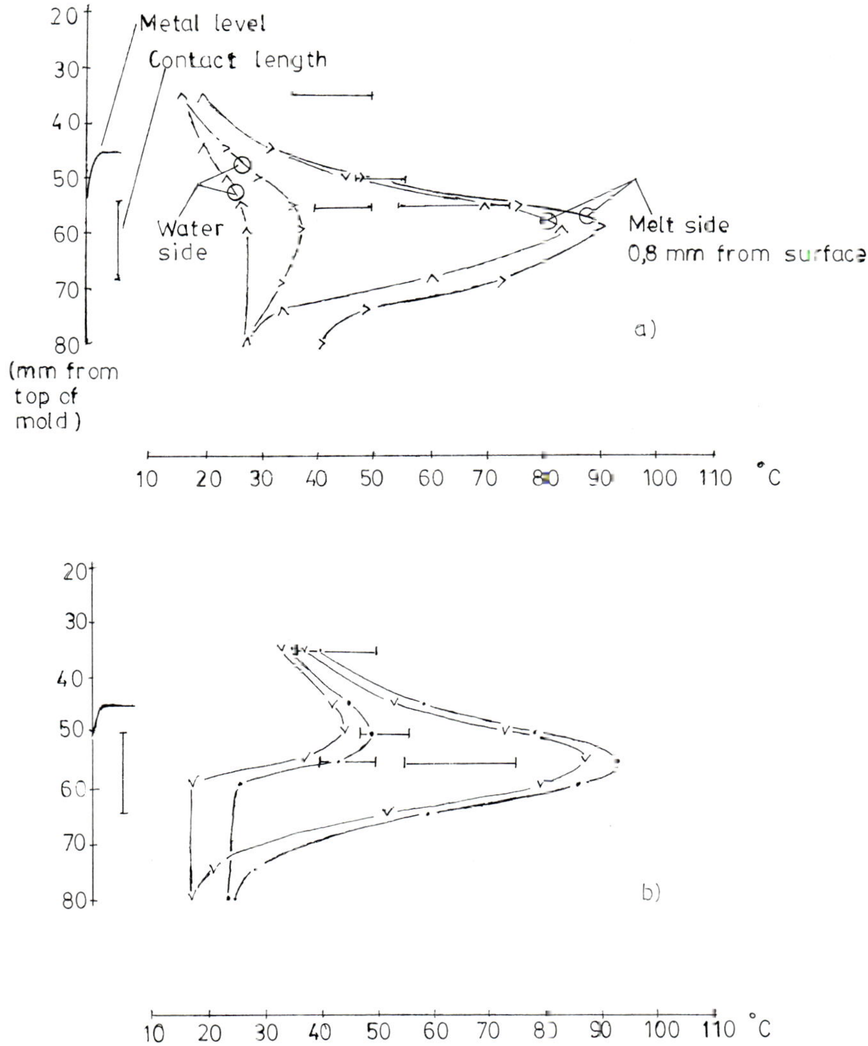

Fig. 9. Effect of water holes in the mould wall and position of contact zone.

		Direct contact (mm below metal level)	Effect of water holes in mould wall
Fig. 9a.	644415	> 9-23	Increasing
	644417B	^ "	
Fig. 9b.	644420	. 5-19	
	644421	v "	

├─┤ Measured temperatures.

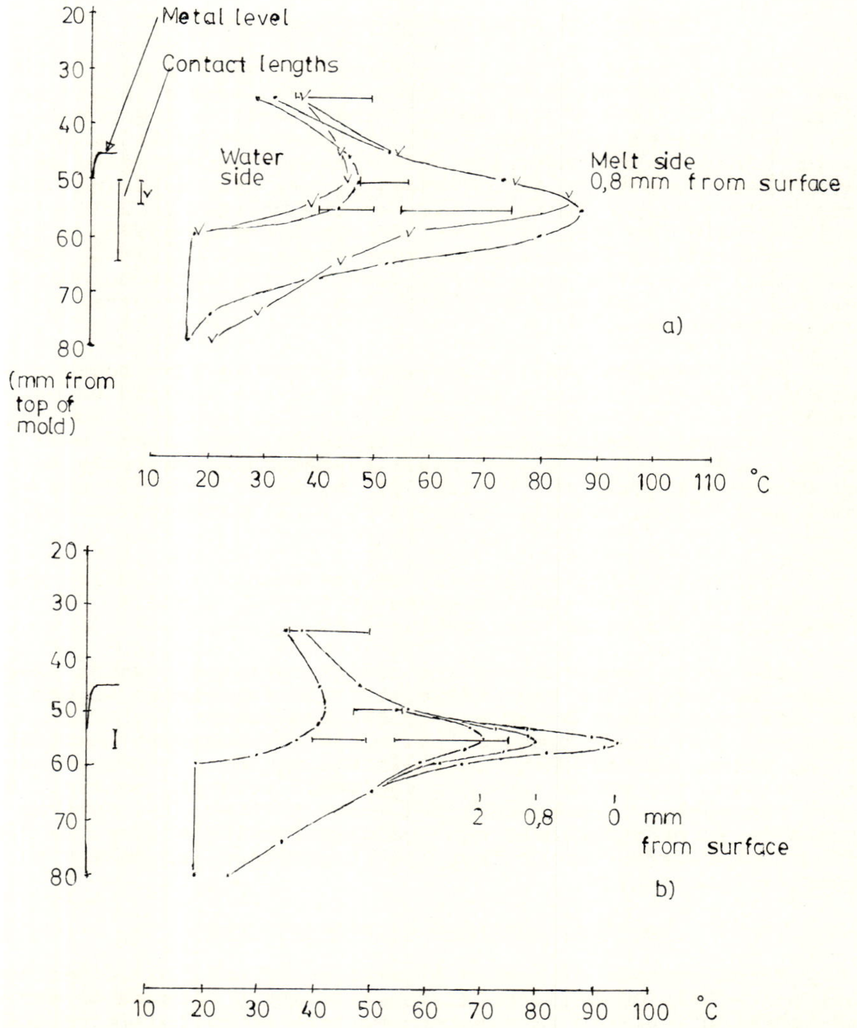

Fig. 10. Effect of varying heat transfer in air gap zone and position of contact zone.

			Direct contact (mm below metal level)	Heat transfer air gap zone
Fig. 10a.	644421	.	5-19	α = 55 W/m2 °C, q=0,03
	644424	v	5-9	" =1000 " ", "=0,52
Fig. 10b.	644428	.	8-12	" =1100 " ", "=0,55 MW/m2

├──┤ Measured temperatures 0,8 mm from mould wall surface.

Effect of water holes in mould wall

It is plausible that the presence of water holes in the lower part of the mould wall (here: present over a length of 20 mm, i.e. 20/35 = 57% of metal level (head) in the mould) can have a marked influence on the temperature profiles in the mould. Thus in our case, the effect cannot be neglected as having been done hitherto calculationwise, only assuming uniform cooling along the water side of the mould. In following modified calculations the effect is simulated by giving a high heat conductivity for the lower right triangel corner of the mould wall (i.e. the area to the right of the water holes), re. fig. 11e , as well as a high heat transfer to the cooling water here.

As an example, the values were chosen to give mould temperatures in the triangle nearly equal to the cooling water temperature. This simulates the physical presence of cooling water of actual cooling water temperature in the mould wall itself. (Thus, in this way, mould temperatures can only be calculated correctly to the left of the water holes in the lower part of the mould). The effect is illustrated in fig. 9.

Naturally, this relatively marked change in cooling will also slightly modify the temperatures in fig. 7 and 8 somewhat,but not alter the conclusions.

The mould temperatures are clearly shifted towards lower values, but still the max. temperature is calculated higher than observed unless a maximum on the temperature curve is accepted, positioned somewhat below the thermocouple position (10 mm below nominal metal level in the mould). Observations when varying the metal level in the mould suggest, however, a rapid decrease in temperature below about 10 mm (1). Further checks have therefore been made calculationwise to see the effects of varying the contact length/position and heat transfer intensities.

Vertical position of contact zone

First contact of liquid Al with mould wall is judged to take place 5-10 mm below metal level in mould because of surface tension/meniscus curvature. Fortina et al (3) indicated 9 mm, Rönigk (1) suggests 5 mm. Exact position for a given contact length is important for correct calculation of position of maximum mould temperatures, especially for short contact lengths.

As illustrated in fig. 4 and 10, calculations seem to give best fit with the available temperature data for a contact zone starting about 8 mm below metal level and a short contact length, re. next paragraph.

The temperature level is also influenced by vertical position of contact zone because of proximity to intense cooling action from water holes in mould wall.

Heat transferred in the "contact" zone versus the "air-gap" zone

In the preceding paragraphs (except in fig. 4), the majority of the heat ($\sim$95%) transferred to the mould is assumed to take place over the short contact length l and thus relatively negligible amounts of heat crossing the "air gap" zone. In order to bring the max. calculated temperatures well below 100°C as observed, a relatively "long" contact length (or at least a high heat transfer zone) of about 10-15 mm has to be assumed, in order not to have too high heat intensities and mould temperatures.

If the "air gap" zone can transfer a larger part of the total heat than hitherto assumed, this minimum length of the direct contact zone can be reduced correspondingly.

A final calculation is made with q in the air gap zone increased from about 0,03 MW/m2 (α = 55 W/m2/°C) to 0,55 MW/m2 (α = 1100 W/m2/°C). In the contact zone of length l (reduced from 13,6 to 4 mm) q = 1,79 MW/m2 and 2,97 MW/m2 respectively. Now the short contact zone and the 6 times longer "air gap" zone transfer each about 50% of the total heat to the mould. Results are shown in fig. 10 with a very good fit with measured data. (See also fig. 11 d, e).

If such relatively high heat intensity in the "air gap" zone is realistic, considerable contributions to the heat transfer must come from other sources than heat radiation, i.e. conductive/convective contributions, possibly also vapourization - condensation of oil and also water vapour from below the mould.

Mould isoterms

In fig. 11 a-c calculated isoterms in the moulds are shown for contact lengths of 5, 2 and 0,5 mm respectively.

Main objective is here to illustrate the changes in the isotherm positions near the direct contact zone in the mould.

Fig. 11 d, e illustrates the isotherms corresponding to the situations mentioned earlier: a long contact length to give low max. temperature when most of the heat is transferred over this zone, or a much shorter contact length when more of the heat passes the air gap zone. In the latter, also the effect of water holes in the mould wall has been taken into account. (The material to the right of the hatched line approximating the water hole line in the mould wall is given a very high conductivity in the calculations).

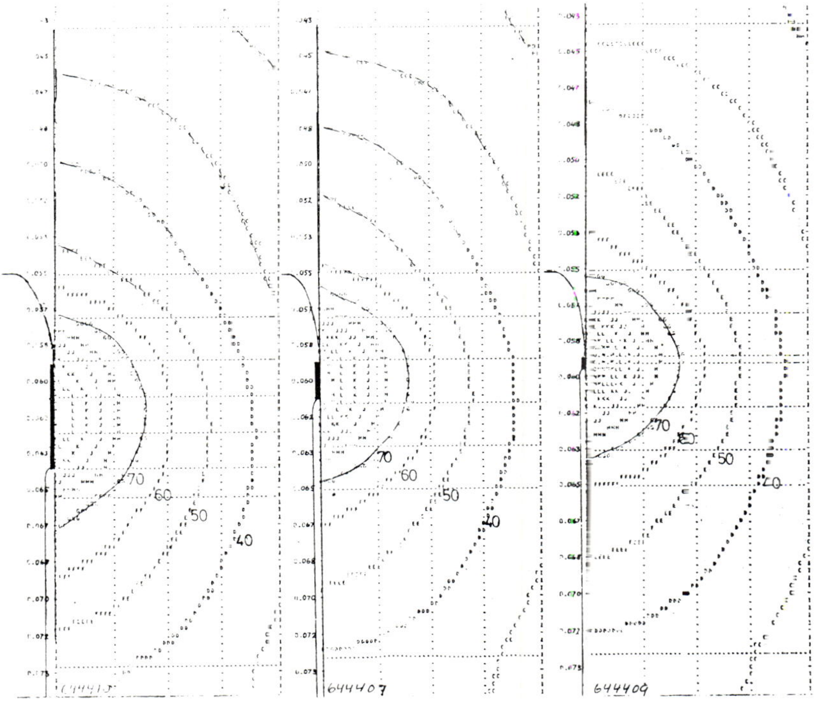

a) b) c)

Fig. 11. Calculated isotherms in mould wall for 5, 2 and 0,5 mm contact lengths respectively.

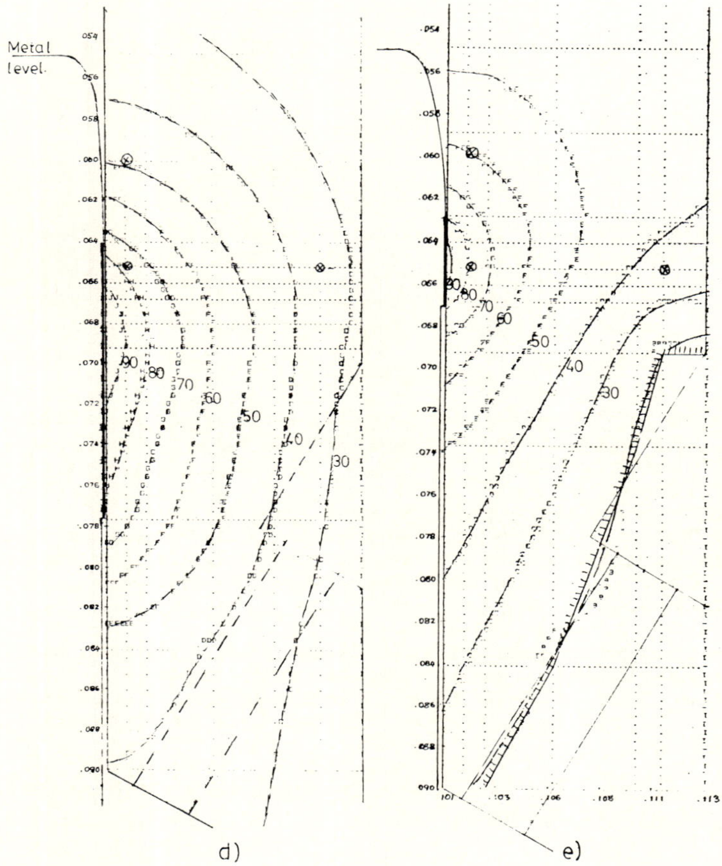

Fig. 11 (continued).
Calculated isotherms in mould wall for 14 and 4 mm contact lengths, ⊗=Thermocouples(1).

Oil film thicknesses

Thicknesses of oil film with "effective" thermal conductivity λ [W/m °C] can be estimated as follows:

$$s = \frac{\lambda}{\alpha} = \frac{\lambda}{q/\Delta T} \sim \frac{0{,}16}{6E6/(600-100)}$$

$$\sim 10\,\mu$$

This result also seems to agree well with practical experience when comparing with normally used oil feeding rates and e.g. also with max. tolerable surface roughness on mould surfaces. For the latter, experience indicates max. roughness of about 5 µm before ingot surface starts to deteriorate.

Conclusions

Recent experimental findings as well as calculations indicate that maximum mould temperatures will normally not reach much above 100°C.

Exact temperature calculations require detailed knowledge about position and length of contact zone as well as heat transfer intensities in the contact and air gap zones. Also the effect of water holes in the mould wall has to be taken into account.

The following seem to give quite good agreement with available experimental data: Maximum heat transfer intensities of the order of 2-10 MW/m2, heat transfer coefficients 5-10 kW/m2 °C, oil film thickness about 5-10 µm, contact lengths 2-4 mm.

Exactly how such heats are transferred to the mould through the oil film is not clear.

References

(1) B. Rönigk, Results of casting experiments presented on the DGM-Fachausschuss Stranggiessen meeting in Geisweid November 11, 1982, also to appear in Zeitschrift für Metallkunde.

(2) E. Madsen, H. Fossheim, "Application of a mathematical model in level pour DC casting of sheet ingots". AIME Annual Meeting 1979.

(3) G. Fortina, F. Gatto, "Thermal distribution during ingot solidification in DC casting". TMS paper no. LM 78-49, presented at the AIME meeting 1978.

(4) J. Baumgarten, M. Stark, "Kokillenwandtemperaturen beim Stranggiessen von Al 99,5 und AlMgSi 0,5". Aluminium 55 (1970) 5, 334.

(5) E.K. Jensen et al, Casting Experiments, level pouring moulds, Mosal Aluminium, Lista Aluminiumverk, Farsund, Norway.

PRODUCTION OF CONTINUOUSLY QUENCHED WIRE ROD OF Al-Mg-Si ALLOYS AND PROPERTIES OF RODS AND WIRES

P.Fiorini [*], L.Gasparotto [**], A.Lana [**], G.Zago [***]

Alluminio Italia S.p.A. - ITALY

The most typical manufacturing cycle of Al-Mg-Si rod and wire, involves the Properzi continuous casting process, discontinuous solubilization in a furnace, water quenching, preageing, drawing, and final ageing of wires. According to the current trends (integrated as much as possible) concerning wrought products aimed at a larger energy saving, a feasibility study has been performed regarding a continuously quenched Properzi rod. Such a development permits to avoid two different high cost, energy-wise, heat treatments, consisting in a solubilization process at 520-540°C for 10-12 hours and in an artificial preageing

Present paper describes the first production parameters modifications and the preliminary results obtained on rods and wires produced by experimental and half-scale tests.

[*] Istituto Sperimentale Metalli Leggeri, 28100, Novara

[**] Stabilimento Bolzano, 39100, Bolzano

[***] Stabilimento Fusina, 30033, Fusina (Venezia)

Introduction

The use of aluminum in the electrical field dates back to the end of the last century but only in the last twenty years the use of light me - tal has reached very interesting levels. Considerations of economic nature have played a determinant role in favour of aluminum with respect to copper, even if the last is considered, from a technical point of view, the most suitable material for electrical conductor.

Starting from this consideration, it is easily explained the parti - cular success of aluminum in the high voltage sector, i.e. where the cost of material is important.

The following aluminum conductors are employed: AAC, All-Aluminum Conductor obtained with EC aluminum wires; ACSR, Aluminum Conductor Steel Reinforced, consisting of steel wire for strength wrapped with EC alumi - num wires; AAAC, All-Aluminum Alloy Conductor, obtained with wires of 6101 or 6201-T8 alloy; ACAR, Aluminum Conductors Alloy Reinforced, which use 6201 or 6101-T8 wire reinforcement, with outer layers of EC Aluminum wires.

As general trend in the world, the use of AAAC cables has progressively increased having these cables, with respect to the classical ones, some important advantages such as lightness, surface hardness, higher electrochemical and mechanical homogeneity.

Furthermore, during the last years, also che mechanical application of Al-Mg-Si alloy wire has been highly developed.

This market evolution has determined important modifications on the production plant of aluminum alloy wire rod with respect to quality and quantity of the product.

The present paper analyzes an interesting innovation which Allumi - nio Italia is introducing in its plant, concerning the production of continuously quenched rod of Al-Mg-Si alloys for electrical and mechanical applications. In particular, the technological solution and the proper - ties of rod and wire, obtained through experimental and half-scale pro - duction, will be examined.

The statement of the problem

Alluminio Italia produces Al-Mg-Si alloys rod by using a T7 Properzi machine; the most significant steps of the process are the following:

a) preparation of the casting alloy in a "waiting" furnace, followed by proper degassing and skimming treatments to obtain a clean molten metal;
b) pouring off of the liquid in the casting furnace, with other cleaning treatments;
c) continuous refinement with Al-Ti-B rod;
d) solidification in the casting wheel with a cross section of the ingot of about 2300 mm^2;
e) hot plastic deformation by rolling to obtain a wire rod with a diameter varying from 15 to 9.5 mm;

f) coiling of the rod;
g) solution heat treatment of the coils in a furnace at 510÷530°C for 12÷15 hours followed by a water quenching process;
h) drying treatment of the coils at 120÷150°C for 2 hours, to avoid corrosion problem; this treatment is useful at the same time to induce on the material a preageing process.

The rod is now ready for drawing operation, followed by final ageing treatment at 160÷180°C to obtain the required electrical and mechanical properties.

The operations from point a) to point f) are performed non-stop, with a high level of efficiency.

On the contrary, points g) and h) represent a sharp discontinuity for the process, which causes apart from moving and storing problems, a considerable consumption of electrical energy as a consequence of both the long solution heat treatment at high temperature and the drying reheating.

In addition, the solution and quenching treatments of big coils (weight 1000÷2000 kg) can produce remarkable scattering of the properties of the rod, as a consequence of a different heating and cooling rate of the internal and external part of the coil.

The realization of particular furnaces, although it strongly reduces the scattering level, doesn't solve the energetic problem.

The continuous quenching of the rod, performed directly on line, seems to represent one of the most interesting way of saving electric energy. The general approach seems to be quite simple, consisting of a drastic cooling treatment of the rod at the exit of the rolling mill.

Actually, the problem is rather complex, involving important transformations of both metallurgical processing parameters and working conditions of the casting apparatus.

Alluminio Italia began some years ago this kind of researches with laboratory experiments; when the energetic problem became dramatic, the results were extended to the industrial production.

As expected, this phase presented many difficulties, but today the continuously quenched rod provides a sufficient reliability.

In the next paragraphs the most important processing parameters and the properties of rod and wire, will be examined, after a short description of Al-Mg-Si alloys.

General properties of Al-Mg-Si alloys

Al-Mg-Si alloys represent, with Al-Cu and Al-Zn-Mg systems, the typical age-hardenable aluminum alloys and are used for electrical application and, more in general, when intermediate strength and high resistance to corrosion and stress corrosion cracking, also after welding, are required.

The strengthening of these alloys is obtained by precipitation hardening induced by thermal treatment at low-medium temperature (< 200°C) after quenching from the solubilization temperature (500–530°C). The first decomposition stage of the solid solution is characterized by needle-like GP zone formation consisting of atom rows all along ⟨100⟩, so that one row of Si-atoms is bounded by two rows of Mg-atoms. In concentrated alloys and in the maximum hardening conditions, the needles can reach a length of about 100 nm, with a 6 nm diameter and a density of about $5 \cdot 10^{15}$ cm^{-3}. Since only a small elastic interaction at the needle-matrix interface has been determined, the high strength of Al-Mg-Si alloys is mainly related to a chemical effect, i.e., to the energy necessary to break Mg-Si bonds when the dislocations cut the needles. The thermal stability of GP zones is very high, even if as a consequence of a prolonged ageing treatment a continuous growth of zones into rods is obtained, the rods being an intermediate form of the equilibrium phase β Mg_2Si. A decrease of the strength accompanies the transformation of GP zones into intermediate or equilibrium phases (overageing). Some 6000 series alloys can be considered as pseudo-binary Al-Mg_2Si alloy. The maximum solubility of Mg_2Si is 1.85% at 595°C and that explains why in this system the age-hardening is less than in Al-Cu and Al-Zn-Mg systems. Additional strength can be obtained with a controlled excess of Mg or, above all, of Si-content, which causes an increase of the supersaturation of the alloys.

Also a thermomechanical treatment can markedly increase the hardening level of these alloys, by combining both the effects of Mg_2Si precipitate and dislocation-induced by cold work. It is important to notice that an artificial ageing treatment after plastic deformation can cause an overageing phenomenon as a consequence of the nucleation process exerted by dislocation on intermediate or equilibrium phases.

The electrical resistivity of Al-Mg-Si alloys varies with Si and Mg content and with the considered temper. In an ideal solid solution, Si and Mg should contribute separately to the resistivity; the contribution of 1 at% of Mg and Si respectively amount to 0.49 and 0.73 $\mu\Omega$ cm. Practically, after solution treatment and quenching, Mg and Si react immediately to form GP zones; during the natural ageing, the electrical resistivity increases till the zones reach a critical dimension, while it decreases subsequently. Artificial ageing, on the contrary, is always characterized by a decrease of electrical resistivity for both the GP zone and the Mg_2Si phase formation.

For the other physical properties it is more difficult to establish the effect of alloying elements, taking into account that Mg and Si react to form three kinds of Mg_2Si phase, consequently with different behaviour.

From a general point of view, the density generally decreases, whereas the thermal expansion coefficient is lower than that expected from a calculation based on the individual Mg and Si additions.

Rod production parameters

To produce continuously quenched rod, the following topics have been mainly investigated:

- liquid metal purity
- solidification and rolling parameters
- quenching conditions

Liquid metal purity

Preliminary experiments clearly showed that the normal cleaning process, i.e. the process needed to produce EC grade aluminum and Al-Mg-Si not quenched rod, were insufficient to obtain the necessary cleanliness. In fact, amounts even quantitatively and dimension-wise reduced of oxide or heterogeneous particles, strongly reduce the workability of the continuously quenched rod, owing to the high strength of this material.

As a consequence, a particular type of ceramic filter has been studied and fitted just before the pouring pot. In addition to filtering properties, an interesting behaviour of the filter is represented by an easy replacement in agreement with the continuity of the casting process.

Solidification and rolling parameters

The determination of solidification conditions represents the most important part of the process, considering the effect on the final properties of wire rod.

A correct choice of the working parameters must guarantee both an homogeneous solidification structure and a high degree of supersaturation of the casting bar. Apparently these results seem to be incompatible, requiring the latest a temperature value of the bar at the entrance of rolling mill the neerest as possible to the solvus limit of the alloys (~ 500°C); nevertheless in this case the risk of having an incomplete solidification of the bar at the exit of the wheel is very high.

With previous laboratory experiments, it has been possible to determine working parameters suitable to satisfy both preliminary assumptions.

In particular, with a rotation speed of the casting wheel of 2.5÷2.7 revs/min, a temperature of liquid metal in the pouring pot of 680÷710°C and with a proper adjustement of the cooling system of the wheel, we can obtain for the casting bar a temperature of 480÷510°C at the entrance of rolling mill. The bar is characterized by a good solidification structure, as shown by macrograph of fig.1 and by micrographs of figs.2 and 3 relative to the most critical zones of cast bar, i.e. the core (fig.2) and a part of flat surface solidified in contact with steel band (fig.3).

Naturally these casting conditions, in particular the casting speed, must permit to maintain, also after rolling, a high solubilization degree of wire rod.

Quenching conditions

The quenching treatment of the rod is strictly connected with the processing parameters previously described and therefore the realization of the cooling system has been carried out starting directly from the most useful speed casting condition.

Different solutions of the cooling facility have been proved to obtain a decreasing rate of the rod temperature, from about 300 to about 100°C, high enough to avoid a decomposition process of the solid solutions.

Table I summarizes an example of working parameters used for continuously quenched rod production. In this condition we have measured a

Figure 1 - Etched cross section of the cast bar, showing grain structure.

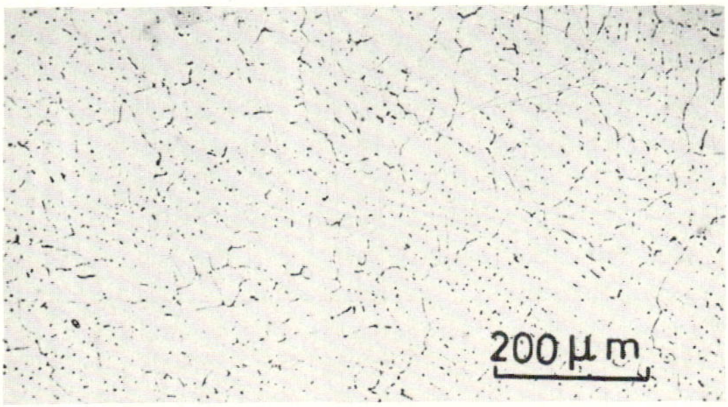

Figure 2 - Optical micrograph of the cast bar core, showing a homogenenous distribution of secondary phases (etching HF 0.5%)

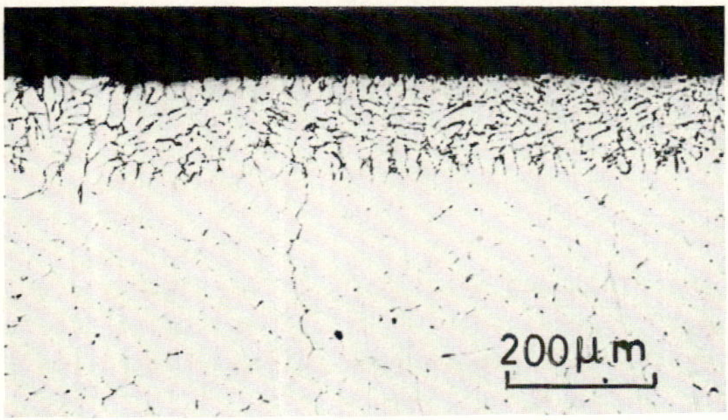

Figure 3 - Optical micrograph of the flat face of cast bar. It is observable a cortical layer of eutectic segregation of about 0.1 mm thickness (etching HF 0.5%)

temperature of about 90°C on the external part of the coil.

Table I - Example of working parameters employed to produce continuously quenched wire rod of Al-Mg-Si electrical conductor alloy

Parameters	Values
Liquid metal temperature in the pouring pot	695°C
Casting wheel speed	2.6 revs/min
Cast bar speed	11.1 m/min
Cast bar temperature at the entrance of rolling mill	495°C
Speed of the rod Ø 9.5 mm	5.8 m/sec
Cooling water flow inside the quenching tunnel	20 m^3/hr
Temperature of cooling water:	
Tunnel entrance	9°C
Tunnel exit	18°C

Finally the cooling process of the coil from 90°C to room temperature determines the first step of the natural ageing process which can be considered concluded after other 10÷20 days at room temperature.

Experimental production of rod and wire

The working parameters exemplified in table I, have been employed for some experimental production of 1500 kg weight coils, from which we have obtained the first information about the properties of the rod and wire.

Tables II and III give, respectively, the chemical composition and the electrical-mechanical properties of the rod produced in one of the experiments.

Table II - Chemical composition of the rod (experimental production)

Element	Si	Mg	Fe	Ti	Mg_2Si	Si in excess
wt %	0.60	0.61	0.21	0.005	0.97	0.24

Table III - Mechanical and electrical properties of the continuously quenched rod, with the composition of table II, after different hours of natural ageing

Time of natural ageing	U.T.S. MPa	E %	ρ 20°C ($\mu\Omega$cm)
24 hrs	185	14.5	3.33
7 days	215	13.5	3.38
30 days	227	13.0	3.42

Figs.4 and 5 shows the microstructure of the rod, determined by electron microscope examination.

We can realize that the rod presents a subgrain structure characterized by the presence of free dislocation inside subgrains and subgrain boundaries with non completely organized dislocations (fig.4).

In addition a very reduced amount of Mg_2Si incoherent particles is observable (fig.5).

These characteristics are connected with the quenching treatment which freezes both the hot deformation structure and the solubilization stage of the rod, and avoids the precipitation process at lower temperatures.

The microstructural evidences well explain the high properties of table III; in fact the resistivity values are essentially connected with the solubilization degree of the rod, and tensile properties are due both to work and solution hardening.

As comparison purpose, figs.6 and 7 shown the microstructure of rods obtained with standard casting and rolling conditions, without and with a final quenching treatment, respectively.

Both cases, are characterized by an important, even if quantitatively different, decomposition process of solid solution.

The drawing experiments have been performed with a laboratory facility, by employing a single die machine with a reduced working speed.

During this operation we did not have any drawing problems, obtaining wire with high surface quality.

Table IV gives the mechanical and electrical properties of the wire, 2.6 mm diameter, in different tempers (tempers are referred to Aluminum Association).

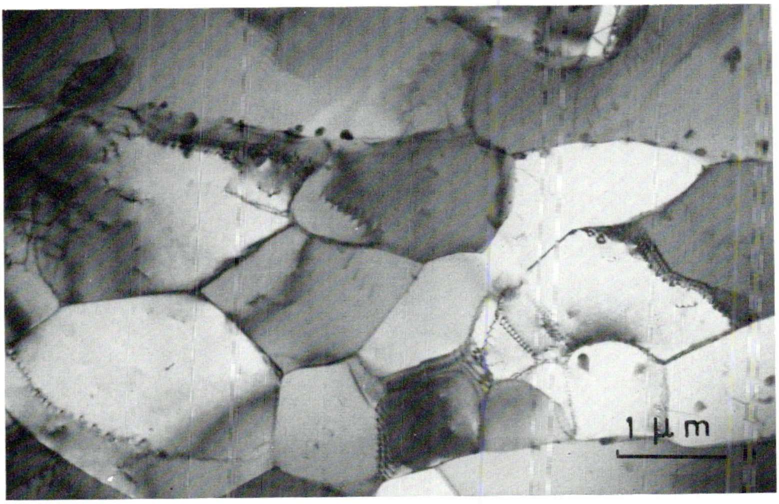

Figure 4 - Transmission electron micrograph of continuous quenched rod produced by using experimental conditions of table I. It is evident the subgrain structure and the presence of a very reduced amount of Mg_2Si particles.

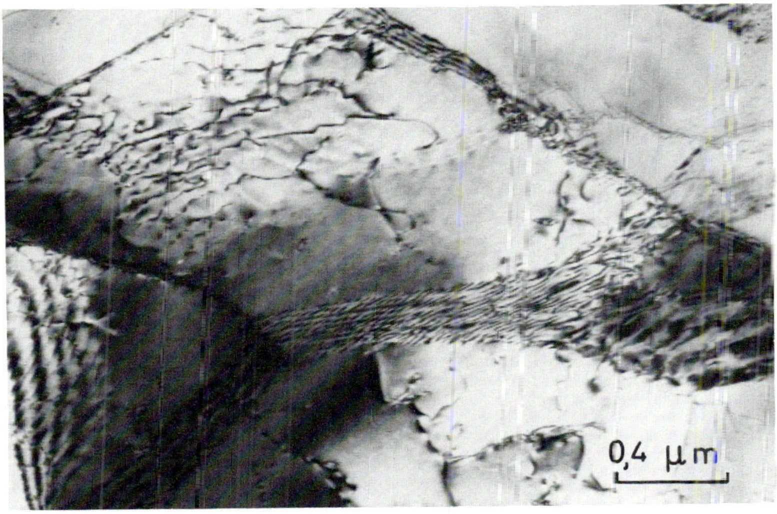

Figure 5 - Transmission electron micrograph of continuous quenched rod produced by using experimental conditions of table I. It is evident the dislocation distribution.

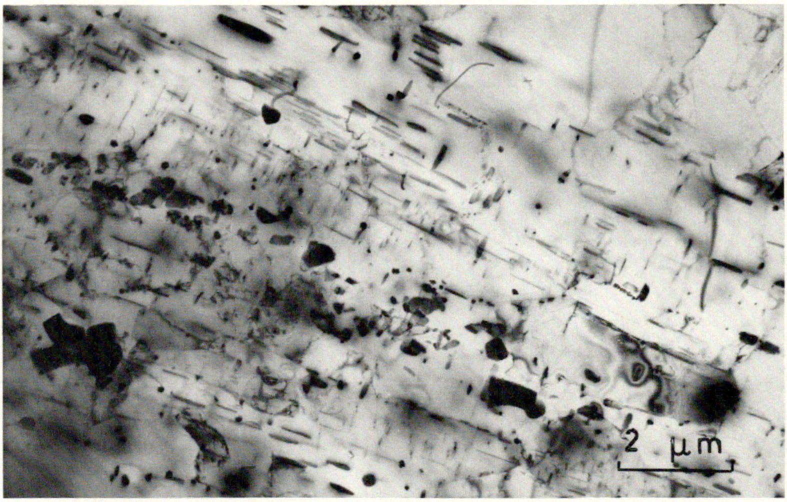

Figure 6 - Transmission electron micrograph of standard production rod, showing the rich and coarse precipitation of Mg_2Si particles.

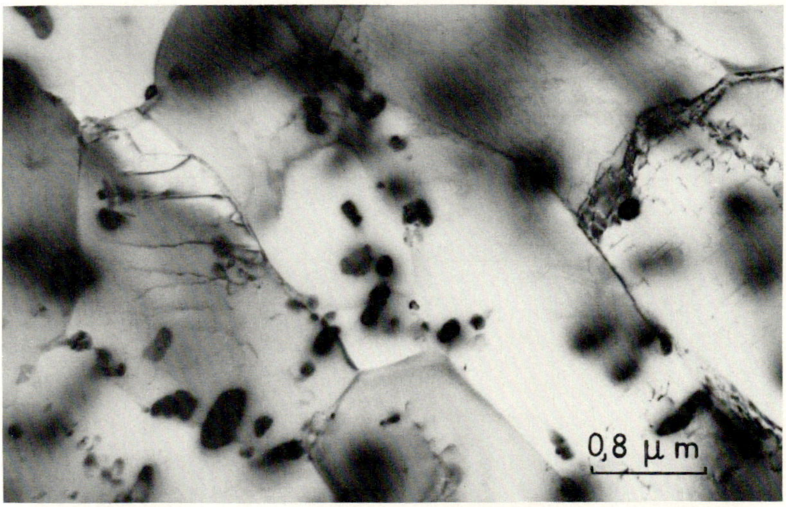

Figure 7 - Transmission electron micrograph of rod produced with standard condition, with a final continuous quenching treatment. An important presence of coarse Mg_2Si particles is observable.

Table IV - Mechanical and electrical properties of wire 2.6 mm diameter in different tempers (laboratory drawing facility)

Temper	U.T.S. MPa	E %	ρ 20°C ($\mu\Omega$cm)
T2 (as drawn)	368	4.0	3.48
T10 (A 160°C x 1 hr)	346	9.0	3.30
T10 (A 160°C x 2 hrs)	338	8.5	3.24
T10 (A 160°C x 4 hrs)	326	6.5	3.17
T10 (A 160°C x 6 hrs)	312	6.5	3.12
T10 (A 150°C x 4 hrs)	338	8.7	3.24
T10 (A 150°C x 6 hrs)	333	7.7	3.19
T10 (A 150°C x 10 hrs)	312	7.5	3.12

As comparison purpose, table V shows the properties of wire obtai - ned through the traditional processing system (water quenching of the coil from the furnace, drying treatment at ~ 120°C for 2 hours, drawing and final ageing).

Table V - Mechanical and electrical properties of wire 2.6 mm diameter, obtained by traditional system, in different tempers (laboratory drawing facility)

Temper	U.T.S. MPa	E %	ρ 20°C ($\mu\Omega$cm)
T3 (as drawn)	361	3.0	3.45
T8 (A 160°C x 1 hr)	360	9.5	3.31
T8 (A 160°C x 2 hrs)	352	8.0	3.27
T8 (A 160°C x 4 hrs)	344	6.5	3.21
T8 (A 160°C x 6 hrs)	333	6.5	3.15
T8 (A 160°C x 8 hrs)	328	5.5	3.11

The value examination essentially shows that the continuously quenched material presents a higher overageing tendency at 160°C.

This behaviour can be explained with two different mechanisms: the first involves a smaller nucleation capability exerted at 160°C by GP zones formed at room temperature in continuously quenched rod, with respect to GP zones formed during drying treatment in standard material; the second, is referring to a higher recovery rate of continuously quenched rod due to a higher dislocation density.

A slight decrease of the ageing temperature (for example 150°C) is nevertheless sufficient to reduce this effect and guarantee mechanical and electrical properties well comparable with the standard material's ones.

Half-scale production of rod and wire

Starting from the satisfactory preliminary results, we have done a half-scale production of quenched rod consisting of about 50 tons subdivided in eight series (6 tons each), to obtain a first evaluation of the reliability of the process (for parameter production see table I).

Table VI shows the chemical, tensile and electrical properties of the different series of the rod, after seven days of natural ageing.

Table VI - Chemical, tensile and electrical properties of 8 production series (6 tons each) of quenched rod (half-scale production)

Series n.	Fe %	Si %	Mg %	U.T.S. MPa	E %	ρ20°C ($\mu\Omega$cm)
1	0.18	0.57	0.61	210	11.6	3.35
2	0.16	0.57	0.60	215	11.9	3.35
3	0.17	0.57	0.60	213	12.5	3.36
4	0.16	0.57	0.60	208	12.4	3.32
5	0.17	0.58	0.62	214	13.8	3.32
6	0.18	0.57	0.60	209	11.9	3.35
7	0.18	0.57	0.60	212	13.9	3.35
8	0.19	0.59	0.60	208	13.5	3.35

Table VII gives the mean properties for 2.6 mm diameter wire, obtained by using the laboratory drawing machine previously mentioned, thermal treated at 150 and 160°C for 4 hours.

Table VII - Mechanical and electrical properties of 2.6 mm diameter wires, in different tempers (laboratory drawing facility)

Series n.	Wire characteristics								
	T2			T10 (A150°Cx4hrs)			T10 (A160°Cx4hrs)		
	U.T.S. MPa	E %	ρ20°C ($\mu\Omega$cm)	U.T.S. MPa	E %	ρ20°C ($\mu\Omega$cm)	U.T.S. MPa	E %	ρ20°C ($\mu\Omega$cm)
1	353	3.7	3.44	344	8.5	3.27	336	8.5	3.20
2	358	4.5	3.43	347	9.0	3.27	330	7.5	3.19
3	357	3.7	3.44	344	9.5	3.26	334	8.5	3.19
4	352	3.5	3.42	340	8.5	3.24	330	8.0	3.18
5	355	3.7	3.42	342	8.0	3.24	331	8.0	3.18
6	368	3.0	3.45	351	8.5	3.27	336	8.5	3.20
7	369	3.5	3.45	352	9.0	3.27	336	8.5	3.21
8	361	3.5	3.43	346	8.0	3.25	336	8.5	3.19

The values are generally comparable both with those obtained in preliminary experiments and with each other.

Part of this production has been also drawn by employing an industrial facility (13 dies no-slip machine) characterized by high drawing speed. In this case we have observed some rupture events of the wire, generally connected with more reduced diameters.

Table VIII gives, as an example, mean properties of 3.15 mm diameter wire, in different tempers, as function of three values of drawing speed.

Table VIII - Tensile and electrical properties of 3.15 mm diameter wire (series n.1), for different drawing speeds (industrial drawing facility)

Drawing speed	Wire characteristics					
	T2			T10 (A160°Cx4hrs)		
	U.T.S. MPa	E %	ρ20°C ($\mu\Omega$cm)	U.T.S. MPa	E %	ρ20°C ($\mu\Omega$cm)
12 m/sec	327	5.0	3.40	332	10.2	3.18
17 m/sec	328	5.2	3.39	333	9.5	3.17
20 m/sec	330	6.0	3.42	337	9.5	3.20

The values appear correct. A statistical analysis of the values obtained for 50 samples for each drawing condition gave values of the standard deviations ranging from 8 to 13 MPa for U.T.S. and from 0.020 to 0.045 x 10^{-8} Ω m for electrical resistivity.

These values can be considered better than the typical ones normally obtained from traditionally quenched coils; nevertheless they don't always satisfy the level that French standardization is introducing.

The mechanical characterization of the wire has been completed by wrapping test performed according to ISO 958. No rupture has been observed and only a few samples have shown little cracks on the surface.

Optical microscope observation clearly revealed that both drawing breaks and wrapping cracks are connected with the segregations evidenced in fig.3. As it is well known, this structure is normally present also in the rod produced by standard method, but for continuously quenched material it can be dangerous as a consequence of the high strength of the rod.

As final consideration, it seems possible to conclude that the production of continuously quenched rod is quite promising even if some problems must yet be overcome. From the casting point of view we are operating to obtain a further reduction of segregation level and the maximum control of production parameters in particular chemical composition and cast bar temperature.

Conclusive remarks

Starting from energy saving requirements, we are introducing in our Al-Mg-Si wire rod production cycle, the continuous quenching process of the rod. This operation avoids the discontinuous thermal treatment of the coil carried out in a furnace at high temperature for a long time. This purpose requires important modifications of traditional working parameters of our Properzi facility, from liquid clean treatment to casting and rolling conditions, and the introduction of a useful quenching apparatus.

In particular the most important point is represented by the solidification step, which must guarantee both a suitable solidification structure, and a high solubilization degree of the rod.

Laboratory researches and a first attempt on Properzi machine have clearly shown the possibility of reaching the goal. Subsequently, a half scale production has confirmed the validity of the choosen solutions, taking into account the properties of the rod.

Drawing tests carried out with both laboratory and industrial facilities have indicated that the properties of the wire are quite satisfactory as far as the actual International Standards are concerned.

Nevertheless Alluminio Italia is still working to improve the quality and reliability level of this promising product, by setting up the complete process automatization, which is necessary to strictly control and reduce the main variations of the production parameters.

CHAIRMEN

E.L. Rooy
Aluminum Company of America
Pittsburgh, Pennsylvania

J.L. Dassel
Kaiser Center for Technology
Pleasanton, California

CORELESS INDUCTION VERSUS ROTARY FURNACE MELTING OF BALL-MILLED DROSSES

Dr. P.J. Bamji

ALCAN International Limited
Kingston Laboratories
P.O. Box 8400
Kingston, Ontario
Canada
K7L 4Z4

Abstract

Efficient metal recovery from dross depends on furnace practices, dross cooling, ball milling and the process used for extracting metal. Remelting of dross is invariably done in rotary salt flux furnaces which produce copious quantities of slag. With increased emphasis on pollution control and stricter dumping legislations, other means of metal recovery are sought.

A coreless induction furnace offers an alternate route to rotary melting. With its inherent electromagnetic stirring the Coreless furnace produces the same level of turbulence as a rotary and, under certain operating conditions, it produces comparable metal recovery using very little salt and generates no slag. Technical and operational aspects of metal recovery from ball milled drosses using both these furnaces is discussed.

Introduction

A common method used to recover aluminium from drosses is to first reduce its non-metallic fraction. This is invariably done by breaking up the raw dross in a crusher and then to process it through a ball mill. The ball milled product hereafter called Metallic Concentrate is then screened to remove and reject dust which is usually below 1 to 1.5 mm. The metal contained in the metallic concentrate is extracted using rotary furnaces in which large amounts of salt flux are used to help release the entrapped metal. This process produces copious quantities of slag which are invariably dumped. With increased emphasis on the control of industrial pollution, stricter Environmental Protection Agency regulations and the advent of anti-dumping state laws, a process which releases metal by other means is imperative. The process itself must be capable of:

(a) Rapidly submerging the metallic concentrates into a bath of molten aluminium.

(b) Reducing the surface tension between the entrapped metal and the neighbouring oxide.

(c) Rupturing the oxide pocket thus releasing the entrapped metal.

(d) Maintaining heat within the bath throughout the melting cycle.

A mains frequency coreless induction furnace with its inherent electromagnetic stirring fulfills most of the above requirements.

This paper briefly outlines the process of melting metallic concentrates using both the rotary and the coreless induction furnaces and makes a comparison of the process performance using both these routes. Some present-day problems associated with the coreless induction process are listed and possible solutions to these problems are given.

Rotary Furnace Melting of Metallic Concentrates

Rotary furnaces have been the work horses of secondary smelters for many decades. The furnace itself is simple to fabricate and, in many ways, very simple to operate. It comprises a horizontally-disposed cylindrical steel container (Figure 1) which is lined with refractory bricks. The container is made to rotate on its horizontal axis by means of steel tyres located at each extremity of the furnace. A burner located at one extremity of the furnace radiates heat on the charge and on the refractory lining. Because the furnace is rotating and the heated lining comes in contact with the relatively cold charge, the furnace refractories operate at a lower temperature than the roof of a reverberatory furnace.

The operating practice is to melt the salt flux first and then to charge in the metallic concentrates. The rotation of the furnace causes the metallics to submerge into a turbulent pool of molten flux (Figure 2). Once submerged, the flux reduces the surface tension between the oxide and the entrapped metal so that it can be released. Dependant upon the type of flux used, the furnace operating temperature is usually between 750 and 850°C. As melting progresses the molten metal separates to the bottom and the salt slag floats to the top. At the end of melting the tap hole is

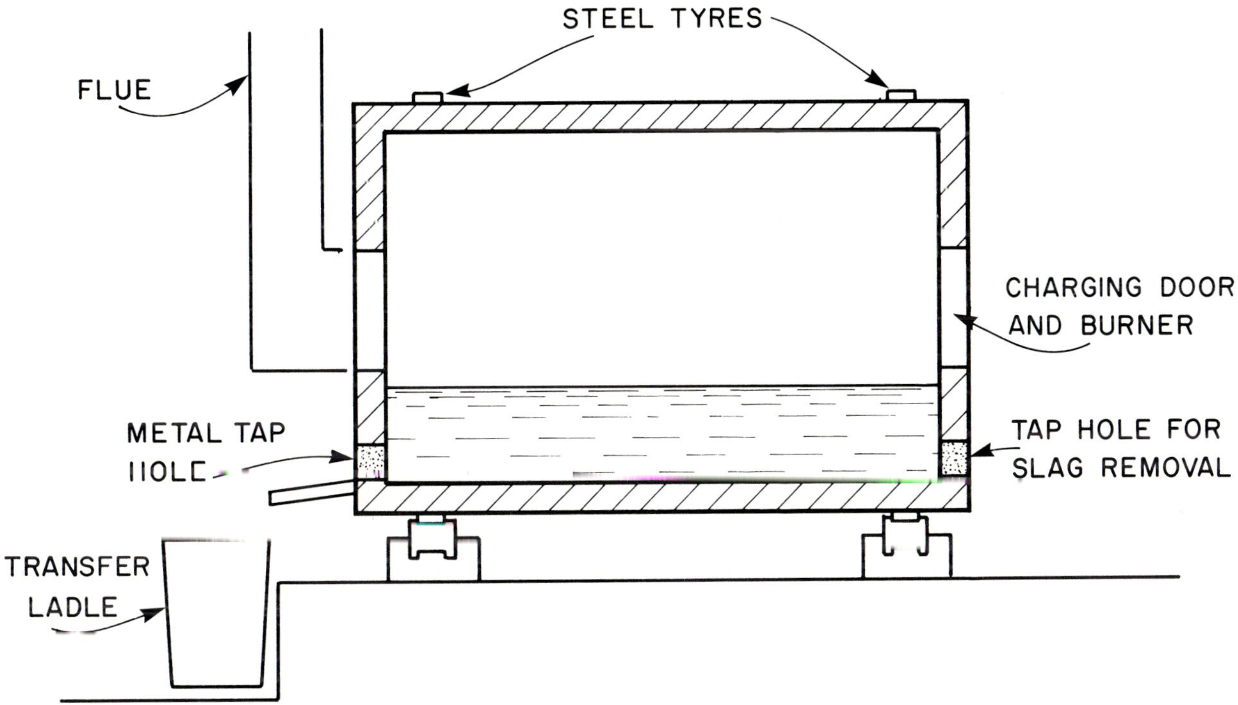

Figure 1 - A rotary salt flux furnace.

brought to its lowest position and molten metal is tapped out first. Slag is next discharged into sow pans and cooled down before it is dumped. Some plants prefer to leave the salt slag in the furnace and process a second charge of metallic concentrates (with limited quantity of fresh salt) before disposing the slag. This practice causes the resultant slag to reduce its fluidity; it becomes sticky and needs to be vigorously stirred before tapping.

Fluxes used in rotary furnaces vary. Either rock salt + 3% fluoride salts (m.p. 790°C) is used or a naturally occurring salt called Montanal (69% NaCl, 29% KCl, 2% CaF) (m.p. 700°C) or a mixture of both is used. The quantity of flux used varies from 25 to 60% by weight of the metallic concentrates depending on their size.

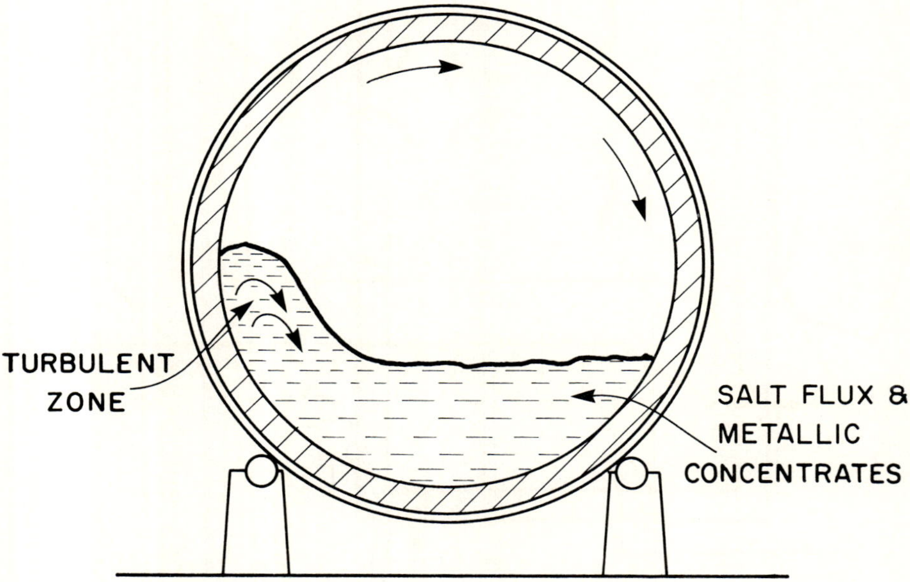

Figure 2 - Mixing caused by rotation of the charge.

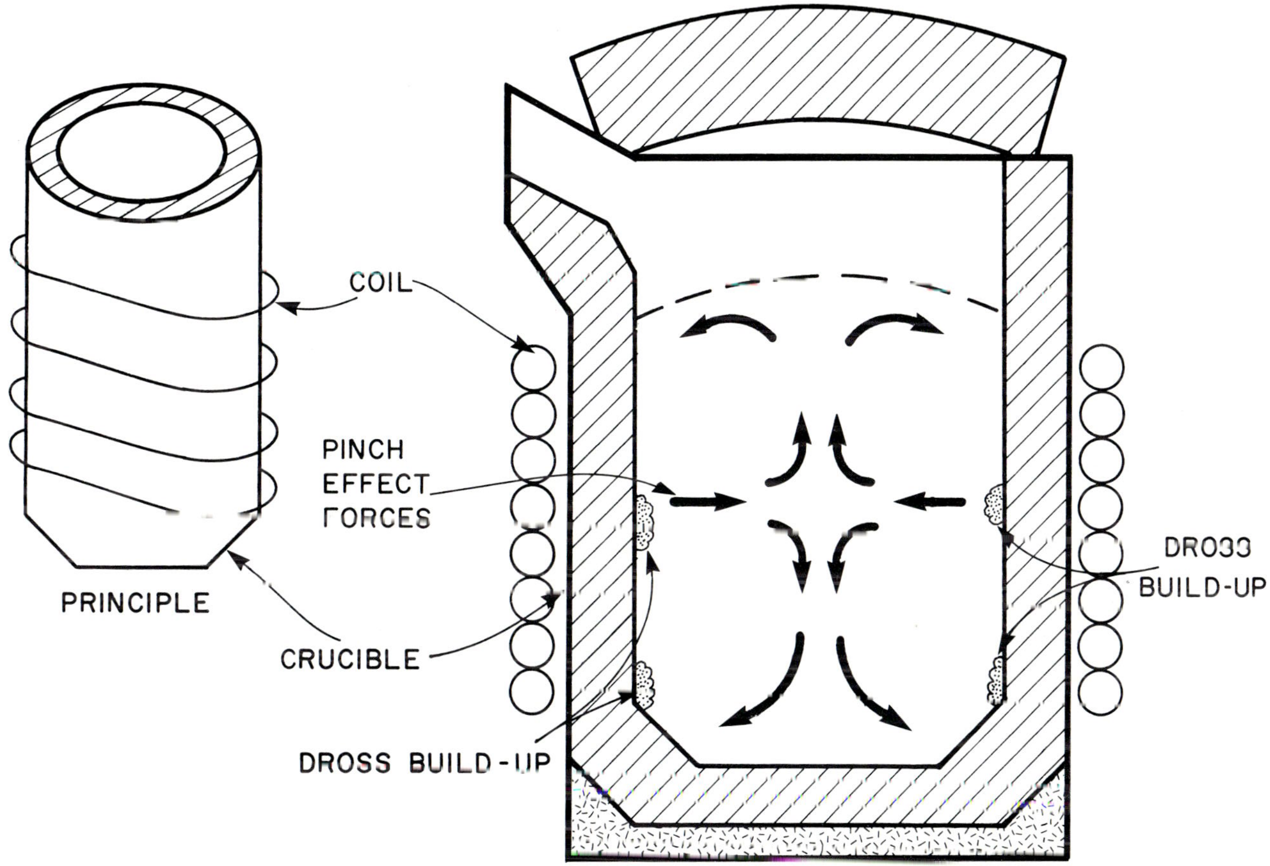

Figure 3 - Stirring Pattern in a mains-frequency coreless induction furnace.

the same charge was melted in a mains frequency coreless furnace, there was total and rapid ingestion of the charged material.

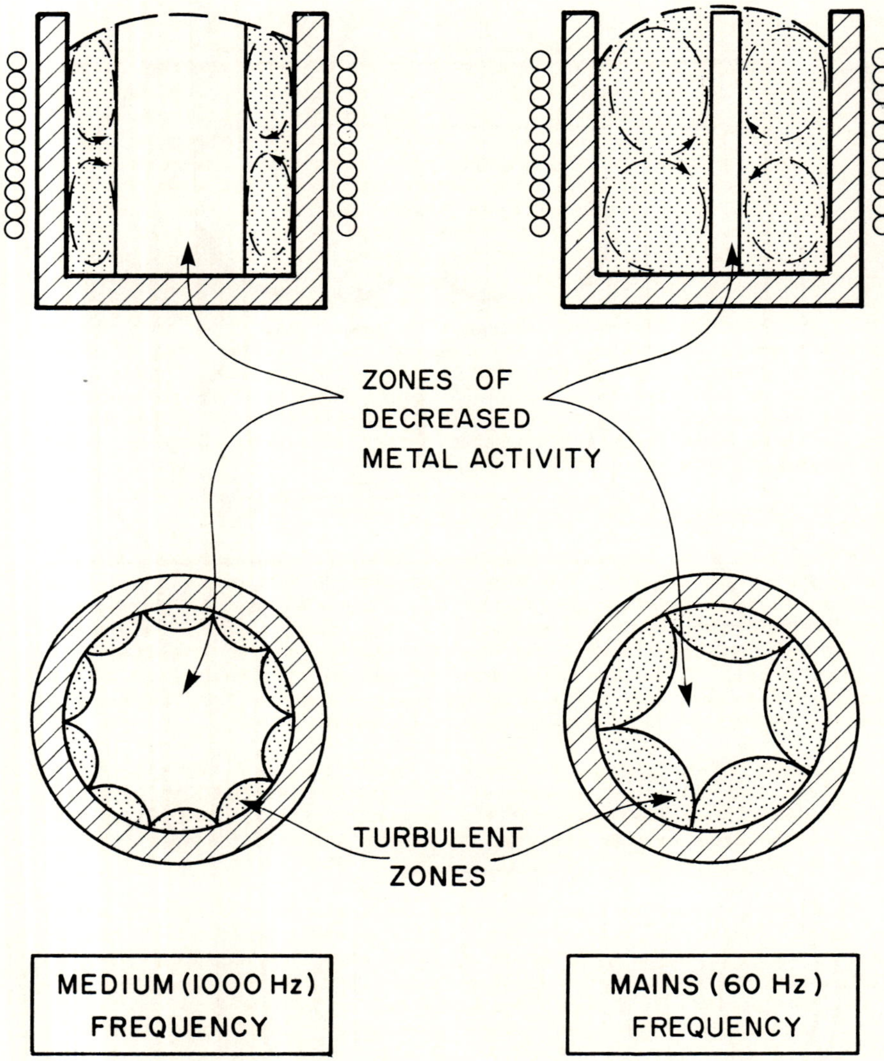

Figure 4 - Stirring Patterns in Mains and Medium Frequency Furnaces

Mains Frequency (60 Hz) Coreless Induction Furnace Melting of Metallic Concentrates

Mechanism of Melting

The mains frequency coreless induction furnace fulfills a number of conditions necessary for recycling of metallic concentrates listed earlier. The mechanism of submerging light scrap in a coreless furnace relies on a hot heel practice, i.e. a heel of metal of 1/3 to 1/2 the overall furnace capacity is in the furnace prior to charging of the scrap. When the furnace coil is energized the pinch effect forces (Figure 3), which are induced in the metal, produce distinct stirring patterns both on the surface of the metal and within. When viewed from the top, the metal surface assumes a mushroom-shaped "dome" with streams of metal moving from the centre of the dome towards the furnace periphery (wall). Light material such as metallic concentrates when charged at the apex of the dome is rapidly taken towards the furnace wall and then down into the molten metal. Rapid submergence is a key feature of the process of melting of metallic concentrates. Once submerged, the metallic concentrates follow the sub-surface stirring pattern and, on impact with the furnace walls, the hot oxide film is ruptured thereby releasing the metal contained within it. The oxide being tenacious gets lodged on the furnace wall. At the termination of the melting campaign a hoop of dross is observed towards the bottom of the furnace and a similar ring of dross also forms towards the central horizontal axis of the furnace (see Figure 3). A preferred practice is to charge the metallic concentrates into the heel at such a rate that there is no build up of solids on the metal surface; slow trickle feed being better than dump charging. When flux is used this is mixed in with the charge.

Choice of Frequency - Mains Versus Medium Frequency Furnaces

A key factor in electric melting of metallic concentrates is rapid submergence of the charge. The ideal flow pattern within the metal in an induction furnace is shown in Figure 3. The intensity of movement is strongly influenced by the frequency of the power supply, the furnace capacity and the power input. The degree of turbulence is proportional to the power input and is inversely proportional to the square root of the operating frequency, i.e., for a given power input, as the frequency is lowered, the stirring action becomes more vigorous.

Frequency also governs the depth of penetration of the electromagnetic field into the pool of metal. The penetration depth is inversely proportional to the square root of the frequency, i.e., for a given power input, as the frequency is lowered, the field penetrates deeper into the pool of metal.

Figure 4 is drawn to elucidate these two features of frequency selection which govern the ultimate feasibility of melting metallic concentrates. To quote an example of an improper choice of frequency, on one occasion metallic concentrates in size range 1.5 to 50 mm were melted in a 100 kg, 300 kw, 1000 Hz medium frequency coreless furnace with very little surface turbulence. Metallic concentrates were charged into a heel as previously discussed. At the end of 20 minutes no melting had occurred and the charge still lingered on the crown of the metal. During this time the bath temperature had increased from 720 to 840°C. In contrast, when

Process Performance

Rotary Versus Coreless Induction Melting

The performance data is based on test work on two rotary and two coreless induction furnaces at different sites. Table 1 gives brief details of the furnaces which were used.

Table I. Furnaces Used to Carry Out Tests

Rotary					Coreless Induction			
Plant	Capacity (Tonnes)		R.P.M.	Fuel	Plant	Capacity (Tonnes)	Rating (KW)	Frequency (Hz)
	Metal	Flux						
A	2.5	1.5	1	gas	B	5.5	1450	50
B	5.0	2.5	2	oil	C (Laboratory)	0.45	200	50

Table II summarises the test results. It is emphasized that the metallic concentrates used during these tests were not always from the same source. The table as such, shows trends, and any comparison made from one test to another should be made with discretion.

Table II. A Summary of Some Test Results

Plant	Type of Furnace	Size Range of Metallic Concentrates (mm)	Gross Metal Recovery (%)	Ratio of Metal Recovered to Flux	Power Consumption kwh/Tonne Recovered	Tonnes Recovered Per Hour
A	Rotary	1.5-50	73.8	1:0.43	1099	0.5
B	Rotary	1-50	74	1:0.56	1638	0.72
B	Coreless	1-50	68.2	1:0	846	0.65
B	Coreless	1-50	72.8	1:0.16	776	0.63
C	Coreless	1.5-50	74.7	1:0.14	852	-

The data in Table II highlights the following trends:

(a) When a small quantity of flux is used, metal recoveries using an induction furnace are equivalent to those using a rotary furnace.

(b) For similar recoveries, the rotary process uses over three times as much salt as the coreless induction process.

(c) The energy consumed by the electric furnace is consistently less than that used by the rotary furnace. However, these figures do not include a correction for the efficiency of electrical power generation.

Effect of Particle Size of Metallic Concentrates on Metal Recovery

In a separate series of tests the effects of different size fractions on the gross metal recovery in both types of furnaces were studied. Table III gives details.

Table III. Effects of Types of Furnace and Particle Size on Gross Metal Recoveries From Metallic Concentrates

Size Range (mm)	Rotary		Induction		Difference in Recovery (%)
	Gross Recovery (%)	Recovered Metal: Flux	Gross Recovery (%)	Recovered Metal: Flux	
1.5-6	70	1:0.54	57	1:0	13
6-50	75	1:0.54	72	1:0	3
1.5-50	75	1:0.49	65	1:0	10

The presence of finer particles in the metallic concentrates produce lower recovery in a coreless induction furnace. From results of several tests it is concluded that if the metallic concentrates contain particles below 6 mm, then they are best suited for rotary melting.

Use of Salt Flux in an Induction Melting Furnace

The results of several extended plant trials have shown that limited flux additions in a coreless induction furnace has improved metal recoveries, made scraping of furnace walls relatively easy and, in general, prevented dross from burning. It is difficult to suggest an optimum quantity of salt flux to be used in coreless induction melting. However, the experimental trends indicate that 10 to 20% flux is adequate whereas rotary melting consumes 25 to 65% flux. Figure 5 is drawn to show these trends. The net recoveries were obtained from fire assays of the metallic concentrates used in each test.

FURNACE	PLANT A	PLANT B	PLANT C
INDUCTION	———	△ ▲	□ ■
ROTARY	○ ●	▽ ▼	———

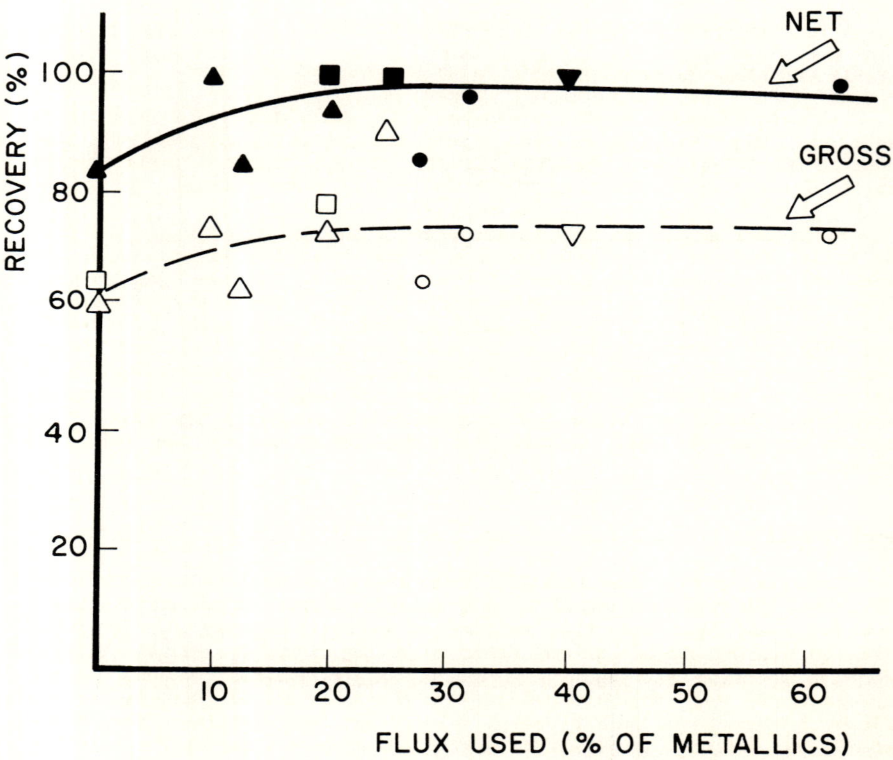

Figure 5 - Metal Recovery versus quantity of flux used - some trends.

Dross Generation in the Coreless Induction Furnace: Problems and Possible Solutions

Melting of metallic concentrates produces copious quantities of dross. The stirring forces in an induction furnace are such that oxide is dragged towards the furnace walls in distinct places (see Figure 3) where it tends to accumulate. Scraping of dross away from the walls is arduous and time-consuming. Further, dross which lodges towards the bottom of the furnace is difficult to dislodge and invariably results in localized build-up of heat. Under such circumstances premature failure of lining can occur. The process therefore necessitates several additional features to make it attractive. These are:

(1) Frequent and meticulous scraping of walls.

(2) Mechanised scraping and removal of dross.

(3) Refractory coatings which ease the removal of dross.

Induction furnace manufacturers now offer mechanised furnace scrapers and dross scoops which would reduce both the labour content and the time required for drossing. Additionally the furnace crucible walls can be coated with refractory coatings which would ease dross removal.

Tests carried out on drosses produced by the induction route and from the rotary furnace slags have shown that the induction furnace-generated dross can be re-processed through the normal route. In fact the quantity of salt flux used during induction melting is about the same as that used by some recycling plants when melting painted and oily scrap in sidewell type furnaces. Rotary slags on the other hand are difficult and uneconomic to process. Recycling of the induction furnace-produced drosses is therefore an extremely attractive feature of the process. As dumping regulations become stricter and environmental pressures increase, this feature on its own will dictate the wide spread use of the induction process.

Metal Quality

Considering that the metallic concentrates themselves contain up to 30% oxide, the resulting metal is surprisingly clean. Metal quality tests have confirmed that induction melting of metallic concentrates produced metal of quality similar to any other recycling furnaces. Metal cleanliness in induction melting is attributed to the stirring pattern in the furnace where oxide migrates towards the furnace wall and adheres to it, thereby rendering the bulk of metal relatively clean and free from inclusions. The presence of excessive dust (less than 1.5 mm) in the metallic concentrates causes the eventual metal quality to deteriorate. This was observed in several tests where excessive dust (~12%) was present in the charged material. In this context, the rotary furnace deals with material with excessive dust better and the resulting metal is comparatively clean.

Concluding Comments

The use of a coreless induction furnace for melting metallic concentrates has several useful features which include:

(1) Rapid submergence capability.

(2) Low salt consumption.

(3) No slag formation.

(4) Elimination of slag disposal.

(5) Ability to recycle dross.

(6) Efficient energy consumption.

(7) Rapid and precise control of temperature.

By far the most important feature of the process is its ability to extract metal from the metallic concentrates using very little salt and thereby eliminating problems of slag disposal. A judicious choice of furnace refractory and refractory coatings, the use of a mechanized way of scraping dross off the walls and of a mechanized way of scooping dross out of the furnace should make this process both technically and economically viable.

Acknowledgement

The experimental work referred in this paper was conducted at Alcan Enfield Alloys Ltd., Bradford, U.K; Alcan Aluminio Italiano S.p.A., Borgofranco, Italy; and at the Electricity Council Research Centre, Capenhurst, U.K. The author wishes to acknowledge the contribution of his colleagues at these three locations.

ELECTROLYTIC REMOVAL OF MAGNESIUM FROM SCRAP ALUMINUM

Basant L. Tiwari and Ram A. Sharma
Electrochemistry Department
General Motors Research Laboratories
Warren, Michigan 48090-9055

An electrolytic process was used to reduce the magnesium content of synthetic aluminum-magnesium scrap from 1.5 w/o to less than 0.1 w/o. Magnesium in the form of salt-coated globules was recovered as a by-product. The electrolytic cell, which used a $MgCl_2$-$CaCl_2$-NaCl-KCl electrolyte at 1025 K, was operated at current densities as high as 1.66 A/cm^2 with current efficiencies near 100%. With further development, this process could become a viable alternative for the removal of magnesium from molten aluminum.

Introduction

In certain aluminum casting alloys such as 380 (a major die-casting alloy) and 319 (a popular permanent mold alloy), magnesium content is specified to be below 0.1 w/o (1). In recent years there has been an attempt to increase this specification (2). But the secondary smelters which supply the charge material to the foundries mostly use the scrap of the mill products for their raw materials. This scrap is generated during the production of such varied products as sheets, sidings, bars and beverage cans, and it contains from 0.5 to 5.0 w/o magnesium. Therefore, removal of the excess magnesium (termed "demagging") from the aluminum melt prior to casting is required.

Chlorination (3-5) is the current approach for removing magnesium. This process is based on the favored reaction of magnesium with chlorine, producing a magnesium chloride dross which separates from molten aluminum. While the process is reasonably efficient at high magnesium content, it has certain drawbacks (5). The present investigation was undertaken to develop an electrolytic method for extracting magnesium from aluminum scrap. In this method, a concentration-type cell, Figure 1, was used in which the heavier molten aluminum-magnesium alloy formed the bottom anode layer, the molten electrolyte the middle layer, and the lighter transferred magnesium the top cathode layer. On applying a potential between the electrodes, magnesium oxidizes and enters the electrolyte at the anode-electrolyte interface and simultaneously deposits at the cathode-electrolyte interface. This electrolytic process recovers magnesium as metal and apparently causes no environmental problems.

Theoretical Consideration

This section presents the thermodynamic principles used to calculate the potential at which magnesium begins to be transferred across the cell. It also addresses the question of stability of the electrolyte and indicates the equilibrium potential at which undesirable transfer of aluminum would take place.

Magnesium will be transferred from the bottom magnesium-aluminum alloy anode to the top cathode by the cell

$$\text{Mg-Al(l)/MgCl}_2\text{-CaCl}_2\text{-NaCl-KCl/Mg(l)} .$$

The theoretical open circuit potential can be estimated by the relation

$$E = \frac{RT}{nF} \ln a_{Mg} \qquad (1)$$

E = cell potential,
F = Faraday constant,
R = gas constant,
T = temperature, K,
a_{Mg} = activity of magnesium in the alloy with respect to pure magnesium, and
n = 2

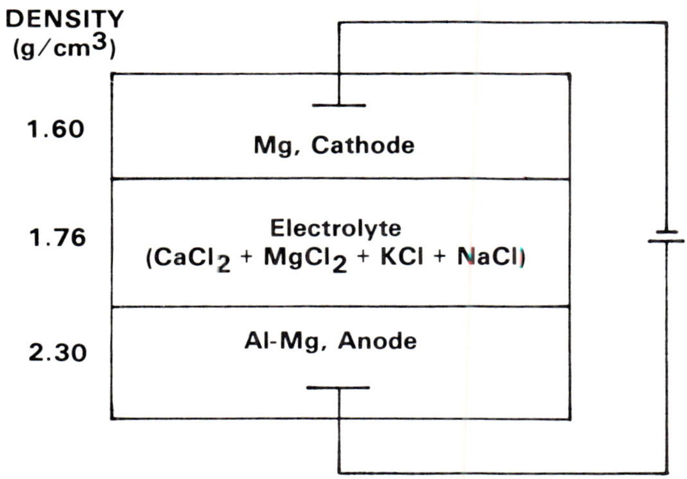

$$Mg\ (in\ Al) \longrightarrow Mg^{+2} + 2e \qquad \text{at Anode}$$

$$Mg^{+2} + 2e \longrightarrow Mg \qquad \text{at Cathode}$$

$$Mg\ (in\ Al) \longrightarrow Mg\ (Cathode)$$

Figure 1. Schematic Presentation of Conventional Three Layer Cell.

The cell potentials at 1.5 and 0.1 a/o magnesium in the alloy (the initial magnesium content and the final magnesium content permissible in casting alloys) are calculated at 1000 K to be 0.20 and 0.32 V, respectively, using an activity coefficient for magnesium of 0.63 (6). These small values suggest that most of the electrical energy will be consumed in overcoming ohmic resistance of the cell, a point clarified in the discussion of the results.

The composition of the electrolyte selected for this investigation (7) is 10 w/o, or 8.6 mole percent $MgCl_2$, 45 w/o $CaCl_2$, 30 w/o NaCl and 15 w/o KCl. Activity of $MgCl_2$ in this electrolyte is assumed to be 0.086, equal to the mole fraction. The reaction by which $AlCl_3$ can form is as follows:

$$\underset{\text{in electrolyte}}{(MgCl_2)} + \underset{\text{in Al-Mg Alloy}}{2/3[Al]} = \underset{\text{in electrolyte}}{2/3(AlCl_3)} + \underset{\text{in Al-Mg alloy}}{[Mg]}$$

The equilibrium constant at nearly unit activity of aluminum in the aluminum-magnesium alloy is

$$K = \frac{a_{AlCl_3}^{2/3}\ a_{Mg}}{a_{MgCl_2}} = e^{-\frac{\Delta G^\circ}{RT}} \qquad (2)$$

where ΔG° is the standard free energy change of the reaction. It is calculated at 1000 K to be 23,720 cal using the values of standard free energy of formation for $MgCl_2$ and $AlCl_3$ of -115,200 and -137,220 cal, respectively (8). Activity of $AlCl_3$ in the electrolyte at 1.5 and 0.1 a/o Mg in the Al-Mg alloy is calculated to be 5.9×10^{-5} and 8.9×10^{-4}, respectively, using the 0.63 activity coefficient for magnesium in the alloy, as above. The concentrations of $AlCl_3$ in the electrolyte at these activities are likely too low to transfer a significant amount of aluminum to the cathode.

The standard cell potential calculated by the relation

$$\Delta G^\circ = -nFE^\circ \qquad (3)$$

is 0.51 V. In the case where the cell potential (excluding resistance losses) approaches 0.51 V, $AlCl_3$ will form at the alloy/electrolyte interface, dissolve in the electrolyte and at some stage react with magnesium at the electrolyte/cathode interface; thus aluminum will be transferred to the cathode.

Experimental Section

Laboratory Cell

The assembled cell is shown schematically in Figure 2. The stainless steel vessel was 7.5 cm in diameter and 46 cm in height, with an insulating lid to position the cathode and the thermocouple. A dense-graphite container was placed inside the stainless steel vessel to hold the aluminum-magnesium alloy, electrolyte, and the recovered magnesium.

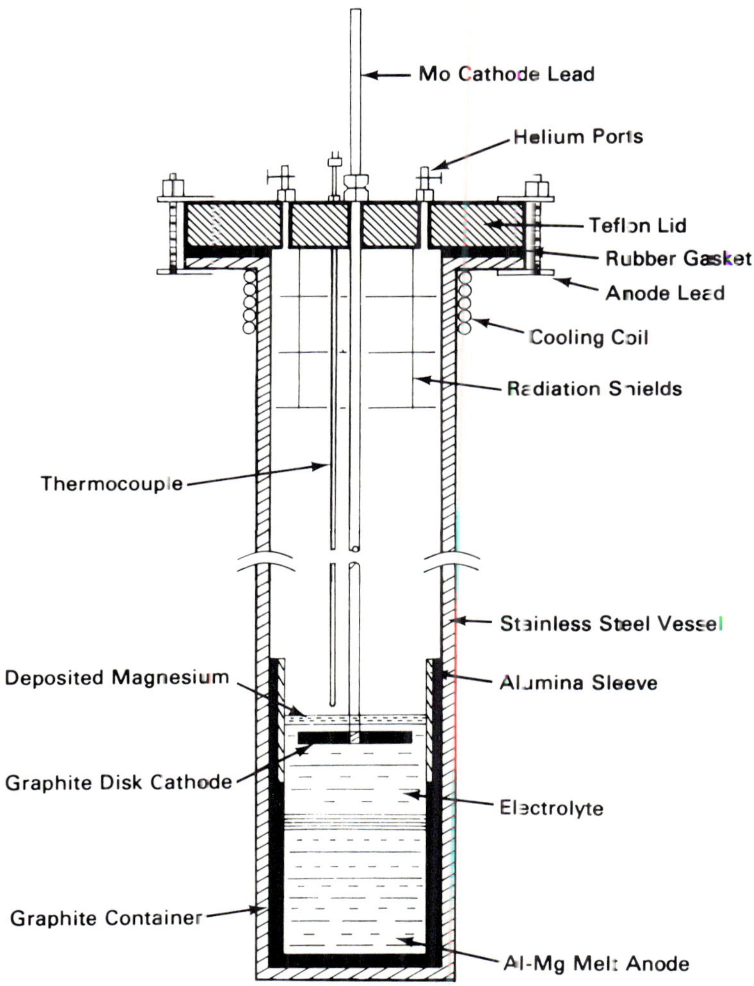

Figure 2. Schematic Diagram of the Electrolytic Cell.

An alumina sleeve was fitted into the upper part of the graphite container to electrically insulate cathode from anode. A molybdenum rod was threaded into a graphite disk, 5 cm in diameter and 0.6 cm thick, which served as cathode current collector. A rubber gasket was used to seal the lid to the stainless steel vessel.

The lid contained ports for helium circulation to prevent oxidation of the graphite container and the other materials. Leak-tight fittings were used for inserting the cathode lead and the thermocouple. The anode connection was made to the graphite container through the stainless steel vessel. The cell was heated in a tubular resistance furnace and a dual-channel recorder was used to record the cell voltage and current.

Electrolyte

The electrolyte was prepared by mixing weighed amounts of anhydrous magnesium chloride (99% purity), anhydrous calcium chloride (98% purity), sodium chloride (99% purity), and potassium chloride (99.9% purity) in a helium-atmosphere glove box. The electrolyte melts at approximately 775 K and has a density of about 1.76 g/cm^3 at 975 K (9). The standard free energies of formation for the component chlorides (Figure 3) indicate that the electrolyte is stable in contact with both magnesium and the aluminum magnesium alloy.

Aluminum-Magnesium Alloy

A representative alloy was prepared by melting 8.9 kg of aluminum (99.5% purity) in a graphite crucible and adding to it 181.4 g pure magnesium. The melt was purged with dry nitrogen for twenty minutes to remove dross and hydrogen, and it was then cast into a slab of 3.8 cm thickness. The slab contained 1.5 w/o magnesium, as indicated by a chemical analysis. Pieces cut from the slab were used as the charge material for the test cell.

Procedure

Calculated amounts of alloy and the electrolyte were introduced in a graphite crucible placed inside the stainless steel vessel. The stainless steel vessel was closed with its lid holding the cathode current collector and the thermocouple, and a helium flow was started through the cell. Next the vessel was heated to about 1025 K, the cathode was lowered into the electrolyte, and the thermocouple was positioned.

After completing the electrical circuit, magnesium was transferred to the cathode at a controlled current. Because of their differing densities, the alloy (2.3 g/cm^3), electrolyte (1.76 g/cm^3), and magnesium (1.6 g/cm^3) formed three separate layers. After a time interval calculated to be approximately that required to remove the magnesium from the alloy, the circuit was opened and the cell was cooled. Refined aluminum and the transferred magnesium, which was in the form of globules, were recovered by washing away the electrolyte with water. A representative sample of the refined aluminum was sent for chemical analysis and another piece was polished and examined under a microscope for inclusions.

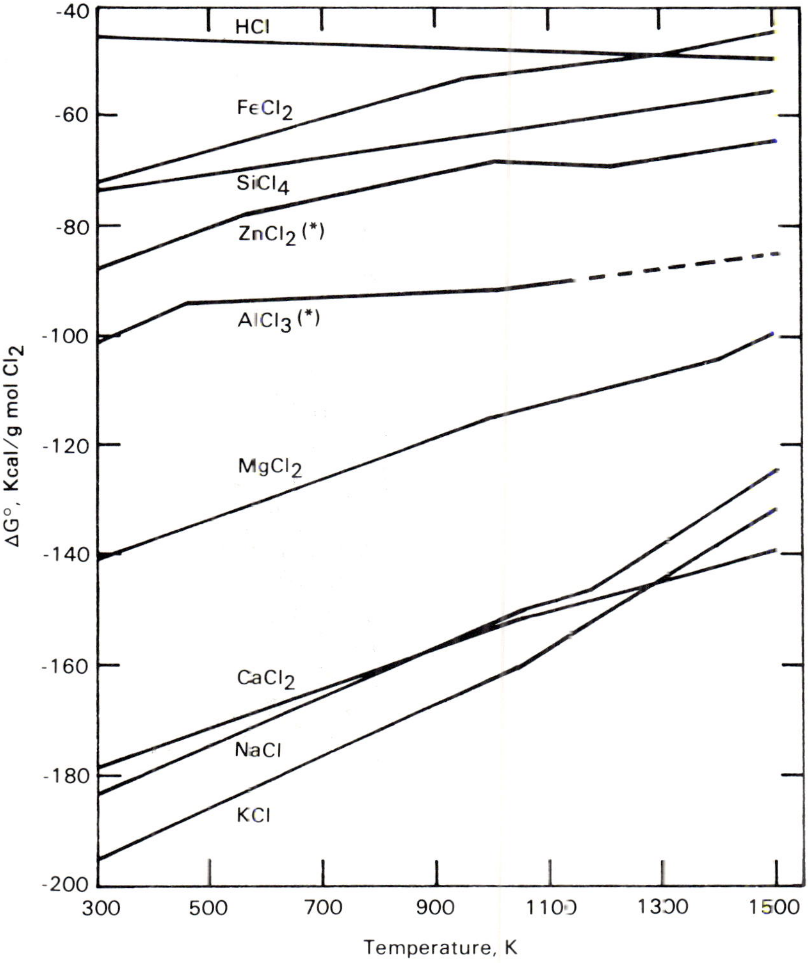

(*) A. Glassner, The Thermochemical Properties of the Oxides, Fluorides, and Chlorides, ANL-5750, 1965.

Figure 3. Standard Free Energy of Formation, ΔG°, of the Chlorides versus Temperature (7).

Results and Discussion

The first stage of feasibility determination was focused on the two component alloy: aluminum-magnesium. Results summarized in Table I indicate that magnesium level as low as 0.005 w/o, with current densities as high as 1.66 A/cm^2, is attainable. In all runs, coulombic efficiency as determined from the alloy composition varied from 85% to 100%. Most of the magnesium was recovered as globules floating on the electrolyte (Figure 4), although some was usually found condensed on the cooler part of the stainless steel vessel.

The refined aluminum showed a surprising uniformity of magnesium distribution throughout the 10.5-cm depth. Mixing during electrolysis is apparently sufficient to keep large concentration gradients from forming. The electrorefined aluminum showed no visible inclusions (Figure 5).

The cell potential at a constant current was observed to remain constant during transfer of most of the magnesium from the alloy anode to the cathode. This was not unexpected because the change in potential due to the change in magnesium concentration was small (E = 0.20 and 0.32 V at 1.5 and 0.1 a/o Mg in the alloy). The magnesium concentrations in the alloy calculated by Equation 1 and from the open circuit potentials measured at the end of the tests are given in Table I. They are in good agreement with magnesium concentrations determined by chemical analysis.

Figure 6 shows that the applied potential varied almost linearly from 0.8 to 3.4 V as the current was increased from 1.8 to 30.0 A (0.1 to 1.66 A/cm^2), yielding a cell resistance of 0.0912 ohm which compares well with the 0.0743 ohm estimated from the electrolyte's conductivity (9). It also means that most of the electrical energy was consumed in overcoming the ohmic resistance, as pointed out before.

The electrolytic process appears to cause no environmental problems and recovers magnesium in the form of salt-coated globules. Perhaps these globules could be used to desulfurize iron and steel melts (10) and alloy aluminum (11). With further development, the electrolytic process could become a viable alternative for the removal of magnesium from molten aluminum.

Table I. Summary of Operating Conditions and Results

Electrolyte = 160 g for Runs 1 and 2 and
202 g for Runs, 3, 4, and 5

Aluminum-Magnesium Alloy = 325 g for Runs 1 and 2 and
600 g for Runs 3, 4, and 5

Magnesium Content of Feed Aluminum = 1.5 w/o.

Cathode Area = 18.1 cm^2. Temperature = 1025 K.

run	applied cell	current	recovered as globules	chemical analysis	calculated from Eq. 1
Run #	Applied Cell Potential (V)	Current Density (A/cm^2)	Percent of the Removed Mg Recovered as Globules	Mg Content of the Refined Aluminum (w/o) Chemical Analysis	Mg Content of the Refined Aluminum (w/o) Calculated from Eq. 1
1	0.8	0.10	84	0.009	0.007
2	2.0	0.46	--	0.009	0.007
3	1.6	0.55	90	0.074	0.060
4	3.4	1.66	82	0.005	0.004
5	2.0	1.10	92	0.122	0.100

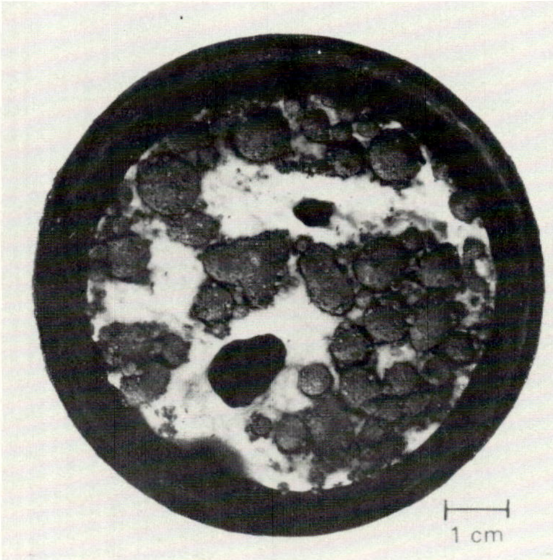

Figure 4. Globules of recovered magnesium.

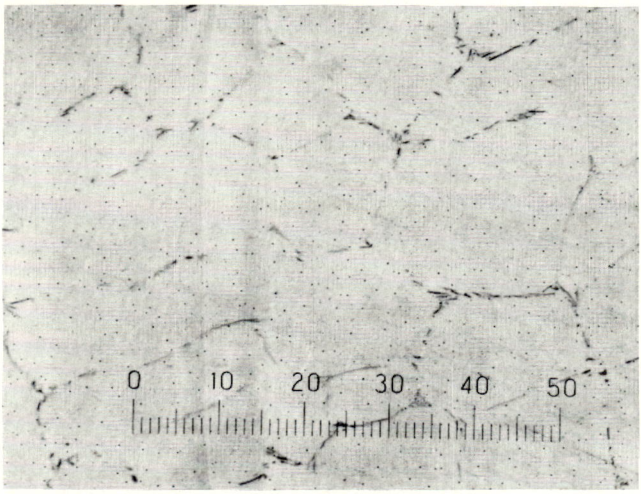

Figure 5. Photomicrograph of the electrolytically refined aluminum showing no inclusions. Magnification 75X (1 div. = 0.02 mm).

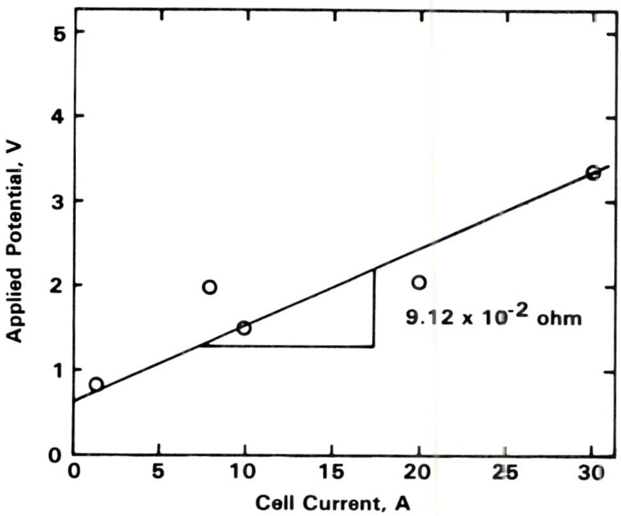

Figure 6. Variation in applied potential of the cell with current, 1025K, 18.1 cm^2 cathode area.

References

1. "Aluminum Standards and Data," 194, Aluminum Association (1978).

2. Private Communication, W. P. Shulhof, GM Central Foundry, Saginaw, Aug. 1983.

3. M. C. Mangalick, Light Metals, The Metallurgical Society of AIME, 2 (1974) p.613.

4. Alcoa Technology Marketing Division, Report No. 1, June 1972, Pittsburgh, PA.

5. B. L. Tiwari, Journal of Metals, 34 (9) (1982) pp. 54-58. Also see Light Metals, The Metallurgical Society of AIME, (1982), pp. 889-902.

6. E. E. Lukashenko and A. M. Pogodayev, Russian Metallurgy (Metally) 5 (1971) pp. 69-72.

7. N. Sevryukov, B. Kuzmin, and Y. Chelischev, "General Metallurgy," translated from the Russian by B. Kuznetsov, Peace Publishers, Moscow.

8. JANAF Thermochemical Tables, Second Edition, NSRDS-NBS-37, June, 1971.

9. G. J. Janz, Molten Salt Handbook, Academic Press, New York, NY, 1967.

10. A. M. Smillie and A. Huber, "Desulfurization of Iron and Steel and Sulfide Shape Control," Edited by W. G. Wilson and A. McLean, Iron and Steel Society of AIME, (1980), pp. 77-89.

11. E. Myrbostad and K. Venas, Light Metals, The Metallurgical Society of AIME, (1983) pp. 921-930.

IMPROVED GRAIN REFINING WITH TiBAl ALLOYS CONTAINING 3% TITANIUM

J. Pearson and M.E.J. Birch

London & Scandinavian
Metallurgical Co Limited
London, United Kingdom

A range of grain refining alloys containing 3% titanium has been developed. The master alloys have been evaluated on commercial aluminium alloys. Results from the laboratory and the casthouse have shown that cost savings can be made without loss of grain refining efficiency.

Introduction

Many advances have taken place in the grain refining of aluminium and its alloys in the last decade. For the production of wrought alloys the major grain refining alloy used is titanium boron aluminium (TiBAl). The alloy was traditionally added in waffle plate or ingot form to the furnace prior to casting. In the early Seventies, the addition of 3/8" (9.5 mm) diameter rod to the casting trough was introduced, and currently approximately 60% of TiBAl additions are added in this form. Rod grain refinement is more efficient than furnace additions and has led to lower addition rates (typically 50%) and a constant grain refining effect during the whole casting cycle.

An important factor in the changeover to rod grain refinement has been the improvement in the quality of TiBAl master alloys over the last ten years. Major advances have been made in cleanliness (reduction in oxide and elimination of TiB_2 agglomerations) and the contact time of the alloys. Contact time is a measure of the alloy's ability to achieve maximum grain refining efficiency immediately after addition to the molten aluminium. In some DC casting systems where high metal flow rates and short troughs are used the contact time available can be as little as 30 seconds. These developments in quality have enabled many critical alloys (can stock, foil stock and anodising qualities) to be cast using TiBAl rod grain refinement with or without the use of filters. Most casthouses using filters add the rod upstream of the filter without significant loss in grain refining efficiency.

TiBAl alloys traditionally had a titanium content of 5% (weight), with boron in the range 0.1% to 1%. In the USA, the 5/0.2 TiBAl alloy in rod form is widely used, whereas in the rest of the world 5/1 TiBAl rod is still the main alloy. In the rest of the world, however, there is a trend towards lower boron contents (5/0.2 and 5/0.6 TiBAl) on critical alloys such as can stock. Lower boron contents in the master alloy generally give a reduced grain refining effect, but improved manufacturing techniques have increased the grain refining efficiency of the lower boron alloys enabling them to be used at economical addition rates. When low boron containing alloys were first available they had long contact times (up to 30 minutes to achieve maximum efficiency), but are now produced with contact times equivalent to 5/1 TiBAl.

Previous papers by Pearson and Birch (1,2) investigated the effect of the titanium:boron ratio on the efficiency of aluminium grain refining alloys. The authors concluded from laboratory experiments that 3/1 TiBAl was the most efficient alloy and offered a lower cost method of grain refining than 5/1 TiBAl. Following this work, 3/1 TiBAl rod was successfully introduced to the industry and is used as the normal rod grain refiner in a number of smelters and semi fabricators. Consequently it was decided to carry out an R&D programme at London & Scandinavian Metallurgical Co Ltd. to develop a family of TiBAl master alloys based on 3% titanium.

Experimental Programme

The main aims of the programme were to:-

(i) Produce a series of TiBAl alloys containing 3% titanium with boron contents of 0.2%, 0.5% and 1% (3TiBAl).

(ii) Measure the grain refining properties of the 3TiBAl alloys and compare their efficiency with equivalent TiBAl alloys containing 5% titanium (5TiBAl).

(iii) Study the metallographic structure of the alloys to determine factors which may be advantageous to the casthouse.

The alloys evaluated are given in Table 1. The alloys were produced on a pilot scale (40 Kg) by a practice similar to full scale production at London & Scandinavian Metallurgical Co Ltd.

TABLE 1 COMPOSITION OF MASTER ALLOYS STUDIED

Master alloy	Composition, wt.% Ti	B	Ti/B ratio
3/0.2 TiBAl	3	0.2	15
3/0.5 TiBAl	3	0.5	6
3/1 TiBAl	3	1	3
5/0.2 TiBAl	5	0.2	25
5/0.5 TiBAl	5	0.5	10
5/1 TiBAl	5	1	5

Table 2 shows the two commercial alloys used to evaluate the master alloys. The alloys were 99.7% Al (representative of pure grades of Al) and 6063 (a common extrusion alloy). The alloys were chosen for the tests because of their commercial importance, and their contrasting properties in that commercially pure 99.7% Al is relatively difficult to grain refine, whilst alloy 6063 is easier because of the silicon and magnesium contents.

TABLE 2 COMPOSITION OF COMMERCIAL ALUMINIUM ALLOYS EVALUATED

	Composition, wt.%
Commercial purity Al	99.7 Al
6063	0.5 Si - 0.5 Mg

The Alcan grain refining test was used to obtain the data. The test, already described by Jones and Pearson (3), is outlined below:-

The base alloy is melted in a resistance furnace to give a batch weight of 10 kg. A calculated grain refining addition is made to the melt at 720°C and stirred for 30 seconds. One minute after the addition a sample is cast into a water quenched conical mould. Further samples are taken at intervals up to 6 hours. After solidification, the samples are sectioned and the surface prepared to allow the grain size to be measured. It is then possible to construct a grain refining curve such as shown in Figure 1 which illustrates the effectiveness of the master alloy over a 6 hour period.

There are three important features to note in the curves: contact time, ultimate grain size, and fade.

Contact time is defined as the time required for the master alloy to achieve maximum efficiency.

The ultimate grain size is the smallest grain size achieved for a given addition rate.

Fade is the decline in grain refining properties and is usually due to the agglomeration and settlement of TiB_2 particles. In this case stirring of the melt will almost completely reactivate the particles. Fade can also be due to a poisoning effect by, for instance, zirconium or chromium. In this case the TiB_2 particles cease to be efficient nuclei, and stirring will not reactivate them.

Metallography was carried out using the technique given in the Appendix, followed by examination on a Reichert Biovar microscope.

Results

For ease of comparison of the results, addition rates have been calculated on the basis of the amount of master alloy added to the test charge.

Figure 1 shows the effect of grain refining commercially pure (99.7 wt.%) aluminium with a range of master alloys containing 3 wt.% titanium. Similar results for 6063 alloy are shown in Figure 2. These curves allow a comparison of the grain refining efficiency of the master alloys with holding time.

Figures 3 and 4 provide data for the range of 5 wt.% titanium master alloys.

In order to show the effect of adding various amounts of the alloys containing 0.2 wt.% and 1 wt.% boron, an experimental program was devised using a range of addition rates normally found in commercial practice. Figure 5 compares the grain refining effect two minutes after making the grain refining addition. A two minute holding time equates to a typical time available in commercial rod addition practice, i.e. between the time of addition and solidification of the metal. Figure 6 shows similar results for a holding time of two hours and represents the effect expected if the grain refiner is added as waffle plate to the furnace.

Photomicrographs are shown in Figures 7 and 8 of typical micro-

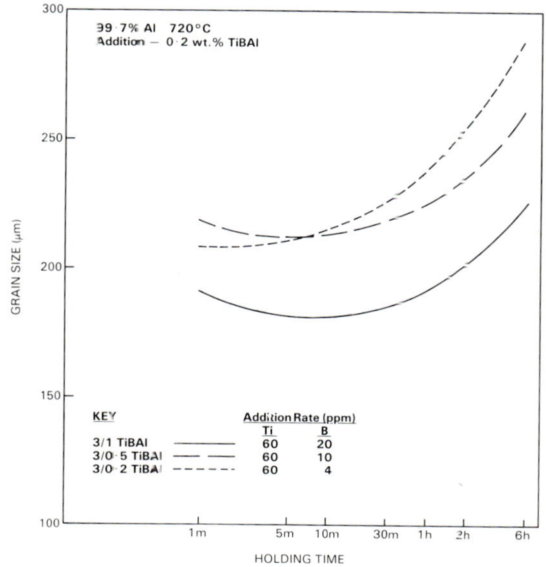

Figure 1 - Effect of holding time on grain size of 99.7Al with 3TiBAl alloys

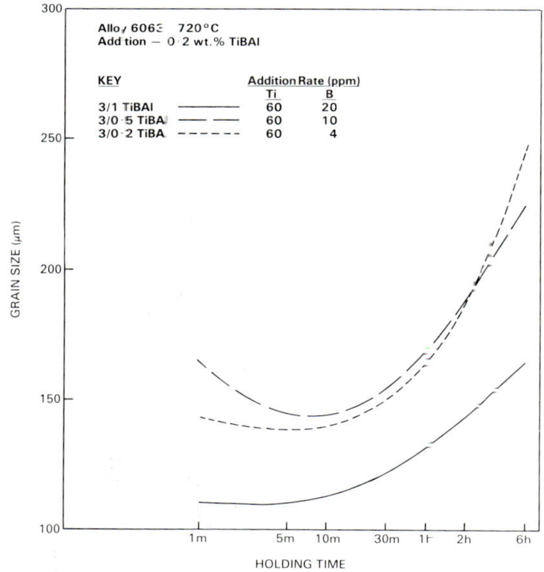

Figure 2 - Effect of holding time on grain size of alloy 6063 with 3TiBAl alloys

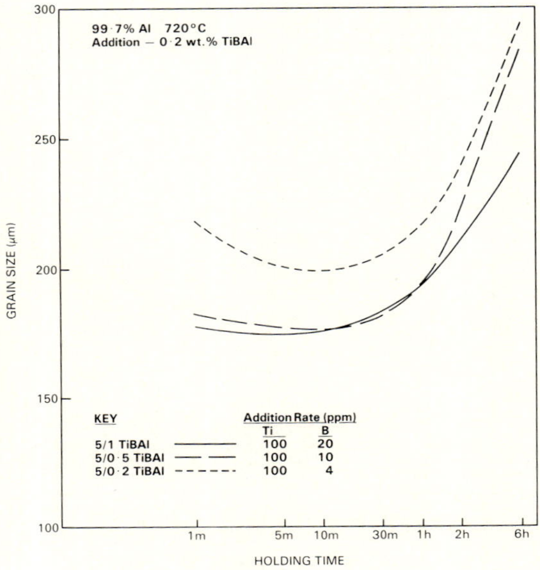

Figure 3 - Effect of holding time on grain size of 99.7Al with 5TiBAl alloys

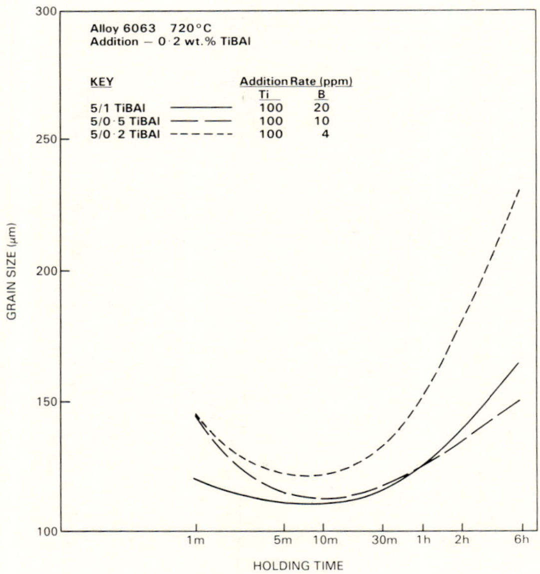

Figure 4 - Effect of holding time on grain size of alloy 6063 with 5TiBAl alloys

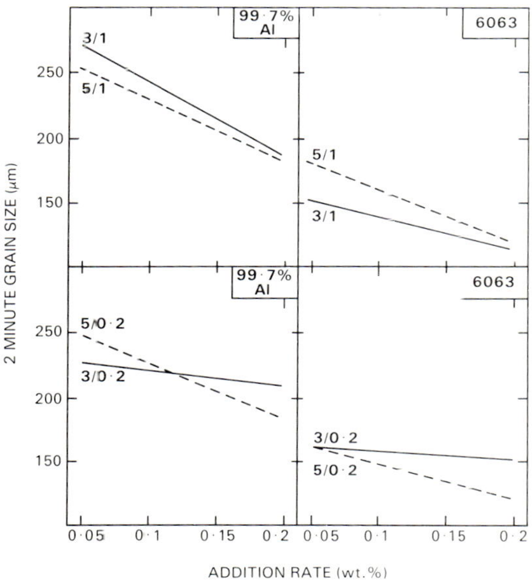

Figure 5 - Effect of addition rate on 2 minute grain size

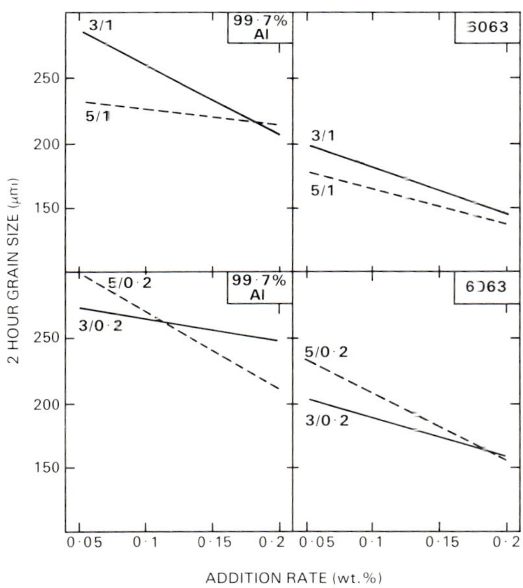

Figure 6 - Effect of addition rate on 2 hour grain size

3/1 TiBAl

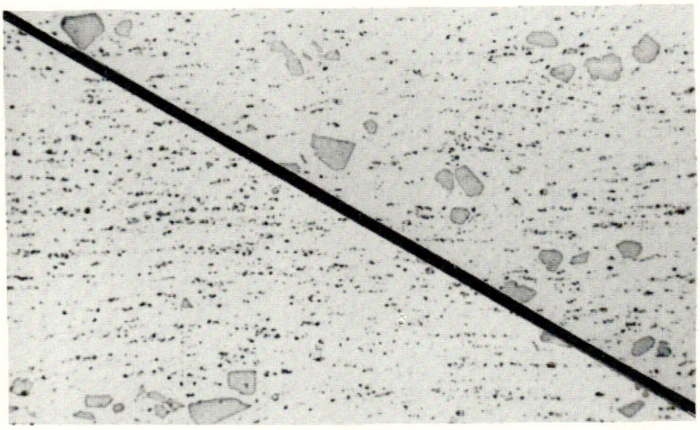

5/1 TiBAl

100 µm

3/0.2 TiBAl

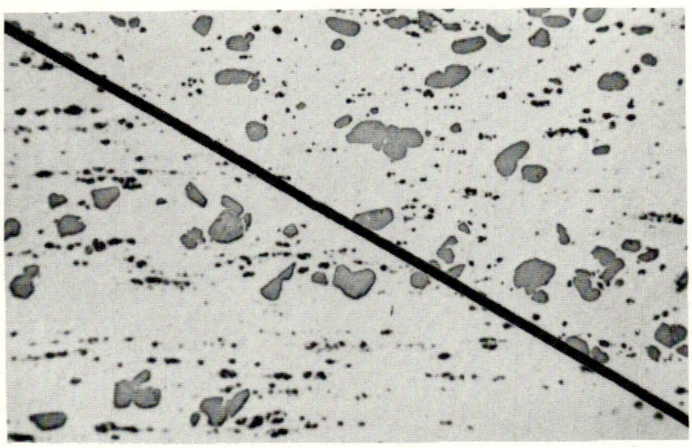

5/0.2 TiBAl

Figure 7 - Micrographs of TiBAl rod

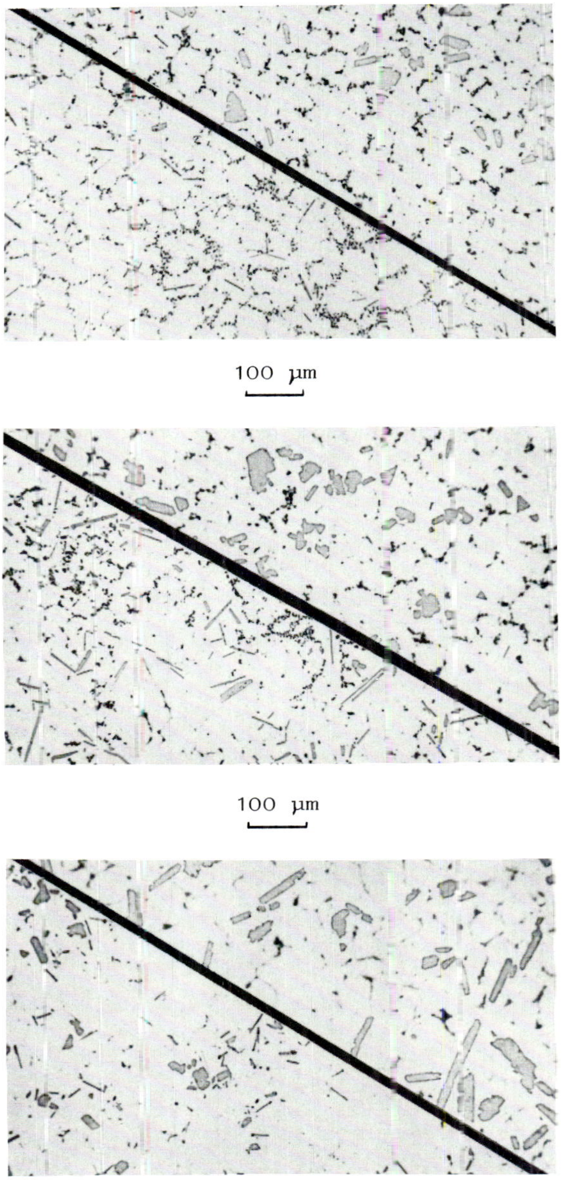

3/1 TiBAl 5/1 TiBAl

3/0.5 TiBAl 5/0.5 TiBal

3/0.2 TiBAl 5/0.2 TiBAl

Figure 8 - Micrographs of TiBAl waffle plate

structures of the 3 and 5TiBAl series alloys. No etch was used in the preparation of these micrographs.

Discussion

The grain refining of 99.7% Al with a 3TiBAl series alloy is achieved most effectively by the use of 3/1 TiBAl, with the 3/0.5 and 3/0.2 alloys giving rather less refinement at the same alloy addition rate (Figure 1). This result is not unexpected as previous work (2) has shown the 3:1 Ti:B ratio to be the most effective. Figure 3, however, shows that the grain refining efficiency of the more dilute boron alloys of the 5TiBAl series is fairly close to that achieved by 5/1 TiBAl. Our test results show 5/0.5 TiBAl to have similar grain refining properties to 5/1 TiBAl up to two hours, although it shows rather more rapid fade. It will be noted that 5/1 and 3/1 TiBAl alloys show very similar grain refining properties.

When 6063 alloy is grain refined with 3TiBAl series alloys (Figure 2), the grain refining efficiency of the whole master alloy range is similar, although the 3/1 TiBAl still performs better than the alloys containing less boron at all times after addition. A similar result is found for the 5TiBAl series alloys (Figure 4). The 5/0.2 TiBAl alloy, however, shows pronouced fade after 30 minutes. Again, the grain refining efficiency of the 3/1 and 5/1 TiBAl alloys is seen to be similar under the test conditions. The grain size achieved in 6063 alloy for a given refinement practice is finer than achieved with 99.7% Al. 6063 alloy may be taken as typical of the easier-to-refine alloys. The effect of alloying elements on grain refining is not straightforward, and optimisation of addition rates is needed for each alloy. In general, elements such as magnesium, silicon, zinc and copper enhance grain refinement by a constitutional undercooling mechanism, whereas zirconium and chromium have a poisoning effect (2). Manganese has little effect on grain refining properties.

The effect of TiBAl addition rates on the grain size achieved is shown in Figure 5, corresponding to a rod addition practice, and Figure 6, which corresponds to a furnace addition practice. The results are as expected, i.e. larger grain refining additions result in smaller grain sizes.

It will be seen that in most cases, the difference in performance between 5/1 and 3/1 TiBAl, or 5/0.2 and 3/0.2, under similar test conditions, is not great, the grain size being in most cases within thirty microns.

Under the test conditions used, the grain size obtained at an addition rate of 0.05 wt.% master alloy on 6063 alloy was of the same order as that achieved using a 0.2 wt.% addition to 99.7% aluminium. This is evidence of the need to optimise the grain refining practice for each alloy in order to obtain the required grain size at the lowest cost.

As the addition rate increases, the tendency is for a decreasing difference in the grain size produced by the 5TiBAl and 3TiBAl alloys.

In nearly all cases, the grain size produced two minutes after the grain refining addition is smaller than that achieved after holding for two hours. This illustrates the improvement in contact times achieved by London and Scandinavian grain refining alloys in recent years. It is interesting to compare the results in Figure 3 with results from a similar series of tests published five years ago (Ref. 1, Figure 2). This further demonstrates how improvements in production techniques over this period have led to considerable reductions in the grain sizes expected for a given addition rate. These improvements are most noticeable in the improved contact times achieved using the more dilute (5/0.2 and 5/0.5) TiBAl alloys.

The results obtained in this investigation demonstrate that it is not necessary to have a large excess of titanium (over the amount required to form TiB_2) to obtain good grain refinement. The 3TiBAl series alloys are as effective as the 5TiBAl series. This can be explained on the basis of the number of nucleating TiB_2 particles present in the alloys; for example, 3/1 TiBAl and 5/1 TiBAl contain the same number of TiB_2 particles.

Figure 7 shows photomicrographs of continuously cast and rolled rod, and Figure 8 shows the as-cast waffle plate microstructure.

5TiBAl series alloys have at least twice as much titanium in the form of $TiAl_3$ compared to the 3TiBAl alloys, when one allows for the stoichiometric amount of titanium required for TiB_2 formation. This explains why the photomicrographs show less $TiAl_3$ present in the 3TiBAl series alloys. It will also be noticed that the size of the $TiAl_3$ particles is generally smaller for TiBAl alloys containing 3% titanium. Recent evidence (4,5) shows that the $TiAl_3$ phase dissolves very rapidly when added to molten aluminium. Dissolution of $TiAl_3$ particles occurs within 30 seconds with 5TiBAl series alloys; in 3TiBAl alloys the smaller $TiAl_3$ particles dissolve even more rapidly. This is of particular interest in those cast-houses where a grain refining addition is made as rod with the filter immediately downstream, as it reduces any risk of undissolved $TiAl_3$ being removed by the filter.

Metallographic examination reveals that the size and distribution of the TiB_2 particles in 3/1 and 5/1 TiBAl are effectively identical. The numbers and distribution of TiB_2 particles are also the same in the 0.2% B and in the 0.5% B alloys. In all cases, the average TiB_2 particle size is less than 1 micron.

The build up of titanium in revert scrap can be a problem when TiBAl addition rates are high, e.g. in the DC casting of anodising quality sheet ingot and sheet caster production of 99.0% Al. In such cases titanium addition rates using 5/1 TiBAl rod can be in the region of 0.015% and, with the continual recycling of scrap, titanium contents in the finished metal can be very high. For example, consider a sheet caster operation based on a charge of 50% virgin ingot and 50% grain refined revert scrap. Grain refining is carried out using 5/1 TiBAl rod at an addition rate of 3 kg per mt, i.e. 0.015% Ti. After three remelts of the scrap part of the charge, the titanium content of the product would rise to 0.024% Ti (excluding Ti present in the virgin ingot). The use of 3/1 TiBAl rod, however, at the same alloy addition rate (3 kg per mt) would result in a total Ti pick up of only

0.015%.

As a result of the reduced level of titanium in the 3TiBAl series alloys, they can be produced and sold at a cost several percent less than 5TiBAl series alloys (at the same boron content). This can give a considerable saving in the cost of grain refining over a period of time.

As already described, 3/1 TiBAl rod has already been successfully adopted by a number of casthouses for the DC casting of sheet ingot and extrusion billet. It is also being used very effectively in sheet casting operations. As a result of the work described in this paper, considerable interest has been expressed in 3/0.5 and 3/0.2 TiBAl alloys, and these are now available commercially.

Conclusion

1. A family of 3TiBAl alloys (containing 0.2% to 1% boron) has been successfully developed and is now available commercially.

2. The grain refining efficiency of 3TiBAl alloys has been shown to be similar to 5TiBAl alloys. This new range of alloys reduces the cost of grain refinement without reducing effectiveness.

3. 3/1 TiBAl is more efficient than 3TiBAl alloys with lower boron contents.

4. The use of 3TiBAl alloys reduces titanium build up in aluminium and its alloys. This is of great benefit in areas where high levels of revert scrap or high addition rates are used, e.g. sheet casting.

5. 3TiBAl alloys contain much less $TiAl_3$ than 5TiBAl alloys. The size of the $TiAl_3$ particles in 3TiBAl reduces the solution time for $TiAl_3$, which is important when rod additions are made immediately before a filter.

References

1. J. Pearson and M.E.J. Birch, "The Effect of the Titanium: Boron Ratio on the Efficiency of Aluminium Grain Refining Alloys", Journal of Metals, 1979, Vol. 31, No. 11, pp 27-31.

2. J. Pearson, M.E.J. Birch and D. Hadlet, "Recent Advances in Aluminium Grain Refinement", paper presented at the conference "Solidification Technology in the Foundry and Casthouse", 15th-17th September, 1980, University of Warwick, U.K.

3. Jones G.P. and Pearson J., "Factors Affecting the Grain Refinement of Aluminium Using Titanium and Boron Additives", Metallurgical Transactions, Vol. 7B, June 1976, pp 223-234.

4. Cappleman G.R., "The Dissolution of Solids in Molten Metals", Ph.D. Thesis, University of Sheffield, U.K., January 1983.

5. T.W. Clyne and M.H. Robert, "The Behaviour of Aluminides in Liquid Aluminium as it Relates to Grain Refinement by Melt Innoculation", paper presented at the conference "Solidification Technology in the Foundry and Casthouse" 15th-17th September, 1980, University of Warwick, U.K.

A P P E N D I X

Preparation of Microspecimens of Aluminium-Titanium-Boron Alloy

1. Mount the microspecimen in cold-setting (acrylic resin) plastic.

2. Turn down the specimen face on a lathe using a round-nose tool. Take the last skim fine and slow. Carried out with sufficient skill, step 3 below can be omitted.

3. Remove lathe marks from the specimen using a worn 800 grit silicon carbide paper fitted to a revolving wheel. Keep the paper well lubricated with running water and linish only as long as is necessary to remove the lathe marks. Wash and dry the specimen.

4. Polish the specimen by hand, using medium/heavy pressure, on a rotating 'Selvyt' pad lubricated with metal polish. Polish only as long as is necessary to remove the 800 grit scratches about 60 seconds. Wash the specimen thoroughly with detergent, then rinse thoroughly under running water and dry.

5. Mechanically polish the specimen on a rotating 'Selvyt' pad mounted in a rotating bowl containing an aqueous suspension of 0.05 micron gamma alumina (40% Al_2O_3 by volume). An adequate polishing time will be of the order of 30 minutes. After polishing, rinse the specimen thoroughly in running water. Wash out the polishing bowl and renew the 'Selvyt' pad before polishing further specimens.

6. Finally, polish the specimen on a similar rotating pad in a similar bowl using an aqueous suspension of light magnesium oxide (5-10% MgO by volume). The polishing time required will be between 30 seconds and 3 minutes depending on the characteristics of each individual batch of magnesium oxide. Wash the specimen thoroughly under running water, dry quickly and examine under the microscope. Equipment used for alumina polishing may not be used for magnesium oxide polishing or vice versa.

7. The microspecimen is best examined unetched.

THE USE OF UNDERCOOLING MEASUREMENTS TO EVALUATE

GRAIN SIZE OF WROUGHT ALUMINUM ALLOYS

S. A. Levy and J. L. Kirby

Metallurgy Laboratory
Reynolds Metals Company
Richmond, Virginia 23219
USA

Equipment has been developed recently, by Metallurgical Products and Technologies, Inc., which makes the analysis of undercooling data an easy task in the cast house, to predict grain size and silicon modification in aluminum casting alloys. The current work reports tests concerning the use of this equipment for a series of wrought alloys. In particular, can stock and foil alloys were investigated. The goal was to determine the reliability of this approach in an effort to provide a rapid quality control tool for use in the cast house to determine the degree of grain refinement. Existing procedures can not provide sufficiently rapid information to allow on-the-spot correction of addition levels.

INTRODUCTION

A rapid on-line method for determining the degree of grain refinement of cast aluminum alloys would be of great utility. Currently either ingot slices or a test casting must be machined and etched. The difficulties and time delays make this procedure less than ideal. In a laboratory environment, various investigations[1-4] have used undercooling to study the grain refining processes. The theory is that if nucleation is difficult, due to an absence of heterogeneous nuclei, then a melt will undercool until a nucleus forms homogeneously. Once nucleation occurs, the melt temperature will increase and growth will occur at the normal equilibrium temperature. Recently equipment was developed by Metallurgical Products and Technologies (MP&T) which appears promising for using undercooling measurements in a production environment. One of the outstanding parts of the MP&T development is a test cup which is accurate[*], inexpensive (∿85¢), and provides a relatively constant cooling rate. It is connected to a dedicated microprocessor which analyzes the cooling curves. It had been used predominently for the Al-Si casting alloys, for relating the degree of modification to the details of cooling curve. It should be noted that for the Al-Si alloys, undercooling of the eutectic transformation is desired. Elements such as sodium or strontium are added which "poison" the silicon phase such that it must continuously renucleate. Thus properly modified Al-Si alloys, show short segmented silicon needles and a relatively large undercooling, compared to unmodified alloys.

The current study was designed to provide additional data concerning the relationship between the details of the cooling curve and the as-cast grain size, for six important wrought alloys. These include two foil alloys (MD230 and 8111), two can stock alloys (3104 and 5182) and two aircraft alloys (7075 and 7050). Initial tests showed that the existing MP&T microprocessor needed modification, such that the temperature range evaluated was raised, to encompass the higher freezing temperatures of these six alloys. Thus for the current work only their thermocouple/cup assembly, plus the cup holder, were employed. The remainder of the system used a laboratory digital data acquisition system.

PROCEDURE

A digital data acquisition system based upon the Hewlett-Packard (HP-85) desktop computer was employed. The computer was connected to a HP model 3456A digital voltmeter and a model 3495A, 20 channel thermocouple scanner. Only three of these channels were employed: (1) the test thermocouple, (2) a room temperature channel, and (3) a calibrator channel. To the calibrator channel was connected an Analogic DIGI-CAL II digital reference/calibrator device. Prior to each molten metal test, at least five calibration tests, in the vicinity of the liquidus temperature of each alloy, were performed. During the molten metal tests, time-temperature data, at .5 second intervals, were: (1) printed, (2) plotted, and (3) stored on floppy

*The thermocouple wire is specified to be accurate to $\pm$ 0.4^{o}C

disks. The raw millivolt data for the type K thermocouples were first converted to temperature using polynomials based upon on data in NBS monograph MN-125. The raw data were then adjusted, by linear interpolation, based upon the five calibrator readings. The cooling rate was also printed, which facilitated the location of the minimum undercooling temperature and the horizontal arrest temperature.

The grain size was determined using a "golf tee" mold described previously.(5) The grain size was determined by comparison to a set of standards produced in our laboratory. Three individuals did the comparisons and their results were averaged, to yield an ASTM grain size number. Chemical analysis was performed using standard spectrographic procedures. Initially, metal from a previously prepared stock pile of laboratory billets was employed as melting stock. For the last half of the work, virgin melts were prepared from 5A aluminum, due to a concern about the starting titanium level in the earlier material. The starting material was melted in a Lindberg laboratory, electric furnace, controlled to ± 10 F. Different commercial 6% titanium and 5:1 titanium-boron hardeners were employed. For the second half of the experiments only a Shieldalloy 6% titanium rod hardener was employed. The general procedure was to start with the base alloy and slowly raise the innoculant level. A few experiments were conducted to evaluate holding times, with no further addition.

Figure 1 shows the MP&T test cup and cup holder. The cup is formed of CO_2 hardened foundry sand, and the bottom is configured such that it is impossible to insert one into the stand with reversed polarity of the thermocouple wires. A typical cooling rate, observed just above the liquidus, is of the order of 10 F/second.

RESULTS AND DISCUSSION

Initial tests, performed with the undercooling apparatus, were made using foil and can stock alloys. The high production volumes and critical grain refinement requirements of these alloys were very attractive for us. Use of the Alu-Delta type undercooling measurement equipment would find maximum utilization under these conditions. Test data obtained for these alloys (degrees F undercooling, horizontal arrest temperature, and ASTM grain size number for each titanium addition) are presented in Table 1.

Typical grain size curves for additions of 6% titanium rod are shown in Figure 2. Similar curves for the can stock alloys 5182 and 3104, using 6% titanium waffle additions, are shown in Figure 3. In each case, "golf tee" grain size samples were taken one minute after the titanium addition was made to the melt. As expected, titanium rod is much more efficient at one minute after the addition than titanium waffle. Note that the ASTM grain size number is logarithmic, and the larger the grain size number, the finer the grain size. As a typical conversion to average grain diameter, in microns, a grain size number of one is equivalent to 254 microns, two equivalent to 180 microns, and three equivalent to 127 microns.

Figure 1a - Thermocouple wire and test cup.

Figure 1b - Test cup and cup holder.

TABLE 1

Alloy	Titanium Addition	ASTM Grain Size #	Horizontal Arrest (°F)	Under-Cooling (°F)	ASTM Grain Size #	Horizontal Arrest (°F)	Under-Cooling (°F)
MD230	0	.01*	1218.62	1.35	.01*	1219.15	1.19
	0	.07	1218.55	4.74	.05	1217.97	0.86
	.001	.13	1218.16	1.79			
	.005	.58	1217.68	2.39	.91	1210.69	1.07
	.010	1.0	1217.81	5.91	.91	1219.25	7.10
	.010	.91	1219.53	0.09			
	.015	1.25	1218.51	3.98			
	.020	1.25	1218.10	0.58	1.25	1217.63	4.15
8111	0	.02*	1211.15	3.45	.25**	1210.12	1.76
	0	.02	1210.36	1.87	.20	1209.77	2.96
	0				.20	1211.05	7.11
	.001	.43	1211.18	2.30	1.5	1210.20	0.81
	.005	1.08	1211.28	2.07	2.0	1210.60	3.46
	.008				2.5	1211.80	3.46
	.010	1.75	1212.00	2.00	2.0	1213.53	1.00
	.010				2.5	1211.68	1.52
	.015	2.08	1211.96	2.96	2.5	1211.62	2.80
	.020	1.41	1212.20	1.22			
	.025	1.83	1212.21	2.10			
	.030	2.16	1212.70	2.43			

*Rod (6% Ti)
**5:1 Tibor Rod

TABLE 1(Continued)

Alloy	Titanium Addition	ASTM Grain Size #	Horizontal Arrest (^oF)	Under-Cooling (^oF)	ASTM Grain Size #	Horizontal Arrest (^oF)	Under-Cooling (^oF)
3104	0	.20*	1207.12	4.90	.07**	1201.10	2.61
	0	.10	1204.30	2.51	.07	1200.44	0.84
	0	.10	1204.50	1.09	.21	1200.85	1.20
	.001	.75	1205.70	2.38	.38	1202.42	1.30
	.005	1.0	1205.76	0.82	.41	1202.17	8.00
	.008	1.0	1206.57	0.05			
	.010	1.5	1207.32	0.86	.70	1202.94	1.00
	.015	2.0	1206.90	0.41			
	.020				1.0	1203.23	1.10
	.030				1.5	1204.28	0.90
5182	0	.43*	1173.22	1.20	.23**	1169.70	1.15
	0	.40	1173.46	1.97	.46	1170.98	1.19
	.001	.43	1175.52	1.16	.46	1172.28	1.04
	.005	2.0	1176.25	0.60	1.0	1172.60	0.57
	.010	2.41	1176.60	0.06	1.41	1174.13	0.41
	.015	2.53	1177.22	0.23	1.83	1174.23	0.23
	.020	2.58	1177.63	0.46			
	.030				2.33	1175.21	0.25

*Rod (6% Ti)
**Waffle (6% Ti)

TABLE 1(Continued)

Alloy	Titanium Addition	ASTM Grain Size #	Horizontal Arrest (°F)	Under-Cooling (°F)	ASTM Grain Size #	Horizontal Arrest (°F)	Under-Cooling (°F)
7075	0	.20*	1163.72	2.89			
	0	.20	1164.13	3.27			
	.001	.43	1163.87	2.68			
	.005	.40	1165.26	2.38			
	.010	1.30	1167.02	3.98			
	.015	3.0	1168.36	0.58			
	.020	3.5	1168.11	0.26			
	.030	3.5	1168.79	0			
7050	0	.23*	1162.49	3.00			
	0	.30	1162.36	2.64			
	.001	.30	1163.93	1.41			
	.005	.58	1162.90	1.56			
	.010	.83	1163.86	1.15			
	.015	1.0	1164.32	0.86			
	.020	1.5	1164.91	0.93			
	.030	1.75	1165.95	0.53			

*Rod (6% Ti)

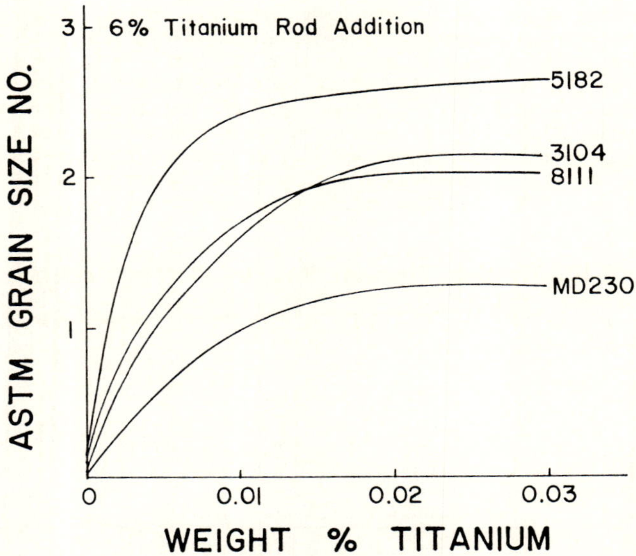

Figure 2. Grain refinement response of foil and can stock alloys for increasing levels of 6% titanium rod.

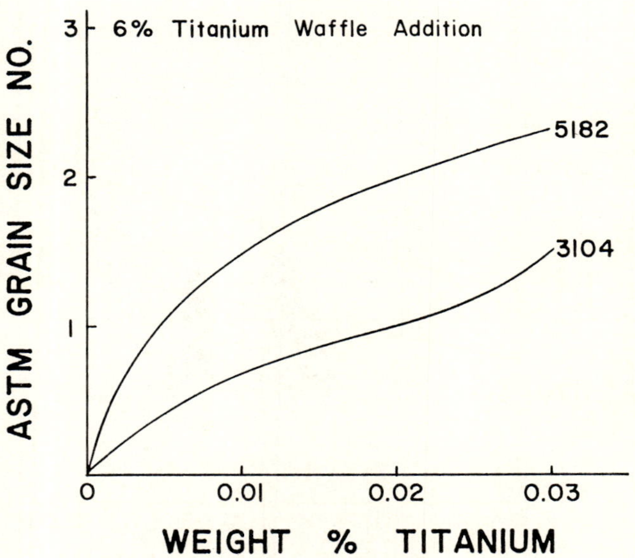

Figure 3. Grain refining response of can stock alloys for increasing levels of 6% titanium waffle.

Alloy MD230 is most difficult to grain refine. It contains small amounts of Si and Fe, and acts similarly to pure aluminum in its grain refining response. The horizontal arrest temperature, of 1218.2 degres F, is only slightly below that of pure aluminum, i.e., 1220 degrees F. Introduction of various weight percentages of titanium, up to .03%, had no effect upon the horizontal arrest temperature. Alloy 8111 is easier to grain refine than MD230. The horizontal arrest temperatures exhibit a tendency to increase slightly, from 1210.7 degrees F for no titanium addition, to 1213.0 degrees F for .03% titanium addition, per Figure 4.

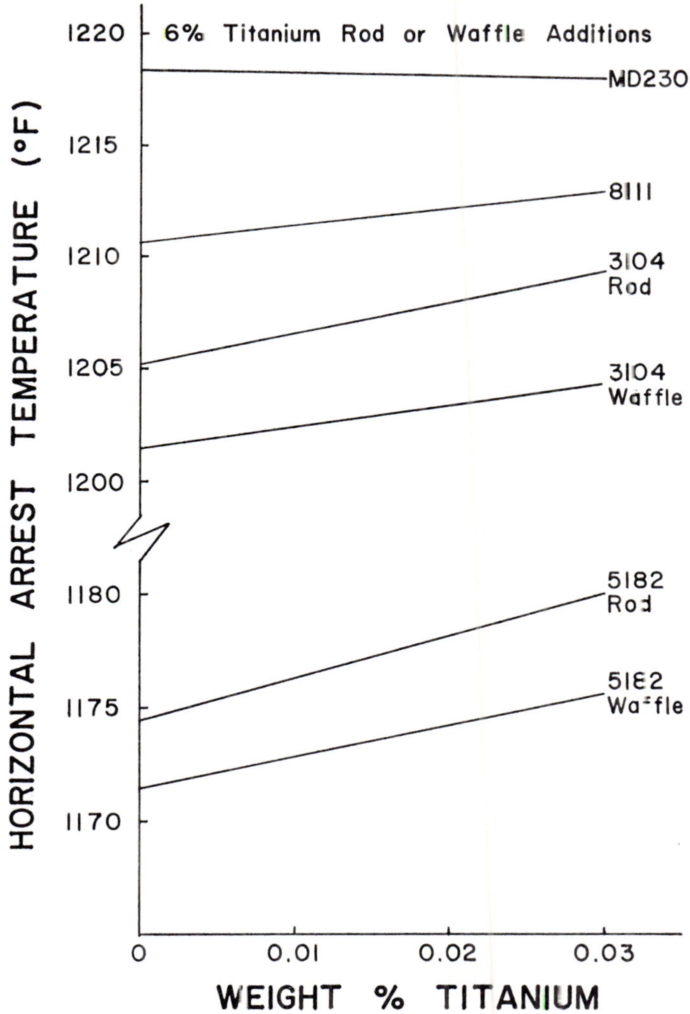

Figure 4. Change in horizontal arrest temperatures of foil and can stock alloys for increasing titanium levels.

Actual degrees of undercooling were plotted for each titanium addition, up to .03%. Alloys MD230 and 8111 had considerable scatter, Figures 5 and 6. No direct correlation of undercooling to percent titanium could be drawn from this data.This is confirmed in Figures 7 and 8, which show no correlation between degrees of undercooling and ASTM grain size number.

Alloys 3104 and 5182 were relatively easy to grain refine, especially 5182. Horizontal arrest temperatures showed more change, with increasing titanium, than for the foil alloys. The data, Figure 4, were slightly different for rod and waffle additions. The two melts used to produce the base stock for the individual tests had slightly different chemistries, and rod grain refiner having better nucleation characteristics than waffle cause the differences in the two curves. The slopes of the curves are similar, but rod additions had slightly more effect on raising the horizontal arrest temperature than waffle additions. The differences in effectiveness for each type of addition, per Figures 2 and 3, show that rod additions should affect nucleation more than waffle.

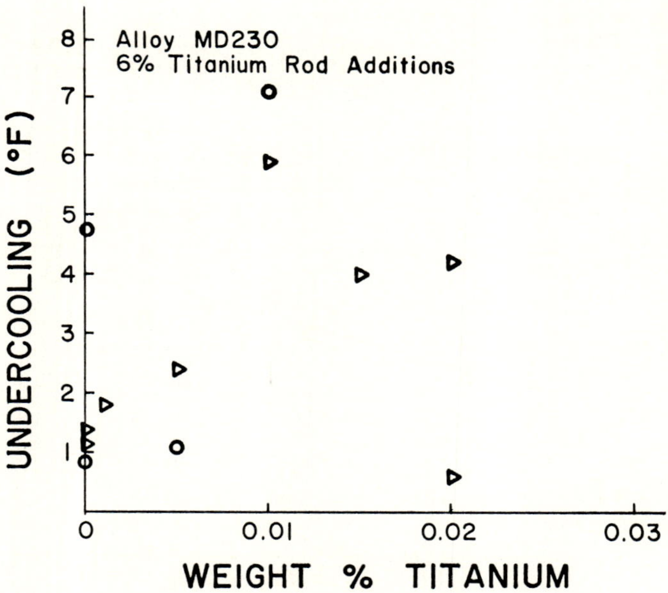

Figure 5. Lack of correlation of degrees of undercooling for increasing levels of titanium in alloy MD230.

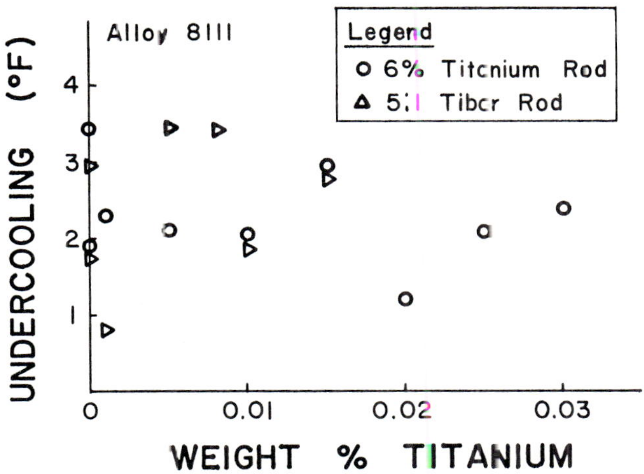

Figure 6. Lack of correlation of degrees of undercooling for increasing levels of titanium in alloy 8111.

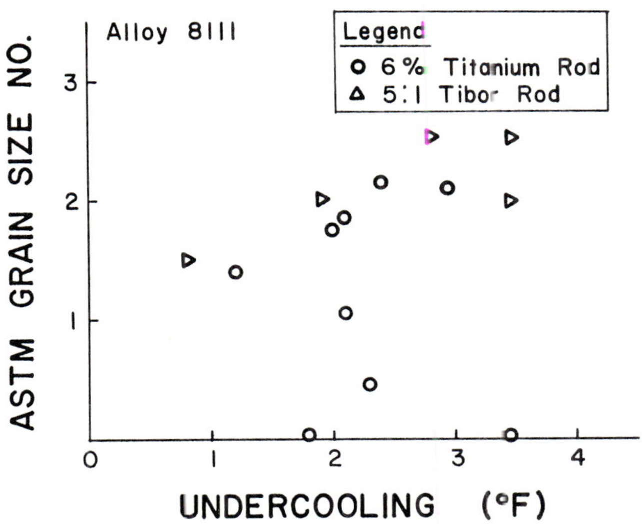

Figure 7. Relationship between undercooling and grain refining response for alloy 8111.

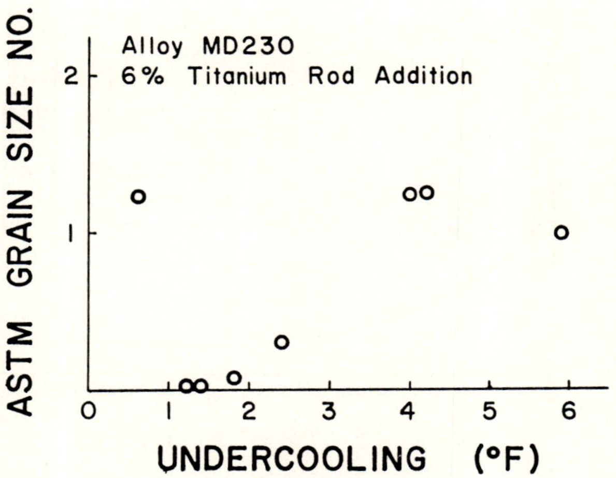

Figure 8. Relationship between undercooling and grain refining response for alloy MD230.

The wider freezing ranges for the can stock alloys, over the foil alloys give a larger systematic change in horizontal arrest temperature and also degrees of undercooling. The degrees of undercooling were again plotted for titanium additions to .03%, Figures 9 and 10. The amount of undercooling decreased for 5182 and 3104 with increasing additions of titanium.

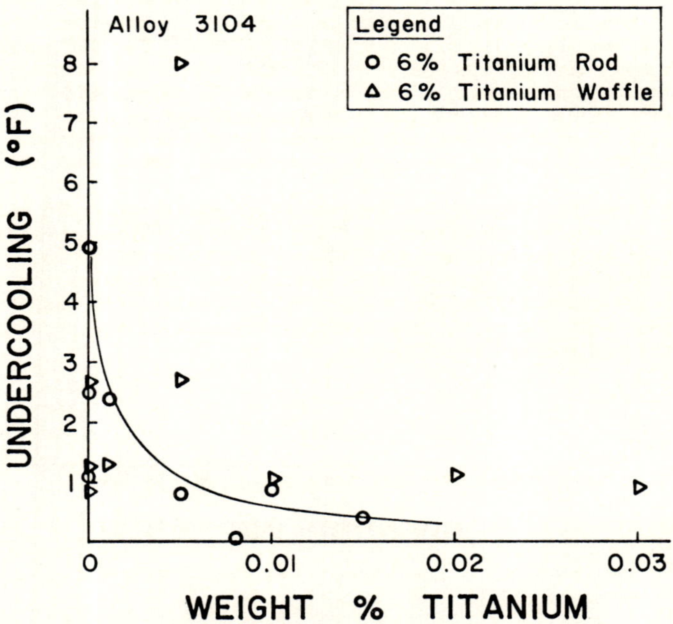

Figure 9. Relationship between undercooling and level of titanium for alloy 3104.

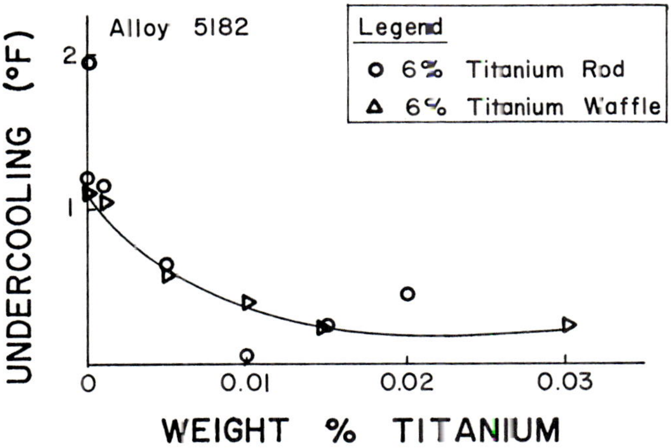

Figure 10. Relationship between undercooling and level of titanium for alloy 5182.

For the 3104 and 5182 the number of degrees of undercooling was plotted versus the ASTM grain size number, Figure 11. As the undercooling decreased, the grain size became finer. A very small level of undercooling reflects a fine grained structure. As a plant quality control tool, it may not be possible to have a highly reliable direct relationship between the number of degrees of undercooling and the grain size for certain alloys. For the purer alloys the best that may be possible is a go/no-go relationship between the maximum acceptable undercooling and a tolerable grain size for each alloy. More data must be obtained before a statistically meaningful level of undercooling can be specified. Other types of analyses of the cooling curves area being examined.

Wider freezing range alloys generally are easier to grain refine because of the nucleation effect of other elements. This is evident in the better test performance for 3104 and 5182 than MD230 and 8111. For this reason 7050 and 7075 were chosen. The chemistry of all alloys are shown in Table 2. Alloy 7075 is relatively easy to grain refine, and 7050 relatively difficult because of the presence of zirconium "poisoning" the titanium.

Table 2

Typical Chemical Analysis of Alloys Used

Alloy	Si	Fe	Cu	Mn	Mg	Cr	Zn	Zr
MD230*	.02	.38	-	-	-	-	-	-
8111	.50	.65	-	-	-	-	-	-
3104	.22	.46	.17	1.12	.96	-	-	-
5182	.05	.34	-	.35	4.65	-	-	-
7075	.11	.25	1.80	-	2.63	.16	5.79	-
7050	.07	.13	2.31	-	2.42	-	6.18	.12

*~A. A. 1235

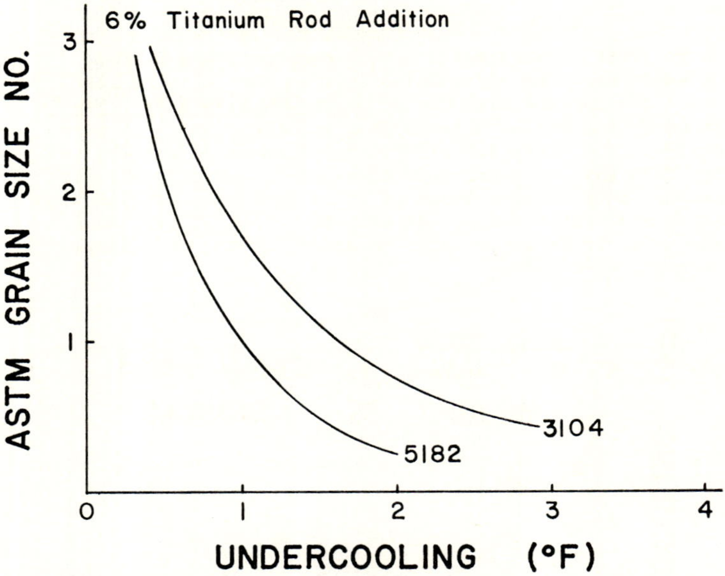

Figure 11. Relationship between undercooling and grain refining response for the two can stock alloys.

Typical grain size curves for these two alloys are shown in Figure 12. Alloy 7075 shows a greater tendency to produce a fine grained structure at lower levels of titanium than 7050. Due to the titanium-zirconium interaction, this would have been expected, especially since all of the sampling was performed one minute after the titanium addition was made to the melt.

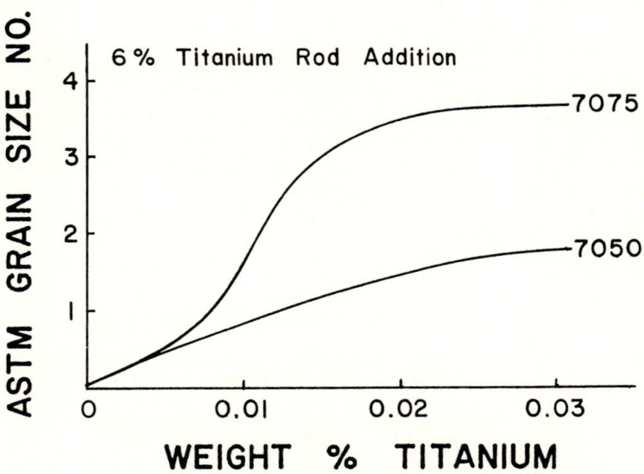

Figure 12. Grain refining response of aircraft alloys for increasing levels of 6% titanium rod.

The horizontal arrest temperatures ranged from 1164.3°F for no titanium addition to 1169.8°F for .03% titanium for 7075. Alloy 7050 had a 1162.8°F arrest temperature for no titanium addition to 1165.9 F horizontal arrest temperature for .03% titanium addition. The correlation between horizontal arrest temperature and percent titanium addition was very good. The correlation of degrees of undercooling and titanium level was also very good for these two alloys, per Figures 13 and 14. Note that 7075 has one extraneous data point, but only one test was performed so it is not known if it is real or a problem with data taking and manipulation. As with the other alloys, much more data needs to be accumulated to establish the confidence levels for the test method.

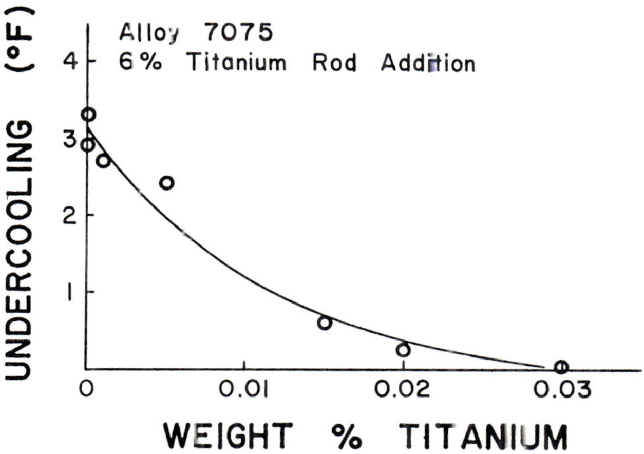

Figure 13. Relationship between undercooling and level of titanium in alloy 7075.

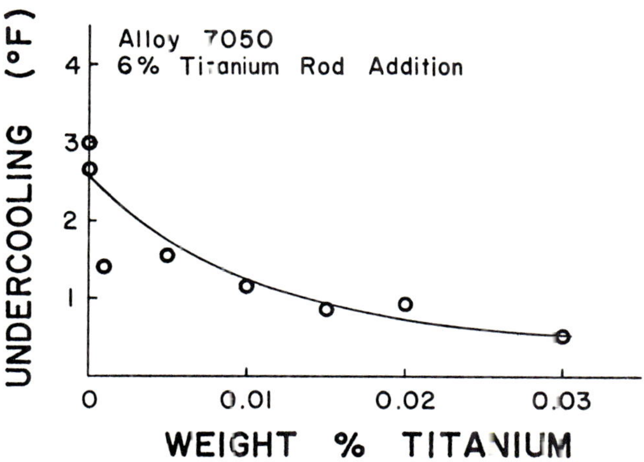

Figure 14. Relationship between undercooling and level of titanium in alloy 7050.

Test data from Figures 13 and 14 were plotted to reflect ASTM grain size numbers versus undercooling, Figure 15. Both alloys had a tendency for finer grain sizes with lower levels of undercooling. The curves were very similar to those of Figure 11, for 3104 and 5182 alloys. In each case, undercooling levels of less than about 0.5°F are required for adequate grain refinement. This level of undercooling seems to be alloy independent for the can stock and aircraft alloys. Therefore, a single test apparatus calibrated to perform at any reasonable arrest temperature should have the same go/no-go field parameters. A single device would not have to be calibrated for each alloy or alloy system to be effective. These laboratory tests have shown the merits of the concept of using undercooling as a plant quality control tool. Further work is required before such an apparatus will become a reality for plant utilization on a routine basis.

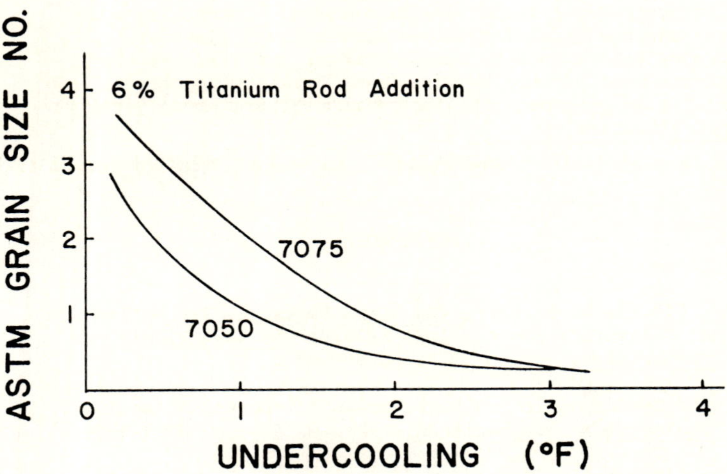

Figure 15. Relationship between undercooling and grain refining response for the two aircraft alloys.

CONCLUSION

The MP&T equipment appears promising to be sufficiently rapid, inexpensive, and rugged to be used in a plant environment for measuring undercooling. For the aircraft alloys 7075 and 7050 a rather good correlation between the grain size determined by the golf tee test and the number of degrees of undercooling was obtained. The correlations for the can alloys 5182 and 3104 was not quite as good. For the two foil alloys considerable scatter occurred. During holds with no further additions, the undercooling returned to a large value at a faster rate than the measured grain size increased. *

The equipment merits further use in studies of the fundamentals of grain refinement. The subjects of fading, contact time, and scrap recycling are only partially understood. Merely having a given level of Ti and B in an alloy does not guarantee a consistent grain refining response. For this study, all of the sampling for grain size and degrees of undercooling were performed one minute after a fresh titanium addition. In practice, residual titanium is present quite often, but it may not be useful as a nucleating agent. The correlation of percent titanium and grain size is excellent because of the controlled laboratory conditions employed. Fundamentals relating the metallographic structure of a hardener and its performance in service are needed. Currently, only lengthy and costly qualification testing in a plant or laboratory can be used to check hardener quality. We are optimistic that further studies using this equipment can materially improve both the theory and the practical quality control situation.

*This is a preliminary observation; to be confirmed in future tests.

REFERENCES

1. J. Cisse', H. W. Kerr, and G. F. Bolling, "The Nucleation and Solidification of Al-Ti Alloys," Metallurgical Transactions, Vol. 5 1974, pp 663-641.

2. J. A. Marcantonio and L. F. Mondolf, "Denucleation," Metallurgical Transactions, Vol. 5 1974, pp 1325-1329.

3. A. Cibula, "The Mechanism of Grain Refinement of Sand Castings in Aluminum Alloys," J. Institute Metals, Vol. 76 1949, pp 321-360.

4. L. Arnberg, L. Backerud, H. Klang, "Grain Refinement of Aluminum, 1: Production and Properties of Master Alloys of Al-Ti-B type and Their Ability to Grain Refine Aluminum," Metals Technology, January 1982, pp 1-6.

5. J. L. Kirby and S. A. Levy, "Grain Refinement of Aluminum Ingots," Light Metals, 1978, Vol. 2, pp 38?-389.

ROLE OF INGOT STRUCTURE IN STRUCTURAL STREAKING

Douglas A. Granger and Craig L. Jensen

Aluminum Company of America
Alcoa Technical Center
Alcoa Center, PA 15069

Structural streaking on anodized 1100 and 5005 aluminum alloy sheet has been a continuing source of product rejections. Attempts have been made to solve this problem during fabrication and finishing, but substantial improvements in recovery were only realized when attention was paid to ingot structure. Evaluation of structural streaking is traced in this paper from the as-cast ingot through surface scalping, preheating and rolling. Association of structural streaks with different constituent phases is identified and impact of the casting process on constituent phase formation is described with examples taken from practice.

Introduction

Aluminum alloys AA1100 and 5005 are widely used for etched and anodized panels for architectural applications such as curtain wall. In this application, an essential requirement is freedom from surface defects that would blemish the buildings finish. Sources for these defects include grease marks, roll pick-up and mechanical damage, all of which can be avoided by paying careful attention to fabricating practices and handling procedures. More troublesome and difficult-to-categorize defects arise from heterogeneities in metallurgical structure such as non-uniform grain structure, variations in cell size, segregation and local differences in constituent type. Variations in constituent type as a source of structural streaking is the subject of this article.

It is well known that cell size and primary intermetallic compound size are inversely proportional to freezing rate (1). There is an increasing body of evidence that constituent type is dependent upon cooling rate during solidification. Considerable progress has been made in AA1XXX series aluminum alloys. In 1962 Hollingsworth, et. al., (2) showed that, in addition to the stable phase Al_3Fe, the metastable phase Al_6Fe could be formed when freezing rate was sufficiently rapid, such as in small diameter direct chill (D.C.) cast ingot. More recently, a noteworthy article by Westengen (3) showed the presence of several other metastable phases in D.C. cast commercial purity aluminum ingot (AA1050 alloy), including Al_mFe and Al_xFe previously reported by Japanese workers (4-7).

For architectural sheet applications, principal interest is in AA5005 which is often used in the non-clad (or bare) condition. It will be shown that in this alloy similar variations in intermetallic compounds form, as a function of freezing rate, in D.C. cast ingot. Furthermore, these phases are metastable, transforming to intermediate or stable forms during ingot preheat which, depending on their distribution, may be responsible for structural streaking in etched and anodized sheet.

Experimental Material

Low-head D.C. cast 5005 alloy ingot, with chemical composition shown in Table I, was used in this investigation.

Table I. Composition of AA 5005 Alloy Ingot Slice

	Si	Fe	Cu	Mn	Mg	Cr	Zn	Ti
AA5005	0.11	0.49	0.09	0.003	0.74	0.001	0.008	0.013

The alloy was grain refined with Al-5% Ti-0.2% B alloy added in rod form to the metal flowing from furnace to casting unit. The grain structure was equiaxed, ranging in size from about 100 μm close to the surface to about 200 μm near ingot center. Ingot size was 20 x 54 in. (508 x 1372 mm) and casting speed was 4 in./min.(100 mm/min). For comparison, ingot of similar composition was cast at ∿ 2 in./min. (50 mm/min.) using conventional D.C. casting practice where the effective mold length was 3 in. (75 mm) as opposed to the 1-1/4 in.(32 mm) mold length used for low-head D. C. cast ingot.

The effect of these differences in casting practices on cell size and cooling rate are illustrated in Fig. 1. It is apparent from these results that the low head D.C. cast ingot has a smaller and more uniform cell size close to the cooled surface than conventionally cast ingot. Small cell size is a direct consequence of the more rapid cooling rate, and the uniformity results from the virtual absence of a reduction in heat transfer due to air gap formation. When cell size measurements are converted into cooling rates (1, 3), it is found that a broad zone of metastable phase is predicted (4-7) for low head D.C. casting in contrast with the narrow zone for conventionally cast material. Crater profiles and predicted zones of metastable Al_mFe phase formation are shown schematically in Figure 2.

Longitudinal ingot slices representative of these two casting practices were surface machined then etched in NaOH (5% solution) at 150°F (66°C) for several minutes. The "fir tree" structure (8), seen in the conventional D.C. cast ingot, is essentially absent in the low-head D.C. ingot where the dark etching zone is deeper and more uniform as illustrated in Figure 3 (a) and (b). The difference in etching response between the light and dark appearing zones is the result of the presence of different types of intermetallic compounds.

Results of Structural Analysis in As-Cast Material

Optical metallography and Guinier X-ray diffraction analysis were employed to quantify cell size and determine constituent type respectively. In addition to identifying intermetallic compounds, Guinier X-ray technique

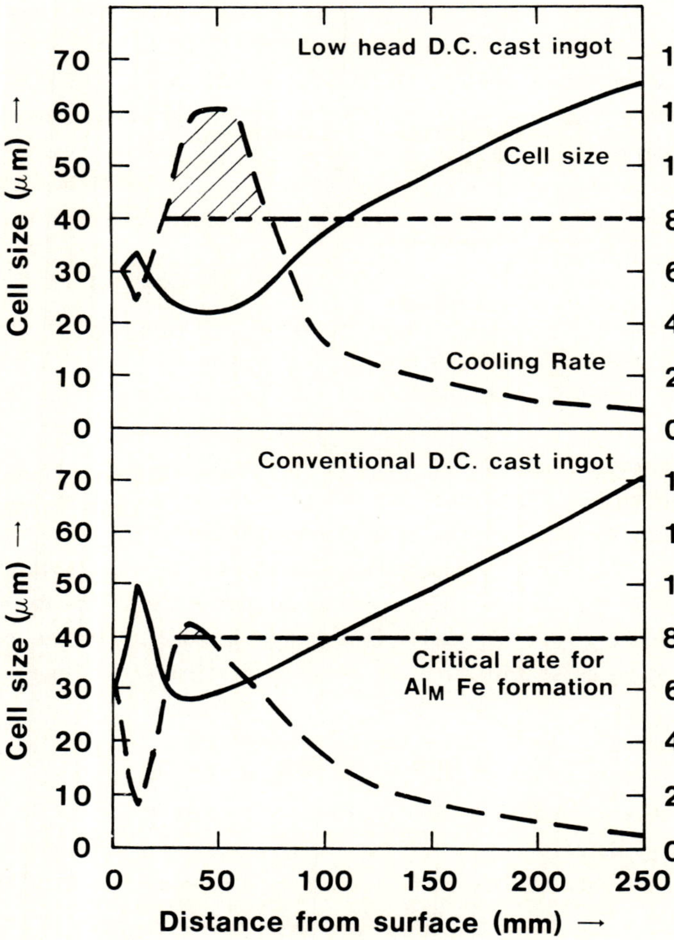

FIGURE 1

Cell size and cooling rate as a function of distance from the surface in low head and conventional D.C. cast ingot.

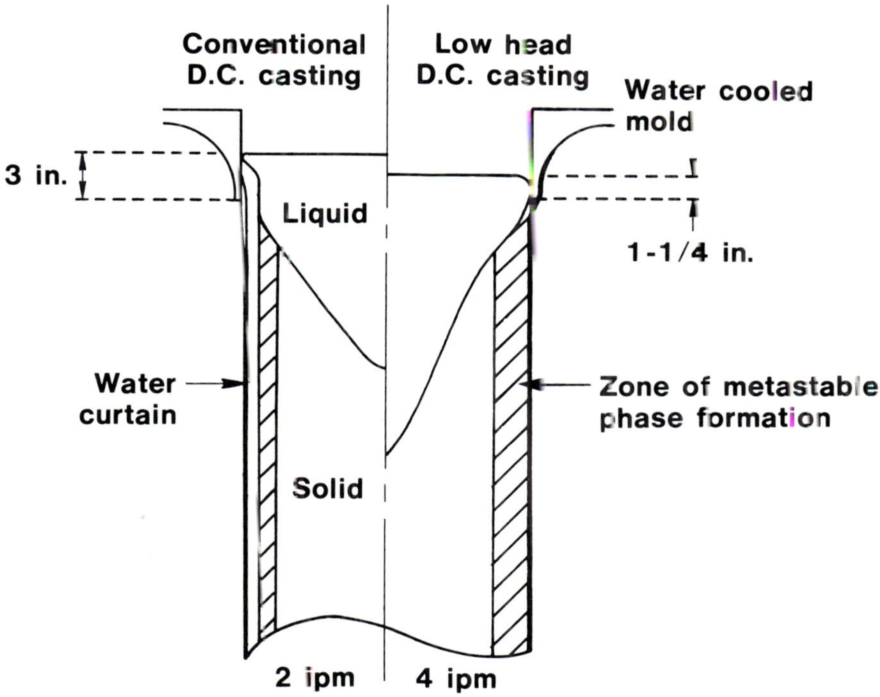

FIGURE 2

Schematic illustration of zones of metastable zone formation and crater profiles in conventional and low head D.C. ingot casting.

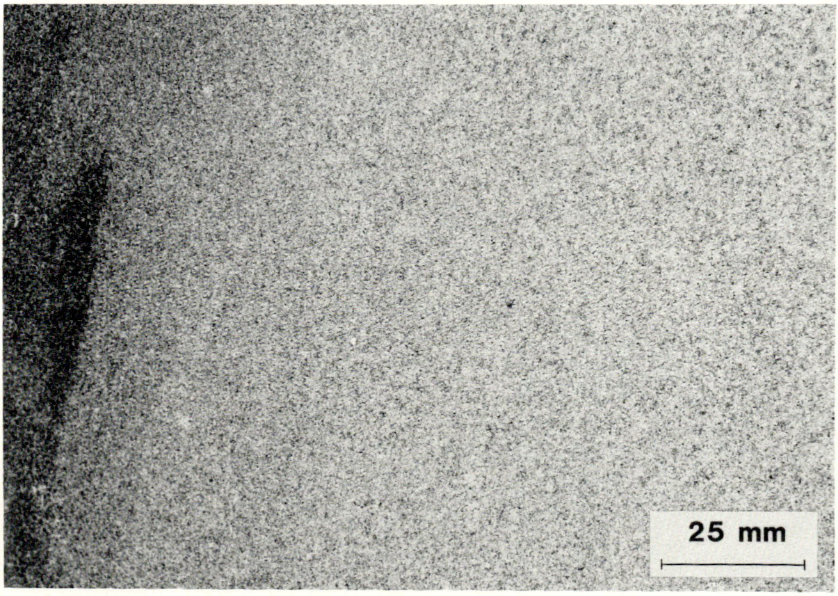

FIGURE 3 **(a)**

Longitudinal sections of 5005 alloy ingot showing region of metastable phase (dark etching) near surface in (a) conventional and (b) low head D.C. cast ingot.

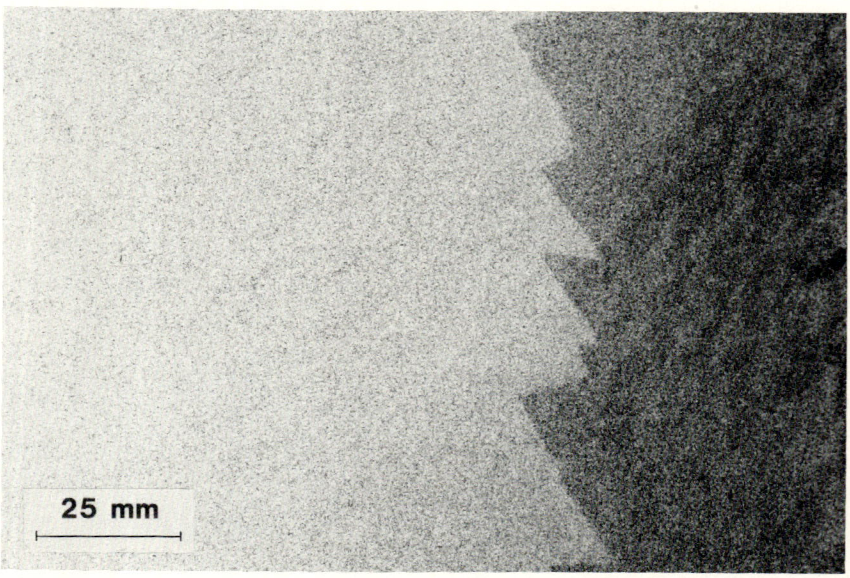

(b)

can be used to estimate sample-to-sample variability in the amount of any particular phase. Sections were taken at 10 mm intervals from the surface of the 5005 alloy low head D.C. cast ingot, made using the practice described. The distribution of intermetallic phases as a function of depth from the surface of the ingot, which is shown in Figure 4, is similar to that reported previously (3). Al_mFe is found in the region near the surface of the ingot where freezing rates were high. The distinct boundary between the dark and light etching zones occurs at the transition from Al_mFe to Al_6Fe. Finally, there is a gradual increase in the amount of Al_3Fe as one moves deeper into the ingot.

Initially, X-ray diffraction analyses were made on sections taken from a longitudinal section cut along the ingot center line, i.e. through the center of the rolling faces of the ingot. However, examination of a transverse slice taken from the same ingot shows that the cooling rate, as judged by the depth and uniformity of the dark etching zone, is not the same around the ingot periphery. Examination of several transverse slices showed that non-uniformities in the depth of the dark zone were particularly prevalent at the ingot quarter points, i.e., at a distance of about 13-1/2 in. (343 mm) in from the ends of the ingot. From time-to-time, "islands" of another constituent type (e.g., Al_6Fe) would appear close to the ingot surface as illustrated in Figure 5.

Influence of Ingot Scalping

The surface of D.C. cast ingot is not sufficiently homogeneous to allow rolling direct to sheet. It is normal to remove surface roughness, and sub-surface segregation, by machining about 5/8 in. (16 mm) from rolling faces prior to further fabrication. After surface machining, or scalping, it is essential that the exposed surface contain a single constituent type and be uniform in both grain structure and cell size. Any non-uniformity at this stage will result in structural streaking in sheet. Small areas of non-uniformity, such as the islands described, will be elongated up to 200x during subsequent rolling operations.

Preheating

A number of transformations involving Fe-bearing phases can occur when AA5005 alloy ingots are heated prior to rolling. The majority of these reactions involve the metastable intermetallic phases which are present

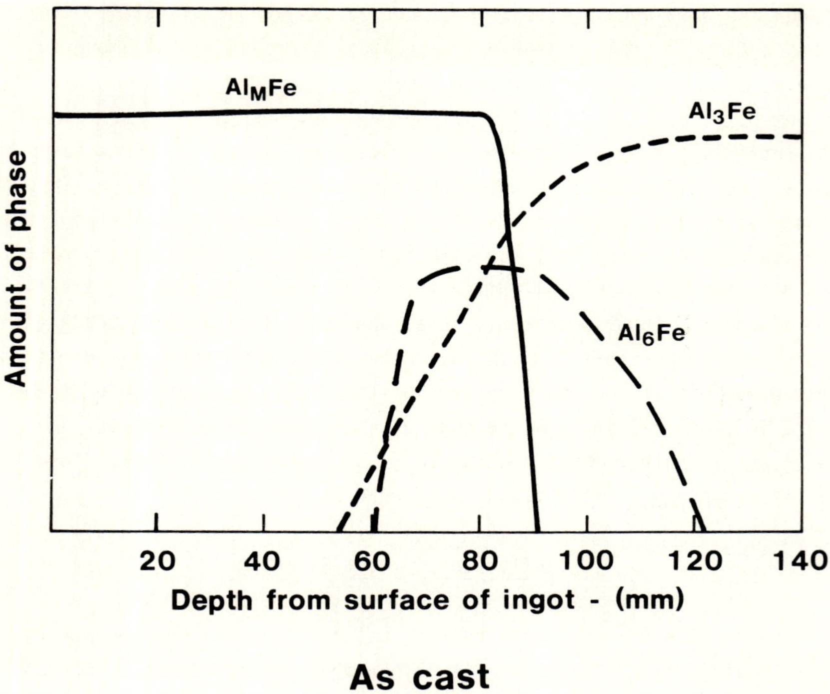

FIGURE 4

Distribution of intermetallic compounds in low head D.C. cast 5005 alloy ingot.

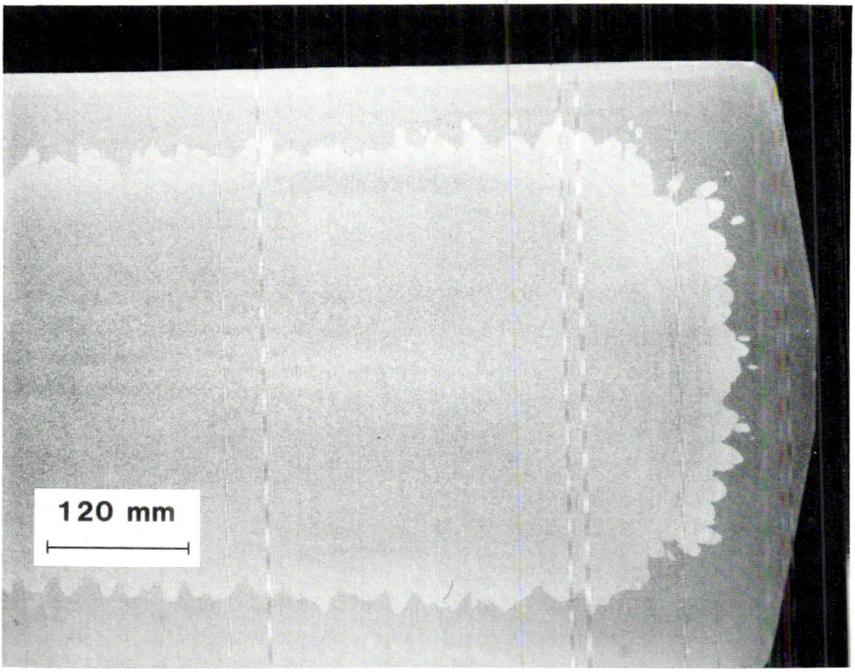

FIGURE 5

Transverse slice from a 20 x 54 in. 5005 alloy ingot cast using the low head D.C. method. Note the "islands" of light etching material in the otherwise dark etching zone.

Etch: 5% NaOH at 150°F

near the surface of the ingot. Because the appearance of sheet products which are etched or anodized is sensitive to the properties of the Fe-bearing phases which are located at the surface of the sheet, it is desirable to understand how the properties of these phases evolve during preheating. Previous investigations (3, 5, 9 & 10) have considered the transformation of the metastable intermetallic phases into Al_3Fe in Al-Fe and Al-Mg-Fe alloys and the formation of Al_3Fe when Al-Fe-Si and Al-Mg-Fe-Si alloys are heated to high temperatures. The current study has examined the transformations which occur when an Al-Mg-Fe-Si alloy is heated to a temperature where Al_3Fe does not form readily.

The distribution of intermetallic phases in the as-cast structure is presented in Figure 4. After initial X-ray analysis, these samples were preheated for 24 hours at 850°F (454°C) and identification of the intermetallic phases was repeated. Figure 6 shows the distribution of Fe-bearing phases as a function of depth below the surface for the preheated material. Comparison of Figures 4 and 6 indicates that significant changes in the distribution of intermetallic phases occurs during preheating. However, the metastable phases, which are responsible for the light and dark etching zones, are still present after preheating. This indicates that structural streaking could still occur.

Additional Al_3Fe forms after 24 hours at 850°F. The Al_3Fe which formed during the thermal treatment is not distributed uniformly throughout the thickness of the ingot. Rather, Al_3Fe formation is most evident at the distance from the surface where Al_6Fe occurred in the as-cast condition. The relatively large decrease in the amount of Al_6Fe and Al_mFe in the region where Al_3Fe is formed is consistent with previous observations that Al_3Fe forms after dissolution of the metastable phases (9, 10).

It was observed that changes in the intensity and position of some lines in the X-ray diffraction pattern of Al_6Fe occurred during thermal treatment. The structural changes in Al_6Fe, which are responsible for the variation in the X-ray pattern, are not known at this time.

A substantial amount of α which is formed decreases slightly with increasing distance from the as-cast surface. The formation of large amounts of α has not been observed in previous studies because the Si

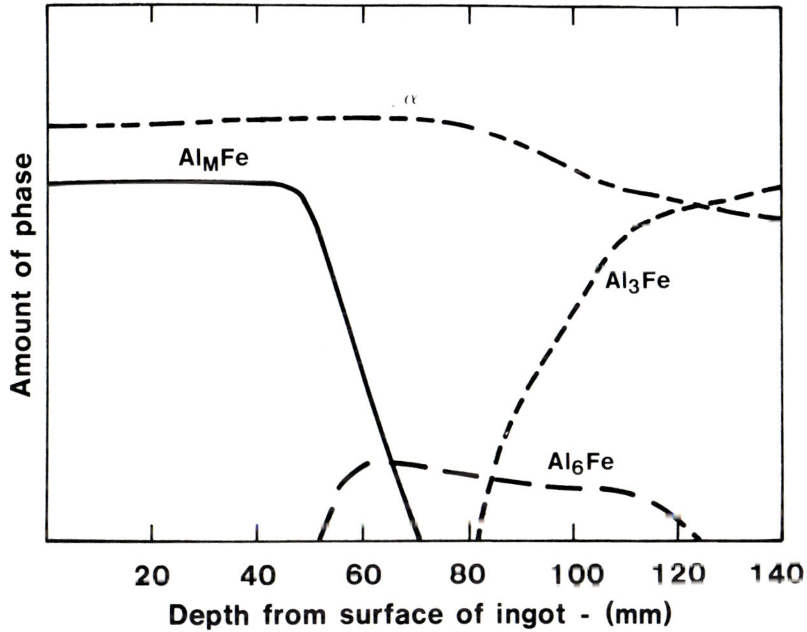

Preheated at 454°C (850°F) for 24 hrs.

FIGURE 6

Transformation of the metastable phases in Al_mFe and Al_6Fe into ∝ and the stable phase Al_3Fe at 850°F.

levels were too low or because the alloys were heated to a high temperature where Al_3Fe forms rapidly.

An additional experiment was conducted to determine the relative rates at which α is formed and Al_mFe dissolves. A number of samples were cut from 20 mm below the as-cast surface of the ingot. These samples were heated to 850°F for times ranging from 1 hour to 24 hours. Figure 7 illustrates the relative amounts of α and Al_mFe as a function of time at 850°F. The presence of α is noted after only 1 hour while the dissolution of Al_mFe proceeds much more slowly. It is not known if α forms by precipitation from solid solution or if α nucleates on existing metastable phases as a result of the relatively high diffusivity of Si compared to Fe at this temperature.

Discussion

There are basic differences in the structure of ingot produced by the low head and conventional D.C. casting methods. Not only are there superficial effects, such as a smoother exudation-free surface on low head ingot, but there are significant differences in the sub-surface cell size and types of intermetallic compounds that are formed. Predictably, architectural sheet produced from conventional D.C. cast 5005 alloy ingot is plagued with structural streaking arising from the non-uniform and narrow zone where metastable phase formation can occur. Surface machining prior to preheating and rolling will inevitably produce a non-uniform array of intermetallic compounds. Although the metastable phase Al_mFe will dominate, patches of Al_3Fe or Al_6Fe will be present; these are elongated during rolling to give the well known, light etching and anodizing, "tiger stripes."

Low head D.C. ingots are less susceptible to the structural streaking discussed in this paper. Higher freezing rates which accompany the low head technique cause the Al_6Fe phase to be formed further from the surface of the ingot. Thus, scalping is less likely to expose areas of Al_6Fe-bearing metal which could result in a streak after etching or anodizing. However, it appears that -- if this phase is exposed -- preheating at 850°F modifies the structure of Al_6Fe; these changes affect the reactivity of Al_6Fe-bearing metal during etching and anodizing and can result in a streak that is dark compared with the matrix.

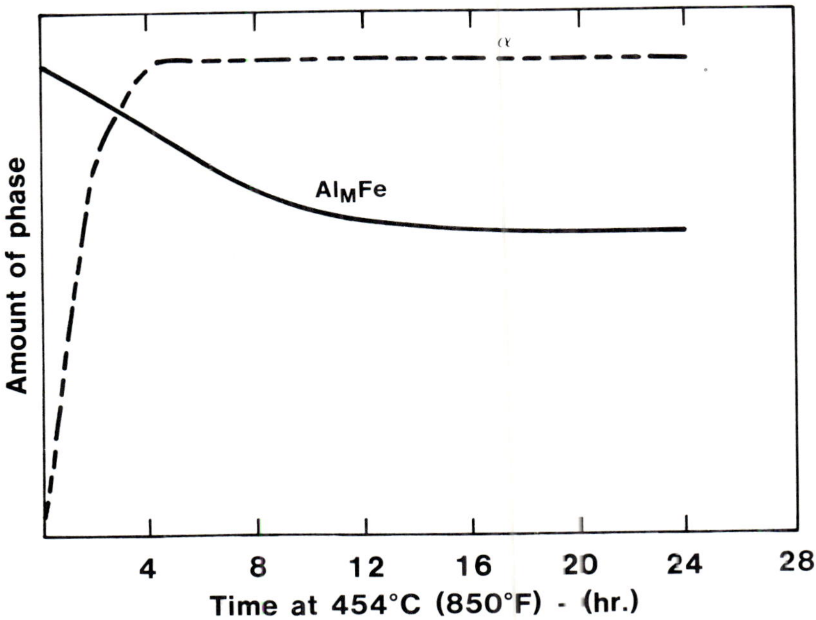

FIGURE 7

Graph illustrating transformation to the metastable Al_mFe phase into ∝ as a function of time at 850°F.

The as-cast distribution of intermetallic phases and the relationship between these phases and structural streaking in 5005 alloy appear to be similar to those results which have been reported for 1XXX series alloys. As previously suggested (7), current work indicates that the addition of Mg extends the range of freezing rates over which Al_mFe can be formed. In contrast to 1XXX alloys, preheating of 5005 alloy at 850°F results in the formation of α Al-Fe-Si. It is apparent that the formation of α does not affect the uniformity of the finishing response of the material.

Conclusions

1. Different ingot casting processes result in different distributions and types of intermetallic compounds in 5005 alloy.
2. Two metastable phases dominate, Al_mFe and Al_6Fe, where Al_mFe is formed at higher cooling rates than Al_6Fe. In Al-Mg-Fe-Si alloys the Al_mFe compound is most common, especially in low head D.C. casting.
3. In low temperature (850°F) preheat, the metastable phases slowly transform: the Al_mFe to α Al-Fe-Si and the Al_6Fe to a phase whose X-ray pattern is similar to Al_6(Fe, Cu). The former compound causes no non-uniformity in etching or anodizing of the surface; the latter is present in dark streaks.
4. The presence of Mg in an Al-Fe-Si alloy extends the freezing rate range over which the metastable phase Al_mFe is formed.

Acknowledgements

We wish to recognize the support and encouragement of our colleagues Messrs. J. E. Jacoby (Alcoa Technical Center), E. L. Rooy, H. G. Reavis (Pittsburgh Office), and R. T. Richter (Davenport Works), throughout this investigation. Our thanks are due to the Alcoa Editorial Committee for permission to publish the work.

References

1. R. E. Spear and G. R. Gardner, "Dendrite Cell Size," Trans. Am. Foundrymen's Soc., 71 (1963), 209-215.

2. E. H. Hollingsworth, G. R. Frank, Jr., and R. E. Willett, "Identification of a New Al-Fe Constituent $FeAl_6$," Trans. Met. Soc. AIME, 224 (1962), 188-189.

3. H. Westengen, "Formation of Intermetallic Compounds During D.C. Casting of a Commercial Purity Al-Fe-Si Alloy," Z. Metallkunde, 73 (1982), 360-368.

4. I. Miki, H. Kosuge and K. Nagahama, "Supersaturation and Decomposition of Al-Fe Alloys During Solidification," J. Jap. Inst. Light Met., 25 (1975), 1-9.

5. H. Kosuge and I. Mizukami, "Behavior of the 'Fir Tree' Structure in Al-Fe-Si Alloy Ingots During High Temperature Heat Treatment," J. Jap. Inst. Light Met. 22 (1972), 437-444.

6. H. Kosuge and I. Mizukami, "Formation of 'Fir Tree' Structure in D.C. Cast Ingot of Al-0.6% Fe Alloys," J. Jap. Inst. Light Met., 25 (1975), 48-58.

7. A. Asami, T. Tanaka and A. Hideno, "Fir Tree Structure in D.C. Cast Ingot of Al-Mg-Fe-Si Alloy," J. Jap. Inst. Light Met., 28 (1978), 321-327.

8. D. Altenpohl, "Investigation of the Structures of Continuously Cast Commercial and High-Purity Aluminum Ingots," Z. Metallkunde, 46 (1955), 535-544.

9. H. Suzuki, M. Kanno, H. Tanabe and K. Iloi, "The Metastable to Stable Phase Change in an Al-1.0% Fe Alloy," J. Jap. Inst. Light Met., 28 (1978), 284-291.

10. H. Kosuge and H. Takada, "Phase Change of the Metastable Phases in Al-Fe Alloys at Elevated Temperatures," J. Jap. Inst. Light Met., 29 (1979), 64-69.

CHAIRMEN

J.E. Dore
Consolidated Aluminum
Hendersonville, North Carolina

N.J. Pappas
Union Carbide-Linde Division
Tarrytown, New York

THE TAC PROCESS: A PROVEN TECHNOLOGY

Bruno Gariépy and Ghyslain Dubé

Alcan International Limited
Arvida Laboratories and
Experimental Engineering Centre
P.O. Box 1250
Jonquière, Québec, Canada
G7S 4K8

Charles Simoneau and Guy Leblanc

Alcan International Limited
Engineering Division
2001 University Street
P.O. Box 6090
Montréal, Québec, Canada
H3C 3H2

The TAC process has been introduced into Alcan casting plants since 1982. Production experience combined with process demonstrations in other plants permitted a full evaluation of metallurgical variables treatment conditions and process and equipment performances. The effect of variables such as metal temperature, presence and type of electrolyte was studied. A new type of treatment (TAC II) involving metal transfer in a treatment crucible was demonstrated and led to improved alkali removal. Recycling of spent AlF_3 through potlines was established. The effect of the treatment on metal cleanliness was substantiated specially for the removal of Al_4C_3 inclusions. Process and equipment performances were evaluated and will be presented.

Introduction

Because of ever growing concerns over energy costs and environmental regulations, the use of lithium electrolytes in smelter operations has become more and more attractive to aluminium producers. Over the past few years in particular, Alcan Smelters and Chemicals has operated partly or in whole several of its Canadian smelters with lithium-containing electrolyte. Alcan International is continuing its development program to provide a more economical, efficient and competitive method to remove contaminants from smelter metal produced thereby.

The TAC process (Treatment of Aluminium in Crucibles) for the removal of lithium and other alkali metal contaminants in primary aluminium was introduced by Alcan at the beginning of 1982. Following Alcan's basic approach (1) to perform metal treatment before furnace transfer, the TAC process has proved highly flexible, energy saving and effective. Apart from advantages related to reduced dross formation as compared with handling of Li contaminated aluminium, the process does not involve toxic gases or waste products often associated with in-line fluxing after furnaces.

Having been designed and engineered on basic principles, the TAC process has easily been automated and incorporated into existing hot metal installations. Production experience at two different locations (above 200,000 mt TAC processed metal), as well as process demonstrations in other plants, enabled us to make a full evaluation of treatment conditions, metallurgical variables and process performances. A new type of TAC production station will also be presented.

Process Parameters

Alkali removal by the TAC process was established to proceed according to the reaction:

$$AlF_{3(s)} + \left\{ \begin{matrix} Li \\ Na \\ Ca \end{matrix} \right\} \longrightarrow (Li_x\ Na_y\ Ca_z)_3\ AlF_6 \qquad (1)$$

In practice, a large excess of AlF_3 powder is required over the stoichiometric quantity.

Reaction products, mixed with residual electrolyte material, either agglomerate on the crucible lining or float to the surface of the melt where they can easily be recovered. The following variables were previously (2) identified to control the removal process:

1) Particle size and quantity of AlF_3 powder
2) Stirring rate
3) Reaction time
4) Metal temperature
5) Presence of bath in crucibles.

TAC production units are operated at constant AlF_3 addition, impeller rpm and reaction time. Crucibles are treated immediately upon arrival at the station in order to minimize disruption of hot metal delivery for furnaces.

Under these conditions, TAC process performances will only be affected by metal temperature and amount of bath in crucibles (not taking into account variation in metal level).

Metal Temperature

Figure 1 describes the effect of metal temperature on lithium reaction constant and removal efficiency. Under similar treatment conditions, the lithium removal rate is inversely proportional to metal temperature. Removal efficiencies varied from 80 to 90% in the normally encountered metal temperature range of 750 to 840°C.

Typical lithium contents measured after TAC treatments, also under production conditions, are shown in Figure 2 for 1.3 and 2.0 kg/mt AlF_3 additions. The results illustrate the importance of metal temperature distribution in the determination of process parameters. Off-specification furnace charges are prevented by appropriate AlF_3 additions and stirring time that will take into account high metal temperature frequency.

Effect of Electrolyte

Figures 1 and 2 also indicate that TAC treatments performed at a given temperature will result in variable alkaline contents. Close monitoring of the process performance in Arvida showed that the presence of a large amount of electrolyte in crucibles during the treatment is detrimental to TAC process efficiencies. Results are given in Table I.

Table I. Effect of Crucible Cleanliness and Metal Temperature on TAC Process Efficiency

Crucible Condition	Metal Temperature °C	Lithium (ppm) Before	Lithium (ppm) After	Efficiency
Clean	T<760°C	17.7	1.0	94%
Clean	760<T<830°C	22.0	3.2	85%
Electrolyte in mushy state	830<T	21.4	5.2	76%

Higher temperatures associated with the presence of mushy state electrolyte shift thermodynamic equilibriums toward the left-hand side of equation 1 and therefore reduce cryolitic compounds formation. Moreover, the AlF_3 powder readily reacts with cryolite (Na_3AlF_6) to form chiolite ($Na_5Al_3F_{14}$), thus reducing the effective AlF_3 concentration for alkali removal.

Metal temperature distribution and conditions of crucibles with regard to bath at the TAC treatment station depend on the hot metal delivery system from potrooms to casting, on siphoning practices in potrooms and on crucible cleaning practices.

Because these are characteristics in plant operation, they can influence TAC efficiency from one plant to another. In cases of frequent treatments

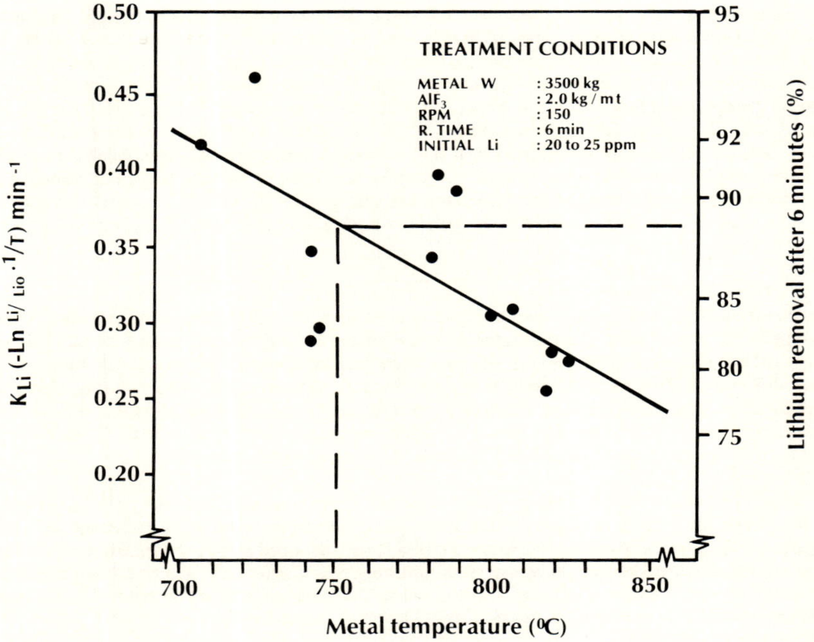

Figure 1 - Effect of Metal Temperature on Lithium Removal

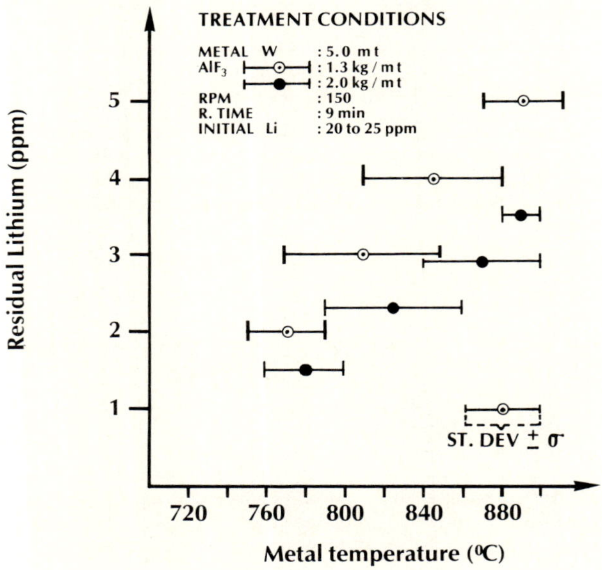

Figure 2 - Residual Lithium Concentration (After TAC) vs Metal Temperature

to be performed at high temperatures and in the presence of large amounts of mushy state electrolyte, the process would have to be carried out with an AlF_3 addition close to or above 3.0 kg/mt in order to achieve reasonable alkali removal efficiencies. In those conditions, however, the process efficiency would be limited to a residual Li content from 2 to 3 ppm minimum. Aluminium fluoride additions larger than 2 kg/mt were not considered acceptable, both from process economics and crucible cleaning operations. For this reason a modified process, involving the treatment after metal transfer, has been developed and tested at the beginning of 1982.

TAC II: TAC Treatment After Metal Transfer

Instead of treating smelter metal directly in potroom siphoning crucibles, metal was first transferred into a designated treatment crucible and then TAC processed (TAC II). Metal temperature dropped after transfer from about 850 to 810°C while most of the liquid electrolyte stayed in the original crucible or solidified and floated on the surface of the metal in the treatment crucible. Lithium residual levels obtained with the TAC II method are shown in Figure 3 and compared to the original TAC process.

The TAC II method led to substantial improvement in the efficiency and reliability of the process. For example, from an initial content of 20 to 25 ppm, a 3 ppm residual lithium level was reached with less than 0.5 kg AlF_3/mt aluminium while 2.5 kg/mt had been needed previously. Furthermore, very low Li contents (0.5 to 2 ppm) could be obtained with the TAC II only. Lithium residual contents after TAC II were also more consistent from one treatment to another.

Associated advantages are reduction in operating costs due to reduction of AlF_3 addition, extended rotor life and easier crucible cleaning operations (1 kg AlF_3/mt is equivalent to $1.40/mt).

TAC Process and Other Electrolytes

The TAC process has been demonstrated in three AS&C plants where potlines are operated under lithium-containing electrolyte. In all cases sodium and calcium were removed faster and at levels lower than those achieved for lithium. For this reason, lithium is preferably used as a measurement of the TAC process performances.

Standard Electrolyte

The TAC process was further evaluated under plant conditions using standard electrolyte (6-7% excess AlF_3, no LiF) with the objective of optimizing the process with respect to sodium removal. Dissolved Na in smelter metal ranged from 40 to 60 ppm upon arrival at the cast house. All crucibles coming from one potline were treated during a 10-day trial period using a portable demonstration TAC unit.

Figures 4 and 5 demonstrate the effectiveness of the TAC process with regard to sodium removal. Treatment parameters set at 6 min and 0.5 kg AlF_3/mt Al reduced sodium concentration down to 1 ppm. Lower sodium concentration could easily be obtained by increasing the AlF_3 addition up to 1.0 kg/mt without increase in treatment time.

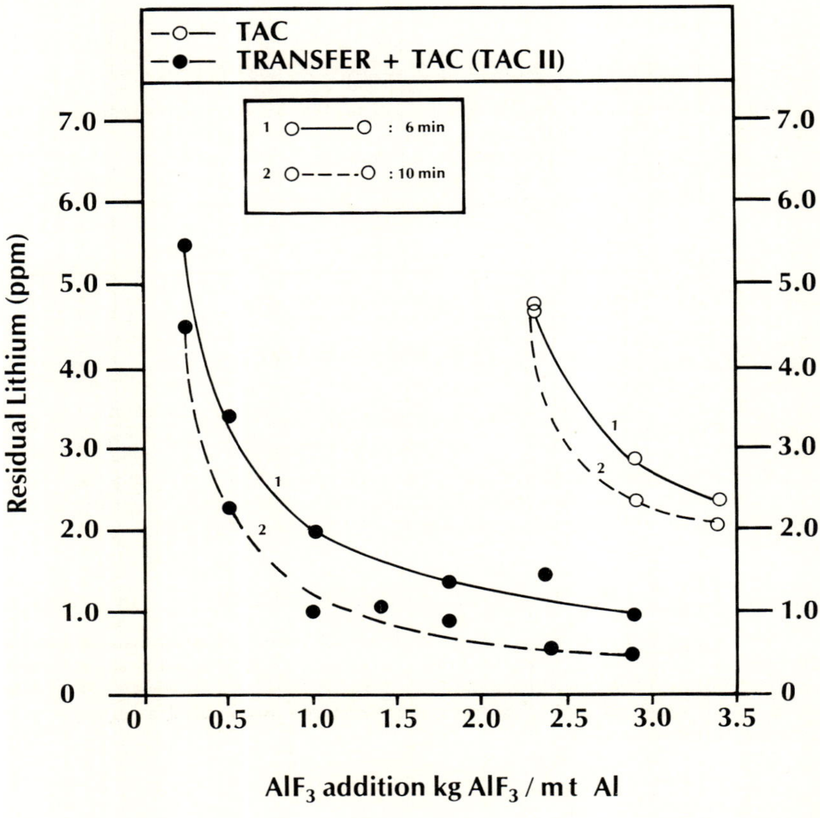

Figure 3 - Residual Lithium Concentration vs AlF_3 Addition
TAC vs TAC II (Transfer + TAC)

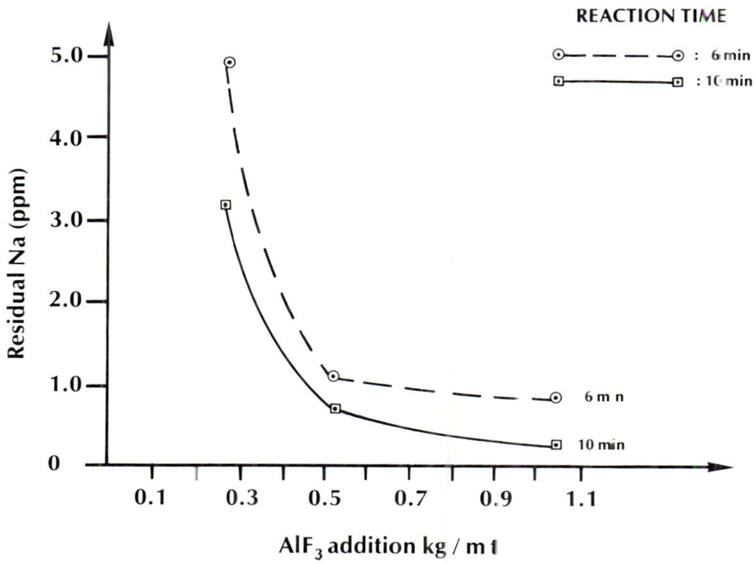

Figure 4 - Residual Sodium Concentration After TAC vs AlF_3 Addition
Standard Electrolyte

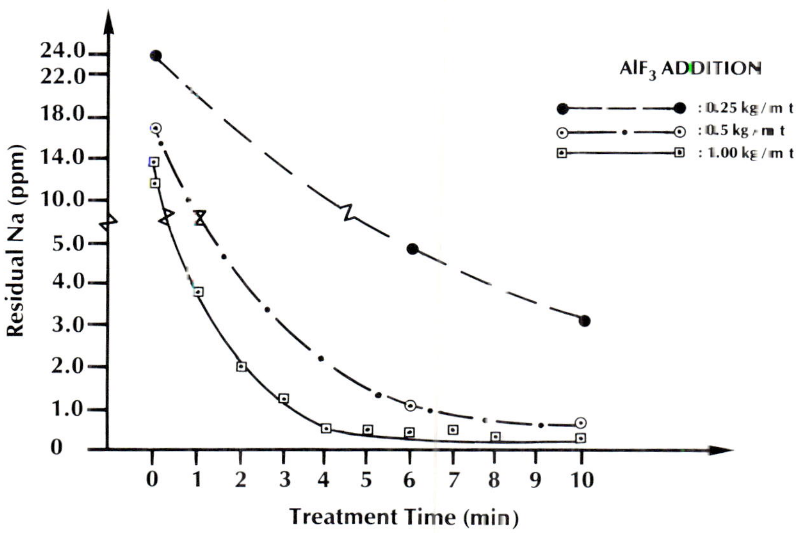

Figure 5 - Sodium Concentration vs TAC Treatment Time
Standard Electrolyte

As mentioned previously, sodium is removed more rapidly by the process than lithium. For example, after only two minutes of stirring, a 1.0 kg/mt addition will remove all but 2 ppm of sodium. It is believed that the higher sodium vapour pressure assists sodium removal both by reaction with aluminium fluoride and oxidation. It is also likely that the high (6-7%) excess AlF_3 in the electrolyte present in the crucibles during the treatment assists sodium removal.

Mixed Electrolyte (LiF/MgF_2)

Mixed electrolytes LiF/MgF_2 were also investigated to measure TAC effectiveness on dissolved Mg removal. Smelter metal of about 20 ppm Li, 40 ppm Na and 100 ppm Mg was treated during continuous plant testing following the TAC II (Transfer & TAC) method. Typical results are shown in Figure 6. Fluoride compound Na_2MgAlF_7 was identified by XRD in reaction products.

TAC Process Additional Benefits

It was previously stated (2) that, by the action of AlF_3 rich flux pellets agglomerated on the metal surface, the process does not suffer any melt loss and could even improve metal/bath separation. An improvement of metal cleanliness was also attributed to the continous recirculation of

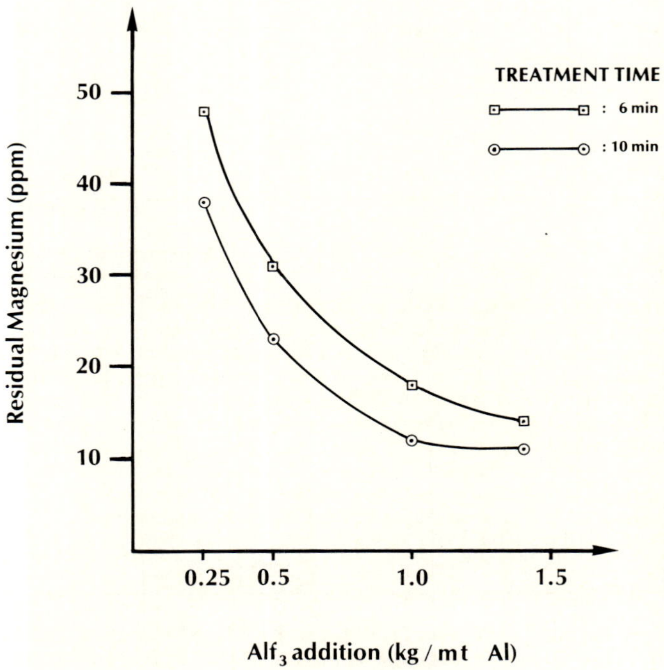

Figure 6 - Magnesium Concentration After TAC (II) Process vs AlF_3 Addition. Mixed Electrolyte LiF/MgF_2

fluoride particles through molten metal during the treatment. The latter was substantiated during plant implementation of both Arvida and Isle Maligne production units. Almost 100% nonmetallic inclusions in smelter metal are aluminium carbide particles. As reported in Table II, Al_4C_3 concentrations (analyzed by gas chromatography) were reduced from an average 14 ppm down to 3 and 4 ppm respectively. It was also observed that aluminium carbide removal increased with decreasing metal temperature during treatment. At this level, metallographic examinations did not reveal any Al_4C_3 inclusion. The residual Al_4C_3 concentration could be associated with dissolved carbides or submicron size particles. Other nonmetallic inclusions such as oxides or nitrides are not generated by the TAC process either.

Table II. TAC Process: Effect on Nonmetallic Al_4C_3

	Arvida		Isle Maligne	
Treatment				
AlF_3 Addition	1.5 kg/mt		2 kg/mt	
Treatment Time	6 minutes		10 minutes	
Metal Weight	3.6 mt		5.6 mt	
Al_4C_3 Concentration (ppm)	Before	After	Before	After
	9	2	14	4
	17	3	16	3
	14	5	15	4
	13	5	12	6
	16	2	15	5
Average	14	3	14	4
% Eff.	79%		71%	

Material Recycling

The Isle Maligne TAC unit has been in operation for more than one year. The excess AlF_3 remaining in crucibles from the TAC treatment is removed along with smelter electrolyte coming from pot tapping operations. A partial rough cleaning is made manually at the end of each shift. Crucibles are thoroughly cleaned after 4 to 6 days of continuous operation. The total material recovery was measured at approximately 11 kg/mt of processed aluminium. From an AlF_3 addition of 1 to 2 kg/mt Al, the TAC operation could account for 1/10 to 1/5 of the total. Because of the excess AlF_3, the material is mostly made up of lower melting point chiolite ($Na_5Al_3F_{14}$). All of it is readily recycled in potroom operations as bath additions.

TAC Process: The Drive-In TAC Station

A second TAC production unit was introduced by Alcan Smelters and Chemicals at its Isle Maligne plant at the end of 1982. This drive-in TAC station is designed to service cylindrical crucibles handled bail-down using a tractor and crucible transport trailer arrangement. A typical TAC station's main components are schematically reproduced in Figure 7. It shows the main

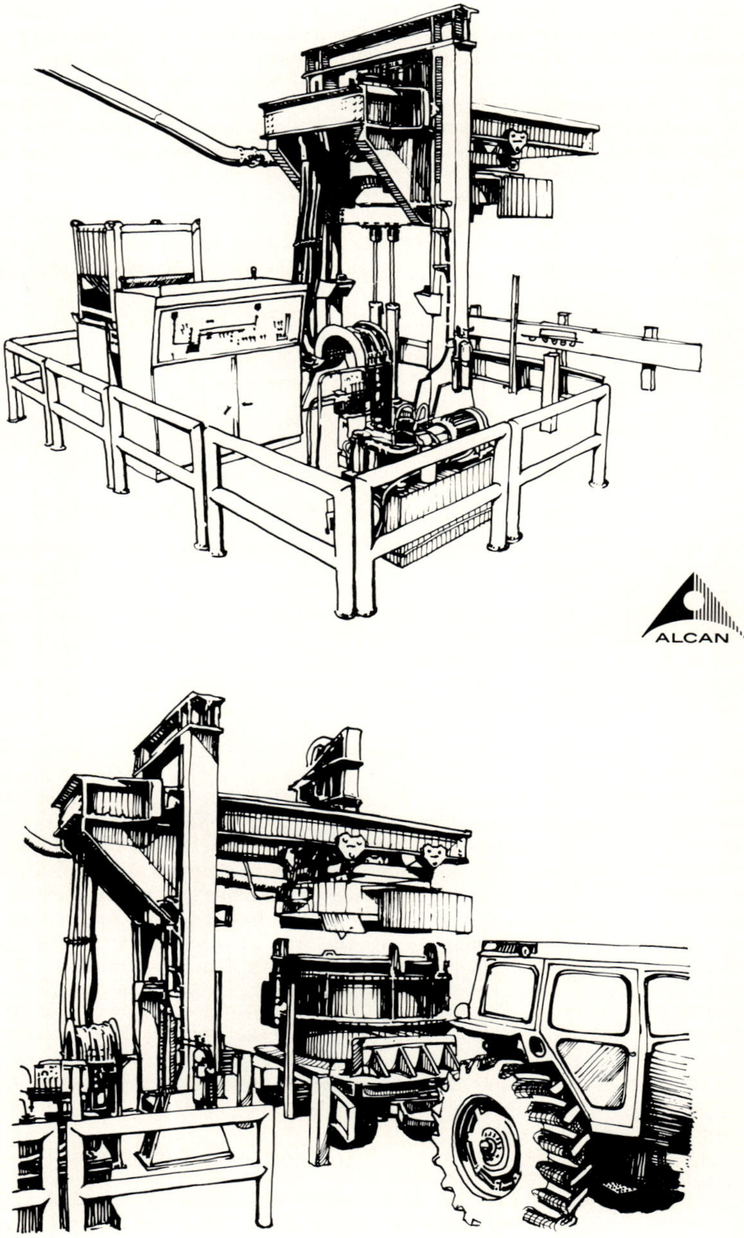

Figure 7 - Alcan Drive-In TAC Station

structure coupled with a programmable controller, hydraulic unit, AlF_3 injection and dust collector systems.

All operations are initiated in sequence or simultaneously by a programmable controller in the automatic mode. The reliable use of this central component has proven to be effective in terms of system flexibility and control of operations. Under its control, the whole station is operated without the need for any additional manpower. A typical "drive-in" TAC station can process up to 32 crucibles per shift for a rated capacity of about 150,000 to 200,000 mt/year.

At this location, crucibles carry an average of 5.6 mt of Al. Because of their size, a specific rotor design had to be developed to create proper vortex characteristics. The unit operates under the following parameters:

Reaction Time	:	9 minutes		
AlF_3 Addition	:	1.33 kg/mt (16 lbs/cruc.)	and	1.0 kg/mt (12 lbs/cruc.)
Average T^oC	:	845^oC		
Initial Li Content	:	20 to 22 ppm		

However, the system is versatile enough to permit rapid adjustments of any of the process parameters.

Residual lithium concentrations after TAC treatments are shown in Fig. 8 for AlF_3 additions of 1.33 kg/mt. The average lithium content after TAC was 2.6 ppm for a process efficiency of 87%. This TAC station was also operated at 1.0 kg/mt. The average efficiency was measured at about 78% for an average residual lithium content of 4.5 ppm.

The wide range may be accounted for by the effect of metal temperature and quantity of bath variations. It is also attributed to variations in metal weights to be treated. In this particular case, an addition of 1.0 kg/mt to 1.33 kg/mt is sufficient to meet cast product alkali specifications. Metal transfer in furnaces and subsequent batch preparation steps are responsible for an additional loss of alkaline contaminants. Figure 9 indicates the range of lithium contents in commercial metal cast ingots.

Conclusions

The TAC process was developed to remove alkali and alkaline earth impurities from primary aluminium without utilization of chlorine gas.

The heart of the process is the rotor designed to achieve efficient contact between aluminium fluoride flux and liquid aluminium. The TAC process is characterized by the following advantages:

- Low operation and maintenance costs
- Limited floor space requirement
- No pollution
- No additional manpower required
- Flexible and easy to incorporate in existing hot metal installations.

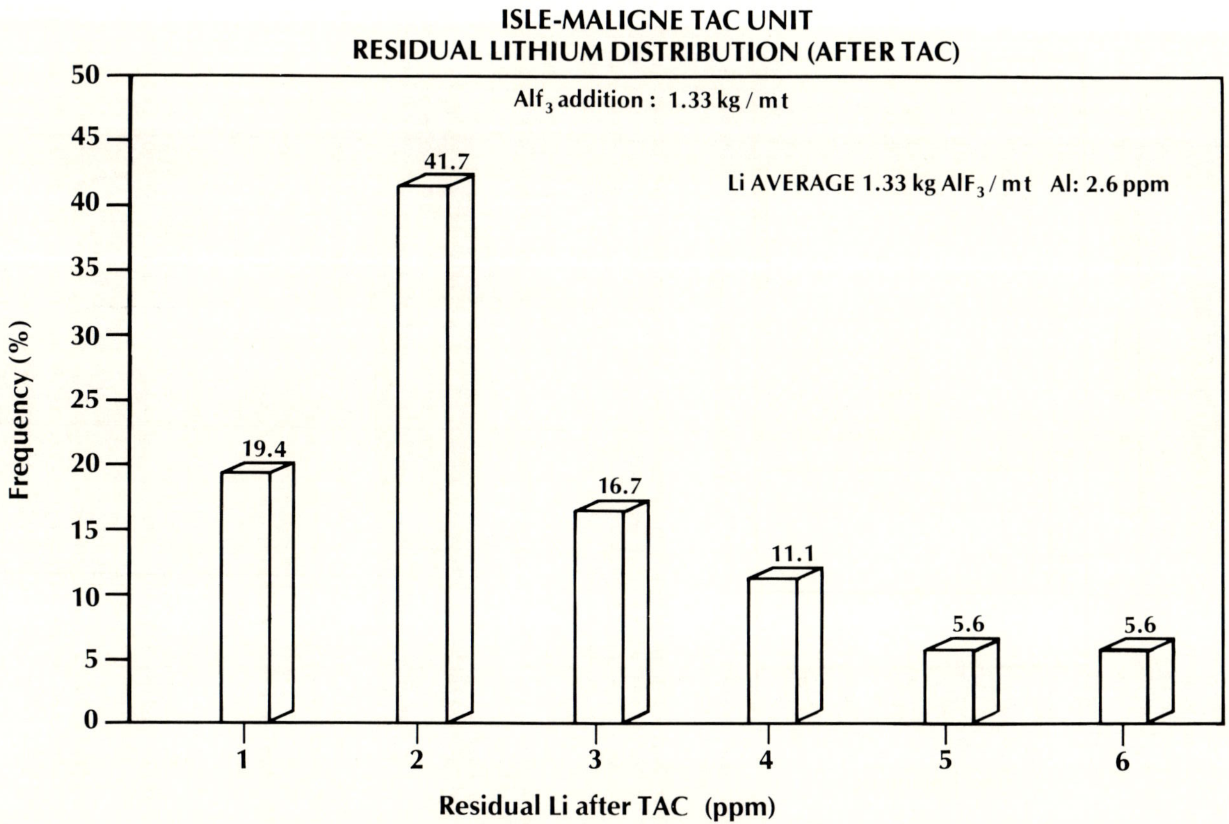

Figure 8 - TAC Process Performances in Production Conditions

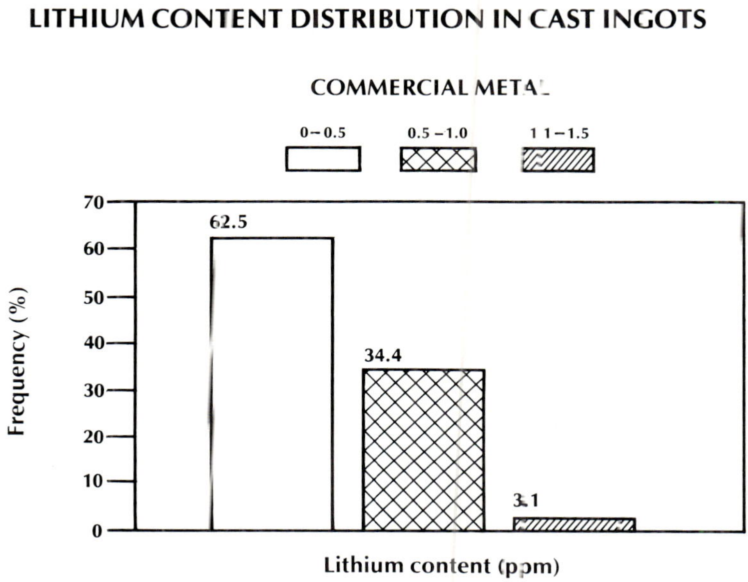

Figure 9 - Lithium Concentration in Cast Ingots - 100% TAC Processed Metal

The process also has extra-beneficial effects on metal cleanliness. Two types of TAC stations are operating to accommodate different hot metal transportation systems.

The TAC process is patent pending in the U.S.A. and elsewhere.

Acknowledgement

The authors are grateful to the Management of Alcan International Limited for the permission to publish this paper. Collaboration of production personnel from Alcan Smelters and Chemicals Limited, Arvida and Isle Maligne casting plants is also acknowledged.

References

(1) Pierre Achim, Ghyslain Dubé, "Removal of Lithium in Commercial Metal", pp. 903-918 in Light Metals, 1982.

(2) Ghyslain Dubé, V.J. Newberry, "TAC - A Novel Process for the Removal of Lithium and Other Alkalis in Primary Aluminium", pp. 991-1003 in Light Metals, 1983.

MOLTEN ALUMINUM FILTRATION:

FUNDAMENTALS AND MODELS

C. E. Eckert and R. E. Miller

Ingot Casting Division
Alcoa Technical Center
Alcoa Center, PA 15069

D. Apelian and R. Mutharasan
College of Engineering
Drexel University
Philadelphia, PA 19104

The removal of inclusions from an aluminum melt usually involves four basic separation processes: sedimentation, floatation, cake mode filtration, and deep bed filtration. In this paper each of the techniques are examined, with an in-depth treatment of cake mode and deep bed filtration. An analytical expression is derived that relates operating parameters of constant volume cake mode filters to performance, and a method of interpreting experimentally obtained constant pressure filtration data is presented. A kinetic expression is also applied to a finite element deep bed filter to demonstrate short term behavior.

Introduction

Historically, aluminum ingot metal quality has been recognized as a critical element in the melting-casting-fabrication sequence. The increasing demands on product integrity have paralleled improvements in nondestructive testing methods, and in turn has stimulated advances in molten metal characterization technology. This trend will continue, and perhaps even accelerate as aluminum based alloys find application in areas with ever increasing property demands. This is certainly the case with aerospace alloys where strength, fracture toughness, corrosion, and fatigue properties are paramount. Similarly, the metal forming criteria for aluminum containers are known to be a direct function of the molten metal quality prior to solidification. In 1970, an acceptable performance level for inclusion related defects in drawn and ironed beverage can sheet was approximately three tear-off failures per ten thousand cans.[1] By contrast, today's performance expectations require more than an eighty percent reduction in inclusion frequency, and a one-third lower inclusion size. Many other products, such as stock for computer memory disks, bright finish trim, etc. are also highly dependent upon internal integrity to satisfy even todays standards.

It is generally recognized that todays product/property limitations will only be extended by markedly advancing the state of metal treatment technology. Whereas much of the progress to date has been achieved through a heavy reliance upon empirical techniques, significant progress in fundamental mechanistic understanding is prerequisite for futher advances.

Background

Metal quality, in a wholistic sense, is a composite of three interrelated components: the control of alkali trace elements, hydrogen reduction, and inclusion removal. Operations utilized for the removal of trace elements and hydrogen are generally referred to as fluxing, and commonly employ inert-reactive gas sparging in batch devices,[2] or flow through back-mix reactors, as a precursor to filtration. A typical treatment scheme is illustrated in Figure 1.

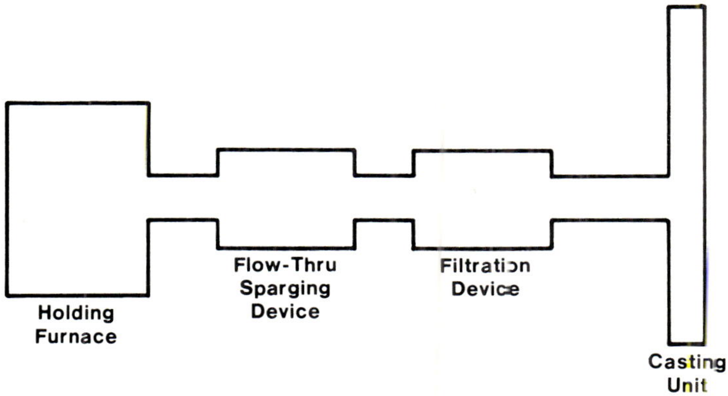

Figure 1 - Generalized Metal Treatment Scheme

The removal of non-metallic and intermetallic particulate suspensions present the most significant challenge to molten metal treatment systems. Inclusion concentrations in an unprocessed aluminum melt range from 0.005 to 0.02 volume fraction, and have been observed in a variety of particulate and film morphologies. Figure 2 depicts several types of inclusions found in such melts.

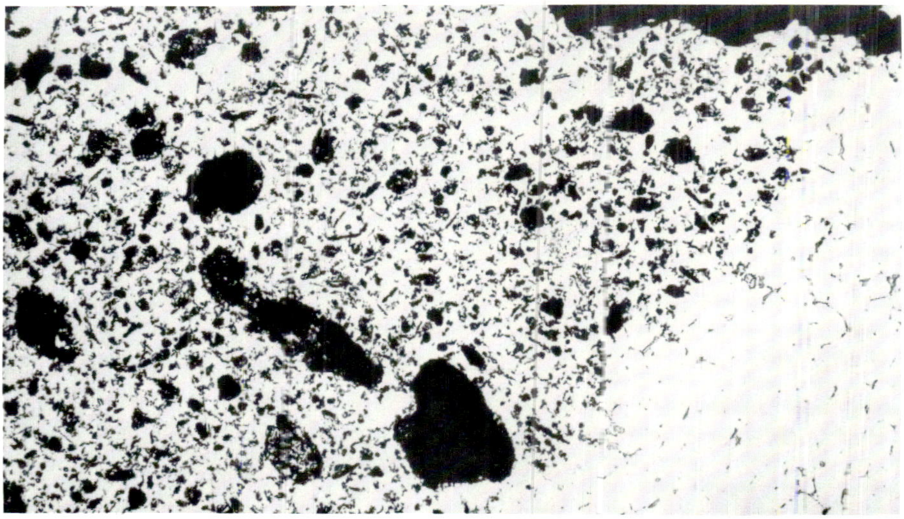

Figure 2 Typical Melt Inclusions (After Blackburn)[3] **50X**

A detailed discussion of the origin and formation mechanics of various inclusion types is beyond the scope of this paper, however, a tabulation of broad inclusion classifications and sources is provided in Table 1:

Classification	Types Observed	Potential Source(s)
Non-metallic exogenous	Various refractory particles, Al_4C_3, etc.	Refractory degradation, potroom metal, refractory/metal reactions
Non-metallic - insitu	M_gO, Al_2O_3 films clusters, and dispersoids; $M_gAl_2O_4$ films and clusters	Melting, alloying: metal transfer turbulence
Homogeneous Halide Salts	M_gCl_2-NaCl-$CaCl_2$, etc.	Poor separation of fluxing reaction products
Particle/Salt	$MgCl_2$-NaCl-$CaCl_2$/MgO, etc.	Salt generated during chlorine fluxing of magnesium containing alloys, filter and metal handling system releases

Table 1 Inclusion Types in Molten Aluminum

The observed inclusion particle sizes in unprocessed aluminum melts range from M_gO dispersoids of a few microns in size to Al_2O_3 films and clusters that extend several millimeters. Most products require the particle size of an inclusion population to be less than 50 , at concentration on the order of several volume parts per million.

The intent of this paper is to consider the four active mechanisms for removing particulates from molten aluminum. These are sedimentation, floatation, cake-mode filtration, and deep bed filtration. For each case the important parameters, pertinent mathematical expressions, and practical significance will be discussed.

Sedimentation

Non-metallic inclusion removal techniques are predicated on body force driven separation methods and filtration. Body force separation entails simple gravity sedimentation, whereas floatation employs a sparging gas, or a combination of both. Gravity sedimentation methods are practically limited to inclusion sizes greater than 100μ, because of inordinately low particle terminal velocities and high drag forces that are encountered. This is illustrated for an aluminum/magnesia system through a simple analysis deriving from Stoke's Law (at 1g force):

$$v_\infty = \frac{2}{9}\,\frac{R_i^2(\rho_s-\rho_1)g}{\eta_{Al}}, \qquad (1)$$

Where R_i is the inclusion radius, ρ_s is the inclusion density, ρM_gO = 3.65 gm-cm^{-3}; ρ_1 = aluminum density $\rho_{Al}(1)$ = 2.32 gm-cm^{-3}; η_{Al} the viscosity of aluminum, η_{Al} = 1.2 x 10^{-2} gm-cm^{-1}-sec^{-1} and v_∞ the terminal velocity of the particle in cgs units. This analysis is valid only for spherical particles in creeping flow ($Re_s < 1$). Since:

$$Re_s = \frac{2\, v_\infty R_i \rho e}{\eta} \quad 1, \qquad (2)$$

and v_∞ will be of the form:

$$v_\infty = KR^2, \qquad (3)$$

where K is a proportionality constant determined by evaluating equation (1).

Substitution of equation (3) into equation (2) yields:

$$R_i,\max = \left(\frac{Re\,\eta_l}{2K\rho_l}\right)^{1/3} \qquad (4)$$

Equation (4) is a first approximation of the maximum inclusion radius for which this analysis is valid (which can be improved by iterative techniques). Substituting for values of K and Re→1, it is found that the analysis is approximately valid for inclusion particle diameters less than 100μ. The terminal velocities and estimated separation times for a particle size range of 1μ to 100μ are tabulated in Table 2:

Particle Diameter	v (cm-sec^{-1})	Separation Time (minutes) for 36 in. Deep Bath
1μ	$6x10^{-5}$(cm-sec^{-1})	$1x5x10^{6}$ (min.)
10μ	$6x10^{-3}$	$1.5x10^{4}$
50μ	0.15	$6.1x10^{2}$
75μ	0.34	$2.6x10^{2}$
10Cμ	0.60	2.52

TABLE 2

These results are plotted in Figure 3. It can be seen that sedimentation as a practically viable separation method is limited to inclusion diameters greater than 90μ, for those operations limited to inverval separation times of 30 minutes or less. Extension of the separation time will obviously result in the sedimentation of correspondingly smaller particles.

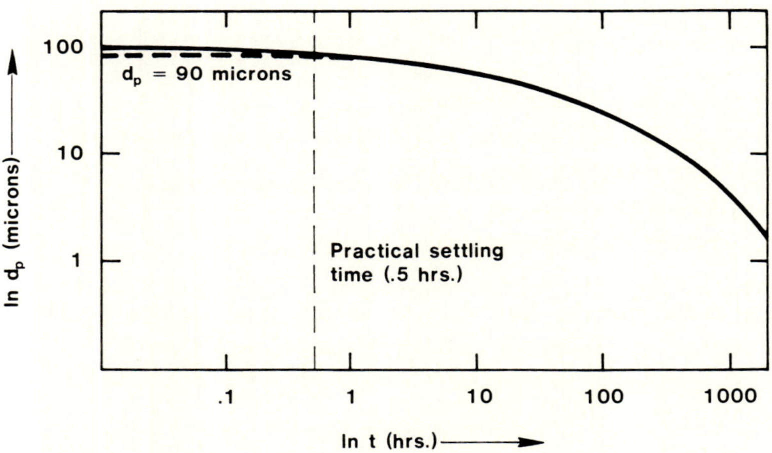

Figure 3 - Particle Separation Time-Sedimentation

Floatation

Particle floatation mechanisms have been comprehensively investigated by Szekely[2] and found to be a result of two elementary capture operations: inertial impaction and peripheral interception. Inertial impaction occurs when the inertia of an inclusion particle exceeds that of a local fluid volume element resulting in departure from fluid flow streamlines around a rising gas bubble. This mechanism, not unlike that responsible for particle transport in deep bed filtration, results in the impaction of an inclusion on the gas bubble surface. If attachment to the bubble occurs and viscous shear forces do not cause detachment, the inclusion is separated from the melt by floatation. Since inertial impaction is gas bubble and inclusion diameter dependent, Szekely concludes that particles larger than 80μ can be removed by these means with bubble diameters as great as 1.0mm.

The second mechanism of particle floatation, peripheral inception, is predicated on problematic equatorial contact of an inclusion and rising gas bubble. An analytical expression relating the fraction of inclusions removed to inclusion radii and bubble diameter (for spherical particles) is:

$$\eta = \exp\ (1.0 \times 10^4\ G\ r_i/r_b^2) \qquad (5)$$

Where: η = percent inclusions removed

G = specific gas sparging rate (volume flowrate gas per unit mass treatment zone, m^3/Kg)

r_i = inclusion radius, m

r_b = bubble equivalent radius, m

For typical normalized gas sparging rates used in single stage in-line sparging devices (corrected for density reduction due to thermal effects), an expression is obtained for 100% removal of inclusion particles:

$$d_p = .222\ d_b^2 \qquad (6)$$

Where: d_p = inclusion diameter, μ

d_b = sparging bubble diameter, mm

Equation (6) is evaluated for $0 \leq d_b \leq$ 20 mm and plotted in Figure 4.

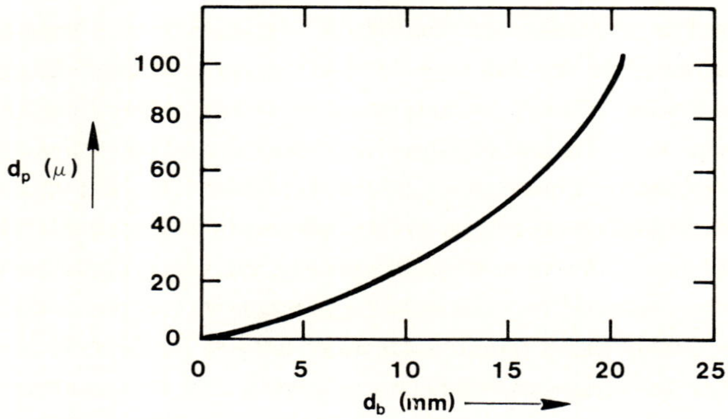

Figure 4 - Inclusion Removal by Floatation

Since the bubble size commonly encountered in back mix type in-line sparging devices is often as large as 15 μ, it can be concluded from Figure 4 that reliable separation of inclusions greater than 30-40μ may occur by floatation. This size range is obviously particle constitution and morphology dependent, but this analysis is in reasonable agreement with practical observation. The reader should recognize that the analysis is based on one hundred percent inclusion removal in a single stage device. Obviously, multiple staging would lower the inclusion fraction removal loading per stage and permit the use of a lower η in developing equation (6). In this manner, the overall floatation efficiency of the multi stage process with respect to inclusion diameter would increase. The efficiency relative to inclusion fraction removed of a diameter less than 40μ would asymptotically approach one hundred percent for n stages, however.

Many high volume aluminum products require the removal inclusions significantly smaller than 40μ in diameter. Failures in beverage container sheet can occur at an inclusion particle size of approximately ten percent of the residual container gage which corresponds to effective inclusion diameters substantially lower than 20μ.[5] Inclusion removal by cake mode and deep bed filtration methods are effectively used in this inclusion size range.

Cake Mode Filtration

Both cake mode and deep bed filtration processes are in widespread use throughout the aluminum industry for melt clarification. Commercial examples of cake mode filters include the use of reticulated ceramic foam plates[6], and consolidated rigid media. Deep bed filters are typified by the Aluminum Company of America's Alcoa 94, Alcoa 181[7], and Alcoa 528 processes; and a tubular cartridge filter comprised of rigid media[8]. It is particularly important to note that although commercial aluminum filters are categorized by the principal operative filtration mode (e.g.: Alcoa 94 process-deep bed), bench scale modeling has demonstrated[9,10] that a secondary filtration mode can indeed occur within certain operating parameter regimes. For example, filtration modeling and commercial scale operating data indicate that it is more probable for a deep bed process to exhibit cake mode behavior than the converse. This can be largely attributed to dissimilar flow dynamics and requisite conditions for particle capture on cake mode filter media[11].

Cake Mode Filtration -- Model Development

A rudimentary example of aluminum melt clarification by cake mode filtration is the fiberglass screen or "trough sock" used for the removing of large inclusions. Here, intrusion of the silica screen by inclusions cannot occur if particle capture is to take place. Referring to Figure 5, $d_i/d_o > 1$ is an essential condition for incipient cake mode filtration:

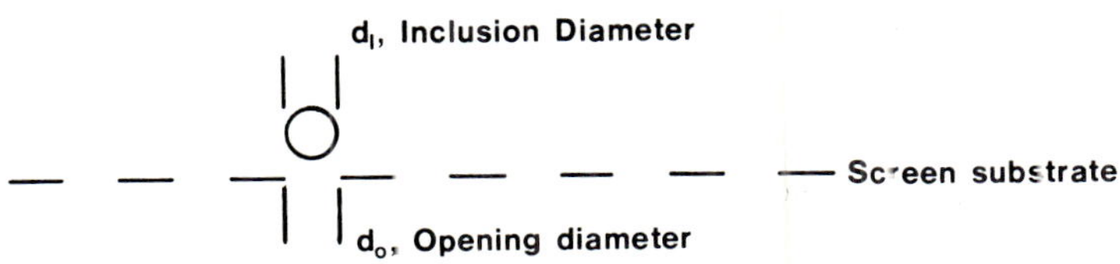

Figure 5 - Mechanical Inclusion Capture

As a layer of separated solids is deposited at the interface of the screen substrate, the effective opening diameter is progressively decreased. Smaller diameter inclusion particles can now be captured in subsequent layers of separated solids, thus forming a cake. The cake effective opening diameter and void fraction decreases, while cake tortuosity increases resulting in improved particle capture. This process is illustrated in Figure 6.

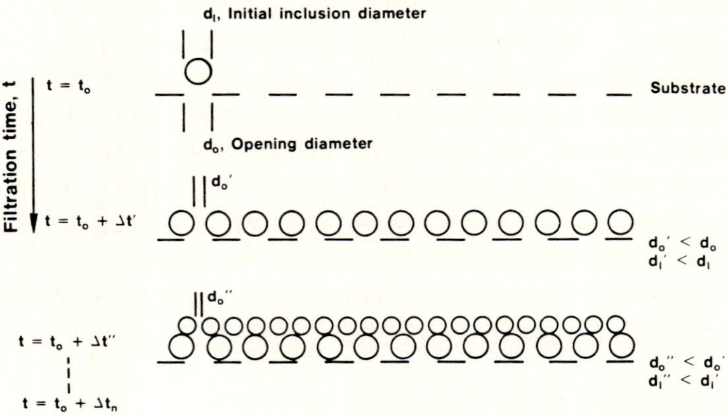

Figure 6 - Progressive Cake Development

The resulting filter cake is characterized by α, the specific cake resistance, and has the dimensions $L\text{-}M^{-1}$. This a particularly important parameter, as it can be used to obtain functional relationships of filtration pressure changes with respect to both filtration time and particle capture efficiency. Intuitively, as the value for α increases in constant volume flow rate filtration (physically corresponding to cake compression) the driving force or filtration pressure also increases; it will soon be shown that the net effect of this is to increase particle capture efficiency. In incompressible cake constant pressure filtration, α may not increase significantly, however the cake thickness builds thereby reducing metal volume flow rate. A common physical example of this behavior is the clogging of an automotive intake air filter, and subsequent reduction of combustion air flow rate and engine performance loss.

The authors have recently developed a model describing the constant rate cake mode filtration of metals. Using this model, inclusion removal characteristics can be related to system parameters such as filtration time, pressure, and entry concentration. The rigorous derivation of this model will be made in a subsequent publication, and only the results will be presented here.

A fundamental study of cake mode filtration must begin with an approximation for α. Such an approximation is made by conducting constant pressure filtration experiments in melts where cake compression effects can be controlled. Although some cake compaction and void fraction reduction will occur, the effect is typically small. Filtrate volume flow rate with respect to time data are collected from constant pressure experiments and graphically used to derive α, for a particular melt-inclusion system. The relationship used by the authors is predicated on D'Arcy's Law at low Reynolds number flow. A concise derivation follows:

D'Arcy's Law is applied to the filter cake and yields:

$$\alpha = \frac{A^2 \Delta P}{\sigma V_o \eta} \frac{dt}{dV_o} \tag{7}$$

Where: A = plan area of cake

ΔP = pressure loss through the cake

σ = dV_c/dQ, the rate change in cake volume with respect to metal volume flow rate

V_o = the total metal volume used through time, t, during the experiment

η = metal viscosity

Equation (7) could be integrated directly to yield an expression relating V_o and t, however fluid flow resistance is only implicit in α. In reality, total fluid flow resistance is a serial resistance of the cake and supporting substrate (media). While integration of (7) may be valid for systems with very low substrate resistance, (e.g.: fiberglass trough socks), most practical filtration pressure drop data represents serial flow, thus, (7) will be modified to include substrate resistance, R_s:

$$R_s = r_s \, dA \tag{8}$$

Where r_s = local substrate resistance

A = substrate plan area, essentially equivalent to A in equation (7)

Insertion of (8) into (7) and rearranging yields:

$$\frac{dt}{dV_o} = \frac{\eta(\sigma V_o \alpha + R_s A)}{A^2 \Delta P} \tag{9}$$

Since the filtration pressure is constant, equation (9) is an ordinary differential equation and is integrated to provide a relationship for t and V_o:

$$t = \frac{\eta \sigma V_o^{\,2}}{2\Delta P A^2} + \frac{\eta R_s V_o}{\Delta P A} \tag{10}$$

Equation (10) may be divided by V_o, and fitted to data obtained from constant pressure experiments. The resulting linear equation is of the Y-intercept form, $t/V_o = KV_o + b$, where $K = \frac{\eta \sigma V_o \alpha}{2\Delta P A^2}$ and $b = \frac{\eta R_s}{\Delta P A}$. Through substitution of experimental parameters for η, V_o, ΔP, and A; values for α and R_s can be obtained from K and b respectively for a characteristic filtrate metal implicit in σ. This procedure can be performed graphically or by a regression analysis curve fit technique. If the cake is truly incompressible, a straight line will result.

A deviation from linearity of α will result if the cake is highly compressible. Melt suspensions comprised largely of particulate M_gO, Al_4C_3, and SiC would be expected to exhibit linearity (except for solids migration effects in the cake) as previously mentioned.

Equation (10) is restricted to use as an investigative tool, and does not describe constant rate melt filtration. Most commercial aluminum cake mode filtration systems are operated at constant rate, and permit ΔP to increase to the maximum available aluminostatic driving force or metal head capacity of the system. Additionally, inclusion types such as M_gO and Al_2O_3 films can form compressible cakes and invalidate equation (10).

A model has been developed by the authors to describe constant rate cake mode melt filtration. In this case, analytical expressions are developed based on:

1) constant rate-variable pressure conditions
2) compressible cake and variable α

A mass balance is constructed over the filter cake to yield:

$$Vm = \frac{M_i(1-f_s/f_{sc})}{\rho f_s} \tag{11}$$

Where: Vm = normalized volume of filtered metal (per unit area of cake)
M_i = normalized mass of inclusions in cake
f_s = fraction inclusions in unfiltered metal
f_{sc} = fraction inclusions in filter cake

Since cake compression results in a variable α, an expression is developed to express an apparent cake resistance, α; as a function of pressure dissipation in the cake. Substituting this relationship into equation (11) results in:

$$\int_0^{\Delta pc} {}^{d}p/\alpha' = \frac{Q_N^2 \eta \rho f_s t}{(1-f_s/f_{sc})} \tag{12}$$

Where: Q_N = normalized volume flow rate of metal relative to filter area
P_c = pressure dissipated in total cake layer

The filter cake void fraction will decrease as filtration progresses, and α' must be expressed as some functional relationship with Δp_c. From empirical observation, it is reasonable to expect a power function relationship of the form:

$$\alpha' = \beta + \alpha (P_c)^n \quad (13)$$

Where α' = bulk specific cake resistance as obtained through constant pressure experiments.

P_c = Pressure associated with the cake

β = constant relating to filter media flow resistance

n = empirical constant

At the onset of filtration, P_c will have a value of zero. Consequently, the value of the integrand in equation (12), d_p/α', will also be zero. To avoid this complication, the constant term β is introduced that will physically relate to media flow resistance. Since β always assumes a positive finite value, α' will always be positive. In the most rigorous sense, a third flow resistance will be presented by the reduction in void fraction at the cake/media interface. This additional resistance will be an assumed component of β and not dealt with separately.

Substituting equation (13) into (12), integrating, and evaluating over the limits 0 to ΔP_c yields:

$$(\Delta P_c)^{(1-n)} = \frac{Q_N^2 \eta \rho (\alpha+\beta)\ f_s t}{(1-f_s/f_{sc}) gc} (1-n) \quad (14)$$

The insertion of g_c preserves dimensional consistency:

$Qn = L^3\text{-}L^{-2}\text{-}t^{-1}$

$\eta = m\text{-}L^{-1}\text{-}t^{-1}$

$\rho = m\text{-}L^{-3}$

fs, fsc - dimensionless

α - $L\text{-}m^{-1}$

Since:

$$f_{sm} = f_s - f_{sc} \quad (15)$$

Where: f_{sm} = fraction solids in filtered metal

Equation (14) can now be used as an analytical expression characterizing filtration performance and indeed demonstrates the exponential relationship between Δpc and t, characteristic of constant rate cake mode filtration. Figure 7 depicts this general behavior.

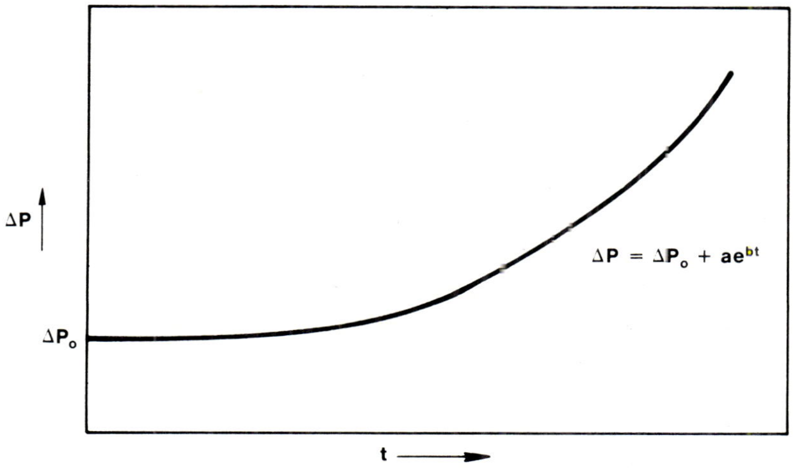

Figure 7 - Typical Pressure-Time Relationship for Constant Rate Cake Mode Filtration

Deep Bed Filtration

Cake mode filtration is particularly useful in aluminum melt clarification when:

a) the inclusion concentration is high.

b) ingot plant floor space is limited.

c) transient performance characteristic can be tolerated.

d) the capture of extremely small inclusions is not required.

The inclusion concentration of the melt is a particularly important consideration, as it gives rise to expedient cake development and improved filtration. As the $(1-f_s/f_{sc})$ term in equation (14) suggests, however, an inordinantly high solid fraction can lead to unacceptably rapid filter head build-up, this warranting pretreatment clarification methods such as sedimentation and floatation.

Deep bed or depth filtration of aluminum melts, on the other hand, operate most effectively when inclusion mean effective diameters are low, and the concentration of suspended inclusions is relatively dilute. Tien[12] cites the following desirable volume concentration regimes for cake mode and deep bed filtration in aqueous systems:

$f_{inc} > 2000$ ppm - cake mode

$f_{inc} < 500$ ppm - deep bed

It is believed that these regimes are approximately correct for aluminum melt clarification, depending on inclusion size and morphology. Inclusion size is an extremely important parameter in the correct operation cf deep bed filters since inclusion intrusion of the bed by must occur for correct performance. Cake formation is undesirable since it results in exponential head increase and necessitates frequent bed replacement. Deep bed filters are particulary useful in capturing of small inclusions ($d_i < 20\mu$) because their effectiveness is predicated on surface effects that are enhanced by high media surface area bed filter.

Inclusion capture is considered to be a result of two sequential events: transport of an inclusion to capture sites on the filter media surface, and subsequent attachment of the particles to these sites. In general, particle transport can occur by impingement, interception, sedimentation, diffusional, and hydrodynamic effects. Particle attachment can be a result of forces developed through pressure, chemical, or Van der Waal effects. The relative dominance of each mechanism is a function of particulate type and size, fluid approach velocity, as well as temperature and media characteristics. It is believed that impingement is a dominant transport mechanism in aluminum melt filtration, and pressure/Van der Waal forces are responsible for particle attachment. Excellent descriptions of potential capture mechanisms in aqueous systems are provided in the literature by Ives,[13] Tien,[14] and Herzig,[15] et. al.; and investigations are currently underway to improve the understanding of deep bed capture mechanisms in aluminum melts.

Accurate mathematic representations of deep bed filtration processes can be powerful system design and use optimization tools. Two types of mathematical models are commonly used to describe deep bed filtration:

1) parametric or functional models based on fluid flow mechanics through porous media. One objective of these models is to develop relationships between filter media and operating parameters, such as filtration pressure and void fraction.

2) Phenomenological or kinetic models predicated on kinetic rate expressions that are developed for a particular filtration system. Kinetic models are useful in predicting filter performance.

The proper treatment and form of differential equations used to develop kinetic models is often made possible through physical insight into a particular filtration system. This understanding can be obtained from knowledge of the elementary particle capture processes gained through the correct functional models. One example of this interaction is the modification of a kinetic model to include bed filter release effect terms. The rigorous development of functional models in molten aluminum filtration systems is particularly difficult, however, due to the intermediate Reynolds number flow ranges encountered. This will be reviewed later.

The literature is abundant with phenomenological models of various investigators in low temperature systems. Of particular interest are the discussions by Ives,[16] Herzig,[15] et. al., Tien,[17] et. al., and Kraj.[18] Herzig and Tien provides simplifying transformations that result in more easily solved ordinary differential equations.

The first application of kinetic models to melt deep bed filtration was made by Apelian, Mutharasan, and O'Malley,[19] and Apelian and Mutharasan.[20] Development of this model begins by constructing a mass balance over a differential volume element of the filter bed:

$$\frac{\partial}{\partial t} (\sigma + \varepsilon C) + V_m \frac{\partial C}{\partial Z} = 0 \qquad (16)$$

Where: σ = normalized volume of retained inclusions in element
ε = bed void fraction
V_m = melt superficial velocity
C = volume concentration of inclusions
Z = flow direction dimension

The kinetic rate expression relating the change of retained inclusion volume with respect to time is:

$$\frac{\partial \sigma}{\partial \tau} = KC^n \qquad (17)$$

Where: τ = normalized time resulting from a transformation used with (16)

K = kinetic parameter

n = rate expression order

Equations (16) and (17) result in a first order partial differential equation that, in conjunction with dimensionless parameters for time, distance, and K; assignment of $\eta = 1$; and use of appropriate boundary conditions, is solved to yield:

$$C_L = Co\ e^{\frac{-KoL}{V_m}} \tag{18}$$

Where: C_L = inclusion conetration at the volume element outlet (Z=L)

Co = inlet inclusion concentration

Ko = the kinetic parameter coefficient, related to K by $K = Ko\ [1 - \frac{\sigma}{\sigma_m}]$ where σ_m is σ at bed saturation

L = volume element length

The group Ko/V_m can be conveniently manipulated through designation as λ, the filter coefficient . Thus, the applicable expression describing short term filter behavior becomes:

$$C_L = Co\ \ e^{-\lambda L} \tag{19}$$

A macroscopic application of equation (19) will begin by considering a deep bed filter that has been partitioned into five discrete volume elements, as indicated in Figure 8.

The fraction of inclusions removed (at constant particle diameter), f_{rem}, by layer 1 is:

$$f_{rem} = \frac{Co - C_{z=1}}{Co} = 1 \quad e^{\lambda L} \tag{20}$$

When $L = 1/\lambda$ (equation 20) reduces to:

$$f_{rem} = 1 - e^{-1} = \ 632 \tag{21}$$

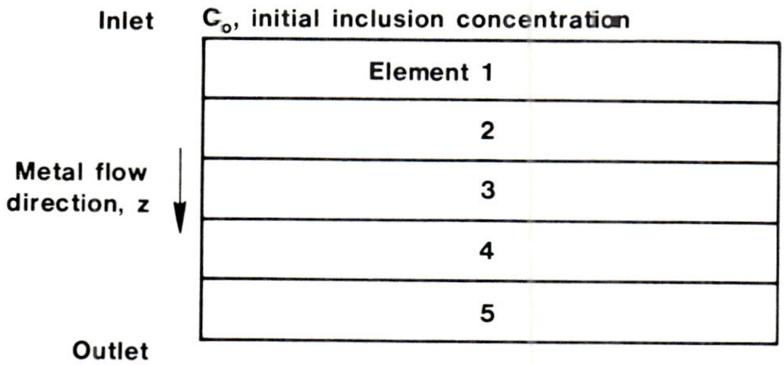

Figure 8 - Finite Element Deep Bed Filter

Consequently, each volume element or layer in Figure 8 will remove 63.2 percent of the incoming inclusions of constant particle size. A differential expression for the local inclusion concentration at any volume element outlet is:

$$dC = (1-.368\ Co)^{Z}\ dZ \tag{22}$$

Where Z increases is the flow direction

Integration of equation (23) and application of the appropriate boundary conditions gives an expression for inclusion concentration, C, at any point along the length (depth) of the filter:

$$C = (1 - .632\ Co)\ (.368)^{Z-1} \tag{23}$$

Equation (23) is evaluated at the outlet of each serial volume element to result in the following concentration (with Co = 1).

Z	C
0	1
1	.368
2	.135
3	.049
4	.018
5	.007

Plotting these results yield the concentration profile in Figure 9.

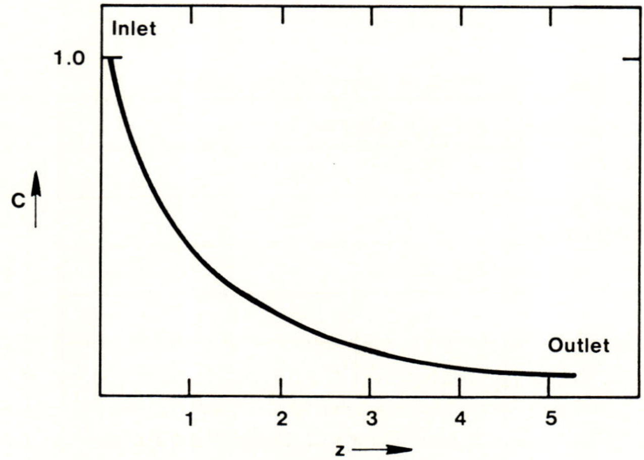

Figure 9 - Inclusion Concentration Profile in Deep Bed Filter According to Equation (19)

The profile in Figure 9 is considered valid for short term filter behavior, i.e.: when the bed filter is new. Equation (19) does not, however, account for the migration of captured solids in the filter bed which later become detached. Systematic detachment of inclusions can result in a break-through" phenomenon or an emergence of a continual inclusion efflux at the filter outlet. Instantaneous detachment may give rise to sporadic slug releases which are of equally serious consequene. A comprehensive deep bed filtration theory must, then, provide for a description cf such conditions giving rise to release events.

The acknowledgement of release events is a controversial subject in the literature and appears to be strongly a function of a particular investigator's system. Some investigators, for example, claim that filter releases do not occur under any conditions, while others maintain that releases may result from interruption or a step increase of fluid flow through the filter. Another belief is that filter releases occur when the filtrate interstitial velocity exceeds a threshold limit. Experimental evidence strongly indicates that the latter is the case for the long term behavior of aluminum melt filters.

Factors leading to bed filter releases can be quantatively described as follows: The melt local interstitial velocity in a bed filter is

$$v_i = \frac{W}{\rho A \varepsilon_L} \qquad (24)$$

Where v_i = interstitial metal velocity
W = mass flow rate of the melt
A = plan area of media
ρ = melt density
ε_L = local media void fraction

It must be pointed out that ε_L is a dynamic parameter that necessarily results from inclusion capture in the filter bed. Thus, as a filter ages, capture occurs at the expense of ε_L and the local interstitial metal velocity increases. A consequence of increasing v_i is that shear or detachment forces on a captured inclusion particle also increase. The effects of viscous drag are illustrated by equation (25) and Figure 10.

$$\tau_{r,z} = \eta \frac{dv_r}{d_r} R \cos \Theta \qquad (25)$$

Where: τ = shear stress in bulk flow direction, Z
η = viscosity of melt
$\frac{dv_r}{d_r}$ = melt velocity gradient, direct r
Θ = inclusion orientation angle, defined in Figure 10.

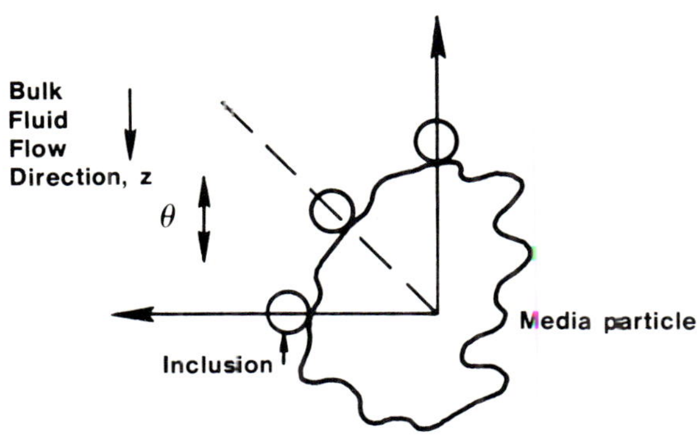

Figure 10 - Viscous Drag on Captured Inclusion

As can be seen, inclusion detachment is a function of orientation, as values of Θ 0° result in maximum shear forces, while as Θ 90°, shear forces vanish.

Attempts are being made to analytically derive parametric expressions to predict the critical interstitial metal velocities at the onset of detachment. This is a difficult task, since the development of a radial metal velocity profile cannot be readily obtained by analytical methods. The problem arises with the regime in which metal flow typically occurs in aluminum bed filters. A typical Reynolds number range is 8-to-50, where Re ib the porous media is defined as:

$$Re = \frac{4 \, Dg \, \rho}{6(1-\varepsilon)\eta} \tag{26}$$

Where: Re = bed Reynolds number
Dg = diameter of filter grain

Simplifications to the Navier-Stokes equation cannot be made through inviscid (creeping flow) or non-inertial (potential flow) considerations at this intermediate Reynolds number regime.

A method of improving the kinetic model is to include a release term in the kinetic rate expression. The differential expression would be of the form:

$$\frac{\partial \sigma}{\partial \tau} = K_c C - K_r \tag{27}$$

Where: K_c = Kinetic parameter association with capture
K_r = Kinetic parameter associated with release

The important concept of bed filter releases will be extensively dealt with in a future publication.

Conclusions

In this paper, each of the four particle removal techniques has been considered in detail. From these discussions, the following salient points have been extracted.

I. Sedimentation

- o Gravity driven sedimentation in aluminum can be effective or inclusion > 90μ.

II. Floatation

- o Most single stage back mix sparging devices cannot reliably remove inclusion <30-40μ.

III. Cake Mode Filtration

- o The specific cake resistance (α) increases with filtration pressure and subsequently results in improved capture efficiency in constant rate filtration.
- o An expression has been derived to interpret constant rate filtration.
- o An expression has been derived to interpret constant rate filtration data from incompressible cake experiments.
- o An apparent cake resistance α′ has been derived to account for pressure dissipation effects through the cake.
- o Filtration pressure change has been shown to be an exponential function of time in constant rate filtration.
- o The particle capture characteristics of cake mode filters are transient and expressions have been derived to describe them analytically.

IV. Deep Bed Filtration

- o Deep bed filters can also exhibit cake mode behavior, depending on operating conditions.
- o Optimizing high media surface area effects make deep bed filters particularly effective at removing dilute inclusion concentrations of a size ≤ 20μ.
- o A kinetic rate expression has been applied to a partitioned deep bed to demonstrate short term filter behavior.
- o Several factors affecting intermittent inclusion releases from deep bed filters have been described.

References

1. H. G. Reavis, Alcoa internal correspondence, 1983.

2. R. E. Miller, L. C. Blayden, M. Bruno, C. E. Brooks, "In Line Fumeless Magnesium Removal," Light Metals, 1978.

3. R. D. Blackburn, "Sampling Molten Aluminum for Non-Metallic Inclusions," Light Metals 1983

4. A. G. Szekely, "The Removal of Solid Particles from Molten Aluminum in the SNIFF Process," Trans AIME, Vol. 7B, June 1976.

5. H. G. Reavis, Alcoa internal correspondence, 1982

6. F. R. Mollard, N. Davidson, Modern Casting, Vol. 69, No. 3, 1979.

7. K. J. Brondyke, P. D. Hess, Trans AIME, Vol. 230, 1964

8. M. C. Mangalick, Light Metal Age, Feb. 1973

9. R. Mutharasan, D. Apelian, C. Romanowski, Light Metals, 1981.

10. Unpublished work, Alcoa, 1982-83.

11. D. Apelian, R. Mutharasan, C. Romanowski, R. E. Miller, C. E. Eckert, Light Metals, 1982

12. C. Tien, A. C. Payatakes, Journal of Alche, September 1979.

13. K. J. Ives, Proceedings of the NATO Advanced Study Institute, July 1973, pp 183-201.

14. K. J. Ives, Filtration and Separation, November-December 1970.

15. J. P. Herzig, D. M. LeClerc, P. LeGoff, Industrial and Engineering Chemistry, Vol. 62 No. 5, May 1970.

16. K. J. Ives, Proceedings of the NATO Advanced Study Institute, July 1978, pp 203-224.

17. C. Tien, R. M. Rurian, H. Penose, Journal of AlChe, May 1979.

18. W. Kraj, Bul. Acad. Pol. Sci Ser. Tech, Vol. 14, No. 8, 1966.

19. D. Apelian, R. Mutharasan, R. O'Malley, "Depth Filtration of Liquid Metals" submitted to Trans AIME, 1981.

20. D. Apelian, R. Mutharasan, Journal of Metals, Vol. 32, Sept. 1980.

21. D. M. Mintz, "Modern Theory of Filtration," INT Water Supply Congress, 1966.

22. J. R. Baylis, Journal of the American Water Works Assn., 1937.

23. D. F. Eben, Journal of the American Water Works Assn., July 1960.

MOLTEN ALUMINUM QUALITY MEASURED WITH REYNOLDS 4M™ SYSTEM

Thomas L. Mansfield

Metallurgy Laboratory
Reynolds Metals Company
Richmond, Virginia 23219
USA

The Reynolds 4M™ System was used to evaluate three levels of molten aluminum quality. Several solid state tests and examinations were performed on the semi-finished product to determine what the 4M™ System was detecting in the molten state. A direct correlation of molten metal results to solid state results confirm that the 4M™ System is indeed a valuable addition to existing quality control methods. The production 4M™ System, tests performed, and results will be discussed.

Introduction

Until the development of the Reynolds 4M™ System (Mansfield Molten Metal Monitor), ultrasonic inspection within Reynolds operations had been confined to solid materials, e. g., plate, extrusions, ingot, and sheet.

The Reynolds Production 4M™ System offers a means for determining the quality of molten aluminum before the casting operation on a routine basis. It is, clearly, a valuable addition to existing quality control methods. This pulse-echo ultrasonic system allows for continuous monitoring of the quality of molten metal processing and/or cleaning operations such as fluxing practices, holding time, and filtering methods.

In the cited literature, Pitcher and Young[1-2] and others [3] show that ultrasonic techniques can be used successfully to determine the presence of nonmetallics, precipitation effects, and velocity of sound in molten aluminum.

For practical applications, we found that a departure from the conventional means of ultrasonic inspection was necessary. Our earlier papers[4,5] describe Reynolds initial work in this new field. Our involvement with the prototype system soon led to the development of a production system.

Production Test Setup

For the author's work, a new approach was taken. A titanium probe was designed and constructed that greatly simplifies the procedures required. The design includes a built-in cavity with top and bottom surfaces machined flat and parallel. The bottom surface (reflector) provides a means for continuously monitoring sound in the melt. This design allows molten metal to fill up and/or flow through the cavity. The patented 4M™ probe is shown in Figure 1.

The success of any molten aluminum inspection method requires that the immersed portion of the probe be "wetted" by the molten metal. Based upon the author's research, a means of coupling ultrasonic energy into molten aluminum quickly and very effectively was developed.[6] After this initial "wetting"of the immersed portion of the probe, no additional treatment is required.

The probe can be moved about, held stationary, raised, and lowered without the occurrence of any misalignment. It is a very stable probe -- thermally, ultrasonically, and physically. Probe "life" is dependent mainly on the velocity of molten aluminum flowing through the probe cavity. Average probe "life" for most of our applications extends from 1 to 3 million lbs. of aluminum inspected before replacement of the probe insert is required. Replacement of the spent probe insert is easily achieved and saves the need for replacement of the water cooling jacket.

Two ultrasonic measurements can be made simultaneously with the production 4M™ System without making any changes in the test setup. They are measurements that do not require any special preparation prior to inspection. Usually, a 15 second waiting period is all that is needed after the probe is immersed in the melt. The system then measures relative attenuation of the melt and also counts and classifies the number of discontinuities detected. These two measurement capabilities are discussed and described in earlier papers.[4,5]

The Reynolds 4M™ System used for making these measurements is shown in Figure 2. This production system is approximately one-third the size of the prototype and contains a complete data acquisition, control, and processing computer to simplify operation and to provide a hard-copy analysis of the inspection run at the end of the casting operation.

This advanced system has been developed jointly by Reynolds Metals Company and General Research Corporation (formerly Adaptronics, Inc.). It collects data from the probe and continually records analysis results. The system features automatic calibration independent of the operator and evaluation of the probe/transducer quality prior to each run. It is a practical and effective, microprocessor based pulse-echo ultrasonic inspection system featuring state-of-the-art high speed arithmetic analog to digital circuitry, and a self-contained tone-burst pulser/receiver. The operator interfaces with the unit through a front panel keypad and cathode ray tube (CRT). External equipment for printing, plotting and temperature monitoring are controlled by the microprocessor. The unit is designed for portability and is shock-mounted to a special equipment cart. This cart moves about in typical cast house environments with ease and can negotiate small steps.

Figure 3 shows the probe placed into a flowing stream of molten metal in the run condition.

Discussion and Results

Figure 4 shows the real time screen display of the production unit in the RUN mode. Notice that the RUN conditions are continuously displayed. For example in this figure we see that we are indeed in the RUN mode collecting data at the F1 frequency with the probe placed UP stream of a filter. The top trace shows real time discontinuity amplitude vs. time information. The bottom bar graph displays the magnitude of the A_1 and A_2 amplitude signals and the resultant attenuation computed in decibels (dB). Also, the magnitudes of detected discontinuities are displayed by the % discontinuity bar graph.

Figure 5 shows the summary of data collected for a particular run segment. In this IDLE/HOLD mode, the system is

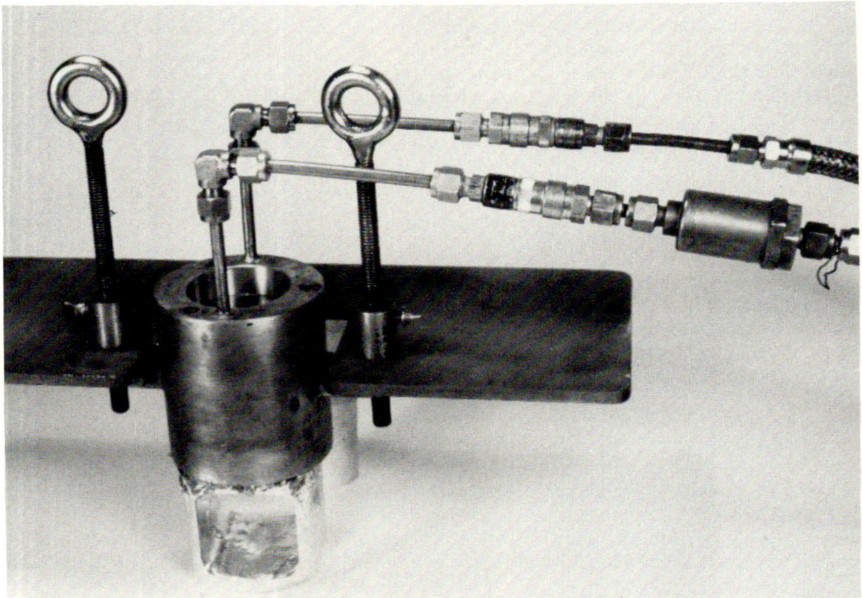

Figure 1 - Reynolds 4M™ Probe

Figure 2 - Reynolds 4M™ System

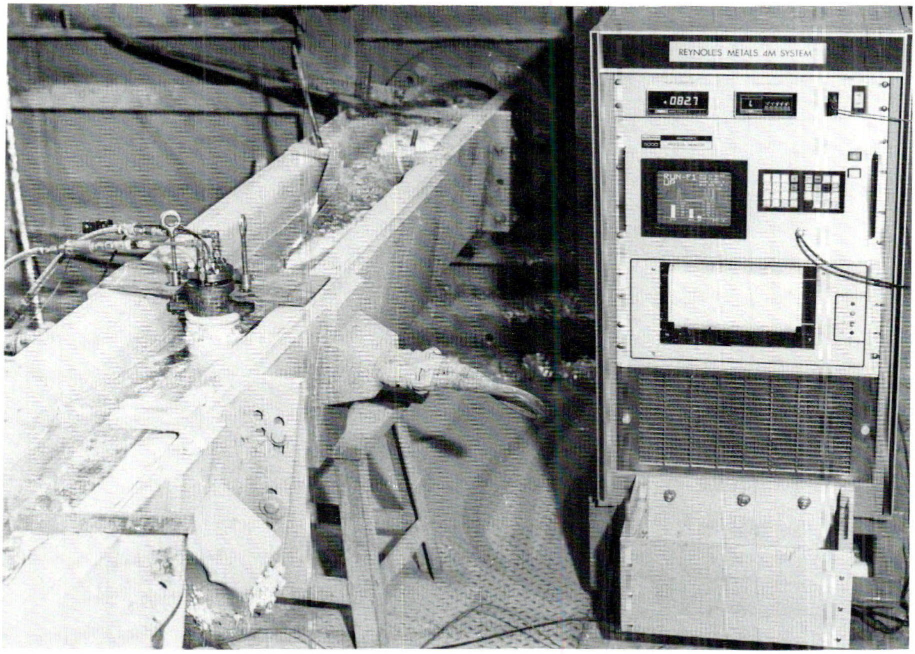

Figure 3 - Probe Placed into a Flowing Stream of Molten Metal in Run Condition.

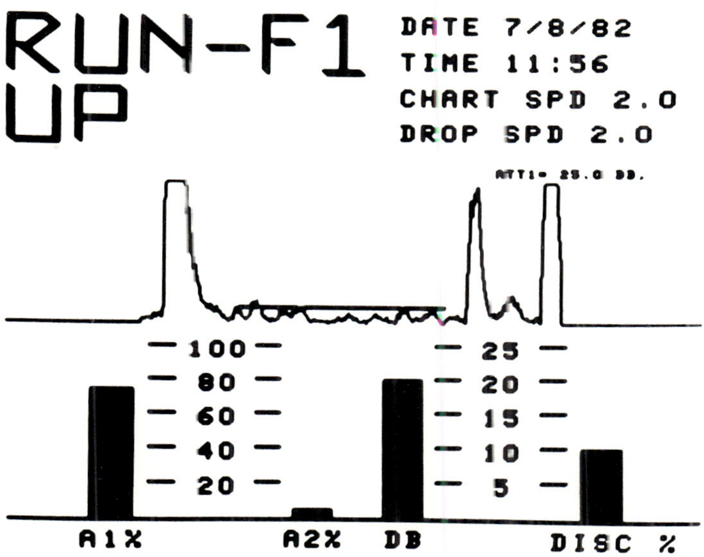

Figure 4 - Real Time Screen Display of the Production Unit in the RUN Mode of Operation.

IDLE/
HOLD

DATE 7/8/82
TIME 11:56
CHART SPD 2.0
DROP SPD 2.0

FREQ 1 UP 7.0 MHZ

AVERAGE = 13.534
STD DEV = 4.155118
HIGH ATTEN = 24.879
LOW ATTEN = 9.842

DISCONTINUITY RESULTS:

5 - 10%	488
10- 15%	39
15- 20%	0
20- 25%	0
25- 100%	0
TOTAL=	527

Figure 5 - IDLE/HOLD Mode - Provides a Summary of Data Collected for a Particular Run Segment

awaiting further instructions from the operator. While in this mode of operation, the system has automatically summarized all data obtained in the previous segment. This information is also printed by the recorder for a permanent record of results as shown in Figure 6. At the end of a casting operation, all "segments" are summarized for final anlaysis. For example, if the system was used to evaluate a filter, both upstream and downstream data would have been collected on several occasions during a casting operation. At the end of a test, the system would summarize all upstream data as well as all downstream data to yield a comparison between the two positions. This type of information can lead to a very efficient evaluation of filters.

Using the Reynolds 4M™ System a successful attempt was made at correlating molten metal results to solid state ultrasonic results and semi-finished product quality.

Several attempts were made at taking "grab" samples with the hope of showing direct correlation to 4M™ results. We had very limited success. Also, we felt that "another grab sampling technique" would just raise additional questions about the validity of the sampling technique. We, therefore, devised a method where actual cast ingot sections could be identified and directly related to the molten metal monitored with the 4M™ Probe. This yields the best possible sample to evaluate.

Thus the discussion of results will center on actual ingot slice evaluations and how they related to 4M™ results from metal that was monitored downstream from a two filter in tandem casting setup. The test was conducted under actual casthouse conditions. This cast metal is the material which would eventually end up being purchased by a customer.

The casting station set up is shown in Figure 7.

Molten metal is poured from ladle 1 and 2 into the transfer trough which feeds the molten metal through a "fiber glass bag." The molten metal then passes into filter A and then into filter B. The metal coming from filter B then goes to the distribution pan at which point the metal is distributed to the D. C. (direct chill) molds where the ingot is cast.

The 4M™ probe was positioned in one "leg" of the distribution pan adjacent to the mold. Thus, the molten metal was monitored just before it went into the mold.

By coordinating the drop speed and the position of the 4M™ probe, we could very accurately mark areas of interest directly on the cast structure. Thus when an event was created or occurred and the 4M™ detected the event, the resulting cast structure material was identified. After the casting practice was completed, the ingot was homogenized. After the homogenization practice, the ingot was sawed into 14" diameter slices 6" to 8" thick at those areas where the 4M™ detected abnormal events taking

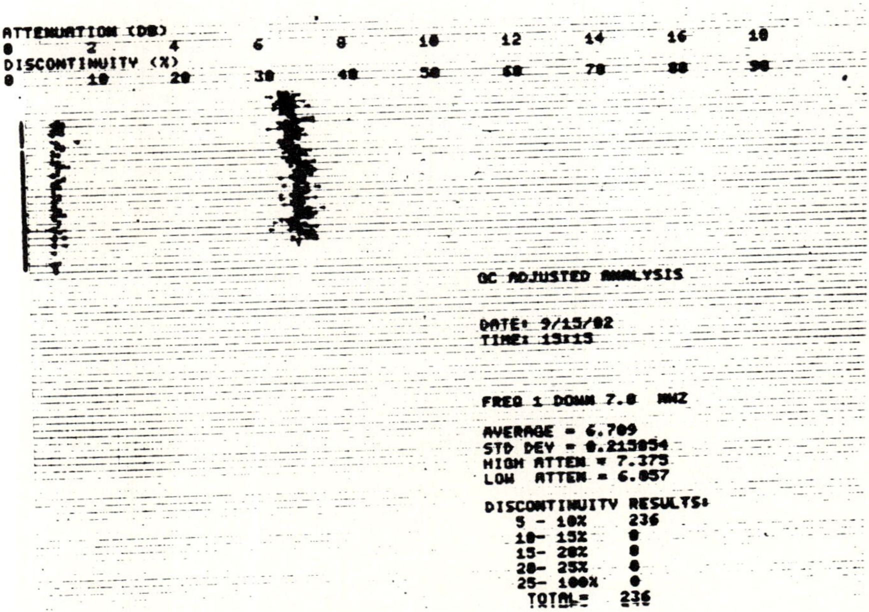

Figure 6 - Hard Copy of Results for Permanent Record

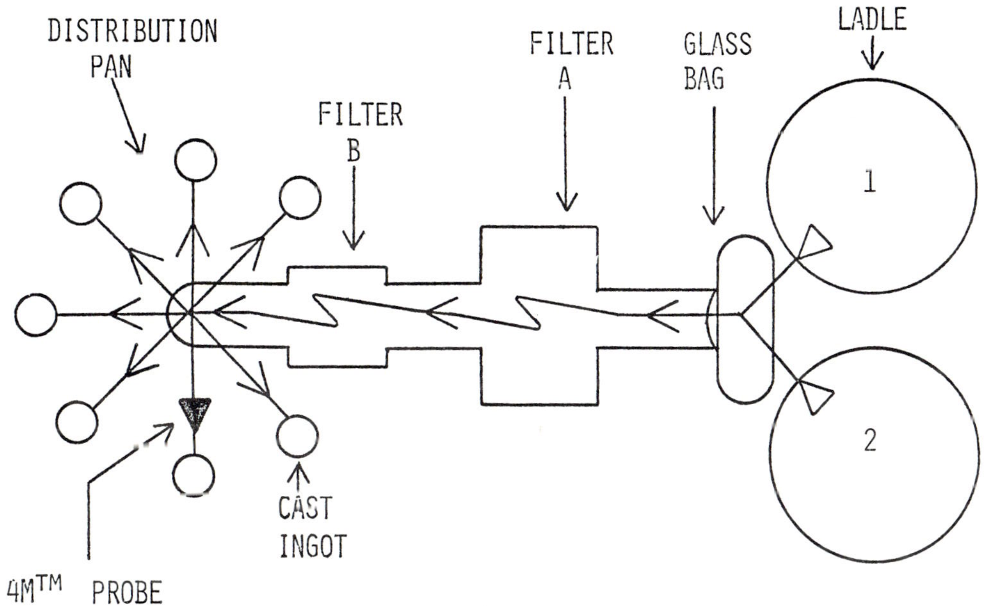

Figure 7 - Casting Station Setup

place. These known events caused the molten metal to display a higher attenuation and/or number of detected discontinuities as recorded by the 4M™ system. The identified ingot slices were then returned to the Metallurgy Laboratory where a thorough examination was performed.

Examples of Practical Effects for Different Molten Metal Conditions:

For this work, three ingot slices were selected for initial examination -- one representing a very high level, one representing an intermediate level, and one representing a lower level of molten metal attenuation and/or discontinuity counts. In each case laboratory techniques have been applied to identify what the 4M™ System was detecting and how it might affect the semi-finished product.

Table I gives the test identifications and the 4M™ results for each test condition. Condition I exhibited very poor ultrasonic molten metal characteristics whereas Condition II exhibited somewhat better ultrasonic characteristics. Condition III represents better ultrasonic characteristics than Condition I or II. However, this condition's characteristics are considered less than desirable.

Table I

Test Identification	Average Attenuation (dB)	Standard Deviation of Attenuation(dB)	Total Counts
Condition I: Filter 'B' Physically Disturbed	21.6	10.9	112
Condition II: Glass bag above filter 'A' was broken prematurely on start of ladle #2. Exogenous material from bag entered filter 'A'.	15.0	2.4	24
Condition III: Filter 'A' adjusted for known molten metal condition as determined by 4M™ System.	7.8	0.3	22

The ingot slices representing Conditions I, II, and III were then diamond machined on all surfaces with the faces being machined flat and parallel. From this material, seven tests were performed to characterize the metal quality level. These tests are as follows:

1 - Ultrasonic examination

2 - Nitrogen Carrier Fusion (NCF) Hydrogen Determination

3 - Fast Neutron Activation - O_2 Content

4 - Fabrication into Sheet for Evaluation by:

a) Caustic Etch Smut Pattern
b) Buffed and Bright Dipped Analysis

5 - Analysis of Filtrate from Chemical Dissolution of Ingot Slice Samples

6 - Metallographic Examination of Fracture Specimens Macro, Microscopic and Scanning Electron Microscope (SEM)

7 - Metallography and SEM of Cross Sections Through Discontinuity Areas

Results from the tests are as follows:

1. Ultrasonic Examination

The immersion method of ultrasonic inspection was employed to determine the solid state quality characteristics of the ingot slices.[7] The sensitivity of the test was such that discontinuities equal to or greater than 1/64" ASTM flat bottom hole reference standards would be detected.

A 1/2" diameter, 5MHz search unit, using a scan index of 0.2" with a water path of 3" was used for this evaluation. Table II summarizes the solid state ultrasonic results

Table II
Solid State Ultrasonic Results

Identification	Number of Discontinuities > 1/64" < 3/64" Reference Standards
Condition I	115
Condition II	112
Condition III	56

The initial observation to be made with this data is that a much higher number of discontinuities are present in the cast structures which represent the higher attenuating molten metal. Also, the largest number of discontinuiies observed in the molten state is carried over into the solid state when filter 'B' was physically disturbed. When Filter 'B' is not disturbed the counts in the molten state were much lower. This indicates the possibility that filter 'B' will allow particles up to a particular size to pass through and not be filtered out of the melt. This solid state ultrasonic analysis also revealed another significant piece of information. The discontinuities which were detected in each of the ingot slices were not randomly located. There was a preferred orientation, and this orientation was such that the discontinuities were concentrated near the center 6 inches of the diameter of each slice. This indicates that the exogenous material may possibly move along the solidifying front toward the center region which is the last metal to solidify.

2. Nitrogen Carrier Fusion (NCF) Hydrogen Determination

Samples were taken from the center area of each ingot slice and submitted for NCF hydrogen determination. The results are given in Table III.

Table III
Nitrogen Carrier Fusion (NCF) Hydrogen Results

Identification	Volume Content Hydrogen (cc/100 grams) Average of Three Samples
Condition I	0.114
Condition II	0.195
Condition III	0.213

From this data we see no correlation with the molten metal or solid state ultrasonic results. Of particular interest is the fact that the highest hydrogen content metal exhibited the better ultrasonic results both in the liquid and solid state. This indicates that the hydrogen content at least in the solid state samples is not a good indicator of metal quality alone.

3. Fast Neutron Activation - Oxygen Content

Samples were taken from the center area of each ingot slice and submitted for oxygen analysis by Fast Neutron Activation. The results are given in Table IV.

Table IV
Fast Neutron Activation - Oxygen Results

Identification	Oxygen Concentration (Average of of Two Samples) PPM
Condition I	16.5 ± 2.8
Condition II	16.0 ± 3.1
Condition III	14.6 ± 3.3

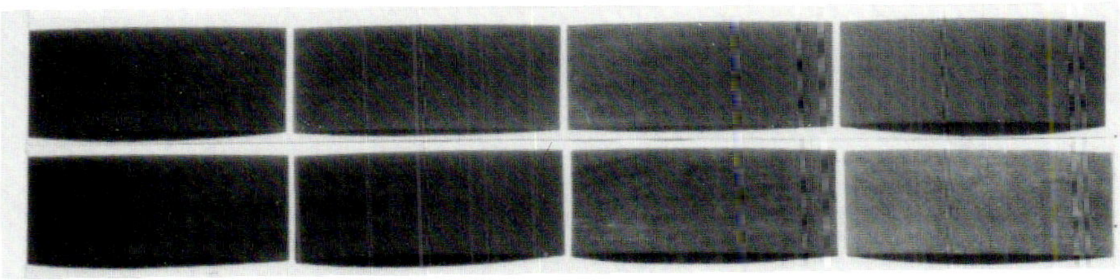

CONDITION III

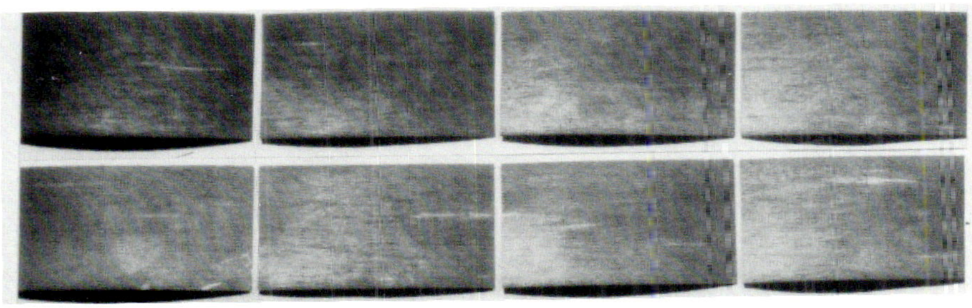

CONDITION II

CONDITION I

Figure 8 - Caustic Etch Smut Pattern

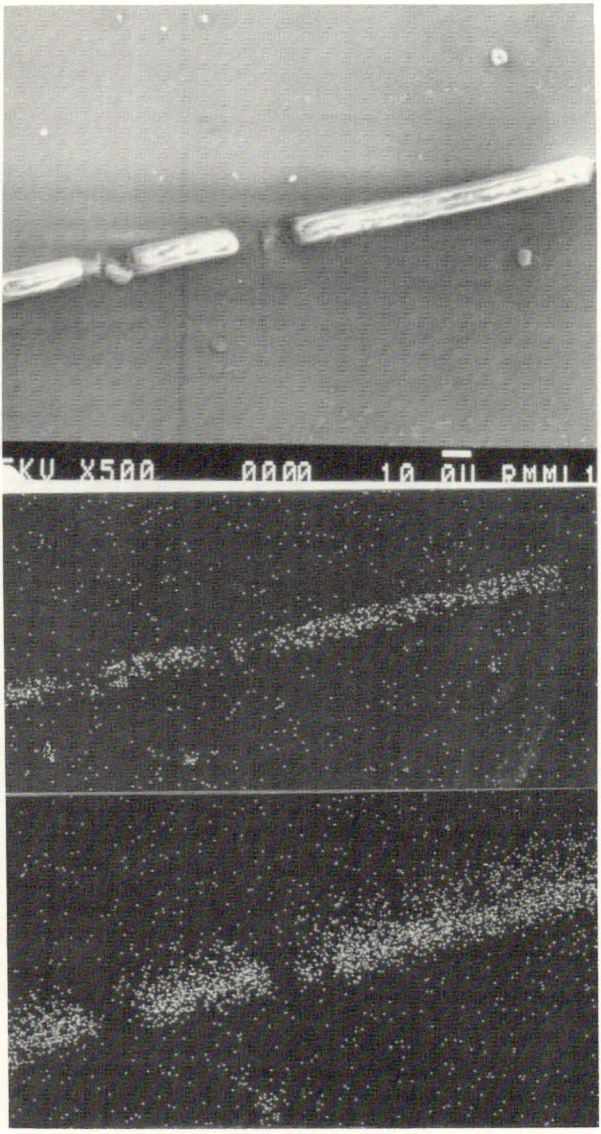

Figure 9 - Glass Fiber Found in Condition II Rolled Sheet Sample with Silicon EDS and Oxygen WDS Dot Maps.

filtrate consisted of silica in all cases. However, a glass fiber of similar size and composition to those found in procedure 5-b was found in the Condition II filtrate. See Figure 10. This is a very strong indication that what appears to be manufactured fibers are present in Condition II structure. They, in all probability, were introduced to the melt when the glass bag was broken on the start up of ladle 2. This is highly significant in that these fibers passed through two "acceptable" filters and ended up in the resultant cast structure. No other exogenous material was identified with this method; it is felt that most items of interest were dissolved in the solution.

6. Metallographic Examination of Fracture Specimens

Samples were sawed from all three conditions in which an ultrasonically identified discontinuity was present. The samples were notched in the region of the discontinuity and then fractured to reveal the structure through the discontinuity area. Five samples from Condition I, eight samples from Condition II, and three samples from Condition III were fractured. All fractured halves contained what would be considered as a defective region.

Through imaging techniques, it was possible to see that there were areas of ductile and brittle fracture in all samples examined. However, no real differences could be found between surface chemistries using Energy Dispersive Spectroscopy (EDS). In scanning the surfaces, there were miscellaneous areas found to be rich in Mg, Cl, Fe, and Ti.

It is usually not a proper procedure to do Wavelength Dispersive Spectroscopy (WDS) on rough samples such as these but they were run anyway for oxygen determination. There were strong indications that the brittle areas in Condition I were higher in oxygen than in the ductile areas. The results on conditions II and III were less revealing. Figure 11, 12, and 13 show macroscopic views of representative defective regions in Condition I, II and III respectively.

To further characterize and identify the discontinuities, additional samples were metallographically cross sectioned, mounted and carefully diamond polished to intersect the ultrasonically located discontinuity area. The samples were first received in the as-polished condition and photomicrographs were taken of the nonmetallic inclusions. They were further identified with the SEM using WDS and EDS, and found to be higher in oxygen and magnesium than the typical areas with no inclusions present.

Note: WDS or Wavelength Dispersive Spectroscopy requires flat, polished samples for accurate analysis. The samples in this last part of Procedure 6 satisfy this criteria. While the WDS analysis in the first part (rough, fractured surfaces) did show oxides on the rough surface, the analysis could not stand alone due to the extreme surface roughness. However, when viewing all data collectively, the overwhelming consensus is that the discontinuities evaluated in Procedure 6 are of the aluminum and/or magnesium oxide type.

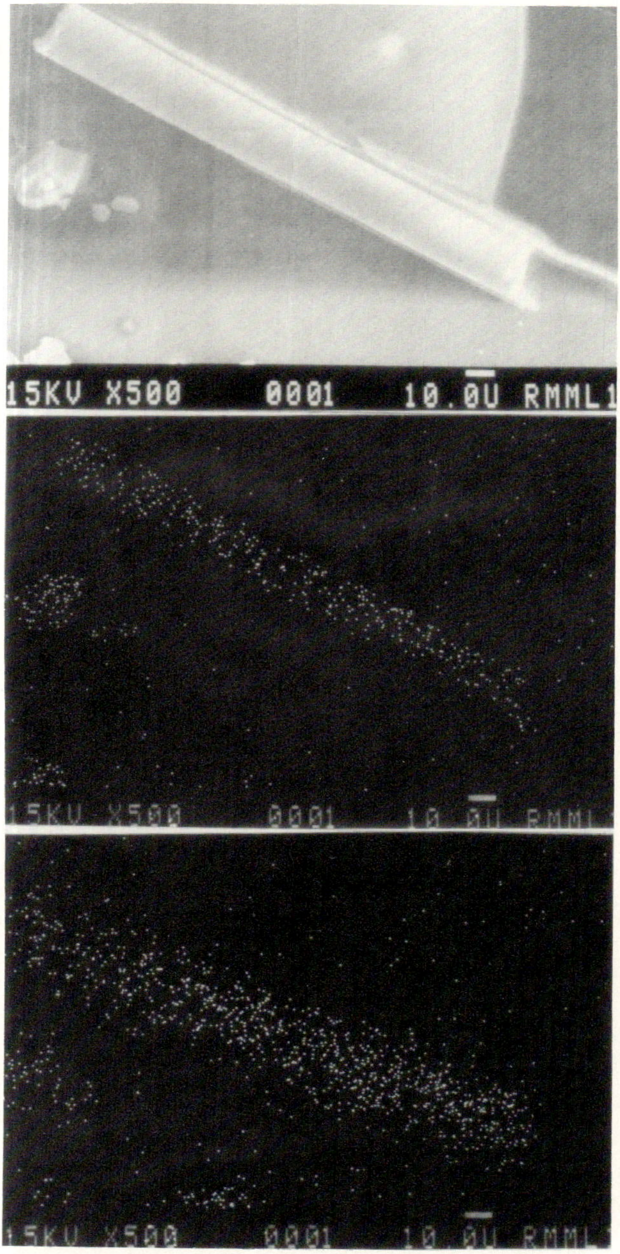

Figure 10 - Glass Fiber Found in Condition II Filtrate with Silicon EDS and O_2 WDS Dot Maps.

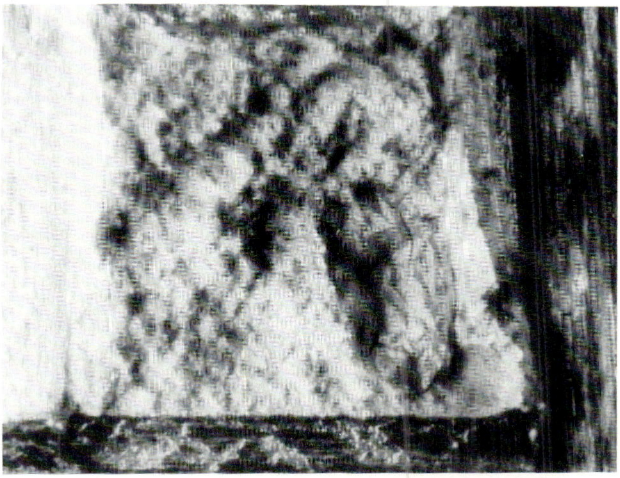

Figure 11 - Condition I - Fracture Through Defective Region

Figure 12 - Condition II - Fracture Through Defective Region

Figure 13 - Condition III - Fracture Through Defective Region

The following figures 14 and 15 are photomicrographs of areas depicting the regions of discontinuity with an analysis describing the nature of the region for sample condition I and II, respectfully. We did not have success in polishing through a region of discontinuity in condition III.

Summary and Conclusions

The Reynolds 4M™ production system, a practical and effective microprocessor based pulse-echo ultrasonic inspection system featuring state-of-the art high speed arithmetic analog to digital circuitry, and a self-contained tone-burst pulser/receiver, was used to characterize three molten metal conditions. The resulting cast ingot structure was rigorously examined and evaluated in an attempt to identify what the 4M™ system was detecting and how it might affect the semi-finished product.

It was felt that the most representative and best sample to evaluate was the actual cast structure, since it is that material which will eventually be sold to a customer. Several attempts at taking and evaluating grab samples have met with very limited success.

The evaluation of the cast ingot structure samples by several laboratory techniques yield positive proof that the 4M™ system can detect adverse conditions in the molten state and that these conditions can create defective semi-finished products.

When the molten metal attenuation and discontinuity counts were high, subsequent test of the solid structure by ultrasonic examination, and examination of sheet produced from the ingot structure show a direct correlation with the 4M™ results. Additional metallographic and scanning electron microscope were used to identify the resultant defective regions. When the molten metal attenuation and discontinuity counts are lower, subsequent tests of the solid structure again show a direct correlation with the 4M™ results in that material contains a much lower level of defects.

The Nitrogen Carrier Fusion (NCF) Hydrogen determination and the Fast Neutron Activation for O_2 content did not show any correlation to 4M™ results or solid sample analysis.

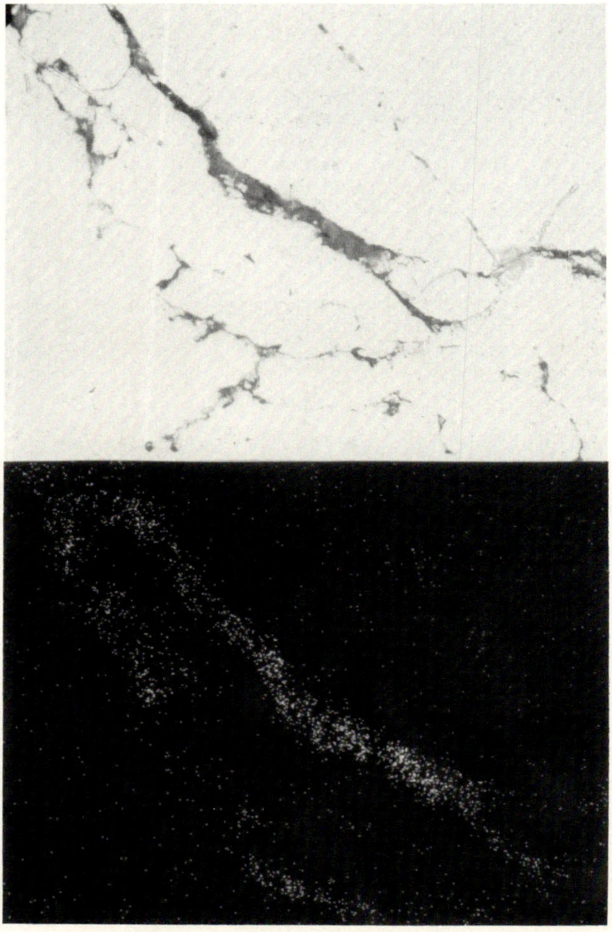

Figure 14 - Condition I - Metallographically Cross Section Mounted and Diamond Polished to Intersect Ultrasonically Located Discontinuity Area with Oxygen Dot Map - 500X.

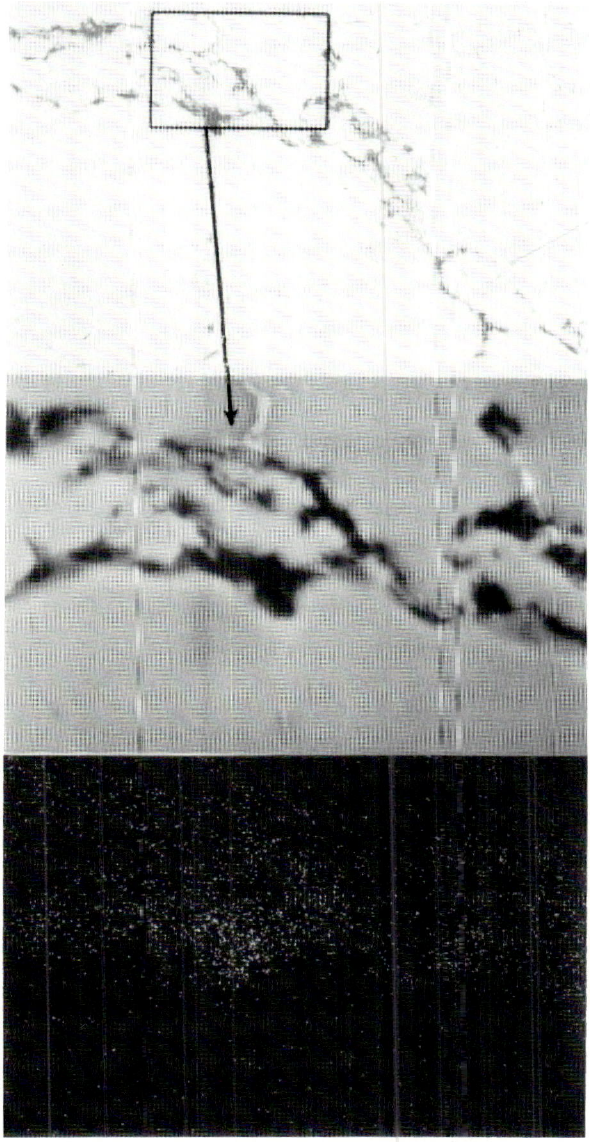

Figure 15 - Condition I - Metallographically Cross Section Mounted and Diamond Polished to Intersect Ultrasonically Located Discontinuity Area - 500X and 1000X with Oxygen Dot Map

From this data we see no correlation with the molten metal or solid state ultrasonic results. In a paper by Levy [8], et al, it was felt that the fast neutron activation technique had little utility as an evaluation technique for solid state material. Based on the evaluation in this work, we would concur with that idea.

4. Fabrication into Sheet for Evaluation

Each ingot slice had a 1" x thickness of ingot slice (approximately 6-7") x diameter (approximately 14") section sawed from it. These ingot slice samples were carefully picked to include some of the larger discontinuities identified by the ultrasonic examination. Each of these sections was preheated, hot rolled to 0.22", then cold rolled to 0.050". The resulting sheet from each condition was then given a standard 6061 solution heat treatment practice and naturally aged to T-4 condition. The results are as follows:

a) Caustic Etch Smut Pattern

Each rolled section, which represented edge to center to edge of the ingot slice, was given a caustic etch and then dried and photographed. As can be seen in Figure 8, the smut patterns revealed discriminating differences between the three conditions. Condition I and II were non-uniform with several streaks present, while Condition III was the most uniform with very few streaks present.

b) Buffed and Bright Dipped

The samples were then desmutted, buffed and bright dipped. The same results obtained with the smut pattern were seen on the bright dipped samples. Since the bright samples did not photograph well, only subjective visual ratings can be documented.

Based on the above results, good correlation exists between the 4M™ data and Condition III with its uniform surface quality and, depending on end use, the non-uniform and rejectable quality of Conditions I and II.

It should be noted that during procedure b) several fibers were found in the surface of Condition II material. These were identified by Scanning Electron Microscope (SEM) techniques as rich in silicon and oxygen and are very probably glass-fibers. See Figure 9.

5. Analysis of Filtrate from Chemical Dissolution of Ingot Slice Samples

Approximately ten grams of metal from each ingot slice, taken from the center region of each slice, were dissolved in 20% HCL, 1% HNO_3 and 79% H_2O solutions. These solutions were then vacuum filtered through 0.5 micron fluopore filter medium, dried and examined on the SEM for analysis of non dissolved solids. The

The equipment described in this paper represents developments of Reynolds Metals Company with U. S. and foreign patents granted or applied for. This publication is not a licence, implied or expressed, to use or operate under and is not intended to suggest infringement of existing patents. Interested parties are advised to contact the Manager of the Planning, Communications, and Resources Operation, Corporate Research and Development, Reynolds Metals Company, Richmond, Virginia 23261.

Acknowledgments

Many people have contributed to this work. The contributions of C. E. Burley, H. A. Lecky, M. E. Thurston, R. E. Hughes and K. Archibald of Reynolds Metallurgy Laboratory; and J. M. Marr and G. N. Chaffin of Reynolds Phoenix Extrusion Plant are particularly acknowledged.

References

1. Pitcher, D. E., and R. S. Young, "Improvements in or Relating to Methods of and Apparatus for Testing Molten Metal," Great Britain Patent, Patent 1 127 231 (Sept. 18, 1968).

2. Pitcher, D. E., and R. S. Young, "Methods of and Apparatus for Testing Molten Metal," U. S. Patent 3 444 726 (May 20, 1969).

3. Stremousov, V. I., and V. V. Tekuchev, "Velocity of Ultrasound in Molten Binary Alloys of Aluminum with Tin and Zinc," Zhurnal Fizicheskoi Khimii, Vol. 50, 1976, pp. 2373-2375.

4. Mansfield, T. L., "Ultrasonic Technology for Measuring Molten Aluminum Quality," Materials Evaluation, Vol 41, Number 6, May 1983, pp. 743-747.

5. Mansfield, T. L., "Ultrasonic Technology for Measuring Molten Aluminum Quality," Journal of Metals, September, 1982, pp. 54-57.

6. Mansfield, T. L., "Probe for the Ultrasonic Inspection of Aluminum," U. S. Patent 4 261 197 (April 14, 1981) and U. S. Patent 4 2878 755 (Sept. 8, 1981).

7. Krautkramer, J., and H. Krautkramer, Ultrasonic Testing of Materials, 2nd edition, 1977. Springer-Verlag, Berlin and Heidelberg, Federal Republic of Germany, and New York City. New York.

8. Levy, et al. "Molten Metal Quality Particulate Tests"; Light Metals, 1977. The Metallurgical Society of AIME, New York, New York.

REMOVAL OF HYDROGEN FROM MOLTEN ALUMINIUM
BY GAS PURGING

T.A. Engh and T. Pedersen

The University of Trondheim
The Norwegian Institute of Technology
Division of Metallurgy
N-7034 Trondheim - NTH,
Norway

Summary.

A model of hydrogen removal from molten aluminium by gas purging is presented. The model includes diffusion of dissolved hydrogen in the melt to the gas bubbles, surface removal, pickup of hydrogen from humidity above the melt and influence of reactive gas.

The theory is compared with measurements taken in laboratory and industrial batch and continuous reactors using lances, porous plugs and spinning nozzles. It is assumed that mass transfer in the melt boundary layer is rate-limiting. The generally accepted mass transfer coefficients have been employed. D_H is taken from a previous paper. Contact areas for the bubbles have been measured in water model experiments and scaled to molten aluminium. The theory gives the ratio end hydrogen/initial hydrogen within limits of $\pm$ 15 %. Thus it is possible to calculate purge gas requirements and removal times.

Introduction

Hydrogen is picked up by molten aluminium mainly due to a reaction with H_2O-vapour. Especially for remelted metal it is necessary to reduce hydrogen levels.

Inert gas purging seems to be well suited for the removal due to the high vapour pressure of hydrogen in equilibrium with dissolved hydrogen in molten aluminium

$$\underline{H} = \frac{1}{2} H_2 \tag{1}$$

A detailed work which compares theory and experimental results will be found in a Ph.D. thesis (1). Here part of this work is presented.

Assumptions

The mathematical description of gas purging is based on the following main assumptions:

a) The bath is well-stirred so that the concentration of $\underline{H}$ is the same throughout the melt.

b) In a horisontal cross-section of the bath hydrogen pressure, p_{H_2}, in the bubbles is the same.

c) Mass transfer of dissolved hydrogen in melt boundary layers (surrounding the bubbles and at the bath surface) is rate limiting.

d) The molten bath is "shallow" (depth less than 1 m) so that we do not have to take the change in total pressure into account. In the calculations we use an average value for the total pressure in the bubbles.

e) The hydrogen partial pressure p_{H_2} is much smaller than the total pressure, p_{total}.

Discussion of assumptions

Assumption a) does not always apply. For instance if there is only one or two porous plugs in a long shallow refining furnace we may not have complete mixing. For a SNIF (Spinning Nozzle Inert Flotation) unit where the impeller diameter is 1/3 of the melt diameter the assumption will be fulfilled. Mixing time (2) will then be very short compared to the time employed in gas purging.

p_{H_2} will depend on bubble diameter (p_{H_2} will be greatest in the small bubbles). Therefore strictly, b) will not be correct if there is a wide distribution of bubble sizes. This case may be worked out introducing the bubble size distribution and bubble rise velocities (for instance obtained from water model experiments). The bubble size distribution is then employed in Eq. 2. The solution becomes rather complicated.

With regard to assumption c) chemical rates may easily be included (for first order kinetics) by introducing total mass transfer coefficients. If diffusion of H_2 in the gas phase must be included, the situation is complicated and it seems, that extensive numerical calculations must be carried out. It seems that variation in total pressure only has to be taken into account for vacuum operation.

Similarly, assumption e) applies for units operating at atmospheric pressure. The reason is that usually $\underline{H}$ is so low that $p_{H_2} \ll 1$ atmosphere.

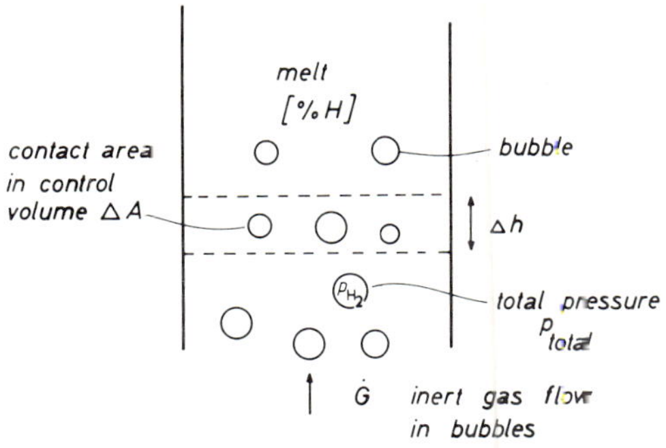

Figure 1 - Control volume of a gas purging reactor.

The hydrogen partial pressure in the bubbles as they escape the bath p_{H_2} is found as follows:

A mass balance for hydrogen over a horisontal slice (see Fig. 1) of the bath gives:

Hydrogen that diffuses into bubbles =
increase of hydrogen content of bubbles

or in mathematical form

$$\rho k \left([\% \underset{\cdot}{H}] - [\% H]_e\right) \Delta A/(100 \cdot m_H) = 2\,\Delta(\dot{G}\, p_{H_2}/p_{inert}) \tag{2}$$

Here $[\% \underset{\cdot}{H}] - [\% H]_e$ is the driving force for diffusion of hydrogen. $[\% \underset{\cdot}{H}]_e$ is the concentration of hydrogen in the melt in equilibrium with the partial pressure of hydrogen in the bubbles

$$[\% \underset{\cdot}{H}]_e = K \sqrt{p_{H_2}}/f_H \tag{3}$$

$$[\% H]_e \text{ (at bath surface)} = K \sqrt{p^o_{H_2}}/f_H \tag{3a}$$

If eq. (3) is introduced into (2), we may integrate from bottom to top of the melt giving $p^o_{H_2}$. For this solution reference is made to a previous paper (3). Here the results are presented graphically in Fig. 2 giving dimensionless variable

$$Z = [\% H]_e \text{ (at bath surface)}/[\% \underset{\cdot}{H}] \tag{4}$$

as a function of a dimensionless group

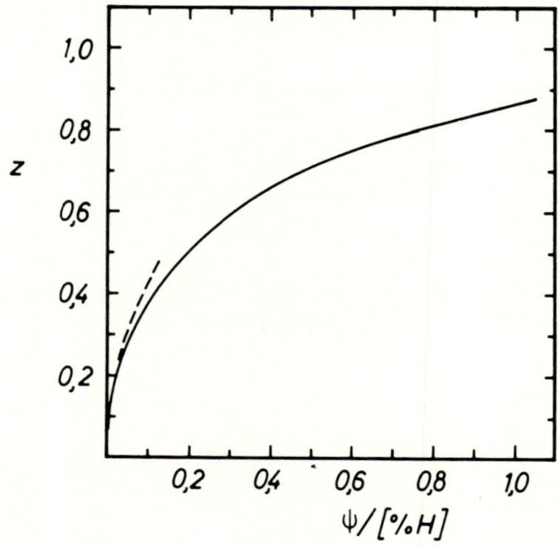

Figure 2 - Dimensionless gas composition, Z, as a function of inverse hydrogen concentration. The limiting solution for diffusion control, $Z^2/2$, is shown by the dashed line.

$$\frac{2\,\psi}{[\%\ H]} = \frac{\text{Maximum hydrogen that can diffuse to bubbles}}{\text{Equilibrium hydrogen content of bubbles}}$$

$$= \frac{k\cdot A\cdot \rho[\%\ H]/(100\cdot m_H)}{2\ \dot{G}\cdot p^o_{H_2}/(p_{inert}\cdot Z^2)} = \frac{k\cdot \rho\cdot A\cdot p_{inert}\cdot K^2}{2\ f_H^2\cdot 100\cdot m_H\ \dot{G}\cdot [\%\ H]} \qquad (5)$$

It is seen that the group is inversely proportional to hydrogen concentration.

If $Z < 0.1$ we have the diffusion controlled case (3). It Z is close to one (A "large"), chemical equilibrium controls the removal rate. For $0.1 < Z < 0.9$ we have a case of mixed control (3). For gas purging with a lance where bubbles size is large, Z has a low value, around 0.1; for a porous plug Z has a somewhat higher values. For a system with a rotating gas impeller $Z \sim 0.7$. These systems are compared on Fig. 5.

Equations for hydrogen removal

Once we know the hydrogen content in the escaping bubbles, hydrogen concentrations can be obtained employing a hydrogen mass balance for the refining reactor.

At the top surface hydrogen removal is given by

$$\frac{k_s A_s \rho \left([\%\,H] - [\%\,H]_v\right)}{100 \cdot m_H}$$

See numerical example for $[\%\,H]_v$.

For a "completely exposed" surface $[\%\,H]_v$ is the concentration of hydrogen in the melt given by the equilibrium

$$1/2\ H_2O + 1/3\ Al(\ell) \ = \ 1/6\ Al_2O_3(s) + \underline{H} \qquad (6)$$

where p_{H_2O} is the water vapour concentration in the air.

Often the atmosphere above the melt will be protected with a lid so that there should be little or no hydrogen pickup from the air. In this case the gas composition is equal to that of the escaping bubbles so that: $[\%\,H]_v \ = \ [\%\,H] \cdot Z$.

By blowing inert gas on the melt surface (combined with a lid), $[\%\,H_v] \approx 0$.

A mass balance equation for a steady-state continuous reactor gives:

> Difference between hydrogen flow into and out of reactor = hydrogen removed by bubbles per unit time + hydrogen removal rate at bath surface

or

$$\frac{\dot{M}\left([\%\,H]_i - [\%\,H]\right)}{100 \cdot m_H} = \frac{2\,\dot{G}\,p^o_{H_2}}{p_{inert}} + \frac{\rho\,k_s A_s\left([\%\,H] - [\%\,H]_v\right)}{100 \cdot m_H} \qquad (7)$$

or if eqs. (3) and (4) are introduced

$$\frac{\dot{M}\left([\%\,H]_i - [\%\,H]\right)}{100 \cdot m_H} = \frac{2\,\dot{G}\,f_H^2\,[\%\,H]^2\,z^2}{K^2 \cdot p_{inert}} + \frac{\rho\,k_s A_s\left([\%\,H] - [\%\,H]_v\right)}{100 \cdot m_H} \qquad (8)$$

Using b and B, in the list of symbols eq. (8) can be written:

$$\left([\%\,H]_i - [\%\,H]\right) \ = \ + \frac{[\%\,H]^2\,Z^2}{B'} + \frac{b}{B'}\left([\%\,H] - [\%\,H]_v\right) \qquad (9)$$

As mentioned previously Z is given in Fig. 2. $[\%\,H]$ is obtained from eq. (9) as the solution of a simple quadratic equation.

The physical meaning of B'and b may be understood from the following groups:

$$\frac{[\% H]}{B'} = \frac{\text{Hydrogen flow in bubbles}}{\text{Hydrogen flowing out in melt}}$$

$$\frac{b}{B'} = \frac{\text{Hydrogen flow out by mass transfer at top surface}}{\text{Hydrogen flowing out in melt}}$$

$$\frac{[\% H]}{B} = \frac{\text{Hydrogen flow in bubbles}}{\text{Hydrogen content of melt}}$$

Similarly the mass balance equation for a batch reactor is:

Reduction of content of hydrogen in bath = hydrogen removed by bubbles + hydrogen removed at bath surface

or replacing the lefthand side in eq. (8) with $-\frac{M\, d\,[\% H]}{100\, m_H\, dt}$;

B' with B where $\dot{M}$ is replaced by M

$$\frac{d\,[\% H]}{dt} = - \frac{[\% H]^2 \cdot Z^2}{B} - \frac{b}{B}\left([\% H] - [\% H]_v\right) \tag{10}$$

where now Z given by Fig. 2 changes with time since [% H] changes with time. [% H] in Fig. 2 is also given by eq. 11:

$$\psi/[\% H] = - Z - \ell n\,(1-Z) \tag{11}$$

Z is solved numerically by Newton-iteration on eq. (11) or from Fig. 2. Now [% H] is integrated over time by using Runge - Kutta on eq. (10). During the integration corresponding values of [% H] and Z are calculated. Some results are shown in Fig. 5.

Mass transfer coefficients

The mass transfer coefficient to the bubbles is taken from Fig. 4, line d.

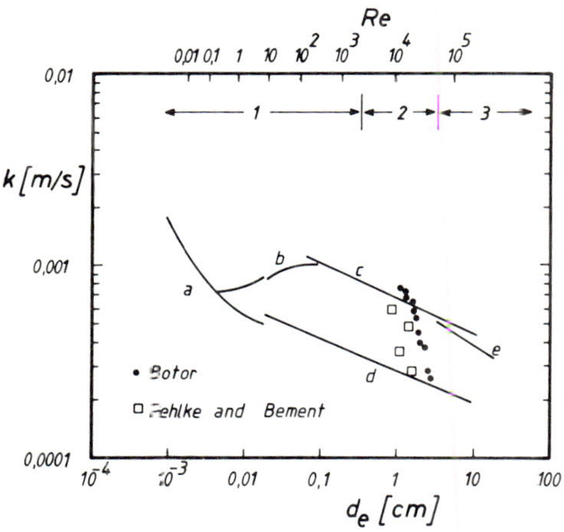

Figure 4 - Predicted and observed mass transfer coefficients for the removal of hydrogen to inert gas bubbles in aluminium at 1000 K. Regimes 1, 2, and 3 correspond to spherical, ellipsoidal, and spherical cap bubbles.
(a) Mass transfer for viscous flow. The upper line is for a freely circulating bubble; the lower is for a rigid bubble. (b) Mass transfer for turbulent flow to a circulating bubble in the transition region. (c) Mass transfer to a circulating bubble in potential flow. (d) Mass transfer to a rigid bubble in potential flow. (e) Mass transfer to the front and rear face of a spherical cap bubble.

The basis for this figure is discussed in a previous paper (3).

Mass transfer to the surface is described for lances and porous plugs using Machlins' equation (4).

$$k_s = \sqrt{\frac{8}{\pi}} \, (U_R \cdot R_c \cdot D_H)^{1/2} / R_c \qquad (12)$$

where U_R is the radial velocity of the metal surface at the crucible edge and R_c is the radius of the crucible.

U_R is obtained using the equation (5):

$$U_R = \left\{ \frac{\dot{Q} \cdot p \cdot \ln \left(1 + \frac{\rho \cdot g \cdot H}{1.01 \cdot 10^5}\right)}{\pi \, R_c^{\,2} \cdot \rho} \right\}^{1/3} \qquad (13)$$

where $\dot{Q}$ is the volumetric gas flow in m_N^3/s; p is the pressure equal to 1 atm; ρ is the melt density; g the acceleration of gravity 9.81 m/s^2 and H the bath depth.

For impeller systems $k_{s_{Na}}$ for the removal of dissolved sodium was determined experimentally (6). For hydrogen k_s was then determined using the relation

$$k_s = k_{s_{Na}} \sqrt{\frac{D_H}{D_{Na}}} \qquad (14)$$

Eq (14) is based on the assumption that mass transfer at a "free interface" is proportional to the square root of the diffusion coefficient.

Numerical example

We take batch - gas purging, Run No. 1.3.1. as a numerical example.

AlSi 7 Mg0.5 $\quad f_H = 1.75$

$\quad (f_H = 1$ for pure Aluminium)

$A = 0.17 \; m^2$

$d_b = 26$ mm (measured in water-models (9) and scaled to Aluminium) (1)

$A_s = 0.20 \; m^2$

$M = 200$ kg

$\dot{G} = 4.46 \cdot 10^{-6}$ kmols^{-1}

$k = 2.3 \cdot 10^{-4} \; m \; s^{-1}$

$k_s = 1.43 \cdot 10^{-4} \; m \; s^{-1}$

$K = 1.2 \cdot 10^{-4}$

$[\% \, H]_i = 0.14 \cdot 10^{-4}$ %

$[\% \, H]_v = 0.14 \cdot 10^{-4}$ % ($[\% \, H]_v$ is $[\% \, H]$ found in melt after long holding time with no gaspurging. Experimental value is much lower than given by eq. (6) and lower than given by eq. (1) with $p_{H_2} = 1$ atm.)

$B = 1.05\ 10^{-3}$

$b = 3.48\ 10^{-7}$

$\psi = 2.4 \cdot 10^{-7}$

Eqs (10) and (11) are solved by Rhunge-Kutta and plotted as Run No. 1.3.1. in Fig. 5. The symbols + are the experimental values.

Experimental

The batch gas purging experiments were carried out in circular vessels, electrically heated. The SNIF-unit is rectangular in shape. Hydrogen analysis was performed only before and after gas purging in the cast-shops.

The Telegas instrument (7) was employed in all the experiments. It was very important to preheat the probe well, let the probe stay for about 5 min. in the metal while bubbling gas through it and use the mean value of the next three readings. The initial one was discarded since it often gave too high values.

Fig. 6 shows the correspondence between measured and calculated values of end hydrogen for batch refining units. Figs 7a, b, c compare three sets of measured and calculated end values for a continuous refining unit. Also the initial measured values are indicated. Fig. 8 shows the comparison for a system with two units in series.

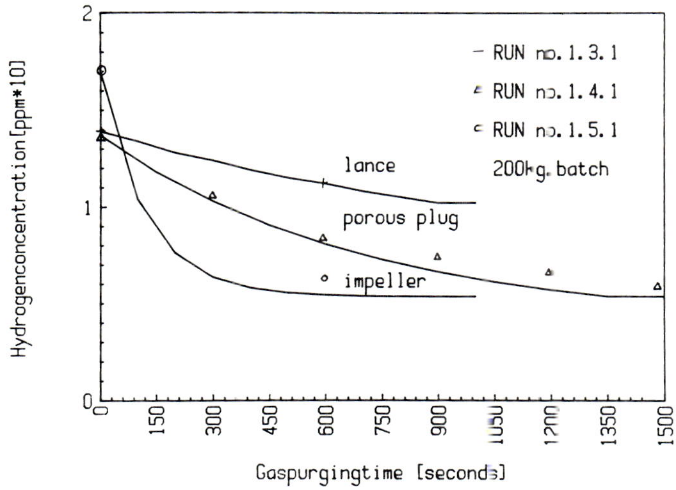

Figure 5 - Gaspurging with lance, porous plug and impeller in 200 kg batch for the same gasflow $\dot{Q} = 0.36\ m_N^3/h$.

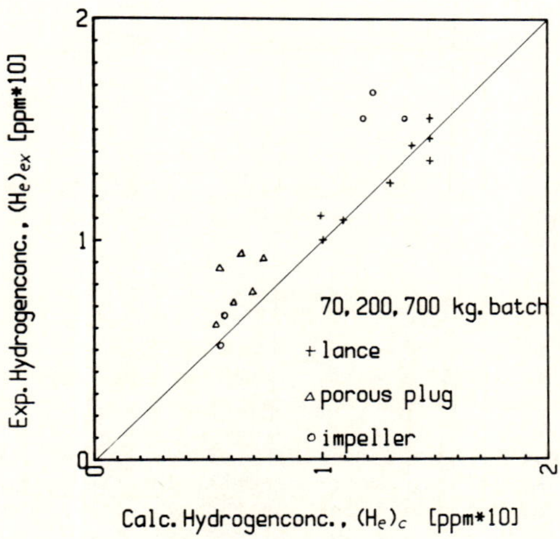

Figure 6 - Experimental and calculated hydrogen concentrations in 70, 200 and 700 kg batch using lances, porous plugs and impellers.

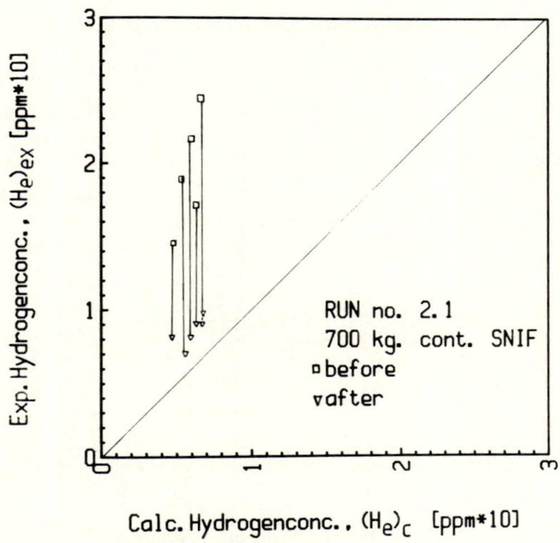

Figure 7a - Experimental and calculated hydrogen concentrations in 700 kg continuous SNIF.

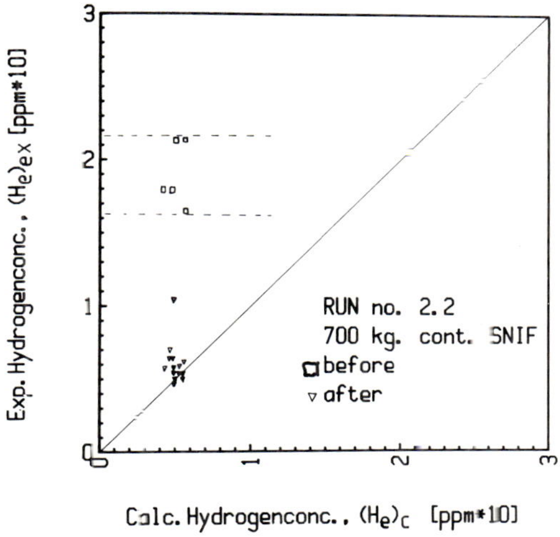

Figure 7b - Experimental and calculated hydrogen concentrations in 700 kg continuous SNIF.

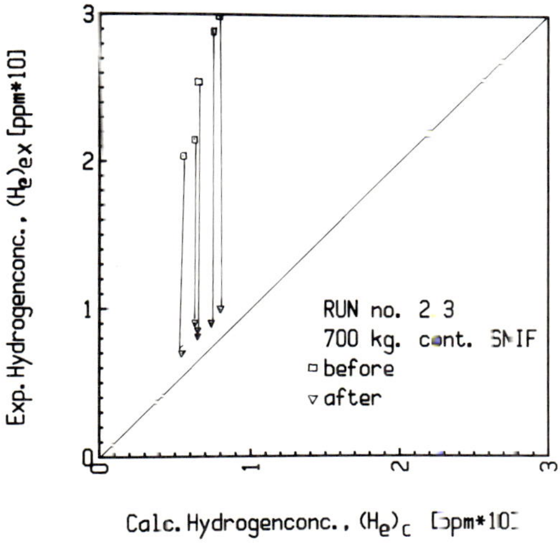

Figure 7c - Experimental and calculated hydrogen concentrations in 700 kg continuous SNIF.

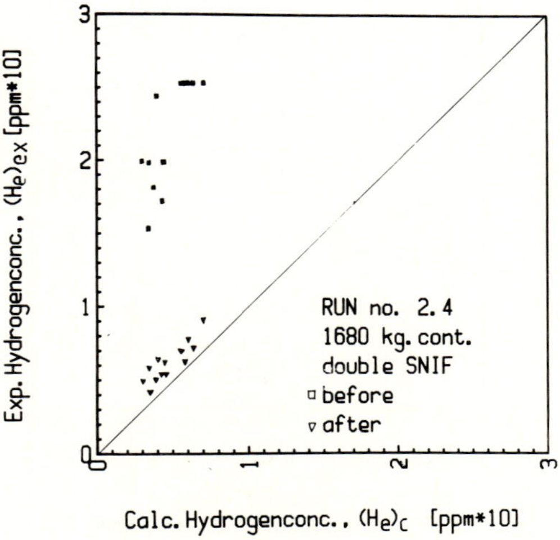

Figure 8 - Experimental and calculated hydrogen concentrations in 1680 kg double continuous SNIF.

Discussion

The measured values tend to be somewhat higher than the calculated values. This may be partly explained by the fact that the telegas instrument tends to give too high readings (1).

From ending the gas purging to reading the telegas instrument there will also be some hydrogen pickup from atmosphere. There will also be some hydrogen pickup from humidity in the refractory-material.

An other factor is that the gas may not be properly dispersed throughout the melt. This occurs if gas flow rates are too high so that "flooding" occurs (9). This effect is illustrated in Fig 9 where the ratio $\frac{H_{exp}}{H_{cal}}$ is given as a function of the total flow rate for a double SNIF.

Furthermore the contact area A inverse proporsional to bubble-radius employed on the calculations is obtained from full scale water-model experiments (9). Our theory (1) indicates that $R_{b,\ melt} \approx 2 \cdot R_{b,\ H_2O}$.

Finally it should be pointed out that the mass transfer coefficients k and k_s are not known very accurately. There are several reasons for this. The mass transfer coefficient is proportional to the square root of the diffusivity of hydrogen, D_H, which is known only roughly. Also k_s is proportional to the square root of the surface velocity which depends on, geometry, position of impeller, etc. Furthermore the surface velocity does not follow the assumptions made in the theory.

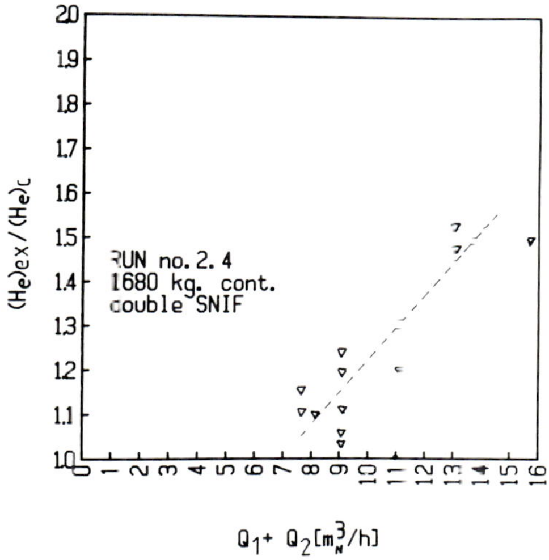

Figure 9 - Ratio between experimental and calculated hydrogen concentrations out of a double continuous SNIF as function of gasflows $\dot{Q}_1$ and $\dot{Q}_2$, showing lower efficiency of the gas at higher flow rates when flooding appears.

"Pickup" of hydrogen from moisture in air may partly be held to low levels by use of a lid.

In Figure 10 is shown theoretical curves for hydrogen removal with (a) surface completely exposed to air with moisture, (b) with a lid and (c) blowing inert gas on the surface. We see from case b in Fig. 10 that a lid greatly improves efficiency. Blowing inert gas on the surface seems to be interesting only if extremely low hydrogen levels are required.

Chlorine in the purge gas will react with Al to $AlCl_3$. It will not react with $\underline{H}$ to give HCl.

Above the melt $AlCl_3$ may react with H_2O to give Al_2O_3 and HCl. However, HCl should in contact with the bath-surface react according to the scheme

$$3\,HCl + Al = AlCl_3 + 3\,\underline{H}$$

Therefore based on thermodynamic reasoning there seems to be no advantage in the use of Cl_2. However, mass transfer may increase marginally according to Botor (10).

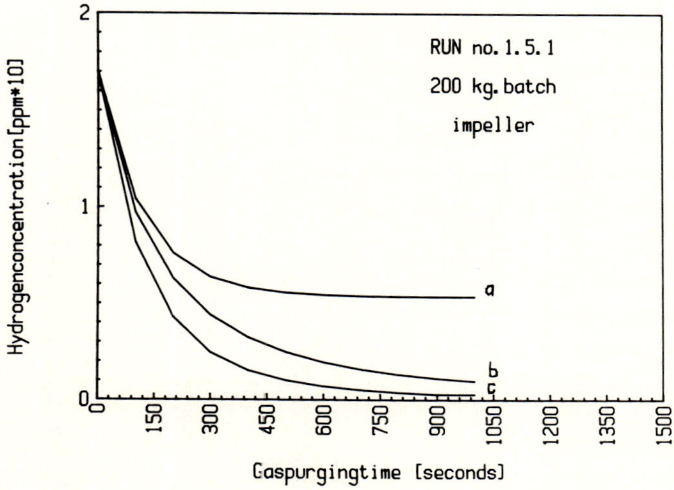

Figure 10 - Calculated curves for hydrogen removal with
a. surface completely exposed against air (as Fig 5)
b. a lid
c. blowing inert gas on the surface (combined with a lid).

Conclusion

A model (1, 3) that gives end or output hydrogen both for batch and continuous inert gas purging units has been checked experimentally both in the laboratory and industrially. Correspondence between calculated hydrogen levels and "Telegas" instrument readings seems acceptable.

For impeller systems operating efficiently hydrogen can be calculated approximately using only the equilibrium between hydrogen in gas and in melt and the inert gas consumption. From the comparison between calculations and measurements or from full-scale water model experiments it is possible to check if "flooding" (large bubbles) reduces efficiency. If flooding occurs ,inert gas consumption must be reduced, impeller rotation speed increased or the impeller system should be redesigned.

For the less efficient lance and porous plug systems also the contact areas between gas and melt and mass transfer coefficients must be known in the calculation of hydrogen levels. The contact areas can be measured in full scale water-model experiments and scaled to molten aluminium.

Using a lid to protect the melt against humidity in the air will improve the efficiency.

List of symbols

A	contact area of bubbles in melt [m^2]
A_s	contact area of melt surface [m^2]
B	$M\, p_{inert} \cdot K^2/(100 \cdot m_H \cdot 2\, f_H^2\, \dot{G})$ [s]
b	$\rho \cdot k_s \cdot A_s \cdot p_{inert} \cdot K^2/(100 \cdot m_H \cdot 2\, f_H^2\, \dot{G})$ [s]
d_e	equivalent diameter of bubble [m]
D_H	diffusivity of hydrogen [m^2/s]
E	$\frac{2\,M}{k\,A\,\rho}$
f_H	Henrian activity coefficient of hydrogen
g	acceleration of gravity [m/s^2]
$\dot{G}$	flow rate of inert gas kg·mole/s
[% H], $[\%\,H]_e$, $[\%\,H]_i$, $[\%\,H]_\infty$	hydrogen concentration, in equilibrium with bubbles, initial and at long times, respectively
k	mass transfer coefficient to bubbles for hydrogen [m/s]
k_s	mass transfer coefficients to bath surface for hydrogen [m/s]
K	equilibrium constant for hydrogen eq (3)
M	mass of aluminium bath [kg]
m_H	molecular weight of hydrogen
p_{inert}	pressure of inert gas [atm]
p_{H_2}	pressure of H_2 in bubbles [atm]
$p^o_{H_2}$	pressure of H_2 in bubbles at melt surface [atm]
p_{total}	total pressure [atm]
ρ	density of melt [kg/m^3]
R	gas constant 1.9873 kcal/kmole K
R_b	radius of gas bubbles [m]
R_c	radius of vessel [m]
t	time from start of gas purging [s]
U_R	surface velocity [m/s]
Z	square root of $p^o_{H_2}$ to the equilibrium value
ψ	$\frac{k \cdot A \cdot p_{inert} \cdot K^2}{4\, f_H^2 \cdot 100\, m_H \cdot \dot{G}}$

B' is found from B by replacing M with $\dot{M}$.

Acknowledgement

The authors wish to acknowledge gratefully the financial support of the Royal Norwegian Council for Scientific and Industrial Research under project MK 0520.09808 which makes this research possible.

References

(1) T. Pedersen, Ph.D. thesis, to be published.

(2) S. Asai, T. Okomoto, J.C. He, and I. Muchi, "Mixing time of refining vessels stirred by gas injection," Trans ISIJ, 23 (1983) pp. 43-50.

(3) G.K. Sigworth, and T.A. Engh, "Chemical and Kinetic factors related to hydrogen removal from Aluminium," Metallurgical Trans. B, 13B, (1982) pp. 447-460.

(4) F.D. Richardson, Physical Chemistry of Melts in Metallurgy, Vol. 2, pp. 477-487, Academic Press, New York, N.Y., 1974.

(5) T.A. Engh, and S.T. Johansen, "Convection and Stirring Induced by Bubbling from a Porous Plug," Proceedings of the Shanghai Symposium on Injection Metallurgy, Nov. 1982, Shanghai, China.

(6) T. Pedersen, "Fjerning av lettelementene Na, Li og Ca ved spylgassbehandling og røring,"STF34 A83014, SINTEF, NTH, Trondheim, 1983.

(7) C.E. Ramsley, and D.E.I. Talbot, "An instrument for measuring the gas content of Aluminium alloy during melting and casting," Journal of the Institute of Metals 86, (1957-58) pp. 212-219.

(8) S. Nagata, "Mixing, Principles and Applications", Kodansha Ltd., 1975.

(9) B. Rasch, "Bestemmelse av boblestørrelse og kontaktareal gass/væske ved hjelp av vannmodell-forsøk", STF34 A82094 SINTEF, NTH, Trondheim, 1982.

(10) J. Botor, "Kinetics of hydrogen-degassing of molten Aluminium by purge gas," Aluminium,56 (8) (1980) pp. 519-522.

DETERMINATION OF HYDROGEN IN ALUMINUM AND ITS ALLOYS BY THE FUSION/CARRIER GAS METHOD USING THE LECO RH-3 DETERMINATOR

W.D. Lamb

ALCAN International Limited
Kingston Laboratories
P.O. Box 8400
Kingston, Ontario
Canada
K7L 4Z4

Abstract

The LECO RH-3 determinator analyses Al and its alloys for hydrogen by the N_2 carrier gas fusion technique. It has been used successfully for commercial purity Al, Al-Mn-Mg, Al-Cu, Al-Si and Al-Mg alloys, the results comparing favorably with the Ransley sub-fusion method. A number of problems were encountered in its evaluation such as ingress of atmospheric moisture and measurement of the so-called surface hydrogen correction. Raising the sample temperature in stages shows that no significant hydrogen is evolved below 400-500°C, where hydrogen may also be diffusing from the interior of the sample. Without surface correction by either method, the LECO RH-3 results are frequently much higher than the Ransley result for alloys such as AA-7075, containing Zn and Mg. This disagreement has not yet been resolved although three explanations are discussed (a) reaction of volatilized Mg or Zn with hydroxyl groups in the silica network of the quartz furnace tube to produce H_2, (b) reaction of H_2O absorbed on oxide inclusions which occurs only when the sample is molten and (c) presence of large amounts of water in cavities on the prepared surface, caused by extensive shrinkage porosity in this alloy.

Introduction

Until 1979 the determination of hydrogen in aluminum and its alloys at Alcan was done by the Ransley sub-fusion technique. This has been considered the standard method for the industry since its publication in 1955 by Ransley and Talbot (1). This method involves the extraction of hydrogen under vacuum by heating a carefully machined cylindrical sample to a temperature just below the commencement of fusion. The hydrogen which diffuses from the sample is collected in a known volume of the equipment and its pressure is measured using a McLeod gauge. The major drawback of this method is that it is time consuming, taking four to eight hours for a single determination of hydrogen. To shorten the analysis time the LECO RH-3 hydrogen determinator was purchased which can do a complete analysis in approximately 20 minutes. The principle of operation of this equipment is identical with that of the Ithac-02 instrument which is based on the work of Degréve (2). In this method the cylindrical sample contained in a graphite crucible is melted in a stream of nitrogen (see Figure 1). The hydrogen is carried downstream to be detected and measured by a catharometer which senses the hydrogen because its thermal conductivity is approximately seven times that of the N_2 carrier gas. The main difference between the Ithac-02 and LECO RH-3 instruments is that the latter is provided with a loading head for the introduction of the sample into the equipment. This loading head permits the sample to be stored in a cavity which has been purged with dry nitrogen gas and sealed, prior to its introduction into the graphite crucible and subsequent fusion while contained in the quartz tube of an induction furnace. This permits the baking out of the crucible and subsequent fusion of the sample to be accomplished without opening the apparatus at any stage of the analytical cycle. With the Ithac-02 instrument it is necessary to open the furnace for sample introduction after baking out the crucible. The LECO RH-3 thus has an advantage since it is more difficult to introduce atmospheric moisture which may produce spurious hydrogen by reaction with metal.

In order to provide confidence in the use of the RH-3 it was necessary to determine its accuracy by comparison with the more conventional Ransley sub-fusion technique.

Theoretical Considerations

Thermal Conductivity

Table I shows the thermal conductivity of various gases relevant to the analytical process. Normally oxygen should be excluded from the analysis since it would react with and damage the graphite crucible but the presence of small amounts would not seriously affect the results since its thermal conductivity is close to nitrogen. The same can be said for carbon monoxide which may be produced by the above reaction. Water and carbon dioxide would produce catharometer signals of opposite polarity to that of hydrogen since their thermal conductivity is slightly below that of the carrier gas. If methane is produced by the reaction of hydrogen with the graphite crucible then low results should occur since its thermal conductivity is much less than that of hydrogen.

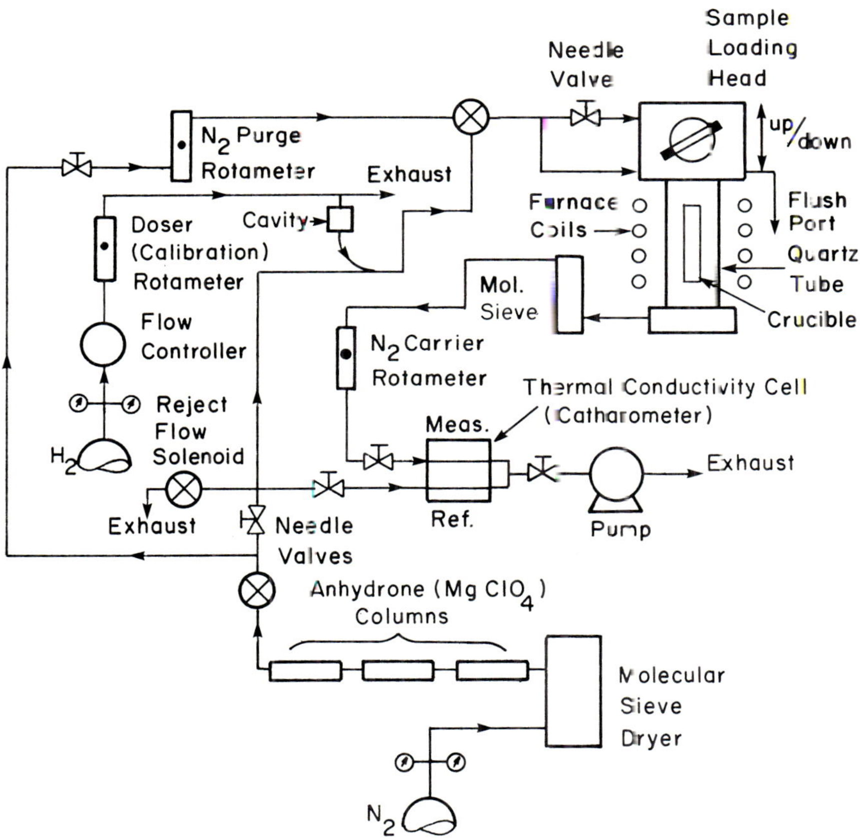

Figure 1 - Schematic Diagram of LECO RH-3

Table I. Thermal Conductivity of Gases at 48.9°C[3]

Gas	Thermal Conductivity cal/(s) (cm^2) (°C/cm) x 10^{-6}
Nitrogen	65.71
Oxygen	68.19
Water	46.70
Methane	89.26
Carbon Monoxide	63.89
Carbon Dioxide	43.81
Hydrogen	471.11

Reaction Equilibria

Figure 2 shows a logarithmic plot of the equilibrium constant (K_p) against reciprocal absolute temperature for reactions which may occur in the analytical process. The free energy argument shows that the reactions that give the highest K_p at a given temperature are the most likely to occur. These are the reactions of metal with water, which produce spurious hydrogen. These reactions are favoured by lower temperature but their progression towards equilibrium will also be governed by kinetic considerations. Reaction of carbon with water will occur during the initial baking of the crucible and Figure 2 shows that this reaction is favoured by higher temperatures and possibly different kinetics. The significance of molecular breakdown of hydrogen will be discussed below.

Metal Volatility

Comparative melting point and vapour pressure data are shown for aluminum, magnesium and zinc in Table II. The question of metal volatility is important since when the sample is melted, the volatile components may be carried out of the crucible and deposited in cooler parts of the equipment such as the furnace tube. These deposits may have an important role to play in the analysis in the sample as described below.

Table II. Comparative Volatility of Alloying Metals[4]

Metal	Melting Point (°C)	Vapour Pressure at 850°C Torr
Al	660	<0.1
Mg	651	47
Zn	419.4	550(approx)

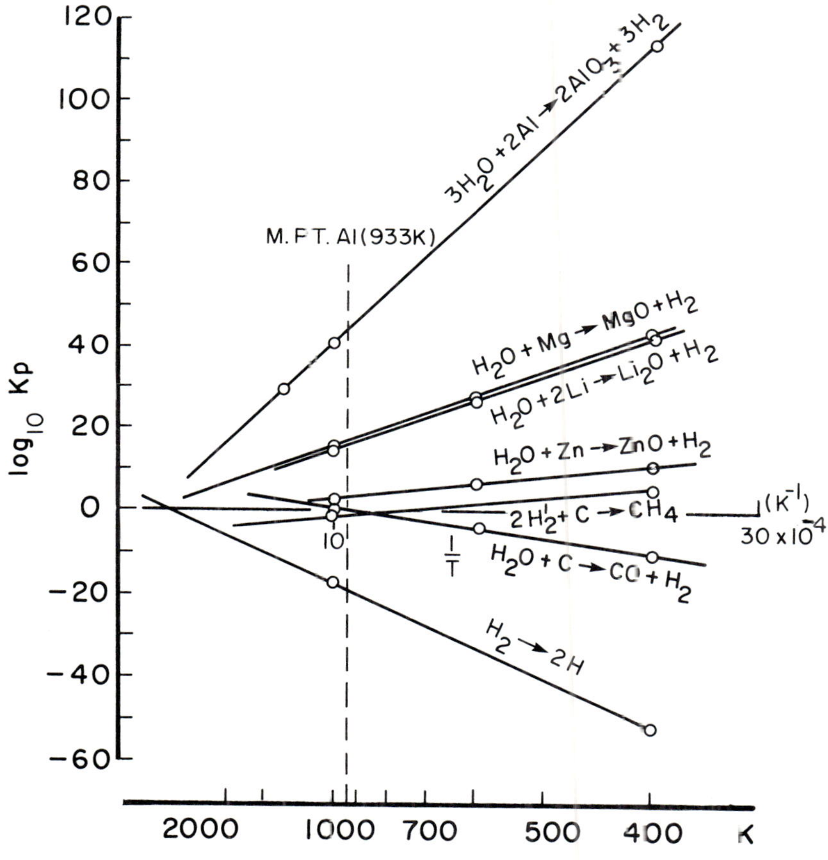

Figure 2 - Reaction Equilibria

Effect of Water

The first effect of water seen during the analytical process is during the initial baking out of the graphite crucible at 1200°C. This produces a large positive displacement from the base line which represents production of hydrogen by reaction of water with graphite as described above and by pyrolysis of hydrocarbons in the graphite. Following this baking out period the crucible blank is done by heating the empty crucible under the same conditions used for fusion of the sample. The instruction manual specifies that this crucible blank must be below 0.05 mL H_2/100g based on a one gram sample. It is important to realize that this crucible blank does not relate to reaction of water with the metal sample, which is governed by entirely different thermodynamics and most probably kinetic considerations as described above. Assuming that the crucible blank is satisfactory then fusion of the sample can proceed. During fusion, spurious hydrogen can be produced by reaction of water with metal vapour. This water may come from two sources (a) the carrier gas and (b) hydroxyl ions in the silica networks of the quartz tube. The former interference can be overcome by ensuring that the carrier gas is dried sufficiently to satisfy analytical requirements. Appendix 1 shows that a dewpoint of -78°C is required to prevent interference greater than 0.01 mL H_2/100g. This assumes that water in the nitrogen can react with metal for the entire fusion/ analytical cycle of approximately 3 minutes. Although the completion of this reaction is thermodynamically favoured the validity of this assumption depends upon metal volatilisation and reaction kinetics. A simple calculation will show that only 7 x 10^{-6} grams of magnesium is required to produce a result of 0.01 mL/100g of sample. Thus the importance of dry carrier gas cannot be overemphasized. The extent of reaction with water in the quartz tube will be dependent on a number of factors such as the transport of metal vapour to the tube, the temperature required for the reaction and the availability of hydroxyl groups. Simensen and Lauritzen, (5) when using the capsule method observed that any magnesium evaporated into the degassing chamber will condense on quartz walls outside the furnace and claimed that such deposits produce hydrogen by reaction with water in the quartz tube. However, they demonstrate this by heating the magnesium-rich layer and moving it along the tube. It may be thus argued that reaction is generated by supplying heat which will aid the difusion of hydroxyl groups towards the deposited magnesium and possibly accelerate the rate of reaction. In the LECO RH-3 induction furnace the quartz tube does not get very hot since it is not directly heated. There is thus some uncertainty concerning the extent of this reaction in the RH-3 but if it does occur then evidently high results will be experienced.

Gettering of Hydrogen

A study of the literature (6) shows that gettering of molecular hydrogen by magnesium and zinc does not occur. Magnesium will getter atomic hydrogen, but as shown by Figure 2, molecular decomposition is not favoured under the conditions of the analysis. Degréve (2) also reports no evidence for a gettering effect.

Experimental Procedure

The Surface Correction

One of the complications in the determination of hydrogen in aluminum is the fact that surface reactions can occur which generate hydrogen additional to that which is present in the bulk of the metal. These reactions are mainly:

(a) Breakdown of hydrocarbons from contaminating oil or grease.

(b) Reaction of adsorbed surface moisture with the metal sample.

The former problem is usually minimized or overcome by dry machining of the sample and avoiding subsequent manual contact. In any method for hydrogen determination therefore, the surface hydrogen must be separated from the bulk hydrogen which is the required measurement.

In the Ransley sub-fusion method this is accomplished by first measuring the total hydrogen, then removing the sample from the apparatus, re-machining the surface and replacing the sample in the equipment for a second hydrogen determination. The latter is subtracted from the total hydrogen to provide a surface correction. This presupposes that the surface contamination on the newly exposed surface is identical with that on the original sample and that the new surface is close to or identical with that presented to the instrument during the total hydrogen determination.

With the fusion method, Degrève (2) maintains that surface hydrogen could be eliminated by a thermal treatment at 400°C prior to melting and Simensen and Lauritzen (5) found that preheating to 500°C is necessary in the case of magnesium/aluminum alloys. Based on the experience of these investigators we found that operation at 0.3 A plate current for 40 seconds produced the temperature time curves given in Figure 3. This shows the sample spends approximately 1 minute above 400°C and the temperature rises to a maximum of approximately 480°C before falling. This is the surface treatment condition used in all of our subsequent analysis.

It was later considered appropriate to determine the temperature range at which the surface hydrogen is released. To do this, experimental conditions were selected which would permit the sample temperature to be increased in stages from room temperature up to 600°C. Further conditions were determined to raise the sample from room temperature to 200, 400 and 500° directly. These conditions are shown in Table III. Figure 4 demonstrates that these temperatures can be maintained for periods of at least 4 to 6 minutes which was considered adequate for the release of surface hydrogen. Since this surface treatment step is done after a crucible blank it was also found that 20 minutes are required for the sample to cool to room temperature after the blank determination.

Using these conditions a number of experiments were done.

In the first experiment a sample of pure aluminum was heated stepwise up to 600°C and then melted. No measurable emission of hydrogen was observed below 500°C where a peak corresponding to 0.014 mL/100g was detected followed by no further emission until fusion which produced a peak corresponding to 0.036 mL/100g. The chart recorder trace for this experiment is shown in Figure 5.

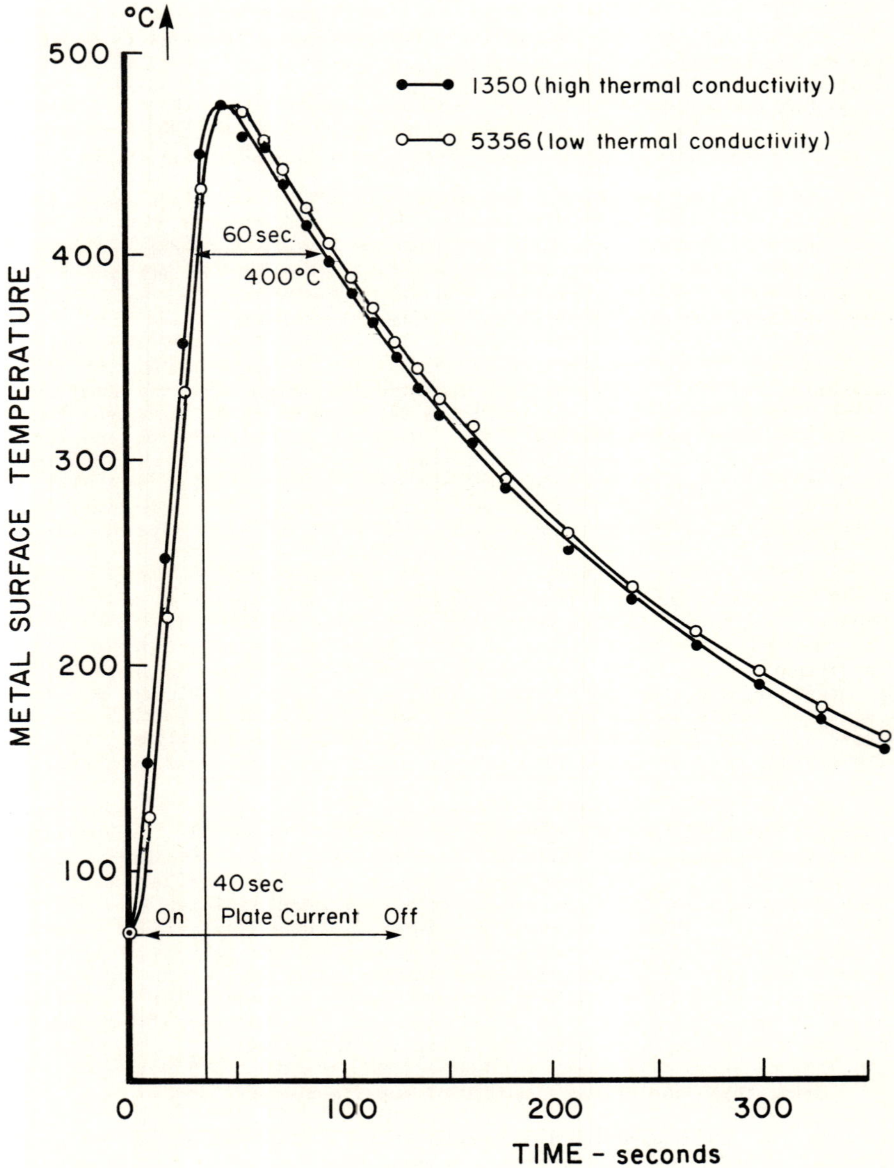

Figure 3 - Surface degassing for Al alloys at 0.3 A plate current for 40 s.

In a second experiment a pure aluminum sample was deliberately contaminated by manual contact and heated in the same way as for the previous sample except that it was not fused. In this case no significant emissions were observed below 500°C and then small peaks were observed at this temperature and at 550 and 600°C. The corresponding recorder trace is shown in Figure 6.

Heating a contaminated pure aluminum sample directly from 25-500°C produced a rather large peak corresponding to 0.32 mL/100g as shown in Figure 7.

In a further experiment the sample of AA-5356 extruded rod of known analysis (0.17 mL/100g by the Ransley method) was analyzed with and without surface degassing at 500°C resulting in the trace shown in Figure 8. This shows good agreement without degassing but a much lower result of approximately 0.03 mL/100g with the degassing step. Evidently in this case the severity of the surface treatment is too great and results in the removal of some of the bulk hydrogen prior to fusion. The measured surface gas concentration appears to be too low to explain the difference. This is because the integrator does not recognize the peak until the base line is displaced by one division. Thus the greater part of the surface gas peak has not been integrated.

In a final experiment two specimens of AA-7029 alloy remaining after a Ransley analysis were analyzed using our normal surface degassing condition of 0.3 A plate current for 40 seconds. One of these samples was deliberately contaminated by manual contact. The resulting recorder traces are shown in Figure 9. The contaminated sample clearly demonstrates significantly larger surface gas and analytical peaks. This shows that the thermal treatment of the surface was inadequate to remove all of the contamination.

Table III. Power/Time Settings Required to Increase and Control Sample Temperatures for Surface Degassing

Temperature Change (°C)	Initial Current* (A)	Time (s)	Control Current (A)
25 → 100	0.20	11	0.040
100 → 200	0.30	9	0.070
200 → 300	0.30	9	0.100
300 → 400	0.30	15	0.130
400 → 500	0.30	18	0.155
500 → 550	0.30	11	0.170
550 → 600	0.30	15	0.185
25 → 200	0.40	8	0.070
25 → 400	0.40	20	0.125
25 → 500	0.40	30	0.155

* Plate Current

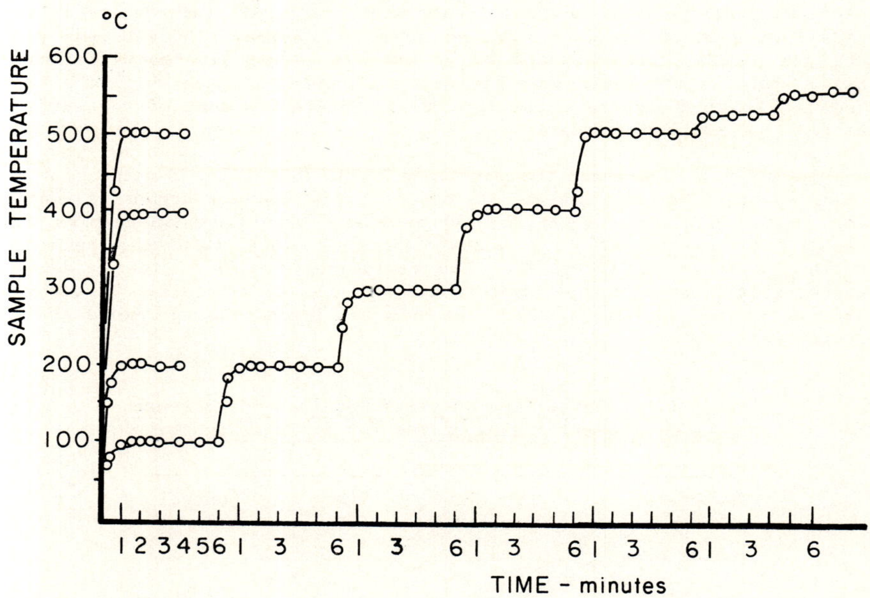

Figure 4 - Raising sample temperature in stages to examine surface degassing

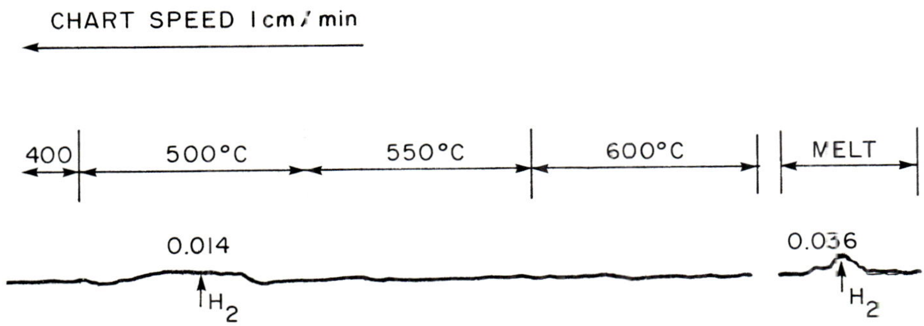

Figure 5 - Stepwise heating of pure Al sample showing H_2 evolution in mL/100g

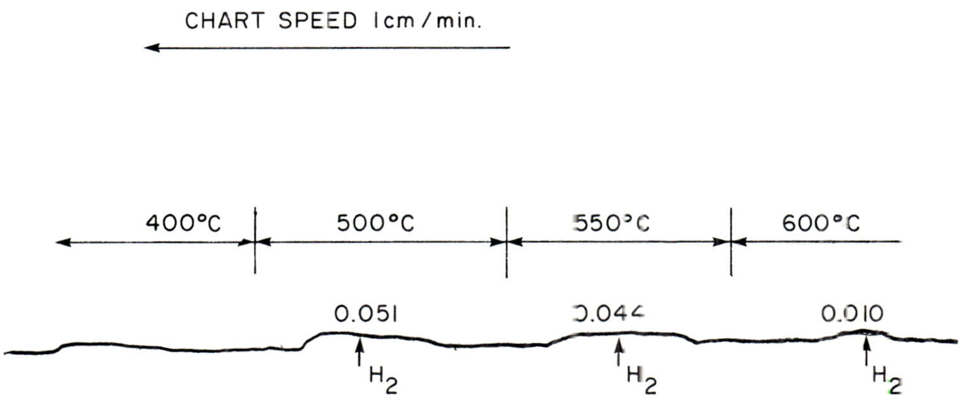

Figure 6 - Stepwise heating of pure Al sample with surface contaminated by manual contact (H_2 in mL/100g)

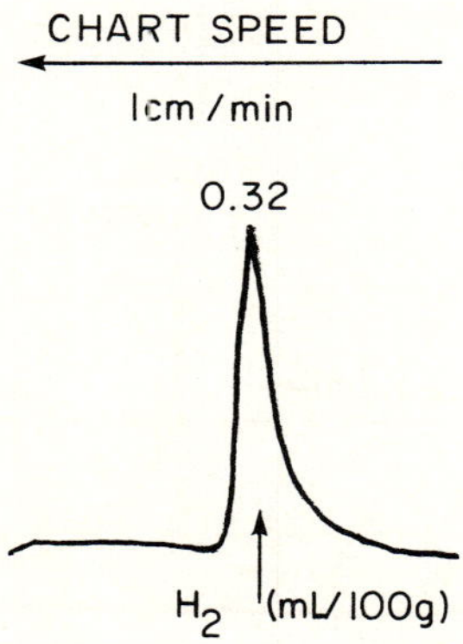

Figure 7 - Heating pure Al sample from 25 - 500°C surface contaminated by manual contact

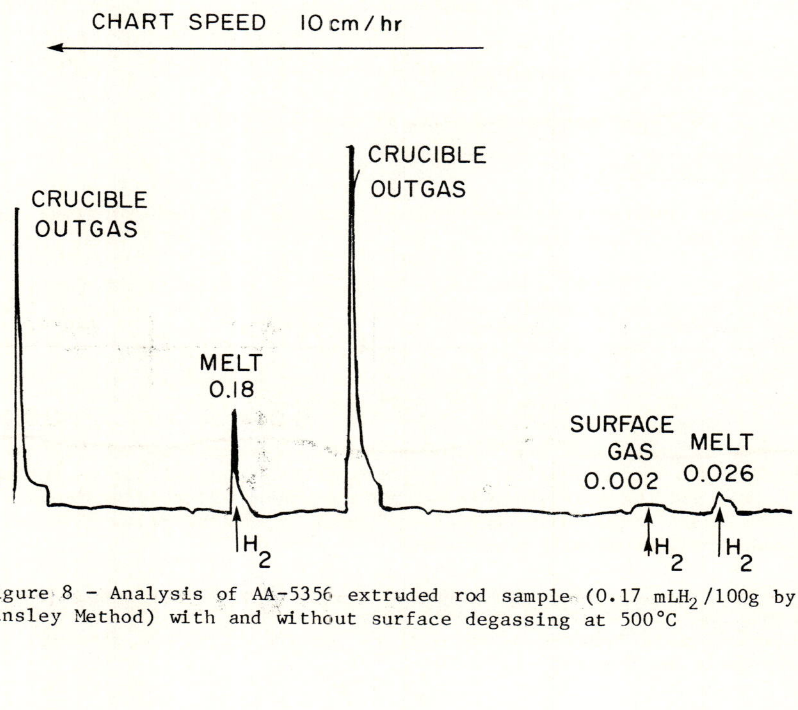

Figure 8 - Analysis of AA-5356 extruded rod sample (0.17 mLH_2/100g by Ransley Method) with and without surface degassing at 500°C

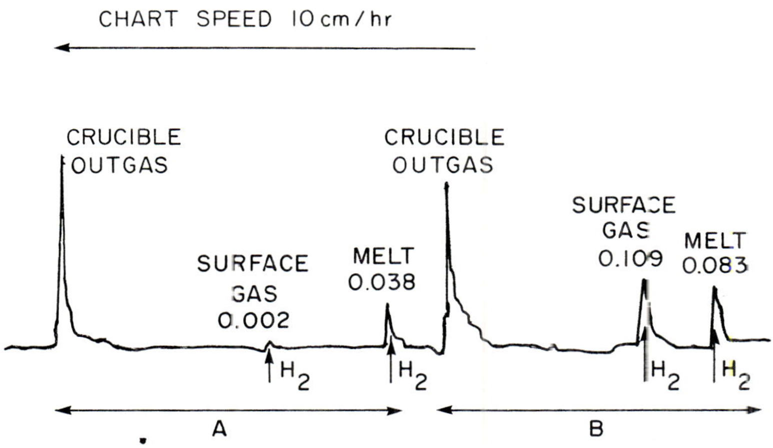

Figure 9 - Analysis of AA-7029 Ransley bar sample after analysis by Ransley method. Surface degassed 0.3 A for 40s.

(A) Uncontaminated surface

(B) Surface contaminated by manual contact

Analytical Procedure

In the interests of brevity full details will not be given but a rough outline follows below.

Step 1 A graphite crucible is placed in the induction furnace and a carefully prepared sample weighing 6.0-6.5g is sealed into the loading head.

Step 2 The crucible is baked out for 3 minutes using a plate current of 0.85 A. Cooling takes place for 5 minutes.

Step 3 With the instrument weight compensators set at 1 g and the furnace plate current at 0.7 A the crucible is heated for a period of 23 seconds and the instrument is set to analyze any evolved hydrogen. The crucible blank measurement usually produces a reading below 0.01 on the instrument's digital volt meter (DVM).

Step 4 The sample is introduced from the loading head into the crucible and its weight is entered into the weight compensator. Surface contamination is removed using a plate current of 0.3 A for 40s. The sample cools for 5 min.

Step 5 Using a plate current of 0.7 A the sample is melted which takes approximately 23 seconds. The furnace power is turned off and the analytical result is read on the DVM after a period of about 3.5 minutes.

Calibration

This is accomplished by passing a known volume of pure hydrogen through the catharometer and correcting for pressure and temperature to reduce the result to a pressure of 1 atmosphere at 0°C.

Drying the Carrier Gas

The three narrow diameter steel tubes containing magnesium perchlorate which are a standard part of the equipment, needed to be supplemented by a larger drying column containing molecular sieve to produce satisfactory drying. The dew point found with this arrangement is approximately -60°C.

Sample Type

Samples were taken from prepared melts for the investigation using a Ransley mould. The sproule was sawn off the mould and the bar was cut in half for the analysis. Half of the bar was analyzed by each of the two methods. The bars were dry machined on a lathe to a final weight of about 6.5g.

Results

These are shown in Table IV in the form of a standard deviation for the LECO RH-3 compared with the Ransley method for 99.8% aluminum and the various alloys examined in this investigation. The associated degrees of freedom are also shown together with a calculated average bias for the group of samples examined. In most cases different ranges of hydrogen concentration were examined and the foregoing statistical information is given for each range.

The importance of adequate drying of nitrogen carrier gas is demonstrated by the first results shown for AA-4043 and AA-2036 alloys where a significant positive bias of 0.05-0.06 mL/100g is found such that the LECO result is higher than the sub-fusion result. When extra drying was added in the form of a molecular sieve, the bias became almost insignificant. The standard deviation of internal precision of the Ransley sub-fusion method is generally considered to be 0.02 mL/100g and assuming that the LECO RH-3 procedure has approximately the same precision then the accuracy can be considered satisfactory for both of these alloys although its magnitude tends to increase at the higher concentration ranges as might be expected. The results are also satisfactory for AA-3003 and AA-3004 alloy although in the former case the samples were not Ransley bars but were taken from a 3" diameter DC ingot and in this case the bias is slightly larger, such that the Ransley result is slightly higher than the sub-fusion result.

In the case of AA-5182 alloy there was concern over the possible deposition of magnesium on the furnace tube and elsewhere. This was

Table IV. Standard Deviation (σ) LECO V. Ransley for Various Alloys

Range (mLH_2/100g)	Degrees of Freedom	σ mLH_2/100g	Bias[1] mL/100g
		Pure Al 99.8%	
0.09-0.25	6	0.019	-0.008
		AA-4043 Alloy 4.5-60% Si	
0.08-0.20	8	0.019	0.059[2]
0.16-0.28	14	0.031	-0.0007
0.41-0.49	13	0.039	-0.005
		AA-2036 Alloy 2.2-3.0% Cu, 0.3-0.6% Mg	
0.10-0.20	12	0.029	0.051[2]
0.05-0.12	10	0.036	0.003[3]
0.14-0.25	7	0.045	0.002[4]
0.35-0.67	13	0.031	0.006
		AA-3003 Alloy 1-1.5% Mn	
0.17-0.26	10	0.021	-0.01[5]
		AA-3004 Alloy 1-1.5% Mn, 0.8-1.3% Mg	
0.16-0.24	5	0.034	0.001
		AA-5182 Alloy 0.2-0.5% Mn, 4-5% Mg	
0.06-0.15	14	0.027	0.004
0.09-0.15	18	0.022	0.006
0.38-0.52	15	0.050	-0.069
0.44-0.51	6	0.028	0.015[6]
0.67-0.80	6	0.052	-0.006[6]
		AA-7075 Alloy 1.2-2.0% Cu, 2.1-2.9% Mg, 5.1-6.1% Zn	
0.06-0.24	16	0.026	0.003[3]

(1) LECO RH-3 - Ransley
(2) Inadequate drying of N_2 carrier gas
(3) 2 Rogue values rejected
(4) 4 Rogue values rejected
(5) Samples from 3" diam DC ingot
(6) No surface correction

checked out by comparing analytical results from the Ransley method with those from the LECO RH-3 method when the furnace tube had been used from 0-5 times. An analysis of variance, see Appendix 2, of the results showed no significant effect due to multiple use of the furnace tube and visual observation showed no significant deposition of magnesium film. Also no signficant bias was observed nor any significant affect on the crucible blank. Therefore, reaction of deposited magnesium with hydroxyl groups from the silica network is not important.

For AA-5182 alloy the internal precision for both methods was measured by using duplicate halves of the same Ransley bar and from Table V this has a standard deviation of approximately 0.03 mL/100g. One might therefore expect the accuracy for this alloy to be somewhat worse. From Table IV poorer accuracy seemed to be associated with the higher concentration levels and the results on samples from one melt in the range 0.38-0.52 mL/100g also demonstrated a strong bias such that the LECO results were significantly lower than the sub-fusion results. We considered that this might be due to removal of some of bulk hydrogen during the surface treatment step thus further samples were analyzed at high concentration levels in which no surface treatment was applied. Comparing the results with Ransley determinations also with no surface correction demonstrated that the bias was practically eliminated and accuracy was within acceptable limits.

Our Ransley determinations usually involve the application of a standard surface correction factor specific for each alloy which obviates the second hydrogen extraction necessary to actually measure the surface gas. However, when we actually measured the surface correction for AA-5182, we found that the average value for 22 determinations was 3.18×10^{-4} mL/cm^2 compared to the customary value of 2.57×10^{-4} mL/cm^2. The relative standard deviation of the measured value was very high at 30%.

For AA-7075 alloy in the concentration range of 0.06-0.24 mL/100g the accuracy is satisfactory and the bias negligible when surface corrections are applied. The precision of both methods for AA-7075 as shown in Table V suggest that the accuracy is somewhat better than might be expected. Using the same logic as for AA-5182 alloy a suite of samples at higher range of H_2 concentration namely 0.43-0.63 mL/100g was analyzed with no surface correction by either method. However in this case the results were much different and they are shown in detail in Table VI. The relevant column is the last one which shows the difference between the two methods without surface correction. In eight cases out of the 18 samples tested the LECO result is significantly higher than the Ransley result. This represents a large percentage of the results and suggests a new analytical problem. At first an instrument problem was suspected and several samples of AA-5356 rod were analyzed after the sixth AA-7075 sample. The AA-5356 rod had previously shown a value of 0.18 mL/100g by Ransley analysis. Results varied from 0.17-0.26 mL/100g, slightly high but not high enough to account for the magnitude of the differences already observed with the AA-7075 alloy. All desiccants were changed and the dew point checked giving a value of -60°C. Analysis was then continued but as can be seen from Table VI high results from the LECO RH-3 continued to appear, erratically interspersed with other results which seemed reasonably acceptable. One of the samples, AWWH-18 was degassed at 500°C but this resulted in the loss of some of the bulk hydrogen producing in this case a result which was much too low. The results also show that the measured Ransley surface correction is higher than the usual standard surface correction factor and its relative standard deviation is again close to 30%.

Table V. Internal Precision* (σ_F) of Both Methods on AA-5182 and 7075 Alloy

Alloy	Method	σ_p ($mLH_2/100g$)	Degrees of Freedom	Range $mLH_2/100g$
AA-5182	LECO RH-3	0.031	17	0.12-0.27
	Ransley	0.027	18	0.08-0.43
AA-7075	LECO RH-3	0.027(1)	13	0.11-0.48
	Ransley	0.022(2)	10	0.09-0.27

(1) 5 Rogue values rejected
(2) 3 Rogue values rejected
* With surface gas correction

At this point a metallographic examination was made of Ransley bar sections for AA-7075 and AA-6063 alloys. A Ransley sample bar taken from a large ingot of AA-5182 alloy was also examined. The photographic results are shown in Figures 10-12 which show extensive and well-distributed shrinkage porosity occurring in the AA-7075 sample extending through to the surface of the machined sample and visible at a magnification of X100. The extent of shrinkage porosity in the AA-6063 sample visible at magnification of X500 is obviously much less and does not extend close to the machined surface. The AA-5182 alloy also demonstrates shrinkage porosity visible at X100 but although the pores visible are quite large their distribution is not as extensive as with the AA-7075 alloy.

Discussion of Results

In general the accuracy of the LECO RH-3 results is satisfactory providing that the carrier gas is adequately dried. Providing this aspect is taken care of high results generally do not occur. The secondary concern of the reaction of volatilised magnesium or zinc with the hydroxyl ions in the quartz network of the furnace tube should be evident with AA-5182 and AA-7075 alloys if this effect is important. The absence of high results with the former alloy suggest that it is not important and although high results were frequently observed at high levels of hydrogen in the latter alloy when it was not surface de-gassed the problem here, is thought to be due to other factors.

The gettering effect would produce significantly lower results and this was not observed except at high levels of hydrogen in AA-5182 alloy where this was shown to be due to the removal of some of the bulk hydrogen due to surface degassing conditions which were too severe. Thus as forecast on theoretical grounds the gettering effect does not constitute a problem.

The surface treatment step has been shown to be a sensitive area. The main problem is that surface hydrogen is evolved at temperatures close to the melting range of some alloys. Thus the danger is that too severe a treatment will remove some of the bulk hydrogen which we want to measure.

Table VI. H_2 Determination in AA-7075 Alloy

Comparison of results without surface correction, 0.43-0.63 mL/100g

Sample	Measured Surface Correction $mL/cm^2 \times 10^4$	H_2 (mL/100g) Ransley (A)	LECO (B)	(B)-(A)
AWWH-1	8.14	0.47	0.60	0.13
AWWH-2	4.66	0.55	0.61	0.06
AWWH-3	5.84	0.49	0.64	0.15
AWWH-4	5.09	0.56	0.59	0.03
AWWH-5	5.23	0.52	0.65	0.13
AWWH-6	-	-	0.58	-
AWWH-7	3.15	0.50	0.82	0.32
AWWH-8	6.11	0.47	0.83	0.36
AWWH-9	6.49	0.52	0.58	0.06
AWWH-10	4.81	0.43	0.84	0.41
AWWH-11	3.90	0.64	0.59	-0.05
AWWH-12	3.89	0.60	0.63	0.03
AWWH-13	5.67	0.52	0.51	-0.01
AWWH-14	3.81	0.63	0.60	-0.03
AWWH-15	6.93	0.44	0.73	0.29
AWWH-16	4.05	0.58	0.48	-0.10
AWWH-17	4.07	0.46	0.80	0.34
AWWH-18[(1)]	3.36	0.60	0.12	-0.48
AWWH-19	2.70	0.53	0.54	0.01

Standard surface correction = 1.92×10^{-4} mL/cm^2

Average measured surface correction = 4.88×10^{-4} mL/cm^2

Standard deviation of surface correction = 1.44×10^{-4} mL/cm^2

(1) Sample given preliminary heating at 500°C before melting.

On the other hand if the surface contamination is very high then too mild a treatment will not remove it all and the result will be too high. If the main contribution to the surface problem is adsorbed moisture then it seems possible that this might be removed by thermal treatment at lower temperatures. If this could be done successfully then fusion of the sample following this step should produce more accurate results since all the water would be gone. The success of this approach would require such absorbed water to be quite labile at low temperatures. The results shown in Figures 6 and 7 show that less hydrogen is evolved from a contaminated surface during step-wise heating up to 600°C than is involved when such a sample is heated directly and rapidly from 25–500°C. Thus longer heating at lower temperature is an approach worth investigating.

The contrast in the manner of the surface correction step between the two methods, i.e., fusion and sub-fusion is worth discussing here. In the Ransley sub-fusion method surface correction is made after the total hydrogen determination by removing the sample, re-machining the surface and replacing it in the apparatus for a second extraction. The assumption of parity between surface quality and contamination before and after remachining is illogical, especially when one considers the presence of extensive shrinkage porosity as in the case of AA-7075 alloy. Here, remachining of the surface may produce a whole new distribution of cracks and fissures compared to that presented to the apparatus for the total hydrogen determination. The removal of surface contamination in a preliminary separate step prior to the bulk hydrogen determination with the sample contained in the apparatus for both steps would appear to be much more logical. If this can be successfully accomplished in both methods, precision and accuracy may be improved.

The behaviour of AA-7075 compared to AA-5182 alloy at high levels of hydrogen was quite different when analyzed without the surface degassing step. The frequently occurring high results were at first thought to be due to an instrument problem such as inadequate drying of the carrier gas. After verifying the dryness of the gas and obtaining satisfactory results on a known sample of AA-5182 alloy, the high results continued to occur. This suggested something different about the nature of the sample itself and thus led to the metallographic examination discussed above. This was expected to reveal the presence of oxide film inclusions especially since there had been no salt cover on the melt and agitation had occurred during its preparation prior to taking the Ransley bar samples. If oxide film inclusions were present they could have acted as traps for water which would react to form spurious hydrogen when the Ransley bar samples were melted. It was conceivable that such reaction would not occur in the sub-fusion method where the sample is not melted. However such oxide film inclusions were not observed; the main visible feature being extensive shrinkage porosity. This shrinkage porosity is caused by inadequate liquid phase filling of the interdendritic voids which appear as the metal solidifies. The extent of this type of porosity depends on such factors as the alloy's freezing range and casting conditions. It is known that additions of magnesium to aluminum impair the feeding ability of the liquid phase and interdendritic microporosity is observed in alloys containing 7-10% magnesium. Zinc additions are soluble in solid state with little or no eutectic formation. Thus feeding ability is poor and the alloys are subject to microporosity (7). Thus the appearance of

100X

Surface shrinkage porosity

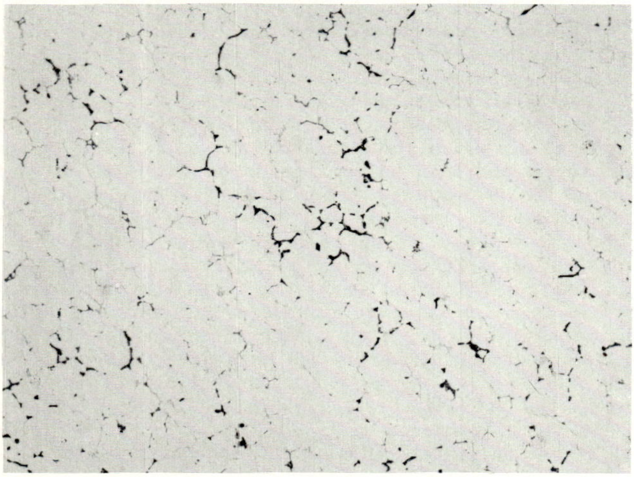

100X

Centre shrinkage and gas porosity

Figure 10 - AA-7075 Alloy (Ransley Bar)

500X

Surface shrinkage and gas porosity in subfusion sample

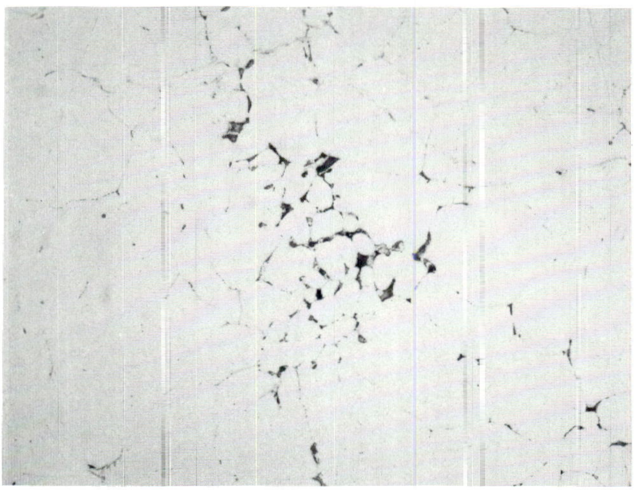

500X

Shrinkage porosity in centre of 6063 sample

Figure 11 - AA-6063 Alloy (Ransley Bar)

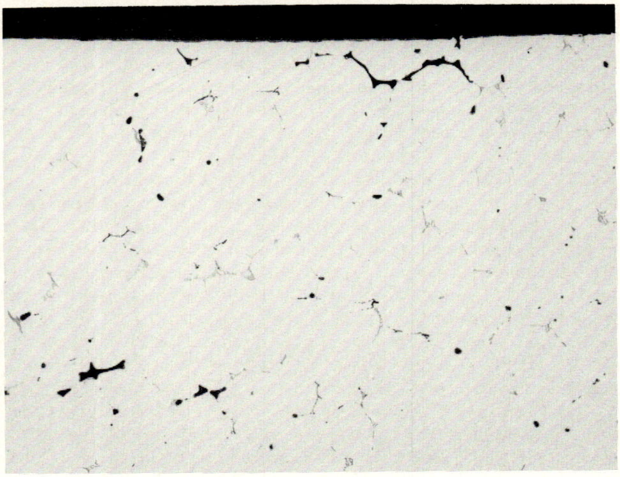

100X

Surface shrinkage porosity

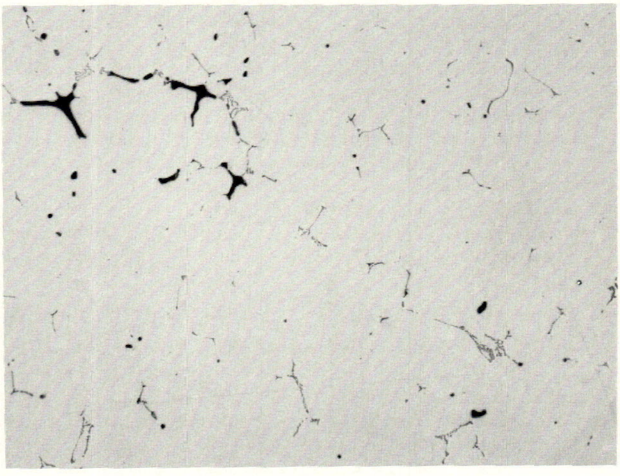

100X

Centre shrinkage porosity

Figure 12 - AA-5182 Alloy (Ingot Sample)

extensive shrinkage porosity in the AA-7075 is not unexpected. These voids can act as traps for hydrogen gas and when present at the surface could also serve as traps for water vapour. In the sub-fusion procedure where the sample is heated slowly to the extraction temperature such entrapment of water vapour may readily escape and either be absorbed into the silica networks of the surrounding glassware or pumped into the gas collection part of the equipment. Such water would not report as hydrogen which is finally measured by allowing it to pass through a heated palladium tube. In the fusion method by contrast the sample is melted within 20 to 30 seconds. Such entrapped water would have little chance to escape and would react to form hydrogen. Thus we have a possible explanation for the frequently observed high values during the analysis of AA-7075 samples without the degassing step. As suggested above, suitable thermal treatment at lower temperature may have succeeded in removing this water and preventing its interference at the fusion stage. Another possible effect caused by such extensive porosity would be produced if the latter is not homogeneously distributed. If we assume the presence of more porosity at one end of the Ransley bar then more hydrogen will be likely trapped in this end and when analyzed, disparate results will occur. This phenomenon was observed when the surface degassing step was included at lower hydrogen levels in AA-7075 alloy and in AA-2036 alloy at both high and low levels. This produced rouge values which were rejected from accuracy calculations on statistical grounds as noted in Table IV. It was again observed within the LECO RH-3 and Ransley methods during the determination of internal precision when the same method was used on the duplicate portions of the bar (see Table IV).

A non-zero value of hydrogen content is often obtained upon fusion analysis of the same sample after a Ransley determination as seen in Figure 9 or upon remelting the sample without removing it from the furnace. This is similar to the so-called surface hydrogen which is measured when a sample is replaced in the Ransley apparatus after the total hydrogen measurement. No definite explanation has been found for this but it may be due to the slower release of hydrogen which is more strongly bound in the metal than the conventional physically entrapped hydrogen which is usually measured. Talbot (8) defines both physical and chemical traps for hydrogen, the latter for example being due to hydrides such as zirconium hydride and sodium hydride. Evidence for the existence of these is presented by Konar and Banerjee (9) and Varhegyi et al (10) who demonstrate its slower release from the metal and suggest a method for its determination. VanLancker (11) suggests that such chemically combined hydrogen can produce deleterious affects due to the formation of porosity during rolling.

Conclusions

(1) The accuracy of the fusion approach as employed by the LECO RH-3 determinator is satisfactory for the measurement of hydrogen in aluminum and common alloys discussed in this paper.

(2) Generation of spurious hydrogen by reaction of water vapour with volatilised metal can be avoided by adequate drying of the carrier gas.

(3) Gettering of hydrogen by volatalised metal is unimportant.

(4) No significant release of hydrogen caused by surface contamination occurs below 400-500°C. This complicates the surface degassing step since depending upon the severity of the thermal treatment, either incomplete removal of the contamination, or release of hydrogen from the bulk of the metal may occur.

(5) Reaction of volatilised magnesium or zinc with hydroxyl groups in the silica network of the quartz furnace tube, producing hydrogen is not considered to be a problem.

(6) Extensive shrinkage porosity in the case of AA-7075 alloy may be linked to frequent high results which occur when surface degassing is not employed but further work will be necessary to verify this.

References

1. W.E. Ransley and D.E.J. Talbot, "The Routine Determination of the Hydrogen Content of Aluminum by the Vacuum Hot Extraction Method", J. Inst. Metals, 84, (1955), 445.

2. Francis Degrève and Christian Jardin, "New Methods for the Determination of Hydrogen Content of Aluminum and its Alloys, Part II Rapid Determination by the Nitrogen Carrier Gas Fusion Method". Metallurgical Transactions B., (December 1975), 6B, 545.

3. Handbook of Chemistry & Physics, 51st Edition, 1970-71, Chemical Rubber Co.

4. Perry et. al., "Perry's Chemical Engineers Handbook", 4th Edition, McGraw Hill Publishing Co.

5. Chr. J. Simensen and J.L. Lauritzen, "Analysis of Hydrogen in Aluminium-Magnesium Alloys", Aluminium, 56, Jahrg., 1980, 2.

6. S.L. Reimann, "Vacuum Technique", pp. 353-354, Chapman and Hall, 1952.

7. K.R. VanHorn, "Aluminum Vol. 1., Properties, Physical Metallurgy and Phase Diagrams", pp. 185-186, American Society for Metals, 1967.

8. D.E.J. Talbot, "Effects of Hydrogen in Aluminium, Magnesium, Copper and their alloys" International Metallurgical Reviews, 1975, Vol. 20, pp. 166-184.

9. J. Konar and N.G. Banerjee, "Vacuum Heating Hydrogen Determination in the Aluminum and Aluminum Alloys N.M.L. Tech. J., Vol. 16, Feb-May 1974, pp. 18-19.

10. G. Varhegyi et al, Method for the Determination of Hydrogen Content of Aluminum or Aluminium Alloys. Hungarian Patent 18,441, 12 January 1978.

11. Marc VanLancker, "Book of Metallurgy of Aluminum alloys" Chapman and Hall, London, 1978.

Appendix 1

Calculation of H_2 Produced by Reaction of H_2O Vapour in N_2 Carrier gas With Al, Mg, Zn, Li During LECO RH-3 Determination

1. Reactions:-

$$2Al + 3H_2O \rightarrow Al_2O_3 + 3H_2$$

$$Mg + H_2O \rightarrow MgO + H_2$$

$$Zn + H_2O \rightarrow ZnO + H_2$$

For all reactions, 1 volume H_2O gives 1 volume H_2.

2. Terminology:-

V_N = N_2 carrier gas flow = 0.28 L/min

p_{H_2O} = partial pressure of H_2O in N_2 (torr)

W = weight of sample = 6.5 g

3. Assumptions:-

(a) furnace pressure = 1 atm (760 torr).

(b) water in N_2 reacts for entire analytical cycle (3 min).

4. H_2 produced:-

$$V_H(\text{mL}/100\text{gAl}) = \left(V_N \frac{LN_2}{\text{min}} \times 3 \text{ min} \times \frac{pH_2O}{760}\right) LH_2O \times \frac{1LH_2}{LH_2O} \times \frac{1000 \text{ mL}}{1L} \times \frac{100g}{Wg}$$

$$= \frac{394.7\ V_N pH_2O}{W}$$

for V_H = 0.01 mL/100g:-

$$p_{H_2O} = \frac{V_H W}{V_N} \times 0.00253$$

$$= \frac{0.01 \times 6.5 \times 0.00253}{0.28} = 0.00059 \text{ torr (dew point of } -78°C)$$

Appendix 2

Table I. Effect of Re-Using Furnace Tube (AA-5182 Alloy)

Number of Previous Uses of Furnace Tube											
0				1				2			
	H_2 (mL/100g)				H_2 (mL/100g)				H_2 (mL/100g)		
	A	B	A-B		A	B	A-B		A	B	A-B
Blank	LECO	Ransley		Blank	LECO	Ransley		Blank	LECO	Ransley	
0.023	0.23	0.27	-0.04	0.013	0.24	0.29	-0.05	0.007	0.20	0.28	-0.08
0.014	0.26	0.25	0.01	0.009	0.24	0.31	-0.07	0.011	0.32	0.32	NIL
0.014	0.34	0.27	0.07	0.001	0.16	0.12	0.04	0.007	0.16	0.16	NIL
0.020	0.21	0.16	0.05	0.020	0.12	0.11	0.01	0.020	0.17	0.12	0.05
0.020	0.18	0.15	0.03	0.004	0.15	0.15	NIL	0.004	0.14	0.18	-0.04
0.004	0.15	0.14	0.01	0.030	0.14	0.16	-0.02	0.004	0.15	0.16	-0.01

Number of Previous Uses of Furnace Tube											
3				4				5			
	H_2 (mL/100g)				H_2 (mL/100g)				H_2 (mL/100g)		
	A	B	A-B		A	B	A-B		A	B	A-B
Blank	LECO	Ransley		Blank	LECO	Ransley		Blank	LECO	Ransley	
0.036	0.20	0.33	-0.13	0.001	0.32	0.24	0.08	0.007	0.22	0.28	-0.06
0.030	0.28	0.28	NIL	0.030	0.38	0.30	0.08	0.030	0.20	0.26	-0.06
0.007	0.18	0.16	0.02	0.017	0.19	0.15	0.04	0.004	0.28	0.15	0.13
0.017	0.14	0.12	0.02	0.011	0.15	0.15	NIL	0.004	0.15	0.12	0.03
0.004	0.20	0.17	0.03	0.001	0.16	0.15	0.01	0.001	0.14	0.12	0.02
0.001	0.13	0.16	-0.03	0.040	0.17	0.22	-0.05	0.040	0.22	0.17	0.05

Table II. Analysis of Variance - LECO V Ransley

	Sum of Squares	Degrees of Freedom	Mean Square
Between tube uses	0.0123	5	0.00246
Within tube uses	0.087	30	0.00272
Total	0.0940	35	-

$F = \frac{0.00246}{0.00272} = 0.9$ (not significant)

Table III. Analysis of Variance - Crucible Blanks

	Sum of Squares	Degrees of Freedom	Mean Square
Between tube uses	0.000252	5	0.00005
Within tube uses	0.004636	30	0.000155
Total	0.00489	35	-

$F = \frac{0.00005}{0.0000155} = 0.323$ (not significant)

SEGREGATION EFFECTS IN ALUMINUM METAL SAMPLES

EFFECT OF SAMPLE MOULD DESIGN

Jean-Luc Fortier, Christian Tremblay

Alcan International Limited
Arvida Laboratories and
Experimental Engineering Centre
Arvida (Jonquière), Québec, Canada
G7S 4K8

More than ever before the metallurgical industry needs to cast alloys with narrow tolerances in their chemical composition in order to reduce the variability of the mechanical properties and fabricating characteristics. Good quality control involves adequate casting practices, proper sampling techniques and finally appropriate analytical methods. Instrumental inaccuracies of modern optical emission spectrometers are no longer the limiting factor in quality control but rather, sampling practices and sample mould design. This communication examines the segregation profiles of the major constituents for several alloys using the most commonly employed sample shapes.

Introduction

More than ever before the metallurgical industry needs to cast alloys with narrow tolerances in their chemical composition in order to reduce the variability of the mechanical properties and fabricating characteristics. This can only be achieved by adequate analytical quality control. Presently most of these control analyses are carried out by optical emission spectrometry (OES).

As with other aspects of the fabricating technology, the quality and performance of analytical instrumentation have been greatly improved during the past decade. Fifteen or twenty years ago a precision of about 5% in the chemical composition of alloys was the goal, 1-2% was very optimistic. Today a precision of 1% or even better can be obtained by OES. However, instrumentation is only one link in the analytical process chain. There are many others which have to be considered.

Quality Control Requirement

Good analytical quality control of molten metal by OES involves many requirements and all are equally important. For example:

- reliable analytical instrumentation
- adequate and good quality spectrochemical standard
- homogeneous and representative samples
- accurate analytical methods
- short analytical time to assure rapid feedback

Many of the quality control requirements can be met. Thanks to recent improvements in emission spectrometry (1) including the use of Nitrogen (2) or Argon (3) atmosphere surrounding the spark, new types of excitation sources (4,5) which reduce the structural and matrix effect of the sample, and the availability of homogeneous spectrochemical standards. But as with any other type of analysis, the quality of the result cannot be better than the quality of the samples (6).

Sampling Requirement

Sampling and sample preparation are key parts of the analytical procedure. Unfortunately, the complexity of sample procedure is usually underestimated. Adequate sampling and sample preparation involve many factors:

- samples must be representative of the melt
- sample composition must be homogeneous
- the metallurgical history of the sample should be identical to that of the standard used
- sampling sequence must test the homogeneity of the melt
- sample preparation must be simple

Using adequate procedure most of these sampling requirements can be achieved, except for assuring the homogeneity of the sample. Controlling the segregation phenomena and thus the homogeneity is critical if we consider the fact that the spark-to-plane analytical technique only involves about 1 mg of the sample. In this study we examined the segregation profiles of 5 different sample shapes cast under various sampling conditions to evaluate the effect of mould design, temperature of the molten metal, temperature of the mould and type of cooling of the sample.

Experimental Procedure

Type of Moulds and Samples

We examined five types of sample moulds: horizontal disk, mini-disk, vacuum disk, small pin and big pin (Figures 1 and 2). The mould for the horizontal disk is an Alcan version of the ASTM type A mould (7). The mini-disk mould is a smaller modified version of the ASTM type A mould with improved sample preparation and handling characteristics. The vacuum disk mould is commercially available. Small pin and big pin are both book type moulds, the former giving a pair of pins of 0.250 inch diameter x 1.25 inch long, the big pin mould is a small pin mould with a drilled orifice to give one pin of 0.75 inch (1.8 cm) diameter x 1.50 inch (3.6 cm) long. The size, weight of sample including the sprue are given in Table I along with the weight of the moulds. All moulds were provided with a thermocouple for monitoring the mould temperature during sampling.

Table I. Physical Characteristics of Sample Moulds and Samples

	Horizontal disk	Mini disk	Vacuum disk	Small pin	Big pin
Length (inch)				1.25	1.5
Width (inch)	.375	.20	.50		
Diameter (inch)	2.25	1.55	1.5	.25	.75
Weight sample (g)	155	40	48	53	63
Weight of mould (kg)	9	3	2.5	3	3

Sampling of Molten Metal

Sampling tests were carried out with 3 alloys representative of the majority of commercial aluminum alloy types AA1000, 3000 and 6000. A strict casting sequence was used to evaluate the effect of molten metal temperature, mould temperature, mould coating and type of sample cooling. For each type of alloy, more than 75 samples were cast from a melt weighing 75 lbs (34 kg) prepared in a 24 kW electric furnace (Bikley, Model T40H) using maximum composition limits as target. Sampling was carried out at two temperatures of the melt: 780°C ±5 followed by 680°C ±5.

In order to evaluate the effect of mould temperature, the samples were cast consecutively without cooling the mould. The mould temperature was allowed to rise from 25°C to 350°C and was monitored with a thermocouple fixed on the mould base. To evaluate the effect of the surface of the mould, moulds with and without a MoS_2 spray coating were used. The samples were removed from the mould as soon as metal had solidified. Samples were alternately water-quenched by submerging in a bucket of water and air-cooled by leaving exposed on a steel bench.

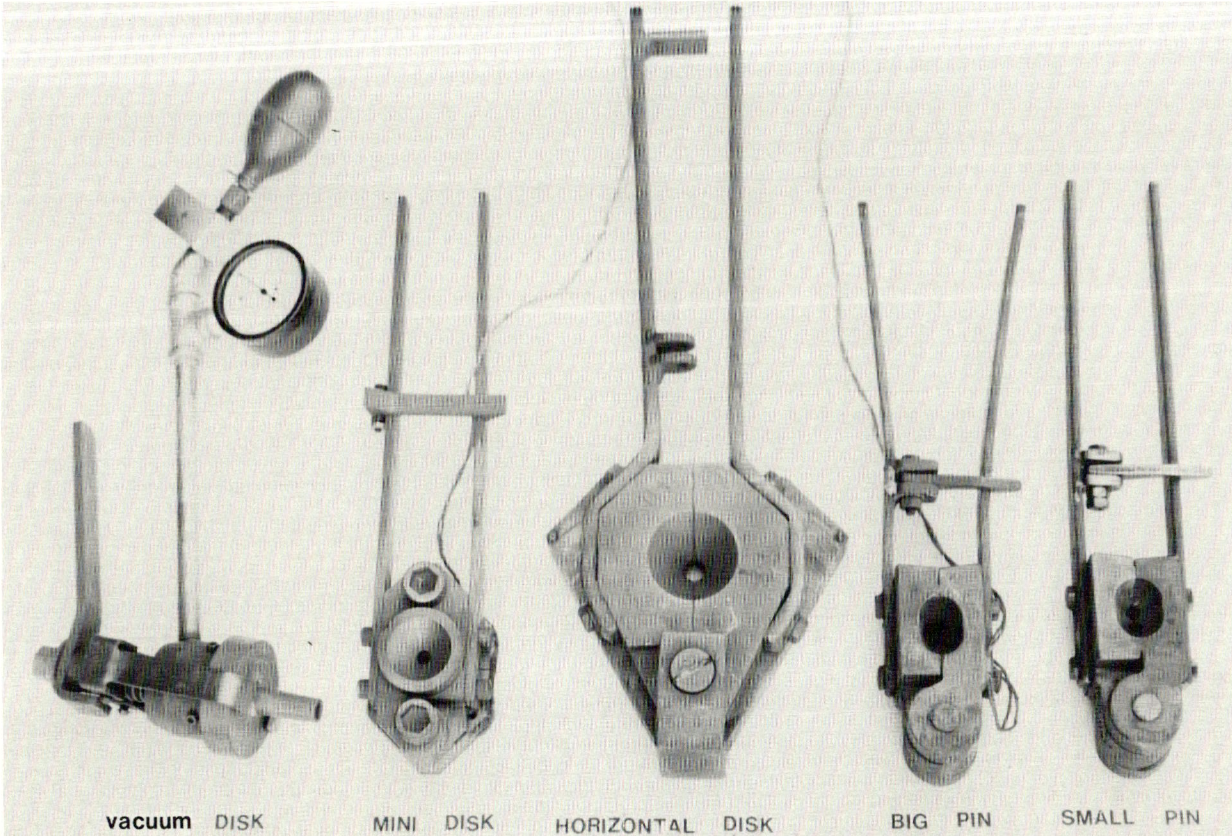

Figure 1 - Spectrochemical sample moulds

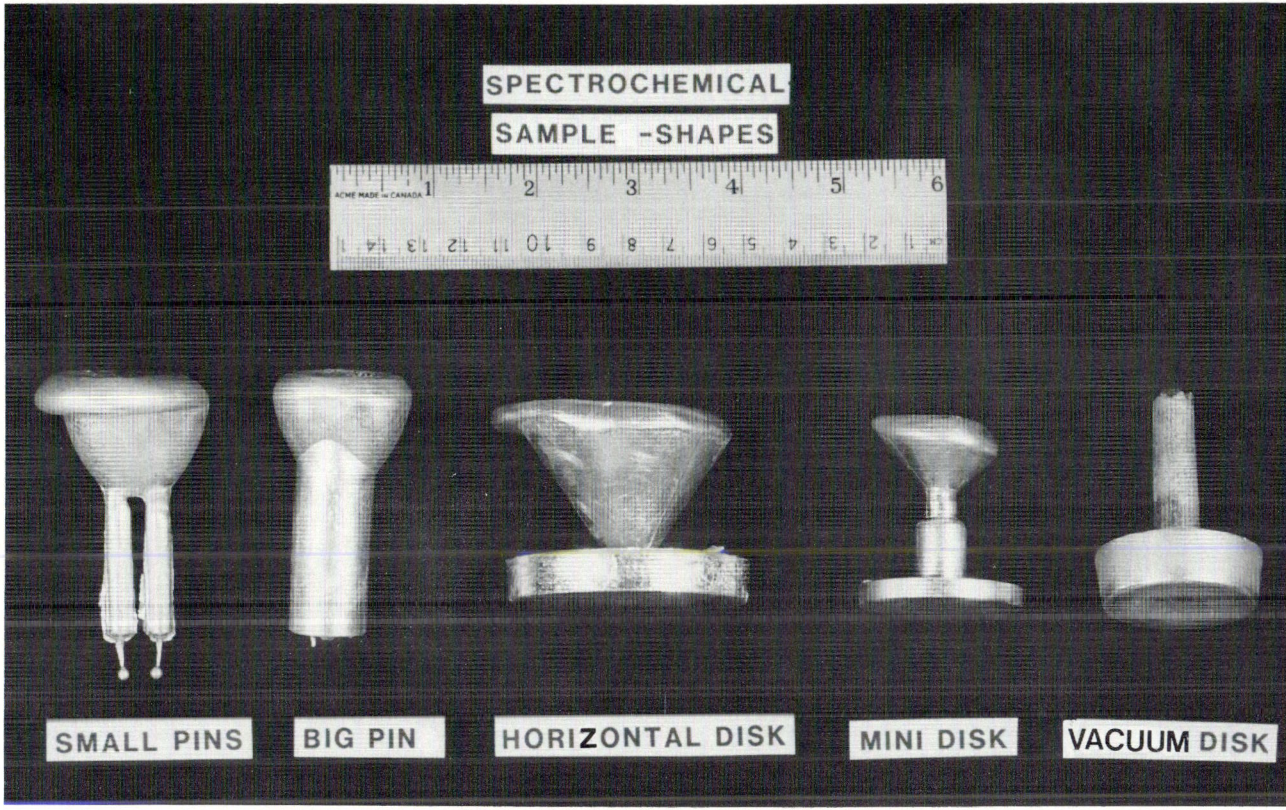

Figure 2 - Spectrochemical sample shapes

OES Analysis

All disk and the big pins samples were analysed with an ARL-34000 Argon atmosphere spectrometer (vacuum type). For the small pins, an ARL-29000 air-gap spectrometer with an Alcan computer console was used. The sparked area and the analysed depth on each type of samples is shown in Figure 3. For disk samples, values at each cutting depth were averaged for 4 sparks, in the case of pins, the average of 2 sparks on the same sample were calculated; the second value being recorded after removing only a very small amount of metal ($\sim$ .005 inch). All depths are given relative to the surface opposite to the sprue of the samples.

Wet Chemical Analysis

Wet chemical analysis of alloying elements were carried out using a Perkin-Elmer, Model 4000 Atomic Absorption spectrophotometer, a Bausch & Lomb, Model 2000 spectrophotometer and a ARL-35000 ICP spectrometer. Essentially two types of analyses were carried out, one to evaluate the true chemical composition of the alloys, the other one to evaluate the segregation profiles of elements in the sample. In the former case, because of the radial and axial segregation in the horizontal disk, special sampling of the sample was undertaken. Chips were obtained using a 1/8 inch milling cutter. But unlike Olson and Macy (8) we took a pie shaped cut instead of a rectangular one. For the segregation profile evaluation, the metal chips recovered from the OES sample preparation were analysed. They represent the metal between each analysed depth of the sample over the entire surface of the disk or pin.

Results and Discussions

Real or Apparent Segregation

It is well known that point-to-plane analysis is dependent of matrix and metallurgical structure in the sample (4,5). When segregation profiles are measured by OES, doubt exists whether this segregation is real or apparent. The profiles measured by wet chemical analysis, see Figure 4, are unequivocal, the segregation is real and significant. As expected, the chemical values do not explain all the segregation measured by OES. So part of the segregation must be apparent. Its relative magnitude is a function of the analytical condition used for OES analysis. Depending on the depth analysed, the OES results may be higher or lower than the true chemical value for the original melt. Biases of 10% relative are not unusual.

As shown in Figure 5, the shape of the segregation profile is dictated by the nature of the element and the phase diagram of the alloys. For elements forming eutectics with aluminium (like Fe, Si, Mg, etc.), there is usually an enrichment in the zones which are cooled first, that is to say, toward the walls of the sample. For peritectic elements (like Ti), the phenomena is inverse, the enrichment is found toward the zone which is cooled last. Other elements, like Mn, which have a flat liquidus curve are less sensitive to segregation.

Other tests showed that radial segregation is significant (Figure 6)

MACHINING REFERENCE AND SPARKING AREA

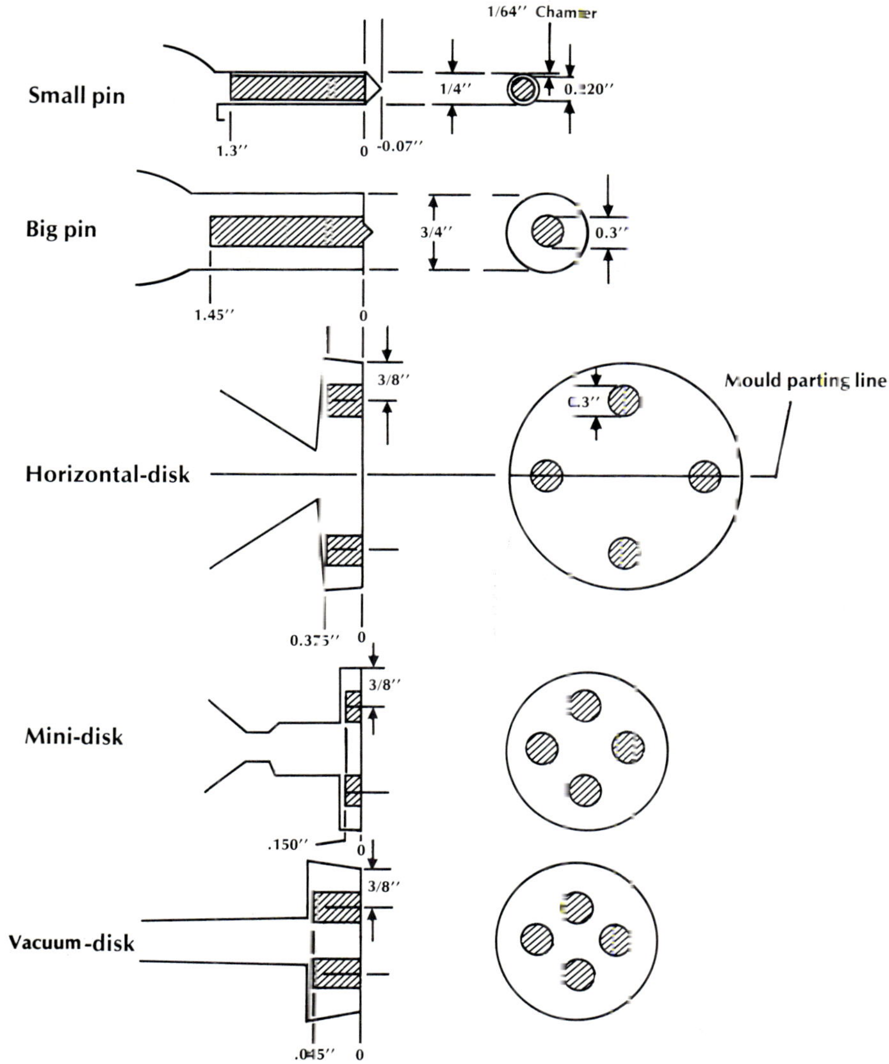

Figure 3 - The sparked position and analysed depth on each of the 5 sample shapes are indicated by the shaded surface. All depth analyses were started from the surface opposite to the sprue and referred to the C value as defined in the drawing.

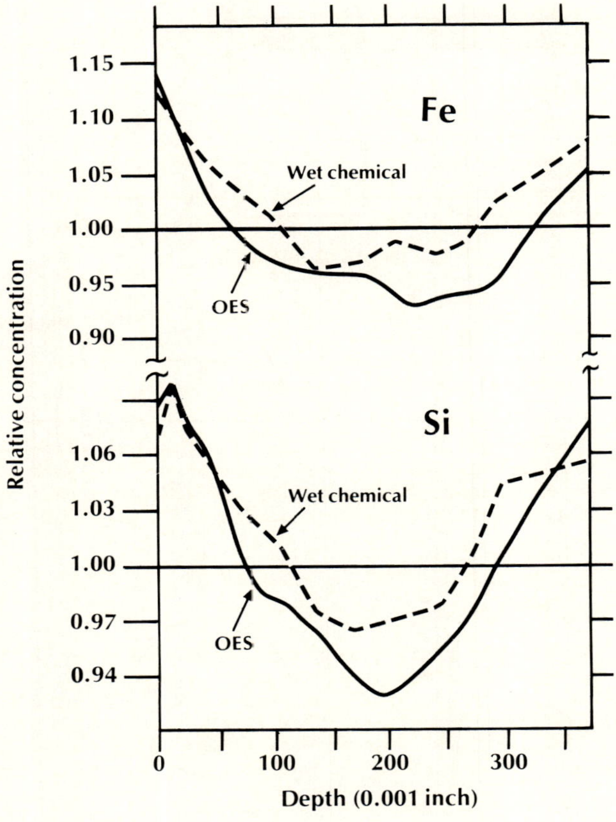

Figure 4 - Depth segregation profile for Fe and Si in horizontal disk measured by OES (——) and wet chemical analysis (----). Concentration (%) variations are expressed in relative value to the true chemical composition of the metal (horizontal lines).

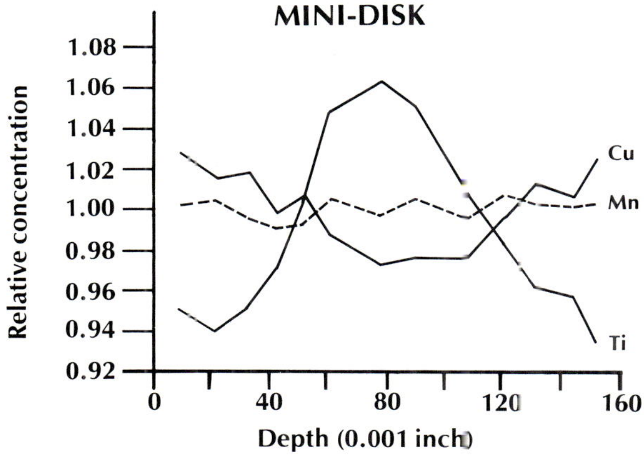

Figure 5 - Depth segregation profile of Cu, Mn and Ti in mini-disk measured by OES. Concentration (%) variations are expressed in relative value to the true chemical composition of metal (1.00).

and that the symmetry of the mould design also has some effect on the apparent and real segregation. It was further found that the segregation profile was very dependent of the sampling conditions: mould type, temperature of the metal, mould temperature and coating, and type of cooling of the sample.

Effect of Metal Temperature

For some alloy types, the temperature of the metal in the holding furnace and during casting may be different by about 100°C. As indicated in Figure 7, the temperature of the metal has a significant effect on the concentration profile. Depending of the sample preparation practice and the attention given to it, different results may be obtained even if the batchmaker made no adjustment in the chemical composition of the molten metal.

Effect of Mould Design

The design of the mould, its weight, type of material, the amount of sampled metal, the shape of the inlet will have an effect on the heat transfer rate. This leads to changes in the concentration profile as shown in Figure 8. Usually, disk type samples showed a concentration plateau slightly lower than the true chemical composition of the metal. In the case of small pins, there is no apparent plateau and values oscillate around true value. For big pins, the situation is worse, the values show a constant decrease in the axial direction of the length, but the initial values are close to the true values.

Depending of the type of alloy and the experimental conditions, all 5 mould types studied show variations in the composition with depth.

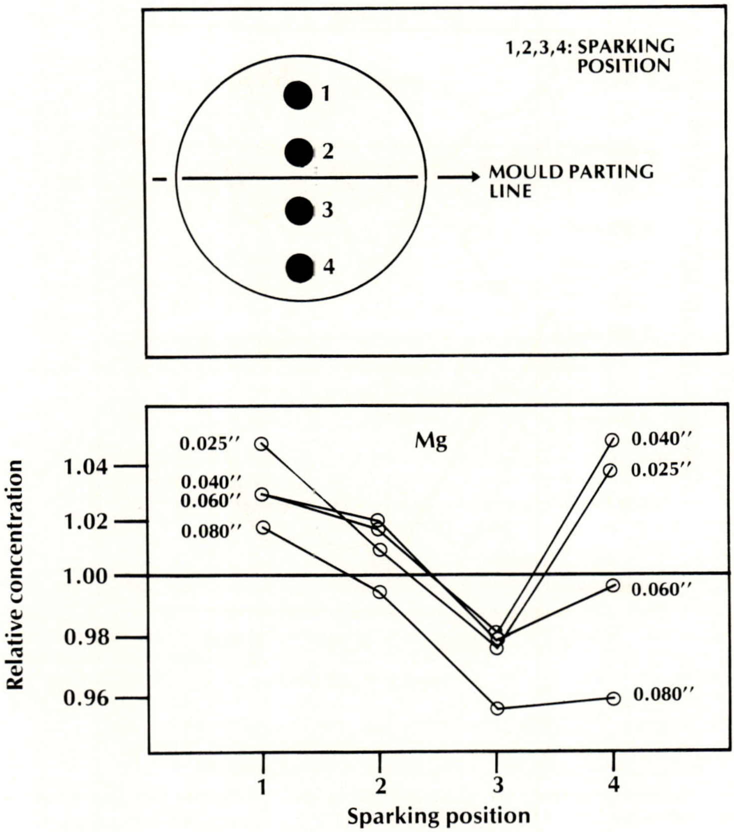

Figure 6 - Radial segregation profiles at various depths of Mg in horizontal disks measured by OES.

Effect of Mould Temperature and Coating

Since the rate of metal cooling affects the concentration-depth profile, the value at a given cutting depth will depend on the initial mould temperature and the rate of heat transfer. As shown in Figure 9, for the analysis of Mg in horizontal disks at a cutting depth of 0.275 inch, the mould temperature can cause a concentration bias of up to 2%. The same effect appears with other mould types although to different extents.

A coating, usually MoS_2, is commonly used to protect moulds and to facilitate removal of sample. The effect of this coating must be taken into account since it modifies the heat transfer rate and, by the same token, changes the segregation profile of the elements as shown in Figure 9.

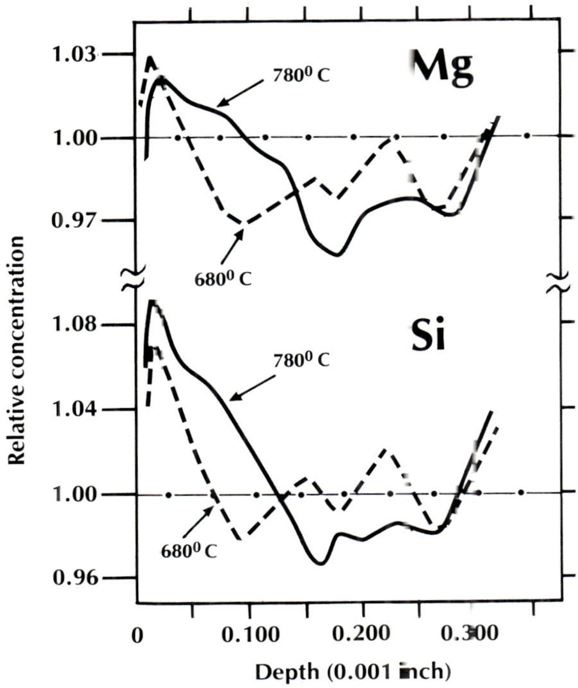

Figure 7 - Effect of sampled metal temperature on concentration depth profile in horizontal disks. Concentration (%) variations are expressed in relative value to the true chemical composition of metal (1.00).

Effect of Type of Cooling

The manner in which the samples are cooled also has an effect on spectrometric analysis. In Figure 10, we present data for Fe and Si obtained with horizontal disk samples, where immediately after the metal had solidified, one series was quenched in water while the other was left on the bench for cooling by the ambiant air. Usually, water-cooled samples gave analytical results slightly higher than those found for air-cooled samples.

Conclusion

To obtain further improvements in the analysis of metal samples, the instrumental inaccuracy of modern optical emission spectrometer is not a limiting condition. As indicated by our results as well as those by Olson and Macy (8), Dugain (9) and Mannweiler (10), the sampling and sample preparation practices should now be considered as the critical step. The magnitude of segregation profiles for different elements, variation with their sampling conditions including the temperature of the metal and the mould and

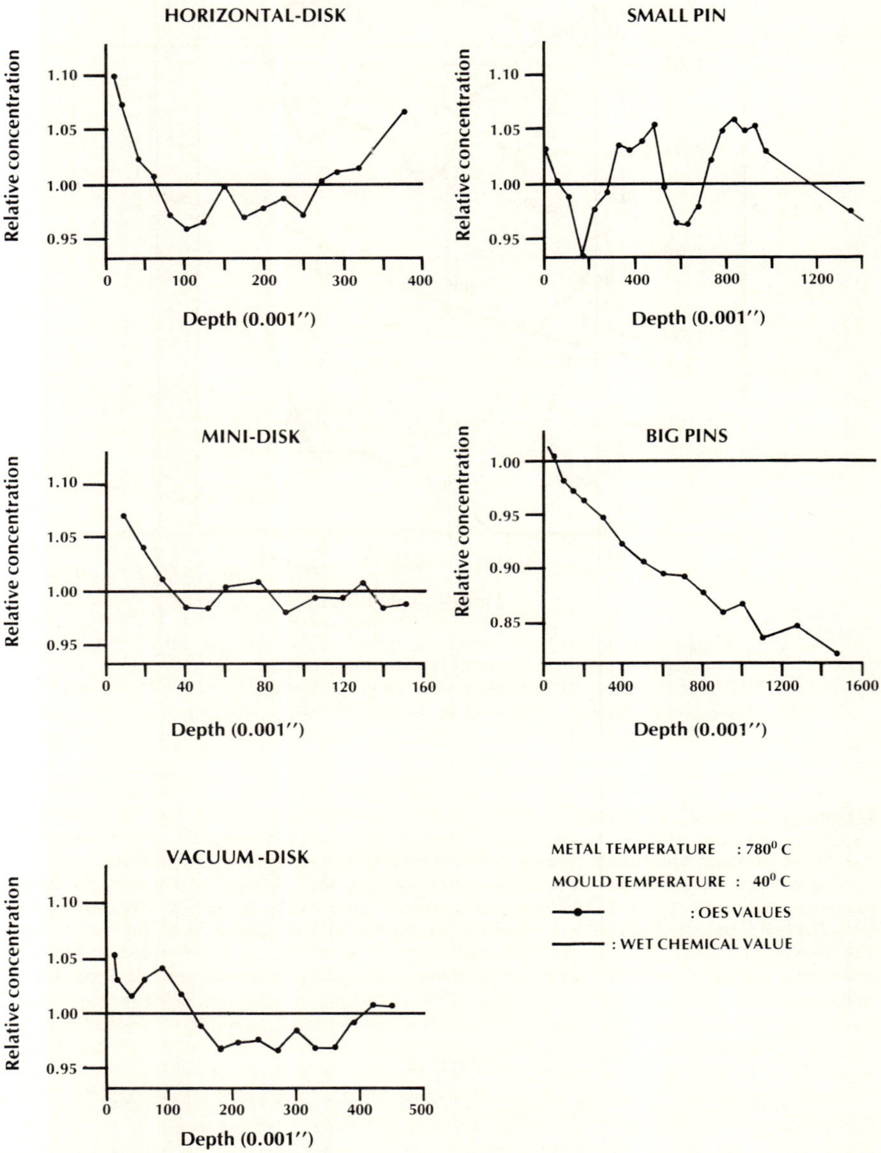

Figure 8 - Depth segregation profile of Fe obtained in 5 different types of mould: horizontal disk, mini-disk, vacuum disk, small pin and big pin. Metal temperature 780°C, mould temperature 40°C.

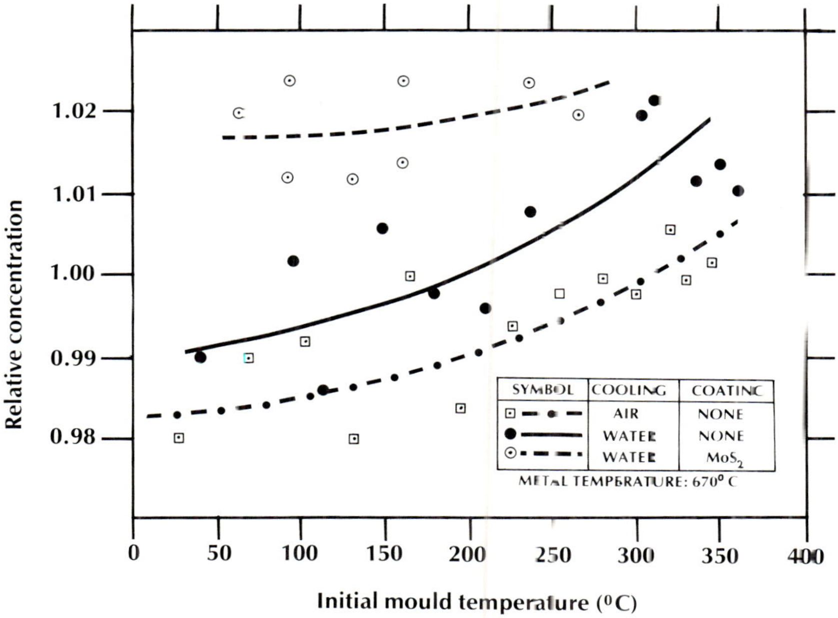

Figure 9 - Effect of mould temperature and coating on the OES analysis of Mg in horizontal disks at a depth of 0.275 inch.

the rate of cooling of the sample, make it very difficult for any improvements to be achieved. Significant gains can be obtained by using more rigid sampling, sample preparation and sample sparking conditions. If the analysis is carried out at a depth where the values are less affected by experimental conditions, it may become necessary to apply a correction to OES values to convert to true chemical composition of metal.

The design of a new mould type whose segregation phenomena will be minimal and less dependent of operational conditions should be considered. Although it has been proposed (8) that wet chemical methods constitute a better alternative, our work has shown that segregation effects are real and the problem lies not with the analytical instrumentation but with sample heterogeneity. Sampling remains the chief hurdle to be overcome in achieving improvement in the analytical results.

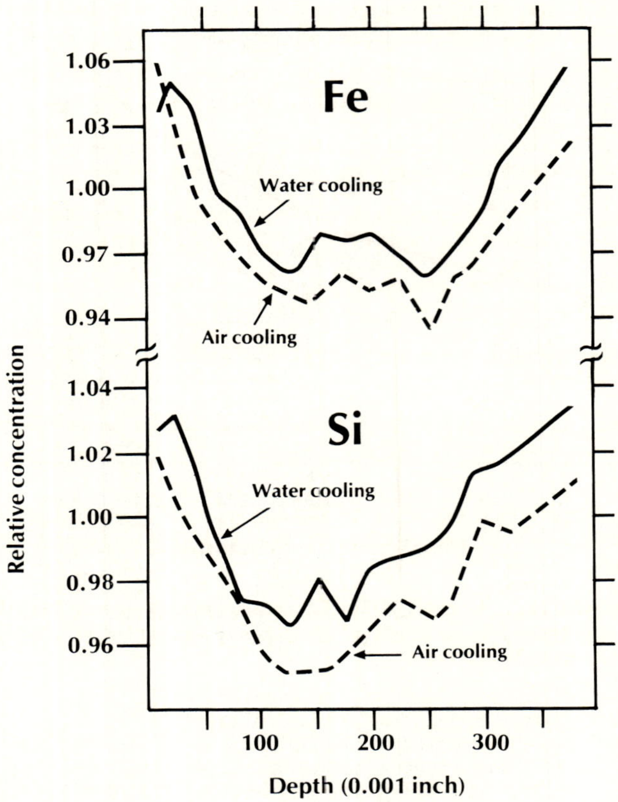

Figure 10 - Effect of type of cooling on the concentration profiles of Fe and Si measured by OES in horizontal disks. Immediately after the metal had solidified one sample was quenched in water (———), the other one left on the bench for cooling by the ambiant air (----).

References

(1) Walter J. Boyko, Peter N. Keliher and James M. Malloy, "Emission Spectrometry", Analytical Chemistry, 52, (1980), pp. 53R-69R.

(2) Charles K. Matocha and John Petit, "Spectrochemical analysis of Aluminium Alloys using nitrogen gas shielding and point-to-plane spark excitation", Applied Spectroscopy, 22, (1968), pp. 562-571.

(3) Karl Slickers, E. Wild and E. Hofman, "Spectrometric Aluminium Analysis with Argon", Aluminium, 50, (1974), pp. 588-591.

(4) Karl Slickers and J.C. Schmitt, "Analyses spectrochimiques des alliages d'aluminium", Analysis, 3, (1975), pp. 317-322.

(5) Karl Slickers, "Automatic Emission Spectroscopy", Applied Research Laboratories, Lausanne, Switzerland, (1980), 243 pp.

(6) Byron Kratochvill and John K. Taylor, "Sampling for Chemical Analysis", Analytical Chemistry, 53, (1981), pp. 924A-938A.

(7) Methods for emission spectrochemical analysis, ASTM, 1969, pp. 59-66, Method E 101-67.

(8) H.A. Olson and D.W. Macy, "Metallurgical Approach to Evaluating Chemical Samples Discs", Light Metals, (1973), Vol. 2, pp. 301-311, AIME.

(9) F. Dugain, "La précision dans l'analyse des métaux massifs, exemples de l'aluminium et de ses alliages", Revue de l'Aluminium, Mars, (1981), pp. 117-123.

(10) Ulrich Mannweiler, "Genauigkert und Reproduzierbarkert Spektrometrischer Analysen Non Aluminium und Seinen Legierungen", Ergmetall, 28, (1975), pp. 386-390.

ENVIRONMENTAL CONTROL I
Disposal and/or Recovery of Spent Potlining: Presentations and Panel Discussion

CHAIRMEN

S.G. Epstein
The Aluminum Association
Washington, D.C.

L.C. Blayden
Aluminum Company of America
Alcoa Center, Pennsylvania

The transcript of the papers and discussion of this session will be published in a 1984 issue of *Journal of Metals*.

CHAIRMAN

L.C. Blayden
Aluminum Company of America
Alcoa Center, Pennsylvania

HYDROCARBON EMISSIONS FROM SODERBERG POTS

D. R. Augood and H. J. Seim

Kaiser Aluminum & Chemical Corporation
Pleasanton, CA 94566

ABSTRACT

This study concerns the emissions obtained using coal tar pitch (CTP) or petroleum pitch (PP) as anode binders for Soderberg cells.

The experiments and discussion range over: Gas trains for collecting hydrocarbon emissions -- particularly benzene solubles (BS); improved analytical methods for measuring polynuclear aromatics (PNAs) -- including benzo(a)pyrene (BaP); the amount and nature of the hydrocarbons emitted; preliminary mouse skin painting tests; the relevancy of such tests to humans; and so on.

Among the many conclusions, it is observed that:

- Anodes made with CTP or PP emit approximately the same amount of BS and, therefore, according to current standard, are no different exposure-wise.
- While (solid) PP contains only about one-fifteenth the BaP in CTP, the corresponding ratio for gaseous emissions is greater -- at one-fifth or one-half.
- A separate, large epidemiological study shows no increase in mortality from lung cancer for U.S. workers exposed to CTP emissions in alumina reduction plants.

INTRODUCTION

At the time this study was made, Kaiser Aluminum & Chemical Corporation (KACC) operated two Soderberg reduction plants. The cells, or pots, at these plants are of horizontal stud design and use coal tar pitch (CTP) anode binder. On line, these pots emit some coal tar pitch volatiles (CTPV). Some of these hydrocarbons -- most notably benzo(a)pyrene (BaP) -- are potential carcinogens.

In 1980, we reported laboratory work aimed at reducing BaP emission by decreasing BaP content in anode binder (1). This was attempted by using petroleum pitch (PP) -- which usually contains less BaP than CTP -- and/or by subjecting the pitch before use to chemical or physical treatment. Overall, we found little benefit in using a binder containing <0.1% BaP -- since, at this level, more BaP was generated than destroyed when the anode paste was baked. It was decided that a plant-scale test using low-BaP-PP-binder was in order. This paper outlines such work carried out during the 1974-78 period.

INITIAL OBJECTIVES

While there were many objectives connected with this overall study, we are involved here only with those concerning hydrocarbon emissions. Our initial objectives were to:

- Develop better understanding of hydrocarbon (and BaP) emissions from Soderberg pots.

- Compare the emissions obtained from pots using CTP versus PP as anode binder.

- Concentrate on the hydrocarbon fraction -- presumably that containing less volatile hydrocarbons -- most relevant to carcinogenicity.

- Use reduction, or decontamination, factors to express results and guide the work.

  The reduction factor, R, represents the weight ratio for a given component in the CTP case to that in the PP case. For example, $[R_p\text{-BaP}]_{PP}^{CTP} = 19$ indicates that the ratio of BaP contents in CTP and PP pitches is 19. Clearly, if the objective using PP is to reduce exposure to potential carcinogens, then R for a given component, or group of compounds, should have a large value. If R=1, of course, the exposure is the same in the PP as in the CTP case.

- Initiate animal studies to assess the carcinogenicity of samples trapped, or extracted, from plant streams.

- Devise a method for obtaining major (gram) quantities of hydrocarbons representing gaseous emissions in the potroom. These samples might then be used in additional, more definitive, animal studies.

- Develop analytical and procedural techniques, as required.

GENERAL APPROACH

Benzene Solubles (BS)

A few words of explanation are necessary concerning the use of the term "Benzene Solubles (BS)". In CTPV-exposure circles, this term means "BS extracted from Total Particulate Matter" -- sometimes referred to as "Particulate Polycyclic Organic Matter (PPOM)". In this study, for reasons which will become apparent, a wider meaning is employed, such that BS also includes:

- The benzene extract of pitch.

- The total hydrocarbons, perforce benzene-(or toluene)-soluble, collected using various gas trains to sample potgas, areas and personnel exposure.

Plant Sampling

The PP employed in these tests was delivered in either flaked or molten condition. Flaked pitch could be loaded directly into the paste mixers, but required more mixing/heating to obtain satisfactory paste. Molten pitch, on the other hand, could be handled through the plant's normal CTP tank and piping system -- but gave rise to problems described later.

The collection of 50 g or more of hydrocarbons representing potroom emissions is not easy. Thus, while these hydrocarbons might be separated from potroom atmosphere, the undertaking is formidable because of their dilute concentration. Alternatively, hydrocarbon solids could be scraped from the ducts, but this might prove to be different material. We decided:

- To use some easily-collected deposits to initiate some animal studies.

- To collect gram quantities of hydrocarbons from potgas in the ventilation duct(s) leaving the pot(s). This potgas is enriched in hydrocarbons relative to potroom atmosphere and these hydrocarbons may well have the same composition.

- To prove, or disprove, this point by:

    - Collecting potroom personnel and area samples.

- Comparing the chemical fingerprints of these samples with the potgas sample.

It took three plant sampling tests and much work to achieve these goals.

As this study began, the chosen Soderberg plant was busy enlarging and modifying its dry scrubber system. A calciner was being installed to burn away the hydrocarbons adsorbed on "enriched ore" returning to the potrooms from the dry scrubber. Because of this, and possible interference from any adsorbed hydrocarbons, we arranged to test only pots receiving "virgin ore" (fresh alumina) feed.

In selecting pot(s) or potroom(s) we chose similar configurations for CTP and PP. In each instance, potgas was sampled from the duct at a point having linear flow. A pitot traverse was made to determine flow rate, but the potgas sample was taken isokinetically from the center of the duct only.

Calculations indicated that it took approximately one month for fresh paste to reach the "volatilizing" level in the anode, and approximately three months to reach the working face in the bath. Sampling was not begun, therefore, until at least two months had elapsed after changing the binder -- sampling first the CTP system and then the PP system.

In general, most anodes remained quiet for about 1.5 hr after flexraise. After about 4 hr, "paste" oozing around the newly-placed studs, was burning. After about 30 hr, the anode appeared fairly quiet. Unless otherwise defined, we use the term "flexraise" to describe the ~48-hr period after actual flexraise, and "quiet period" to describe the quiet time between flexraises.

All sample components -- filters, cyclone catches, washings, etc. -- were carefully packed and shipped to the Center for Technology (CFT) for workup and analysis.

ANALYTICAL

Gravimetric Measurements

Gravimetric procedures for BS, etc., are sufficiently well known and/or are adequately defined within this text that no further explanation is required (see **Benzene Solubles** above).

Gas Chromatographic Method (OV-7 Column)

Gas chromatography (GC) was very useful for characterizing BS extracts. This was especially true for CTP samples.

The main GC workhorse featured an OV-7 column and conditions summarized in Table 1. The GC determinations are normally made on the BS extracts. Because benzene is not a suitable solvent for GC, an aliquot of said extract is evaporated just to dryness at 35°C under a gentle flow of nitrogen. The residue is then dissolved in a measured volume of CS_2 to give a concentration in the range given in Table 1.

Table 1. Main GC Method: OV-7 Column

Instrument	Perkin-Elmer 900 Flame ionization detector Integrator*
Column	Stainless steel
Length	3.7 m
Diameter	3 mm
Support	Chromosorb G-HP
Mesh size	80-100
Coating	3 wt % OV-7
Carrier Gas	Helium
Rate	40 ml/min
Sensitivity	8 x 10^{12} amp/mV
Temperatures (Isothermal)	
Column	285°C
Detector	305°C
Injection port	290°C
Sample size	3 μl
Concentration	0.3 to 0.5 mg/ml
Chart speed	38 cm/hr

*Electronic integrator set to pick off fingerprint peaks above background.

PNA Fingerprint

Volatiles collected from thermal treatment of CTP produce a characteristic GC fingerprint that varies little with the source of the CTP. Thus, Figure 1 shows fair separation of benzo(a)pyrene (BaP), benzo(e)pyrene (BeP), and perylene (Per), and also of benz(a)anthracene (BaA) and chrysene (Ch). For CTP-derived samples, all nine PNAs, or groups of PNAs, identified in this fingerprint are usually present in sufficient quantity such that they can be quantified using appropriate standards.

In general, the fingerprint PNAs account for 95% of the "GC measurables", but only 45% of the BS. The remaining 55% of the BS comprises a mixture of heterocyclic and high molecular weight compounds.

Samples derived from PP produce a different fingerprint from that obtained using CTP. This PP fingerprint, moreover, changes more with the source, the manufacture and/or the thermal treatment of the pitch. The PP used in this study was specially made at relatively low temperature to minimize BaP content.

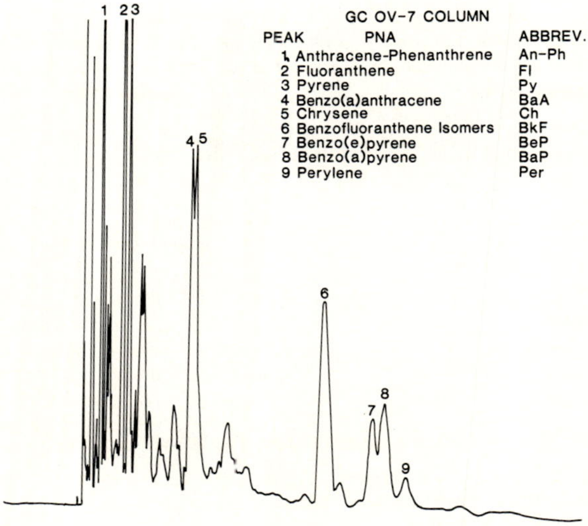

Figure 1. Typical GC Fingerprint for CTPV

Chromatograms for two early samples of PP are shown in Figure 2. These scans are virtually identical and exhibit many of the peaks found in the CTPV fingerprint (Figure 1). The corresponding concentrations are, however, very much smaller.

Determination of BaP by GC

The determination of BaP by GC posed problems, particularly for PP samples. Thus, as shown in Figure 2, the BeP and BaP peaks are confounded, although the form of the double peak suggests that BeP and BaP are present in approximately equal amounts.

Analytical resolution and sensitivity certainly left something to be desired. Sensitivity could not be improved by increasing the GC sample size because this would quickly foul the GC column. Our early technique, therefore, was only capable of describing PP as containing <0.1% BaP.

Gradually, our GC procedure improved. For example, we compared unknown sample chromatograms with chromatograms for known mixtures of BeP and BaP. The overall ratio of BeP to BaP observed ranged from 70:30 to 30:70.

These problems were eventually mitigated, as described below, with the development of the HPLC method.

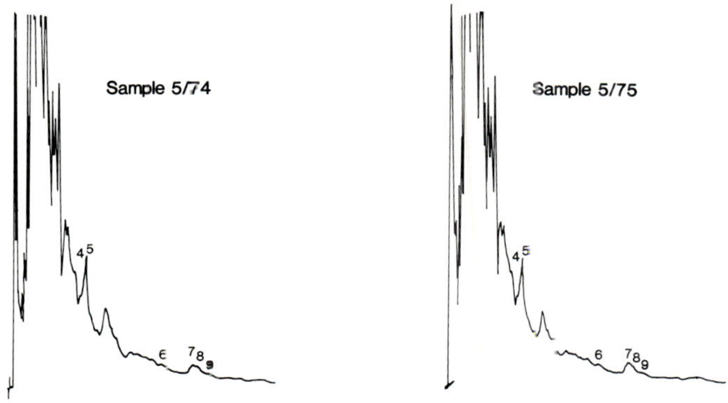

Figure 2. GC Chromatograms for Early PP Samples

(OV-7 Column. See Figure 1 for CTPV and key to peaks. HLPC work showed these PP samples to contain 0.07% BaP.)

Other Analytical Methods for BaP

In focusing upon BaP, we were concerned about analytical sensitivity using PP samples. Indeed, the GC method, just described, was only marginally acceptable.

Another method for determining BaP was the GC-UV method proposed by NIOSH as P&CAM 183 (2). This procedure uses GC and a 10:1 column splitter to trap selected fraction(s) of the larger stream in a cold glass tube. The trapped fraction containing BaP, BeP and Per is dissolved in cyclohexane and BaP is determined by UV spectroscopy. The quoted limit of detection is 1.2 μg BaP per sample, which is also marginal for PP samples. The procedure, moreover, is time consuming.

The most sensitive method available when this study began was NIOSH P&CAM 186 (2) using thin layer chromatography (TLC). Here, the spot containing BaP is scraped off the TLC plate, extracted, and the BaP determined by fluorescence spectroscopy. The quoted limit of detection is 4 ng per sample or 300-times more sensitive than the foregoing GC-UV method. Using the TLC method, and meticulous care, we obtained satisfactory results with CTP-derived samples, but values high by more than a factor of two with PP-derived samples. Better results were obtained for PP using the GC-trapping technique (above) to "clean up" the samples before TLC. This procedure, however, was much too cumbersome for routine use.

Determination of BaP by High Pressure Liquid Chromatography (HPLC)

A method for determining BaP by HPLC was developed. H. Boden reported an early HPLC approach to the Aluminum Association task group for PPOM and published his method in 1976 (3). In 1977, the technique was proposed as P&CAM 251 by NIOSH with a reported detection limit of 6 ng per sample (2).

The HPLC method developed at CFT involved separation of PNAs on a column of 5 μm Spherisorb silica using cyclohexane as the mobile phase. Measurements for BaP were made using a variable UV detector set at 383 nm. At this wavelength, interference from BeP and Per is eliminated. The operating conditions are summarized in Table 2.

The HPLC method was applied later in this study to selected samples as a check on the other procedures, particularly the main GC separation method.

Table 2. HLPC Method for BaP

Instrument	Varian 4100
Precolumn	3.2 mm Ø 5 cm long stainless steel tube filled with 5 μm Ø Spherisorb silica
Analytical Column	4.6 mm Ø 25 cm long tube filled as above
Temperature	40°C
Mobile phase	Cyclohexane at 0.33 ml/min
UV Detectors	254 nm for high mol. wt. material 383 nm for BaP
Sample size	25 μl

Determination of BaP in PP

In an attempt to better define PP, some analyses were made using larger samples in both GC and HPLC methods. As mentioned, larger samples are generally avoided in order not to foul the columns. For GC, concentrations of 5.5 mg/ml were injected, and for HPLC, concentrations of 0.55 mg/ml were used -- both concentrations being about 10-times higher than usual. Typical scans are presented in Figure 3, along with a 50:50 BeP:BaP standard. These scans show BeP and BaP, often confounded in PP GC scans, to be present in nearly equal amounts.

In fact, the analyses presented in Figure 3 shows that this sample of PP-2 contains 0.062% BaP (by GC) and 0.075% (by HPLC). After analyzing other samples and rounding, it was established that PP-2 contained 0.06% BaP (by GC) and 0.07% (by HPLC).

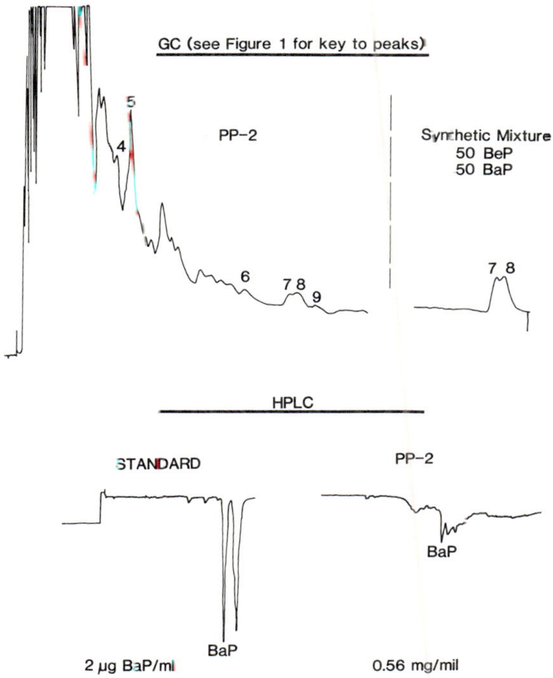

Figure 3. GC and HPLC Scans of PP-2

POTROOM TEST I

Experimental

Flaked PP-1 was used in this test. The pot configuration, potgas sampling apparatus, and some experimental details are given in Figure 4.

There were no obvious differences between the physical properties of small laboratory anodes made using CTP or PP-1 binders. Likewise, at least initially, there were no obvious operating differences between corresponding pots on-line.

Potgas samples were taken for periods of 45 min. at 90-min. intervals for the day of the actual flexraise, and daily thereafter. The sampling apparatus worked well except for some faltering of the gas meter towards the end of each run. After each sampling, the "Teflon" lines and condenser were washed down with benzene.

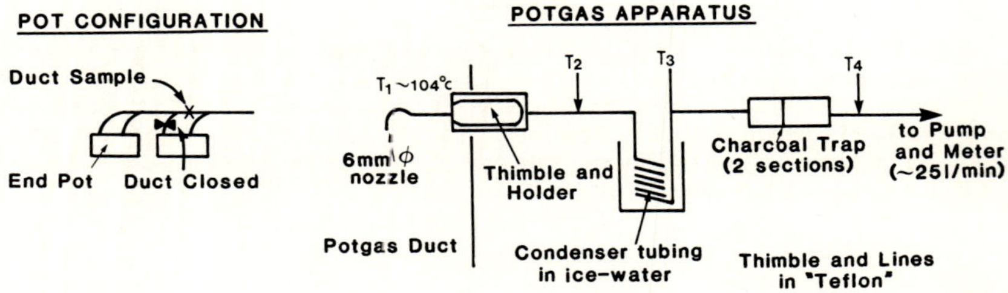

PITCH		
Type	Soft Pt. °C	% in Anode
CTP	98	30.0
PP-1	107.5	27.1

POTGAS SAMPLING	NUMBER
45 min samples:	
Every 90 min after actual flexraise	8
Daily thereafter	~15

Figure 4. Test I: Apparatus and Sampling (1975)

At CFT:

- The "Teflon" thimbles were extracted with benzene for ~4 hr in a Soxhlet extractor and small amounts of fine carbon were removed from the extracts using fine fritted glass filters.

- The condenser washings were passed through Millipore glass fiber filters to remove fine water droplets.

- Benzene solubles (BS) were determined on the benzene extracts by carefully evaporating aliquots in tared aluminum pans and weighing on a 5-place Mettler balance.

- The BS values for thimble and condenser extracts were combined for each sample set to give the concentration of BS in mg/m^3 gas sample. These data were then converted to pot emission rates, expressed in g/min, using potgas flow rates.

- The benzene extracts were analyzed for BaP by GC. The extracts for the thimble and condenser were analyzed separately for the first half of each series of samples and combined for the balance. In all cases, the total BaP emission rate, expressed in mg/min, was calculated in a way similar to that described above.

- The BaP contents in the pitches were determined via benzene extraction and GC analysis.

In related experiments, BaP could not be detected in the delayed and fluid coke used in making anode pastes.

Results

The BS data for CTP and PP-1 potgas emissions are given in Figures 5 and 6, respectively. It should be noted that these data represent BS in the condensates alone and do not include any organics adsorbed in the activated charcoal traps of the sampling train. Later work showed that this adsorption loss might be significant.

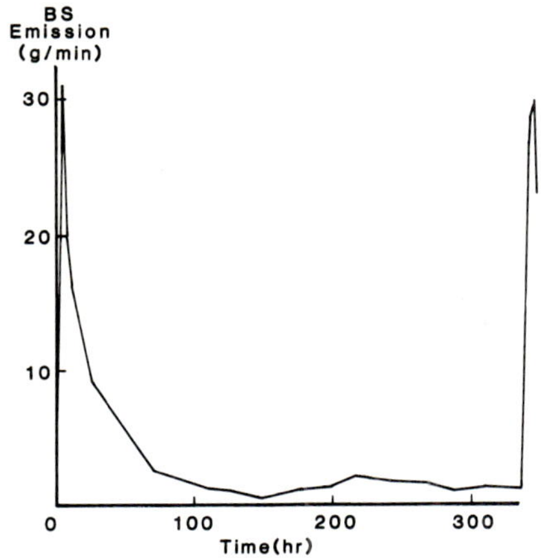

Figure 5. Test I: BS in CTP Potgas

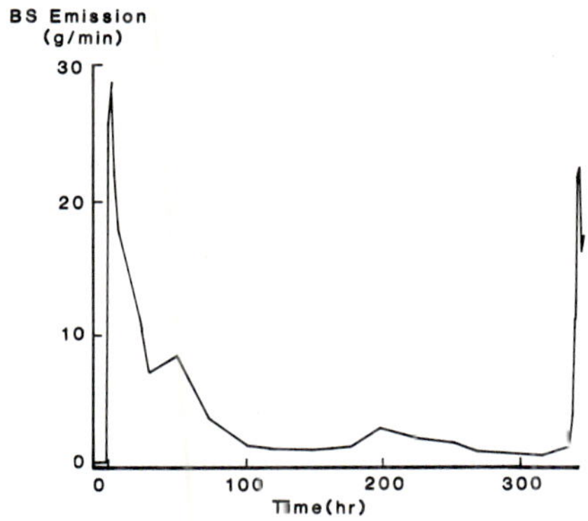

Figure 6. Test I: BS in PP-1 Potgas

The impressive observation from Figures 5 and 6 is the magnitude (concentration and time) of the fume surge arising from flexraise. It is unfortunate that more samples weren't taken through the flexraise -- thereby providing more information about this surge.

The BaP data for CTP and PP-1 potgases are given in Figure 7. The smaller amount of BaP for PP-1 compared with CTP is evident.

By integrating the areas under these curves, the relative emission rates for CTP and PP-1 can be compared. These data -- rounded to whole numbers -- are given in Table 3 along with some other information.

These results, together with operating data, indicate that:

- Separately, the pitches and potgases contain approximately the same amount of BS. Thus, $[R_P\text{-BS}]^{CTP}_{PP\text{-}1}$ and $[R_{PG}\text{-BS}]^{CTP}_{PP\text{-}1} \approx 1$.

- While the pitches possess a BaP-reduction ratio $[R_P\text{-BaP}]^{CTP}_{PP-1} = 19$, the potgases show $[R_{PG}\text{-BaP}]^{CTP}_{PP-1} = 13$.

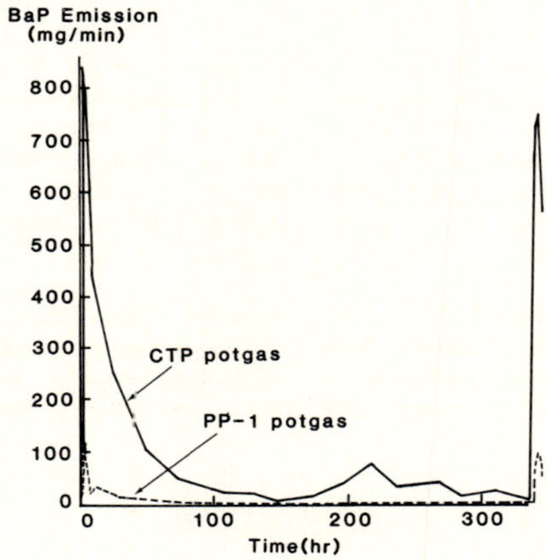

Figure 7. Test I: BaP in CTP and PP-1 Potgases

Table 3. Test I Results

Sample	Case	Component	Reference or Concentration		R-Value (Rounded)
Pitch (P)	CTP	BS	72	wt% in P	$[R_P\text{-BS}]^{CTP}_{PP-1} = 1$
	PP-1		76		
	CTP	BaP	1.3		$[R_P\text{-BaP}]^{CTP}_{PP-1} = 19$
	PP-1		0.07		
Potgas (PG)	CTP	BS	Figures 5 & 6 (a)		$[R_{PG}\text{-BS}]^{CTP}_{PP-1} = 1$
	PP-1				
	CTP	BaP	2.4	wt % in BS (b)	$[R_{PG}\text{-BaP}]^{CTP}_{PP-1} = 13$
	PP-1		0.15		

(a) By integration.

(b) By integration of Figures 5 through 7.

Later, we noted that the BaP in the BS emitted by PP-1 averaged ~0.4% during the fume surge and ~0.1% during the quiet period.

In reflecting on these results:

1. We were concerned about taking "spot" samples and integrating data over a long period.

2. We were worried about our GC method for determining low levels of BaP.

3. We were intrigued by the fact that both CTP and PP-1 anodes emitted approximately the same amount of BS. This BS must be related to the BS upon which the gravimetric standard for exposure is based. Given the same amount of BS, the exposure from CTP and PP-1 is the same. This, however, may say little about relative risk and/or carcinogenic potency.

4. We were encouraged by the fact that the gaseous reduction ratio for BaP of 13, since this was about four times greater than that derived from the laboratory baking tests (1). Due to Items 1 and 2, however, we were hesitant to accept this result.

TEST II

Updated Objectives

Following Test I and a series of meetings -- some with a consultant -- it was decided that:

- To obtain a representative sample from potgas, we would be forced to average the gas composition in some way -- perhaps by sampling the duct serving many pots and/or by using long sampling times.

- We would be reasonably sure of collecting all potentially carcinogenic compounds from a gas stream if we trapped and recovered all PNAs having >4 rings. The more volatile, lower molecular weight hydrocarbons were considered innocuous. These hydrocarbons may be described as "vaporous polycyclic organic matter" (VPOM) comprising PNAs having <4 rings.

  In order to provide information and guarantee collection of PNAs >4, we decided that it would be necessary to collect some VPOM as well as the less volatile material.

- We would examine adsorbents other than activated charcoal for trapping VPOM.

- Additionally, we would:

  - Develop an apparatus capable of collecting 50 g to 1.5 kg of potgas hydrocarbons (PNA >4).

  - Simultaneously collect area and personnel samples.

  - Compare all of these samples and, hopefully, show them to be chemically the same for PNA >4.

Analytical

The examination of various analytical techniques and the early development of the HPLC method provided confirmation of the general GC method (see **ANALYTICAL**).

Search For Alternate Adsorbent

Experiments with activated charcoal carrying adsorbed hydrocarbons confirmed our suspicion that PNAs having >3 rings were strongly held and only reluctantly desorbed under special extraction or vacuum distillation conditions. Fortunately, we learned of a new absorbent (Tenax-GC) and, in separate adsorption-desorption experiments using PNAs ≥4 we found this to be a viable replacement for charcoal.

As stated earlier, the GC determinations are made using CS_2 as the solvent for sample injection. However, because Tenax-GC is soluble in CS_2, this adsorbent cannot be desorbed directly into CS_2 and a milder solvent must be used. Separate experiments were carried out to prove a quantitative procedure. Thus, the Tenax is extracted in 2 ml of n-pentane for 30 minutes. A 1 ml aliquot of this extract is slowly evaporated with CS_2 repeatedly, without going to dryness, until the pentane is removed. The final volume is then adjusted with CS_2 to give the desired concentration for injection.

As an example of some experiments conducted with Tenax, Figure 8 shows what happened when a synthetic mixture of PNAs in an oven was swept with a gentle stream of heated air. The curves show that:

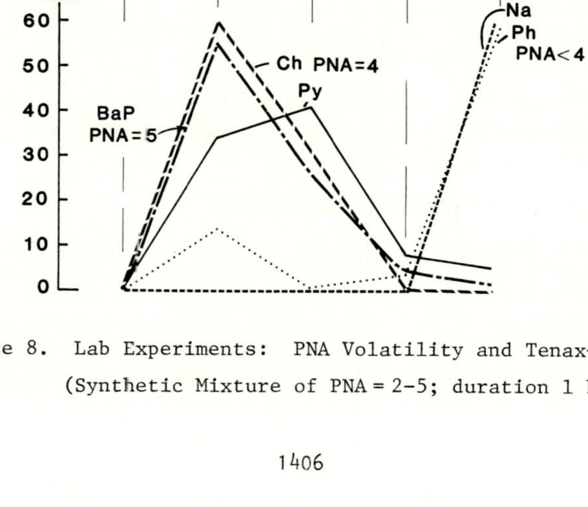

Figure 8. Lab Experiments: PNA Volatility and Tenax-GC (Synthetic Mixture of PNA = 2-5; duration 1 hr.)

- Tenax is an effective trapping agent.

- In spite of high melting and boiling points, the PNAs in the oven are sublimed into cooler parts of the apparatus. None of the initial PNAs remained in the oven.

- Similarly, we found that the rate at which PNAs sublimed from the oven was substantially reduced if these PNAs were initially adsorbed on particulate.

These findings have implications to gas sampling systems.

General Plant Operation

An extensive test was planned to evaluate the potroom performance of PP and to obtain more emission data. A room containing 72 pots, therefore, was converted to PP. Because of the amount of PP required to service this room, the petroleum pitch (PP-2) was delivered in molten form.

This potroom test lasted for 17 months, during which emission Tests II and then III were carried out. Also during this time, some minor compositional changes were made in PP-2 for pitch property and potroom operating reasons.

During Test II, a problem surfaced related to the use of PP-2. Although PP-2 was stored in a dedicated tank at the paste plant, some CTP was inadvertently dumped into this tank contaminating the PP-2. It was difficult, also, to purge the plant's lines so that they were free from CTP when PP-2 paste was to be made. The paste-mixing operations were, therefore, carefully monitored and clearly improved with time. However, due to this uncertainty, it was elected to designate both PP potroom and potgas during Test II as "contaminated". Thus, though PP-2 was shipped to the plant, contaminated PP (CPP) was actually tested.

Figure 9 provides evidence for the contamination of PP-2. Not only is the BaP in CPP increased relative to PP-2, but the other fingerprint PNAs are similarly increased. There is good agreement between chromatograms for CPP and a "Synthetic Mixture" of PP-2 and CTP. The degree of contamination of PP-2 varied but generally decreased with time.

Later for Test III, though there was much less uncertainty, we also designated the potroom "contaminated".

Potgas Sampling

In sampling potgas in Test II we tested two gas trains, first for the CTP and then for the CPP system. Doubting that we would be able to sample continuously, we selected sampling positions in ducts serving sets of 9 pots -- reasoning that, given relatively short-sampling times and only 1 pot of the given set on flexraise, we would still obtain a fairly well-averaged sample. Some of these details are given in Figure 10.

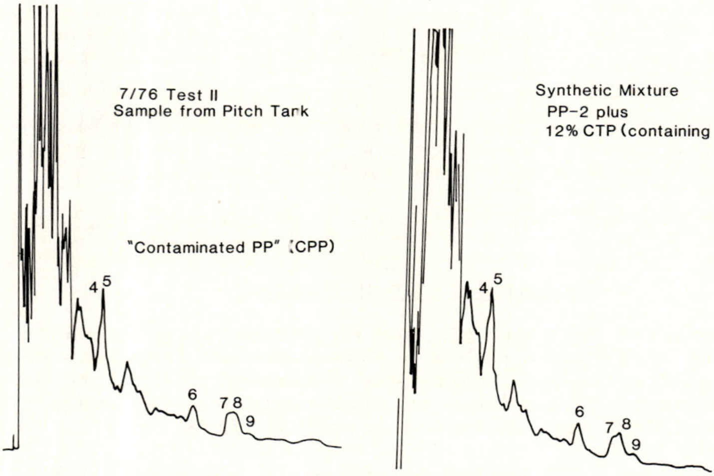

Analysis of PP-2 delivered to plant is shown in Figure 2. Said PP-2 contained 0.07% BaP compared with the above samples which contained 0.21% BaP.

Figure 9. Contamination of PP-2 During Test II
(see Figure 1 for key to peaks)

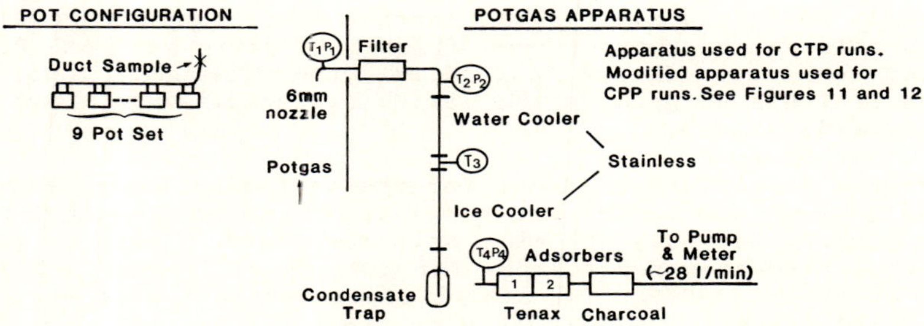

PITCH

Type	Soft Pt°C	% in Anode
CTP	98	29.0
PP-2 } CPP }	~123	26.7

POTGAS SAMPLING

Potrooms: CTP: Four~6hr runs
CPP: One~6 and one~34hr run
In all cases, 1 pot in 9 pot set was on flexraise
Area: 6 multihead runs at 3 locations in CTP potroom.

Figure 10. Test II: Apparatus and Sampling (1976)

The first runs were made with the CTP system. The capacity of the filter -- the first component of the gas train (Figure 10) -- limited the sampling time to ≤6 hr, and the ice-water cooler proved troublesome. The component samples collected from these runs were worked up and analyzed in a manner similar to that used for the laboratory oven tests. Figure 11 shows that the ice-water cooler contributed little to the collection of PNAs ≥4 and might be eliminated.

Changes were now made in the sampling train as it was transferred to the CPP system. Figure 12 shows that the cyclone in this train displaced the filter to a later position and took some load off the filter. Indeed, we found it possible to run this train for ~34 hr before being forced to shut down.

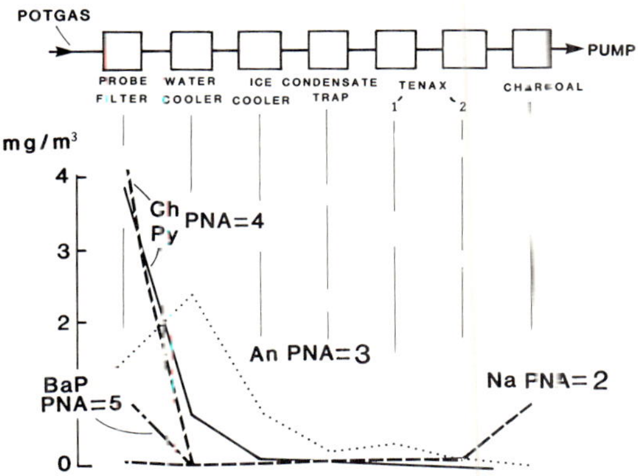

Figure 11. Test II: PNA s in Apparatus Sampling CTP Potgas (Average from four ~6 hr runs)

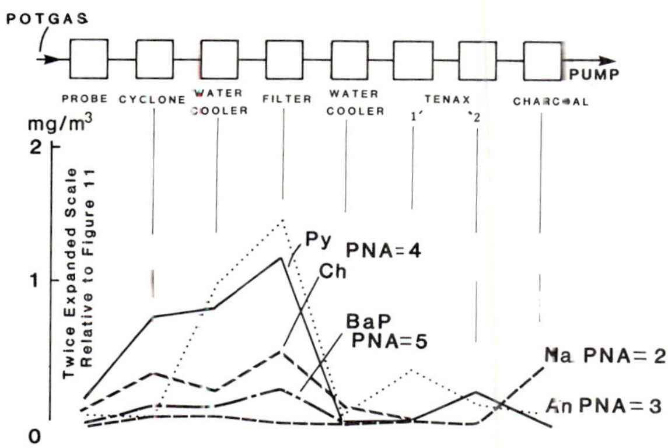

Figure 12. Test II: PNA s in Apparatus Sampling CPP Potgas (Average from ~6 and 34 hr runs)

Further evaluation of these data indicated that we might include a water cooler ahead of the cyclone and dispense with the adsorbent traps.

Potgas Results

The potgas samples were worked up, using benzene, very much along the lines described for Test I.

As shown in Table 4, the consolidated results for Test II differ from those obtained in Test I. This is not surprising, perhaps, in view of the different test configurations and sampling procedures employed. Notably in Test II more gaseous BaP was observed -- particularly in the CPP case.

Table 4. Test II Results

Sample	Case	Component	Reference or Concentration (a)		R-Value (Rounded)
Pitch (P)	CTP	BS	70.5	wt % in P	$[R_P\text{-BS}]^{CTP}_{PP-2} = 1$
	PP-2		70.0		
	CTP	BaP	1.2		$[R_P\text{-BaP}]^{CTP}_{PP-2} = 18$
	PP-2		0.07		
Potgas (PG)	CTP	BS	73.0 (b)	mg/m^3 in PG	$[R_{PG}\text{-BS}]^{CTP}_{CPP} = 1$
	CPP		78.0 (b)		
	CTP	BaP	1.59 (b)		$[R_{PG}\text{-BaP}]^{CTP}_{CPP} = 4$
	CPP		0.40 (c)		

(a) Average data for sampling runs -- see Figure 10.

(b) These concentrations are approximately twice those obtained in Test I -- showing the greater trapping efficiency and possibly some decrease in pot ventilation rate during Test II.

(c) This concentration is about seven times that obtained in Test I.

- In quantitative terms, the R-values in Table 4 for Test II show no difference between pitches or emissions for BS, in agreement with Test I. Thus $[R_{P,PG}\text{-BS}]^{CTP}_{PP-2,CPP} \approx 1$.

- For BaP, while pitches exhibit $[R_P\text{-BaP}]^{CTP}_{PP-2} = 18$, potgases give $[R_{PG}\text{-BaP}]^{CTP}_{CPP} = 4$. Though this latter result differs from that obtained in Test I, it tends to the $[R_{PG}\text{-BaP}]^{CTP}_{PP}$ = 3 result obtained in the laboratory baking tests (1).

Area Sampling

Area gas samples were taken as part of Test II. Multiple personnel-type cassette samplers were used. In an extension of normal cassette practice, a two-section Tenax adsorber was added to the filters to collect VPOM. Three of these samplers were mounted in a multihead unit -- as shown in Figure 13.

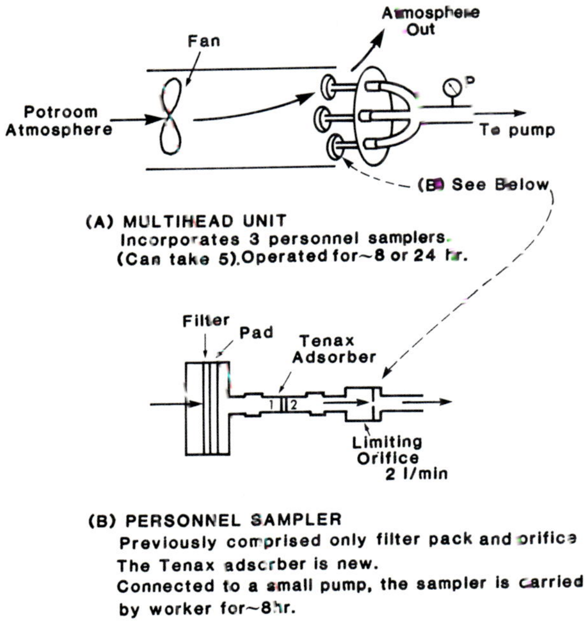

Figure 13. Multihead (Area) and Personnel Sampler

The personnel sampler, when attached to a potroom operator, is usually operated for an 8 hr shift. The multihead unit, freed from such constraint, was operated for 8 hr during flexraise and for 24 hr during the quiet period.

Details of the multihead sampler, originally designed to collect five simultaneous samples, have been reported in a paper by Seim and Dickeson (4). The precision data for TP and BS in Table II of said paper were taken from results obtained in this study. Prior to the use of multihead samplers in this study, the reproducibility of data obtained in their use was proven in other plant tests. Both GC and HPLC analyses showed good agreement between replicate samples.

Some consolidated data -- for tests using additional charcoal traps -- are shown in Figure 14. Given the criterion to collect all PNAs >4, Figure 14 shows that it is important to add Tenax to the conventional personnel sampler. The marked difference in the distribution of 3- and 4-ring PNAs collected using this personnel sampler (Figure 14), compared with that obtained using the Test II duct sampling train (Figures 11 and 12) should be noted. In the latter case, the function of the cooled condensers (Figure 10) in retaining these PNAs is evident -- with relatively small amounts of PNAs collected on the Tenax. With the personnel samplers, however, there is no provision for cooling the sample stream and appreciable quantities of 3- and 4-ring PNAs pass through the filter and support pad into the Tenax. To simplify sample workup in this program, we elected to eliminate the support pads in these samplers. It may be observed that where we employed two sections of Tenax, there was never more than 0.2% relative of any PNA >4 compound in the second section of the Tenax trap.

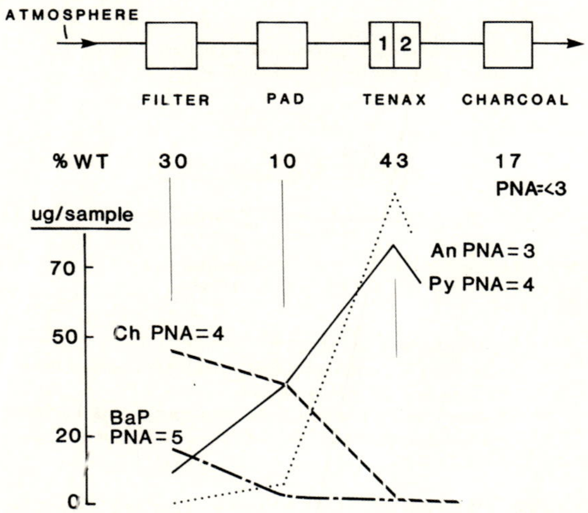

Figure 14. Test II: Area Data

Chemical Fingerprints (Potgas cf. Area)

It was particularly interesting to compare the chemical fingerprints of duct potgas and area (potroom atmosphere) samples.

Table 5 presents PNA >3 data for CTP, Test II. There is good agreement within experimental error, between potgas and area compositions.

Test II work, therefore, qualified an improved potgas sampling train and personnel and multihead samplers and we were ready for Test III.

Table 5. Fingerprint Comparison (Duct cf. Area Samples)

CTP: Normalized data (PNA ⩾3) = 100%

PNA		Duct Samples			Area Samples (a)	
Abbrev.	Rings	1	2	3	1	2
An	3	26	24	23	22	32
Fl	4	24	24	24	26	26
Py	4	24	24	23	28	24
Ch	4	19	20	21	20	13
BaP	5	7	8	8	4	6

(a) Multihead unit on top of cell during flexraise.

TEST III

Updated Objectives

In approaching Test III -- it was decided that:

- We would practice all plant sampling methods simultaneously -- one potroom (or pitch) at a time.
- We would eliminate the possibility, now slight, of contaminating PP anode binder by basing 3 pots within the CPP potroom on flaked PP-2 pitch. This flaked pitch would be mixed directly (in contrast to bulk liquid PP-2 going through the plant's tank and lines) and would, therefore, retain its integrity.
- We would take potgas samples from the given 3-pot set operating continuously for two flex cycles with each pot contributing equally.
- We would take area samples using multihead units (Figure 13). In the CPP potroom, however, to further minimize contamination, these samples would be collected around the middle cells of the 3-pot set using flaked PP-2.
- We would collect personnel samples using potroom operators in both CTP and CPP cases. The work schedule of these operators would be adjusted to eliminate cross-room exposure. These personnel samplers would carry Tenax adsorbers -- just like those used in the multihead units.
- We would continue analytical development. In working up samples, however, we would switch to toluene -- now regarded as more acceptable than benzene in carcinogenic studies.

Potgas Sampling

Using information obtained in Test II, a new potgas sampling train (Figure 15) was constructed. In sampling continuously, we aimed to remove the cyclone catch and change the filter on-line, as needed. Some plugging and reduced flow problems were experienced.

As shown in Figure 15, three samples were collected. Note that the second PP-2 sample (ii) was obtained over a 4-day period while two pots were on flexraise. This sample might be expected to be richer in BaP than average [or the PP-2(i) sample]. It might also be noted that during this test, the plant encountered problems with the hooding fans -- decreasing the ventilation rate for CPP pots.

The components of all samples were worked up carefully, with toluene being removed under reduced pressure, to obtain final solutions of BS (in toluene). The amounts collected for the CTP, PP-2(i) and (ii) cases were 55, 63 and 27g, respectively. Clearly, the collection of kg-quantities would require a much larger apparatus and/or longer sampling times.

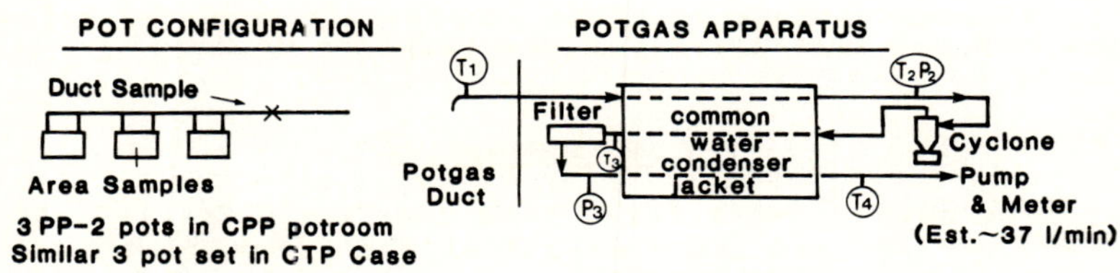

PITCH

Type	Soft Pt. °C	% in Anode
CTP	98	29.0
PP-2 / CPP	~123	26.7

POTGAS SAMPLING

Potrooms: Continuous
CTP~3 week (55g)
PP-2(i)~3 week (63g)
PP-2(ii)~4 days (27g)
2 cell flexraise

Area: Any CTP cell in 3 pot set
Middle cell of 3 PP-2 pot set
Top, aisle and side of pot
3 8hr samples during flexraise period. Repeated later on.
2 24hr samples during quiet period.

Personnel: Potroom operators for both CTP and CPP rooms.
4 8hr samples in each case.

Figure 15. Test III: Apparatus and Sampling (1977)

Chromatograms for these potgas samples are presented in Figure 16:

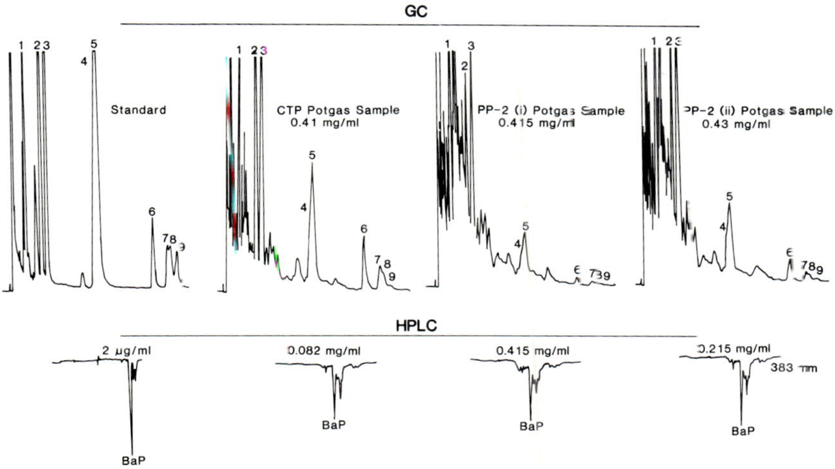

Figure 16. GC and HPLC Scans for Test III Potgas Samples (See Figure 1 for key to GC peaks)

The compositions of these samples are compared below with those of area and personnel samples.

Area Sampling

Area samples were taken from the "top", "aisle" and "end" of the CTP pot and the middle PP-2 pot. Figures 17 and 18 show how the amounts of TP and BS varied during the flexraise cycle for CTP and PP-2, respectively. The surge of fume emitted from the pot during the first 48 hr after actual flexraise is clearly illustrated by the "top" samples. In contrast, very little surge is evidenced by the "front" and "end" samples taken close to floor level. For PP-2, one "top" sample exhibited a high emission level.

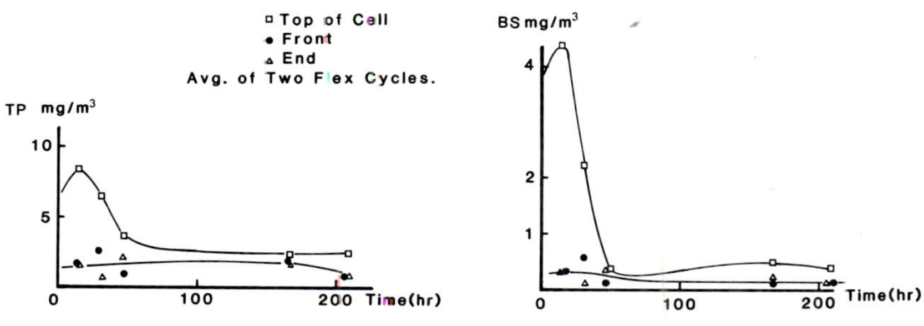

Figure 17. Test III: TP and BS in CTP Area Samples

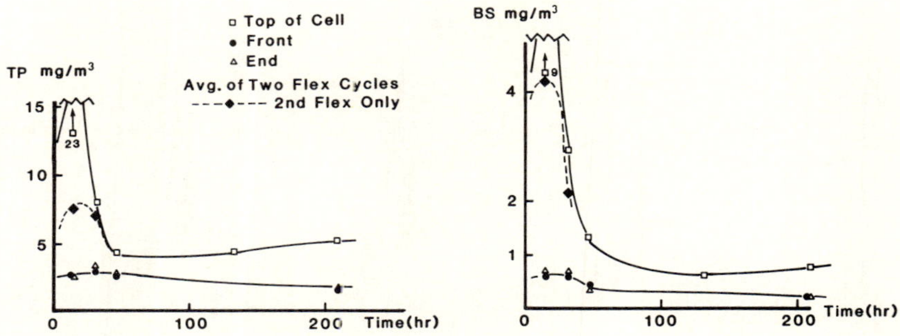

Figure 18. Test III. TP and BS in Area Samples Around PP-2 Pot

The observed emission concentrations reflect not only a possible difference between binders but problems in positioning the samplers, the number and nature of the operations performed on or by the pot, and a difference in ventilation rates between potrooms. These things highlight the difficulty of obtaining average samples from Soderberg systems. They also indicate that different workers may experience different emission exposures.

While the amounts of sample collected varied according to location and time relative to flexraise, etc., the chemical fingerprints -- plotted in Figures 19 and 20 for CTP and PP-2, separately -- show the corresponding PNA >4 compositions to be similar.

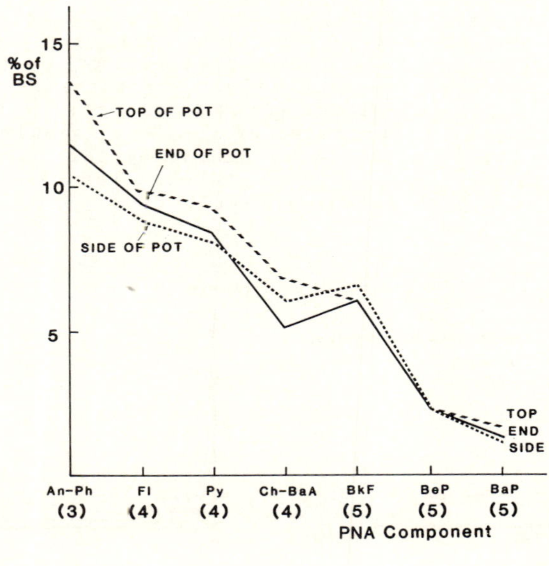

Figure 19. Test III: Fingerprints for Area Samples Around CTP Pot (Average data)

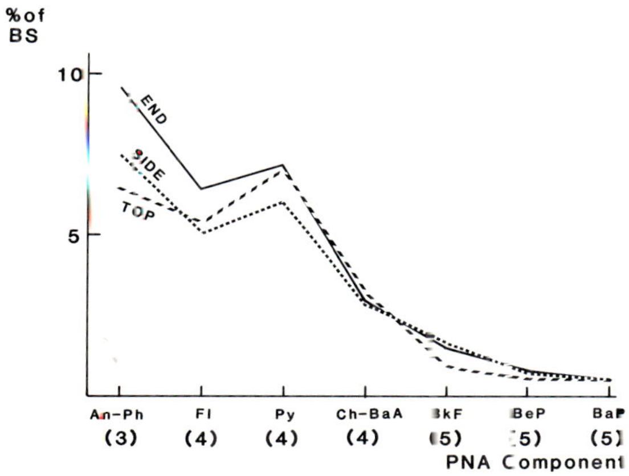

Figure 20. Test III. Fingerprints for Area Samples Around PP-2 Pot (Average data)

Personnel Sampling

Personnel samples were collected for pot operators in both CTP and CPP potrooms -- replicating four times in each case. For the CTP potroom, additional (routine) personnel sampling data were available.

The data obtained from these samples are shown in Table 6. This was the first time we had obtained VPOM data in addition to the usual BS, or PPOM, data. It is observed that the TP, BS and VPOM values for CTP are approximately equal to those for CPP. The TP and BS values, moreover, are close to standard. Again, therefore, these samples show little difference in the amount of BS emitted by CTP or PP pots. The R-values are close to unity. The amounts of VPOM collected by the Tenax absorber are notable.

Table 6. Test III: Personnel (Pot Operator) Samples

Sample Species	Organics (mg/m³) BS	VPOM	Particulate TP
Standard	0.2	(a)	10
CTP Routine	0.17	(a)	11.1
Test III	0.11	0.09	10.1
CPP Test III	0.15	0.08	10.2

(a) There is no standard for VPOM and routine samples are not taken.

The fingerprints of the CTP and CPP samples are shown on a normalized basis in Figure 21 and on a concentration basis in Figure 22. There's a considerable difference in composition between these samples.

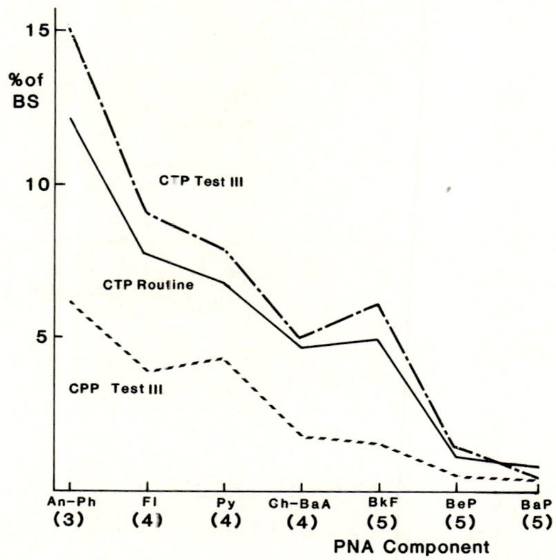

Figure 21. Test III: Personnel Samples: CTP and CPP Potrooms Pot Operator - Normalized Basis

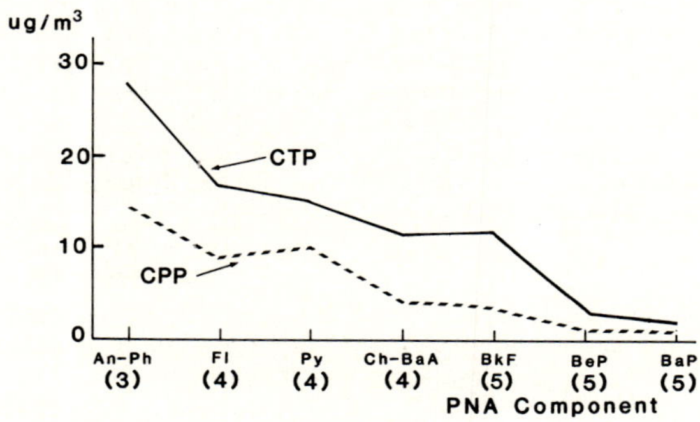

Figure 22. Test III: Personnel Samples: CTP and CPP Potrooms Pot Operator - Concentration Basis

Comparison of Potgas, Area and Personnel Results

The chemical PNA ≥4 fingerprints for samples collected in Test III may now be compared. Figures 23 and 24 present normalized data for CTP and PP-2/CPP, respectively:

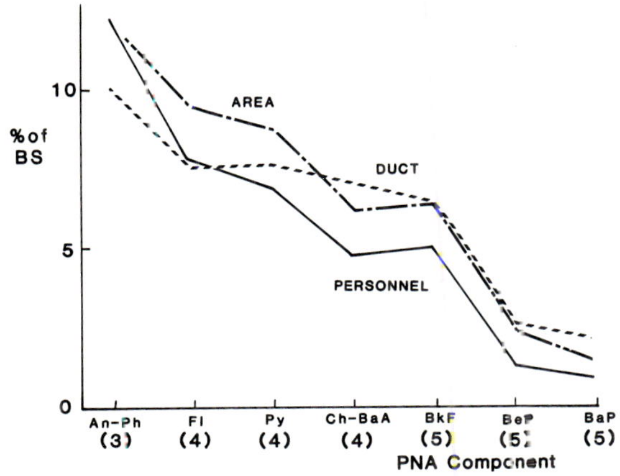

Figure 23. Test III: Potgas, Area and Personnel Fingerprints for CTP

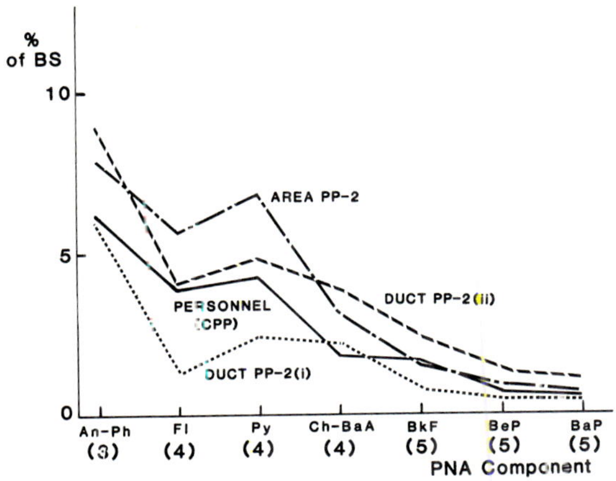

Figure 24. Test III: Potgas, Area and Personnel Fingerprints for PP-2/CPP

- For CTP, there is fairly good agreement between duct, area, and personnel samples.
- For PP-2/CPP, there is more dispersion -- but we expected PP-2(ii) to contain more high molecular weight material (see **Potgas Sampling** above).

Table 7 summarizes the reduction ratios obtained in Test III. It is apparent that:

Table 7. Test III Results

Sample	Data Source or Reference		R-Value (Rounded)
Pitch (P)	Pitch extractions	$\left[R_P\text{-BS}\right]^{CTP}_{PP-2}$	= 1
		$\left[R_P\text{-BaP}\right]^{CTP}_{PP-2}$	= 13
Potgas (PG)	BS analyses	$\left[R_{PG}\text{-BS}\right]^{CTP}_{PP-2}$	= 1
	Figure 16	$\left[R_{PG}\text{-BaP}\right]^{CTP}_{PP-2}$	= 5(i); 2(ii)
Area (A)	Figures 17 and 18	$\left[R_A\text{-BS}\right]^{CTP}_{PP-2/CPP}$	= 1 (a)
	Figures 19 and 20	$\left[R_A\text{-BaP}\right]^{CTP}_{PP-2/CPP}$	= 2
Personnel (PO)	Table 6	$\left[R_{PO}\text{-BS}\right]^{CTP}_{CPP}$	= 1
	Figure 21	$\left[R_{PO}\text{-BaP}\right]^{CTP}_{CPP}$	= 2

(i)-value probably more representative than (ii)-value. See text.

(a) Discounting the high emission in one PP-2 case.

- There is fair agreement between the various samples for corresponding R-values.
- There is little difference in BS emissions for CTP and PP-2 pots: $[R_{PG}\text{-BS}]^{CTP}_{PP-2} = 1$. Also, the initial pitches show little, if any, difference.

  These results are essentially the same as those obtained previously through Test II.
- There is less BaP emission from PP-2/CPP pots than from CTP pots. The reduction ratios for potgas, area and personnel samples: $[R_{PG,A,PO}\text{-BaP}]^{CTP}_{PP-2/CPP} = 2\text{-}5$: are similar to those obtained in the lab baking experiments (1) and in Test II. The initial pitches, of course, are considerably different: $[R_P\text{-BS}]^{CTP}_{PP-2} = 13$.

The promise of obtaining the same reduction ratio in the emissions as observed in the pitches, therefore, was not achieved.

- These results are comparable to some obtained in Russian work (5).

OTHER PLANT STUDIES

Trend of "Smoke Index"

For many years, a smoke index was measured and recorded for the potrooms. This index is measured on every shift and averaged for each month. An operator walks along the center passageway between potrooms and, by experience, rates the smokiness of the rooms -- using 3 = "Good", 2 = "Fair" and 1 = "Poor". Figure 25 shows a plot of the smoke index for 1968 through 1976. Potroom conditions improved slightly through 1970 -- when increased ore blanket, and, later, tighter maintenance brought about further improvements.

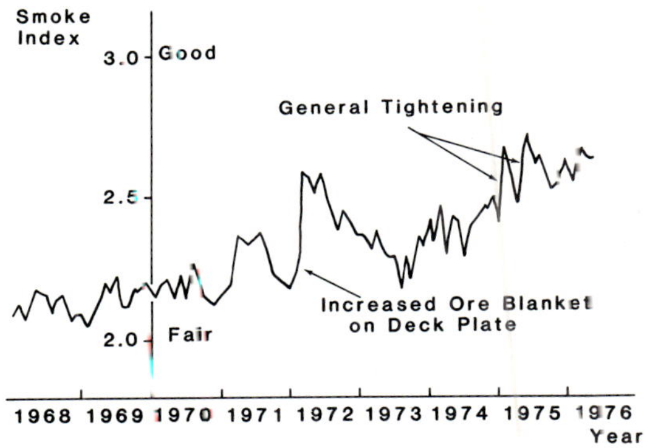

Figure 25. Average Potroom Smoke Index 1968-1976

Improved smoke index would probably correlate with decreased personnel exposure.

Assessment of Relative Conditions

Wanting to relate general potroom conditions some twenty years ago with those for 1976 -- and recognizing the difficulty of trying to recreate those conditions, we structured a survey for "old hands" at the plant. This was done as follows:

- Stress (s) was defined as a combination of smoke, heat and labor, etc., contributing to worker fatigue. (This approximately parallels exposure to PNAs.)

- The "old hands" were now asked to rate selected work zones for stress -- on an increasing scale of 1 to 10 -- for 1951/4 and 1976.

- Various jobs were broken down to time (t) spent in these zones.

- The summations $\Sigma(s)x(t)$ for typical average work days were then calculated to provide semi-quantitative measures of stress for workers in these situations.

The results of this exercise indicated tht the average stress level for potroom workers at the plant in 1976 was approximately 33% less than in 1951/4.

Decreased stress is synonymous with decreased personnel exposure.

Potroom Performance of Petroleum Pitch

Concurrent with Tests I through III above, the performance of the Soderberg pots using PP binder were being studied by others. Some problems were encountered and, while it is not our intent to detail these problems, a few comments are in order. Thus: The relatively high softening point of the PP made mixing difficult; the geometry of baked material within the anodes affected stud setting; the anodes suffered increased airburn and tended to "chunk"; and, finally, there was an indication of reduced metal productivity. Given more development time some of these difficulties might be eliminated.

ANIMAL EXPERIMENTS

Mouse Skin Painting Tests

Our initial objectives included some animal tests. Early in this program, with the counsel and advice of a consultant, we decided to have some mouse skin painting tests carried out. As a first step, we decided to study CTP and PP pitches and some condensed (CTP) hydrocarbons scraped from plant ducts. The preliminary nature of this work must be emphasized: Our objective was to learn something about methodology and the materials we were handling. We expected to undertake some definitive studies later on using materials collected from potgases.

The preliminary mouse study was completed in mid-1976. Solutions of the above-mentioned samples in toluene were tested, along with some BaP standards. For each case, an Average Latent Period was determined representing the average number of weeks for the mice to exhibit a tumor. In 1977, we were able to define the fingerprints of the materials we had tested. By tenuous argument around the few pieces of experimental data, we concluded that while the hydrocarbon material emitted from PP anodes (a material we had not tested) might have a slightly longer Average Latent Period than its CTP counterpart, the difference was probably within experimental error.

Extrapolation

As our studies progressed, we became increasingly doubtful of our ability to prove -- in terms of human exposure and response -- the relative, or absolute, value of process modifications in the reduction plant.

The literature encompassing the carcinogenicity of pitch-related compounds is voluminous and confusing. It is inappropriate to review this field here, but we offer a few observations:

- The foregoing work highlights the difficulty of collecting representative samples that can be used in animal tests. Consider just a few items:
  - The source of the sample.
  - The need to collect over a long period to obtain an average sample.
  - The makeup of the sample -- whether or not VPOM should be included; the exclusion of particulate by using solvent-extracted samples; etc.
- Mouse (or rat) skin painting tests may be questioned. Is the protocol sound; the number of animals large enough; what about strain; dose rate; biology; evaluation; statistics; etc?
- Animal skin tests may also be questioned in terms of relevance to human lung cancer.
- It would be more appropriate, perhaps, to instill samples into hamster-throats. Still further, respiration tests could be structured using dogs or monkeys. Such studies are very expensive, however, and still suffer problems. For example, consider the generation of an atmosphere which, to obtain an effect, must be loaded with BS. This BS condenses on the fur of the animals who, in preening, ingest additional (spurious) quantities of BS.
- Strictly speaking, the test atmosphere should also contain other substances present in the potroom. Various gases like HF, SO_2 and CO can be provided fairly easily -- although they introduce questions of level and/or synergism. The provision of particulate, however, is not so simple. Respirable dust may well play a role in irritating human lungs -- facilitating biochemical reaction with BS. Alternatively, this particulate may adsorb the BS -- rendering it less reactive.
- We've focussed upon BS, and even more narrowly upon BaP. The action, interaction and mechanisms whereby these and other compounds engender carcinogenic response are unclear.

- Overall, this subject is complex and incompletely understood. Because of the enormous difficulties of testing pitch-derived compounds in animals and correlating the results to humans, approximations are often assumed and/or are justified and are then converted into standards for human exposure. In spite of great effort, therefore, we may still be dealing with gross, or superficial, aspects of the subject.

This rationale was disquieting with respect to our objectives and we were wondering what our next step should be when we learned preliminary results from a study now to be described.

Human Epidemiological Studies

In the above context, the best experiment is to examine the effect of pitch-derived compounds on humans. This can only be done in retrospect, as it were, by conducting an epidemiological study. Indeed, over the years several such studies have been made and, while certain of these suggest that hazards exist, others suggest that many of the earlier results are not statistically significant.

It happens that shortly after we commenced this work, a massive epidemiological study of U.S. reduction plant workers was begun. The results of this study were eventually published in September, 1981 as, "The Tripartite Study" (6). This investigation covered: ~23,300 men having ≥5 years employment between 1946 and 1974; 14 reduction plants; Soderberg and prebake systems; 59 causes of death; exposure to BS, PPOM, BaP, TP, F, and SO_2; race, time, plant, job function, cumulative employment/exposure; and comparison against both U.S. and local populations. It was concluded that the data do not support the finding of excess mortality from lung cancer due to working in the potrooms (or carbon departments) of reduction plants.

It appears, therefore, that while some animal tests indicate that hazard may exist, the large-scale experiment shows the effect to be non-significant. The Tripartite result is heartwarming, but no one is signaling "stop" to continuing improvements in reduction plants.

With these results in hand, we decided to at least temporarily close our study of hydrocarbon emissions from Soderberg cells.

To our knowledge nobody has yet suggested that the Tripartite result provides an example of a "threshold effect". Perhaps we might be excused for doing so. Contributing to such an effect, the presence of particulate may well decrease the potency of potroom hydrocarbon emissions (see **Search For Alternate Adsorbent**, and **Extrapolation**).

CONCLUSIONS

- Improved methods are developed for trapping hydrocarbons from Soderberg potroom and duct gases. In the later instance, multigram quantities of BS are obtained.
- Improved analytical methods are developed for determining PNAs (and thence PNA fingerprints) for the trapped hydrocarbon mixtures.
- It is shown that, within experimental error, the BS trapped from potgas (in duct close to the pot) are chemically equivalent in terms of PNA ≥4 to material present in the potroom atmosphere represented by area and personnel samples.
- An area sampling device, using multiple personnel samplers, is described. The replicates provided by this apparatus are generally in excellent agreement.
- In addition to BS (PPOM) collected by the usual personnel samplers, almost equal additional amounts of vaporous, lower molecular weight hydrocarbons (VPOM) are trapped.
- Tenax-GC adsorber is useful in this work and, due to the ease of recovering PNAs, is preferred over charcoal.
- Soderberg cells emit a large surge of fume for ~48 hr after actual flexraise. The amount of this fume varies -- but represents >50% of the total fume emitted over the 11-14 day flex cycle.
- Anodes made with CTP and PP emit essentially the same amount of BS. With the current emission standard based on BS there is, therefore, no difference in exposure using these binders.
- As supplied, CTP contains ~15 times more BaP than PP. The gaseous emissions obtained using these binders, however, show a 2-to-5 fold difference. The lowering of BaP content in pot room atmosphere in switching from CTP to PP is, therefore, only one-to-three tenths of that expected from the BaP contents of the starting pitches.
- During anode carbonization in Tests II and III, the BaP emitted corresponds to ~25% and ~100% of the BaP initially present in the CTP and PP binders, respectively. These figures represent the net balance of BaP destruction, formation and emission in the anode.
- Certain operating and anode performance problems are apparent during a 17-month test of PP in 72 pots. If these problems could not be solved, the use of PP in place of CTP would involve considerable cost penalty.
- Very preliminary mouse skin painting tests, coupled to fingerprint work, suggest that PP (both in original and gaseous emission form) may be a slightly weaker tumorogen than its CTP counterpart.

- Many difficulties are involved in conducting animal carcinogenic studies relevant to human exposure, particularly the lungs. Indeed, the exercise virtually demands that assumptions be made.

- As an alternative, the results of a massive epidemiological study covering ~23,000 reduction plant workers have recently been published. Covering a wide range of variables, including local as well as total U.S. populations, this study shows no increase in mortality of said workers due to lung cancer.

- This conclusion is interesting in that BaP (or BS) is carcinogenic to animals under certain circumstances. The absence of an effect with humans suggests that either a threshold exists for humans and/or animal results may be inapplicable. In this regard, the presence of particulate in the potroom may be important.

- A review of smoke index data and an assessment of relative stress conditions for a given Soderberg plant indicate that potroom exposure to smoke and stress has lessened considerably over the 1951 through 1976 period. Because these variables include hydrocarbon emissions, this result also indicates a decrease in potroom hydrocarbon exposure.

ACKNOWLEDGMENT

This study would not have been possible without the support and enthusiasm of many colleagues. In particular:

- At the plant, we thank J. E. Andersen, J. Frish, J. Manghine, R. B. Murzyn and W. C. Selassi for their assistance with plant sampling.

- At CFT, we thank G. H. Walden for building the potgas apparatus and organizing the sampling work, R. K. Ng for adsorbent studies and sample work up, and W. W. Hanneman and J. Benjamin for organic analytical development.

- At miscellaneous locations, we thank W. H. Cundiff, R. L. Humphrey, J. E. Todd, T. J. Walker, many members of KACC Management, and Eula Bingham (Kettering Laboratory, Cincinnati) -- who provided counsel and encouragement.

ABBREVIATIONS, ETC.

A	Area
An-Ph	Anthracene & Phenanthrene (PNA=3)
BaP	Benzo(a)Pyrene (PNA=5)
BeP	Benzo(e)Pyrene (PNA=5)
BkF	Benzo(k)Flucranthene (PNA=5)
BS	Benzene Solubles. Used more widely than in CTPV-exposure circles. See **GENERAL APPROACH**,

	Benzene Solubles.
CFT	Center for Technology, Pleasanton, CA 94566
Ch-BaA	Chrysene & Benz(a)Anthracene (PNA=4)
CPP	Contaminated PP (containing some CTP)
CTP	Coal Tar Pitch
Fl	Fluoranthene (PNA=4)
Flexraise	See **GENERAL APPROACH, Plant Sampling**
GC	Gas Chromatography
HPLC	High Pressure Liquid Chromatography
KACC	Kaiser Aluminum & Chemical Corporation
NA	Naphthalene (PNA=2)
P	Pitch. Sometimes Pressure.
PG	Potgas
PNA	Polynuclear Aromatics. Integer indicates number of aromatic rings.
PO	Personnel (Pot Operator)
PP	Petroleum Pitch. 1 or 2 denote different lots.
PPOM	Particulate Polycyclic Organic Matter
Py	Pyrene (PNA=4)
Quiet Period	See **GENERAL APPROACH, Plant Sampling**
R	Reduction factor expressing ratio of given component, or group of compounds, in CTP system to that in PP system. For example: $[R_{PG}\text{-BaP}]_{PP-1}^{CTP} = 13$ means that the ratio of BaP contents in CTP to PP-1 potgases = 13.
T	Temperature
TLC	Thin Layer Chromatography
TP	Total Particulate
VPOM	Vaporous Polycyclic Organic Matter
ϕ	Diameter

REFERENCES

1. D. R. Augood, S. S. Jones and H. J. Seim, "Hydrocarbon Emissions From Pitch," Light Metals, 1981, 963.

2. NIOSH Manual of Analytical Methods, U.S. Dept. of Health, Education and Welfare, Second Edition (1977) Cincinnati, OH.

3. H. Boden, "The Determination of Benzo(a) Pyrene in Coal Tar Pitch Volatiles Using HPLC with Selective UV Detection," J. Chromatog. Sci., 1976, (14), 391.

4. H. J. Seim and J. A. Dickeson, "A Device For Collecting Replicate Industrial Hygiene Samples in the Workplace," Am. Ind. Hyg. Assoc. J., 1983, 44, (8), 562.

5. V. G. Konstantinov et al, "Hygienic Assessment of the Use of Petroleum Pitch as a Binder for Anode Masses in the Electrolytic Production of Aluminum," Gig. Sanit., 1973, (3), 27.

6. H. E. Rockette and V. C. Arena, "The Tripartite Study." "Mortality Patterns of Workers in Aluminum Reduction Plants." 2 vols. Graduate School of Public Health, Dept. of Biostatistics, Univ. of Pittsburgh, PA 15261, Sept. 1981.

MATHEMATICAL MODEL OF FLUORIDE EVOLUTION FROM HALL-HEROULT CELLS

W.E. Haupin

Alcoa Laboratories
Alcoa Center, Pennsylvania 15069
USA

Fluoride evolution from Hall-Heroult cells can be predicted with reasonable accuracy using only thermodynamics, kinetics, and the physical properties of the electrolyte. Kinetics, fortunately, so limits the hydrolysis reaction responsible for the formation of HF that only about 28% of the equilibrium partial pressure is achieved. The pot gas appears to achieve the equilibrium partial pressure of $NaAlF_4$. This condenses to form most of the particulate fluoride. The calculated particulate fluoride is, however, always less than measured. This difference was assumed to represent entrained bath and was correlated with bath composition and temperature. The combined mathematical relationships for volatilized, hydrolyzed, and entrained electrolyte predicted the measured fluoride evolution from Alcoa cells with a standard error of 1.8 kg per metric ton (7.5% relative standard error) for 1.10 to 1.30 bath ratio, 3 to 6% Al_2O_3 and 940 to 990°C.

Introduction

In this paper, fluoride evolution will be defined as the F content of the HF and particulate that leaves the cell with the cell gas. Most of this will be captured by the cell's hooding system. That which escapes capture will be called fluoride emissions. Hence, fluoride emissions can be calculated by multiplying the evolution by the factor: one minus the hooding efficiency. Hooding efficiency generally lies between 0.95 and 0.975.

Fluoride evolution from smelting cells is often classified as gaseous and particulate. Gaseous fluorides are those that continue to be gases at ambient temperature: namely, HF, CF_4, SiF_4, and C_2F_6. Entrained and volatilized electrolyte, on the other hand, becomes particulate at room temperature. In addition, alumina and carbon dust will sorb HF and, therefore, contribute to particulate fluoride. The classification into gaseous and particulate fluorides is somewhat arbitrary since the relative amounts depend upon sampling technique, sampling location, and the analytical procedure. Particulate fluorides may react with water vapor in the air to form hydrogen fluoride. Conversely, HF will sorb on nonfluoride particulates and be reported as particulate fluorides. Hence, total fluoride evolution is a more useful value.

Other sources of fluoride emission include the fluoride from spent anodes, fluoride from tapping ladles, and fluoride from carbon baking furnaces. These other sources of fluorides and the CF_4, C_2F_6, and SiF_4 from the cell are not included in this model.

Gaseous Fluoride Evolution

The major gaseous fluoride is HF, although small amounts of CF_4 and C_2F_6 may also be formed. HF is generated by hydrolysis of the bath. The water contributing to this hydrolysis is introduced into the cell by (a) the water content of the alumina, (b) burning of hydrocarbons from the anode, (c) moisture in the anode, (d) moisture in aluminum fluoride and other bath additions, and (e) perhaps from moisture in the air drawn through the pot crust. Fortunately, only a small percentage of the total moisture entering the pot is converted to HF. Alcoa measurements (1) indicated that only 8.6% of the water in the alumina, 8.7% of the moisture plus hydrogen in the anodes, and 0.3% of the moisture in the air sweep reacted to form HF.

Thermodynamic calculations indicate that the HF evolution will increase markedly at low alumina concentrations:

$$1/3\ AlF_3(\text{in bath}) + 1/2\ H_2O(g) \rightarrow HF(g) + 1/6\ Al_2O_3(\text{in bath}) \qquad (1)$$

$$P_{HF} = \mathrm{Exp}\,(-\Delta G/6RT)\; P_{H_2O}^{1/2}\; a_{AlF_3}^{1/3}/a_{Al_2O_3}^{1/6} \quad (2)$$

Where:

P_{HF} = partial pressure of HF (atmospheres)
ΔG = Gibbs free energy for Reaction 1
R = gas constant
T = temperature of electrolyte (K)
P_{H_2O} = partial pressure of H_2O in pot gas (atmos)
a_{AlF_3} = activity of AlF_3 in electrolyte
$a_{Al_2O_3}$ = activity of Al_2O_3 in electrolyte

It was assumed that moisture reacts with AlF_3 because the equilibrium constant for H_2O reacting with AlF_3 is 10^3 higher than that for H_2O reacting with Na_3AlF_6 and 10^8 higher than for H_2O reacting with NaF.

Expressing ΔG for Reaction (1) as a linear function of temperature over the range 1200 to 1300 K gave:

$$\Delta G/6RT = (8444/T) - 7.533 \quad (3)$$

Activity data (2,3) for AlF_3 was fitted by the expression:

$$\ln a_{AlF_3} = 1.32597 - 9.5199(R_b) + 2.3481(R_b)^2 \quad (4)$$

Where:

R_b = bath ratio (weight ratio NaF/AlF_3)

Activity data (4,5) for Al_2O_3 was fitted by the expression:

$$a_{Al_2O_3} = (\%Al_2O_3/\%Al_2O_3^*)^{2.77} \quad (5)$$

Where:

$\%Al_2O_3$ = weight percent Al_2O_3 in electrolyte
$\%Al_2O_3^*$ = weight percent Al_2O_3 at saturation

Using these values in Equation (2), a match with the Alcoa test data (1) was obtained when the assumption was made that the reaction proceeded between 26.6 and 28.8% of the way toward completion, increasing with decreasing bath ratio, or alternately that only 3% of the available water entered the bath. This led to Equation (6). Figure 1 compares the Alcoa test data obtained over the range of 3-6% alumina with the predictions of Equation (6):

$$F_G = \frac{2310000}{(\%CE)} - \frac{955000}{(P_b)} R_b \left[EXP \left(7.533 - \frac{8444}{T} \right) \right] \left[\frac{\%H_2O \text{ in } Al_2O_3}{37.44} + \frac{\%H \text{ in Anodes}}{21.5} \right]^{1/2} [EXP\ (0.44199 - 3.1733R_b + 0.78127R_b^2)]/ \left[\frac{\%Al_2O_3}{\%Al_2O_3 \text{ sat}} \right]^{0.462} \quad (6)$$

Where:

F_G = Gaseous fluoride per metric ton
%CE = % current efficiency
R_b = bath ratio, weight NaF/AlF_3
P_b = barometric pressure, kPs

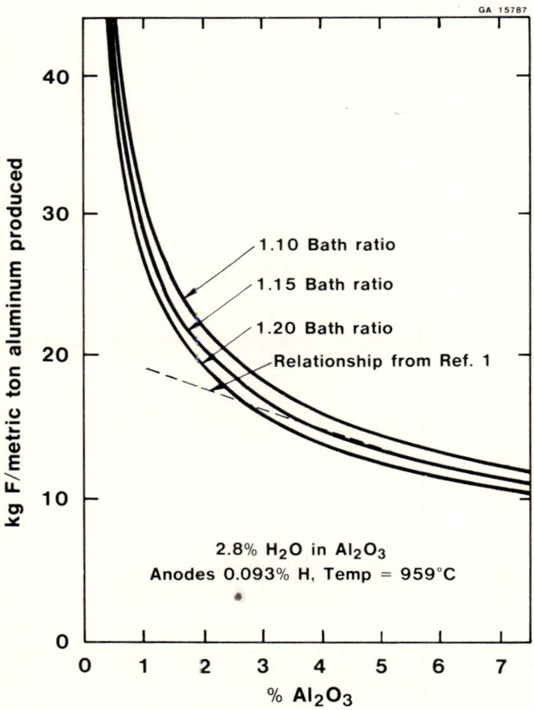

Figure 1 - HF evolution calculated by Equation (6)

Particulate Evolution by Volatilization of Electrolyte

E.H. Howard (6) demonstrated that the vapor over Hall bath has the composition $NaAlF_4$ (sodium tetrafluoroaluminate). He identified the structure of this compound by X-ray diffraction of rapidly quenched samples. Howard also reported that the condensate formed by slow cooling gave diffraction patterns for chiolite ($Na_5Al_3F_{14}$) and aluminum trifluoride (AlF_3). These original observations have since been confirmed by many other investigators. Mass spectographic measurements by Sidorov et al (7) showed a small, but significant, amount of dimerization:

$$NaAlF_4 \rightleftarrows Na_2Al_2F_8 \qquad (7)$$

Their measurements of the equilibrium constant can be correlated:

$$K = P_d/P_m^2 = \mathrm{EXP}\ (-21414/T + 15.6) \qquad (8)$$

Where:

K = Equilibrium constant for dimerization
P_d = Partial pressure of $Na_2Al_2F_8$ (kPa)
P_m = Partial pressure of $NaAlF_4$ (kPa)

The vapor pressure of molten cryolite with additives has been studied extensively and there is good agreement among recent investigators (8-12). We have combined these data into the following formula giving the vapor pressure in kPa:

$$P = \mathrm{Exp}\ (B - A/T) \qquad (9)$$

Where:

$$A = 7101.6 + 3069.7(R_b) - 635.77(R_b)^2 + 51.22\ (\%LiF) - 24.638(R_b)(\%\ LiF) + 764.5\ (\%Al_2O_3)/[1 + 1.0817\ (\%Al_2O_3)] + 13.2\ (\%CaF_2)$$

$$B = 7.0184 + 0.6844(R_b) - 0.08464(R_b)^3 + 0.01085\ (\%LiF) - 0.005489(R_b)(\%LiF) + 1.1385\ (\%Al_2O_3)/[1 + 3.2029\ (\%Al_2O_3)] + 0.0068\ (\%CaF_2)$$

% = wt %

Bath vapor pressure increases with decreasing bath ratio and increasing temperature. Additions of calcium fluoride, magnesium fluoride, lithium fluoride, and alumina all decrease vapor pressure. Alumina has the greatest effect. The partial pressures of the monomer and the dimer were calculated from the vapor pressure, Equation (9), and the dimerization equilibrium constant, Equation (8).

$$P_m = [-1 + (1+4K \cdot P)^{.5}]/2K \quad (10)$$

$$P_d = P - P_m \quad (11)$$

Kilograms of volatilized fluoride per metric ton of aluminum produced were calculated by assuming that the pot gas (9.3277 g mols/kAh) is saturated with $NaAlF_4$ and $Na_2Al_2F_8$ at their calculated vapor pressures. For equal partial pressure the dimer gas contains twice as much F as the monomer. This leads to Equation (12):

$$F_{VP} = (204000/\%CE)\,(P_m + 2P_d)/P_b \quad (12)$$

F_{VP} = Particulate fluoride (from vaporized electrolyte) per metric ton of aluminum

Figures 2, 3, and 4 present calculations made at an assumed current efficiency of 90%. Particulate fluoride evolution at other current efficiencies can be obtained by multiplying the indicated evolution by 90 and by dividing by the true current efficiency. Figure 2 shows the variation in particulate evolution and vapor pressure as bath ratio and bath temperature are varied. Figure 3 shows the same relationships at constant superheat temperatures above liquidus. Figure 4 shows the change with alumina concentration.

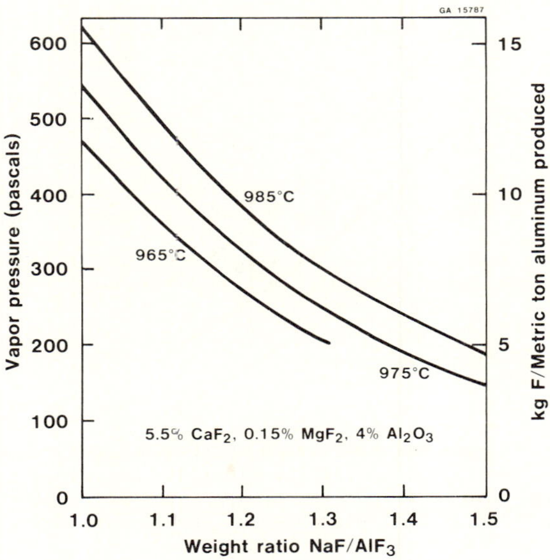

Figure 2 - Vapor pressure as function of bath ratio and temperature

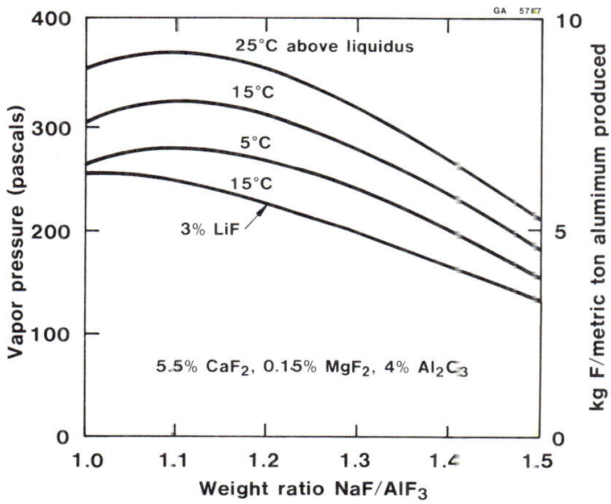

Figure 3 - Vapor pressure as function of bath ratio and super heat

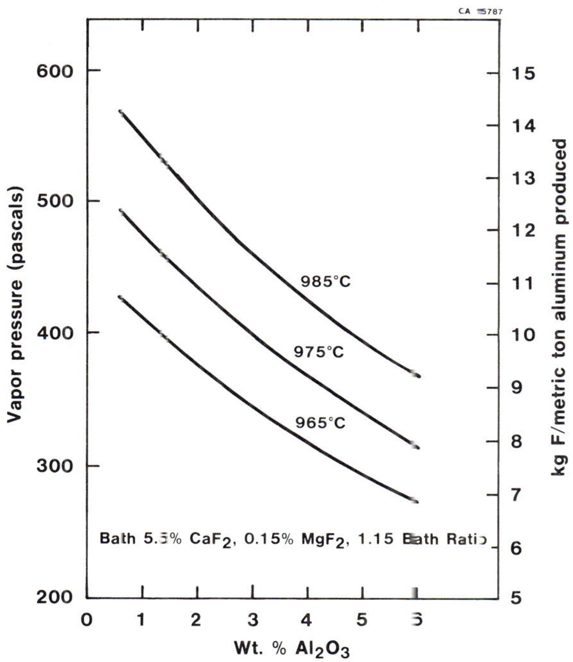

Figure 4 - Vapor pressure as function of % Al_2O_3 and temperature

Particulate Evolution by Entrainment of Liquid Droplets

Entrainment of electrolyte in the pot gas as it leaves the bath has not been subjected to nearly the quantitative investigation that hydrolysis and volatilization of the bath have. Newitt, Dombrowski, and Knellman (13) described two mechanisms by which bubbles rising through any liquid entrain the liquid in the gas. When a bubble reaches the surface, its dome breaks and tiny droplets, generally under 0.025 mm in diameter, form and are entrained in the gas. Azbel and Lee (14) present equations to calculate this source of entrainment. The calculated value falls far short of accounting for the observed entrainment. Because of their small size, dome particles contribute little on a weight basis, although they dominate on a numerical basis. In the second mechanism described by Newitt et al, liquid rushing in to fill the cavity made by the breaking bubble forms a jet. Bubbles larger than 0.5 to 0.8 cm in diameter yield jets that are too thick and too short to break; hence, only the dome produces entrainment. As bubbles get smaller, the jet thins, gets taller and breaks, ejecting droplets. These droplets range from 0.1 to 1 mm in diameter and account for the major mass of ejected liquid available for entrainment.

It is not clear whether the mechanism (ballistics) of ejection of liquid electrolyte into the gas phase or the transport by the sweep gas of the entrained solidfying liquid controls the amount of entrainment. If the amount of air sweep were reduced, would entrainment be markedly reduced? Unfortunately, there is little published data on the entrainment of electroyte in Hall cell gas and the authors have not attempted to correlate the measured entrainment with bath properties, temperature or operating conditions. Evidence for significant amounts of entrained liquid include the fact that the NaF/AlF_3 ratio of the particulate fluoride is intermediate between that of the bath vapor ($NaAlF_4$) which has a bath ratio of 0.5, and that of the electrolyte itself. Also, calcium fluoride is detected in the particulate. Since the calcium species in the bath have negligible vapor pressure, calcium fluoride in the particulate must arise from entrained droplets of bath.

Less and Waddington (15) investigated the size distribution and chemical composition of dust in unburned gas from prebake cells and burned gas from Soderberg cells. In both cases, the dust could be separated into two fractions; one with particles smaller than one micrometre and the other with particle size greater than five micrometres. The coarser material consisted principally of glassy appearing, frozen droplets of bath -- presumably entrained electrolyte -- plus solid particles of carbon and alumina. The fine dust consisted primarily of chiolite and aluminum fluoride, and presumably was formed by condensation of the vapor $NaAlF_4$. It was deficient in aluminum fluoride and this was assumed to indicate that the vapor had been hydrolyzed. Their data lead to the conclusion that somewhat over 35% of the particulate arises from entrainment.

Grjotheim et al (11) compared the measured particulate evolution with that calculated from vapor pressure measurements and concluded that approximately half of the particulate resulted from entrainment. Miller (17) estimated from microscopic examination of the particulate that 20 to 40% of the particulate came from entrainment. In the Alcoa test (1) entrainment was determined by analyzing the calcium content of the particulate. This indicated that only about 6% (0.7 kg/t) resulted from entrainment; however, the vapor pressure of the bath accounted for only 53% of the observed particulate. Adding 6% for entrainment still left 41% of the particulate unexplained. Wahnsiedler assumed this additional particulate was generated by air being drawn under the crust. This latter mechanism would have to extract nearly as much bath vapor as the anode gas which is generated under the bath. An alternate explanation would be that the calculation of entrained bath, based on calcium content of the particulate, was in error and that the unexplained 41% is largely entrained bath plus some fluoride sorbed on alumina and carbon dust. On this basis the particulate would be approximately 53% voltatilized bath and 47% entrained bath, putting it in much better agreement with other investigators. Using this alternate explanation, the differences between the calculated and measured particulate was considered to be entrained electrolyte. These data were correlated empirically with bath ratio, $\%Al_2O_3$, and temperature by multiple regression analysis:

$$F_E = [-17030 + 29800R_b - 13000R_b^2 + 67(\%Al_2O_3) - 173t - 0.3896t^2 + 141.6R_bt]/\%CE \quad (13)$$

Where:

F_E = Entrained electrolyte (kg F/metric ton Al)
t = T - 1243K

The sum of HF (gaseous), vapor pressure generated particulate, and entrained electrolyte gives total fluoride evolution:

$$F_{tot} = F_G + F_{VP} + F_E \quad \text{kg F/ton Al} \quad (14)$$

Description of Fluoride Evolution Tests

In the Alcoa tests (1) previously referenced, an attempt was made to get 100% hooding efficiency by using new hoods and sealing all gaps with aluminum foil. Cell volts, amps, bath temperature, exhaust gas flow, exhaust gas temperature, HF concentration in the exhaust gas, ambient temperature, relative humidity and barometric pressure were determined continuously. Isokinetic 15-minute pot gas particulate samples were taken. Electrolyte samples were taken hourly. Every four hours an alumina feeder dump was caught, weighed, and sampled. An attempt was made to control bath ratio and anode hydrogen content at various levels. Other variables, although measured, were allowed to fluctuate normally. Fluoride evolution was analyzed by multiple regression analysis giving the following equations:

$$F_{tot} = [-13930 - 5275R_b + 21.98t + 0.352_{hum} + 312V_{ore} - 125\ (\%Al_2O_3) + 63\ (\%CaF_2) + 200(LSO) + 1200(AE) - 214\ (TR) + 745H_{an}]/\%CE \quad (15)$$

$$F_{gas} = [-8994 - 3364R + 14.44t + 304P_{hum} + 251W_{ore} - 66.3\ (\%Al_2O_3) + 26.9\ (\%CaF_2) + 200(LSO) + 809(AE) - 261(TR) + 1066H_{an}]/\%CE \quad (16)$$

Where:

F_{tot} = Total fluoride (kg/t Al)

F_{gas} = F of the HF evolved

t = Bath temperature, °C

P_{hum} = Vapor pressure of H_2O in air (kPa)

W_{ore} = % H_2O in ore

LSO = Low surface area ore. If HF on surface of reacted ore exceeds monolayer, enter 1; otherwise, enter 0.

AE = If on anode effect, enter 1; otherwise 0.

TR = Track. If Al_2O_3 feeding has been stopped, enter 1, otherwise 0.

H_{an} = % H in anode

Discussion

The multiple correlation coefficients adjusted for degrees of freedom were 60% for Equation (11) and 58% for Equation (12). A difficulty with this approach lies in the linear relationship assumed between fluoride evolution and each variable. Figure 1 illustrates what can happen when one attempts to extrapolate a relationship derived by linear regression. Notice how a straight line fits the data quite well over the normal range of 3-6% Al_2O_3, but there is large deviation from linearity at low alumina concentrations. The new model based on thermodynamics and kinetics, except for entrainment, predicted total measured fluoride evolution with a standard error of 1.8 kg/metric ton of aluminum or a 7.5% relative standard error. A disadvantage of these revised equations is that they do not take into account operating variables such as the condition of the pot crust, anode effects, use of ore having insufficient surface area in a dry scrubber, etc. Anode effect could be handled by knowing the temperature rise and duration of the anode effect. Low surface area ore could be handled by assuming all HF exceeding one monolayer would flash off as it is added to the cell. Calculation of entrained bath is particularly weak. Being derived by difference, this correlation contains all the error of measurement plus the predictive error of the other models. It predicts impossible negative entrainment above 990°C for electrolytes having bath ratios greater than 1.4. This demonstrates again the danger of extrapolating an empirical correlation. Further investigation of entrainment is needed.

References

1. W. Wahnsiedler, R. Danchik, W. Haupin, D. Backenstose, and J. Colpitts, Light Metals 1978, v. 2, p. 407.

2. K. Yoshida and E.W. Dewing: Met. Trans. 3, 683 (1972).

3. H. Kvande, Dr. Thesis NTH, Univ. of Trondheim, Norway (1979).

4. M.M. Vetyukov and N. van Ban, Tsvet. Met., 44:31 (1971).

5. M. Rolin, Rev. Int. Hautes Temp. et Refract., 9:333 (1972).

6. E.H. Howard, Journal of American Chemical Society, v. 76, p. 2041 (1954).

7. L.N. Sidorov, E.N. Kolosov, and V.B. Shol'ts, Russ. J. Phys. Chem. 42:1382-1386 (1968).

8. M. Rolin and J. Houriez, Bull. Soc. Chem., France, p. 891 (1964).

9. U. Kuxmann and U. Tillessen, Erzmetall, 20: 147 (1967).

10. F. Mesrobian, M. Rolin, and H. Pham, Rev. Intern. Hautes Temp. Refract. 9 (1) 139-146 (1972).

11. J. Gerlach, V. Hennig, and M. Mucke, Erzmetall. 26 (10) 496-501 (1973).

12. K. Grjotheim, H. Kvande, and K. Motzfeldt, Light Metals 1975, v. 1, pp. 125-137.

13. Newitt, Dombrowski, and Knellman, Trans. Inst. Chem. Engr., 32, 244 (1954).

14. D.S. Azbel and S.L. Lee, Two-Phase Momentum Heat Transfer Chem., Process Eng. Syst., v. 1, p. 159 (1978).

15. L.N. Less and J. Waddington, Light Metals 1971, p. 499.

16. J. Grjotheim, H. Kvande, K. Motzfeldt, and B J. Welch, Can. Metal. Quart., 11 (4) 586-599 (1972).

17. J. Miller, Proceedings: Engr. Aspects of Pollution Control, Metals Soc. London, p. 165, November 1974.

METHODS OF MONITORING OCCUPATIONAL EXPOSURE IN ALUMINIUM SMELTERS

A. Steinegger,
Swiss Aluminium Ltd.,
Zurich/Switzerland

Introduction

To determine airborne contaminants at the workplace in Al smelters, various methods can be applied: stationary monitoring, personal monitoring and biological monitoring. The methods mentioned all have advantages and disadvantages[1]) and should be applied specifically, therefore,

In the following a few examples from practice are given:

1. Comparison of Stationary Measurement/Personal Sampling in Aluminium Reduction

Both methods are used to check the influence of two different technologies of aluminium reduction on the workplace. Plant A is equipped with open pots, in which the gases and dusts arising in the work process, are removed openly by large air circulation supported by the air convection of the scrubber system in the roof. Plant B has hooded, sideworked pots.

In the following Tables 1 and 2, the measurements are compiled for stationary and personal sampling. The standard deviation (σ) is given in brackets.

Comparison of Plants A and B by Stationary Sampling:
Potroom air = macro climate TABLE 1

	Dust (mg/m³)			Fluoride (mg/m³)		
	Total	Fine	% of Total	Gaseous	Particle	Total
Plant A	3.00 (±1.26) n = 13	0.80 (±0.16) n = 13	26.7	0.31 (±0.08) n = 10	0.35 (±0.09) n = 10	0.66 (±0.10) n = 10
Plant B	1.13 (±0.37) n = 7	1.01 (±0.39) n = 7	89.4	0.37 (±0.12) n = 7	0.30 (±0.12) n = 7	0.67 (±0.22) n = 7
Ratio A / B	2.65	0.79	0.30	0.84	1.16	0.99

Comparison of Plants A and B by Personal Sampling:
Exposure = micro climate TABLE 2

	Dust mg/m³			Fluoride mg/m³		
	Total	Fine	% of Total	Gaseous	Particle	Total
Plant A	9.71 (±3.68) n = 8	2.76 (±0.95) n = 8	28.4	0.64 (±0.40) n = 7	0.92 (±0.37) n = 7	1.56 (±0.65) n = 7
Plant B	4.8 (±0.36) n = 3	0.87 (±0.30) n = 3	18.1	0.30 (±0.08) n = 10	0.62 (±0.14) n = 10	0.92 (±0.17) n = 10
Ratio A / B	2.02	3.17	1.57	2.13	1.48	1.70

o Stationary Measuring (Table 1)

- for the total dust, the value in plant A is approx. 2.5 times higher than in plant B

- the fraction of fines is percentagewise considerably higher in Plant B, however,

- for the fluorides, no great differences are ascertained.

o Personal Sampling (Table 2)

- All values from personal sampling measurements of plant A are higher both for dust and for fluoride.

- The ratio of plant A to plant B is highest for fine dust (~3 fold).

- In the case of fluorides, the ratio of plant A to plant B is highest for gaseous (~2 fold).

In this comparison, it has to be taken into account that a few other parameters typical of operation such as the kind of alumina, the mode of working etc. influence the measurements and can thus make a direct comparison more difficult. Nevertheless, the following conclusions can be drawn:

- the macro climate (= general potroom air) is approximately the same for both potrooms with the exception of the total dust.

- The employee doing the work on the pots (micro climate) is exposed to a higher degree to contaminants than in plant B, however. The reason is probably that no mechanical ventilation is effective in the direct vicinity of his work.

- the lower values of the macro climate for the fine dust and the gaseous fluoride in the stationary measurement can be explained by the fact that in the potrooms without covered pots (plant A), there is a much greater air convection and the fine fractions and gases are more efficiently removed as a result, although they, as personal sampling shows, arise in work at the pot in larger quantities in the direct vicinity. In the case of the hooded pots (plant B), the discharge of volatile contaminants into the potroom air is smaller. Gaseous substances and fine dust, however, stay longer near the bottom because of the smaller air convection.

- due to the determination of the macro climate in the example shown, only limited conclusions can be drawn as to the expected exposure of the workers.

2. Comparison of Stationary Measuring/Personal Sampling - Anode Bake Furnace

Both methods, stationary sampling and personal sampling, were also used to determine the polycyclic aromatic hydrocarbons in a building for the production of anodes. Here attention is paid that the flow speed through the filters was roughly the same in both cases. The sole difference was that in personal sampling a silver membrane filter was used in which, according to previous examinations[2], approx. 10 % PAH (= polycyclic aromatic hydrocarbons) are retained according to the glass fibre filter. Altogether 14 different PAH's were determined. The results are listed in Table 3.

Comparison of Personal Sampling, Stationary Sampling

TABLE 3

	CSM $\mu g/m^3$	PAH $\mu g/m^3$	% PAH of CSM	% BaP of PAH
Personal sampling	77	26	33.8	2.4
Stationary	27.5	10.5	38.2	2.9
Personal/ stationary	2.8	2.5	0.9	0.8

- Due to personal sampling, higher substance concentrations are obtained than with stationary sampling. This applies to the totality of the compounds soluble in cyclohexane (CSM) and the polycyclic aromatic hydrocarbons (PAH).

- The percentage fractions on the other hand are not very different either for the PAH's or for the benzo(a)pyrene (BaP) fraction.

- This indicates that in the example shown, the contaminant situation is generally recorded by both methods but that the concentrations depend on the place of sampling. That does not apply in every case, however.[3]

3. <u>Comparison Personal Monitoring / Biological Monitoring</u>

The data from various aluminium reduction plants were checked with a view to a possible correlation of the F' concentrations measured by personal monitoring with the fluoride content of the urine. The results are entered in Fig. 1.
10 - 20 analyses by personal monitoring and 60 - 100 in the urine at the end of a shift period correspond to each value. The fluoride values were not corrected. As can be seen from the curve, a high degree of correlation was found which generally agrees with the results of Dinman et al[4]. For the exposure to HF, a correlation factor for the 24 hour urine of 0.8615 was published by Zober et al[5].

The good correlation is probably favoured both by the good suitability of biological monitoring for fluoride and by the large number of measurement data. If only limited populations and/or very different groups of employees form the basis, the result is negatively influenced. Data of this kind can also be found in literature for example in a paper by Morgan et al[6] in the case of exposure to nickel.

Linear regression between F' concentrations at the workplace and in the urine (aluminium smelter)

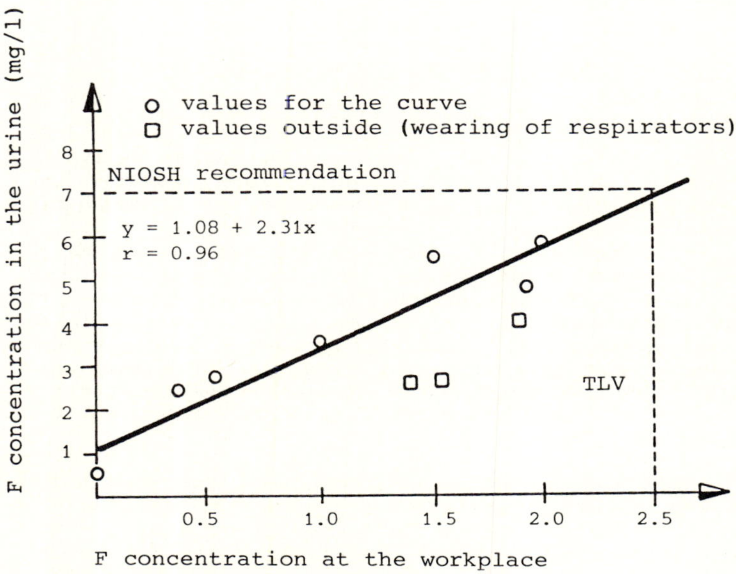

Fig. 1

In Fig. 1 values from an aluminium smelter have also been entered in which the wearing of respirators is very widespread. It is a matter of light one-way masks which, however, nevertheless, as can be seen from the location of these dots, have a good protective effect.

Conclusions

- In the interpretation of the measurements obtained with personal sampling compared to stationary sampling, the specific technical conditions have to be taken into account. Knowledge of the fractions of exposure due to the macro climate (= potroom atmosphere) determined by stationary sampling and the micro climate (= directly in the vicinity of individual sources of emission) determined by personal sampling allow purposeful measures to reduce the exposure. This applies not only to the examples shown when determining dust and fluoride concentrations but is valid generally.

- In the case of relatively consistent emissions over large areas e.g. from the anode bake furnace, above all the distance to the source plays a role for the extent of the concentrations measured. On the other hand the fractions of PAH in cyclohexane soluble matter and of the BaP in PAH are not very different. At workplaces with varyingly heavy dust loads, these statements do not apply, however[3]).

- Personal monitoring and biological monitoring show a good correlation as regards fluoride in the potrooms. This applies for the time being, however, only for the mean value of a relatively large number of measurements of fluoride in urine (>60). For an individual control, the biological monitoring no doubt offers advantages. However if discipline in wearing respirators is high, too low values are obtained with biological monitoring. Personal sampling continues to be important in clarifying special cases such as for example individual, too high F' values, influence of different raw materials etc.

Bibliography

1. A. Steinegger

   Control of Workplaces as Regards Worker's Hygiene Symposium on Guidelines for Industrial Toxicology Research Topic 3: Applied Processes, Future Processes, p. 129
   Nancy, 27th - 29th September, 1982,

2. A. Steinegger and R. Glaus

   The Determination of Polycyclic Aromatic Hydrocarbons at the Workplace in Carbon Plant and Reduction Plant. Light Metals 1981, p. 977 (Proceedings 110th AIME Meeting)

3. R. Glaus and A. Steinegger

   Contribution to The Interpretation of Coal Tar Pitch Measurements at the Workplace
   AIME 1984

4. B. D. Dinman, M.D.; W.J. Bovard, M.D.; T.B. Bonney, B.S.; J. M. Cohen, B.S., M.P.H.; and M.O. Colwell, M.D. 'Absorption and Excretion of Fluoride Immediately after Exposure - Pt. 1'
   JOM 18 (1976) p. 7 - 13

5. A. Zober, M. Geldmacher v. Mallinckrodt and K.H. Schallter 'Renal Fluoride Excretion as a Useful Parameter for Monitoring Hydrofluoric Acid-Exposed Persons'
   Int. Arch. Occup. Environ. Hlth 40, (1977), p. 13 - 24

6. L.G. Morgan and P.J.C. Rouge
   Inco Europe Ltd., Clydach, Swansea
   'A Study into the Correlation between Atmospheric and Biological Monitoring of Nickel in Nickel Refinery Workers'
   Ann. Occup. Hyg. 22, (1979), p. 311 - 317

REDUCTION CELL HOODING EFFICIENCIES WITH VARIOUS CELL,

SUPERSTRUCTURE, AND SHIELD DESIGN FACTORS

R. J. Schlager and R. K. Ng

Center for Technology
Kaiser Aluminum & Chemical Corporation
Pleasanton, California 94566
USA

Reduction cell design parameters can significantly affect the efficiency of containing emissions generated by the aluminum smelting process. Hooding efficiencies of various experimental prebake reduction cells were measured with a helium tracer gas technique. Environmental aspects relating to hooding efficiency are discussed along with correlating the hooding efficiency relationships to various fume control design factors. The effects of the various design factors on hooding efficiency become readily apparent when cell evacuation requirements are compared on a normalized basis.

Introduction

The smelting of primary aluminum produces a number of pollutants, of which particulate and gaseous fluorides have received the most attention from an environmental control viewpoint. A number of scrubbing systems have been developed to recover the fluoride from reduction cell gases, but little has been published concerning the factors that affect the efficiency of containment of cell-generated fumes. These fumes are contained by shielding a cell with hoods that can be moved or removed from cells. The efficiency of fume containment is a function of various cell design parameters such as cell superstructure and shielding. The shielding practice, shield maintenance, and cell exhaust gas flowrate also affect the degree of containing cell emissions. Since cell gas scrubbing is a very efficient process for recovering fluoride from cell-generated fume, the major fluoride losses from a reduction plant result from fluorides escaping cell hooding.

Experimental prebake reduction cells were tested in order to determine hooding efficiency characteristics. The cells tested were different in size, shield design, superstructure design, and general cell operating practices. Results of a study such as described in this paper can be used to predict emissions from reduction cells, provide design criteria for the operation of new reduction cells, and provide data for economic analysis of the most suitable fume containment strategy for a reduction plant.

Experimental

The experimental procedure for determining cell hooding efficiency was a helium tracer gas technique, as described by Johnson, et al. (1). Briefly, a constant volumetric flow of helium was released into a test cell. The helium was picked up along with cell-generated gases by the fume control system and transported away by ductwork. Samples of this potgas were extracted from a sampling port located in the cell evacuation duct. They were collected in glass syringes and were referenced as "cell samples". At the completion of cell sampling, the same volumetric flow of helium gas was released directly into the cell evacuation ductwork at a port next to the cell superstructure. Samples collected at this time were designated as "duct samples" which were used as the reference concentration. The relative helium concentrations in the gas samples were determined by a portable gas chromatograph (G.C.). The hooding efficiency (H. E.) was defined as:

$$\text{H. E.} = \frac{\text{G. C. response of cell sample}}{\text{G. C. response of duct sample}} \times 100\% \qquad (1)$$

This test method is unique in that it allows hooding efficiency measurements to be performed on individual cells. Since only one cell is involved, neighboring cells in a potroom are not distrubed during the testing and manipulation of the experimental cell. Under ideal testing conditions the procedure allows as many as 18 individual determinations of efficiency to be performed in a workshift. The standard deviation of a number of measurements typically falls within ±1% of the measured value. Hooding efficiency values of experimental cells determined by roof ventilator and main duct fluoride measurements were within ±1% of the hooding efficiency values predicted from previously obtained helium tracer gas test data.

Results

Shield Design

Two 190 kA prebake test cells were evaluated, each with different side shield designs. Shield design differences are shown schematically in Figure 1. Design A is a conventional shield, whereas design B incorporates a step for worker access to the cell superstructure. Both designs are essentially the same in all other respects. Results of hooding efficiency measurements at different cell exhaust rates are given in Figure 2. The hooding efficiency values listed on the Figure are for a cell-day; that is, the hooding efficiency of the cell including effects of tapping and anode change operations on the overall hooding efficiency. The hooding efficiency characteristics of the cell with shield design A shows a design which is more efficient in containing cell emissions than design B. In order to maintain the same level of fume collection from the cells tested, approximately 10 M^3/min more evacuation airflow is required if shields with design B are used on the cells. Since design A is more streamlined than design B, it allows a shorter path for cell-generated gases to be picked up at the superstructure exhaust points. It is also conceivable that "dead spaces" exist under shield design B, making it a less efficient design for transporting the fume to the superstructure.

The 190 kA test cells underwent another series of tests in which the effects of sealing various openings on the cell was studied. This cell was fitted with shields of design A.

The sources of openings in a reduction cell from which emissions can potentially escape are 1) spaces between anode rods and the superstructure, 2) spaces between side shields, and 3) spaces around the end doors and view slots. The total open space area of the test cell was measured, and was found to be distributed approximately 18%, 65%, and 17% for sources 1, 2, and 3 respectively. Duct tape was used to seal openings to varying degrees around the cell in order to arrive at the following test conditions:

1) Cell with normal open space area.
2) Totally closed cell, all openings sealed.
3) One-half of all openings sealed.
4) All openings sealed except the space around the end doors.

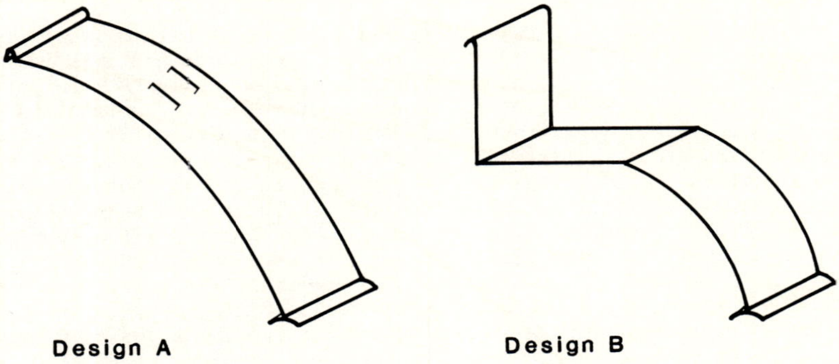

Figure 1: SCHEMATIC DIAGRAM OF DIFFERENCE BETWEEN SHIELD DESIGNS USED ON 190 kA CELLS.

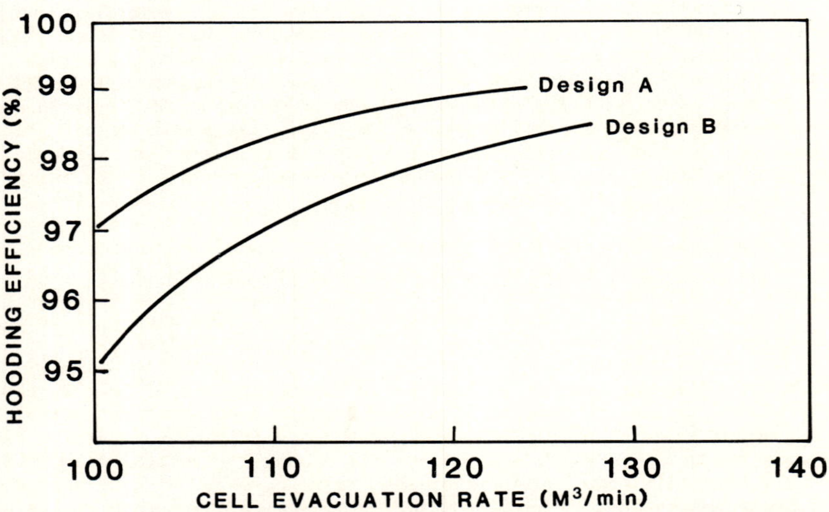

Figure 2: HOODING EFFICIENCY CHARACTERISTICS OF 190kA CELL WITH DIFFERENT SHIELD DESIGNS.

Figure 3 shows the hooding efficiency curves for the various test conditions at evacuation rates between 80 and 120 M^3/min (2825 - 4250 ACFM).

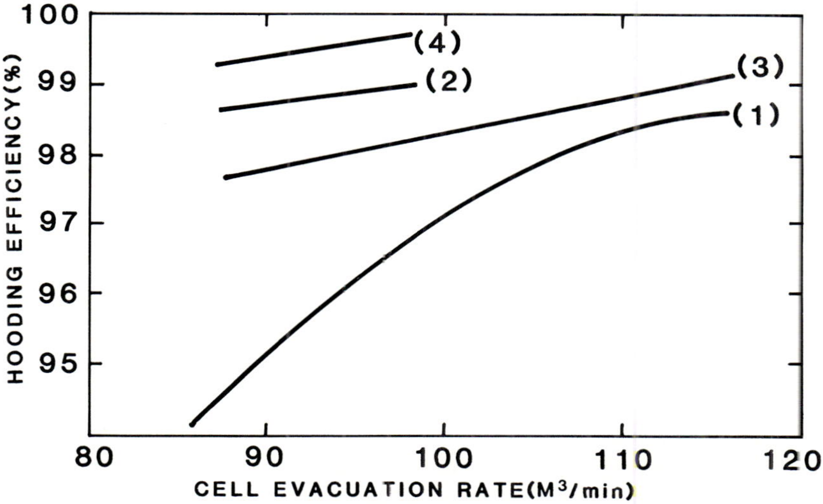

Figure 3: HOODING EFFICIENCY EFFECTS AS A RESULT OF SEALING OPENINGS ON THE CELL.

The cell was tested at two evacuation flow rates in the case of the totally sealed cell (Curve 2, Fig. 3). The hooding efficiency reached 99.0% at a flow rate of 97.8 M^3/min (3450 ACFM). It increased to 99.7% by opening the spaces around the end doors (Curve 4, Fig. 3). Apparently, gases generated near the ends of the cell require make-up air to sweep them into the superstructure pickup points. Otherwise these "stagnant zones" may eventually allow gases to diffuse outside the cell.

Since the hooding efficiency characteristics represented by Curves 2 and 4 are within experimental uncertainty of the test method (± 1%), the differences in efficiencies between test conditions 2 and 4 are not statistically significant. It is felt, however, that the curves show realistic trends in hooding efficiency behavior at the given test conditions.

For practical purposes it is impossible to operate a cell that is completely sealed. With the thought of allowing sufficient space for normal cell operation, such as shield wear and anode rod deformation, a series of tests were conducted in which one-half of all openings were sealed and the other half were left open. Results of this test (Curve 3, Fig. 3) show a considerable improvement in hooding efficiency over the base case cell of Curve 1, Fig. 3.

Dual Flow Concept

The 190 kA test cells were equipped with dampers which allowed a greater amount of evacuation air to be drawn from the cells during those operations requiring shields to be removed from the cells. Activating the "dual flow" dampers increased the evacuation rate from a cell by approximately 75%.

Tests show that under simulated anode change conditions (removal of the required number of side shields to perform an anode change) a hooding efficiency of 85.8% is obtained at normal cell evacuation rates. When the dual flow damper was activated, the hooding efficiency increased to 93.0%. Increasing the cell evacuation rate with use of the dual flow damper in effect reduced cell emission losses by 50% during the simulated anode change.

A similar series of tests were performed to determine the hooding efficiency effects of a simulated tapping operation by use of the dual flow damper. In this case an end door was removed from the test cell and hooding efficiencies determined under normal and high flow rate conditions. The hooding efficiency difference between the two evacuation rates was negligible.

The impact of the use of the dual flow damper on the cell-day hooding efficiency can be calculated knowing the hooding efficiencies and the length of time the cell undergoes various cell operations (1). This calculation shows that the use of the dual flow damper increases the cell-day hooding efficiency by only 0.2%. For example, a cell not equipped with the damper may have a cell-day hooding efficiency of 98.3%, whereas an identical cell which does have the damper will have a cell-day hooding efficiency of 98.5%.

Superstructure Design Considerations

Hooding efficiency measurements were taken on two 75 kA experimental cells, in which one cell had an upgraded superstructure design. The upgraded design differed from the conventional design in two respects: 1) fume pickup points in the superstructure were modified to minimize the accumulation of solids, and 2) a modification was made to allow a tighter seal with the side shields. Hooding efficiency characteristics of the two 75 kA experimental cells are shown in Figure 4. The upgraded design shows a considerable improvement in the cell fume collection capability. In maintaining a cell hooding efficiency of 98.0%, it can be seen that the modified cell requires approximately 30% less evacuation flow rate than the conventional cell. A visual observation of the conventional cell showed a noticeable entrainment of alumina feed in the fume pickup system at cell evacuation rates greater than 105 M^3/min.

Figure 4 shows two curves of different shape. The curve for the modified superstructure shows the hooding efficiency asymptotically approaching 100% at higher evacuation rates. It would be expected that the curve for the conventional superstructure would also asymptotically approach 100% at higher evacuation rates. The curve for the modified superstructure shows a change in slope at an evacuation rate of about 75 M^3/min. At evacuation rates higher than 75 M^3/min, the incremental improvement in hooding efficiency is much less than at lower evacuation rates. The economic considerations of increasing evacuation rate versus the benefits gained by such an increase can be evaluated and included as part of the overall assessment of fume collection design factors.

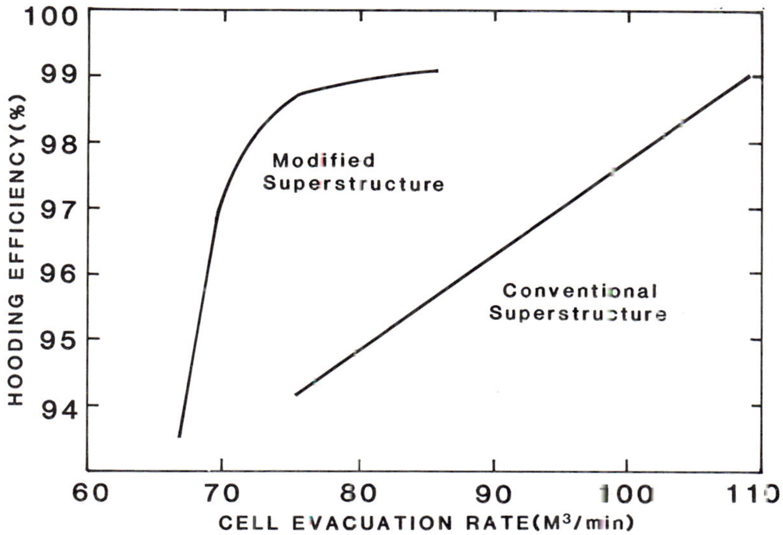

Figure 4: 75 kA CELL HOODING EFFICIENCY CHARACTERISTICS.

Shield Maintenance

A normal production cell was tested in an operating plant to determine the effects of shield maintenance on hooding efficiency. Testing showed a hooding efficiency of 91.7% for a cell which was typical of the plant shield maintenance and placement practices. When the shields on the same cell were replaced with new shields of the same design, and obvious sources of improper shield placement were eliminated (bath on deckplates), the hooding efficiency increased to 97.7% at the same cell evacuation rate. Results of this test indicate the potnetial for hooding improvements using existing cell and shield designs. Since emissions escaping reduction cells are directly related to hooding efficiency, shield placement and maintenance practices can significantly affect overall potroom emission levels.

Discussion of Results

A number of different cells and design parameters have been tested for hooding efficiency performance. This information can be calculated on a normalized basis in order to compare fume collection performance of cells, even though they may differ in size, production rate, and design. In the case of the cells described in this paper, the normalized cell evacuation rates required to achieve a hooding efficiency of 98.0% are shown in Table I. The Table leads to several interesting points about the cells tested. A comparison of the 190 kA cells with the 75 kA cells shows that the smaller cells require a greater amount of evacuation airflow per unit weight of aluminum produced in order to maintain the same level of fume collection as the larger cells. Comparisons of Table I information for the various 190 kA design possibilities show those design variables most influential in affecting cell hooding efficiencies. Any design variable that is contemplated for a reduction cell must be weighed against the practical and economic considerations that also affect the ultimate design and operating parameters prescribed for cell operation.

Table I - Normalized Evacuation Rates Required to Maintain Hooding Efficiencies at 98.0%

Cell Confirguration*	M^3/kg Al Produced
190 kA, Shield Design A	102.5
190 kA, Shield Design B	120.1
190 kA, Normal Cell Openings	102.5
190 kA, Half-Sealed Openings	89.1
190 kA, Completely Sealed Openings	63.1
190 kA, End Seals Open	54.8
75 kA, Conventional Cell	249.6
75 kA, Modified Superstructure	181.1

* *See text for detailed descriptions.*

Conclusions

As a part of reduction cell design procedures, those factors pertaining to the containment of cell-generated fumes must also be considered. Cell designers who are equipped with a knowledge of hooding efficiency and cell emission characteristics can incorporate functional and efficient designs while maintaining compliance with appropriate emissions regulations.

Acknowledgements

We are grateful for the support, assistance, and cooperation from those reduction plants who have participated in the described measurements. We also want to thank the management of Kaiser Aluminum & Chemical Corp. for permission to publish these findings.

References

1. A. R. Johnson, T. A. Lowe, W. W. Hanneman, and R. J. Schlager, "A Study of Reduction Cell Fluoride Emissions", Light Metals, pp. 749-757, C. J. McMinn, ed., AIME, New York, N. Y., 1980.

CONTROL OF GAS COLLECTION EFFICIENCY IN VSS-POTLINES BY MONITORING CO_2 AND CO IN THE PRIMARY POT GASES

L. Wikström
Gränges Aluminium Metall Aktiebolag
S-851 76 Sundsvall, Sweden

Satisfactory working conditions require high collection efficiency of primary pot gases. Compared to other pot types, it is most difficult to obtain high primary gas collection efficiency on the VSS-pot due to such factors as gas-skirt and burner conditions, operating schedules and, last but not least, the labour skill. Hence, it should be important to continuously monitor collection efficiency in order to implement quick corrective actions when required. This paper describes a method for continuous monitoring of collection efficiency based on comparing the amount of carbon oxides in primary pot gases with a standard carbon consumption. Infra-red photometry is used. The system is calibrated against fluoride measurements. For testing purposes the system is installed at Gränges Aluminium Metall, Sweden, where it has proved to respond adequately under different conditions. It should be a useful instrument to obtain the highest possible collection efficiency.

Introduction

Improving the collection efficiency η of the electrolysis gases is the most direct way to improve the pot-room atmosphere in aluminium-smelting pot rooms and to decrease the emissions from them to the surrounding environment. This is of special importance for VSS pots, for which η is lower than for hooded pots.

When working with the improvement of η, it is valuable to have a technique for measuring it. On a long-time basis, the consumption of AlF_3 is a measure of η, if there is fluoride recovery of the pot gases. Methods that give values on a shorter time basis are, however, necessary for direct evaluation of the effect of different actions taken. Such methods could involve sampling of fluorides in the pot gases as well as in the ventilation air, and the subsequent analyses in the laboratory. In this way, values for a time length of down to one hour are obtainable, but the results are often not available until the following day. Furthermore, the sampling involves difficulties of a considerable degree, especially in the ventilation air.

These circumstances have made us look for other methods. An ultimate method would be one that can continuously record η. A possible solution to this would be to monitor the contents of carbon dioxide and carbon monoxide in the primary pot gases, assuming the total evaluation of these compounds to be constant. This is the method being under evolution with an experimental system, installed for one third of the pots in the VSS plant.

Theoretical considerations

When talking about collection efficiency it is often implied that it is the collection efficiency of fluorides that is referred to, as fluorides are among the most important of the pollutants. When attempting to measure η by monitoring other compounds, such as CO and CO_2 it is desirable that η for these compounds equals, or at least correlates well with η_F. Another important consideration is whether carbon oxides must be monitored both in the pot-gases and in the ventilation air, or if the total evolution of them can be assumed to be constant for large groups of pots.

Monitoring carbon oxides in the ventilation air was over-ruled as impractical, when literature data revealed that the concentration of CO_2 in ambient air was of the same magnitude as the contribution from the pots. Therefore the total specific consumption of carbon producing CO and CO_2 has to be obtained through simultaneous measurements of carbon oxides and F_t (total fluorides) in the primary pot gases and of F_t in the ventilation air. These measurements would also show the correlation between η_C and η_F.

Plant lay-out and gas cleaning facilities

The VSS-plant at the Sundsvall smelter comprises two pot-lines with 120 and 142 pots, respectively. The pots are run at appr 105 kA, and the bath-temperature averages 965°C.

There is one burner per pot and the suction rate is 450 m^3_n/h/pot. The pot gas is cleaned in a dry-recovery plant.

The ventilation air is also cleaned and the cleaning takes place in wet-scrubbers, located between the two pot rooms, see fig 1. The scrubbers are connected to 70 m high chimneys as shown in the figure.

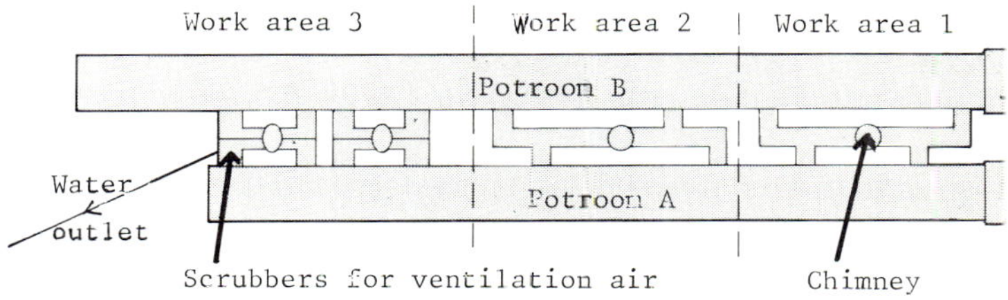

Figure 1 - The Sundsvall VSS-smelter

The plant is divided into three work areas across the pot-rooms. Within each work area the pot operators form a team with job-rotation. The measurement system for the collection efficiency (in the following referred to as the IR - infra-red - system) is installed in work area 3.

Measuring equipment

Equipment for η_C

Sampling. The pot-gases are sampled in the main ducts of the two groups of pots as shown in fig 2. The sampled gas stream is first filtered in a coarse-filter assembly (Hartmann & Braun AG, West-Germany), attached direct to the duct. It consists of a ceramic tube-filter with a pore-width of 60-90 μm mounted in a stainless steel housing. Then the gas is sucked through 2 m of polyethene tubing into a home-made cold-trap of stainless steel. The cold-trap is cooled with an ordinary household freezing-box, maintaining the cold-trap at < -20°C. After the cold-trap, the gas is led through appr 20 m of polyethene tubing into the instrument shed. There it is filtered through solid calcium-chloride to further decrease the content of HF and SO_2 and possibly of other corrosive compounds. Then follows a system of magnetic valves by which it is possible to sample gas from the A or B side or from the calibration gas cylinder. Then follows the pump and a rotameter with an electrical sensor. Finally the gas is split up in two parallel streams which enter the measuring instruments.

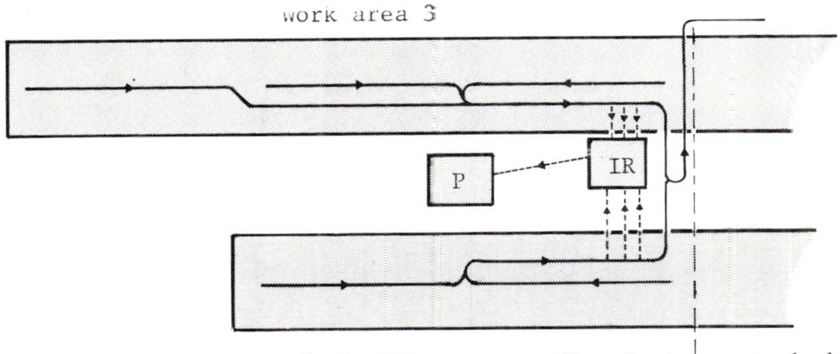

Figure 2 - Lay-out of the IR system. IR = Instrument shed with IR-meter. P = Control room with printer for η_C

Determination. The concentration of CO_2 and CO are continuously monitored by non-dispersive infra-red photometry. Two instruments of type Beckman Model 864 are used for the two gases. They are of double-beam type. The light from two equal-energy cermet-sources is first simultaneously chopped with a mechanical chopper at 10 Hz. Then it is filtered so that only one band of light enters the cells. (For CO_2: silica passing 2 - 4.5 μm; for CO: sapphire passing 2 - 6 μm and a high-pass filter for > 4.5 μm). The cells are a flow-through sampling cell and a sealed reference cell filled with dry nitrogen. The detector is pneumatic and of the so called Luft-type (1). It has two chambers, separated by a diaphragm and filled with the gas to be detected. The detector acts as a capacitor and forms part of an oscillator system and amplitude-modulates a 10 MHz carrier signal, which is then processed to a DC-voltage. Because of the light-filters and the detector-type, the instruments are selective.

Calibration is done each 4th hour with calibration gas from Alfax AB, Sweden, with appr 9% CO_2 and 1.7% CO in dry nitrogen. Nitrogen is used instead of air because air causes the composition to change. When new calibration-gas is bought, this is compared with the old to ensure continuity.

Flow measurement. The flow is continuously monitored by an S-pitot tube inserted in the gas-duct. The use of an ordinary pitot tube was not considered feasable due to the risk of dust clogging the small holes for the pressure measurement. The S-pitot tube remains open for weeks. The only problem was clogging of the plastic tubing for the total pressure, but this was solved by connecting a bottle to collect the dust entering the pitot tube. The whole assembly is cleaned weekly with compressed air.

The dynamic pressure is converted to a 4 - 20 mA current with a differential pressure transmitter of type Rosemount 115 1 DR which is of the capacitive type.

The flow measurement system has several times been calibrated against manual pitot tube traversings.

Pot-line current. The pot-line current for each of the two lines is measured via the same 4 - 20 mA loop as the process-computer uses.

Data-correction and computer system. This consists of an Analog Devices μMAC-4000 measurement and control system, a Hewlett-Packard 9915A computer, a Sanyo DM 5109 CX video monitor and a Seiko EPSON MX-80 F/T TYPE II printer.

The μMAC-4000 has analog inputs to which the 8 measuring signals are connected. The signals are multiplexed to a programmable amplifier, and then follows a 13-bit dual slope A/D converter. An Intel 8085 A microprocessor with 8 kbytes ROM and 1 kbytes RAM scales and conditions the signals to engineering units.

The μ-MAC communicates with the HP-9915A host computer via an RS-232 C interface. The HP-9915A is equipped with 32 kbytes RAM and 16 kbytes ROM for the HP-BASIC-interpreter. The mass-memory is a 3M DC 100 A magnetic tape cartridge of 280 kbytes capacity. This was preferred to a floppy-disc because tape is less sensitive to dust. An 18 kbytes BASIC-programme controls the measuring and calibration sequences and does the calculations of η_C. The η_C-values are sent via a 20 mA current loop to the EPSON-printer.

This printer is located in the control room for the pot operators of work area 3 (see figure 2).

Equipment for fluorides

Sampling of gas. Sampling of the pot-gases was performed with the conventional method utilizing a filter and impingers with sodium-hydroxide solution.

The ventilation air had, as mentioned above, to be sampled in horizontal ducts with large cross-sections. Utilizing impingers here would have been difficult. The development of a dry absorption technique for HF as described in (2) for measurements with personal sampling in hygienic measurements, where there also is a need for light-weight equipment, was useful in our case. This dry absorption technique, utilizing cellulose pads impregnated with sodium-formiate immediately after the dust-filter, was developed to suit the sampling in the ducts. With ordinary impingers the measurements would have been hard to carry out.

Sampling of water. Water was sampled in the main outlet from the 8 scrubbers (see figure 1) with an automatic sampler (Terag 7, AB Innomatic, Sweden), working with compressed air.

Flow measurements. The pot-gas flow was measured manually, using pitot-tube and inclined-tube gauge. The flow of the ventilation air was measured in each scrubber with calibrated vane anemometers once per each sampling occasion. The water flow is indicated in each scrubber by means of a measuring-venturi and an electric indicator, which converts the differential pressure to flow-units. The indicators were read once per measurement.

Analyses. All the fluoride samples were analysed using the computerized Sintalyzer system developed by SINTEF, Trondheim, Norway. The system employs a special buffer, and the determination is performed with a fluoride-sensitive electrode with the aid of the multiple standard addition technique.

Testing the response of the IR-system

A test of the system's response to artificial changes of η_C was performed when the system had been in operation for some months. On one group of pots the suction from some of the pots was blocked by obstructing the burners with mineral fiber (there is one burner per pot). On the test occasion 37 pots out of 40 were in operation and the result is shown in fig 3.

The test began with all pots well covered with alumina after crust-breaking. η_C was then 90%. (A value of 510 kg C/ton Al was used as carbon factor.) A successive number of burners were blocked and, as can be seen, the result was a very linear decrease of η_C. With 19 burners blocked η_C was 33%. The burners were then successively unblocked and η_C returned to its original value of 90%. It should be observed, however, that the decrease in η_C, 2.7% per blocked burner, is slightly higher than expected. This could be due to leaks in the gas ducts between the points of flow measurements and sampling. The result was, however, considered quite satisfactory indicating that the system worked properly.

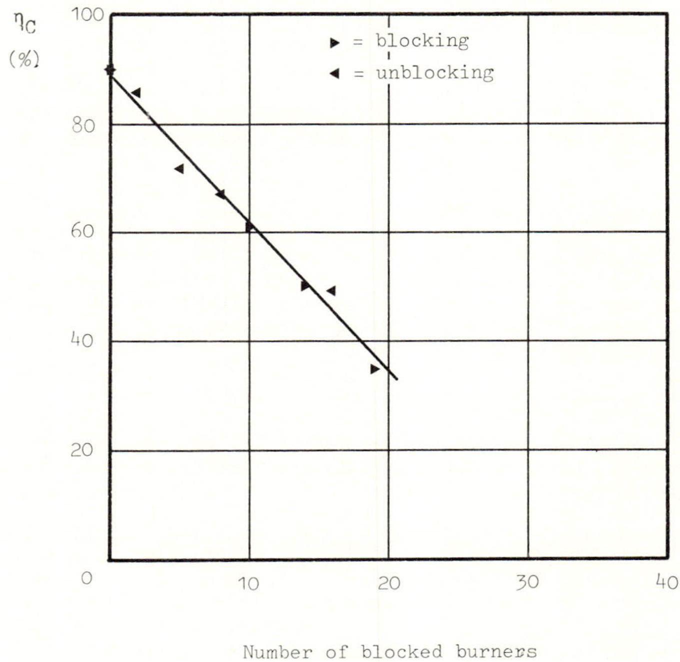

Fig. 3 - η_C for 37 pots as a function of number of artificially blocked burners

Comparing η_C and η_F

Measuring strategies

Once established that the system worked and responded properly to changes in η_C a carbon factor had to be established, that is the total amount of elementary carbon converted to CO and CO_2 had to be quantified. Secondly the relationship between η_C and η_F using a constant carbon factor had to be studied.

A basic condition to measure η_F is that there are possibilities to measure fluorides both in the pot-gases and in the ventilation air. Is this condition fulfilled at the Sundsvall VSS-plant? Concerning the pot-gas, manual measurements are quite straight-forward, for example in the group-ducts. Regarding the ventilation air two conditions have to be fullfilled. First all the air has to be collected in ducts and no air should escape through fugitive leakages. Secondly circulation of emitted fluorides back to the pot-rooms should not occur. This last observation is well documented from another smelter in a tracer study (SF_6) done by the Norwegian Institute for Air Research (NILU) (3). At the Sundsvall smelter the pot-rooms are not ventilated by natural draught but by fan-batteries. Furthermore the air is (after scrubbing) emitted through stacks at a height of 70 m. This last fact should rule out the possibility of back circulation of fluorides. In another tracer study made by NILU at the Sundsvall smelter(4) it is shown that fugitive releases do not occur but all pot-room emissions are brought to the scrubbers.

Thus it can be stated that the basic conditions for measuring η_F are fulfilled.

Sampling of the uncleaned ventilation air is best performed in the ducts between the fan-batteries and the scrubbers. Because of the large cross-section and the short length of these ducts the sampling and also the flow-measurement are susceptible to errors. An additional difficulty lies, however, in another area. A scrubber nominally collects air from a part of the pot-room corresponding to 10 pots or up to 26 pots for the largest scrubber. Due to drift of the air along the pot-rooms and between them, one cannot be sure from how many pots the fluorides in the sampled duct originates. It is also uncertain where the turning point in the longitudinal roof-channel is.

Complete measurements. In spite of the described difficulties complete measurements of η_F for the comparison with η_C were started. The first series was performed in a group of 20 pots. This was before the IR-system was computerized. Later two other series were carried out and these were done on 40 and 60 pots respectively to minimize the effect of the described drift of the air, but also because the IR-system had now been permanently installed for these two groups. The results are summarized in the table below.

TABLE I

Three measuring series for comparing η_F and η_C

F_t = total fluorides, F_p = particulate fluorides

Series No.	Measuring location	Total evolution of F_t, kg/ton Al Av	Min	Max	η_F (%)	η_C % carbon-factor= = 510	Number of meas.	Meas time (h)	Fp/Ft in pot-gas %
1	Group 6 A	22	15.6	31.4	86	86.5	14	-	33
2	Group 5-7 B	26	22.4	30.1	80	76.0	7	∿4	11
3	Work area 3	27	26.0	29.6	86	79	6	∿4	12

The first series is probably impaired by systematic errors. As can be seen the F_p in the pot-gas constitutes 33% av F_t and this is unusually high for VSS-potgas. It can be explained if gaseous fluorides (F_g) were not totally trapped in the sampling impingers. This will also explain the low value of the total evolution of F_t as compared with the other two series. For the first series the carbon factor must be set to 510 kg C/ton Al to obtain η_C equal to η_F.

The other two series have a more normal F_p/F_t ratio, ∿11%, and the total F_t-evolution is higher, or 26-27 kg/ton. The carbon factor turned out to be 490 and 475, respectively, a relative difference of 3%. When the computer system was started up in July 82 the value 510 was, for different reasons, assigned to the carbon factor. The three series had, however, given no reliable value for the factor, and furthermore the correlation between η_C and η_F seemed to be weak, as is shown in fig 4.

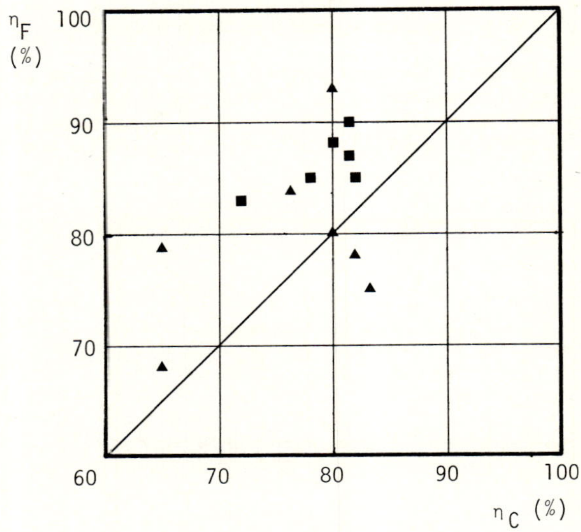

Figure 4 - Correlation between η_C and η_F. Carbon factor: 510. η_F determined through complete measurements of F_t emitted via the pot gases and ventilation air.
▲ = April 1982 ■ = November 1982

The poor correlation could be due to several factors or combinations of them:

- a varying carbon factor
- the poor correlation obtained could be real
- measurement errors.

To answer these questions another approach was tried.

Simplified measurements of η_F

This approach relies on the total evolution of F_t and the cleaning efficiency of the ventilation air scrubbers being known. The two last series described above gave approximately the same total evolution rate, namely 26 and 27 kg F_t/ton Al. From several efficiency measurements of the scrubbers it is known that their cleaning efficiency is appr 80% for F_t. Then the quantities of fluorides discharged via the scrubber water would give η_F according to:

$$\eta_F = \frac{(F_{out} - F_{in}) \cdot Q}{0.8 \cdot 26 \cdot P \cdot n} \quad (1)$$

where Q is the water flow, P is the specific production and n the number of pots. n is precisely known and P is known within appr 2%. Q is measured as described earlier and the method should be quite accurate.

A series of measurements was now carried out according to the assumption made above. They were performed as 1/2-hour measurements during the day-time, after the main operations had been done. These fluoride measurements were carried out under carefully controlled conditions and the result is shown in figure 5.

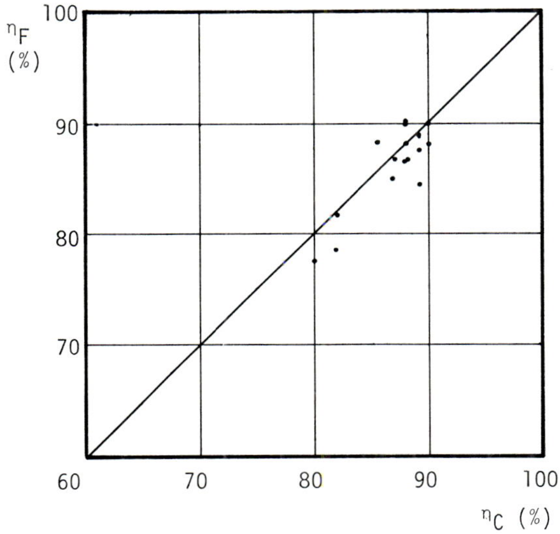

Figure 5 - Correlation between η_C and η_F for 100 pots. Carbon factor: 510. η_F determined through measurements of F_t in the scrubber water
· = 1/2-hour values (from June-July 1983)

It can now be seen that the correlation between η_C and η_F measured via water is good. The poor correlation in figure 4 can thus be attributed to measurement errors in the ventilation air (and possibly also in the pot-gases). Furthermore the carbon-factor 510 seemed to hold.

The value 510 should be compared with the anode-paste consumption which is appr 550 kg/ton Al. Then the difference of 40 must be accounted for. From the pitch-content of the paste and its hydrogen-content the amount of hydrogen can be calculated to ∿18 kg/ton. Other constituents of the pitch and coke amount to appr 1.6% which gives 0.016 · 550 = 9. The amount of unburned tar and coal (tar and coal dust emission to the environment, coal-dust in the electrolyte and amounts of points) has been estimated to 12. Then we have 510 + 18 + 9 + 12 = 549 kg anode material/ton Al.

The good agreement between this number and the value 550 obtained from paste-production data gives further support to a carbon-factor of around 510.

The estimation of the carbon-factor can, together with measurements of the side-burning of the anodes, also give an idea of the quantities of carbon consumed by the Boudouard reaction. So far, the measurements indicate an amount of >40 kg/ton, but further work is needed (cf (5), p. 58).

Results obtained during September 1982 to September 1983

Long-time trends. The IR-system was first started in September 1982 with print-out of η_C in the control room. Continuous sampling of water for fluoride analysis twice a week had been started in May 1982. This was intended both for monitoring of the effluents and for comparison with the IR-results. In September 1982 η_C, calculated using the carbon-factor 510, and η_F, calculated from fluorides in the water, agreed well, as was expected from the results in figure 5. Successively the agreement was poorer. This is illustrated in figure 6. In January 1983 the IR-measurements were discontinued mainly because of problems with the cold trap, but also because of suspected leakage in the culverts. During the spring these and other problems were solved and the IR-system was started again in July 1983. The agreement with the water measurements were now considerably better, but the systematic difference had now changed sign, as can be seen in the figure. The most important thing is, however, that the two curves follow each other. It is also interesting to notice the decrease of both η_C and η_F during July to September. The reason is an increase of carbon dust in the bath. Because of this many pots had to be opened and cleaned from dust. The curves illustrate well how very sensitive η is for problems with the pots, at least under the present practices of handling sick pots.

The still more pronounced dip in η during the summer 1982 is attributed to several problems originating both from the alumina and from the anodes.

It should also be noticed that during the winter and especially in April and May, η was high. During these last two months it climbed up to around 90% and this was a period of very stable operation.

Day to day variations. In figure 7 a part of the η_C-curve is shown with higher time-resolution. Day averages have been used and a very striking pattern is revealed. During the week-ends beginning on Fridays η_C begins to drop and this continues until Monday. This is especially pronounced over the week-end 35/36. On Monday mornings there is always a steep rise in η_C except on one occasion. At the weekly meetings among the pot-room people, a lot of attention has been paid to these observations, and it seems as if the dip has almost disappeared during week-end 37/38. The dips are probably connected with the fact that only two burner-cleaners (for the whole plant) are in work during week-ends, whereas there are 8 working from Monday to Friday. But, it can also be that the pot-operators pay less attention to the pots during week-ends and it is their job to keep the burner inlet open. It has now been decided to keep a close record of different observations during the week in order to get explanations for the week-end dips and other phenomena.

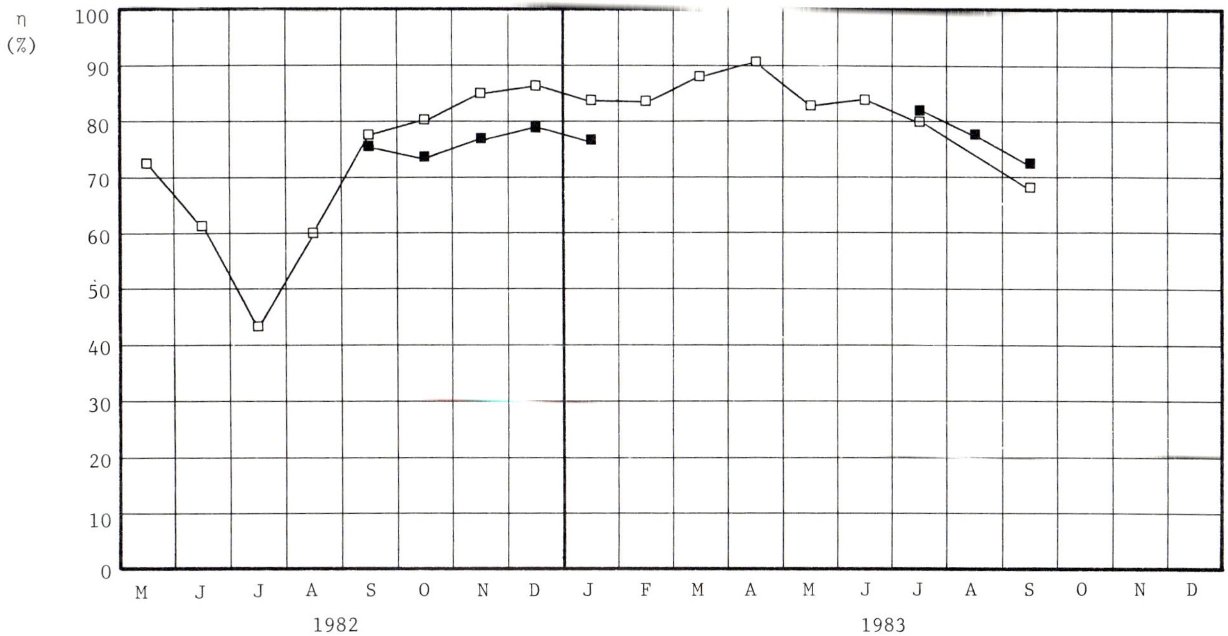

Figure 6 - Collection efficiency for work area 3. Monthly averages.

■ = η_C. Carbon-factor used: 510

□ = η_F, obtained from F_t in water-discharge. Total F_t evolution used: 26 kg/ton Al

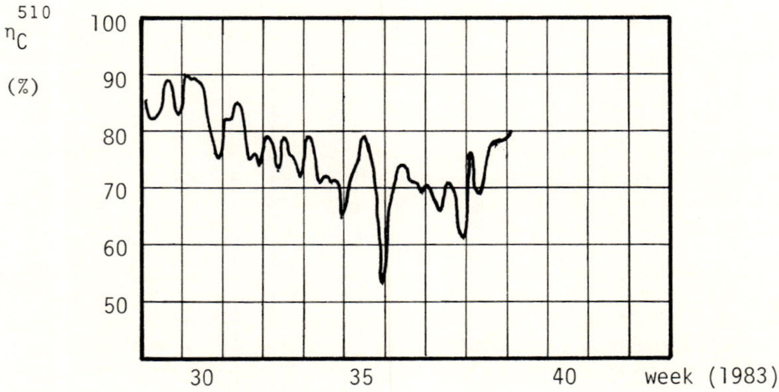

Figure 7 - η_C for the A-group (40 pots). The diagram shows that η_C drops over the week-ends.

Within-day variations. Within-day variations can be observed on the actual print-outs from the printer. In figure 8 a sample copy is shown (redrawn and translated) which examplifies the effect upon η of some activities in the pot-rooms. At 12.30 the B-group is rolled and this is seen as a decrease of η_C immediately followed by an increase when alumina is charged (from truck). At 13.00 the same operation takes place on the A-group and here only an increase is seen due to the charging. This may be illustrative of a short time-lag between rolling and charging. At shift-changes the shift-averages are printed. Then it can be seen that η_C for the A-group is slightly decreasing whereas η_C for the B-group increases. This is due to the fact that tapping has begun on the A-group. On the B-group the new shift have fed extra alumina on open pots and done manual covering as well.

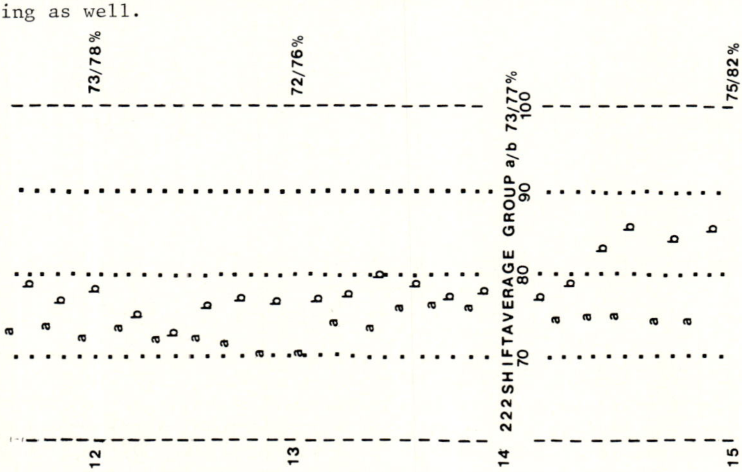

Figure 8 - Sample copy from printer showing η_C for the A- and B-group

The IR-system used in an attempt to quantify different factors affecting η_C

Under ideal conditions η should be close to 100%. An average value during 1983 at the Sundsvall VSS-plant can roughly be set to 80%. What factors have off-set η from the ideal 100%?

Experiments with complete coverage of pots have shown that η can be raised to at least 97% with existing gas-skirts and suction-rates. Simultaneous measurements in the work-place atmosphere gave 0.08 mg/m^3 of F_t and 0.25 μg/m^3 of b(a)p. Emissions to air were 0.2 kg F_t/ton Al.

It is obvious that many routine operations such as crust-breaking and tapping decrease η_C. To get an estimate of this the IR-system has been used to evaluate the effect on η_C of such operations. A large scattering of the data was obtained but nevertheless they can be used as guidelines. To evaluate the effect of a certain operation it is also necessary to know for how long the pot is open. This time is fairly well known by the pot-room people and as will be seen below, the time factor is most critical. Table II gives rough estimates of the importance of each operation.

TABLE II

Estimates of the decrease in η_C due to different routine operations

Operation	Time of open pot (min)	Frequency Times/ /pot-day	Percentage open time % per pot	Decrease in η_C during the operation % per pot	Average decrease in η_C %
Crustbreaking, sides	0.5	5	0.2	20-40	
Crustbreaking, ends	30	0.14	0.3	"	
Tapping	10	1	0.7	"	0.3-0.6
Measurement of bath- and metal depth	10	0.3	0.2	"	
Bath sampling	2	0.28	0.02	"	
Anode-effect	30	1.8	3.7	40-70	1.5-2.6

Sum 1.8-3.2

According to the table the deterioration of η due to routine operations including anode-effects is as small as appr 3%. (It should be noted that anode-effects are quenched by conventional means, i.e. by crust-breaking followed by pooling with a wood-rod and subsequent alumina-feeding.) As mentioned before a crude average of η_C can be set to 80%. If η_C ideally is 100% and the routine-operations only lowers it to 97%, what factors do then cause the remaining large loss of ~17%?

Below, an attempt is made to rank these different factors in order of decreasing importance with the present state of knowledge.

	Factor	Remarks
a)	Sick pots	This is obviously the most important factor. Preventing it has been constantly worked on but further work is needed such as: better and faster methods of quality control of raw materials (coke, pitch, alumina), adjusting paste production and pot-room practices to the properties of the raw-materials, improving the handling of sick pots.
b)	Insufficient cleaning of gas-skirts and burner inlets	The IR-system print-outs and weekly meetings around these results seem to have evoked an increasing concern for these factors. This will be closely followed up.
c)	Insufficient sweeping of burners and ducts	
d)	Anode effects	The open time of the pot has been estimated to 38 min. Even if this is an overestimation, it is important to decrease this time.
e)	Routine operations (crust-breaking, tapping, etc)	Of minor importance. See also below.

The very small effect that the routine operations have on η_C is important to emphasize. Some words may also be said about crust-breaking. This has for appr 10 years been done by wheel-breakers and the alumina feeding by trucks. It has been thought that this operation caused large dips in η. This might also have been the case earlier. Then the time-lag between breaking and feeding could be considerable, due to such factors as poor communication between people or too few alumina trucks. These and other related problems have during the last years been worked on, and a great awareness has been implanted in the pot-operators. Now the alumina-truck follows immediately after the crust-breaker and the open-time is as low as 0.5 min (see table II).

Due to pressure from environmental authorities it has, however, also been going on work with automatic crust-breaking and alumina feeding. Both knife-breakers and point-breakers have been tried. Mechanical problems with the constructions have so far prevented implementation of these techniques. It seems that this is a more difficult task to introduce on VSS-pots than on pre-bake pots. From the estimation in table II it is clear, however, that automatic breaking and feeding is in no way of any significant importance for η. The present practice with the short-time lag between wheel-breaker and truck is quite sufficient. From this point of view the factors that are most important are a, b, and c in the ranking list.

Conclusions

- Collection-efficiency can be continuously monitored by measuring CO and CO_2 in the primary pot gas.

- Simultaneous measurements of fluorides in the scrubber water indicate that the total amount of elementary carbon generating CO and CO_2 is appr 510 kg C/ton Al.

- Routine operations including anode-effects decrease η_C by as little as appr 3%.

- Sick pots are responsible for the largest decrease of η_C with the present routines for handling them.

- The η_C-curves have revealed significent dips over week-ends and other patterns that are being worked on.

References

(1) K.V. Luft, Z. tech. Phys. 24 (1943) p 97

(2) J. Jahr, "Eine neue Doppelfiltermethode zur separaten Bestimmung von Fluorwasserstoff und staubförmigen Fluoriden in der Luft", Stab-Reinh. Luft 32 (1972) pp 248-252.

(3) B. Lamb, O. Skogvald, "A tracer investigation of the wake down-wind of an aluminium smelter hall", Norwegian Institute for Air Research, Oslo, Norway, 1978.

(4) B. Sivertsen and V. Vitols, "A tracer study of dispersion patterns from Gränges Aluminiums smelter in Sundsvall", Norwegian Institute for Air Research, Oslo, Norway, 1980.

(5) K. Grjotheim and B.J. Welch, "Aluminium Smelter Technology", p 58, Aluminium-Verlag GmbH, Düsseldorf, W Germany (1980).

CHAIRMAN

L.C. Blayden
Aluminum Company of America
Alcoa Center, Pennsylvania

THE EFFLUENT LIMITATIONS GUIDELINES

FOR THE PRIMARY ALUMINUM INDUSTRY

J. H. Goldman, Ph.D.

The Aluminum Association
Washington, D.C. 20006
USA

The effluent limitations guidelines for the nonferrous metals manufacturing point source category, including regulations for the primary aluminum industry, were proposed by EPA in February 1983. Final regulations are expected to be issued by the Agency in January 1984. The history and mechanism used by EPA in developing the guidelines are discussed; requirements of primary aluminum plants are outlined in detail. A hypothetical compliance example is presented. Technology variations and options available to individual plants in negotiating a permit are also discussed.

I. Introduction

In February 1983, the U.S. Environmental Protection Agency proposed regulations covering treatment of wastewater from aluminum reduction plants. As of October 1983, final regulations were expected in January 1984.

The regulations mandate maximum levels of contaminants that can be released, and indirectly what in-process and end-of-pipe treatment technologies are to be used, in achieving compliance. Pending Congressional action, parts of the regulation must be implemented by July 1, 1984. The regulations make use of a "building block" approach that combines projected water use for specific operations with targeted pollutant concentrations. The concentrations are based on treatment effectiveness of existing technologies, often borrowed from related or other industries.

The regulations are quite complex. Environmental control specialists at each plant, as well as permit writers, will need to be educated on the structure and implementation of the regulation. This paper is intended to assist them in learning about the regulation.

II. Legislative Mandate and Regulatory History

The Federal Water Pollution Control Act Amendments of 1972 established a comprehensive water pollution control program for the nation. Five sets of standards were established for different kinds of point sources of pollutants.

Sources are generally separated into "direct" and "indirect" dischargers. Direct dischargers send effluent directly into bodies of water, such as streams, lakes and rivers; indirect dischargers send wastewater into sewers, and ultimately into municipal sewerage plants, also known as publicy owned treatment works (POTWs).

The 1972 Amendments empowered the EPA to set effluent limitations guidelines for existing direct dischargers in two steps. The first step, best practicable technology currently available (BPT), is an intermediate step and is interpreted by the Agency to be the average of the best existing technologies; it must be in place by July 1, 1977.[1] The second step, best available technology economically achievable (BAT), is more stringent and is interpreted by EPA as the best available demonstrated technology. It was to be in place by July 1, 1983[2].

Direct discharge facilities constructed or expanded after the issue date of the proposed regulation are considered to be new sources; they were to be covered by new source performance standards (NSPS) developed by EPA[3]. The standards were to reflect the newest, most effective control technology.

Indirect dischargers were to be regulated by EPA under pretreatment standards for existing sources (PSES)[4] and pretreatment standards for new sources (PSNS)[5]. The regulations were to cover pollutants that would "pass through" or interfere with the operations of POTWs.

The EPA could not produce standards by the dates set in the legislation, and was sued in 1976 by several environmental groups. A court-approved settlement agreement was reached, setting a schedule for effluent limitations guidelines, pretreatment standards, and new source performance standards for 21 major industries, covering 65 "priority" pollutants and pollutant classes (totalling 129 priority pollutant substances)[6].

Congress incorporated some of the provisions of the settlement agreement in 1977 Amendments to the Clean Water Act. BAT standards were to cover the priority, or toxic, pollutants, and to be implemented by July 1, 1984, one year later than originally established[7]. Congress is presently considering legislation to extend the deadline to three years after EPA promulgates, or issues, final regulations.

The 1977 Amendments also established best conventional technology (BCT) to cover conventional pollutants (biological oxygen demand, total suspended solids, fecal coliform, and pH) in lieu of BAT[8].

EPA has thus been mandated to develop BPT, BAT, BCT, NSPS, PSES, and PSNS for 21 industries, or point source categories.

III. Review of the Regulation

One of the twenty-one point source categories that EPA is empowered to regulate is the nonferrous metals manufacturing industry. The category includes aluminum smelting (Subcategory B).

The Agency originally promulgated standards for primary aluminum - for BPT, BAT, NSPS, and PSNS - in March 1974[9]. Pretreatment standards for existing sources were not promulgated because there were (and still are) no plants discharging effluent into municipal systems.

EPA proposed revised regulations for aluminum smelting on February 17, 1983[10], with corrections published on April 18, 1983[11]. Final regulations were expected in January 1984.

The regulation builds on data and logic applicable to all industries falling under the label of nonferrous metals manufacturing, but sets standards unique for each component of the point source category. All standards for aluminum smelting will be in 47 CFR Part 421 Subpart B - Primary Aluminum Smelting Subcategory. The Agency has estimated compliance costs of $46.7 million in capital costs and $25.1 million in annual costs (1982 dollars) for the aluminum smelting industry[12].

A. Description of the Regulation

The effluent limitations guidelines, pretreatment standards, and new source performance standards (abbreviated as "effluent guidelines") for primary aluminum are "mass-based" standards. For specific waste streams in the industry, specific loadings of pollutants are allowed per mass of aluminum processed (see Tables 1-5). The figures are based on a "building-block" approach used by the Agency, combining water use limits for specific processes with pollutant concentrations achievable by best practicable or best available technologies. The mass-based approach may limit both water flow and pollutants discharged by a plant. EPA will use the approach whenever possible, because they fear that concentration-based standards - setting only limits on concentrations of pollutants discharged by the whole plant - could be met by plants by diluting their wastewater streams.

1. Production - Normalizing Parameters. The Agency identified the waste streams in the primary aluminum industry based on a survey taken in 1978. Those streams considered common to aluminum smelting were designated independently of each other for tailoring the effluent guidelines to individual plants. The common processes in the industry that contain waste streams were identified as: anode paste plant wet air pollution control, anode bake plant wet air pollution control, cathode manufacturing, cathode reprocessing, anode contact cooling, potline wet air pollution control, degassing wet air pollution control, direct chill casting contact cooling, continuous rod casting contact cooling, and stationary casting contact cooling[13].

Theoretically, when a National Pollutant Discharge Elimination System (NPDES) permit is developed, only the waste streams present at the plant would be included. Any other waste streams not categorized by EPA, such as pot repair and coke plant contact cooling wet air pollution control, must be considered on a case-by-case basis by the permit writer.

For each operation with a waste stream, the Agency surveyed primary aluminum plants for water use. With the goal of reducing industry water usage, EPA set water use limits for operations in terms of a production-normalizing parameter, usually the product produced.

The average of the best (or lowest) industry water consumption was used for BPT and BCT (Table 6). For example, potline wet air pollution control is allowed 201 gallons per ton (gpt) (838 liters per metric ton (kkg)) of aluminum produced.

Table 7 presents the water flow allotments based on lowest ("best") industry water usage, used for BAT. For example, anode contact cooling is allowed 149 gpt (621 liters per metric ton) of anodes cast. Many flows are based on recycling process water or making process changes.

Table 1

BPT EFFLUENT LIMITATIONS
(in pounds per billion pounds [or mg/kkg] of aluminum processed)[24]

	Pollutant or Pollutant Property		
Standard	Fluoride	Total Suspended Solids	pH
One-day Maximum	2.0	3.0	(1)
Monthly Average	1.0	1.5	(1)

(1) Within range of 7.5 to 10. at all times.

Table 2

BCT EFFLUENT LIMITATIONS
(in pounds per billion pounds [mg/kkg] of aluminum processed)[25]

Waste Stream	Standard	Pollutant or Pollutant Property: Oil & Grease	Total Suspended Solids	pH
Anode Paste Plant Wet Air Pollution Control				
	One-day Max.	20,560.0	42,148.0	(1)
	Monthly Average	12,336.0	20,560.0	(1)
Anode Bake Plant Wet Air Pollution Control				
	One-day Max.	12,340.0	25,297.0	(1)
	Monthly Average	7,404.0	12,340.0	(1)
Cathode Manufacturing				
	One-day Max.	1,550.0	3,177.5	(1)
	Monthly Average	930.0	1,550.0	(1)
Cathode Reprocessing				
	One-day Max.	19,040.0	39,032.0	(1)
	Monthly Average	11,424.0	19,040.0	(1)
Anode Contact Cooling				
	One-day Max.	29,800.0	61,090.0	(1)
	Monthly Average	17,880.0	29,800.0	(1)
Potline Wet Air Pollution control				
	One-day Max.	16,760.0	34,358.0	(1)
	Monthly Average	10,056.0	16,760.0	(1)
Potroom Wet Air Pollution Control				
	One-day Max.	26,100.0	53,505.0	(1)
	Monthly Average	15,660.0	26,100.0	(1)

Table 2 Cont'd

Waste Stream	Standard	Pollutant or Pollutant Property		
		Oil & Grease	Total Suspended Solids	pH
Degassing Wet Air Pollution Control				
	One-day Max.	52,320.0	107,256.0	(1)
	Monthly Average	31,392.0	52,320.0	(1)
Direct Chill Casting Contact Cooling Water				
	One-day Max.	39,980.0	81,959.09	(1)
	Monthly Average	23,988.0	39,980.0	(1)
Continuous Rod Casting Contact Cooling				
	One-day Max.	20,840.0	42,722.0	(1)
	Monthly Average	12,504.0	20,840.0	(1)
Stationary Casting Contact Cooling				
	One-day Max.	0	0	(1)
	Monthly Average	0	0	(1)

(1) Within the range of 7.5 to 10 at all times.

Table 3

BAT EFFLUENT LIMITATIONS
(in pounds per billion pounds [mg/kkg] of aluminum processed)[26]

Waste Stream	Standard	Pollutant or Pollutant Property					
		Benzo(a)Pyrene[(1)]	Antimony	Cyanide	Nickel	Aluminum	Fluoride
Anode Paste Plant Wet Air Pollution Control							
	One-day Max.	1.03	198.70	20.60	56.85	312.90	4,089.10
	Monthly Average	--	88.58	8.24	36.11	127.72	1,812.80
Anode Bake Plant Wet Air Pollution Control							
	One-day Max.	0.49	95.34	9.88	27.17	149.68	1,961.18
	Monthly Average	--	42.48	3.95	18.28	61.26	869.44
Cathode Manufacturing							
	One-day Max.	--	149.38	15.48	42.57	234.52	3,072.78
	Monthly Average	--	66.56	6.19	28.64	95.96	1,382.24
Cathode Reprocessing							
	One-day Max.	9.52	1,837.36	190.40	523.60	2,884.56	37,794.40
	Monthly Average	--	818.72	76.16	352.24	1,180.48	16,755.20
Anode Contact Cooling							
	One-day Max.	--	1,196.72	124.22	341.61	1,881.93	24,657.67
	Monthly Average	--	534.15	49.69	229.81	770.16	10,931.36
Potline Wet Air Pollution Control							
	One-day Max.	8.38	1,617.34	167.60	460.90	2,539.14	33,268.60
	Monthly Average	--	720.68	67.04	310.06	1,939.12	14,748.80
Potroom Wet Air Pollution Control							
	One-day Max.	13.05	2,518.65	261.0	717.75	3,954.15	51,808.50
	Monthly Average	--	1,122.30	104.40	482.85	1,618.20	22,968.0

Table 3 Cont'd

Waste Stream	Standard	Pollutant or Pollutant Property					
		Benzo(a)Pyrene[1]	Antimony	Cyanide	Nickel	Aluminum	Fluoride
Degassing Wet Air Pollution Control							
	One-day Max.	--	0	0	0	0	0
	Monthly Average	--	0	0	0	0	0
Direct Chill Casting Contact Cooling Water							
	One-day Max.	--	3,858.07	399.80	1,099.45	6,056.97	79,360.30
	Monthly Average	--	1,719.14	159.92	739.63	2,478.76	35,182.40
Continuous Rod Casting Contact Cooling							
	One-day Max.	--	201.30	20.86	57.37	316.03	4,140.71
	Monthly Average	--	89.70	8.34	38.59	129.33	1,835.68
Stationary Casting Contact Cooling							
	One-day Max.	--	0	0	0	0	0
	Monthly Average	--	0	0	0	0	0

(1) At the Source

Table 4

NSPS EFFLUENT LIMITATIONS
(in pounds per billion pounds [mg/kkg] of aluminum processed)[27]

Waste Stream	Standard	Benzo(a)pyrene[(1)]	Antimony	Cyanide	Nickel	Aluminum	Fluoride	Oil & Grease	Total Suspended Solids	pH
		Pollutant or Pollutant Property								
Anode Paste Plant Wet Air Pollution Control										
	One-day Max.	0	0	0	0	0	0	0	0	(2)
	Monthly Average	0	0	0	0	0	0	0	0	(2)
Anode Bake Plant Wet Air Pollution Control										
	One-day Max.	0	0	0	0	0	0	0	0	(2)
	Monthly Average	0	0	0	0	0	0	0	0	(2)
Cathode Manufacturing										
	One-day Max.	-	149.38	15.48	42.57	234.52	3,072.78	774.0	1,161.0	(2)
	Monthly Average	-	66.56	6.19	28.64	95.48	1,362.24	774.0	928.80	(2)
Cathode Reprocessing										
	One-day Max.	9.52	1,837.36	190.40	523.60	2,884.56	37,794.4	9,520.0	14,280.0	(2)
	Monthly Average	-	818.72	76.16	352.24	1,180.48	16,755.20	9,520.0	11,424.0	(2)
Anode Contact Cooling										
	One-day Max.	-	1,196.72	124.22	341.61	1,881.93	24,657.67	6,211.0	9,316.50	(2)
	Monthly Average	-	534.15	49.69	229.81	770.16	10,931.36	6,211.0	4,453.20	(2)
Potline Wet Air Pollution Control										
	One-day Max.	0	0	0	0	0	0	0	0	(2)
	Monthly Average	-	0	0	0	0	0	0	0	(2)
Potroom Wet Air Pollution Control										
	One-day Max.	0	0	0	0	0	0	0	0	(2)
	Monthly Average	-	0	0	0	0	0	0	0	(2)
Degassing Wet Air Pollution Control										
	One-day Max.	-	0	0	0	0	0	0	0	(2)
	Monthly Average	-	0	0	0	0	0	0	0	(2)
Direct Chill Casting Contact Cooling										
	One-day Max.	-	3,858.07	399.80	1,099.45	6,056.97	79,360.30	19,990.0	29,985.0	(2)
	Monthly Average	-	1,719.14	159.92	739.63	2,478.76	35,182.40	19,990.0	23,988.0	(2)
Continuous Rod Casting Contact Cooling										
	One-day Max.	-	201.30	20.86	57.37	316.03	4,140.71	1,043.0	1,564.50	(2)
	Monthly Average	-	89.70	8.34	38.59	129.33	1,835.68	1,043.0	1,251.60	(2)
Stationary Casting Contact Cooling										
	One-day Max.	-	0	0	0	0	0	0	0	(2)
	Monthly Average	-	0	0	0	0	0	0	0	(2)

(1) At the source
(2) Within the range 7.5 to 10 at all times

Table 5

PSNS EFFLUENT LIMITATIONS
(in pounds per billion pounds [mg/kkg] of aluminum processed)[28]

Waste Stream	Standard	Benzo(a)pyrene[(1)]	Antimony	Cyanide	Nickel	Fluoride
			Pollutant or Pollutant Property			
Anode Paste Plant Wet Air Pollution Control						
	One-day Max.	0	0	0	0	0
	Monthly Average	-	0	0	0	0
Anode Bake Plant Wet Air Pollution Control						
	One-day Max.	0	0	0	0	0
	Monthly Average	-	0	0	0	0
Cathode Manufacturing						
	One-day Max.	-	149.38	15.48	42.57	3,072.78
	Monthly Average	-	66.56	6.19	28.64	1,362.24
Cathode Reprocessing						
	One-day Max.	9.52	1,837.36	190.40	523.60	37,794.40
	Monthly Average	-	818.72	76.16	352.24	16,755.2
Anode Contact Cooling						
	One-day Max.	-	86.95	124.22	341.61	24,098.68
	Monthly Average	-	37.27	49.69	229.81	9,813.38
Potline Wet Air Pollution Control						
	One-day Max.	0	0	0	0	0
	Monthly Average	-	0	0	0	0
Potroom Wet Air Pollution Control						
	One-day Max.	0	0	0	0	0
	Monthly Average	-	0	0	0	0
Degassing Wet Air Pollution Control						
	One-day Max.	-	0	0	0	0
	Monthly Average	-	0	0	0	0
Direct Chill Casting Contact Cooling Water						
	One-day Max.	-	3,858.07	399.80	1,099.45	79,360.30
	Monthly Average	-	1,719.14	159.92	739.63	35,182.40
Continuous Rod Casting Contact Cooling						
	One-day Max.	-	201.30	20.86	57.37	4,140.71
	Monthly Average	-	89.70	8.34	38.59	1,035.68
Stationary Casting Contact Cooling						
	One-day Max.	-	0	0	0	0
	Monthly Average	-	0	0	0	0

(1)At the source

Table 6

WATER FLOW ALLOTMENTS FOR BPT, BCT[29]

Wastewater Stream	BPT,BCT Normalized Discharge Rate		Production Normalizing Parameter
	1/kkg	gal/ton	
Anode Paste Plant Wet Air Pollution Control	1,028	247	Paste Produced
Anode Bake Plant Wet Air Pollution Control	617	148	Anodes Bakes
Cathode Manufacturing	77.4	18.6	Cathodes Produced
Cathode Reprocessing	952	228	Aluminum Produced from Electrolytic Reduction
Anode Contact Cooling	1,490	358	Anodes Cast
Potline Wet Air Pollution Control	838	201	Aluminum Produced from Electrolytic Reduction
Potroom Wet Air Pollution Control	1,305	314	Aluminum Produced from Electrolytic Reduction
Degassing Wet Air Pollution Control	2,616	628	Aluminum Produced from Electrolytic Reduction
Direct Chill Casting Contact Cooling	1,999	479	Aluminum Produced from Direct Chill Casting
Continuous Rod Casting Contact Cooling	1,042	250	Aluminum Produced from Direct Rod Casting
Stationary Casting Contact Cooling	0	0	Aluminum Produced from Direct Stationary Casting

Table 7

WATER FLOW ALLOTMENTS FOR BAT[30]
(from EPA Development Document)

Wastewater Stream	BAT Normalized Discharge Rate l/kkg	BAT Normalized Discharge Rate gal/ton	Production Normalizing Parameter
Anode Paste Plant Wet Air Pollution Control	103	24.7	Paste Produced
Anode Bake Plant Wet Air Pollution Control	49.4	11.9	Anodes Bakes
Cathode Manufacturing	77.4	18.6	Cathodes Produced
Cathode Reprocessing	952	228	Aluminum Produced from Electrolytic Reduction
Anode Contact Cooling	621	149	Anodes Cast
Potline Wet Air Pollution Control	838	201	Aluminum Produced from Electrolytic Reduction
Potroom Wet Air Pollution Control	1,305	314	Aluminum Produced from Electrolytic Reduction
Degassing Wet Air Pollution Control	0	0	Aluminum Produced from Electrolytic Reduction
Direct Chill Casting Contact Cooling	1,999	479	Aluminum Produced from Direct Chill Casting
Continuous Rod Casting Contact Cooling	104	25.0	Aluminum Produced from Direct Rod Casting
Stationary Casting Contact Cooling	0	0	Aluminum Produced from Direct Stationary Casting

Table 8 presents the water flow allotments for NSPS and PSNS. The EPA has prohibited discharges from anode paste plant wet air pollution control, anode bake plant wet air pollution control, potline wet air pollution control, and potroom wet air pollution control for new sources; their allowed discharge is zero.

2. Pollutant Concentrations. In its effluent guidelines, the EPA will be regulating levels of antimony, nickel, cyanide, and fluoride for all sources. Aluminum, total suspended solids, oil and grease, and pH will be limited for existing and new direct dischargers. Benzo(a)pyrene will be limited for all direct and indirect dischargers, existing and new sources, for the following waste streams: anode paste plant wet air pollution control, anode bake plant wet air pollution control, cathode reprocessing, potline wet air pollution control, and potroom wet air pollution control.

The Agency's method of choosing pollutant parameters and setting limits involved surveys of wastewater concentrations in the primary aluminum industry, taken in 1978, augmented by wastewater samples taken by a contractor for the Agency in 1978 through 1981. Samples were analyzed for all listed toxic pollutants. Up to two plants, for up to three days each, were sampled for each waste stream[14].

When all of the data were assembled, the Agency divided the pollutants into five groups: pollutants never detected, pollutants never found above their analytical quantification level, pollutants detected below levels achievable by treatment, pollutants detected (above levels achievable by treatment) in a small number of sources, and pollutants selected for consideration in establishing regulations. Pollutants in the last group were found above detectable and treatable limits.[15]

In assigning the pollutants to one of the five groups, the Agency used detectable levels as determined by its own lab results and as reported in the published literature. Treatable levels may come from technology demonstrated in the smelting industry or other industries considered to have similar wastewaters. Data were used for the primary aluminum industry from a "Combined Metals Data Base", which used data on chemical precipitation and sedimentation performance from the aluminum forming, copper forming, battery manufacturing, porcelain enameling, and coil coating industries. A statistical analysis by EPA purports to show that effluent data from the industries can be pooled, and is similar to raw effluent data from the primary aluminum industry and other industries in nonferrous metals manufacturing.[16]

Each of the pollutants in the fifth category need not be specifically limited by the Agency. Several classes of pollutants, such as metals, might be removed by a common treatment technology, so only one or two indicator metal pollutants were chosen by the Agency. In addition, if a pollutant in the last group is not too common, and/or treatment is very expensive, it may not be chosen for control.

Table 8

WATER FLOW ALLOTMENTS FOR NSPS,PSNS[31]
(from EPA Development Document)

Wastewater Stream	NSPS,PSNS Normalized Discharge Rate		Production Normalizing Parameter
	l/kkg	gal/ton	
Anode Paste Plant Wet Air Pollution Control	0	0	Paste Produced
Anode Bake Plant Wet Air Pollution Control	0	0	Anodes Bakes
Cathode Manufacturing	77.4	18.6	Cathodes Produced
Cathode Reprocessing	952	228	Aluminum Produced from Electrolytic Reduction
Anode Contact Cooling	621	149	Anodes Cast
Potline Wet Air Pollution Control	0	0	Aluminum Produced from Electrolytic Reduction
Potroom Wet Air Pollution Control	0	0	Aluminum Produced from Electrolytic Reduction
Degassing Wet Air Pollution Control	0	0	Aluminum Produced from Electrolytic Reduction
Direct Chill Casting Contact Cooling	1,999	479	Aluminum Produced from Direct Chill Casting
Continuous Rod Casting Contact Cooling	104	25.0	Aluminum Produced from Direct Rod Casting
Stationary Casting Contact Cooling	0	0	Aluminum Produced from Direct Stationary Casting

Concentration numbers were based on mean treatability averages, from the data described above, multiplied by variability factors from the Combined Metals Data Base, to yield daily maximum limits and average monthly limits. Table 9 shows each of the three values for the pollutant parameters chosen by the Agency for BAT (antimony, nickel, cyanide, fluoride, aluminum), NSPS (all parameters listed), and PSNS (antimony, inckel, fluoride, cyanide) limits; Table 10 shows values used for BPT (fluoride, total suspended solids) and BCT (total suspended solids, oil and grease) standards. The benzo(a)pyrene limit of 10 ug/l is at the quantification limit. Under proposed EPA regulations, BCT can be set equivalent to BAT if two cost tests are passed.[17] In the case of aluminum reduction, both failed, so BPT level technology and its removal efficiencies were used.

The levels for cyanide are based on treatment performance of ferric/ferrous sulfate precipitation systems from the coil coating industry, plus filtration. Antimony, nickel, total suspended solids, and indirectly, aluminum levels are derived from chemical precipitation and sedimentation (lime and settle) data from the Combined Metals Data Base, plus filtration, which is assumed to account for an additional 33% removal over lime and settle. Fluoride removal was based on data from the electronics and electrical components industry using lime and settle technology, plus filtration for BAT, PSNS, NSPS. Treatable levels of oil and grease were derived from performance data on oil skimming/chemical emulsion breaking from the aluminum forming and other industries. Data on toxic organics removal using activated carbon columns was used for benzo(a)pyrene removal.[18]

Recent joint research between EPA and the Aluminum Association may result in an upward revision in the cyanide treatability levels; in addition, aluminum levels are also expected to be raised. Final limits will be published in the promulgated regulation in January 1984.

3. Effluent Limitations as Expressed in the Regulations. Because the regulations are mass-based, the concentration limits discussed in Section IV.A.2 cannot be directly used by a plant to determine allowable mass loadings. The Agency used the combination of concentrations and flow allowances to limit both pollutant concentrations and water use.

To calculate the total mass of pollutants allowed in waste streams for BAT, for example, the total flow for the waste stream allowance is multiplied by the allowable concentration for each pollutant, expressed as mg/l, to yield a value expressed as milligrams of pollutant per metric ton (kkg) of product (mg/kkg). The figure is equivalent to the value expressed as pounds of pollutant per billion pounds of product. For example, the BAT daily maximum limit for nickel for the potline wet scrubber waste stream is calculated by multiplying the stream flow allowance, 838 l/kkg (Table 6), by the BAT daily maximum concentration for nickel, 0.55 mg/l (Table 9), to yield a daily maximum of 460.9 mg nickel per kkg of aluminum reduced, or 460.9 lbs. per billion pounds reduced (see Table 3). Similarly, the BCT allowable monthly average loading of oil and grease for anode paste plant wet scrubber liquor was calculated by multiplying the flow allowance, 1,028 l/kkg (Table 6), by

Table 9

SUMMARY OF TREATMENT EFFECTIVENESS (mg/l)[32]

Pollutant Parameter	Treatment Performance for BAT,NSPS,PSNS: Mean	One-Day Maximum	Monthly Average
Antimony	0.47	1.93	0.86
Nickel	0.22	0.55	0.37
Cyanide	0.047	0.20	0.08
Aluminum	0.74	3.03	1.24
Fluoride	9.67	39.7	17.6
Oil and Grease	-	10.0	10.0
Total Suspended Solids	2.6	15.0	12.0

the monthly average concentration limit, 12.0 mg/l (Table 10), to yield a limit of 12,336 mg oil and grease per kkg of anode processed, or 12,336 pounds per billion pounds (see Table 2). In this fashion, all limits for all waste streams have been calculated for BAT, BCT, NSPS, and PSNS (see Tables 2-5). BPT limits were simply carried over from the original BPT standards of 1974, expressed in terms of product produced (Table 1).

4. <u>What mass loadings will be allowed at your operations</u>. The final step in calculating what mass loadings will be allowed at your plant is to multiply the quotient of the mass of product actually processed in pounds divided by 10^9 pounds processed, by the allowable mass loading for a particular pollutant. The product has been specifically identified for each process stream at the production normalizing parameter (see Tables 6, 7, and 8). You should divide your production by one billion pounds, and multiply that quotient by the allowable mass loading for the pollutant. For example, if you produce 6,000 tons of aluminum per month, then the allowable monthly average of nickel for the potline wet scrubber waste stream under BAT is calculated by dividing 6,000 tons (12,000,000 pounds) by 1,000,000,000 pounds, and multiplying the result, 0.012, by 310.06 lbs/billion lbs. (Table 3) to yield a monthly average of 3.72 pounds of nickel. A full example for a plant will be given later.

5. <u>Treatment Technologies</u>. Although the EPA does not mandate specific treatment technologies, the limits for pollutant loadings are based on specific technologies. In complying with the effluent guidelines, any control technology - not just those used to justify the regulation - can be used if it will result in compliance. In addition, only those pollutants that occur over the limits must be controlled; for example, if benzo(a)pyrene is not found above the effluent limitations, an activated carbon column or other removal technology need not be installed. Monitoring must still take place, however.

The Agency selected chemical precipitation and sedimentation (lime and settle) as the control technology to serve as the basis for BPT and BCT limitations[19]. The following control technologies were selected by EPA to serve as the basis of BAT, NSPS, and PSNS limitations: preliminary treatment with oil skimming, cyanide ferrous sulfate precipitation, and activated carbon adsorption (for toxic organics removal), where necessary, followed by chemical precipitation, sedimentation, and multimedia filtration for control of toxic metals and fluorides, and approximately 90% flow reduction[20].

For BAT, an overall in-process water flow reduction of 90% was envisioned by reducing water use for casting contact cooling water and scrubber liquor from anode paste plants, anode bake plants, potlines, and potrooms. Zero water discharge has been proposed for the degassing wet air pollution control (by replacing chlorine degassing with inert gases) and for stationary casting contact cooling waste streams. Comparing Tables 6 and 7 show the flow reductions in terms of product processed for each stream.

Additional flow reduction for NSPS and PSNS was set by EPA as zero discharge by the use of dry air pollution (alumina) scrubbing on potlines and anode bake plants, and elimination of scrubber discharges from potrooms with 100% recycle (with blowdown recycled to casting). A comparison of Tables 7 and 8 shows the additional flow reduction that EPA believes is achievable.

6. Summary of Regulations. A summary of the regulations for BPT, BAT, BCT, NSPS, and PSNS is presented in Table 11. As of October 1983, the compliance deadline for BAT will be July 1, 1984. Congress is currently considering legislation to set the deadline at three years after promulgation. Since promulgation is expected in January 1984, the proposed legislation would set a BAT deadline of January 1987. EPA is also considering administrative relief to the BAT deadline problem. BCT will be effective three years from promulgation. NSPS and PSNS will be installed on all plants built or expanded after February 1983.

B. Application of Regulations to a Hypothetical Situation

A hypothetical primary aluminum plant discharges treated wastewater into a nearby river. Its annual production rate is 160,000 tons/yr. The plant is a prebake plant, using water for anode cooling and wet scrubbers on its anode paste and anode bake plants; the anode bake plant scrubber recycles 33% of its water. The potline uses wet scrubbers with 95% recycle. The plant also recycles its direct chill casting contact cooling water. Other processes producing wastewater are noncontact cooling water for compressors and cathode storage leachate. Stormwater is combined with treated wastewater before discharge. Under its present discharge permit, the plant has an oil skimming and lime and settle system. The plant management will decide what additional steps must be added to comply with BAT and BCT standards.

1. Inventory of Water Flows. The plant currently discharges approximately 979,250 gallons of wastewater per day (gpd), with stormwater surges of 200,000 gpd. In terms of production, the plant's waste streams currently discharge aapproximately 2440 gallons of wastewater per ton of aluminum produced (See Table 12, column 1). Approximately 0.5 pounds of anode are consumed per pound of aluminum produced; the waste streams anode paste scrubber, anode bake scrubber, and anode contact cooling water are multiplied by 0.5 to estimate water use per ton of aluminum produced. Anode paste plant wet air pollution control uses 289.5 gpt of anode paste produced; anode bake plant wet scrubbers use 434 gpt of anode baked with 33% recycle. Anode contact cooling uses 2171 gpt of anode cast. The potline wet scrubber uses 419.7 gpt of reduced aluminum, with 521.1 gpt used in direct chill casting (98% recycle). Noncontact cooling water, used to cool the compressors, uses about 0.1 million gallons per day (mgd). Spent cathode is stored on concrete pads on the plant grounds; rainfall leaches 600,000 gallons per year (1644 gallons per day), which must be treated. Finally, stormwater runoff contributes a surge of about 0.2 mgd; such flows occur about 20 times per year. The plant uses inert fluxing agents for degassing, and does not recover cryolite from spent cathode.

Table 10

SUMMARY OF TREATMENT EFFECTIVENESS (mg/l)[33]

Pollutant Parameter	Treatment Performance for BPT and BCT		
	Mean	One-Day Maximum	Monthly Average
Fluoride	14.5	59.5	26.4
Oil and Grease	-	20.0	12.0
Total Suspended Solids	12.0	41.0	20.0

Table 11

SUMMARY OF EFFLUENT GUIDELINES
FOR THE PRIMARY ALUMINUM INDUSTRY

Category	Item	BPT	BAT	BCT	NSPS	PSNS
Pollutant Parameters	Antimony		X		X	X
	Nickel		X		X	X
	Cyanide		X		X	X
	Fluoride	X	X		X	X
	Aluminum		X		X	
	Benzo(a)pyrene[1]		X		X	X
	Total Suspended Solids	X		X	X	
	Oil & Grease			X	X	
	pH	X		X	X	
Treatment Technologies	Chemical Prep. & Sed.	X	X	X	X	X
	Multimedia Filtration[2]		X	X	X	X
	Cyanide precip.[2]		X		X	X
	Oil Skimming[2]		X	X	X	X
	Activated Carbon[1,2]		X	X	X	X
	Water Flow Reduction[3]		X		X	X
	Schematic:Figure		1	1	1	1
Effluent Limitations						
a. Concentrations	From Table:	10	9	10	9	9
b. Flow Allowances	From Table:	6	7	6	8	8
c. Limitations	From Table:	1	3	2	4	5

Notes
1. Control and monitoring at the source.
2. If necessary.
3. Including use of dry scrubbers, recycling, and alternate fluxing methods for degassing.

Table 12

WATER USAGE

Waste Stream	Flow As Per Ton Of	Current Discharge (gpt)	Regulatory Allowance BAT Allowed Discharge (gpt)	Compliance Discharge Reduced Discharge Under BAT (gpt)
1. Anode Bake Plant Wet Air Pollution Control	Anode Baked	434.2	24.7	26
2. Anode Paste Plant Wet Air Pollution Control	Anode Paste Produced	289.5	11.9	6
3. Potline Wet Air Pollution Control	Aluminum Produced	419.7	201	168
4. Anode Contact Cooling Water	Anode Cast	2,171.0	149	0
5. Direct Chill Casting Contact Cooling Water	Aluminum Cast	521.1	479	521
6. Noncontact Cooling Water	-	22.8 (1)	-	0
7. Cathode Storage Runoff	-	3.75 (2)	-	3.75
8. Stormwater Runoff	-	25 (3)	-	25

(1) 0.01 mgd - expressed as gallons per ton of aluminum produced.
(2) 0.002 mgd - expressed as gallons per ton of aluminum produced and averaged over the year.
(3) 0.2 mgd on 20 days - expressed as gallons per ton of aluminum produced and averaged over the year.

2. Calculation of Allowable Mass Loadings. The first step in deciding on a compliance strategy for BAT and BCT is to calculate allowable mass loadings of the pollutant parameters. For each set of limitations, the daily mass of product processed for each waste stream should be multiplied by the monthly limitation for that waste stream.

Since annual production is 160,000 tons per year, daily production is 438.4 tons, or 876,712.3 lbs/day. Approximately 450,000 pounds of anode are produced and consumed daily. For each waste stream at the plant, the daily production in pounds should be multiplied by the BAT and BCT monthly average limits. For example, the BAT monthly average limitation for the pollutant parameter fluoride from anode contact cooling is 10,931.36 lbs/billion pounds of anode cast (from Table 3). On a daily basis, therefore, 10,931.36 pounds of fluoride per 10^9 pounds of anode cast is multiplied by 450,000 pounds of anode produced per day, to yield a limitation of 4.92 pounds of fluoride in the effluent per day. The BCT limit of oil and grease in the effluent, from anode contact cooling, based on the monthly average, is 17,880.0 pounds of oil and grease per 10^9 pounds of anode cast (Table 2) extruded per day, to yield 8.05 pounds of oil and grease per day.

To calculate total allowed effluent from the plant, the monthly limits must be calculated from all waste streams for all pollutant parameters (using Tables 2 and 3), and summed for each pollutant parameter. The calculations have been made in Table 13. Daily exceedances cannot be greater than the calculated daily maximum limit, while maintaining the monthly average. Daily maxima serve as upper limits to an allowable range; they have not been calculated for this paper. The total allowed daily effluent for the hypothetical plant under BAT, based on the average monthly limit, is: antimony, 2.54 lb/day; nickel, 1.09 lb/day; cyanide, 0.24 lb/day; aluminum, 3.66 lb/day; fluoride, 52.0 lb/day; and benzo(a)pyrene, 0.0077 lb/day. Benzo(a)pyrene limits are summed for the anode paste wet scrubber, anode bake wet scrubber, and potline wet scrubber flows only.

Daily limits under BCT (for conventional pollutants), based on the average monthly limits, can be calculated in a similar fashion, using the daily production rate and Table 2. The results are presented in Table 13. Allowed levels of BCT are: oil and grease, 63.9 lb/day; and total suspended solids, 82.1 lb/day.

Calculations of the total allowed plant effluent in Table 13 include calculated limits for three unregulated (or "nonprocess") waste streams, noncontact cooling water, cathode storage leachate, and stormwater runoff. Although not specifically included in the regulation, negotiations with the permit writer can result in calculating limits from those streams, if the flow is significant and the water needs treatment. In this example, the limit of each unregulated stream is calculated as in Note 1, Table 13.

Table 13

CALCULATION OF MONTHLY AVERAGE PLANT EFFLUENT
LIMITS UNDER BAT, BCT (in pounds per day)

		Wastewater Stream								
Standard	Pollutant Parameter	Anode Paste Scrubber	Anode Bake Scrubber	Potline Scrubber	Anode Contact Cooling	Direct Chill Casting Contact Cooling	Non-Contact Cooling Water[1]	Cathode Storage[1]	Storm Water Runoff[1]	Plant Total
1. BAT	Antimony	.044	0.019	0.63	0.26	1.51	0.033	0.005	0.036	2.54
	Nickel	0.018	0.008	0.27	0.11	0.65	0.014	0.002	0.016	1.09
	Cyanide	0.004	0.002	0.059	0.025	0.14	0.003	0.001	0.003	0.237
	Aluminum	0.063	0.028	0.91	0.38	2.17	0.048	0.008	0.052	3.66
	Fluoride	0.90	0.39	12.93	5.41	30.8	0.68	0.11	0.74	52.0
	Benzo(a)pyrene(2)	0.0005	0.0002	0.007	-	-	-	-	-	0.0077
2. BCT	Oil and Grease	6.10	3.33	8.82	8.85	21.0	0.46	0.076	0.51	63.9
	Total Suspended Solids	10.2	5.55	14.7	14.8	35.1	0.77	0.12	0.85	82.1

(1) Note listed in regulations; limit calculated in negotiations with permit writer as quotient of flow of unregulated stream divided by flow of potline wet scrubber, times limit calculated for plant's potline wet scrubber stream.

(2) For specific regulated waste streams only.

3. Bubbling. Although limits are based on particular waste streams under the building-block approach, the total effluent from the plant is the key to determining compliance with the regulations, with the exception of benzo(a)pyrene, which must be monitored and controlled at the source (before combination with other process streams). As a result, the effluent level from one waste stream can be above the calculated limit if another is sufficiently below the limit. The theoretical combination of streams to yield an average value is termed bubbling. Only the plant effluent will be monitored. For example, if the effluent of oil and grease from potline wet scrubbing is 13 pounds per day, 4.2 pounds over the calculated BCT limit for that waste stream, but the total plant effluent of oil and grease is 60 lb/day, the plant is still in compliance with the BCT limits for oil and grease. In essence, the plant operator need only be concerned with total plant effluent. The building block approach is used primarily for calculating the total plant discharge, but may be used as a guide in designing the plant's approach to compliance.

4. Comparison of the Limits with the Plant's Effluent. The next step is to compare the BAT and BCT limits, expressed on a daily basis, with the actual average daily plant effluent. If effluent levels are below limits for any pollutant, then it will have to be monitored under terms of the permit, but no control measures beyond those already in place need to be taken for it.

5. Control Strategies. Depending on the source of pollutant loadings that exceed BAT or BCT limits, the plant will have three control strategies available. Water flow can be reduced in such a manner as to decrease raw waste loading, thus bringing the plant into compliance. Control technologies can supplement existing oil skimming and lime and settle technology, without reducing water flow (and without diluting). The third approach is a combination of the first two.

What is important in these mass-based regulations is that the mass loading limits calculated for the plant (Table 13) are achieved. The concentrations and water flow allowances used in calculating the mass loading limits need not be specifically met. So under BAT, for example, the allowed plant discharge of cyanide is 0.24 lb/day. Whether the limit is achieved by reducing the effluent concentration below 0.08 mg/l (Table 9) and/or by reducing the water flow of the streams that are the major sources of cyanide is at the plant's discretion and not mandated by the regulations.

In all actuality, a combination of water treatment and flow reduction will probably be needed; for the sake of simplicity, examples of compliance using only water flow reduction and only water treatment will be discussed.

a. Water Flow Reduction. The plant currently discharges 2440 gpt. Under BAT, the discharge flow allowance used to calculate the mass limits would be 801.55 gpt (see Table 12). Unregulated flows of 28.75 gpt were added to the regulated flow allowance of 772.8 gpt; this is assumed to be the result of negotiations with the permit writer, as described below. Remembering that the total effluent is what counts, not the distribution, the plant could elect to recycle anode contact cooling water to the anode paste plant wet scrubber, increase the recycle rate of anode bake wet scrubbers from 33% to 98%, recycle anode paste plant scrubber waters 98%, and increase potline wet scrubber recycle from 95% to 98%. Direct chill casting contact cooling water use would not be decreased. Of the nonprocess, or unregulated, streams, noncontact cooling water would be recycled to direct chill casting; cathode storage and storm water runoff would not be decreased. The result is a new discharge rate of 734 gpt. Note that although the direct chill casting flow, 521 gpt, exceeds the BAT discharge of 479 gpt, the plant discharge is in compliance by reducing other flows more than necessary to compensate. This is an example of a "bubble" approach. Because cathode storage and stormwater flows are episodic, if distributed throughout the year, the monthly average flows may be achievable; negotiations with the permit writer may be needed to increase daily maxima limits to allow compliance during episodic surges in flow. If the reduced water flow results in reducing mass loading below the BAT and BCT limits (Table 13), the plant has done all it needs to do to comply with BAT and BCT limitations.

b. Addition of Control Technologies. The other compliance approach is to add treatment technology to the plant's existing treatment train. In our hypothetical example, the plant already has installed oil skimming for direct chill casting contact cooling water, and chemical precipitation and sedimentation (lime and settle) for all wastewater streams to comply with its existing discharge permit. Data on current pollutant loadings show that the plant is already in compliance with daily BAT limits for antimony, nickel, and aluminum. The plant must still achieve compliance with limits for fluoride, cyanide, benzo(a)pyrene, total suspended solids, and oil and grease.

To achieve compliance, the plant could install a cyanide precipitation system as a pretreatment before cathode storage water is combined with other wastewaters. Because meeting the limits with EPA's suggested treatment technology may still be a problem, treated wastewater should be analyzed for cyanide. If EPA raises the cyanide limit, the problem may be eliminated.

Multimedia filtration would also be added after the chemical precipitation and sedimentation clarifier; fluoride levels should be analyzed in the wastewater to see if compliance can be achieved with EPA's proposed treatment technology.

Water from potline wet scrubbing, anode bake plant wet scrubbing, and anode paste plant wet scrubbing must be tested for benzo(a)pyrene. An activated carbon system has been identified by EPA as the basis for setting a concentration limit of 10 ug/l for benzo(a)pyrene. Recent research undertaken by EPA and the Aluminum Association investigated the removal efficiencies of chemical precipitation and sedimentation, multimedia filtration, and activated carbon on potline scrubber water using a pilot plant. Results showed sufficient removal of benzo(a)pyrene without using activated carbon treatment.[20,21] If activated carbon would have been necessary, the plant would have had two treatment options. The effluent going to the carbon column could be pretreated with chemical precipitation, sedimentation, and multimedia filtration to avoid fouling the column, followed by activated carbon treatment, followed by 1)routing into the regular treatment train or 2)discharging the effluent directly into the receiving waters. Because of the research results, it is possible that the effluent from the potline and anode bake and paste plant scrubbers can be combined with other effluent and sent diretly to the mixer for chemical precipitation, sedimentation, and filtration.

It is believed that this treatment will bring the plant into compliance for BAT and BCT limits. The choice of which of the three options to choose - flow reduction only, water treatment only, or a combination, depends on many factors, including current mass loading levels, cost of equipment and maintenance, plant configurations and other engineering constraints, compliance deadlines, additional local standards, and other factors. Some of these issues will be discussed in the following section.

IV. Factors to Consider in Complying with the Effluent Guidelines

No primary plant will be able to design its approach to compliance as easily as in the hypothetical example. Some of the factors were mentioned above; some are discussed in more detail below.

A. Uncategorized Wastewater Streams

Some primary plants may have wastewater streams, such as cathode storage leachate, not considered by the Agency to be a process wastestream. Under the regulations, process and nonprocess wastewaters should be treated separately. No flow allowances have been made; the permit writer could require segregation of the streams for separate treatment, which may be costly. Not including certain waste streams may be significant in determining allowable mass loadings for your plant. The omitted source contributes to the plant discharge, and possibly to the plant's mass loading, and must be considered by the plant in its approach to compliance. Yet if not included in the effluent guidelines, the stream contribution is not taken into account in calculating loading limits. As a result, the plant will have to control its effluent water flow more stringently than intended in the regulations.

Solutions to this problem are to reduce the flow, use the bubble approach as discussed earlier to "cancel out" the discharge, or to negotiate with the permit writer for a flow allowance for the waste stream to be included in the calculation of a mass loading limit. If the flow is large, the last option would provide the most relief. In our hypothetical example, the last solution was used. The permit writer has discretion to tailor permits to individual cases and should be encouraged to do so when the situation warrants.

B. Combining Several Sets of Effluent Guidelines

Several primary plants will share an industrial site, and often wastewater treatment systems, with other plants that may or may not be covered by other sets of effluent guidelines. If there is combined treatment of wastewater, all applicable sets of effluent guidelines must be considered.

Difficulties may arise if the guidelines are incompatible. An example is the need to combine mass-based standards (as are the primary aluminum guidelines) with concentration-based standards, as for example the metal finishing effluent guidelines. Another case is when different regulations control different pollutant parameters. The metals proposed for control in the primary aluminum guidelines are antimony, nickel, and aluminum; in the aluminum forming regulations, zinc, chromium, and aluminum are proposed for limitation.[22] Even when the same parameters are limited, the concentrations used in calculating mass limits may be different.

Two solutions are possible. The permit writer has the discretion to make choices so that a comprehensive set of limits is in effect for the central outfall from the plants. Such situations are aided by the presence of another facility also feeding into the central treatment facility that is not controlled by any set of effluent guidelines. Effluent from the unregulated facilities can be used as a "wild card" to help develop a permit from the incompatible regulations.

Another solution is for the Agency to specify solutions in its regulations, which it has done for one situation in the aluminum forming guidelines. Anodizing and conversion coating operations normally will be regulated under the metal finishing effluent guidelines. The metal finishing regulations are concentration-based, control different pollutants, and are controlled at higher levels than the forming guidelines. If the anodizing or conversion coating operation is "conducted as integral parts of the aluminum forming (extrusion) process" it is considered as etch lines and will be regulated as such in the forming guidelines, not the metal finishing guidelines[23]. The result is that only one set of guidelines need be used in developing a permit, although independent andozing operations will have a less stringent set of guidelines than combined extrusion-anodizing operations.

C. State Standards

A final factor to consider is state water quality standards. It is possible that the water quality standards for the receiving stream may necessitate installation of control technology exceeding BAT, BCT, NSPS, or PSNS. The requirement may be the result of existing or revised water quality standards. Plans for achieving compliance must take the local water quality standards into account.

V. Conclusions

The EPA has promulgated effluent guidelines for the primary aluminum industry. The standards control both water use and mass loadings of pollutants by smelters. Limitations are expressed as pounds of pollutant per billion pounds of aluminum (or other product) processed. Various options for complying with the regulations, using the reduction of water use and the reduction of pollutant concentrations, are available to each plant. Because of the regulation's complexity, the permit writer will have to use his or her discretion to settle several plant-specific issues.

Acknowledgements

I would like to thank Margaret Convey for her technical assistance, and Judy Snyder, Richard Peeks, and Shelly Lankin for their invaluable assistance in arranging and producing the manuscript.

REFERENCES

1. Federal Water Pollution Control Act Amendments. Sections 301 (b)(1) (A), 304 (b). (1972)

2. Ibid. Sections 301(b) (2) (A), 304(b).

3. Ibid. Sections 304(c), 306.

4. Ibid. Sections 304(f), 307(c).

5. Ibid. Sections 304(f), 307(b).

6. Natural Resources Defense Council, Inc. v. Train, 8 ERC 2120 (D.D.C. 1976), modified 12 ERC 1833 (D.D.C. 1979).

7. Federal Water Pollution Control Act Amendments. Sections 301 (b) (2) (A), 301 (b) (2) (C), 307 (a). (1977)

8. Ibid. Section 301(b) (2) (E).

9. 39 Federal Register 12822.

10. 48 Federal Register 7032. (February 17, 1983)

11. 48 Federal Register 16508. (April 18, 1983)

12. 48 Federal Register 7056.

13. Proposed 47 CFR Section 421.23 through 421.27

14. U.S. EPA. Proposed Development Document for Effluent Limitations Guidelines and Standards for the Nonferrous Metals Manufacturing Point Source Category - Supplemental Development Document for: Primary Aluminum. EPA Report No. 440/1 - 83/019-b. Washington, D.C. March 1983. pp. 41 - 96.

15. Ibid. pp. 97-112

16. EPA Memo. Henry D. Kahn to James Berlow Washington, DC. (March 9, 1983).

17. 47 Federal Register 49176.

18. U.S. EPA. Proposed Development Document for Effluent Limitations and Standards for the Nonferrous Metals Manufacturing Point Source Category - Supplemental Development Document for: Primary Aluminum. EPA Report No. 440/1 - 83/019 - 6. Washington, D.C. (March 1983). pp. 113-120.

19. Ibid. pp. 139, 195-196.

20. Ibid. pp. 141-156, 173-176, 185-189.

21. U.S. EPA. Physical/Chemical Treatment of Aluminum Plant Potline Scrubber Wastewater - Preliminary Project Summary, prepared by Rexnord, Inc. for EPA (Contract 68-02-3928). Washington, D.C. (October 1983).

22. Aluminum Association. Report on Analysis of Split Samples from EPA Feasibility Study. Prepared by Am Test, Inc. Washington, D.C. (September 6, 1983).

23. 47 Federal Register 52626. (November 22, 1982)

24. Ibid. 52655; Proposed 47 CFR 467.01

25. Proposed 47 CFR 421.22

26. Proposed 47 CFR 421.27

27. Proposed47 CFR 421.23

28. Proposed 47 CFR 421.24

29. Proposed 47 CFR 421.26

30. James P. Berlow. EPA Effluent Guidelines Division. Personal Communication.

31. U.S. EPA. Proposed Development Document for Effluent Limitations and Standards for the Nonferrous Metals Manufacturing Point Source Category - Supplemental Development Document for: Primary Aluminum. EPA Report No. 440/1-83/019-b. Washington, D.C. (March 1983). P.162.

32. Ibid. p. 177

33. U.S. EPA Proposed Development Document for Effluent Limitations and Standards for the Nonferrous Metals Manufacturing Point Source Category. EPA Report No. 440/1-83/019-b. Washington, D.C. (March 1983). p. 307. (As amended by personal communication from J.P. Berlow).

ST. JUDE INDUSTRIAL PARK'S SANITARY WASTEWATER

ZERO DISCHARGE SYSTEM

Robert E. Hurt

Noranda Aluminum, Inc.
P. O. Box 70
New Madrid, Missouri 63869
USA

Technologies for the control of pollutants from wastewater discharges are diverse and can be extremely complex. The majority of these adopt the principle of removing as many of the "contaminants" as possible prior to discharge to a receiving body of water. St. Jude has made use of one of the oldest concepts of water utilization, irrigation of farmland, to achieve the Clean Water Act's goal of zero discharge and at the same time make beneficial use of many of the constituents of the wastewater.

This paper discusses all phases of the project including the background development work, feasibility testing, equipment selection considerations, land requirements, application rates, and sample well installation and monitoring. The economics of such a system are also presented in addition to the steps followed in obtaining the necessary environmental permits from the regulatory control agencies.

Introduction

The use of wastewater for irrigation of farmland is neither a new concept nor a new practice. One of the oldest systems in the United States is in Lubbock, Texas. It began operation in the 1920's and by 1937 was distributing 1.5 million gallons per day on only 200 acres. In 1977, this project was distributing 14 million gallons per day over 5 000 acres (1).

However, "zero discharge" permits are relatively new. NPDES permits developed over the last 15 years are, by definition, permits that regulate the amount of discharge to waters of the United States. Further, the Environmental Protection Agency's mandate from Congress, via the Clean Water Act, is to promulgate effluent guidelines for this nation's major point source categories. These guidelines are giving the states direction in developing limitations on the discharge of pollutants from those sources. With few exceptions, the regulations written to date have determined a maximum concentration level of the particular contaminant which is considered safe if discharged to the environment.

Noranda Aluminum, Inc. operates a primary aluminum reduction facility in New Madrid, Missouri. Plants like ours have been identified by the EPA for control under the nonferrous metals manufacturing point source category. The guidelines for this category, like many others, have been in the development stage since the late 1970's. Noranda has always been in compliance with all federal and state water pollution control laws. However, the uncertainty over what the new regulations would require led to some innovative thinking to insure continued compliance.

A number of factors led to the decision to use land application as the desired method of wastewater treatment.

1. Level of treatment necessary was unknown since regulations were not final.
2. No guarantee of continuous compliance since the history of environmental process has documented a tendency to pass more stringent regulations, as analytical techniques and equipment allow detection of pollutants at lower and lower levels.
3. Availability of land - St. Jude Industrial Park still leases over 2 500 acres to local farmers while it waits for additional industrial tenants.
4. Availability of an existing oxidation pond sanitary wastewater system for use in a pilot project that was also suitable for expansion if all tests were positive.
5. Realization of a return on the capital investment. Most conventional treatment systems do not achieve a commercial benefit and are a continuous economic liability due to loan repayments and operating and maintenance expenses.

A test plot with one center-pivot system (4 towers, each 125 feet long), covering 18 acres, was installed in 1978. Monitoring was conducted on a monthly basis for the first 12 months, then reduced to a quarterly basis. In 1982, a monitoring frequency of twice per year was established. The Park's 2-stage (4-acre) lagoon system was expanded to a total of four cells, covering 27 acres, late in 1981. Since that time, there has been no discharge from the system to waters of the State. Two new pivot systems, one of 33 acres (6 towers, each 105 feet long) and one of 133 acres (12 towers, each 105 feet long), were added early in 1982. All water distributed to date has been sanitary type wastes (toilets, wash basins, drinking fountains) created by the Park tenants' 1 000 to 1 500 employees. Future plans call for construction of a new 150-acre pivot system, installation of a second 33-acre system, and inclusion of the aluminum plant's process wastewater.

In the pages that follow, I have presented details of various aspects of the project. For convenience and quicker reference to information most pertinent to different individuals, I have organized the text of this presentation into the following categories:

1. Soil Characteristics and Site Selection
2. Equipment Options and Special Considerations
3. Area Requirements and Application Rates
4. Effluent Quality and Land Treatment Capabilities
5. System Design and Project Installation
6. Sample Wells and Long-term Monitoring
7. Economics of Cost and First Year Return
8. Environmental Permitting and Future Plans

Soil Characteristics and Site Selection

The New Madrid site has several different soil types, and special considerations were required. There are areas of heavy clay soils where the permeability is low. These do not absorb water readily. The ground had to be put to grade to provide adequate retention time and also to prevent surface runoff. These areas must be watered at a slower rate, which can be accomplished by increasing the speed at which the irrigation system rotates. Conversely, in the areas of lighter and sandy soils, excess water is required because of the soil's tendency to pass the water through too rapidly to provide full benefit to the crops.

The final locations for the pivots were determined by their proximity to the oxidation ponds. The soils further away were of the same quality in our case, but any length of piping system could be installed if it were determined that increased yields would be achieved at a greater distance. The extra capital expense of the pipe would be quickly offset by the higher annual yields, and over the life of the system a better rate of return would be realized.

Equipment Options and Special Considerations

The St. Jude system uses center-pivot type irrigators. A fixed sprinkler type system was rejected, even though it requires minimal maintenance, because it is not suitable for planting and harvesting row crops. A linear traveling system would have required fewer "towers," but our intentions to introduce process wastewater may have required us to pave or line the distribution ditch. The center-pivot type systems are adaptable to relatively rough terrain, and only one small field required being put to grade.

A four-tower system, covering 18 acres, was installed in the first part of 1978 to serve as a test plot. The unit was all aluminum, manufactured by Reinke, Inc., and distributed 200 gallons per minute at 25 psig. Even with the lighter aluminum structure, during extended operation for several days, we observed the wheels on each tower sinking into the ground as they continued to revolve in the same track. Our solution on the test plot was to install steel, military surplus, aircraft landing mats. They are approximately 1½ feet wide by 12 feet long, and were connected end-to-end to form circular runways for each tower.

This problem was eliminated on the new units by specifying greatly over-sized tires and shorter spans to minimize the ground pressure loads. Further, the ratio of gallons per minute to acres covered was reduced from 11.1 to an average of 7.2. All the new units were made by Valmont, Inc.

Another feature of the newer units is sprinkler nozzles with a minimum orifice diameter of 3/8 inch. The wastewater is so clean in the final cell that algae, fish and turtles thrive and would quickly plug the smaller nozzles of the pilot system. A fine mesh net has been installed in the corner of the final cell to protect the pump inlet and provide a place to treat the water to kill the fish, if necessary.

Area Requirements and Application Rates

A total of 130 million gallons per year was estimated to be the maximum amount of wastewater that could be generated by the present park tenants. This was to be distributed over a six-month period. Under normal conditions, Southeast Missouri soils can absorb an average of 1 inch per week per acre. Using 26 weeks as the irrigation period, a minimum of 185 acres was necessary.

$$\left(\frac{130\text{ MM gallons}}{\text{year}}\right)\left(\frac{\text{year}}{26\text{ weeks}}\right)\left(\frac{\text{week}}{1\text{ inch}}\right)\left(\frac{1\text{ acre-inch}}{27\ 154\text{ gallons}}\right) = 185\text{ acres}$$

We chose to set this up in three separate systems, covering a total of 200 acres. The existing 18-acre test system was to be expanded to 33 acres. The second system is also 33 acres, and the third is 133 acres. The two smaller units are designed to distribute 275 gallons per minute each, and the large system operates at 900 gallons per minute.

Actual wastewater quantities have not reached the 130 million gallon level yet. Lesser amounts of water can be handled by reducing (a) the number of weeks of irrigation, (b) the acres to be irrigated, and/or (c) the average application per week. The 1983 growing season was the first full year of operation. We began irrigating on May 4 with 50.9 million gallons in storage. Influent plus rain, during the following 23-week period to October 14, added 53.4 million gallons. During this time, a total of only 53.8 million gallons was put through the irrigators to reach our minimum storage level of 2.0 feet (16.8 million gallons). This means that nature took care of 33.7 million gallons for us through evaporation over the hot and dry summer, due to the large surface area of the cells.

The annual rainfall and average evapotranspiration are the key factors in the amount of water that can be applied through irrigation. Maximum flexibility was designed into the system to take advantage of dry weather, yet still insure the ability to distribute the wastewater during wet periods.

Effluent Quality and Land Treatment Capabilities

The water in the oxidation pond at the discharge point in cell number four is measured routinely at frequencies of not less than once per month. The table below lists the range and conservative average of all parameters used in calculations concerning soil and crop removals and uptakes.

Table I. St. Jude Park Effluent Concentrations

PARAMETER	RANGE (mg/l)	VALUE USED FOR CALCULATIONS (mg/l)	LBS/ACRE-YR BASED ON 26"/ACRE-YR
Total N	2 - 16	10	58.9
Phosphorus	1 - 8	5	29.5
Potassium	1 - 7	5	29.5
Calcium	20 - 70	35	206.2
Sodium	10 - 18	13	76.6
Magnesium	6 - 10	8	47.1
Lead	<.01 - <.05	0.02	0.12
Zinc	<.01 - .07	0.03	0.13
Copper	<.01 - .04	0.02	0.12
Nickel	<.01 - .04	0.02	0.12
Cadmium	<.005 - <.01	0.005	0.03
Chromium	<.01 - <.05	0.01	0.05

Calculations are made to check the nutrient removal capabilities against the amount applied to the soil and crops via the wastewater. Crop rotation will be utilized, as necessary, to prevent buildup of any one parameter.

The main concern of the regulatory personnel is the potential for buildup of heavy metals which are neither removed nor broken down by the soil-plant filter system. The table below lists the total amount of each metal that is recommended to be applied to an acre of ground. Once this level has been accumulated, no more water should be applied to the land.

Table II. Maximum Cumulative Amounts for Heavy Metals (2)

PARAMETER	CUMULATIVE AMOUNTS (POUNDS/ACRE) FOR CATION EXCHANGE CAPACITIES OF		
	0-5	5-15	15+
Cadmium	4.5	9	18
Chromium	500	1 000	2 000
Copper	125	250	500
Lead	500	1 000	2 000
Nickel	125	250	500
Zinc	250	500	1 000

Unless the concentration of these elements in our wastewater increases, we could apply 130 million gallons per year for over 150 years before the least of the above values would be exceeded.

System Design and Project Installation

Expansion of the oxidation ponds from a 2-cell to a 4-cell system was completed during the last half of 1981. Installation of the irrigation system, including piping, was completed in the first part of 1982.

Burns and McDonnell, an engineering/architectural firm in Kansas City, Missouri, was engaged for consulting and preparation of bid documents on the expansion portion. The chief design engineers on the total project were William Swann, Manager of Planning and Development at Noranda Aluminum, and this author. A flow schematic of the entire project is provided on the following page as Figure 1.

Cell number 3 was purposely constructed at 9.0 acres to conform with regulatory requirements for 3-stage oxidation pond systems. The wastewater flow is from cell 3 to cell 1 to cell 2. At this point, the wastewater has been sufficiently treated and could be discharged. However, our "discharge" is to cell number 4, where it is held until irrigation is needed.

Sample Wells and Long-term Monitoring

Monitoring of the water quality may be the single most important aspect of the project. We wanted as much information as possible for our own information and protection. In addition, our monitoring has assured the control agencies of our intention to maintain a properly operated system and insure protection of the environment.

One deep (35-foot) and one shallow (6-foot) well were installed before the first gallon of wastewater was distributed in 1978. The initial sampling data is used to provide background information on the groundwater quality. Monthly sampling was conducted for one year. Sample frequency was reduced to a quarterly basis because no unusual deviations or trends in the water quality were observed. The deep well was located to the southeast of the test pivot area in the expected direction of underground water movement. The shallow well was installed to the northwest, corresponding to the surface drainage direction.

The expansion of the irrigation system in 1981 added five more deep wells and three more shallow wells. During 1982, the frequency of monitoring was reduced to twice per year. It is performed in the spring prior to start-up of the system, and again in the fall just after the crop harvest.

In 1982, annual sampling of the soil and plant tissue was begun. A total of 15 sites was selected under the three irrigators, and they are checked for the six metals previously mentioned to detect if any buildup occurs. All monitoring locations are depicted in the figure below. Each square is 1 000 feet x 1 000 feet.

Figure 2. Monitoring Locations

□-deep well △-shallow well ○-soil & vegetation

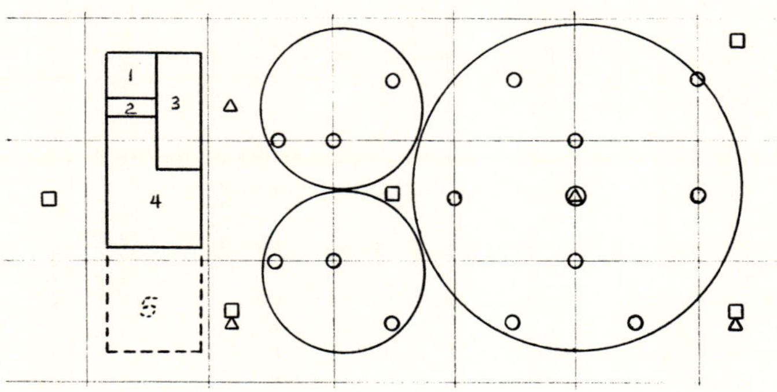

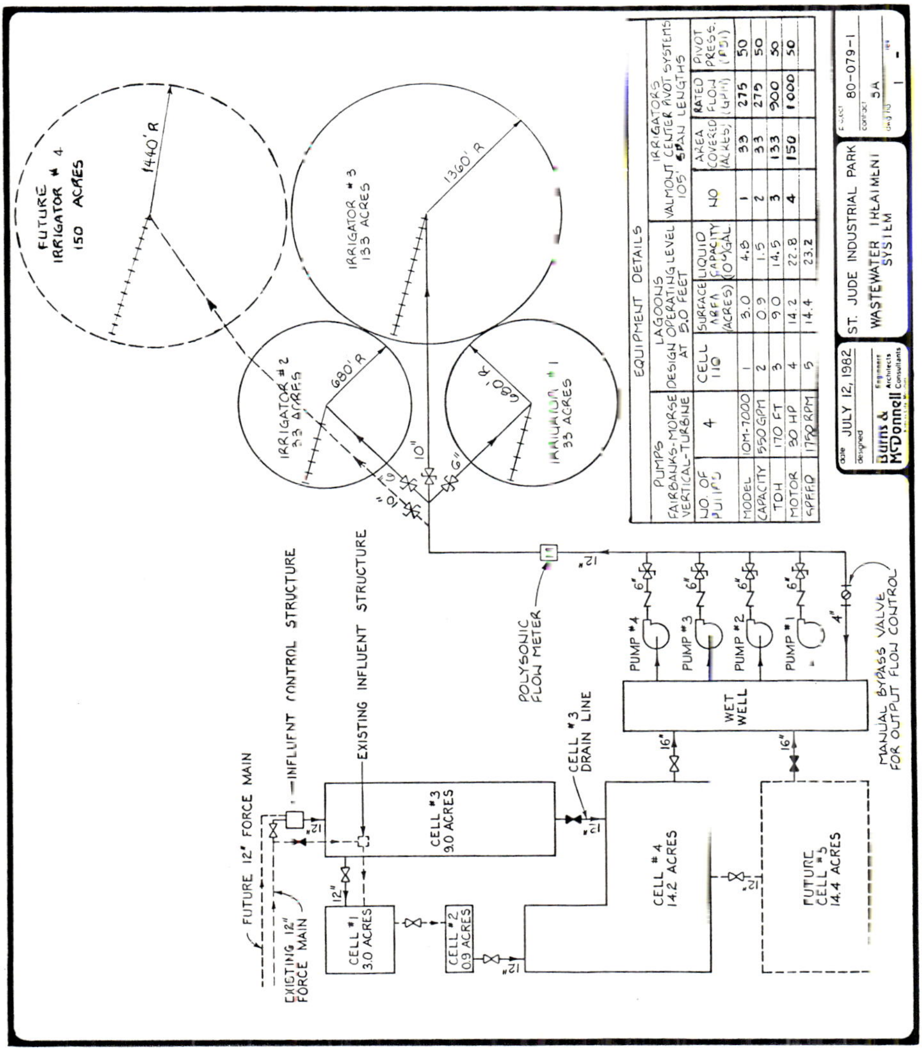

PUMPS FAIRBANKS-MORSE VERTICAL-TURBINE	
NO. OF PUMPS	4
MODEL	10M-7000
CAPACITY	550 GPM
TDH	170 FT
MOTOR	30 HP
SPEED	1770 RPM

LAGOONS DESIGN OPERATING LEVEL AT 5.0 FEET		
CELL NO	SURFACE AREA (ACRES)	LIQUID CAPACITY (10^6 GAL)
1	3.0	4.8
2	0.9	1.5
3	9.0	14.5
4	14.2	22.8
5	14.4	23.2

IRRIGATORS VALMONT CENTER PIVOT SYSTEMS 105' SPAN LENGTHS			
NO	AREA COVERED (ACRES)	RATED FLOW (GPM)	PIVOT PRESS. (PSI)
1	33	275	50
2	33	275	50
3	133	900	50
4	150	1000	50

Figure 1. Irrigation System Schematic

Economics of Cost and First Year Return

The total installed cost of the expansion project was approximately \$834 000. The earthwork and piping for the new ponds consumed the major share of this, at \$529 000. The two center-pivot systems, the irrigation piping and electrical controls, and the pumping station amounted to \$263 000. Additional expenses of \$42 000 were incurred for land leveling and installation of the monitoring wells.

Operating and maintenance expenses of the pivots are about equal to the expenses previously incurred to treat the wastewater in oxidation pond number 2 prior to discharge. The permit specified a maximum limit of 30 mg/l for total suspended solids. This could not be met during the warm months of the year without chemical treatment and settling to kill and remove the algae.

As of this writing, 70 percent of the irrigated ground has been harvested. The average yield for field number 3 was 7 270 pounds of milo per acre. The average yield for nonirrigated milo in adjacent fields, which was planted, fertilized, and cultivated identically, was only 3 150 pounds per acre. The irrigated portion of field number 2 had a reduced yield of only 4 156 pounds per acre because of the land leveling performed last year. It is anticipated that this field would not have produced even 2 000 pounds per acre without the irrigation.

These differentials of over 100 percent are mainly due to the extremely hot and dry summer experienced throughout the Midwest during 1983. Only 0.86 inches of rain fell during the period June 30 through October 3. The average benefit normally resulting from irrigation is 25 percent in Southeast Missouri.

Due to the drought conditions, the price of milo was exceptionally high, at \$5.25 per hundred weight, during our harvest time. A total of 95 acres of irrigated milo was harvested at an average improved yield of 3 115 pounds per acre. The total net benefit of the irrigation was \$16 354 (\$172 per acre). Approximately 40 acres of soybeans will be harvested in late October. An improved yield of 15 bushels per acre is expected. At a conservative price of \$8 per bushel, this acreage will net a benefit of \$4 800 (\$120 per acre). Only 135 of the 166 acres were planted in 1983 because the extremely wet spring prevented access to one low area in field number 3.

Environmental Permitting and Future Plans

The State of Missouri has primacy for enforcement of all water quality regulations. Contacts with the agency personnel were periodically made to inform them of our progress and plans. The level of monitoring and thoroughness of all submitted documents were the major factors in obtaining approvals without any delay. An appendix to this report is provided for reference, and includes [1] the table of contents for the original application for authorization to construct the land application system, and [2] and [3] monitoring forms listing all parameters analyzed in each sample.

Only one deviation was ever requested by the State. Their calculations showed that 100 million gallons per year was the maximum application for our soils and crops to insure that no wastewater could leach through the soil to the groundwater. Their authority is limited to discharges to waters of the State, and as long as there is no runoff from the property, we can legally distribute any volume of water. However, we voluntarily agreed to limit our system output to 100 million gallons over the 200 acres (500 000 gallons/acre ≅ 18 inches/acre). On that basis, the State agency approved and issued a zero discharge operating permit for the St. Jude Industrial Park's sanitary wastewater system on October 8, 1982.

Future plans call for expansion of the irrigated area by addition of a 150-acre center-pivot system and installation of the second 33-acre unit. The resulting annual capacity will then be 175 million gallons over 450 acres. This is expected to be more than enough to offset the increased load due to Noranda Aluminum's third potline, which tapped first metal on July 29, 1983. The excess capacity is reserved for future expansions or new Park tenants. If approval is granted, a groundwater well and pump system will also be installed. This will mainly supply the new 150-acre pivot area with fresh water until the wastewater volumes exceed the capacity of the existing acreage. By keeping the waters separate, it will be possible to calculate the increase in crop yields attributable to the nutrients in the wastewater, over and above the benefit received from irrigation alone.

In addition, the process contact wastewater from the reduction plant will be introduced into the system. This was the driving force behind the entire project. Conversations have been held with State personnel, and monitoring of both quality and quantity are being conducted on a basis of 10 consecutive samples per month. Item [4] of the appendix is the form developed specifically for this.

The Industrial Park will require Noranda to collect and treat, if necessary, this process wastewater to insure compliance with minimal pretreatment standards. This is to insure that the Publicly Owned Treatment Works status of the Park will not be jeopardized. Since a zero discharge system removes 100% of all pollutants, it is technically correct that an infinite amount of pollutants could be sent to the system.

Proper design will balance the aluminum plant's desire to avoid costly wastewater treatment equipment expenses against its commitment to guarantee long-term integrity of the soils under the center-pivot system.

References

(1) Golueke, Clarence G.; "Wastewater - A Solution in Land Application," Compost Science/Land Utilization, Nov./Dec. 1978, pp 13-15.

(2) Cavender, David, et al; Agricultural Use of Municipal Wastewater Sludge - A Planning Guide, pp 16; Missouri Department of Natural Resources, Division of Environmental Quality, June 1982.

Appendix

[1] Table of Contents of Application to Construct a Zero Discharge Land Application System

[2] Irrigation System Analyses Form - Metals

[3] Irrigation System Analyses Form - Other Parameters

[4] Process Wastewater Analyses Form

TABLE OF CONTENTS OF APPLICATION TO CONSTRUCT A ZERO DISCHARGE WASTEWATER TREATMENT SYSTEM

A. Introduction and Application for Approval 1

B. Overall System Design Considerations 3

1. Effluent Details 3
   a. Quality 3
   b. Quantity 3
2. System Details 5
   a. Capacity 5
   b. Water Balance 6
3. Preliminary Site Information 8
4. Field Investigations 12
   a. Percolation Tests 12
   b. Soil Analyses 19
   c. Soil Evaluation by ASCS 31
5. Effluent Renovation by Soils and Crops 36
   a. Nutrient Removal 36
   b. Potential Toxicity 37
   c. Alkalinity-Salinity Hazards 38
6. Effects on Livestock 39
7. Method of Application 40
   a. Type of System 40
   b. Equipment Details 40
   c. Positive Containment 40
8. Monitoring System 44
   a. Test Wells 44
   b. Effluent Analyses 45
9. System Management 50
   a. Design Considerations 50
   b. Schedule of Operations 50
   c. Routine Inspections 55
   d. Maintenance 55

[2]

IRRIGATION SYSTEM ANALYSES FORM - METALS

City of New Madrid

St. Jude Industrial Park

PARAMETER	UNITS	SAMPLE PERIODS				AVERAGE	
Arsenic*	µg/l						
Barium*	µg/l						
Beryllium*	µg/l						
Boron*	µg/l						
Cadmium*	µg/l						
Chromium (Total)*	µg/l						
Cobalt*	µg/l						
Copper*	µg/l						
Lead*	µg/l						
Magnesium*	mg/l						
Manganese*	µg/l						
Mercury*/ Mercury**	µg/l						
Nickel*	µg/l						
Potassium*	mg/l						
Selenium*	µg/l						
Sodium*	mg/l						
Zinc*	µg/l						
Aluminum*/Aluminum**	mg/l						

* Analyses performed by St. Louis Testing Laboratories

** Duplicate sample analysis performed by Noranda Aluminum laboratory

[3]

IRRIGATION SYSTEM ANALYSES FORM - OTHER PARAMETERS

City of New Madrid

St. Jude Industrial Park

PARAMETER	UNITS	SAMPLE PERIOD				AVERAGE	
Total Suspended Solids	mg/l						
Total Dissolved Solids	mg/l						
B.O.D.	mg/l						
C.O.D.	mg/l						
Nitrogen (Total)*	mg/l						
Nitrogen (Nitrate)	mg/l						
pH	units						
Calcium	mg/l						
Chloride	mg/l						
Chlorine	µg/l						
Cyanide*/ Cyanide**	µg/l						
Fluoride / Fluoride**	µg/l						
Iron*	mg/l						
Linear Alkylate Sulfonates	mg/l						
Phosphorus (Total)	mg/l						
Phenols*	µg/l						
Sulfates	mg/l						
Fecal Coliform	col/100 ml						
Fecal Streps	col/100 ml						
Oil & Grease	µg/l						
Odor (Threshold No.)	TON						

* Analyses performed by St. Louis Testing Laboratories

** Duplicate sample analysis performed by Noranda Aluminum laboratory

[4]

PROCESS WASTEWATER ANALYSES FORM

NORANDA ALUMINUM, INC.

WATER SAMPLES FROM ______________________ MONTH OF __________

SAMPLE DATE	pH (units)	PARAMETERS (mg/ℓ)										TOTAL DISCHARGE (gallons)
		TOTAL SUS. SOLIDS	OIL AND GREASE	FLUORIDE	CYANIDE			METALS				
					FREE	COMPLEX	TOTAL	Al	Ni	Sb		
AVERAGES												

COMMENTS: ______________________ DATE: __________

COPIES TO: __________ __________ __________

__________ __________ __________

CONTROL OF PITCH VOLATILES IN GREEN ELECTRODE PRODUCTION

Thomas J. Robare
Maurice W. Wei

Aluminum Company of America
Pittsburgh, Pennsylvania 15219

Introduction

The production of carbon electrodes is a major activity within the primary aluminum industry. The Hall process typically consumes about one-half a pound of carbon anode for every pound of aluminum produced. Anodes are made from sized aggregates of petroleum coke with 15 - 20 percent by weight coal tar pitch as a binder. Therefore, 150 - 200 pounds of coal tar pitch are used per ton of aluminum produced.

Coal tar pitch is a solid at ambient temperatures, melts over a range of $70^{o}C$ to $110^{o}C$, and boils at about $600^{o}C$. It is commonly used at 170-185°C in electrode mix. In that temperature range, a vapor phase exists whose composition depends on the partial pressure of the volatile components. These substances may be irritating to the skin, eyes and respiratory tract. They condense to aerosols to produce a characteristic yellow-brown particulate matter and are highly visible as either fugitive or stack emissions.

Pitch is received at the green electrode facilities in hot liquid and/or solid form, at ambient temperature, with most of the newer plants using liquid. It is added to sized coke aggregates and mixed in either batch type or continuous mixers. The mix is then conveyed to forming equipment such as bolters, vibratory formers, extruders or presses. The single largest emission source of pitch volatiles is the mixer. The second largest occurs at spreader devices or belt transfer points where the mix is agitated. There may also be a significant fume emission around the forming equipment, particularly when mix is loaded into a weigh hopper or mold.

The control of pitch volatiles involves the capture of the fumes with adequate exhaust air volume followed by a treatment to remove aerosols. Most electrode facilities have addressed the problem of fume capture by using large volumes of air which only increases the cost of subsequent fume treatment. This paper will discuss two systems which were developed to effectively treat the pitch volatiles from both batch and continuous mixers, using a minimum amount of exhaust air. These systems not only improve the ambient air quality, but a better worker environment results from improved collection of fugitive emissions.

Process Description

Batch Type Mixer Systems

There are several operating modes commonly used in batch mixing plants. These include:

1. Preheated dry aggregates and liquid pitch.

2. Preheated dry aggregates and solid pitch.

3. Ambient dry aggregates and liquid pitch.

4. Ambient dry aggregates and solid pitch.

In all four of these operating modes the dry aggregate is charged to the mixer on a batch basis. A fume is generated when the displaced air entrains a portion of the fine coke dust contained in the charge. This smoke either disperses into the the work area or is captured by exhaust hooding. Many exhaust systems are retrofit designs rather than integral parts of the mixer design. Retrofit designs allow fugitive dust releases through packing glands, mixer ports, or other openings in the mixer itself. In addition, retrofit designs often include a ductwork system through which the fume must be conveyed. Retrofit designs generally have the disadvantages of:

a. Large air volume requirements with attendant make-up air energy costs.

b. Ductwork plugging problems.

c. Loss of fines from the mix which results in a lower quality mix.

d. Atmospheric particulate emission and opacity problems.

In liquid pitch systems hot liquids are added to the aggregate in the mixer. The vapor pressure of coal tar pitch at 350°F is about 200 mm. Hg so a significant amount of vapor is released during the pitch addition. The aggregate is quickly wetted and dusting subsides. However, the pitch volatiles continue to be released during the mixing cycle. The intermeshing action of the mixer paddles creates an effect similar to a positive displacement pump and generates an internal positive pressure which forces pitch fumes through any openings in the vessel.

The solid pitch systems present a different problem. The pitch is heated and melted in the mixer over a time-frame that depends on whether preheated or ambient-temperature aggregates are used. Until melting and wetting occurs, a combination of dust and pitch fumes may be emitted.

Emission Control

An integrally mounted small bag collector (Figure 1) offers a solution to the mixer emission problem and to the disadvantages of retrofit hooding and exhaust systems. When the dry aggregate is charged, the displaced dust is captured and deposited on the bags as a precoat. The baghouse fan maintains a slight negative pressure inside the mixer so there are no fugitive leaks. The air volume required is very small because the fan only requires a capacity sufficient to overcome the displacement volume and any

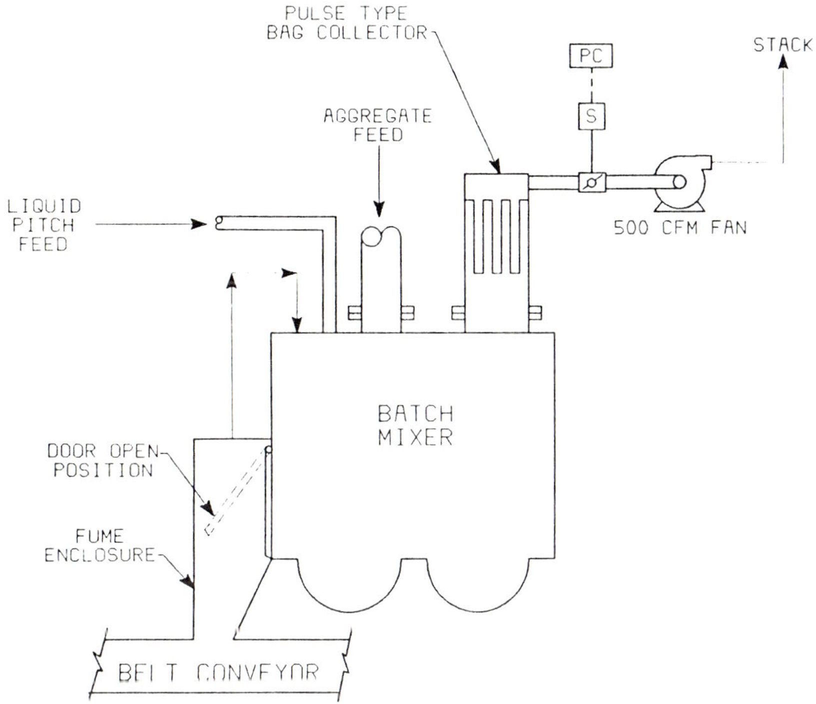

FIGURE #1

SCHEMATIC OF BATCH MIXER VENT COLLECTOR

miscellaneous air inleakage through small openings. Following the initial surge resulting from solids additions to the mixer, the baghouse fan can be throttled back to maintain a slight internal negative pressure on the mixer. The pitch volatiles are pulled through the precoated bags where the condensed phases are trapped on the dust and some vapors are adsorbed. After the mixer is dumped and prior to starting a new charge cycle, the bags are pulsed directly back into the mixer. This returns the fines fractions that in most other control schemes would have been lost.

The cleaned air is exhausted to the atmosphere with a near zero opacity. The constant negative pressure inside the mixer maintains a clean work area and significantly reduces worker exposure as well as maintenance and housekeeping problems.

The total installed cost of a retrofit system will vary with space and structural support requirements, but is typically about $15,000 per mixer. A greenfield installation would be less because headroom access and duct routing considerations would be part of the initial design.

Continuous Mixer Systems

The continuous aggregate mixer uses a helical screw to mix the dry aggregate and liquid pitch which are continuously charged at the feed end. The finished mix is discharged to a conveyor system for continuous transport to forming equipment. The screw type mixer causes a pulsating positive pressure which is relieved at both ends of the mixer.

The dry aggregate and hot liquid pitch produce both coke dust and pitch volatiles simultaneously at the feed end. The discharge end generates primarily pitch volatiles as the wetted mix falls through a chute to a conveyor. There are many designs for localized hooding and capture of these emissions to remove them from the work place. In many cases, these exhaust systems vent directly to the atmosphere which can create a particulate emission and/or an opacity problem.

Emission Control

A solution to the atmospheric emission problem is a bag filter unit installed on the exhausted air. The dust generated at the feed end of the continuous mixer provides a coating of coke to trap the pitch volatiles. No supplemental dust feed is required for a system (Figure 2) which collects smoke from the feed and discharge ends of the mixer, the conveyor transfer points and the forming equipment. The collected dust is recovered and transported back to the aggregate make-up system and is recycled into the process. After three years of operation, no difficulties have been encountered in handling the dust nor have adverse effects been observed on anode quality from recycling this dust. The stack exhaust has near zero opacity.

The total installed cost of this system is about $14/ACFM or $210,000 for a typical 15,000 ACFM system.

Testing Results

Alcoa has commercial-scale installations of both the batch mixer bin filter system and the continuous mixer filter system. These systems have been evaluated on the basis of total particulate, hydrocarbon particulate, and hydrocarbon vapor phase collections. Of particular importance is the

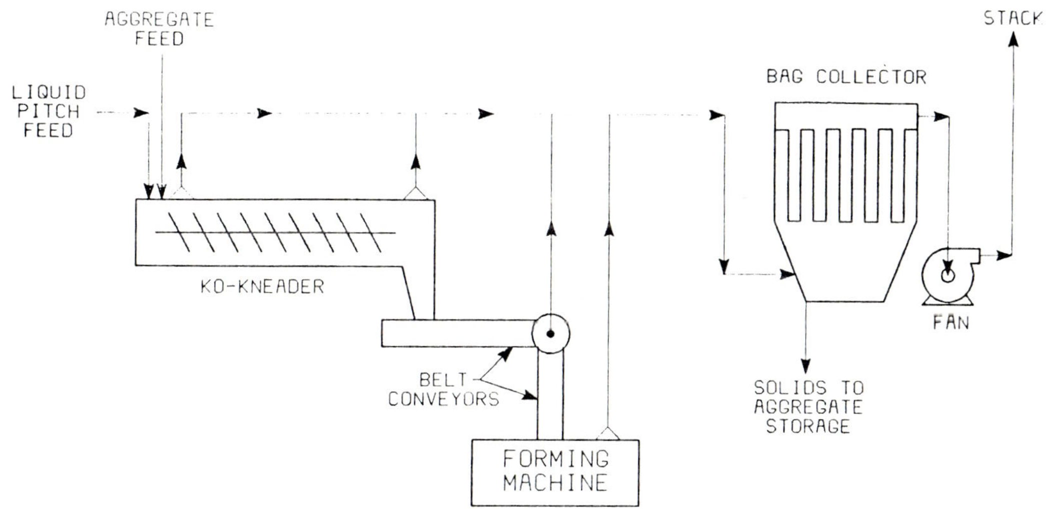

FIGURE #2

SCHEMATIC OF CONTINUOUS MIXER BAG COLLECTOR

performance of the filter systems with respect to different hydrocarbon components in the fume. The performance has been evaluated on the basis of the following parameters:

1. The portion of the particulate and vapor phases that is soluble in benzene. This is commonly referred to as the benzene soluble fraction or BSF.

2. The portion of the benzene soluble fraction, both particulate matter and vapor phases, that consists of polynuclear aromatics.

3. The specific concentration of benzo-alpha-pyrene (BaP) in the polynuclear aromatics.

Tests have been conducted at three different installations -- two batch mixers with different operating conditions and a continuous mixer. One batch mixer uses preheated aggregate and liquid pitch with a cycle time of about 10 minutes per batch. The second batch mixer uses ambient aggregate with the solid pitch contained in the charge and a cycle time of about 50 minutes per batch. The continuous mixer uses 120°C preheated aggregate and liquid pitch. The following tables present the performance of these three units:

TABLE I

BATCH MIXER, PREHEATED AGGREGATE, LIQUID PITCH
Average Exhaust Gas Temperature = 180°F (82°C)
Average Exhaust Volume = 80 SCFM
Temperature of Coke Precoat = 185°C

PARTICULATE MATTER	Inlet #/Hr.	Outlet #/Hr.	% Removal
Total Part.	0.250	0.01	96.0
Insoluble Part.	0.232	0.004	98.3
BSF	0.018	0.006	66.7
PNA	0.016	0.0016	90.0
BaP	66.3×10^{-6}	$<2.8 \times 10^{-6}$	>96.0
VAPOR PHASE	Inlet #/Hr.	Outlet #/Hr.	% Removal
Tot. Condensibles	0.118	0.050	57.6
Insol. Fraction	---	---	---
BSF	0.118	0.050	57.6
PNA	0.055	0.046	16.4
BaP	$<5 \times 10^{-5}$	$<4 \times 10^{-5}$	0.0
Total BSF	0.136	.056	58.8
Total PNA	0.071	.048	33.0
Total BaP	116.3×10^{-6}	42.8×10^{-6}	63.2

TABLE II

BATCH MIXER, AMBIENT AGGREGATE, SOLID PITCH
Average Exhaust Gas Temperature = 155°F (68°C)
Average Exhaust Volume = 400 SCFM
Temperature of Coke Precoat = 75°F

PARTICULATE MATTER	Inlet #/Hr.	Outlet #/Hr.	% Removal
Total Part.	7.64	0.053	99.3
Insoluble Part.	7.34	0.052	99.3
BSF	0.30	0.001	99.6
PNA	0.21	0.0003	99.9
BaP	0.004	<5×10^{-6}	>99.9
VAPOR PHASE	Inlet #/Hr.	Outlet #/Hr.	% Removal
Tot. Condensibles	0.011	0.009	20.0
Insol. Fraction	0.0	0.0	---
BSF	0.011	0.009	20.0
PNA	0.004	0.004	0.0
BaP	<1×10^{-5}	<1×10^{-5}	0.0
Total BSF	0.311	0.010	96.8
Total PNA	0.214	0.0043	98.0
Total BaP	0.004	15×10^{-6}	99.6

TABLE III

CONTINUOUS MIXER, PREHEATED AGGREGATE, LIQUID PITCH
Average Exhaust Gas Temperature = 75°F (24°C)
Average Exhaust Volume = 15,000 SCFM
Temperature of Coke Precoat = 120°C

PARTICULATE MATTER	Inlet #/Hr.	Outlet #/Hr.	% Removal
Total Part.	256.0	3.4	98.7
Insoluble Part.	250.0	1.6	99.4
BSF	6.0	1.8	70.0
PNA	1.5	0.005	99.7
BaP	0.007	<1×10^{-4}	>99.0
VAPOR PHASE	Inlet #/Hr.	Outlet #/Hr.	% Removal
Tot. Condensibles	0.11	0.11	0.0
Insol. Fraction	0.0	0.0	---
BSF	0.11	0.11	0.0
PNA	0.11	0.11	0.0
BaP	<0.0006	<0.0002	67.0
Total BSF	6.11	1.91	68.7
Total PNA	1.61	0.115	92.9
Total BaP	0.0076	0.0003	96.1

Discussion of Results

In order to understand these results more completely, it is necessary to explain the significance of the terms used in the stack sample analyses. Total particulate matter includes coke dust and condensed aerosols from the coal tar pitch volatiles. Benzene soluble fraction (BSF) is the hydrocarbon portion of the total particulate matter. Polynuclear aromatics, or PNA, are the C_{10} and higher molecules which exist in the benzene soluble fraction. Finally, BaP is the fraction of PNA compounds that is specifically benzo-alpha-pyrene. Total condensibles are the pitch volatiles which are in the vapor state at the exhaust gas temperature. They are also totally soluble in benzene and contain PNA and BaP found in the condensed aerosols.

The results indicate that coke dust is an effective filter aid for removing coal tar pitch volatiles, and that hydrocarbon removal efficiency varies with the temperature of the coke dust. The highest total hydrocarbon (particulate plus vapor) removal efficiency of 96.8 percent was achieved on a batch mixer with a cold dust precoat, even though the gas temperature was higher than that from a continuous mixer using a preheated aggregate. The lowest total hydrocarbon removal of 58.8 percent was on a batch mixer with a hot coke dust precoat. This behavior could be anticipated because the cold coke particles present a more suitable condensation site for the pitch volatiles.

Further analysis of the data show that the system is very effective on the particulate polynuclear aromatic compounds which are frequently referred to as particulate polycyclic organic matter or PPOM. PNA removal efficiency varied from 90 to 99.9 percent. Similarly, the removal efficiency of the BaP fraction was 96 to 99.9 percent. Again, the removal efficiency varies inversely with the coke dust temperature and seems relatively independent of the exhaust gas temperature. As would be expected of a fabric filter, it is 99+ percent efficient on removing the coke dust which is recovered and recycled to the process.

Another variable which affects the hydrocarbon efficiency is the amount of coke dust relative to the hydrocarbon level in the gas stream. Higher efficiencies would be expected if additional coke is added over and above the process dust associated with the pitch fumes. Our results indicate that hydrocarbon removal efficiency increases with increasing coating levels.

In summary, the results of these tests demonstrate that coke dust is an effective coating material for controlling pitch fumes. Practically all of the polynuclear compounds are removed and the recovered coke dust is reused in the process.

Conclusions

1. Coal tar pitch volatile emissions from batch mixers can be effectively controlled by an integrally mounted bag filter unit. This system offers the following advantages:

    a. Low total air flow rates to the bag filter of 500 CFM or less per mixer. This means low horsepower and less energy usage for heated make-up air.

    b. Improved anode mix quality by retention of fines.

c. Elimination of ductwork and attendant plugging.

d. Minimized worker exposure levels both in the workplace and in the ambient air.

e. Improved plant housekeeping.

f. Reduced overall maintenance

2. Coal tar pitch volatile emissions from continuous mixers, conveyors and forming equipment can be effectively controlled by a bag filter system precoated with process coke. The coke dust is recycled within the plant, effecting a raw materials cost savings.

3. Atmospheric emissions of particulate polynuclear aromatic hydrocarbons from coal tar pitch processing can be reduced by more than 90 percent using a bag filter system precoated with coke dust. The benzo-a-pyrene portion can be reduced by 95 percent.

GRADIENTS OF FLUORIDE ACCUMULATION IN SOUTHERN PINE FORESTS NEAR A NEW PRIMARY ALUMINUM SMELTER

R. C. Dickie[1], D. C. McCune[2], D. C. MacLean[2], and J. A. Laurence[2]

[1]Alumax of South Carolina
P. O. Box 1000
Goose Creek, SC 29445
and
[2]Boyce Thompson Institute at Cornell University
Tower Road
Ithaca, NY 14853

The chemical analysis of foliage of loblolly pine (*Pinus taeda* L.) for fluoride has been part of an environmental monitoring program before, during, and after start-up of a new primary aluminum smelter. One series of sites was established at intervals along a transect and sampled periodically during the growing season to determine the spatial and temporal patterns of fluoride accumulation in the forest canopy. Height above ground (10 vs 20 m) had no consistent effect. Washing of foliage removed up to 50% of the accumulated fluoride at the nearest sites. The concentration of fluoride within needles decreased with distance through the canopy (about a fourfold decrease in 1000 m). Although the general trend was for the concentration of foliar fluoride to increase with time, growth and maturation of the foliage, climate, and meteorological factors influenced the course of accumulation at any site.

Introduction

The accumulation of airborne fluoride by vegetation and the direct and indirect effects caused thereby have long been associated with the smelting of aluminum. However, smelters constructed in the last decade and equipped with the most effective state-of-the-art emission control systems, or older smelters retrofitted with modern air handling systems, can achieve significant control that results in total fluoride emissions of less than 1 kg F per metric ton of metal produced.

Even with this degree of control there are concerns about siting smelters in areas of commercial agricultural and forestry operations. For this reason, modeling the dispersion of airborne fluoride is a component of planning and siting, and monitoring of fluoride is a component of operations. A systematic program for monitoring fluoride in foliage serves a variety of purposes, such as determining compliance with an air quality standard based on foliar fluoride levels; estimating the distribution of airborne fluoride around a source; or measuring the "dose" received by plants in relation to the occurrence of possible effects. Any explanatory or predictive model must consider many factors and their possible interactions; however, not all can be economically included in a monitoring program but some can serve more effectively than others. Those chosen for this study were: washing of tissue by a standard procedure; height above ground at which samples were obtained; age or developmental stage of foliage; time of year; and, distance from the smelter.

Experimental Procedures

The Smelter

Alumax of South Carolina operates a modern pre-bake primary alumina reduction plant 32 km north of Charleston, SC, on a flat coastal plain. It is the first Greenfield smelter constructed in the U.S. since 1973 and the adoption of the Clean Air Act and its Prevention of Significant Deterioration (PSD) and New Source Performance Standards (NSPS) requirements. The smelter consists of two potlines, each having 180 center-work pre-bake reduction cells. The cells are operated at 180,000 amperes and utilize Alcoa 697 technology.

The production capacity of the smelter is about 182,000 metric tons of metal per year. Operations began in June of 1980, and the facility has run at full capacity since mid-December 1980. Fluoride emissions average 0.45-0.5 kg ton^{-1} of metal produced with a range of 0.375 - 0.505 kg ton^{-1}. The major sources of fluoride-containing emissions are five high level (60 m) point sources (primary dry scrubber and bake oven dry scrubber stacks) and eight low-level (18 m) line sources (potline roof monitors).

The Forest

The smelter is surrounded by a 2400 ha plantation, the majority of which is wooded. The vegetation near the plant consists of southern pines (loblolly, longleaf, shortleaf, and pond) and, in low-lying swampy areas, hardwoods (oaks, gums, maples, and a variety of undergrowth).

The growing season begins in late March and continues into October. Warm temperatures and adequate moisture support a lush growth of vegetation. Pines generally have two flushes of growth annually, one in early April, and another in late June or July. However, during the period of this study, abnormally dry conditions prevailed and secondary flushes of growth were infrequent or absent althogether.

Routine management of the forests includes controlled burning on an annual basis, regular thinning after 20 years, and harvest rotations of between 40 and 80 years depending on the intended use of the trees.

Experimental Site

In May 1981, a transect was established following an old logging road that leads southwest from the plant site to an ambient air monitoring station approximately 2 km from the center of the pot rooms, as shown in Figure 1.

FIGURE I

LOCATION OF TRANSECT IN RELATION TO PRIMARY ALUMINUM SMELTER

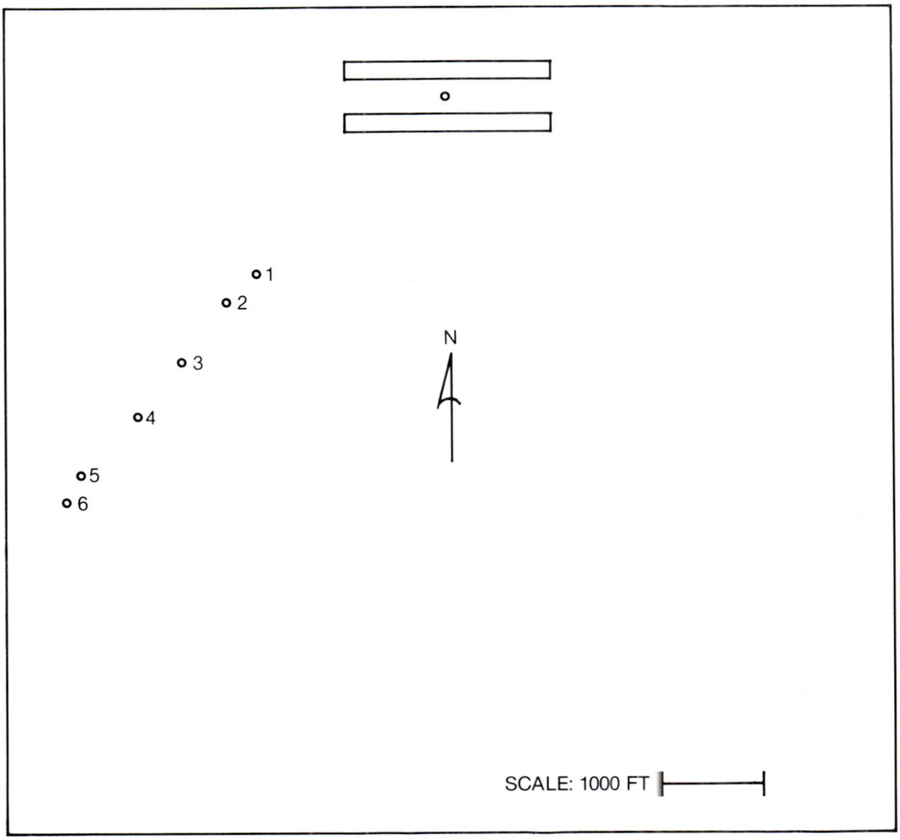

Large (> 25 m tall and 30 cm diameter) loblolly pines were selected at six locations: 865, 990, 1230, 1450, 1740, and 1830 m from the center of the smelter. The edge of the forest is 700 m from the center of the pot rooms. Trees were permanently identified to allow repeated sampling of the same individual each time, i.e., in March, May, July, and September.

Sampling and Analysis

Samples of foliage were collected at heights of 10 and 20 m above the ground using a pole pruner and a Pitman "Hot-Stik" boom truck. After removal from the tree, needles were photographed, examined for symptoms of disease, insect infestation, fluoride toxicity, and abiotic stresses. Needles were removed from the branch, segregated by year of their production, divided into two lots, and stored in paper bags. After completion of the transect (approx. 2 hr.) the samples were returned to the laboratory and one fraction of each sample was immediately dried in a forced draft oven at 80 C for at least 24 hours, while the other was washed in Alconox and EDTA (2) and then dried as above. Dried samples were ground to 40-mesh and the concentration of fluoride content of the samples was determined using the semi-automated method (2).

Meteorological Measurements

Measurements of wind speed and direction were collected at 10 and 55 m above the ground at a location on the smelter site. Summaries of wind data were prepared by calculating the frequency of wind direction and average wind speed for each sampling period (the time between foliage collections). An estimate of relative fluoride exposure was calculated for each sampling period (assuming constant source strength) as the product of the direction frequency and the average wind speed for each height.

Results & Discussion

Washing of tissue

There was no consistent effect of washing on the concentration of fluoride found in needles. However, in samples collected from the site nearest the smelter in March, May, and September of 1982, about one-third of the total fluoride was removed by washing the 1981-flush of needles. About one-half of the total fluoride was removed by washing the 1982-flush of needles in May 1982 (when they were about 50% elongated) but there was no apparent effect of washing in September.

Differences in the concentration of fluoride between washed and unwashed foliage are customarily used as estimates of superficial fluoride on foliage and the contribution of particulate *vis a vis* gaseous forms of emissions. Because superficial fluoride may be eluted by precipitation, the fraction of foliar fluoride removed by washing at any one time will be the resultant of both emissions and climate. Because washing had an effect only at the nearest site, one could conclude that lower sources of emissions (i.e., roof monitors) were responsible for accumulation and the particulate fraction decreased more rapidly than the gaseous with distance from the source.

Height of sampling

There was no overall difference between the heights (10 m and 20 m) at which samples were obtained on the concentration of fluoride in foliage. Although the idealized model expects the concentration of an airborne pollutant to decrease with height within the canopy, and thereby the accumulation by the plant, many different factors and processes can interact to determine concentration profiles (4, 8). In the forest sampled, the canopy was neither regular nor closed and its structure, in addition to micrometeorolgy, could have given the observed uniformity of accumulation. It should also be noted that after a controlled burn of the understory (a standard forest management practice), the concentrations of fluoride in needles at 10 m tended to be greater than at 20 m. Such would be concordant with the estimates of combustion mediated release of fluoride from forest litter (7) and its sorption by the canopy.

Temporal variations

The temporal patterns observed for the concentration of fluoride in needles is shown in Figure 2 for each site (with washed samples and 10 and 20-m values pooled). There are three interesting features to these temporal variations.

The first is that the concentration of fluoride in needles fluctuates from one sampling period to another; although increasing concentrations with time might be expected, decreases also occur. Because these decreases are found in fully expanded needles, they are explained more by a removal of fluoride from the tissue than dilution by growth. This loss of fluoride could be caused by weathering of and loss of the foliar cuticle with fluoride bound to it (and not removable by washing) or leaching of fluoride from the foliage.

The second feature is that the variations in accumulation of fluoride are associated with variations in the persistence and average velocity of winds during the periods between sampling dates (also shown in Figure 2 for 10 and 55 m wind measurement heights). Whereas fluctuations in the concentration of foliar fluoride at the site closest to the smelter appear to be associated with lower level winds and sources, those at further distances appear to respond to contributions from both lower- and upper-level winds and sources. Fluctuations in meteorological variables that are associated with variations in foliar fluorine may be indications of the relative contributions of the two kinds of sources to the accumulation of fluoride by the forest canopy at each site.

A third feature of temporal variation is the age or developmental stage of the tissue. Temporal variations tend to be parallel although older needles have greater concentrations of fluoride.

Distance

The effect of distance from the smelter on the concentration of fluoride in needles is shown in Figure 3 for three sampling periods. The general pattern shows about a fourfold decrease in the foliar concentration of fluoride over the length of the transect. Because two kinds of sources are present and differ in effective height of release and because of the intervention of meteorological factors, the gradient

FIGURE II

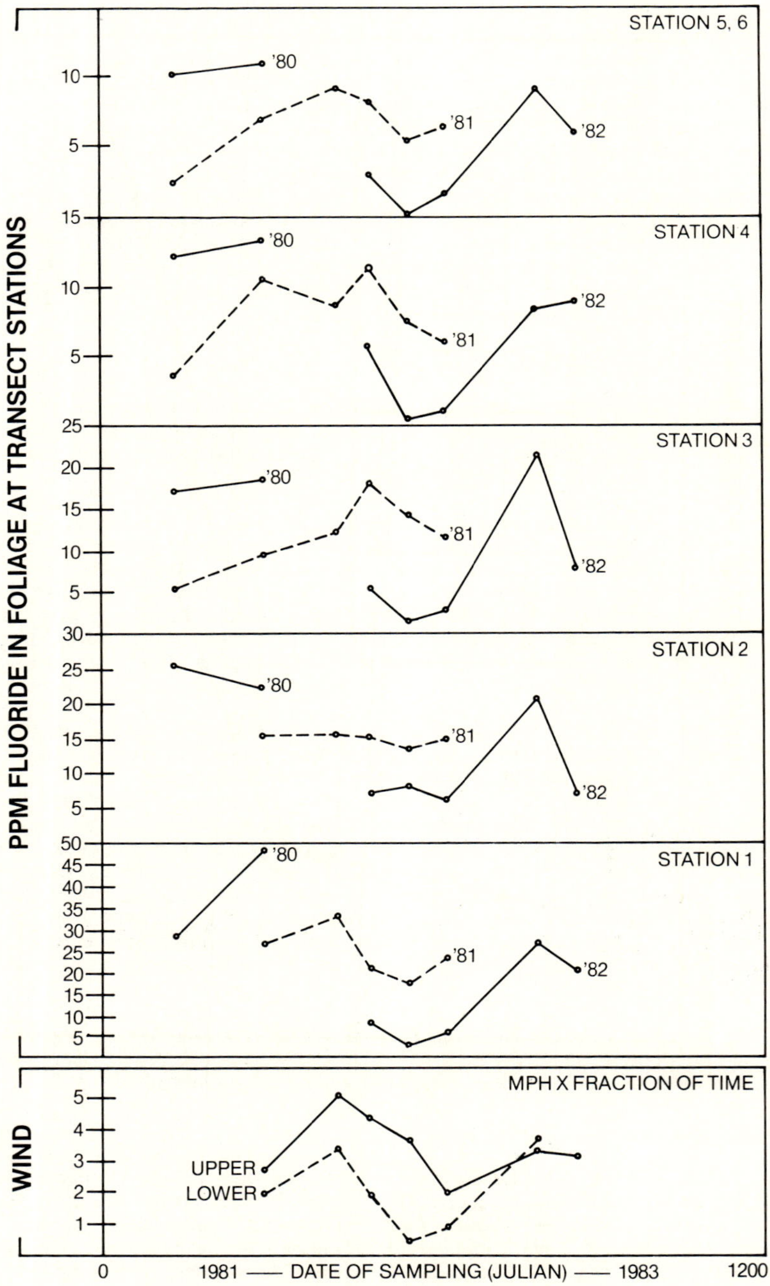
PPM FLUORIDE IN FOLIAGE AT TRANSECT STATIONS
STATION 5, 6
STATION 4
STATION 3
STATION 2
STATION 1
'80
'81
'82
WIND
MPH X FRACTION OF TIME
UPPER
LOWER
0
1981 —— DATE OF SAMPLING (JULIAN) —— 1983
1200

FIGURE III

FLUORIDE UPTAKE - FOREST TRANSECT

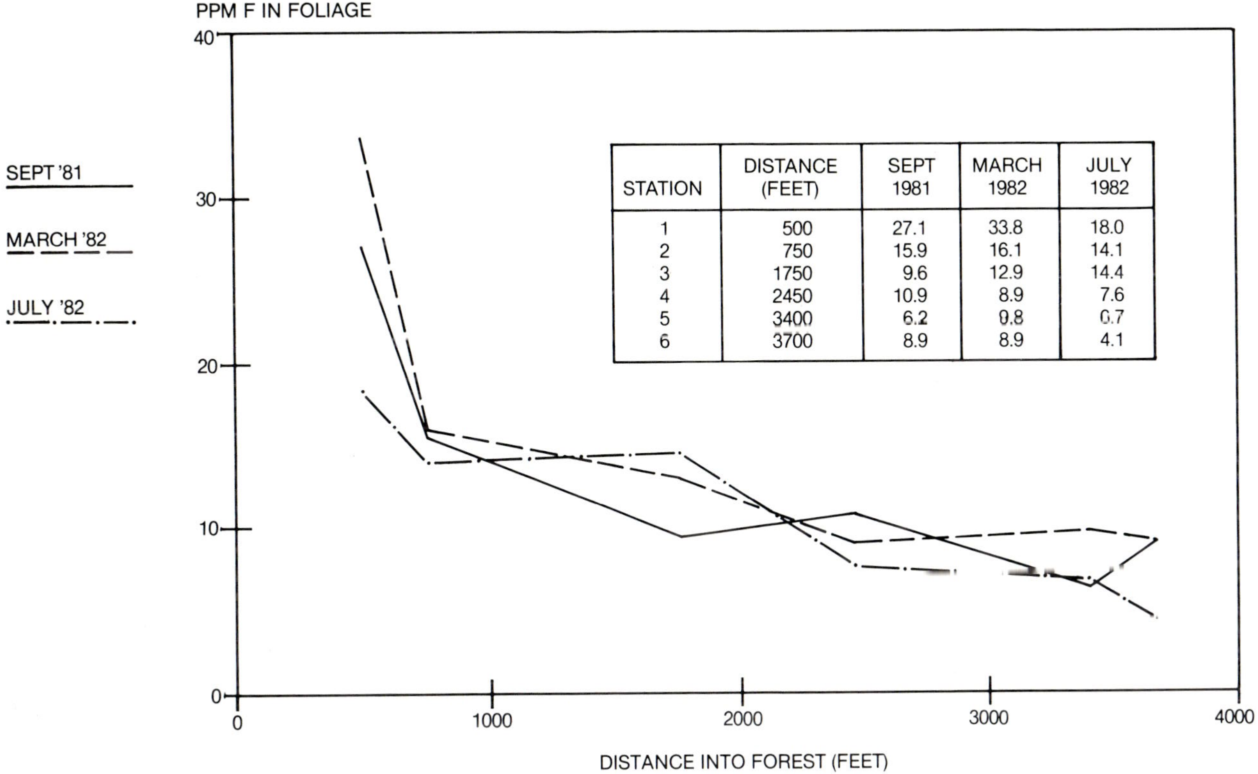

STATION	DISTANCE (FEET)	SEPT 1981	MARCH 1982	JULY 1982
1	500	27.1	33.8	18.0
2	750	15.9	16.1	14.1
3	1750	9.6	12.9	14.4
4	2450	10.9	8.9	7.6
5	3400	6.2	9.8	6.7
6	3700	8.9	8.9	4.1

does not follow a single exponential decrease that could be expected from dilution and depletion of a plume. Instead it is a composite of rather rapid decline (with distance) from the lower-level (18 m) sources and a touchdown and then decline (at the further sites) from the higher-level (55 m) sources. In more extreme situations than those depicted, the concentration of fluoride decreased from the first to the second site and then increased at the third site with a decrease to the sixth site.

Conclusions

Many factors can influence the accumulation of airborne fluoride by vegetation (5) and they comprise the general classes of: physical and chemical characteristics of emissions; environment, which includes meteorologic factors that effect dispersion and climatic factors that determine the response of the plant; biologic factors, such as age of tissue, species of plant, etc.; conditions of exposure, such as concentration of pollutant and duration and recurrency of exposure. Different models have been proposed (3, 6) for the uptake of gaseous and particulate pollutants by forests. However, the observed concentrations of fluoride and other pollutants in the foliage of plants are the resultants of the processes of redistribution and elimination as well as accumulation of airborne materials and are affected by sampling and analytical error.

It is apparent that in a southern pine forest, the accumulation of fluoride by foliage is a dynamic system and the resultant of continuous processes of uptake and loss. Consequently, additional measurements of environmental variables as well as knowledge of the emissions must be incorporated into any explanation of the temporal and spatial patterns. One limitation is that the magnitude of the spatial and temporal variations (as well as the concentrations of foliar fluoride) are low with reference to analytical and sampling errors and baseline (background) values. A remedy would be to sample more intensively in space and time. At present, however, a more economic solution would appear to be a revision of the location of some sites and sampling protocols along the present transect and the initation of a sampling program along two or three new transects.

Several models have been used to predict the dispersion of atmospheric fluoride around a potential source, but most do not take into account the removal of airborne fluoride through adsorption/absorption by vegetation. The accumulation of fluoride in loblolly pine foliage at various distances downwind from a modern, well-controlled primary aluminum smelter is presented in this paper to illustrate the efficacy of vegetation as a sink for fluoride and to emphasize the need to consider uptake by vegetation in dispersion models for fluoride. The results reported herein represent what has been learned from an exploratory effort to study the transfer of fluoride in a specific ecosystem and derive a model that could predict what environmental changes might be associated with present or future operations.

REFERENCES

1. J. O. Ares, "Fluoride Cycling Near a Coastal Emission Source," _Journal of the Air Pollution Control Association_, 29 (1978) pp. 344-349.

2. ASTM, "Analysis for Fluoride Content of the Atmosphere and Plant Tissues (Semiautomated Method)," pp. 713-726 in 1976 Annual Book of ASTM Standards, Part 26, American Society for Testing Materials, Philadelphia, PA, 1976.

3. Y. Belot, A. Baille, and J. L. Delmas, "Modele Numerique de Dispersion des Polluants Atmospheriques en Presence de Couverts Vegetaux," Atmospheric Environment 10 (1976) pp. 89-98.

4. R. P. Hosker, Jr., and S. E. Lindberg, "Review: Atmospheric Deposition and Plant Assimilation of Gases and Particles," Atmospheric Environment 16 (1982) pp. 889-910.

5. D. C. McCune, and A. E. Hitchcock, "Fluoride in Forage: Factors Determining Its Accumulation from the Atmosphere and Concentrations in the Plant," pp. 289-292 in Proceedings from the International Clean Air Congress, 2nd, H. M. Englund & W. T. Berry, eds; Academic Press, New York, 1971.

6. C. E. Murphy, Jr., T. R. Sinclair, and K. R. Knoerr, "An Assessment of the Use of Forests as Sinks for the Removal of Atmospheric Sulfur Dioxide," Journal of Environmental Quality 6 (1977) pp. 388-396.

7. F. Murray "Ecosystems as Sinks for Atmospheric Fluoride," pp. 191-205 in Fluoride Emissions: Their Monitoring and Effects on Vegetation and Ecosystems, F. Murray, ed.; Academic Press, Sydney, Australia, 1982.

8. G. A. Sehmel, "Particle and Gas Dry Deposition: A Review," Atmospheric Environment 14 (1980) pp. 983-1011.

EFFECTS OF STAGED COMBUSTION ON NO_x FORMATION DURING COKE CALCINATION

Alvin F. Bopp

Martin Marietta Laboratories
Baltimore, Maryland 21227
USA

John F. Bent

Martin Marietta Aluminum, Inc.
Bethesda, Maryland 20817
USA

A laboratory-scale combustion apparatus was used to measure nitric oxide formation over a range of combustion conditions. The formation of nitric oxide was controlled by regulating oxygen availability during the early phases of combustion. The laboratory results confirm that staged combustion is a realistic strategy for controlling NO_x from the volatile nitrogen-containing compounds evolved during coke calcination. Reductions approaching 50% of the total NO_x formed were achieved in plant tests.

Introduction

Metallurgical-grade coke is used worldwide in the production of aluminum metal by the Hall-Heroult process. In most commercial processes, petroleum or pitch coke is heated to 1200°C in a rotary kiln, driving off volatiles and carbonizing the coke (Figure 1). Martin Marietta operated such a facility in Carson, CA, and in the late 70's elected to construct a state-of-the-art calciner in Wilmington, CA. This new plant features higher capacity (300,000 tons per year calcined coke) and electrical power cogeneration utilizing normally wasted heat. As a requirement for the operating permit, a limit of 90 lb NO_x/hr (as NO_2) was imposed by the South Coast Air Quality Management District compared to typical emissions of 120 lb NO_x/hr at the Carson site. It was evident that to achieve the stringent NO_x limit at higher production, an effective NO_x control strategy was needed. After reviewing many options, the staged combustion concept, in combination with burner operation under reducing conditions, was incorporated into the design to control NO_x. To demonstrate the feasibility of this concept, a detailed laboratory study was conducted in 1981. Once the plant came on-stream in February 1983, operational testing was carried out on site to verify the concept and define optimum parameters.

NO_x Formation Considerations

A brief description of the mechanism for NO formation is given below as a background for later discussions of staged combustion. NO_x is initially formed as nitric oxide (NO) and is slowly oxidized in the atmosphere to nitrogen dioxide (NO_2). In combustion processes, NO is produced via two general pathways: 1) thermal NO and 2) fuel-bound-nitrogen NO.

The mechanism of the formation of thermal NO from the oxidation of atmospheric nitrogen has been studied extensively. The formation of thermal NO occurs in the post-flame region and can be described mechanistically by the extended-Zeldovich mechanism (1).

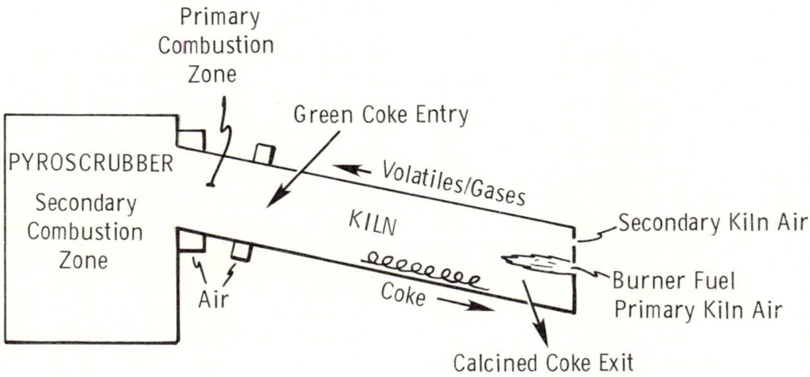

Figure 1. Block diagram of Martin Marietta's coke calciner, at Wilmington, CA.

$$O + N_2 \longrightarrow NO + N \qquad (1)$$
$$N + O_2 \longrightarrow NO + O \qquad (2)$$
$$N + OH \longrightarrow NO + H \qquad (3)$$

The rate limiting step in this mechanism is Equation (1), which has a high activation energy of 78.5 kcal/mole (1). Therefore, this step and the formation of thermal NO are highly temperature sensitive. Hence, to control the formation of thermal NO, it is necessary to minimize the temperature and/or O atom concentration in the post-flame region.

While the nitrogen source for thermal NO is atmospheric nitrogen, it may also be the nitrogen-containing compounds in the fuel itself. NO formed from this nitrogen source is termed fuel-bound-nitrogen NO. Although the mechanism for the production of NO from fuel nitrogen is not as well defined and accepted as is the extended-Zeldovich mechanism for thermal NO, a qualitative model can be discussed for the homogeneous gas phase formation of NO. Prior to entry into the combustion zone, the nitrogen-containing compounds undergo pyrolysis reactions as do hydrocarbons. Low molecular weight compounds and radicals, such as NH_3, HCN, NH_2, etc., are formed that oxidize very rapidly under fuel-lean conditions to form NO. The chemistry behind the oxidation reactions is exceedingly complicated and involves the reactions of atoms and radicals. Since the activation energies for reactions involving these species are usually low, 10 kcal/mole or less, the activation energies for the oxidation of these nitrogen-containing pyrolysis products are also expected to be low. This conclusion has been verified experimentally (1). The formation of NO from fuel-bound-nitrogen sources does not exhibit a strong temperature dependence but does show a strong dependence on oxygen availability and residence time. Localized oxygen availability is critical. In an oxygen-deficient atmosphere, the nitrogen-containing pyrolysis products decompose to molecular nitrogen as a function of time. Given fuels with a high nitrogen content, an effective NO control strategy would be to minimize available oxygen until the fuel-bound nitrogen converts to N_2 and only then add air to complete the combustion process. This rationale forms the basis for the staged combustion approach to reducing NO_x emissions.

Staged Combustion

In staged combustion, a fuel-rich primary combustion zone is promoted initially, and the air that is necessary for complete combustion is added further downstream. Because the primary combustion zone is at less than maximum temperature, thermal NO_x formation is reduced. In the primary zone, the nitrogen-containing fuel components pyrolyze to form, among other things, molecular nitrogen. The residence time in the primary zone must be long enough to permit the pyrolysis reactions to take place because any remaining organic nitrogen will rapidly oxidize in the secondary combustion zone. In the secondary combustion zone, air is added rapidly to bring the system to a fuel-lean condition to complete combustion while maintaining low flame temperature. Rapid mixing is important in order to maintain a low combustion temperature through stoichiometry control. Since both temperature and oxygen availability are regulated, both thermal and fuel-bound-nitrogen NO_x can be affected.

Staged combustion as an approach to the reduction of NO_x emissions has been studied extensively, focusing on gas-, coal-, or oil-fired utility boilers. The combustion modification literature has been recently summarized in an EPA Project Summary (EPA-600/57-82-034, Sept. 1982). The staged combustion of calcination by-product volatiles, as reported here, is the first such application for waste-heat boilers.

The object of this work was to demonstrate the applicability of staged combustion in reducing NO formation from fuel nitrogen sources both laboratory experiments and plant trials. Laboratory experiments centered on the simulation of plant processes and conditions with the resulting data being used for concept validation. Following the development of the design criteria, the concept was verified in commercial application through field testing at the Martin Marietta Carbon calciner in Wilmington, CA.

Experimental

A laboratory-scale apparatus was built to simulate the kiln and pyro-scrubber sections of the coke calcining process. The apparatus is shown in Figure 2. The rotary kiln was represented by the combustor, whose function was to provide a hot gas stream for the introduction of the volatile combustible material (VCM) simulants. The combustor was constructed of inconel tubing connected to and fired by a modified Meeker burner. The exhaust from this tube flowed into a ceramic bucket used as the primary afterburner, sized so that the residence time of the gas stream containing the VCM was about 6 ms. The hot gases exiting this zone were mixed with air in the last section -- the secondary afterburner -- to achieve the desired overall stoichiometry. The air flows into the primary and secondary combustion afterburners were carefully regulated; therefore, the VCM:air ratio in the afterburner could be controlled.

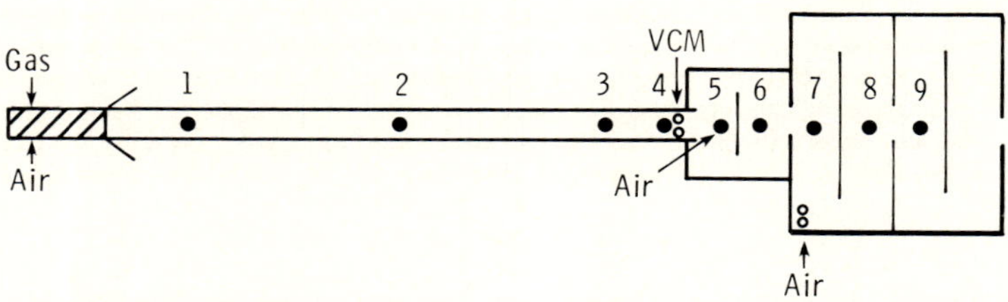

Combustor	76 x 3.2 cm	Thermocouple ports	- Ports 1, 2, 4, 6, 7, 8
Primary Afterburner	20.3 x 10.3 cm	VCM Inlet	- Between Ports 4 & 5
Secondary Afterburner	67 x 29.5 cm	Primary Staged Air	- Port 5
		Secondary Staged Air	- Between Ports 6 & 7
		Nitric Oxide Sampling	- Port 9

Figure 2. Block diagram of the laboratory combustion apparatus.

Natural gas and air were used to heat the combustor, and these flows were held constant during the staged combustion experiments. Previous analyses of coke volatiles indicated that the main components are H_2 and CH_4 (85% and 10% respectively) and the main nitrogen-containing species are NH_3 and HCN (2). The VCM was simulated by H_2 with NH_3. In one experiment, small amounts of CH_4 were added to the VCM to be more representative of actual volatiles. In these tests, the NH_3 and air flows were chosen to provide a direct comparison of the VCM:air ratio between this experiment runs on NH_3/H_2 VCM. The VCM simulant was injected into the combustor near the discharge into the primary afterburner (see Figure 2). All gas flows were metered independently at each entry location.

Temperatures were measured with chromel/alumel thermocouples. Temperatures in the primary combustion zone ranged from 1250–1350°C depending on NH_3 flow and the VCM:air ratio in the primary combustion zone. The temperature at Port 2 was typically 1120–1130°C. The temperature at Port 4, just prior to VCM introduction, registered VCM combustion and varied with VCM composition; it ranged from 1040–1150°C.

A Theta Sensors US 5000 NO_x/SO_2 Monitor was used to measure NO. The principle of operation involves a combination of semipermeable membranes and selective redox reactions in a sealed transducer. Gas was drawn continuously from the secondary afterburner, and passed over an ice water trap to remove moisture prior to introduction into the transducer. The NH_3 and air flows were set to a precalculated value, and after a period of time, typically 4 min, the NO level was recorded. The instrument was calibrated at the beginning of each day and checked at the end of each day with a standard NO in N_2 mixture.

The laboratory experiments measured NO formation as a function of: 1) NH_3 level in the VCM, 2) the overall VCM:air stoichiometry, and 3) the fraction of stoichiometric air added into the primary combustion zone. The experiments are summarized in Table I. Included in Table I are flow-rate data for air and VCM simulants. Excess air was controlled between 2 and 7%. V_p/V_{st} is defined as the ratio of the air flow to the primary zone to the air requirement for stoichiometric VCM combustion. The maximum V_p/V_{st} was fixed by the VCM flow rates and composition while the minimum was limited to maintain stable combustion. The maximum V_p/V_{st}, reported in Table I, represents the case in which the total staged combustion air flow was added to the primary afterburner; hence, it is also the overall VCM:air ratio.

Table I. Summary of Laboratory-Staged Combustion Experiments

Exp't No.	NH_3 (l/min)	H_2 (l/min)	CH_4 (l/min)	V_p/V_{st} Range max	min
1	0.88	28.3	0	1.02	0.40
2	0.88	28.3	0	1.06	0.40
3	1.18	28.3	0	1.05	0.65
4	1.18	28.3	0	1.05	0.40
5	1.18	28.3	0	1.07	0.40
6	1.57	28.3	0	1.02	0.40
7	1.57	28.3	0	1.05	0.40
8	1.18	28.3	1.27	1.07	0.50

Laboratory Results and Discussion

The results of the laboratory-staged combustion experiments are shown in Figures 3, 4, and 5. All plots show the same basic shape. As less air is added into the primary combustion zone (decreasing V_p/V_{st}), NO formation, from both burner and VCM sources, shows a dramatic decrease followed by a gradual increase. The shape of these curves can be explained by recalling the prior discussion on staged combustion and NO formation. The initial dramatic decrease is evidence of the effects of reducing oxygen availability in the primary zone. However, after air flow is decreased to a certain point, insufficient air is being added to the combustion zone to maintain temperatures needed for NH_3 pyrolysis. The NH_3 passes through this zone intact only to oxidize rapidly in the secondary combustion zone. These plots are interpreted as evidence that NO_x can be reduced by staged combustion.

A comparison of the effects of fuel N content on NO formation and staged combustion at a given overall VCM:air ratio is shown in Figure 3. The amount of NO formed is clearly a function of the NH_3 content in the VCM simulant. However, even with stoichiometric air flow in the primary afterburner, the conversion of NH_3 to NO is much less than stoichiometric. When a simplified rate law was written, the NH_3 reactant order was determined to be > 1, indicating a complex reaction mechanism. No effort was made to elucidate the reaction mechanism. In addition, the efficiency of the staged combustion reduction process, calculated from maximum and minimum NO levels, increases with increasing NH_3 content in the VCM. Lastly, the minimum NO production appears to occur at higher V_p/V_{st} with increasing N content. This latter observation can also be explained qualitatively by the temperature/air flow effect discussed in the preceeding paragraph and the high heat capacity and ignition temperature of NH_3.

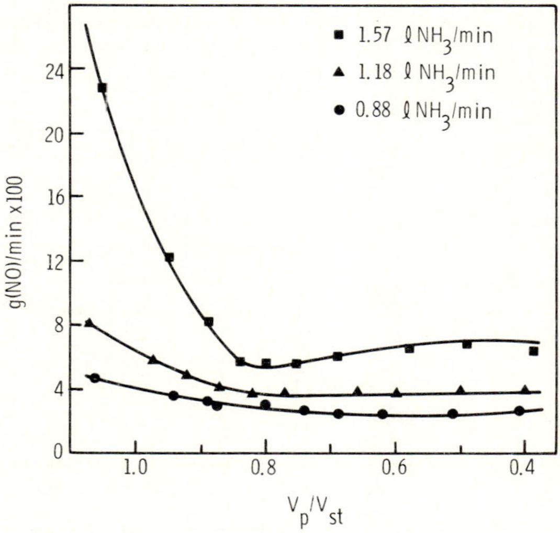

Figure 3. Staged combustion reduction of NO_x at various NH_3 flow rates (V_p/V_{st} = the ratio of air flow into the primary afterburner to the air requirements for stoichiometric VCM combustion).

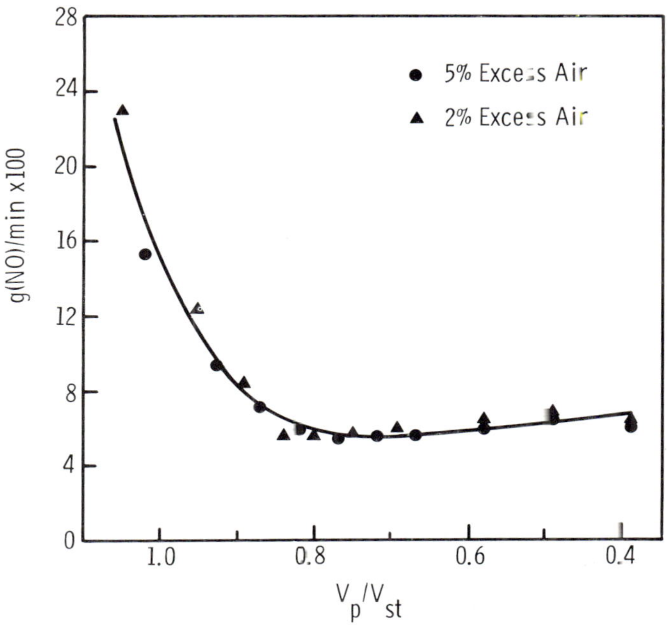

Figure 4. The effect of VCM:air ratio on staged combustion reduction of NO_x.

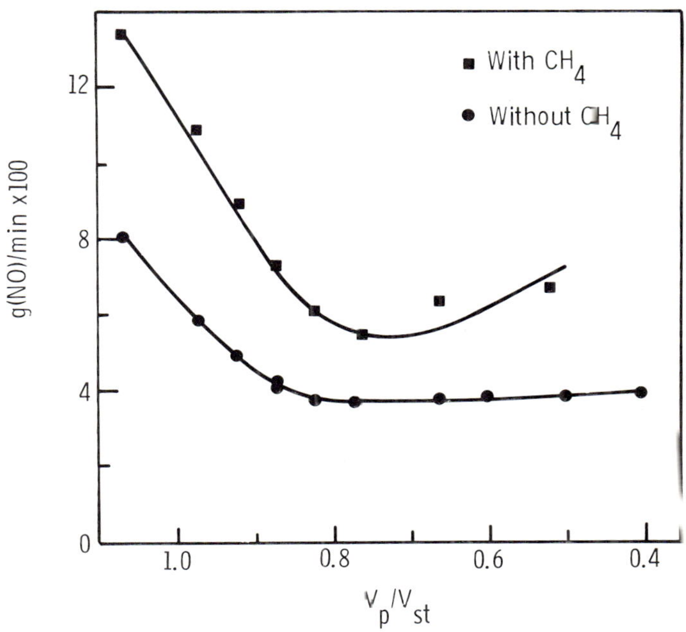

Figure 5. Effect of CH_4 and primary zone stoichiometry on NO formation and control through staged combustion.

Figure 4 shows the effects of VCM:air ratio on NO formation at a constant NH_3 flow. This plot clearly shows that the overall VCM:air ratio does not have an impact on NO formation, but the ratio is critical in the primary combustion zone. Note the steep increase in NO formation as the combustion zone approaches and exceeds stoichiometric VCM:air. The reduction efficiency may appear greater for 5% vs 2% excess air; however, more importantly, the minimum NO production is the same for both stoichiometries.

The effect of adding CH_4 to the VCM is shown in Figure 5. This Figure includes data from the equivalent experiment without CH_4. Based on limited data, the results indicate that NO formation and the efficiency of staged combustion are enhanced by the CH_4. By reducing the air flow in the primary afterburner from 1.05 to 0.8 of stoichiometric, NO reduction of 57% with CH_4 present in the VCM compared to 53% with NH_3/H_2 VCM was achieved. It can be inferred from this experiment that CH_4 promotes the formation of an intermediate species that competes favorably with NH_3 for O_2 but is probably less thermally stable than NH_3 because of the greater effect of staged combustion.

The nitrogen level in the plant VCM was estimated by elemental analysis of green and calcined cokes. Laboratory results extrapolated to this VCM composition (lower VCM nitrogen) indicate that NO reductions of 40-45% are possible by reducing V_p/V_{st} to 0.8. However, since this VCM nitrogen is expected to be less than the simulants, a V_p/V_{st} lower than 0.8 may realize even greater NO reductions. Recall that the optimal V_p/V_{st} decreased with decreasing VCM nitrogen.

No attempt was made to separate temperature effects on NO formation from effects due to the VCM:air ratio. Since the primary zone temperature is a function of VCM:air, it like NO, is a dependent variable. This temperature was taken as an experimental parameter. In these experiments, the air flows into the primary and secondary afterburner are the independent variables. Excessive air in the combustion zone promotes high NO formation through both high temperature and high oxygen availability.

The combustor contribution to NO_x was approximately 0.7 g (NO)/min under the conditions of the staged combustion experiments. This contribution to NO formation was not subtracted from the measured NO levels shown in Figures 3, 4, and 5.

Plant Sampling Tests

During coke calcination, the volatiles, which are evolved in the kiln, are swept by the gas stream through the transition section, where staged combustion takes place, and into the pyroscrubber for complete combustion. This process is shown in Figure 1. The staged combustion air is provided through two dampers that can be manipulated to regulate the flow of air into the kiln exhaust stream. During the plant tests, the air flow through the first of these dampers was varied to observe the effects of the primary combustion zone VCM:air ratio on NO. The opening of the first damper was adjusted from completely open to completely closed, while the second damper was held constant at 3% open. Throughout these tests, kiln throughput was maintained at 20 tons per hour. Other process conditions are listed in Table II.

TABLE II. Process Conditions During Sampling Test

Time	Back End Temp (°F)	Pyro. Temp (°F)	Kiln Fuel (CFM)	Kiln Prim. Air (CFM)	Kiln Sec. Air (CFM)	Back End ϕ
1300	1498	1995	738	6754	3719	3.69
1400	1482	2138	548	5394	2905	4.43
1530	1485	2114	578	5609	3589	4.03
1605	1520	2186	539	5353	3302	4.24

The Theta Sensors VS 5000 was used to measure the NO level in the stack gas. The ppm NO was converted to lb NO_2/hr, which was normalized to the maximum NO_2 production for the test. From the stoichiometry of the VCM and burner combustion reactions and measured burner fuel and kiln air flows, the equivalence ratio, ϕ, and the air necessary for stoichiometric VCM combustion could be estimated (see Table II). It was assumed that the coke contained 11% volatiles consisting of 85% H_2 and 10% CH_4. The air flow through each damper in the transition section was calculated from the damper position and known fan characteristics. With this information, a V_p/V_{st} analagous to the laboratory parameter was determined for each test point. The results of these measurements are listed in Table III and illustrated in Figure 6. The data plotted in Figure 6 show that as less air is added to the primary zone, thereby decreasing V_p/V_{st}, NO production decreases and then subsequently increases (as was characteristic of the laboratory studies). A reduction approaching 50% of the total NO was observed by reducing V_p/V_s from 1.05 to .50. As in the laboratory studies, the burner contribution to NO was not separated from the measured NO levels, which represent NO of both burner and VCM origin

These plant tests are viewed as a single data point During these tests, the plant was operated below design capacity and no attempt was made to optimize gas flows for NO_x minimization.

TABLE III. NO_x Reductions through Staged Combustion in Commercial Application

Primary Air Damper Position (% open)	V_p/V_{st}	NO_x/NO_x (max)
0	0	0.69
10	0.22	0.41
20	0.45	0.40
25	0.56	0.29
30	0.66	0.33
30	0.66	0.35
40	0.84	0.52
50	1.05	0.49
60	1.20	0.54
70	1.35	0.56
80	1.43	0.39
90	1.44	0.39
100	1.44	1.00

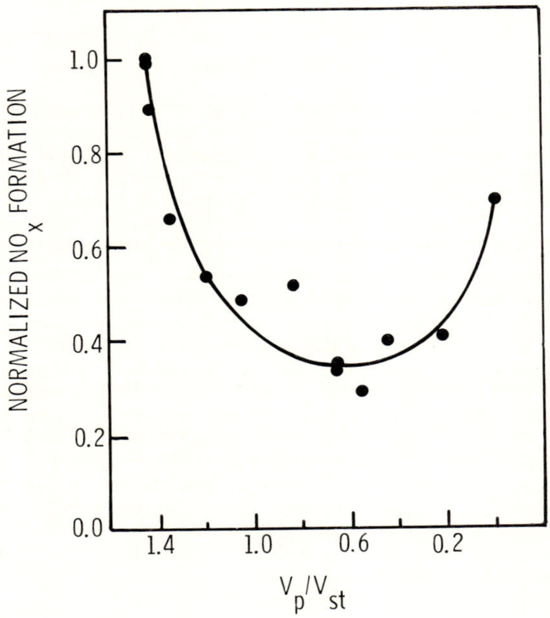

Figure 6. NO_x reductions from commercial staged combustion application.

Conclusions

The results of these tests can be summarized as follows:

- The laboratory results showed the criticality of controlling air flow into the primary zone. The minimum NO production in the laboratory apparatus occurred for V_p/V_{st} in the region 0.7 to 0.8 and reductions in NO production of 50% were achievable when compared to V_p/V_{st} of 1.05. Extrapolation to anticipated plant VCM compositions indicated that 40-45% reductions were possible by decreasing V_p/V_{st} from 1.05 to 0.8 and that lowering V_p/V_{st} below 0.8 might further enhance staged combustion efficiency.

- The combination of operating the kiln burner under reducing conditions and removing VCM through staged combustion is an effective way to limit the formation of NO. In the Martin Marietta application NO reduction approaching 50% of the total NO was achieved by reducing the airflow into the primary combustion zone from 105% to about 50%-60% of stoichiometric without any further optimization.

- Though unrefined, the laboratory experiment proved to be an excellent predictor of plant performance. Both the NO reduction and the V_p/V_{st} for minimal NO production followed the trends predicted from the laboratory data extrapolated to plant conditions.

Acknowledgements

The authors would like to thank both the Martin Marietta Aluminum management for their support throughout this project and permission to publish the results as well as the plant management at Martin Marietta Carbon for their cooperation during the sampling tests. Members of Martin Marietta Environmental Center provided a valuable critique on the project and helpful comments on the manuscript. Lastly, the authors would like to thank David Creighton at Theta Sensors for his support work on our instrumentation.

References

1. I. Glassman, Combustion, p. 213; Academic Press, New York, NY, 1977.

2. M. McCarty, "Thermodynamics of the Martin Marietta Carbon Kiln," Report, Martin Marietta Laboratories, Baltimore, 1975.

CONTRIBUTION TO THE INTERPRETATION OF COAL TAR PITCH MEASUREMENTS AT THE WORKPLACE

R. Glaus
Swiss Aluminium Ltd.
Research & Development
CH-8212 Neuhausen
Dr. A. Steinegger
Swiss Aluminium Ltd.
Feldeggstrasse 4
CH-8034 Zurich

1. Introduction

Generally, measurements of the workplace atmosphere to determine the exposure of employees are influenced by various factors, namely workplace procedures and conditions together with variations in measuring techniques.

Procedures and conditions may vary as a function of technical installations, raw materials (e.g. material causing a great deal of dust), air temperature and time of year.

Measuring techniques depend on the choice of analytical methods and sampling procedures e.g. personal and stationary, where fluctuations in the absolute values and their composition can occur[1]).

For the measurements of coal tar pitch concentrations, additional problems arise in the interpretation of the data determined, since correlations between different kinds of values and possible health hazards have not been sufficiently well established[2]).

In this investigation, the samples were examined for the fraction soluble in an extraction agent, for individual polycyclic aromatic hydrocarbons (= PAH) and their sum.

Altogether more than 1'000 samples were taken during various measuring campaigns to monitor the situation and check technical improvements. Further series of measurements were conducted to obtain additional information for interpreting the results.

Based on the test material this paper attempts to clarify the complex correlations in the practical measurements of coal tar pitch at the workplace.

2. Sampling and Analysis

2.1 Personal Sampling

Sampling is carried out throughout an entire shift (approx. 8 hours) on a glass fiber filter (dia. 3.7 cm) and a silver membrane (8 μm pore size). A flow of 2.0 l/minute is set with electronically controlled pumps (Casella 3137/80).

2.2 Stationary Sampling

For the stationary sampling, filter holders with glass fiber filters (dia. 6.7 cm) and Reciprotor pumps (flow 6.5 l/minute) are used.

2.3 Extraction

The filters and the membranes are normally extracted with cyclohexane in an ultrasonic bath for 10 minutes. For special tests, benzene and toluene were also added.

2.4 Gravimetric Determination

The evaporated extract is weighed on a microbalance with an accuracy of 0.1 μg.

2.5 High Performance Liquid Chromatographic Separation

Separation took place on a 25 cm long RP 18 column. The polycyclic aromatic hydrocarbons were recorded with a UV detector (254 nm) and a fluorescence photometer (355/430 nm or 285/410 nm) and evaluated on a data system (Spectra Physics 8000).

The following substances are routinely determined:

Phenanthrene	Ph	Perylene	Per
Anthracene	A	Benzo(a)pyrene	BaP
Fluoranthene	Fl	Benz(g,h,i)perylene	BPer
Pyrene	Py		
Benzo(c)phenanthrene	BcPh	(The abbreviations are used in Figs. 1-3)	
Benz(a)anthracene	BaA		
Chrysene	Chry		
Benzo(b)fluoranthene	Bbfl		
Benzo(e)pyrene	BeP		

The relative standard deviation is better than 6.8 % for all the substances recorded based on mean values from 10 determinations (usual sample concentration in the range of 10-100 µg PAH/ml).

3. Influence of the Extraction Agents Used

3.1 Basic Situation

In the literature[3-7], mainly benzene and cyclohexane are suggested as solvents. Other extraction agents are also used, for example ethanol, toluene, propylene carbonate etc. For historical reasons, benzene was initially used almost exclusively. For reasons of workers' hygiene, cyclohexane was frequently suggested in place of benzene in the laboratory. However, some doubts arose due possibly to the inadequate extraction effect of cyclohexane compared to benzene.

In an earlier publication[8], certain deviations in the differently extracted polycyclic aromatic hydrocarbons were established. However, cyclohexane could still be used successfully.

To confirm these results, in the following investigation, stationary sampling was applied with a view to small deviations compared to personal sampling (individual factors).

In addition the results were compared with those using toluene as the extraction agent. The following abbreviations are used:

BSM = benzene soluble matter
CSM = cyclohexane soluble matter
TSM = toluene soluble matter.

3.2 Comparison - Gravimetric Method

Initial investigations were carried out to determine to what extend values measured gravimetrically are influenced by the extraction agent.

Compared to the BSM values, the following percentages were obtained:

for CSM 68 % in the paste plant
71 % in the prebake reduction plant

for TSM 61 % in the paste plant
62 % in the prebake reduction plant.

The values do not differ very much in the range of 10 - 300 μg/m^3. Single tests revealed that even at values higher than 1000 μg/m^3 there is no basic change.

3.3 Comparison of the PAH Contents

The percentages of the PAH contents compared with benzene as solvent are somewhat lower:

for CSM 52 % in the paste plant
57 % in the prebake reduction plant

for TSM 58 % in the paste plant
44 % in the prebake reduction plant.

Table 1 lists the percentages of individual substances evaluated routinely. In addition to the listed polycyclics 10 - 15 % of PAH substances (e.g. methyl-phenantrenes, benzofluorenes; methylpyrenes, benzo - j + k - fluoranthene) were determined as a group. Although the analyses were conducted with stationary measurements, a relatively large standard deviation was ascertained as shown in Table 2. Even greater fluctuations were obtained with personal sampling. This applies both for the carbon plant and for the reduction plant.

On the other hand, in the percentage distribution per solvent, only small differences are ascertained, as can be seen from the figures in Table 3.

Apart from the larger standard deviation, the relations for personal sampling essentially remain unchanged as presented above.

Table 1

Content of Individual Components in Benzene, Cyclohexane and Toluene Extracts

	Paste Plant			Reduction Plant [1)]		
	BSM $\mu g/m^3$	CSM vs.BSM %	TSM vs.BSM %	BSM $\mu g/m^3$	CSM vs.BSM %	TSM vs.BSM %
Phenanthrene	21,0	61	64	1,45	65	46
Anthracene	2,2	59	64	0,14	64	43
Fluoranthene	23,9	50	56	1,49	58	49
Pyrene	17,0	56	67	0,98	82	55
Benzo (c) - phenanthrene	1,8	78	83	0,11	73	73
Benz (a) - anthracene	8,1	64	65	0,49	65	49
Chrysene	12,1	56	56	0,81	61	38
Benzo (b) - fluoranthene	6,5	45	43	0,43	49	37
Benzo (e) - pyrene	2,2	41	46	0,20	45	40
Perylene	1,3	42	46	0,08	53	50
Benzo (a) - pyrene	2,3	39	41	0,22	41	40
Benz (g,h,i) - perylene	1,2	46	50	0,07	43	57

1) Prebake

BSM = Benzene soluble matter
CSM = Cyclohexane soluble matter
TSM = Toluene soluble matter

Table 2

Relative Standard Deviations in %
(to values used in table 1)

	Paste Plant			Reduction Plant		
	BSM	CSM	TSM	BSM	CSM	TSM
Phenanthrene	18	27	22	25	31	27
Anthracene	18	25	25	30	35	35
Fluoranthene	12	21	15	15	15	19
Pyrene	15	15	19	20	12	19
Benzo (c) - phenanthrene	22	27	29	13	27	25
Benz (a) - anthracene	18	14	22	15	22	19
Chrysene	16	14	20	7	15	17
Benzo (b) - fluoranthene	18	12	20	10	9	11
Benzo (e) - pyrene	14	17	15	15	12	15
Perylene	25	28	26	18	22	25
Benzo (a) - pyrene	17	17	19	13	10	14
Benz (g,h,i) - perylene	24	22	25	20	18	15

Table 3

Percentage Composition of Benzene, Cyclohexane and Toluene Extracts (Stationary Sampling)

	Paste Plant			Reduction Plant		
	BSM %	CSM %	TSM %	BSM %	CSM %	TSM %
Phenanthrene	21,1	23,5	22,8	22,4	22,7	21,9
Anthracene	2,2	2,4	2,4	2,2	2,2	2,0
Fluoranthene	24,0	21,5	22,5	23,0	21,4	24,2
Pyrene	17,1	17,3	19,2	15,1	19,7	17,9
Benzo (c) - phenanthrene	1,8	2,6	2,6	1,7	2,5	2,6
Benz (a) - anthracene	8,1	9,5	8,9	7,6	7,9	7,9
Chrysene	12,1	12,4	11,5	12,5	12,1	0,3
Benzo (b) - fluoranthene	6,5	5,3	4,8	6,6	5,2	5,3
Benzo (e) - pyrene	2,2	1,6	1,6	3,1	2,2	2,6
Perylene	1,3	1,0	1,0	1,2	1,2	1,3
Benzo (a) - pyrene	2,3	1,6	1,6	3,4	2,2	2,6
Benz (g,h,i) - perylene	1,2	1,0	1,0	1,1	0,7	1,3
Σ PAH	99,9	100,1	99,9	99,9	100,0	99,9
% PAH in the extract	54,6	49,5	46,4	26,8	24,5	20,9
Standard deviation	±17,3	±16,8	±15,2	± 5,4	± 5,7	± 6,3

3.4 Conclusions

With cyclohexane, the extracted fraction of soluble matter (61 - 68 %) and PAH (52 - 57 %) are smaller compared with benzene, however, the percentage composition of the individual substances is not much changed. It must be taken into account nevertheless that for BaP, the percentage referring to absolute values is approx. 45 %; for the small molecules, the percentage contents are higher. However, in our opinion the values, which are based on cyclohexane as solvent, can be used as a basis for determining PAH. By using correlation factors resulting from the above-mentioned figures, the values will come close to those where benzene is used as solvent. Toluene as dissolving agent behaves similarly in its extraction effect to benzene and cyclohexane but offers no advantages. By using cyclohexane or toluene, the health risk for the laboratory is considerably smaller.

4. Influence of Kind of Sampling on the Result

4.1 Basic Situation

Large deviations in the CSM values determined by personal sampling can, as previously shown[8], be due to the work procedure even at the same workplace. The large deviations in the PAH fractions, however, can only be partially explained in this way. It has been ascertained before[9] that, with a few exceptions, the PAH fractions are low for a specified working area in the case of high CSM values, and vice versa i.e. high PAH fractions with low CSM values. A probable reason for this phenomenon ist that the high CSM values are caused by dust with a low PAH content, while the lower CSM values correspond to the components formed directly in the process. From another reference[10], it was seen that the high PAH fractions are due to evaporated products. On the other hand, it was noted that in stationary sampling, the deviations in the PAH fractions were much smaller. For this reason the correlation was investigated systematically.

4.2 Difference in the PAH Fraction in the case of Stationary and Personal Sampling

If one compares the PAH fraction of stationary measurements with those of personal samples taken in the same working area, it can be found for example that the percentage in stationary sampling at 58 % PAH is more than twice that of personal sampling with 27 % PAH. The relevant CSM values were considerably higher in the case of personal samples. As expected, the deviations were particularly marked for the CSM values but were also higher in the case of the PAH values from personal sampling. The average percentage distribution of the individual components on the other hand is much the same for both kinds of sampling.

These results were compared with tests on pure coal tar pitch with the following PAH fractions:

coal tar pitch: solid: 20 % PAH of total CSM

coal tar pitch: vapor: 34 % " " " "

This comparison shows that the major share of the CSM in particulate particles does not only consist of PAH but also of aliphatics, sulfur and nitrogen compounds etc. If on the other hand a sample with more volatile and later recondensed compounds is collected, the PAH fraction in the CSM is considerably greater. In the example mentioned, the PAH fraction found is essentially determined by the tar-containing dust.

4.3 Conclusions

Taking the above-mentioned correlations into account, the total PAH concentration is a more suitable criterion than the CSM values for assessing a health hazard from exposure.

5. PAH Profiles

5.1 Basis

It was suggested in the literature[11-13]) that PAH measurements of the individual components in the form of PAH profiles, allows a better overview. The expediency of such a presentation is presented below.

5.2 Some examples

With a percentagewise presentation of the values (cf. Fig. 1, bar chart) different workplaces can be compared with one another as regards the distribution. Certain differences are then ascertained. A completely different picture is obtained in this case if the absolute values are plotted (solid and broken lines in Fig.[1]). The large differences observed are mainly due to the different absolute quantities. For assessing workplaces it is often important to know the absolute and relative distributions.

If only a few measurements e.g. three, are available, the reliability of this procedure is doubtful because the variations in the individual substances is considerable from one measurement to the next even in the same plant. Our measurements have shown that often 3 samples, taken one after the other, show rather homogeneous profiles (cf. Fig. 2).
The results of all samples have a different profile with a higher standard deviation. For some compounds, the mean values of the three individual measurements are outside the standard deviation of the total determination.

If one compares the profiles of different kinds of operations, for example paste plant and bake furnace in anode production, and Soederberg and prebake elecrolysis (Fig. 3), typical differences can be ascertained. The large differences in the phenanthrene are partially caused by temperature both as regards the emission at the workplace as well as in the actual sampling. The phenanthrene content is not typical for the type of plant. More characteristic differences can be observed for fluoranthene and pyrene, specially for benz(a)anthracene, chrysene and benz(b)fluoranthene.

Figure 1

Content and Percentage Composition of PAH at Two Different Workplaces in the Pate Plant

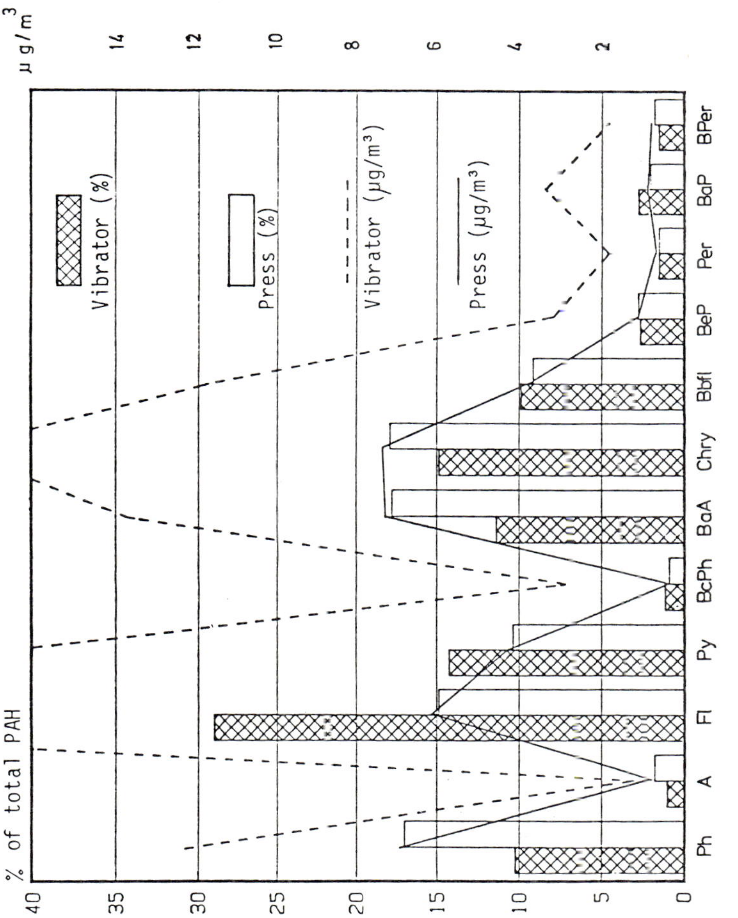

Figure 2

Profile and Standard Deviation of PAH in an Anode Baking Plant

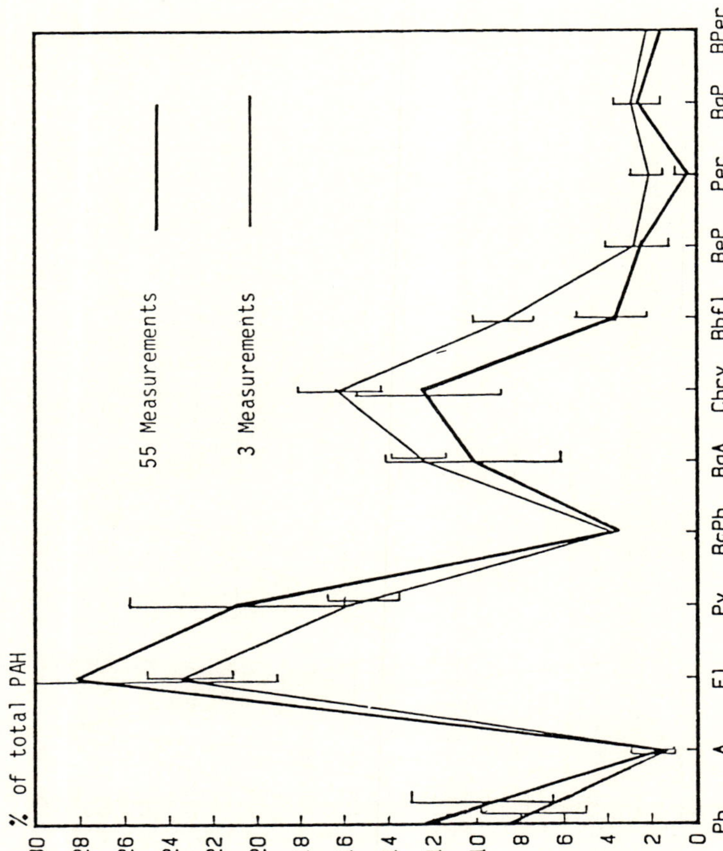

PAH Profile from Different Type of Plant

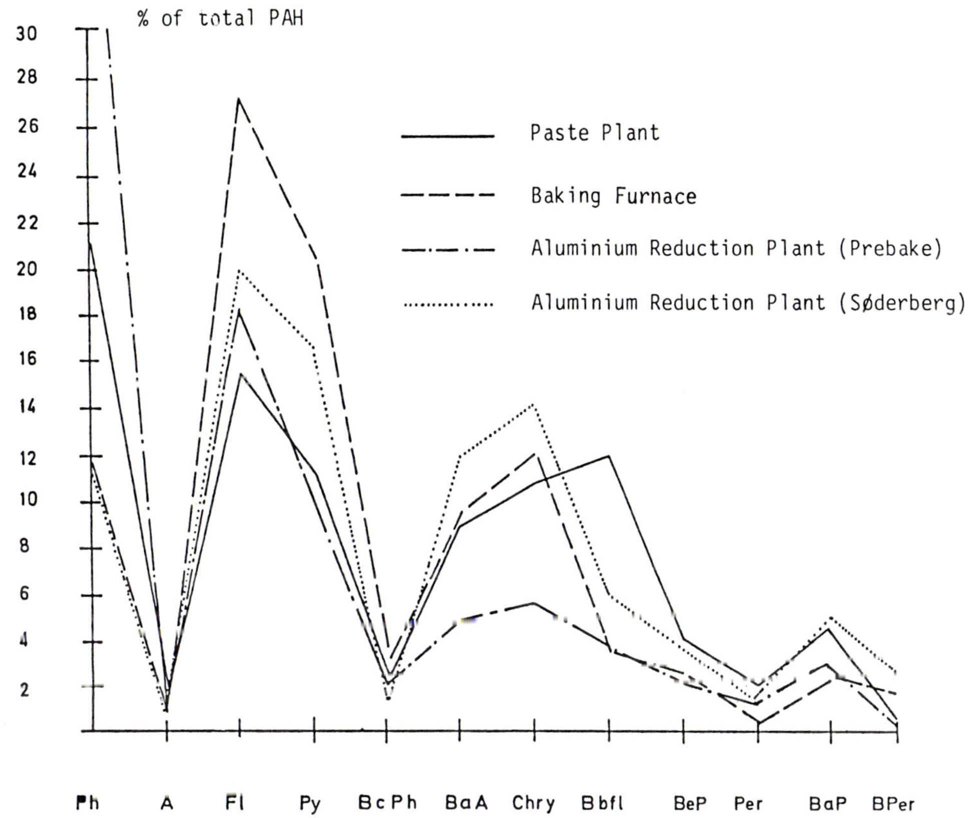

Figure 3

There are also marked differences as regards the benzo(a)pyrene content. For the other substances listed, the contents measured are not characteristic of the various operations.

5.3 Conclusions

The use of PAH profiles for assessing an exposure is only applicable to a limited extent. One important factor is the number of measurements. Only when more than 10 measurements are available, can the profile reliably be assumed to be correct.

For differentiating individual workplaces in a sector of a plant, the use of PAH profiles is often very difficult due to the influence of different sources of emission, the time spent at different places, the kind of work and the temperature and wind conditions.

Furthermore the question has to be asked whether the necessary high expense is worthwile in view of the modest results. A look at the overall PAH also reveals a useful criterion for assessment. Knowledge of the BaP content might be very useful additional information.

Bibliography:

1) A. Steinegger, Paper 1984 AIME Annual Meeting
Methods of Monitoring Occupational Exposure in Aluminium Smelters

2) Th. B. Bonney, Second IPAI Health Protection Seminar, Montreal 1981, International Primary Aluminium Institute, London (in press)
Measurement of Polycyclic Aromatic Hydrocarbons in The Aluminum Industry - North America

3) U.R. Stenberg and T. E. Alsberg, Anal. Chem. 53, 2067 (1981) Vacuum Sublimation and Solvent Extraction of Polycyclic Aromatic Compounds Adsorbed on Carbonaceous Materials

4) A. Bjørseth et al, Scand. J. Work Environ. & Health 4, 212 (78) PAH in Working Atmospheres. Determination of the PAH Content in an Aluminum Reduction Plant.

5) D.T. Koschani Staub-Reinhalt. Luft 43, 72 (8[illegible])
Determination and Recovery Rate of PAH in Samples with Diesel Soot Content

6) A. Liberti et al, Ann. Chim. 65, 573 (1975)
Comparative Determination of Polynuclear Hydrocarbons in Atmospheric Dust by Gas Liquid Chromatography and Spectrophotometry

7) W. Coutreels and K. Van Cauwenberghe, Water, Air and Soil Pollution 6, 103 (76) Extraction of Organic Compounds from Airborne Particulate Matter

8) A. Steinegger, R. Glaus, Light Metals 1981, 9[illegible]7
(Proceedings 110th AIME Meeting)
The Determination of PAH at the Workplace in Carbon Plants and Reduction Plants

9) A. Steinegger, Second IPAI Health Protection Seminar, Montreal 1981, International Primary Aluminium Institute, London (in press), Practical Experience of PAH in the Aluminium Industry; Switzerland

10) W.L. MacEachen et al Light Metals 1978, 509
(Proceedings 107th AIME Meeting)
Composition of Benzene Soluble Matter Collected at Aluminum Smelter Workplaces

11) G. Grimmer VDI Bericht 358 39/50 (1980)
Luftverunreinigungen durch polycyclische aromatische Kohlenwasserstoffe. Analytik und Vergleich der PAH-Profile aus Umweltproben

12) A. Bjørseth
Polynuclear Aromatic Hydrocarbons edited by P.W. Jones, Ann Arbor 1979 (p. 371)
Determination of PAH's in the Working Environments

13) H. Blome and K. Baus, Staub-Reinhalt. Luft 43, 367 (1983)
Konzentrationen polycyclischer aromatischer Kohlenwasserstoffe bei Herstellung und Verwendung von Pyrolyseprodukten aus organischem Material

ADDENDUM

Papers presented at the 112th AIME Annual Meeting in Atlanta, Georgia, March 6-10, 1983.

ADDENDUM

OVERVIEW OF ALCOA DIRECT REDUCTION PROCESS TECHNOLOGY

M. J. Bruno

ALUMINUM COMPANY OF AMERICA
Alcoa Technical Center
Alcoa Center, PA 15069

Summary

From 1977 to 1982, Alcoa and the Department of Energy conducted a cost sharing contract to determine the technical feasibility of producing Al and Al-Si alloy by carbothermic reduction of alumina-silica ores. Previously, Alcoa established reaction mechanisms and optimum silica to alumina ratios and had expertise in applicable refining technology for recovery of alloy and commercial grade aluminum. Three process concepts were evaluated--a combustion heated blast furnace, a combination blast-arc and a submerged arc. Burden preparation methods and formulations were developed for producing reactor feedstocks. Mathematical models were developed and utilized to calculate reactor heat and mass balances. The capabilities of alloy refining by fractional crystallization were defined in terms of limiting concentrations of key impurities. Recovery of commercial grade Al was demonstrated in diaphragm electrolytic cells at high current efficiencies and current densities for relatively low power consumption. Cost estimates and a profitability analysis were determined for a commercial Al-Si submerged arc process.

From 1977 to 1982, Alcoa and the Department of Energy conducted a cost sharing contract to determine the technical feasibility of producing aluminum and aluminum-silicon alloys by carbothermic reduction of alumina-silica ores (1). Potentially, a carbothermic process potentially provides a number of advantages over electrolytic processes, including reduced utilization of net energy and electrical energy, lower capital and operating costs, reduced dependence on imported materials such as petroleum oil and bauxite, greater flexibility in use of raw materials such as coal refuse or oil shale tailings, and improved environmental control.

The preferred, optimum process (Figure 1) was identified as an oxygen blown aluminum blast furnace to produce Al-Si-Fe alloys, followed by fractional crystallization to separate eutectic Al-Si alloy from ferro-silicon type compounds, and further purification of the eutectic alloy to commercial grade aluminum and silicon in a diaphragm transport cell.

Energy Requirements

Calculated energy requirements of various aluminum producing processes, shown in Table I, indicate the potential total net energy and electrical power savings for carbothermic reduction compared to the alternatives. The blast furnace case was for high iron bauxite, requiring a sub-halide type purification, since the Alcoa crystallization-diaphragm cell method was limited to low iron alloys. Alcoa purification was applied to the blast-arc and arc estimates since alumina was the aluminum source. The net thermal energies of the direct reduction processes are significantly lower than the Hall process. The net electrical energies of the process utilizing combustion of coke, blast furnace and blast-arc, are also the lowest. The importance of the by-product values from direct reduction are clearly indicated.

Table II summarizes similar calculations for production of Al-Si alloy with an 85-15 composition. Again, the carbothermic processes have great potential for reduction of net and electrical energies, taking credit for by-product values.

By-Products

The impact of by-products from direct reduction is shown in Table III for a commercial plant with a capacity of 273,000 annual metric tons of aluminum as metal or alloy. The blast furnace has the highest carbon to metal ratio and produces the most carbon monoxide. The CO energy equivalents are shown for burning to CO_2 or conversion to electrical power at 34% efficiency. A 1977 study by Battelle Laboratories predicted that industrial by-product CO for methanol synthesis will be competitive with conventional technology when the price of natural gas exceeds \$0.82/mm BTU (2). The study was based on highly concentrated CO, similar to the quality that would be generated from an aluminum direct reduction process.

Scope of Project

Prior to the contract, Alcoa had established the primary reduction reaction mechanisms and optimum silica to alumina ratios in the raw materials (3, 4). Also, Alcoa had experience in design and operation of crystallizers and diaphragm cells applied to certain aluminum alloy systems.

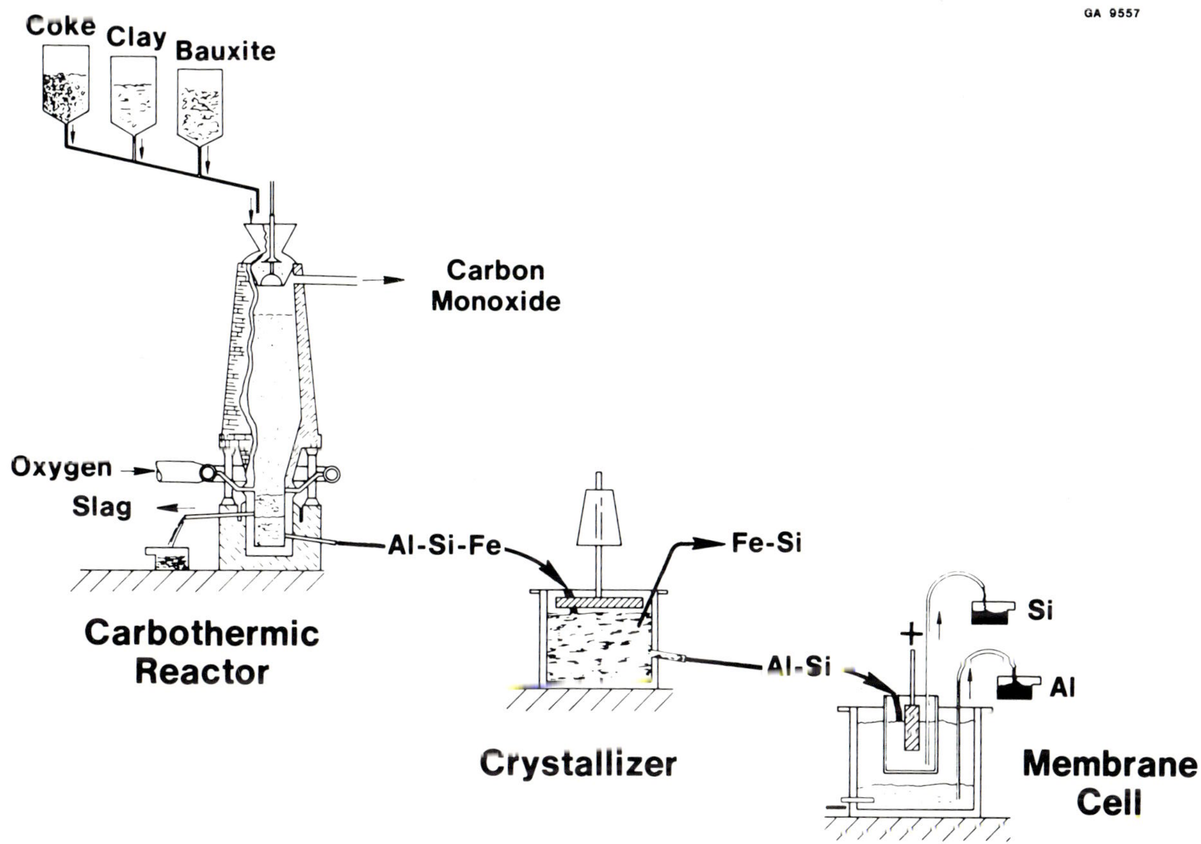

Figure 1. Blast Furnace Process Flow Sheet

TABLE I

PROCESS ENERGY REQUIREMENTS, mm BTU/TON ALUMINUM

Process	Gross Energy			By-Product Energy		Net	Net Elec
(Al Source)	Bayer	Smelting	Purification	CO	Intermetallics	Energy	kWh/lb
Blast Furnace (Bauxite-Clay)	0	321	56	203	69	105	2.0
Blast-Arc (Alumina-Silica)	42	229	30	75	104	122	3.5
Arc Furnace (Alumina-Clay)	29	206	30	28	104	133	5.7
Alcoa Chloride (Alumina)	42	130	0	0	0	172	4.5
Subhalide (Bauxite)	0	180	0	0	0	180	8.6
Hall (Alumina)	42	190	0	0	0	232	8.0

TABLE II

PROCESS ENERGY REQUIREMENTS, mm BTU/TON
ALUMINUM-SILICON (85-15) ALLOY

Process (Al-Si Source)	Gross Energy			By-Product Energy		Net Energy	Net Elec kWh/lb
	Bayer	Smelting	Purification	CO	Intermetallics		
Blast-Arc (Alumina-Silica)	36	194	6	64	71	101	2.7
Arc Furnace (Alumina-Clay)	25	175	6	24	71	111	4.5
Hall + Si Arc (Alumina + Silica)	36	157	0	0	0	255	7.8

TABLE III

BY-PRODUCT BALANCE FOR COMMERCIAL PLANT
(273,000 Tonnes/Year Al as Al or as 85% of Al-Si Alloy)

Process	Intermetallics tonnes/yr	CO, tonnes/yr	CO, SCF/yr	CO Energy MW	Equivalents BTU/yr
Blast Furnace	699,000 (77%Fe-19%Si-4%Al)	6.37×10^6	197×10^9	695	61×10^{12}
Blast-Arc	196,000 (85%Si-15%Al)	2.37×10^6	73×10^9	262	23×10^{12}
Arc Furnace	196,000 (85%Si-15%Al)	0.89×10^6	28×10^9	103	9×10^{12}

Under the contract, three basic carbothermic reduction concepts were evaluated: the combustion-heated blast furnace; variations of blast-arc systems using combustion heat for prereduction of silica to silicon carbide, followed by electric heat for the alloy producing step; and the submerged arc furnace. Two purification processes were evaluated: fractional crystallization of alloy product from reduction to recover Al-Si hypereutectic alloys and Al-Si-Fe-Ti intermetallics; separation and recovery of aluminum from hypereutectic alloy by electrolysis in a diaphragm transport cell, also producing a high Si-Al alloy for further processing to recover silicon. An economic study was made for the production of Al-Si alloy by the submerged arc process, including crystallizer refining to marketable eutectic alloy. Mathematical modeling of the reduction processes and fundamental phase equilibria studies on the Al-Si-O-C and Al-Si-C systems were also included in the contract.

Areas of Investigation

Technical progress on the Alcoa Direct Reduction Process was reported in a series of quarterly and annual reports from December 1977, through April 1981. The final report covering the entire contract period was submitted in March 1983. The following areas of investigation will be briefly summarized:

1. Burden Preparation
2. Blast Furnace
3. Blast-Arc Reactor
4. Submerged Arc Furnace
5. Fractional Crystallization
6. Diaphragm Transport Cell
7. Phase Equilibria
8. Mathematical Modeling
9. Process Economics

Burden Preparation

Most of the burden for reactor operations was 3/8 inch x 6 mesh pellets made in a pilot sized drum pelletizer. Typical burden formulations and properties are shown in Table IV. Average compressive strength for both bauxite and alumina hydrate burdens containing kaolin clay binders was 2.2 mega pascals (319 psi); abrasion resistance measured as percent breakdown through a 6 mesh screen was 4 percent for a Coal Tumbler Test, which translated to 11 percent for an Iron Ore Pellet Tumbler Test - indicating adequate resistance for a shaft reactor process based on the iron blast furnace.

TABLE IV

BURDEN PREPARATION: DRUM PELLETIZER

Bauxite, %	20	0
Alumina Hydrate, %	0	30
Kaolin Clay, %	58	48
Met. Coke, %	22	22
Compressive Strength, MPa	2.2	2.2
Abrasion Resistance, % Breakdown*	---	4.1 [11.0]

*Routinely Measured by Coal Tumbler Test
[ ] is comparable value from Iron Ore Pellet Tumbler Test

The compressive strength of clay bonded pellets was derived from the formation of amorphous glass and mullite upon calcination. Bonding strength increased with calcination temperature, faster calcining rates and higher clay concentrations. Activated alumina was also used successfully as a binder for non-clay type pellets.

Blast Furnace Reduction

Both bench and pilot scale vertical shaft reactors were operated. Bench units, as shown in Figure 2, ranging from 63 to 102 mm inside diameter and up to 1118 mm high were externally heated by induction systems and were run either batch or semi-continuously. The pilot reactor was 210 mm inside diameter by 2400 mm high and was equipped with both a 100 kW staged induction coil for external heating and three water-cooled oxygen tuyeres for internal combustion heat (Figure 3). The pilot unit was operated semi-continuously as a blast furnace with alloy product cooled to solid ingots in the hearth.

It was determined that a low pressure combustion-heated Al-Si blast furnace was not feasible due to severe bridging and interruption of burden movement caused by total reflux of Al, Al_2O and SiO vapors. Experimental results were verified by thermodynamic calculations using revised thermochemical data for the activity of Al-Si alloy, the vapor pressure of Al_2O, and the phase and temperature of formation for Al_4O_4C.

The addition of iron improved operations by reducing reflux of volatiles and fuel rate requirements, and could result in a technically feasible combustion-heated process for dilute Al-Si alloy. However, an innovative refining method must be developed for separating the iron from the alloy.

Details of the research on the aluminum-silicon blast furnace were presented in another paper (5).

Blast-Arc Reduction

To overcome the bridging and reflux problems of the combustion-heated blast furnace, the blast-arc process was evaluated in bench scale and pilot units. In this process, the high temperature heat requirements of the metal producing phase were supplied by electrical energy in the form of a submerged arc operation. The pilot shaft reactor was modified

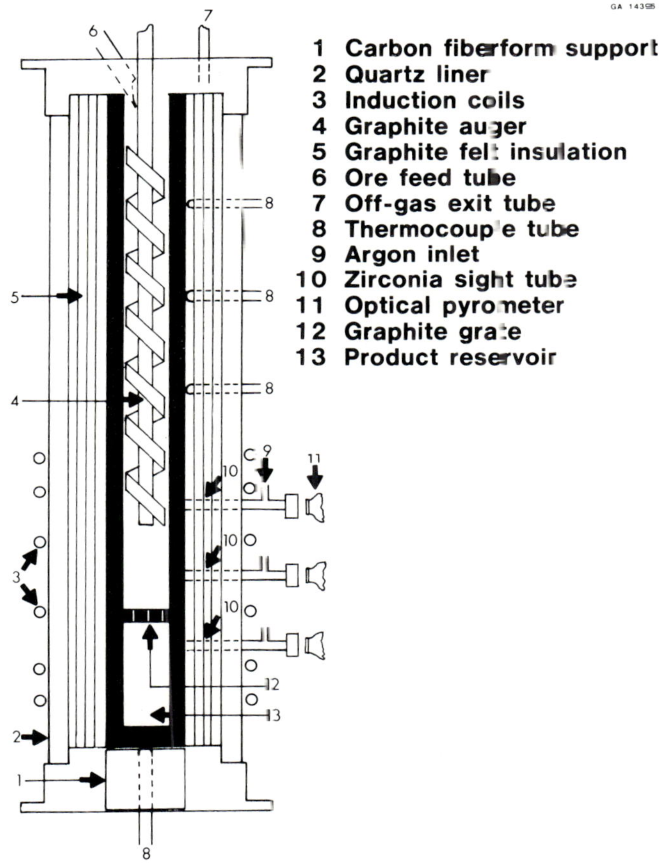

Figure 2. Bench Scale Vertical Shaft Reactor

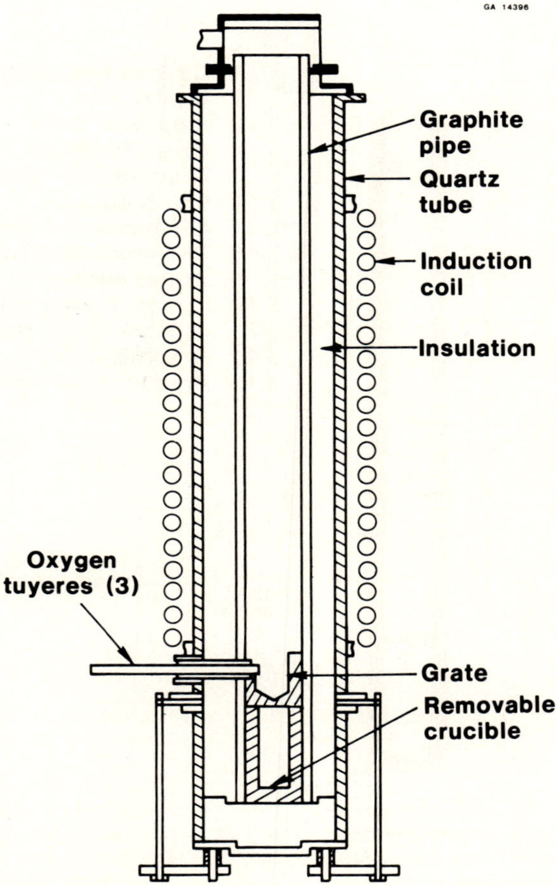

Figure 3. Pilot Scale Vertical Shaft Reactor

as shown in Figure 4, to discharge solid products, and was operated continuously as a prereducing step for silica to silicon carbide.

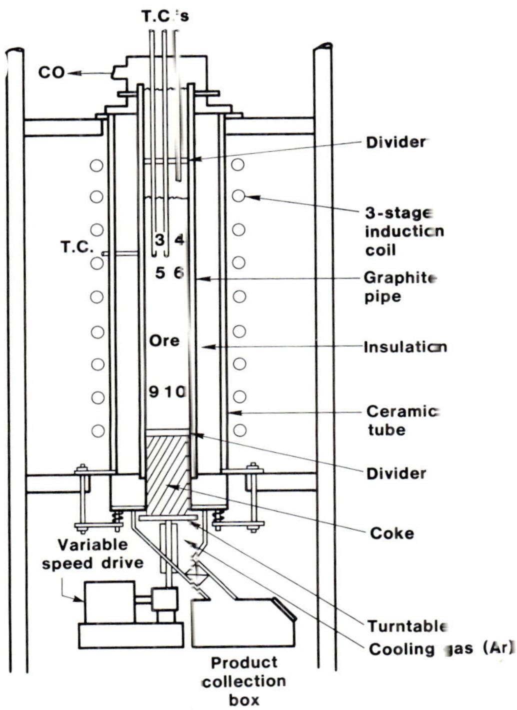

Figure 4. Vertical Shaft Reactor With Solids Discharge

An integral, one-step blast-arc reactor could potentially reduce the electrical energy requirement by up to 38% compared to the submerged arc furnace. However, engineering design of the integral process was complex. Also, the kinetics determined for silicon carbide formation required temperatures greater than 1700°C to attain practical conversion rates. At such temperatures, simultaneous reduction of alumina occurred, causing disruption of burden movement in the shaft from fusion products. Therefore, a two-step blast-arc process was evaluated, with silica and coke charged to the blast step and the resulting silicon carbide plus alumina containing ore and coke fed to the arc furnace. This process would require 32% less electrical energy than the submerged arc furnace.

Details of the aluminum-silicon blast-arc process are reported elsewhere (6).

Submerged Arc Furnace

A pilot scale submerged arc furnace was designed, built and operated continuously to evaluate carbothermic reduction of prereduced material for blast-arc as well as clay-alumina-coke burdens for total reduction to Al-Si alloys. Although straight arc furnace processing offered limited electrical energy gains, potential total net energy savings were significant. The pilot unit contained a cavity 200 mm wide by 400 mm long by 270 to 470 mm deep, and was powered by a 150 kVa A.C. supply through two 89 mm diameter electrodes. The furnace was designed to simulate one third of a six electrode three-phase system with each phase connected between a pair of electrodes (Figure 5).

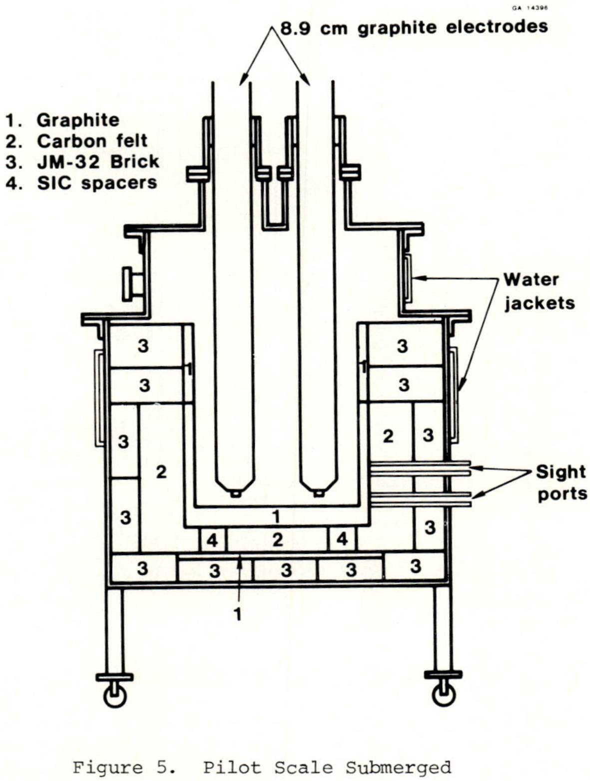

Figure 5. Pilot Scale Submerged Arc Furnace

Typical operating conditions and results are shown in Table V for the pilot submerged arc furnace on clay-alumina-coke burden. The unit was run at 1600 amps and 60 volts, with an average burden feed rate of 20 kg/hr. Approximately 16 kg of alloy was tapped during the six hours of continuous feeding. The product alloy contained 62% Al, 29% Si, 3.5% Fe, 2.1% Ti and 2.8% C.

TABLE V

PILOT ARC FURNACE OPERATIONS

Typical Burden:

47% Clay - 30% Alumina - 23% Met. Coke

-3/8 inch x 4 mesh pellets

2.6 MPa Compressive Strength

3.5% Breakdown, Abrasion Resistance

0.6 SiO_2/Al_2O_3

81% Theoretical Carbon

Operating Conditions:

Amps	1600	Alloy Tapped, kg	15.6
Volts	60	Run Time, Hrs.	5.7
Feed Rate, kg/hr	20		

Product Alloy:

62.2% Al	2.1% Ti
29.4% Si	2.8% C
3.5% Fe	

Production of Al-Si alloy was demonstrated in the arc furnace on clay-alumina-coke pellets containing 85-90% of theoretical carbon for optimum operation. Operation was smooth and controlled by careful attention to burden feed rates and adjusted carbon concentration during startup. Molten alloy with a carbon content of 1.3 to 3.3% was routinely tapped.

Arc smelting of prereduced burdens resulted in rapid formation of liquid slag, which inhibited smooth operation of the arc furnace. Further experimentation is required to develop control of slag formation rates through burden formulation and sizing.

Fractional Crystallization

The primary product from reduction is an Al-Si alloy containing 30 to 40 percent silicon and lesser amounts of iron and titanium. Upon cooling, the alloy partially solidifies, forming crystals of silicon and intermetallics of Fe-Si-Al-Ti. The concentration of silicon and other impurities in the melt are reduced, resulting in a useful casting alloy containing 12.6 (eutectic) to 17 percent silicon, less than one percent iron, and less than 0.1 percent titanium.

The operating principles are illustrated in Figure 6. Molten alloy from reduction is charged to the crystallizer and force cooled at the melt surface by jets of compressed air. Cooling and crystal formation are aided by tamping the crystal bed with a blade intermittently positioned to contact the entire surface. Silicon platelets and intermetallic crystals form as the melt temperature drops from about 1000°C to 590°C, the eutectic point for Al-Si. The eutectic alloy product is then drained through a bottom taphole to ingot molds. Part of the remaining solids are remelted through several temperature levels by sidewall induction coils, drained and added to the eutectic as usable casting alloy up to 17 percent cumulative silicon concentration. Liquid drained with higher silicon levels is recycled to the crystallizer, and the crystal bed remaining at the end of remelt is recovered for use as a reductant in iron or magnesium processing.

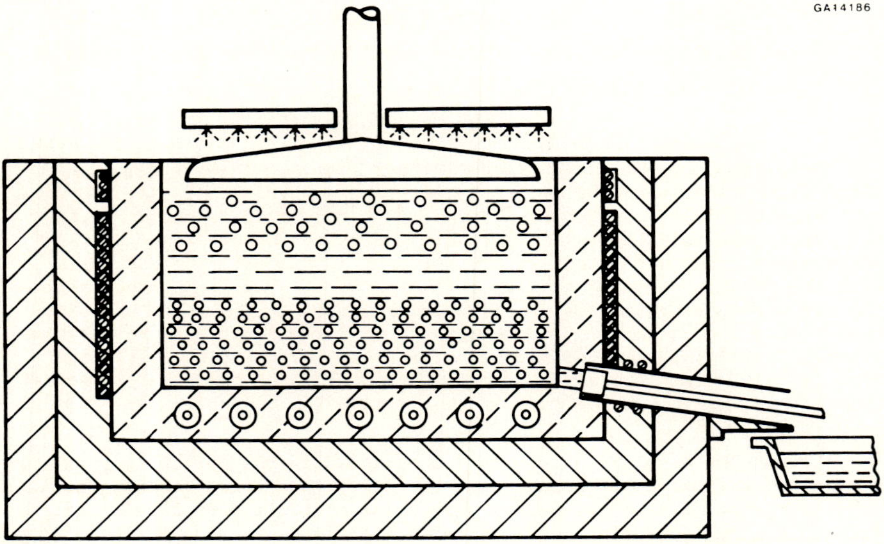

Figure 6. Pilot Scale Fractional Crystallizer

The fractional crystallization process was evaluated in 2 kg bench scale units and a 1000 kg pilot unit. The pilot crystallizer was 1020 mm inside diameter by 765 mm high. The sidewall induction heating system for remelt was 250 kW.

Due to the high aluminum content of the Fe-Si-Al intermetallics formed ($FeSiAl_4$ and $FeSi_2Al_5$), alloy purification by fractional crystallization was limited to processing alloys with iron at 3 percent or less. Even then, staged crystallization was required to attain acceptable casting alloy yields. Results of a two-stage pilot run are shown in Table VI.

TABLE VI

TWO-STAGE FRACTIONAL CRYSTALLIZATION

First Stage Inlet	64.8 Al - 31.2 Si - 3.2 Fe - 0.8 Ti
Second Stage Inlet	74.0 Al - 22.2 Si - 3.4 Fe - 0.4 Ti
Product Alloy	85.3 Al - 13.6 Si - 1.0 Fe - 0.1 Ti

Product Alloy Yield - 58%

Product Al Yield - 74%

An Al-Si direct reduction process would be complete after crystallization.

Diaphragm Transport Cell

Recovery of commercial purity aluminum from Al-Si alloy requires a third processing step--the electrolytic diaphragm transport cell. Similar to the principal of the Hoopes process, direct current is applied from the impure anodic alloy to the pure metal cathode through a molten salt electrolyte. The net effect is transport of aluminum to the cathode, with the impurities concentrating at the anode. Unlike the Hoopes process, the diaphragm cell utilizes a physical permeable barrier, such as porous carbon or graphite cloth, to separate the metallic electrodes rather than a thick layer of molten salt. The barrier facilitates a thin film of high conductivity electrolyte, i.e. a narrow anode-cathode (A-C) spacing, resulting in minimum power consumption (Figure 7).

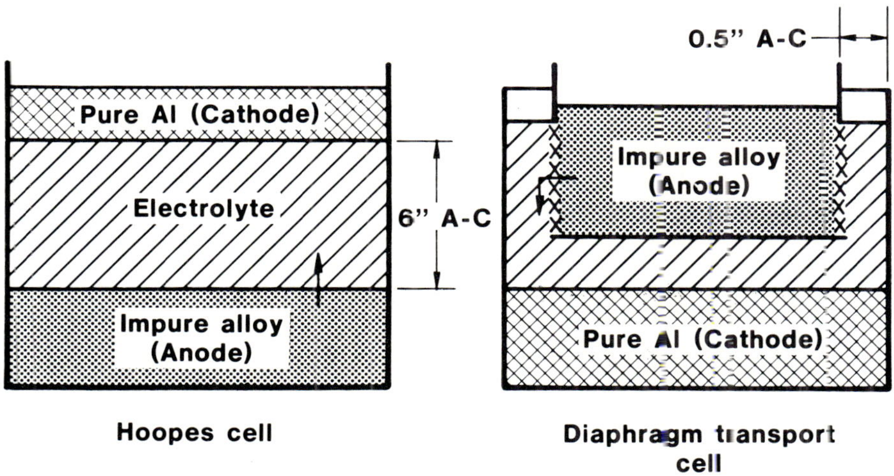

Figure 7. Electrolytic Transport Cells

Purification to commercial grade aluminum was evaluated in bench scale cells ranging in capacity from 0.8 kg to 6.4 kg of molten metal. Typical results for a graphite cloth diaphragm are shown in Table VII.

TABLE VII

DIAPHRAGM TRANSPORT CELL OPERATION

Graphite Cloth	0.5 mm Thick 90-95% Void Volume
Initial Alloy	84.2 Al - 14.1 Si - 1.5 Fe - 0.2 Ti
Product Purity	99.94% Al
Current Efficiency	97%
Current Density	3.3 A/cm^2 (20 A/in^2)
Power Consumption	5.5 kWh/kg Al (2.5 kWh/lb Al)

The major commercial advantages of the diaphragm cell are: high current densities, up to 40 $amps/in^2$ as compared to 2.5 for the Hoopes cell and 6.5 fcr Hall cells; power consumption as low as 1.9 kWh/lb Al compared to 7 for Hoopes and 6 to 8 for Hall; and high current efficiencies of greater than 95 percent.

Phase Equilibria

Fundamental phase equilibria experiments were conducted on the Al-Si-O-C and Al-Si-C systems. The former study indicated that SiC, when present in a ratio of one to two moles SiC per mole Al_2O_3, effectively isolates portions of the charge material from carbon, resulting in a carbon activity less than unity. Thus, aluminum carbides become unstable and Al-Si alloys can be produced at temperatures 100°C lower than reported for carbon saturated systems, starting at about 1980°C.

The tests on the Al-Si-C system indicated that Al_4SiC_4 does not have a high degree of stability above 2000°C. The estimated appearance of the Al-Si-C ternary diagram at 2100°C is shown in Figure 8. The empirically determined equilibrium composition of Al-Si alloy with carbon at 2185°C is 59 Al - 36 Si - 5 C.

Mathematical Models

In addition to the experimental programs on reduction and purification, a parallel effort was made to develop mathematical models for gas/solid countercurrent shaft reactors. Models based both on chemical equilibrium and on chemical kinetics limited by equilibrium were developed (7). They describe a one-dimensional reactor in which height is determined for a given unit cross section. Plug flow of gas upward and condensed phases downward are assumed, thus avoiding the necessity of detailed three-dimensional flow modeling. Provision is made for the introduction of thermal energy directly (as with an electric arc) or indirectly by combustion (as in a blast furnace).

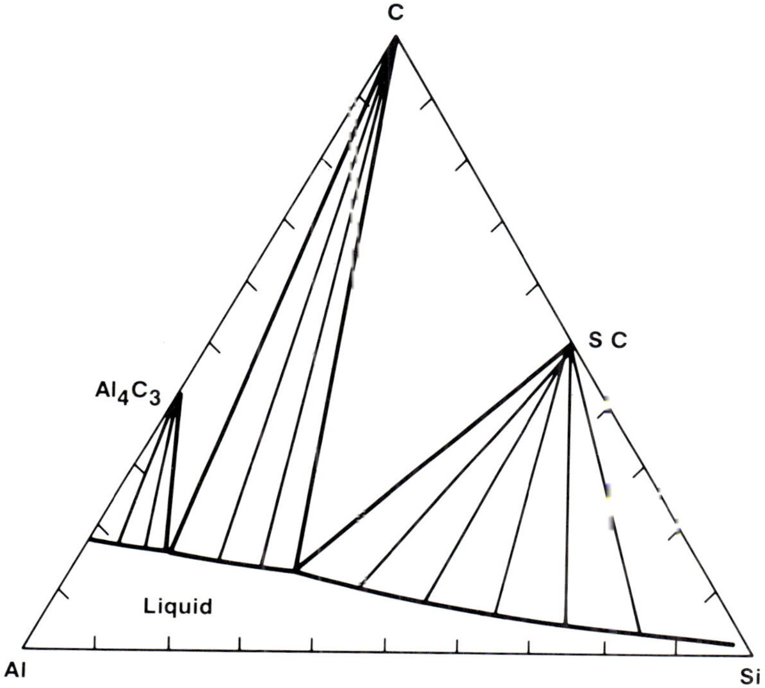

Figure 8. Al-Si-C Estimated Phase Diagram

Capabilities of the heat and mass balance models are indicated in Table VIII.

TABLE VIII

SHAFT REACTOR MODEL CAPABILITIES

	Heat Transfer	Kinetics	Recycle
FLWEQUIL	X		
FLOWREAC	X	X	
SOLGASMIX-REACTOR	X	X	X
RCYCRERC	X	X	X

The models are useful in predicting process heat and mass balances as well as effects of operating conditions on recycle loads, energy requirements and reaction temperatures. Impurity concentrations can be traced and the interactions between heat requirements, feed requirements, and impurities determined. A large number of side reactions can be considered.

Process Economics

The models were utilized to calculate heat and mass balances for conceptual commercial scale submerged arc processes. The calculations, for both closed and semi-closed furnaces, were the basis for cost estimates of Al-Si alloy production. Results are shown in Table IX. For a 353,000 ton plant, the closed furnace process capital cost estimate was $2235 per annual ton of 85 Al - 15 Si, 6 percent lower than the semi-closed process. Closed furnace operating costs were estimated as $.81 per pound of alloy, assuming a power cost of $.06 per kWh and a by-product credit of $.42 per pound of Al-Si-Fe intermetallic.

TABLE IX

Al-Si ARC FURNACE COST ESTIMATES (1981 Dollars)

Plant Capacity - 353,000 tons/yr
85 Al - 15 Si Alloy

	Closed Furnace	Semi-Closed Furnace
Capital Cost	$2235/ton-year	$2366/ton-year
Cash Cost @ $.06/kWh	$1.01/lb Al-Si Alloy	--
Net Mfg. Cost, Credit		--
For CO and Inter metallics	$.81/lb Al-Si Alloy	--

The discounted cash flow rate of return (DCF-ROR) ranged from 6 percent at $.06 per kWh power cost to 12 percent at $.04 per kWh based on a 1981 selling price of $.95 per pound of Al-Si alloy. A 15 percent DCF-ROR would require an alloy selling price of $1.09 per pound in the best case calculated (Figure 9).

Summary

In summary, results of research and development carried out under the contract have indicated that a low pressure, low iron blast furnace is not feasible; that a significant portion of the process energy requirements must be supplied by electrical power; that an integral blast-arc process using conventional electrodes will be very difficult to engineer; that a two-step blast-arc process reduces the engineering complexities at the expense of some power savings; and that a submerged arc process is technically feasible but offers no significant power or cost advantages for producing aluminum. It has been speculated that a high iron blast furnace, possibly at high pressures, may overcome the operating difficulties related to volatility but will certainly necessitate innovative purification technology to economically separate and recover the metal products. It has been empirically shown that prereduction of silica to

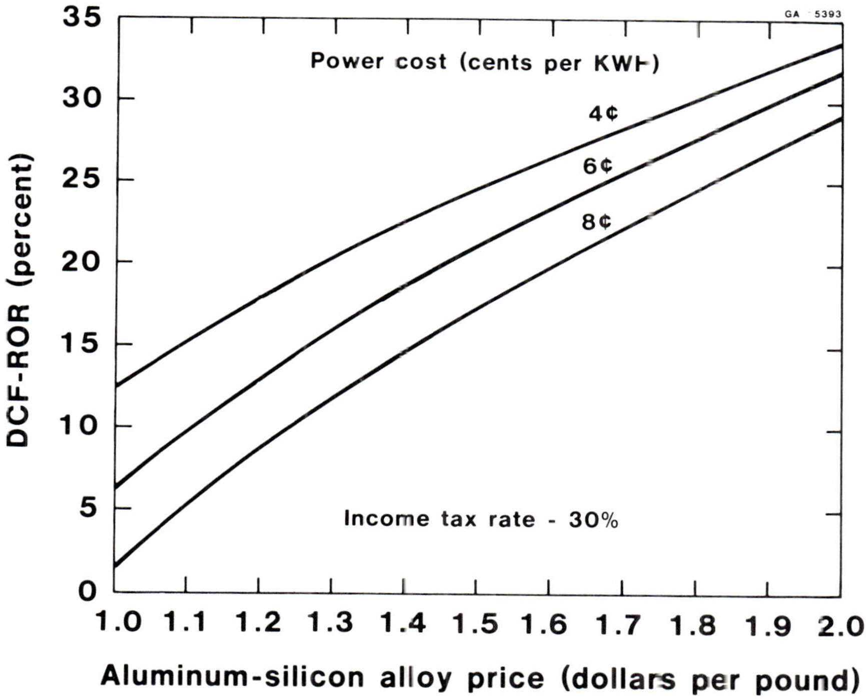

Figure 9. Profitability Analysis for Closed Arc Furnace

silicon carbide in a blast-arc concept will require control of reaction kinetics for slag formation. Therefore, a major technical effort remains before a commercial direct reduction process can become a reality, not even considering the challenge of economically competing with alternative aluminum processes.

Acknowledgment

During the contract period many engineers, scientists, and technicians contributed to the research and development effort on the direct reduction processes. In addition to the Alcoa staff, the following subcontractors, consultants, and monitors should be recognized: Koppers Company, Incorporated, H.F. Leonard, Project Consultant; Professors W.O. Philbrook, G.K. Sigworth, and graduate student M. Riley, Carnegie Mellon University; Dr. J.C. Agarwal, Charles River Associates, Incorporated; Professor Julian Szekely, MIT; Professor J.F. Elliott, MIT; Professors Ketil Motzfeldt and T.A. Engh, University of Trondheim. Also, the Department of Energy technical project managers provided support and encouragement throughout the program.

REFERENCES

1. M.J. Bruno, Production of Aluminum-Silicon Alloy and Ferrosilicon and Commercial Purity Aluminum by the Direct Reduction Process, Final Technical Report, CONS-5089-16, UC-95f, February 1983.

2. C.A. Rohrmann, et al, Chemical Production From Waste Carbon Monoxide - Its Potential for Energy Conservation, BNWL-2137, UC-95f, November 1977.

3. U.S. Patent No. 4,046,558 (September 1977), S.K. Das and R.A. Mileto, "Method for the Production of Aluminum-Silicon Alloys".

4. U.S. Patent No. 4,053,303 (October 1977). C.N. Cochran, et al, "Method of Carbothermically Producing Aluminum-Silicon Alloys".

5. R.L. Troup, "Feasibility of an Aluminum-Silicon Blast Furnace", presented at the 112th AIME annual meeting, Atlanta, GA, March 9, 1983.

6. D.T. Stevenson, "Feasibility of an Aluminum-Silicon Blast-Arc Process", presented at the 112th AIME annual meeting, Atlanta, GA, March 9, 1983.

7. W.E. Wahnsiedler, et al, "Mathematical Modeling of Countercurrent Shaft Reactors", Light Metals 1982, pps. 405-431.

FEASIBILITY OF AN ALUMINUM-SILICON BLAST FURNACE

R. L. Troup

ALUMINUM COMPANY OF AMERICA
Alcoa Technical Center
Alcoa Center, PA 15069

Summary

Alcoa and DOE jointly sponsored a multi-year project to develop an energy-efficient alternative to the Hall Process. The new process involved carbothermic reduction of alumina-silica ores in a combustion-heated reactor followed by alloy refining and purification to commercial grade aluminum. Prior to the joint project, batches of Al-Si alloy were produced using external heat. A thermochemical model indicated that combustion heat might be feasible. Combustion heating was simulated by sweeping excess CO through an externally heated shaft reactor. At CO rates well below those required by a self-heated process, severe bridging occurred due to total reflux of Al_2O, Al(v), and SiO. Combustion with O_2 in a larger shaft reactor failed to produce alloy. These experimental results are consistent with recent thermochemical data and revised assumptions. It may be possible to produce a dilute aluminum alloy with combustion heat, but feasibility requires an innovative aluminum collection process.

Introduction

While aluminum companies are vertically integrated to a substantial degree, they are vulnerable to outside forces which could affect cost and availability of bauxite, petroleum coke and coal tar pitch for anodes, and electric power. One strategy for increasing security is to develop an alternative reduction process based on domestic ores with coal as the reductant and energy source. A carbothermic reduction process in which alumina-silica ores are reduced with coal or coke satisfies the materials availability criteria. But, up to this time, all demonstrated carbothermic processes have used electricity--now an expensive, embattled form of energy. A process based on a combustion-heated, shaft reactor, similar to an iron blast furnace, would satisfy the energy criteria provided the energy balance was favorable.

In 1977, Aluminum Company of America (Alcoa) and the U.S. Department of Energy (DOE) entered into a contract to determine the feasibility of a three-step process for producing aluminum by carbothermic reduction. The three steps involved production of an Al-Si alloy with combustion heat in a shaft reactor followed by fractional crystallization and electrolytic purification. Prior to the contract, small batches of Al-Si alloy had been produced in an electrically heated, bench scale reactor. An optimum range for the SiO_2/Al_2O_3 ratio in the ore had been established, and it was found that the various constituents of the ore should be finely ground and pelletized with carbon for reduction. Three primary reaction stages were identified by CO evolution.

$$\text{I:}\quad 3SiO_2 + 9C = 3SiC + 6CO$$

$$\text{II:}\quad 2Al_2O_3 + 3C = Al_4O_4C + 2CO$$

$$\text{III:}\quad Al_4O_4C + 3SiC = 4\underline{Al} + 3\underline{Si} + 4CO$$

A thermochemical model (1) was used to calculate staged equilibrium heat and material balances for a process heated by combustion of carbon with oxygen. The flow of reactants between stages is shown schematically in Figure 1. Alloy yield was predicted when 100% of the heat was supplied by combustion. The model also predicted a high reflux ratio but could not predict whether the process could be operated at such a high ratio, or whether reoxidation and carbide formation could be avoided.

Blast Furnace Processes

The internal structure of an iron blast furnace is shown in Figure 2 (2). While there is a great deal known about the iron blast furnace, there are a number of significant differences between it and a hypothetical Al-Si blast furnace. Temperatures required for reduction to liquid Fe are about 1500°C, while Al-Si requires 2000-2100°C. The reduction of Al_2O_3 and SiO_2 is much more endothermic than the reduction of FeO. While CO can reduce Fe_2O_3, Fe_3O_4, and FeO, it cannot reduce Al_2O_3 or SiO_2. The reduction to Al-Si takes place via gaseous intermediates; relatively high partial pressures of SiO, Al(v) and Al_2O exist at the high reduction temperatures. These species reflux by condensing as solids, primarily Al_2O_3 and SiC, and must be returned to the higher temperature zone by solids flow. The condensation of these solid phases can act as a cement between particles. The condensation also liberates heat so that the reactor behaves as a heat pipe, removing heat from the reaction zone. At

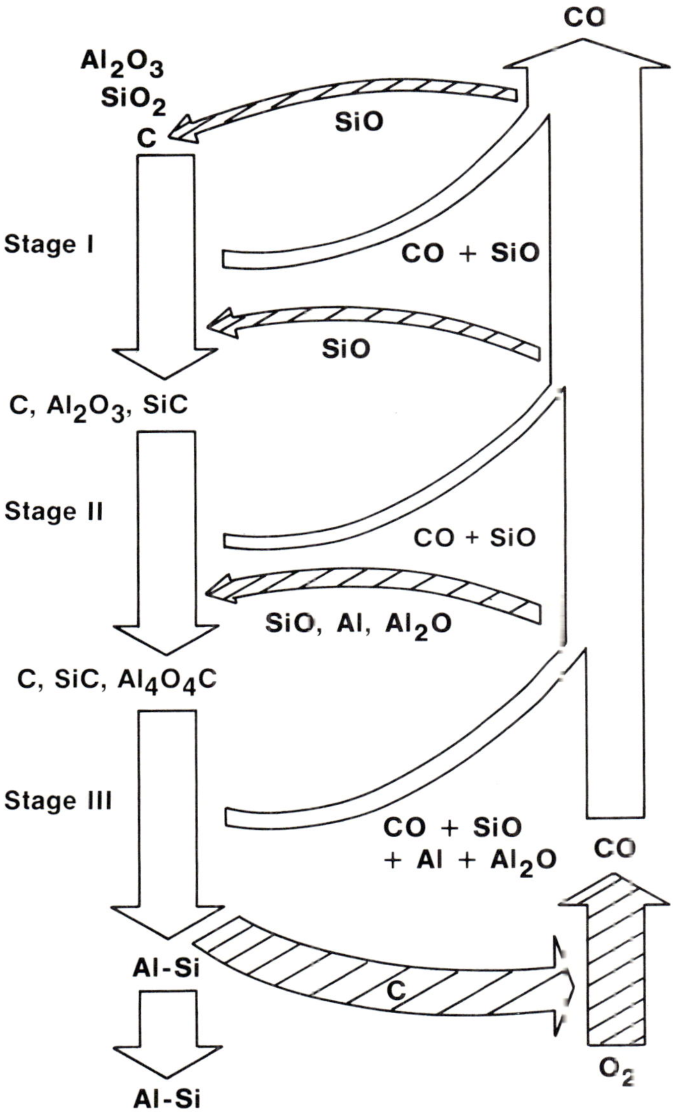

Refluxing Mass Balance

Figure 1

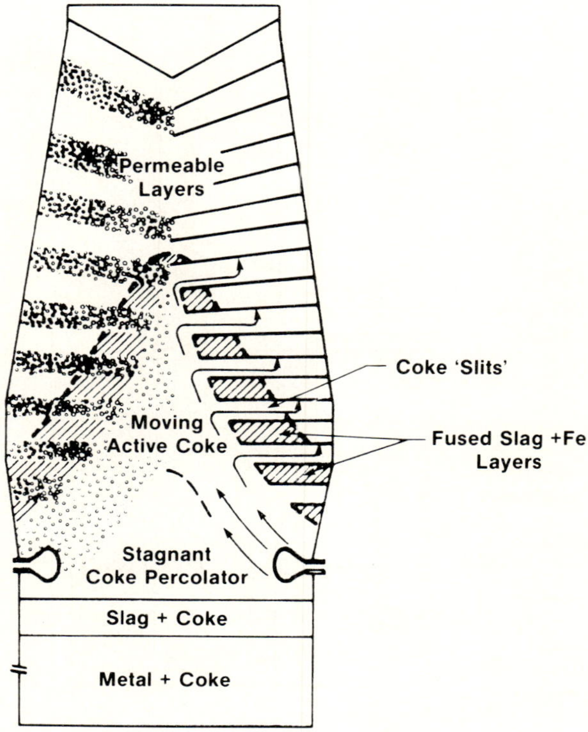

Sketch of the Internal Structure of a Large Blast Furnace (Nakamura, 1978)

Figure 2

1500°C, Fe alloy containing about 4.5% C is in equilibrium with excess carbon, while Al-Si alloy is in equilibrium only with carbides below their decomposition temperatures. Previous attempts to make ferrosilicon in blast furnaces encountered severe problems above 20% Si.

Pilot Scale: The Raceway

The initial phase of pilot scale tests was directed at the raceway zone where combustion occurs. The primary goals were: (1) develop a tuyere design which could withstand the high temperature oxidizing/reducing environment; (2) reach alloy formation temperature with coke-O_2 combustion; (3) determine optimum tuyere size, coke size, and O_2 flow rate; and, (4) react batches of ore in the vicinity of the raceway. The first pilot reactor is shown in Figure 3. The reactor vessel was constructed from a set of carbon-bonded SiC crucibles. Between the inner 19 cm I.D. crucible and an outer crucible, the annular gap was filled with alumina for insulation. To further reduce heat loss, the Crucible Reactor was located inside a gas-fired furnace operated at 1200°C. A perforated

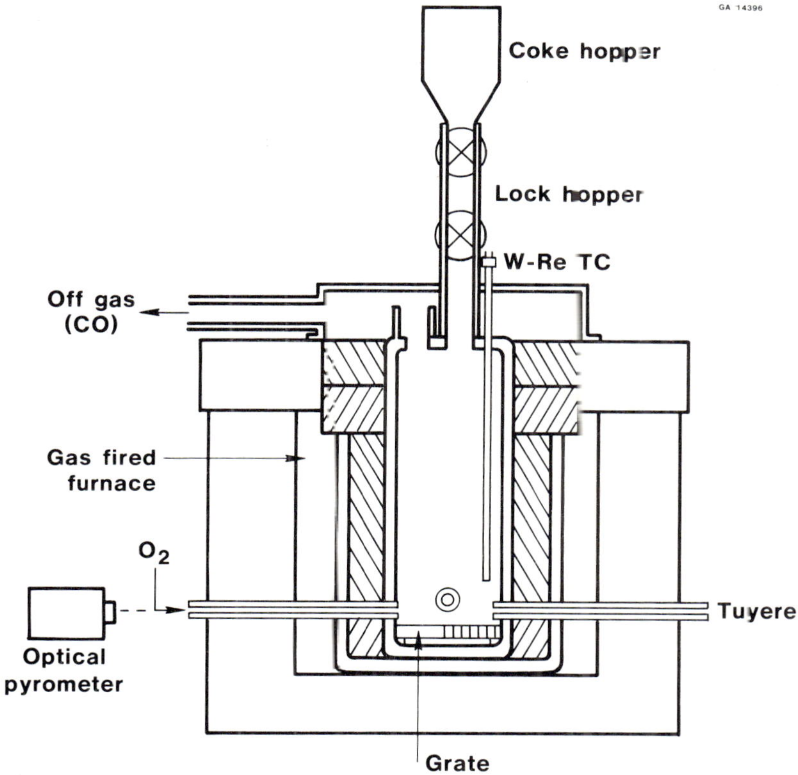

Crucible Reactor with 3 Side Entry O_2 Tuyeres

Figure 3

graphite plate spaced about 2 cm above the inner crucible bottom acted as a grate. Three tuyeres, spaced 120°C apart, were located just above the grate. During a run, coke was fed from a hopper through a double ball valve feeder.

When attempts to develop a ceramic oxygen tuyere were unsuccessful, a water-cooled design, shown in Figure 4, was adopted. The tuyere was fabricated from three concentric stainless steel tubes welded to a machined stainless steel tip. Oxygen was blown through the inner 4.6 mm tube and discharged from a diverging 9.5 mm nozzle. The design proved troublefree in terms of integrity, but the tuyeres acted as major heat sinks.

Initial combustion tests were terminated by bridging of the coke bed which caused the raceway to expand and consume the reactor. To control bridging, petroleum coke was used instead of higher-ash metallurgical coke. It was also necessary to use a graphite liner, 15 cm I.D., which extended from the grate to about 5 cm above the tuyere. This liner protected the crucible wall which had been a source of SiO.

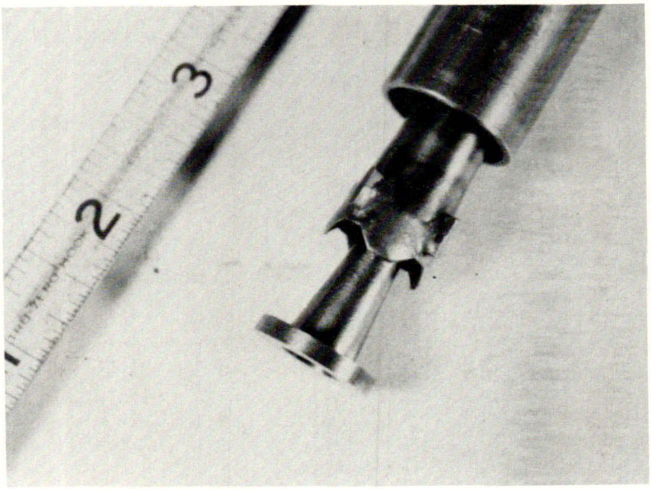

WATER COOLED O_2 TUYERE
(Scale in Inches)

Figure 4

To determine the size of the combustion zone, the burn-off pattern of vertical graphite rods was observed. The rods were set in the grate and spaced about 2 cm apart along a diameter extending from the axis of each tuyere (Figure 5). Based on burn-off patterns, a range of coke sizes and O_2 flow rates was established which gave reasonable combustion zone sizes. During a typical experiment, nearly complete burn-off occurred just in front of the tuyeres while little, if any, burn-off occurred between the center of the reactor and the far wall (Figure 6).

Pilot Scale: Smelting of Ore - Indirect Heating

In the initial combustion-heated smelting runs, ore was heated indirectly by placing a barrier between the ore and the combustion zone. In these runs, 20 mm I.D. graphite tubes containing agglomerated ore were placed in locations adjacent to the raceways. The tubes isolated the ore from combustion CO to reduce volatility and from excess carbon to reduce carbide formation. The ore was typically formulated from low iron bauxite, clay, and metallurgical coke with SiO_2/Al_2O_3 = 0.7. The raw materials were briquetted and crushed into -3/8 inch + 1/4 inch particles.

Oxygen was blown into a bed of petroleum coke at flow rates of 4.2 m^3/hr (150 cfh) to 12.6 m^3/hr (450 cfh) in a series of runs. The optimum O_2 flow rate and coke size were 4.6-5.0 m^3/hr (165-180 cfh) and -3/8 inch x + 4 mesh, respectively. A mid-radius position between the tuyere centerlines was the optimum ore tube location. At these locations, the ore tubes survived the reactor conditions. Although tubes were partially burned under non-optimal conditions, metallic product was found in each run (Figure 7).

LAYOUT OF GRAPHITE RODS IN CRUCIBLE REACTOR
BEFORE COMBUSTION RUN

Figure 5

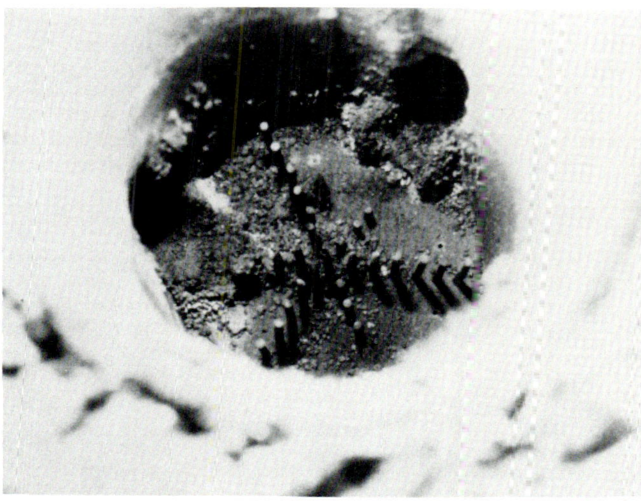

BURN-OFF PATTERN OF GRAPHITE RODS IN CRUCIBLE
REACTOR AFTER COMBUSTION RUN

Figure 6

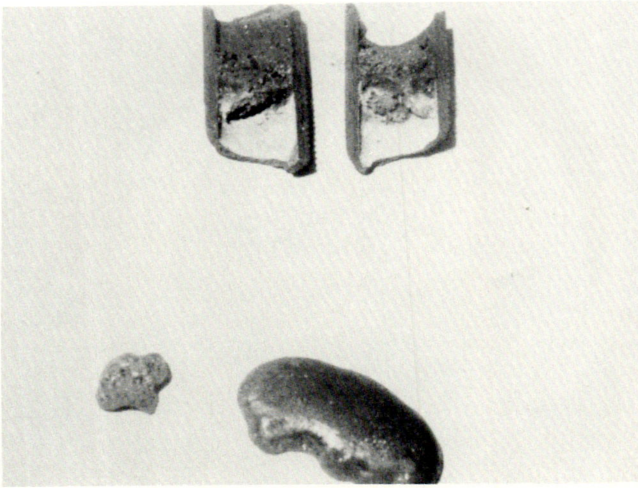

METALLIC PRODUCT FROM COMBUSTION RUN IN CRUCIBLE REACTOR USING ORE TUBES

Figure 7

In two runs, two long ore tubes with sealed bottoms were extended through the furnace top and were connected to a wet test meter. This made it possible to track the course of the reactions by CO evolution. In the first run, CO evolution from SiC formation overlapped the CO evolution from Al_4O_4C and alloy formation. Therefore, in a second run, a powdered charge of SiC, Al_2O_3, and C was used to eliminate CO from SiC formation. Carbon monoxide evolution indicated that significant reduction to Al_4O_4C and Al-Si alloy was occurring. A coalesced product resulted in both runs. Considerable amounts of refluxed material condensed on the charge in all ore tube runs, and ore bridged the tubes.

Microprobe analyses of coalesced product from two runs made at nearly optimal conditions are shown in Table I. Phases containing Al-Si-Fe and Al-Si-C were identified. Although the theoretical iron content based on ore composition was about 5%, the noncarbide phases contained significant amounts of iron due to the volatility of Al and Si species.

Pilot Scale: Smelting of Ore - Direct Heating

The Crucible Reactor was limited by the height and temperature of the furnace chamber. A new unit, the Vertical Shaft Reactor (VSR), shown schematically in Figure 8, was designed to overcome those limitations. The VSR was heated by a 160 cm high induction coil with three separately controlled zones powered by a 100 kW motor generator. This was sufficient power to reach 2100°C. The outer shell of the VSR was a nominal 510 mm O.D., 460 mm I.D., 2000 mm high glazed ceramic pipe. The susceptor, which also served as the reactor lining, was a 250 mm O.D., 210 mm I.D., 2400 mm high graphite pipe. The susceptor was wrapped with about 50-75

TABLE I

Short Ore Tube
Coalesced Material in Graphite Tube

Appearance	Probable Phase	%	Al	Si	Fe	C	Total
Dark Gray	Si-Al-C	wt	6.9	66.0	--	31.7	105
		at	4.8	44.8	--	50.3	100
Gray	Al-Fe-Si	wt	41.6	21.3	40.6	--	104
		at	50.9	25.1	24.0	--	100
Gray	Al-Fe-Si	wt	21.4	27.6	56.2	--	105
		at	28.5	35.3	36.2	--	100

Long Ore Tube
Coalesced Material in Graphite Tube

Appearance	Probable Phase	%	Al	Si	Fe	C	Total
Dark Gray	(Al-Si-C) (plus possible) (Ti)	wt	48.5	19.1	--	20.1	88
		at	43.3	16.4	--	40.3	100
Gray	Al-Fe-Si	wt	43.6	32.0	30.4	--	106
		at	49.0	34.6	16.5	--	100

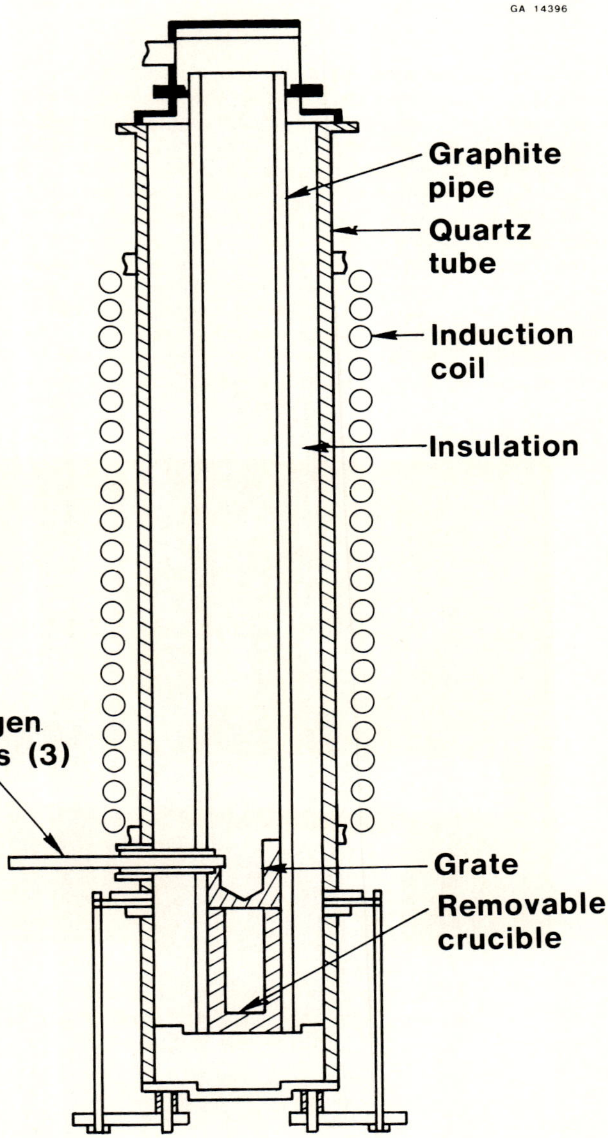

Schematic of Vertical Shaft Reactor

Figure 8

mm of graphite or carbon felt and 25-50 mm of alumina-silica felt insulation. The three tuyeres from the Crucible Reactor were reused in the VSR.

After a series of runs to determine a satisfactory combination of coke size and O_2 flow rate in the VSR, two runs were made in an attempt to produce alloy by a combination of direct heating and combustion. VSR-8 was started on a bed of petroleum coke and, after establishing bed movement at an O_2 flow rate of 2.5 m^3/hr (88 cfh) per tuyere, a constant ratio of ore-to-metallurgical coke equal to 0.6 was charged. The coke and ore, which consisted of extrudates containing 10% Fe_2O_3, were charged in alternating layers. Problems with bed depth measurement, due to a bridge, resulted in a final bed depth of only 90 mm above the tuyeres. This low bed height prevented further bridging since volatiles tended to blow out of the bed. About 1200 gm of coalesced product consisting of metal, slag, and embedded coke had solidified around the graphite pegs above the grate (Figure 9). Apparently, the zone above the grate was too cold for drainage of alloy and the product flooded the tuyeres.

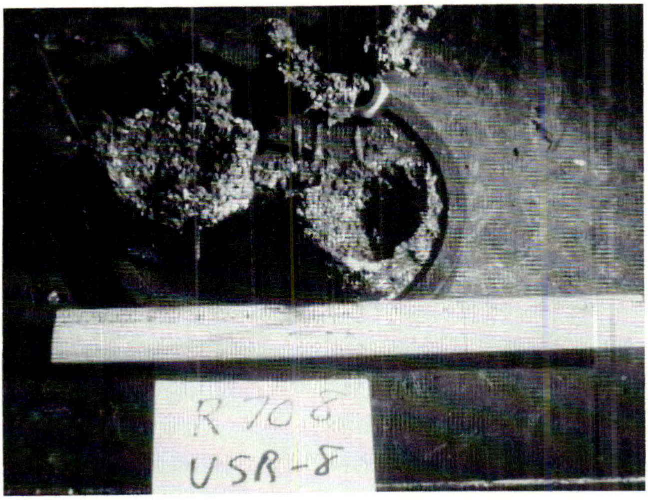

COALESCED PRODUCT CONSISTING OF METAL, SLAG AND EMBEDDED COKE FROM VSR COMBUSTION RUN
(Scale in Inches)

Figure 9

In VSR-9, better control of bed height was obtained. Steady-state bed movement was achieved on a petroleum coke start-up bed with 16 kW input to maintain a 2000°C wall temperature. Addition of ore required 35 kW to maintain temperature. About 760 gm of coalesced product were found on the grate. A bridge was found at the top of the bed.

Product samples from the two runs were remelted and liquid was separated from unreacted solids. The composition of the samples before and after remelt are shown in Table II. The alloy product collected below the grate was FeSi with 7-8% Ti plus C and 2-14% aluminum. The low aluminum content indicated sensitivity of the alloy formation to CO sweep and/or carbide formation.

TABLE II

Remelt Analysis, %Wt

Run	Sample	Al	Si	Fe	Ti	C
VSR-1-8	Before Remelt	21.9	15.7	28.7	5.42	12.70
VSR-1-8	Above Grate After	12.2	20.8	31.8	10.20	20.00
VSR-1-8	Below Grate After	14.5	24.5	50.2	3.47	4.32
VSR-1-9	Before Remelt	1.78	25.5	41.0	5.21	23.40
VSR-1-9	Above Grate After	1.65	26.0	34.7	5.50	29.40
VSR-1-9	Below Grate After	2.61	23.7	59.3	3.94	3.05

Then the reactor was modified by moving the grate up to a location between the middle and bottom induction coil in an attempt to reduce heat loss. An auger was installed to maintain bed movement. The auger was essentially a double-fluted, right hand screw and was rotated clockwise. It was designed to lift the bed near the shaft and spill it into the annulus between the auger and the wall. By keeping bed particles in constant motion, and by increasing the void fraction through lifting, it was possible to accumulate condensed material in the bed without causing agglomerates. The auger was constructed of graphite with a 64 mm diameter shaft and 130 mm outer diameter.

In the next series of runs, a constant ore:coke ratio of 1:1 was charged after establishing bed movement on a 100% coke bed. The runs were all terminated when either the auger shear pin broke or a bridge formed below or within the auger and charging could not be continued.

Figure 10 shows the results of a typical run at an oxygen rate of 3.4 m^3/hr (120 cfh). The bridge and a mass of agglomerates in front of the tuyeres contained SiC, Al_2O_3, C (graphite) and Al_4SiC_4. In some runs Al_4O_4C was found in the bridge. The only alloy collected at the grate was a thin metallic coating on the graphite lumps which covered the grate. This was the same location where considerably larger accumulations were found in VSR-8 and VSR-9 with high iron charges.

Effect of Excess Carbon

Assuming that alloy is produced, the presence of excess carbon in the form of combustion coke introduces the possibility of forming carbides (Al_4C_3, Al_4SiC_4, and SiC). Since a small amount of Al_4SiC_4 was obtained in most combustion runs, it appears that the kinetics of carbiding are sufficiently fast that some alloy formed carbides before it could vaporize. Schmitt and Wittner (3) reported that carbide formation could be minimized by removing the alloy from the reduction zone as quickly as possible. In order to investigate the drainage of alloy through, and reaction with layers of carbon-containing materials, a batch run was made. The reactor

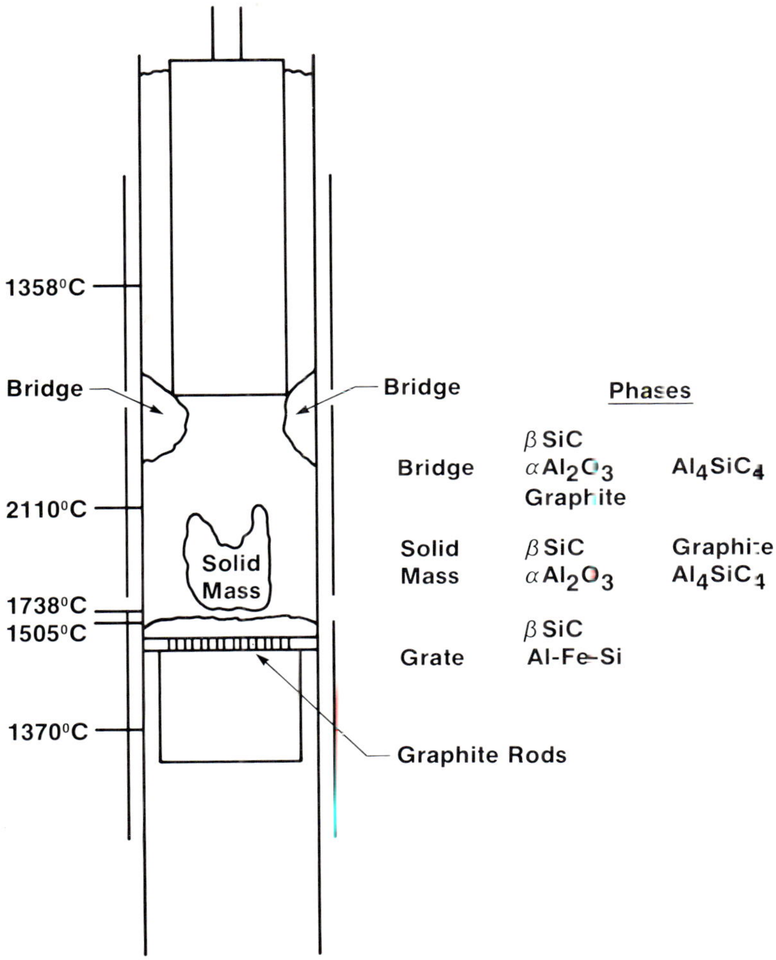

AUTOPSY OF TYPICAL O_2 COMBUSTION RUN
IN VSR WITH AUGER

Figure 10

contents shown schematically in Figure 11, contained alternating layers of ore and carbonaceous material--graphite, SiC or petroleum coke. The particle size of the materials decreased from bottom to top. The top layer was a mixture of ore and coke equivalent to the ratio in ore-coke layers.

No auger was used and no burden material or oxygen was introduced during the run. The reactor was heated to 2060°C and held until the CO in the off-gas dropped to a low level which indicated completion of the

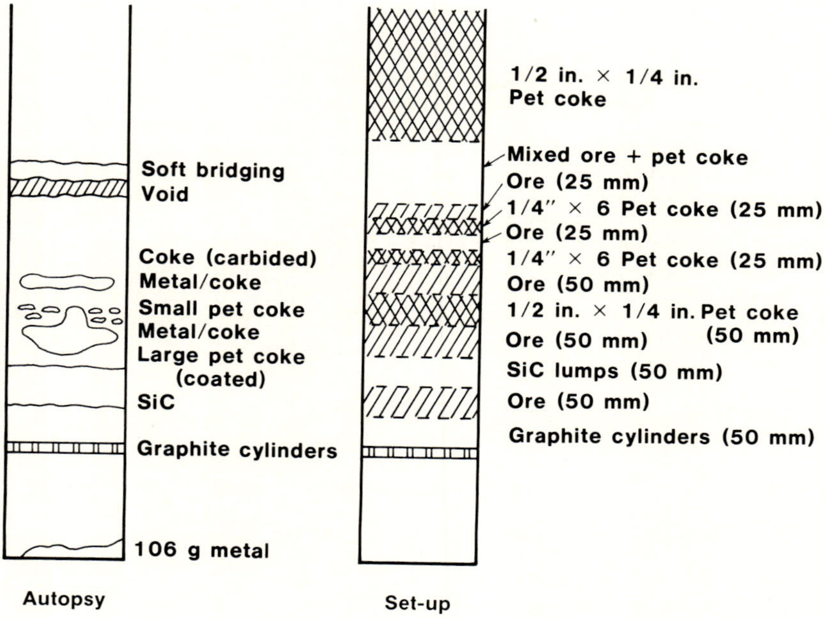

SET-UP AND AUTOPSY OF EXCESS CARBON RUN
IN VSR WITHOUT COMBUSTION

Figure 11

reaction. An autopsy (also shown in Figure 11) revealed that alloy product which had formed in the ore layers above the graphite and SiC lumps had drained through those layers. Most of the product coated the graphite cylinders and the grate while 106 gm flowed to the hearth crucible. The 50 mm and 25 mm thick ore layers apparently reacted to alloy, which penetrated the lower coke layers and reacted with them to form carbides. The only trace of reaction products within the mixed ore-coke layer was carbide-coated coke. This run demonstrated recovery of alloy from a layered charge where the particle size of the coke or carbide exceeded 1-2 cm. Neither the mixture of ore with combustion coke nor the layers of ore and standard size combustion coke (-1/4 +6 mesh) produced metal.

Bench Scale: Effect of CO on Reflux

The earliest bench scale experiments were carried out in a low bed isothermal batch reactor from which volatiles escaped without reflux. The goal of the bench scale experiments was to achieve continuous operation in a refluxing reactor with a CO sweep rate that simulated 100% combustion heat. In the first version of a refluxing reactor, a long column of ore was charged before each run. The column effectively condensed all volatiles, but a bridge formed in the middle of the column in every run. About 50-60% of the charge reacted, leaving a void below the bridge. Increasing CO sweep from 0.12 to 1.8 m^3/hr (2 to 30 Lpm) reduced alloy recovery at constant run time by about 50%.

To simulate a continuous refluxing reactor, some means was required to prevent bridging and achieve bed movement. Therefore, a smaller version of the pilot VSR was designed to introduce preheated CO at the level of the grate. This reactor, designated AF, had an I.D. of 102 mm with a 76 mm O.D. auger, and was operated semicontinuously using an ore feeder. About 8-10 kg of ore pellets were fed during a successful run, producing an ingot which filled the cavity below the grate. CO sweep rate, SiO_2/Al_2O_3 ratio, and per cent carbon in the ore were varied. The AF reactor was operated successfully at 0 and 0.9 m^3/hr (15 Lpm) CO sweep with ores containing SiO_2/Al_2O_3 ratios from 0.5 to 1.22. Table III shows data for a series of runs at 0.9 m^3/hr (15 Lpm) in which the CO/<u>Al</u> ratio was calculated from an average aluminum production rate for the time the reactor was above 2000°C. The CO/<u>Al</u> ratio varied from 4.2 to 8.2 and was generally higher at higher SiO_2/Al_2O_3 ratios. Average alloy analyses are listed in Table IV.

TABLE III

Summary of AF Reactor CO Sweep Runs At Various Burden SiO_2/Al_2O_3 Ratios

Run Number	AF13	AF15	AF18	AF20	AF14	AF12
SiO_2/Al_2O_3 Wt Ratio	.50	.78	.99	1.22	.99	.73
Total Wt% Stoichiometric C in Ore	80.1	80.0	80.0	78.5	80.0	80.5
Wt% Fe_2O_3	3.0	2.9	2.7	1.9	2.7	2.1
Time >2000°C min	215	245	280	130	155	291
Time CO on min	299	305	385	290	265	365
T_{max}, °C	2110	2120	2115	2155	2115	2105
Partially Reacted Charge, gm	3068	1640	2373	757	1949	2381
Total Reacted Charge, gm	11484	9826	9370	370[illegible]	9418	11572
Al Yield, gm	915	935	676	287	613	773
Si Yield, gm	413	504	530	263	508	572
Actual Wt Metal / Max Wt Metal	.34	.45	.40	.46	.40	.35
Reaction Zone Phase	Oxide	Oxide	Oxide	Carbide	Oxide	Oxide
CO/Al	4.2	4.7	7.5	8.2	4.6	6.8

TABLE IV

Analyses of Alloy Product From AF Reactor
CO Sweep Runs

Run Number	AF13	AF15	AF18	AF20	AF14	AF12
Average Metal Analysis Wt Basis						
%Al	56.0	52.4	45.8	42.5	41.1	45.9
%Si	23.7	26.5	29.5	34.0	31.9	30.8
%Fe	5.5	5.8	5.5	5.0	5.1	3.7
%Ti	3.4	2.9	2.7	2.2	5.9	3.0
%O	3.1	2.9	7.3	4.8	2.9	3.9
%C	7.6	4.7	6.1	6.7	7.5	8.6
TOTAL %	99.3	95.2	96.9	95.2	93.9	95.9

When CO sweep was increased from 0.9 to 1.35 m^3/hr (15 to 22.5 Lpm), no yield was obtained although some alloy was found in the partially reacted material above the grate. Apparently, increased vaporization which accompanied the increase in CO sweep reduced the alloy collection rate below a critical level for coalescence. Based on the thermochemical model calculations which follow, alloy production was found to be impossible at a CO sweep rate well below that required for blast furance operation.

Thermochemical Model of Al-Si Blast Furnace

Failure to produce Al-Si alloy at the required CO sweep rates led to a review of the original thermochemical calculations. The original data base contained 1972 JANAF data for Al_2O (4), a series of hypothetical Al_xSi_x compounds with properties derived from Batalin (5), and estimated data for Al_4O_4C(S, ℓ). Because only Al_4O_4C and SiC were allowed to react in Stage III, a unique alloy composition was calculated for a given CO sweep rate and pressure.

Whenever significant CO sweep was introduced, the model calculated that the heat balance was controlled by Stage III with the heat for Stages I and II supplied by reflux. Simplified heat and material balance equations for Stage III are shown in Equations (1-4). The basis is 1 mol of Al and 0.67 mols of Si.

Carbon Balance: (1)

$$n_C + n_{SiC} + n_{Al_4O_4C} = n_{CO}$$

Silicon Balance: (2)

$$n_{SiC} = 0.67 + c \cdot n_{CO}$$

Aluminum Balance: (3)

$$4n_{Al_4O_4C} = 1 + (2a + b) \cdot n_{CO}$$

Heat Balance: (4)

$$\Delta H_{alloy} + (a\ \Delta H_{Al_2O} + b\ \Delta H_{Al(v)} + c\ \Delta H_{SiO})n_{CO} - \Delta H_{CO} \cdot n_C = 0$$

where: ΔH_i = heat of reaction to form species i (see Appendix)

n_i = mols of species i

$a = n_{Al_2O}/n_{CO}$

$b = n_{Al(v)}/n_{CO}$

$c = n_{SiO}/n_{CO}$

The second term in Equation (4) will be referred to as the reflux reaction term. Solution of Equations (1)-(3) gives the relationship between n_C, the fuel rate or CO sweep rate, and n_{CO}, the total of combustion and reduction CO, as shown in Equation (5).

$$n_{CO} = (n_C + .92)/(1 - .5a - .25b - c) \qquad (5)$$

The aluminum reflux ratio, defined as the mols of aluminum gasified as Al_2O and Al(v) per mol of aluminum in the alloy, is given by Equation (6).

$$R_{Al} = (2a + b)\ n_{CO} \qquad (6)$$

By combining Equations (1)-(4), the fuel rate and aluminum reflux ratio were calculated, as shown in Table V, using values for a, b, and c from the thermochemical model.

$R_{Al}(n_C = 0)$ is the predicted reflux ratio for an externally heated process, for example, a submerged arc furnace. When combustion heat is used, the table shows that R_{Al} increases by a factor of about 20. Although n_C and R_{Al} are large, the feasibility of the process is not precluded.

TABLE V

Reflux Ratio and Fuel Rate Based on Original Data

a	0.0187
b	0.0447
c	0.0627
R_{Al}	1.73
n_C	8.3 mol C/mol Al
R_{Al} (n_C = 0)	0.082 (external heat)

The principal weakness of the thermodynamic calculation in this system is the uncertainty of some of the data. A review of the literature revealed that Batalin's data on activities in Al-Si alloys, which predicted significant negative deviations from ideality, were questionable. More recent data by Loseva (6) at 1700 and 1900°C indicated that the alloy approached an ideal solution at high temperatures. Data on the free energy of formation of Al_2O at 2300°K was revised downward from -63.898(50) to -67.469(54) kcal/g mol according to a revision of JANAF (7). Finally, data for both solid and liquid Al_4O_4C were revised to correspond to Gitlesen and Motzfeldt data for the solid (8), and a hypothetical melting point at 2173°K corresponding to peritectic decomposition.

Using the revised data, the items in Table V were recalculated and are shown in Table VI.

TABLE VI

Reflux Ratio and Fuel Rate Based on Revised Data

a	0.1801
b	0.0888
c	0.1108
R_{Al} (n_C= 0)	0.53
R_{Al}	---
n_C	---

It was calculated that both recycle ratio and fuel rate were negative. This implies that the reflux reactions which form Al_2O, Al(v) and SiO, consume too much heat for the heat balance to close. An adiabatic reactor would not be able to reach alloy producing conditions.

Comparison of Bench Experimental Results With Calculations

A direct comparison of experimental CO sweep results with heat and materials balance calculations is difficult due to high experimental heat losses. However, it was useful to estimate what percentage of combustion heat had been simulated by passing CO gas through the bench scale "AF" reactor. To this end, the heat balance, Equation (4), was used to devise Equation (7), using the revised data.

$$\%\Delta H^{III} = \frac{20.298\ n_C}{177.41 + 64.022\ n_C} \qquad (7)$$

Table VII shows the calculated values for $\%\Delta H^{III}$ and the aluminum recycle rate, R_{Al}, as a function of fuel rate, n_C. According to this calculation, and the revised data, less than 31.7% of the heat can be supplied by combustion.

TABLE VII

Effect of Fuel Rate on Per Cent Heat Supplied by Combustion and Recycle Ratio

Fuel Rate mol C/mol Al	$\%\Delta H^{III}$	R_{Al}
1	8.4	1.1
5	20.4	3.4
10	24.8	6.3
20	27.8	12.1
100	30.8	58.5
∞	31.7	∞

At a CO sweep rate of 15 Lpm, alloy was produced in the AF reactor as shown in Table III. The average CO/Al ratio for two runs with SiO_2/Al_2O_3 = 0.78, was 5.75. Based on Equation (7), this corresponded to about 21% combustion heat and a recycle ratio of 3.85.

Process Modifications

Three modifications can be applied to increase the heat input or decrease the heat requirement for reflux reactions: preheat the oxygen, increase pressure, and dilute the alloy with a solvent. Preheating the oxygen to the Stage III reaction temperature (2355°K) does not close the heat balance since the enthalpy per mol CO is increased by only 43%. Increasing pressure reduces the Al(v)/CO, Al_2O/CO, and SiO/CO ratios (a, b and c) in the gas. More heat is available at lower ratios although the advantage is partially offset by higher reaction temperature. The upper limit on pressure is assumed to be 5 atmospheres, the maximum blast pressure for modern blast furnaces. By itself, or in combination with O_2 preheat, pressurization is unable to correct the heat deficiency.

Dilution offers the most flexibility since the activities of Al and Si can be reduced to any desired level. A lower activity in the alloy reduces the partial pressures of Al(v), Al_2O and SiO which lowers the reflux heat requirements. Possible solvents include Fe, Cu, Ni, and Sn. Table VIII shows the calculated fuel rate for various Fe-Si-Al alloy compositions at various pressures. In these calculations the alloy was assumed to be an ideal solution. Alloy formation was assumed to occur in two stages (IIIa and IIIb) to approximate continuous alloy formation over a temperature range and take advantage of heat reflux from stage IIIb to IIIa. The calculated Al content increases from 30 to 45% as pressure and temperature are increased. Fuel rate decreases with pressure but increases with aluminum content at constant pressure.

TABLE VIII

Al-Si-Fe Blast Furnace: Calculated Alloy Compositions

	T, °K					Fuel Rate
P, atm	IIIb	IIIa	% Fe	% Al	% Si	mols C/mols Al
1	2290	2270	56	30	14	18
5	2470	2435	43	38	19	10
5	2480	2455	38	42	20	14
5	2490	2465	33	45	22	18

Summary

Oxygen-coke combustion generated temperatures that produced high Fe, Al-Si alloy from ores in an indirectly heated batch process. While an indirectly heated process reduces the problems associated with CO sweep and reaction with the combustion zone, problems with bridging remain unless volatiles are not refluxed. Scale-up of an indirectly heated process would involve serious difficulties.

Direct combustion heated runs were unsuccessful in producing significant amounts of alloy. High iron ores yielded small amounts of alloy with very low Al content while low iron ores yielded only a trace of alloy. Since the selection of O_2 rate and/or coke ratio was based on physical considerations such as raceway size and charge descent, it was not possible to reduce O_2 flow sufficiently to avoid complete alloy vaporization.

Alloy was produced at a maximum CO/Al ratio of 5.75. Efforts to run a higher CO flow were unsuccessful since no alloy was recovered.

Using the best available data on Al-Si alloy activities and free energies of Al_2O and Al_4O_4C, equilibrium calculations showed that a 100% combustion heated process is not possible. The highest percentage of combustion heat that was simulated in CO sweep runs was 21%. The calculated limit is 32%. By adding a solvent such as iron, it was calculated that the heat balance could be closed for an alloy containing 30 to 45% Al depending upon fuel rate and pressure.

Pilot experiments indicated that, even if volatility had not been so severe, other problems would make operation difficult. It was found that alloy would not drain through layers of carbon-containing material with a particle size smaller than 1-2 cm. Combustion experiments indicated that a much smaller coke size, -1/4 inch +6 mesh, was required for smooth operation. Therefore, the pilot reactor would probably have to be scaled up by a factor of 3 in diameter in order to achieve satisfactory combustion and drainage.

Although the auger was partially successful in preventing bridging, the disruption of layers and back-mixing would probably prevent alloy recovery since the combustion carbon would be intimately mixed with the alloy. A low bed height which would permit excess volatiles to excape might allow bridging control, but would also create major reactor top problems because of high temperatures and dust loading.

Al and Si activity reduction by the addition of a solvent such as iron could result in a technically feasible combustion-heated process if a refining technique can be developed for separating the solvent from the alloy. Fractional crystallization is not an option since yield decreases drastically above 3-5% iron.

References

1. C.N. Cochran and N.M. Fitzgerald, "Computer-Aided Model of Coke-Fired Reduction of Alumina-Silica," Light Metals 1980, p. 191.

2. J.G. Peacey and W.G. Davenport, The Iron Blast Furnace, Theory and Practice, Pergamon Press, 1979.

3. H. Schmitt and Wittner, Zeitschmitt Fver Erzbergbau ard Metallhuettewesen 13 (1960) No. 10, pp. 471-77 (German).

4. M.W. Chase et al., "JANAF Thermochemical Tables, 1974 Supplement", J. Phys. Chem. Ref. Data 3 (1974), pp. 311-480, $Al_2C(g)$, 1972 June 30.

5. G.I. Batalin, E.A. Beloborodova, and V.A. Stukulo, "An Examination of the Thermodynamic Properties of Al-Si Metals." Russian Metallurgy, No. 2, 1971, pp. 44-48.

6. A.F. Loseva, et al., "Thermodynamic Properties of Liquid Aluminum-Silicon Alloys," Russian Journal of Physical Chemistry 51 (2), 1977.

7. M.W. Chase et al., "JANAF Thermochemical Tables, 1978 Supplement," J. Phys. Chem. Ref. Data 1 (1978), pp. 793-940. $Al_2O(g)$, 1975 June 30.

8. G. Gitlesen, O. Herstad, and K. Motzfeldt, "Selected Topics in High Temperature Chemistry," Ed. T. Foreland, et al., Universitets Forlaget, Oslo 1966, pp. 179-196.

APPENDIX

Heats of Reaction

Alloy Formation for Al-40% Si:

$$.25Al_4O_4C(\ell) + .67SiC = \underline{Al} + .67Si + CO \qquad \text{(A-1)}$$

$$\Delta H_{alloy} = +118.51 \text{ kcal/mol Al}$$

Suboxide and Vapor Formation:

$$.5Al_4O_4C(\ell) + .5C = Al_2O + CO \qquad \text{(A-2)}$$

$$\Delta H_{Al_2O} = +181.43 \text{ kcal/mol } Al_2O$$

$$.25Al_4O_4C(\ell) + .75C = Al(v) + CO \qquad \text{(A-3)}$$

$$\Delta H_{Al} = +169.55 \text{ kcal/mol Al}$$

$$SiC + CO = SiO + 2C \qquad (A-4)$$

$$\Delta H_{SiO} = +17.997 \text{ kcal/mol SiO}$$

<u>Combustion</u>:

$$C + 1/2O_2 = CO \qquad (A-5)$$

$$\Delta H_{CO} = -20.298 \text{ kcal/mol CO}$$

FEASIBILITY OF AN ALUMINUM-SILICON BLAST-ARC PROCESS

D. T. Stevenson

ALUMINUM COMPANY OF AMERICA
Alcoa Technical Center
Alcoa Center, PA 15069

Summary

The joint Alcoa-DOE project to develop a carbothermic aluminum smelting process originally focused on a combustion-heated shaft reactor to produce Al-Si alloy. Experimental evidence and revised thermodynamic calculations determined that a totally combustion-heated Al-Si process was not feasible. Therefore, a process for combining combustion heat and electrical energy was evaluated. A one-step process using combustion heat to produce SiC from an Al_2O_3-SiO_2-C pellet which was fed directly to a submerged arc metal producing stage offered the greatest prospect for electrical energy savings. Numerous technical and engineering design problems with the one-step process led to the evaluation of a two-step concept where SiC was produced in a separate combustion-heated shaft reactor from SiO_2-C burdens and then charged along with Al_2O_3-C agglomerates into a submerged arc furnace to make Al-Si alloy. The kinetics of SiC formation was a very important factor in the evaluation of these processes.

Introduction

The most economic carbothermic reduction process for producing aluminum rich alloys is based on a blast furnace heated by combustion of coal and coke. However, a 100% combustion heated process is not possible because of the high reflux of volatiles created (1). Alternative concepts involve a combination of combustion and electrical heat in a blast-arc process.

Previous studies indicate that reduction reaction takes place in three stages (2). Silicon carbide formation can be complete at 1550°C, oxycarbide slag formation takes place between 1900-1950°C, and metal production between 2000-2050°C. Thermodynamic calculations show that utilization of combustion heat is practical only for SiC production. The reasons for this are threefold. Any CO sweep from combustion in the two higher temperature stages greatly increases the formation and recycle of Al, Al_2O, and SiO vapors; this is the main reason for failure of the blast furnace. Second, the slag and metal production stages are so close in temperature that separation of these reactions is unlikely. Thirdly, the sensible and chemical heat in the gas generated from an electrically heated metal production stage is sufficient to supply all of the energy needed to form the oxycarbide slag and preheat the burden exiting the SiC production stage to the metal producing temperature.

To take fullest advantage of heat released by backreaction of the volatiles, it is necessary to combine all three chemical stages into one reactor. This one-step blast-arc process is shown schematically in Figure 1a. Heat for SiC formation is supplied by combustion, while electric heat is used for slag and metal production. The electric heat could come from a submerged arc using carbon electrodes or from a plasma torch. In place of an oxygen tuyere for burning coke charged to the top of the furnace, a burner could be employed where carbonized powdered coal is conveyed pneumatically to an oxygen-char burner. This burner would allow better control of the O_2-C balance in the burden as it passes to the metal producing zone. Carbon balance control is necessary to prevent carbiding of the metal alloy which causes yield loss and operating difficulties in the hearth zone. However, in this case, the preheat for the combustion carbon that would otherwise come from the rising hot gases as in the tuyere case would come from the consumption of extra oxygen and carbon. The series one-step concept offers the potential to use 38% combustion heat to produce a 60% Al-40% Si alloy.

Another possibility for partial use of combustion heat is to supply some electrical heat in the highest temperature zones in a superimposed arrangement as shown in Figure 1b. This decreases the volatile generation that is experienced in the blast furnace, but not to as low a level as in the series one-step process. The combustion heated trials conducted to test a blast furnace operation (1, 2) were combinations of electric and combustion heat in a parallel configuration. Calculations showed that for 31% combustion heat (1) there was no aluminum yield. CO sweep runs simulating 21% combustion heat demonstrated the practical limit for operating a refluxing furnace. Mechanical means were used to prevent bridging of the burden during volatile backreaction in these runs.

A third blast-arc concept for producing Al-Si alloy involves the production of SiC in a separate combustion heated reactor using an Al_2O_3

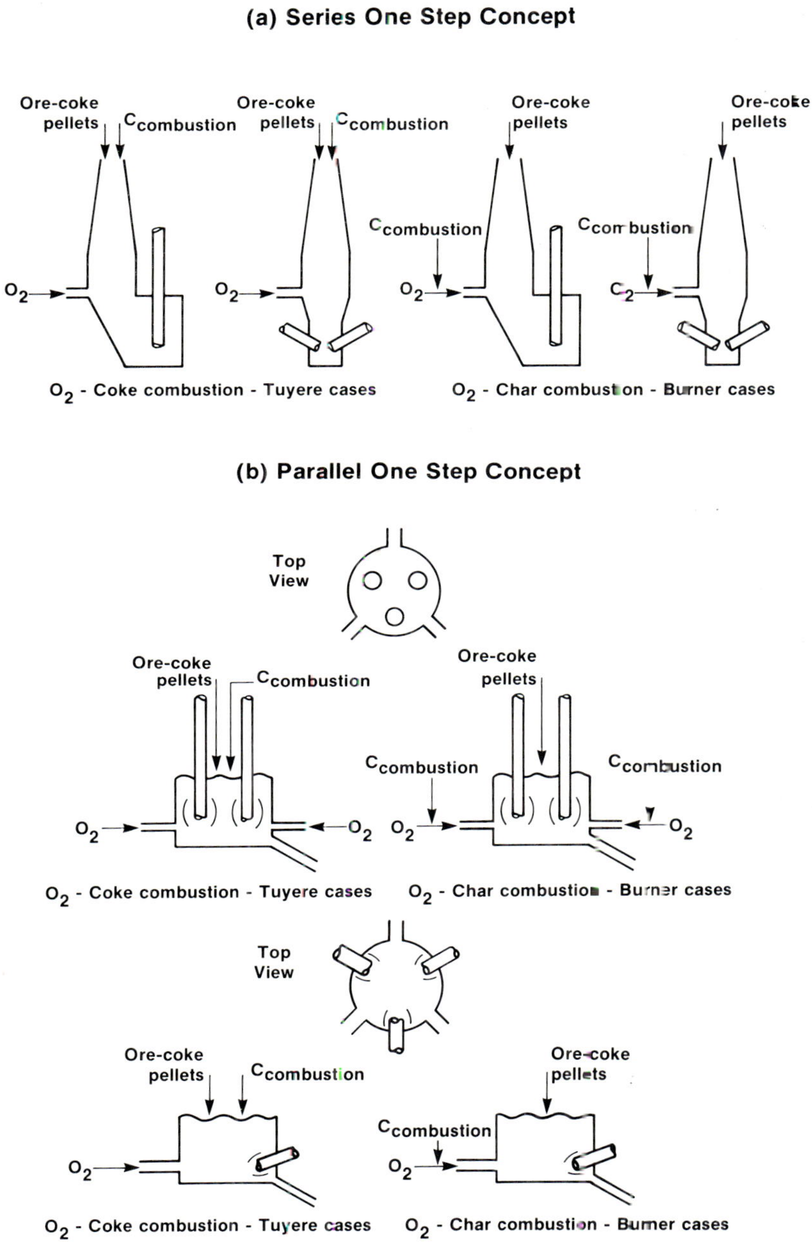

Blast-Arc Concepts

Figure 1

free source of silica. The SiC product is then combined with Al_2O_3, and any additional required reduction carbon, and fed to a submerged arc furnace to make metal. This two-step process cannot take full advantage of the volatile backreaction and superheat from the metal and slag production stages, but at least some combustion heat could be used. Separation of the combustion stage from metal production precludes the overlap of these two stages as may occur in the series one-step process.

As before, the tuyere and burner cases shown in Figure 1c are considered where reoxidation of SiC is less likely to occur in the burner case due to improved $C-O_2$ control during combustion. Separation of the SiO_2 feed from the moving bed prevents fusion of silica prior to reduction. This allows smoother furnace operation and takes advantage of the gas-solid reaction mechanism predominant in the formation of SiC. Hot SiO vapor generated in an oxygen-char burner stage backreacts in a bed of coke to form SiC in a second stage.

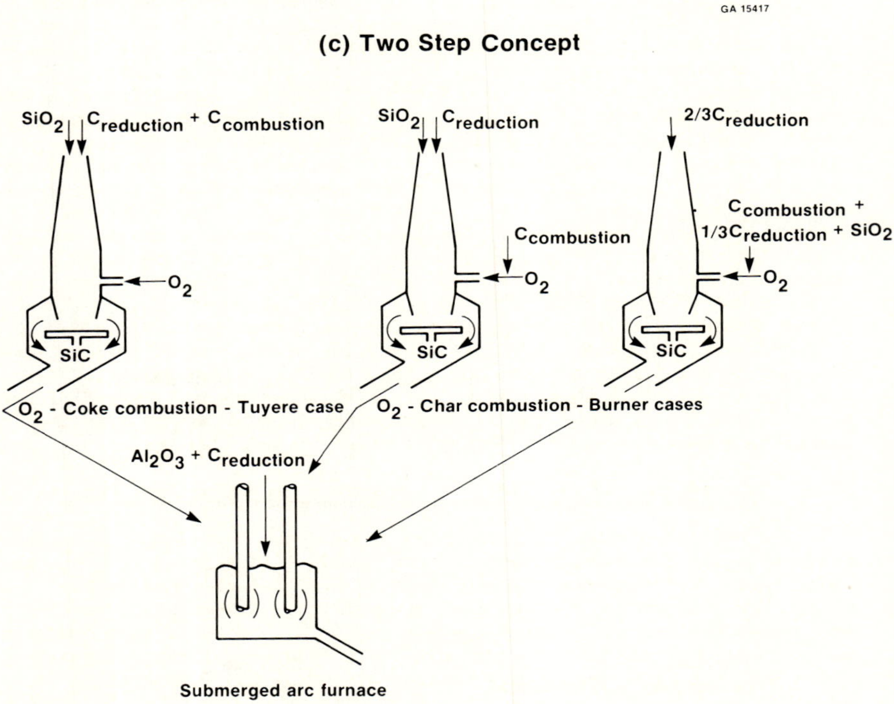

Blast-Arc Concepts

Figure 1

Experimental Studies for the Al-Si Blast-Arc Process

Based on earlier attempts to make metal with and without CO sweep in electrically-heated bench-scale and pilot reactors (2), SiC formation always proceeded as an intermediate step without any difficulty. It appeared that prereduction and metal production could be easily accomplished in the same reactor in a staged fashion along an axial temperature gradient. Furthermore, pellet strength was adequate to support a packed bed above the metal-producing zone, and CO sweep seemed to have no effect on SiC production.

To determine whether this intermediate step could be separated from higher temperature stages, tests were run in a 210 mm I.D. induction heated pilot shaft reactor shown in Figure 2. This reactor was fitted with a slowly revolving table to regulate solids discharge from the reactor through an isolation valve into a sealed box. Argon flow introduced below the turntable was used to cool the product prior to discharge. The first runs focused on SiC production from Al_2O_3-SiO_2-C pellets to demonstrate the first stage of the series one-step blast-arc concept. Results from these runs at 1600°C showed that SiC formation was slow and that pellet strength decreased with increasing SiC formation.

Tests of the SiC reaction kinetics were conducted in an induction heated shallow bed isothermal batch reactor as shown in Figure 3. The 64 mm I.D. graphite susceptor also served as the crucible to contain the pellets used in these tests. One liter per minute of argon flowed through the reactor, with most of the gas passing through the bed and a small amount through the insulation used to seal the moderately gas tight reactor against air inleakage. The Ar fed to the reactor and the total gas flow of CO from reduction plus Ar were measured by mass flowmeter and a wet test meter, respectively. The CO flow, obtained by subtracting argon from the total flow, was correlated with reaction time and temperature. The percentage of prereduction was based on the CO produced as compared to the total evolution expected from the initial SiO_2 analysis of the sample. These flow measurements lead to an error of less than ±5% of the absolute percent prereduction, and hence the calculated time averaged reaction rates. Physical properties of the sample pellets were measured in a compression test and a 1560 revolution tumble test. This was done to determine the relative attrition loss expected during materials handling and flow through the blast-arc reactor. The sample from the tumble test was screened to 10 mesh to predict fines generation.

Experimental Results for the Series One-Step Blast-Arc Process

The results of the physical tests on prereduced pellets from this study shown in Table I indicate the unreduced pellets are fairly strong and durable, but that the fully prereduced burdens are considerably weaker and susceptible to breakdown by abrasion. As expected, partially prereduced burdens are generally stronger and more durable than completely reduced pellets. It is not surprising that reduced pellets are weak, since SiO_2 and Al_2O_3 supply the bonding strength, and formation of SiC causes a considerable decrease in the bonding phase.

The reaction rates at various levels of prereduction are shown in Figures 4 and 5, and the time-averaged reaction rates in Table II. For both burdens, there is a significant increase in reaction rate between 1650°C to 1700°C. Some of these tests show a peak in the reaction rate in the early stages of reduction. While it is believed that the maximum reaction rate occurs before 50% prereduction, these peaks are an artificial occurrence related to slight overheating of the samples as they are heated to the desired reaction temperature.

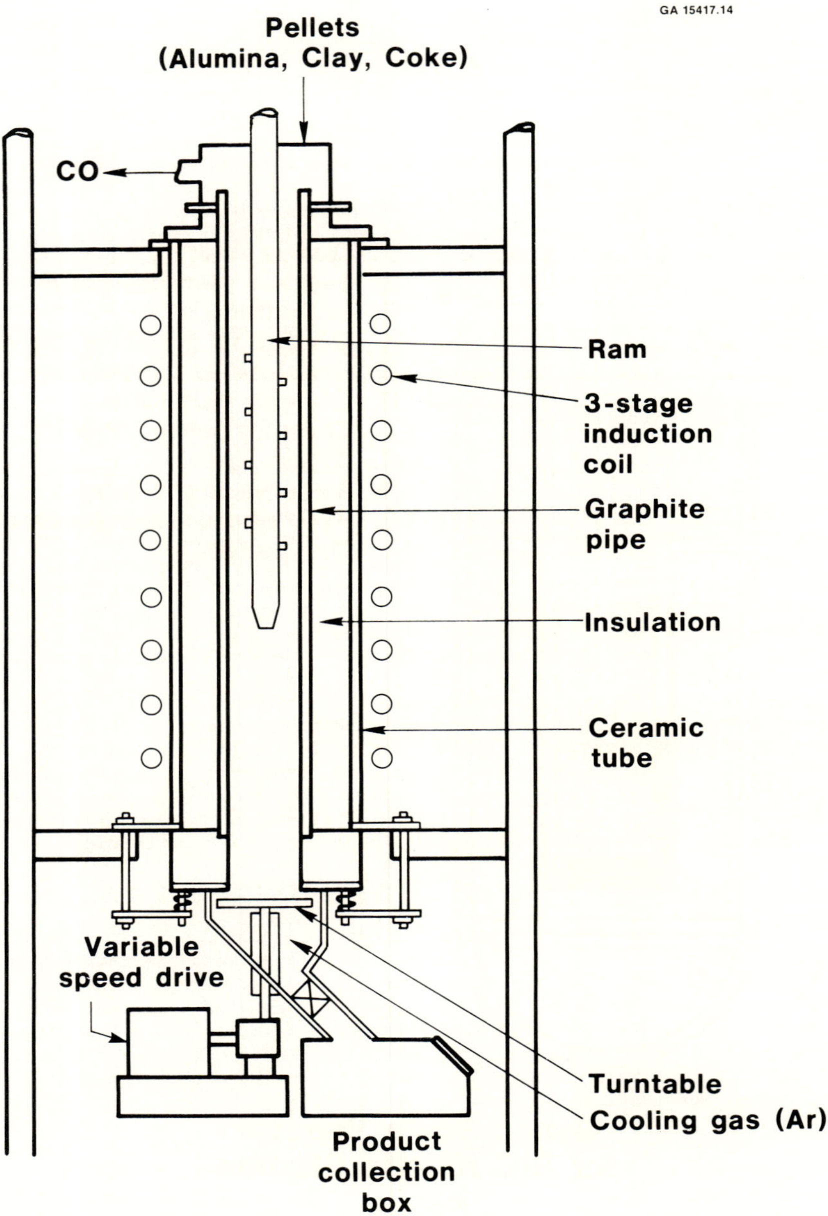

Vertical Shaft Reactor with Solids Discharge

Figure 2

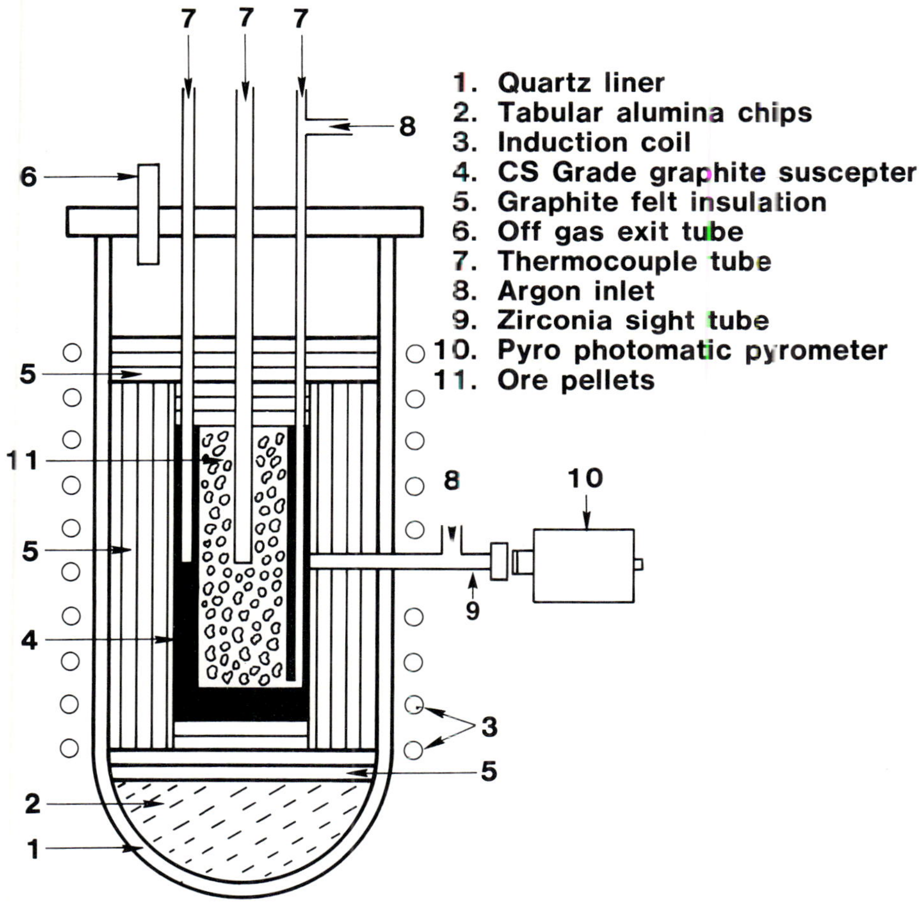

Bench Scale Reactor for SiC Kinetic Studies

Figure 3

GA 15417.26

Table I

Physical Properties of Prereduced Pellets After SiC Formation

	Prereduction Temperature	% Prereduction Calculated From CO Evolution	Compressive Strength 1/4" Pellet	Weight Loss To Abrasion
Clay-Al_2O_3	—	0%	12.7 kg	2%
Hydrate-Met. Coke	1600°C	99.7	2.3	74
	1650	106.5	2.4	74
3/8"+6 Mesh	1700	105.8	3.0	71
	1650	76.7	12.4	28
	—	0%	19.0 kg	11%
Fused SiO2 - Activated	1600°C	101.8	9.3	43
Al_2O_3-Pet. Coke	1650	102.6	4.1	49
3/8"+1/4"	1700	105.7	4.0	54
	1650	68.5	6.6	33

I³ 1006-295

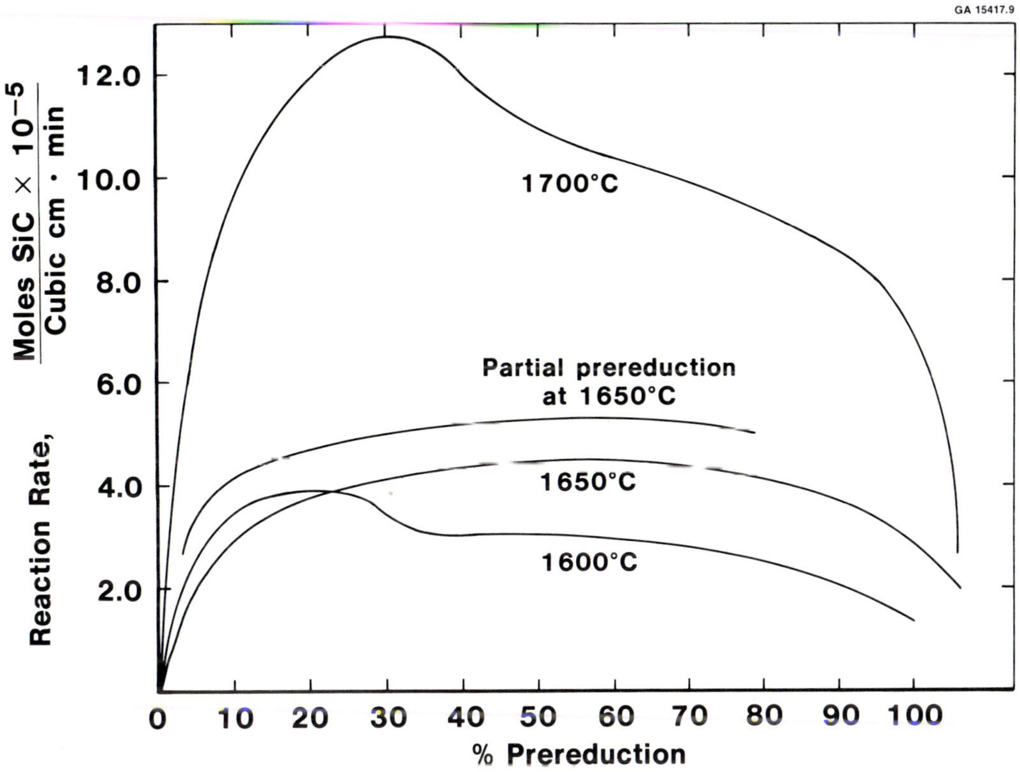

Reaction Kinetics of SiC Formation in Clay-Alumina Hydrate-Met Coke Pellets

Figure 4

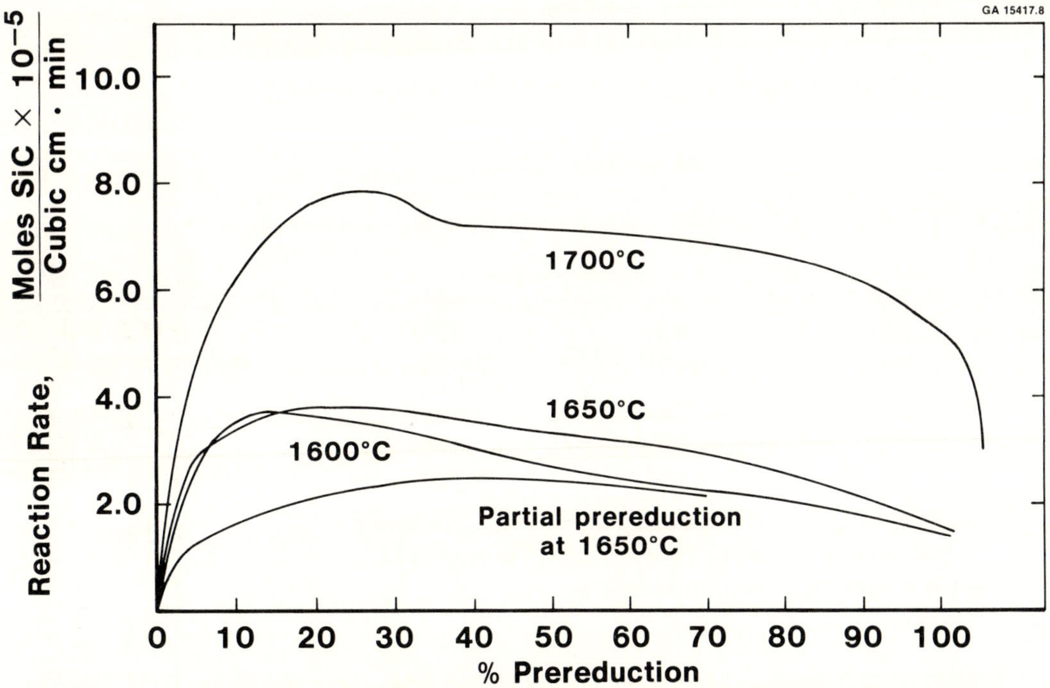

Reaction Kinetics of SiC Formation in Fused SiO_2 - Activated Al_2O_3 - Pet Coke Pellets

Figure 5

GA 15417.29

Table II

Reaction Rates for the Prereduction of Pellet Burdens

	Prereduction Temperature	% Prereduction Calculated From CO Evolution	Reaction Rate $\frac{\text{Moles SiC}}{\text{Cubic cm min}} \times 10^{-5}$
Clay-Al_2O_3 Hydrate-	1600°C	100%	1.4
Met. Coke	1650	100	2.9
3/8" +6 Mesh	1700	100	7.0
Fused SiO2-Activated	1600°C	100%	1.4
Al_2O_3-Pet. Coke	1650	100	1.5
3/8" +1/4"	1700	100	5.2

I³ 1006-295

Experimental Results for the Two-Step Blast-Arc Process

Specific burdens that could be used in the two-step process are outlined in Table III. The first burden is of carbon bonded pellets containing the required carbon for SiC formation. These pellets could be used in the tuyere or burner case, and localized formation of SiC would occur in each pellet due to the intimate mixture of the reactants. Pellets containing the carbon required for reduction and combustion could be used only with a tuyere operation. Lump burdens of quartzite and coke, like those used in the Acheson process, (3) could be chosen. A two stage process of SiO generation by the injection of C/SiO_2 in a 1/1 ratio into the combustion zone and backreaction of SiO in a moving coke bed could be considered. Use of a burner where additional carbon is consumed to supply the preheat for the combustion reactants results in a consumption of two more moles of carbon per mole of silica reacted as compared to the tuyere case.

Table III

GA 15417.27

Concepts for a Two-Step Countercurrent Shaft Reactor Blast-Arc Process

Case	Burden Composition	Tuyere	Burner	Heat Source
SiO_2 + C Pellets	C/SiO_2 = 3/1	X		O_2 blast burning separate coke fed along with the pellets
	C/SiO_2 = 3/1		X	O_2 blast burning char fed with the O_2 through a burner
	C/SiO_2 = 10/1	X		O_2 blast burning excess carbon in pellet
Quartzite + Coke Chunks	C/SiO_2 = 3/1	X		O_2 blast burning separate coke fed along with the quartzite chunks and reduction coke
	C/SiO_2 = 3/1		X	O_2 blast burning char fed with the O_2 through a burner
Two Stage Process				
SiO Generator	10/1 = C/SiO_2 powders + O_2		X	SiO_2 + char powders + O_2 injected through a burner at bottom of hot zone to supply SiO and heat
Backreaction	Coke Bed C/SiO_2 = 2/1			SiO backreaction in bed to form SiC

In the C/SiO_2 = 3/1 pellet, all of the carbon comes from metallurgical coking coal and in the C/SiO_2 = 10/1 pellet, half of the carbon comes from coking coal and half from petroleum coke. The results, shown in Figure 6 and Table IV, indicate that the higher carbon content does not improve the prereduced pellet strength, but does increase the reaction rate by accelerating the backreaction of SiO with excess carbon. This phenomenon was also observed in another study (4).

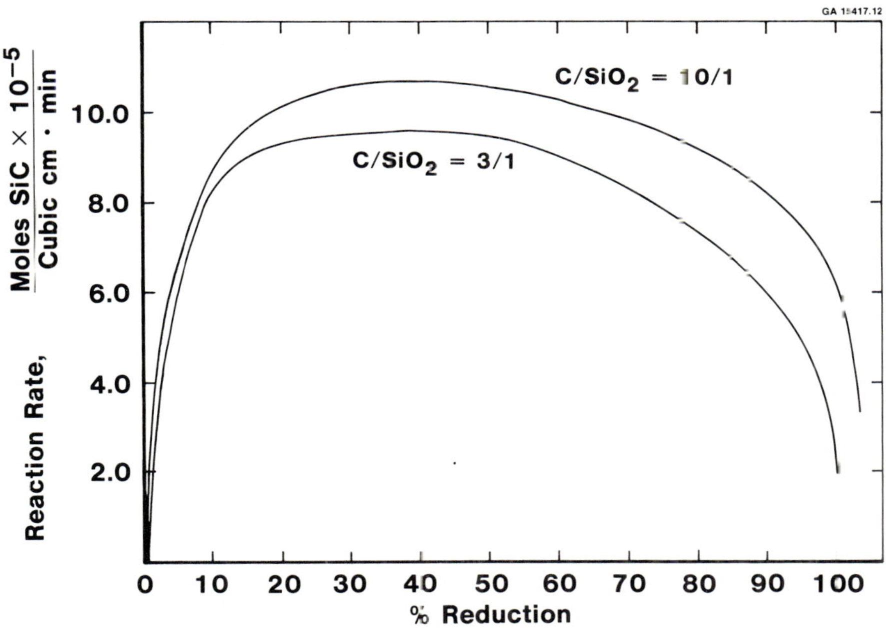

Reaction Kinetics of SiC Formation In SiO_2 + C Pellets at 1700°C

Figure 6

Table IV

Production of SiC from SiO_2 + C Pellets

	Reduction Temperature	% Prereduction Calculated From CO Evolution	Reaction Rate Moles SiC Cubic cm min x 10^{-5}	Compressive Strength 1/4" Pellet	Weight Loss To Abrasion
Stoichiometric Carbon C/SiO_2 = 3/1 3/8" +6 mesh	1700°C	98.5%	3.4	.22 kg	93%
Excess Carbon C/SiO_2 = 10/1 3/8" +6 mesh	1700°C	103.4	6.6	.65 kg	74%

Lump burdens in particle sizes that would not fluidize in a combustion heated shaft furnace react slower than pellets containing SiO_2 and C. The data shown in Figure 7 and Table V verify this, and indicate that smaller particle sizes and commercial SiC production temperatures (3) give good results.

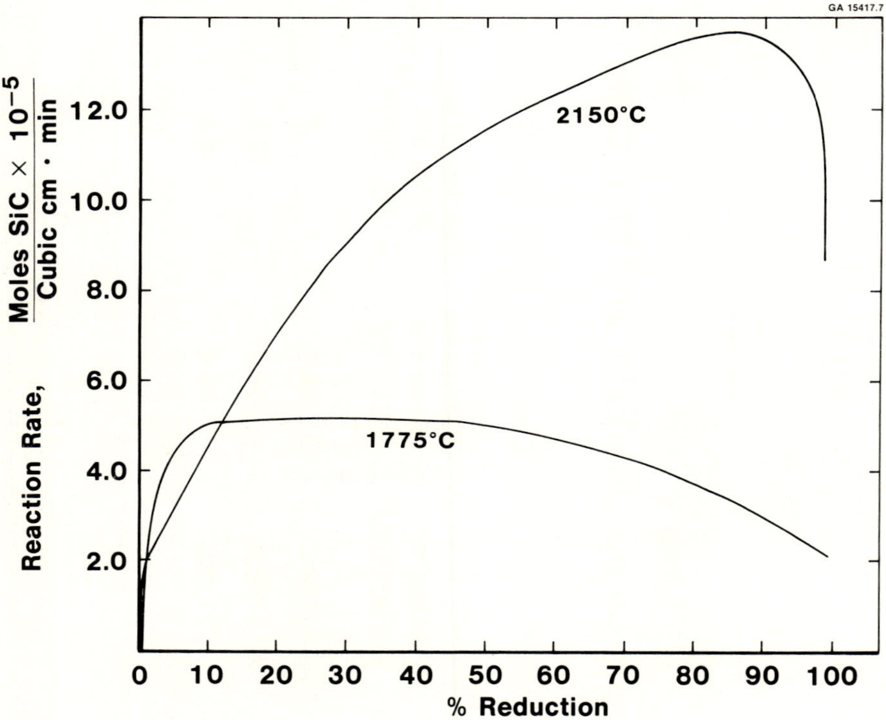

Reaction Kinetics of SiC Formation from−10+20 Mesh Lump Quartzite and Met Coke

Figure 7

To test the two-stage concept, fused SiO_2 and petroleum coke powders in a 1/1 ratio were placed below a grate in the isothermal batch reactor and twice as much coke was placed on top of the grate to capture the SiO as it was evolved. At 1775°C, SiO generation was very slow. The SiO_2 melted and the products in the generator below the grate were SiO_2 glass, some SiC, and unreacted carbon. There was little or no SiC in the bed above the grate. Charcoal was used to increase the carbon reactivity for SiO back reaction in a second run, but there was no improvement in the results (Table VI). This indicates that SiO generation is the rate limiting step. An increase in temperature to 2150°C gave the highest SiC reaction rate measured in these tests, and demonstrated the two-stage concept (see Table VI and Figure 8).

Table V

SiC Production from Lump Quartizite and Met. Coke

Quartz Size	Met. Coke Size	Reduction Temperature	% Prereduction Calculated From CO Evolution	Reaction Rate $\frac{\text{Moles SiC}}{\text{Cubic cm min}} \times 10^{-5}$
-1/4" +7 mesh	-1/4" +6	1700°C	30.5	.7
-10 +20	-10 +20	1775°C	98.5	2.2
-10 +20	-10 +20	2150°C	98.4	8.8

GA 15417.30

Table VI
Two Stage Production of SiC

SiO Generator	Carbon Bed Material	Reduction Temperature	% Prereduction Calculated From CO Evolution	Reaction Rate $\frac{\text{Moles SiC}}{\text{Cubic cm min}} \times 10^{-5}$
$SiO_2/C = 1$	activated coke	1775	—	—
$SiO_2/C = 1$	charcoal	1775	—	—
$SiO_2/C = 1$	charcoal	2150	106.7	16.1

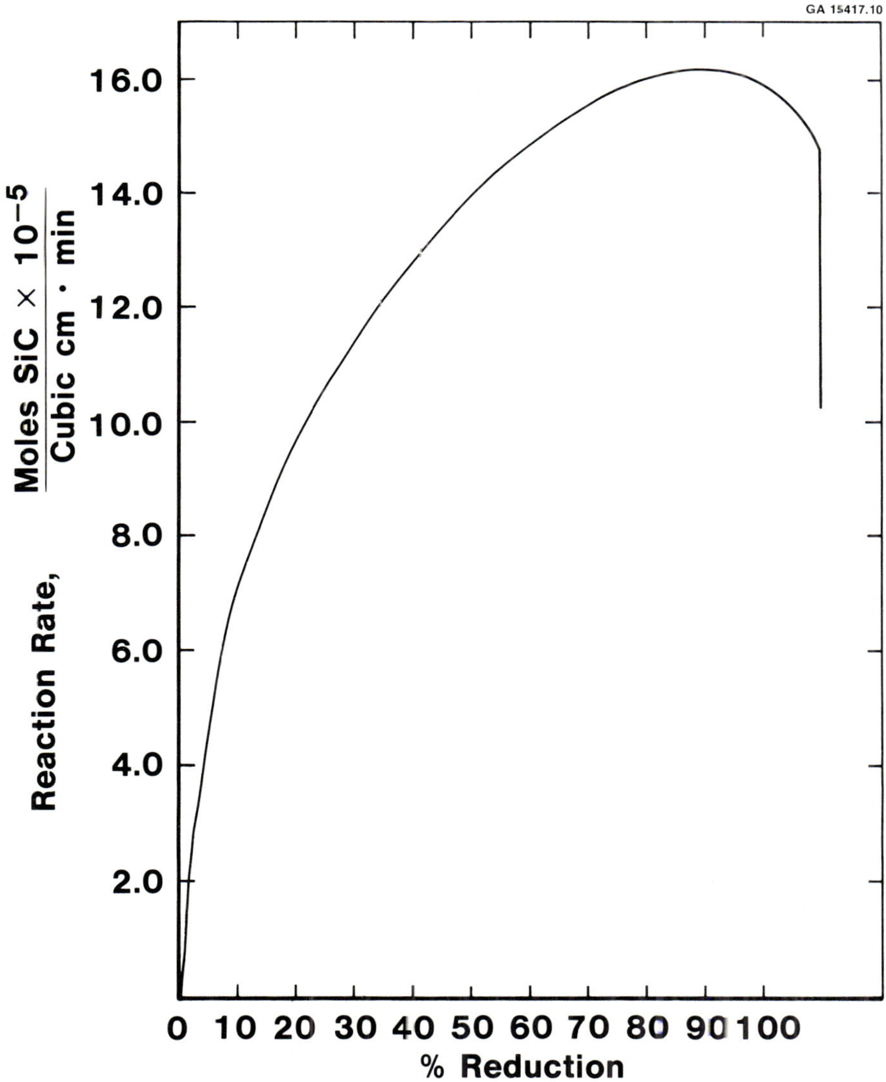

Two Stage SiC Formation at 2150°C
$SiO_2/C = 1$ in SiO Generator
Charcoal in the Bed

Figure 8

Conclusions

One-Step Blast-Arc Process

Operation of a combustion heated prereduction step in a series one-step blast-arc process is not practical for SiC production from agglomerated ores containing alumina. The prereduction reaction rates are too slow to attain complete SiC formation without heating to temperatures at which Al_2O_3 reduction begins. This would be advantageous in preventing degradation of the burden as predicted by the SiC kinetics tests, but would make it very difficult to separate the SiC production step from Al_2O_3 reduction. For these reasons, the series one-step blast-arc process is rejected in favor of a two-step process.

Two-Step Blast-Arc Process

The best SiC formation rates result from reacting pellets of mixtures of fine powders of C and SiO_2 in a ratio of C/SiO_2 >3/1, but agglomerates of these powders turn to dust after reduction. Reaction rates for mixtures of lumps of SiO_2 and C can be increased by lowering the lump size, but increases in temperature result in much greater increases in reaction rates. The two-stage method of SiC formation, where a superheated mixture of CO and SiO backreacts with a moving bed of carbon at temperatures above 2000°C, offers the highest reaction rate of SiC formation. The two-stage burner case listed in Table III is the most feasible concept. Most importantly, the two-stage SiC production step of the two-step blast-arc process has been demonstrated.

Production of metal from non-prereduced burdens in electrically heated pilot and bench scale reactors was in general very successful. Decrepitation of pelletized burdens did not occur, and a moving packed bed for reflux could be maintained above the slag and metal producing zones. Metal readily coalesced to form a pool, and slag accumulations in the hot zone were in general not large. The results of the kinetic experiments, however, indicated that an overlap between SiC and slag formation stages could be expected, due to the slow kinetics of SiC formation below 1800°C. Attempts to make metal from 100% prereduced burdens in both the pilot arc furnace and bench scale shaft reactor resulted in a large accumulation of slag, indicating a high rate of slag production. In addition, no refluxing bed above the hot zone could be maintained with the burdens that were used.

From the standpoint of heat transfer, it is not surprising that the slag formed very fast because the ratio of the heats of reaction of metal/slag production is greater than 3/1. The heat energy supplied to these reactors to make metal under normal circumstances went to producing slag as long as reactants were available. More work must be done on this step of a two-step blast-arc process.

Summary

The parallel one-step blast-arc concept uses the least combustion heat, but results in the greatest reflux. It is the least desirable of the concepts considered. The series one-step process offers the greatest opportunity to utilize combustion heat, but separation of the combustion heated SiC stage from an electrically heated slag production stage located in the same reactor is not practical because the kinetics of SiC formation are too slow below 1800°C. It is reasonable to assume that the kinetics of SiC formation should be equal to or greater than those of the Acheson process to successfully operate a countercurrent reactor for SiC production. The two-step process offers a high utliization of combustion heat and the two-stage burner case is the preferred first step. However, more work is required on the metal production stage of totally prereduced burdens.

References

1. R.L. Troup, "Feasibility of an Aluminum-Silicon Blast Furnace," presented at the 112th AIME annual meeting, Atlanta, GA, March 9, 1983.

2. M.J. Bruno, "Production of Aluminum-Silicon Alloy and Ferrosilicon and Commercial Purity Aluminum by the Direct Reduction Process," Final Technical Report, CONS-5089-16, UC-95f, February 1983.

3. H. Fuchs, "Silicon Carbide,' Chemie-Ing.-Techn., 46 No. 4, 1974, pp. 139-142.

4. B.C. Bechtold, "Carbothermal Reduction of Koalin Clay and Resulting SiC Formation," PhD. Thesis, University of Utah, 1973.

THE LIFE SPAN AND MARKET DEMANDS OF SOME BAUXITE MINES*

V.G. Hill and Slavko Ostojic

International Bauxite Association
67 Knutsford Boulevard
P.O. Box 551, Kingston 5,
Jamaica, (W.I.)

The bauxite-production growth curves for the producing countries, which supply the principal aluminum markets, are examined, differentiating between the countries involved on the basis of the market which they supply, and their domestic bauxite resources. The relationships between the shape of the bauxite production growth curves, the trend in the grade of bauxite produced over time and its competitiveness vis-à-vis other bauxites traded internationally are used as the basis for identifying two basic types of bauxite mine life-span curves.

The data suggest that when a bauxite-producing country is industrialised, it develops a significant domestic market for its bauxite, and consequently is able to produce progressively lower grades of bauxite thereby increasing its economic bauxite resource base. The conclusion, therefore, is that the economic bauxite resource base of a country changes to reflect the development of its aluminum industry, whether or not smelting is done locally. It is, therefore, suggested that the goal of developing a local or regional aluminum industry should be the preferred long-term option of bauxite-producing countries, if they are to optimise the benefits from the exploitation of bauxite.

*The opinions expressed in this paper are those of the authors and not necessarily those of the International Bauxite Association.

INTRODUCTION

The modern aluminum industry, from its beginning towards the close of the last century, has developed from small operations in North America and Europe into the present system with four principal market centers in North America, Japan, Western and Eastern Europe. The bauxite resources to supply these demands have expanded from domestic deposits, except in the case of Japan, to the present large-scale international trade in bauxite and alumina.

Two factors helped determine the bauxite supply linkages between mines and Bayer plants. Firstly, there are the characteristics of the bauxite resources supplying the plants and secondly, the transport costs for bauxite, alumina and (to some extent) metal which became important factors in the industry (Hill and Robson, 1982). The need for aluminum to maintain its competitive edge over alternative materials, which it progressively replaced in specific target markets, is still an important requirement for the continued growth of the industry. To this must now be added the need for reliable supplies of relatively low-cost electricity. Thus, the industry has been increasingly characterised by significant inputs of bauxite and recently, by energy from countries outside of the major market areas, so the resulting mutual interest between the suppliers of these inputs is now a characteristic of the industry (Hill, 1980 and Davis, 1982).

The objectives of this paper are: (a) to compare the long-term bauxite production growth trends for representative bauxite-producing countries; (b) to use these results as the basis for identifying possible long-term options, and prospects which the bauxite-producing countries face. This information may then be used as one of the basic factors for the development and assessment of policies of the bauxite-producing countries, and which will form the background against which investment decisions are made.

This is our second paper on this general subject. The first, (Hill, Robson and Ostojic, 1982) highlighted the world bauxite supply pattern with the supply trend changing first from domestic to neighbouring deposits and then to the present world-wide bauxite trade. This paper is a preliminary study of the economic bauxite resource base of some bauxite-producing countries, and of their development options and prospects.

The locations of the major bauxite provinces in which the most important identified bauxite resources and mines occur are identified by Hill and Ostojic (1981, p. 81) and by Bardossy (1982, Figure 5).

APPROACH

The approach of this paper involves, firstly, the plotting of the annual bauxite production data for the major aluminum consuming countries and for some major and minor bauxite-producing countries which supplied the major markets for the period 1900 to 1981. Long-term production growth curves were then fitted by eye to the plots for the bauxite-producing countries, smoothing out the effects of such events as the depression of the thirties, the Second World War and the present deep recession. The inclusion of such events at this stage tends to obscure or at least unduly complicate the long-term bauxite production growth curves. In a future work, we intend to use other curve-fitting techniques, and quantitative interpretations.

The second step involves the demonstration of the relationships between the changes in the grade of bauxite produced over time and the size of the economic bauxite resources of the following categories of bauxite-producing countries:

(i) the Arkansas bauxite region in the United States is used as the example from the major aluminum market countries;

(ii) the major bauxite suppliers to the aluminum market countries, differentiating according to their level of industrialisation:

(a) industrialised: Australia, Hungary and Yugoslavia[1], and

(b) under-industrialised: Jamaica;

(iii) minor bauxite suppliers to major aluminum market countries, using Haiti, Sarawak (Malaysia) and Indonesia as examples.

Finally, typical bauxite production life span curves are identified, reflecting the relationships between the size of the economic bauxite resources base of a country, the annual bauxite production, and changes in the grade of bauxite produced over time. The probable options and prospects for each category of bauxite-producing countries set the limits within which these countries may establish mutually acceptable arrangements. Bauxite producing countries in Africa are absent from the list of countries to be studied because, firstly, all but Ghana are relatively newcomers to the industry, and secondly bauxite production in the countries selected has reached the stage where confident prognosis can be made.

DATA BASE

Bauxite production growth curves

Figure 1 shows the domestic bauxite production growth curve for France, the United States and the Soviet Union between 1900 and 1981. The domestic bauxite production curve for France shows a general increase up to 1930, and then levelled off until the mid-forties, when it again increased until the mid-sixties, levelling off again in the early seventies, finally decreasing when massive imports of high-quality bauxite began. Bauxite production began in the Soviet Union in 1931. This production growth curve shows a rapid increase until the early forties, and rejuvenating from 1944 to 1970, and levelling off. The bauxite production growth curve for the United States is more complex and suggests the resolution of composite growth curves. It shows a gradual increase in production between 1900 and 1920 and a decline from then until the early thirties. The decrease in domestic bauxite production was later followed by an increase until 1950, since, when it has remained steady. Since 1960, the domestic production in each of the three major aluminum market areas has matured, as is indicated firstly, by their need to import higher grades of bauxite, and secondly, by the levelling off of the bauxite production growth curves.

1. Yugoslavia is listed in the upper quarter of the middle-income countries in the World Bank, World Development Report-1981, pp. 134-135.

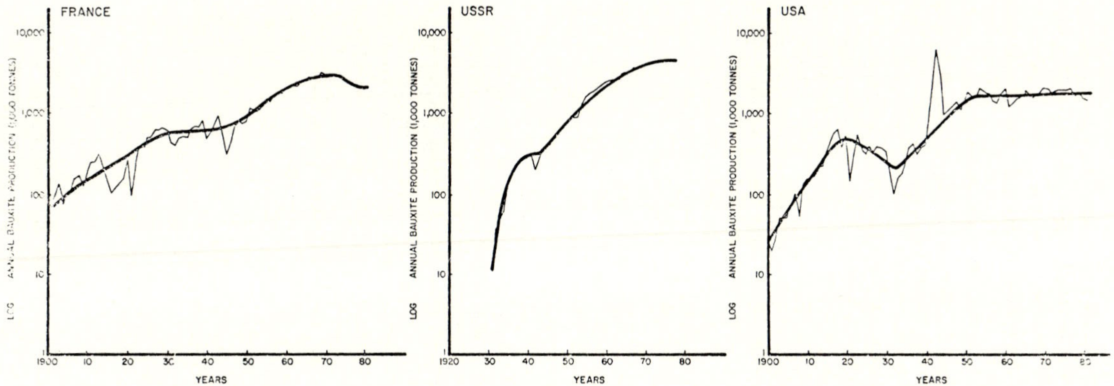

Figure 1: Bauxite Production Growth Curves for Major Aluminium Market Countries with significant bauxite resources

Figures 2 and 3 show the bauxite production growth curves for some of the overseas suppliers to the United States market. Suriname and Guyana were the first two and produced Guyana type bauxite which is the type to which Arkansas bauxite also belongs. The similarity of the bauxite production growth curves for Suriname and Guyana is striking. With the entry of Reynolds Metals Company and Kaiser Aluminum (then Permanente Metals) into the aluminum industry in the United States[2], Jamaica became a major supplier from 1952, with Haiti and the Dominican Republic later becoming secondary suppliers (Figure 3). The bauxite production growth curve for Jamaica shows a sharp increase from the commencement of mining until the sixties, followed by a slightly slower growth rate until the mid-seventies, when production decreased and levelled out. The cases of Haiti and the Dominican Republic are similar and suggest managed supplementary patterns, which appear to change in a step-wise fashion, declining after the mid-seventies. The production growth curve for Brazil (Figure 2, Brazil) the sixth country in this group, shows minimal growth in bauxite production until 1978 when the export of bauxite from Trombetas began. In the North American market, bauxite from the Sangaredi Mine in Guinea and Trombetas in Brazil are now preferred for use in the sweetening stage of US Gulf Coast plants using this process modification. This is so because on the one hand, their higher grade permits them to compete with bauxite from the Guianas, which although of the same type, now has a higher silica content. On the other hand, imports of these bauxites are also replacing some bauxites from Jamaica. The effects of the commencement of bauxite mining in Jamaica in 1952 on the production in Suriname and Guyana as against the effects of the coming on stream of Trombetas and Sangaredi can be noticed.

Bauxite production in Europe has been complicated by political factors following the Second World War. These resulted in the establishment of separate Eastern and Western European markets. Among the European bauxite-producing countries (other than France and the Soviet Union), there is firstly, the group comprising Yugoslavia, Hungary, and Greece (Figure 4), countries which have major bauxite resources; a second group includes Italy and the Federal Republic of Germany, countries which have only minor bauxite resources. The bauxite production growth curves for all three members of the first group of European countries, that is those with major bauxite resources, show an initial steady increase in bauxite production, and then a discontinuity due to the Second World War. This is followed by a slower growth rate as these countries changed emphasis from being major exporters of bauxite to developing their local aluminum industries. The timing and implementation of this new policy reflect the relative stage of industrialisation, availability of energy and emphasis on the development of the local aluminum market of these countries. On the other hand, Italy and the Federal Republic of Germany, as industrialised countries with only minor bauxite resources, rapidly exhausted their meagre economic bauxite resources and had to rely on imports, initially, from Yugoslavia, Hungary and Greece. However, at the beginning of the seventies, Italy and the German Federal Republic began importing bauxite from Australia and Guinea, which then emerged as major bauxite-producing countries.

2. Alcan played a vital role in the development of the bauxite resources of Jamaica. However, because the alumina which Alcan produced in Jamaica since 1952 was mainly exported to Canada and Europe, it is not directly relevant to the arguments here.

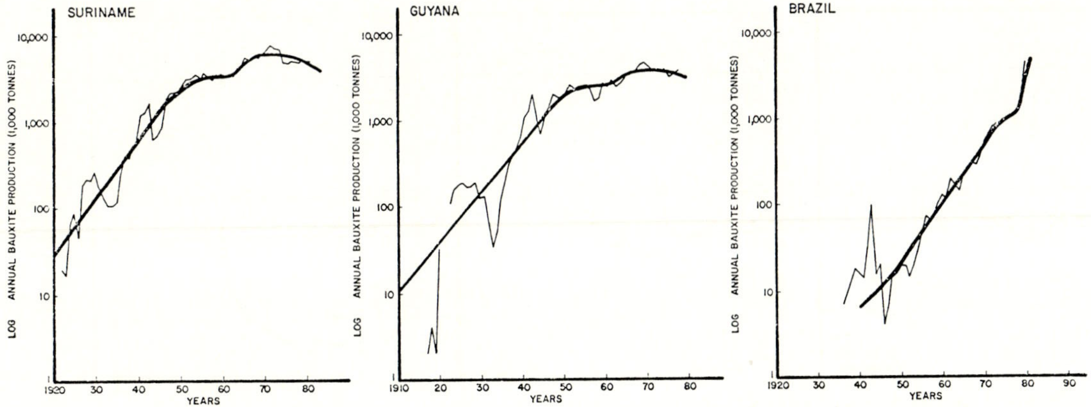

Figure 2: Bauxite Production Growth Curves for Overseas Guyana-type Bauxite Suppliers to the United States Market

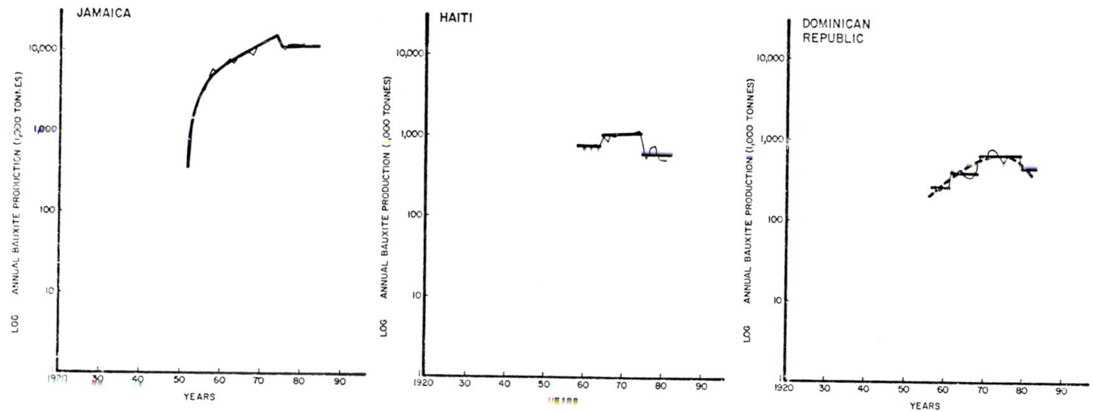

Figure 3: Bauxite Production Growth Curves of Jamaica Type Bauxite for the United States Market

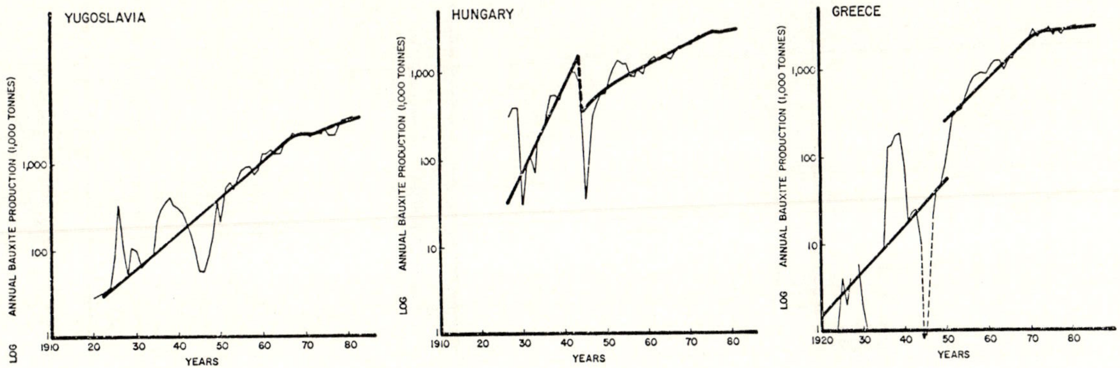

Figure 4: Bauxite Production Growth Curves for the Primary Bauxite Suppliers to Europe

Japan initially depended almost entirely on imports of bauxite, mainly from Malaysia, Indonesia and India to supply its aluminum industry. During the Second World War, alunite from Korea was used for aluminum production (Bracewell, 1962, p. 198). However, Australia became the primary supplier to this market shortly after 1963.

Figure 5 illustrates the bauxite production growth curves of Australia and Indonesia, two of Japan's important suppliers. Indonesia as a small producer, plays a secondary role in this market, but has the advantage of being an energy-endowed country, and as such it is now developing further down-stream processing in co-operation with Japanese, American and West German companies. The bauxite-production growth curve for Australia shows distinct stages of development. There is, firstly, the stage when only minor bauxite mining occurred in Victoria, Queensland and New South Wales for use chiefly in their domestic chemical and steel industries (Bracewell, 1962, p. 128). However with the commencement of bauxite mining in Weipa and the Darling Ranges in 1963 and later at Gove, production rose rapidly until the late seventies when it appears to have levelled off at approximately 29 million tonnes per annum, if we ignore the present short-term reduction due to current market conditions.

Bauxite Resources and Production Grades

(i) Major Aluminum Market Countries

The relationship between the bauxite resources of the major aluminum market countries is illustrated by the Arkansas bauxite deposits. Table I illustrates the data on the resources.

Table I: Changes in Bauxite Resources in Arkansas
(in million tonnes)

Year	Economic Resources	Sub-economic Resources	Source
1901	50.0		Hayes (Gordon et al., 1958 p. 168)
1941	17.5	9.7	Mionen & Bourchard, (Gordon et al., 1958, p. 6)
1946	39.4	23.7	US Bureau of Mines
1949	47.7	24.6	US Bureau of Mines
1963	36.4	43.1	Min. Res. of Arkansas, 1964

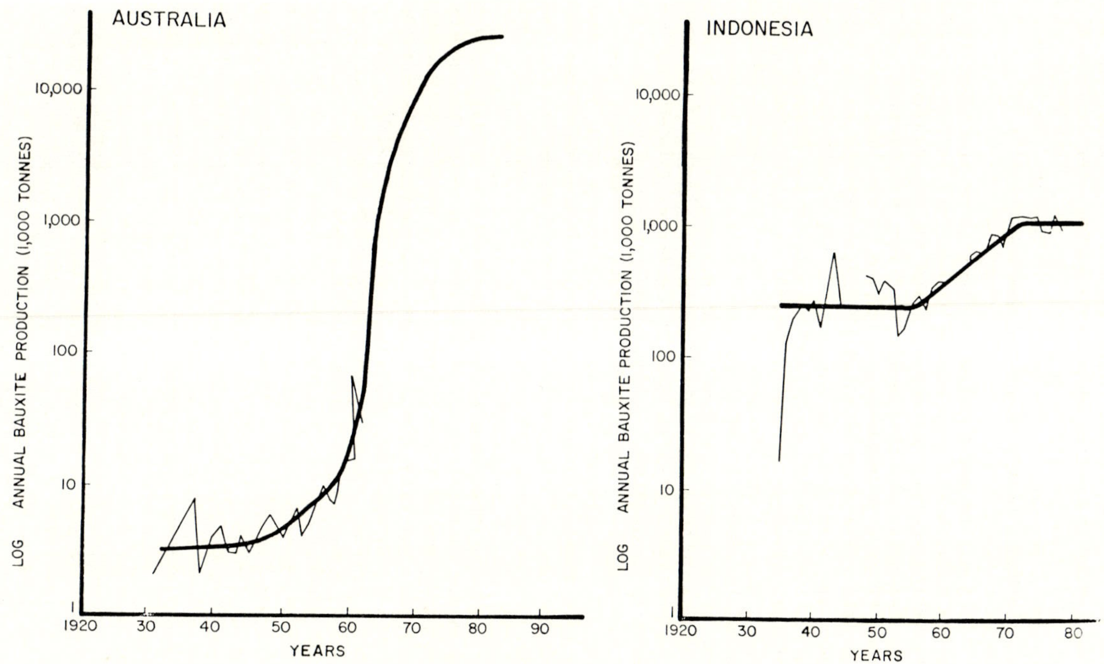

Figure 5: Bauxite Production Growth Curves for Some Suppliers to the Japanese Market

The authors understand that since 1963, the reserves available in the Arkansas area have exceeded 30 years' supply in spite of continued mining. This reflects the effects of improved mining methods, the mining and processing of lower-grade bauxite, (up to 18 percent silica) and the blending of local bauxite with imported bauxite, notably from Guyana and Suriname. Before the Second World War, bauxite produced from the Arkansas mines contained approximately 58 percent alumina and 6.5 percent silica. At that time, bauxite containing more than 7 percent silica was either discarded or left unmined except where the iron content was sufficiently low to permit its use as chemical grade bauxite (Gordon et al, 1958).

(ii) The Cases of the Major Bauxite Suppliers

Yugoslavia and Hungary are typical of the countries which originally exported bauxite to neighbouring industrialised countries with primary aluminum industries, but which, after the Second World War, developed and integrated their own aluminum industries. Figure 6 (A and B) shows the changes in bauxite grade over a period of time for these countries. The associated effects of improved digestion efficiency in Bayer plants in Hungary are also shown in Figure 6B. Yugoslavia has a long association with bauxite. Breithaupt identified bauxite in the territory which is now Yugoslavia as early as 1847 and trial mining started in Istria in 1912. However, significant production did not begin until 1919 and by 1938 it reached 405 thousand tonnes to make it the third largest producer in the world, after France and the United States. All of the bauxite produced in Yugoslavia[3] before 1952 was exported as crude ore. Between 1952 and 1972, the export level of crude bauxite decreased slightly but was still over 90 percent of production. However, as a result of the construction of integrated facilities, most of the bauxite produced has been processed locally since 1976. Consequently, Yugoslavia is now able to produce lower-grade bauxite economically and its economic bauxite resources have remained at approximately 300 million tonnes since 1970 despite increased mining (Besic et al., 1965; Buric, 1966; Chadwick, 1982; Marusic, 1954 and Zeravica, 1954).

The relationships between the bauxite production growth rate, growth of economic bauxite resources and level of down-stream processing and development are well demonstrated by the growth of the industry in Australia. The large Darling Ranges deposits contain approximately 30 percent available alumina, and 24 percent total silica, but only 1.5 percent reactive silica. However, because the deposits are located in the Perth area, which has well developed infrastructure and because most of the gangue mineral can be rapidly separated and washed immediately after digestion, thereby minimising dilution of the liquor, the bauxite is economically beneficiated to alumina in <u>in situ</u> dedicated Bayer plants (Sibly and Buckett, 1981, pp 130-131). At present, the Kwinana, Pinjarra and Wagerup plants have a total operating capacity of 4.1 million tonnes of alumina per annum and, including the Worsley Plant, the projected total capacity is approximately 6 million tonnes of alumina per annum. Similar developments occurred almost concurrently in Queensland, based on the large high-grade Weipa bauxite deposits and later, at Gove in the Northern Territory (Anon., 1981, pp 4-8). There is now a projected total metal grade alumina capacity of slightly under 10 million tonnes per annum. This capacity is associated with mines, Bayer plants, smelters and fabrication plants. Of course, most of the alumina and metal produced will be exported for the short to medium term. There are in addition, large bauxite deposits in the Kimberley Area.

3. Before the Second World War, Yugoslavia had only a very small alumina and smelting capacity near Lozovac.

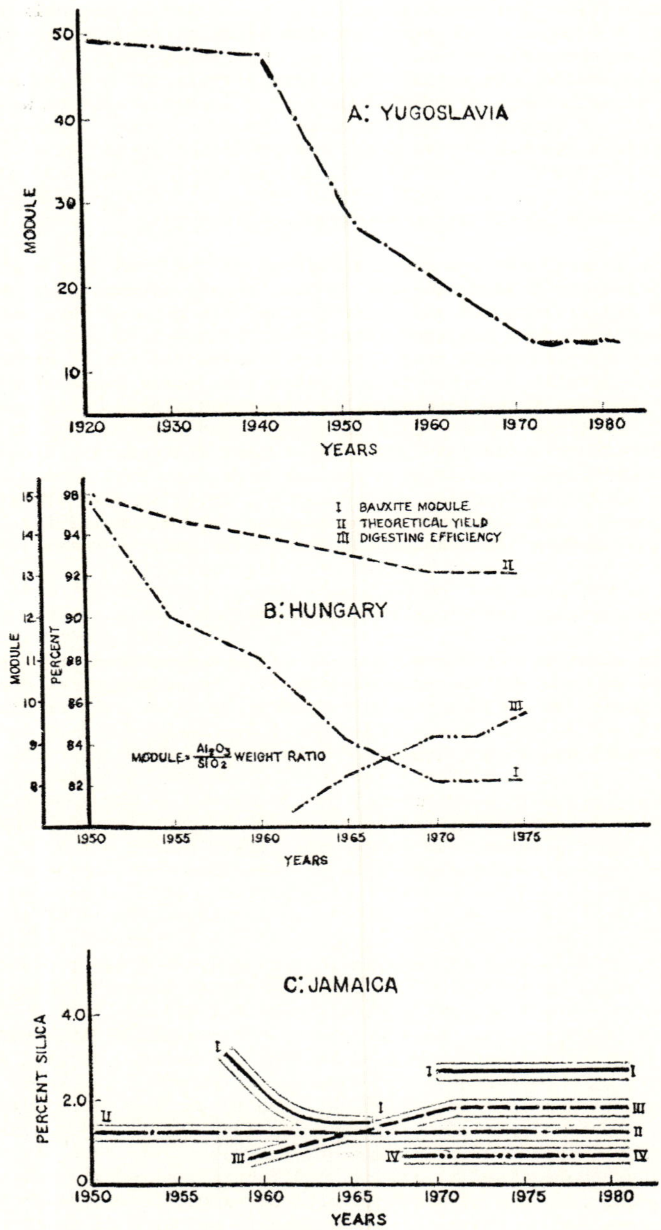

Figure 6: Generalised bauxite quality trend for some producing countries. (8B after Juhasz, 1978, p. 363)

West Australia, giving Australia measured identified developed economic bauxite resources, of 2,500 million tonnes and, in addition over 4,000 million tonnes of identified inferred and demonstrated undeveloped bauxite resources. Given these bauxite resources, coupled with abundant coal resources, and taken in conjunction with their present expertise in the aluminum industry, the rise of the industry in Australia is understandable, and the projections are optimistic. The proximity of Australia to Japan and also to the west coast of North America is another advantage.

Jamaica developed into a major supplier of crude bauxite and alumina to the nearby North American market, but to date, it has had only minor involvement in the other stages of the industry, despite abundant bauxite resources and the development potential of the regional market.

The bauxite grade-time production trends for the operations in Jamaica (Figure 6C) show that one of the operations exporting crude bauxite (Figure 6C-III) decreased the grade of the bauxite produced until 1971, after which the grade produced levelled off. Another operation exporting crude ore, and two others which produce alumina locally, mined bauxite of different but relatively constant grades. The grade trend of another operation appears to be more variable than that of the other cases.

The economic bauxite resources of Jamaica increased from an initial US Bureau of Mines estimate of 5 million tonnes in 1943 to 100 million tonnes in 1953 (Anon., 1971, pp 7 and 10). In 1973, the estimate was 1,000 million tonnes, and is currently estimated to be 2,000 million tonnes (Lyew-Ayee, 1980, p. 28) reflecting the results of increased exploration.

(iii) The Cases of the Minor Bauxite Suppliers

The total silica content of bauxite produced from the small operations in Haiti has been progressively reduced below the average grade of the deposits in the Rochelois Plateau since 1977. The result has been that the mineable bauxite was prematurely depleted in mid-1982. Table II gives our estimate of this trend.

Table II: Bauxite Resources Grade Relationship for Haiti

Year of Estimate	Maximum Average Silica	Estimated Mineable Resources
1974	3.3%	9.5 million tonnes
1978	2.8	9.5 million tonnes
1979	2.2	2.2 million tonnes
1980	2.0	1.3 million tonnes

Sarawak, Malaysia is another documented case of a minor bauxite mine supplying the Japanese market. Bauxite was first recorded as occurring there as small concretions in clays in 1911, and by 1955 twelve bauxite occurrences were identified in the Kuching areas. Table III gives the then estimates of the reserves.

Table III: Estimates of Bauxite Reserves in Sarawak, Malaysia

Deposit	Area in acres	Average thickness in feet	Tonnage of washed ore	SiO_2 Percent	Al_2O_3 Percent
Munggu Belian	.. 220	10	2,600,000	3.42	56.8
Gunon Gebong	.. 340	5 - 7.5	1,900,000	4-7	48-55
Tanjong Serabang		5.4	1,100,000	6	47
Total			5,600,000		

(Source: Bracewell, 1962, p. 117)

Mining operations commenced at Munggu Belian in 1958 and about four tonnes of crude material were required to produce one tonne of washed bauxite for export. Between 1958 and 1965, the Munggu Belian Mine produced approximately 1.6 million tonnes of washed bauxite averaging 3 percent silica of an initially estimated 2.6 million tonnes with an average of 3.42 percent silica. Mining operations ceased at this juncture[4].

The development of the bauxite resources of Indonesia contrasts with the cases of Haiti and Sarawak. Bauxite was discovered in Indonesia in 1924 and the exploitation of the largest deposits which were on the island of Bintan, began in 1935. Their measured and indicated economic bauxite resources were then estimated to be 17.5 million tonnes of washed bauxite with an average silica content of 4 percent. The grade of bauxite exported in 1935 was 53.0 percent alumina by difference and 2.5 percent total silica. However, the average grade shipped before the commencement of the Second World War was 3.0 percent total silica. This grade was changed to 4 percent after the war (Bemmelen, 1941, pp 136-137), and in recent years the export grade has been increased to slightly less than 5 percent total silica. In 1977, P.T. Aneka Tambang estimated the economic bauxite resources to be over 19 million tonnes of washed ore, of which 8.8 million tonnes contain an average of 5.07 percent total silica, which their current export grade approximates. The remainder contains an average grade of 7.5 percent total silica. In addition, it is estimated that the bauxite resources in West Kalimantan amount to approximately 800 million tonnes of washed bauxite, averaging less than 3 percent total silica (Simandjuntak, 1979, pp 12-17). Current development plans for the industry involve the construction of a 600,000-tonne per annum American Bayer alumina plant in Bintan designed to process bauxite containing up to 13.3 percent total silica, but with a re-active silica content of under 7.0 percent (Anon., 1981A, pp 18-21). This plant will complete the integration of the aluminium industry in Indonesia, since the Asahan smelter which shipped its first ingot in mid-October 1982 and some fabrication facilities already exist.

4. Letter dated November 11, 1982, from Principal Geologist, Goelogical Survey of Malaysia, Sarawak.

DISCUSSION

The bauxite production growth curves and grade data presented demonstrate that economic bauxite resources like those of other mineral commodities are dynamic rather than static. They further show that the situations of major aluminum market countries with significant bauxite resources (e.g. the United States, France and USSR) and the technologically advanced and the most advanced developing countries which were initially primary bauxite exporters (e.g. Australia, Brazil, Greece, Hungary and Yugoslavia) are similar in that they now have aluminum industries and established markets. The idealised bauxite-production life-span curves for these countries as shown in Figure 7 1A & 1B demonstrate that the annual bauxite production levels off rather than declines. Their economic bauxite also levels off in spite of continued mining. The difference is a matter of the time frame within which these countries develop their own aluminum industries and, obviously, the size of their bauxite resources. On the other hand, there is a major difference between the technologically advanced and the under-industrialised primary bauxite suppliers. The economic bauxite resources of the latter increase with exploration and development to a maximum and then decrease until they are no longer competitive, unless this situation is changed by down-stream developments, which would provide the impetus for these countries to become technologically advanced primary bauxite suppliers. The eventual decrease in bauxite production of exporting countries is often aggravated by competition from new bauxite mines coming on stream in other countries, making the older deposits increasingly uncompetitive and, in the long run, uneconomic. On the other hand, if industrialisation is successful, then the producer switches from being a class 2a producer to becoming a class 1b producer. Indonesia typifies this at present (Anon. 1981, p. 18). The bauxite production growth curves of the countries with small bauxite resources appear to be similar superficially, but the difference is that the industrialised bauxite-producing countries (e.g. Germany and Italy) have the capacity to develop aluminum industries based on bauxite imports, while the under-industrialised countries eventually inherit only ghost mining camps.

It should be clearly pointed out that neither the life-span curves nor the production growth curves are suitable for forecasting the bauxite production rate of a country or region because they are subject to the effects of the policies applied to the industry. Their advantage is that their use facilitates the understanding of the options and prospects of a bauxite-producing country on the basis of its bauxite resources, the maturity of its industry and its industrialisation policies. In other words, small bauxite-producing countries such as the Dominican Republic, Ghana, Haiti, Malaysia and Sierra Leone require different policies from those of the larger producers such as Australia, Brazil, Guinea and Jamaica. Further, the effects of specific policies or changes in policies can be assessed with improved reliability once the appropriate life-span grouping and maturity of the industry in a particular country are correctly recognised (see Figure 7).

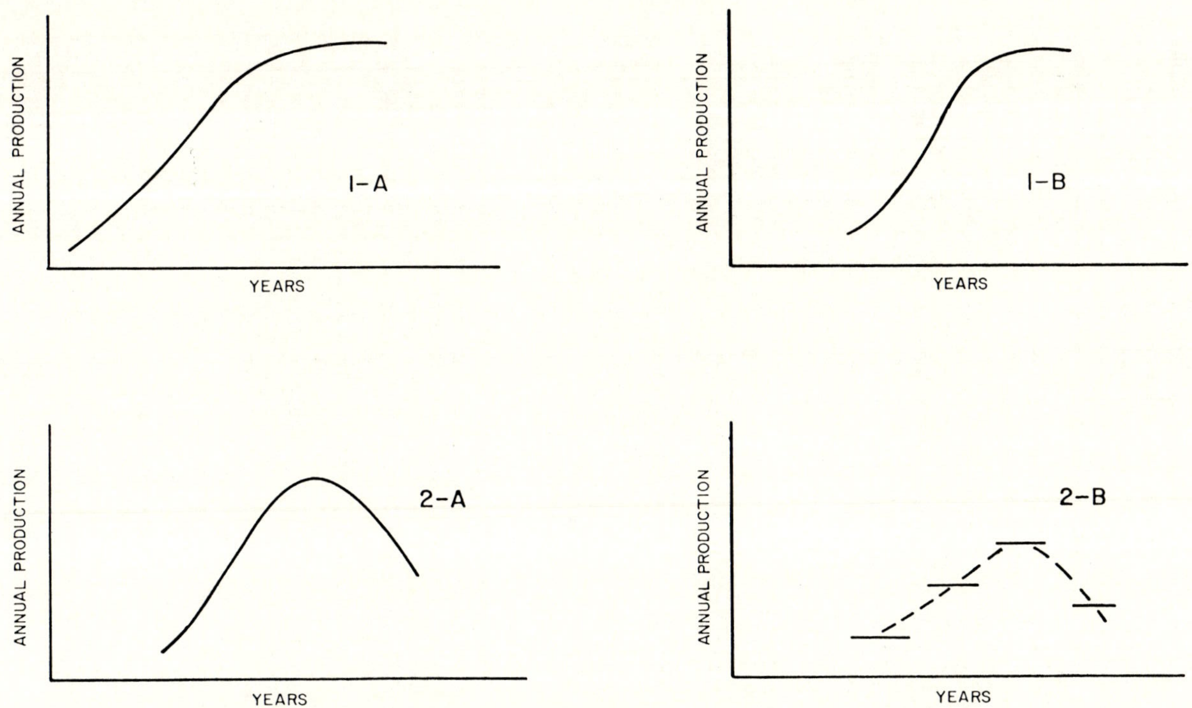

Figure 7: Idealised Bauxite Production Life Span Curves for the Four Types of Bauxite-producing Countries Identified

1A: Major Aluminum Market Countries with Significant Bauxite Resources
2A: Technologically Advanced Primary Bauxite Producers

2A: Technologically Disadvantaged Primary Bauxite Suppliers
2B: Secondary Bauxite Suppliers

The major aluminum companies have a responsibility in influencing the industry, particularly when long-term national security considerations of their home countries are considered. They can play a crucial and constructive role in the development of the aluminum industries and increase aluminum consumption in the under-industrialised bauxite-producing countries by supplying technology and scarce skills, undertaking training programmes for host-country nationals and initially, providing market off-take arrangements for part of their production. To be effective, this important consideration will need to be matched by the willingness on the part of the international finance institutions to play the role of financing of the required project or the required project infrastructure, and bauxite-producing countries must guarantee the security of the investments (Touré and Hill, 1981, p. 31).

The major under-industrialised bauxite-producing countries may prefer the options of the development of the aluminum industry as against the medium to long-term export of crude bauxite which lead to the depletion of this mineral resource, possibly without adequate compensation. The development of either a local or regional aluminum industry in bauxite-producing regions or countries deserves evaluation.

In the case of bauxite-producing countries, particularly those with inadequate energy resources, the evaluation of the merits of a product-sharing system to allocate the metal produced between the suppliers of energy and technology may be considered. At that stage, a mutually agreed product-sharing formula must be designed to allocate the metal between the contributors of inputs, and all steps taken to ensure that the metal is sufficiently competitive with alternative materials that it can continue to penetrate new markets, thereby achieving continued growth. The use of secondary aluminium is one approach which has proven useful in increasing the competitiveness of aluminium because its net effect is to significantly lower the energy content of the aluminium supply. It is recognised that this lowers the demand for primary aluminium, but on the other side the make-up primary aluminium needed for the production of secondary aluminium alloys may represent a market which would not be were recycling absent.

CONCLUSIONS

The conclusions from this study are:

(i) The options and prospects of bauxite-producing countries depend on the extent of their bauxite resources, the maturity of their industry, and their industrialisation policies. In other words, small bauxite producers such as the Dominican Republic, Ghana, Haiti, Malaysia and Sierra Leone require different policies from those of the larger producers such as Australia, Brazil, Guinea and Jamaica.

(ii) The policy adopted affects the quantum of the economic bauxite resources of a country, and so bauxite-producing countries need to tailor their strategy to their requirements, in order to optimise the benefits which they derive from the exploitation of their bauxite resources.

The market competition arising from the present international trade in bauxite compels most bauxite-exporting countries to exploit their higher grades of bauxite. The long-term continuation of this policy will lead to the early depletion of the economic bauxite resources in some bauxite-producing countries. The mines in the poorer countries, whose bauxite resources are small, will usually be closed first. This is not in the long-term interest of either the bauxite producing or consuming countries, and a change of policy appears to be urgently necessary. It is suggested that this should involve the beneficiation of the bauxite to alumina near the mines and the participation of the bauxite-producing countries in further down-stream processing with emphasis on the production of semi-fabricated and aluminum products, whether or not the country produces metal. This will promote an expansion of the aluminum market and the increased industrialisation of these countries. It is suggested that if this is done, either on a national or regional basis, the development and the expansion of the economic bauxite resources, rather than the depletion and closure of bauxite mines, will occur in many bauxite-producing countries. In the final analysis, this will also enure to the benefit of the major consuming countries.

ACKNOWLEDGEMENTS

The bauxite production data for the period 1900 to 1969 are from US Bureau of Mines publications, but the IBA Secretariat statistics were used for the period 1970 to 1981. The authors wish to thank their colleagues of the IBA Secretariat for critically reviewing the manuscript, but they alone are responsible for the opinions expressed.

REFERENCES

1. Anon., 1956: Materials Survey: Aluminium; Compiled for Office of Defence Mobilization, the Business and Defence Services Administration, US Department of Commerce.

2. Anon., 1971: Reynolds in Jamaica, Reynolds Jamaica Mines Ltd., 85 pp.

3. Anon., 1981: Submission by the Department of Trade and Resources, Senate Committee on National Resources; Australia, 34 pp.

4. Anon., 1981A: Pertambangan Bauksit dan Proyek Alumina Bintan, Aneka Tambang, Jakarta, 21 pp.

5. Bardossy, Gyorgy, 1982: Karst Bauxites; Elsevier Scientific Publishing Company, Amsterdam, Oxford, New York.

6. Bemmelen, R.W. Van., 1949: The Geology of Indonesia, Vol. II, Economic Geology, 265 pp., Government Printing Office, The Hague.

7. Besic, Z.; Vukovic, V. and Cicovic, B., 1965: Boksiti Crne Gore, Rudnici Boksita, Niksic, 164 pp.

8. Bracewell, Simth, 1962: Bauxite, Alumina and Aluminium, Overseas Geological Survey, H.M. Stationery Office 235 pp.

9. Buric, P., 1966: Geologija Lezista Boksita Crne Gore. Posebna Izdanja Geoloskog Glasnika. Sarajevo. VIII (1966) 276 pp.

10. Chadwick, J.R., 1982: Yugoslavia: Mining Industry with Considerable Potential. World Mining, Vol. 35, pp. 52-55.

11. Davis, C.E., 1982: The Changing Geography of Bauxite Production and its Impact on the World Aluminium Industry. Metal Bulletin (in press).

12. Gordon, MacKenzie, Jr.; Tracey, J.I., Jr. and Ellis, M.W., 1958: Geology of the Arkansas Bauxite Region: US Geol. Survey Prof. Paper 299, 268 pp.

13. Grubic, A., 1975: Geology of Yugoslav Bauxites. The Serbian Academy of Sciences and Arts, Monographs.

14. Hill, V.G., 1980: The Rational Management of Bauxite Resources, J. Geol. Soc., Jamaica, Special Issue, Bauxite Symposium V., pp. 1-19.

15. Hill, V.G. and Ostojic, Slavko, 1981: The Bauxite Supply Potential for the Aluminium Industry, ICSOBA; Tihany, Hungary, pp. 71-89.

16. Hill, V.G. and Robson, R.J., 1982: Transport Costs and Refinery Sizing; J. Geol. Soc. Ja., Special Issue, Bauxite Symposium V., pp. 257-264.

17. Hill, V.G., Robson, R.J. and Ostojic, Slavko, 1982: Some Implications of the Changing Bauxite Supply Pattern; Light Metals, pp. 3-13.

18. Marusic, R., 1954: Reserven und Qialitsbauxit in Jugoslawien, Aluminium, Vol. 30; Dusseldorp. r/a 399 pp.

19. Juhasz, Adam, 1978: Complex Development of Alumina Processing; Hungary, ICSOBA, Athens Meeting, Vol. 1, pp. 342-369.

20. Sibly, John M. and Buckett, Leslie N., 1981: Alcoa of Australia Limited, Western Australian Alumina Refining Operations; Light Metals, pp. 129-137.

21. Simandjuntak, H.R. Wilher, 1979: Some Notes on Ore Deposits of Indonesia, DMR-Indonesia Report, 20 pp.

22. Touré, Lamine and Hill, V.G., 1981: The Development of Energy Resources in Bauxite-Producing Countries; in Towards Increased Co-operation in Bauxite Development; Proceedings of the International Conference on Bauxite, Ocho Rios, Jamaica.

23. Zeravica, M., 1954: Boksiti Jugoslavie S Narocitim Osvrtom Na Ekonomicnost Rroizvodnje. Rudarstvo i metarlugija BR. 3, Beograd.

THE USE OF SYNTHETIC FLOCCULANT POLYMERS IN SETTLING RED MUDS DERIVED FROM HIGH GOETHITE BAUXITE ORES

S.E. Sankey and R.J. Schwarz
Hercules Incorporated
Wilmington, Delaware

Introduction

A previous paper(1) discussed the substitution of synthetic water soluble polymers for starch in the separation of red mud from process liquor in digester blow-down and in the washing of this mud to remove valuable aluminate and caustic.

This paper continues that discussion with emphasis on the use of synthetic water-soluble polymers for the settling of red muds derived from yellow bauxite ores in which the iron oxide component contains high goethite concentrations. These are recognized as difficult-to-settle muds.

In this paper, we show how the mineralogy of the high goethite ores and the Bayer process chemistry control the resulting digester red mud slurry properties, and how these in turn affect the performance of typical synthetic water-soluble polymers used in settling these slurries.

Included in this discussion is a review of flocculation principles to help the reader better understand how synthetic water-soluble polymers interact with the red mud particles.

Data obtained from laboratory digestions are used to illustrate the performance of synthetic flocculant polymers in the settling of red mud slurries.

Bauxite and the Bayer Process

Although the major emphasis in this paper is on the performance properties of synthetic water-soluble polymers used in red mud settling, a brief review of bauxite mineralogy is given for the unfamiliar reader since this mineralogy has a major effect on the settling characteristics of the red muds obtained in the Bayer process.

The major components of Jamaican bauxite are listed in Table 1. The general classifications of red and yellow bauxite, based on composition, are given in this table. The classification of the bauxite is dependent upon the relative hematite and goethite content of the ore. The red bauxites have a relatively high hematite and low goethite content, while the yellow bauxites have a relatively high goethite and low hematite content. Additionally, there is a higher substitution of AlOOH in the goethite of yellow bauxite compared to red bauxite. Another very important difference is that the yellow bauxites contain a much higher concentration of secondary phosphates which are later shown to be severe antagonists in the red mud settling process.

The major alumina components in these ores consist of gibbsite and boehmite, the trihydrate and the monohydrate, respectively. Boehmite requires much more severe digestion conditions in temperature and alkalinity to form the soluble sodium aluminate.

Bayer Process Chemistry

The basic Bayer process chemistry for gibbsite is:

$$\underset{\text{(Gibbsite)}}{Al_2O_3\cdot 3H_2O} + 2NaOH \rightarrow \underset{\text{(Sodium Aluminate)}}{2NaAlO_2} + 4H_2O$$

The alumina trihydrate and monohydrate are reacted with caustic to form the soluble sodium aluminate. The valuable component, that is the aluminum, is thus dissolved in the process liquor; while the waste iron oxides, silica, etc., (red mud) are present as particulates in the slurry. The separation of the liquor from the red mud is a classical liquid-solid separation. A typical synthetic liquor used for a low temperature laboratory digestion of low boehmite ores consists of the following: 140 g/l NaOH, 50 g/l Na_2CO_3, giving 235 g/l (as Na_2CO_3) total soda. The digestion is run at an alumina to total alkalinity ratio of about 0.65, a digestion temperature of 140°C. and a digestion time of 1/2 hour.

The major process steps of the Bayer process are outlined in Figure 1. The digestion portion of this process has been covered above. In general, the blow-off from the digester is cooled in a flash tank and then diluted before being sent to the settler. The settler represents the first stage of liquid-solid separation, that is, separation of the aluminate liquor from the red mud. In the settler, sufficient flocculation of the red mud fines must occur in order to adequately reduce the suspended solids level of the liquor being sent to the filters. Inadequate removal of suspended solids may cause blinding of these filters. After filtration, the subsequent stages of precipitation and calcining recover alumina which will eventually be electrolytically reduced to produce aluminum.

The red mud which is removed from the settler contains valuable aluminate and caustic. The function of the washers, which may be single or multi-staged countercurrent decantation units, is to concentrate this mud and return the valuables back to the process. The washers or thickeners must be operated so that red mud levels and compaction do not impose severe loadings on the rakes within these units. Too rapid settling and/or severe compaction could also promote difficulty in removing (pumping) the mud from these units.

Synthetic Water-Soluble Polymers

In the flocculation of red mud, the synthetic water-soluble polymers which have found significant application are copolymers of acrylamide and sodium acrylate. The chemical representation of these polymers is shown in Table 2. The majority of polymers in use are high functionality types which range from 50-100% anionic in character. Typical application ranges of these polymers are also shown in Table 2.

TABLE I

JAMAICAN BAUXITE MINERALOGY

MAJOR COMPONENTS

COMPONENT	TYPICAL COMPOSITION (%) RED BAUXITE	YELLOW BAUXITE
GIBBSITE ($Al_2O_3 \cdot 3H_2O$)	72	66
BOEHMITE ($Al_2O_3 \cdot H_2O$)	2	4
HEMATITE (Fe_2O_3)	12	3
GOETHITE (FeOOH)	6	19
(% AlOOH in GOETHITE)	(11)	(17)
MINOR COMPONENTS	8	8
(SECONDARY PHOSPHATES)	(LOWER)	(HIGHER)

TABLE 2

ACRYLAMIDE - SODIUM ACRYLATE COPOLYMERS

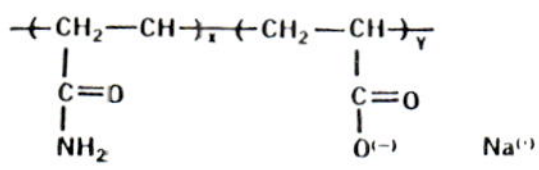

	SYSTEM ALKALINITY (G/L AS Na_2CO_3)	POLYMER FUNCTIONALITY X (ACM)	Y (SA)
SETTLING DIGESTER BLOW-OFF (210°F, 3% SLURRY)	250	0-5	100-95
WASHING RED MUD (100-75°F, 6-15% SLURRY)	70 ↓	0-20	100-80
	2	40-50	60-50

FIGURE I

BAYER PROCESS

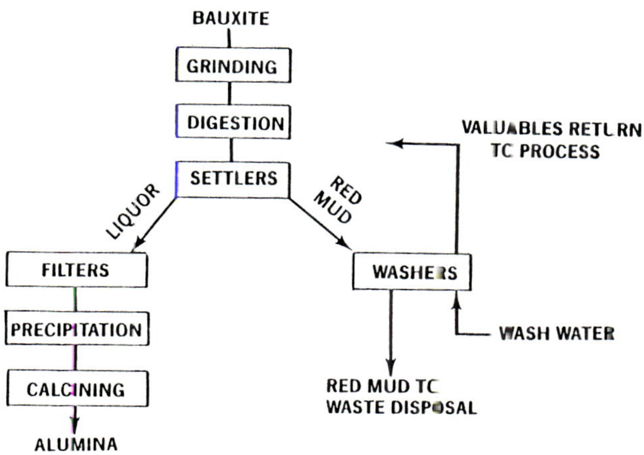

In settling red muds derived from digester blow-offs where the system alkalinity can be as high as 250 g/l, the high anionic functionality polymers such as poly(sodium acrylate) are typically used. Copolymers containing a small amount of acrylamide have also been shown to be extremely effective for this application. The reason for this will be discussed in the section of this paper that covers the bonding forces of polymers with red mud particles.

In the washing of red mud to remove valuable sodium aluminate and caustic, the lower anionic functionality copolymers are found to be most effective. The copolymers most effective in settling these washer muds will depend on the alkalinity of the mud. Typical ranges for these polymers are shown in Table 2. Higher anionic copolymers will be most effective with the more alkaline slurries. The data in Table 2 show that the bauxite processor could use copolymers of at least two different functionalities in the settling of red mud in his plant. The highest anionic types would be used in the digester settlers and possibly in the early washer stages. A lower anionic type would be used in the latter red mud washer stages. These lower functionality copolymers would also be used in the final thickening stages prior to pumping the red mud to disposal ponds.

Polyacrylamide copolymers useful in red mud settling are produced in three basic product forms: dry powders, gels and emulsions. Dry powders contain approximately 90% polymer solids and 10% volatiles. Gel type products contain approximately 33% polymer solids and 67% water. Emulsions will contain approximately 33% polymer solids, 30% hydrocarbons and surfactants, with the rest being water.

The product form is controlled by the process technology which the manufacturer of the acrylamide copolymer is practicing. Irrespective of product form, acrylamide copolymer manufacturers can alter the anionic functionality of their polymers to meet the specific needs of the flocculation application.

Flocculation Mechanisms

In Figure 2, two mechanisms of flocculation are illustrated. In both mechanisms, the adsorption of a positively charged cationic polymer occurs on a negatively-charged substrate. In a later section of this paper, we describe why red mud particles have negative charges on them. We also discuss why anionic polymers are used in red mud flocculation in preference to cationic polymers. But for the discussion of these flocculation mechanisms, it is useful to illustrate the adsorption of a positively charged cationic polymer on a negatively-charged substrate because the attraction of oppositely charged substances is well recognized.

In the top illustration of Figure 2, the adsorption of a low molecular weight cationic polymer occurs on the negatively-charged particles. Negative charges on these particles normally cause them to be repelled, inhibiting the self-coagulation mechanism. The chain lengths of the low molecular weight polymers are insufficient to cause the polymer to bridge between two particles. The net effect of the polymer absorption is the reduction of the effective negative charge on the particles. This reduction of charge in many cases allows the particles to coagulate; that is, to flocculate. These coagulated particles normally exist as very small flocs.

In the bottom illustration, the bridging of particles by a high molecular weight cationic polymer is shown. In this mechanism, large flocs can form which will settle very rapidly. In this illustration, the bridging polymer could actually be a very high molecular weight polymer or an aggregate of polymers; that is, entangled polymers which could give a very large molecular structure. Later on in this paper, we discuss the bonding mechanisms which give these long aggregated polymeric structures.

Polymer effectiveness as a function of dose level is illustrated in Figure 3. Either an underdose or overdose of polymer will result in poorly flocculated or restabilized particles. The optimum polymer dosage will cause large flocs to form which will settle rapidly.

There are several bonding forces which can come into play in inter-molecular bonding as well as the bonding of molecules to suspended particles. These forces are illustrated in Figure 4. Van der Waal's forces which occur through electron orbit synchronization are extremely weak and are easily broken at the thermal energies associated with flocculation. Hydrogen bonding forces are much more powerful and can play a significant role in the bonding of two molecules together as well as in the bonding of flocculant molecules to suspended particles. The role of ionic bonding forces, also illustrated in Figure 4, has already been described when the charge neutralization and bridging mechanisms in flocculation were discussed.

Anionic acrylamide copolymers, used in red mud flocculation, can therefore bond via two mechanisms. These include hydrogen bonding through the amide group ($-NH_2$) on the acrylamide portion of the chain as well as ionic bonding through the carboxylate group ($-COO^{(-)}$) on the anionic portion of the chain.

Particles in Suspension

The flocculation performance of a polyelectrolyte on a given substrate (suspended particle) will depend in part on the charge type and charge density on the substrate, the charge type (functionality) and charge density (functionality level) of the polyelectrolyte, and the ionic strength of the slurry medium. As previously discussed, chemical adsorption forces such as hydrogen bonding, determined by the functional groups present on the substrate and the polyelectrolyte, will also control the interaction of flocculant and substrate.

The suspended particles which are found in red mud slurries can be extremely small, ranging from 10 microns and below. These particles will have very long sedimentation times; that is, very slow settling velocities. In addition to the gravitational settling forces, which are moderated by hydrodynamic and thermal forces, ionic forces will play a significant role in stabilizing the particles in these red mud slurries.

Suspended particles in water will have ionic charges on their surfaces. A typical clay platelet in water will have opposite ionic charges on its surface and edge as illustrated in Figure 5. There is some argument at the present time as to what gives rise to the negative charge on the platelet surface. The silanol mechanism is illustrated in this Figure. Atomic vacancy mechanisms have also been proposed. Considering the platelet structure, one can easily envision stacking mechanisms where the positively charged edges stack upon the negatively charged surfaces forming tight agglomerates which can easily hold a significant quantity of adsorbed and entrained water or process liquor. Iron oxide particles in red muds will have negative charges randomly distributed on the particle surface. The mechanism leading to this is discussed below.

FIGURE 2

FLOCCULATION MECHANISM

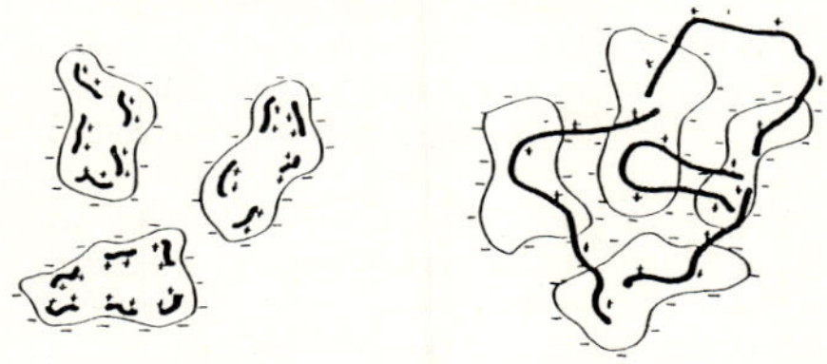

COAGULATION (CHARGE NEUTRALIZATION) — FLOCCULATION (BRIDGING)

FIGURE 3

FLOCCULATION MECHANISM

POLYMER UNDERDOSED — POLYMER OPTIMUM DOSE — POLYMER OVERDOSED

FIGURE 4

FLOCCULATION BONDING FORCES

VAN DER WAALS 1 Kcal/MOLE

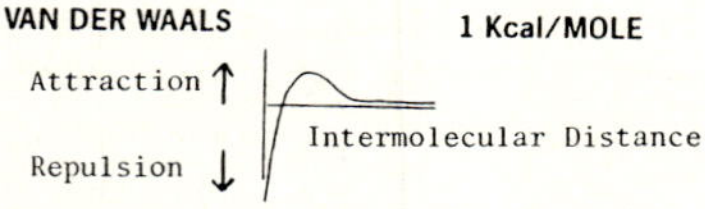

HYDROGEN BONDING 5 Kcal/MOLE

$R - H - - - - O^{(-)} - R$

IONIC BONDING 8 Kcal/MOLE

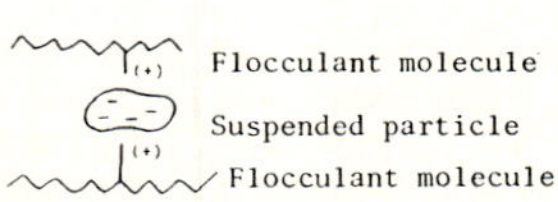

(KE THERMAL ~ 0.5 Kcal/MOLE)

FIGURE 5

CLAY PLATELET
SURFACE CHARGE IN WATER

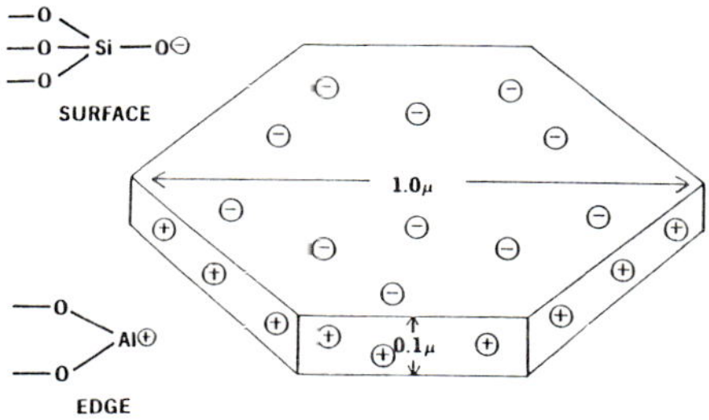

Zeta potentials are typically used to characterize the ionic charges on particles in suspension. Simply stated, the zeta potential of a particle is its effective ionic potential in the medium in which it is located. Slurries whose particles have zeta potentials greater than 40 millivolts (mV) are extremely stable, the particles tending to repel each other and not consolidate. Particles with zeta potentials of 10 mV or below can be considered fairly unstable and their slurries easily coagulated.

The electrical charge on the iron oxides present in red mud will be pH dependent. The isoelectric pH has been reported by a number of investigators[2] to be around 8.5. The particles assume an increasing negative charge as pH increases because of the increased concentration of $Fe\text{-}O^-$ groups on the iron oxide surface in preference to Fe-OH groups.

H. Hirosue, et al[3] report the zeta potential for red mud particles to be in excess of -40 mV at a pH of 12. These authors additionally show that low concentrations of sodium hexametaphosphate, a well-known dispersant for red mud, can increase (negatively) the zeta potential of these particles.

Although the iron oxide particle will have a high negative potential at the pH's encountered in red mud flocculation, the alkaline media encountered in the settlers and washer circuits will have extremely high ionic conductivities. Even low alkalinity solutions of around 3 g/l will have conductivities in excess of 10^4 mho/cm. Under these conditions the charges on the particles will be shielded causing the "effective" potential on them to be essentially zero. It is well known in flocculation technology that negatively-charged particles in a high conductivity slurry medium can be efficiently flocculated with anionic polyelectrolytes, i.e., negatively-charged polymers. This is all brought about by the shielding effect of the high ionic strength slurry medium.

An anionic polymer, or any polyelectrolyte for that matter, will also be affected by the high ionic strength of the highly alkaline slurries encountered in red mud settling. In effect, the anionic charges on the polymer will be shielded by the ionic charges in the surrounding medium, causing the polymer to assume a more random coil configuration rather than an expanded open structure. This will cause the polymer to have a lower solution volume (hydrodynamic volume) and consequently lower solution viscosity. The decrease in solution viscosity of some red mud flocculants with increasing alkalinity of the solution medium is illustrated in Table 3.

TABLE 3

SOLUTION VISCOSITY (0.03% RSV) OF RED MUD FLOCCULANTS IN ALKALINE WATERS

WATER ALKALINITY G/L NaOH)	POLYMER VISCOSITY (RSV) DL/G POLYMER FUNCTIONALITY 60%	80%	90%	100% (PSA)
0	419.1	498.1	346.2	395.2
2	100.2	84.1	90.6	119.6
10	54.5	48.2	47.3	47.6
30	39.6	34.7	32.7	27.6
75	31.2	26.2	22.5	16.6
140	29.1	23.8	19.8	16.1

At relatively low alkalinities encountered in red mud flocculation, there is an initial rapid viscosity decrease caused by contraction of the molecule, followed by a slower decrease. The data show that the polymers under consideration have molecular sizes which are inversely proportional to the level of functionality, i.e., the higher the functionality, the smaller the molecular size. In this range of functionalities, this is typical for acrylamide-sodium acrylate copolymers.

Considering the particle, polymer, and medium interactions discussed above; one could anticipate that all polymers described in Table 3 would show some efficacy in settling red mud across a broad range of alkalinities. However, poly(sodium acrylate) does appear to have a more extended structure at low alkalinities, a property not entirely compatible with a flocculation mechanism dependent on hydrogen or ionic bonding forces where polymer conformation to the particle surface is required. Lower functionality polymers are thus preferred at low alkalinities. Conversely, low functionality polymers may excessively hydrogen bond to red mud particles at the higher temperature and alkalinity conditions found in settlers, and thus not provide sufficient bridging between particles. In this application, high functionality polymers are preferred.

Red Mud Flocculation

In previous sections of this paper, the chemical and physical properties of red mud flocculants and the nature of suspended red mud particles were reviewed. Before illustrating the actual performance of these flocculants in settling digester slurries, the important interactions between polymers and suspended particles which play a significant role in the flocculation of red mud slurries will be reviewed.

The acrylamide copolymers under consideration are extremely high molecular weight polymers. These polymers will have aggregation properties resulting from molecular entanglements or from chemical bonding forces which can cause the aggregate to have large physical dimensions.

The polyelectrolyte flocculants under consideration as red mud flocculants have anionic functionality which is stable under the high pH conditions encountered in red mud slurries. The polymers also have flexible chains which permit them to conform to substrate surfaces and bridge through the suspension medium to effectively capture other particles.

The ionic strength of the alkaline medium causes sufficient ionic shielding to permit a negatively-charged polymer to be used to flocculate a negatively-charged suspended particle in the environment found in red mud slurries. Hydrogen bonding mechanisms can be effective in bonding polymer molecules to the oxide functionality found on the red mud particle. Therefore, polymers with slightly lower functionality than poly(sodium acrylate) may find applicability in the flocculation of red mud in digester slurries. However, significantly lower functionality copolymers will tend to be over-adsorbed on the red mud particles at the high temperature and alkalinity conditions found in these slurries and thus will not be effective as red mud flocculants.

Laboratory Evaluation of Red Mud Flocculants

Laboratory evaluation of red mud flocculants involves the determination of performance-dosage profiles for the polymers in question. The performance factors will include settling rates of the red mud, the suspended solids found in the liquor supernatant, and the compaction level of the mud after a specific period of settling. Column settling tests are traditionally performed in 1,000 ml. cylinders which contain simulated rakes revolved at low speed ca 3 RPM. The factors that control the performance of the red mud flocculation system will include the mineralogy of the bauxite being used and the parameters of the Bayer process; slurry alkalinity, temperature, and percent solids; as well as polymer type, which would include functionality, molecular weight, and dosage level.

In this section, laboratory data are presented to illustrate several of the performance factors that have been discussed previously.

We illustrate the settling of red mud slurries which are derived from the digestion of Jamaican bauxites in which the iron oxide component has a high concentration of goethite. Two types of bauxites are considered. The first is one containing a low boehmite content that was digested at relatively low temperature, 140°C. The other is one containing a high boehmite content that was digested at a relatively high temperature, 230°C.

The mineralogy of the low boehmite ore and the red mud derived from the digestion of this ore at 140°C is given in Table 4. We see that the hematite/goethite ratio of the iron oxide component did not change in the digestion process as was to be expected under the relatively mild conditions of digestion. The phosphate content in this ore could be considered relatively high. As previously discussed, phosphates tend to be antagonistic to the red mud settling process.

The second type of Jamaican bauxite of interest is a high boehmite, high goethite type. The mineralogy of this bauxite and the red mud mineralogy derived from processing this ore at 230°C. is given in Table 5. In this bauxite, the iron oxides are predominately in the goethitic form. After digestion at 230°C, a portion of the goethite has been converted to the hematite form.

In a previous paper(1) the relative effectiveness of starch and poly(sodium acrylate) in the settling of red muds derived from digester blow-offs was illustrated. The synergistic effect of combining starch with poly(sodium acrylate) in this settling process was demonstrated. These data are reproduced in Table 6. In essence, the use of starch and synthetic flocculant polymers in these settling processes can lead to very rapid settling rates and good clarity of the process liquor.

For the purposes of the present discussion, only synthetic water-soluble polymers are being considered in the flocculation process. The settling experiments were conducted on the blow-offs of laboratory digestions where either synthetic liquor or plant process liquor was used in the digestion.

TABLE 4

JAMAICAN BAUXITE DIGESTION
LOW BOEHMITE - HIGH GOETHITE
SYNTHETIC LIQUOR - 140°C - A/C = 0.62

COMPONENT	BAUXITE MINERALOGY	RED MUD MINERALOGY[1]
Al_2O_3 TOTAL	48.0	14.9
AVAILABLE AT 230°C	45.5	
140°C	44.6	
Fe_2O_3	18.5	48.2
HEMATITE/GOETHITE	35/65	36/64
TiO_2	2.1	7.0
SiO_2	1.2	2.5
P_2O_5	0.8	1.7

[1]POLY(SODIUM ACRYLATE) AS FLOCCULANT

TABLE 5

JAMAICAN BAUXITE DIGESTION
HIGH BOEHMITE/HIGH GOETHITE
SYNTHETIC LIQUOR - 230°C - A/C = 0.65

COMPONENT	BAUXITE MINERALOGY	RED MUD MINERALOGY[1]
Al_2O_3 TOTAL	47.8	16.8
AVAILABLE AT 230°C	43.2	
140°C	40.7	
Fe_2O_3	18.8	41.9
HEMATITE/GOETHITE	6/94	14/86
TiO_2	2.5	6.1
SiO_2	1.6	4.0
P_2O_5	0.5	1.3

[1]POLY(SODIUM ACRYLATE) AS FLOCCULANT

Lime is traditionally used by alumina refiners in the digestion process to remove phosphorus as carbonate apatite. Phosphorus has been discussed as an antagonist in the red mud settling process. Its absorption onto red mud particles serves as an effective dispersant for these particles. Conventional flocculants do not compete very effectively with phosphorus in this absorption process. Consequently, those muds which contain high concentrations of phosphorus available for absorption onto red mud particles will be extremely difficult to settle.

In all the laboratory digestions that are discussed in this section of the paper, lime has been used to remove phosphorus. Figure 6 shows that adding lime to the digester made poly(sodium acrylate) much more effective in settling the red mud. The effect on settling rate is most apparent at the higher polymer dosages. These data can be rationalized on the basis that those red mud particles obtained from the digestion process with lime addition have more adsorption sites available for the synthetic flocculant polymer.

Questions always arise as to the settling behavior of red muds which are derived from synthetic liquors versus actual plant process liquors. It is known that plant process liquors, which are recycled, will develop a buildup of organic components which can seriously compete with flocculant polymers in the absorption on red mud particle surfaces. Flocculant polymers would then be expected to be less effective in settling red muds derived from plant process liquors than those from synthetic liquors.

The above principles are shown in Figure 7. In this illustration we see that poly(sodium acrylate) settled red mud more effectively in the synthetic liquor than in the plant liquor, especially at higher dosage levels. It is interesting to note that the red mud derived from the plant liquor settles much more rapidly than the red mud derived from synthetic liquors when low dosages of flocculant are used. This is not uncommon. In general, red muds derived from laboratory digestions incorporating synthetic liquor have little tendency to settle. Most red muds produced in plant operations do have an inherent settling tendency with no flocculant addition. Of course, these settling rates are much too slow for effective process utilization.

The effectiveness of a copolymer of sodium acrylate and a small quantity of acrylamide - termed a high functionality copolymer - in settling red muds is illustrated in Figure 8. In comparing the performance of this copolymer against poly(sodium acrylate), whose performance in a high and low temperature digestion was illustrated in Figures 6 and 7, we see the following. The high functionality copolymer is much more effective than poly(sodium acrylate) in settling red muds derived from the high-temperature digestion system. On the other hand, the high functionality copolymer and poly(sodium acrylate) perform very similarly in settling red muds derived from the low-temperature digestion. The data in Figure 8 indicate that the high functionality copolymer may lose some effectiveness if it is in use and process conditions are changed towards a lower temperature. The above point illustrates that bauxite processors should expect a shift in the performance of the synthetic red mud flocculant if the bauxite mineralogy changes and the digestion conditions are altered in the process.

FIGURE 6

RED MUD SETTLING
JAMAICAN BAUXITE, HIGH BOEHMITE - HIGH GOETHITE
DIGESTION - SYNTHETIC LIQUOR, 230°C
POLYMER - POLY(SODIUM ACRYLATE)

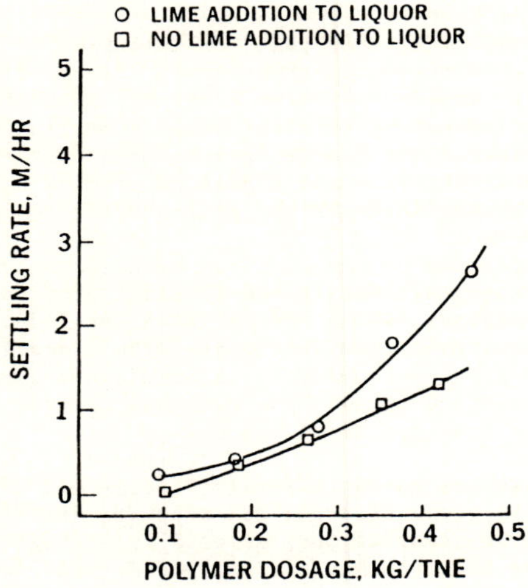

FIGURE 7

RED MUD SETTLING
JAMAICAN BAUXITE, LOW BOEHMITE - HIGH GOETHITE
DIGESTION - 140°C
POLYMER - POLY(SODIUM ACRYLATE)

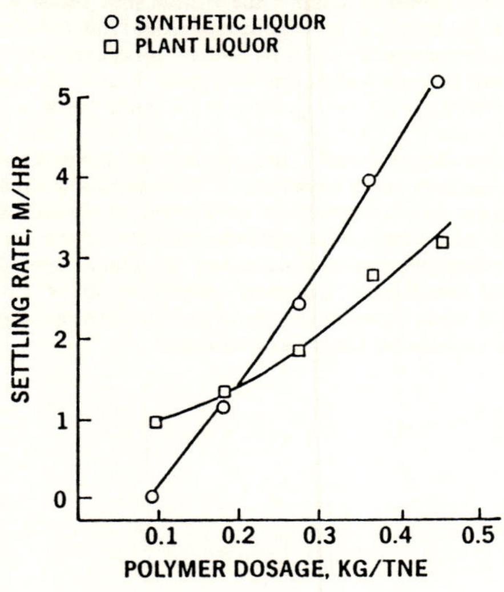

In an earlier part of this paper, we indicated that much lower functionality copolymers of sodium acrylate and acrylamide find application in the settling of red muds found in low alkalinity washer systems. We thought it useful to demonstrate that these much lower functionality copolymers are not very effective in settling red muds derived from digesters. This point is vividly illustrated in Figure 9. The relatively flat response of the copolymer containing 80% sodium acrylate and 20% acrylamide suggests that the amide functionality is extremely effective in absorption onto the surface of the red mud particle. The absorption could be so effective that an overdose situation is simulated and there is little bridging effectiveness for satisfactory particle flocculation. As the functionality of the copolymer is lowered as in the case of the 60% sodium acrylate and 40% acrylamide copolymer, the polymer becomes ineffective through loss of bridging efficiency. Data in Figures 8 and 9 show that slight amide functionality in the red mud flocculant polymer may be desirable but too much of the amide functionality will lead to rendering the polymer completely ineffective for flocculating digester muds.

Thus, copolymers used in flocculating digester muds will contain from 90-100% anionic functionality. Actual laboratory tests will be required to determine which of the copolymers are most effective in a given settling operation.

Conclusions

This paper has illustrated the effectiveness of high functionality synthetic water-soluble polymers in settling red muds derived from bauxite ores in which the iron oxide component contains a high concentration of oxide in the goethite form. We have shown that the higher functionality copolymers of sodium acrylate and acrylamide are the most effective ones in settling these red muds.

The performance of these synthetic flocculant polymers will be a function of the bauxite mineralogy and the process conditions used in the digestion process. In a given plant situation, changes in either the bauxite or the process conditions can be expected to affect the performance of the synthetic flocculant polymers in use at that plant.

Synthetic polymer manufacturers have changed polymer functionality in order to design polymers which are effective under specific red mud slurry conditions. A bauxite processer will be able to effect good performance trade-offs by incorporating two different types of flocculant polymers in his total process.

This paper discussed the role of surface-active agents, either phosphates or organic moieties present in the process liquor, in competing with flocculant polymers for adsorption onto the surfaces of red mud particles. Removal of the antagonistic absorbants from the red mud particle surfaces leads to more effective utilization of the flocculant polymers. Future improvements in flocculant polymer performance will be attained by additional removal of these agents or by structuring polymers to compete more effectively with them in adsorption on particle surfaces.

FIGURE 8

RED MUD SETTLING
JAMAICAN BAUXITE
POLYMER - 95/5 SODIUM ACRYLATE/ACRYLAMIDE
DIGESTION - AS INDICATED, SYNTHETIC LIQUOR

O HIGH BOEHMITE, HIGH GOETHITE, 230°C DIGESTION
□ HIGH BOEHMITE, HIGH GOETHITE, 140°C DIGESTION

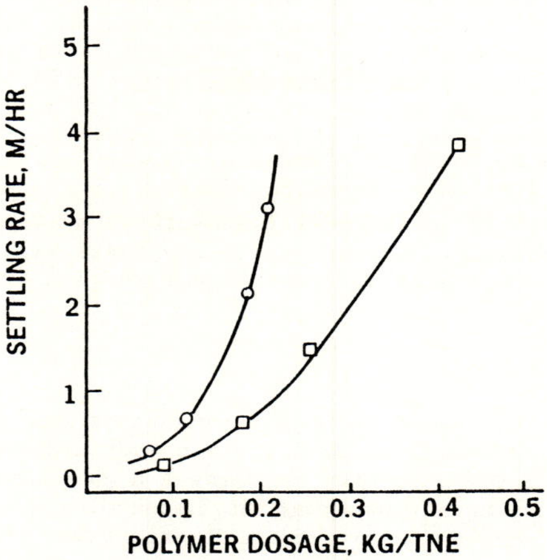

FIGURE 9

RED MUD SETTLING
JAMAICAN BAUXITE, LOW BOEHMITE - HIGH GOETHITE
DIGESTION - SYNTHETIC LIQUOR, 140°C

COPOLYMER O 100 SODIUM ACRYLATE
□ 80 SODIUM ACRYLATE/20 ACRYLAMIDE
△ 60 SODIUM ACRYLATE/40 ACRYLAMIDE

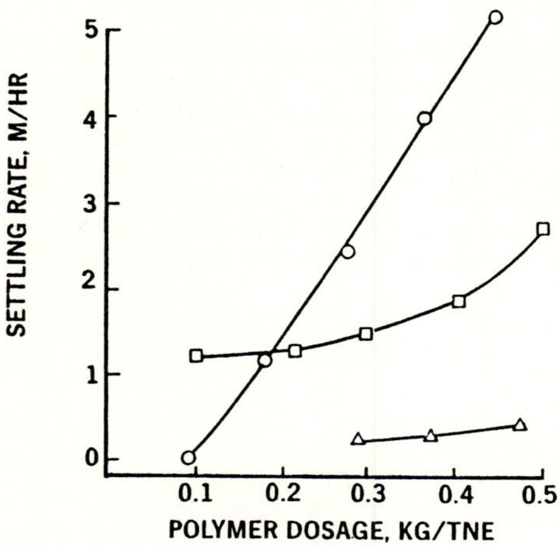

These conclusions show that synthetic water-soluble polymer manufacturers have a unique opportunity for making considerable improvements in the performance of red mud flocculant polymers, and thus a contribution in the economics of bauxite processing.

ACKNOWLEDGEMENT

The authors wish to thank Dr. Lynden C. Nugent and Dr. Conrad Douglas of the Jamaica Bauxite Institute, Kingston, Jamaica, who carried out the research program for Hercules Incorporated that provided the data used in this paper.

REFERENCES

1. S. E. Sankey and R. J. Schwarz, "The Structure-Property Relationship of Polymers used in Red Mud Flocculation. Presented at the 111th AIME Annual Meeting, Dallas, Texas - February 16, 1982.

2. G. A. Parks and P. L. DeBruyn, J. Phy. Chem., 66 967 (1962).

3. H. Hirosue, et al, Powder Technology, 27 197-202 (1980).

FLASH CALCINER

BY FIVES-CAIL BABCOCK

J. P. HENIN
FIVES-CAIL BABCOCK RESEARCH CENTER
59105 - LILLE-FRANCE

The world's first industrial alumina suspension calciner, based on the flash calcination process, is under setting up at Aluminium de Grèce's (ADG) St Nicholas plant in Greece. The 900 mt/d installation, which will feed sandy alumina to modern batteries of electrolysis tanks ("point tapping"), has evolved through research efforts carried out by Fives-Cail Babcock (FCB) and pilot tests conducted jointly by FCB, Aluminium Pechiney and ADG.

BACKGROUND OF FLASH CALCINATION

Until now, three main types of kilns existed for ores : shaft kilns, rotary kilns and fluid-bed kilns. The suspension flash calcining process is the outgrowth of improvements to rotary kilns in the cement-making industry to reduce heat consumption and increase productivity.

The first breakthrough was the short-kiln and dry-process preheater design. The preheater is made up of four stages of exchange ducts and cyclones. The heat exchange in each stage takes place in parallel flow between the ground material carried by the gases and the carrying gases (whence the terms suspension-exchange and suspension-preheater SP). The stacking of several parallel flow stages (fig. 1) achieves an overall counter-current heat exchange. This process has reduced heat consumption from 5,000 KJ to 3,300 KJ/Kg of clinker. Shortening the rotary kiln, which is still the only device permitting tacky particles mass-flow required

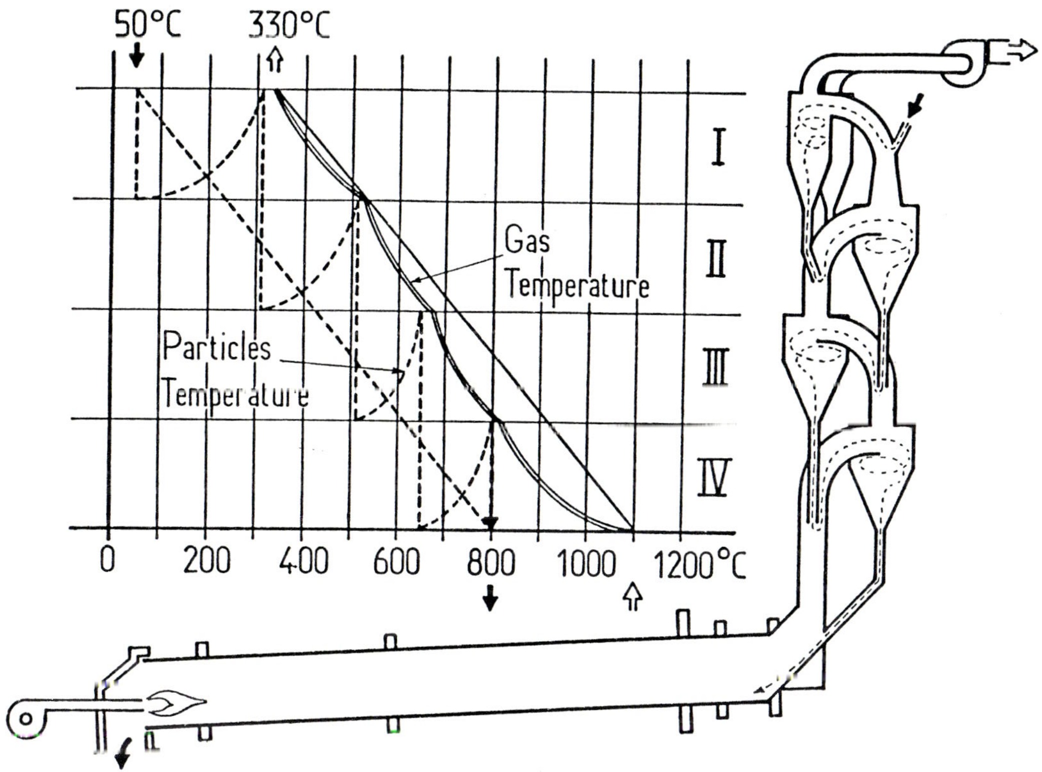

FIG.1 Short rotary Kiln with 4 stages suspension preheater (SP)

by clinkerisation, has also reduced maintenance costs.

The second breaktrough has been the Japanese development of precalcination. This SPP (suspension preheater and precalciner) process involves the burning of fuel at the lower stage of the suspension preheater (SP) in a special furnace called precalcining chamber (PC), to increase the exchange rate in the preheater without increasing the gas temperature at the exit of the rotary kiln. Since the rotary kiln is sized by its fuel flow, its clinker output with precalcination can be increased up to a factor of 2.4, depending on the adopted flow-sheet (fig. 2).

FCB has set up many dry-process cement plants with SP kilns and has also developed its own precalcining chamber design (fig. 3).

For other ores, rising energy costs have let users to choose shaft kilns or fluid-bed kilns. Rotary kilns, whenever they remain necessary, are often fitted with head preheaters such as the vertical shaft preheaters for lime kilns.

Considerable development work is currently underway by FCB on the improvement of heating performance of rotary kilns.

ALUMINA ROTARY KILN CONVERSION

The combination of an intermediate-carried burner pipe, for reducing the length of the hot zone and increasing the

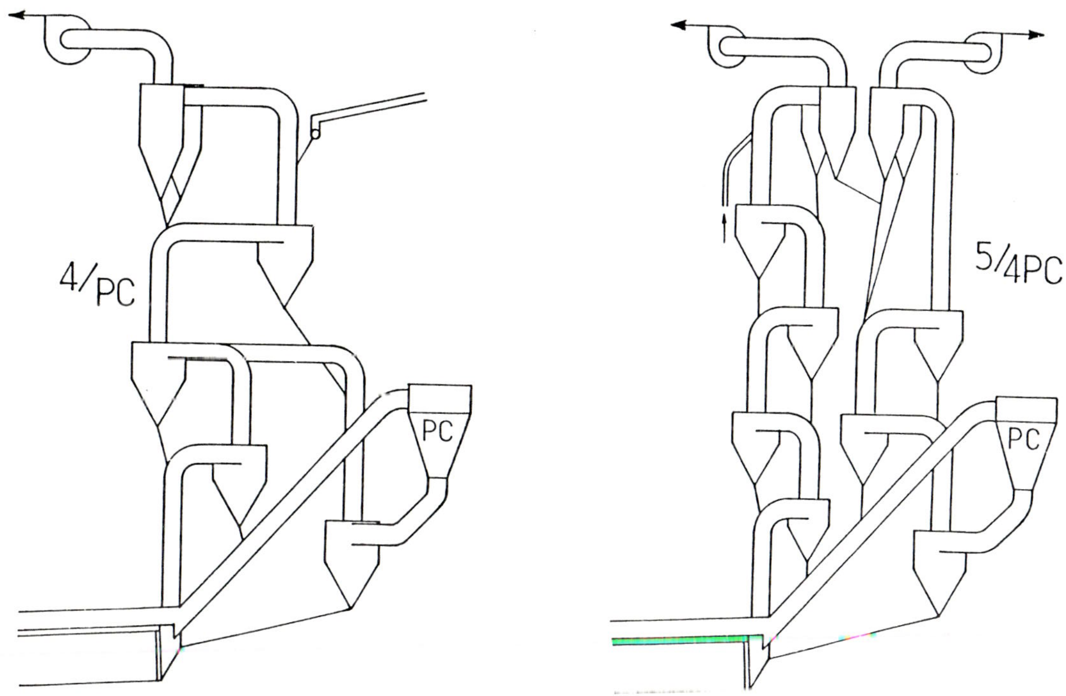

FIG. 2 SPP Flow-sheets

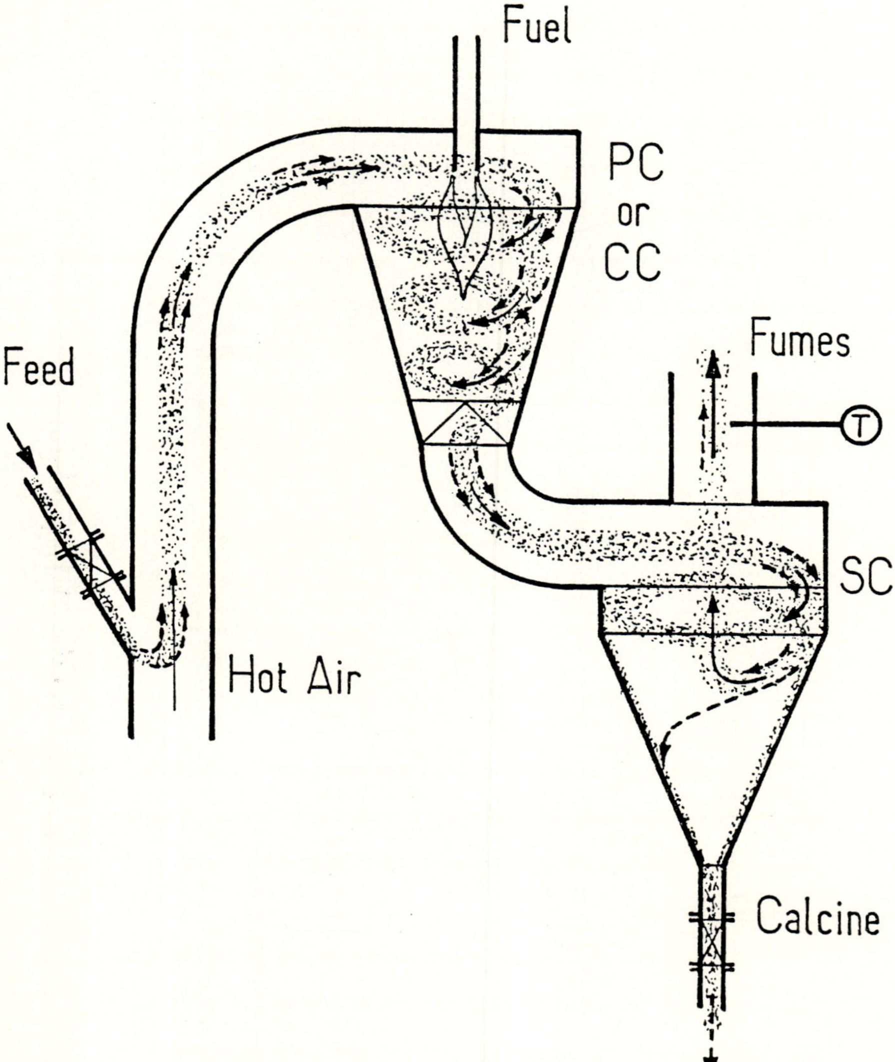

FIG. 3 Precalcination (cement) or Flash Calcination stage by F. C. B.

length of the cooling zone, and of a two-stage SP, decrease consumption to less than 88 kg of Bunker C fuel-oil/m of calcined alumina. Fig. 4 shows the first modified alumina kiln, which, with its patented process, set the stage for more kiln-conversions (5 operating, 2 under construction). According to the decrease of fuel consumption, kiln output can be rated at 20-30 % more than its former value.

Such kiln modifications have had no substantial effect on the quality of the calcined aluminas. Thus, it is possible to obtain either floury aluminas (with alpha-alumina contents higher than 90 %) or more reactive sandy aluminas, the same as in conventionnal rotary kilns, depending on the degree of calcination.

For reactive aluminas, the same α-content/BET surface relationship is obtained as in a nonmodified kiln. For example, for a given alumina the following results were obtained.

Table 1.

If the addition of cyclones brings about a certain attrition in the largest and most fragile particles, however there is no generation of fines smaller than 15 μ harmful for electrolysis.

Experience gained in starting the first modified kilns and all along the running of flash calcination pilot tests, have pointed out to the importance of gas flow rate upon

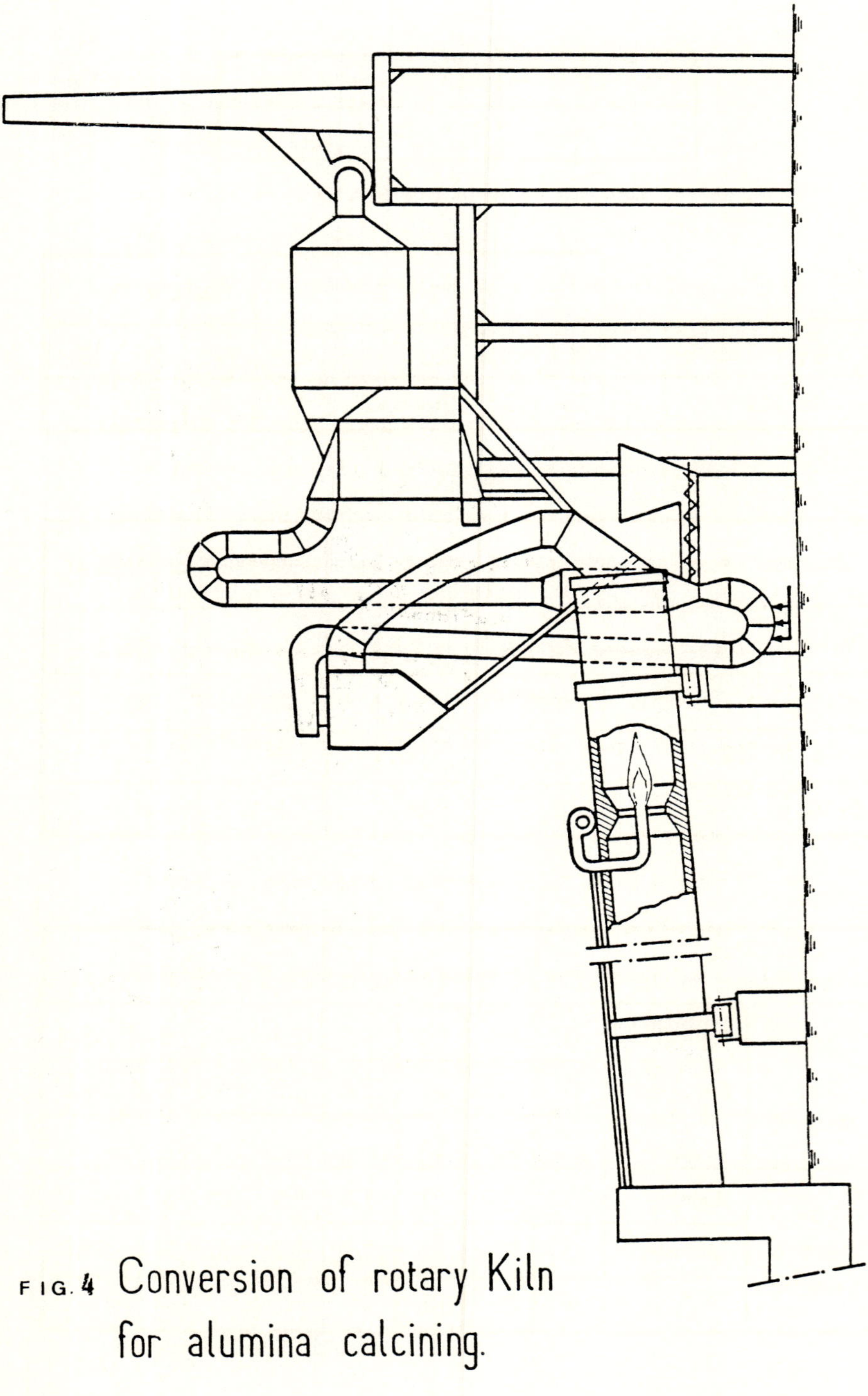

FIG. 4 Conversion of rotary Kiln for alumina calcining.

Table 1

% α	18	21	24	26
BET (m2/g)	77,5	74,5	66,0	52,5

Table 2

	Hydrate A					Hydrate B		
BET (m2/g)	32,2	44,3	58,8	68,3	81,5	55	60	82
% α	28	10,5	0,8	0	0	2	0,8	0

Table 3

Average samples	Hydrate A calcined at 60 m2/g		Hydrate B calcined at 33 m2/g	
Particle-size category	% α	BET	% α	BET
<40 µ	1,5	53,6	42,7	30,0
40 to 100 µ	0	60,2	23,9	31,3
>100 µ	0	65,4	16,1	41,3

Table 4

Cumulated Percentage minus		<15	<30	<45	<60	<80	<100	<125	<145
High Gas Velocity	Hydrate	0	0,4	1,9	4,0	12,1	41,0	84,8	95,6
	Calcined	0	1,3	8,0	25,6	59,5	80,8	97,3	99,5
	Δ	0	0,9	6,1					
Low Gas Velocity	Hydrate	0	0,9	1,9	3,1	13,5	42,5	81,7	96,4
	Calcined	0	0,9	2,8	11,8	35,2	68,5	94,8	99,0
	Δ	0	0	0,9					

generation of particles under 45 μ (325 mesh). With proper design of SP flues, attrition expressed as the increase of the percentage minus 45 μ (wet sieving) resulting of calcination can be limited to 2 points.

FLASH CALCINING FLOW-SHEET

At the same time, FCB has developed a calcining flow-sheet for fine sized ores or materials (minus 1 mm), in which a suspension calcining chamber of the type used in cement manufacture is utilized. The rotary kiln is completely eliminated,and the cooling of the calcined product takes place in the same manner as its preheating (i.e. in a dry-process exchanger).

The resulting flow-sheet (fig. 5) indicates that the calcining zone, the heart of a suspension or flash-calcining installation, consist of a cyclone calcining chamber (CC) and of an associated separation cyclone (SC). The preheated particles are dragged upwards by the hot air coming from the cooler, and this suspension is introduced tangentially into the calcining chamber CC (fig. 3). The fuel is introduced at one or several points of the chamber roof. The presence of material absorbing the calories released by the combustion is such that there is no high flame temperature within the chamber, hence there is less risk of overheating. The centrifuging effect of the material on the walls increases residence time (especially for the largest particles). At the outlet a round-square transformation of the flue, homogeneizes the suspension for completing the heat exchange in parallel flow. The associated cyclone SC, separates the calcined product from the fumes. Since the exchange is of the parallel-flow type, the temperature of the fumes is very close to the temperature of the solid particles, and consequently, it is easy to control the degree of calcination by adjusting the fuel flow to a set fume temperature. A suspension preheater makes it possible to use the hot fumes from the calcining zone to dry and precalcine the solid particles, while a suspension cooler allows the use of heat from the calcined product to preheat the combustion air entering the calcing zone.

In fig. 5, the preheater and the cooler have two exchanger stages. The number of stages, however, should be optimized according to the material processed. Account must be taken mainly of the desired chemical reactions, thermal requirements and temperature ranges of these reactions.

PROCESS APPLICATION TO ALUMINA

The flow-sheet of an alumina flash-calcination installation, is only an application of the general flow-sheet (fig. 6).

The preheater consists of two stages (C1 and C2). The introduction of hydrate into the upper stage, the separating element of which is a multicyclone (C1), takes place through

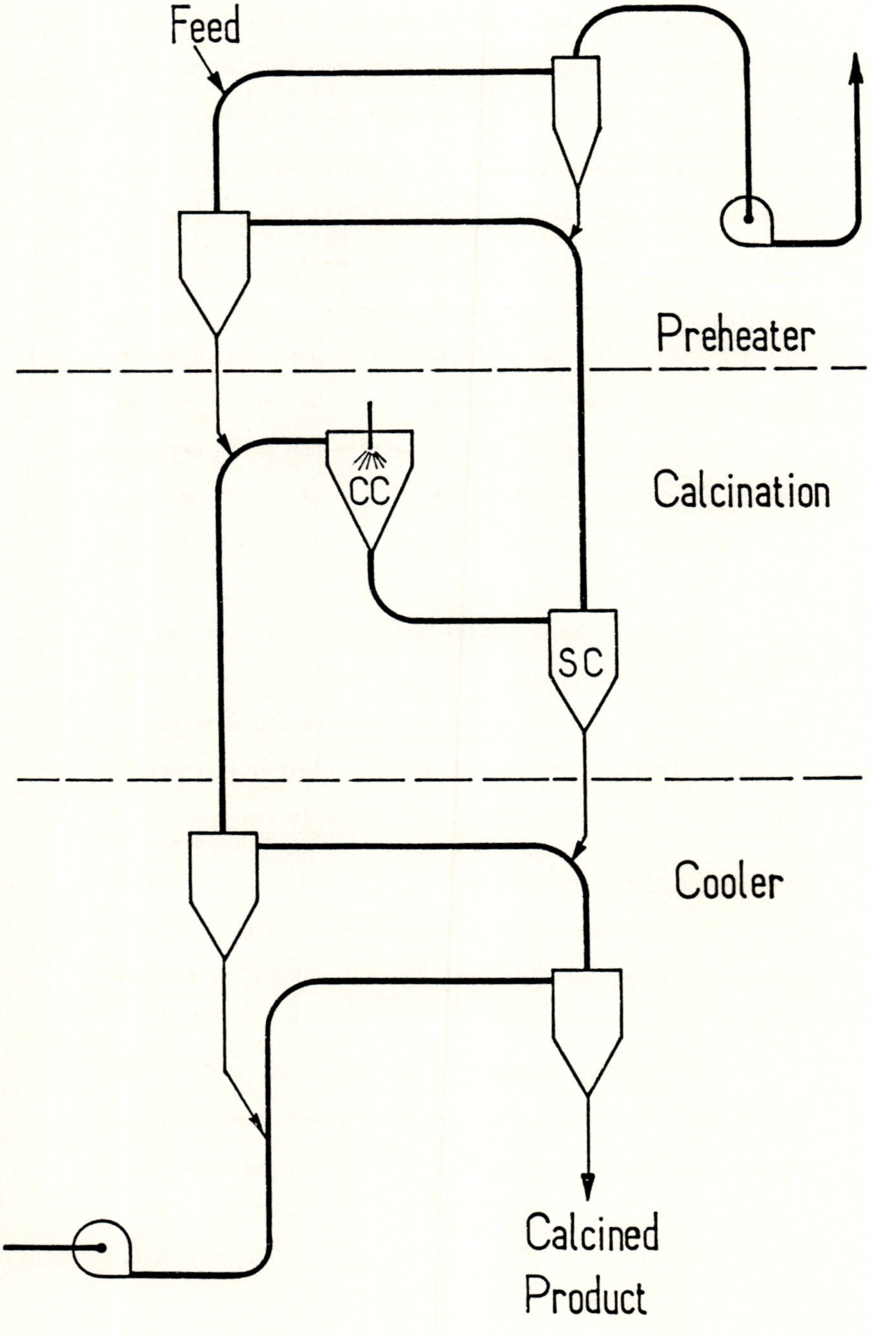

FIG. 5

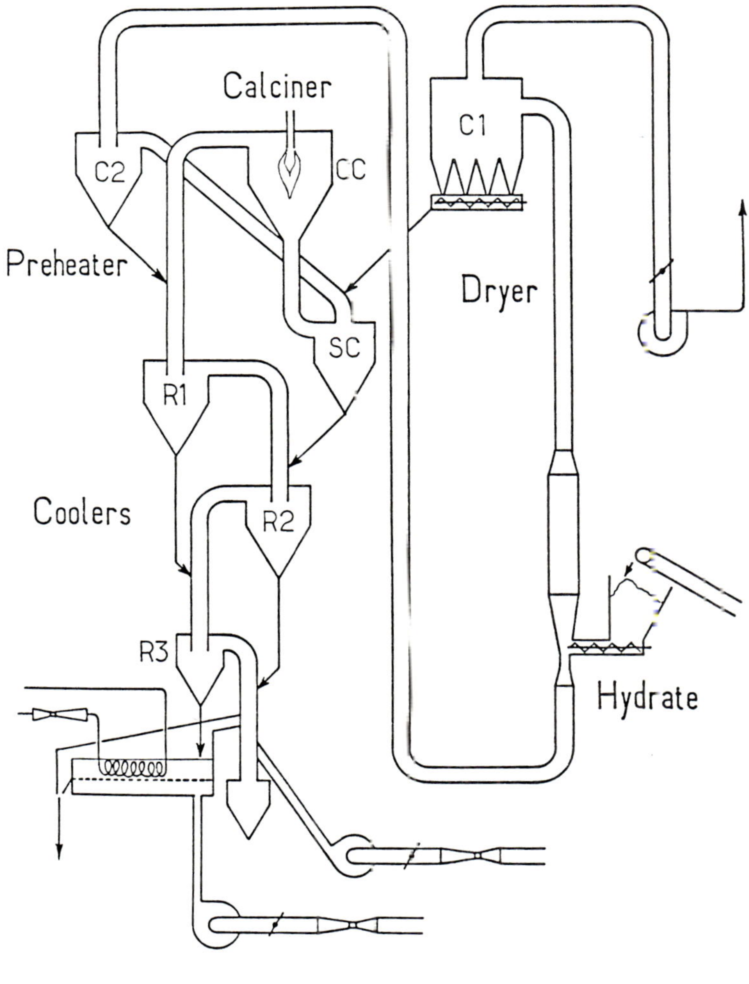
Calciner
C2
CC
C1
Preheater
Dryer
SC
R1
Coolers
R2
R3
Hydrate

FIG. 6

a drying flash venturi, which accepts the hydrate with a 9-13 % moisture content. The alumina leaves the preheater (C2) at 320-350 °C and with a loss in water content of 85-90 %.

The cooler is made up of three stages. The air off the cooler is preheated at about 800 °C, while the calcined alumina exits the last cyclone (R3) at about 280-300 °C. Hence, a final fluid-bed cooler with water circulation coils is needed to reach the required 100 °C temperature.

The air-particles flow enters the calcining chamber (CC) at 400-500 °C. The calcined product leaves the calcining zone (SC) at temperatures between 1,050 °C and 1,200 °C depending on the aimed degree of alumina calcining.

Heat consumption is about 3,100-3,200 kJ/kg of alumina and power consumption about 13 kWh/kg.

Although the calculation of heat consumption and the sizing of the different devices did not raise any difficulties, the main aspect to be dealt with was the quality of such a calcined alumina. FCB initiated a laboratory flash-calcination test program, including alumina, lime and phosphates calcination tests on a 100 kg/h unit. While the results were satisfactory only testing on a high-capacity pilot installation could provide FCB the required assurances.

PILOT PLANT DESCRIPTION

The pilot installation was designed to allow testing of different solids having different calorific requirements and different reaction temperature ranges (thus with different output) (fig. 7).

The main component of the pilot installation is the calcining zone which must be representative of real industrial conditions.

The preheater must be flexible enough in order to simulate any number of exchanger stages, from which, the choice of a single stage preheater but with an auxiliary furnace making possible to reach any preheating temperature.

The cooler has only two stages and the supporting structure is of a limited nature. Like flues and cyclones, floors are of unit construction to make erection and transportation more easy.

PILOT TESTING

The pilot installation was set up at Aluminium Péchiney in the Gardanne factory in June 1980, and testing was performed by the personnel of the FCB Research Center with the cooperation of the Industrial Testing Department and the Alumina Center Laboratories of Gardanne.

Operation of the pilot plant was generally at a flowrate of 1.2 - 1.5 mt/hr of calcined alumina, with the capability of increasing to 1.7 mt/hr.

A first set of calcination tests carried out on a given alumina hydrate was run to determine the different qualities of sandy alumina obtainable by varying the calcination

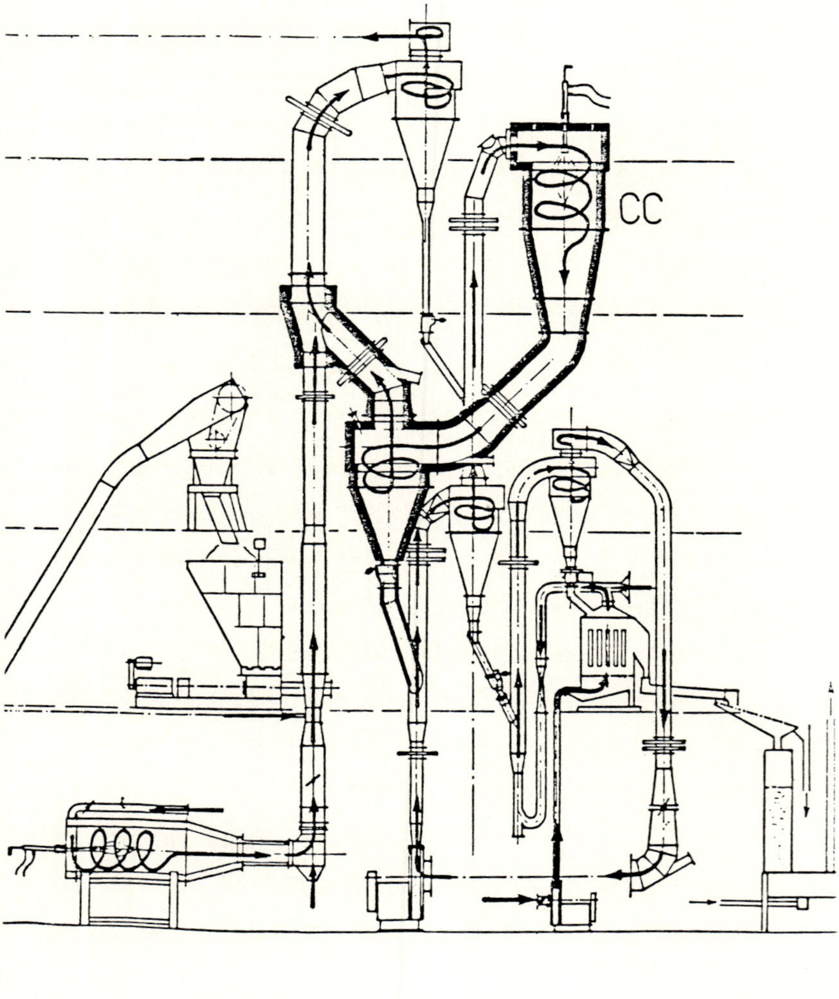

FIG. 7

temperature. From checking the results it was found, that for a given BET surface, the α content of flash-calcined alumina is much lower than by using other calcining devices. Also demonstrated was the easy operation and the flexibility of the installation.

A second set of calcination runs from the same hydrate was carried out to study the problems of grain breakage under different operating conditions.

Based upon the good results, ADG requested FCB to transfer the pilot installation to St Nicholas for a long-duration test feeding a serie of electrolysis tanks (fig. 8)
Fig. 8 is the photography of the pilot in ADG's plant.

At St Nicholas, the pilot installation fed a buffer silo of about 50 mt. The alumina was then sent to an independant tank gaz purification loop before feeding a serie of 26 electrolysis tanks at a rate of 30 mt/d for two periods of six months without interruption. The BET surface aimed for was 60 or 70 m^2/g, corresponding to a calcination temperature of 1120 or 1080 °C.

Following satisfactory electrolysis results, ADG awarded FCB a contract for a 900 mt/d flash calciner. Since then, the pilot installation at St Nicholas has been used for additionnal tests. In agreement with ADG, it is now available to alumina producers wishing to test this new calcination process on their hydrates. Moreover, FCB has built a second pilot installation to pursue its suspension calcina-

F I G. **8** FCB PILOTE PLANT IN SAINT NICOLAS - GREECE

tion test program on other ores, particularly phosphates.

FLASH CALCINED ALUMINA CHARACTERISTICS

BET surface

We plotted the curve of BET surface of calcined aluminas as a function of calcinating temperature expressed as the temperature of gases at the exit of cyclone SC (fig. 9). This temperature is only slightly different of that of the calcined particles (parallel flow exchange), and its measurement is much more reliable.

From this relation, to obtain a required BET surface and hence a certain degree of calcination, one need only keep the calcining temperature constant, while adjusting either the fuel flow or the hydrate flow. The ease and accuracy of this adjustement make it possible to obtain an alumina of very constant quality. The BET/calcining temperature curve, plotted from the first test runs showed to be also valid for calcined aluminas from other hydrates.

Alpha-alumina content

Alpha-alumina is the least reactive, most fragile, and least soluble allotropic form of calcined alumina. The most striking characteristic of flash-calcined aluminas from the pilot installation was their low α-alumina content for a given BET surface. With the same hydrate, a 14.3 % α content was obtained for a BET of 74.5 with a rotary kiln, and a 0.2 % α content for a BET of 50.6 with the pilot flash-calciner.

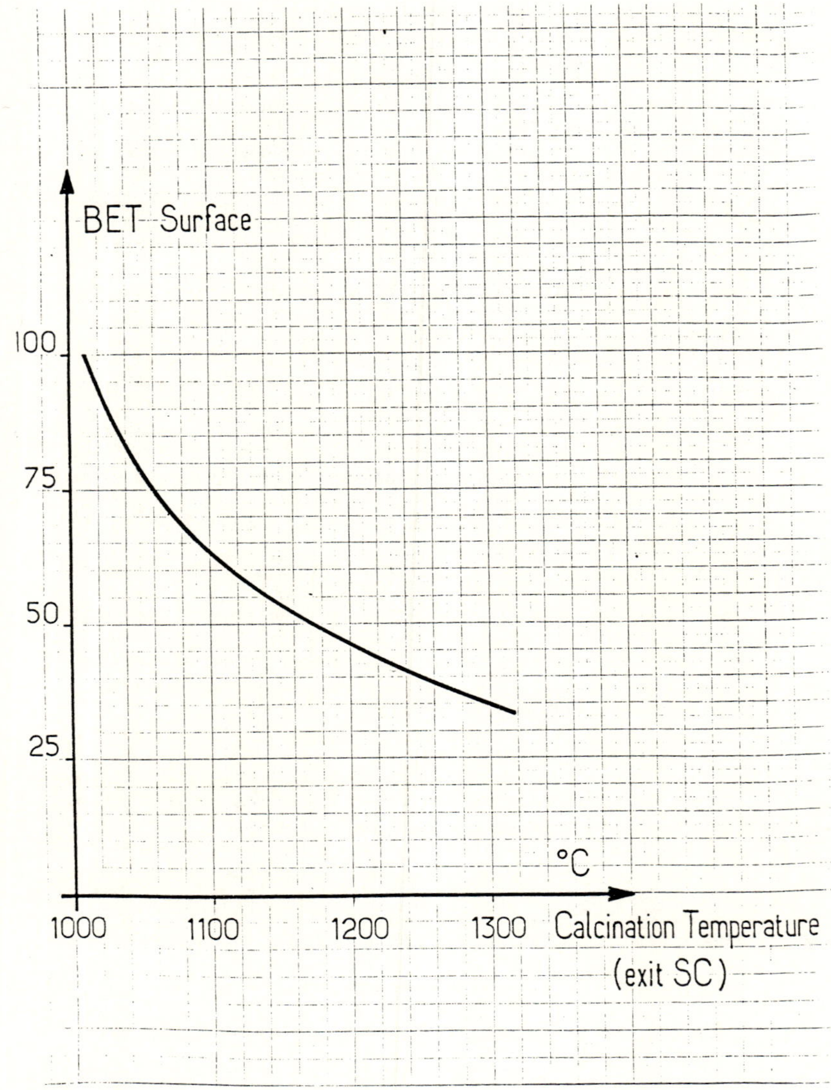

BET Surface
100
75
50
25
°C
1000
1100
1200
1300
Calcination Temperature
(exit SC)

FIG 9

Fig. 10 show the ranges of variations in α content as a function of BET surface for different calcinating devices. Alumina from the pilot flash-calciner countains less α alumina than the aluminas calcined in rotary kilns or fluid-bed kilns. Notice particularly that for BET higher than 60 m2/gm there is practically no α alumina. Figures are given in the table below for flash-calcined aluminas from two different hydrate A and B.

Table 2.

It was observed (see table below) that the different particle-size categories of the flash calcined aluminas were of very homogeneous quality, particularly compared with aluminas calcined in rotary kilns.

Table 3.

Loss of ignition and water recovery

Fig. 11 and 12 show the curve of LOI and water recovery as functions of the BET surface of flash-calcined Aluminas.

For a given BET, the corresponding LOI is rather low which can also be explained by the good homogeneity of calcination (no under-calcined coarse particles).

The water recovery curve show no significant differences with other calcining devices.

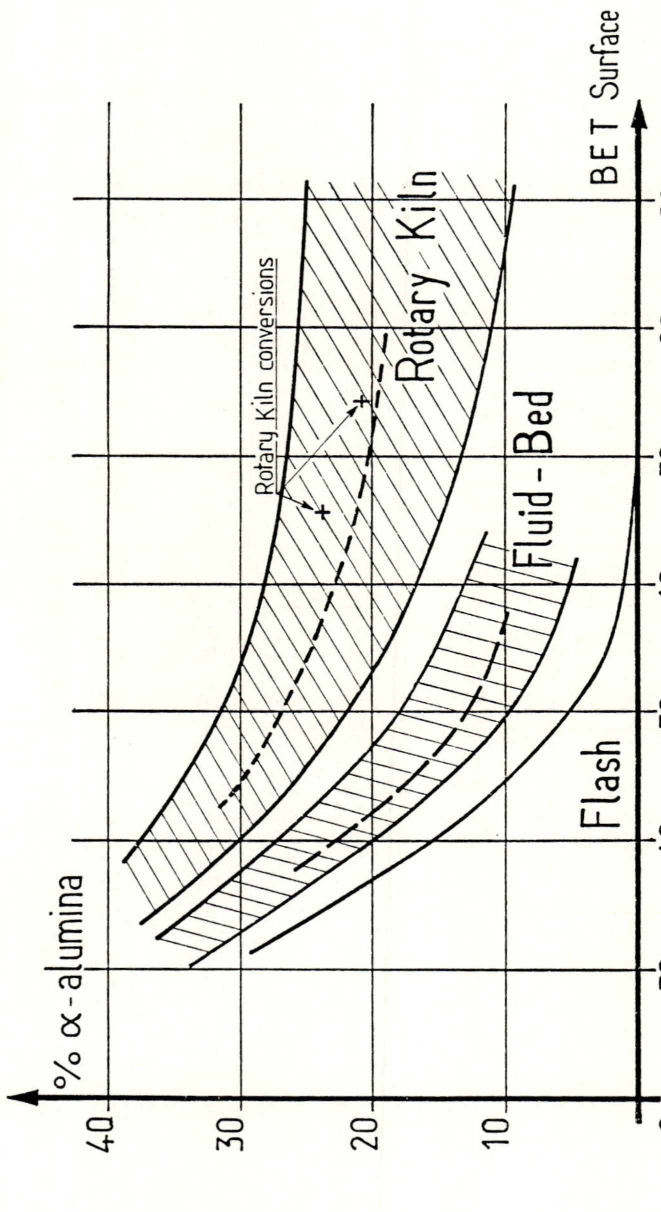

% α-alumina
40
30
20
10
0
30
40
50
60
70
80
90
BET Surface
Flash
Fluid - Bed
Rotary Kiln
Rotary Kiln conversions

FIG. 10

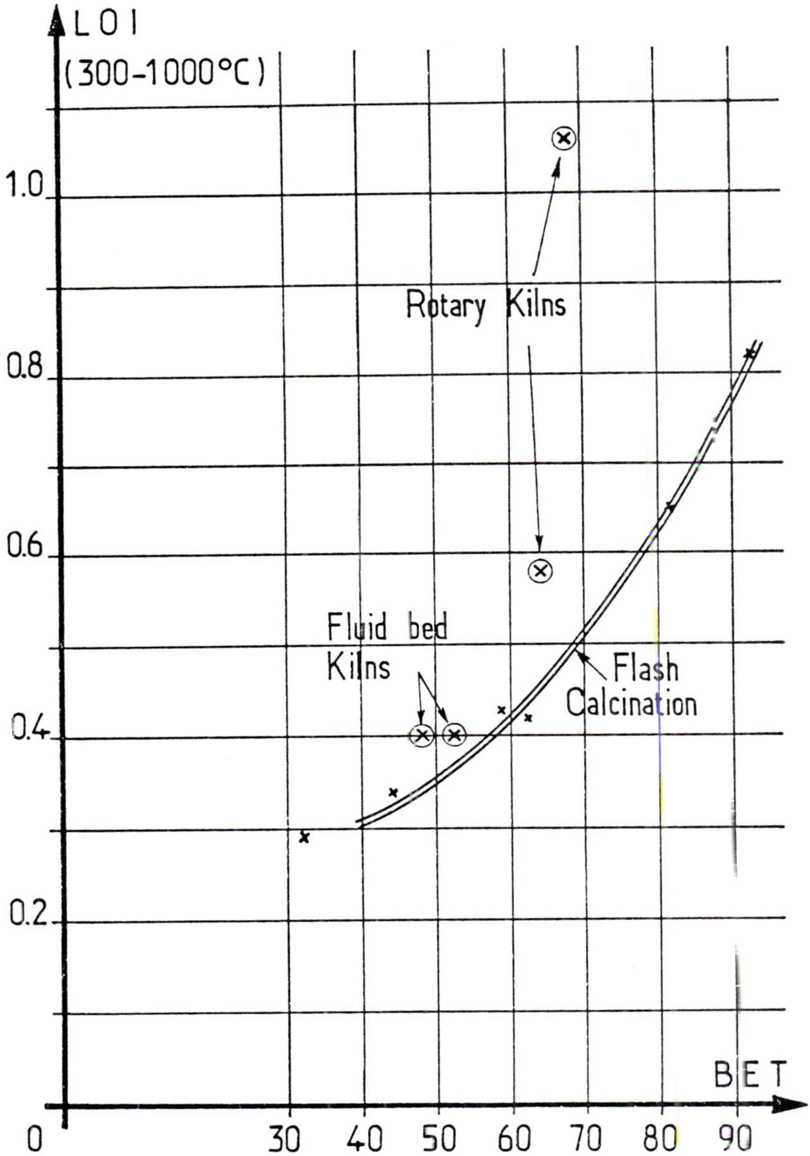

LOI
(300-1000°C)
1.0
0.8
0.6
0.4
0.2
Rotary Kilns
Fluid bed
Kilns
Flash
Calcination
BET
0
30
40
50
60
70
80
90

FIG. 11

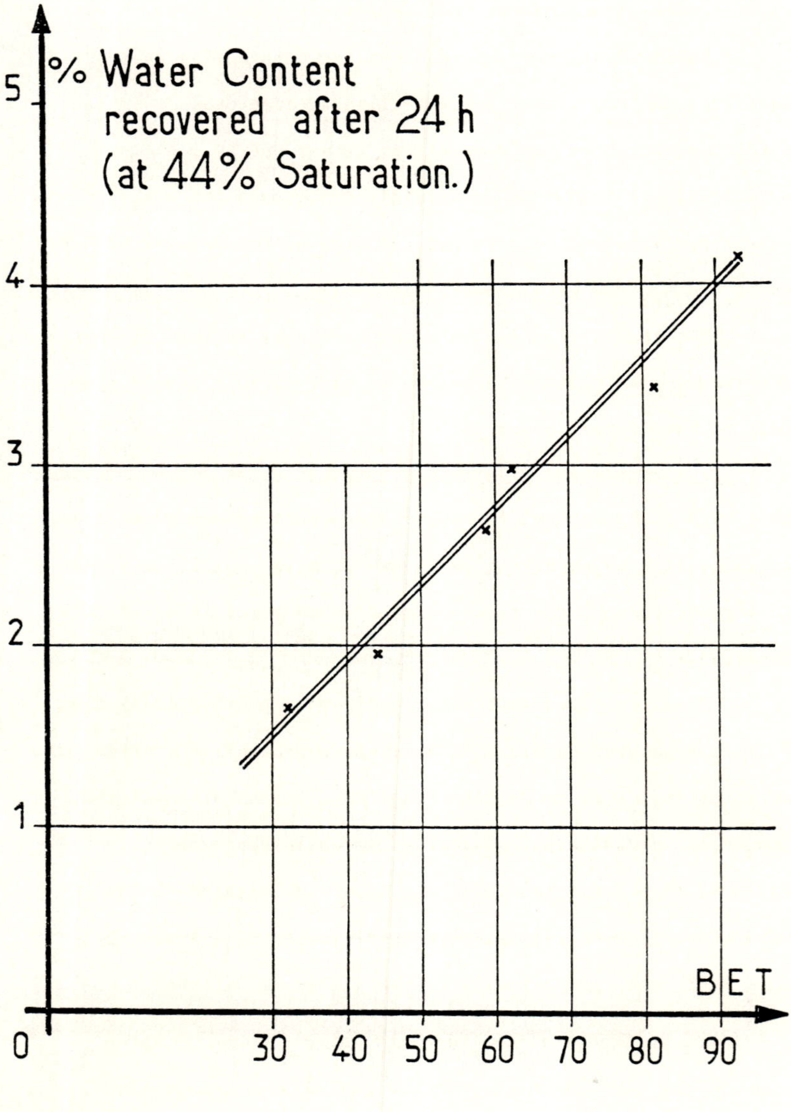

% Water Content
recovered after 24 h
(at 44% Saturation.)
5
4
3
2
1
0
30
40
50
60
70
80
90
BET

FIG. **12**

Attrition

During calcination, the coarse grains which are the most fragile, break into several fragments that enrich the lower particle-size fractions without generating superfines.

Explanations for the relative weakness of the coarse particles include : they are sometimes agglomerates of smaller particles, which become dissociated rather easily ; they are more sensitive to thermal shocks, either because the stresses resulting therefrom are larger, or because the removal of water vapor takes place with greater difficulty.

The result of these particle-breakages is that grain-size distribution of calcined aluminas is more closely graded than that of the initial hydrates.

But commercial requirements ask for a low amount of minus 45 μ (325 mesh), so calcinating process ability has also to be characterised by an attrition index which is usually defined as the increase of the percentage minus 45 μ resulting of calcination.

From the first test runs, we found attrition index from 0 up to 5 points. We had then to study thoroughly the effect of parameters such as hydrate fragility, amount of impurites, degree of calcination, temperature in preheating zone...
We found that attrition index was mainly correlated to two parameters : initial grain-size distribution of hydrate and, most important, gaz velocity in the flues and cyclones.

The influence of gaz velocity is well shawn in table 4 and by fig. 13. The higher is the gas velocity, more important is the breakage rate of the coarse particles and higher is the probability to get fragments minus 45 μ.

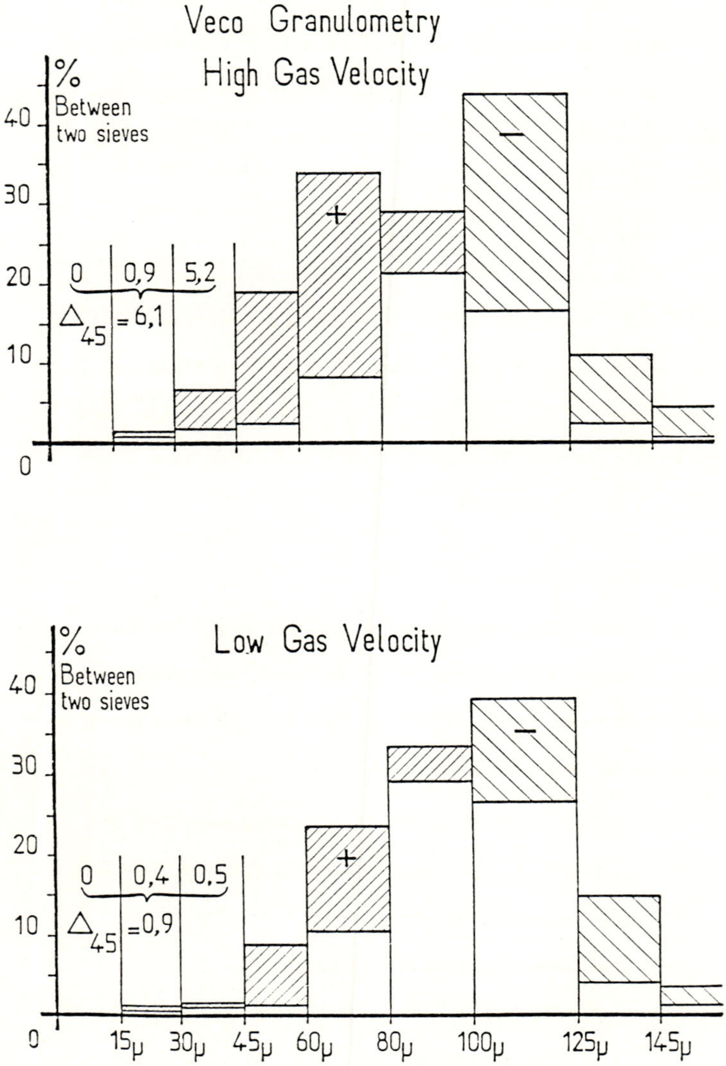
Veco Granulometry
High Gas Velocity
%
Between
two sieves
40
30
20
10
0
0
0,9
5,2
Δ_{45} = 6,1
+
−
Low Gas Velocity
%
Between
two sieves
40
30
20
10
0
0
0,4
0,5
Δ_{45} = 0,9
+
−
15μ
30μ
45μ
60μ
80μ
100μ
125μ
145μ

FIG. **13**

So, as for rotary kiln conversion, for almost all hydrates, proper design of the plant allows to get rid of this problem with attrition Index under 2 points.

Fragility

From comparative laboratory tests of fragility on both aluminas from the same hydrates, but flash-calcined on pilot installation or calcined in rotary kilns, we did not notice any significant difference of the behaviour in relation with the calcining process.

Solubility

The alumina calcined in the flash-calciner has an excellent solubility which is related to the low α amount and to the closely graded particle-size distribution. The steadiness of the calcined alumina characteristics improves the efficiency of the cell's control. Result is a raise in the Faraday efficiency.

ADVANTAGES OF THE NEW PROCESS

Currently, for new plants, fluid-bed kilns have overcome rotary kilns due to their lower heat consumption. The new type of flash calcining kiln offers the same advantage and in addition :

- lowers the initial capital outlay by 15-20 %
- reduces electric power consumption, owing to the low pressure drop
- increases utilisation flexibility
- yields a better calcined alumina quality : lower α alumina content and lower attrition index.

THE USE OF REFRACTORY LIFTERS IN COKE CALCINATION

M. F. Vogt, G. R. Jones, G. A. Tyler

Carbon Division
Kaiser Aluminum & Chemical Corporation
Chalmette, Louisiana 70044
USA

Summary

Experimental refractory tumblers were installed in the rotary kiln at Kaiser's Gramercy Coke Plant in May 1981. Results showed an improvement of VBD of 1.6 lb/ft^3 (vibrated bulk density, -8 +14 mesh) in the product calcined petroleum coke. These results were noted with similar medium sulfur delayed cokes at identical calciner production rates.

In March 1982, an improved tumbler installation, based on the Gramercy experience, was completed at Kaiser's Chalmette Coke Plant. Here, results showed a VBD increase of 2.1 lb/ft^3 under identical feed quality and rate conditions.

Uniformity of calcination between fine and coarse particles, as measured by specific resistivity, improved 20-30%. Tumblers expose the calcining coke bed to a more uniform temperature, which promotes product homogeneity.

These coke quality improvements result in improved anode quality, thereby reducing aluminum reduction plant raw material and energy costs.

Additionally, tumblers improve kiln operating stability. At Gramercy a 7% production rate increase was experienced after installation.

How Tumblers Work

Rotary Kiln Calcination

During the progression through a rotary kiln, reduction grade delayed petroleum coke experiences:

1) Moisture removal at 100°C followed by uneventful heat-up to about 400°C.

2) Volatile Material (VM) removal between 400°C and 1000°C.

3) Densification from 800°C to a final calcined temperature of 1300°C.

The event of moisture removal is obvious. Most coke kilns are fed with coke containing 6 to 8% moisture. Between 100°C and 400°C the coke is returned to the temperature it experienced during coking (about 420-440°C). During this process the coke softens and swells.

Volatile (VM) removal for most anode quality cokes begins at 400°C and is complete at 1000°C (Ref. 1,2). The profile of the emission of VM is shown in Figure 1. This volatile emission spectrum is for a typical anode grade coke. VM emission proceeds slowly at 400°C, peaks at 600 to 700° and is essentially complete at 1000°C. The quality critical heat-up zone during VM removal is between 400°C and 800°C as determined by Kaiser research and that of others. The porosity of the calcined coke is adversly affected by heat-up rates which are in excess of 50 to 100°C/min, according to Rhedey (Ref. 2).

Between 800°C and the final temperature, densification of the coke occurs (Ref. 2,3). During this transition, all measures of density-real density, apparent density and vibrated bulk density (VBD) - increase with increasing temperature, as shown in Figure 2. However, the final apparent density and VBD are fixed by the temperature increase rate experienced between 400 and 800°; if the heat-up rate is too high, excessive porosity is generated, causing densification to start from a lower point on Figure 2. The increase in apparent density would then parallel the graph given, at a constant value less at each T point.

How Tumblers Can Improve the Calcination Process

The Kaiser tumblers are strategically placed in the kiln to provide a more uniform heat-up in the critical temperature zone, 400 to 800°C. Minimum heat-up rate for all coke particles (50°C/min or less) is the key to maximum bulk density (minimum porosity). Mechanically, the tumblers stir the bed, equalizing bed temperature and eliminating two major causes of excessive heat-up rates:

1. The "isolated core"

2. Coke "slides"

The Isolated Core. Delayed coke is a heterogeneous material. Particle sizing of green coke is typically 2 inches (5 cm) by less than 0.01 in (0.025 cm). The density of the coke in the green state is also variable, depending on the volatile (VM) content (usually a function of position in the coke drum during production). In a 10 foot (3 m) diameter rotating kiln, the coke bed is about 18 inches (46 cm) deep at its maximum, as shown in Figure 3. Coarse particles of coke in this non-homogeneous

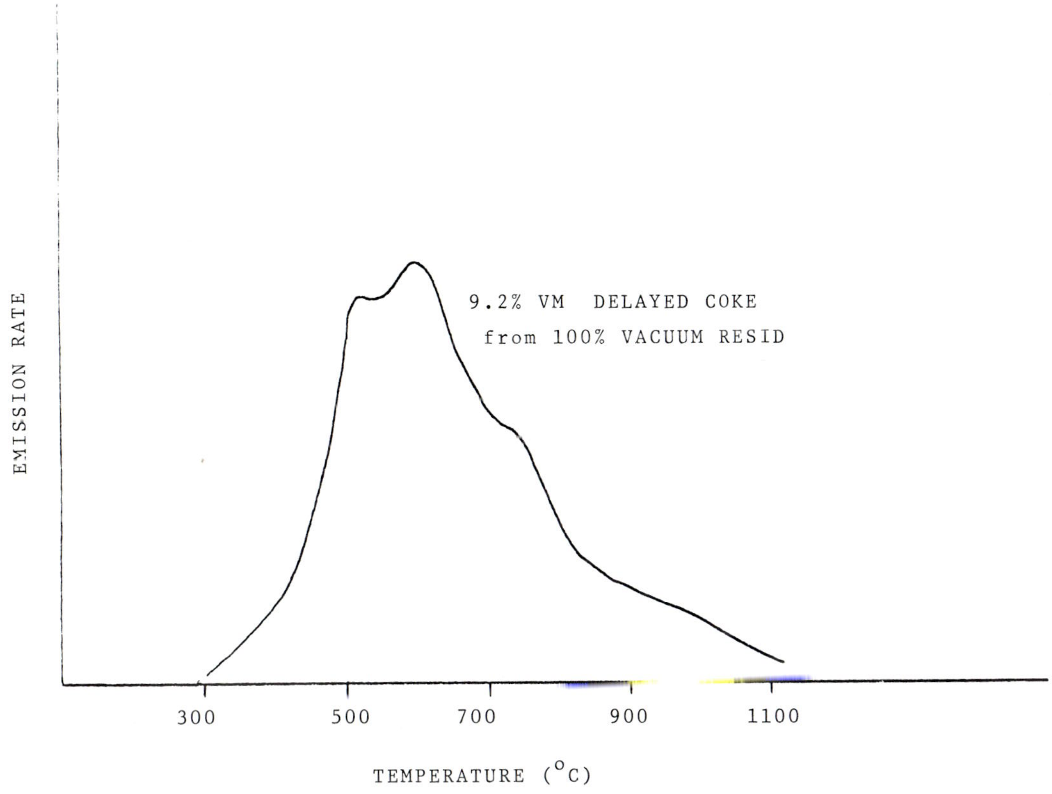

FIGURE 1

VOLATILE EMISSION SPECTRUM

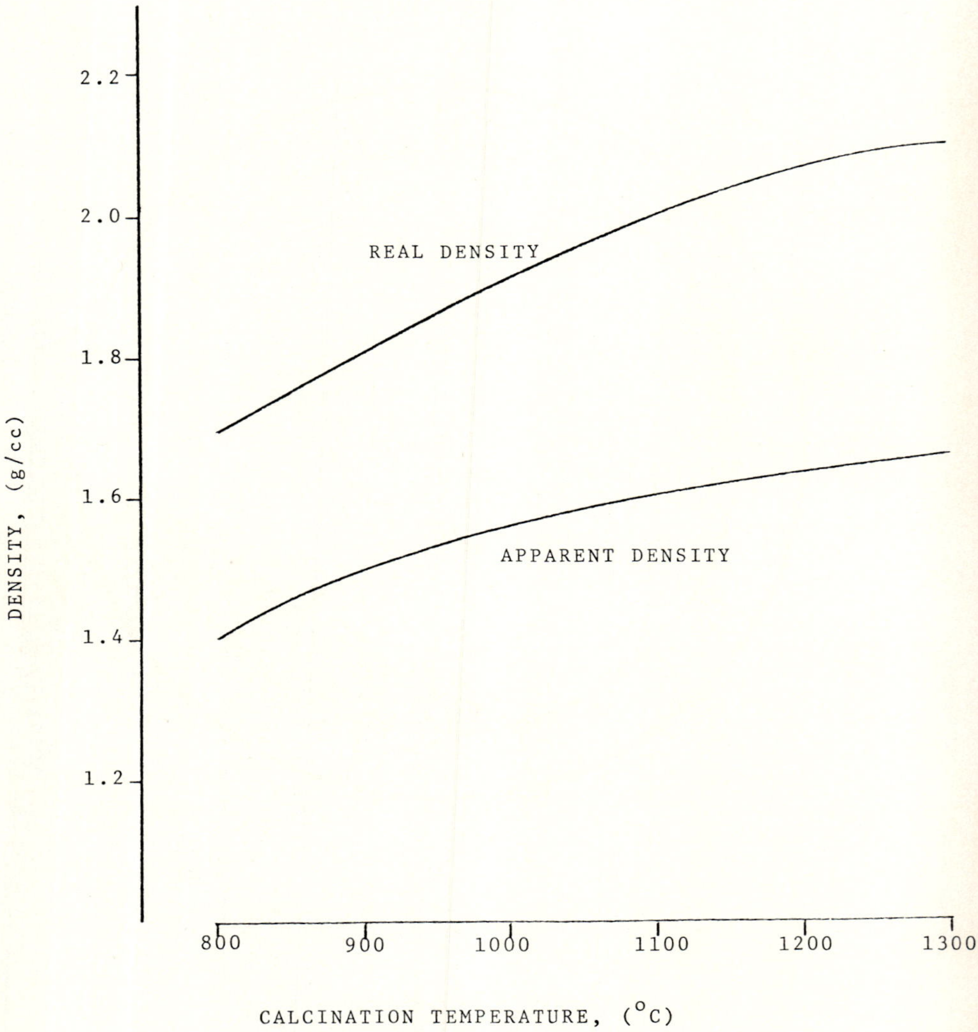

FIGURE 2

REAL and APPARENT DENSITY
vs. CALCINATION TEMPERATURE

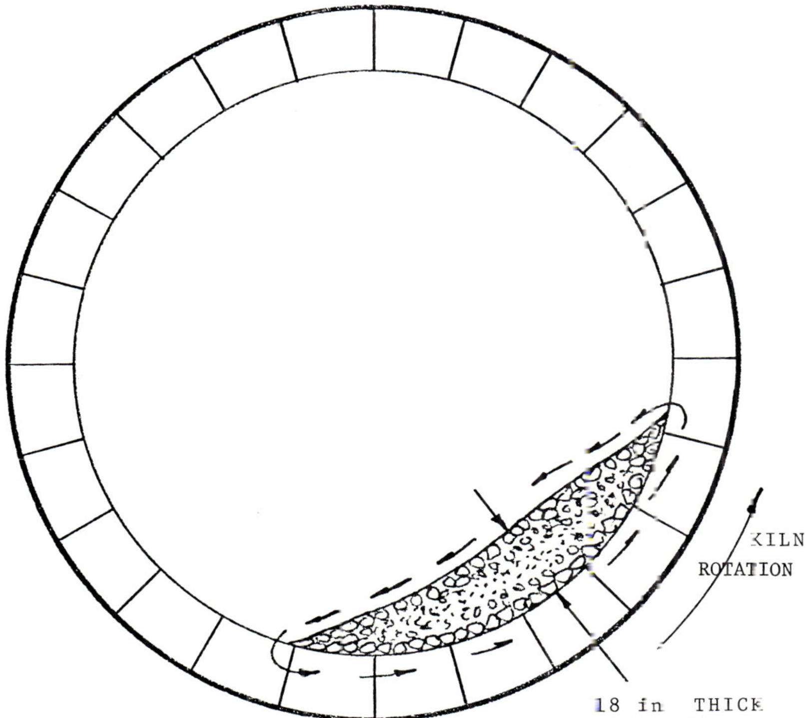

FIGURE 3

COKE BED IN A ROTATING KILN

bed will move to the outer periphery. The inner core of the bed remains composed of finer (generally higher VM) particles. The coke bed becomes a rotating envelope (as indicated by the directional arrows in Figure 3) surrounding a relatively stagnant core, the "isolated core." This situation is similar to that which occurs in fine grind feed kilns, such as in the lime and cement industries (Ref. 4).

As the coke progresses down the kiln, the outer envelope will heat up due to radiation from the countercurrent hot gas stream and conduction from the refractory lining. It was theorized and verified by mathematical model that the finer, denser inner core experienced a lower heat-up rate. The rate of heat conduction from outer envelope to inner core is relatively slow. The result, again theorized, is a lower temperature core such as shown in Figure 4. As VM evolve rapidly from the coke at 600^oC, the outer envelope will fluidize. This phenomenon has been reported by Alcan (Ref. 5). The colder, inner core coke will be stirred by this fluidization phenomenon. The inner core particles will be then exposed to rapid heat-up rates creating porosity. The net effect is reduced VBD of the product.

Coke Slides. Figure 5 illustrates how the coke bed expands during VM removal between 400^oC and 1000^oC. For some cokes (depending on VM content and volatile emission profile) and/or high kiln production rates, this fluidization, if excessive, can cause the coke bed in Zone 2 (Figure 5) to cascade down the kiln into Zone 3. These "slides" are caused by excessive fluidization seen with rapid VM emissions. This unstable condition is a periodic phenomenon noticed at high kiln feed rates or with excessively high VM content coke. The shortened residence time results in under-calcination of the coke.

Tumbler Installation

Installing the tumblers in the zone of VM emission significantly improves heat transfer. Gas temperature and coke bed temperature, determined by mathematical model are shown in Figure 6. Heat exchange in the area of VM removal (and burning) is due largely to radiation. This includes direct radiation from the gases to the bed surface and to the refractory surface. In turn, the refractory surface radiates heat to the bed or conducts heat to the bed as the coke makes contact with the refractory.

A tumbler installation similar in principle to that in lime and cement kilns (shown in Figure 7) will:

a) Increase the surface area of the bed, by retaining portions of the bed higher up the rotating side;

b) expose the "inner core" cooler material to the "outer envelope" by stirring;

c) transport heat more efficiently from gas stream to bed by the increased available surface area (both refractory and bed area).

The net effect of the coke kiln tumbler installation is the elimination of the non-uniformity of the coke bed temperature. Therefore, excessive heat-up rates are avoided and increased stability of the bed prevents coke "slides."

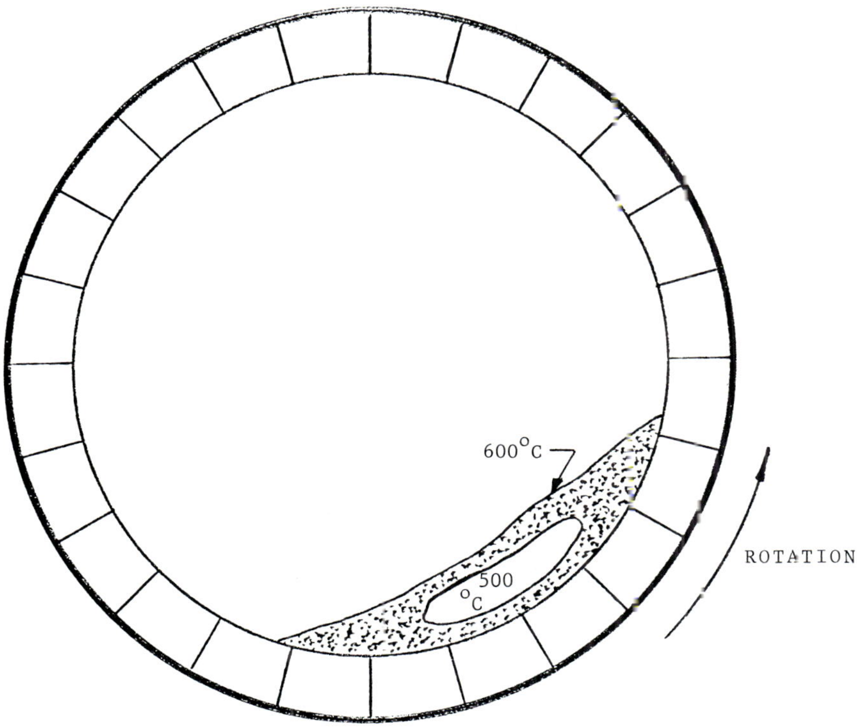

FIGURE 4

THE "ISOLATED CORE"

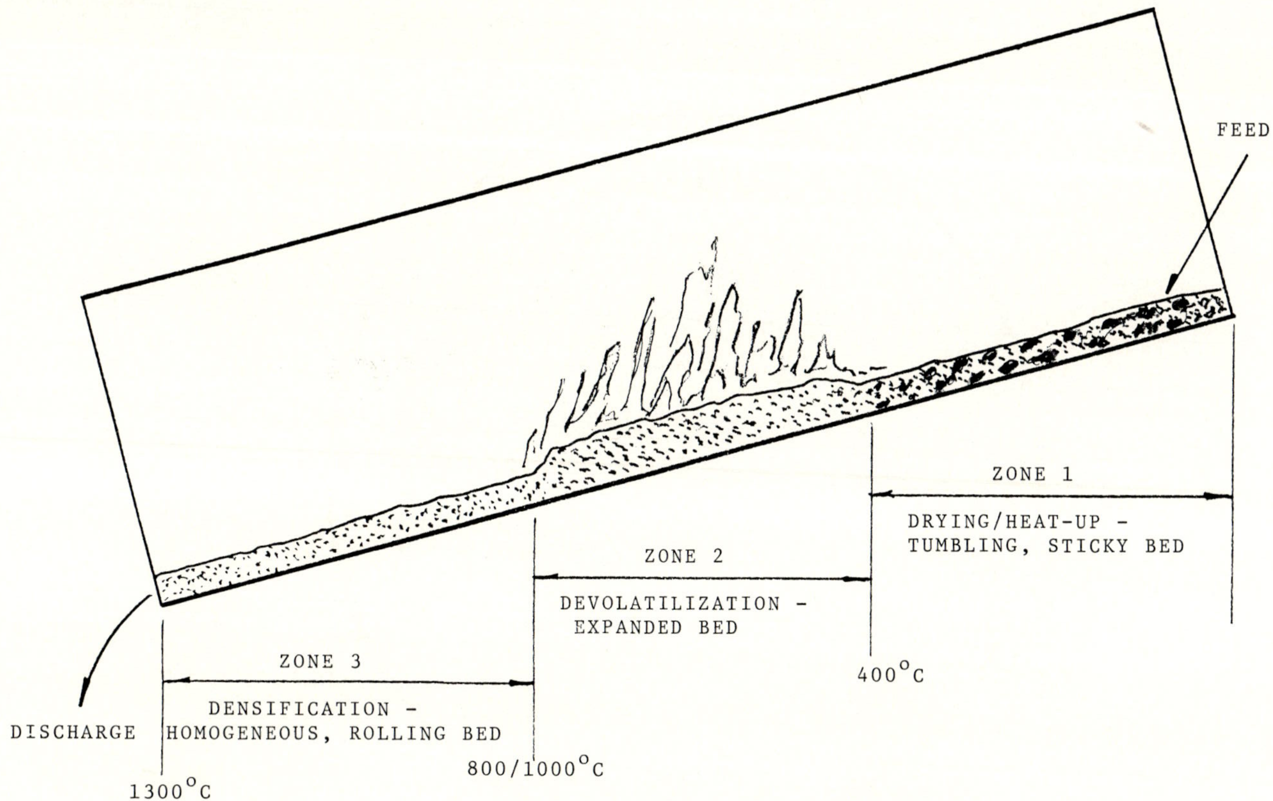

FIGURE 5

KILN SCHEMATIC DIAGRAM
SHOWING ZONES OF COKE CALCINATION

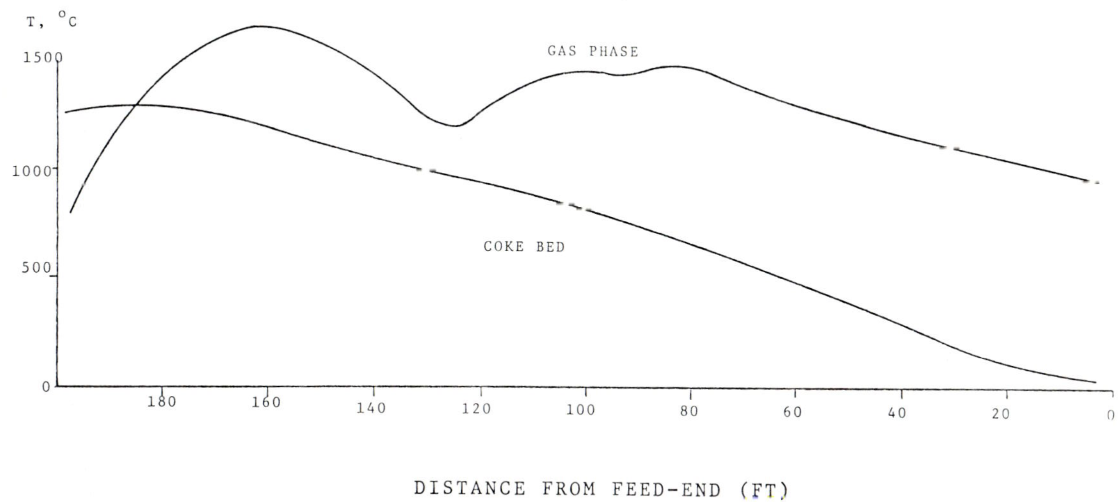

FIGURE 6

KILN TEMPERATURE PROFILES

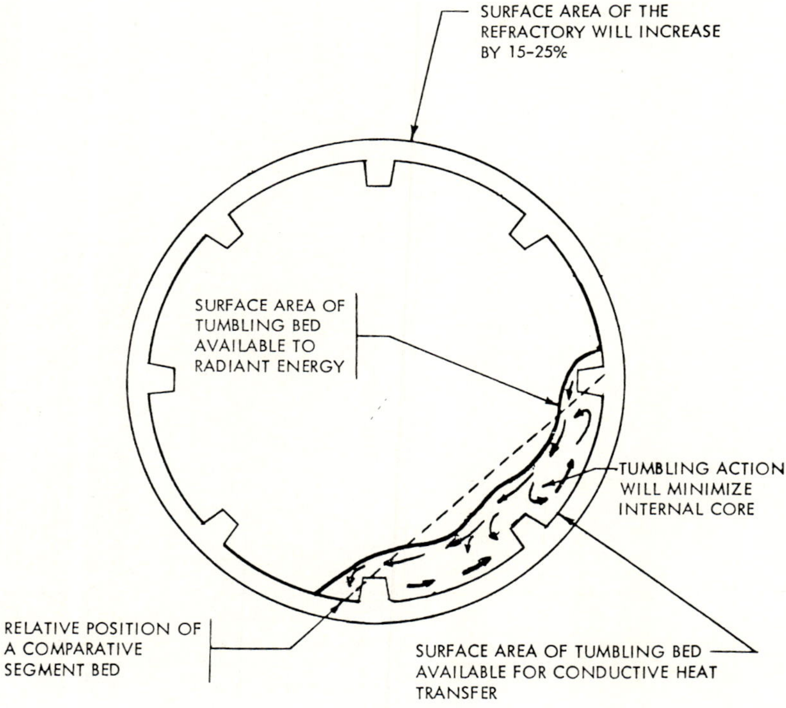

FIGURE **7** RELATIVE PHYSICAL CONDITION OF THE TUMBLING BED VS. THE SLIDING BED

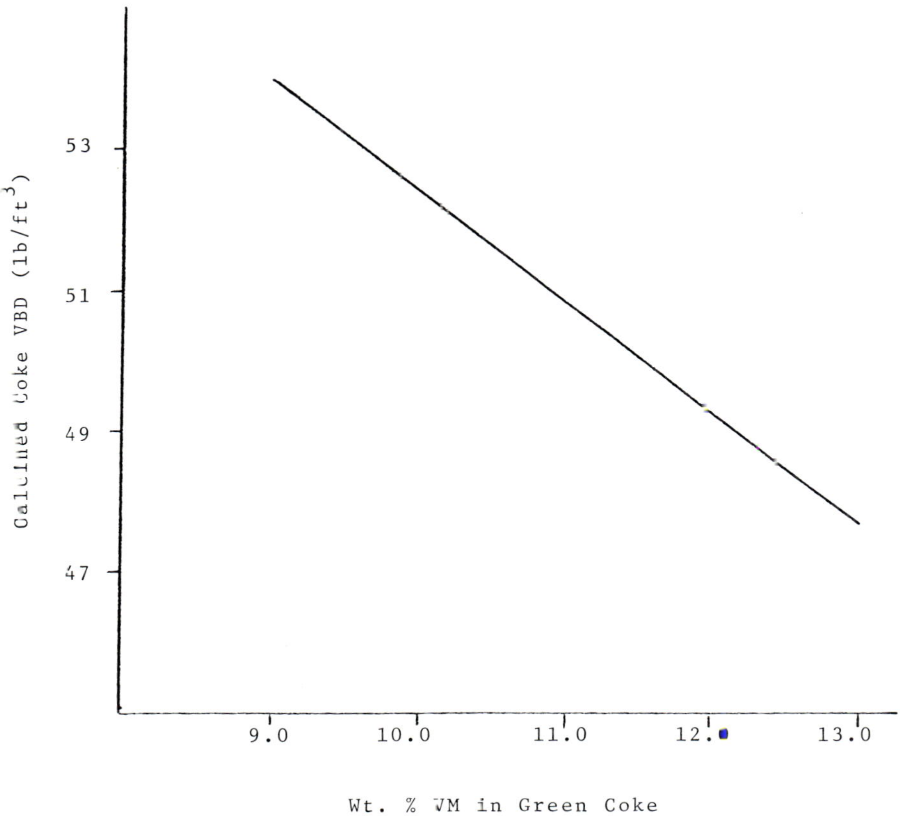

FIGURE 8

LABORATORY CALCINED VBD
VS. GREEN VOLATILE CONTENT

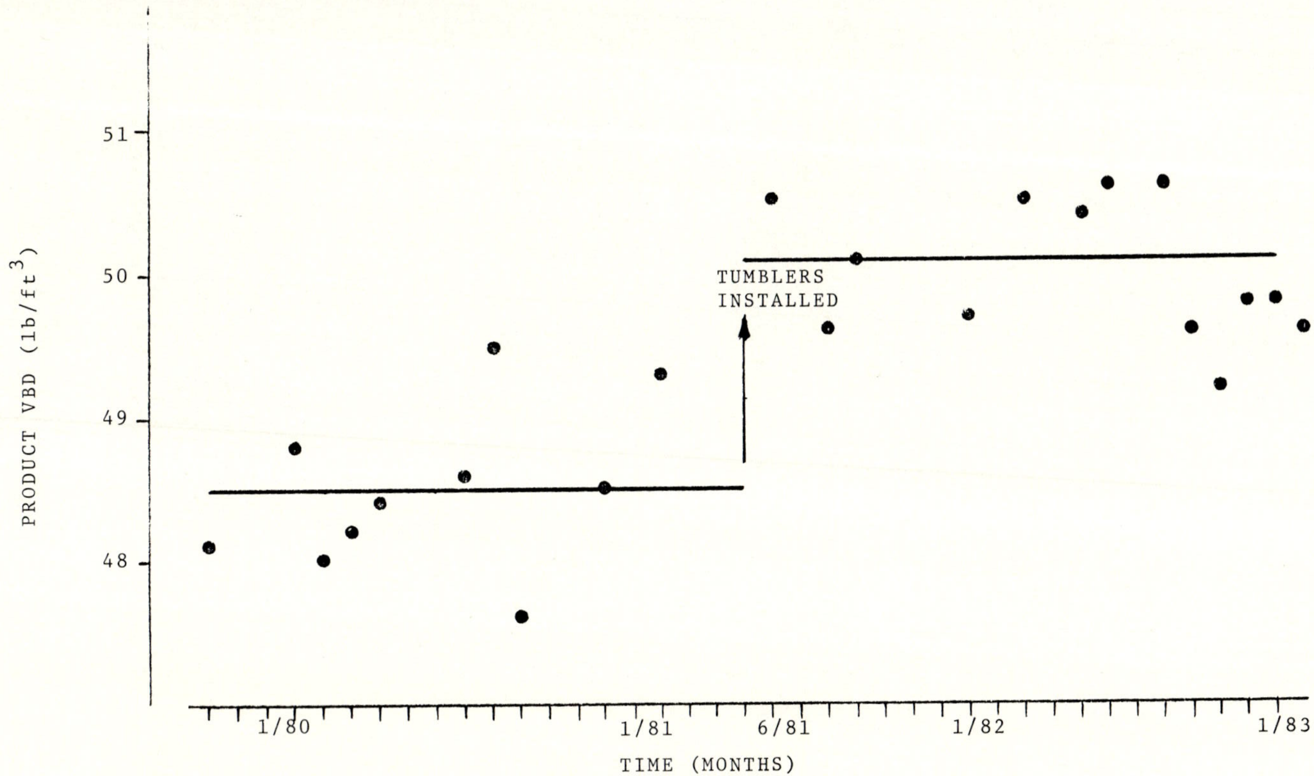

FIGURE 9

GRAMERCY COKE CALCINER
BULK DENSITY HISTORY

The initial tumbler installation at Gramercy uncovered some basic design deficiencies. Better placement and design were implemented at Chalmette and the results have improved. Location is specific to the kiln design, operation and feedstock. Tumbler detail is still under investigation.

Results

Importance of Volatile Material to Vibrated Bulk Density

Paul Rhedey (Ref. 2) defined the effect of heat-up rate on porosity and apparent density in carefully-controlled laboratory experiments. Vibrated bulk density (VBD) is another measure of porosity. The Kaiser VBD test is run with coke sized between 8 and 14 Tyler mesh fractions. For most anisotropic cokes in the Kaiser system, there is a relationship between volatile content of green delayed petroleum coke and the calcined VBD. A typical graph of such is shown in Figure 8. For this relationship the coke was calcined in a carefully controlled laboratory environment. Optimum commercial kiln calcination conditions can duplicate, but not improve, these results. Often, commercial kiln calcination results in a somewhat lower VBD. This is due to higher heat-up rates, as found by Rhedey. The most important consideration is the green coke volatile (VM) content. While the use of tumblers has shown that VBD can be improved in commercial kiln calcination, this should not be considered an acceptable alternative to low volatile content in the green coke raw material. The objective of the calciner quality control system must remain the procurement of low VM green coke.

Effect on Vibrated Bulk Density

Figure 9 shows a graph of Gramercy's VBD normalized to a constant green coke blend and quality (see Note 1). Prior to tumbler installation in May 1981, average VBD was 48.5 lb/ft^3. After tumbler installation, VBD averaged 50.1 lb/ft^3. The periods represented correspond to the following green coke quality:

	Pre-tumbler	With tumblers
coke % S	2.08	2.38
coke actual % VM	10.5	10.7
normalized % VM	10.5	10.5
production, short tons/hr	32.2	23.6

(Note 1: As shown in Figure 8, there is a relationship between VBD and VM content for anisotropic cokes. Also, each green coke can demonstrate a slightly different slope or intercept. Therefore, in order to eliminate the effects of differing VM content and the changing blends of green cokes used, the mathematical correllations relating VBD to VM and blend ratios were normalized. If this had not been done, the change in VBD after tumbler installation would have been even greater due to an improved feed blend.)

While the average production rate for the post-tumbler period is lower than pre-tumbler, tests run after installation at identical pre-tumbler rates showed a 1.7 lb/ft^3 VBD increase. These tests verify the 1.6 lb/ft^3 VBD increase on the average is a valid result. More on production rate is reported below:

Figure 10 shows a graph of Chalmette's VBD normalized to a constant green coke blend and quality (see Note 2). An improved design and installation position for the tumblers at Chalmette resulted in a VBD improvement of 2.1 lb/ft^3 since March 1982. The periods represented correspond to the following green coke quality:

	Pre-tumbler	With tumblers
coke % S	2.1	2.1
coke actual % VM	10.8	9.9
normalized % VM	10.8	10.8
production, short tons/hr	26.0	25.6

(Note 2: As per Note 1, a mathematical adjustment was again performed to eliminate VM and blend effects. Without this adjustment, the improvement in VBD would have been even higher.)

A word should be said about the sensitivity of VBD to production rate. Any sensitivity of VBD to rate for stable kiln operation has not been discernable over the range of rates examined. This does not imply that, in the extreme, a sensitivity does not exist. However, for the small changes in rates during this study, the operation has been in the relatively flat portion of the heat-up rate curve presented in Rhedey's work (Ref. 2).

Uniformity of Calcination

Degree of calcination is measured by the real density (RD) and specific resistivity (SR) of the product. SR is measured in the Kaiser system on the 35x65 Tyler mesh fraction using all product which can be crushed to this size.

SR tests were run before and after installation of the tumblers on various particle size distributions: +4, 8x14 and 35x65 Tyler mesh naturally occurring in the product. It was found that, before tumblers, the difference in SR values was 0.0040 to 0.0070 at a mean of 0.0380 ohm-in. After tumblers, the difference was reduced to 0.0030 to 0.0050 at a mean of 0.0380 ohm-in.

The above improvement in uniformity is attributed to more constant exposure of the particles to calcination temperature via isolated core elimination.

Improved Product Performance

A previous study indicated the sensitivity of reduction anode quality to calcined coke properties (Ref. 6). Baked density is an important factor which is strongly influenced by calcined coke bulk density. Generally, as bulk density increases in anisotropic cokes, baked density also increases. This, in turn, leads to improved anode performance in aluminum reduction cells. Jones et al. (Ref. 7) showed, in full size reduction cell tests, the correllation of carbon efficiency was a function of both baked density (BD) and airburn (550 AB):

$$\text{Net C} = A - (b \times BD) + (c \times 550\ AB)$$

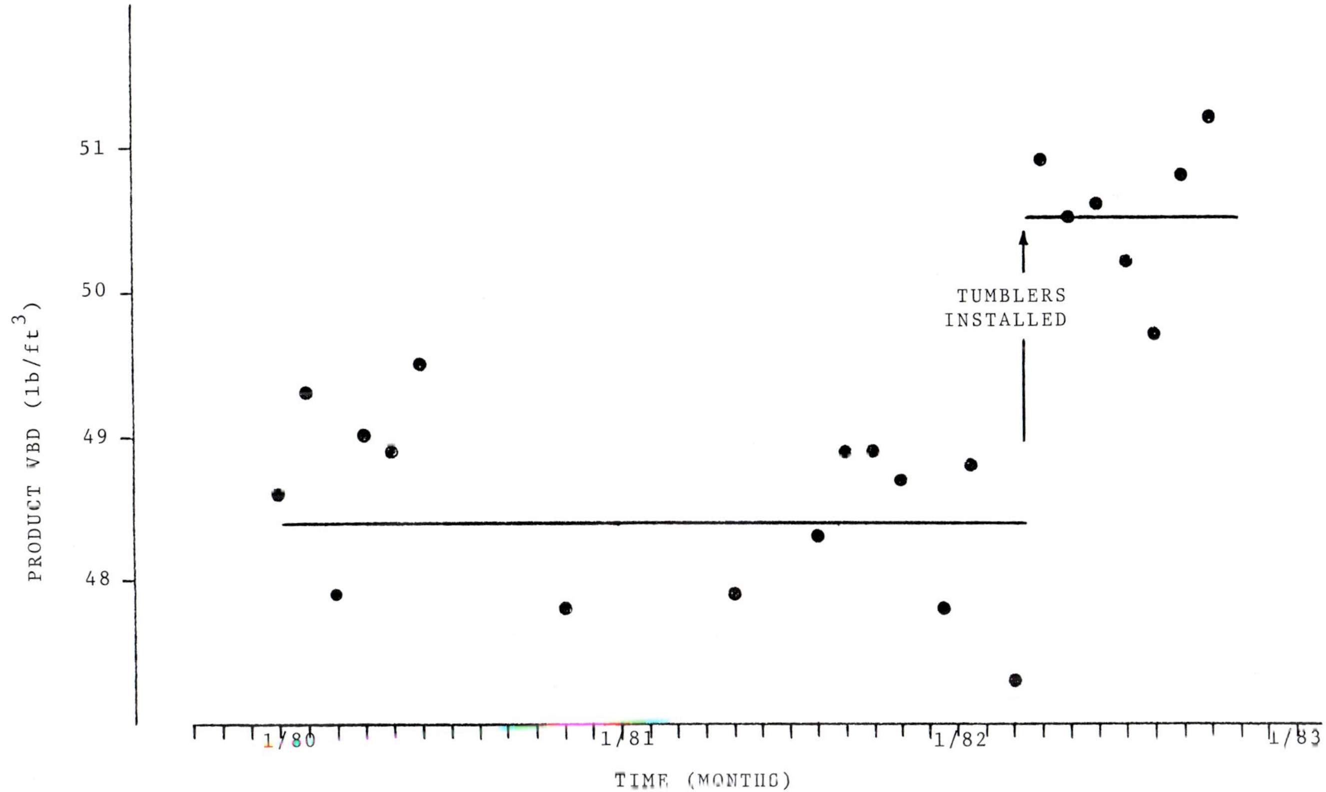

FIGURE 10

CHALMETTE COKE CALCINER
BULK DENSITY HISTORY

Where Net C = net carbon consumption

A, b and c are constants

In fact, actual plant performance bears this correllation out. Chalmette plant performance data comparing the operation before and after tumbler-improved coke was received shows significant improvment in production rate, power consumption and carbon efficiency during comparable periods. This data is shown in Figure 11. Chalmette is a horizontal-stud Soderberg plant. It is estimated that 2/3 of this performance improvement is due to tumbler-improved VBD.

At an average power cost of 20 mils/KWH, these improved efficiencies would translate into the following annual savings for a 100,000 metric ton/year reduction plant:

Carbon Cost Savings	=	\$540,000
Energy Cost Savings	=	\$530,000

These cost savings illustrate the importance of improved coke quality on the overall operation of aluminum reduction.

Improved Kiln Performance

The installation of tumblers has resulted in a significant improvement in kiln operation. Historical demand for calcined coke resulted in the operation of the Kaiser kilns at maximum production rates consistent with quality requirements. Prior to tumbler installation at Gramercy, this rate was about 32.7 short tons calcined coke per hour. After tumbler installation, this rate was increased to 35 tons/hr with no adverse effects.

The above-noted production rate increase is due to the improvement in kiln stability at maximum throughput. As mentioned earlier, during the VM removal phase of coke calcination, fluidization of the bed will occur in Zone 2. If this fluidization is excessive, coke slides will result. A coke slide is the single most serious problem limiting throughput. The tumblers, by virtually eliminating slides, increase stability allowing rate increases to the point where coke quality (degree of calcination) dictates production. VBD at the higher rate is identical to that at the previous maximum rate.

Conclusions

1. Refractory tumblers installed in coke calcining rotary kilns improve VBD by 2 lb/ft^3 when calcining reduction grade anisotropic delayed coke.

2. Consistency of calcination (as measured by specific resistivity) between fine and coarse particles was improved 20 to 30% by tumbler installation.

3. By installing tumblers, kiln operating stability was substantially improved, resulting in a 7% increase in attainable production rate.

FIGURE 11

PRODUCTION DATA

CHALMETTE REDUCTION PLANT

	Pre-Tumbler June - Dec. 1981	After Tumblers June - Dec. 1983	Difference 1982-1981
Al Production - lb	69,128,650	69,980,200	851,550
Pot Days	61,174	61,011	-163
lb/Pot Day	1,130	1,147	17
KWH/lb Al	7.81	7.69	-0.12
Amps	72,441	72,468	27
Current Efficiency	87.9	89.2	1.3
lb Carbon/lb Al	0.526	0.498	-0.028

References

1. Jones, S. S., Hildebrandt, R. D., "Green Coke Volatile Emission Spectra," from Light Metals 1981, AIME, pp. 423-440.

2. Rhedey, P., "Structural Changes in Petroleum Coke During Calcination," Transactions of the Metallurgical Society of AIME, Vol. 239, pp. 1084-1091, July 1967.

3. Whittaker, M. P., Miller, F. C. and Fritz, H. C., "Structural Changes Accompanying Coke Calcination," Industrial and Engineering Chemistry Product Research and Development, Vol. 9, pp. 187-190, June 1970.

4. Tyler, G.A., "The Utilization of Refractory Tumbling Ledges to Improve the Fuel Efficiencies and Production Rate in Rotary Cement Kiln," a paper presented at the Rock Products Eleventh International Cement Industry Seminar, Chicago, Illinois, 1975

5. Farago, F. J. and Sood, R. R., "Advances in Rotary Kiln Calcination of Petroleum Coke," Light Metals 1976, AIME, Vol. 1, pp 351-362.

6. Jones, S. S., Hildebrandt, R. D. and Hedlund, M. C., "Variation of Anode Performance with Coke Quality," Paper No. A 77-97, presented at the 106th AIME Annual Meeting, Atlanta, Georgia, 1977.

7. Jones, S. S., Hildebrandt, R. D., and Hedlund, M. C., "Influence of High Sulfur Cokes on Anode Performance," Light Metals 1979, AIME, Vol. 2, p. 553.

LOSS OF WEIGHT SYSTEMS IN CONTINUOUS CARBON PRODUCTION - A COMPETITOR TO TRADITIONAL BELT SCALES:

Steinar Frosta and Tormod Naterstad,
Aardal og Sunndal Verk a.s.,
5875 Aardalstangen, Norway

Belt scales, the traditional metering for the continuous production of carbon paste, are influenced by a number of factors (temperature variations, fluidizing and dust etc.) which can result in inaccurate weighing.

With the aim of overcoming this difficulty, and also to facilitate the accurate metering of binder , a weighing system based on the loss of weight principle has been developed.

This report deals with the principle of the Simplex Loss of Weight system for continuous metering, the development of the system and practical experience from operations. Finally the question is discussed as to whether this system of weighing fulfils expectations and whether, in practice, it can compete with belt scales in the continuous manufacture of carbon products.

INTRODUCTION

The consept of large scale carbon production for the Al-industry usually includes the following 5-8 separate unit operations.

- Raw material preparation, cleaning, drying
- Crushing/grinding
- Classifying (4-6 aggregate fractions deviding a typical $<10^{-3}$ to 20 mm particle range)
- Metering (separate fraction metered in fixed ratios to the binder)
- Mixing
- Forming
- Baking
- Machining/rodding

(Mixing, Forming, Baking, Machining/rodding: Prebaked products only.)

For a given raw-material, crushing and grinding characteristics are the bases for obtaining well defined, consistant aggregate fractions through classifying.

In the overall process the achievement of a uniform consistant product through fixed total aggregate distribution and binder content is very much dependant on the metering precision.

PRINCIPLE

Metering systems

A number of different metering systems are currently applied in the carbon industry. In batch systems cumulative or fraction by fraction weighing may be carried out in a single hopper combined with mechanical or electronic weighing. By principle, however, a continuous process has to include separate, continuous metering of each fraction as well as the binder.

So far belt weights of mainly two different designs have been market leaders in the continuous carbon aggregate formulation as well as for solid pitch metering. In Fig. 1 the principle of a common belt weight is shown.

For liquid pitch metering volumetric pumps have been favoured. In Fig. 2 an example of a total metering concept in a more or less general flow sheet of a continuous paste production is shown.

Further discussion is mainly concerned with the continuous process.

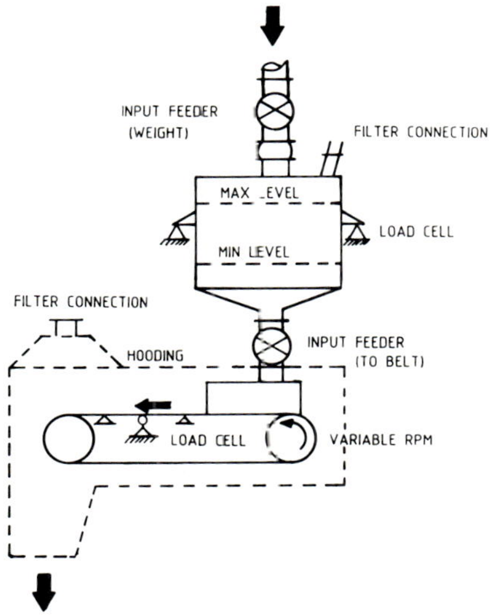

Fig. 1. Principle of a common belt weight.

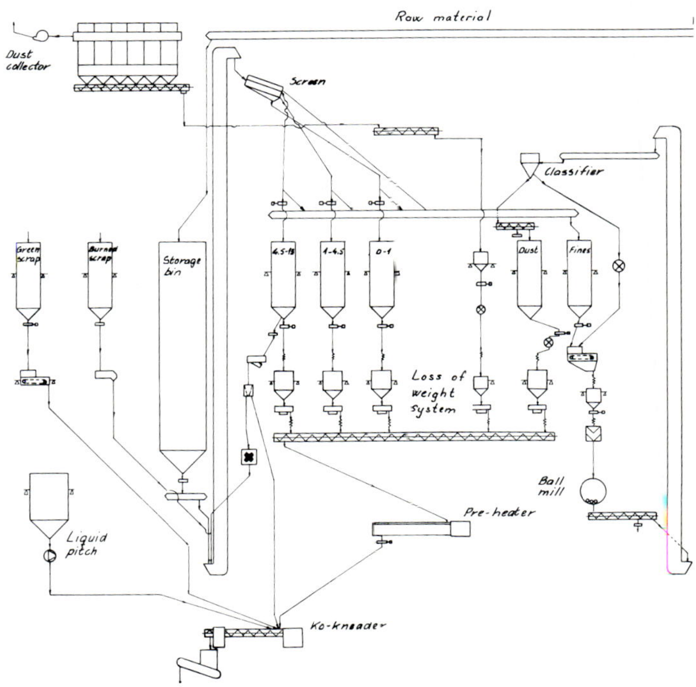

Fig. 2. Flow sheet of a continuous paste production.

Metering precision

Although difficult to state in a general way independent of fraction ranges, product specifications, mix formulation as well as process capacity, the following terms may serve as approximate guidelines for metering device precision (σ denoting standard deviation).

Repeatability: $2\sigma < +$ 1% measured by 10 consecutive samples over 1 min.

Linearety: $\sigma <$ 0,5% of set rate measured by 10 consecutive samples taken over 3 min.

Stability: $\sigma < +$ 1% over 3 months.

Generally, lower precisions in coarce and medium fractions are more tolerable than for the finer fractions due to drastic change in binder demand by roughly the inverse square of the particle size.

Aside of the metring device precision, fluidization and suck-off of the typical bagfilter fraction (< 63 μ m) subsequent to metering may present a particular problem.

Depending on rawmaterials origin and desintegration characteristics through transportation and comminution processes the filter return fraction easily accumulates to a 3-8% level of the continuous processed aggregate. Thus, contributing a typical 9-20% to the total fines (< 63 μ m) of the total aggregate, the return of collected dust if not well controlled, may present a quality consistency problem as indicated by Fig. 3.

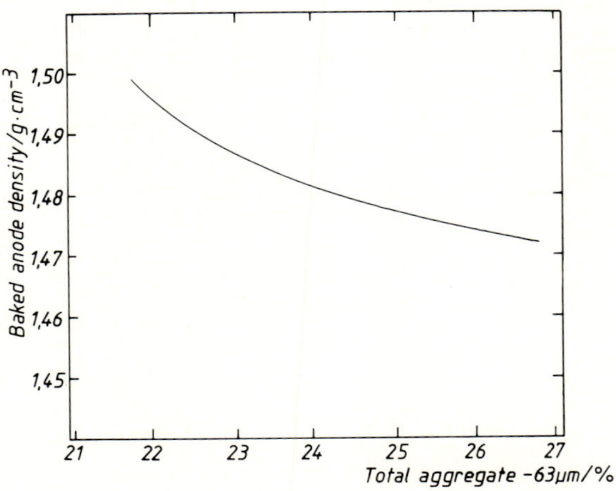

Fig. 3. Variation in baked density versus -63 μm in the total dry aggregate with constant pitch content.

Design of "the loss of weight" system for continuous metering of carbon fractions.

With the increased push on improved working and environmental condition in the late 1960's ÅSV decided to persuade a development of well closed aggregate processing and careful control on return of collected dust as a separate fraction in paste plants. Based on previous experience of "loss of weight" feeders in batch mixing, a modified continuous prototype-feeder was built in 1968 for filter dust metering.

Supported by the convincing long term tests obtained, continuous "loss of weight" feeders were built for metering all dry aggregate fractions as well as liquid pitch in a complete production line in 1972. Major design factors were:

- High capacity range for each unit independent of particle size.
- Minimum dead weight to feed rate capacity.
- Closed feeder system to minimize necessary draught through the bag filter system (minimum subtraction of already metered aggregate).
- Design of weighing hopper to avoid obstructions and segregation in the mass flow of aggregate to the outlet feeder.
- Un-bridging device to face changes in particle flow characteristics (i.e. by change in moisture and volatiles).
- High precision of load cells with triangle suspension to the hopper.
- Short refilling sequence for the hopper.

In Fig. 4 the principle of the "loss of weight" system is shown, while Fig. 5 shows in some details of the outlet feeder arrangement.

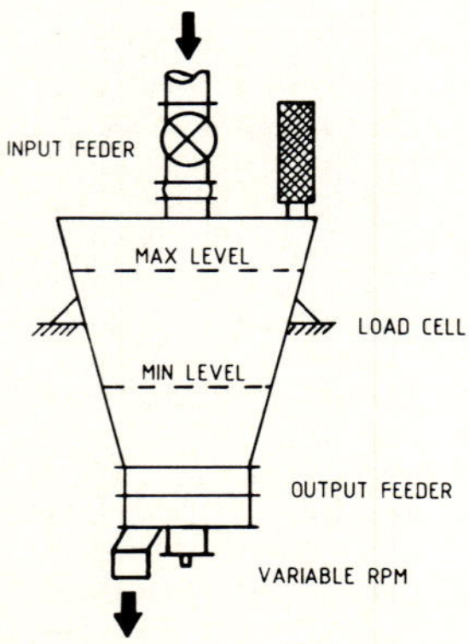

Fig. 4 Principle of a continuous "loss of weight" system

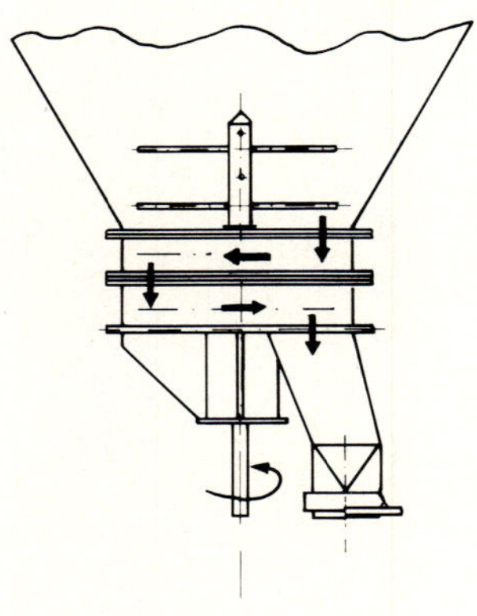

Fig. 5 Outlet feeder design

The second generation feeders was further improved by adapting a superior control/monitoring system. In principle the control system is based on a pre-amplified signal from 3 load cells, which through a derivative unit enters a PID integrated circuit controller with adjustable characteristics and reference capacity settings. The maximum and minimum level for stop/refilling functions are adjustable by reference values which are compared to the actual load. By a variable speed motor the velocity of the outlet feeder is adjusted according to the error signal from the PID controller. Besides built in operation cycle-indicators, overflow alarm systems for most vital function as well as malfunction error signals are indicated. Further, continuous process monitoring of actual feed rate, deviation in feeding rate, aggregate load in the hopper as well as the integrated feeder output are shown on front panel meters.

Characteristic refilling period for the hoppers vary between 10-60 seconds. During this period a fixed feeder speed may be kept constant - or can be changed through specified adjustable steps.

In our latest microprocessor version a self-updating logic system defines a step-up or step-down function during refilling according to accumulated characteristics in some proceeding cycles. In this way the influence of variations in bulk density, static pressure and wearance of the feeder characteristics over time are continuously compensated for.

In Fig. 6 a typical recording of the loss of weight and refilling sequence between upper and lower set limits is shown.

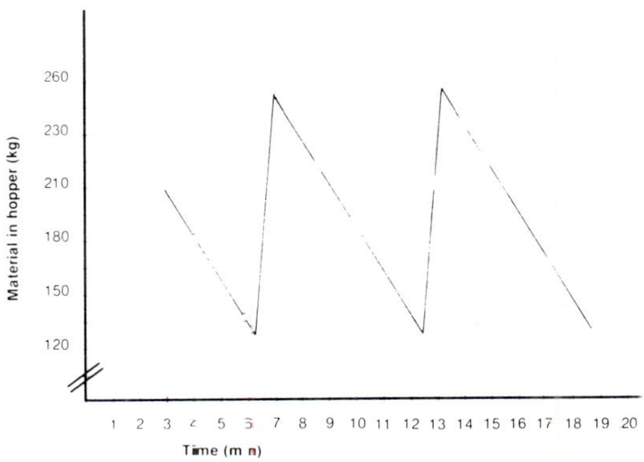

Fig. 6. Actual weight in hopper over time.

Table 1: Comparison of different characteristics of "loss of weight" and belt weight feeders in usual carbon production.

Characteristics	"Loss of weight" feeder	Belt weight feeder
Metering reproducibility [x]	$\pm$ 0,3 - 0,8%	$\pm$ 0,8 - 1%
Ordinary linear capacity range	1:20	1:20
Stability [xx]	$\pm$ 0,9	$\pm$ 1%
Sensivity towards aggregate temperature - 25 to + 150°C	none	essential
Calibration routine (mass to signal)	quick	slow
Sensivity towards foreign ductile particles	some	none
Restrictions in particle size	none below 50 mm	Dusting/material loss/frequent cleaning for fine fractions (< 0,63 μm)
Maintenance	low	low

[x] Standard deviations in repeated measurements every 10 min. for 3 h.

[xx] Standard deviation for set capacity over 3 months measured by arbitrary sampling.

"Loss of weight" principle applied in liquid pitch metering

Volumetric pumps with no feed-back information on density, viscosity and temperature are likely to introduce problems in exact meteric. Furthermore the characteristics of most pumps are more or less sensitive to abrasion by hard particles (5-14% by weight in the range 0,5-50 μm in the pitch.

The "loss of weight" principle easily applies to liquid pitch metering also by using a common type rotary pump. Combined with a self-updating control system, high precision and long service life of pumps is made possible.

DISCUSSION

Experience with "the loss of weight" system.

From our first version off in 1968, the "loss of weight" concept has been introduced in all of our new or rebuilt carbon process lines. At the present the number of units in operation count 15 on solid fractions and 3 on liquid pitch. Furthermore the system is applied in coal firing, lime, dolomite and quarts aggregate metering covering a bulk density range of approximately 0,2-1,3 $g.cm^{-3}$ and capacity range of 0,1-30 m^3/h.

In Table 1 characteristics particularly related to carbon paste production is compared with belt weight feeders.

Both systems are reliable within precision limits specified, which are somewhat in favour of "the loss of weight" system. The real essential advantage of the latter system is however, in precise fine dust metering and practical maintenance routines.

With a straight flow-through of aggregate from efficient drying (i.e. on butts) through classifying to the fraction metering, the aggregate temperature may easily raise from ambient to 80-100 ^{o}C. Most belt weights are sensitive to much lower temperature variations, while by principle a whole-steel "loss of weight" feeder is unaffected.

Further characteristics of the loss of weight system are:

- Compact, uniform, closed design

- Interchangeable control/monitoring and software design

- Non-sensivity towards variations in tare of the hopper.

A disadvantage of "the loss of weight" system has been a non feed-back control during the refilling period, particularly if combined with variations in flow characteristic, static pressure and change in characteristics of the feeder itself (wearance etc). Practically these limitations are reduced by introducing a safe, mechanical unbridging device inside the aggregate hopper and continuous self-updating logics in the control function. The sensivity of "loss of weight"feeders towards over-sized foreign or ductile particles may be solved by modified tolerances of the outlet feeder or by a different design.

For pasty products (i.e. green paste of low softening point) a regular mass flow through the hopper has to be secured by forced screw movement in the lower part of the hoppper.

SUMMARY

Loss of weight metering is developed to an approved technology in carbon paste production.

Design characteristics:

- Compact and uniform.

- Steel and light metal parts.

- Compatible control/monitor/software.

Precision:

- Typical short term standard deviation: $\sigma < \pm 0,5\%$.

  - Equal or higher than for most belt weights in coarce aggregate (0,1 - 20 mm) metering.

  - Superior to belt weights in fines and dust (< 0,1 mm) metering.

- More precise than non feed-back volumetric system in liquid pitch metering. Non-sensivity towards density, viscosity and usual temperature variations.

- Non-sensivity towards variations in tare and in aggregate temperature.

Maintenance:

- Compares to belt weights for solid aggregates.
- Less than for volumetric liquid pitch metering systems.

RECRYSTALLIZATION OF COMMERCIAL BERYLLIUM SHEET

James M. Marder
Brush Wellman Inc.
17876 St. Clair Avenue
Cleveland, Ohio 44110

A study was conducted to determine the recrystallization behavior of commercial beryllium sheet rolled from vacuum hot pressed S-65 impact ground powder. Microstructural response to rolling and subsequent annealing was determined using optical metallography, and transmission electron microscopy. Crystallographic texture studies were conducted, utilizing {0002} X-Ray pole figures.

It was found that upon rolling to 75% reduction at 1425 F (773 C), a high temperature anneal was required to produce a uniform recrystallized grain size. Low temperature anneals, below 1550 F (843 C), gave rise to a duplexed grain structure. Subsequent rolling at 1250 F (677 C), to 30% reduction gave rise to a fairly uniform, recrystallized grain size. The activation energy for recrystallization after 1250 F (677 C) rolling was calculated at 173 Kcal/mole. This value is contrasted to the 18-22 Kcal/mole determined for ultrahigh purity ingot. The relationship between the oxide content and recrystallization kinetics probably accounts for the variation in recrystallization energies.

Introduction

Rolled beryllium products are currently employed in a variety of applications for their structural strength and transparency to electromagnetic radiation. Energy dispersive analysis of X-Rays (EDAX units), Computer Aided Tomography (CAT Scanners) and other precision X-Ray devices use beryllium as a window material to isolate the X-Ray source or detector, generally kept under vacuum, from the rest of the environment. In addition, weight-constrained applications such as satellite structural members, also use beryllium to advantage. The beryllium sheet must therefore have considerable strength and ductility as well as X-Ray transparency. In recent years, new grades of beryllium have evolved which can better meet the needs of current technology requirements.

A recent major advance in the production of improved grades of beryllium has come about with the introduction of impact grinding (I.G.) as the final step in reducing beryllium vacuum-melted ingot to the powder required as input for consolidation. Impact grinding, as opposed to attrition milling, results in a blocky, rather than flake-like powder giving rise to a higher degree of property isotropy in the consolidated product. Consolidation is generally carried out by vacuum hot-pressing (VHP), although Hot Isostatic Pressing (HIP) has also been employed and is finding wider application. In addition impact grinding (I.G.), schematically shown in Figure 1, is done at relatively low powder particle temperatures due to adiabetic expansion of the gaseous impact grinding medium. Therefore, a lower beryllium oxide content is possible with I.G. powder. The lower BeO inclusion content in the current grades of impact ground structural material is exemplified by the 1.0% maximum BeO content of S-65 in comparison to the 2.0% maximum allowed in the S-200 attritioned powder. The S-65 chemistry specification is presented in Table I.

Table I
Chemical Composition of S-65

Constituent	wt.%
Beryllium Assay, % Min.*	99.0
Beryllium Oxide, % Max.	1.0
Iron, % Max.	0.08
Aluminum, % Max.	0.06
Carbon, % Max.	0.10
Magnesium, % Max.	0.06
Other Metallic Impurities (% each, Max) as determined by normal spectrographic methods.	0.04

*Includes Be content of BeO present.

The difference in oxide content between impact ground and attritioned powder implies a different response to deformation and recrystallization. It was the objective of this work to establish the recrystallization and grain growth behavior of S-65 material.

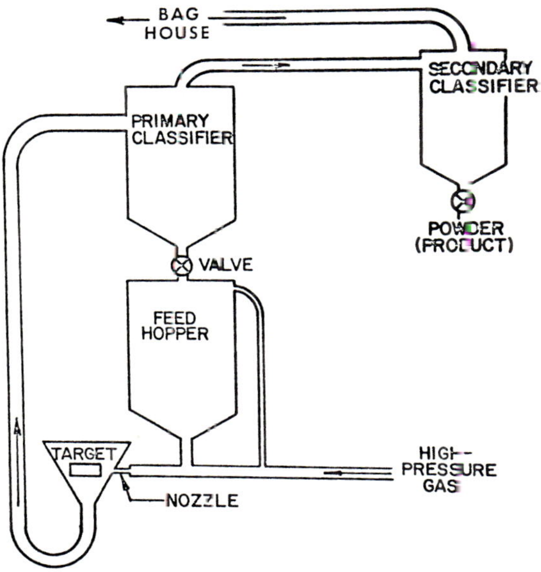

Figure 1 - Schematic Illustration of Impact Grinding Unit

Experimental Procedure

Samples of S-65 vacuum hot-pressed blocks, 1 in. thick were canned in 7/16 in. thick 1010 low carbon steel. These compacts were oriented such that the plane of the plate was perpendicular to the hot pressing direction. The compacts were undirectionally rolled at an initial temperature of 1400-1425 F (760-773 C), to a total reduction of 75%. Approximately 20 passes were used to achieve this reduction with reheating accomplished between each rolling pass.

At the completion of this stage of rolling, the steel cans were removed. Final rolling was then carried out at 1250 F (677 C) to achieve 30% additional reduction at this temperature. The rolling direction was the same as that in the 1425 F (773 C) operations, so that an undirectional rolling schedule was used throughout.

Subsequent to rolling, heat treatment was undertaken to investigate the kinetics and microstructural changes associated with recrystallization. These studies were conducted upon both 1250 F (677 C) and 1425 F (773 C) rolled material. In addition to optical metallography, X-Ray pole figure analysis and electron metallography were performed to characterize both the as-rolled and recrystallized material.

The results of the program consist of a concise statement of the recrystallization behavior of the S-65 grade of beryllium as a function of working temperature and rolling reduction.

Results and Discussion

The As-Rolled Structure

The optical microstructure of vacuum hot pressed S-65 is shown in Figure 2. The typical grain diameter is approximately 12 microns and the relationship between the coarse (>0.25μm) BeO particles and the grain boundaries is evident. Figure 3, shows the microstructure after rolling at 1425 F (775 C) to 75% total reduction. The extent of rolling deformation is apparent ini the high aspect ratio grains which result. The oxide particles themselves remain undeformed, but there is some tendency to produce elongated "stringers" of oxide in the rolling direction. The transmission electron micrograph shown, Figure 4, was taken under 2-beam dynamical conditions, near the $\{\bar{1}2\bar{1}3\}$ zone axis. A complex network with a very high dislocation density is visible in the bright-field micrograph. Tilting experiments indicated that the majority of the dislocations had Burger's vectors parallel to $\{0002\}$. The grain boundaries are generally free of oxide although oxide can sometimes be associated with them. A typical low angle boundary structure is also evidenced in Figure 4. Figure 5, a dark field electron micrograph taken with the $\{1\bar{1}01\}$ reflection of the $\{\bar{1}2\bar{1}3\}$ zone axis, reveals dislocations being pinned by intragranular, fine BeO particles. It should be noted that the grain boundaries are relatively free of BeO particles, and high angle grain boundaries are somewhat difficult to distinguish.

A very marked crystallographic texture was also produced by the rolling operation. The $\{0002\}$ X-Ray pole figure, Figure 6, shows the c-axis poles aligned in a plane perpendicular to the sheet plane direction. The tendency, therefore, is for the basal planes to lie in the sheet plane, with some preferred orientation along the rolling direction.

This is similar to the occurrence of sheet texture in other hexagonal close-packed materials, such as titanium[1]. This preferred orientation is not affected by such processing variables as rolling temperature, within the 1250-1425 F (677-773 C) range of this study. The ineffectiveness of a post-rolling heat treatment is documented by the next $\{0002\}$ X-Ray pole figure for material which has been fully recrystallized, Figure 7. The persistence of the rolling texture, even after a full recrystallization anneal is not unusual[2]. The ineffectiveness of annealing in altering the texture of rolled beryllium is an important consideration when fabricating or performing other operations with this material. In titanium, for example, full recrystallization, which is accompanied by the α β allotropic transformation, does not destroy rolling-induced texture[1]. It is not surprising, therefore, than in beryllium, in which no analogous potentially texture-randomizing process exists, that the strong texture persists through the recrystallization and grain growth heat treatments.

After rolling at 1425 F (773 C) and 75% reduction, further rolling was carried out at 1200 to 1250 F (649-677 C). The maximum reduction possible was 25-30% before transverse cracking was noted. Therefore, this single level of reduction was used for the low temperature rolling portion of the program.

Figure 8, a transmission electron micrograph demonstrates the very highly strained nature of the beryllium subsequent to the low temperature rolling sequence. The dislocation density is very high, and there is really very little in the structure that can be discerned. Again, the

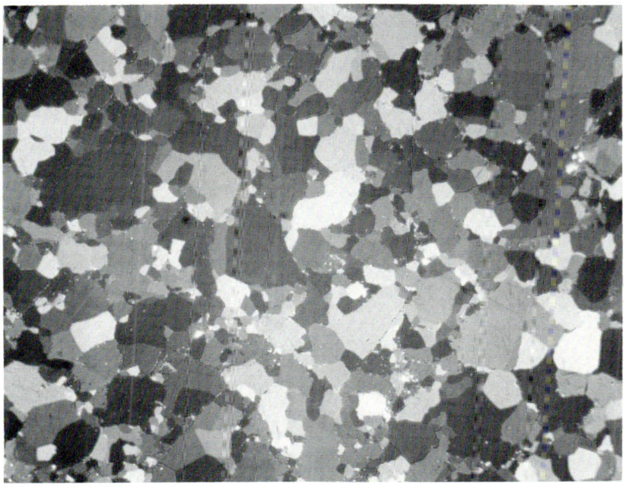

Figure 2 - Optical Micrograph of Vacuum Hot Pressed S-65. Note Grain Size of Approximately 12μ and a Random Distribution of Coarse BeO. Polarized Light 250X.

Figure 3 - Optical Micrographs Showing Structure of S-65 After Rolling at 1425 F (773 C) to 75% Reduction. Polarized Light: Top 100X, Bottom 500X. RD Indicates Rolling Direction.

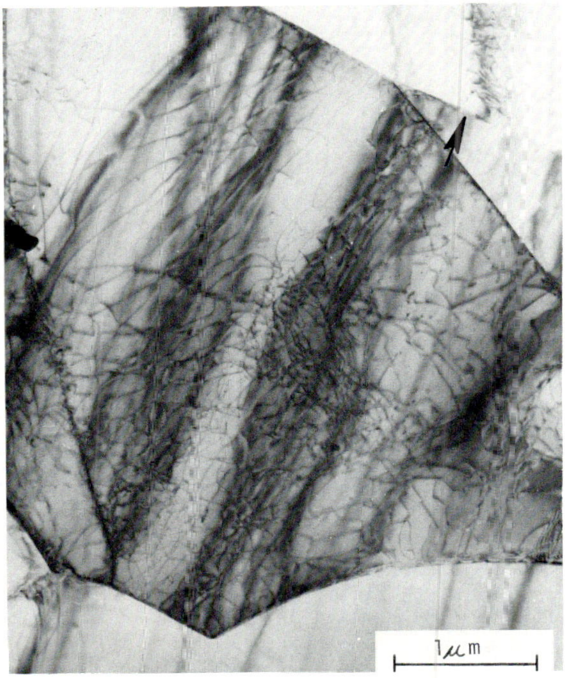

Figure 4 - Bright Field Transmission Electron Micrograph of Material Rolled 75% at 1425 F (773 C). Note the High Dislocation Density. Note the Lack of BeO Particles at the Grain Boundaries. Note the Low Angle Grain Boundary (arrow).

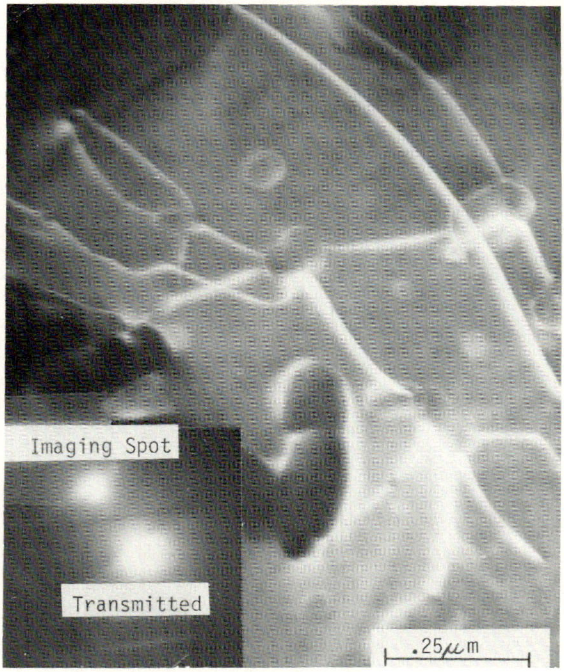

Figure 5 - Dark Field Electron Micrograph Showing Dislocation Pinning by Fine Intragranular Oxide Particles. Their Micrograph was taken with the {1101} Reflection of the {1213} Zone Axis, S Approximately equal to 0. The Associated Diffraction Pattern and Dark Field Reflection are shown in the Inset.

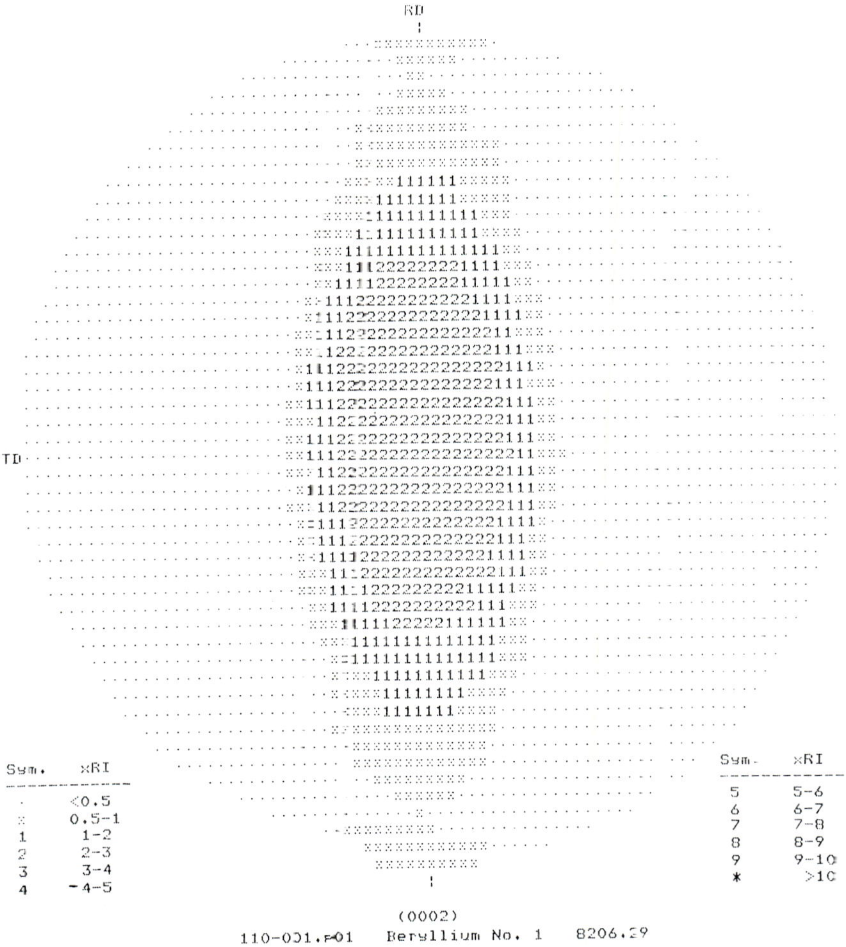

Figure 6 - Digitized X-ray Pole Figure for S-65 Rolled at 1425 F (773 C) Showing Density of {0002} Plane Poles. Numerical Values are Proportional to the Random Pole Density.

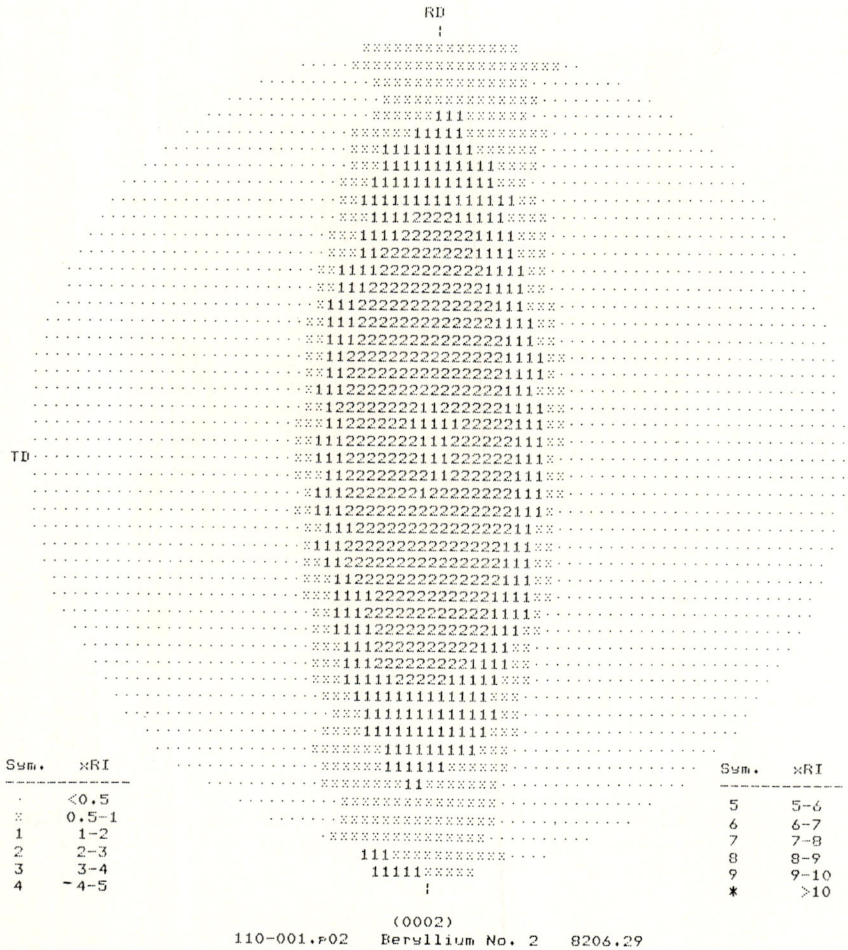

Figure 7 - Digitized X-ray Pole Figure for S-65 Rolled at 1425 F (773 C) and Recrystallized at 1600 F (871 C). Note the Minimal Change in Comparison with the As-Rolled Condition.

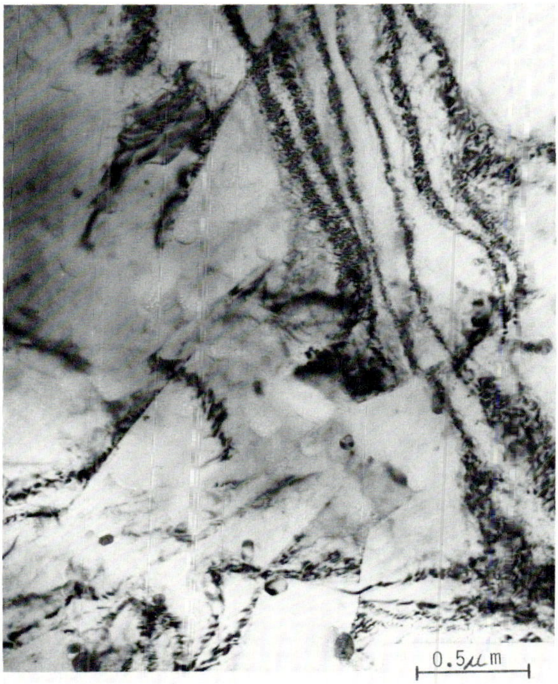

Figure 8 - Bright Field Transmission Electron Micrograph Illustrating The Very Highly Strained Condition of S-65 Sheet After Rolling at 1250 F (677 C).

{0002} axis pole figure, Figure 9, shows a concentration of c-axis poles perpendicular to the sheet plane and elongated elliptically in the rolling direction.

It is interesting to note that the degree of texture and the maximum pole density has remained the same as after the 1425 F (773 C) operations, so it appears that we have either saturated the texture-producing mechanism, or the 25-30% additional reduction was insufficient to alter the texture.

Having examined the structure of the as-rolled beryllium, we can now turn our attention to the kinetics of recrystallization and microstructure produced as a result of subsequent heat treatments.

Recrystallized Structures

Annealing treatments for S-65 rolled as 1250 F (680 C) were carried out in the temperature range of 1350 to 1650 F (732-898 C). Optical metallography was used to determine the extent and structure of recrystallized material. The results of a series of isothermal anneals at 1450 F (787 C) are presented in Figures 10a-10d.

Recrystallization is not in evidence after 1 hour (10a), has begun to occur after 3 hours (10b), is largely complete after 8 hours (10c) and is certainly complete at 16 hours (10d). Sequences such as this were carried out at 1400 F (760 C) and 1500 F (815 C). Figure 11 shows the results of this study. The results were plotted as log t for completion versus the inverse of the absolute temperature. The calculated activation energy for the recrystallization process Q, is 173 Kcal/mole°K, in the Arrhenius equation:

$$\left(\frac{1}{t} = A \exp\left(-Q/RT\right)\right)$$

Where t is the time for recrystallization A is a constant and R and T have their usual meaning.

The fit of the line to the data is good, and the curve was used to predict the recrystallization time at the 1550 F (843 C). The calculated value was 280 seconds, about 4-1/2 minutes. Experimentally, recrystallization occurred in 5-8 minutes. We therefore see that after substantial cold work at 1250 F, recrystallization is well behaved, and quite predictable.

Rolling reduction at 1425 F (773 C) with similar recrystallization experiments produced some parallel behavior, and some areas in which behavior differed significantly. As in the 1250 F (676 C) rolled material, the preferred orientation of {0002} poles was not affected by recrystallization. Figures 12 and 13 however, show a distinct difference in behavior. The recrystallization was fairly uniform when rolling was accomplished at 1250 F (676 C). This is not always the case for high temperature reductions. For high temperature anneals, 1550 F (843 C) and above, one does not get relatively uniform grain sizes, at 1600 F (871 C) for 30 minutes and 1650 F (898 C) for 12 minutes as shown in Figure 12. At 1550 F (843 C) and below, however, there does not seem to be sufficient energy available to initiate uniform recrystallization, and we get anomolous grain growth instead. Figure 13 illustrates the behavior after a 1550 F (843 C) anneal, for 3 and 31 hours. If uniform recrystallization had occurred, we would expect to complete recrystallization in 2-5 hours. It was seen that

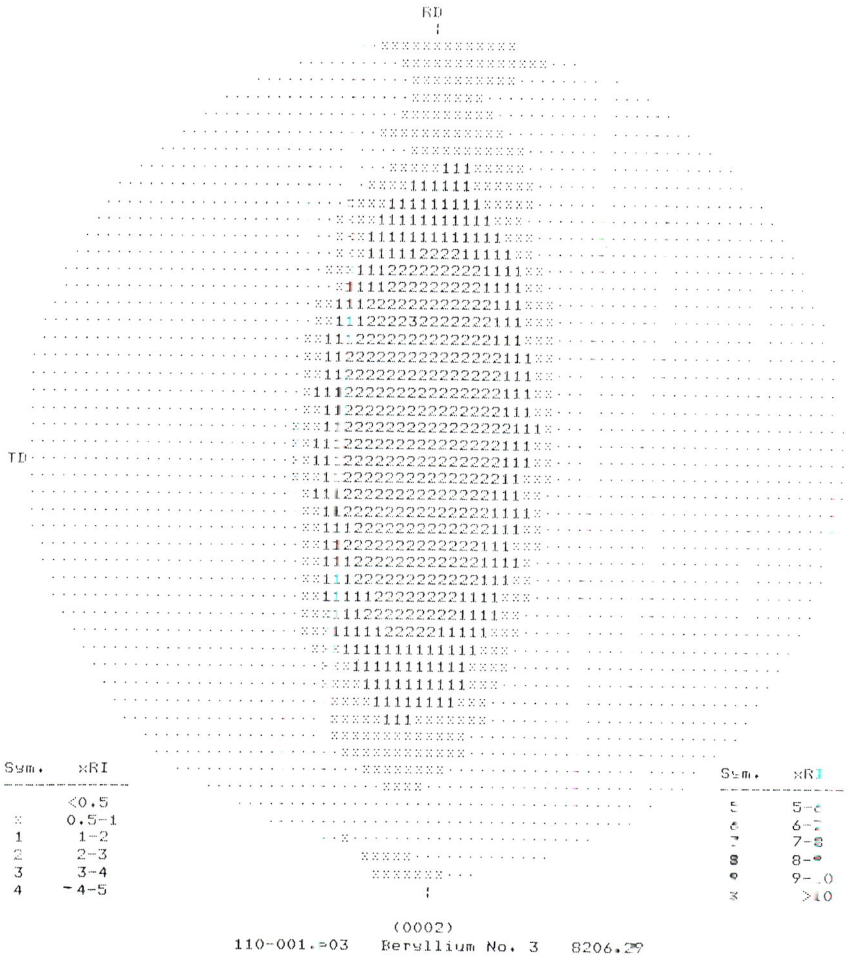

Figure 9 - Digitized {0002} X-ray Pole Figure for S-65 Rolled at 1425 F (773 C) Followed by 30% Additional Work at 1250 F (677 C). Note the Similarity to Pole Figure Obtained After Rolling at 1425 F (773 C).

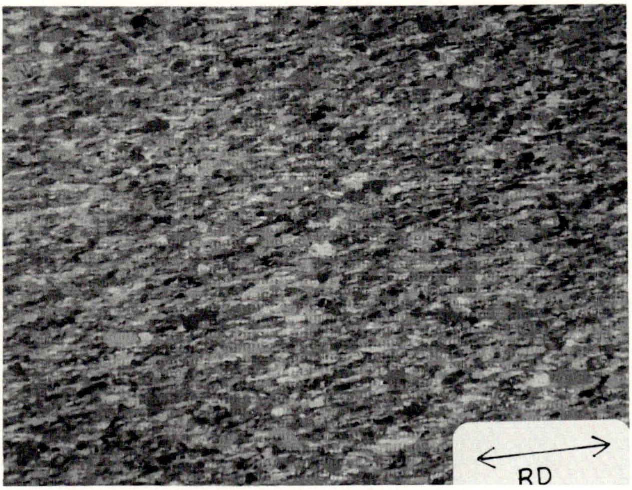

a. 1 hour

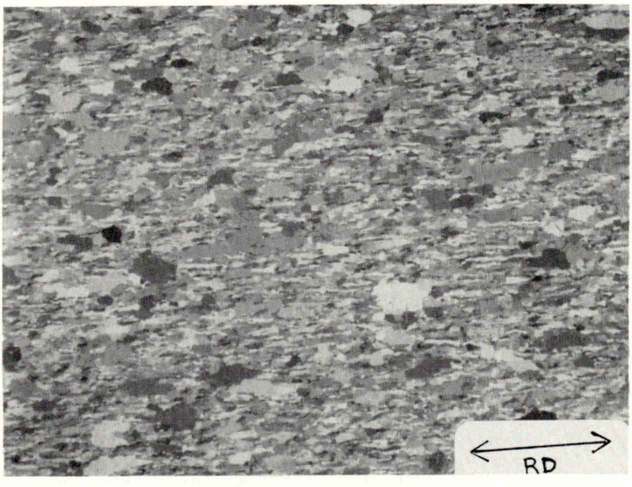

b. 3 hours

Figure 10 - Sequence of Optical Photomicrographs Showing Progress of Recrystallization After Rolling at 1250 F (677 C) and Annealing at 1450 F (787 C). All Photomicrographs are at 100X, taken in Polarized Light.

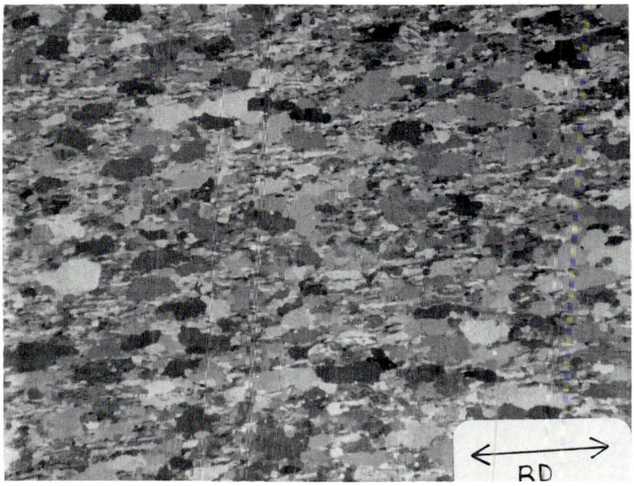

c. 8 hour

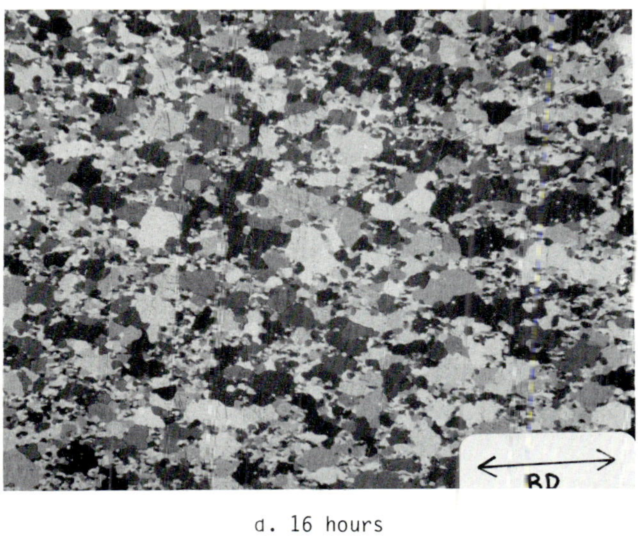

d. 16 hours

Figure 10 - Sequence of Optical Photomicrographs Showing Progress of (cont'd) Recrystallization After Rolling at 1250 F (677 C) and Annealing at 1450 F (787 C). All Photomicrographs are at 100X, taken in Polarized Light.

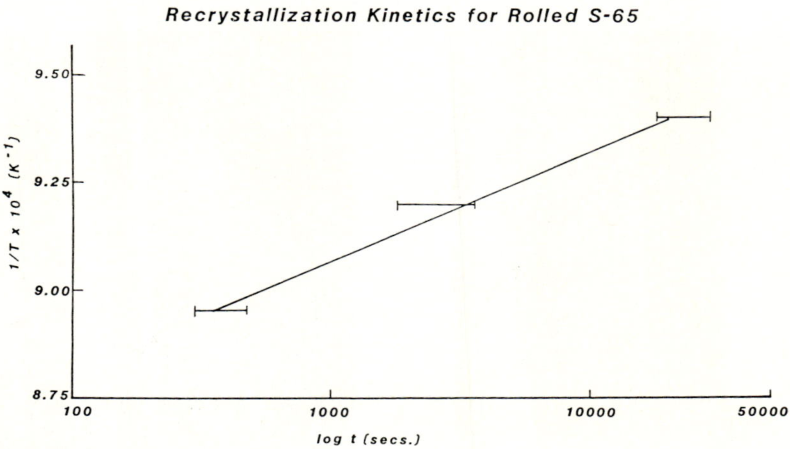

Figure 11 - Summary of Recrystallization Kinetics for S-65 Sheet Rolled at 1250 F (677 C). Graph Shows Log t for Completion Versus 1/T.

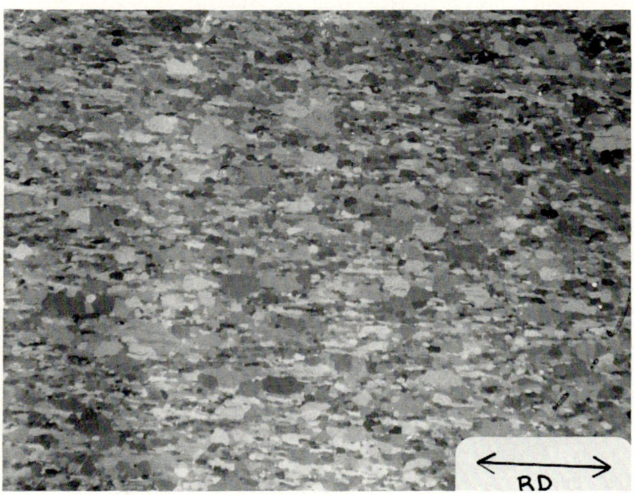

Figure 12 - Optical Micrograph Showing Structure of S-65 After Rolling at 1425 F (773 C), followed by a 1650 F (898 C) Anneal for 15 minutes. Note the Relatively Fine, Uniform Grain Size. 100X Polarized Light.

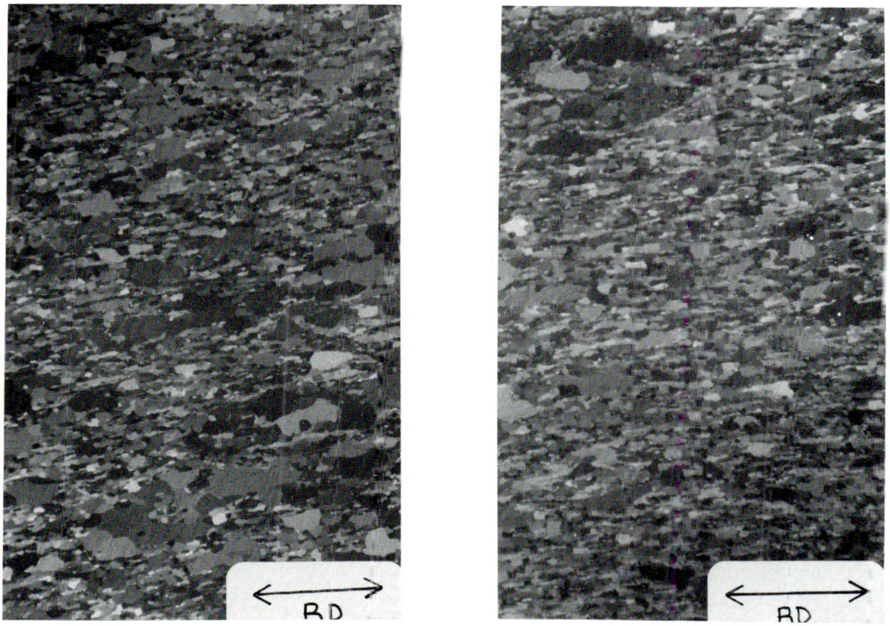

Figure 13 - Optical Photomicrographs Illustrating Structure of S-65 After Rolling at 1425 F (773 C), followed by a 1550 F (843 C) Anneal. Left-3 hours, Right-31 hours. Note Highly Duplexed Grain Size, and Lack of Uniformity of Grains. 100X, Polarized Light.

between 3 and 31 hours, little has occurred in the way of recrystallization. Selected grains have grown however, giving a severely duplexed structure. The high temperature rolling therefore gave rise to an undesirable microstructural condition. The kinetics after the 1425 F (773 C)/75% reduction were more sluggish than those at 1250 F (767 C)/30% reduction shown in Table II.

TABLE II

COMPARISON OF RECRYSTALLIZATION KINETICS
AFTER HIGH AND LOW TEMPERATURE ROLLING

	t	
	1425 F/75%	1250 F/25%
1450	----------	8 hours
1500	----------	45 minutes
1550	*	8 minutes
1600	30 minutes	----------
1650	12 minutes	----------

* Not recrystallized in 31 hours.

The kinetics could not be defined to any greater precision than the comparison because of the rapidity of the process at higher temperatures than 1600 F (871 C), and the lack of uniform response at lower temperatures.

In all cases, the fully annealed structure was one of a relatively dislocation-free matrix, with some low angle grain boundaries in evidence, as shown in Figure 14, a bright field electron micrograph. A low-angle grain boundary is easily visible. Although dislocations are in evidence, the structure is quite clean compared to the as-rolled material.

The changes in recrystallization behavior which occur as a result of oxide dispersions have been previously investigated by Webster, et.al.[3]. It was experimentally determined that the most effective agent in retarding the recrystallization of rolled beryllium was a dispersion of fine (<600Å) BeO particle. Coarser particles, 1650 Å and above, were not as effective as the fine dispersed BeO. The conclusion is consistent with the observation made herein based upon TEM examination that fine intragranular BeO appeared to be an effective dislocation pinning agent. Coarser grain boundary BeO was not as intimately involved with immobilizing dislocations.

Recrystallization studies performed in the Soviet Union[4,5], utilizing very high purity beryllium ingot (99.98% Be), indicated that the activation energy for recrystallization was 18 $\pm$ 3 Kcal/mole. This value, approximately 10% of the value measured in the current work, is indicative of the magnitude of the changes in the activation energy which can be occasioned by a decrease in BeO content, and an increase in reduction (1400%).

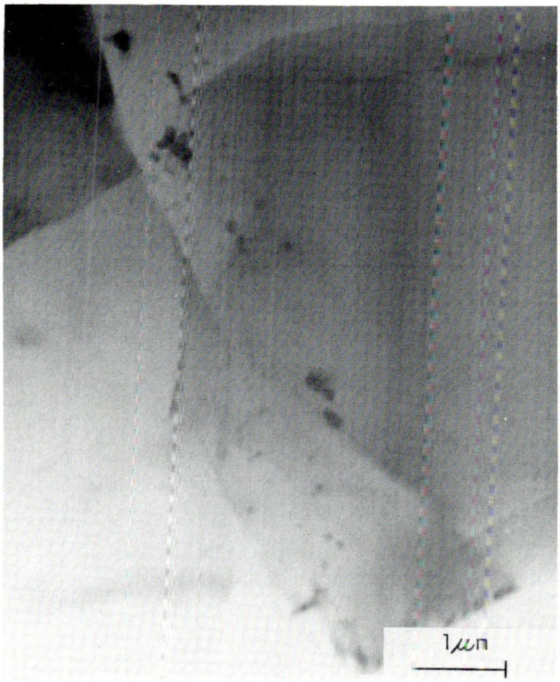

Figure 14 - Bright Field Electron Micrograph showing the Presence of Sub-Grain Boundaries, and a Relatively Dislocation-Free Structure After a Full Recrystallization Anneal.

Conclusions

The following conclusions can be drawn regarding the microstructure and kinetics of recrystallization of rolled S-65 sheet.

1. Low temperature working, below 1425 F (773 C), and high temperature annealing is required to produce a uniform recrystallization microstructure in rolled S-65 beryllium.

2. Recrystallization does little to alter the crystallographic texture of rolled beryllium sheet.

3. Rolling and recrystallization give rise to a substructure of low angle grain boundaries.

4. Fine intragranular oxide appears to be a potent barrier to dislocation motion. Large grain boundary oxides did not appear to be as effective in pinning dislocations.

References

1. F. R. Larson, A. Zarkades and D. H. Avery, "Twinning and Texture Transitions in Titanium Solid Solution", Titanium Science and Technology, V. II, p. 1169 (1973).

2. M. J. Harrigan, A. W. Sommer, P. G. Reimers and G. H. Alers, "The Effect of Rolling Texture on the Fracture Mechanics Properties of Ti-6Al-2Sn-4Zr-6Mo Alloy", Titanium Science and Technology, V. II p. 1297 (1973).

3. D. Webster, D. O. Crooks, and A. E. Vidoz, Met. Trans., V. 4, p. 2841, December 1973.

4. L. A. Kornienko, A. A. Nikolaenko and I. A. Taranenko, Kharkov Physico Technical Institute, Academy of Sciences of the Ukranian SSR. No. 4, p. 78, April 1978.

5. L. A. Kornienko and A. A. Nikolaenko, Kharkov Physico Technical Institute, Academy of Science of the Ukranian SSR, No. 8, p. 56, August 1980.

Author Index

Adjemian, A., 3
Aeschbach, P., 475, 1053
Apelian, D., 1281
Arakawa, N., 197
Assakura, S., 351
Augood, D.R., 1393
Austrheim, O.G., 825
Auton, J., 1027

Bamji, P.J., 1193
Bartlett, R.W., 397
Becker, A.J., 757
Beisswenger, H., 357
Belitskus, D., 923
Bent, J.F., 1541
Bertaud, Y., 667
Birch, M.E.J., 1217
Bjørgo, R., 501
Bockris, J. O'M., 759
Bolles, T.R., 397
Bopp, A.F., 869, 1541
Bourgeois, T., 1033
Boxall, L.G., 555, 573, 649
Bradley, B.F., 541
Brandt, H.H., 883
Braun, D.J., 257
Broad, J.C., 1015
Brooks, D.G., 961
Brown, J.J., Jr., 841
Bruno, M.J., 1571
Bui, R.T., 1033
Bullough, V.L. 961, 989

Careatti, D.R., 757
Castonguay, L., 901
Charette, A., 1033
Chaubal, M.V., 307
Chaudhry, O.S., 461
Chin, E.J., 1027
Chinloy, D.R., 13
Colombe, P., 3
Cooke, A.V., 573
Crawford, R.F, 385
Creehan, J.T., 39
Cresswell, P.J., 211
Crowell, W.F., 1027
Cukier, S., 889

Dahlstrom, D.A., 385
Da Roit, S., 601
Denisenko, V.I., 943
Dernedde, E., 1033
Dewing, E.W., 541
Dickie, R.C., 1531
Dubé, G., 1267

Ebersole, R.R., 1027
Eckert, C.E., 1281
Ellison, W.W., 1027
Elsner, D., 411
Engh, T.A., 1329
Entner, P., 701
Etzel, K., 475, 1053

Fan, L., 789
Farmer, B.D., 1015
Faunce, J.P., 1145
Fellner, P., 805
Fiorini, P., 1177
Fjørnes, B., 701
Förster, L., 1041
Fortier, J.L., 1373
Fortin, S., 721
Franke, A.J., 1041
Fremstad, D,. 501
Frosta, S., 1715

Gariépy, B., 1267
Gasparotto, L., 1177
George, A.L., 989
Gesing, A.J., 721
Gerhardt, M., 721
Glaus, R., 1553
Goldman, J.H., 1475
Granger, D.A, 1249
Grojtheim, K., 805
Groff, G.B., 869
Gruber, U., 701
Guthrie, J.D., 127

Hafeez, A., 1071
Hamano, S., 157
Harato, T., 157
Haugen, Ø., 743
Haupin, W.E., 1429
Henin, J.P., 1669
Hester, R.B., 519
Heuillard, P., 1103
Hill, V.G., 1633
Holzwarth, R.K., 13
Horinouchi, E., 157
Howard, B.H.. 869
Hsieh, H.P., 287
Hsu, M.B., 977
Hurt, R.E.. 1507
Husøy, A.H., 825
Hyland, W.W., 711

Imbrogno, W.D., 127
Iricibar, R , 1129
Itoh, M., 63
Itskov, M.L., 943

Jeltsch, R.C., 493
Jenkins, D.H., 411
Jensen, C.L., 1249
Jensen, E.K., 1159
Jin, I., 1129
Jones, G.R., 1697
Jones, S.S., 933

Kaiser, J., 1027
Kapner, R.S., 883
Kawai, Y., 197
Keller, R., 513
Kibby, R.M., 589
Kirby, J.L., 1231
Kitahara, T., 351
Kohno, T., 63
Kolodin, E.A., 943
Kremer, H., 889
Kvande, K., 805

Lama, A., 1177
Lamb, W.D., 1345
Landi, M.F., 601
Laurence, J.A., 1531
Lavastre, F., 343
Leblanc, G., 1267
Lectard, A., 667
Lee, S.S., 841
Lei, K.S., 841
Levy, S.A., 1231
Lid, O., 743

MacLean, D.C., 1531
Macmillan, D.W., 443
Malito, J.T., 147
Mansfield, T.L., 1305
Marchessaux, Ph., 325
Marder, J.M., 1727
Marshall, R.J., 237
Martin, E.S., 619
Matocha, C.K., 101
Matsushima, T., 689
McCune, D.C., 1531
Miller, R.E., 1281
Milne, D.J., 211, 223
Minear, G., 385
Mitrovic, Z., 701
Monita Benito, J.L., 53
Moriceau, J., 1103
Mutharasan, R., 1281

Nadkarni, S.K., 859, 901
Nagy, K., 501
Naterstad, T., 1715
Nes, K., 1111
Ng, R.K., 1449
Nordmo, F., 825
Nunn, R.F., 373
Numata, H., 759

Ohto, T., 689
Ostojic, S., 1633
Øye, H.A., 1059

Parker, T.L., 81
Peacy, J.G., 1027
Pearson, J., 1217
Pearson, R.M., 81
Pedersen, T., 1329
Perra, S., 269
Piras, L., 601

Qiu, Z., 789

Ratcliffe, D.S., 433
Rhedey, P.J., 859
Robare, T.J., 1521
Rogers, C.J., 443
Rogers, J.N., 541
Roussel, R., 115
Ruch, L.M., 589
Rufer, D., 701
Russell, E.R., 989
Russell, P.P., 933

Saavedra, A.F., 589
Sankey, S.E., 1653
Sato, C., 197
Saulnier, R., 343
Schlager, R.J., 1449
Schmidt, H.W., 357
Schmidt-Hatting, W., 701
Schwarz, R.J., 1653
Seim, H.J., 1393
Sharma, R.A., 1205
Shaw, H., 1145
Shibue, Y., 197
Simoneau, C., 1267
Sinha, H.N., 411
Sivilotti, O.G., 1085
Sorlie, M., 909, 1059
Stankovich, I., 635
Steemson, M., 223, 237
Steinegger, A., 1441, 1553
Stevenson, D.T., 1613
Sturm, A., 1041
Sulzer, J., 1085
Svoboda, R.V., 943

Tabereaux, A.T., 519
Talcott, R.M., 373
Tanabe, T., 351
Tarapore, E.D., 455
Tauscher, J.E., 433
The, K., 115
The, P.J., 127
Thonstad, J., 825
Tiwari, B.L., 1205
Townsend, D.W., 555, 649

Tremblay, C., 1373
Troup, R.L., 1591
Tyler, G.A., 1697

Vadla, J.J., 483
Vedernokov, G.F., 943
Vee, K.Ø., 825
Vogt, M.F., 1697

Wagstaff, F.E., 1145
Wefers, K., 619
Wei, M.W., 1521
Weisbrod, K.R., 933
Weng, T.L., 977, 1005
Wesely, R.J., 397
Westengen, H., 1111
White, E.T., 223, 237
Wikström, L., 1457

Xu, P., 841

Yamada, K., 157
Yoshida, O., 351

Zabreznik, R.D., 455
Zago, G., 1177
Zwicker, J.D., 173